ENVIRONMENTAL ENGINEERS' handbook

BÉLA G. LIPTÁK, Editor

ENVIRONMENTAL ENGINEERS' handbook

Volume II

Air Pollution

CHILTON BOOK COMPANY
Radnor, Pennsylvania

Published in Radnor, Pa., by Chilton Book Company
and simultaneously in Ontario, Canada,
by Thomas Nelson & Sons, Ltd.

Designed by Carole L. DeCrescenzo

Manufactured in the United States of America

Library of Congress Cataloging in Publication Data

Main entry under title.

Environmental engineers' handbook.

Includes bibliographies.
CONTENTS: v. 1. Water pollution.
1. Environmental engineering—Handbooks, manuals, etc. 2. Pollution—Handbooks, manuals, etc.
I. Lipták, Béla G., ed.
TD145.E57 628 72-14241
ISBN 0-8019-5692-7 (v. 1.)

To the Framers of the
United States of Central Europe

CONTRIBUTORS

IRVING M. ABRAMS, BCh, PhD
Manager, Technical Development, Diamond Shamrock Chemical Company
(Sections I.6.16, I.7.17)

CARL E. ADAMS, JR., BSCE, MSSE, PhDCE, PE
Technical Director, Associated Water & Air Resources Engineers, Inc.
(Section I.5.16)

ROBERT C. AHLERT, BChE, MSEng, PhD
Professor of Chemical & Biochemical Engineering, Rutgers University
(Sections I.2.10, I.2.15, III.5.10)

DAVID J. ALLEE, BS, MS, PhD
Professor of Resource Economics, Cornell University
(Section III.6.4)

SEYMOUR B. ALPERT, BSChE, MS
Senior Industrial Economist, Stanford Research Institute
(Section II.6.1)

ELMAR R. ALTWICKER, BS, PhD
Associate Professor of Bio-Environmental Engineering, Rensselaer Polytechnic Institute
(Sections II.3.1, II.3.16)

ROLLIN A. ARMER, AEM
Senior Design Associate, Lawrence Radiation Laboratory
(Sections II.5.46, II.5.47, II.5.48)

DONALD B. AULENBACH, BSCh, MS, PhD-S
Associate Professor of Bio-Environmental Engineering, Rensselaer Polytechnic Institute
(Sections I.6.31, I.7.4)

STEPHEN M. AYRES, MD
Director, Cardiopulmonary Laboratory, St. Vincent's Hospital and Medical Center
(Section II.3.7)

Contributors

RICHARD C. BAILIE, BSChE, MSChE, PhDChE
Professor of Chemical Engineering, Chemical Engineering Dept., West Virginia University
(*Section III.2.16*)

RICHARD L. BENNETT, BMetE
Combustion Engineer, The North American Manufacturing Company
(*Sections II.5.25, 26, 27, 32, 38, II.7.7, 7.8, 7.10, 7.15, 7.18, III.2.21, 25*)

HAROLD BERKSON, BA, MA, PhD
Specialist in Environmental Policy, Congressional Research Service, Library of Congress
(*Sections I.1.1, I.3.1*)

JAIME BIGU DEL BLANCO, ChE, NE
Research Associate, National Research Council of Canada
(*Section II.3.19*)

EDWARD C. BINGHAM, BSCh, MBA
Technical Assistant to General Manager, Farmers Chemical Association, Inc.
(*Sections I.3.8, 6.25, 7.20, II.7.6*)

JAMES H. BLACK, BSCh, MSCh, PhDCh, PE
Professor of Chemical Engineering, University of Alabama
(*Section II.5.2*)

CHRISTOPHER P. BLAKELEY, BSChE
Market Manager, Water Management, Honeywell, Inc.
(*Sections I.4.2, 8, 14, 16, 24*)

L. JOSEPH BOLLYKY, PhD
President, Pollution Control Industries Ozone Corp.
(*Sections I.2.22, 25, 4.18, 6.33, 36, 7.15, 21, 24, 27, 33, 45, II.4.14, 6.20, 21, 23*)

JERRY L. BOYD, BSChE
Chief Process Application Engineer, Eimco Corporation
(*Section I.5.9*)

AUGUST BRODGESELL, BSEE
Chief Instrument Engineer, Crawford & Russell, Inc.
(*Sections I.4.9, 23, II.4.9*)

THOMAS F. BROWN, JR., BSAE, EIT
Assistant Director Environmental Engineering, Commercial Solvents Corporation
(*Sections I.5.13, 7.2, 8.12, III.3.27*)

BARRETT BRUCH, BSME, BSIE
Oil Spill Control Project Leader, Lockheed Missiles & Space Company
(*Section I.7.40*)

ROBERT D. BUCHANAN, BSCE, MSCE, PE
Chief Sanitary Engineer, Bureau of Indian Affairs
(*Sections I.2.1, 5.1, 12, 23, 7.23, 8.1, III.3.17, 24, 6.5*)

DON E. BURNS, BSCE, MSCE, PhD-SanE
Senior Research Engineer, Eimco Corporation
(*Sections I.5.11, 6.5, 14*)

RICHARD D. CADLE, BA, PhD
Head, Atmospheric Chemistry Dept., National Center for Atmospheric Research
(*Sections II.3.2, 3, 12, 20*)

JOHN CAIRNS, JR., AB, MS, PhD
Director, Center for Environmental Studies and Research Professor, Biology Dept., Virginia Polytechnic Institute & State University
(*Sections I.2.19, 20, III.5.2*)

LARRY W. CANTER, BE, MS, PhD, PE
Director, School of Civil Engineering & Environmental Science, University of Oklahoma
(*Sections I.2.13, 7.31, II.2.10, 3.10, 7.1*)

PAUL J. CARDINAL, JR., BSME
Manager, Sales Development, Envirotech Corporation
(*Section I.5.20*)

CHARLES A. CASWELL, BS-Geology, PE
Vice President, University Science Center, Inc.
(*Sections I.1.6, 2.6, 16, 7.5, 10, 8.14, 15*)

MIKIO CHIBA, MSc, DSc
Research Scientist, Tokyo Univ. of Agriculture, Japan.
(*Sections II.3.17, 4.16, III.5.15, 18*)

NICHOLAS P. CHOPEY, BSChE, MA-Econ.
Senior Associate Editor, McGraw-Hill, Inc.
(*Section II.6.2*)

DAVID E. COLE, BSME, BS-Math, MSME, PhDME
Associate Professor, University of Michigan
(*Section II.5.50*)

RICHARD A. CONWAY, BS, MSSE, PE
Group Leader, Union Carbide Corporation, Research & Development Dept.
(*Sections I.7.7, 11, 32, 35, III.5.17*)

GEORGE J. CRITS, BSChE, MSChE, PE
Technical Director, Cochrane Div., Crane Company
(*Sections I.6.2, 17, 7.30*)

ERNOE CSÁKY, BS-Physics
Manager, The Dow Chemical Company
(*Section I.4.17*)

DONALD DAHLSTROM, PhDChE
Vice President & Director of Research & Development, Eimco Corporation
(*Sections I.5.9, 11, 15, 6.5, 14*)

Contributors

STACY L. DANIELS, BSChE, MSSE, MSChE, PhD
Development Engineer, The Dow Chemical Company
(*Section I.6.6*)

ERNEST W. J. DIAPER, BSc, MSc
Manager, Municipal Water & Waste Treatment, Cochrane Div., Crane Company
(*Sections I.6.13, 7.1*)

KENNETH L. DICKSON, BSEd, MS-Biol, PhD-Zoology
Assistant Director, Center for Environmental Studies, Virginia Polytechnic Institute & State University
(*Sections I.2.20, III.5.2*)

BASIL DIMITRIADES, ChE, MSPhCh, PhD, PhCh
Assistant Research Supervisor, Bureau of Mines, U.S. Department of the Interior
(*Section II.5.49*)

FRANK W. DITTMAN, BSChE, MSChE, PhD, PE
Professor of Chemical Engineering, Rutgers University
(*Section I.5.19*)

LAURENT DUBOIS, BS
Research Scientist, Government of Canada
(*Section II.4.4*)

PATRICK R. DUGAN, BS, MS, PhD
Professor of Microbiology, Ohio State University
(*Sections I.2.3, 9, 3.2, 3, 5*)

BERNARD J. EASTLUND, BS, PhD
Physicist, U.S. Atomic Energy Commission
(*Section III.3.21*)

WAYNE F. ECHELBERGER, JR., BSCE, MSE, MPH, PhD
Associate Professor of Civil Engineering, University of Notre Dame
(*Sections I.2.12, 5.28, 6.22, 24, 8.16, III.3.22, 25*)

JOHN E. EDINGER, BCE, MSE, PhD
Associate Professor Civil Engineering-Water Resources, Towne School, University of Pennsylvania
(*Sections III.5.12, 14*)

ROBERT H. ESSENHIGH, BANS, MANS, PhDChE
Professor of Fuel Science, Pennsylvania State University
(*Section II.5.35*)

ROBERT P. FARRELL, JR., BSEE, PE
Manager, Water Products Engineering, Environment/One Corporation
(*Sections I.6.1, III.2.11*)

J. W. TODD FERRETTI
President, The Bionomic Systems Corporation
(*Sections I.7.14, II.5.1, 3, 7.5, 17*)

RONALD G. GANTZ, BSChE
Senior Process Engineer, Continental Oil Company
(*Sections I.5.6, 7.36, II.5.39, 40, 41*)

WILLIAM C. GARDINER, BA, MA, PhD, PE
Director-Electrochemical Development, Crawford & Russell, Inc.
(*Sections I.7.28, II.7.4, III.5.20*)

FERRUCCIO GERA, D-Geol
Functionary, Comitato Nazionale per l'Energia Nucleare, Italy
(*Section III.5.8*)

WALTER F. GERDES, BSEE, PE
Technical Specialist, The Dow Chemical Company
(*Section I.4.21*)

JAMES E. GERMAIN, BS, MS, PE
Vice President-Concept Technology Division, Roy F. Weston, Inc.
(*Section I.5.14*)

LOUIS C. GILDE, JR., BSSE
Director-Environmental Engineering, Campbell Soup Company
(*Sections I.5.29, 7.9, III.3.28*)

JOHN L. GILLILAND, BSChE, PE
Technical Director, Ideal Cement Company, Div. of Ideal Basic Industries, Inc.
(*Sections II.1.1, 3, 7.3*)

RICHARD GODER, PE
President, Joseph Goder Incinerators
(*Section III.2.21*)

BRIAN L. GOODMAN, BS, MS, PhD
Director, Technical Services, Ecodyne Corporation, Smith & Loveless Division
(*Sections I.5.17, 18*)

ROBERT J. GORDON, PhD
Assistant Professor, U.S.C. School of Medicine
(*Sections II.2.8, 3.5, 9, 4.5, 7, 10, 12, 21*)

WILLIAM C. GOUGH, BSE, MSEE
Project Manager in the Office of Controlled Thermonuclear Research, U.S. Atomic Energy Commission
(*Sections III.3.21, 6.1*)

CHARLES L. GREISER
Senior Liaison Engineer, Lockheed Missiles & Space Company, Inc., Research Div.
(*Section I.8.2*)

CHARLES E. HAMILTON, BSChE
Project Leader, The Dow Chemical Company
(*Sections I.4.4, I.4.13*)

Contributors

CHARLES W. HART, JR., BA, BS, MA
Administrator of Consultant Programs Dept. of Limnology, Academy of Natural Sciences of Philadelphia
(*Section I.2.17*)

WILLIAM J. HATCHER, JR., BSChE, MSChE, PhDChE, PE
Associate Professor of Chemical Engineering, University of Alabama
(*Sections II.5.4, 22, 23, 24, 28, 29, 30, 33, 34, 37, 6.5, III.2.20*)

JOHN A. HAYDEN, BS, MS, PhD
Director, European Operations, Hayden, Harding & Buchanan, Inc., West Germany

JARED E. HAZLETON, BBA, PhD
Vice Chairman & Associate Professor, Economics Department, University of Texas at Austin
(*Sections I.1.2, 4, 5*)

JOEL W. HEDGPETH, PhD
Professor, Dept. of Oceanography, Oregon State University
(*Section I.2.4*)

ROBERT A. HERRICK, BSChE, PE
Consulting Engineer
(*Sections II.4.2, 6, 15, 20*)

DANIEL A. HIRSCHLER, BSME, AB
Director, Automotive Research, Ethyl Corporation
(*Section II.5.51*)

ELBERT HOOPER, BA, JD
Attorney at Law, Hooper & Robinson, Attorneys at Law
(*Sections I.1.3, 7, III.1.4*)

DAVID L. HOYLE, BSChE
System Design Engineer, The Foxboro Company
(*Section I.6.37*)

JAY S. JACOBSON, BS, MA, PhD
Plant Physiologist, Boyce Thompson Institute for Plant Research
(*Sections II.2.12, 4.8*)

FREDERICK W. KEITH, JR., BSChE, PhDChE, PE
Manager, Applications Research, Pennwalt Corporation
(*Section I.5.21*)

EDWARD G. KOMINEK, BS, MBA, PE
Manager, Industrial Water & Waste Sales, Eimco Processing Machinery Division, Envirotech Corporation
(*Sections I.7.3, 6, 13, 16, 19, 26, 29*)

P. KRISHNAN, BSCE, MSc, PhDSE, PE
Senior Project Engineer, Roy F. Weston, Inc.
(*Section I.5.14*)

CHARLES KURKER, BSChE
Senior Staff Advisor, Solid Waste Program, State of Connecticut Department of Environmental Protection
(*Reviewed Section III.3.18*)

HELMUT E. LANDSBERG, CCM, PhD
Research Professor, University of Maryland
(*Section II.2.7*)

ROBERT C. C. LAO, BS, MS, PhD
Research Chemist, Department of Environment, Government of Canada
(*Sections II.4.3, 4, 22*)

TRUMAN S. LIGHT, BSCh, MSCh, PhDCh
Senior Research Scientist, The Foxboro Company
(*Section I.4.19*)

BÉLA G. LIPTÁK, ME, MME, PE
Staff Consultant, Crawford & Russell, Inc.
(*Preface, Glossaries, Abbreviations, Sections I.4.1, I.5.27, II.4.1, III.2.15, 3.10, Appendices and Miscellaneous*)

JÁNOS LIPTÁK, CE, PE
Senior Partner, János Lipták & Associates
(*Sections I.5.2, 3, 5*)

PÉTER LIPTÁK, BS
Chemist, Presbyterian Hospital
(*Sections I.8.11, III.3.26*)

DAVID H. F. LIU, BSc, MS, PhDChE
Research Associate in Charge of Chemical Engineering, Pollution Abatement and Applied Mathematics, UniRoyal Chemical Division, of UniRoyal, Inc.
(*Sections I.7.42, II.4.18, 5.45, III.2.3*)

IHOR LYSYJ, MS
Project Development Scientist, Rocketdyne, Division of North American Rockwell Corporation
(*Section I.4.5*)

BRYCE I. MacDONALD, BSChE, PE
Director of Engineering, Kennecott Copper Corporation
(*Reviewed Section II.6.12 and contributed II.7.14*)

WOLFGANG A. MACK, ChE, DEng Econ.
President, Werner & Pfleiderer Corporation
(*Section III.3.9*)

L. CHARLES MATSCH, DSc
Assistant Director, Linde Division, Union Carbide Corporation
(*Section I.6.10*)

MICHAEL J. MATTESON, DChE
Assistant Professor, Georgia Institute of Technology, School of Chemical Engineering
(*Sections II.6.3, 4, 6 to 11*)

PETER G. MAURIN, BSME, MBA
Assistant to the Manager Air Pollution Control Systems, Combustion Engineering, Inc.
(*Section II.6.12*)

FOSTER LEE MAYER, JR., BSBiol., BSCh, MSBiol., PhDEcology
Fishery Biologist, U.S. Department of the Interior, Bureau of Sport Fisheries and Wildlife, Fish-Pesticide Research Laboratory
(*Section I.3.4, III.5.16*)

CARL A. MAZZOLA, BS & MS Meteorology
Meteorologist, Scott Research Laboratories
(*Section II.2.6*)

ANDREW F. McCLURE, Jr., BSChE
Manager, Industrial Concept Design Div., Betzon Environmental Engineers
(*Section I.7.22*)

JOHN L. MONKMAN, BS, MS
Chief, Chemistry Division, Air Pollution Control Directorate, Dept. of Environment, Government of Canada
(*Sections II.4.4, 22*)

DAVID C. MORRIS, BSChE
Assistant Technical Director, Mill Division, Green Bay Packaging, Inc.
(*Section I.6.12*)

THOMAS J. MYRON, JR., BSChE
Senior Systems Design Engineer, The Foxboro Company
(*Sections I.6.27 to 30, 37*)

SUEO NISHI
Chief Chemist, Government Chemical Industrial Research Institute, Tokyo, Japan
(*Section I.4.12*)

RICHARD T. OLIVER, BSCh, MSCh, PhDCh
Senior Analog Systems Design Engineer, The Foxboro Company
(*Sections I.4.11, 15*)

JUDITH PAPP, ChE, PE
Senior Process Engineer, Crawford & Russell, Inc.
(*Sections I.6.9, 26*)

FRANK L. PARKER, BA, MS, PhD, PE
Professor of Environmental and Water Resources Engineering, Vanderbilt University
(*Sections I.1.8, 2.14, 3.9, 6.35, 7.46, II.3.18, III.1.3, 5.4, 5, 6, 7, 11, 13*)

WILLIAM H. PARTH, BS, MS in Physics
Senior Research Physicist, The Dow Chemical Company
(*Section I.4.20*)

GRACE PAUL, MT(ASCP)
Research Associate, The Institute for Advanced Environmental Studies and Center for Environmental Information and Evaluation
(*Sections III.2.1, 3.1 to 6, 8, 5.3*)

PETER N. B. PEACOCK MA, ChB, MD, DPH, DIH, DTMH
Professor and Chairman, Department of Public Health and Epidemiology, University of Alabama in Birmingham
(*Section II.2.11*)

NORMAN PEARSON, BA, MTPIC
Professor of Political Science, Talbot College, The University of Western Ontario
(*Sections I.2.5, II.2.3, III.2.2*)

JOSEPH G. RABOSKY, BSChE, MSE, PE
Senior Project Engineer, Calgon Corporation
(*Sections I.2.18, 5.25, III.2.18, 23*)

GEORGE W. REID, BSCE, MSCE, PE
Professor and Director of Water Research Laboratory, University of Oklahoma
(*Sections I.2.11, 21, 24, III.6.2, 6*)

LeROY H. REUTER, MS, PhD, PE
Consultant
(*Section I.5.32*)

BERNARDO RICO-ORTEGA, BSCh, MSSE
Product Specialist, Pollution Control Dept., Nalco Chemical Company
(*Section I.6.8*)

HOWARD C. ROBERTS, BAEE, PE
Professor of Engineering (Retired)
(*Sections III.1.2, 4.1 to 9, 5.21*)

REED S. ROBERTSON, BSChE, MSEnvE, PE
Senior Group Leader, Nalco Chemical Company
(*Section I.7.39*)

DAVID M. ROCK, BSChE, MSChE, PE
Staff Engineer, Environmental Control American Enka Company
(*Section I.7.34*)

CESAR A. ROMERO, MD, DSc
Associate Professor (Anatomy), Queen's University, Kingston, Ontario
(*Section II.3.19*)

J. ELDON RUCKER, JR., BSCE, MSSE, PE
Project Engineer, Research and Development, Mobil Oil Inc.,
(*Section I.2.8*)

Contributors

JAMES D. RUSHTON, BSChE, PE
Chemical Process Design Engineer, J. E. Sirrine Company
(*Sections II.2.9, 7.12, III.6.3*)

CHAKRA J. SANTHANAM, BSChE, MSChE, ChE, PE
Senior Environmental Engineer, Crawford & Russell, Inc.
(*Sections I.6.7, 18 to 21, 32, 34, 7.43, II.3.14, 5.5 to 11, 36, 42, 43, 44, III.2.13, 14, 19*)

E. STUART SAVAGE, BSChE, PE
Manager, Research and Development, Water & Waste Treatment Dept., Dravo Corporation
(*Sections I.6.4, 23*)

LETITIA S. SAVAGE, BS-Biol.
North Park Naturalist, Latodami Farm Nature Center, Allegheny County Dept. of Conservation
(*Section I.7.38*)

FRANK P. SEBASTIAN, MBA, BSME
Senior Vice President, Envirotech Corporation
(*Sections I.5.24, 26, 6.38, 8.7, III.2.17, 22, 3.14*)

JAMES L. SETSER, BS, MS
General Manager, Measurement Services Div., Environment/One Corporation
(*Section I.4.3*)

INDRAVADAN S. SHAH, BSChE, MSChE
Vice President, Combustion Equipment Associates, Inc.
(*Sections II.3.15, 5.12 to 21, 31, 6.25, 26, 28, 7.2, 9, 16*)

GERALD L. SHELL, MSCE, PE
Director of Sanitary Engineering, Eimco Corporation
(*Sections I.5.9, 11, 15, 6.5, 14*)

ROBERT J. SHERWOOD, MSEnvE
Director of Wastewater Treatment Division, Envirotech Europe, Inc. Paris, France
(*Section I.8.10*)

DONALD J. SIBBETT, AB, MA, PhD
Vice President, Geomet, Inc.
(*Sections II.3.11, 4.11, 6.27*)

ROBERT SIEV, BSChE, MBA, CE
Engineering Specialist, Bechtel Corporation
(*Section I.4.10*)

S. FRED SINGER, PhD
Professor of Environmental Science, University of Virginia
(*Sections I.2.2, II.2.1, 4, III.5.9*)

FRANK G. SMITH, JR., BSChE, MSCh, PE
Manager, Environmental Control, Celanese Coatings Company
(*Section II.7.11*)

RALPH G. SMITH, BS, MS, PhD
Professor, University of Michigan, Department of Environmental & Industrial Health
(*Sections II.3.4, 6, 8, 13, 4.17*)

STUART E. SMITH, BChE, MSChE, MSSE, PE
Manager, Industrial Wastewater Operation, Environment/One Corp.
(*Sections I.7.18, 25, 37*)

JOHN R. SNELL, BECE, MSSE, DSSE, PE
President, John R. Snell Engineers
(*Sections I.5.30, 8.4, 5, 6, 8, 13, 17, III.2.10, 12, 3.11, 12, 13, 15, 16, 18, 19, 20, 23*)

PETER E. SNOEK, BSCE, MSCE, PE
Senior Engineer, Bechtel, Inc.
(*Sections I.5.31 and III.2.9*)

MICHAEL J. STANKEWICH, JR., MSSE
Staff Engineer, Union Carbide Corporation, Linde Division
(*Section I.6.10*)

PAUL L. STAVENGER, BSChE, MSChE
Director of Technology, Process Equipment Division, Dorr-Oliver, Inc.
(*Sections I.5.4, 6.3*)

HERMAN D. STENSEL, PhD
Senior Research Engineer, Eimco Processing Machinery, Division of Envirotech Corp.
(*Sections I.8.9, III.3.7*)

GEORGE P. SUTTON, BSME, MSME, PE
Executive Vice President, Sumitomo Jukikai Envirotech, Inc., Japan
(*Section II.7.13*)

JOSEPH A. TANNER, BSc, MEng, PhD
Head, Control Systems Laboratory, National Research Council of Canada
(*Section II.3.19*)

FLOYD B. TAYLOR, BSSE, MPH, PE
Regional Representative, Division of Water Hygiene, Environmental Protection Agency
(*Sections I.2.23, 26, 27, 3.6, 7, 7.41, II.1.2, III.1.1, 5.19*)

MARK W. TENNEY, SB, SM, Sc.D, PE
Associate Professor, Dept. of Civil Engineering, University of Notre Dame
(*Section I.2.7*)

TERRY L. THOMPSON, BSME, MSBA, PE
Chief Slurry Engineer, Bechtel, Inc.
(*Sections I.5.31 and III.2.9*)

CARL C. TINSTMAN III, BS, MBA
Senior Quantitative Analyst, City of New York Environmental Protection Administration.
(*Section III.2.5*)

Contributors

AMOS TURK, BS, MA, PhD
Professor, Dept. of Chemistry, The City College of the City University of New York
(*Sections II.4.13, 6.22, 24*)

DAVID R. VAUGHN, BSCE, MSCE, PE
Development Engineer, Dorr-Oliver, Inc.
(*Section I.5.4*)

P. AARNE VESILIND, BSCE, MSCE, MSSE, PhD, PE
Assistant Professor of Civil Engineering, Duke University
(*Sections I.5.7, 8, 10, 22*)

CECIL C. WALDEN, BA, MA, PhD
Associate Director, B. C. Research, Canada
(*Section I.7.8*)

JAMES V. WALTERS, BSCE, MSCE, PhD, PE
Professor of Civil Engineering, University of Alabama
(*Section III.2.4*)

EDWARD J. WASP, BSChE, MSChE, MS-Math, PE
Manager of Slurry Systems for Pipeline and Production, Service Division, Bechtel, Inc.
(*Sections I.5.31 and III.2.9*)

GEORGE P. WHITTLE, BSChE, MSChE, PhDChE, PE
Associate Professor, University of Alabama
(*Section I.4.7*)

JOHN I. YELLOTT, BSME, MME, PE
Director, Yellott Solar Energy Laboratory
(*Section II.4.19*)

IRVING G. YOUNG, BS, MS, PhD
Chemist, Advanced Technology Staff, Honeywell, Inc.
(*Sections I.4.6, 17, 22*)

ROBERT R. ZALL, BS, MS, PhD
Associate Professor, Cornell University
(*Section I.8.3*)

ROGER H. ZANITSCH, BSCE, MSSE
Senior Project Engineer, Calgon Corporation
(*Sections I.6.15, 7.12, 44*)

WILLIAM C. ZEGEL, ME, MS, ScD
Senior Research Associate, Ryckman, Edgerley, Tomlinson & Associates, Inc.
(*Sections II.1.4, 2.2, 5, 6.13 to 19, III.5.1*)

MATTHEW M. ZUCKERMAN, BCE, MSCE, PhD
Manager of Development, Z-M Process for Wastewater Treatment, Envirotech Corporation
(*Section I.6.11*)

PREFACE

CONTENTS OF THE PREFACE

The three volumes of the *Environmental Engineers' Handbook* contain technical and scientific information needed to utilize fully our modern technology in cleaning up the environment. The first volume is devoted to the many aspects of water pollution; the second deals with air pollution; and the third discusses various forms of land pollution including solid waste, noise, pesticides, radiation, thermal and others. Recognizing the integrated nature of our environment, where various forms of pollution are only interrelated symptoms and therefore cannot be treated separately, the last volume devotes a full chapter to the broader context of environmental engineering.

This reference book, in addition to giving complete coverage to all aspects of pollution causes, effects and remedies also emphasizes the recycling, recovery, salvaging and reuse potentials. Thereby, the reader is provided with information on the technology available for making resources out of wastes. In these three volumes, each section is built upon and supported by the others through cross-references and subject indexes. Since many of the discussed processes are biological in nature, special emphasis is given to the bioengineering aspects.

A large number of environmental scientists and engineers have participated in the preparation of this reference set. The preliminary list of contents for the three volumes was reviewed and discussed by over one thousand environmental experts. The authors who were invited to contribute to this reference book include American, European, Japanese and Canadian scientists and representatives of the nontechnical professional communities, who prepared the texts on medical, biological, agricultural, meteorological and legal aspects of environmental engineering. In order to keep this reference source up to date, future editions and updated supplements are planned.

An engineer answers society's needs through technical innovation. The tools of engineers are the basic sciences, such as physics or chemistry. It is conceivable that members of the traditional engineering disciplines might be attracted to the tools themselves and so might work on, instead of with, these tools. This may contribute to the misuse of technology by allowing the engineer to concentrate on narrow, specific tasks without having the opportunity, or sometimes even the desire, to question his goals and their broader consequences. The members of the new discipline called "environmental engineering" are in a privileged and challenging position. Their tools are the totality of man's scientific knowledge, and their target is nothing less than making man's peace with nature.

The Condition of the Environment

The biosphere is of fixed dimensions. It is a ten-mile thick layer extending over the 200 million square mile surface of the earth. This thin crust of earth, air and water which sustains life is already near its limit. One may note for example that the rate at which mammals alone have disappeared from the world wildlife family during the last 200 years is more than ten times the rate for the previous 2,000 years. According to some scientists, if the present growth rate persists on a worldwide basis, the total demand on the environment will increase by a factor of *32 over the next 66 years.*[*] The gravity of the ecological crisis and the speed at which it is coming justify the concern of many. Without corrective action, the disruption of the ecosystem can be anticipated together with an exhaustion of resources that could lead to the failure of food supplies and to the collapse of society. Such irreversible disruption of the life-support system may be the price of inaction within the lifetimes of our children.

Life on earth has been governed by laws of nature for three billion years. For millions of years, man too had accepted these laws, but in recent history man has succeeded in "conquering" nature and in developing social structures which can respond to desires that conflict with the laws that govern nature. Man finds it difficult to "govern" nature because he does not have nature's most abundant commodity: time. For example, the replacement of the waters in deep oceans takes 1000 years; the half life of plutonium is 24,000 years. The result of such conflict can become a crisis of survival.

On the other hand, concentrated action in this area can do more than just abate pollution. It can have a unifying and sobering effect on a sated and weary society which, if properly informed and motivated, will welcome the challenge. Curing the environmental crisis will necessitate changes in man's consciousness and his value system, and such changes will, in effect, change the course of history.

The Condition of the Waterways

The present daily water withdrawal rate of the United States is near 400 billion gallons (consumption is about 100 bgd). If expressed on a daily per capita basis, this is about 2,000 gallons of water withdrawn per person. These withdrawal figures represent about one-third of the average daily runoff, but

[*] A Blueprint for Survival, *The Ecologist*, London, January 1972.

in any given geographic area, the supply/demand ratio can be drastically different.

The extent of water pollution is also a function of the geographic area (Fig. 1). For centuries it was believed that great rivers, such as the Mississippi which drains 41% of the continental United States, were too large to be destroyed by pollution. Recent reports indicate, however, that the waste dumped into this river is threatening both the health of humans in southern Louisiana and the aquatic life in the Gulf of Mexico. Untreated sewage, industrial dumping, fertilizers and pesticides are the major sources of contamination. As a result, it has been estimated that as much as 60% of the discharged pollutants reenter water supplies because of tidal effects on reservoirs and the close proximity of wastewater discharge and water supply points in the waterways. The deterioration of the quality and safety of the water sources continuously increases the cost of water supply treatment.

The increasing cost of municipal water is useful insofar as it motivates recycling. If it costs more for an industry (or other water user) to purchase fresh water than it does to clean and reuse the wastewaters it generates, water recycling will be more widely practiced and less pollutants will be discharged into the waterways. As the the "closed circuit" (recycled) use of water is increased and the "once through" use reduced, the restoration of water bodies will become a reality.

The present economic picture is such that 1000 gallons of fresh municipal water costs about 25¢. To treat the wastewater effluent so that it will meet the prevailing discharge quality regulations, it costs another 10¢ to 80¢. Therefore it costs from 35¢ to 105¢ to obtain and discharge 1000 gallons of water on a "once through" basis. In many instances, it is already more economical to reuse treated water than to discharge wastewater. Sea water desalting (60¢ to 100¢ per 1000 gallons) seems to be another potential economic alternative to the once through use of fresh water.

If organic wastes are not recycled into the soil but are burned, buried or dumped into the ocean, they create pollution. As an example of practices in 1972, five million tons of sewage sludge and six million tons of dredged material were still being dumped every year into the New York Bight twelve miles off of Sandy Hook. Jacques Cousteau, the underwater explorer, has said: "In 30 years of diving, I have seen this slow death everywhere under the water. If it continues, I predict that man has only another 50 years on this planet."

Some pollutants poison and kill plants and animal life directly. Others, such as oil, make so great a demand on the oxygen supply that marine life suffocates. Some pollutants, such as fertilizers and detergents, produce an excessive algae growth, and the decomposition of algal remains deprives the water of its life-supporting oxygen content. Waste is the wrong material in the wrong place at the wrong time in the wrong quantity.

DDT and other chlorinated hydrocarbons accumulate in those types of

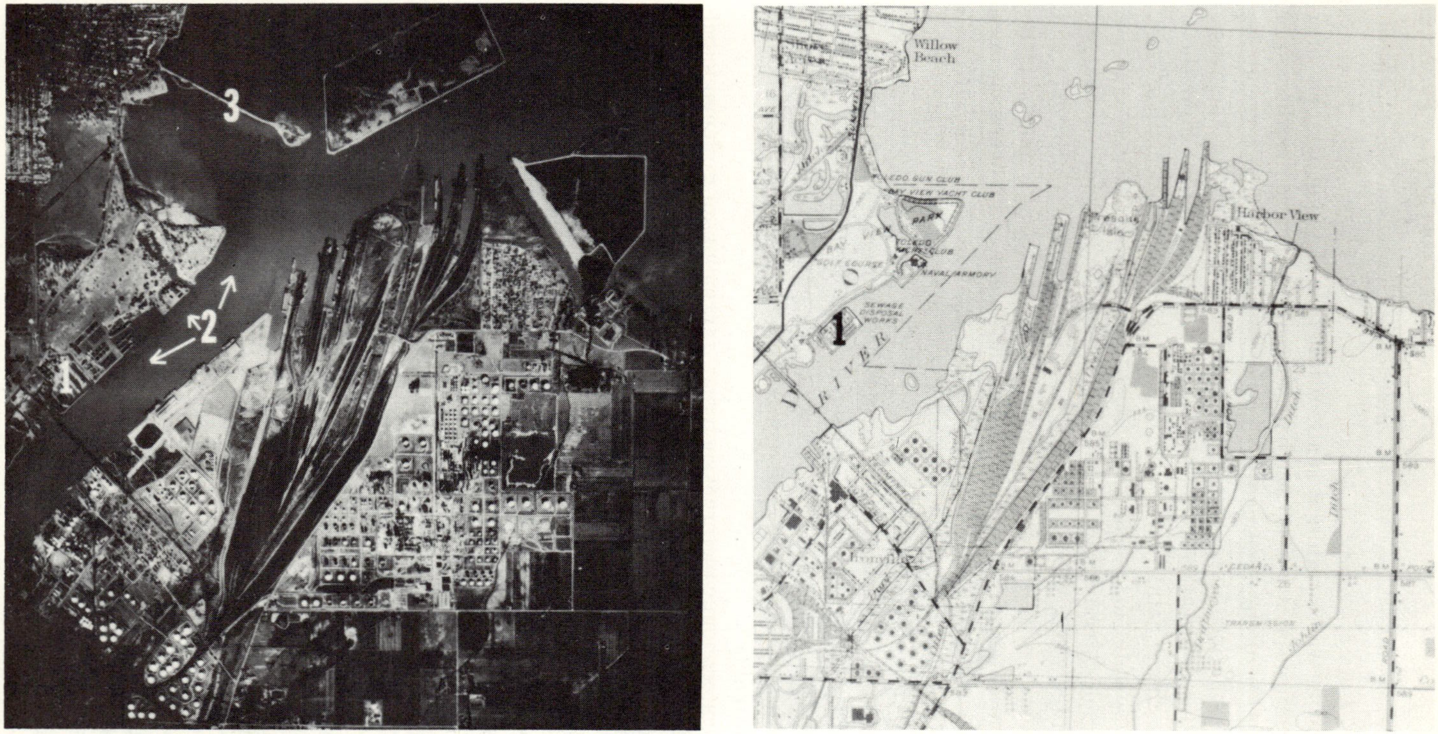

Figure 1 Mouth of Maumee River at Lake Erie, showing water and cultural features associated with water pollution.
(*Courtesy of NASA*)

marine life that have an affinity for them, and they eventually pass into the food chain with dangerous or lethal effects. The production of organic chemicals in the United States has increased in the last 25 years by 1,200%. In terms of biochemical oxygen demand, the amount of industrial wastes released in the United States is equivalent to the municipal wastes from a population of 400 million people. Potentially toxic chemicals are produced on a large scale, and traces of them enter the environment and the food supply. Fertilizers or sewage which could be food for the soil if carefully applied in limited quantities will become poison for the rivers and lakes if carelessly used. Some 10,000 miles of rivers in the United States are polluted by acid mine drainage.

The effort to clean up the rivers and lakes has been rather slow and ineffective to date. In the United States since 1965, the concept behind water legislation has been that the setting of water quality standards for lakes and rivers, along with the regulation of municipal and industrial discharges will in itself clean up the waterways. This concept was based on the estimated "assimilative capacity" of the lakes and streams—a kind of pollution-dilution—which would achieve the desired water quality. It has been found that this approach, largely administered by the states, is not effective. A more recent concept is that to clean up the nation's waters effectively, pollution must be controlled at the source through a permit program administered by the states under federal approval. The Water Pollution Control Act of 1972 requires all industries to use "the best practicable" control technology by 1977, and if they are unable to eliminate all discharges at a reasonable cost by 1983, they must be able to demonstrate that they are at least using "the best available technology." A "no discharge" goal is proposed for 1985.

This goal of zero discharge by 1985 is criticized by some representatives of industry who claim that the goal is both unrealistic and unattainable. They refer to the high costs, to the "unbalancing" of the uses of the nation's waters (they would prefer the "assimilative capacity" concept, because it allows for discharges into the yet unpolluted waterways). They also would prefer to maintain the primary responsibility for pollution control in the hands of local authorities, who are more intimately familiar with local conditions. Other opponents of the no discharge goal (by 1985) feel it would deceptively arouse expectations that could not be fulfilled.

In other aspects of water pollution, the progress has been even slower. The problems related to ground water pollution (due to the water leaching of solid waste landfills, strip mine sites and deep well disposal of wastes) are just beginning to be recognized. Damage due to oil pollution is still increasing.* In 1971 in the Long Island Sound region alone, 64 oil spills occurred. The offenders (when found) were billed for the cleanup cost only and not for the cost of the damage caused. The consequences of introducing 1.5 million tons

* The Coast Guard predicts that one out of every nine tankers will suffer an accident each year.

of oil every year into the oceans are largely unknown, but there are theories that it could increase the albedo of the earth, thereby possibly causing a new ice age.

As of this writing, the water pollution abatement effort is just beginning, and in most areas the levels of pollution are still increasing, but a serious effort is now underway to reverse this trend. In 1972, an important precedent was set when a federal judge invoked the 1899 Federal Refuse Act and, by ruling that sewage is refuse, ordered 15 communities along the New Jersey shoreline to stop pumping sewer sludge into the Atlantic.

Unfortunately, unregulated disposal beyond the three-mile limit continues and can imperil the waters, resources and beaches. Therefore, specific legislation is imperative if the resources of the oceans are to be protected.

The Quality of the Air

It has been predicted that by the year 2,000, the carbon dioxide content of the atmosphere will have increased by from 14 to 30%. Table 2 lists the yearly amounts of pollutants discharged into the atmosphere in the United States. Figure 3 is a visual presentation of the sources of pollution in the county of Los Angeles. It can be deduced that in spite of air pollution abatement efforts to date, the contamination of the atmosphere is still increasing.

Air pollution problems are at their worst during thermal inversions (Fig.

Table 2
ESTIMATED EMISSIONS OF AIR POLLUTANTS BY WEIGHT, NATIONWIDE, 1969
(In millions of tons per year)

Pollution Source	Pollutant						
	CO	Particulates	SO_x	HC	NO_x	Total	Change, % 1968–1969[a]
Transportation	111.5	0.8	1.1	19.8	11.2	144.4	−1.0
Fuel combustion in stationary sources	1.8	7.2	24.4	0.9	10.0	44.3	+2.5
Industrial processes	12.0	14.4	7.5	5.5	0.2	39.6	+7.3
Solid waste disposal	7.9	1.4	0.2	2.0	0.4	11.9	−1.0
Miscellaneous	18.2	11.4	0.2	9.2	2.0	41.0	+18.5
Total	151.4	35.2	33.4	37.4	23.8	281.2	+3.2
Change, % (1968–1969)	+1.3	+10.7	+5.7	+1.1	+4.8		

[a] Computed by the 1969 method from the difference between 1969 estimates and 1968 estimates. The new method results in higher values for 1968 than those computed by EPA for 1968.

SOURCE: The Mitre Corp. MTR-6013. Based on Environmental Protection Agency data.

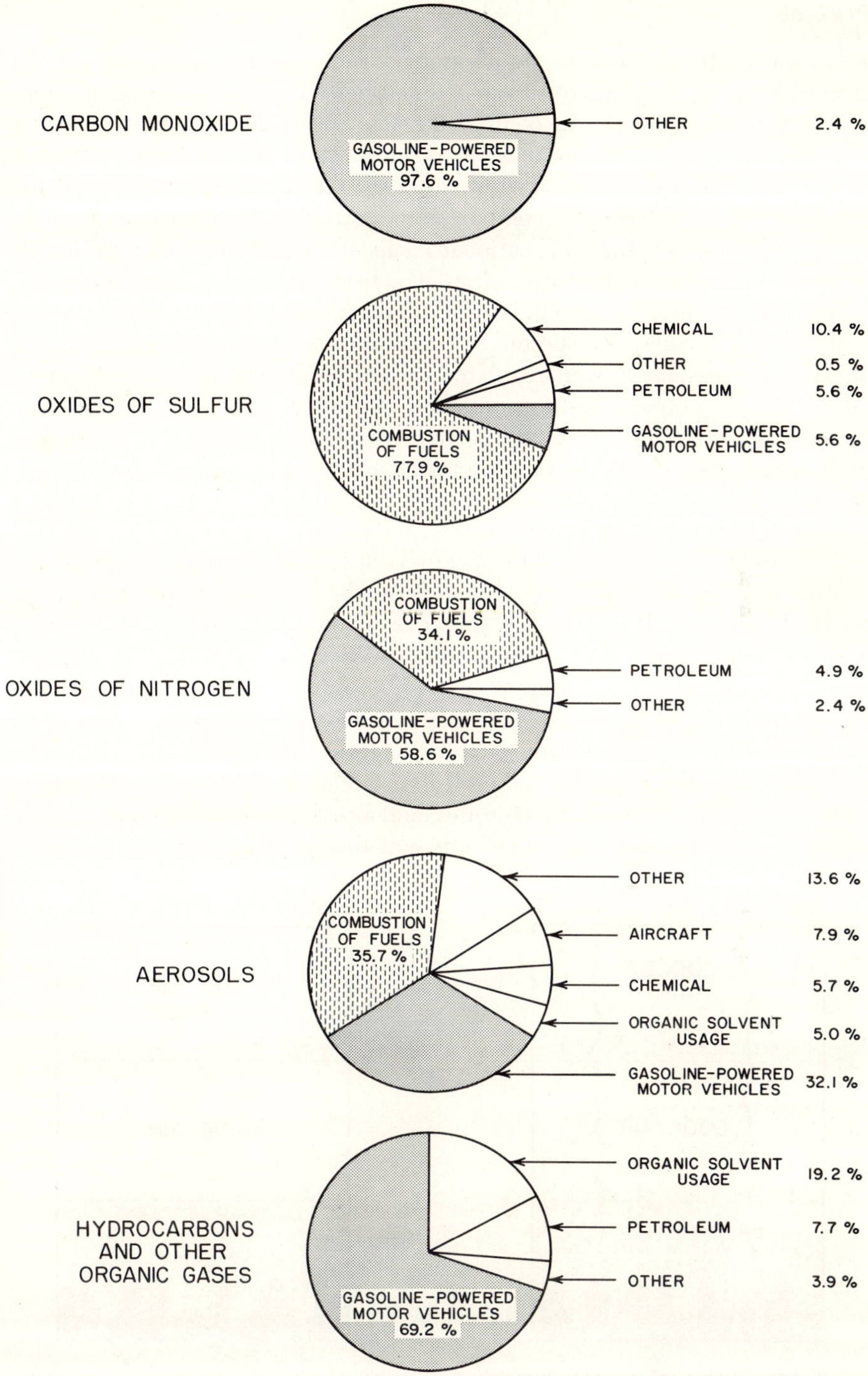

Figure 3 Percentage contributions of air contaminants from major sources in Los Angeles County. (*Bulletin Los Angeles County Air Pollution Control, District, January 1967*)

4). In the inverted situation, the denser cool air remains motionless on the ground, keeping the pollutants from being dispersed by normal air circulation. The lower the atmospheric ceiling, the less space the pollutants have to disperse into. Such a situation occurred in 1971 in Birmingham, Alabama, where the emergency provisions of the 1970 Clean Air Act were used for the first time in ordering 23 companies to restrict their operations.

It now appears that the national clean air requirements for 1975 will not be met in some of the larger cities. The best chance they have is compliance with the sulfur oxide standards (0.03 parts of sulfur dioxide per million parts of air), because the stationary combustion sources and their fuels are likely to be reasonably well regulated by that date. The recently proposed tax of 15¢ per pound of sulfur emitted is also likely to help in meeting this target. On carbon monoxide and particulates, drastic action will be needed in the large cities if the 1975 requirements are to be met. The federal target is 75 micrograms of particulate matter per cubic meter of air. The average reading in New York City was 98 in 1969, and it rose to 104 in 1971. Corrective action will have to be directed toward fuels and transportation. The pollution control problem is further complicated by the shortage of natural gas (a relatively clean fuel) and by the presence of used and foreign cars (which will not be equipped with pollution control devices by 1975). Yet another problem is that as of 1972, some states had not submitted for federal approval their plans for meeting the 1975 air quality standards, while other plans from states who had submitted them were rejected as insufficient.

The national air quality and emission standards now cover particulates, carbon monoxide, the oxides of sulfur and nitrogen, hydrocarbons, photochemical oxidants, asbestos, beryllium and mercury. In addition, for new

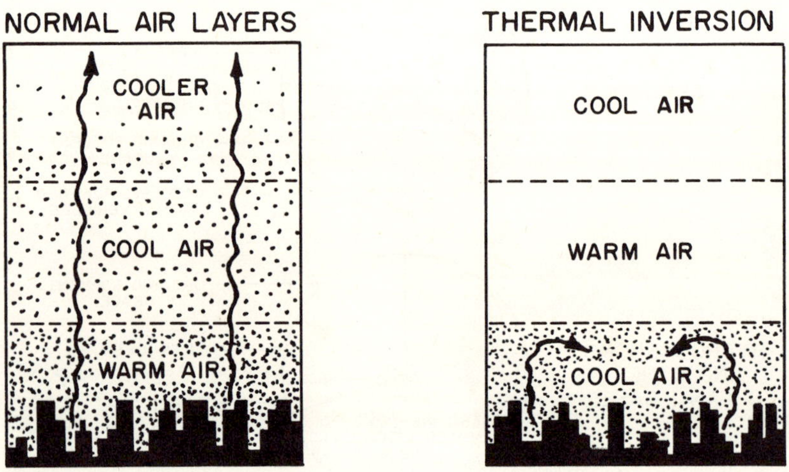

Figure 4 Thermal inversion stops air circulation.

plants, there are national air pollution performance standards. The categories of plants for which limits on the emission of pollutants were set are fossil-fueled steam-generating plants, sulfuric and nitric acid manufacturers, portland cement plants and large incinerators.

The pollutants to be controlled are particulate matter (smoke and soot), sulfur dioxide, nitrogen oxides and sulfuric acid mist. Similar standards are being developed for some 40 other industries.

Some 75 new steam generators are built annually in the United States, and 96% of them are used by the electric industry. The emission standards for such new plants are summarized in Table 5.

Table 5
EMISSION IMPROVEMENT REQUIRED AND COSTS RESULTING FROM PERFORMANCE STANDARDS[d]

Source category	Emissions		Cost of Added Control, %		Affected size
	Existing installations	NSPS[b]	Investment	Operating	
Fossil Fuel-Fired	(lb/10⁶ Btu)	(lb/10⁶Btu)			Over 250 million
Steam Generators					Btu/hr heat
Particulates	0.04–4.0	0.1	3[a]	0.01¢/kW-hr[a]	input
Sulfur dioxide					
liquid fuel	3.0 (av.)	0.8	10[b]	7–30[b]	
solid fuel	5.0 (av.)	1.2	12[a]	0.12¢/kW-hr[a]	
Nitrogen oxides					
gaseous fuel	0.4 (av.)	0.2	5[b]	4[b]	
liquid fuel	0.7 (av.)	0.3	7[b]	4[b]	
solid fuel	1.4 (av.)	0.7	1[a]	0.01¢/kW-hr[a]	
Visible emissions		#1 Ringelmann or 20% opacity			
Incinerators	(gr/SCF)	(gr/SCF)	9–15[b]	[c]	Over 50 tons/ day charging rate
Particulates	1.0	0.08			
Portland Cement			12[b]	5–7[b]	
Particulates	(lb/ton feed)	(lb/ton feed)			None
kilns	45	0.3			
clinker coolers	30	0.1			
Visible emissions					
kilns		#½ Ringelmann or 10% opacity			
others		<10% opacity			
Nitric Acid	(lb/ton acid)	(lb/ton acid)	4–8[b]	0–2[b]	None
Nitrogen oxides	43	3.0			
Visible emissions		<10% opacity			
Sulfuric Acid	(lb/ton acid)	(lb/ton acid)	14–38[a]	0	None
Sulfur dioxide	22–85	4			
Acid mist	0.4–0.9	0.15			
Visible emissions		<10% opacity			

[a] Costs from Federal Register (37 FR 55), March 21, 1972.

[b] Costs from Background Information for Proposed New Source Performance Standards (NSPS).

[c] For refractory wall, $29,000/year for 100-ton unit; $65,000/year for 300-ton unit. For water wall, $13,000/ year for 100-ton unit; $23,000/year for 300-ton unit.

[d] R. E. Jenkins, G. A. McCutchen, "New Source Performance Standard." *Environmental Science and Technology*, *6*(10):884, October 1972.

The air pollution control effort has been the most successful in developing quality and emission standards and in establishing target dates and enforcement procedures. This is partially due to the nonbiological nature of air (land and water are complex biosystems), which is most compatible with present technology. Yet as of this writing, the contamination of the air in heavily populated industrial areas is still increasing (Fig. 6).

The Condition of the Land

The third medium through which pollution is transmitted is the land itself. The largest pollutant of the land is solid waste: 365 million tons of waste are generated annually in the United States, and the present cost of disposing of it without recovery is from 5 to 6 billion dollars. Solid waste pollution abatement has received very little attention to date. Because this is a big city problem and large cities seem to have less and less political influence and financial resources, the likelihood of progress is not good.

There are very few national standards or regulations dealing with solid waste collection, transportation, processing or disposal. This is particularly unfortunate as solid waste holds the greatest potential for the recovery and utilization of resources. Without the recovery of its material and power content, the handling and the disposal of solid wastes is very expensive and a source of other forms of pollution. A concentrated effort toward recovering the valuable constituents would not only pay for the collection costs, but could reduce the overall need for raw materials. At the present rate of usage, the reserves of some important metals will be exhausted in 50 years (Table Vol. III, 6.1h). The technology for the recovery of valuable materials is already available; some of it is in daily use in the mining and minerals industries. All that is needed to solve problems of solid waste pollution is the use of existing technology.

The other forms of land pollution have been similarly neglected. For example, there is no national regulation of the timber industry, which follows the wasteful and land-damaging practice of clear cutting (cutting down all trees regardless of age) for reasons of efficiency. An even worst violator of the land is strip mining. As the demand for metals and coal soared and the grade of available raw materials declined, the only way to reduce costs was to increase the scale of the mining operations. The result was that mining came above ground, and today 90% of metal mining is done by the strip mining technique. In these giant-sized operations, the fertile topsoil is mixed with the lower layers of overburden (the layers covering the coal or mineral), and even if the topsoil is later returned to the open pits, it can no longer support plant life.

As of this writing, there are no nationwide environmental controls on strip mining. Some large coal users, such as the Tennessee Valley Authority, have developed their own policy in this area. The TVA's policy includes the

Figure 6 Smog encroachment at the base of the San Gabriel Mountains. (*Courtesy of Professor Leonard W. Bowden, Department of Geography, University of California, Riverside*) A, 5:15 P.M., February 29, 1968—smog from Los Angeles moves inland; *B*, 5:25 P.M., same day—smog advancing at the rate of one mile per minute.

Figure 6 (*continued*) *C*, 6:00 P.M., same day—combined haze and smog engulfed Riverside; *D*, 12:15 P.M., same day—atmospheric conditions before smog moved in.

withholding of 2% of the contract payments until reclamation is completed (reclamation does not include the restoration of fertility). This 2% figure appears to be insufficient motivation because, according to the president of one coal company, it would cost $6000 an acre to restore the land to its original fertile state.

There has been some progress in establishing national standards relative to noise pollution, but their enforcement is yet to take place. The first statewide noise control law was enacted by the state of New Jersey. The Department of Environmental Protection can now levy fines of up to $3000 for each violation. Previously excessive noise was a local violation of disturbing the peace with an associated fine of about $25.

There is no national regulation of the nonaesthetic or glare pollution caused by many forms of advertising and other eyesores. In some areas, due to the changing employment market, the population has become so mobile that there is no time to develop a sense of community pride (this also contributes to land pollution).

The Causes and the Remedies

Population and Its Distribution

The world population is nearing 4 billion and increasing at a yearly rate of 2% (Fig. 7). The present population is about the most that can be fed equitably without forced agricultural fertilizing. The world's food production between 1951 and 1966 increased by 33%, while in the same period, the worldwide use of fertilizers and pesticides increased 146% and 300% respectively. There is danger in this tendency toward transforming the ecosystem into a world consisting solely of man and a chemically sustained food factory. Assuming that the populations of the developed nations of the world were stabilized by the year 2000 and those of the other countries by 2040—according to some British scientists—the world population would still exceed 15 billion.

The distribution of the population has also changed because of industrialization and the better economic opportunities in big cities (Fig. 8). The drive to congregate in the urban areas has resulted in 70% of the people living on 2% of the land in the United States today, and the trend is continuing toward the complete abandonment of spacious rural living.

It seems that society as a whole has a responsibility in reversing this trend. By the use of zoning, taxation and good mass transportation, society could effect a more natural distribution of the population. It is important that the coalescing urban masses be reduced and that the cities themselves be surrounded by green belts. This would not only bring the food supplies and sport and recreational facilities closer to the city, it would also reduce air pollution,

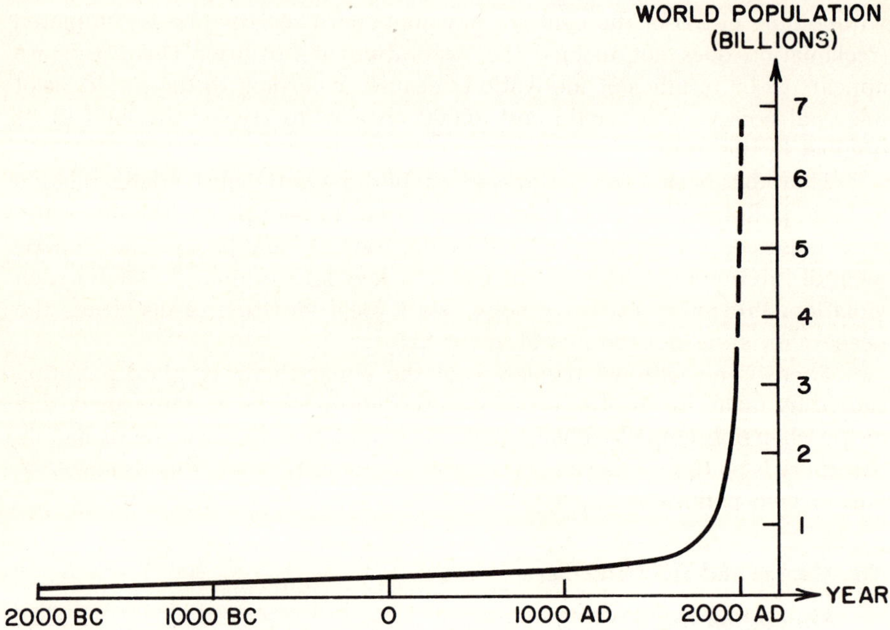

Figure 7 Growth of world population. *(U.S. Department of Agriculture projections)*

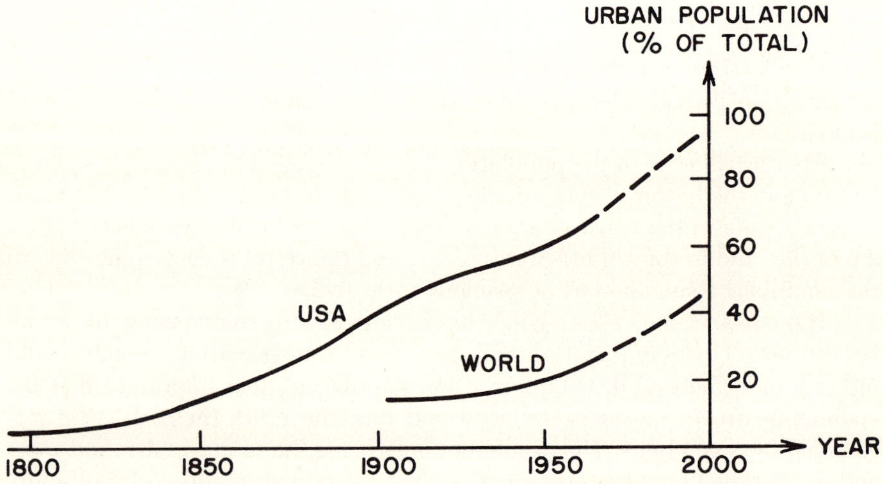

Figure 8 Increase in percentage of population living in urban area. *(Encyclopaedia Britannica)*

for plants and the soil itself are the most natural and efficient cleansers of pollution.

In addition to controlling and stabilizing the population, it would appear desirable to utilize the presently available facilities (cultural, recreational, transportation, sport and others) at a better overall efficiency in the densely populated areas. This could be accomplished by staggering the workhours and workweeks so that everybody will not be on the road attempting to be at work or the ball games at the same time.

Such basic changes cannot be made without a complete national commitment; they will require the development of long range national and international plans to unify and coordinate the many fragmented and as of yet ineffective local efforts.

The Culture and Value System

It is a widespread impression that the main causes of pollution are affluence and the population explosion. Statistics indicate that, while the gross national product of the United States has increased since the end of the Second World War by only 100%, the population by only 50% and the per capita consumption by only 25%, pollution has increased by 2,000%. This seems to imply that there are other major contributors to the environmental crisis besides the increases in population and affluence. These other contributors include our careless technology and our outdated concept of progress.

The Runaway Technology

The first machines were applied to replace the human muscle. Then others were developed to substitute the routine functions of man's mind. It is important to prevent the evolution of a society in which machines can influence man's decisions and life style. Technology itself is neither humane nor inhumane, but its use can be either. Technology and production can be the great benefactors of man, but they are at the same time mindless instruments. If undirected, they will roll along with a momentum of their own and will pulverize all that is in their path, including the landscape and the natural environment.

Historically, technology has been developed to conquer, regulate and control nature. As such, it has a tendency to use nature as a commodity to be exploited and turned into profit.

Past generations of engineers were taught to function as human components in production oriented organizations. The next generation of engineers will learn that all of us are responsible for the world we live in. The new generation of environmental engineers will have an education enabling them to make use of technology, to control it, to give it direction and to cause it to serve those values which society has chosen.

The Concept of Progress

There is in our culture a deeply rooted belief in continuous growth. There is also a fear of a steady state economy because, according to our concepts, that would mean deadly stagnation. According to this value system, progress can be defined in a one-dimensional, materialistic manner as the continuous expansion of a production oriented economy. Such a definition of progress also implies the need for a continuously increasing per capita usage of raw materials and energy.*

If progress requires an economy that is continuously expanding—without at the same time making the domestic poor or the underdeveloped nations into users of the goods produced—then it also requires the creation of artificial markets for its goods. These artificial markets may include not only wars and space exploration but also wasteful production (wasteful both in the sense that the products are designed to last only for a short time and in the sense that the products are designed to be thrown away without the possibility of being repaired or traded in). This outdated concept of progress contributes to the cultivation of the prevailing disposable and wasteful mentality which accepts a low quality of life in order to have a "high standard of living."

There are many nonmaterial and nondestructive paths of growth. Man must be the master of the technology he has developed. The fact that faster planes or space shuttle systems can be built does not automatically mean that they should be put into service.

It would seem desirable to concentrate not only on the improvement of society but also on the improvement of man and his quality of life. In some ways, it is more important that work be a form of self-expression and a source of satisfaction than it is that work be efficient quantitative production. Excessive technology can also contribute to unemployment. In the United States, for example, it already takes $30,000 in machinery and chemicals to support the job of one farm worker, and this figure is continuously rising.

In facing up to and solving the present environmental crisis, we must not regard it as just another challenge to be overcome by old methods but must realize that our concepts and values will have to change so that thoughtful, humane and measured steps can be taken toward a stable society.

Power Consumption

While the world population presently doubles every 30 years, the doubling period for energy consumption is 15 years, for electric energy use 10 years** and for the consumption of uranium only 3 years. The energy consump-

*The United States presently imports one-third of its oil or iron consumption and 90% of its aluminum needs.

**This can not continue for long, because if one assumes this rate of increase to continue for one or two hundred years, the corresponding electric power usage by the end of these periods will have increased 2,000 and 2,000,000 fold respectively.

tion of the world is growing at an even faster rate than does its population. The slopes of the curves in Figure 9 are so steep that one needs to use a logarithmic scale for the illustration. The rate at which the world energy usage is increasing exceeds that of the United States, and by the early part of the next century, the amount of energy released by man will equal the total energy supply of the plant and animal kingdoms on earth. The amount of energy released by processes on earth in some urban and industrial areas is already equivalent to the energy reaching the earth from the sun ($\simeq$100 watts per square meter). The consequences of such energy releases upon the climate are not yet well understood.

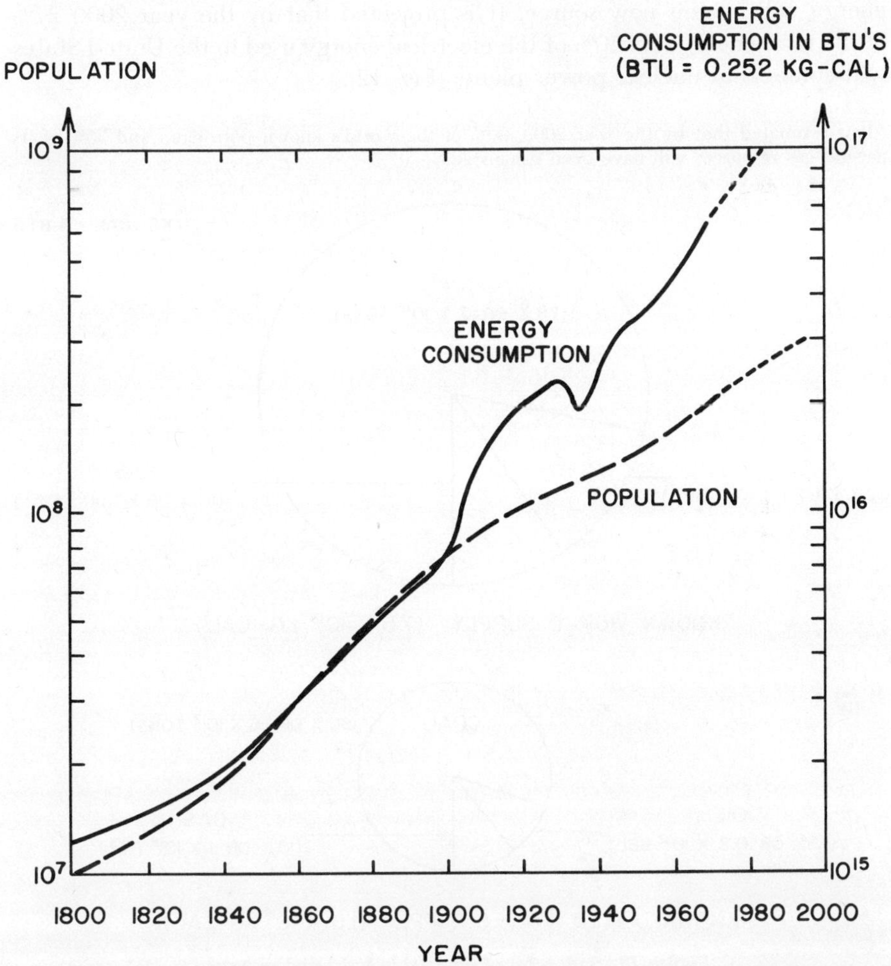

Figure 9 Growth of population and energy consumption in the United States.

Depletable Energy Sources

A recent study of southern California has concluded that that area can double its electric power production just one more time without deadly consequences to marine and human life because the air and water could not absorb the heat produced by a greater increase.

Excepting the small contribution of hydraulic energy, practically all the energy used today comes from the exhaustable fossil sources (coal, oil and gas). The total fossil deposits of the world amount to about 7.5×10^{18} kg-cal (Fig. 10). This amount of energy, considering the exponentially increasing rate of usage, is sufficient to supply the world for less than the life span of one man* (Table 11). Therefore mankind, for the first time in its history, is faced with the decision of selecting a new energy source. It seems now that nuclear energy will be this new source. It is projected that by the year 2000 27% of the total energy and 50% of the electrical energy used in the United States will come from nuclear power plants (Fig. 12).

*It is estimated that by the year 2000, 90% of the world's known petroleum and 75% of its natural gas resources will have been exhausted.

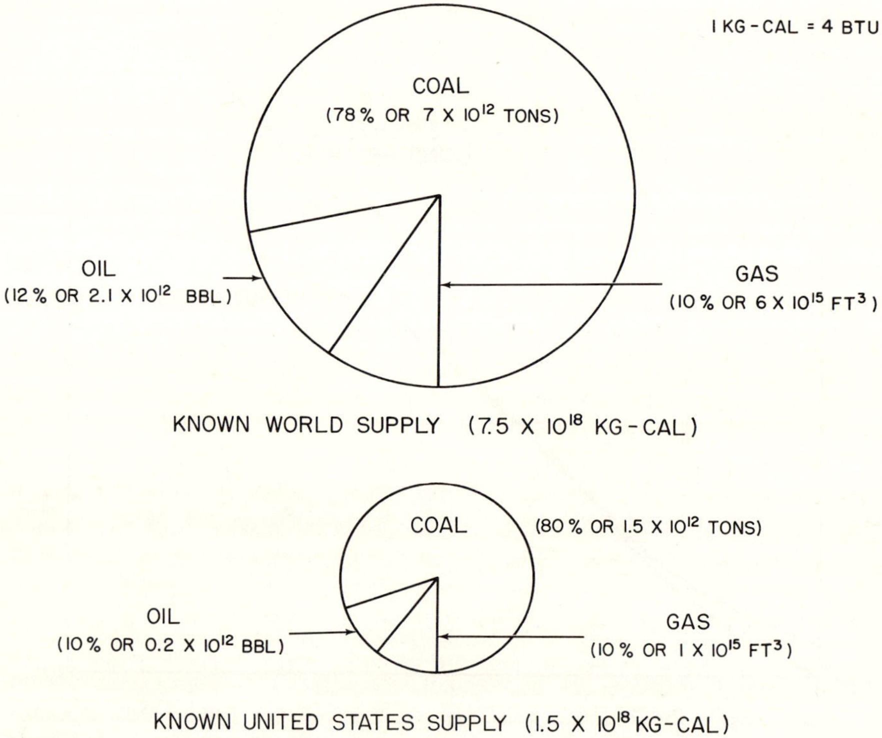

Figure 10 Sum total of depletable fossil fuel reserves.

Table 11
DEPLETABLE ENERGY SUPPLIES OF THE WORLD

Type of Energy Reserve	*Sufficient for the Period of (years)*[a]
Fossil Fuels (Coal, oil, gas)	50
Nuclear Fission (Conventional Reactor)	150
Nuclear Fission (Breeder Reactor)	5–15,000
Nuclear Fusion (Thermonuclear)	Practically unlimited

[a] These numbers were arrived at by using the estimated world energy consumption in the year 2000 (15×10^{16} kg-cal/year) as the unit. In other words, the noted sources could be the sole source of energy for the noted number of years if consumed at the rate estimated for the year 2000.

The trend in this direction was not initiated by the judgment of a well-informed public. Those who question the wisdom of this trend point to the safety and pollution aspects of nuclear power generating stations and to the fact that nuclear fission is also a depletable energy source and therefore does not represent a permanent solution. The only safe, nonpolluting and inexhaustable alternative to nuclear energy is the direct conversion of solar energy.

In the immediate future, the trend will be toward a reduced use of coal and an increased utilization of oil and gas. The United States is an exporter of coal and an importer of both natural gas and oil. Partially for this reason and partially because of the desirability of using less polluting fuels (liquified petroleum gas and natural gas) in the large cities, it appears desirable to perfect the technology of the gasification of coal. The domestic gas demand in the next 25 years will more than double, and some cities will not be able to meet the federal air quality standards unless their gas supply is increased. Yet according to the American Petroleum Institute, the proven gas reserves of the United States are sufficient for less than 20 years. Therefore the power supply situation in some locations is much worse than what the overall worldwide energy balance would imply. In some areas, there already is the danger of a power crisis being used to overrule and create a blackout on environmental considerations.

Nuclear Energy. In 1970 there were 109 nuclear power reactors in 15 countries of the world, 27 of them in the United States. By 1976 the United States will have 107 nuclear reactors in operation (Fig. 13). All but a few of the new reactors planned for 1976 will be the conventional water-cooled fission type reactor in which uranium 235 undergoes a division of the atomic nucleus accompanied by the release of one or more neutrons and large amounts of energy. The efficiency of these nonbreeding reactors is below 2.5%. They use low grade uranium at such a rate that its world supply would be exhausted in a century, and they are a major source of thermal pollution (Fig. 14). There is also a problem with radioactive waste disposal. Man has not lived with

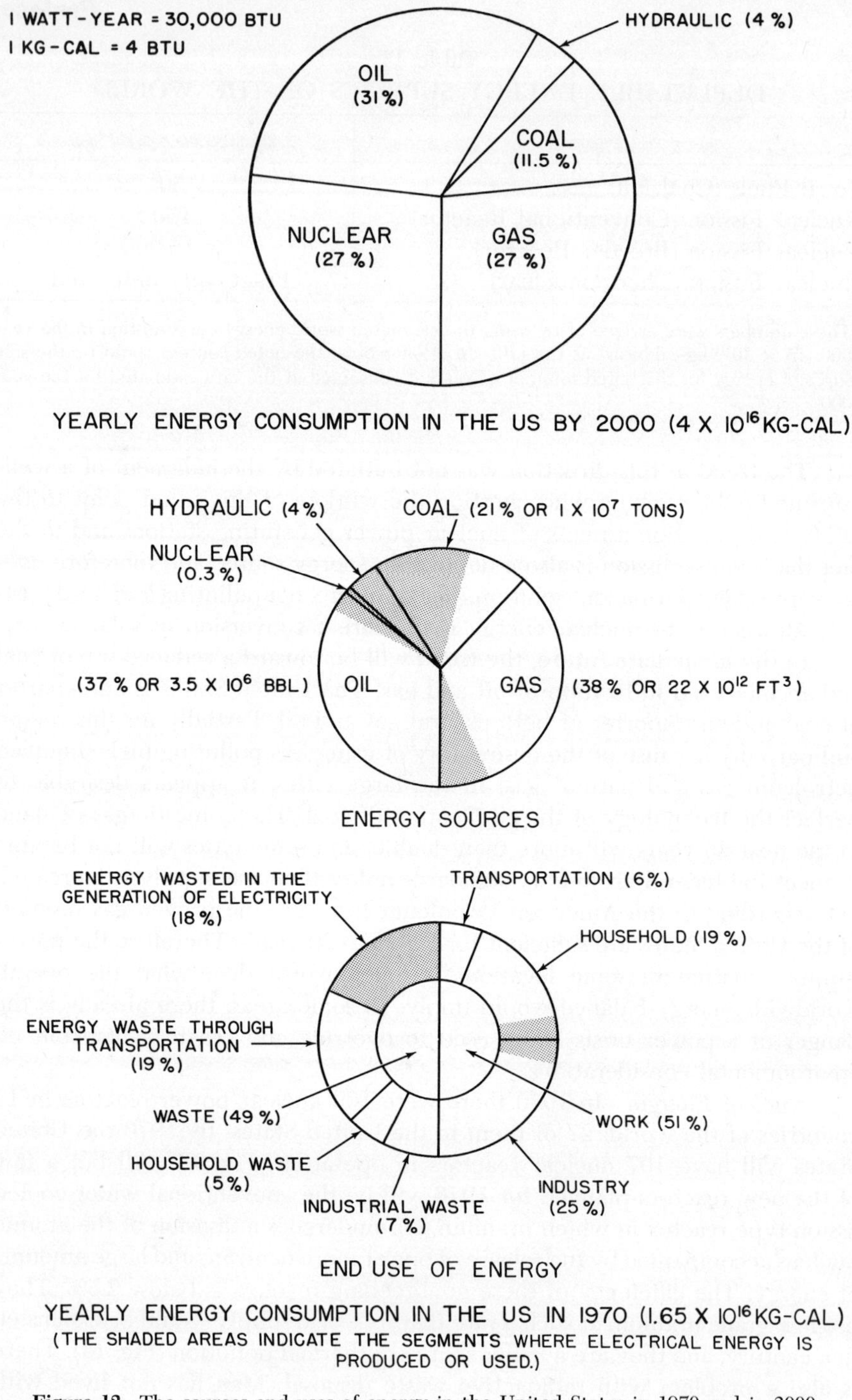

Figure 12 The sources and uses of energy in the United States in 1970 and in 2000.

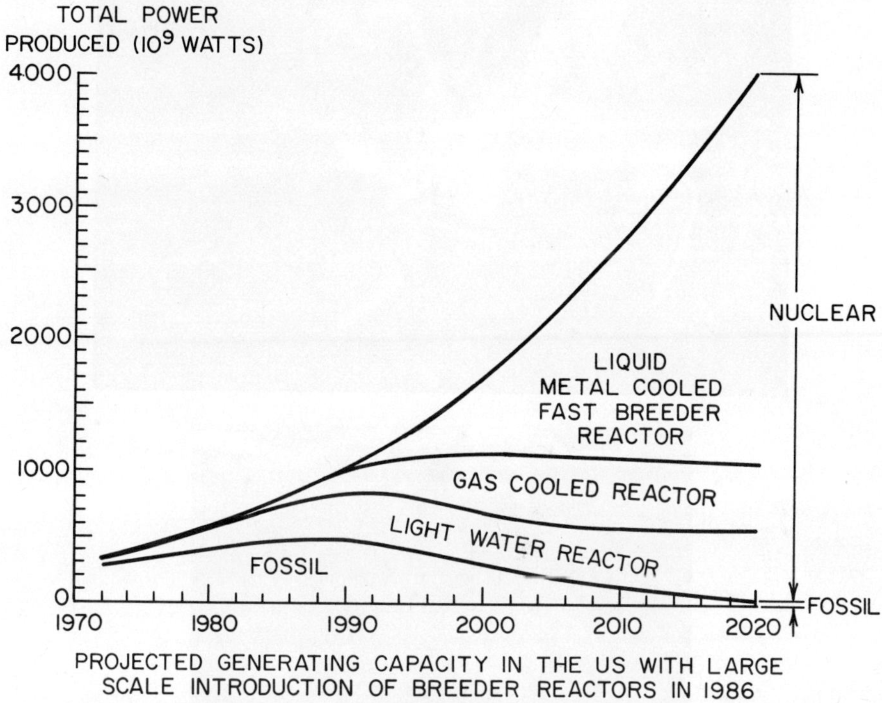

Figure 13 Projections on nuclear power production in the United States.

radiation long enough to know if storing up to a million cubic feet of nuclear plant wastes yearly in various burial sites is safe. There is some danger involved in the safeguarding of this highly toxic material for a 1000 years or longer. The strontium 90 alone accumulated in the United States by 1980 would contaminate the fresh waters of the world to a level 10,000 times exceeding the limits of standards for drinking water.

In the United States, environmental reviews are required before a nuclear power plant is given its operating license. This recently instituted practice along with the public resistance to accept nuclear power plants has resulted in halting the construction of some 68 nuclear plants. As of this writing, few of the environmental impact statements were yet approved, and only two of the construction projects were restarted or given operating licenses.

By locating these units offshore in the ocean some of the objections relative to safety and thermal pollution could be reduced and the energy generated could also be used for desalinating the sea water in areas where the fresh water supply is limited.

Nonbreeding reactors are provided with an emergency safety control system which floods the core with water if the regular cooling water is lost.

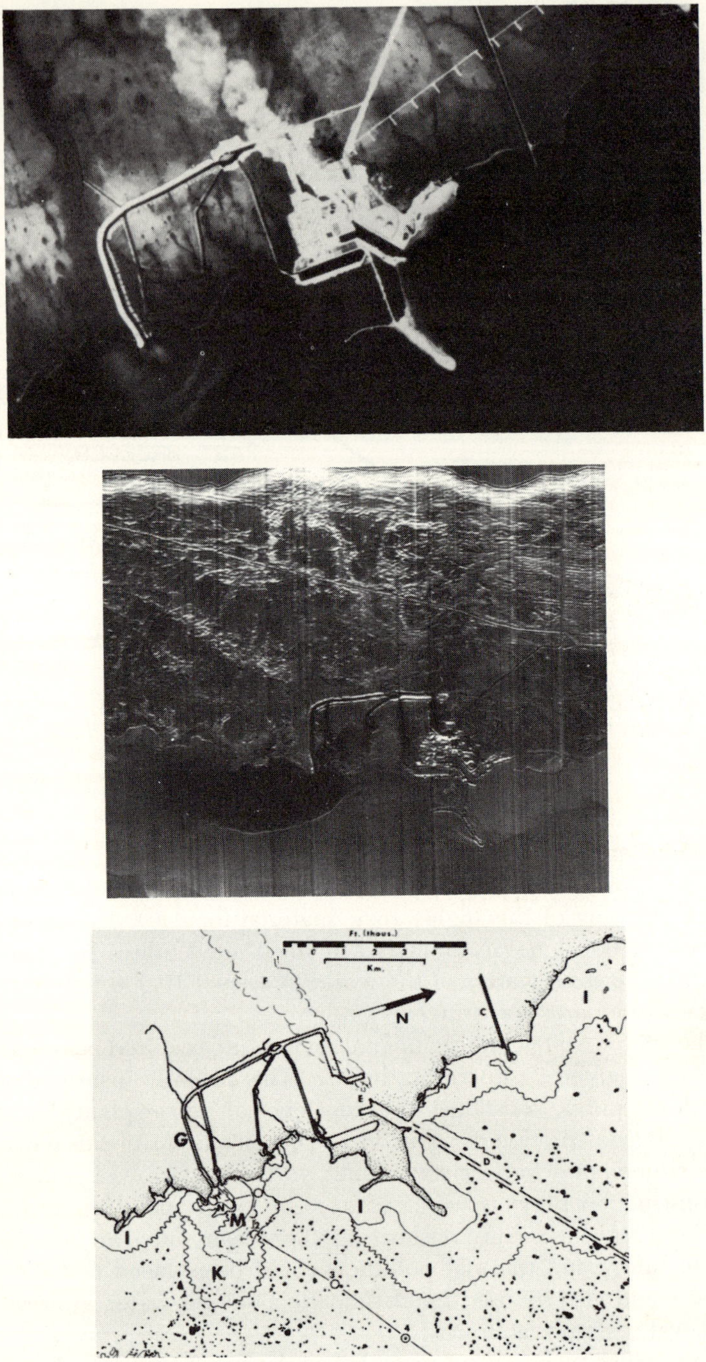

Figure 14 Destruction of bottom biota by thermal pollution. Concept: Use of aerial photography and thermal IR imagery to trace extent of thermal pollution. Test site: Turkey Point power plant, Biscayne Bay, Florida. Top: Color photo. Middle: Thermal IR scan. Dark area is 7°C warmer than ambient water. (Bottom) Surface truth analysis. M—visible inner area of bottom damage; K—outer zone of bottom damage; I—coastal grass zone; J—mixed bottom cover zone.

(Courtesy of NASA)

Otherwise the reactor core would quickly melt and release a fallout cloud of radioactivity. Unfortunately, this emergency system has not yet been tested. Such tests are not scheduled until 1974–1975, and until these tests are completed, the effectiveness of the safety systems will not be known.

The Breeder Reactor. Breeder reactors utilize surplus neutrons from the fission of uranium 235 to create other nuclear fuels such as plutonium 239 and uranium 233. They produce a neutron radiation sufficiently intense and energetic to convert uranium 238, which cannot be used as a reactor fuel, into plutonium, which can, and thus they manufacture more fuel than they consume. Breeder reactors can be fueled with a mixture of fissionable plutonium 239 and the abundant isotope of uranium (uranium 238, which makes up 99 percent of all the natural uranium deposits) with a small admixture of uranium 235. This process opens the prospect of making full use of the world's nuclear fuel resources.

In a breeder reactor, 1 gram of uranium releases the same energy as would 2.7 tons of coal or 1.9 tons of oil in a fossil fuel power plant. The experimental breeder reactors are cooled mostly by liquid sodium, although gas cooling is also being considered and does appear to have fewer technical problems.

Construction of the first large commercial fast breeder reactor has been completed in the Soviet Union. The French (jointly with the Germans) and the British are both planning to start up their units in 1973. The first breeder reactor in the United States is planned for 1980. American and Japanese units will use the "loop" design, the British, French and German reactors will be of the "pot" design with four times the sodium in the system. The Soviet Union has both but is shifting toward the "pot" concept.

Breeder reactors are much more efficient than their conventional counterparts. Their use can prolong the world's uranium 235 supply for several thousand years. In a breeder reactor, it is estimated that after 10 years of operation the atomic core will have produced 120 percent more plutonium than was loaded into it. This plutonium could be extracted and used to fuel other nuclear power plants.

Breeder reactors produce less thermal pollution than do the nonbreeder types. They create no air pollution at all. On the other hand, plutonium is the most toxic material known to man, (Section III.5.8) and to make this material the essential basis of fulfilling the world's energy needs for the next several thousand years would be understandable only if there were no alternatives available. It appears that, by the inertia of prevailing technological trends (the technology to build nuclear reactors does exist, while the technical and economic details of the use of solar energy have not yet been resolved), an all-out commitment in favor of breeder reactors has been made which has shortchanged the search for alternative power sources.

Fusion Reactors. Fusion involves the combining of light nuclei into larger ones, the various forms, for example, of heavy hydrogen (deuterium and tritium) into helium. Large quantities of energy are released in this process;

thermonuclear fusion occurs in the sun and also in the hydrogen bomb. Control of the fusion process involves many scientific phenomena that are not yet fully understood, and its engineering feasibility has not yet been seriously studied. In order for thermonuclear fusion to occur, a temperature of 100,000,000 °F has to be reached by the use of an intense magnetic field (60,000 times that of earth's) which squeezes and heats the reacting gas in the center of a tube. Depending on the process used, controlled fusion might open up not only an important, additional energy resource but also a virtually unlimited one, since its fuel is the abundantly available hydrogen.

The fusion process is inherently safer than the fission process because it does not produce radioactive wastes (the tritium can be completely reused and recycled). Given the advantages of this process, it would appear justified to increase research efforts in this area.

Nondepletable Energy Sources

The continuously available energy sources include geothermal, tidal, wind, solar radiation and hydropower sources. Out of these, only the hydroelectric power generating stations are used today. One tidal power plant is in operation in France, and there are some small users of geothermal (Fig. 15) and wind energy around the world. The total energy received by solar radiation *in one year* is nearly 200 times the total depletable fossil fuel supply of the world. Of this radiation, 30% is reflected back into space by the atmosphere, 47% is absorbed by the atmosphere and 23% is consumed in the hydrologic cycle. Only one five-thousandths of the total solar energy is captured by the chlorophyll in plant leaves to provide the plant and animal kingdoms with energy (Fig. 16).

If the sunlight that falls on only 1% of the land area of the United States (or the world) could be converted to electricity, it could satisfy the total energy demand projected for the year 2,000. The solar conversion cells could be located in unused desert areas or the moon's surface. They would create no pollution, and their energy supply would be inexhaustible. The land area required for the commercially significant collection of solar radiation is large necessitating a high capital investment. About 3,000 square miles would be required to match the present electric power generating capacity of the United States. (Paved roads cover 15,000 square miles.) Coupled with the cost of the necessary energy conversion systems and storage facilities, this makes solar power very expensive, at the present. On the other hand, the direct conversion of solar energy is the only significant long-range and inherently safe alternative to the use of nuclear power, and as such it would justify the initiation of the largest research and development effort ever mounted.*

*The 1974 federal budget for all types of energy research is 800 million. 75% of this sum is for nuclear R & D. In 1972 the federal funding for fusion research was 28 million, for synthetic fuel development, 20 million, and for terrestrial solar research, 2 million. This is not sufficient because the results of the Manhattan and Apollo projects prove that high levels of funding are required to reduce the development periods of major new technologies.

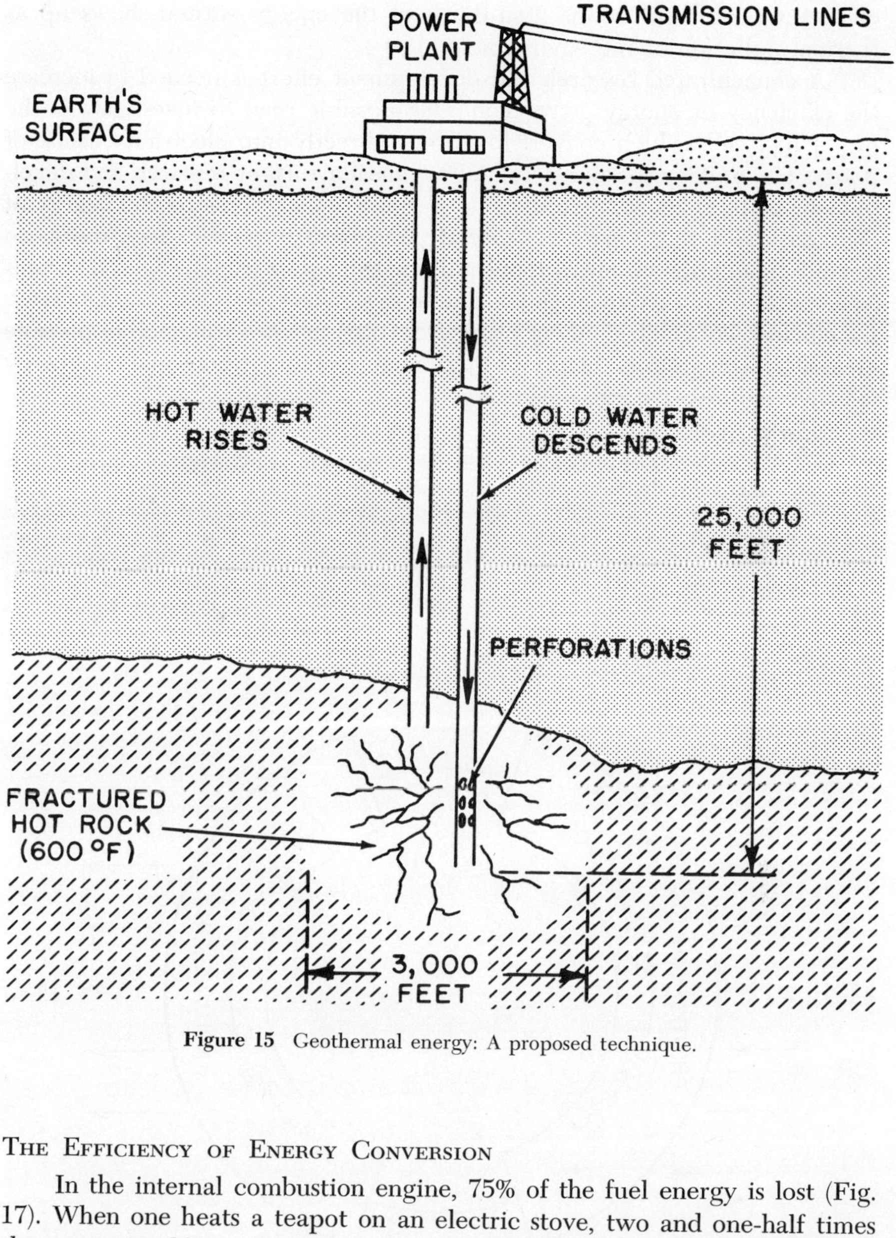

Figure 15 Geothermal energy: A proposed technique.

The Efficiency of Energy Conversion

In the internal combustion engine, 75% of the fuel energy is lost (Fig. 17). When one heats a teapot on an electric stove, two and one-half times the energy used by a gas stove is required. The wasteful and inefficient nature of various energy conversion systems is illustrated in Figure 12. One way of reducing the total energy demands of society is to increase the efficiency at which energy is being converted into useful forms such as electricity. In the most modern fossil fuel-operated electric generating stations, the conversion efficiency is still only 33%. Two-thirds of the energy wasted shows up as thermal pollution of the environment.*

A concentrated research and development effort is needed to increase the efficiency of energy conversion. One possible road to investigate is the use of fuel cells which convert fossil fuels directly into electricity inside of porous electrodes. Another potential avenue for further research and development is the magnetohydrodynamic turbine, which converts the energy of charged liquid or gas particles directly into electric power. Such units can operate in a closed circuit without polluting the environment. One such unit is in operation in the Soviet Union, and its designers feel that it could increase the efficiency of electric power plants from 33% to 50%. Still another possible area of investigation is the use of thermionic reactors, where nuclear energy is converted directly into electricity.

*Presently for each unit of electricity generated two units of heat energy are disposed of in the form of thermal pollution. If utilities marketed both electricity and heat (as hot water and steam) in a total energy conservation scheme, both thermal pollution and the heat island effect on cities could be eliminated. Steam and hot water can serve the energy needs of many industries in addition to municipal space and water heating. These heat users presently consume 30% of the total energy use in the U.S.A., which is four times the total energy consumed as electricity. (Hot water can be stored in confined underground aquifiers for several months without high costs or loss of heat content and can be withdrawn as needed to match the seasonal demand variations).

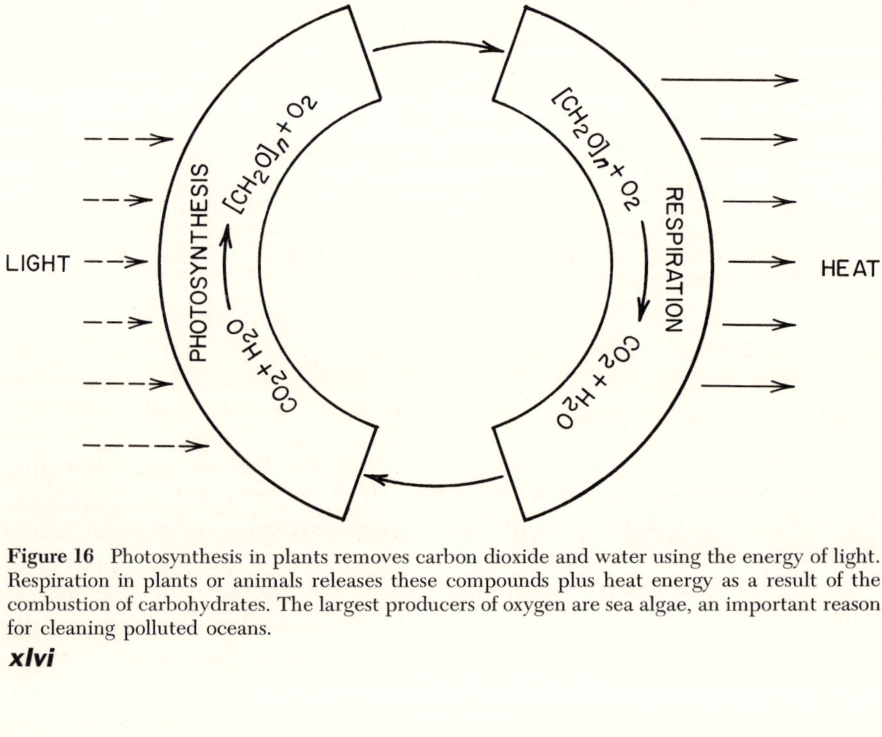

Figure 16 Photosynthesis in plants removes carbon dioxide and water using the energy of light. Respiration in plants or animals releases these compounds plus heat energy as a result of the combustion of carbohydrates. The largest producers of oxygen are sea algae, an important reason for cleaning polluted oceans.

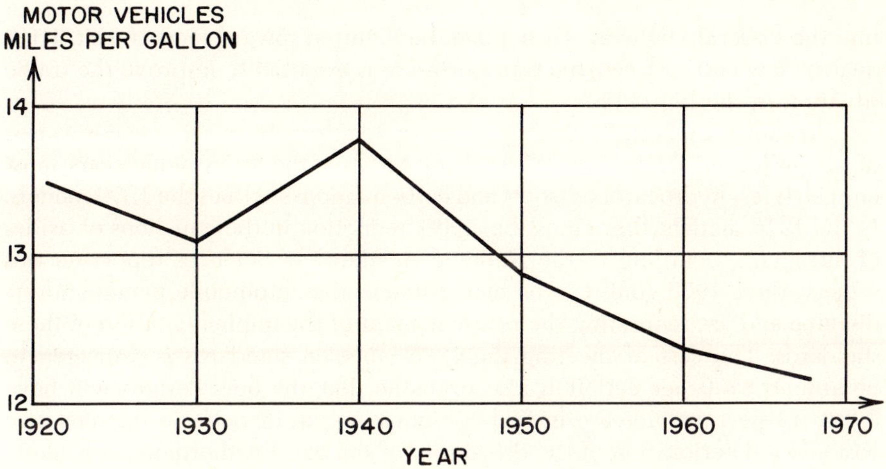

Figure 17 Decreased efficiency of internal combustion engine.

WASTEFUL USE OF ENERGY

It would seem desirable to curtail the ever-increasing per capita use of energy. Besides the obvious method of increasing the cost of commercial energy, this can be accomplished through a better awareness of the "energy content" of various commercial products. For example, it takes six times the energy to produce a pound of aluminum as it takes to make a pound of steel. In the area of housing and office buildings, constructions which necessitate simultaneous heating and cooling should be avoided. Generally one should avoid the substitution of aluminum for steel, plastics for wood products, detergents for soap, synthetic materials for cotton and wool and throw-away bottles and cans for returnable bottles. A higher level of consciousness regarding power problems could substantially reduce the rate of energy use.

Transportation

In densely populated areas, the automobile is not a practical means of transportation but a major source of air pollution. In the United States between 1940 and 1970, the total automobile horsepower increased from 2.5 to 18 *billion,* and the total railroad horsepower declined from 92 to 53 *million.* Highways in the United States presently cover an area as large as New England, and still more would be needed to speed up traffic. Each year, 420,000 acres are turned into highways and airports. Even representatives of the car, oil and road construction industries have realized that more of the same is not a solution to the transportation problem. The head of one major automobile manufacturing company has recently recommended that the yearly five to six billion dollars collected from gasoline taxes, which presently go into the Federal Highway Trust Fund, be diverted toward mass transit. High quality, low cost or free mass transportation is essential to improve the traffic situation in the big cities.

As shown by both Table 2 and Figure 3, automobiles are a major source of air pollution. Under the 1970 Clean Air Act, the 1975 model cars must emit 90% less hydrocarbons (soot) and carbon monoxide than the 1970 models. In the 1976 models, there must be a 90% reduction in the emissions of oxides of nitrogen. According to the National Academy of Sciences report on this subject, these 1975 goals can be met. Some of the automobile manufacturers disagree and are requesting the postponement of the implementation of these standards. The cost of meeting the 1975 emission standards is projected to be around $200 per car. It is also probable that the new designs will have a 3 to 12 percent increase in fuel consumption, an increase in maintenance cost and a deterioration of the drivability of the car. Furthermore, it is likely that if catalytic pollution control devices are used, they will not last the life of the car. In short, attempts to make the automobile compatible with big city conditions are far from being promising.

There are many partial solutions to the transportation problem under consideration including: (1) the use of smaller cars; (2) engines, such as the Wankel, which pollute less; (3) the use of electric cars (the intensive research underway by AEC and others to develop new batteries is likely to improve their potential); (4) the use of fuels, such as liquified petroleum gas, which are less polluting; (5) the banning of all old or foreign cars not provided with pollution control devices from the cities; (6) the control of truck traffic, which presently contributes 50% of the pollution in the cities (the average loading of these trucks is only about 10%); (7) the creation of taxi stands so that no empty cabs would be cruising the streets; (8) the elimination of toll charges if there are three or more passengers in the car. Each of these proposals would alleviate the situation, but the complete solution lies in the recognition that automobiles and large cities are not compatible. Good (preferably free) mass transit is the only full solution.

Pollution Abatement Expenses and Savings

There are very few reliable estimates on the expenses of cleaning up the environment and even fewer on the costs of inaction, but it does appear that pollution costs more than does its abatement. The White House Council on Environmental Quality has estimated that in order to meet the air and water standards which were in effect as of 1971 by the year 1976, a total of 105 billion dollars would have to be spent in a six-year period (Fig. 18). On a yearly basis, this amounts to less than 20 billion dollars, which is less than 1.5% of the estimated gross national product for the year of 1975, or about 25% of the yearly defense budget (Fig. 19).

According to the third annual report of the Council on Environmental

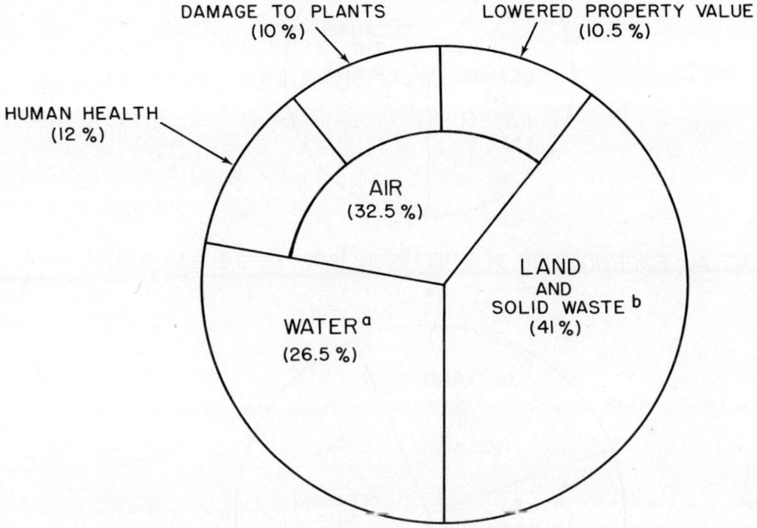

THE APROXIMATE COST OF POLLUTION IN THE US IN A 6-YEAR PERIOD
($ 300 BILLION)

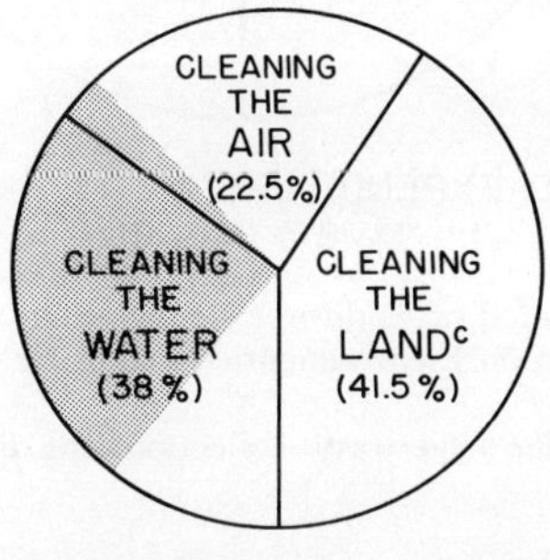

TOTAL ESTIMATED COST FOR CLEANING UP THE ENVIRONMENT IN A 6-YEAR (1970-1975) PERIOD ($ 105 BILLION)

Figure 18 Pollution control costs, 1970 to 1975. *Note:* The shaded areas represent the estimated state and federal contributions of the public sector; unshaded areas represent the costs borne by both municipalities and private sector of the country.

[a]This estimate excludes intangibles such as lost recreational values.
[b]This approximation is for lost energy, raw material and damaged or used land costs.
[c]This expenditure is primarily for solid waste collection and disposal, not for reduction of land pollution.

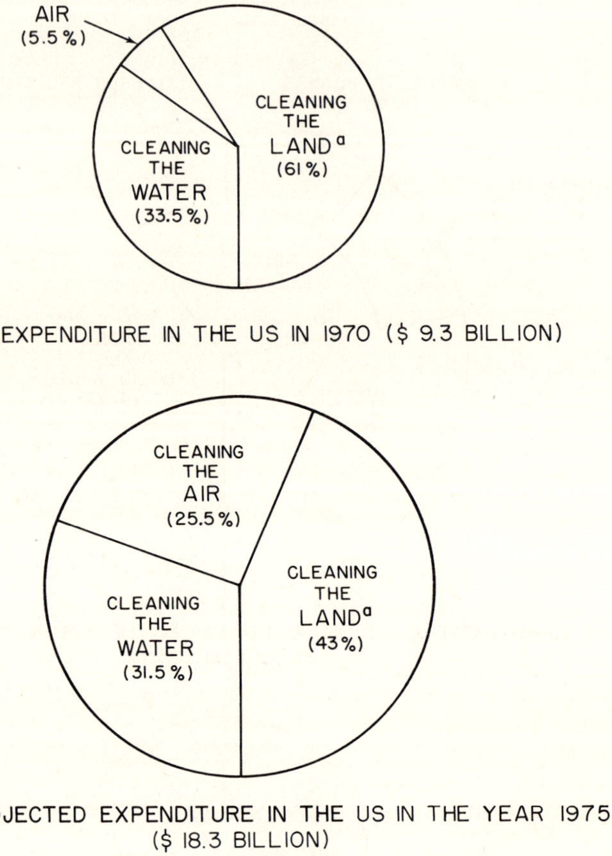

Figure 19 Current and projected expenditures for pollution control in the United States. *Note:* The projection of the Council on Environmental Quality for 1980 is $33.3 billion.

[a]This expenditure is mostly for solid waste collection and disposal, not for reduction of land pollution.

Quality, it would cost at least 287 billion dollars to solve the country's major pollution problems over a ten-year period. They estimate that this sum would be approximately evenly distributed between air pollution (37.5%), water pollution (30.5%) and other pollution (32.5%) expenses.

As pollution standards become more stringent and enforcement increases, further expenses can be anticipated. According to some estimates, the cost of eliminating the discharge of all the pollutants in the United States might total two trillion dollars. Another estimate is an annual cost of 40 billion dollars for the next 25–30 years. (The North River Treatment Plant for Manhattan alone is projected to cost 0.8 billion dollars.) Regardless of which figure one prefers to accept, it is quite clear that man has no choice but to make his peace with nature. Whatever the costs, the savings and benefits are much greater.

Opinions are diversified on how these pollution abatement costs should be shared between the public and private sectors of society. There is a general agreement that industries can no longer regard the nation's air, land and water as free "dumps." It is also recognized that property rights do not extend to the right of individuals and firms to pollute the air and the water. Yet due to the prevailing power distribution in society, public financing is mostly in the areas of water pollution (federal expenditure for this alone is proposed by Congress at the level of from 20 to 25 billion dollars for a three-year period). In the areas of solid waste and air pollution, which are the main problems of the big cities, most of the expense is left to the municipalities and the private sector (Fig. 18). The present funding of solid waste-handling programs covers only collection and disposal transportation expenses and does not include anything for salvaging, reuse, recycling or recovery programs. It seems that a more balanced distribution of expenses between the public and private sectors would be desirable, as would a taxation and subsidy program which gives the motivation to reclaim resources and to clean up the environment. The yearly 5–6 billion dollars, for example, which are collected from gasoline taxes could be used to pay for some of the air pollution control and mass transportation expenses of the urban centers.

The immediate cost of pollution abatement to the public is not yet excessive (Table 20), and it is probable that as expenses increase so too will contributions from the public sector. The federal expenditure toward pollution abatement in 1971 was $2.72 billion and in 1972 $4.38 billion. In both years, less was spent than was authorized, and almost all the expenditures were in the area of water pollution control. Only 1% of the federal budget submitted for the year 1973 is for environmental purposes, while this proportion in the

Table 20
SPECIFIC COSTS TO THE PUBLIC FOR POLLUTION ABATEMENT IN THE IMMEDIATE FUTURE

Pollution Abatement Programs	*Estimated Costs*
1975 Emission standards for new automobiles	$200/car
Standards for new steam generators would increase average electricity bill by	$1/year
Meeting the standards on new municipal incinerators would cost	$1/capita

budgets of the larger cities is 3–5% for the same year. The federal expenditure on research and development and on solid waste control has been practically nil in recent years.

Another very important aspect of costs is the funding of research programs. These should be much increased, particularly in the area of solid waste handling and recovery. Funding for energy research is also too small and out of proportion with the needs. For example, while the breeder nuclear reactor program is being funded by several billion dollars (as of 1972), only $28 million is being devoted to nuclear fusion research, only $10 million is being spent on coal gasification pilot projects and practically nothing is being spent on the development of solar power plants.

The Role of the Nontechnical Community

A well-informed and concerned public is an essential prerequisite of progress. With the backing of public opinion, simple, clear-cut and realistic laws can be enacted which will properly define the role of the agencies involved, and these laws must be enforced with sufficient vigor so that potential violators are not tempted to consider inaction as a viable alternative.

The Need for a Well-informed Public

The worst enemy of progress is an uninformed and confused public. On the subject of the environment, there unfortunately exists an unusually high degree of confusion. This is partially caused by the overstatements of well-intentioned but unqualified antipollution activists, who by their emotional exaggerations tend to do damage to the credibility of scientists and professionals working in this area. An even more important cause of the prevailing confusion is the contradictory statements and announcements which tend to leave the public with the impression that nobody really knows the facts. It is hoped that this handbook will make a contribution by showing that while there might not be full agreement on how clean is clean, there certainly is a knowledge of what the causes of contamination are and of what technology is needed for pollution abatement.

One example of the prevailing confusion is the topic of detergents and their phosphate content. Announcements on the harmful nature of detergents have probably contributed to the decision of the representatives of Suffolk County, Long Island to ban the sale of all detergents. Other areas of the United States such as the city of New York have interpreted the same information as meaning that it is only the phosphate content that should be limited or abolished. (Phosphate is a nutrient that has been linked to the unchecked growth of algae, which in turn choke the other life forms out of lakes, bays and rivers.) Later announcements have left the impression that phosphate detergents are not really the worst choice because other types, such as the enzymes, represent an even worse health hazard. The National Research

Council after a one-year study could find no evidence of major health hazards from enzyme detergents. Even the best-motivated housewife or representative of the public will be at somewhat of a loss after such a series of claims and denials and will tend to reach the conclusion that the best thing to do is to wait "for the dust to settle." This, of course, is just the opposite result to what is needed to clean up the environment.

Trial and error techniques or careless experimentation cannot be allowed when dealing with nature or with human health. Yet there are a number of examples of it taking place today in the nuclear industry, the oil industry (offshore drilling, Alaskan pipeline) and many others. A small example of the results of such experimentation can be seen in Grand Junction, Colorado where during 1953–1966, from 50,000 to 200,000 tons of sand (called "tailings") left over from the uranium-processing operation at the mill were spread around the town. The fine gray sand was used as a base for laying concrete basements and building slabs in private homes and schools. A study of birth and death records in the area shows increased genetic problems, higher cancer rates and lower birth rates than those of the rest of the state. These radioactive tailings are also carried by the wind into rivers. Drinking water should not contain more than 3 picocuries of radium per liter of water; each gram of tailings contains 100–1000 picocuries of radium. The chairman of the Atomic Energy Commission has acknowledged that the use of radioactive sands in building houses in Grand Junction, Colorado was a mistake and that action is needed to remove these tailings. He indicated that the federal government might be willing to pay some of the removal cost. He said that there was a lack of knowledge about the problem in the early days of uranium milling, and that authority in those days was fragmented. One would hope that such mistakes will serve at least an educational purpose.

One could make quite a list of areas in which man is still using the trial and error approach to learn the consequences of his acts. For example, it is known that ozone in the upper atmosphere protects the earth from ultraviolet radiation that would otherwise kill all plant life. We do not know for sure what the nitrogen oxides dumped by supersonic airplanes or methane from man-related activities will do to the ozone concentration. Yet many seem willing to experiment.

Another area where the public is justifiably confused relates to the use of pesticides in general and to the use of DDT and other chlorinated hydrocarbon pesticides, such as deldrin and aldrin and the weedkiller 2,4,5-T, in particular. The communication media report that, since there is no acceptable alternate for DDT, it should be used against the insects that carry malaria, yellow fever, elephantiasis, sleeping sickness and cholera. It is also reported that DDT is essential to feed the growing population of the world, 65% of which is now malnourished even with the practically unrestrained use of pesticides and other agricultural chemicals. Yet the public also reads in their newspapers that there is a relationship between DDT and cancer in laboratory animals and that all humans are contaminated with the pesticide in ways that could have long-term effects on human health and genetics. In the United States, the use of DDT has been practically outlawed (as of the end of 1972). There are half a million man-made chemicals in use today, and their properties and effects (individually or in combination) are largely unknown.

It is damaging to public confidence when, for example, the Administrator of the Environmental Protection Agency argues against the use of returnable cans and bottles on the basis of the findings of a "California study" which he later admits was never conducted. One of the most important tasks ahead is in the area of education and communication. The public on its own cannot be expected to reach the right conclusions in some of these highly technical areas, yet it is essential to have a well-informed and properly motivated public. A highly respected organization which is beyond the reach of pressure groups or daily politics should be given the task of advising the public on factual matters relating to the environment and pollution. One such body is the National Academy of Science. If it cooperates with the academies of other nations (most likely under the auspices of the United Nations), it can provide the respected guidance which the presently sometimes confused public needs for responsible action. The establishment of the Governing Council for Environmental Programs (by the 1972 UN Conference in Stockholm) was an important step toward this goal.

In striving for a well-informed public, one cannot overlook the area of advertising. Researchers have found that of the 289 pages of environmentally oriented advertising, costing about $6 million, in the 1970 issues of Time, Newsweek and Business Week magazines, more than half was from the five industries pinpointed in independent studies as having the most pollution cleanup to do—electric utilities, steel, oil, paper and chemicals. Examples of false advertising included depicting a lobster, which could have come from nowhere near the plant, as a contented resident of power plant wastewater (by a southern California power supplier) and an advertisement (by an oil company) representing the Palm Springs courthouse as a company research center.

Enforcement of the Laws

In the areas of pollution abatement, law enforcement is not yet sufficient. Court action takes so long and fines are frequently so small that pollution levies can be shrugged off by corporations as a cost of doing business. As an example, the Clairton Works (Allegheny County, Pa.) of the U.S. Steel Corporation daily emits about 225 tons of pollutants into the air. As of early 1972, this had resulted in total fines of less than $20,000, and even these are being appealed. Similarly in the case of an oil spill, it is insufficient motivation to have the responsible corporation pay only for the costs of cleaning up with no consideration of the damage done (Fig. 21).

Government agencies should also become more sensitive to environmental considerations. It should not be necessary, for example, for the U.S. Circuit Court of Appeals in Washington, D.C. to order the Atomic Energy Commis-

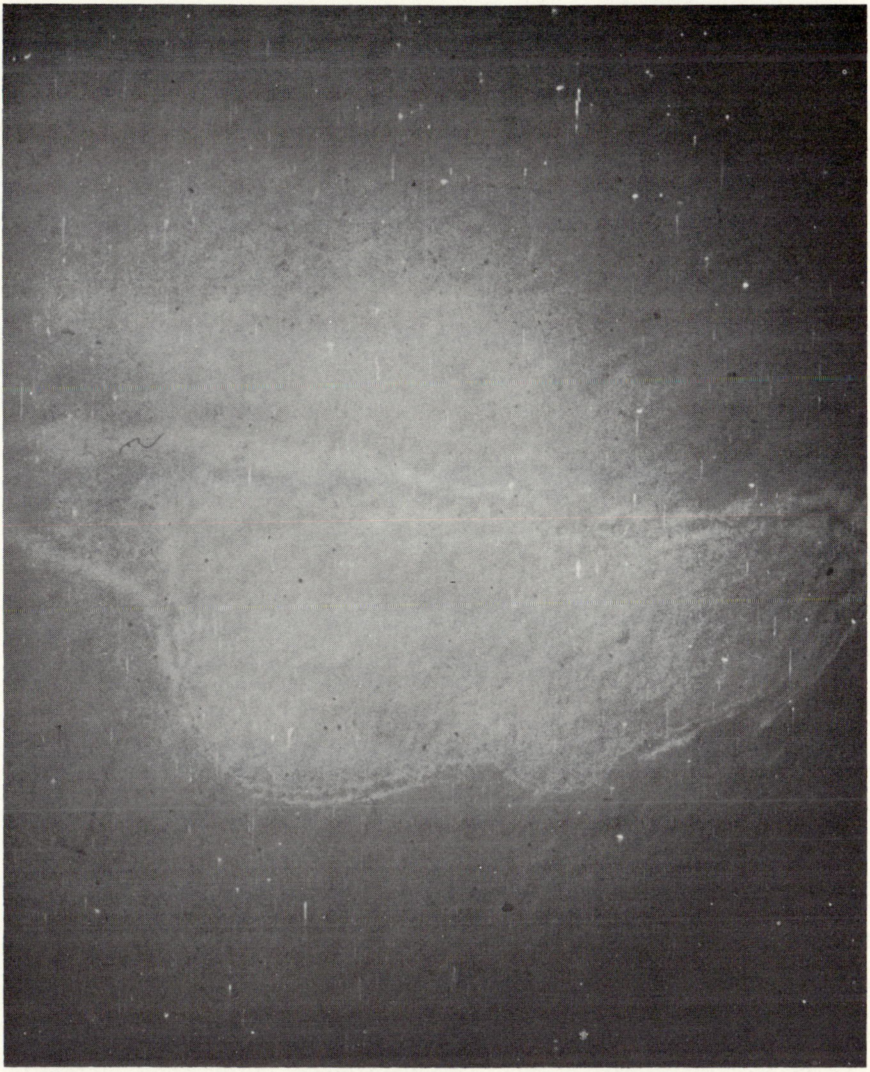

Figure 21 Sensors have been developed which can be mounted on satellites for spotting and tracking oil slicks on a daily basis. (*Courtesy of NASA*)

sion (Calvert Cliffs, Md.) to follow the Environmental Protection Act in licensing nuclear plants.

Almost certainly with the passage of time, law enforcement and regulation will become more energetic. As of yet, although the signs of this trend are few, they are definitely present. The slow progress is partially due to the feeling of elected officials that those on whom a regulation falls will remember it in the polling booths.

In the case of the coal burning power plant at Arthur Hill in Staten Island, the Consolidated Edison Company maintained for some time that it was impossible to meet the city's air pollution abatement demand and at the same time provide reliable power. Yet due both to the charges of dozens of $100-a-day violations of the city's air pollution control code and even more importantly to the pressure of public opinion, the company finally agreed to convert this plant to the use of low pollution fuels.

In the area of air pollution, the first injunction was made against polluters under the emergency provisions of the Clean Air Act of 1970 in 1971 in Birmingham, Alabama.

In the area of water pollution, the enforcement effort has been slow and ineffective to date. Most of the changes were based on the seldom-enforced Federal Refuse Act of 1899, which makes it a criminal misdemeanor to discharge "any refuse matter of any kind or description" into navigable waters. In 1972 a federal judge ruled that sewage is also covered by this act and ordered some New Jersey communities to stop their sewer sludge discharges into the Atlantic. Under the law, violators are subject to one year's imprisonment and $2,500 in fines for each count. While this act is a horse-and-buggy means of dealing with a jet-age problem, it was used to fine Anaconda some $200,000 on a hundred-count indictment for washing copper fragments into the Hudson River. The size of the fine in this as yet isolated case is sufficient to motivate serious pollution abatement efforts.

Another direction undergoing experimentation by pollution control agencies is the holding of corporate executives criminally responsible for water pollution caused by their plants. This approach, it is hoped, will be more productive than the old practice of bringing criminal and civil charges against the company, which itself can hardly be thrown in jail. Mr. J. J. O'Donnell of the J. J. O'Donnell Woolens Company in Massachusetts was the first executive to be fined after the federal government brought criminal charges against him for not stopping (for some seven years) the discharging of dyes into the Blackstone River.

It is hoped that federal involvement in pollution abatement efforts will increase with the passage of time. This involvement began when the Federal Environmental Protection Agency moved against two large New York State paper companies in the first step of a major new program that will take over pollution control activities from lagging state efforts. Another example of such federal involvement occurred when the administrator of the Environmental Protection Agency asked the Justice Department to file suit against the Reserve Mining Company for their pollution of Lake Superior (a daily discharge of 67,000 tons of taconite tailings).

Violations of the pollution laws are so widespread today that pollution control by law enforcement alone would be hopeless, but law enforcement as a part of a serious national commitment is an essential tool of success.

Conclusions

While the changes and adjustments needed to make man's peace with nature are very basic, they will still have to be made gradually. The first corrective steps will recognize the fundamental "incentive structure" of the profit and production oriented society and will concentrate on developing a structure of laws and taxes that will discourage pollution. These taxes will have to recognize that property rights do not extend to the right of treating the air, land and waters as free dumps and that environmental costs must be added to production costs. The first steps in this direction have been made by the introduction of legislation increasing the taxes on leaded gasoline and sulfur dioxide emissions from power plants and requiring environmental impact statements from nuclear power plants before licensing them. This latter practice should be extended to all major industrial plants.

Taxation policy will have to be gradually extended to tax the use of power and raw materials and to penalize the resource-intensive industries and favor the labor-intensive ones. It will also be necessary to introduce amortization taxes which will increase the tax burden on those products which are designed to last for only a short time. On the other hand, industries which participate in recycling, recovery and salvaging programs to make resources out of wastes will have to be rewarded by tax benefits. That this is not yet understood by some officials* results in steps such as the recently imposed federal tax on reused oil. This new tax eliminates the until now existing economic motivation to reclaim used oil.

In some regions, it will be found impractical to have each industry treat its own waste separately, and service industries will have to be established to collect and treat these wastes. But regardless of who does the actual cleaning of plant effluents, it is very important to maintain the goal of no discharge of any pollutants as the ultimate target.

A well-informed public is essential in the long-range pollution abatement effort. With public opinion behind it, a national commitment can be made in favor of such reforms as a safer energy policy, a better mass transit system

* Another example of poor judgement is the policy of the Interstate Commerce Commission setting higher freight rates for recycled materials than for raw materials. Relative costs of shipment for a ton of raw iron ore and a ton of scrap steel compare as $1.64 to $4.12.

or a zoning policy requiring such improvements as green belts around the cities, just to mention a few.

The road of pollution abatement will gradually lead to the abandonment of some basic human ideas: the instinct for fertility, the worship of economic growth and the tendency to become more industrialized, urbanized, centralized. In moving toward "a stable society" characterized by a steady or declining population, decentralized living and limits on the use of resources, it will become clear that, by making man's peace with nature, human history can be changed for the better.

ACKNOWLEDGMENTS

I wish to thank my brother Péter, my son Ádám and my daughter Ágnes for helping me with the proofreading and indexing tasks; Pamela Smith for patiently typing all the correspondence and manuscripts; Alma Coppola for helping with the reproduction of the text; Samuel Russell for providing me with much needed secretarial and other help; and all the contributors for their outstanding spirit of cooperation and for their patience in submitting to modifications which were needed in order to produce a uniform handbook. Finally, I wish to thank my wife Mártha for tolerating the consequences of this project during the last few years.

Béla G. Lipták
Stamford, Connecticut

CONTENTS

LIST OF IMPORTANT TABLES AND CHARTS

List of Important Tables and Charts

List of Important Tables and Charts

List of Important Tables and Charts

Chapter 1

AIR POLLUTION LAWS AND REGULATIONS

J. L. Gilliland F. B. Taylor
W. C. Zegel

CONTENTS OF CHAPTER 1*

* For a glossary of terms, refer to the back portion of this volume; and for a list of abbreviations used in this chapter, refer to the listing given after these contents, preceding the first section.

Abbreviations

EPA	Environmental Protection Agency
HEW	Department of Health, Education and Welfare
NAPCA	National Air Pollution Control Administration
PL	Public Law
USPHS	United States Public Health Service

1.1 FEDERAL LAWS[1]

Air Pollution Control Research and Technical Assistance Act, 69 Stat. 322

The original federal air pollution control legislation (P.L. 84-159), adopted in 1955, authorized the Public Health Service of the Department of Health, Education and Welfare (HEW), in cooperation with other public and private agencies as well as with industry, to recommend research programs on methods for controlling air pollution. Upon the request of state or local agencies, the Surgeon General could conduct investigations and surveys on air pollution problems. Research, training and demonstration projects could be conducted by grants-in-aid or by contract with state and local agencies, private agencies and individuals.

Extension of 1955 Act, 73 Stat. 646

In 1959, the 1955 act was extended (P.L. 86-365), with specific instructions for other federal agencies to cooperate with HEW and the state, interstate and local air pollution control agencies with respect to discharges from federal facilities.

Study of Motor Vehicle Discharges, 74 Stat. 162

In 1960, specific emphasis (P.L. 86-493) was placed on the motor vehicle exhaust problem.

Clean Air Act, 77 Stat. 392

The 1963 Clean Air Act (P.L. 88-206) forms the basis for many of the laws which followed. It declared "that the prevention and control of air pollution at its source is the primary responsibility of state and local government." It provided for a research program to develop low-cost techniques for extracting sulfur from fuels. The secretary of HEW was instructed to compile and publish, for information only, criteria reflecting the latest scientific knowledge on the effects of pollutants. Up to 30% of the total authorized program was available for grants to air pollution agencies for developing, establishing or improving programs. These grants could cover up to two-thirds of the cost of the program, except that 75% grants were available for intercity or interstate programs.

The 1963 act provided for abatement conferences upon the request of a state or municipality in which the discharge was originating. If the pollutant originated in another state, an abatement conference could be called after consultation with all affected states. After the conference, the secretary would recommend the necessary remedial action. If this action was not taken, a hearing board was convened to abate the pollution.

The automotive and fuel industries, through technical liaison committees with HEW, were encouraged to develop devices and fuels to reduce discharges from automotive exhausts.

Motor Vehicle Air Pollution Control Act, 79 Stat. 992

The 1965 amendments (P.L. 89-272) required the secretary of HEW to prescribe emission limits for *new* motor vehicle engines. Prototype testing was required before certification. Research programs on emissions of hydrocarbons, nitrogen oxides and aldehydes from motor vehicle engines were authorized. The secretary was instructed to develop low-cost techniques to reduce sulfur oxide emissions from fuels. He was given authority to call conferences for problems of international significance.

Clean Air Act Amendments of 1966, 80 Stat. 954

The 1966 amendments (P.L. 89-675) provided that 50% grants could be made to control agencies for *maintaining* programs and authorized 60% grants for maintaining *regional* programs.

Air Quality Act of 1967, 81 Stat. 485

In the Air Quality Act of 1967 (P.L. 90-148), adopted November 21, 1967, many specific provisions of earlier acts were removed and the entire concept was broadened. For the first time, the law called for the enhancement of the nation's air resources, as a purpose of the act. The act required the secretary to give special emphasis to research and development into methods, having industry-wide application, for the prevention and control of *all* air pollution resulting from the combustion of fuels, rather than limiting the studies to sulfur oxides. It provided for federal grants to study and demonstrate these methods, as well as for the operation of federal facilities for this purpose. Studies were authorized on the recovery and marketing of by-products resulting from the removal of pollutants. Special attention was given to the development of low-cost instrumentation for measuring emissions.

Whereas the 1963 act had specified that the criteria on air pollution, published by the secretary, would be for informational purposes, the 1967 act instructed the secretary to issue to the states ". . . such criteria of air quality as in his judgment may be requisite for the protection of the public health and welfare. . . ." Simultaneously, the secretary would issue information on recommended control techniques and on the costs for achieving the levels of air quality defined by the criteria. The secretary was instructed to consult with appropriate advisory committees and with other federal departments in establishing these criteria and control technology.

After consultation with state and local authorities, the secretary was to designate air quality control regions. When the criteria and control technology had been made available, the *states* would hold public hearings, then *adopt* air quality *standards* applicable to their regions and an implementation plan to achieve these standards. Both the standards and the implementation plan were subject to the approval of the secretary. However, the agencies were not to be prevented from adopting programs to achieve a higher quality of ambient air than approved by the secretary. The question of national emission standards was to be studied for two years, both for stationary sources and for aircraft emissions.

Grants to air pollution control agencies could now be made for *planning* control programs and for the implementation of the air quality standards required by the act. Special provision was made for grants to *interstate* agencies.

Authority was given to the secretary to abate an emission source which was presenting an imminent and substantial danger to health.

The 1963 act provision for technical liaison committees with manufacturers was replaced by an Air Quality Advisory Board, composed of the secretary and fifteen public members appointed by the President. The Board advised the secretary on matters of policy and made recommendations to the President.

No states or their political subdivisions were permitted to adopt or enforce any standards controlling emissions from *new* motor vehicles. They could not require certification, inspection or any other control as a condition of initial sale or registration. However, the secretary could hold public hearings, then waive application of this section to any state which had adopted standards for new motor vehicles prior to March 30, 1966. California was the only state qualifying under this exception. Grants of up to 75% could be made to states to develop uniform programs for inspecting and testing motor vehicle emission control devices.

It was required that all fuel additives must be registered, with information on the chemical composition of the additive and the purpose of its use. The penalty for noncompliance was $1,000 per day.

The secretary was required to make comprehensive economic cost studies. These included the cost of carrying out the provisions of the act, the cost of program implementation by governmental units, the economic impact of the air quality standards on communities and industries and the cost of controlling emissions to national standards.

Under the 1967 act, HEW issued criteria and control technology documents for sulfur oxides and particulates on February 11, 1969. Manuals for carbon monoxide, nitrogen oxides and hydrocarbon emissions from both stationary and mobile sources were issued in 1970. Manuals for fluorides and photochemical oxidants were issued in 1971.

National Environmental Policy Act of 1969, 83 Stat. 852

The act of 1969 (P.L. 91-190) established the Council on Environmental Quality consisting of three members in the Executive Office of the President. This council appraises the federal programs and formulates national policies to promote the improvement of the quality of the environment.

The act requires that every recommendation or report on proposals for legislation and other major federal actions significantly affecting the human environment must include a detailed statement on the environmental impact of the proposed action. These statements are reviewed by the other affected federal, state and local agencies and are made available to the public.

The council submits an annual "Environmental Quality Report" to the President for transmittal to the Congress.

Environmental Quality Improvement Act of 1970, 84 Stat. 114

The act of 1970 (P.L. 91-224) declares a national policy of enhancing the environmental quality, with the primary responsibility for implementing the policy resting with the state and local governments.

The act provides for an Office of Environmental Quality in the Executive Office of the President. The chairman of the Council on Environmental Quality is the director of the office. The function of the Office on Environmental Quality is to provide the professional and administrative staff for the council.

Clean Air Amendments of 1970, 84 Stat. 1676

The 1970 amendments (P.L. 91-604) to the Clean Air Act made many major revisions, shifting much of the responsibility from the states to the federal government. On December 2, 1970, the Environmental Protection Agency (EPA) was established in the Executive Office of the President to administer the major environmental activities of the Departments of Agriculture, HEW and Interior, the Atomic Energy Commission and the Federal Radiation Council. All references in the Clean Air Act were revised to give these responsibilities to the administrator of EPA.

Research

Special emphasis is placed on the short and long-term effects of air pollutants on public health and welfare. (See Section 2.11.) Research is to be conducted on improving the efficiency of fuel combustion and the production of synthetic or new fuels whose use would decrease atmospheric emissions.

Grants are made available for programs to develop low-emission alternatives to the internal combustion engine.

Air Quality Control Regions

All regions designated under the 1967 act retain their status. The administrator is required to designate any other areas by March 31, 1971. Finally, all portions of a state which had not been previously designated within a control region are made into one or more regions.

New Stationary Source Standards

The Administrator will publish lists of categories of stationary sources to which new source standards will apply. Within 120 days after the publication of the list, proposed regulations establishing the federal standards will be published. They are to be promulgated within 90 days after publication. This "standard of performance" for emissions of air pollutants shall reflect ". . . the degree of emission limitation achievable through the application of the best system of emission reduction which (taking into account the cost of achieving such reduction) the administrator determines has been adequately demonstrated." The first stationary source categories listed,[2] on March 31, 1971, were fossil-fuel fired steam generators, incinerators, plants for the manufacture of portland cement, sulfuric acid and nitric acid.

Hazardous Air Pollutants

Provision is made for national emission standards for hazardous air pollutants to which ambient air quality standards are not applicable. Asbestos, beryllium and mercury were the first pollutants listed.[2]

National Primary and Secondary Ambient Air Quality Standards

The act requires the administrator to establish national primary and secondary ambient air quality standards. *Primary standards* are ambient air quality levels, based on the criteria and allowing an adequate margin of safety, which, in the judgment of the administrator, are necessary to protect the public health. *Secondary standards* are the levels required to protect the public welfare from any known or anticipated adverse effects associated with the presence of the pollutants in the ambient air. Public welfare ". . . includes, but is not limited to, effects on soils, water, crops, vegetation, man-made materials, animals, wildlife, weather, visibility, and climate, damage to and deterioration of property, and hazards to transportation, as well as effects on economic values and on personal comfort and well-being." These standards must be proposed within 30 days and promulgated within 120 days for those pollutants for which criteria have been issued. For those pollutants for which criteria are issued after December 31, 1970, the proposed standards will be published simultaneously with the criteria, with adoption not more than 90 days later. Standards for sulfur oxides, particulates, photochemical oxidants, hydrocarbons and carbon monoxide were published[3] on April 30, 1971.

Implementation Plans

Within nine months after the promulgation of the National Ambient Air Quality Standards, the states are required to hold public hearings and submit a plan which provides for implementation, maintenance and enforcement of the standards. The administrator must approve or disapprove the plan within four months. The plan must show that the primary standards will be attained within three years and the secondary standards within a reasonable time. The plan must include emission limitations and provision for monitoring ambient air quality. The states must have the authority for pre-construction review of the *location* of new sources to which the standards of performance will apply.

Federal Enforcement

Violations of an implementation plan, new source standards or hazardous emissions standards are subject to a fine of not more than $25,000 per day or imprisonment for not more than one year, or both. After a conviction, the second offense is subject to a fine of $50,000 or imprisonment for not more than two years, or both. Knowingly making false statements or records or tampering with monitoring equipment is subject to a fine of $10,000 or imprisonment for not more than six months, or both.

Inspection, Monitoring and Entry

The administrator has the right of entry and may inspect records and monitoring equipment and sample any emissions. Except for trade secrets, all records and reports must be available to the public.

State Authority

Except for the specific restrictions on regulations for motor vehicles, any state or local subdivision may adopt and enforce any standard or limitation on emissions, provided that they are not less stringent than the federal standards or limitations.

Control of Pollution from Federal Facilities

All federal departments and agencies must comply with federal, state and local requirements. The President may exempt a federal source if he determines it to be in the interest of the United States to do so, except that no exemption may be granted from the "new source" standards of performance, and only under special restrictions for hazardous pollutants. Each January, the President must report to Congress on the reasons for all exemptions granted during the year.

1.1 Federal Laws

Motor Vehicle Emission Standards

The administrator will prescribe emission limits applicable to all new motor vehicle engines. These standards are applicable for five years or 50,000 miles, whichever occurs first. For light-duty vehicle engines, a reduction of 90% (as compared with 1970) is required by model year 1975 for emissions of carbon monoxide and hydrocarbons. A reduction of 90% (as compared with 1971) is required by model year 1976 for emissions of nitrogen oxides. One year extensions may be granted by the administrator. All motor vehicles offered for sale must carry a certificate of compliance. It is unlawful to remove any pollution control device. Violators are subject to a civil penalty of up to $10,000. The administrator will conduct tests on new vehicles or engines submitted by manufacturers and may conduct assembly line tests. The manufacturer must warrant to all purchasers that the engines meet the standards and must provide instructions for proper maintenance.

Regulation of Fuels

The administrator may control or prohibit the sale of any fuel or fuel additive if he finds that emissions from such fuels endanger public health or impair the performance of emission control devices. Manufacturers may be required to conduct tests to determine potential public health effects. Violations are subject to a penalty of $10,000 per day.

Development of Low-Emission Vehicles

The Low-Emission Vehicle Certification Board will certify vehicles for government procurement.

Aircraft Emission Standards

The administrator is required to study the effects of aircraft emissions on air quality and to establish standards to be enforced by the secretary of transportation. No state may adopt or enforce aircraft emission standards unless they are identical with the federal standards.

Emergency Powers

If the state or local agency does not act, the administrator may bring suit in the appropriate U.S. District Court to stop the emission of air pollutants endangering public health.

Citizen Suits

Citizens may bring suit against violators of the standards or against the administrator for failure to perform a nondiscretionary act. A 60-day notice must be given before suit is filed. The administrator may intervene in any suit where he is not a party.

Table 1.1a
FEDERAL FUNDING FOR
AIR POLLUTION CONTROL[a]
(MILLIONS OF DOLLARS)

Fiscal Year	*Authorized*	*Appropriated*
1955		0.2
1956	5	1.7
1957	5	2.7
1958	5	4.0
1959	5	3.9
1960	5	4.2
1961	5	5.9
1962	5	8.8
1963	5	11.1
1964	5	13.0
1965	25	21.0
1966	30.4	26.7
1967	46	40.1
1968	109	64.2
1969	185	88.7
1970	134.3	108.8
1971	200	107.8
1972	350	
1973	450	

[a] Reprinted by permission of the American Bar Association and its Section of Natural Resources Law.

Federal Procurement

No federal agency may contract for goods or services with any person convicted of violating the act until the administrator certifies that the condition has been corrected.

Mandatory Licensing

Owners of patents may be required to license them on reasonable terms, as defined by the district court, to other persons who require them in order to comply with the act. Table 1.1a[4] lists the sums which the federal government provided for air pollution abatement purposes over the years.

REFERENCES

1. Schroeder, Stanley W.: "Pollution in Perspective: A Survey of the Federal Effort and the Case Approach," vol. 4, no. 2, American Bar Association April 1971.
2. Federal Register, 36 F. R. 5931, March 31, 1971.
3. Federal Register, 36 F. R. 8186, April 30, 1971.
4. Schroeder, op. cit.

1.2 FEDERAL AGENCIES

The origin of the National Air Pollution Control Administration can be found in the U.S. Public Health Service where most of the federal environmental programs had their start. Under P.L. 410, the Public Health Service Act, the Surgeon General of the United States has long been charged with the responsibility of overseeing and protecting the national health. The act gives him broad powers to prevent and contain the spread of communicable disease and to conduct research. As new threats to public health emerged, the Public Health Service introduced new programs designed to meet these threats. Air pollution control was first handled by a small group in the Public Health Service Division of Industrial Hygiene.

Development of the Organization

The Donora, Pennsylvania, incident in 1948 was the first to focus public and official attention on the need for doing something about air pollution. Beginning on October 27, 1948 and continuing for five days, this Monongahela River Valley town of 14,000 people was subject to an atmospheric inversion which trapped industrial smoke and fumes over the community. By the end of this episode, 5,910 persons were sick and there were several deaths. A Public Health Service field team of physicians, nurses and engineers, under the direction of George D. Clayton, Division of Industrial Hygiene, USPHS, was dispatched to the town.

Following Donora, officers of the Public Health Service accumulated facts and conducted limited research for several years and pressed for the enactment of federal legislation which would provide the resources needed to cope with air pollution on a national scale. These efforts together with those of concerned citizens, political leaders and scientists culminated in the passage of Public Law 159 by the eighty-fourth Congress and signed by President Eisenhower on July 14, 1955. (See Section 1.1.) This was the first Air Pollution Control Act. The Public Health Service efforts that began in 1955 were not large but included research on the causes and control of air pollution, encouragement of state and local activities and gathering and disseminating information and training.

To administer the act in the Public Health Service Bureau of State

Services, an Air Pollution Engineering Program under Mr. Frank Tetzlaff and an Air Pollution Medical Program under Dr. Wilton Fisher were established. These divisions had engineering, medical and research branches, the latter being located at the Robert A. Taft Sanitary Engineering Center in Cincinnati, Ohio. When the next significant legislation in the form of the Clean Air Act (P.L. 88-206) was passed in 1963, the Division of Air Pollution carried on a program to generate greatly expanded air pollution control by state and local governments through program grants (Table 1.1a).

A series of amendments were added to P.L. 88-206 to increase its scope and provide more funds. (Section 1.1.)

The federal administrative machinery had been expanded to the National Center for Air Pollution Control in 1968 and thereafter to the National Air Pollution Control Administration (NAPCA) (1967), Public Health Service, Department of Health, Education and Welfare. Reorganization Plan No. 3 of December 2, 1970, transferred this and other environmental programs to the new Environmental Protection Agency (EPA). Although NAPCA has kept its basic organizational components more or less intact, it has lost its name and become first the Air Quality Office of EPA and most recently the Office of Air Programs, EPA.

Present Organization

As of this writing, it is not clear what the final pattern of the federal air pollution control organization will be within the EPA. The underlying philosophy of many of the leaders of that administration is that the whole EPA should be organized on a functional rather than on a categorical basis. (Fig. 1.2a). However, it has not been possible to entirely eliminate the environmental categories, and most of the program elements persist until the present. For air pollution control, these program elements are shown by Figure 1.2b. Figure 1.2a also illustrates the organizational position of the EPA regional offices.

Regional Offices

Federal activities for many years have been conducted at the field level through a series of regional or (as they were first called) district offices. Each EPA region is directed by a regional administrator who has been given broad powers including the approval of grants (Fig. 1.2c).

Grant Mechanisms

There are five EPA grant programs relating to air pollution control: (1) Air Pollution Research Grants, (2) Air Pollution Survey and Demonstration Grants, (3) Air Pollution Manpower Training Grants, (4) Air Pollution Fellowships and (5) Air Pollution Control Program Grants. The fiscal year 1971 funding level for these was: (1) $7,028,500; (2) $803,000; (3) $3,494,366; (4)

Figure 1.2a Organization of the EPA.

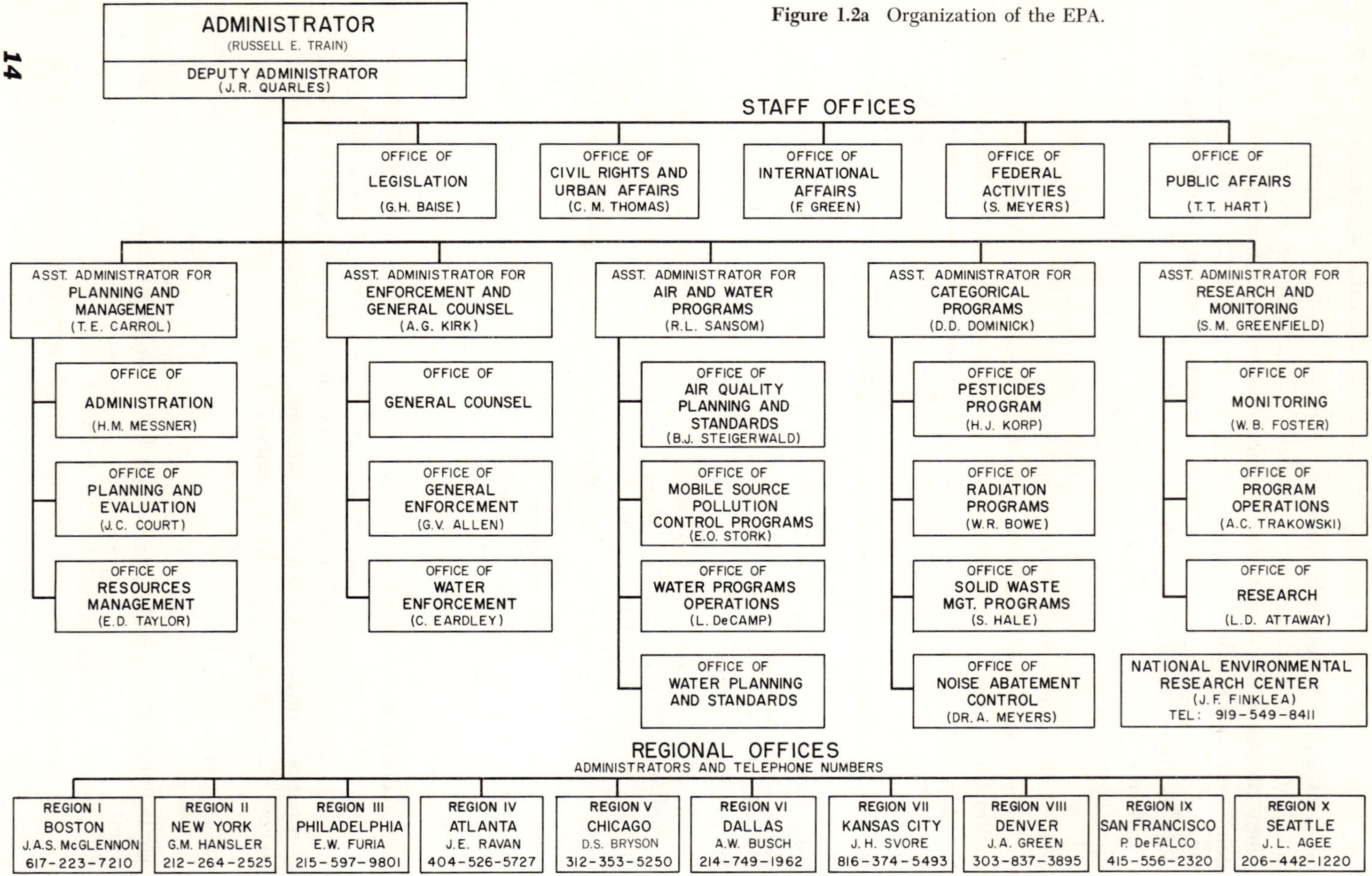

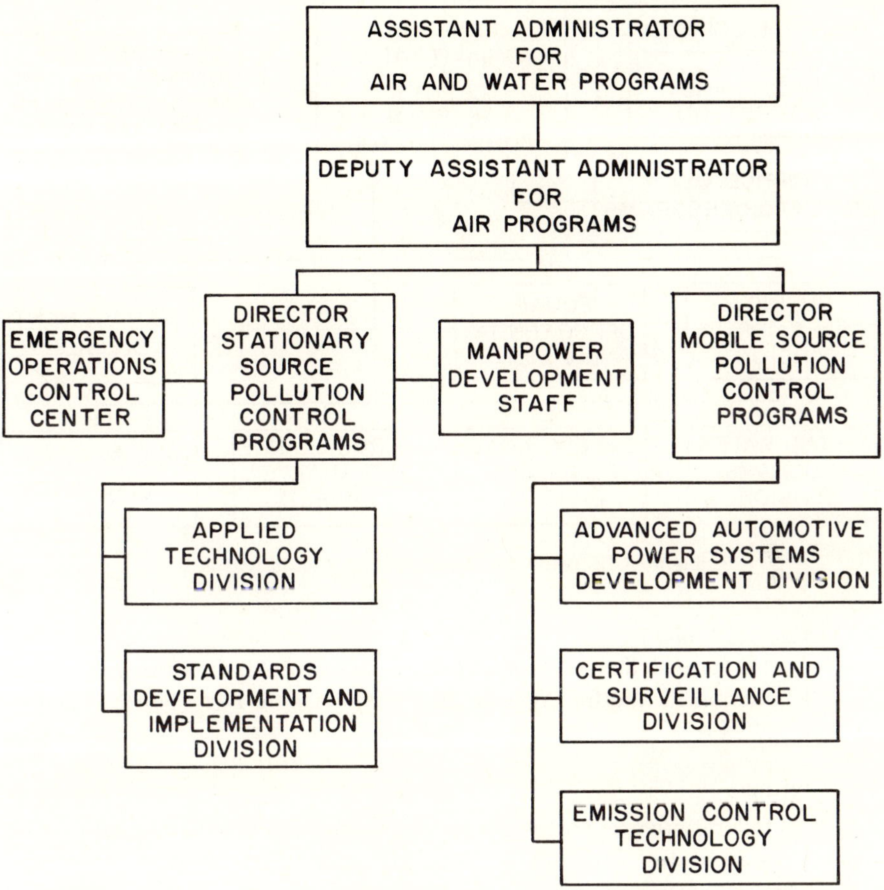

Figure 1.2b Organization of the Office of Air Programs, EPA (July 14, 1972).

$332,539 and (5) $29,446,996, respectively. Statutory authority for these grants is found in Sections 103, 103(b), 104, 105 and 106 of the Clean Air Act, 42 U.S.C. 1857 and the proper contacts for applications are as follows:

Research, Survey and Demonstration Grants.
Chief, Project Processing Section, Resources Management Branch, Office of Research and Monitoring, Environmental Protection Agency, Washington, D.C. 20460

Manpower Training and Fellowship Grants.
Director, Office of Manpower Development, Office of Air Programs, Environmental Protection Agency, Research Triangle Park, North Carolina 27711

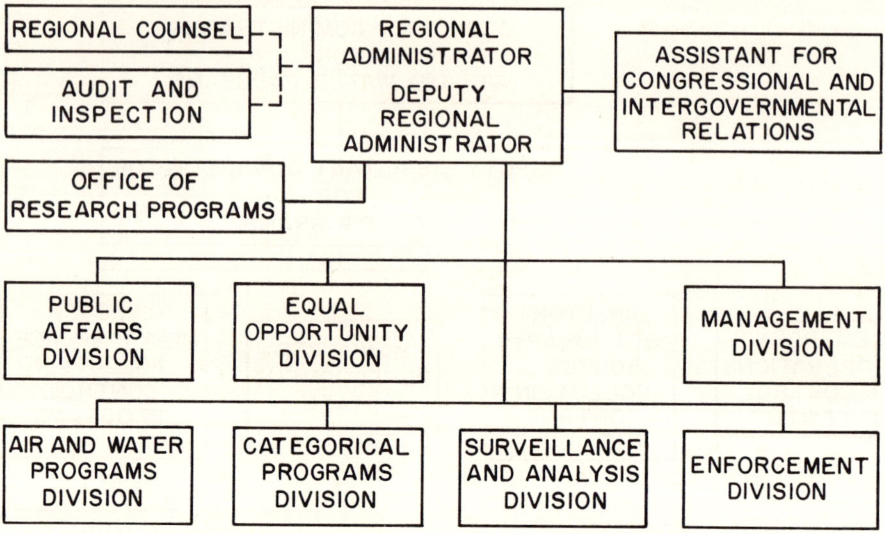

Figure 1.2c Structure of EPA organization in Region I.

Program Grants.

Director, Office of Grants Policy, Office of Air Programs, Environmental Protection Agency, Research Triangle Park, North Carolina 27711

Conclusions

The federal government is the largest single air pollution control entity in the United States. It exercises broad programs of research, enforcement, standards setting, information dissemination and encouragement of state and local programs through a complex administrative structure in the Environmental Protection Agency. The level of funding in fiscal year 1971 was $112 million (Table 1.1a). Much of this money was spent through a comprehensive grant program.

Headquarter's mailing address: U.S. Environmental Protection Agency, Washington, D.C. 20460
For information on grant applications see publication: "Catalog of Federal Domestic Assistance," Public Documents Department, Government Printing Office, Washington, D.C. 20402
For information on survey and demonstration grants: Water 703-557-9338; Air 919-688-8146/345; Solid waste 513-684-4333
For information on research: 703-557-9338

1.3 STATE AND PROVINCE AGENCIES*

State Agency Addresses

Alabama	Department of Public Health, State Office Building, Montgomery, Alabama 36104 Alabama Air Pollution Control Commission (Same address as above)
Alaska	Department of Environmental Conservation, Pouch 0, Juneau, Alaska 99801
Arizona	Division of Air Pollution Control, 4019 N. 33rd Ave., Phoenix, Arizona 85017
Arkansas	Department of Pollution Control and Ecology, 1100 Harrington Ave., Little Rock, Arkansas 72202 Arkansas Pollution Control Commission (Same address as above)
California	Air Resources Board, 1025 "P" St., Sacramento, California 95814
Colorado	Air Pollution Control Division, Colorado Department of Health, 4210 E. 11th Ave., Denver, Colorado 80220 Colorado Air Pollution Control Commission (Same address as above)
Connecticut	Air Pollution Control Section, Department of Environmental Protection, 79 Elm St., Hartford, Connecticut 06115
Delaware	Department of Natural Resources and Environmental Control, Division of Environmental Control, Tatnall Building, Capitol Complex, Dover, Delaware 19901 Water and Air Resources Commission (Same address as above)
District of Columbia	Air Pollution Division, Department of Human Resources, 25 K St., N.E., Washington, D.C. 20002

* Names and addresses for the state and province agencies have been taken from the 1971–1972 directory, "Governmental Air Pollution Agencies," published by the Air Pollution Control Association, 4400 Fifth Avenue, Pittsburgh, Pennsylvania 15213. This directory, which is published annually, includes federal and local agencies. The directory may be purchased for $2.50 from the Association.

1.3 State and Province Agencies

Florida	Department of Pollution Control, Suite 300, Tallahassee Bank Building, 315 S. Calhoun St., Tallahassee, Florida 32304 Florida Air and Water Pollution Control Board (Same address as above)
Hawaii	Air Sanitation Branch, Division of Environmental Health, P. O. Box 3378, Honolulu, Hawaii 96801 State Board of Health (Same address as above)
Idaho	Air Pollution Control Section, Environmental Improvement Division, Idaho Department of Health, Statehouse, Boise, Idaho 83707 Idaho Air Pollution Control Commission (Same address as above)
Illinois	Environmental Protection Agency, 2200 Churchill Road, Springfield, Illinois 62706 Pollution Control Board, Illinois Environmental Institute, 189 W. Madison St., Chicago, Illinois 60602
Indiana	State Board of Health, 1330 W. Michigan St., Indianapolis, Indiana 46206 Air Pollution Control Board of the State of Indiana (Same address as above)
Iowa	Environmental Engineering Service, Iowa State Department of Health, Lucas State Office Building, Des Moines, Iowa 50319 Iowa Air Pollution Control Commission (Same address as above)
Kansas	Kansas State Department of Health, State Office Building, Topeka, Kansas 66612
Kentucky	Kentucky Air Pollution Control Commission, 275 E. Main Street, Frankfort, Kentucky 40601
Louisiana	Air Control Section, Bureau of Environmental Health, Louisiana State Department of Health, P. O. Box 60630, New Orleans, Louisiana 70160 Louisiana Air Control Commission (Same address as above)
Maine	Department of Environmental Protection, State House, Augusta, Maine 04330 Bureau of Air Pollution Control (Same address as above)
Maryland	Bureau of Air Quality Control, Maryland State Department of Health and Mental Hygiene, 610 N. Howard Street, Baltimore, Maryland 21201 Air Quality Control Advisory Board (Same address as above)

Massachusetts	Bureau of Air Use Management, Division of Environmental Health, Department of Public Health, 600 Washington St., Boston, Massachusetts 02111 Public Health Council (Same address as above)
Michigan	Air Pollution Control Section, Division of Occupational Health, Michigan Department of Public Health, 3500 N. Logan Street, Lansing, Michigan 48906 Michigan Air Pollution Control Commission (Same address as above)
Minnesota	Division of Air Quality, Minnesota Pollution Control Agency, 717 Delaware St., S.E., Minneapolis, Minnesota 55440
Mississippi	Mississippi Air and Water Pollution Control Commission, P. O. Box 827, Jackson, Mississippi 39205
Missouri	Missouri Air Conservation Commission, P. O. Box 1002, 112 W. High St., Jefferson City, Missouri 65101
Montana	Montana State Department of Health, Cogswell Building, Helena, Montana 59601 Air Pollution Control Advisory Council, Montana State Board of Health (Same address as above)
Nebraska	Division of Air Pollution Control, State Department of Environmental Control, 411 S. 13th Street, Lincoln, Nebraska 68508 Nebraska Environmental Control Council (Same address as above)
Nevada	Bureau of Environmental Health, Nye Building, 201 South Fall Street, Carson City, Nevada 89701 State Commission of Environmental Protection (Same address as above)
New Hampshire	New Hampshire Air Pollution Control Agency, 61 S. Spring St., Concord, New Hampshire 03301 Air Pollution Control Commission (Same address as above)
New Jersey	New Jersey State Bureau of Air Pollution Control, Division of Environmental Quality, Department of Environmental Protection, P. O. Box 1390, Trenton, New Jersey 08625
New Mexico	Environmental Improvement Agency, PERA Building, College and W. Manhattan, Santa Fe, New Mexico 87501 Environmental Improvement Board (Same address as above)

1.3 State and Province Agencies

New York	New York State Department of Environmental Conservation, 50 Wolf Road, Albany, New York 12205 State Environmental Board (Same address as above)
North Carolina	Department of Water and Air Resources, P. O. Box 27048, Raleigh, North Carolina 27611 Board of Water and Air Resources, P. O. Box 206, Broadway, North Carolina 27505
North Dakota	North Dakota State Department of Health, State Capitol, Bismarck, North Dakota 58501 Air Pollution Control Advisory Council (Same address as above)
Ohio	Air Pollution Unit, Ohio Department of Health, 450 E. Town Street, Columbus, Ohio 43216 Ohio Air Pollution Control Board (Same address as above)
Oklahoma	Air Pollution Control Division, Environmental Health Services, Oklahoma State Department of Health, 3400 N. Eastern Avenue, Oklahoma City, Oklahoma 73105 Oklahoma Air Pollution Council, Room B-201, 205 N.E. 28th Street, Oklahoma City, Oklahoma 73105
Oregon	Department of Environmental Quality, 1400 S.W. Fifth Avenue, Portland, Oregon 97201 Environmental Quality Commission (Same address as above)
Pennsylvania	Bureau of Air Quality and Noise Control, Department of Environmental Resources, Commonwealth of Pennsylvania, P. O. Box 2351, Harrisburg, Pennsylvania 17120 Environmental Quality Board (Same address as above)
Puerto Rico	Department of Health, Ave. Ponce de Leon, Santurce, P.R. 00908 Environmental Quality Board, 40 Ave. Ponce de Leon, Stop 22 ½, Santurce, P.R. 00910
South Carolina	South Carolina Pollution Control Authority, Owen Building, 1321 Lady Street, P. O. Box 11628, Columbia, South Carolina 29211
South Dakota	South Dakota State Department of Health, Division of Sanitary Engineering and Environmental Protection, Air Quality Control Program, Office Building #2, Pierre, South Dakota 57501 South Dakota Air Pollution Control Commission (Same address as above)
Tennessee	Division of Air Pollution Control, Tennessee Depart-

	ment of Public Health, C2-212 Cordell Hull Building, Nashville, Tennessee 37219
	Tennessee Air Pollution Control Board (Same address as above)
Texas	Air Pollution Control Services, Texas State Department of Health, 1100 W. 49th, Austin, Texas 78756
	Texas Air Control Board, 820 E. 53rd Street, Austin, Texas 78751
Utah	Utah State Division of Health, 44 Medical Drive, Salt Lake City, Utah 84113
	Utah Air Conservation Committee (Same address as above)
Vermont	Agency of Environmental Conservation, Air Pollution Control, Montpelier, Vermont 05602
Virginia	State Air Pollution Control Board, Room 1106, Ninth Street Office Building, Richmond, Virginia 23219
Virgin Islands	Division of Environmental Health, Department of Health, P. O. Box 1442, St. Thomas, Virgin Islands 00801
	Air Pollution Advisory Commission, St. Thomas, Virgin Islands 00801
Washington	Washington State Department of Ecology, P. O. Box 829, Olympia, Washington 98504
	Ecological Commission (Same address as above)
West Virginia	West Virginia Air Pollution Control Commission, 1558 Washington St. East, Charleston, West Virginia 25311
Wisconsin	Bureau of Air Pollution Control and Solid Waste Diposal, 4610 University Avenue, Madison, Wisconsin 53705
	Air Pollution Advisory Council (Same address as above)
Wyoming	Industrial Hygiene Services, Division of Health and Medical Services, Department of Health and Social Services, State Office Building, Cheyenne, Wyoming 82001
	Wyoming Air Resources Council (Same address as above)

Province Agency Addresses

Alberta	Air Pollution Control Section, Division of Environmental Health, Alberta Department of Health, Administration Building, 109th Street and 98th Avenue, Edmonton 6, Alta
British Columbia	Pollution Control Branch, Department of Lands, For-

	ests and Water Resources, Parliament Buildings, Victoria, B.C.
Manitoba	Clean Environment Commission of the Province of Manitoba, Room 304, 402 York Ave., Winnipeg 1, Manitoba
Nova Scotia	Nova Scotia Department of Public Health, P. O. Box 488, Halifax, Nova Scotia
Ontario	Air Management Branch, Department of the Environment, 880 Bay Street, Toronto, Ontario
Quebec	Environmental Health Branch, Ministry of Health, 1075 Chemin Ste-Foy, Quebec, Quebec
Sasktachewan	Air Management Division, Occupational Health Branch, Department of Public Health, 3211 Albert Street, Regina, Saskatachewan

1.4 LOCAL ORDINANCES

Historically, local air pollution ordinances have been based upon the classification of air pollution as a public nuisance because it causes inconvenience, annoyance, discomfort or harm to the general public. In these ordinances, the definition of air pollution and what constitutes an emission subject to abatement is usually rather vague and based upon a sensory response to the emission. The three types of emissions that evoke a sensory response are fine particulates (smoke), coarse particulates (soot) and malodors. These ordinances take little or no account of less obvious contaminants that may be emitted by various sources. Thus, most of the local ordinances in existence are nothing more than smoke and/or odor control regulations.

Visible Emissions

The prohibition of smoke denser than a specified value requires a density or opacity standard and some means for its measurement. The Ringelmann smoke chart is generally used for this purpose (Fig. 1.4a). This chart gives shades of gray with which the density or opacity of columns of smoke can be compared. The Ringelmann system is a series of six cards that presents graduated shades of gray from white to black. It utilizes a rectangular grill of black lines of definite width and spacing on a white background, which can be accurately reproduced. In use, the chart is held at eye level, at a distance such that the lines on the chart merge into shades of gray. The observer then glances from the smoke to the chart and notes the number of the chart most nearly corresponding with the shade or opacity of the smoke. In dealing with a smoke that is not gray in color, the chart most nearly obscuring the observer's view of the white background as the smoke obscures the view of the background is used as the reading.

The Ringelmann system, while not an elegant approach to the measurement of air pollution emissions, has several important advantages for a local official:

1. It is simple in its application.
2. The emissions can be measured from a remote location.
3. Ordinances can be written using the Ringelmann standards.

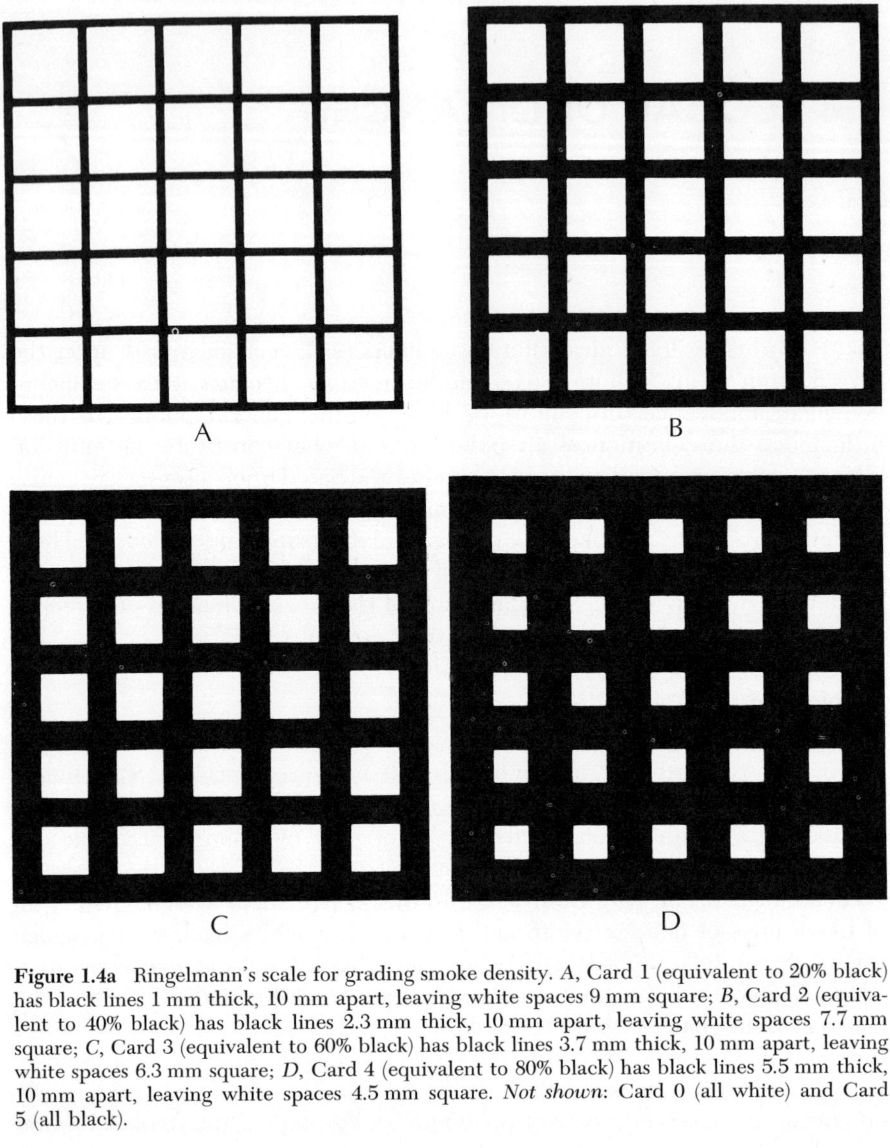

Figure 1.4a Ringelmann's scale for grading smoke density. *A*, Card 1 (equivalent to 20% black) has black lines 1 mm thick, 10 mm apart, leaving white spaces 9 mm square; *B*, Card 2 (equivalent to 40% black) has black lines 2.3 mm thick, 10 mm apart, leaving white spaces 7.7 mm square; *C*, Card 3 (equivalent to 60% black) has black lines 3.7 mm thick, 10 mm apart, leaving white spaces 6.3 mm square; *D*, Card 4 (equivalent to 80% black) has black lines 5.5 mm thick, 10 mm apart, leaving white spaces 4.5 mm square. *Not shown*: Card 0 (all white) and Card 5 (all black).

Odorous Emissions

By way of contrast, ordinances dealing with emissions of odorous materials (Sections 6.19 to 6.24) are awkward to apply and are frequently ineffective because of:

1. A lack of definite emission standards. Because small quantities of contaminants may create strong odors in some situations and not

in other situations, no definite emission standards can be established.

2. Difficulty in relating a source, a contaminant and a complaint. A great deal of rather expensive "detective work" may be required to prove a contaminant, or class of contaminants, caused a complaint and then to establish the exact source of the contaminants.
3. Disagreement among complainants as to quality, severity, time of occurrence, and the like, obscuring the point of whether a nuisance exists.

Thus, the enforcement of odor control and other nuisance ordinances has been rather ineffectual except in very obvious and extreme cases.

Ordinances for Specific Pollutants

Many communities find that the public nuisance approach to air pollution ordinances is adequate for their air pollution problems. However, most large communities have enacted specific air pollution control ordinances.

These ordinances have been designed to deal with a specific type of pollutant or pollutant source. They may include ordinances on particulate emissions from fuel burning plants, particulate emissions from refuse burning equipment, particulate emissions from manufacturing processes, sulfur compound emissions, organic solvent emissions, hydrocarbon emissions, fluoride emissions, and motor vehicle emissions, as well as ordinances on smoke and odor emissions. These types of regulations require a fairly large, sophisticated and, hence expensive, program and staff in order to be effective, thereby restricting them to the larger municipalities.

The leader in the development of air pollution emission control regulations and ordinances has been Los Angeles County. In response to its ever worsening smog problem, the county board of supervisors passed a strong air pollution control ordinance in 1947. This ordinance evolved into the most comprehensive and extensive set of regulations and ordinances for the control of air pollution emissions yet devised for a local level of government. Many of the regulations and ordinances across the country are based upon the Los Angeles regulations, in some cases rather inappropriately because they were developed for needs and processes of Los Angeles.

Projections on Future Local Ordinances

The Clean Air Act (Section 1.1) as amended in December of 1970 charges the state government with the enforcement of air pollution laws, with the federal government acting as backup for the states and as a source of the basic ground rules for air pollution laws. The approach recognizes the limitations of local control efforts and tends to limit local attempts at control to general nuisance ordinances, special local problems, operation of monitoring stations that are part of a large network and monitoring easily checked emissions such as smoke, soot and odors. The more extensive staffs of federal, state and large metropolitan agencies will perform the more sophisticated air quality determinations and stack sampling, as well as the implementation and enforcement of control regulations.

Chapter 2

CONSERVATION OF THE ATMOSPHERIC ENVELOPE

L. W. Canter R. J. Gordon
J. S. Jacobson H. E. Landsberg
C. A. Mazzola P. N. B. Peacock N. Pearson
J. D. Rushton S. F. Singer W. C. Zegel

CONTENTS OF CHAPTER 2*

*For a glossary of terms, refer to the front portion of this volume; and for a list of abbreviations used in this chapter, refer to the listing given after these contents, preceding Section 2.1.

Contents of Chapter 2

Abbreviations

Å Angstrom (10^{-10} meter, 10^{-4} micron)
bbl Barrel
BTU British Thermal Unit

°C	Centigrade
cal	Calorie
cm	Centimeter
COHb	Carboxyhemoglobin
EKG	Electrocardiogram
FAA	Federal Aviation Administration
ft	Foot
g, gm	Gram
gal	Gallon
h	Planck's Constant
HC	Hydrocarbons
$h\nu$	One quantum of radiant energy
°K	Kelvin (°C + 273.2)
Kg	Kilogram
Km	Kilometer
Kw	Kilowatt (1000 watts)
Kwt	Kilowatt Thermal
LTO	Landing-Takeoff
lys	Langley (cal/cm^2)
m, M	Meter
mgm	Milligram
ml	Milliliter
mm	Millimeter
PAN	Peroxyacetyl nitrate having the following form: $CH_3\overset{\overset{\displaystyle O}{\|}}{C}OONO_2$
ppb	Parts per billion
ppm	Parts per million, ppm = μg/m^3(molecular weight/24,500) for gases
SMR	Standardized Mortality Ratio
SST	Supersonic Transport
yr	Year
μ	Micron (10^{-6} meter)
μg	Microgram (10^{-6} gram), μg/m^3 = ppm (24,500/molecular weight) for gases
ν	Vibration frequency

2.1 THEORIES ON THE GLOBAL CONSEQUENCES OF POLLUTION

Before discussing how the earth's climate can be modified by pollution, it is first necessary to summarize our understanding of the present surface and atmospheric temperatures.

The only heat source of any consequence is the sun's radiation which corresponds to that of a black body of about 6,000°K temperature, emitting its maximum energy in the visible region of the spectrum near 0.47 microns. At the distance of the earth, one astronomical unit from the sun (1.5×10^{13} cm), the energy flux s is 1400 watt per m^2. The earth reflects a portion of this energy back into space corresponding to its optical albedo. (See Glossary.) The rest is absorbed and turned into heat and reradiated as infrared or thermal radiation with a maximum wavelength in the vicinity of 10 microns. If the earth had no atmosphere but a good surface conductivity of heat, then its surface temperature would be given by Equation 2.1(1):

$$T = [(1 - A)S/4\sigma]^{1/4} \qquad 2.1(1)$$

where T = the planetary temperature
A = the albedo of the earth
σ = the Stefan-Boltzmann constant.

(The factor 4 appears because the earth absorbs according to its cross section which is a circle, but radiates according to the surface of the sphere.) For an average planetary albedo of approximately 33%, the temperature would be 253°K (−20°C).

The Role of the Atmosphere

The earth's atmosphere changes this simple picture considerably. A clean atmosphere is reasonably transparent to solar visible radiation hence does not absorb much of the incident solar radiation. But even a clear atmosphere contains water vapor (H_2O), carbon dioxide (CO_2) and ozone (O_3) gases which are strong absorbers and radiators in the infrared region of the spectrum. This makes the atmosphere essentially opaque to the infrared radiation which is re-emitted from the earth's surface, except for a "window" which extends approximately between 8 and 12 microns.

In case of a clear atmosphere containing an average amount of dust, about 65% of the incoming solar radiation reaches and is absorbed in the earth's surface. However, only about 10% of the total (infrared) radiation leaving the surface is directly transmitted to space.* The remainder is absorbed by the radiative gases in the atmosphere and is re-emitted successively at the (lower) temperature of the gas layer. Part of it is re-emitted downward to the ground and part of it upward to the next layer and finally to space. The downward radiation helps to keep the surface temperature relatively high by what is termed the "greenhouse" effect. The radiation emitted back into space can be thought of as having two components: (1) that emitted from the very cold outermost layer of the atmosphere and (2) the component directly emitted from the earth's surface and reaching space through the "window" (Fig. 2.1a).[1]

The atmosphere also has the important function of distributing the energy from the sun, cooling the equatorial region and heating the polar regions, in a complicated process of energy transfer which is made more difficult by the rotation of the earth.

The distribution of the radiating gases is largely responsible for the layering of the earth's atmosphere. The troposphere extends from sea level up to approximately 12 kilometers at which point the temperature has dropped to approximately −60°C. Water vapor is about ten times more

* For an average cloudy atmosphere the values are 45% and 5%, respectively.

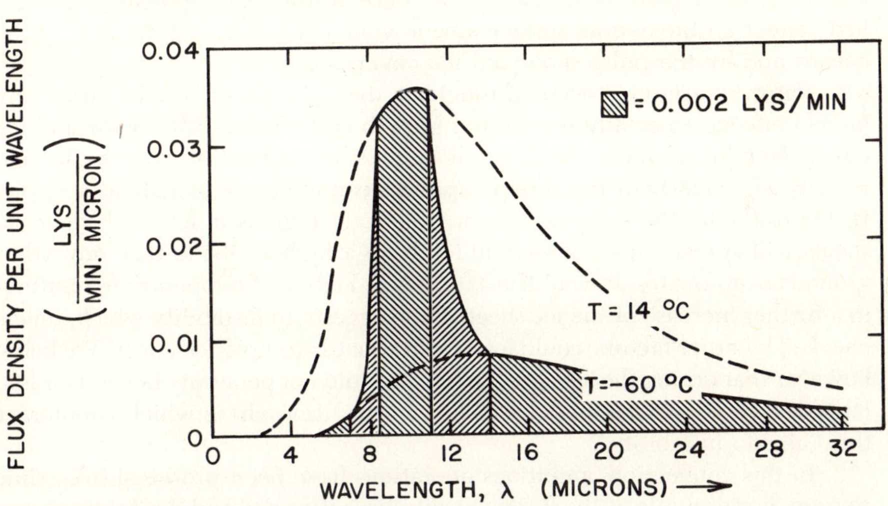

Figure 2.1a The two components of infrared energy flux leaving the earth's atmosphere. The flux in the "window" region corresponds to an average surface temperature. The second component can be thought to originate from an upper and colder atmospheric layer, essentially from the highest 0.3 mm of precipitable water vapor (1 langley per minute = 1 cal per cm^2 per minute).

important than carbon dioxide, both for radiative heating by absorbing solar radiation and for radiative cooling by emitting more infrared than is absorbed. In the stratosphere however, radiative heating by the absorption of solar radiation by ozone is dominant. The radiative cooling that does occur is carried on primarily by CO_2, with water vapor playing a minor role above about 20 kilometers.

This brief discussion sets the stage for understanding how changes in global climate can occur due to changes in the earth's environment. There are four general causes:

1. A change of the earth's surface albedo.
2. A change in the albedo of the atmosphere.
3. A change in the chemical composition of the atmosphere.
4. Releases of heat which are large in relation to solar heating.

(This last mechanism is discussed in Section 2.4.)

Changes in Surface Albedo

The natural albedo of the earth's surface varies from extremely low values (almost black or totally absorbent), like the ocean surface, to extremely high values (very white and therefore nearly completely reflecting), such as freshly fallen snow. In between, there are low-albedo irrigated crop lands and high-albedo deserts. From this preliminary discussion, it can be seen that human activities, principally urbanization and agricultural development, will have little effect on the overall surface albedo which is largely determined by the oceans and by the polar snow and ice covers.

Natural variations occur throughout the year which would dwarf any human effects, especially the natural growth and retreat of the snow and ice cover. Herein, however, lies a point of possible instability whereby human activity, if properly or improperly applied, can influence climate as a whole. If, for example, the temperature in the arctic regions is lowered, then ice sheets will spread. Since snow and ice have a higher albedo than any other ground cover, the result would be a further reduction of temperature, resulting in a further increase of the ice sheet. This suggests an instability which, unless checked by other factors, could lead to an equator-to-pole ice cover. We know however that during the ice ages the ice sheet did not penetrate beyond middle latitudes. Hence, there must be factors in the atmosphere which counteract this "albedo instability."

In this connection, ambitious operations have been proposed from time to time. For example, if the Bering Straits were dammed and the Arctic Ocean were to become ice free, then it might be possible to start another ice age (if, as Ewing and Donn suggested in 1956, the principal requirement for another ice age is the supply of water made available by an ice-free arctic ocean).[2,3] It is believed that once the arctic ice sheet is removed, the arctic ocean would remain ice free. Some intriguing suggestions have been made by J. O. Fletcher for removing the ice sheet.[4] One of them simply requires an albedo change of the ice, spraying it with soot or another dark material which could absorb more solar radiation.

Changes in Atmospheric Albedo

The reflecting properties of the earth's atmosphere with respect to incoming solar radiation are determined mainly by the amount and distribution of water and ice clouds, and less so by particulate material. Water clouds form at lower altitudes, ice clouds (cirrus) in the upper troposphere. Water droplets and ice crystals mainly scatter the visible radiation rather than absorb it. On the other hand, dust and aerosols partially reflect and partially absorb incident solar radiation, depending on their exact composition. It is possible that the small absorptivity of clouds is produced by "dirtiness" introduced by a dust nucleus.

The average cloud albedo varies from 10 to 25% for cirrus to about 70 to 80% for thick cumulus clouds. It should be pointed out that "white" clouds, i.e., with a high albedo in the visible, can be "black", i.e., efficient absorbers and therefore radiators, in the infrared region of the spectrum.

The effect of atmospheric particles in the size range of 0.1 to 2.5 microns is complicated by their optical properties and their exact location. Particles which absorb in the infrared part of the spectrum (but not in the visible) may cool the upper atmosphere, while particles containing absorbing material in the visible may heat it. For example, it is believed that high-altitude cirrus will cool the upper troposphere, while the eruption of the volcano Agung, which put a great deal of dust into the stratosphere, produced worldwide heating of the layer.

Conceivably, human influences can be important here. High-altitude aircraft laying down condensation trails produce artificial cirrus clouds, while at the same time ejecting smoke particles which simulate volcanic dust. At lower altitudes, much particulate material is produced in industrialized areas by stationary power plants, metal smelters and the like and in the tropical agricultural areas by slash-and-burn agricultural procedures.

At present, the injection of particulate material (dust, smoke and aerosol particles) by human activity is small compared to the natural injection of particles which come on a steady basis from forests in the form of terpenes, from oceans in the form of salt particles, from deserts and other wind-blown dust and from occasional eruptions by volcanoes. The latter can, and do, put large quantities of particulates into the stratosphere where they remain for several years and add to the already present stratospheric aerosol layer.

Changes in Atmospheric Composition

The major concern is the steady increase in the carbon dioxide content of the atmosphere produced by the rapid burning of fossil fuels. Also of concern are changes in the water vapor concentration and ozone concentration in the stratosphere. Of lesser concern for global climate change are sulfur dioxide, nitrogen oxides, carbon monoxide, methane and other minor gases, although they may produce intense local problems.

Since late in the nineteenth century, man has released carbon dioxide to the atmosphere at an ever-accelerating rate through the combustion of fossil fuels. About 50% of the CO_2 released has remained in the atmosphere, the rest being absorbed into the ocean. The present rate of CO_2 production by man is estimated at 15×10^9 metric tons per year. The average planetary atmospheric CO_2 concentration in 1970 was very close to 320 ppm, or 11% above the generally accepted nineteenth century base level of 290 ppm. From a variety of observations, one may conclude that the secular rate of increase of CO_2 is now about 0.2% per year, corresponding to a doubling time of about 23 years in the excess above 290 ppm. By 2000 A.D., the increase will be nearly 30% over base level.

According to the numerical investigations of Manabe and collaborators,[5] increases in CO_2 result in a warming of the entire lower atmosphere, with the amount of warming dependent on the humidity, but with cloud distribution, surface albedo and so on remaining constant. Under those conditions, a 10% CO_2 increase can lead to a warming of 0.2 to 0.3°C.

Mitchell[6] concludes that probably not more than one-third of the planetary temperature disturbance of the past century is attributable to variations of atmospheric CO_2. Other mechanisms are evidently required to account for part of the warming observed between 1880 and 1940, as well as for the cooling observed since 1940. It is generally believed that atmospheric turbidity (due to particulates, mostly natural but partly manmade) could be responsible for the difference. On this basis, Mitchell has projected temperature trends for the next 100 years, attributable to anticipated future increases of CO_2 and particulates, both from human activities (Fig. 2.1b). The different curves correspond to different estimates of the future rate of increase of particulates and their effect on temperature.

It should be emphasized that such extrapolations cannot be considered realistic unless the complete atmosphere is modeled at the same time. Increased temperature will lead to increased evaporation from the ocean and therefore an increased water content in the atmosphere. If, therefore, cloudiness increases, then this will offset temperature increases. For the time being, we must rely on careful monitoring over the complete earth's surface by means of weather satellites and on improvements in the numerical modeling of the atmosphere by using high-speed computers.

Another area of concern is a possible general change in the ozone content of the atmosphere. Since ozone has important radiative properties as it controls the heating in the stratosphere, a general concentration change could have further consequences on a global scale. Interest in this problem has been focused recently through discussions on the environmental effects of the

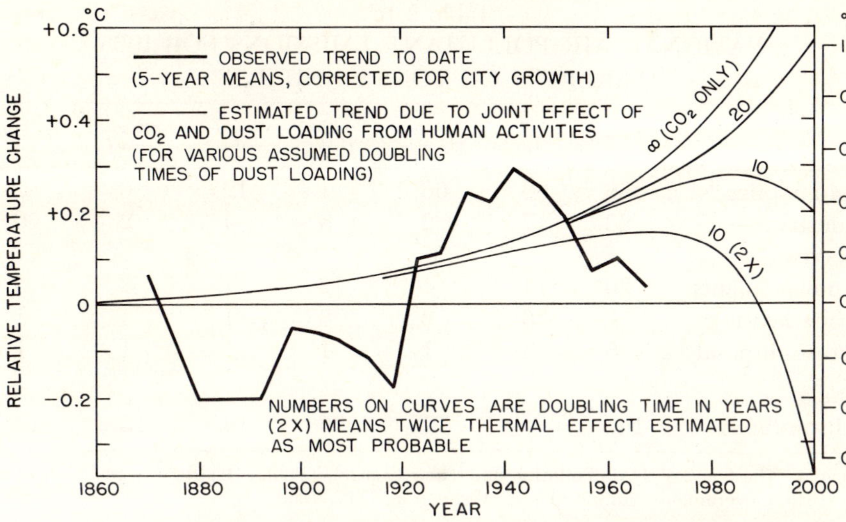

Figure 2.1b Trend of annual mean temperature (Northern Hemisphere, 5-year means, corrected for city growth) 1860–2000 AD. Projected temperature trends in the next 30 years attributable to anticipated future increases of global atmospheric contamination by both CO_2 and particulates from human activities. Different curves correspond to different estimates of the future rate of increase of particulates and their effect on temperature. Numbers on curves are doubling time in years; "(2x)" means twice the thermal effect estimated as most probable. Added contribution of thermal pollution, and all natural sources of climatic change have been neglected.

supersonic transport. Water vapor ejected by the burning of aircraft fuels, and especially nitrogen oxides produced in high-temperature combustion, can remove some of the stratospheric ozone, or possibly even a large fraction. Since the water vapor content in the stratosphere is low, only two parts per million, injection from SST's or from other human activities could lead to important consequences.

Other Chemical Emissions

The emission of other gases may have important local effects and may lead to intense air pollution problems; but on a global scale, they are not considered to be serious. The breakdown shown in Table 2.1c gives the emissions in the United States of various air pollutants. (More recent data is presented in Table 2.10a.) The major aspects of the circulation through the atmosphere of SO_2, NO_x and CO have been discussed by Robinson and Robbins.[7] They estimate the source magnitude, the residual atmospheric concentrations, and discuss the scavenging processes.

One-third of the sulfur reaching the atmosphere comes from pollutant sources, namely as SO_2 (Table 2.1d). Within the atmosphere, there is a net transfer of sulfur from land to ocean areas. The average tropospheric concen-

Table 2.1c
NATIONAL AIR POLLUTANT EMISSIONS FOR 1965
(MILLIONS OF TONS ANNUALLY)

	Totals	*% of Totals*	*Carbon Monoxide*	*Sulfur Oxides*	*Hydro-carbons*	*Nitrogen Oxides*	*Particles*
Automobiles	86	60	66	1	12	6	1
Industry	23	17	2	9	4	2	6
Electric power plants	20	14	1	12	1	3	3
Space heating	8	6	2	3	1	1	1
Refuse disposal	5	3	1	1	1	1	1
Totals in millions	142		72	26	19	13	12

SOURCE: "The Sources of Air Pollution and Their Control," Public Health Service Publication No. 1548, Government Printing Office, Washington, D.C., 1966.

trations of sulfur compounds are shown in Table 2.1e. Most of the scavenging takes place through precipitation and dry deposition as shown in the flow diagram of Figure 2.1f.

The circulation of nitrogen compounds through the earth's environment is quite complex. Of the seven significant atmospheric forms of nitrogen, the only important pollutants emitted by human activities are NO and NO_2 (Table 2.1g). The atmospheric concentrations are shown in Table 2.1h, and a simplified flow diagram is given in Figure 2.1i.

Table 2.1d
ESTIMATED GLOBAL EMISSIONS
OF SULFUR COMPOUNDS

Compound	*Source*	*Estimated Emissions (tons/yr)*	*Emissions as Sulfur (tons/yr)*
SO_2	Coal combustion	102×10^6	51×10^6
	Petroleum refining	6×10^6	3×10^6
	Petroleum combustion	23×10^6	11×10^6
	Smelting operations	16×10^6	8×10^6
H_2S	Industrial emissions	3×10^6	3×10^6
	Marine emissions	30×10^6	30×10^6
	Terrestrial emissions	70×10^6	70×10^6
SO_4	Marine emissions	130×10^6	44×10^6
	Total emissions		220×10^6

Table 2.1e
AVERAGE TROPOSPHERIC CONCENTRATIONS OF SULFUR COMPOUNDS

Compound	*Average Concentration*	*Average Concentration as Sulfur* ($\mu g/m^3$)
SO_2	0.2 ppb	0.25
H_2S	0.2 ppb	0.14
SO_4	2 $\mu g/m^3$	0.7

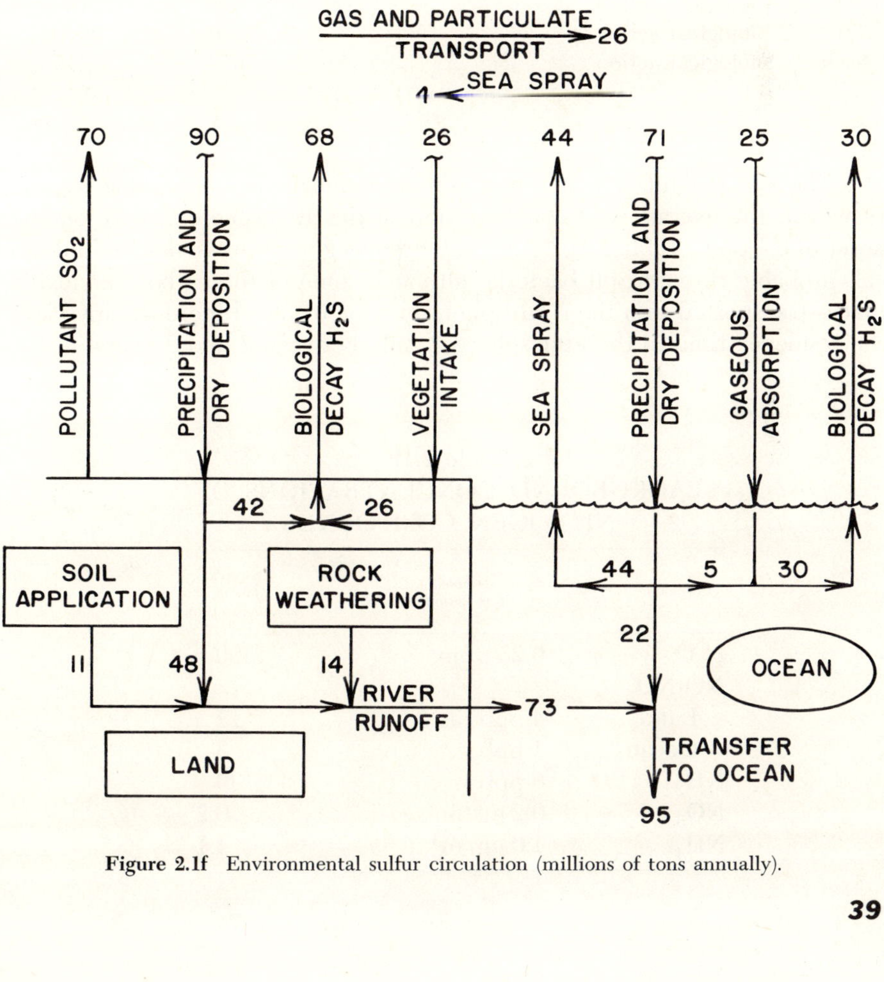

Figure 2.1f Environmental sulfur circulation (millions of tons annually).

Table 2.1g
ESTIMATED ANNUAL GLOBAL EMISSIONS OF NITROGEN COMPOUNDS

Compound	*Source*	*Source Magnitude (tons/yr)*	*Estimated Emissions (millions of tons annually)*	*Emissions as Nitrogen (millions of tons annually)*
NO_2	Coal combustion	3074×10^6	26.9	8.2
	Petroleum refining	11317×10^6 (bbl)	0.7	0.2
	Gasoline combustion	379×10^6	7.5	2.3
	Other oil combustion	894×10^6	14.1	4.3
	Natural gas combustion	20.56×10^{12} (ft^3)	2.1	0.6
	Other combustion	1290×10^6	1.6	0.5
	Total NO_2		52.9	16.1
NH_3	Combustion		4.2	3.5
NO_2	Biological action		500	150
NH_3	Biological action		5900	4900
N_2O	Biological action		650	410

The major significant source of carbon monoxide is the gasoline engine. However, the average CO concentration in the atmosphere is only on the order of 0.1 ppm (Table 2.1j).[8] The major scavenging process may be biological, probably through soil bacteria, although some of the carbon monoxide that is transported into the stratosphere may be oxidized there. In any case, the residence time in the atmosphere is only between 2 and 3 years.

Table 2.1h
BACKGROUND CONCENTRATIONS OF NITROGEN COMPOUNDS

Compound	*Ambient Concentration*	*Atmospheric Mass as N (millions of tons)*
N_2O	0.25 ppm	1500
NO-NO_2		
Land	4 ppb (as NO_2)	12
Ocean	1 ppb	3
NH_3	6 ppb	30
NO_3	0.2 $\mu g/m^3$	0.2
NH_4	1.0 $\mu g/m^3$	4.1

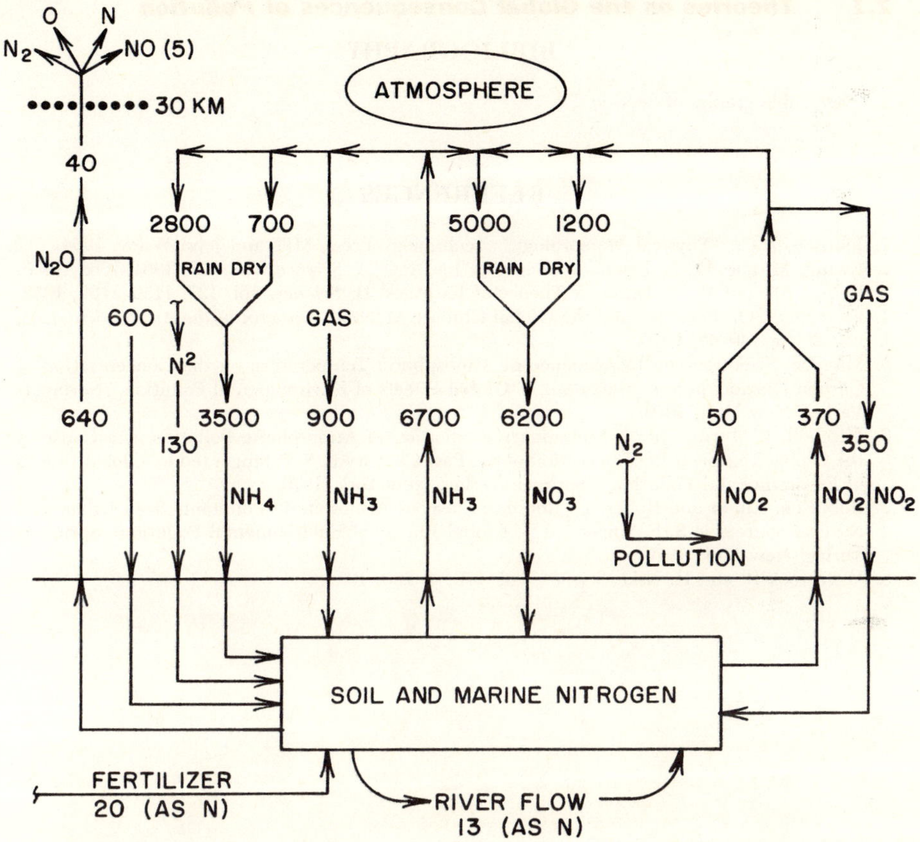

Figure 2.1i Simplified nitrogen compound circulation (millions of tons annually).

Table 2.1j
ESTIMATED GLOBAL EMISSIONS OF CO

Source	*Consumption (millions of tons annually)*	*CO Emission (millions of tons annually)*
Gasoline	379	193
Coal	3074	12
Wood (fuel)	466	16
Incineration	500	25
Forest fires	18[a]	11.3
Total		257

[a] Millions of acres annually.

NOTE: A recent study shows that naturally produced carbon monoxide, by the oxidation of atmospheric methane, amounts to about 5×10^9 tons per year, or about 25 times the amount produced by combustion.

2.1 Theories on the Global Consequences of Pollution

BIBLIOGRAPHY

See Bibliography of Section 2.4.

REFERENCES

1. Johnson, J. C.: "Physical Meteorology," Technology Press, MIT and John Wiley, 1954.
2. Ewing, M. and W. L. Donn: A Theory of Ice Ages, I, *Science,* vol. 123, 1061–1066, 1956.
3. Ewing, M. and W. L. Donn: A Theory of Ice Ages, II, *Science,* vol. 127, 1159–1162, 1958.
4. Fletcher, J. O.: Polar Ice and the Global Climate Machine, *Bulletin of the Atomic Scientists,* vol. 26, pp. 40–47, 1970.
5. Manabe, Syukuro: The Dependence of Atmospheric Temperature on the Concentration of Carbon Dioxide, in S. F. Singer (ed.), "Global Effects of Environmental Pollution," Springer-Verlag, New York, 1970.
6. Mitchell, J. Murray, Jr.: A Preliminary Evaluation of Atmospheric Pollution as a Cause of the Global Temperature Fluctuation of the Past Century in S. F. Singer (ed.), "Global Effects of Environmental Pollution," Springer-Verlag, New York, 1970.
7. Robinson, Elmer and Robert C. Robbins: Gaseous Atmospheric Pollutants from Urban and Natural Sources in S. F. Singer (ed.), "Global Effects of Environmental Pollution" Springer-Verlag, New York, 1970.
8. Weinstock B. and H. Niki: *Science,* vol. 176, p. 290, 1972.

2.2 THE POLLUTION CYCLE

This section investigates some of the interactions and exchanges that may occur between the land, water and air. Its purpose is to make the reader aware of these interactions and exchanges and to illustrate that the somewhat arbitrary lines that have been drawn between air, water and land pollution are really nonexistent. Each type of pollution affects the other carriers, due to the integrated nature of the environment.

The Evolution and Nature of Life

Earth is unique among the planets of the solar system because of its water, free oxygen and life. From its origin as a bit of stellar material, it has evolved and so has the life it harbors. The life, which started somehow in its anaerobic seas, evolved to an autotroph and an organism capable of manufacturing its own food and releasing oxygen in the process. The oxygen built up in the oceans and began to escape into the atmosphere. The harsh ultraviolet light from the sun, which restricted life to certain select sites, dissociated the oxygen and formed ozone. The ozone acted as a shield against the ultraviolet rays and allowed a proliferation of life throughout the seas and a great increase in the production of oxygen. The existence of free oxygen allowed the development of life consisting of different tissues and organs. The development of specialized organs allowed the invasion of the land and resulted in the development of still higher forms of life. This whole pyramid of life is supported by a complex set of interactions and exchanges between water, air, rock and life (Fig. 2.2a).

At the top of the pyramid of complexity, at this point in time, is humankind. As a result of their conditioning through millions of years of evolution, human beings have come to regard nature as "the enemy." They have endeavored to conquer nature, tame the wilderness, make war on pests and vermin and control rivers. They consider life as a battle against the elements in which only the fittest survive. This behavior was essential for this form of life to have survived for as long as it has. However, the crude tools necessary for the survival of the species have proliferated into the tools of the industrial revolution. It is now possible for human beings to produce all the material

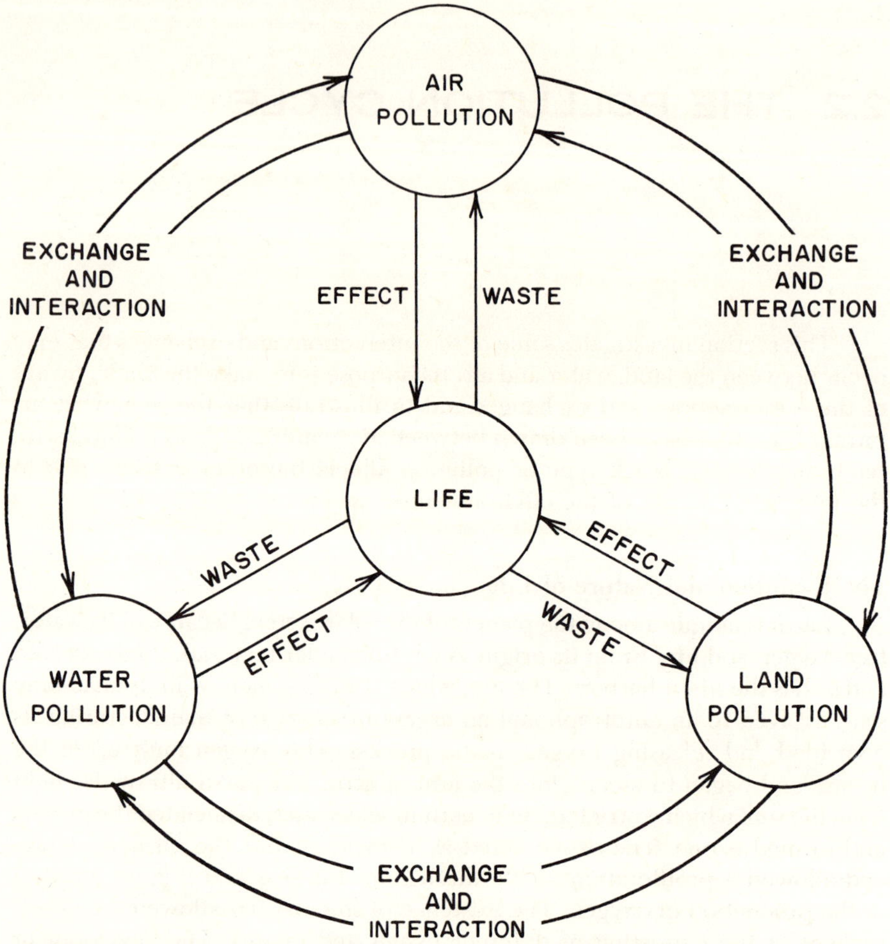

Figure 2.2a Pollutant exchanges and interactions between air, water, land and life.

things needed for survival, and the competition with nature is no longer necessary.

However, changing tens of thousands of years of conditioning is not easy; and the technological civilization that has developed is continuing to treat nature as "the enemy" and is using earth as the dumping ground for its wastes. The contamination of the water, air and rock with humankind's wastes has raised the threat of the elimination of all life.

Just as the pyramid of life is supported by a complex set of interactions and exchanges between water, air, rock and life, so too is the pollution of the environment complicated by interactions and exchanges between water, air, rock and life. Many of the interactions and exchanges are unknown, as is their effect on life. Many are subtle and their true impact has yet to be determined.

Air and Water

Most of the exchanges of pollutants that may occur between the air and the water (Fig. 2.2b) involve the primary air pollutants, which include:

1. Sulfur oxides
2. Particulate matter
3. Carbon monoxide
4. Photochemical oxidants, which result from a series of photochemical reactions involving oxygen, nitrogen oxides and organic material
5. Nitrogen oxides
6. Hydrocarbons

One of the recommended methods of controlling emissions of sulfur oxides, particulates and nitrogen oxides from an industrial plant is to pass the gases through a water scrubber. The water in the scrubber dissolves the acid gases

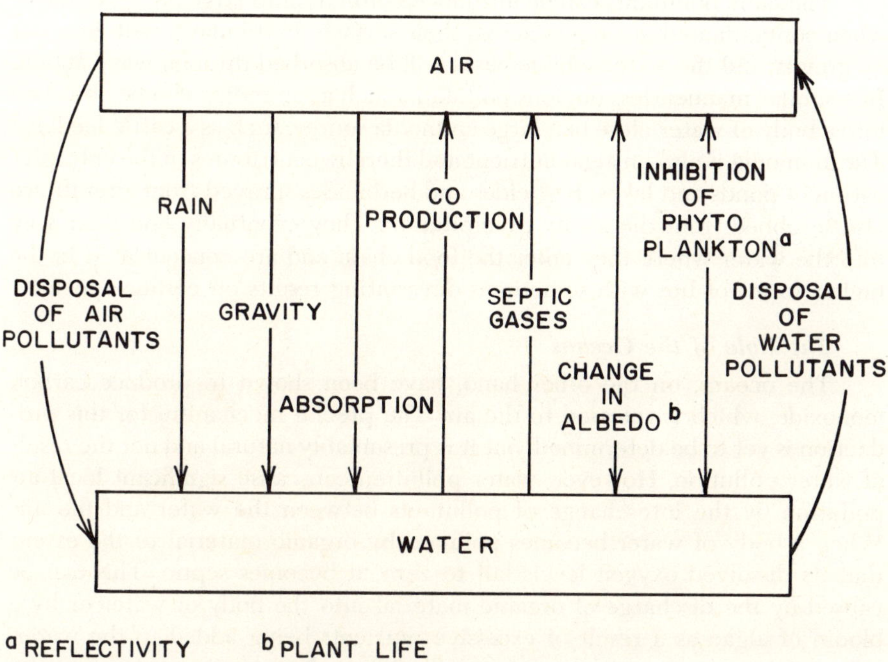

Figure 2.2b Some exchanges and interactions of pollutants between air and water.

and retains some of the particulates as suspended matter and others as dissolved matter. The process requires a large surface area of water in intimate contact with the gases.

Scrubbing Effect of Rain

In nature, this scrubbing process is accomplished by precipitation. Rain presents a large surface area of water to contaminated air. It dissolves the acid gases and retains particulates by impaction or, more importantly, by using them as nuclei for the formation of the rain. Thus, the air is cleaned by the natural scrubbing action of the rain.

Unfortunately, the quantities of pollutants carried by the rain can be sizable. In some areas where rainwater was collected as a source of drinking water, the practice has been discontinued because of the contaminants in the water. The pH of rainwater in the Netherlands and Sweden has been measured to be as low as 4. These air pollutants are ultimately carried into the water causing water pollution. The nitrogen, sulfur and trace minerals act as nutrients for algae in ponds and lakes, hastening the death of these bodies of water. The lead, cadmium, mercury and other pollutants are introduced into the food chain of ponds, lakes and seas to ultimately poison the higher forms of life.

The same pollutants can be introduced directly into large bodies of water when contaminated air moves across their surface. Particulates will drop out by gravity and the water soluble gases will be absorbed through wave action. In a similar manner, less obvious pollutants such as ammonia can be absorbed into a body of water close to a large ammonia source, such as a cattle feedlot.[1] The ammonia is also an algal nutrient and thereby contributes to the eutrophication of ponds and lakes. Pesticides and herbicides sprayed from aircraft are also scrubbed from the air by precipitation. They eventually find their way into the water where they enter the food chain and are concentrated in the higher forms of life with sometimes devastating results on certain species.

The Role of the Oceans

The oceans, on the other hand, have been shown to produce carbon monoxide, which is released to the air. The precise mechanism for this production is yet to be determined, but it is presumably natural and not the result of water pollution. However, water pollution can cause significant local air pollution by the interchange of pollutants between the water and the air. When a body of water becomes polluted by organic material to the extent that its dissolved oxygen levels fall to zero, it becomes septic. This can be caused by the discharge of organic material into the body of water or by a bloom of algae as a result of excessive nutrients being added to the water. When it becomes septic, it emits to the atmosphere noxious gases such as hydrogen sulfide as well as strongly odorous compounds. These can cause sickness and psychological discomfort as well as discoloration of lead-based paints and other physical damage.

More subtle effects on the atmosphere may result from other forms of water pollution. The contamination of the oceans with various pesticides could have an as yet undiscovered effect on the ability of ocean phytoplankton to convert carbon dioxide into oxygen. This would affect the carbon dioxide–oxygen balance in the atmosphere causing an increase in the concentration of the "pollutant" carbon dioxide. This increase may in turn raise the temperature of the earth, melting the polar caps (somewhat) and decreasing the solubility of carbon dioxide in the oceans. The oceans would then release more carbon dioxide, and less heat would be reflected by the smaller polar caps; both of which would further increase the temperature of the earth in a vicious cycle (Section 2.1).

Similarly, the effect on the atmosphere of the 1.5 million tons of oil introduced yearly into the oceans as water pollution from shipping, offshore drilling and accidents is unknown.[2] Aside from direct damage to marine life and possible inhibition of the phytoplankton, a molecule-thick film of oil on the sea could increase the albedo of the earth, lowering the temperature. This could also result in an upset of the carbon dioxide–oxygen balance and another ice age.

Industry as it tries to control its contribution to the pollution problem frequently converts one form of pollution to another. For example, if a water scrubber is employed to control an emission containing large amounts of sulfur dioxide, what can be done with the acidic water saturated with sulfur dioxide? Similarly, the incineration of sewage sludge is complicated by the large amounts of water retained in a filtered sludge. This water can cool the combustion so that products of partial combustion are emitted to the atmosphere as air pollution.

Air and Land

Figure 2.2c illustrates some of the interactions and exchanges of pollutants that may occur between the air and land. The land has been found to be a sink for many air pollutants through microbial or chemical means. Experiments[3] indicated that soil can remove carbon monoxide from the air. The uptake of carbon monoxide is apparently due to a biological rather than a physical mechanism. It is estimated that the soil of the continental United States has a capacity for carbon monoxide of 569 million metric tons per year, which is 6.5 times the annual estimated production of carbon monoxide. In the absence of any other significant carbon monoxide sinks, the soil must be considered the major sink for carbon monoxide.

Later experiments[4] also demonstrated that soil may also be a significant sink for hydrocarbons, sulfur dioxide and nitrogen dioxide from the air. The experiments indicated that the hydrocarbons were removed by a microbio-

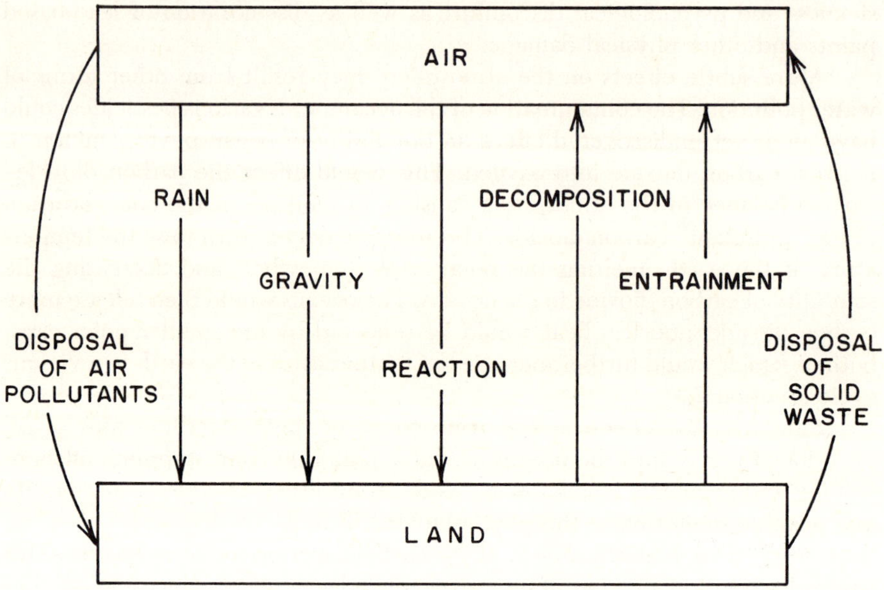

Figure 2.2c Some exchanges and interactions of pollutants between air and land.

logical mechanism, whereas the nitrogen dioxide and sulfur dioxide were primarily removed by a chemical reaction. Seven million tons of ethylene may be removed from the air in the United States by microbial degradation. The capacities of the soil in the United States for nitrogen dioxide and sulfur dioxide are 600 million and 40 billion tons per year, respectively. These compare with estimated United States productions of 15 million tons of ethylene, 33 million tons of nitrogen dioxide and 32 million tons of sulfur dioxide each year.

Particulate matter in the air is also transferred to the soil, as it is to the water, through the action of gravity and precipitation. Other pollutants scrubbed out of the air by precipitation may also be incorporated into the soil. A particular problem around certain industries is the deposit of fluorides in the soil and on the plants growing in the soil. This fluoride can poison grazing animals, particularly cattle, by attacking their teeth and bones.

Waste products discarded on or in the land can also contribute to various forms of air pollution. Fine material placed on the land can become entrained in the air, creating a particulate problem downwind. The decomposition of material in an improperly operated landfill can produce odorous compounds that in turn create a local air pollution problem. Even in a properly operated landfill, methane produced by the decomposition of organic matter can diffuse into the local atmosphere, possibly creating an explosive mixture with the air in low-lying areas.

The problem of disposing of fine particulate matter removed from industrial atmospheric emissions by electrostatic precipitators and other control equipment is another example of converting one form of pollution into another. Some other interactions between the land and the air are the consumption of oxygen by rocks containing reduced compounds and the emission of various gases and particulate matter by volcanic action.

Land and Water

There are several interactions and exchanges of pollutants between the land and water (Fig. 2.2d). The transfer of pollutants from the land to the water is primarily due to leaching. Leaching is the term applied to the process of extracting a water soluble material from a solid by filtering the water through the solid. An excellent example of the process is a coffee pot. In a coffee pot, the soluble coffee is dissolved from the insoluble grounds by the water, and the resultant solution is separated from the grounds. In a similar manner, soluble organic matter is extracted from landfills by the percolation of rainwater through the soil. This organic contaminated water is a particularly potent source of water pollution when it finds its way into streams and wells around the landfill.

Sources of algal nutrients such as phosphates and nitrates can also be leached from the soil and deposited in streams, ponds and lakes. The primary

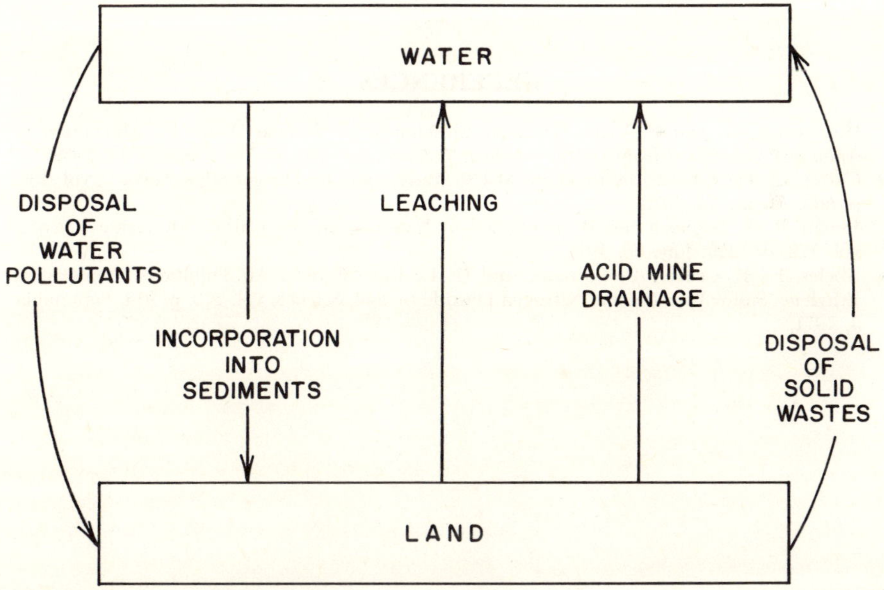

Figure 2.2d Some exchanges and interactions of pollutants between water and land.

sources of these nutrients are agricultural fertilizer and animal feedlots. In ponds and lakes, they stimulate the growth of algae, which in turn hastens the eutrophication of these bodies of water. Other elements, such as lead, mercury, and cadmium, can be leached from deposits of industrial wastes creating other forms of water pollution. As little as 1 to 3 ppm of copper or less than 1 ppm of zinc in a stream may be poisonous to aquatic life.

One of the oldest sources of water pollution is acid mine drainage. This acid water is produced by both working and abandoned mines. It is the result of oxidation of sulfur bearing minerals by air and water to form sulfurous and sulfuric acid. This acid water enters the streams and lakes and quickly destroys plants and fish and the wildlife that depend on them as a food source. The Environmental Protection Agency estimates that over 10,000 miles of streams and rivers are polluted by acid mine drainage in the United States. The only solution seems to be either sealing the mine from groundwater or treatment of the acid water as it comes from the mine.

Pollutants in the water can be removed by various mechanisms involving filtration through the rocks or the production of new rocks by sedimentation. For example, acid water can be neutralized by exposure to basic rocks such as limestone.

When inorganic water pollutants finally reach the sea, reactions such as the carbon dioxide–bicarbonate–carbonate equilibrium can quickly destroy acidic materials and cause heavy metal ions to precipitate and become incorporated into the sea sediments.

REFERENCES

1. Hutchinson, G. and F. Viets: Nitrogen Enrichment of Surface Water by Absorption of Ammonia Volatilized from Cattle Feedlots, *Science*, vol. 166, p. 514, October 24, 1969.
2. Carter, L.: The Global Environment: M.I.T. Study Looks for Danger Signs, *Science*, vol. 169, p. 660, August 14, 1970.
3. Inman, R., R. Ingersoll and W. Levy: Soil: A National Sink for Carbon Monoxide, *Science*, vol. 172, p. 1229, June 18, 1971.
4. Abeles, F., L. Craker, L. Forrence and G. Leather: Fate of Air Pollutants: Removal of Ethylene, Sulfur Dioxide, and Nitrogen Dioxide by Soil, *Science*, vol. 173, p. 914, September 3, 1971.

2.3 PLANNING FOR CLEAN AIR

Planning for clean air is a comprehensive regional planning problem. The uses of our air must be planned along substantially the same lines as the use of entire river-basin units in the conservation of land and water resources. The major problem is to establish a jurisdictional, administrative and financial context for the application of already existing techniques.

The World Context

The atmospheric ocean is a nonlinear system: any input may cause an adverse amplification.[1] All regions are directly or indirectly interdependent: a given air parcel in mid-latitudes circles the earth about once a month, and a complete interchanging of all air masses between latitudes and hemispheres takes place once in every 2 years. For example, urban air pollution studies show that environmental "highways" within the biosphere cause automobile exhaust emissions in concentrated metropolitan areas to affect agriculture hundreds of miles away.

International mechanisms are needed to: (1) take emergency action on imminent and substantial dangers to public health; (2) establish standard international air quality and emission objectives applicable on a regional basis; (3) ensure standard inventory techniques which supply supporting monitoring data to the first global network set up by the World Health Organization (with regional centers in Moscow, Nagpur, Tokyo and Melbourne, fed by twenty other related laboratories, all feeding into international centers in London and Washington); (4) conduct research and disseminate information on control and prevention; (5) conduct surveys and issue abatement directives requiring remedial and preventive measures.

Five fundamental and urgent areas of action stand out: (1) metrology, standardizing and adoption of measurement terms to express various pollutant concentrations; (2) the coordination and sharing of research; (3) declarations of aims and attitudes in regional work; (4) formulation and adoption of standards and procedures for air pollution control; (5) the machinery for enforcement of concluded environmental agreements. Within this world context, action depends on an effective regional approach.[2,3,4]

2.3 *Planning for Clean Air*

Regional Planning

Contaminants from nonregulated point sources beyond local jurisdictions make local control useless. Local enforcement is difficult: experience shows that the larger the region legally affected, the more varied and complex is the socioeconomic base and the less susceptible the planning and control process is to individual and group pressures. Planning for clean air is an integral part of regional planning which treats causes, not symptoms, and which can provide social benefits to offset the expense of compliance. Only in this way can the abuse of common goods such as air be realistically tackled. The solutions lie in regional land use management and in redesigning transport systems, public power supply methods, industrial and domestic processes and systems.[5]

Regional Control

If in all regions of the earth air pollution abatement was initiated, then within 25 years, control would catch up with emissions and would stabilize the situation at present levels, while simultaneously buying time for the environmental managers to pursue strategies of correction and abatement.[6,7]

Three main regional strategies are used, either singly or in combination: (1) treating or reducing the generated pollutants before or after they enter the environment; (2) regulating the location of the causal activity, or of the recipients, or both; (3) modifying or eliminating the causal activity. The main principle is to assign presumptive "source" responsibilities, tackle matters on a priority basis by stages and coordinate the redesigning of future environment systems to generate fewer problems. The major tasks are: (1) the control and prevention of photochemical smog; (2) the control and prevention of sulfur dioxide smog; (3) the removal of present major sources such as automobile exhausts, public power plants, domestic heating systems, "point" sources of industrial emissions, heavy-metals contaminators, airborne pesticide residues and "heat islands," major urban areas and large industries which can affect rainfall and local climate.

The Regional Planning Process

To these ends, the regional planning process can be directed. Regional inventory will identify the particular mix of problems, causal relationships and point sources. A series of computer simulations can produce approximate models of the metabolism of the region and give evidence of systems needing correction. The planners of land-use, transport, services and water-resource systems can produce integrated regional planning alternatives to which models of the simulated metabolism of the region can be applied. This, in turn, leads to the selection of the optimum regional plan[8] including air management strategies, standards and specific objectives. It is essential that the regional plan be fully supported by all levels of government and given statutory force and strong financial commitment. The regional plan then will provide a realistic context for action.[9]

Implementation

The means of implementation is a land-use plan locating residential areas in healthful places and collecting functions into environmental units which can be serviced by a variety of modes of transport and by district heating and servicing systems. They should be intersected by reserved open-space areas allocated to give proper air drainage, to allow the creative use of solid wastes and to permit new communities to be built on the concept of nonpolluting, closed systems.

Implementation methods should include: (1) enforcement of minimal standards[10] and the removal of offensive point-sources by public redevelopment; (2) allocating a definite "life" to problem sources; (3) using "clean air zoning" for progressive upgrading of problem areas; (4) legal supports such as increased enforcement of stricter laws; emission charge systems in the form of taxes, ensuring that such "air pricing" does not become a license to pollute; incentive payments under organized financial subsidization; and compensation to those who were phased out under the plan. The regional plan can be for a 20-year period, updated and revised every 5 years, and developed in 5-year stages of implementation.

REFERENCES

1. Callender, G. S.: Temperature Trends and Fluctuations over the Earth, *Journal of Royal Met. Soc.*, 87, 1967.
2. Commission for Air Chemistry and Radioactivity of the International Association of Meteorology and Atmospheric Physics of the International Union of Geodesy and Geophysics: "Report to the World Meteorological Organization on Station Networks for Worldwide Pollutants," World Meteorological Organization, 1970.
3. Study of Critical Environmental Problems (SCEP): "Man's Impact on the Global Environment: Assessment and Recommendations for Action," The M.I.T. Press, 1970.
4. Citron, R.: "National and International Environmental Monitoring Programs," Smithsonian Institution, Cambridge, Massachusetts, 1970.
5. Peterson, J. T.: "The Climate of Cities: A Survey of Recent Literature," NAPCA, U.S. Government Printing Office, Washington, D.C., 1969.
6. Crocker, T.: "The Structuring of Atmospheric Pollution Control Systems: Seminar on the Economics of Air Pollution," U.S. Public Health Service and the American University, Washington, D.C., April 21, 1965.
7. Stern, A. C.: "Air Pollution, Vol. I: Air Pollution and Its Effects"; "Vol. II: Analysis, Monitoring and Surveying"; "Vol. III: Sources of Air Pollution and Their Control," 2d. ed., Academic Press, New York and London, 1970.
8. Hillhorst, J. G. M.: "Regional Planning," Rotterdam University Press, Rotterdam, 1970.
9. United Kingdom White Paper: "The Protection of the Environment: The Fight Against Pollution," conf. 4373, London, May 1970.
10. N.A.P.C.A.: "Air Quality Criteria for Particulate Matter," Raleigh, North Carolina, 1970.

BIBLIOGRAPHY

Borgstrom, G.: "Too Many: The Earth's Biological Limitations," Macmillan, New York, 1969.

Dansereau, P. (ed.): "Challenge for Survival: Land, Air and Water for Man in Megalopolis," Columbia University Press, New York, 1970.

Herfindahl and Kneese: "Quality of the Environment," Baltimore, 1965.

Lamb, H. I. J.: "The Changing Climate," Methuen Books, London, 1966.

Lowry: "Weather and Life: An Introduction to Biometeorology," Academic Press, New York, 1969.

Scientific American: "Cities," Knopf, New York, 1965.

Shapley, T.: Climatic Change: Evidence, Causes and Effects, Cambridge, Massachusetts, Harvard University Press, 1953.

Singer, S. F. (ed): "Global Effects of Environmental Pollution, Reidel Publishing Company, Dordrecht, 1970.

de Vos, Pearson, Silveston and Drynan: "The Pollution Reader," Harvest House, Montreal, 1968.

Wagner, R. H.: "Environment and Man," Norton, New York, 1971.

2.4 THE GLOBAL HEAT BALANCE

Sections 2.1 and 2.7 discuss the possible changes in global climate with the sun being the sole source of energy. Here, the consequences of the introduction of energy on the earth by human activities will be covered. This heat source will be compared to the solar heat source, in order to estimate its importance.

The worldwide effects of energy consumption can be estimated as follows: as of 1970, the United States population of 204 million was estimated to consume 68.5×10^{15} BTU per year (2.2×10^{12} watts or 17.2×10^{15} Kg-cal/year): the installed electrical power generating capacity was approximately 3×10^{11} watts. Thus, the United States energy consumption per capita was a little over 10 Kw (3×10^{8} BTU/year) while for most of the world it was still just barely above the food intake level of 100 watts. At that time, the United States consumed roughly one-third of all the energy, so that total world consumption was on the order of 6.6×10^{12} watts (Table 2.4a).[1]

Future projections depend very much upon assumptions. If it is assumed that in 50 years the rest of the world will have reached an energy consumption level equivalent to that of present United States consumption, and if a population of 10 billion is assumed, then the total man-made energy production rate would be 1.1×10^{14} watts. This rate of energy production would be distributed, however, in some patchy manner which depends on the actual distribution of population centers as well as on the distributing effects of the atmosphere and oceans.

Human versus Solar Heat Input

These numbers must be compared with the energies which the earth receives from the sun. The solar constant of 1400 watt/m^2 applies to solar radiation outside of the atmosphere. Going from a cross section to a spherical surface introduces a factor of $\frac{1}{4}$. Roughly one-third of this energy is reflected back into space and an additional small amount is abstracted in the upper atmosphere. Therefore, over the earth's total surface, about 220 watt/m^2 is absorbed and converted into infrared energy which is then radiated back out into space, thus preserving the earth's radiation balance (Fig. 2.1a). Over the

Table 2.4a
GLOBAL ENERGY BALANCES

Present and Future Use of Energy	*Units of Energy Consumption Rates*		
	Watt	*BTU/year*	*Kg cal/year*
U.S. energy consumption rate (1970)	2.2×10^{12}	6.6×10^{16}	1.65×10^{16}
World consumption (1970) (estimated)	6.6×10^{12}	1.98×10^{17}	5.0×10^{16}
World consumption assuming population of 10 billion and 10 Kw (thermal) per capita (projection to 2020)	1.1×10^{14}	3.3×10^{18}	8.3×10^{17}
Energy Densities	*Watt/m²*	*BTU/m²-year*	*Kg cal/m²-year*
Local energy dissipation rate in a city (assuming 20,000 people per square mile and 10 Kw per capita)	78	2.34×10^{6}	5.8×10^{5}
Solar constant	1,400	4.2×10^{7}	1.05×10^{7}
Solar energy absorbed (S) (averaged over entire earth)	220	6.6×10^{6}	1.65×10^{6}
Infrared energy emitted (E) (averaged)	220	6.6×10^{6}	1.65×10^{6}
Energy storage (typical value, either positive or negative) computed at 1/2 (F − Fy), where F is radiation balance (half-yearly average), and Fy is yearly average; radiation balance is (S − E) computed over latitude zones of the earth rather than averaged over the whole surface.	30	9×10^{5}	2.25×10^{5}

NOTE: The total area of the earth is 5.1×10^{14} m² and the land area is 1.5×10^{14} m².

whole earth, the heat loss is about 1.2×10^{17} watts or about 1,000 times the energy which would be dissipated by human activity in the year 2020.*

It would be premature, however, to assume that the energy input from human activities can be neglected—just because the average heat input is so

* A preferred estimate, and at the same time the lowest flux figure, would be 30 watt/m² which is the energy storage, computed as one-half the difference between the half-yearly and the yearly average of the radiation balance computed over latitude zones of the earth. Use of this value would lead to an extreme projection for 2020, whereby the "solar effect" would be from 140 to 40 times the human energy dissipation, depending on whether the total earth surface or just land area is considered.

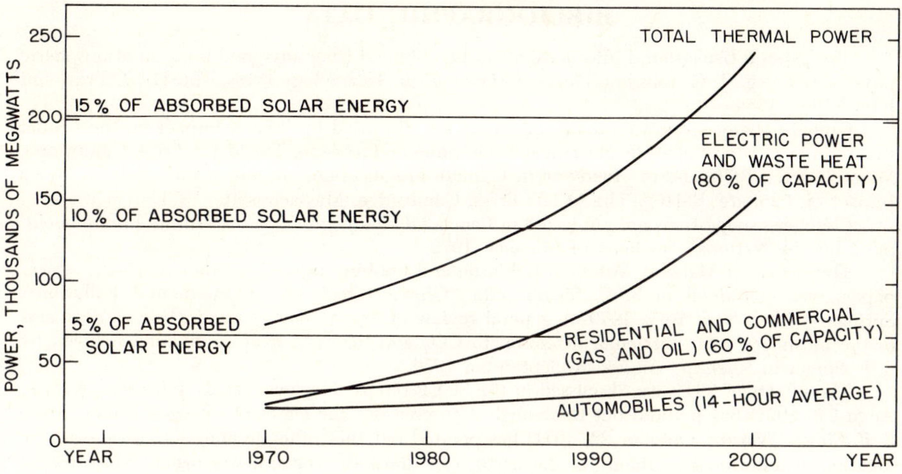

Figure 2.4b Thermal power generation in the Los Angeles basin.

much less than the solar heat input. The atmospheric engine is very subtle and energy inputs in particular places can change local weather patterns at the very least.

Energy Inputs and the Weather

The situation in a particular area (4,000 square miles in the Los Angeles basin) is shown in Figure 2.4b. This area generates the thermal power equivalent of more than 5% of the solar energy absorbed at the ground. By the year 2000, the value is expected to be 18%, based on extrapolating electrical power consumption with a doubling time of 10 years and extrapolating other consumption of energy with a lower rate. Even if this extrapolation is not quite correct, it can be seen that the waste heat loads are very large and can certainly lead to changes in local climate.

Waste thermal power is available to the atmosphere either as sensible or latent heat. The latent heat is produced primarily in the generation of electrical power and is dissipated by deposition in water bodies or by circulation through cooling towers. It is therefore made available at higher altitudes and probably at some distance from the source as compared to sensible heat. From Figure 2.4b,[2] it appears that approximately 30% of the total waste heat will enter through the evaporation-latent heat route by the year 2000.

REFERENCES

1. Johnson, J. C.: "Physical Meteorology," Technology Press, The M.I.T. Press and John Wiley, p. 171, 1954.
2. Lees, L.: SCEP Report, p. 65.

BIBLIOGRAPHIC DATA

The general background information can be obtained from any good book on atmospheric physics, see., e.g., J. C. Johnson, *Physical Meteorology,* Technology Press, The M.I.T. Press and John Wiley, 1954.

Climate changes due to human influences are discussed in "Man's Impact on the Global Environment," report of Study of Critical Environmental Problems, The M.I.T. Press, Cambridge, Massachusetts, 1970; and in "Inadvertent Climate Modification," report of the Study on Man's Impact on Climate (SMIC), The M.I.T. Press, Cambridge, Massachusetts, 1971.

Fletcher's suggestions are published in *Beneficial Modifications of the Marine Environment,* published by National Academy of Sciences, 1972.

The papers of Manabe, Mitchell, Robinson and Robbins, as well as other pertinent review papers, are contained in S. F. Singer (ed.) "Global Effects of Environmental Pollution," Springer-Verlag, New York, 1970. A general review of human energy production as a process in the biosphere with particular attention to CO_2 and to waste heat production is given by S. F. Singer in *Scientific American,* September 1970.

The effects of SST's are discussed in the SCEP report. However, H. L. Johnston (*Science,* August 6, 1971) has pointed out their neglect to consider the effect of nitrogen oxides; while S. F. Singer (*Nature,* October 22, 1971) has pointed out that other human activities, e.g., the creation of additional methane on the earth, can diminish stratospheric ozone. More recent work by R. Ashby, T. Shimazaki and J. A. Weinman (*Bull. Am. Meteor. Soc.,* vol. 53, p. 219, February 1972) indicates that the effects of an SST fleet on ozone are likely to be very minor.

2.5 OXYGEN–CARBON DIOXIDE BALANCE ON A GLOBAL SCALE

The factors contributing to the natural balance of carbon dioxide and oxygen in the atmosphere include both biological and nonbiological processes. Both observed and projected influences of human activities on this balance are presented and discussed in this section.

The balance between oxygen and carbon dioxide in the earth's atmosphere is extremely important to all life on the planet. It has been hypothesized that an increase in carbon dioxide levels could raise the global temperature enough to raise the level of the ocean because of melting of the polar icecaps (Section 2.1). The oxygen requirements of all higher forms of life are such that a significant decrease in its levels would seriously affect these organisms. Similarly, a decrease in carbon dioxide levels could inhibit photosynthesis and thus the production of food; and a significant increase in oxygen levels would no doubt have a strong effect on the biosphere.

Nonbiological Processes

Fortunately, there are a number of processes, both biological and nonbiological in nature, that act to buffer the concentrations of carbon dioxide and oxygen (Fig. 2.5a). Perhaps the most important buffer for carbon dioxide is the sea. Studies of the decline of carbon 14 in the atmosphere, after the cessation of atmospheric nuclear tests, have indicated that the characteristic time for the residence of carbon dioxide in the atmosphere before it is dissolved in the sea is between 5 and 10 years.[1] This means that every year something like 100 billion tons of carbon dioxide is exchanged between the atmosphere and the sea in a dynamic equilibrium. The carbon dioxide dissolved in the sea maintains an equilibrium, through the bicarbonate ion, with the carbonate sediments of the sea. These sediments have been estimated to contain 30,000 times the quantity of carbon dioxide contained in the atmosphere.[1,2] Figure 2.5b illustrates the relative sizes of the various carbon reservoirs. Thus, if the atmospheric levels of carbon dioxide change, the seas have the capability to absorb or release carbon dioxide to maintain a constant concentration in the atmosphere. In reestablishing equilibrium, the bottom sediments and the deep oceans have ultimate control because their exchange of carbon dioxide is the

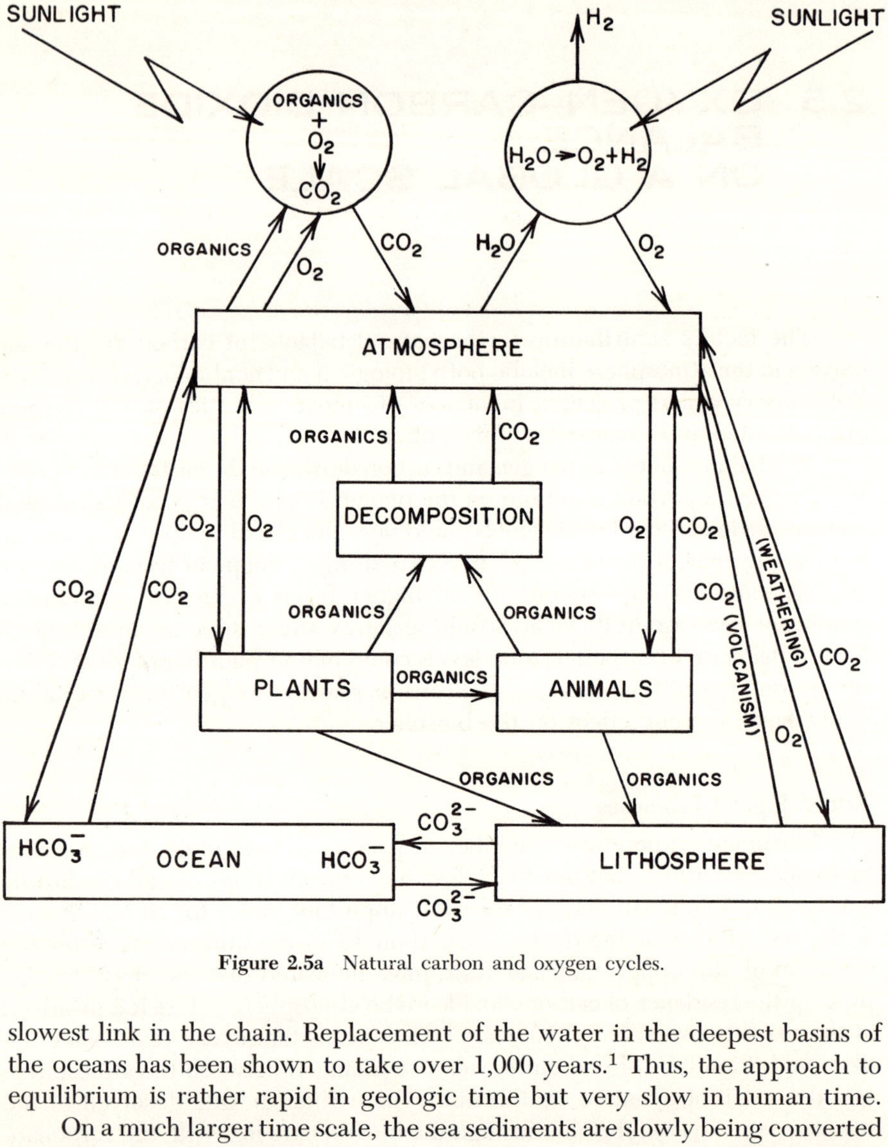

Figure 2.5a Natural carbon and oxygen cycles.

slowest link in the chain. Replacement of the water in the deepest basins of the oceans has been shown to take over 1,000 years.[1] Thus, the approach to equilibrium is rather rapid in geologic time but very slow in human time.

On a much larger time scale, the sea sediments are slowly being converted to carbon dioxide and vented to the atmosphere by the movement of ocean-floor plates under the continents, and their subsequent metamorphism and volcanism. Volcanic gases are also a result of outgassing from the interior of the planet and represent a source of both carbon monoxide and carbon dioxide. Other slow processes that have an effect on the oxygen–carbon dioxide balance in the atmosphere are mountain building and the subsequent weathering of the exposed rocks. Ocean sediments thus exposed to the atmosphere gradually release their stored carbon dioxide. Exposed ferrous oxides, sulfides and other reduced minerals take up oxygen to form oxidized minerals such as ferric oxides and sulfates. All of these are generally felt to be relatively small factors in the overall balance of carbon dioxide and oxygen.

Oxidation of organic gases and the dissociation of water vapor, both of which occur under the influence of ultraviolet radiation in the atmosphere, are additional factors in the carbon dioxide–oxygen balance. Organic gases are reacted with oxygen to carbon dioxide through a complex series of reactions involving sunlight, nitrogen oxides, oxygen and several classes of organic intermediates. These same reactions are responsible for the Los Angeles variety of smog. The photolysis of water vapor in the upper atmosphere to hydrogen and oxygen, with the subsequent escape of the hydrogen from the earth's gravitational field,* is a source of oxygen for the atmosphere. Both of these

* The loss of hydrogen from the earth by this mechanism is generally considered to be negligible when compared with the quantity of hydrogen available on the planet, primarily in the form of water.

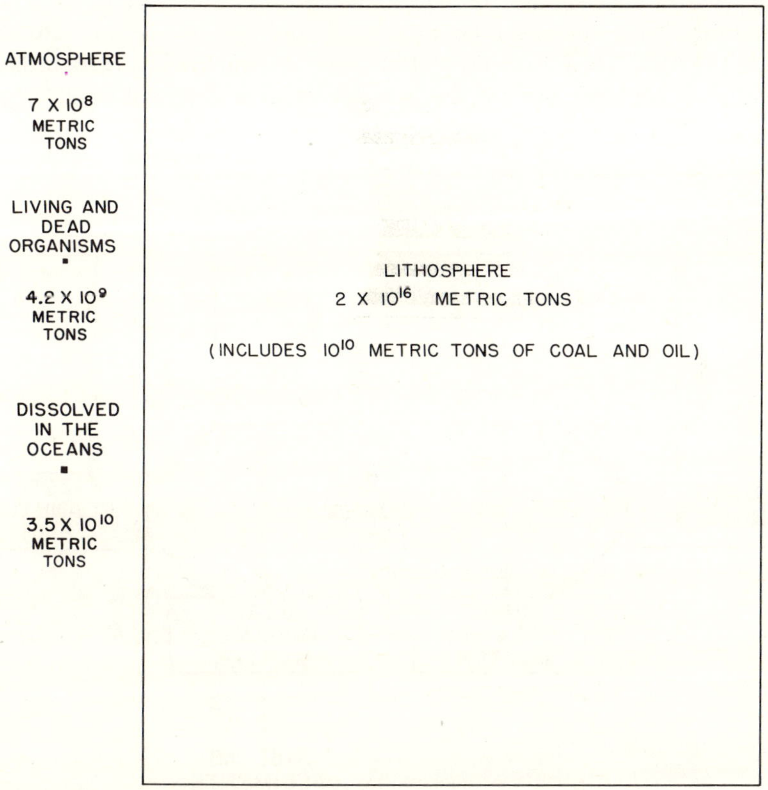

Figure 2.5b Natural carbon reservoirs.

processes make a small contribution to the global oxygen–carbon dioxide balance. The three great systems, the atmosphere, the sea and the continents, are therefore all closely coupled to each other in maintaining the oxygen–carbon dioxide balance through the transfer of carbon dioxide and oxygen across their interfaces. The exchange rates are between decades and thousands of years; consequently, they are stable reservoirs of carbon dioxide and oxygen that ponderously maintain the equilibrium.

Biological Processes

The natural processes with the most immediate impact upon the oxygen–carbon dioxide balance are the biological processes (Fig. 2.5c). Living things with lifetimes of weeks to years operate within the huge oxygen and carbon dioxide reservoirs found in nature to produce short-term deviations from the equilibrium.

The beginning of the cycle, both in the sea and on land, is the fixing of the carbon dioxide and the production of oxygen by plants. The plants are then used as food by the animal life, which in turn may be food for other animal life. These animals take in oxygen and release carbon dioxide in utilizing the food. Dead organic matter is either degraded to carbon dioxide and water or becomes part of the lithosphere as a sediment or an organic

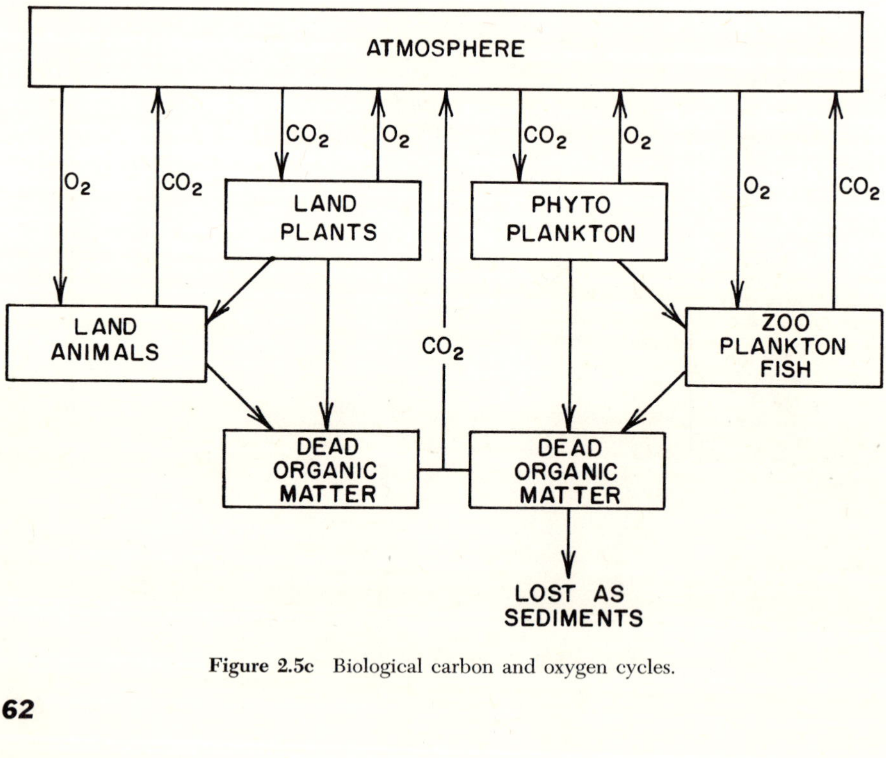

Figure 2.5c Biological carbon and oxygen cycles.

deposit such as peat. In the sea, carbon dioxide may also be incorporated into the sediments as a carbonate shell associated with an organism.

The Effect of Human Activities

Since the onset of the industrial revolution, human beings have had an increasingly strong effect on the oxygen–carbon dioxide balance. So far, this effect has been primarily on the levels of carbon dioxide. Currently, enough fossil fuel is being burned to increase the amount of carbon dioxide in the air by 2.3 ppm per year. Over the last century, the levels have increased from 290 to 320 ppm, and the current rate of increase is about 1.35 ppm per year.[1] The difference between the expected rate of increase and the measured rate can be primarily attributed to the buffering action of the sea. The use of fossil fuels is expected to continue to increase rapidly, pushing the carbon dioxide levels to perhaps 400 ppm by the year 2000.[4]

The effect of human activities on the oxygen levels has been, and is projected to be, negligible.[2] Activities such as clearing and paving large areas of forests will probably have little effect on the oxygen levels since most of the oxygen is produced by the marine phytoplankton. However, the effect of oil spills, pesticides and waste disposal at sea on the phytoplankton is unknown; as is the effect of exhaust from a large fleet of high altitude aircraft upon the ozone layer and upon the albedo of the earth (Section 2.1).

It is a sobering thought that if the pH of the oceans were decreased by a couple of units, huge quantities of carbon dioxide would be released to the atmosphere, eventually eliminating life as we know it. Similarly, the destruction of the ozone layer could inhibit photosynthesis and disrupt the carbon dioxide–oxygen balance. The important point to remember is that the carbon dioxide–oxygen balance is maintained by complex and poorly understood exchanges between the biosphere, hydrosphere, lithosphere and atmosphere. Man cannot control these exchanges; thus, the balance should be left close to the natural state that existed before the industrial revolution. Activities contributing to a major shift in the balance of carbon dioxide and oxygen should be pursued with caution.

REFERENCES

1. Bolin, B.: The Carbon Cycle, *Scientific American,* vol. 223, September 1970, p. 124.
2. Broecher, W.: Man's Oxygen Reserves, *Science,* vol. 168, 3939, June 26, 1970, p. 1537.
3. Cloud, P. and A. Gibor: The Oxygen Cycle, *Scientific American,* vol. 223, 3, September 1970, p. 110.
4. Rasool, S. and S. Schneider: Atmospheric Carbon Dioxide and Aerosols: Effects of Large Increases on Global Climate, *Science,* vol. 173, 2992, July 9, 1971, p. 138.

2.6 ATMOSPHERIC INTERACTIONS, DISPERSION, METEOROLOGY

Underlying Meteorological Principles

Atmospheric interactions include all chemical and physical phenomena occurring to a mixture of airborne pollutants. Immediately upon their emission, this foreign matter undergoes a mass redistribution and a (photo)chemical metamorphosis, varying over many temporal and spatial scales. An understanding of these interactions can be found within the theoretical framework of the atmospheric motions responsible for these changes of pollutant state and distribution.

Planetary Boundary Layer

Figure 2.6a shows the planetary boundary layer, where the frictional effects of the underlying terrain modify the general circulation of the atmosphere and where all significant atmospheric interactions on pollutants occur.[1,2] This layer is the region of small-scale atmospheric turbulence, and these perturbations exert a profound effect on the emitted mixture of pollutants. Both boundary conditions vary; the upper depends on the wind speed, and the lower on the geographic distribution. Generally, 1,000 meters is chosen for the upper boundary.

Mechanical and Convective Turbulence

During the day, a heat transfer process (convection) occurs by the expansion and subsequent rising of the air in response to solar heating. At night, the earth's surface quickly cools by radiation which cools the air in contact with the earth's surface and causes *negative* convective turbulence (negative is oriented *toward* the ground from above). The irregular terrain breaks atmospheric wave motion (analogous to a more dense oceanic wave breaking upon and scraping the ocean floor) into random three-dimensional swirls called eddies. These mechanical and convective eddies are classified as *turbulence,* which has the capacity to thoroughly mix any number of compounds. Pollutant dispersion depends highly on the total amount of turbulent energy. The intensity of atmospheric turbulence varies diurnally. Mechanical turbulence which is always upwardly directed is in phase with the large convective

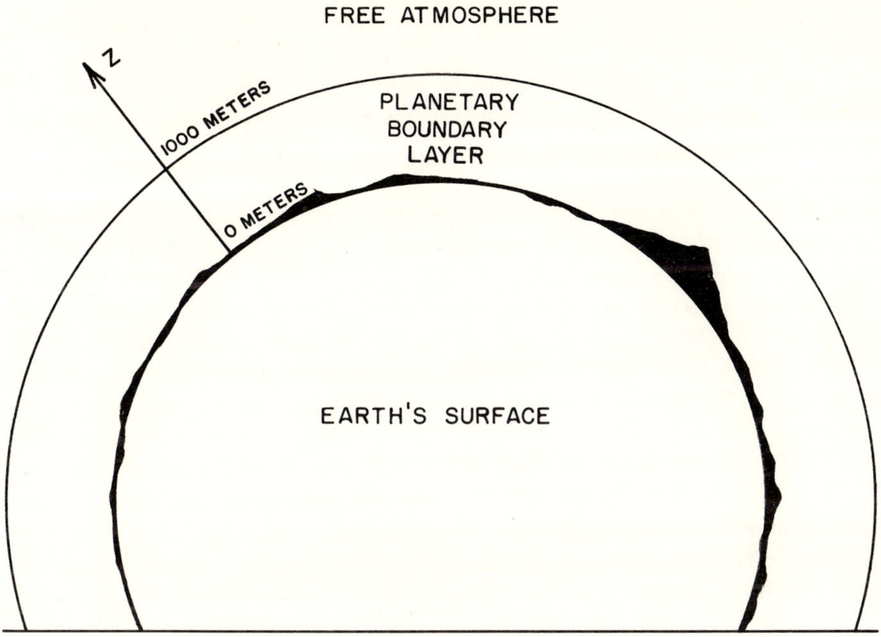

Figure 2.6a Planetary boundary layer in relationship to free atmosphere and the surface of the earth.

turbulence production during the day. At night, negative convection depletes the existing mechanical turbulence. Mechanical turbulence depends on the wind speed and the underlying terrain, while convective turbulent energy is primarily a function of the vertical temperature structure of the planetary boundary layer.[3] This structure can be empirically represented by stability criteria.

Stability Considerations, Inversions

A hypothetical air parcel, given initial upward momentum by a mountain, will expand and cool at the adiabatic lapse rate* of 5.5°F per 1,000 feet in dry air. In the heat of the day, a *super*adiabatic lapse rate occurs, characterized by a large degree of mixing and an unstable atmosphere. A *sub*adiabatic lapse rate occurs at night, with restricted mixing under stable conditions, while neutral conditions (adiabatic) are characteristic of windy, cloudy days and nights. An extensive treatment on stability can be found in standard meteorological texts.[4,5]

To summarize, sunny days exhibit large turbulent energy and strong mixing in unstable air. Cloudy days and nights are of neutral stability and

* For definitions, see Glossary.

of moderate mixing, while evening conditions under fair skies exhibit poor mixing in stable air.

During many cloudless evenings, strong radiative cooling causes the formation of a nocturnal or radiative inversion. This condition, when coupled with light and variable winds, is conducive to high pollutant concentrations, since pollutants are unable to penetrate the inversion *lid*. Inversions also occur over areas where a semipermanent high-pressure cell causes air to subside and, subsequently, to warm by adiabatic *compression*. This *subsidence* inversion occurs around a height of 2,000 feet. Frontal inversions occur during a warm frontal passage, due to the warmer air sliding up over the ground-based cooler air. Both the subsidence and frontal inversions are caused by large-scale phenomena and are not characteristic of planetary boundary layer meteorology.

Other Atmospheric Scales of Motion

Figure 2.6b shows the comparative sizes of a pollutant cloud with various meteorological phenomena.[6] Note the relative size of the motions related to pollutant fields and "weather," which is the scale of motion found on conventional weather maps.

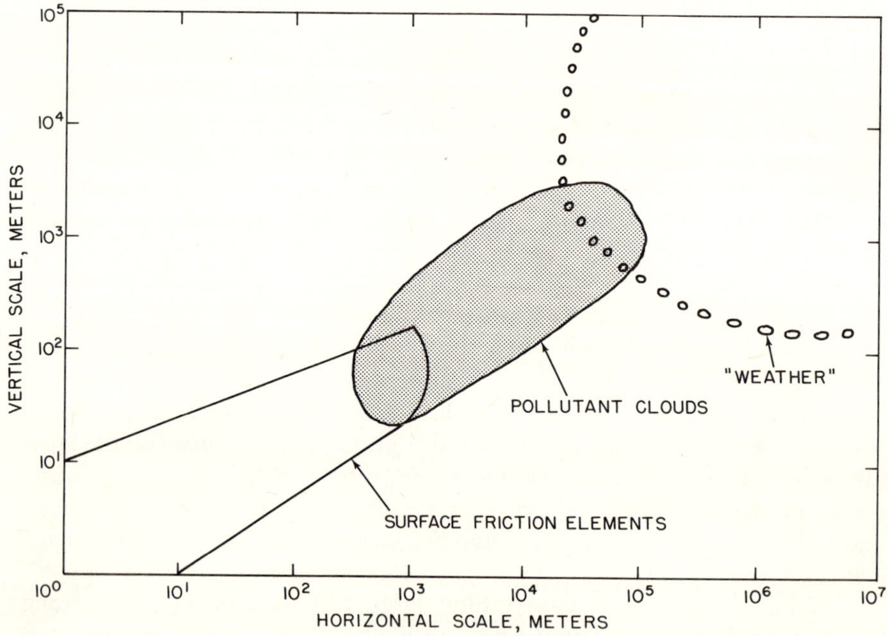

Figure 2.6b Three-dimensional space scale comparison of pollutant clouds with largest and smallest meteorological scales of motion.

Pollutant Entry: Source and Emission Factors

Source Types and Composition

Pollutant sources can be either fixed or moving. Fixed sources include power plants, incinerators, chemical refineries, specialized industrial processing plants and, during the winter season, domestic space heating units. All fixed sources generally emit SO_2 from the combustion of high sulfur coal, and they also emit particulate matter. Additionally, power plants emit large quantities of nitric oxides; refineries release various hydrocarbon products from their cracking processes; and specialized industrial plants emit oxides of their particular process products.

Moving sources include vehicular traffic and airplanes. All other sources contribute a negligible amount to the entire pollutant mixture. Vehicular traffic is generally limited to automobiles and trucks, and their emissions depend upon the driving mode and the specific parameters of the vehicle. Automobile emissions arise from direct evaporation of the fuel before combustion, crankcase blow-by and exhaust. The emissions vary with the driving mode of the vehicle, and their distribution is a function of commuting routes and traffic density. Aircraft emissions occur during ground operations of taxiing, landing and takeoff and during airborne operations of approach and climb out. Carbon monoxide, hydrocarbons, nitric oxides and particulates characterize the pollutant composition from moving sources.

Atmospheric Entry

Pollutant entry into the ambient atmosphere can occur at a point, along a line or over an entire area. Under special conditions, sources may be considered the volume or the virtual point type. In addition, temporal variations exist as sources can emit on a continual basis or at an instant. Instantaneous sources emit *puffs*, while continuous sources discharge *plumes*. An example of each type is listed as follows:

Continuous point source: Strong power plant
Continuous line source: Main urban traffic artery
Continuous area source: An entire urban complex
Continuous volume source: An entire urban complex
Virtual point source: A series of power plants whose emissions are entrained within the plant volume, can be extrapolated behind the emission complex to a virtual point
Instantaneous point source: Explosion, volcanic eruption
Instantaneous line source: A plane emitting pesticides

Once the pollutants enter the atmosphere, their life histories depend entirely on the local atmospheric processes occurring within the emission area.

2.6 *Atmospheric Interactions, Dispersion, Meteorology*

Atmospheric Interactions

Transport and Diffusion

The instantaneous wind vector mathematically describes the complete atmospheric motions entirely responsible for the transport, dilution and dispersion of atmospheric pollutants. At any given time, the instantaneous wind vector ($\vec{V}$) can be expressed as the sum of the prevailing large-scale motions and the superimposed turbulent pertubations:

$$\vec{V}(t) = u\vec{i} + v\vec{j} + w\vec{k} \qquad 2.6(1)$$

where $u = \bar{u} + u'$ (longitudinal component) 2.6(2)
$v = \bar{v} + v'$ (lateral component) 2.6(3)
$w = \bar{w} + w'$ (vertical component) 2.6(4)

The mean wind speed components ($\bar{u}$, $\bar{v}$, $\bar{w}$) exert a twofold effect. They transport the pollutant mix between source and receptor over a given travel time and simultaneously dilute them by mixing in a given volume of air. The stronger the wind speed, the larger the volume of air that will be mixed per unit time, with correspondingly greater dilution. The orientation of the wind vector will determine the direction of transport. The instantaneous wind direction may vary significantly over a given time period.

To simplify the mathematics, the lateral and vertical components are disregarded ($\bar{v} = \bar{w} = 0$), while the abscissa is oriented to the mean flow ($\bar{u}$). These assumptions do not produce any appreciable error since the larger-scale meteorological motions tend to be horizontal. Within the planetary boundary layer, the wind speed increases with height while the wind direction turns clockwise (in the Northern Hemisphere). However, for most diffusion problems the wind vector is assumed to remain constant with height throughout the planetary boundary layer.

The turbulent perturbations (u', v', w') occur randomly in both space and time and are responsible for the diffusion of pollutants about the plume centerline. Turbulent eddies vary over many space scales (1cm to 10m). Therefore, if the eddy is larger than the plume or puff, it will transport the entire mass; while if it is smaller, it will spread or diffuse the mass in space. Turbulent motions vary widely in space and time, never repeating a previous pattern. Due to their lack of time and space history, turbulence must be treated by stochastic methods. The three turbulent energies can be statistically represented by the Equations in 2.6(5).

$$\sigma u = \sqrt{(u - u')^2} \qquad \sigma v = \sqrt{(v')^2} \qquad \sigma w = \sqrt{(w')^2} \qquad 2.6(5)$$

where σu, σv and σw represent the standard deviation of the orthogonal velocities.

The mathematical representation of $\vec{V}(t)$ varies with solar heating, the

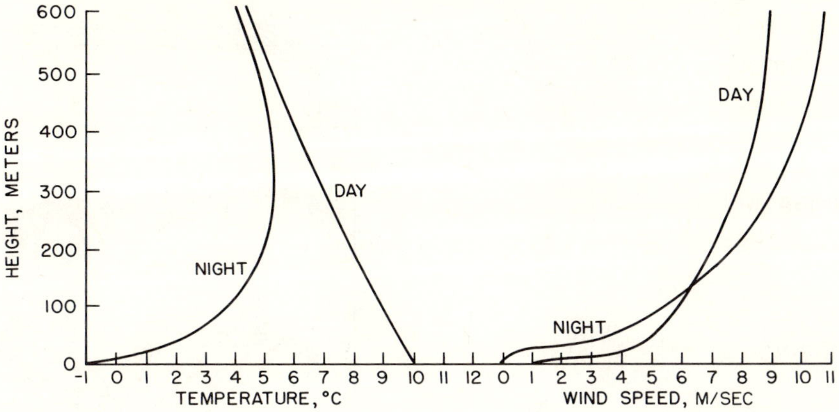

Figure 2.6c Vertical temperature and wind fields in the planetary boundary layer.

underlying terrain distribution and other factors that affect the turbulent structure that is being superimposed on the mean wind field.

Diurnal Effects

Figure 2.6c shows the representative wind and temperature structures during the afternoon and the nighttime hours. During the day, strong solar heating induces a large amount of convective turbulence; while at night, the radiational cooling of the terrain causes low-based temperature inversions. Note the day–night temperature gradient at 500 meters, showing that thermal effects generally occur below that level. There are also transient atmospheric states during the sunrise and sunset periods. Figure 2.6d displays the diurnal effects on a continuously emitting point source.[7]

Looping — In the afternoon, strong solar radiation producing an excessive amount of convective turbulence causes the plume to *loop* in the vertical in response to the strong eddy motion.

Coning — On cloudy days or nights with strong winds, mechanical turbulence prevails and "normal" dispersion occurs, forming a *cone* to the eye of an observer looking upwind into the plume.

Lofting — Over rural areas, just after sunset, a superadiabatic layer exists over the newly forming ground-based inversion. The elevated plume cannot disperse downward due to the inversion, but strongly disperses above it. Urban areas do not have inversion conditions at this time due to heat release from buildings (the "heat island" effect).

Fanning — As radiational cooling progresses, the inversion reaches up to the plume height. Since most of the vertical dispersion is suppressed by the extremely stable conditions, the plume *fans out* in the

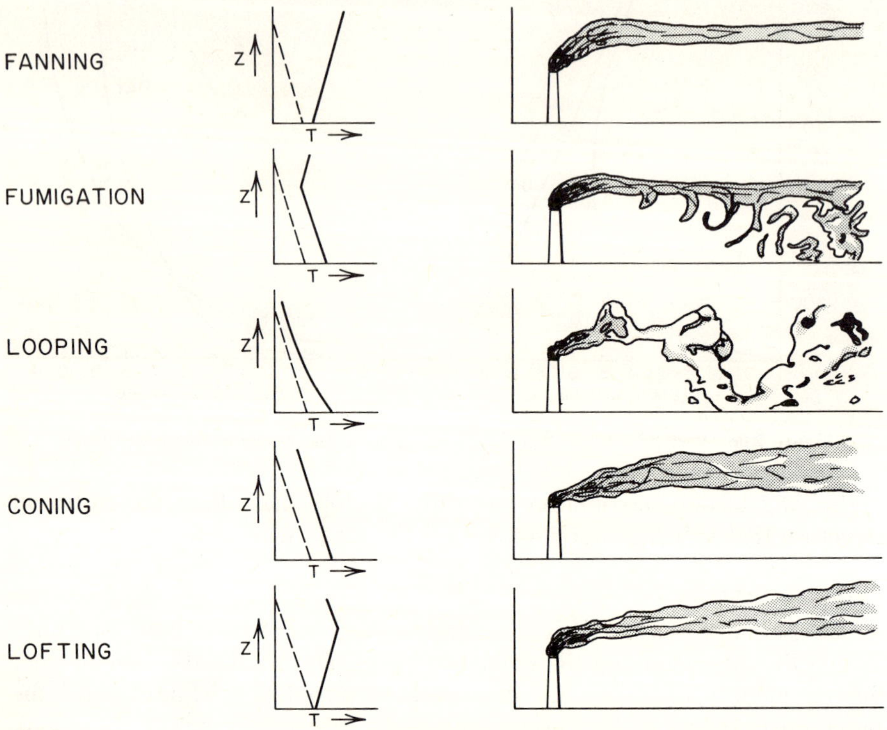

Figure 2.6d Plume behavior as a function of atmospheric stability (dashed lines are dry adiabats).

horizontal. Strong concentrations at plume height are exhibited downwind of the stack.

Fumigation — Sunrise initiates solar heating, and the gradual warming of atmosphere from the ground upwards "burns off" the inversion. As the strong convective mixing reaches the fanned plume, it immediately mixes the large pollutant concentrations towards ground level, *fumigating* that area.

Trapping — This usually occurs during the daylight hours. Under subsidence and warm frontal inversion conditions pollutants are trapped below the inversion level.

TOPOGRAPHICAL EFFECTS

Pollutant events usually occur on days with a diminutive, constantly varying instantaneous wind vector with poor dilution throughout the area. Under these conditions, characteristic of pollutant episodes, the structure of the underlying terrain plays a profound role on the local wind field.[6]

For example, an upslope wind in a valley will significantly affect the pollutant distribution. By the shore, the density gradient caused by the unequal heating coefficients of land and water introduces a now significant onshore breeze; while in the center of the city, the increased absorption of sunlight by the tall buildings alters the temperature and wind structure. Therefore, the structure of the underlying terrain should be evaluated before the wind field can be accurately assessed in terms of its micrometeorological effects.[1]

Plume Buoyancy

Most industrial processes release waste gases at a higher temperature than that of the ambient air. The higher the stack gas temperature, the more the plume is convectively lifted. Plume buoyancy magnitude depends on ambient temperature, waste gas temperature, efflux velocity and wind speed (Section 5.45). The greater the efflux velocity, the larger the initial plume momentum.[8] During periods of light winds, this momentum can be realized to its maximum.

Deposition Processes

Over a long enough period of time, the atmosphere will filter itself of any foreign matter, by a combination of removal processes.

Particulate matter is attracted by the earth's gravitational field and is deposited on the surface. The descent rate depends on the balance between the aerodynamic drag force on the particles and on the gravitational field of the earth.

The second deposition process is precipitation scavenging which can be divided into three subcategories:

1. Delivery of the material to the scavenging site by fog
2. Incloud scavenging as rainout or snowout and
3. Precipitation scavenging below the cloud (washout)

The scavenging process consists of the conversion of particulate matter into condensation nuclei for potential raindrops.[9] Natural precipitation processes cause most of this deposition. The atmospheric interaction research effort concentrates on scavenging.[10]

Other processes, both physical and chemical, that cause the deposition and depletion of pollutants in the atmosphere are aerodynamic downwash from buildings and the radioactive decay and chemical interactions of the particular pollutant species. These areas are very much open to future research.

Modeling Atmospheric Interactions

Purpose

The main purpose of meteorology in the area of environmental pollution is to specify on all scales of time and space the capacity of the atmosphere to absorb, disperse, modify and deposit pollutant materials under any arbitrary

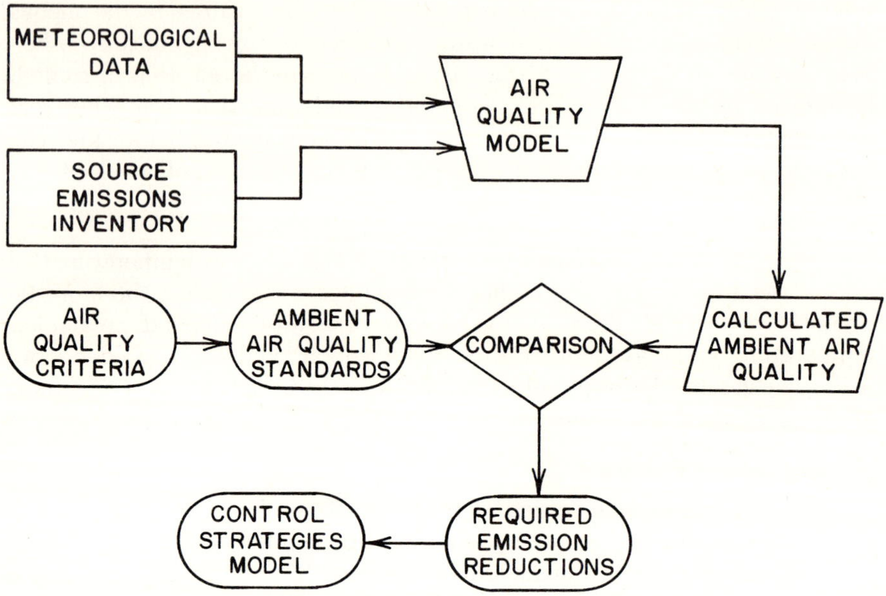

Figure 2.6e Interrelationship between air quality models, air monitoring and air quality criteria.

geographic arrangement and discharge rates of multiple pollutants. Comparing these determinations with the acceptable risk levels will provide the basis for choosing the acceptable levels of pollutant emissions to the atmosphere and will enter as significant criteria in urban, regional and general land-use planning and development[11] (Section 2.3). Figure 2.6e shows an implementation scheme for air quality modeling. There are two "sources" that must be modeled under these criteria, urban and industrial.

Urban Modeling Techniques

There are three complexity levels of urban modeling. The box model is the simplest. An urban area is divided into a checkerboard grid; and it is assumed that all atmospheric interactions occur simultaneously for each box, so that the box is treated as a well-mixed vessel.[12]

The puff and plume models employ the Gaussian statistical assumption for instantaneous and continuously emitting sources respectively, then integrated over the entire urban area. These make up the second level of complexity.[13-20]

The highest level model involves the solution of the mass continuity equations of the turbulent planetary boundary layer. Recently developed models include modules for calculations involving photochemical kinetics, topographical interactions and deposition processes.[21-26] Table 2.6f summarizes the important models.[7,21]

Table 2.6f
WORKING EQUATIONS OF CERTAIN DIFFUSION MODELS DURING THEIR EVOLUTION PERIOD

Gaussian Type Models

Pooler (1961):

$$\frac{X}{Q} = \frac{1}{\pi u(\sigma y + 80)(\sigma z + 30)} \text{ EXP } - 1/2\left[\frac{y^2}{(\sigma y + 80)^2} - \frac{H^2}{2(\sigma z + 30)^2}\right]$$

Clarke (1961):

$$\frac{X}{Q} = \frac{2}{u\sqrt{2\pi}\,\sigma z\,[(\pi/8)X]} \text{ EXP } - 1/2\left[\frac{z^2}{\sigma z^2}\right] \text{ EXP } - 0.693\,(t/T_{1/2})$$

Turner (1964):

$$\frac{X}{Q} = \frac{1}{u\pi(\sigma y + 402)\sigma z} \text{ EXP } - 1/2\left[\frac{y^2}{(\sigma y + 402)^2} - \frac{(z - H)^2}{\sigma z^2}\right] \text{ EXP } - 0.693\,(t/T_{1/2})$$

Koogler et al. (1967):

$$\frac{X}{Q} = \frac{1}{u\pi(\sigma y + 402)\sigma z} \text{ EXP } - 1/2\left[\frac{y^2}{(\sigma y + 402)^2} - \frac{(z - H)^2}{\sigma z^2}\right] \text{ EXP } - 0.693\,(t/T_{1/2})$$

Hilst et al. (1967):

$$\frac{X}{Q} = \frac{1}{\pi\sigma y\sigma zu} \text{ EXP } - 1/2\left(\frac{y^2}{\sigma y^2} - \frac{H^2}{2\sigma z^2}\right)$$

Shieh (1969):

$$\frac{X}{Q} = \frac{1}{(2\pi)^{3/2}\sigma x\sigma y\sigma z} \text{ EXP } - 1/2\left[\left(\frac{x - ut}{\sigma x}\right)^2 + \left(\frac{y - vt}{\sigma y}\right)^2 + \left(\frac{z - wt}{\sigma z}\right)^2\right]$$

Solution of Mass Continuity Equations

Lamb (1968):

$$\frac{\theta C}{\theta t} + \vec{V}_h \cdot \vec{\nabla}_h C_i = Kx\frac{\theta^2 Ci}{\theta x^2} + Ky\frac{\theta^2 Ci}{\theta y^2} + Kz\frac{\theta^2 Ci}{\theta z^2} + F(t)C + S$$

Eschenroeder and Martinez (1969):

$$\frac{\theta C_i}{\theta t} = -u\frac{\theta C_i}{\theta x} + \frac{\theta}{\theta z}\left(Kz\frac{\theta C_i}{\theta z}\right) + R_i$$

Sklarew (1970):

$$\frac{\theta C_i}{\theta t} = -\vec{\nabla} \cdot (\vec{V}C_i) + S_i + R_i$$

Key:
C_i: Concentration of *i*th pollutant species; H: Effective stack height (stack height + plume buoyancy); Kx, Ky and Kz: Eddy diffusivities in the x-, y-, and z-directions respectively; Q: Source emission strength; R_i: Photochemical production of *i*th pollutant species; S_i: Emission strength of *i*th elevated source; t: Time; $T_{1/2}$: Chemical half-life; u: Longitudinal velocity component, mean wind speed; v: Lateral velocity component; w: Vertical velocity component; x: Longitudinal (downwind) direction; y: Lateral (crosswind) direction; z: Vertical direction, height; π: 3.141592; X: Ground concentration at a point downwind; σx, σy and σz: Empirical diffusion coefficients; $\vec{V}_h$: Horizontal components of the instantaneous wind; $\vec{\nabla}_h$: Two-dimensional Hamiltonian operator; $\vec{\nabla}$: Three-dimensional Hamiltonian operator.

REFERENCES

1. Munn, R. E.: "Descriptive Micrometeorology," Academic Press, New York, 1966.
2. Sutton, O. G.: "Micrometeorology," McGraw-Hill, New York, 1953.
3. Lumley, J. L.: and H. A. Panofsky: The Structure of Atmospheric Turbulence, "Interscience Monographs and Tests in Physics and Astronomy," vol. XII, 1964.
4. Byers, H. R.: "General Meteorology," McGraw-Hill, New York, 1959.
5. Donn, W. L.: "Meterorology," McGraw-Hill, New York, 1965.
6. Andersen, G. E.: Mesoscale Influences on the Wind Field, *Journal of Applied Meteorology*, vol. 10, no. 3, pp. 377–386, June 1971.
7. Slade, D. H. (ed.), "Meteorology and Atomic Energy (1968)," July 1968.
8. Briggs, G.: "Plume Rise," U.S. Atomic Energy Commission, 1969.
9. Fletcher, N. H.: "The Physics of Rain Clouds," Cambridge Press, Cambridge, Massachusetts, 1967.
10. Engelmann, R. J. and W. G. N. Slinn (coordinators), "Precipitation Scavenging," USAEC and Pacific NW Laboratory, June 2–4, 1970.
11. Fleagle, R. (chairman), "The Atmospheric Sciences and Man's Needs: Priorities for the Future," National Academy of Sciences, 1971.
12. Reiquam, H.: An Atmospheric Transport and Accumulation Model for Airsheds, *Atmospheric Environment*, vol. 4, no. 3, pp. 233–247, May 1970.
13. Turner, D. B.: A Diffusion Model for an Urban Area, *Journal of Applied Meteorology*, vol. 3, no. 2, pp. 83–91, 1964.
14. Clarke, J. F.: A Simple Diffusion Model for Calculating Point Concentrations from Multiple Sources, *Journal of the Air Pollution Control Association*, vol. 14, pp. 347–352, 1964.
15. Pooler, F.: A Prediction Model of Mean Urban Pollution for Use with Standard Wind Roses, *Int. Journal of Air and Water Pollution*, vol. 4, pp. 199–211.
16. Koogler, J. B., et al.: A Multivariable Model for Atmospheric Dispersion Predictions, *Journal of the Air Pollution Control Association*, vol. 17, pp. 211–214.
17. Croke, E. J., et al.: "Chicago Air Pollution Systems Model," Argonne National Laboratories, 1968.
18. Shieh, L. J., et al.: "*A Multiple Source Model of Turbulent Diffusion and Dispersion in Urban Atmospheres*," New York University, 1969.
19. Bowne, N. E.: A Simulation Model for Air Pollution over Connecticut, *Journal of the Air Pollution Control Association*, vol. 19, pp. 510–514, 1969.
20. Turner, D. B.: "Workbook of Atmosphere Dispersion Estimates," U.S. Department of Health, Education and Welfare, 1967.
21. Neiburger, M.: Meteorological Aspects of Air Pollution and Simulation Models of Diffusion, Transport and Reactions, Task Force 4, *Project Clean Air*, UCLA, September 1, 1970.
22. Sklarew, R. C.: "Systems, Science and Software, a New Approach: The Grid Model for Urban Air Pollution," 63rd Annual Meeting of the Air Pollution Control Association, nos. 70–79, St. Louis, Missouri, June 3–9, 1970.
23. Wayne, L. G., and T. E. Ernest: "Photochemical Smog, Simulated by Computer," USC, no. 69-15, 1969.
24. Fortak, H. G.: Numerical Simulation of the Temporal and Spatial Distributions of Urban Air Pollution Concentrations, *Symposium on Multiple Source Urban Diffusion Models*, U.S. Department of Health, Education and Welfare, 1969.
25. Lamb, R. G., and M. Neiburger: Mathematical Model of the Diffusion and Reactions of Pollutants Emitted over an Urban Area, Task Force 4, *Project Clean Air*, UCLA, September 1, 1970.
26. Eschenroeder, A. Q., and J. R. Martinez, Mathematical Modeling of Photochemical Smog, General Research Corporation, IMR-1210, December 1969.

2.7 IS POLLUTION ALTERING THE WEATHER AND CLIMATE?

Weather designates the instantaneous state of the atmosphere over a place or area. In contrast, climate encompasses the statistical ensemble of all weather conditions over a long interval of time (generally comprising at least several decades), over a place, an area or the whole globe. Weather and climate are usually characterized by measurable elements, such as atmospheric pressure, temperature, wind, precipitation and events such as fog, thunderstorms, freezes. However, climate can also be qualitatively established by its effects on fauna and flora, glaciers and icesheets and geologic formations. Hence, general climatic conditions have been substantiated for long eras prior to instrumental weather records.

Changes in the atmospheric constituents (gaseous, liquid and solid) can affect the energy exchanges that take place and also alter precipitation processes. Such changes have occurred naturally throughout the earth's history and have continued into the present era. Principal among their causes have been volcanic eruptions which bring large quantities of water vapor, carbon dioxide, sulfur compounds and dust into the air. These admixtures can bring about climatic fluctuations. But they are only one of several causes for natural climatic fluctuations. There are other factors, geological and extraterrestrial, which influence the climate. The joint effects of these natural forces are only incompletely understood and a quantitative theory of climate is as yet lacking. Thus, climatic fluctuations constitute a "noise" pattern; and a difficult problem arises when attempting to distinguish effects caused by man's activities from other effects (Table 2.7a).

Global Effects

Carbon Dioxide

Although it is only a minor constituent of the atmosphere in the present geological epoch, carbon dioxide (CO_2) can have an effect on air temperature. This is caused by the interception of outgoing long-wave radiation from the earth (Section 2.1). CO_2 has notable absorption bands in the infrared region where the absorption spectrum of the other long-wave interceptor, water vapor, permits heat loss to space. Much CO_2 is produced naturally by decay and sporadically by volcanic activity (Section 2.5). Combustion of fossil fuels

Table 2.7a
ATMOSPHERIC POLLUTANTS AND CONSTITUENTS, IN METRIC TONS, GLOBAL BASIS

Constituent	Quantity in Natural Composition	Quantity of Annual Injections		Quantity of Residue Remaining in Atmosphere One Year or Longer
		Natural	Anthropogenic	
Nitrogen	4×10^{15}			
Oxygen	1×10^{15}			
Water vapor	1×10^{13}			
Carbon dioxide	2.3×10^{12}		1×10^{10}	5×10^{9}
Dust[a]	1×10^{8}	1×10^{9}	1×10^{8}	6×10^{7}
Sulfur	—	1.4×10^{7}	0.8×10^{7}	[b]

[a]The dust injected per case of major volcanic activity is about 1×10^7 metric tons.
[b]Large percentage is converted to particulates and included in residual dust load.

by man adds about 1×10^{10} tons per year. A considerable part of the CO_2 is absorbed in the water of the oceans. However, at present, 40 to 50% of the manmade CO_2 appears to stay in the atmosphere. Generally accepted figures indicate that the amount of atmospheric CO_2 has increased from 290 ppm in the nineteenth century to 330 ppm at present with a current annual increase of 0.6 ppm.[1] Calculations of the interceptive effect show that a global temperature rise of about 0.3°C can possibly be attributed to the CO_2 increase (Fig. 2.1b). Projections show that at the present rate of increase of atmospheric CO_2, a global temperature gain of 0.8°C could result by the year 2000.

Particulates

Man produces substantial amounts of particulates. Most of them are very short-lived in the atmosphere and confined to the lower layers. They are eliminated by impact, fallout and washout by precipitation. Average residence time varies from a few minutes to 30 days, with an overall mean residence of 10 days. Substantial sources are industrial effluents, car exhaust, tire wear and soil blowing resulting from poor agricultural practices. The smallest particles ($<2\mu$) are convectively carried to higher levels where airplane exhaust also plays a role. Not all the man-made contributions to atmospheric dust loads are of industrial or vehicular origin. Poor farm practices lead to soil blowing, and slash burning of vegetation causes smoke. These two sources are an order of magnitude larger than industrial processes.

Interception of Radiation

Fine particles have the capability of scattering incoming short wave radiation from the sun. A small percentage of this scatter is back into space.

This, in the absence of clouds, causes an energy loss for the earth. Part of the incoming energy is absorbed by the particles thereby warming the layers of the atmosphere where they reside, but causing a net heat loss at the earth's surface. Increasing dust contents of the atmosphere at high levels, provided the earth's cloud cover does not change, should lead to cooling of the earth. This could reach 2 to 3°C in a century. Near-surface dust could, by heat absorption, cause some warming in low layers where most man-made dust stays. However, large volcanic eruptions inject megatons of dust into the stratosphere, which usually stays there several years.

Albedo Changes of Glaciated Areas

Man-made dust rapidly finds its way to all parts of the globe by turbulent atmospheric transport. Some of the dust is eventually deposited on ice and snowfields. Analyses of glacier samples show dusts of undoubted anthropogenic origin, such as lead from car exhaust, in remote polar regions. In some areas, this dust load has increased by a factor of ten in the current century. As long as the dust stays near the surface of the ice, it absorbs solar heat; while clean snow and ice are excellent reflectors of incoming solar radiation. Thus, the dust reduces the planetary albedo in polar regions—a factor that could add slightly to global temperature rise. No quantitative estimates of this effect are available.

Local Effects

In local areas, man's activities have caused observable and measurable atmospheric changes. Some of them are reversible but others, especially in heavily urbanized and industrialized areas, have become quasi-permanent.

Thermal Influence

Notable changes in the energy balance near the earth's surface have been brought about by replacing vegetation with stone and asphalt in urban areas. This leads to a substantial reduction in evaporation, a higher heat absorption because of lowered albedo and ready storage of solar heat in walls and ground because of the better heat conductivity and the higher heat capacity of the artificial surfaces. These factors alone lead to increased temperatures in urban areas. They become particularly pronounced during periods of clear sky and low wind speeds (Section 2.6). Added to this is the heat rejection into the atmosphere from combustion processes, air conditioning and metabolism. The surface changes are the more prominent factors because the buildings also reduce the surface wind speeds.

About 80% of the so-called urban "heat island" effect is attributable to these factors, while the direct heat additions due to human activities presently account for about 20%. The average annual temperature rise in metropolitan areas is 1 to 2°C; but on calm, clear evenings, cities can be 6 to 8°C warmer

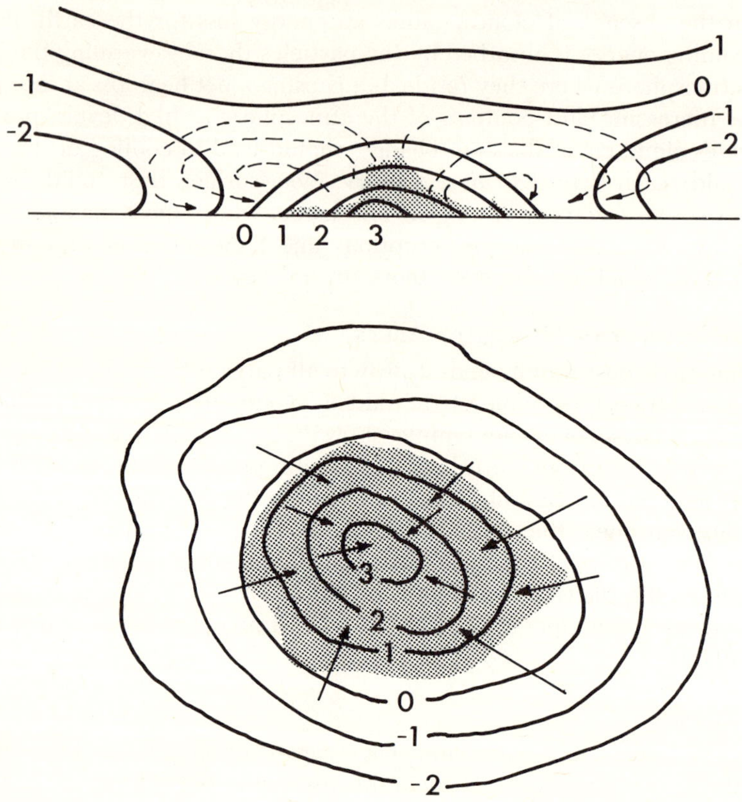

Figure 2.7b Nocturnal heat island produced by city. Upper half shows vertical cross section with heavy lines representing iso-lines of temperature departure at point on city edge and dashed lines with arrows showing streamlines of the wind; hatched area is city silhouette. Lower half shows isotherms as departure from rural value at city edge (heavy lines) with arrows showing directions of surface winds; hatched area represents built-up area.

than the surrounding countryside. Figure 2.7b shows a typical urban heat island and its associated wind field.

Heat rejection is gradually increasing as a factor (Table 2.4a). In some northern cities in winter, it approaches 30% of the solar heat received. On an average, for the year in the megalopolitan area of Washington–Boston, it now reaches 10% and is projected to reach 30% by the year 2000. The amounts of energy rejected into the atmosphere, of anthropogenic origin, *while significant on a local scale,* are less than $^1/_{1,000}$ of the solar energy received by the earth and hence unlikely to produce measurable global climatic changes.

The urban heat island has several consequences. It increases cloudiness, induces or aggravates rain showers and, when general winds are weak, causes

Table 2.7c
CHANGES IN ATMOSPHERIC ELEMENTS CAUSED BY URBANIZATION

Element	*Average Change from Natural Environment*
Contaminants	
Condensation nuclei and particulates	10 times more
Gaseous admixtures	5–25 times more
Cloudcover	5–10% more
Fog, winter	100% more
Fog, summer	30% more
Total precipitation	5–10% more
Days with <5mm	10% more
Snowfall	5% less
Relative humidity	
Winter	2% less
Summer	4% less
Total radiation	15–20% less
Ultraviolet, winter	30% less
Ultraviolet, summer	5% less
Sunshine, duration	5–15% less
Visibility	20–50% less
Annual mean temperature	0.5–1.0°C more
Winter minima, average	1–2°C more
Extreme difference	6–8°C more
Heating degree days (see Glossary)	10% less
Annual mean wind speed	20–30% less
Extreme gusts	10–20% less
Calms	5–20% more

an airflow toward the city from the surroundings. Table 2.7c gives an overview of local climatic changes by urbanization.

Vaporization

Many man-made combustion processes yield water vapor as one of the end products. Large amounts of water are also vaporized in the operation of wet cooling towers, industrial quenching and cooling processes and warm-water discharges into lagoons, lakes and rivers. In some industrial areas, this extra evaporation amounts to a 10^9 cubic meters per year and reaches 30% of the natural surface evaporation. This can reduce surface water runoff by

10%. Runoff is reduced because stream or river water is used up by evaporative cooling. The same water would otherwise run down the streams rather than be directly returned to the atmosphere. Again the local effect is large, but on a global scale it is still negligible.

Wet cooling towers have the most notable meteorological effects on the environment. The plumes readily saturate the air and with high atmospheric humidity can lead to local fog formation. In temperatures below freezing, they cause icing. Drizzle and light snow are not infrequent occurrences in the vicinity. On other occasions, clouds are formed and can be traced for tens of kilometers downwind from the towers.

Condensation and Freezing Nuclei

Industrial effluents and automobile exhaust contain innumerable particles that cause condensation of atmospheric water vapor. Sulfuric acid, a daughter product of sulfur dioxide, will result in droplet formation even before water vapor saturation is reached. Sulfur dioxide content governs visibility. At 70 to 90% relative humidity, a concentration of 0.01 ppm SO_2 will not seriously affect visibility, but at 0.1 ppm it will reduce visibility to 8 to 12 km, and at 1 ppm to 1 km.

Large numbers of condensation nuclei also tend to cause many small droplets to form in clouds. This stabilizes the cloud and limits its potential to precipitate. Thus, overproduction of this type of particles can reduce natural rainfall. Other types of particles can act as freezing nuclei (such as lead iodide forming from exhaust of cars using leaded gasoline) and "seed" subcooled clouds producing added rainfall downwind. There is evidence that such effects exist to the lee of cities prompted by their effluent plumes.

Conclusions

There is no evidence that human activities, including various forms of pollution, have caused any global climatic changes. However, there are measurable, and in part substantial, local climatic changes in urbanized and industrialized areas. As the degree of urbanization progresses, so will these atmospheric changes. Alterations of atmospheric composition by added carbon dioxide and dust as yet have had only minor global influences that disappear in the natural climatic fluctuations. However, the accelerated injection of these anthropogenic additions, projected for the future, as well as the anticipated increases in heat rejection, could be of major concern in the next century unless controlled.

REFERENCE

1. Mitchell, J. M.: A Preliminary Evaluation of Atmospheric Pollution as a Cause of the Global Temperature Fluctuations of the Past Century, in S. F. Singer (ed.), "Global Effects of Environmental Pollution," New York, 1970.

BIBLIOGRAPHY

Landsberg, H. E.: Man-Made Climatic Changes, *Science,* vol. 170, pp. 1265–1274, 1970.

Löf, G. O. G. and J. C. Ward: Economics of Thermal Pollution Control, *J. Water Pollut. Control Federat,* vol. 42, no. 12, pp 1202–2116, 1970.

Matthews, W. H., W. W. Kellogg, and G. D. Robinson (eds.): "Man's Impact on Climate," The M.I.T. Press, 1971.

Singer, S. F. (ed.): "Global Effects of Environmental Pollution," Springer-Verlag, New York, 1970.

Wilson, C. L. and W. H. Matthews (eds.): "Inadvertent Climate Modification," The M.I.T. Press, 1971.

2.8 PHOTOCHEMICAL SMOG

Photochemical smog results from the action of sunlight with oxides of nitrogen and vaporized organics to produce ozone and a variety of noxious products. While the whole system of reactions occurring is complex and not yet completely understood, its major causes and effects are well known.

The Nature of Photochemical Smog

Two basic types of smog occur. "Reducing" smog results from the combination of sulfur-containing fuels which lead to high levels of sulfur oxides and particulate matter in air. The second *"oxidizing"* type smog requires sunlight along with oxides of nitrogen and vapor-phase hydrocarbons or other organics to be evident. Its severity is commonly estimated by its capacity to oxidize iodide. Sulfur dioxide is a *reducing* interference in iodide detectors used to monitor oxidants in the *second* smog type.

Most of the objectionable components in the reducing type of smog are emitted directly from the combustion process as *primary* pollutants. On the other hand, none of the three required ingredients in oxidizing smog is especially noxious in itself: sunlight, nitric oxide and volatile hydrocarbons separately are relatively innocuous at levels ordinarily observed even in heavy smog. These primary pollutants, however, when allowed time to interact, will produce a variety of objectionable *secondary* pollutants as a result of photochemical reactions. These reactions occur in the air and involve a very complex system of reactions of both inorganic and organic molecules.

Inorganic Reactions in Photochemical Smog

Nitric oxide (NO) is the oxide of nitrogen ordinarily emitted as a minor product of high-temperature combustion. It is a colorless gas which is not highly toxic, but it reacts with oxygen (as in air) to form the highly toxic and reactive brown gas nitrogen dioxide (NO_2):

$$2\,NO + O_2 \longrightarrow 2\,NO_2 \qquad 2.8(1)$$

At concentrations present in urban air, this reaction is fairly slow. The NO_2

thus formed, however, is rapidly photolyzed by sunlight (or any ultraviolet light below 4100 Å wavelength) to yield NO and atomic oxygen (O):

$$NO_2 + h\nu \longrightarrow NO + O \qquad 2.8(2)$$

(For a definition of hν see Glossary.) Atomic oxygen is extremely reactive and short-lived because (besides hydrocarbon reactions described later) it rapidly combines with molecular oxygen, present in air in large excess, to yield ozone:

$$O + O_2 + M \longrightarrow O_3 + M \qquad 2.8(3)$$

(M is any third molecule capable of carrying away the energy of bond formation.) Ozone is also a reactive species. In particular, it reacts with NO to form NO_2:

$$O_3 + NO \longrightarrow O_2 + NO_2 \qquad 2.8(4)$$

As a result of reactions 2.8 (1) through 2.8 (4), a steady state is set up where NO_2 is photolyzed to NO and O; O combines with O_2 to give O_3; and O_3 oxidizes NO back to NO_2. The reaction of NO with O_3 is so rapid that both of these species cannot coexist unless one is at "trace" levels. Thus, an approximate equilibrium applies at a given sunlight intensity:

$$K = \frac{(NO)\ \ (O_3)}{(NO_2)} \qquad 2.8(5)$$

From this relationship, it appears that the ozone level will not rise much until the NO/NO_2 ratio becomes small. This occurs when reactive hydrocarbons are present.

Organic Smog Reactions

A photochemical reaction sequence is shown in idealized form in Figure 2.8a. An organic reactive with nitric oxide in air under ultraviolet irradiation disappears as shown. The nitric oxide decreases even more rapidly, while nitrogen dioxide increases to a maximum and then declines. When NO reaches its low levels, ozone and other products begin to rise and eventually reach broad maxima. The specific products and their ratios are highly dependent on the type of organic compounds reacting.

Organics are not only transformed to end products (some of them obnoxious) but may also serve to alter somewhat the inorganic reaction sequence described earlier. There are hundreds of hydrocarbon types in a polluted atmosphere of widely varying reactivity. Table 2.8b includes an abbreviated set of reactions for a single "average" hydrocarbon type, RR. Its interaction with nitrogen oxide and oxygen types can have great bearing on the latter. In particular, the equilibrium given in Equation 2.8(5) is altered to allow greater formation of ozone and other products.

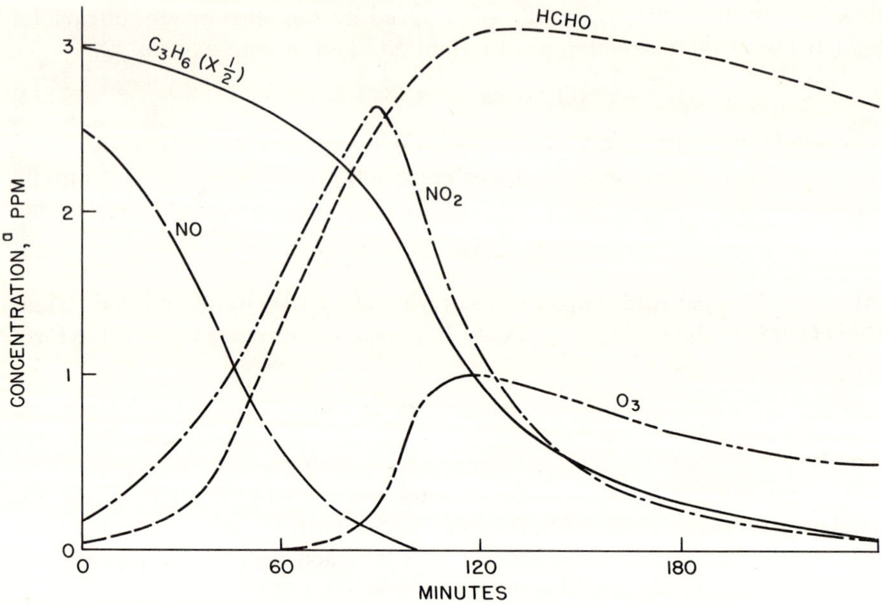

Figure 2.8a Photochemical reaction pattern (propylene + nitric oxide).
[a]The propylene (C_3H_6) concentrations are twice those marked on the ordinate scale.

The photochemical smog sequence thus produces a wide variety of highly reactive intermediates as well as more or less stable products. The reactive species are all *oxidizing* in nature, e.g., O (atomic oxygen), O_3 (ozone), RO_2, NO_3 or NO_2.

Effects of Photochemical Smog

These effects are evident in several ways. The casual observer sees a sometimes dense haze resulting from condensation of reaction products to form aerosol particles large enough to affect visibility. A frequent companion of

Table 2.8b
PHOTOCHEMICAL SMOG REACTIONS

$NO_2 + h\nu = NO + O$	$RO + O_2 (+M) = RO_3 (+M)$
$O + O_2 (+M) = O_3 (+M)$	$RO_3 + NO = NO_2 + RO_2$
$NO + O_3 = NO_2 + O_2$	$RO_2 + NO_2 = RO_2NO_2$
$RR + O = R + RO$	$RO_3 + NO_2 = RO_3NO_2$
$RR + O_3 = RO + RO_2$	$NO_2 + O_3 = NO_3 + O_2$
$RO_2 + NO = NO_2 + RO$	$NO_3 + NO_2 = N_2O_5$
$R + O_2 (+M) = RO_2 (+M)$	$N_2O_5 + H_2O = 2HNO_3$

this type of haze is eye irritation; many of the people exposed to this type of atmosphere suffer smarting or burning eyes with involuntary tears and sometimes nose or throat irritation as well. The known products PAN and some aldehydes have been shown to produce these effects. The effects of ozone and other oxidants are discussed in Section 3.14.

BIBLIOGRAPHY

"Air Quality Criteria for Photochemical Oxidants," Publication No. AP-63, Environmental Protection Agency, U.S. Department of Health, Education and Welfare, March 1970.

Leighton, P. A.: "Photochemistry of Air Pollution," Academic Press, New York, 1961.

2.9 CLEAN AIR COSTS MONEY—DIRTY AIR COSTS MORE

It has been estimated that the damage to individuals and property directly attributed to air pollution is approximately $65 per capita per year, or in excess of $13 billion annually. It is against these figures that the costs for pollution control must be balanced (Fig. 18).

Industrial Expenditures

Stricter air pollution regulations necessitate that companies make sizable investments for pollution control in order to stay in business.

Clean air is, in the long run, less expensive than dirty air. Such a viewpoint can usually only be afforded by the large companies that have considerably more latitude for capital funding. Although the smaller companies may also prefer clean air to dirty air, they are more limited by their sales volume and production margin. It is these smaller companies which find it difficult to comply with emission controls.

The magnitude of air pollution expenditures has, in the past, been large and will become even larger in the future. A summary of total capital expenditures, total pollution control expenditures and air pollution control expenditures, both past and future, in 5-year intervals from 1958 through 1977 is included in Table 2.9a for the chemical, petroleum and the pulp and paper industries.[1]

The percentage of expenditures directed toward pollution abatement as well as the actual volume is expected to increase sharply for the petroleum and the pulp and paper industries. Although the volume of spending for pollution is also expected to increase for the chemical industry, the percentage will probably remain fairly constant.

The chemical industry, in 1970, spent 56% of its pollution control dollars on air and 44% on water, while the pulp and paper industry spent only 12% on air and 88% on water.[2] This apparent anomaly can be explained by the fact that the pulp and paper industry uses much more water per dollar of product produced, and this large volume of water makes it necessary to heavily invest in water pollution abatement. By 1977, the total pollution expenditures of the pulp and paper industry are expected to exceed those of the chemical

Table 2.9a
INDUSTRIAL EXPENDITURES FOR POLLUTION CONTROL, IN MILLIONS OF DOLLARS

	1958	*1963*	*1967*	*1972*	*1977*
Chemical and Allied					
Total capital Expenditures	1,116	1,505	2,900	3,850	5,200
Total pollution Expenditures	33	51	87	125	177
Air pollution Expenditures	12	19	34	54	83
Petroleum Refining					
Total capital Expenditures	647	386	685	730	850
Total pollution Expenditures	39	35	47	78	111
Air pollution Expenditures	18	17	24	43	63
Paper and Allied Products					
Total capital Expenditures	618	675	1,550	1,850	2,400
Total pollution Expenditures	13	22	57	98	182
Air pollution Expenditures	4	7	17	31	60

industry; and the air pollution expenditures will approach those of the chemical industry, even though the total capital expenditures of the chemical industry are expected to be more than double those of the pulp and paper industry.

As the level of allowable emissions is reduced, the operating cost of air pollution control equipment increases.

Expenditures by the Chemical Industry

Two surveys on pollution control spending of major chemical firms conducted in 1970 and 1971[3], leave little doubt that the cost of air pollution can only increase (Table 2.9b). Although a small percentage of the increase could be attributed to inflation, the major cause for the increase is the more stringent pollution control regulations. Tightening the allowable emission standards means that more and/or bigger equipment must be added. The additional equipment means that more power will be needed and usually additional operators will be required, both of which increase operating costs.

Expenditures by Pulp and Paper Companies

The bulk of the expenditures for pollution control to date by the pulp and paper industry has been for water pollution abatement. However, the day is fast approaching when equal emphasis will have to be given to air pollution control. In fact, many kraft mills added abatement systems in 1972

Table 2.9b
COMPARISON OF ANNUAL EXPENDITURE ESTIMATES FOR AIR POLLUTION CONTROL BY SOME CHEMICAL COMPANIES

	Expenditure Estimates									
	1970 Estimate ($millions)					*1971 Estimate ($millions)*				
Company	*1968*	*1969*	*1970*	*Spent Total 1965–1969*	*Projected Total 1970–1974*	*1969*	*1970*	*1971*	*Spent Total 1966–1970*	*Projected Total 1971–1975*
Allied	2.0	2.6	3.0	17.5	35.0	2.6	4.5	6.1	26.6	50–60
Hercules	0.5	0.35	0.45	10.0	20.0	0.98	1.57	4.5	10.0	30.0
Diamond Shamrock	0.9	1.7	. . .	13.4	. . .	0.6	0.66	2.02	18.0	19.0
W. R. Grace	. . .	. . .	. . .	9.0	9.5	. . .	1.5	1.4	12.0	
Monsanto	. . .	. . .	. . .	40.0[a]	55.0[b]	. . .	. . .	. . .	. . .	110.0[b]

[a]Cumulative spending for 1962–1969.
[b]Projected spending for 1970–1972.

Table 2.9c
YEARLY POLLUTION CONTROL SPENDING BY PULP AND PAPER FIRMS, $MILLIONS

Company	*Before 1970*		*After 1970*	
Boise Cascade	11.37	(1967–1969)	49.67	(through 1975)
MacMillan Bloedal Limited (Canada)	18–19		40	(through 1975)
International Paper	23		101	(through 1974)
Scott Paper	15	(1960–1969)	85	(1970–1979)
Kimberly Clark	6.0	(1966–1969)	12[a]	(1970–1979)
Mead Corporation	31	(1960–1969)	50[b]	
Hammermill	4	(1960–1967)	34	(after 1968)

[a]Includes $1.5 million per year for operating costs.
[b]Does not include $5 million per year for operating costs.

and 1973 to treat odor compounds which were then being emitted to the atmosphere as the characteristic "kraft mill" odor. In the future, most of the new recovery boilers installed will be of the new "odor-free" type (Section 7.12).

Table 2.9c indicates the spending by seven paper companies prior to and after 1970. As emission standards are tightened, the capital and operating expenditures will likely exceed the figures now projected.

REFERENCES

1. Spending for Clean Air to Climb, *Steel*, vol. 164, p. 38, February 3, 1969.
2. Pollution: The Cost to Industry, *Modern Manufacturing*, vol. 10, p. 163, October 1970.
3. Pollution Cost Spending Survey—Who, When and How Much, *Chemical Week*, vol. 107, p. 23, July 8, 1970; *see also* It Costs More to Run a Clean Chemical Plant, *Chemical Week*, vol. 109, p. 10, September 8, 1971.

BIBLIOGRAPHY

Beizer, James: Who Pays for Pollution Solution?, *Iron Age*, vol. 198, pp. 27–28, December 22, 1966.

It Costs More to Run a Clean Chemical Plant, *Chemical Week*, vol. 109, p. 10, September 8, 1971.

Kneese, Allen V.: "The Economics of Air Pollution," Harold Volozin (ed.), Norton, New York, 1966.

Mills, Edwin S.: "The Economics of Air Pollution," Harold Wolozin (ed.), Norton, New York, 1966.

Pollution Cost Spending Survey—Who, When and How Much, *Chemical Week*, vol. 107, pp. 23–24, July 8, 1970.

2.9 Clean Air Costs Money—Dirty Air Costs More

The Cost to Industry, *Modern Manufacturing*, vol. 10, pp. 163–164, October 1970.

Ridker, Ronald G.: "Economic Costs of Air Pollution," Praeger Press, New York, 1968.

Spending For Clean Air to Climb, *Steel*, vol. 164, pp. 38–39, February 3, 1969.

Train, Russell E., Robert Cahn and Gordon J. MacDonald: "Environmental Quality," Government Printing Office, Washington, D.C., August 1970.

U.S. Department of Interior, Federal Water Pollution Control Administration: "The Cost of Clean Air and Its Economic Impact," Government Printing Office, Washington, D.C., January 10, 1969.

2.10 EMISSION SOURCES—COMBUSTION, INDUSTRIAL, MOBILE

The sources of air pollutant emissions are manifold; and they can be grouped according to a variety of means, including type of source, number and spatial distribution of sources and type of emissions.[1]

Type of Source

Source type refers to natural and man-made sources, as well as to additional subclassifications within each group. Natural air pollution sources include windblown dust, pollen and other aero-allergens, sea salt nuclei, smoke and flyash and gases from forest fires, microorganisms, gases and odors from swamps and marshes, fog, volcanic ash and gases, natural radioactivity, ozone from lightning, meteoritic dust and natural hazes. Man-made sources of air pollutant emissions include dust-producing processes such as crushing, grinding and demolition; combustion processes such as fuel burning and refuse burning; manufacturing processes such as in metallurgical plants and chemical plants; and agricultural activities such as crop dusting, plowing and field burning.

Number and Spatial Distribution of Sources

Air pollution sources can be grouped into three categories according to number and spatial distribution. The first category includes single or point sources such as a power plant, steel mill or petroleum refinery. The second category includes area-wide or multiple sources, representing a large number of smaller sources distributed over a well-defined area. Examples of this category include an entire residential area, or a block of apartments or commercial buildings concentrated in one section of a community. The third category can be designated as line sources, including freeways, highways and arterials carrying moving vehicles (Section 2.6).

Types of Emissions

Air pollution source grouping can be by the type of emissions, with particulate and gaseous emissions being the two major divisions. Particulates may be defined as any dispersed matter, solid or liquid, in which the individual aggregates are larger than single small molecules (about 0.0002 microns in diameter), but smaller than about 500 microns. Particles in this size range have a lifetime in the suspended state in air ranging from a few seconds to several months. Additional terms used to further describe particulate matter include dust, flyash, smoke, soot, aerosols, droplets, mist, fog and fumes. Each of these terms could be further subdivided into specifics, for example, dust can be cement dust, coal dust or sawdust.

Gaseous emissions include both gases and vapors. Gaseous emissions can be subdivided by type into oxides of sulfur, oxides of nitrogen, oxides of carbon, fluorine, chlorine, hydrocarbons, mercaptans, oxidants, ozone and many others.

Table 2.10a contains the estimated emissions of the major air pollutants, by type of source and type of emissions, in the United States in 1969.[2] The total increase in pollutants by weight in a one year period was 3.2% above the estimated 1968 emissions. Emissions from fuel combustion in stationary sources, from industrial processes and from "miscellaneous" sources all increased. The large increase under "miscellaneous" was due primarily to the increased number and size of forest fires. The slight decline in solid waste emissions is due to a reduction in open burning at municipal dumps.

The greatest single source of air pollution, when measured by pollutant weight, is the automobile. Transportation contributed more pollution than all other sources together; and carbon monoxide alone, primarily from automobiles, outweighed the other four major pollutants combined. However, it is

Table 2.10a
ESTIMATED EMISSIONS OF AIR POLLUTANTS BY WEIGHT, NATIONWIDE, 1969
(In millions of tons per year)

Pollution Source	*Pollutant*						
	CO	*Particulates*	*SO_x*	*HC*	*NO_x*	*Total*	*Change, % 1968–1969*[a]
Transportation	111.5	0.8	1.1	19.8	11.2	144.4	−1.0
Fuel combustion in stationary sources	1.8	7.2	24.4	0.9	10.0	44.3	+2.5
Industrial processes	12.0	14.4	7.5	5.5	0.2	39.6	+7.3
Solid waste disposal	7.9	1.4	0.2	2.0	0.4	11.9	−1.0
Miscellaneous	18.2	11.4	0.2	9.2	2.0	41.0	+18.5
Total	151.4	35.2	33.4	37.4	23.8	281.2	+3.2
Change, % (1968–1969)	+1.3	+10.7	+5.7	+1.1	+4.8		

[a]Computed by the 1969 method from the difference between 1969 estimates and 1968 estimates. The new method results in higher values for 1968 than those computed by EPA for 1968.

SOURCE: The Mitre Corp. MTR-6013. Based on Environmental Protection Agency data.

significant that the contribution of transportation sources declined slightly in 1969. The peak level of emissions from automobiles has apparently been reached and as older cars are replaced by newer ones with pollution control devices, which were first installed in the 1967 models in California and in the 1968 models nationwide, automobile-related pollution can be expected to further decline. The 1975 and 1976 emission standards should accelerate this decline. However, in the long run, if the number of automobiles continues to increase at the present rate, even the effect of emission controls could eventually be negated.

It is somewhat misleading to consider air pollutants in terms of their aggregate weight. Air pollutants are of concern because of their effects (Chapter 3) and, other things being equal, one pound of sulfur oxides has a greater effect on health and welfare than one pound of carbon monoxide. Measuring pollutants by tons emitted also disregards the geographical distribution. Effects are related to the concentration of pollutants in a particular place, but expressions of total weight of emissions do not indicate concentration.

One significant atmospheric reaction to note is that under the influence of sunlight, nitrogen oxides combine with hydrocarbons to form a complex variety of secondary pollutants called photochemical oxidants (Section 2.8). These oxidants, together with solid and liquid particles in the air, make up what is commonly known as smog. The photochemical oxidant pool includes, among others, ozone, nitrogen dioxide, peroxyacetyl nitrates, aldehydes and acrolein.

Carbon Monoxide Emissions

Specific delineations of the carbon monoxide emission estimates[3] of 1968 are included in Table 2.10b. Emission factors information can be utilized to estimate point, area or line source emissions; or it can be used to determine emissions by type of source, that is, mobile combustion, fuel combustion in stationary sources, industrial processes, solid waste combustion and miscellaneous combustion. Information of this type, including emission factors, is also available for particulates, oxides of sulfur, hydrocarbons and oxides of nitrogen.[4,5,6,7]

Mobile Emission Sources

Mobile emission sources include automobiles, trucks, buses, planes and trains. A summary of the emission factors for gasoline and diesel engines is included in Table 2.10c.[1] Carbon monoxide and hydrocarbon emissions are greater, on a pound per 1,000 gallon basis, from gasoline engines than from diesel engines; however, the reverse is true for particulates.

Table 2.10b
CARBON MONOXIDE EMISSION ESTIMATES, 1968

Source	Emissions, millions of tons annually	Emission Factor
Mobile combustion sources		
Motor vehicles		
Light-duty vehicles		
–urban	35.6	
–rural	12.8	
Heavy-duty vehicles		
Class II (6,000–10,000 lb[a])–urban	1.7	Based on method described in reference 3
–rural	0.9	
Class III (10,000–19,500 lb[a])–urban	1.4	
–rural	0.8	
Class IV (19,500 lb[a])		
–urban	3.0	
–rural	2.8	
Diesel	0.2	
Aircraft		
FAA Controlled	1.1[b]	Based on LTO[d] cycle (see reference 3)
Military	1.3[b]	
Railroad	0.1	Distillate fuel oil: 60 lb CO/1,000 gal
Vessels	0.3	
Nonhighway fuel uses	1.8	Distillate fuel oil: 60 lb CO/1,000 gal gasoline: 2,300 lb CO/1,000 gal
Stationary combustion sources		
Coal		
Electrical power	0.1	0.5 lb CO/ton
Industrial	0.2[c]	3.0 lb CO/ton
Domestic and commercial	0.5	50.0 lb CO/ton
Fuel oil		
Domestic, industrial and miscellaneous	0.1	84.0 lb CO/1,000 bbl
Electric power	negligible	1.7 lb CO/1,000 bbl
Wood	1.0	45 lb CO/ton
Industrial processes		
Foundries		
Controlled (with afterburners)	0.2	10 lb CO/ton of charge
Uncontrolled	3.1	250 lb CO/ton of charge

Table 2.10b
(Continued)

Source	Emissions, millions of tons annually	Emission Factor
Petroleum refineries		
Fluid catalytic crackers	2.0	13.7 lb CO/bbl of fresh feed
Fluid coking	0.2	30 lb CO/bbl of fresh feed
Moving-bed catalytic crackers	0.2	3.8 lb CO/bbl of fresh feed
Kraft pulp mills	2.6	215 lb CO/ton of product
Carbon black		
Furnace	0.30	560 lb CO/ton of product
Channel	0.05[e]	
Thermal	0.01	47 lb CO/ton of product
Steel mills		
Beehive coke ovens	0.02	4.5% of exhaust gas by volume
Basic oxygen furnaces	0.1	3.2% of exhaust gas by volume
Sintering	2.4	500 ft^3/ton
Formaldehyde	0.03	100 lb CO/ton of product
Solid waste combustion		
Incineration		
Municipal	0.01	1 lb CO/ton
Onsite	0.8	27 lb CO/ton
Open burning		
Municipal	1.7	85 lb CO/ton
Industrial	0.9	85 lb CO/ton
Commercial, backyard, etc.	0.8	85 lb CO/ton
Metal conical burners	3.6	260 lb CO/ton
Miscellaneous combustion		
Building fires	0.2	45 lb CO/ton
Forest fires	7.2	60 lb CO/ton
Coal banks	1.2	50 lb CO/ton
Agriculture	8.3	60 lb CO/ton

[a] Gross vehicle weight.
[b] Under 3,000 feet.
[c] Does not include 96,400 tons from coke plants.
[d] Landing-Take-Off.
[e] Gross estimate.

Table 2.10c
EMISSION FACTORS FOR GASOLINE AND DIESEL ENGINES (lb/1,000 gal)

Pollutant	Gasoline Engines[a]	Diesel Engines
Aldehydes	4	10
Benzo(a)pyrene	0.3 gram per 1,000 gal	0.4 gram per 1,000 gal
Carbon monoxide	2,910	60[a]
Hydrocarbons	524[b]	180
Oxides of nitrogen	113	222
Oxides of sulfur	9	40
Ammonia	2	not available
Organic acids	4	31
Particulates	11	110

[a] Includes blow-by emissions, but not evaporation losses.
[b] Includes 128 lb/1,000 gal blow-by emissions.

REFERENCES

1. Rossano, A. T., Jr.: "Air Pollution Control Guidebook for Management," E.R.A., Inc., Stamford, Connecticut, 1969.
2. "Environmental Quality—The Second Annual Report of the Council on Environmental Quality," pp. 212–213, U.S. Government Printing Office, Washington, D.C., August 1971.
3. "Air Quality Criteria for Carbon Monoxide," NAPCA Publication No. AP-62, pp. 4–3 and 4–4, U.S. Government Printing Office, Washington, D.C., March 1970.
4. "Air Quality Criteria for Particulate Matter," NAPCA Publication No. AP-49, U.S. Government Printing Office, Washington, D.C., January 1969.
5. "Air Quality Criteria for Sulfur Oxides," NAPCA Publication No. AP-50, U.S. Government Printing Office, Washington, D.C., January 1969.
6. "Air Quality Criteria for Hydrocarbons," NAPCA Publication No. AP-64, U.S. Government Printing Office, Washington, D.C., March 1970.
7. "Air Quality Criteria for Nitrogen Oxides," NAPCA Publication No. AP-84, U.S. Government Printing Office, Washington, D.C., January 1971.

2.11 POLLUTION EFFECTS ON HUMAN HEALTH

Measuring Health and Disease

Health is a subjective and objective state of physical and mental well being, and disease is a continuous variable reflecting any departure from optimum physical or mental health or any state which may shorten the individual's life-span in the future.

At any stage of a disease process, the diagnosis is an expression of probability. The likelihood of obtaining agreement on a diagnosis depends on the disease concerned; how it manifests itself; the knowledge of the examiners; the facilities, support staff and time available; the age, sex, understanding, ability to communicate and ethnic-socioeconomic-geographic situation of the patient; and the fads, prejudices and wishful-thinking of the examiner.

The amount of disease in a community can be assessed at various levels of measured need, perceived need and serviced need; and what information may be made available can be broken down:

1. The need for medical care found by interview and examination of a sample of the population (preferably a random sample including persons who do not present themselves for care) and assessed in terms of some accepted "norm" for that particular population subgroup.
2. The need (demand) for medical care which is perceived by the patient and which may or may not be transferred by him into some form of action (taking some self-prescribed medicine, asking a neighbor's or druggist's advice, staying away from work or school, consulting a physician, going to a clinic, hospital out-patient department or emergency room, calling an ambulance), which may or may not be appropriate when reviewed objectively in light of the need found upon examination.
3. The amount and level of medical care which the patient receives once admitted to the medical system (office or out-patient care, home care, hospitalization, surgery and special procedures, referrals, ancillary services) which will depend on individual factors (such as race and income) in addition to the services (manpower and facilities) available in that particular locality.

4. Recorded disability or death resulting (within some particular time period) from disease (including injury and premature aging).

Each of the approaches may give a different result and the difficulty of arriving at some rigid definition of disease which covers all these concepts should be apparent. Comparisons between the amount of disease in different populations should always be highly suspect.

In relating a disease to some suspect cause (a factor in the absence of which the disease does not occur) or determinant (a factor whose presence or absence influences the distribution or frequency of the disease), it is generally preferable to use incidence figures (the number of new cases over a defined period of time) rather than prevalence figures (the number of cases, new or old, present at a point in time). In a prevalence study, an ambulant population will show a relative excess of mild chronic cases of disease, while hospitalized cases will have an overrepresentation of severe cases (other than sudden deaths) and of patients with multiple diseases. On the other hand, it is often difficult to know in incidence studies whether the individual concerned was healthy before the recorded onset of disease. No matter which form of study is used, the importance of selective recall and other subjective biases may be very difficult to estimate. For acceptable conclusions to be drawn, a control population (one differing from the test population as little as possible except in the variable being tested) is nearly always essential and a historical control (the same population viewed at another point in time) is rarely satisfactory in this regard.

The Multifactoral Approach to Human Disease

The distribution, nature and severity of human disease are almost always determined by a number of factors or determinants operating together. These determinants include inborn characteristics (such as heredity, age, sex and physique), habits (such as smoking, use of alcohol and diet), behavioral features (such as the complaint threshold or unwillingness to seek medical care), environmental exposures (such as to bacteria or noxious agents) and socioeconomic factors (such as the availability of medical care). Many determinants (such as serum cholesterol, blood pressure, the presence of other disease and poverty) are themselves the outcome of the interplay of other more basic determinants. These determinants may influence the probability of developing a given disease independently of others (with an uncommon disease the effects would then be approximately additive) or by interaction.

As a general rule, individual determinants should be thought of as contributory or precipitating factors rather than as causes in producing disease.

The Effects of Air Pollutants Under Experimental Conditions

Under experimental conditions, with exposure to single air pollutants at levels not remarkably higher than those which have been found in urban environments, it has generally proved impossible to reproduce the effects that have been ascribed to natural exposures.

Sulfur Dioxide

Cities in the United States with relatively high SO_2 concentrations in the ambient air, such as Chicago, Cincinnati, Philadelphia[1] and Pittsburgh,[2] have annual mean concentrations of below 0.1 ppm; while in New York, monthly means may go up to 0.3 ppm and 24-hour maximums up to 1.2 ppm.[3] In the London smog of December 1952, a 2-day average value of 1.34 ppm was recorded.[4] (This author was in London at that time as a graduate student and wrote in his diary, "The fog was almost incredibly thick, black and dirty—visibility about 2-yards—there was no road transport and the fog had a strong acid taste.") In volunteers inhaling SO_2 at 1 ppm, a small increase in airway resistance and a decrease in maximum expiratory flowrates have been reported[5,6] with some further increase in airway resistance at 5 to 28 ppm;[7,8] but contradictory findings have also been published.[9,10] It may be that only a small proportion of the normal population is hyperreactive to SO_2[10] or that bronchitic patients would respond more.[3] Animal species vary widely in susceptibility to SO_2 with guinea pigs being the most susceptible. At near-normal exposure levels, rats exposed to 1 ppm SO_2 and 1 mgm/M^3 of graphite dust 12 hours a day for more than 100 days[11] and guinea pigs exposed to 1 ppm of SO_2 for 22 hours a day for 12 months[12] have shown no harmful effects.

Particulates

For most large industrialized cities, annual means for suspended particulates (with some 20% in the respirable range 0.3 to 5 μ) vary round 140-μg/M^3. In Birmingham, Alabama, several sites have measured values over 200-μg/M^3 for 80 or more days a year with occasional daily peaks going up to 600 to 700-μg/M^3. At levels some fifteen times higher (10-mg/M^3) than these naturally occurring peak values, volunteers have shown an increase in airway resistance.[13,14]

Nitrogen Dioxide

The concentration of NO_2 in urban air is usually less than 0.1 ppm,[1] though 24-hour maximums may go up to 2 ppm. Human volunteers exposed to 3 ppm have shown no defined clinical pathology.[15] A few volunteers, at 13 ppm, have complained of eye and nasal irritation;[16] but workers accustomed to the vapors have withstood 20 ppm for 18 months without ill effects.[17] Most experimental animals die when exposed to concentrations above 100 ppm[18] (188-μg/M^3), but concentrations below 3 ppm have only a questionable effect. Many milder cellular effects appear to be reversible though preemphysematous changes have been reported in mice exposed to 0.5 ppm for 3 to 12 months.[19] Lifetime exposures of rats to 2 ppm showed only some ciliary loss and a slight increase in lung weight.[20] Several investigators have shown a reduced resistance to infections after NO_2 exposures and mice exposed continuously to 0.5 ppm for 12 months showed an increased mortality when challenged with an aerosol of *K. pneumoniae.*[21]

Total Oxidant

Peak "total oxidant" levels of up to 1 ppm (with ozone and the PAN compounds the best known constituents)[22] have been reported in photochemical smogs,[23] but even in the Los Angeles basin, 24-hour maximums of 0.3 to 0.5 ppm are more usual.[24] Volunteers exposed to ozone at 0.5 ppm for 2 hours have shown substernal soreness, a cough and laryngitis with reduced pulmonary compliance[25] and at 0.2 ppm for 3 hours reduced visual acuity.[22] Young males exposed to peroxyacetyl nitrate (PAN) at 0.3 ppm had a reduced oxygen uptake when given exercise.[26] Irradiated mixtures of propylene and NO_2 in concentrations of 0.1 to 1.0 ppm produced eye irritation,[27] but irradiated auto exhaust at moderate levels had no effect on fine neuromuscular tasks.[28] Among animals, mice exposed to 0.2 ppm of ozone for 6 hours have shown reduced activity levels;[29] and pneumonitis was more prevalent in mice exposed to ambient air (average oxidant concentration 0.057 ppm) in the Los Angeles basin than among controls.[30] On the other hand, beagles exposed to photochemically reacted automobile exhaust for 18 months showed no significant pulmonary changes.[31]

Carbon Monoxide

Eight-hour averaging time concentrations of CO (a time-span chosen because of the time taken for carboxyhemoglobin levels in the human body to reach equilibrium) along heavily travelled highways usually peak at about 15 ppm;[32] though values of 36 ppm or higher may be reached[33] with as much as 115 ppm inside the vehicles. In road tunnels, values up to 500 ppm have been recorded.[34] An approximation to the equilibrium value of carboxyhemoglobin (COHb) after exposure to concentrations of CO below 60 ppm can be obtained[35] from Equation 2.11(1)

$$[\text{COHb}] \text{ in } \% = 0.181 \times [\text{CO}] \text{ in ppm} \qquad 2.11(1)$$

COHb can be estimated from expired air after breath-holding for 20 seconds.[36] One hundred and eight males exposed for 72 days in a nuclear submarine to 44 ppm CO showed no ill effects;[37] while up to 100 ppm CO on two series of runs for exposures of 4.5 hours had no effect on the critical tracking test (driver simulation) of healthy nonsmoking, male college students.[38] On the other hand, among volunteers exposed to 111 ppm CO after 70 minutes, the nonsmokers (but not the smokers) showed a decrease of visual vigilance with an average COHb of 6.6%.[39] Four percent COHb has been shown[40] to produce a significant decrease in the oxygen debt of exercise. One study on dogs, rabbits and rats using continuous exposure to 50 ppm CO for 3 months[41] showed no differences in EKG or pathological findings between exposed and control animals. In contrast, another study on dogs using continuous exposure to 50 ppm CO for 6 weeks did find some EKG changes.[42]

Experimental work on the combination of various air pollutants at concentrations somewhat above those naturally encountered has shown that when coal dust (at 15-$\mu g/M^3$) and SO_2 (at 10-ppm) are given together to guinea pigs, there is both a reduction in bactericidal activity and a decrease in mechanical elimination.[43]

Air Pollutant Effects Under Natural Conditions

In a report[44] by the Committee on Public Health of the New York Academy of Medicine (1966) entitled "Air Pollution and Health," the conclusion was reached: "Although chronic effects have not yet been demonstrated, it is reasonable to presume that since episodes brought acute sickness and death, exposure to lesser concentrations for prolonged periods will have effects." This presumption cannot be accepted at its face value without some form of proof. It is possible that the human body can deal with lesser concentrations of air pollutants for a lifetime of exposure without effects (adverse or otherwise); or that if adverse effects are found, they are the cumulative result of a series of minor episodes each exceeding some individually determined "threshold level." We must also determine whether any "effects" that are measurable are realistic indicators of potentially harmful disease.

In measuring minimal effects, a continuing problem is to establish limits of normality. According to the available literature, Russian standards[45] are not based on evidence that disease has been produced, but rather on functional changes in animals (e.g., higher nervous activity, cholinesterase activity, excretion of coproporphyrins) and human volunteers (e.g., olfactory threshold, respiratory and visual reflexes, changes in light sensitivity, activity of the cerebral cortex). Attempts to relate pulmonary flowrates with indices of air pollution have been generally inconclusive. The very substantial experimental error (variability) associated with these measurements should be noted. One study, following two groups of Los Angeles school children for 11 months, failed to find any relationship between peak expiratory flowrate and levels of oxidant air pollution.[24] In contrast, a study of four groups of children from Chattanooga, Tennessee,[46] found children from the high NO_2 (0.1 to 0.25 ppm), low particulate area to have a reduced ventilatory performance ($FEV_{0.75}$).* A comparison of children from two districts of Rotterdam[47] showed lower peak flowrates in the district with more air pollution; but this was the poorer part of the city and other parameters such as height and weight were also lower, suggesting that this was primarily a socioeconomic effect. Of interest is a

*FEV means forced expiratory volume and the 0.75 means the volume of air expired in 0.75 seconds.

report from Los Angeles[48] finding that oxidant levels in the air before a race correlated well with track performance over 21 high school cross-country track meets between 1959 and 1964.

Longitudinal studies (see Glossary) of reasonably representative volunteer groups using diaries in which symptoms are reported, backed up by weekly or biweekly interviews to collect this information, have been tried in several areas. A typical report from New York[49] found a runny nose, cough, sore throat, headache and eye irritation were aggravated all during episodic increases in SO_2 levels. The difficulty in coordinating the location of measuring stations and the air breathed by the respondents in this type of study should be appreciated. Studies of this nature in Birmingham, Alabama, have emphasized the role played by such variables as newspaper publicity (increasing response) and the forced shutdown of key industries (decreasing response among those whose relatives are thrown out of work) in eliciting subjective records of this type. The increase from 18 to 33% between 1956 and 1960 in the "percent bothered by air pollution" in the San Francisco Bay area[50] suggests the same subjective bias.

A series of studies on absenteeism from work (in New York[51] and Philadelphia[52]) and from school (in Los Angeles[53]) have shown a strong time dependence (day of the week, season of the year) and dependence on meteorological conditions (temperature, humidity, wind velocity) with little, if any, residual independent effect attributable to air pollution. Using the number of prescriptions for antitussive drugs dispensed to outpatients by the two largest hospitals in Dallas, Texas, over a 6-year period as an indirect method of estimating the prevalence of coughing in the community,[54] similar findings were obtained to those recorded in the cited absenteeism studies. None of these findings discount a possible air pollution effect, but, "there are many variables and—their behavior follows certain regularities and many of these variables which can have an effect are acting in conjunction with the air pollution variable. . . . Isolating in a one-dimensional analysis the effects of . . . one from another is nearly impossible—they are part of the same causal system."[55]

The relationship between hospital admissions and concentration of air pollutants has been looked at by several investigators. One major study, covering admissions from Blue Cross member groups to larger hospitals within 5 miles of a pollution monitoring station in the Los Angeles area[56] from March 17 to October 26, 1961, corrected for industrial activities and admissions related to the day of the week by using for analysis the difference between the daily "observed" values and the "expected" values for that day divided by the standard deviation for that day. In a stepwise multiple regression, various measures of air pollution on the day of admission (oxidants, SO_2, particulates) and 3 days before (including second-order terms for these same variables) were found to have a significant ($p < 0.01$)* though low correlation with the number of "relevant" admissions (allergic, respiratory and cardiovascular disorders), even with temperature and humidity included in the regression. An investigation in Boston[57] failed to show any relationship between two episodes of air pollution and the number of patient visits to the emergency room of a hospital serving patients from the city. (The same investigation did show a relationship between air pollution and the number of clinic visits made by a cohort of patients with chronic, nonspecific respiratory disease.)

Air Pollution Episodes

Historically, the three best known air pollution episodes associated with excess mortality were those involving the Meuse River valley in Belgium in 1930; Donora, Pennsylvania in 1948; and London, England in 1952. The Meuse River episode[58] was associated with some fifty-four excess deaths largely among people who were elderly or who had preexisting cardiopulmonary disease. Autopsy findings suggested these deaths were due to some irritant chemical. The Donora episode[59] was associated with some eighteen excess deaths again often based on preexistent cardiovascular disease. This episode was also associated with a great deal of morbidity and 60% of persons aged 65 years or older reported some illness, with one-half of these illnesses being severe. A follow-up on this population 10 years later[60] found that persons with severe acute illness during the episode had more morbidity and mortality than did those who had mild or no illnesses during the episode. The London episode[61] was associated with over 3,500 excess deaths over the week of the episode and the succeeding week. The greatest excess was among persons 45 years of age and older with chronic cardiorespiratory disease. All three episodes were associated with a temperature inversion; and in each case, more than one irritant pollutant was probably responsible with sulfur dioxide associated with particulate solids (smoke) playing a major role.

Many attempts have been made to correlate excess mortality in other cities with lesser episodes of air pollution, but these have mostly been unable to overcome the analytic (i.e., chance effects alone would dictate that some air pollution peaks would occur synchronously with peaks in total mortality) and logical (i.e., excluding other causes of excess mortality such as an influenza epidemic) difficulties involved. Of several studies from New York which have been reported, one looked at a period of intense air pollution (SO_2 levels averaged 0.46 ppm) from January 29 to February 12, 1963, when there were 809 excess deaths compared with the same 15-day period in 1961, 1962, 1964 and 1965.[62] Unusually cold weather and influenza were also found during this

*$p < 0.01$ means that the probability of this effect occurring by chance is less than one in a hundred.

15 day period in 1963 and estimates of the effect of the air pollution alone ranged from 200 to 400 deaths. During November and December 1962, not only was London affected[61] but similar, though less severe, episodes were reported from the Eastern United States, Rotterdam, Hamburg and Osaka.[63]

Bronchitis and Emphysema

One major difficulty with chronic bronchitis and emphysema has been terminology. The British Medical Research Council[64] has defined chronic bronchitis as "a chronic or recurrent increase above the normal in the volume of mucoid bronchial secretion sufficient to cause expectoration when this is not due to localized broncho-pulmonary disease." (Chronic means[65] on most days for 3 months on at least 2 consecutive years.) The World Health Organization[66] has defined emphysema as "an increase in the size of the air spaces with destructive changes in their walls."

With chronic bronchitis, there is thickening of the glandular component of the bronchial wall (measured using the Reid index), but this is also found in emphysema[67] and the two conditions shade into each other. The term nonspecific chronic respiratory disease is applied to both conditions, to intermediate conditions and (sometimes) to lesser grades of pulmonary disability which do not meet the strict criteria for chronic bronchitis (cough and sputum) or emphysema (shortness of breath, altered lung function, X-ray changes). It appears that while international differences exist (Great Britain has a particularly high prevalence rate) these relate more to the frequency and severity of the disease than to its essential nature.[68]

Several surveys have been done to establish the prevalence of nonspecific chronic respiratory disease and to relate this to personal and environmental variables in an attempt to identify determinants. A listing of more recent American field studies of the prevalence of bronchitis has been published by Stuart-Harris.[69] These surveys have consisted substantially of the administration of a standardized questionnaire (usually the British Medical Research Council questionnaire[70] or a slightly modified form thereof), the performance of a number of lung function tests and the use of X-rays. The amount of overlap between these three indicators of chronic respiratory disease varies with the population being studied. In a Mayo Clinic population of 993 males aged 40 years or older, 50.3% of those with pulmonary symptoms had an abnormal lung function test, compared with an expected value of 19.5%.[71] On the other hand, in a study done on a random sample of 2,718 white males aged 20 years or older from 14 cities in Alabama, only 41.4% of those with pulmonary symptoms had an abnormal lung function test and/or abnormal X-ray, compared with an expected value of 32.8%. Similar difficulties in relating symptoms to lung function to X-ray findings have been reported by other investigators.[72,73]

It has been shown that disagreements on the answer to any one question on the questionnaire can run as high as 20%, with 61% of the disagreement due to the observer, 21% due to the subject and 16% to the question.[74] For lung function tests, sizable instrument and interviewer effects can also be found.[75] These comments may serve as a warning of the danger of making uncritical comparisons between prevalence studies done by different observers at different times on different populations and often using somewhat different criteria.

Wynder has listed fourteen studies in which cough could be clearly related to cigarette smoking.[76] In his own study comparing Seventh Day Adventists with a New York group and a Los Angeles group, the following stands out. Among nonsmokers, the New York group coming from the area with the heaviest air pollution also had the lowest cough record. This finding that the effect of smoking on the symptomatic manifestations of chronic bronchitis greatly overshadows any air pollution effect (if such exists) has bedevilled numerous investigations.

Almost every investigator has found an urban excess (with a concomitant correlation to air pollution) in mortality rates[77,78,79] from chronic bronchitis. A morbidity study[80] by the Respiratory Diseases Study Group of the College of General Practitioners (England) using standardized criteria found the prevalence rates of chronic bronchitis among males aged 40 to 65 years to be: rural, 3%; urban, 7%; and metropolitan, 8%. Studies on the prevalence of chronic bronchitis comparing three areas in New Hampshire,[81] the Vale of Glamorgan with Leigh,[82] the island of Bornholm with the Vale of Glamorgan and Annandale[83] and postal workers in Detroit[84] have, however, all suggested (with some reservations) that the contribution of air pollution to this clinical syndrome was negligible compared with that of smoking. Similar findings emerged from an investigation in Norway[85] and another in Chilliwak, British Columbia,[86] with regard to cough and phlegm; but both these studies suggested that the reduction in lung function found in communities with more air pollution was greater than could be accounted for by smoking alone and air pollution might be the factor responsible for this. Here, however, two other variables (ethnic group and socioeconomic status) also associated with reduced lung function may have been involved.

Confirmatory evidence of a somewhat shaky variety suggesting that air pollution does have an independent effect has been provided by a recent postal survey in Great Britain with 9,975 replies (74% of those eligibles).[87] This survey found that in all air pollution classes smokers had a standardized mortality ratio for bronchitis some 2.7 times that of nonsmokers; but there was also a stepwise increase as the amount of air pollution increased, with (comparing smokers with smokers and nonsmokers with nonsmokers) the SMR among persons exposed to high pollution levels some 1.7 times greater than among persons exposed to very low levels of air pollution.

An alternative approach to overcome some of the problems of comparing different communities which are difficult to define is to view bronchitis in a large industrial complex where personal and environmental measurements can be made with more precision. A recent study on 10,449 men in two integrated steel works in South Wales again demonstrated the overriding importance of cigarette smoking in the etiology of chronic bronchitis, but failed to show any relationship between respiratory disability and atmospheric pollution even though in the worst working areas SO_2 levels averaged 0.5 to 1.0 ppm and respirable particulates averaged 1000 to 1800 $\mu g/M^3$.[88]

Since investigations on adults have given such negative or equivocal results, it was natural to turn to studies on children, where the effect of smoking might be expected to be less; and several such studies are underway at the present time. A recent paper on childhood bronchitis in England and Wales showed a very clear social gradient from low rates in the children from professional and managerial families to the highest rates in those of semiskilled and unskilled workers.[89] Only in this last group (social classes IV and V)* was there also a clear rural–urban gradient with the highest rates being found in the cities of Newcastle and Bolton which have high levels of air pollution.

Another approach to the study of chronic bronchitis has been the use of longitudinal studies, comparing a population with itself over time. These studies are generally not truly comparable with those discussed previously because they mostly deal with those factors which produce clinical exacerbations in patients who already have pulmonary disease. Typical examples are two studies in Chicago which interestingly (and almost typically in this field of study) produced contradictory results. One study covering 41,917 patient days over a 2-year period found that when both season and daily temperatures were held constant only hydrocarbon levels (of minor importance) showed an independent correlation with symptoms.[90] The other study, with the analyzed data based largely on 561 patients followed over a 3-month period and no correction made for temperature effects, showed an increase in acute morbidity in older patients (55 years and older) with more severe bronchitis, especially on the day after exposure in association with increased levels of SO_2.[91]

In summary, it appears that air pollution by itself is not an important determinant of nonspecific chronic respiratory disease; but in combination with smoking, unfavorable living conditions and inclement weather conditions, it may assume a more important role.

Lung Cancer, Asthma and Other Pulmonary Diseases

In addition to the well known and fully proven relationship between cigarette smoking and lung cancer, lung cancer also shows an urban excess, particularly among males.[92] This urban excess has been found for smokers

* See Glossary.

on both sides of the Atlantic[93,94,95] and, with relatively few exceptions,[96] also among nonsmokers. Most investigators have reported that the urban–rural difference increases with the number of cigarettes smoked, suggesting that the two factors have a multiplicative rather than a simply additive effect. The impression that the urban factor is most probably air pollution has been based on the finding in polluted atmosphere of carcinogens (notably 3, 4-benzopyrene) which produce cancer in experimental animals[97] (though not by inhalation) and which, in a few studies, correlate with the distribution of lung cancer.[98]

Lung Cancer

The evidence linking air pollution to lung cancer is, however, far from conclusive. Some of the urban excess, and the puzzling sex difference, may be explained by the greater duration of cigarette smoking (at a given level) among urban dwellers and males. The remaining urban excess could be accounted for by some other variable (such as a subviral infection related to overcrowding and hence to urban conditions) having little or nothing to do with air pollution at the levels usually encountered. Certain arguments against considering the urban factor to be air pollution have been summarized by Goldsmith[63] and deal with the poor correlation (for cities of each size within countries) between air pollution and lung cancer; the higher rates of lung cancer found among migrants to urban areas compared with lifetime urban residents and the lesser impact of the urban factor on women. To these, one can add the better correlation found between population density and lung cancer than between air pollution and lung cancer,[99] the lack of excess cancer among garage workers and policemen exposed to heavy concentrations of motor vehicle exhaust[100] and the low death rate from lung cancer among coal miners.[101] This last is particularly important in view of the experimental work on animals suggesting that lung trauma is an important contributory factor to producing lung cancer when virus infections (PR8, influenza B, Sendai) and a carcinogen are combined.[102]

To the above generalizations dealing with air pollutants mostly found and recorded (particulates, sulfur oxides, nitrous oxides, oxidants, carbon monoxide), certain exceptions should be made. Arsenic is a known carcinogen,[103] but nonoccupational air pollution exposures appear to be very limited. Certain primitive gas generator plants producing large amounts of tarry fume have been associated with an excess of lung cancer,[104] but this does not appear to be a general problem today. One rare type of cancer of the pleura known as mesothelioma has been associated with asbestos[105] as an air pollutant and obviously demands close attention (Section 3.4). As a general community problem, however, the question of any causal relationship between air pollution and lung cancer must be regarded as unsettled.[106,106a] It should be noted that the problems of chronic bronchitis and lung cancer are related since persons who develop bronchitis appear to run a higher risk of developing lung cancer.[107]

Asthma

Among children, several studies have suggested a positive relationship between air pollution and asthma. In Erie County, New York,[108] after standardizing for social class, hospitalization rates for asthma and eczema (especially in boys under the age of 5 years) increased stepwise across the four air pollution levels used. (An obvious difficulty in studies of this type is establishing denominators.) In Philadelphia,[109] a longitudinal study showed more asthma on days of high pollution. Several populations of asthmatics, covering a wider age span, are being followed prospectively in the United States. An early study of this type in Pasadena, California,[110] followed 137 asthma patients on a daily record basis for 98 days and found 8 patients who were called "smog reactors" because their attacks corresponded to days showing plant damage, together with a significantly greater number of attacks on days with oxidant values sufficient to cause eye irritation. In evaluating these and similar reports, the subjective nature of the precipitating factor* for many asthmatics should be remembered. In addition, without adequate controls, one can never be sure whether a relationship with air pollution means that the air pollution is responsible for the asthma or whether the conditions that increased air pollutant levels were directly or indirectly (e.g., through effects on pollen concentrations) of more importance. Good explanations still remain to be found for the special forms of asthma found in the Tokyo-Yokohama area of Japan and in New Orleans, neither of which appear to fit with any of the better recognized sources and types of air pollution.

For the environmentalist, it is easy to overlook the trees for the forest. While community-wide pulmonary effects of general air pollution may be of prime interest, the possibility of localized effects, extending beyond the workers in a particular industry, should not be forgotten. Asbestos has already been mentioned, but others such as beryllium[111] (Section 3.13), hydrogen sulfide (Section 3.10), chlorine (Section 3.14) and some of the phthalates (to mention but a few) need to be considered.

Eye Irritation and Skin Diseases

During a period of temperature inversion in Buffalo, New York, some 48% of persons in one area of the city who were exposed outdoors for 10 minutes or more between 8:30 and 10:30 A.M. on September 18, 1963, complained of burning eyes.[112] The attack rate was only 2% for those remaining outdoors less than 10 minutes. In a longitudinal study in New York of 1,090

* For asthmatics, one requires both an underlying susceptibility to the disease and some precipitating factor such as ragweed pollen to which the person is sensitive and which triggers an actual attack.

adults over the age of 15 years which covered 35,400 person-weeks of observation with data obtained by weekly interview, the question was asked, "Did you have itching, burning or tearing of your eyes?" Time-series analysis[113] of the data showed that SO_2 levels (or something paralleling these levels), but not particulate density, did contribute with almost no lag period to the presence of eye irritation. Principal components analysis[114] on somewhat expanded data (61,000 person-weeks of observation) found eye irritation to be associated with pollutants such as oxidants and organic acids and to occur in warmer weather with stable barometric pressures.

Apart from the above studies, relatively little has been published since 1960 in the English language on community-wide eye irritation produced by air pollution; but prior to this, a series of surveys, especially in California (dealing mostly with oxidants), had documented the relationship. Experimental work has suggested that at concentrations approaching those found naturally, most of the irritant effect found with oxidant pollution can be accounted for by the formaldehyde and peroxyacetyl nitrate present;[115] while a variety of olefins and not sulfur dioxide could also be implicated.[116] Deleterious long term effects, if any, of air pollution on visual performance have not been documented.

It has been suggested[117] that in view of the effect of air pollutants on inanimate objects, it would be logical to look for harmful effects on the skin. No acceptable evidence of such harmful effects has however been presented.

Heart Disease and Lead Poisoning

It has been shown[118] that cholesterol-fed rabbits exposed to small concentrations of carbon monoxide develop more aortic atheromatosis than do controls.* In human cigarette smokers who also have increased levels of carboxyhemoglobin,[119] a similar increase in the frequency and severity of atheroma of the coronary arteries is found.[120] Cigarette smokers not only have an increased risk of developing clinical coronary artery disease[121] (presumably related to the underlying atherosclerosis), but stand a greatly increased risk of sudden death[122] which reverts to something approaching a normal risk shortly after cigarette smoking is discontinued.[123] Persons with some cardiac ischemia show electrocardiographic abnormalities[124] on smoking even one cigarette, which carries a bad prognosis;[125] they resemble those described among persons exposed to carbon monoxide,[126] and may well account for the excess of sudden deaths. It is unknown how far this model, suggesting a possible role for carbon monoxide in increasing the risk of coronary artery disease, can be applied to real experience because of the large number of other, more obvious determinants (e.g., diet, smoking, exercise) involved; but it should not

* A control is a human or animal for which all the other variables are the same as those for the experimental animal excepting the one variable under test.

be ignored. It has been found in one study that an increased myocardial infarction case fatality rate was associated with periods of relatively increased CO pollution.[127] An alternative mechanism for this in persons with borderline circulatory disturbances might be the interference with oxygen transport that has been shown with relatively low levels of carboxyhemoglobin.[128]

Another air pollutant which might conceivably relate to cardiovascular disease is cadmium (Section 3.13). The evidence which has been presented in this context is inconclusive. A positive correlation has been reported between cardiovascular disease and cadmium in the air,[129] while cadmium-fed rats have been made hypertensive.[130] It has also been reported[131] that persons dying of hypertension* had in their kidneys increased concentrations of cadmium as compared to subjects dying of other major diseases, but this has not been confirmed.[132] This author, in a study of a random sample of 127 persons aged 50 to 69 years from Birmingham, Alabama, found the mean serum cadmium level for his 37 hypertensive patients (systolic blood pressure greater than 156) to be 1.40 μg/100-ml compared with 1.31 μg/100 ml for his 90 normotensives;* these two means being less than one standard deviation apart.[133]

From time to time, the risk of lead poisoning as an environmental contaminant has received widespread publicity. While atmospheric levels of lead are low, blood lead levels in exposed persons do correlate with these atmospheric levels.[134] These blood levels are, however, below those values which have been associated with clinical disease though early slight injury could be missed.†

While the evidence relating general air pollution to chronic disease in humans is still inconclusive in many respects, this is no reason for "going slow" on air pollution abatement programs. Health is not just the absence of disease, but also the ability to enjoy life. Any pollutant which is detectable by our senses can interfere with this enjoyment of life.

REFERENCES

1. Goldfield, E. D.: Statistical Reports Division, U.S. Department of Commerce: Statistical Abstract of the United States, 1967.
2. Tebbens, B. D.: Gaseous Pollutants in the Air, in A. C. Stern (ed.), "Air Pollution," 2d ed., Academic Press, New York, 1968.
3. Ayres, S. M. and Meta E. Buehler: The Effects of Urban Air Pollution on Health, *Clin. Pharmacol. Ther.*, vol. 11, p. 337, 1970.

* Definitions of hypertension vary; but in experimental work on animals, levels of blood pressure more than two standard deviations from the expected mean for that species, sex and age constitute hypertension; while normotension means that the blood pressure is within normal limits.

† Minor disturbances of the blood making system of kidney and of the intestinal system will not be picked up by an examining physician unless he has been alerted to the possibility of these changes.

4. Drinker, P.: Deaths During the Severe Fog in London and Environs, *Arch. Ind. Hyg.*, vol. 7, p. 275, December 5 to December 9, 1952.
5. Amdur, M. O., W. W. Melvin and P. Drinker: Effect of Inhalation of Sulfur Dioxide by Man, *Lancet*, London, vol. 2, p. 758, 1953.
6. Snell, A. E. and P. C. Luchsinger: Effects of Sulfur Dioxide on Expiratory Flowrates and Total Respiratory Resistance in Normal Human Subjects, *Arch. Envir. Health*, Chicago, vol. 18, p. 693, 1969.
7. Frank, N. R., M. O. Amdur, J. Worcester and J. L. Whittenberger: Effects of Acute Controlled Exposure to SO_2 on Respiratory Mechanics in Healthy Male Adults, *J. Appl. Physiol.* Bethesda, vol. 17, p. 252, 1962.
8. Speizer, F. E. and N. R. Frank: A Comparison of Changes in Pulmonary Flow Resistance in Healthy Volunteers Acutely Exposed to SO_2 by Mouth and by Nose, *J. Industr. Med.*, vol. 23, p. 75, 1966.
9. Wright, G.: Air Pollution, *Public Health Rep.*, Washington, vol. 75, pp. 11–79, 1960.
10. Barton, G. G., M. Corn, J. B. L. Gee, C. Vasallo and A. P. Thomas: Response of Healthy Men to Inhaled Low Concentrations of Gas–Aerosol Mixtures, *Arch. Environ. Health*, Chicago, vol. 18, p. 681, 1969.
11. Battigelli, M. C., H. M. Cole, D. A. Fraser and R. A. Mah: Long-term Effects of Sulfur Dioxide and Graphite Dust on Rats, *Arch. Environ. Health*, Chicago, vol. 18, p. 602, 1969.
12. Alarie, Y., C. E. Ulrich, W. M. Busey, H. E. Swann and H. N. MacFarland: Long-term Continuous Exposure of Guinea Pigs to Sulfur Dioxide, *Arch. Environ. Health*, Chicago, vol. 21, p. 769, 1970.
13. McDermott, M.: Acute Respiratory Effects of the Inhalation of Coal-dust Particles, *J. Physiol.*, vol. 162, 1962.
14. Toyama, T.: Air Pollution and Its Health Effects in Japan, *Arch. Environ. Health*, Chicago, vol. 8, p. 153, 1963.
15. Rokaw, S. W., H. E. Swann, R. L. Keenan and J. R. Phillips: "Human Exposures to Single Pollutants: NO_2: In a Controlled Environment Facility," preprint, California State Department of Public Health, Berkeley, 1968.
16. Myers, F. H. and C. H. Hine: "Some Experiences of NO_2 in Animals and Man," 5th Air Pollution Med. Res. Conf., Los Angeles, 1960.
17. Patty, F. A. (ed.): "Industrial Hygiene and Toxicology," Wiley, New York, 1962.
18. LaTowsky, L. W., E. L. MacQuiddy and J. P. Tollman: Toxicology of Oxides of Nitrogen: Toxic Concentrations, *J. Ind. Hyg. Toxicol.*, vol. 23, p. 129, 1941.
19. Blair, W. H., M. C. Henry and R. Ehrlich: Chronic Toxicity of Nitrogen Dioxide: II, Effects on Histopathology of Lung Tissue, *Arch. Environ. Health*, Chicago, vol. 18, p. 186, 1969.
20. Freeman, G. and G. B. Haydon: Emphysema After Low-level Exposures to NO_2, *Arch. Environ. Health*, Chicago, vol. 8, p. 125, 1964.
21. Purvis, M. R. and R. Ehrlich: Effect of Atmospheric Pollutants on Susceptibility to Respiratory Infection: II, Effect of Nitrogen Dioxide, *J. Infect. Dis.*, Chicago, vol. 113, p. 72, 1963.
22. Jaffe, L. S.: The Biological Effect of Photochemical Air Pollutants on Man and Animals, *Amer. J. Public Health*, New York, vol. 57, p. 1269, 1967.
23. Tebbens, B. D.: Residual Pollution Products in the Atmosphere, in A. C. Stern (ed.), "Air Pollution," vol. 1, Academic Press, New York, 1962.
24. McMillan, R. S., D. H. Wiseman, B. Hanes and P. F. Wehrle: Effects of Oxidant Air Pollution on Peak Expiratory Flowrates in Los Angeles School Children, vol. 18, p. 941, 1969.
25. Bates, D. V., G. Bell, C. Burnham, M. Hazucha and J. Mantha: Problems in Studies of Human Exposure to Air Pollutants, *Canad. Med. Ass. J.*, Toronto, vol. 103, p. 833, 1970.
26. Smith, L. E.: Cardiorespiratory Effects of PAN Inhalation During Exercise, preprint, California University Riverside Department of Physical Education, 1964.
27. Jones, J. L., N. Endon, E. A. Schuck, R. G. Caldwell and C. J. Doyle: A Progress Report on the Chemistry of Community Air Pollution, California Research Institute, South Pasadena, 1962.
28. Holland, G. J., D. Benson, A. Bush, G. Q. Rich and R. P. Holland: Air Pollution Simulation and Human Performance, *Amer. J. Public Health*, New York, vol. 58, p. 1684, 1968.

29. Murphy, S. D., C. E. Ulrich, S. H. Frankowitz and C. Xintaras: Altered Function in Animals Inhaling Low Concentrations of Ozone and Nitrogen Dioxide, *Amer. Industr. Hyg. Ass. J.*, vol. 25, p. 246, 1964.
30. Emik, L. O., R. L. Plata, K. I. Campbell and G. L. Clarke: Biological Effects of Urban Air Pollution, *Arch. Environ. Health*, Chicago, vol. 23, p. 335, 1971.
31. Vaughan, T. R., L. F. Jennelle and T. R. Lewis: Long-term Exposure to Low Levels of Air Pollutants, *Arch. Environ. Health*, Chicago, vol. 19, p. 45, 1969.
32. Colucci, J. M. and C. R. Begeman: Carbon Monoxide in Detroit, New York, and Los Angeles Air, *Environ. Sci. and Technology*, vol. 3, p. 41, 1969.
33. Ramsey, J. M.: Concentrations of Carbon Monoxide at Traffic Intersections in Dayton, Ohio, *Arch. Environ. Health*, Chicago, vol. 13, p. 44, 1966.
34. Waller, R. E., B. T. Commins and P. J. Lawther: Air Pollution in Road Tunnels, *Brit. J. Industr. Med.*, London, vol. 18, p. 250, 1961.
35. McIlvaine, P. M., W. C. Nelson and D. Bartlett: Temporal Variation of Carboxyhemoglobin Concentrations, *Arch. Environ. Health*, Chicago, vol. 5, p. 308, 1969.
36. Ringold, A., J. R. Goldsmith, H. L. Helwig, R. Finn and F. Schuette: Estimating Recent Carbon Monoxide Exposures, *Arch. Environ. Health*, Chicago, vol. 5, p. 308, 1962.
37. Schulte, J. H.: The Medical Aspects of Closed Cabin Atmosphere Control, Milit. Med., Washington, vol. 126, p. 40, 1961.
38. Hanks, T. G.: Human Performance of a Psychomatic Test as a Function of Exposure to Carbon Monoxide. *Ann. N.Y. Acad. Sci.*, vol. 174, p. 421, 1970.
39. Horvath, S. M., J. F. O'Hanlon and T. E. Dahms: Carbon Monoxide and Vigilance Potential Danger from Existing Urban Concentrations, Project Clean Air, Vol. 2, Int. of Environmental Stress, Santa Barbara, California, 1970.
40. Chevalier, R. B., R. A. Krumholz and J. C. Ross: Reaction of Nonsmokers to Carbon Monoxide Inhalation, *J. Amer. Med. Assoc.*, vol. 198, p. 1061, 1966.
41. Musselman, N. P., W. A. Groff, P. P. Yevitch, F. T. Wilinski, M. H. Weeks and F. W. Oberst: Continuous Exposure of Laboratory Animals to a Low Concentration of Carbon Monoxide, *Aerospace Med.*, St. Paul, vol. 30, p. 524, 1959.
42. Lindenberg, R., D. Levy, T. Preziosi and M. Christianson: An Experimental Investigation in Animals of the Functional and Morphological Changes from Single and Repeated Exposure to Carbon Monoxide, presented at the American Industrial Hygiene Conference, Washington, D.C., May 13–17, 1962.
43. Rylander, R.: Alterations of Lung Defense Mechanisms Against Airborne Bacteria, *Arch. Environ. Health*, Chicago, vol. 18, p. 551, 1969.
44. Committee on Public Health, New York Academy of Medicine: Air Pollution and Health, *Bull. N.Y. Acad. Med.*, vol. 42, p. 588, 1966.
45. Rjazanov, V. A.: Criteria and Methods for Establishing Maximum Permissible Concentrations of Air Pollution, Bull. WHO, vol. 32, p. 389, 1965.
46. Shy, C. M., J. P. Creason and M. E. Pearlman: The Chattanooga School Children Study: Effects of Community Exposure to Nitrogen Dioxide, I: Methods, Description of Pollutant Exposure and Results of Ventilatory Function Testing, *J. Air Pollut. Contr. Ass.*, vol. 20, p. 539, 1970.
47. Biersteker, K. and P. Van Leeuwen: Air Pollution and Peak Flowrates in Schoolchildren, *Arch. Environ. Health*, Chicago, vol. 20, p. 382, 1970.
48. Wayne, W. S., P. F. Wehrle and R. E. Carroll: Oxidant Air Pollution and Athletic Performance, *J. Amer. Med. Assoc.*, vol. 199, p. 901, 1967.
49. McCarroll, J. R., E. J. Cassell, W. Ingram and D. Walter: Health and the Urban Environment: Health Profiles versus Environmental Pollutants, *Amer. J. Public Health*, New York, vol. 56, p. 266, 1966.
50. Hausknecht, R.: Population Surveys in Air Pollution Research, Bur. Chronic Diseases; California State Dept. Public Health, Berkeley, 1962. (Quoted in A. C. Stern (ed.), "Air Pollution," Academic Press, New York, 1968.)
51. Verma, M. P., F. J. Schilling and W. H. Becker: Epidemiological Study of Illness Absences in Relation to Air Pollution, *Arch. Environ. Health*, Chicago, vol. 18, p. 536, 1969.
52. Ipsen, J., M. Deane and F. E. Ingenito: Relationships of Acute Respiratory Disease to Atmospheric Pollution and Meteorological Conditions, *Arch. Environ. Health*, Chicago, vol. 18, p. 462, 1969.

53. Wayne, W. S. and P. F. Wehrle: Oxidant Air Pollution and School Absenteeism, preprint, Div. of Health Effects Research, Southern California University at Los Angeles, 1968.
54. Loudon, R. G. and J. F. Kilpatrick: Air Pollution, Weather and Cough, *Arch. Environ. Health,* Chicago, vol. 18, p. 641, 1969.
55. Goldsmith, J. R.: Air Pollution Epidemiology, *Arch. Environ. Health,* Chicago, vol. 18, p. 516, 1969.
56. Sterling, T. D., S. V. Pollack and J. Weinkam: Measuring the Effect of Air Pollution on Urban Morbidity, *Arch. Environ. Health,* Chicago, vol. 18, p. 485, 1969.
57. Heimann, H.: Episodic Air Pollution in Metropolitan Boston, *Arch. Environ. Health,* Chicago, vol. 20, p. 230, 1970.
58. Firket, J.: Sur les Causes des Accidents Survenus Dans la Vallée de la Meuse Lors des Brouillards de Décembre 1930, *Bull. Roy. Acad. Belg.,* (5 serie), vol. 11, p. 683, 1931.
59. USPHS: Air Pollution in Donora, Pa., Preliminary Report, *Public Health Bulletin,* No. 306.
60. Ciocco, A. and D. J. Thompson: A Follow-up of Donora Ten Years After: Methodology and Findings, *Amer. J. Public Health,* New York, vol. 51, p. 155, 1961.
61. Ministry Health: Mortality and Morbidity During London Fog of December 1952, Report on Public Health and Medical Subjects, No. 95, London, 1954.
62. Greenburg, L., F. Field, C. L. Erhardt, M. Glasser and J. I. Reed: Air Pollution, Influenza and Mortality in New York City, *Arch. Environ. Health,* Chicago, vol. 15, p. 430, 1967.
63. Goldsmith, J. R.: Effects of Air Pollution on Human Health, in A. C. Stern (ed.), "Air Pollution," 2d Ed., Academic Press, New York, 1968.
64. British Medical Research Council: Report of the Committee on the Etiology of Chronic Bronchitis, H.M.S.O., 1965.
65. Fletcher, C. M.: Chronic Bronchitis, *Amer. Rev. Resp. Dis.,* Baltimore, vol. 80, p. 483, 1959.
66. World Health Organization: Report of Expert Committee, Chronic Cor. Pulmonale, WHO Tech. Report Series 213, 1961.
67. Thurlbeck, W. M. and G. E. Angus: The Relationship Between Emphysema and Chronic Bronchitis as Assessed Morphologically, *Amer. Rev. Resp. Dis.,* Baltimore, vol. 87, p. 815, 1963.
68. Fletcher, C. M., N. L. Jones, B. Burrows, and A. H. Niden: American Emphysema and British Bronchitis, *Amer. Rev. Resp. Dis.,* Baltimore, vol. 90, p. 1, 1964.
69. Stuart-Harris, C. H.: Chronic Bronchitis, Part II, *Abstracts World Med.,* vol. 42, p. 737, 1968.
70. (British) Medical Research Council: Chronic Bronchitis in Great Britain, *Brit. Med. J.,* London, vol. 2, p. 973, 1961.
71. Hepper, N. G., R. E. Hyatt and W. S. Fowler: Detection of Chronic Obstructive Lung Disease, *Arch. Environ. Health,* Chicago, vol. 19, p. 806, 1969.
72. Fairbairn, A. S., C. H. Wood and C. M. Fletcher: Variability in Answers to a Questionnaire on Respiratory Symptoms, *Brit. J. Prev. Soc. Med.,* London, vol. 13, p. 175, 1959.
73. Burrows, B., A. H. Niden, C. M. Fletcher and N. L. Jones: Clinical Types of Chronic Obstructive Lung Disease in London and in Chicago, *Amer. Rev. Resp. Dis.,* Baltimore, vol. 90, p. 14, 1964.
74. Gorke, T. M. and B. J. Daffy: Epidemiology of Chronic Bronchitis in Jersey City, *Arch. Intern Med.,* Chicago, vol. 110, p. 606, 1962.
75. Stebbings, J. H.: Chronic Respiratory Disease Among Nonsmokers in Hagerstown, Maryland, Problems in the Estimation of Pulmonary Function Values in Epidemiological Surveys, Environmental Research preprint, ER-205, 1971.
76. Wynder, E. L., F. R. Lemon and N. Mantel: Epidemiology of Persistent Cough, *Amer. Rev. Resp. Dis.,* Baltimore, vol. 91, p. 679, 1965.
77. Pemberton, J. and C. Goldberg: Air Pollution and Bronchitis, *Brit. Med. J.,* London, vol. 2, p. 567, 1954.
78. Daly, C.: Air Pollution and Causes of Death, *Brit. J. Prev. Soc. Med.,* London, vol. 13, p. 14, 1959.
79. Stocks, P.: Cancer and Bronchitis Mortality in Relation to Atmospheric Deposit and Smoke, *Brit. Med. J.,* London, vol. 1, p. 74, 1959.
80. Resp. Diseases Study Group, College of General Practitioners: Chronic Bronchitis in Great Britain, *Brit. Med. J.,* London, vol. 2, p. 973, 1961.

81. Anderson, D. O., B. G. Ferris and R. Zickmantel: Levels of Air Pollution and Respiratory Disease in Berlin, New Hampshire, *Amer. Rev. Resp. Dis.*, Baltimore, vol. 90, p. 877, 1964.
82. Higgins, I. T. T.: Respiratory Symptoms, Bronchitis and Ventilatory Capacity in Random Sample of an Agricultural Population, *Brit. Med. J.*, London, vol. 2, p. 1198, 1957.
83. Olsen, H. C. and J. C. Gibson: Respiratory Symptoms, Bronchitis and Ventilatory Capacity in Men: an Anglo-Danish Comparison with Special Reference to Difference in Smoking Habits, *Brit. Med. J.*, London, vol. 1, p. 450, 1960.
84. Coates, E. O., G. C. Bower and N. Reinstein: Chronic Respiratory Disease in Postal Employees, *J. Amer. Med. Assoc.*, vol. 191, p. 161, 1965.
85. Mork, T.: A Comparative Study of Respiratory Disease in England and Wales and Norway, The Norwegian Research Council for Science and the Humanities, Norwegian Universities Press, 1962.
86. Anderson, D. O., B. G. Ferris and R. Zickmantel: The Chilliwack Respiratory Survey 1963: Part IV, The Effect of Tobacco Smoking on the Prevalence of Respiratory Disease, *Canad. Med. Ass. J.*, Toronto, vol. 92, p. 1066, 1965.
87. Lambert, P. M. and D. D. Ried: Smoking, Air Pollution and Bronchitis in Britain, *Lancet*, London, vol. 1, p. 853, 1970.
88. Lowe, C. R., H. Campbell and T. Khosla: Bronchitis in Two Integrated Steel Works, III, Respiratory Symptoms and Ventilatory Capacity Related to Atmospheric Pollution, *Brit. J. Industr. Med.*, London, vol. 27, p. 121, 1970.
89. Colley, J. R. T. and D. D. Ried: Urban and Social Orgins of Childhood Bronchitis in England and Wales, *Brit. Med. J.*, London, vol. 2, p. 213, 1970.
90. Burrows, B., A. L. Kellogg and J. Buskey: Relationship of Symptoms of Chronic Bronchitis and Emphysema to Weather and Air Pollution, *Arch. Environ. Health*, Chicago, vol. 16, p. 406, 1968.
91. Carnow, B. W., M. H. Lepper, R. B. Shekelle and J. Stamler: Chicago Air Pollution Study: SO_2 Levels and Acute Illness in Patients with Chronic Bronchopulmonary Disease, *Arch. Environ. Health*, Chicago, vol. 18, p. 768, 1969.
92. Haenszel, W. and K. E. Taeuber: Lung Cancer Mortality as Related to Residence and Smoking Histories: II, White Females, *J. Natl. Cancer Inst.*, Washington, vol. 32, p. 803, 1964.
93. Hammond, E. C. and D. Horn: Smoking and Death Rates: Report on Forty-four Months of Follow-up of 187,783 Men, *J. Amer. Med. Assoc.*, vol. 166, p. 1294, 1958.
94. Stocks, P.: On Relations Between Atmospheric Pollutions in Urban and Rural Localities and Mortality from Cancer, Bronchitis and Pneumonia, *Brit. J. Cancer*, London, vol. 14, p. 397, 1960.
95. Buell, P. and J. E. Dunn: Relative Impact of Smoking and Air Pollution on Lung Cancer, *Arch. Environ. Health*, Chicago, vol. 15, p. 291, 1967.
96. Doll, R.: Mortality from Lung Cancer Among Nonsmokers, *Brit. J. Cancer*, London, vol. 7, p. 303, 1953.
97. Hueper, W. C., P. Kotin, E. C. Tabor, W. W. Payne, H. Falk and E. Sawicke: Carcinogenic Bioassays on Air Pollutants, *Arch. Path.*, Chicago, vol. 74, p. 89, 1962.
98. Stocks, P.: Recent Epidemiological Studies of Lung Cancer Mortality, Cigarette Smoking and Air Pollution with Discussion of a New Hypothesis of Causation, *Brit. J. Cancer*, London, vol. 20, p. 595, 1966.
99. Ashley, D. J. B.: The Distribution of Lung Cancer and Bronchitis in England and Wales, *Brit. J. Cancer*, London, vol. 21, p. 243, 1967.
100. Lawther, P. J.: Air Pollution, Bronchitis and Lung Cancer, *Postgrad. Med. J.*, London, vol. 42, p. 703, 1966.
101. Goldman, K. P.: Mortality of Coal Miners from Carcinoma of the Lung, *Brit. J. Industr. Med.*, London, vol. 22, p. 72, 1965.
102. Kuschner, M.: The J. Burns Anderson Lecture, The Causes of Lung Cancer, *Amer. Rev. Resp. Dis.*, Baltimore, vol. 98, p. 573, 1968.
103. Buechley, R. W.: Epidemiological Consequences of an Arsenic Lung Cancer Theory, *Amer. J. Public Health*, New York, vol. 53, p. 1229, 1963.
104. Kawai, M., H. Amamoto and K. Harada: Epidemiologic Study of Occupational Lung Cancer, *Arch. Environ. Health*, Chicago, vol. 14, p. 859, 1967.

105. Selikoff, I. J. and E. C. Hammond: Community Effects of Nonoccupational Environmental Asbestos Exposure, *Amer. J. Public Health,* New York, vol. 58, p. 1658, 1968.
106. Rimington, J.: Smoking, Chronic Bronchitis and Lung Cancer, *Brit. Med. J.,* London, vol. 2, p. 373, 1971.
106a. The discovery of fatal liver cancer among vinyl chloride workers has been reported in 1974.
107. Oettle, A. G.: Lung Cancer, Cigarette Smoke and Smog, *J. Roy. Soc. Health,* vol. 1, p. 27, 1968.
108. Sultz, H. A., J. G. Feldman, E. R. Schlesinger and W. E. Mosher: An Effect of Continued Exposure to Air Pollution on the Incidence of Chronic Childhood Allergic Disease, *Amer. J. Public Health,* New York, vol. 60, p. 891, 1970.
109. Girsh, L. S., E. Shubin, C. Dick and F. A. Schulaner: A Study on the Epidemiology of Asthma in Children in Philadelphia: The Relation of Weather and Air Pollution to Peak Incidence of Asthmatic Attacks, *J. Allerg.,* St. Louis, vol. 39, p. 347, 1967.
110. Schoettlin, E. C. and E. Landau: Air Pollution and Asthmatic Attacks in the Los Angeles Area, *Public Health Rep.,* Washington, vol. 76, p. 545, 1961.
111. Chesner, C.: Chronic Pulmonary Granulomatosis in Residents of Community near Beryllium Plant: Three Autopsied Cases, *Ann. Intern Med.,* Philadelphia, vol. 32, p. 1028, 1950.
112. Partridge, R. A., J. H. Stebbings, W. R. Elsea and W. Winkelstein: Outbreak of Acute Eye Irritation Associated with Air Pollution, *Public Health Rep.,* Washington, vol. 81, p. 153, 1966.
113. McCarroll, J., E. J. Cassell, D. W. Walter, J. D. Mountain, J. R. Diamond and I. M. Mountain: Health and the Urban Environment: Air Pollution and Illness in a Normal Urban Population, *Arch. Environ. Health,* Chicago, vol. 14, p. 178, 1967.
114. Cassell, E. J., M. D. Lebowitz, I. M. Mountain, H. T. Lee, D. J. Thompson, D. W. Wolter and J. R. McCarroll: Air Pollution, Weather and Illness in a New York Population, *Arch. Environ. Health,* Chicago, vol. 18, p. 523, 1969.
115. Schuck, E. A., E. R. Stephens and J. T. Middleton: Eye Irritation Response at Low Concentrations of Irritants, *Arch. Environ. Health,* Chicago, vol. 13, p. 570, 1966.
116. Doyle, G. J., N. Endow and J. L. Jones: Sulfur Dioxide Role in Eye Irritation, *Arch. Environ. Health,* Chicago, vol. 3, p. 657, 1961.
117. Lubowe, I. I.: The Effect of Air Pollutants on the Skin, *General Practice,* May 10, 1964.
118. Astrup, P., K. Kjeldsen and J. Wanstrup: Enhancing Influence of Carbon Monoxide on the Development of Atheromatosis in Cholesterol-Fed Rabbits, *J. Atheroscler. Res.,* Amsterdam, vol. 7, p. 343, 1967.
119. Ringold, A., J. R. Goldsmith, H. L. Helwig, R. Finn and F. Schuette: Estimating Recent Carbon Monoxide Exposures, *Arch. Environ. Health,* Chicago, vol. 5, p. 308, 1962.
120. Auerbach, O., E. C. Hammond and L. Garfinkel: Smoking in Relation to Atherosclerosis of the Coronary Arteries, *New England J. Med.,* Boston, vol. 273, p. 775, 1965.
121. Doyle, J. T., T. R. Dawber, W. B. Kannel, S. H. Kinch and H. A. Kahn: The Relationship of Cigarette Smoking to Coronary Heart Disease: Second Report of the Combined Experience of the Albany, N.Y., and Framingham, Mass. Studies, *J. Amer. Med. Ass.,* vol. 190, p. 886, 1964.
122. Kannel, W. B.: Habits and Coronary Heart Disease: The Framingham Heart Study, *Public Health Service Publication No. 1515,* Bethesda, Maryland, National Heart Institute, 1966.
123. Doll, R. and A. B. Hill: Mortality in Relation to Smoking: Ten Years Observation of British Doctors, *Brit. Med. J.,* London, vol. 1, p. 1460, 1964.
124. Sen Gupta, A. N. and B. P. Ghosh: Observations on Some Cardiovascular and Biochemical Effects of Tobacco Smoking in Health and in Ischemic Cardiacs, *Bull. Inst. Post-Grad. Med. Education Res.,* vol. 9, p. 45, 1967.
125. Kannel, W. B. and P. M. McNamara: The Evidence for Excess Risk in Coronary Disease, *Minn. Med.,* St. Paul, p. 1197, August 1969.
126. Goldsmith, J. R.: Carbon Monoxide Research—Recent and Remote, *Arch. Environ. Health,* Chicago, vol. 21, p. 118, 1970.
127. Cohen, S. I., M. Deane and J. R. Goldsmith: Carbon Monoxide and Survival from Myocardial Infarction, *Arch. Environ. Health,* Chicago, vol. 19, p. 510, 1969.

128. Schulte, J. H.: Effects of Mild Carbon Monoxide Intoxication, *Arch. Environ. Health*, Chicago, vol. 7, p. 524, 1963.
129. Carroll, R. E.: The Relationship of Cadmium in the Air to Cardiovascular Disease Death Rates, *J. Amer. Med. Assoc.*, vol. 198, p. 267, 1966.
130. Schroeder, H. A.: Cadmium, Chromium and Cardiovascular Disease, Circulation, New York, vol. 35, p. 570, 1967.
131. Schroeder, H. A.: Cadmium as a Factor in Hypertension, *J. Chronic. Dis.*, St. Louis, vol. 18, p. 647, 1965.
132. Morgan, J. M.: Tissue Cadmium Concentration in Man, *Arch. Intern. Med.*, Chicago, vol. 123, p. 405, 1969.
133. Peacock, P. B.: The Association of Serum Cadmium Values with Indices of Cardiovascular and Other Disease and Certain Environmental Factors, Final Report to the Environmental Health Service. D. HEW, Contract PH 86-67-39, 1970.
134. Division of Air Pollution, USPHS: Survey of Lead in the Atmosphere of Three Urban Communities, *PHS Publication 999 AP-12*, 1965.

2.12 POLLUTION EFFECTS ON PLANT LIFE

The purpose of this section is to describe some of the effects of air pollutants on vegetation and to provide ample source material for additional reading. The section contains a brief survey of the types and sources of plant-damaging air pollutants, their economic effects on agriculture, the diagnosis of plant effects and suggestions for procedures which can warn of or help avoid impending difficulties.

Types and Sources of Plant Damaging Air Pollutants

Injury to vegetation by emissions from smelting operations was widespread in North America during the first third of the twentieth century.[1] Large areas of vegetation were destroyed by sulfur oxides and components of particulate emissions near Ducktown, Tennessee; Anaconda, Montana; Salt Lake City, Utah; Crockett, California; Sudbury, Ontario; and Train, British Columbia.[2] In the past 30 years, the reduced intensity of sulfur oxide emissions and increased number of sources have caused a change in the pattern of effects on vegetation from acute damage to chronic effects. Instead of areas being denuded of plant life, decreased growth and yield of susceptible species now occur as a result of exposure to sulfur dioxide concentrations in the low parts per million (ppm) by volume range.

Emissions of atmospheric fluorides from the manufacture of aluminum phosphate fertilizer and other products have also caused considerable injury to agricultural crops in the past 30 years. Concentrations of hydrogen fluoride and silicon tetrafluoride in the low parts per billion range have injured susceptible plant species.[3]

A new type of pollution, associated with large urban centers, was identified in Los Angeles and has been found more recently in or near many other cities in the United States. Atmospheric reactions between nitrogen oxides and hydrocarbon compounds from vehicular and industrial sources, catalyzed by sunlight, produce ozone, peroxyacetyl-nitrates and other oxidizing vapors and aerosols. The formation of these substances in moving air masses results in plant damage far from the sources of the original reacting compounds.[4,5] Photochemical oxidants cause the most widespread damage to vegetation in the United States, partly because they are produced in the greatest quantity during the growing season.

Other pollutants, such as chlorine, hydrogen chloride and ammonia, have injured plants usually in limited, well-defined locations as a result of accidents or leaks in storage containers. Organic vapors, insecticides, herbicides such as 2,4-D[6] and ethylene[7] may have effects more widespread than formerly realized. Ethylene is emitted by manufacturing processes and vehicular exhaust in amounts sufficient to affect vegetation.[8]

Economic Effects of Air Pollutants on Agriculture

Reduced growth, yield and quality of forest trees, native vegetation, horticultural and agronomic crops occur each year as a result of exposure to polluted air masses.[9,10,11] A recent estimate in the United States indicates that about one-third of the states have crop losses of more than $0.5 million a year from air pollution.[12] Total loss estimates for the United States have ranged from $150 million to $500 million per year depending on the method of calculation and whether ornamental plants, forest lands and so on are considered. These estimates have been based mainly on reduction in yield and commercial value of crops. Economic loss due to indirect effects of air pollutants on vegetation may never be assessed because of the difficulty in expressing the total value of a natural resource. There may also be an occasional beneficial consequence of pollution on agriculture. Plants growing on sulfur-deficient soil can satisfy part of their sulfur requirement by absorbing atmospheric sulfur dioxide.[13] Farmers in one part of the United States may benefit from higher commodity prices if crops grown in other states are damaged by air pollutants. However, these benefits are uncommon and small in magnitude when compared with the demonstrated incidence of foliar injury, growth inhibition and yield reduction.

Indirect Effects of Air Pollutants on Vegetation

Plants make important contributions to the preservation of the atmosphere, soil and natural waters. Plants improve the atmosphere by replenishing oxygen, providing water vapor, removing carbon dioxide, removing particulate and gaseous impurities, modifying extremes of temperature, reducing wind speed, increasing rainfall, absorbing solar radiation and muffling noise. They increase water-holding capacity, reduce erosion and improve soil structure and composition. Natural waters are provided with oxygen, nutrients which might foster undesirable vegetation are removed and the plants themselves provide food for organisms living in the waters. Consequently, injury to large areas of vegetation will have important ramifications for the environment. For example, significant effects on reflectivity of solar radiation, ambient temperature, humidity, wind speed and rainfall resulted from destruction of vegetation in Tennessee by smelter emissions.[14] Vegetation, after exposure to pollutants, may be more susceptible to the stresses imposed by insects, disease and extreme climatic or soil conditions. Adverse effects on flowering, pollination or fruiting may occur when pollutants are present at the proper stage of plant development. Many of these effects do not become manifest until days, weeks or even months after the period of exposure.

Plants as Agents for Transmission of Potentially Harmful Substances

Plants accumulate substances from the soil and from the atmosphere and vegetation may transfer these substances in concentrated form when used as food for animals and man. In some cases, concentrations can increase to toxic levels. Fluoride, accumulated by forage plants from the atmosphere, causes fluorosis, a serious bone disease in animals.[3] Particles containing lead or other metals are deposited on vegetation and on soil,[15,16,17] where they may be absorbed by plants and transferred to animals. These substances are cumulative toxicants in plants and animals and their effects may not become manifest until long after the initial exposure period.

Air Quality Criteria and Standards

The federal government has promulgated secondary air quality standards designed to prevent the aforementioned injuries to plants and animals. These standards are based on published criteria which provide indications of the relationships between atmospheric concentrations of pollutants and harmful effects.[18] A consistent set of standards for the whole country would be desirable; however, there are several reasons why regulations may differ from state to state. Susceptibility of different species of plants varies greatly for individual pollutants. The occurrence of injury and the type and severity of symptoms often depend on the stage of development of the plant and on soil and climatic conditions before, during and after the periods of exposure. The types of agricultural crops and native plants, the kinds of pollutants and the climatic and soil conditions vary widely from one region to another; consequently, air quality standards for agriculture may reflect these regional differences.

Air quality criteria for vegetation attempt to relate plant response and air pollutant concentrations and durations of exposure. Foliar injury appears to be more closely associated with the frequency and duration of peak concentrations and the time between episodes than with long-term average pollutant concentrations.[19] Atmospheric monitoring methods with sample averaging times of 12 hours or longer are probably of little value for predicting or assessing vegetation injury because short-term peak concentrations are masked. In addition, recent evidence indicates that plants respond to the complex of atmospheric pollutants and not necessarily to individual contaminant concentrations. Combinations of pollutants interact and cause responses which may be more or less than the sum of their individual effects depending on the receptivity of the plant at the time of exposure.[20]

Diagnosis of Air Pollution Injury to Vegetation

The detection of injury to plants by air pollutants is seldom a straightforward matter, and many different lines of inferential reasoning are used. Thorough knowledge of the symptoms of foliar injury on plants by air pollutants and of the relative susceptibilities of plant species are essential.[21] Some effects of air pollution on vegetation are growth reduction, diminished yield and reduced quality of crops, premature senescence, discoloration and malformation of leaves, death of leaf tissue and a shift in population density among native species. It is frequently difficult to determine whether the observed effects are caused by pollutants, disease organisms, insects or nutritional deficiencies or climatic conditions. Chemical analyses of plant leaves can be helpful in diagnosing injury by fluorides, metals, chlorine and sulfur oxides since only small quantities of these elements occur in plants growing in unpolluted atmospheres. Atmospheric monitoring provides extremely valuable information when attempting to interpret the cause of plant injury. Although data on the substances emitted, actual atmospheric concentrations and wind directions are valuable; the most important information relating to plant injury is obtained by visual examination of the vegetation shortly after symptoms are first noticed. Techniques used for obtaining and integrating observations have been extensively described.[22,23,24,25] Ideally, surveys and atmospheric analyses should be made before industrial operations begin as well as afterwards. This procedure will provide comparative data essential to rational decision-making.

Plants as Pollution Indicators

By chemical analysis of leaves, plants can indicate whether potentially toxic substances are transferred to animals or humans via the food chain. They have also proved useful in solving pollution problems by providing indications of the presence of air pollutants, helping to identify the type of contamination and the location of emission sources.[26] Vegetation used for these purposes must have a known response and a high degree of susceptibility to one or more air pollutants, and they must be carefully tended and exposed in specific locations. Although atmospheric monitoring gives more exact information than observation of indicator plants, sufficient funds and equipment for atmospheric analyses are not always available. In addition, plant injury cannot be faithfully predicted from knowledge of atmospheric concentrations of contaminants. As indicated previously, biological, climatic and soil factors influence plant response to such a degree that severe injury may occur after one episode and not after another even though atmospheric concentrations are similar.

Table 2.12a
METHODS OF PREVENTING INJURY TO PLANTS

At Source:

Reducing rates of production of pollutants; removing pollutants from effluents; and changing processes or fuels to avoid sensitive stages of plant growth (daylight periods, flowering and fruit set).

In Pollutant Transfer and Dilution:

Relocating emission source taking into account terrain, meteorological factors and direction from agricultural operations; timing emissions in relation to wind speed and mixing depth; and increasing height of stacks to decrease ground-level concentrations.

At Receptor:

Filtering air (greenhouses); adjusting cultural practices to decrease plant susceptibility; selecting resistant species or varieties; breeding resistant varieties; and applying chemicals to prevent entry by sorption or reaction with pollutants on or near leaf surfaces, to limit pollutant intake by leaves and to enhance physiological resistance to injury.

Plants and Purification of the Atmosphere

Plants accumulate both gases and particles from the air by sorption and deposition. Consequently, large areas of vegetation may reduce the concentration of pollutants in an air mass. Green-belts or shelter-belts are used in England and in Europe to aid in purifying the atmosphere.[27] Obviously, plant species resistant to injury by air pollutants must be used to obtain the maximum benefit. It should also be recognized, however, that plants and agricultural operations can be significant sources of air pollutants.[28]

Preventing Injury to Vegetation by Air Pollutants

Some methods of preventing injury to vegetation are listed in Table 2.12a. Many of these approaches are not applicable or feasible in individual situations. Selection and breeding of resistant varieties have produced tobacco varieties less susceptible to injury by photochemical pollutants.[29] However, the most successful approach to avoiding air pollution problems and to preventing effects on vegetation may be through land-use planning and site evaluation. Exhaustive study of all relevant considerations prior to site selection coupled with source emission controls provides the best hope for protection of botanical and agricultural resources.

REFERENCES

1. Swain, R. E.: Smoke and Fume Investigations: A Historical Review, *Industrial and Engineering Chemistry*, vol. 41, pp. 2384–2388, 1949.
2. Weinstein, L. H. and D. C. McCune: "Implications of Air Pollution for Plant Life," Proceedings of the American Philosophical Society, vol. 114, pp. 18–21, 1970.
3. National Academy of Sciences, Committee on Biologic Effects of Atmospheric Pollutants: "Fluorides," Washington, D.C., 1971.
4. Cobb, F. W., Jr. and R. W. Stark: Decline and Mortality of Smog-Injured Ponderosa Pine, *Journal of Forestry*, vol. 68, pp. 147–149, 1970.
5. Miller, P. R. and A. A. Millecan: Extent of Oxidant Air Pollution Damage to Some Pines and Other Conifers in California, *Plant Disease Reporter*, vol. 55, pp. 555–559, 1971.
6. Sherwood, C. H., J. L. Weigle and E. L. Denisen: 2,4-D as an Air Pollutant: Effects on Growth of Representative Horticultural Plants, *Hort. Science*, vol. 5, pp. 211–213, 1970.
7. Stephens, E. R. and R. F. Burleson: Analysis of the Atmosphere for Light Hydrocarbons, *Journal of the Air Pollution Control Association*, vol. 17, pp. 147–153, 1967.
8. Abeles, F. B., L. E. Forrence and G. R. Leather: Ethylene Air Pollution: Effects of Ambient Levels of Ethylene on the Gluconate Content of Bean Leaves, "Plant Physiology," vol. 48, pp. 504–505, 1971.
9. Scurfield, G.: Air Pollution and Tree Growth, *Forestry Abstracts*, vol. 21, pp. 339–347, 1960.
10. Gorham, E. and A. G. Gordon: Some Effects of Smelter Pollution upon Aquatic Vegetation near Sudbury, Ontario, *Canadian Journal of Botany*, vol. 41, pp. 371–378, 1963.
11. Linzon, S. N.: Economic Effects of Sulfur Dioxide on Forest Growth, *Journal of the Air Pollution Control Association*, vol. 21, pp. 81–86, 1971.
12. Benedict, H. M. and R. E. Olson: "Economic Impact of Air Pollutants on Plants," vol. 1, Stanford Research Institute Project PHS-8115, 1970.
13. Olsen, R. A.: Absorption of Sulfur Dioxide from the Atmosphere by Cotton Plants, *Soil Science*, vol. 84, pp. 107–111, 1957.
14. Hursh, C. R.: "Local Climate in the Copper Basin of Tennessee as Modified by the Removal of Vegetation," USDA Circular No. 774, pp. 1–38, 1948.
15. Goodman, G. T. and T. M. Roberts: Plants and Soils as Indicators of Metals in the Air, *Nature*, vol. 231, pp. 287–292, 1971.
16. Lagerwerff, J. V.: Uptake of Cadmium, Lead and Zinc by Radish from Soil and Air, *Soil Science*, vol. 111, pp. 129–133, 1971.
17. Costescu, L. and T. C. Hutchinson: Soil Contamination by Airborne Metallic Dust Particles and Its Effect on Plants in the Sudbury Basin, Ontario, *Abstract in American Journal of Botany*, vol. 58, p. 481, 1971.
18. McCune, D. C.: "On the Establishment of Air Quality Criteria with Reference to the Effects of Atmospheric Fluorine on Vegetation," Air Quality Monograph, No. 69-3, American Petroleum Institute, New York, 1969.
19. Guderian, R. and H. van Haut: "Detection of SO_2-Effects upon Plants," Staub-Rheinhalt, Luft (Eng. ed.), vol. 30, no. 1, pp. 22–35, 1970.
20. Tingey, D. T., R. A. Reinert, J. A. Dunning and W. W. Heck: Vegetation Injury from the Interaction of Nitrogen Dioxide and Sulfur Dioxide, *Phytopathology*, vol. 61, pp. 1506–1511, 1971.
21. Jacobson, J. S. and A. C. Hill (eds.): "Recognition of Air Pollution Injury to Vegetation: A Pictorial Atlas," Air Pollution Control Association, Pittsburgh, Pennsylvania, 1970.
22. Cole, G. A.: Air Pollution with Relation to Agronomic Crops: III, Vegetation Survey Methods in Air Pollution Studies, *Agronomy Journal*, vol. 50, pp. 553–555, 1958.
23. McKee, H. C. and F. W. Bieberdorf: Vegetation Symptoms as a Measure of Air Pollution, *Journal of the Air Pollution Control Association*, vol. 10, pp. 222–225, 1960.
24. Hendrickson, E. R.: Dispersion and Effects of Airborne Fluorides in Central Florida, *Journal of the Air Pollution Control Association*, vol. 11, pp. 220–225, 232, 1961.
25. McHenry, C. R. and H. Charles: Monitoring of Fluoride Content of Air, Water and Vegetation, *Journal of the Air Pollution Control Association*, vol. 11, pp. 66–70, 1961.
26. Guderian, R. and H. Schönbeck: Recent Results for Recognition and Monitoring of Air

Pollution with the Aid of Plants, Proceedings of the International Clean Air Congress, 2d, Washington, D.C., pp. 266–273, 1970.

27. Bernatzky, A.: Climatic Influences of the Greens and City Planning, *Anthos*, vol. 11, pp. 26–29, 1966.
28. Weinstein, L. H.: "Agriculture's Responsibility in Establishing Air Quality Criteria," Proceedings, Nineteenth Annual Meeting of the Agricultural Research Institute, pp. 16–22, 1971.
29. Feder, W. A.: Modifying the Environment, *Hort. Science,* vol. 5, pp. 247–249, 1970.

Chapter 3

TYPES OF AIR POLLUTANTS, THEIR SOURCES AND EFFECTS

E. R. Altwicker S. M. Ayres
J. Bigu del Blanco R. D. Cadle L. W. Canter
M. Chiba R. J. Gordon F. L. Parker
C. A. Romero C. J. Santhanam
I. S. Shah D. J. Sibbett J. A. Tanner

CONTENTS OF CHAPTER 3*

*For a glossary of terms, refer to the back portion of this volume and for a list of abbreviations used in this chapter refer to the listing given after these contents, preceding Section 3.1.

Abbreviations

Å	Angstrom (10^{-10} meter)
ACGIH	American Conference of Government Industrial Hygienists
ALA	Delta-Aminolevulinic Acid
AM, am	Ante Meridiem
AM	Amplitude Modulation
BBL, bbl	Barrel
BTU	British Thermal Unit
°C	Degree Centigrade
Ci	Curie
cm	Centimeter (10^{-2} meter)
COHb	Carboxyhemoglobin
EEG	Electroencephalogram
EHF	Extremely High Frequency
EPA	Environmental Protection Agency
ev, EV	Electron Volt (ev-23.06 K cal/mole)
FAO	Food and Agricultural Organization
g, gm	Gram
GHz	Gigahertz (10^9 Hertz)
gr, Gr	Grain (1 gram equals 15.4 grains)
h	Plancks' Constant
HbCO	Carboxyhemoglobin
HbO_2	Oxyhemoglobin
HF	High Frequency (3-30 MHz)
Hg	Mercury
$h\nu$	Quantum of Radiant Energy (see Glossary)

hr	Hour
Hz	Hertz
IR	Infrared
KH_z	Kilohertz (10^3 hertz)
°K	Degree Kelvin (°C + 273)
Km	Kilometer (1,000 meters, 0.62 miles)
lb	Pound
LC_{50}	Lethal Concentration to 50% of test animals
LD_{50}	Lethal Dose to 50% of test animals
LF	Low Frequency (30 to 300 KHz)
m, M	Meter
mb	Millibar (1 atmosphere = 1,013 mb)
MEV	Million Electron Volts
MF	Medium Frequency (0.3 to 3 MHz)
mg	Milligram (10^{-3} gram)
MHz	Megahertz (10^6 hertz)
ml	Milliliter
MM, mm	Millimeter
MMD	Median Size Distribution, Mass Median Diameter
mph	Miles per Hour
mrad	Millirad (see Glossary for rad)
mrem	Millirem (see Glossary for rem)
mw	Milliwatt (10^{-3} watt)
Mw	Microwave
mwh	Milliwatt hour
NAPCA	National Air Pollution Control Administration
NCAR	National Center for Atmospheric Research
ng	Nanogram (10^{-9} gram)
NO_x	Nitrogen Oxides, all types
OSHA	Occupational Safety and Health Administration
O_2Hb	Oxyhemoglobin
PAN	Peroxyacetyl Nitrate having the following form: $CH_3\overset{\overset{O}{\|}}{C}OONO_2$
PBN	Peroxybenzoyl Nitrate
PM, pm	Post Meridiem
ppb	Parts per Billion
pphm	Parts per Hundred Million
ppm	Parts per Million
PTD	Percent Toxic Dose
RF	Radio Frequency (3KHz to 300GHz)
RH	Relative Humidity
SCF	Standard Cubic Feet
scfm, SCFM	Standard Cubic Feet per Minute
SHF	Super High Frequency (3 to 30 GHz)

STP	Standard Temperature and Pressure
TLV	Threshold Limit Value
TV	Television
UHF	Ultra High Frequency (0.3 to 3 GHz)
UV	Ultraviolet
V	Visible
VHF	Very High Frequency (30 to 300 MHz)
VLF	Very Low Frequency (3 to 30 KHz)
WHO	World Health Organization
WLM	Working Level Months
wt	Weight
yr	Year
μ	Micron (10^{-6} meter)
μgm, μg	Microgram (10^{-6} gram)
μw	Microwatt (10^{-6} watt)
v	Vibration Frequency

3.1 PHYSICOCHEMICAL NATURE OF AIR POLLUTION

Perhaps it is not inappropriate to open a chapter on the physicochemical nature of air pollution with a recent quote from Lovelock that, "even after 4×10^9 years, the air still consists mainly of nitrogen which acknowledges the overwhelming victory of biological processes over inorganic chemistry and equilibrium thermodynamics."[1]

However, the physics and chemistry of air pollution can be assessed adequately only by a consideration of the trace components and their behavior in the atmospheric setting. The pollutants themselves may lead to changes in circulatory patterns and thus affect the climate and thereby complete a physical-chemical cycle, which may be the cause for detrimental changes in biological processes. For data on threshold limit values, refer to Section 5.3.

The Atmosphere

Earth is surrounded by a life supporting gas mantle which decreases in temperature and pressure with altitude (Fig. 3.1a).[2] The atmospheric layer below the tropopause is known as the troposphere. It is the region of turbulent and large-scale vertical motions of the air masses where the prevailing westerlies (in the middle latitudes) blow in both the Northern and Southern Hemispheres. North–south flow also occurs by a number of processes (including the stratosphere) and thus leads to air exchange between hemispheres. Most of the heat in the troposphere is received from the ground rather than directly from the sun. In the stratosphere, little vertical mixing takes place.

The troposphere contains about 90% of the mass of the atmosphere. This gas mass is composed of a few major and numerous minor chemical compounds (Table 3.1b).[3] Many constituents of the atmosphere can be regarded as pollutants. These compounds have natural origins and produce a "background," although the actual concentrations may vary from location to location. In background air, carbon dioxide concentration remains constant with decreasing altitude, while water vapor concentration decreases. Local variations from such an "average" composition are brought about by changes in humidity, biological cycles, volcanic eruptions, forest fires and release of terpenes from forests, to mention but a few. Any of these combined with mean wind and

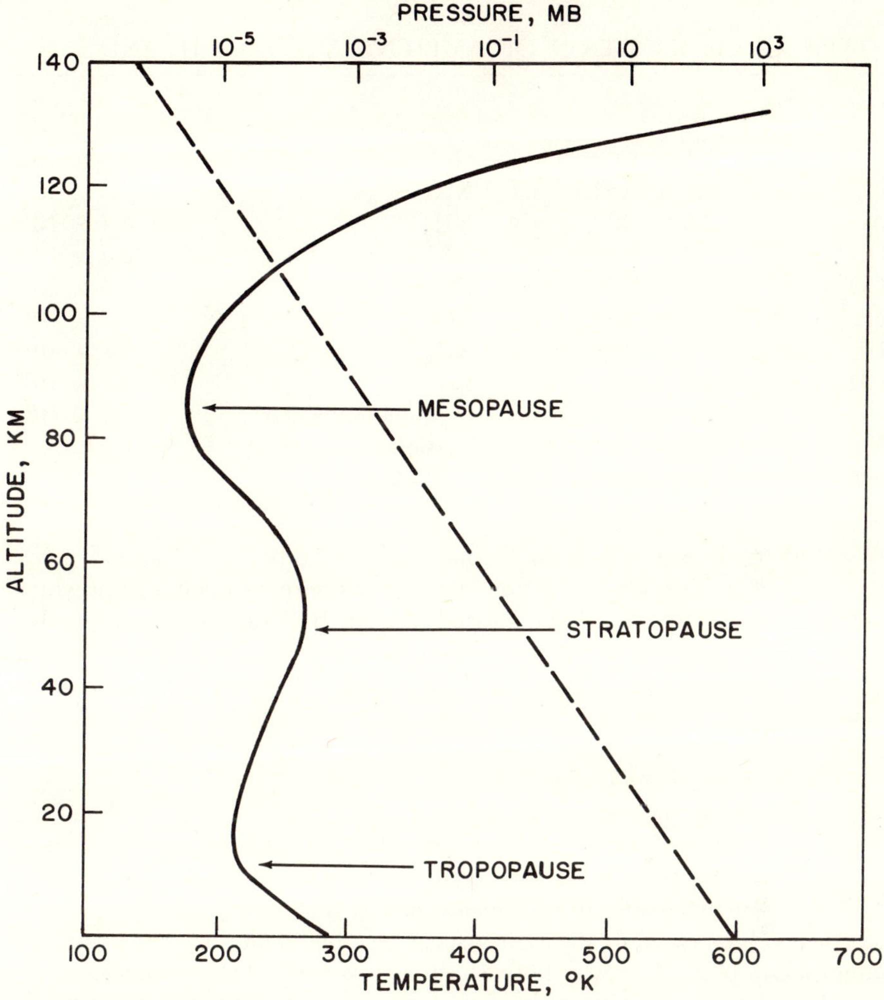

Figure 3.1a Schematic diagram of air temperature (solid line) and pressure (dashed line) as a function of height.

mixing depth (the ventilation index*) determine the actual amounts of trace constituents.

Some of these sources release both gases and particles; therefore, ambient air in the troposphere is not homogeneous but contains particles (aerosols) in the order of a few to several thousand per cubic centimeter. The term particle (aerosol) here denotes dry and wet particles, liquid droplets containing

* Ventilation index is the product of mean wind speed and mixing depth.

Table 3.1b
TYPICAL COMPOSITION OF AMBIENT AIR, 63% R.H. AND 25°C

Gas	Concentration (molecules/cm³)	Gas	Concentration (molecules/cm³)
N_2	2.1×10^{19}	H_2	1.3×10^{13}
O_2	5.6×10^{18}	N_2O	1.3×10^{13}
Ar	2.5×10^{17}	NH_3	5.4×10^{12}
H_2O	5.3×10^{17}	Xe	2.3×10^{12}
CO_2	8.5×10^{15}	O_3	1.5×10^{12}
Ne	4.9×10^{14}	NO	6.7×10^{11}
He	1.1×10^{14}	NO_2	5.4×10^{11}
CH_4	5.4×10^{13}	I_2	2.7×10^{11}
CO	5.3×10^{13}	O (monomolecular)	7.1×10^{4}
Kr	3.0×10^{13}	Particles	$10^8/m^3$
SO_2	2.7×10^{13}	(85% $< 2\mu$)	

dissolved solids; and all of these may carry adsorbed or absorbed gases. Background ambient air thus defined serves as a reference point against which pollution measurements can be carried out. Air pollution is the undesirable alteration in the composition of the atmosphere, undesirable in terms of health or other effects. Both natural and man-made pollution fit under these definitions.

Atmospheric Background Gases

Atmospheric trace gases become part of the major element cycles of the biosphere (such as C, O, N and S cycles).[4] The quantitative aspects of these cycles for several trace gases have been summarized in Table 3.1c.[5,6] Sulfur oxides, carbon monoxide, nitrogen oxides and hydrocarbons are major primary gaseous pollutants, while ozone is a secondary pollutant, being the result of atmospheric photochemistry between nitric oxide and hydrocarbons.

Aerosols can be primary or secondary pollutants. Combustion, volcanic eruptions, dust storms and sea spray are but a few examples of processes emitting particles. The roles played by particles in the atmosphere are primarily a function of their size and shape as well as their chemical composition. Coarse dusts (greater than 10 μ in radius) are only kept in suspension by atmospheric turbulence, while those below 1 μ are not significantly affected by gravity. Particles below 0.1 μ are known as Aitken nuclei; these are capable of serving as condensation nuclei for clouds and fog. Although not completely understood, the chemical composition and crystal structure of particles appear to play a major role in determining the effectiveness of condensation nuclei under certain conditions. Secondary effects are the result of gas phase chemistry and photochemistry which lead to aerosol formation.

A volcanic eruption can be considered a point source which leads to local effects (settling of particles and fumes) and to global effects, since the emissions may enter the circulation in the upper atmosphere, i.e., the stratosphere, and increase the stratospheric aerosol content. There are both short and long-range effects of pollution (Section 2.1).

Polluted Atmospheres

Unless one looks at the emission of a single toxic substance from a single stack, the definition of air pollution remains complex. Certain standards have been set by the Environmental Protection Agency for the most important pollutants, but from a physical and chemical point of view, some of these definitions are poor. "Hydrocarbons" not only denotes compounds containing carbon and hydrogen (be they reactive or nonreactive) but also includes partially oxidized compounds (such as alcohols, aldehydes, acids). The pollutants of Table 3.1d are emitted at certain locations only, and by atmospheric measurements, it has been shown that urban and industrial areas greatly exceed the natural background for particles, sulfur oxides, carbon monoxide and nitrogen oxides (Sections 3.7, 3.12, 3.15 and 3.20). The local effects produced by these emissions are important both to the receptors and to those instituting controls. Thus, carbon monoxide is largely emitted at low levels from vehicles; numerous home heating units combine to form an area of low-level emissions of particles and nitrogen oxides; sheer human activity of movement is an important source of particles in the urban setting. On the other hand, high stacks from power plants lead to increases of sulfur oxides, nitrogen oxides and particles at some distance away from these sources.

Particulates

Many of the particulates in the air are metal compounds which can catalyze secondary reactions in the air or gas phase to produce aerosols as secondary products. The increase in the number of solid particles may be by a factor of 1,000 or greater over background concentrations, and Aitken nuclei counts of $10^6/cm^3$ and more have been measured in industrial atmospheres. Dustfall measurements, i.e., those particles which settle by gravity, give readings of 500 to 1,000 tons per square mile per month in polluted regions versus 10 to 100 tons in relatively unpolluted areas. Although large particles tend to settle rapidly and medium-sized particles (approximately $5\,\mu$) are removed with considerable efficiency by control devices, an accounting of particles strictly on a mass basis is misleading both from the point of view of the total problem and from the proper assessment of the effects. Numerous emissions, such as daily urban activity and quarrying, mining and farming, are difficult to control; and the size range from about 0.1 to $2\,\mu$ may be more important than larger particles where long-range effects are involved. Data on particulate emissions are given in Figure 3.1e.[13]

Table 3.1c

SOURCES, CONCENTRATIONS AND MAJOR REACTIONS OF ATMOSPHERIC TRACE GASES

Contaminant	*Major Pollution Sources*	*Natural Sources*	*Estimated Emissions, 10^6 tons*		*Atmospheric Background Concentrations (ppb)*	*Calculated Atmospheric Residence Time*	*Removal Reactions And Sinks*	*Remarks*
			Pollution	*Natural*				
SO_2	Combustion of coal and oil	Volcanoes	146	None	0.2	4 days	Oxidation to sulfate (ozone, absorption by solid and liquid aerosols, $h\nu$)	Photochemical oxidation with NO_2 and hydrocarbons, heterogeneous reactions involving metal oxides may be important
H_2S	Chemical processes, sewage treatment	Volcanoes, biological action in swamp areas	3	100	0.2	2 days	Oxidation to SO_2	Only one set of background concentrations available
CO	Auto exhaust and other combustion	Forest fires, ocean	220	11	100	<3 yr	Soil? oceans? stratospheric reaction with OH?	May not be as unreactive as previously supposed
NO-NO_2	Combustion	Bacterial action in soil[a]	53	500	1	4 days	Oxidation to nitrate after sorption by solid and liquid aerosols, hydrocarbon photochemical reactions	Very little work done on natural processes

NH_3	Waste treatment	Biological decay	4	5,900	6–20	2 days	Reaction with SO_2 to form $(NH_4)_2SO_4$, oxidation to nitrate	No quantitative rate data on oxidation of NH_3 to NO_3 (seems to be dominant process in atmosphere)
Hydrocarbons	Combustion exhaust, chemical processes	Biological processes	88	480	CH_4: 1,500 non CH_4: 1	16 yr (CH_4)	Photochemical reaction with NO-NO_2, O_3; large sink necessary for CH_4	"Reactive" hydrocarbon emissions from pollution = 27×10^6 tons
CO_2	Combustion	Biological decay, release from oceans	1.3×10^4	10^6	320,000	4 yr	Biological absorption and photosynthesis, absorption in oceans	Atmospheric concentrations increasing (0.06 ppm/month)
O_3	Photochemical smog (NO + hydrocarbons + $h\nu$)	O_2 + O in stratosphere and downward transport	small	2×10^3	10	—	Reaction with organics and reducing agents	High diurnal concentration changes locally

[a] It is not certain that soil bacteria are of a natural source.

Table 3.1d
1968 ESTIMATES OF PRIMARY POLLUTANT EMISSIONS FOR UNITED STATES,[a] MILLIONS OF TONS ANNUALLY

Source	Particulates	Sulfur Oxides	Nitrogen Oxides	Carbon Monoxide	Hydrocarbons
Transportation	1.1	0.7	7.3	66.0	15.1
Stationary combustion sources	8.1	22.1	9.1	1.7	0.6
Industrial processes	6.8	6.6	0.2	8.8	4.2
Refuse disposal	0.9	0.1	0.5	7.1	1.5
Miscellaneous	9.6	0.6	1.7	16.9	8.2
Totals	26.5	30.1	18.8	92.4	29.6

[a] 1968 NAPCA data. Interestingly (and curiously), with the exceptions of emissions of NO_x from stationary combustion sources and industrial hydrocarbon sources, these values are all equal to or lower than 1966 NAPCA data (but see Table 2.10a for more recent data).

Sulfur Dioxide and Smog

Sulfur dioxide is the predominant oxide of sulfur emitted. Although hydrogen sulfide oxidation is a major source of sulfur dioxide (Table 3.1c), the man-made sources (combustion of sulfur-containing fossil fuels) can lead to undesirable local concentrations which have on occasion exceeded 1 ppm. Sulfur oxides are transformed to sulfate aerosols by more than one mechanism; they can be removed via rainfall and coagulation (Section 3.20). Sulfur dioxide is a reducing agent and its chemistry tends to dominate in the atmosphere of a city such as London.

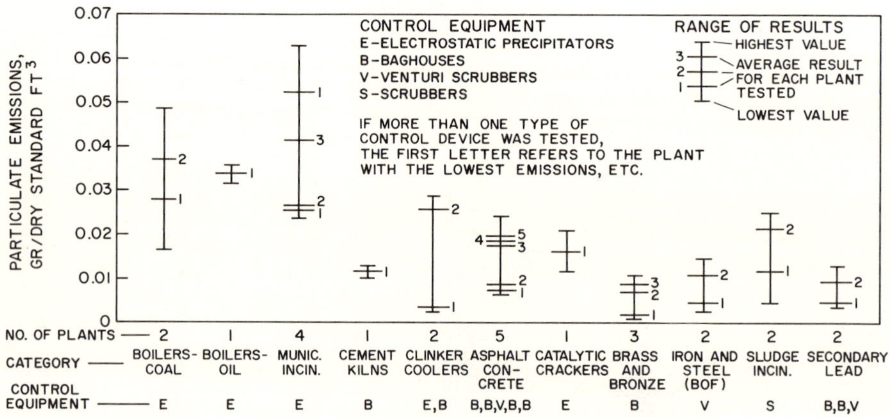

Figure 3.1e Measured particulate emissions from a variety of well controlled plants.

The two oxides of nitrogen of principal importance in polluted atmospheres are nitric oxide and nitrogen dioxide. The former is largely emitted from combustion sources, and atmospheric conversion to the latter form follows.

Hydrocarbons include organic compounds containing other elements in addition to carbon and hydrogen. The combination of hydrocarbons and nitrogen oxides is of special interest as the ingredients of Los Angeles or photochemical smog[7,8] (Sections 2.8 and 3.14). Typical concentrations of pollutants in a smoggy atmosphere are (in pphm): NO_x, 20; CO, 4,000; O_3, 50; CH_4, 250; reactive hydrocarbons, 50; aldehydes, 60; although occasionally much higher concentrations of one or more of these materials have been measured (Fig. 2.8a). The characteristics of three types of smog are:

London	*Los Angeles*	*Tokyo*
Peaks early morning	Peaks midday	Peaks midday
Low temperature (30 to 40°F)	High temperature (75 to 90°F)	High temperature
High humidity	Low humidity	High humidity
Light rain falling or fog	Little or no cloud cover	Little or no cloud cover
Chemically reducing	Chemically oxidizing	Chemically oxidizing
Radiation inversion	Subsidence inversion	Subsidence inversion
Bronchial irritation	Eye irritation	Both

Rapid mixing and dispersion of pollutants normally tend to lessen the rates and effects of atmospheric reactions of the primary pollutants near their sources. In the Los Angeles case, emissions, topography and meteorology combine to generate a huge photochemical reactor; and the effects produced are due in large measure to the formation of secondary pollutants (oxidants, aldehydes, aerosols). It is now recognized that photochemical smog could occur wherever reactive hydrocarbons and oxides of nitrogen and appropriate seasonal parameters (temperature, sun intensity) are present.

Carbon monoxide is emitted at low levels from car exhausts and is the most abundant gaseous pollutant present in urban atmospheres. Instantaneous concentrations have exceeded 100 ppm (Section 3.7).

Transport and Removal of Pollutants

Once a gaseous (or fine aerosol) pollutant has left the source, it is dispersed vertically and horizontally by mechanical and thermal turbulence (Section 2.6). Mechanical turbulence is influenced by terrain roughness and obstacles. Thermal turbulence (Fig. 3.1a) promotes vertical mixing of pollutants. A temperature decrease at a rate of 1°C/100m altitude is known as the dry adiabatic lapse rate. When the decrease is greater, the atmosphere is unstable and good dispersion is provided. At a lesser or even negative lapse rate (thermal inversion), the air is said to be stable and one of two types of inversion can occur; both of these trap pollutants.

A radiational inversion traps pollutants near the ground due to an increase in temperature from ground level to some altitude, while a subsidence inversion denotes a temperature increase aloft. Pollutants emitted underneath rise to the base of the inversion layer (mixing depth) but cannot penetrate it. Larger scale subsidence inversions result from rotating semipermanent anticyclonic (high-pressure) air systems where a pressure gradient toward the center of the downward airflow is set up. The air below is compressed and its temperature rises. This meteorological state serves as the lid on the Los Angeles basin and gives enough time for atmospheric reactions to occur and produce their undesirable manifestations.

Mixing depths for several urban areas indicate A.M. readings of several hundred meters, but P.M. readings for New York and Pittsburgh are much higher (1,500 to 1,800 M) than for Los Angeles (600 to 900 M). Ventilation conditions, thus, are less favorable for Los Angeles.[9] These observations serve to emphasize the fact that a group of pollutants emitted at two locations which are dissimilar in terrain and climatic conditions may follow different removal paths and rates and produce various effects.

While pollutant clouds are transported and dispersed to varying degrees, they also "age," which is a complex combination of homogeneous and heterogeneous reactions and physical processes (such as nucleation, coagulation and Brownian motion). Chemically unlike species may make contact and further branch the complex pattern. The major removal reactions and sinks have been summarized in Table 3.1c. Most of these are not understood in great detail. The effect of water vapor, for example, on atmospheric photochemistry, such as the hydrocarbon-NO_x reaction, is poorly understood even in the lab.[8]

The necessity to predict behavior of pollutant clouds and ground level concentrations has generated a large number of theoretical and working mathematical dispersion models (Table 2.6f). These do not, however, at present, treat adequately the chemical reactions mentioned previously, although efforts in this direction[10] are in progress (Section 2.6).

Effects of Polluted Atmospheres

The effects of air pollution take many forms: health, aesthetics, vegetation damage, accelerated decay of man-made or natural materials. The term "polluted atmosphere" is somewhat elusive. Air quality criteria,[6] for example, are based on relevant health effects (Section 2.15).

Particles in the 0.1 to 2 μ range tend to remain airborne; they are in the range that decreases visibility through the atmosphere and, in the lower ranges, are effectively retained in the lungs. These solid particles are not removed efficiently by pollution control devices; and because this size range contributes little (on a mass basis), efficiency rankings and effect rankings of removal methods do not correlate. The difficulty is compounded by the fact that such aerosols are also produced by atmospheric gas-phase reactions.

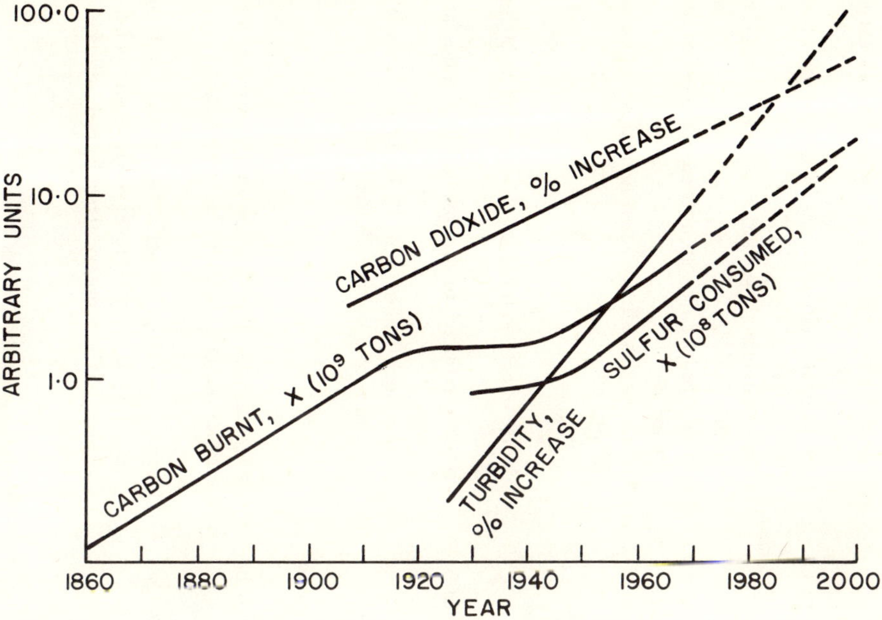

Figure 3.1f Trends in combustion of carbon and sulfur from fossil fuels and relation to carbon dioxide content and atmospheric turbidity.

Moreover, the natural particle inputs vary from location to location and biological sources of gaseous pollutants (Table 3.1c) make their contribution to atmospheric aerosols. Figure 3.1f shows trends in the discharge of combustion products and atmospheric turbidity increases.[1] While the combustion curves are man-made and the extrapolations will change, the relative contribution of man-made particles to the overall decrease in visibility remains difficult to assess.

Urban and industrial heat islands produce local climate changes, although not all of these changes are necessarily undesirable. Annual mean temperature, dust particles, fog and precipitation have all been increased, while relative humidity, radiation and wind speed have been lowered.[11]

Under Section 110 of the 1970 Air Quality Act (Section 1.1), the states must submit plans to the federal government on the implementation of standards for the six major pollutants. The proposed standards are reasonably similar to the suggested levels contained in the earlier criteria documents. The earlier criteria levels of concentration were used as guidelines and did not have the weight of standards. A comparison between the two is given in Table 3.1g. Implementation plans for lead, fine particles, gas phase aerosols and additional pollutants will be required in the future.

Table 3.1g
AIR QUALITY CRITERIA STANDARD REQUIREMENTS

Pollutant	Proposed Standards	Superseded Criteria
Sulfur oxides, primary standards	80 $\mu g/m^3$, annual arithmetic mean	Adverse health effects when annual mean exceeded 115 $\mu g/m^3$
	365 $\mu g/m^3$, maximum 24-hr concentration not to be exceeded more than once per year	Adverse health effects when 24-hr average exceeded 115 $\mu g/m^3$
Sulfur oxides, secondary standards	60 $\mu g/m^3$, annual arithmetic mean	Adverse effects on vegetation at annual mean of 85 $\mu g/m^3$
	260 $\mu g/m^3$, maximum 24-hr concentration not to be exceeded more than once per year	Visibility reduced to approximately 5 miles at 285 $\mu g/m^3$
Particulates, primary standards	75 $\mu g/m^3$, annual geometric mean	Adverse health effects when annual mean exceeded 80 $\mu g/m^3$
	260 $\mu g/m^3$, maximum concentration not to be exceeded more than once per year	
Particulates, secondary standards	60 $\mu g/m^3$, annual geometric mean	Adverse effects on material at annual mean 60 $\mu g/m^3$
	150 $\mu g/m^3$, maximum 24-hr concentration not to be exceeded more than once per year	Visibility reduced to approximately 5 miles at 150 $\mu g/m^3$
Carbon monoxide, primary and secondary standards	10 $\mu g/m^3$, maximum 8-hr concentration not to be exceeded more than once per year	Adverse health effects at 12 to 17 $\mu g/m^3$ with an exposure of 8 or more hours
	15 $\mu g/m^3$, maximum 1-hr concentration not to be exceeded more than once per year	
Photochemical oxidants, primary and secondary standards	125 $\mu g/m^3$, maximum 1-hr concentration not to be exceeded more than once a year	Adverse health effects at an hourly average concentration ranging from 60 to 100 $\mu g/m^3$
Hydrocarbons, primary and secondary standards	125 $\mu g/m^3$, maximum 3-hr concentration (6 to 9 A.M.) not to be exceeded more than once per year	Hydrocarbon concentration of approx. 130 $\mu g/m^3$ corresponds to lowest level at which photochemical oxidant has been observed to adversely affect human health
Nitrogen dioxide	100 $\mu g/m^3$, annual arithmetic mean	None
	250 $\mu g/m^3$, 21-hr concentration not to be exceeded more than once per year	None

Conclusions

Millions of sources pour out millions of tons of both simple and complex chemicals around the globe. The "infinite dilution sink," the earth's atmospheric layer, has shown changes in climate and visibility and man has noted adverse effects on health. Measurements are made but they are yet too few to permit cause–effect analysis and to clarify decision-making for control approaches. The magnitude of the real and potential global effects must be matched by investigations on a global scope, otherwise the triumph of biological processes may lose certainty.[12]

REFERENCES

1. Lovelock, J. E.: *Atm. Environment*, vol. 5, pp. 403–411, 1971.
2. "U.S. Standard Atmosphere, 1962," Government Printing Office, Washington, D.C., 1962.
3. Mohnen, V. A.: *J. Geophys. Res.*, vol. 75, p. 1718, 1970.
4. "The Biosphere," A Scientific American Book, W. H. Freeman and Co., San Francisco, 1970.
5. Robinson, E. and R. C. Robbins: "Sources, Abundance, and Fate of Gaseous Atmospheric Pollutants," Final Report, Stanford Research Institute Report, Project PR-6755, February 1968.
6. Jungo, C. E.. "Air Chemistry and Radioactivity," Academic Press, New York, 1963.
7. Cadle, R. D. and E. R. Allen: *Science*, vol. 167, pp. 243–249, 1970.
8. Altshuller, A. P. and J. J. Bufalini: *Env. Sci. and Techn.*, vol. 5, pp, 39–64, 1971.
9. Bryson, R. A. and J. E. Kutzbach: Commission on College Geography, Resource Paper no. 2, Association of American Geographers, Washington, D.C., 1968.
10. Atkisson, A. and R. S. Gaines: "Air Quality Standards," Charles E. Merrill Publishing Company, 1969.
11. Peterson, J. T.: "The Climate of Cities," National Air Pollution Control Administration Publication No. AP-59, 1969.
12. Singer, S. F.: "Global Effects of Environmental Pollution," Springer-Verlag, New York, 1969.
13. Jenkins, R. E. and G. A. McCutchen: New Source Performance Standard, *Env. Sci. and Techn.*, vol. 6, No. 10, October 1972.

3.2 ALDEHYDES

Although not present in as high a concentration as hydrocarbons (largely unburned gasoline), aldehydes are important constituents of smog because they contribute to eye irritation, and possibly to other physiological effects, and participate in the chemical reactions that produce unpleasant products.

A typical concentration of total aldehydes in moderately severe smog is 60 pphm, most of which is formaldehyde. Other aldehydes in smog include the very eye-irritating acrolein, acetaldehyde and various higher molecular weight aldehydes some of which have two carbonyl groups per molecule.

The automobile is by far the major source of aldehydes in urban pollution. Not only does it emit aldehydes directly, but it also emits the hydrocarbons which are converted to aldehydes in smog by photochemical reactions. However, essentially all burning of organic fuel also introduces aldehydes into the atmosphere. Thus, aldehydes are given off by backyard and municipal incineration, home heating, power plants and a multitude of industrial operations.

Aldehydes are also introduced into the atmosphere by nature. Some are the result of forest and brush fires started by lightning. Other aldehydes are emitted by plants as "essential oils."

Aldehyde Reactions

Aldehydes can participate in smog chemical reactions in at least two ways. Probably, the more important way is involvement in secondary photochemical reactions, i.e., those following the initial photolysis or photoexcitation [Equation 2.8 (2)]. One of these is the reaction with atomic oxygen produced by the photolysis of nitrogen dioxide:

$$NO_2 + \text{sunlight} \longrightarrow NO + O \qquad 3.2(1)$$

$$O + H_3C{-}\overset{\displaystyle O}{\overset{/\!/}{C}}{-}OH \longrightarrow H_2C{-}\overset{\displaystyle O}{\overset{/\!/}{C}}{-}OH + OH \qquad 3.2(2)$$

Both of the products of the reaction shown in Equation 3.2(2) undergo additional reactions, the OH probably being important as the carrier in subsequent chain reactions. Table 3.2a shows the rate constants at room temperature for

Table 3.2a
SECOND ORDER RATE CONSTANTS AT 300°K
FOR THE REACTION OF ATOMIC OXYGEN
WITH SOME ALDEHYDES AND HYDROCARBONS

Organic Compound	*Rate Constant,[a] $cm^3\ mole^{-1}\ sec^{-1}$*
Formaldehyde	2.1×10^{10}
Acetaldehyde	2.7×10^{11}
Propionaldehyde	1.4×10^{11}
Acrolein	1.6×10^{11}
Ethylene	7.2×10^{11}
Isobutene	9.6×10^{12}
Cis-2-pentene	1.1×10^{13}

[a]These are constants which when multiplied by the concentrations of the two reactants yield the rate of reaction.

the reaction of O with several aldehydes and hydrocarbons. These constants, when multiplied by the products of the reactant concentrations, yield the actual reaction rates.

Mixtures of NO_2, aldehydes and air at concentrations that might occur in city smog react chemically in natural or artificial sunlight at much higher rates than would be predicted from the rate constants shown in Table 3.2a.[1-6] These surprisingly high rates may be caused by chain reactions occurring once appreciable quantities of OH radicals have been formed.

Aldehydes, as a result of the carbonyl group, weakly absorb ultraviolet radiation in sunlight to a maximum wavelength of about 3400 Å. As a result, they undergo excitation and decomposition. In the presence of oxygen, photo-oxidation occurs; and if olefinic or aromatic hydrocarbons are also present, the intermediates react with the hydrocarbons. The reaction rates are slower than those induced by oxides of nitrogen, but they are significant and suggest that the control of nitrogen oxides alone will not control air pollution. Table 3.2b[7] indicates the rate of photooxidation of a few such mixtures. The reaction products include alkyl hydroperoxides and hydrogen peroxide, which are oxidizing agents, but not ozone.

Formaldehyde photolyzes with a quantum yield* near unity in two ways:

$$H_2C = O + \text{sunlight} \longrightarrow H_2 + CO \qquad 3.2(3)$$

$$H_2C = O + \text{sunlight} \longrightarrow HC = O + H \qquad 3.2(4)$$

* For definition see Glossary.

Table 3.2b
RATE OF PHOTOOXIDATION OF
ALDEHYDE-AIR MIXTURES IN SUNLIGHT

Aldehyde and Hydrocarbon	*Initial Concentration, ppm by volume*	*% Reacted with Irradiations by Hours*	
		5	6
Formaldehyde	15	61	
2,3-dimethylbutene-2	10	36	
Acetaldehyde	5	12	
2-methylbutene-1	10	4	
Propionaldehyde	19		25
2-methylbutene-1	10		10

This reaction is followed by

$$H + O_2 + M \longrightarrow HO_2 + M^* \qquad 3.2(5)$$

$$HO_2 + NO \longrightarrow NO_2 + OH \qquad 3.2(6)$$

$$OH + CO \longrightarrow CO_2 + H \qquad 3.2(7)$$

which are probably very important reactions in connection with smog formation.

Probably the only important physiological effect of the aldehydes is the eye irritation to which they contribute. Even so, the concentrations of aldehydes in smog are not sufficiently great to account for all the eye irritation it produces. Possibly chronic effects of the aldehydes exist; but if so, they are unknown.

At higher concentrations than what usually occurs in the atmosphere, the mucous membranes of the nose and throat are also irritated. Acrolein, crotonaldehyde and, at somewhat higher concentrations, formaldehyde are especially irritating.

Generally accepted air quality standards have not been established for aldehydes in the United States but they have been established in West Germany and the U.S.S.R.[8] for acetaldehyde and acrolein. In Germany, the permissible standards for a 30-minute averaging time for acetaldehyde and acrolein are 6 and 0.01 ppm by volume, respectively. In the U.S.S.R., for a 20-minute averaging time they are 0.005 and 0.12 ppm by volume, respectively.

* *M* stands for any available atom or molecule which by collision with the HO_2 carries off the excess energy of the reaction and prevents the HO_2 from flying apart as soon as it is formed.

REFERENCES

1. Niki, H.: *J. Chem. Phys.*, vol. 45, p. 2330, 1966.
2. Cadle, R. D. and J. W. Powers: *J. Phys. Chem.*, vol. 71, p. 1702, 1967.
3. Cadle, R. D.: in "Chemical Reactions in Urban Atmospheres," C. S. Tuesday (ed.), American Elsevier, New York, 1971.
4. Elias, L. and H. J. Schiff: *Can. J. Chem.*, vol. 38, p. 1557, 1960.
5. Elias, L.: *J. Chem. Phys.*, vol. 38, p. 989, 1963.
6. Ford, H. W. and N. Endorn: *J. Chem. Phys.*, vol. 27, p. 1277, 1957.
7. Altshuller, A. P., I. R. Cohen and T. C. Purcell: *Science*, vol. 156, p. 937, 1967.
8. Stern, A. C. (ed.), "Air Pollution," 2d ed., Academic Press, New York, 1968.

3.3 AMMONIA AND VOLATILE AMINES

Ammonia and the volatile amines* are bases and therefore react with gaseous and particulate acids in the atmosphere to form salts. At relatively high concentrations, their odor can be highly objectionable. For example, the amine cadaverine is partially responsible for the odor of decaying animal flesh.

Worldwide, most ammonia emitted to the atmosphere is produced by the bacterial breakdown of amino acids in organic waste material. Urban emissions are largely from combustion (Table 3.3a),[1] although industrial operations such as catalytic cracking of petroleum also give off some ammonia. Volatile amines are introduced into the air largely as a result of protein decomposition during putrefaction. The meat and meat products industry and fish processing operations, produce odiferous amines that are very difficult to control.

Ammonia reacts with sulfur dioxide, sulfuric acid droplets and nitric acid vapor in smog. Unpublished studies at the National Center for Atmospheric Research show that at concentrations existing in smog, ammonia rapidly reacts with sulfur dioxide in humid air to form particles of ammonium sulfite and sulfate. The proportion of ammonium sulfate increases with increasing relative humidity.

The reaction of ammonia with sulfuric acid droplets has been studied.[2] The rate of reaction of a particle in a gaseous mixture containing a reactive gas may be controlled by gas phase diffusion of the reactive gas, by the rate of reaction at the particle surface or by diffusion of reactants in the particle. When the particle is a droplet of concentrated sulfuric acid and the reactive gas is ammonia, the reaction rate is controlled by diffusion of the reaction product in the droplet and reaction is rather slow. However, when the droplets are dilute, as is usually the case in city smog or fog, the reaction is extremely fast and is limited by the rate of the gas phase diffusion of the ammonia. The reaction product, ammonium sulfate, is an important constituent of most city smog, contributing to the visibility decrease. The reaction of ammonia with nitric acid vapor in air at concentrations existing in smog has not been studied, but presumably it gives rise to ammonium nitrate particles. The nitric acid

*Those amines which are in vapor form in the atmosphere.

Table 3.3a
WORLDWIDE URBAN EMISSIONS OF AMMONIA

Fuel	*Source*	*NH_3 Emission, millions of tons*
Coal	Power generation	1.2
	Industrial and other	1.8
Petroleum	Refinery catalytic cracking operations	0.02
	Fuel oil	0.8
Natural gas	All uses	0.2
Others	Incineration	0.08
	Wood	0.06
	Forest fires	0.05
		4.21

vapor results from reactions of nitric oxide (NO) and nitrogen dioxide (NO_2) in smog such as described by the reaction in Equation 3.3(1).

$$NO_2 + OH \longrightarrow HNO_3 \qquad 3.3(1)$$

Ammonia may play an important role in sulfur dioxide oxidation. When sulfur dioxide dissolves in fog or cloud droplets, it first forms sulfurous acid:

$$SO_2 + H_2O \rightleftharpoons H_2SO_3 \qquad 3.3(2)$$

The sulfurous acid is slowly oxidized by dissolved oxygen to sulfuric acid; but in the presence of ammonium ion (NH_4^+), the rate of oxidation is greatly increased.

Presumably, amines undergo neutralization reactions in smog that are analogous to those of ammonia. However, essentially no direct knowledge exists concerning such reactions.

The most important physiological effect of ammonia and the volatile amines is their odor. The United States does not have ambient air quality standards for ammonia and the amines. The Ontario basic standard is 3.5 mg/M^3 STP averaged over 30 minutes. The U.S.S.R. basic standard is 0.2 mg/M^3 STP averaged over 24 hours. The odor of ammonia is detectable at about 0.5 mg/M^3 STP.

REFERENCES

1. Robinson, E. and R. C. Robbins: "Sources, Abundance, and Fate of Gaseous Atmospheric Pollutants," Stanford Research Institute Report Project PR-6755, Menlo Park, California, 1968.
2. Cadle, R. D. and R. C. Robbins: *Disc. Faraday Soc.*, vol. 30, p. 155, 1961.

3.4 ASBESTOS

Asbestos is defined by Webster's as "a grayish mineral, a silicate of calcium and magnesium." But in reality, the term asbestos is not limited to any one substance and instead refers to a class of minerals having certain chemical and physical features in common. They are all mineral silicates and are fibrous and noncombustible. Table 3.4a lists some of the more frequently encountered forms of asbestos and their chemical composition. The most widely used form is chrysotile, with more than 90% of world production devoted to this particular mineral. Deposits of asbestos minerals are found throughout the world, but significant mining activity is limited to a relatively few locations where it is commercially feasible. Canada is the principal producer of chrysotile on the North American continent and is probably the leader in world production. Other major producers include the U.S.S.R., South Africa and China. Production in the United States has steadily increased during recent years, with California mines accounting for about two-thirds of the total United States output.

Sources of Asbestos in Urban Air

Inasmuch as asbestos mineral deposits are dispersed throughout the world, fibers could be expected in the air as part of the natural background of particulate matter; but except in the immediate vicinity of asbestos-rich soils, it is doubtful if the natural levels could be measured. However, asbestos fibers have been found in urban air, and there is little doubt that they originate from the many uses made by man of asbestos products. These uses include insulating materials for ducts and pipes, asbestos cement, shingles for home construction, various kinds of fabric materials made with asbestos fibers, brake linings, clutch facings and many similar materials. Although the use of any asbestos-containing materials could result in asbestos air pollution, it is probable that the construction and demolition industries account for most of the asbestos found in urban areas. The spraying of asbestos onto steel girders of buildings under construction is a notable example of appreciable environmental pollution by asbestos, while the demolition of old buildings may also give rise to dust containing asbestos. It might be predicted that the wearing of brake linings and clutch plates on the millions of automobiles in daily use

Table 3.4a
PRINCIPAL ASBESTOS MINERALS

Chrysolite	Hydrous magnesium silicate, fibrous serpentine
Anthophyllite	Magnesium-iron silicate, amphibole
Amosite	Iron-rich anthophyllite, amphibole
Crocidolite	Sodium-iron silicate, amphibole
Tremolite-Actinolite	Calcium-magnesium silicate and calcium-magnesium iron silicate, amphibole

would constitute a major source of asbestos fibers in urban areas. But studies have shown that the heat generated by friction appears to destroy the fibers, so that the resulting dust is chemically similar to asbestos but does not retain the crystalline structure of chrysotile.

Asbestos fibers and dust may be released in considerable quantities in areas where the mineral is mined or in factories where it is processed and made into other products. In many cases, however, such plants have installed control devices to prevent or reduce emissions to the atmosphere; and little environmental contamination by asbestos can be found.

Effects of Asbestos on Health

It has long been known that the inhalation of asbestos dust or fibers can give rise to a disabling lung disease, pneumoconiosis, termed asbestosis. The causes and symptoms of asbestosis are complex. In general, it is a disabling lung disease which may result in death after characteristic lung changes have occurred and are confirmed by roentgenographic examination. The risk of developing asbestosis is directly related to the severity of the exposure and its duration. It is common for the disease to take as long as 20 years or more to be diagnosed. Asbestosis may also develop even though exposure to asbestos terminated a number of years prior to its diagnosis.

Another very serious consequence of overexposure to asbestos is increased incidence of cancer of the lung, and there is conclusive evidence that this disease is much more prevalent in asbestos workers than in the general population. The reason for the carcinogenic activity of asbestos is not clear. Various theories have proposed that the fibers themselves are somehow responsible for tumor production or that impurities associated with asbestos, such as certain metals, or the known carcinogen (benzo-a-pyrene) are responsible. Although some uncertainties exist concerning the mechanisms by which lung cancer is produced, the relationship itself has been conclusively demonstrated; and in addition, certain other unusual forms of cancer have been found in more recent times to be related to excessive asbestos exposure. These include the occurrence of tumors in the pleura, in which the lung is enclosed, and termed mesotheliomas of the pleura. Similar mesotheliomas are also found in the peritoneum, and both types are believed to be related to asbestos exposure.

It should be emphasized that asbestosis and increased incidence of cancer are diseases related to occupational exposure, and there is at present no evidence that exposure to asbestos concentrations present in ambient air has caused these diseases in urban populations. There have been reports of mesotheliomas occurring in persons whose exposures resulted from living or working near an asbestos industry or who lived in households with an asbestos worker; so it cannot be stated that mesotheliomas are uniquely produced by occupational exposure. So-called "asbestos bodies," or "ferruginous bodies," which are frequently found in the lungs of individuals with asbestosis, have also been found in the lungs of persons not known to be exposed to asbestos. The interpretation of these findings is controversial, but it has been argued that their presence demonstrates that asbestos fibers have been inhaled. Other investigators have shown that similar ferruginous bodies are formed by fibers of substances other than asbestos; so it cannot be stated that the presence of asbestos bodies unequivocally demonstrates exposure to asbestos. Nevertheless, it is universally agreed that exposure to asbestos should be minimized. To achieve this objective, the current proposed OSHA standard is an 8-hour time-weighted average of 2.0 fibers greater than 5 μm in length per cubic centimeter of air, with 15-minute excursions to 10.0 fibers/cc > 5 μm allowed. All determinations must be made by the membrane filter method at 400 to 450X* with phase contrast illumination. (For threshold limit values see Section 5.3.)

BIBLIOGRAPHY

American Conference of Governmental Industrial Hygienists, *Documentation of the Threshold Limit Values,* 3d ed. pp. 17–19, 1971.

Litton Systems, Inc., Report, "Air Pollution Aspects of Asbestos," prepared for the National Air Pollution Control Administration, Consumer Protection and Environmental Health Service, Department of Health, Education and Welfare, September 1969. Compiled by Ralph J. Sullivan and Yanis C. Athanassiadis.

Selikoff, I. J. and E. C. Hammond; Community Effects of Nonoccupational Environmental Asbestos Exposure, *American Journal of Public Health,* vol. 58, no. 9, 1658, 1968.

Stern, A. C. (ed.); "Air Pollution," 2d ed., vol. III, Sources of Air Pollution and Their Control, pp. 138–140, Academic Press, New York, 1968.

* A magnification in the range of 400 to 450 times.

3.5 CANCER PRODUCING AGENTS—CARCINOGENS

It has been shown many times in experimental animals that various chemical substances injected or applied to the skin can cause cancer. In some instances of accidental exposure, cancer in humans has been traced to similar causes. Although several broad classes of chemical substances include individual carcinogenic compounds, most individual members of those classes do not show such activity. Many theories have been proposed to explain chemical carcinogenesis, but none yet is completely satisfactory. In the broadest sense, a carcinogen has the capacity to intervene in some intimate cellular process so as to lead to the abnormal growth of an organ or organ system. This is known as cancer.

Any listing of carcinogens must be considered tentative, since there are without doubt others which have not yet been tested adequately to be recognized. (For threshold limit values see Section 5.3.) This is because a given carcinogen may be active only toward certain parts of the organism and perhaps only in certain forms of exposure. Added to this is the variation in susceptibility among the exposed subjects, so that some individuals develop the disease and others, under the same conditions do not. It is prudent to regard any substance which has shown carcinogenic activity in any animal as a potential hazard to man. Those which have not done so can only be considered as unproven, although repeated and varied testing without response increases confidence in their lack of carcinogenicity.

Inorganic Carcinogens

Table 3.5a lists some inorganic substances which are known or suspected by some investigators to be carcinogenic. Exposures to these can occur in industrial situations, in water, as insecticide residues on food, as airborne gases or dusts and from improperly manufactured goods. A particular example is asbestos (Section 3.4) used in insulating pipes and ductwork. The airborne fibers are hazardous occupationally to those handling them and, if not properly controlled when first applied, could endanger others in the vicinity. The main purpose in listing these materials is to call to the attention of design or operations personnel their potential hazards.

3.5 Cancer Producing Agents—Carcinogens

Table 3.5a
POSSIBLE INORGANIC CARCINOGENS

Arsenic (As)	Iron (Fe)	Silicon (Si)
Beryllium (Be)	Lead (Pb)	Silver (Ag)
Cadmium (Cd)	Mercury (Hg)	Zinc (Zn)
Chromium (Cr)	Nickel (Ni)	Asbestos
Cobalt (Co)	Selenium (Se)	

Table 3.5b
ORGANIC CARCINOGENS:
A PARTIAL LIST OF KNOWN OR SUSPECTED AGENTS

Polynuclear Aromatic Hydrocarbons (and derivatives)	
Dimethyl anthracenes	Benzo(e) pyrene
Cholanthrene	Dibenzanthacenes
Benz(a) anthracene	Dibenzophenanthrenes
Benzo(b) phenanthrene	Dibenzopyrene
Chrysene	Dibenzofluorenes
Benzo(a) pyrene	Benzofluoranthenes
Heterocyclic Aromatics	
Benzacridines	Dibenzocarbazoles
Dibenzacridines	
Alkylating Agents	
Nitrogen mustards	Ethylene imines
Diepoxides	Nitrosamines
Propiolactone	
Aromatic Amines	
Napthylamines	Auramine intermediates
Benzidine	Magenta intermediates
4-Aminodiphenyl	
Azo Compounds	
Azobenzenes	Trypan Blue
Naphthyl and pyridinyl azobenzenes	
Miscellaneous Types	
Pyrrolizidine (*Senecio*) Alkaloids	Estrogens
Thiourea, Thioacetamide, Acetamide	Aflatoxin

Organic Carcinogens

There is an even broader collection of known or suspected carcinogens of organic type. A partial listing of these is given in Table 3.5b. The polynuclear hydrocarbons have been very extensively studied, mainly with a view to developing a general theory of carcinogenesis. Even so, since there are hundreds of such hydrocarbons, most have not been tested and a generally accepted theory is still to appear. Such substances are found, for example, in coal tar, asphalt and in combustion effluents. The analysis of such materials is difficult and very incomplete at best. Since polynuclear hydrocarbons as a class have roughly similar properties, the detection of one or a few of them usually suggests the presence of many others.

Closely related to these hydrocarbons are certain heterocyclic polynuclear compounds (Table 3.5b). These are frequently observed to accompany the polynuclear hydrocarbons and may well include members with similar properties.

There is a class of compounds which are able to *alkylate* other compounds which include carcinogens. Among these are nitrogen mustards, epoxides, nitrosamines and others. A number of compounds in these classes are used in chemotherapy against cancer; there seems to be some relationship between the *initiation* and the *treatment* of cancer. Some of these are also listed in Table 3.5b.

Another class of organic carcinogens are those leading to bladder cancers, including mostly aromatic amines. Other carcinogens include certain steroid hormones and other biological metabolites. A notable example is the aflatoxins, products of certain molds. The aflatoxins have been found in moldy peanuts.

BIBLIOGRAPHY

Clayson, D. B.: "Chemical Carcinogenesis," Little, Brown, Boston, 1962.

Hueper, W. C. and W. D. Conway: "Chemical Carcinogenesis and Cancers," Thomas, Illinois, 1964.

3.6 CARBON DIOXIDE

Carbon dioxide is a normal constituent of the atmosphere whose concentration has varied over a tremendous range throughout geological time. It has been theorized that several billion years ago carbon dioxide levels were very much higher than at present and gradually lowered as carbon dioxide was absorbed by the earth and seas, fixing much of it as carbonate rocks. Gradually, plant life displaced carbon dioxide with oxygen by the process of photosynthesis; and in more recent geologic time, carbon was stored in sediments as fossil fuels. In very much more recent time (geologically speaking), the earth's atmosphere has come to be composed largely of nitrogen and oxygen with a fairly stable carbon dioxide content of only 0.03%, or approximately 300 ppm. Nevertheless, there is evidence that even this concentration is not constant and that it has varied within certain limits due to natural causes. Presently, it is increasing as a result of man-generated inputs.

Completely reliable determinations of background carbon dioxide levels cannot be found for years past. But on the basis of an increasing number of such measurements in modern times, particularly in the last decade or so since the geophysical year 1958, it has been calculated that there has been a small but significant increase in carbon dioxide levels. The levels fluctuate considerably throughout any given year, but a steady and increasing upward trend is apparent (Fig. 3.6a).[1] The rate of increase is equivalent to an average rate of about 0.2% per year; and this increase is believed to represent about one-half of the amount of carbon dioxide discharged into the atmosphere, with the other portion having been taken up by the oceans and the biosphere.

Trends in Atmospheric Concentrations

Many predictions have been made about future levels of carbon dioxide and the probable effects on the biosphere. Historically, the carbon dioxide input into the atmosphere was of natural origin and varied widely as a function of volcanic activity and other natural phenomena. The present rise in concentration, however, is believed attributable to man's greatly increased use of fossil fuels for generating heat and power. All estimates of future concentrations of carbon dioxide in the atmosphere are based upon estimates of the rate at which the burning of fossil fuels will continue to accelerate, accom-

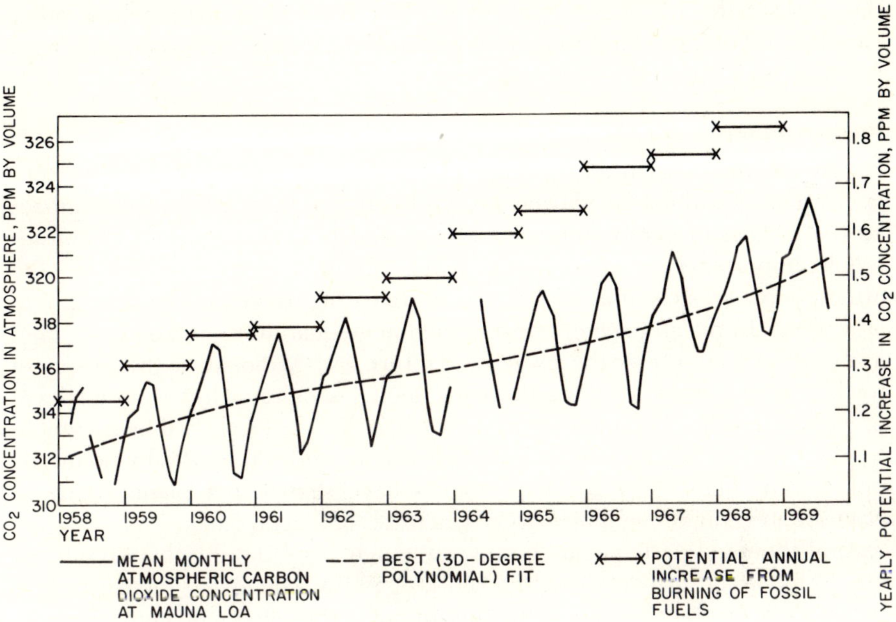

Figure 3.6a Changing CO_2 concentration in atmosphere due to burning of fossil fuels.

panied by assumptions concerning the partitioning of the carbon dioxide once emitted. On the assumption that the growth in the rate of fossil fuel consumption will be similar to that of the recent past and that partitioning will take place in basically the same fashion as in the past, it has been estimated that carbon dioxide in the atmosphere will increase nearly 20% by the year 2000, increasing from the present concentration of 320 to 380 ppm. The President's Science Advisory Committee Environmental Pollution Panel predicted in 1965 that a 25% increase in carbon dioxide content would occur by the year 2000; various other estimates can also be found in the literature (Section 2.1). All such estimates are based on projections of the extent of increase in the use of fossil fuels and are contingent upon the role of development of nuclear power facilities and other technological innovations in satisfying world demands for power.

The implications of changing carbon dioxide levels are profound and exceedingly complex. The most obvious effects relate to the impact of increasing concentrations on the earth's climate (Sections 2.1 and 2.7). Carbon dioxide is transparent to the shortwave energy emitted by the sun but absorbs a substantial amount of the infrared energy radiated by the earth, thus tending to bring about a warming of the atmosphere. Increasing levels of carbon dioxide are usually predicted (Fig. 2.1b) then, to bring about a warming of the atmosphere; but the actual effect is far from clear inasmuch as other variables related to cloud formation, particulate matter and so on may tend to offset such changes by opposing forces.

Photosynthesis and CO_2

No matter what the long-range climatological impact of changing concentrations of carbon dioxide may be, it is likely that there will be some impact on the process of photosynthesis; and any change in this regard is of considerable importance to man. Living plants absorb carbon dioxide from the atmosphere and store it as carbohydrates, releasing oxygen to the atmosphere as a result. In the absence of burning such huge quantities of fossil fuels, man and other animals utilize the oxygen and return carbon dioxide to the biosphere so that, over periods of time, the net effect is a more-or-less constant level of carbon dioxide. Any increase in the carbon dioxide levels will be favorable to plant life, for the present concentration of around 300 ppm is by no means optimal for plant growth. Many studies have shown that plants living in atmospheres containing much larger quantities of carbon dioxide, for example, several thousand parts per million, grow much more rapidly than in normal air. Commercial growers use this fact to good advantage to increase plant production in greenhouses; so in this respect, the effects of high levels of carbon dioxide are beneficial. In actuality, of course, projections of increased crop yields and the like are very uncertain, for the climatological consequences of carbon dioxide build-up may easily overshadow the simple increase in photosynthetic activity predicted by the concentration rise.

Health Effects

Atmospheric carbon dioxide is, in a sense, a contaminant placed into the air by man; but at prevailing levels, it is without any known effect on the health of man or animals. The increased levels which have been projected over the next 30 years are also without known effects on human health, with the exception perhaps of indirect effects related to changes in climate or in photosynthetic processes. Concentrations considerably higher than any level ever likely to be found in the atmosphere are presently believed to be tolerable to man, and only at several thousand parts per million are initial physiological effects observed. At much higher concentrations, carbon dioxide does exhibit toxicity. In combination with an oxygen deficient atmosphere, the effects are likely to be even more pronounced.

The principal impact of changes in carbon dioxide levels should, then, be in relation to plant growth; but the estimated 15% worldwide increase in CO_2 concentration since 1860 has not had any appreciable impact on plant growth. Projections have been made, though, that should the carbon dioxide concentration increase 30% by the year 2000 and an additional 30% by the year 2020, worldwide plant growth rates would also increase by about 5%

and 10%, respectively. All such projections are highly uncertain, for they are inevitably based on many assumptions which are very difficult to prove. The tendency for increased carbon dioxide levels to promote greater plant growth can be offset (to an unknown extent) by future land clearing activities of man.

Most projections of the impact of increased carbon dioxide levels by the end of this century are based upon assumptions concerning the probable effect on average temperature. The effects of a several degree rise in temperature would far outweigh the changes likely to be caused by photosynthetic considerations alone and would include altered ocean temperatures, with resultant changes in the capability of the ocean to retain carbon dioxide, the probability of the forest fire hazard increasing, alterations in marine life, altered rain fall, changing water levels due to melting of the ice cap and many other major phenomena.

It is ironical that modern pollution control is increasing the amount of carbon dioxide emitted to the atmosphere by promoting evermore efficient combustion of all sorts of fuels, particularly gasoline. It would impose an enormous technological and economic challenge to attempt to remove carbon dioxide from industrial emissions. In all probability, it would be nearly impossible to remove an appreciable fraction of the carbon dioxide; so any long-range attempt to minimize carbon dioxide levels in the future must concern itself with the larger issues of population control, fuel consumption practices and the generation of energy by means other than the combustion of fossil fuels.

REFERENCE

1. Report of the Study of Critical Environmental Problems, *Man's Impact on the Global Environment*, Sponsored by The Massachusetts Institute of Technology, pp. 11–12, 46–55, The M.I.T. Press, Cambridge, Massachusetts, 1970.

BIBLIOGRAPHY

Landsberg, H. E.: Man-made Climatic Changes, *Science*, vol. 170, p. 1265, December 18, 1970.

Peterson, E. K.: Carbon Dioxide Affects Global Ecology, *Environmental Science and Technology*, vol. 3, p. 1162, November 1969.

3.7 CARBON MONOXIDE

Carbon monoxide is an ideal model for the study of the biological behavior of air pollutants. The sources of carbon monoxide are known with considerable accuracy; its uptake may be described by simple pulmonary diffusion equations; the body burden may be estimated from analysis of blood carboxyhemoglobin concentrations; the physiological effects are due to competition with oxygen for iron binding sites on hemoglobin and other iron containing proteins; and the intensity of its physiological effects is related to the level of carboxyhemoglobin concentrations. It is a matter of some cosmic irony that, in an energy system in which carbon and oxygen combine to produce carbon dioxide, water and energy, incomplete combustion produces an intermediate product—carbon monoxide—which interferes with oxygen transport in many living systems. Similarly, it is a matter of cosmic consolation that biological oxidation is considerably more efficient than nonbiological oxidation; otherwise, cellular production of carbon monoxide might lead to cellular self-destruction.

Sources and Distribution of Carbon Monoxide

Carbon monoxide is the most abundant and widely distributed air pollutant (Table 2.1j) in the environment: 94 million tons were emitted from technological sources in 1968 and an additional 7 million tons from natural sources such as forest fires. Transportation accounts for 63% of total emissions and at least 90% of these are from gasoline engines. Population density and patterns of transportation account for the large variability in carbon monoxide emissions among geographical areas in the United States. The greater New York City region with a population of 15.4 million in 6,930 square miles produces 5.3 million tons each year; Los Angeles with half the population but 41,000 square miles produces almost as much (5.0 million tons per year), while Denver (population 1.2 million, land area 10,300 square miles) produces only 0.6 million tons each year.[1]

Diurnal, weekly and seasonal variations in carbon monoxide concentration are closely related to vehicular patterns and meteorological factors. Since emission rates are related to engine speed, traffic congestion increases ambient concentrations both by increasing the number of vehicles in a given area and by increasing the emission rates from each vehicle. A decrease in average route speed from 30 to 15 mph, for example, doubles emission rates.[2]

A statistical analysis of 30 technical papers prepared by the National Air Pollution Control Administration revealed that the maximum annual 8-hour average in highly polluted areas would approximate 40 ppm in commercial areas, 115 ppm in motor vehicles in downtown traffic and 75 ppm in vehicles on expressways. This marked variability in atmospheric carbon monoxide concentrations complicates the interpretation of ambient air measurements and underscores the need for study of blood levels.[1]

Biochemistry of Carbon Monoxide–Oxygen–Hemoglobin Interactions

Both oxygen and carbon monoxide combine reversibly with hemoglobin to form oxyhemoglobin and carboxyhemoglobin respectively. Douglas, Haldane and Haldane[3] first demonstrated that the toxicity of extremely low concentrations of carbon monoxide was due to its great affinity for hemoglobin relative to oxygen since the partial pressure of carbon monoxide required to fully saturate hemoglobin is only $\frac{1}{200}$ to $\frac{1}{250}$ of the partial pressure of oxygen required for complete saturation with oxygen. When exposed to a mixture of both gases, equilibrium is gradually reached according to the Haldane equation:

$$\frac{\text{COHb}}{\text{O}_2\text{Hb}} = M\frac{P_{CO}}{P_{O_2}} \qquad 3.7(1)$$

where COHb, O_2Hb, P_{CO} and P_{O_2} represent hemoglobin saturations and partial pressures of carbon monoxide and oxygen respectively, and *M* is an association constant which ranges from 200 to 250 in human blood. Saturation measurements are commonly used in physiological studies and represent the percentage of hemoglobin combined with oxygen or carbon monoxide. Each gram of hemoglobin will bind 1.34 ml of either gas when fully saturated. The product of saturation times hemoglobin concentration (grams per 100 ml) times 1.34 equals carbon monoxide or oxygen content in ml per 100 ml.

Equation 3.7(1) is an equilibrium relationship which would occur rather quickly in vitro (outside the living body). At equilibrium, for example, environments containing 25, 50 and 100 ppm would lead to carboxyhemoglobin saturations of 4.7, 9.1 and 16.1% if arterial oxygen tension were 80 mm Hg. The time required to reach equilibrium between blood and ambient air varies widely and is related to minute ventilation, dead space volume, diffusion capacity of the lung, total red cell mass, rate of blood flow and carboxyhemoglobin concentration in venous blood. The time course of carboxyhemoglobin during the steady state breathing of various mixtures of carbon monoxide[4,5,6] follows complex and frequently exponential relationships. Agreement was found[6] with experimental data using the linear Equation 3.7(2):

$$\text{COHb} = 0.05\ (\text{P}_{\text{CO}}\ \text{in inspired air})\ tR \qquad 3.7(2)$$

Where t is exposure time in minutes and R is ventilatory rate in liters per minute. Solution of this equation reveals that carboxyhemoglobin saturation in a lightly exercising man in an environment containing 50 ppm of carbon monoxide would be 1.4, 2.3, 3.4, 4.6 and 5.7% saturation at 1, 2, 3, 4 and 5 hours, respectively.

Ambient Air and Blood Carbon Monoxide Concentrations

The prediction of carboxyhemoglobin concentrations resulting from various ambient air levels is confused not only by the physiological variables listed but also by the fact that individuals do not breathe a constant concentration of carbon monoxide. A person may breathe 5 ppm for several hours, spend 15 minutes driving through a tunnel containing 150 ppm and then remain in traffic for 30 minutes breathing an atmosphere containing from 20 to 50 ppm. These relationships would be even further confused if he smoked, because cigarette smoke contains 3 to 4% carbon monoxide and each diluted puff is equivalent to about 500 ppm of carbon monoxide.

The problem of fluctuating exposures has been studied utilizing a computer approach with rapidly calculated carboxyhemoglobin saturations from certain assumed constants and a changing environmental carbon monoxide concentration. Figure 3.7a shows that the blood concentration is a damped representation of the air concentration with a definite lag between the two systems.[7]

Blood Carboxyhemoglobin Concentrations in Population

Measurement of carboxyhemoglobin concentrations provides a significantly better estimate of total body carbon monoxide burden than measurement of ambient concentrations. Sampling of blood from large populations is cumbersome and difficult; but, fortunately, alveolar air carbon monoxide is very close to blood levels and may be used to approximate carboxyhemoglobin concentrations. Air within the lung may be considered in two compartments: alveolar air is air which has mixed and equilibrated with pulmonary capillary blood; dead space air is air within conducting tubes which has not equilibrated with blood and is unchanged from ambient air. During exhalation, dead space air is exhaled first and may be discarded. The terminal portions of the exhalation may be collected and considered to represent alveolar air. A number of important studies have shown the relationships between alveolar air and blood carbon monoxide levels and justify the use of alveolar sampling in large-scale epidemiological studies.[8,9,10]

The average carboxyhemoglobin concentration in 511 nonsmokers is

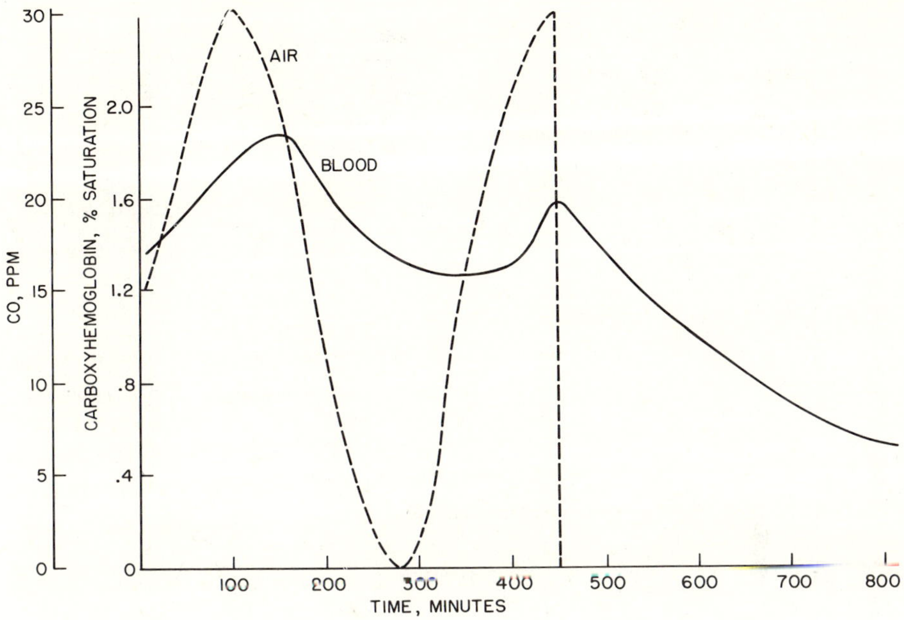

Figure 3.7a Relationship between ambient air (dotted line) and blood (solid line) carbon monoxide concentration. Note that blood levels do not respond immediately to changes in ambient air levels but constitute a damped response to ambient air events. Blood levels are thus a better indicator of the total body exposure to carbon monoxide.

reported to be 1.0 and ranges from 0 to 4.0% saturation.[11] The average carboxyhemoglobin in 509 smokers is 3.9 and ranges to levels as high as 19% saturation. Figure 3.7b shows alveolar carbon monoxide measurements made in subjects working in downtown Manhattan and in another group of subjects working in the Queens Midtown Tunnel. Ambient air concentrations in lower Manhattan ranged from 1 ppm to 19 ppm on an hourly basis and ambient air concentrations ranged from 5 ppm to 19 ppm in various locations in and around the Midtown Tunnel.

Physiological Effects of Carboxyhemoglobinemia

The physiological effects of carbon monoxide resemble the effects of breathing lowered oxygen concentrations, and most investigators have considered that the primary action of carbon monoxide is exerted through its competition with oxygen for binding sites on the hemoglobin molecule. While undoubtedly this is the major site of action, certain recent observations suggest that carbon monoxide may also have a direct effect on other iron-containing compounds (cytochrome oxidase, cytochrome a_3, cytochrome P-450).[12,13]

The net effect of combination of hemoglobin with either carbon monoxide

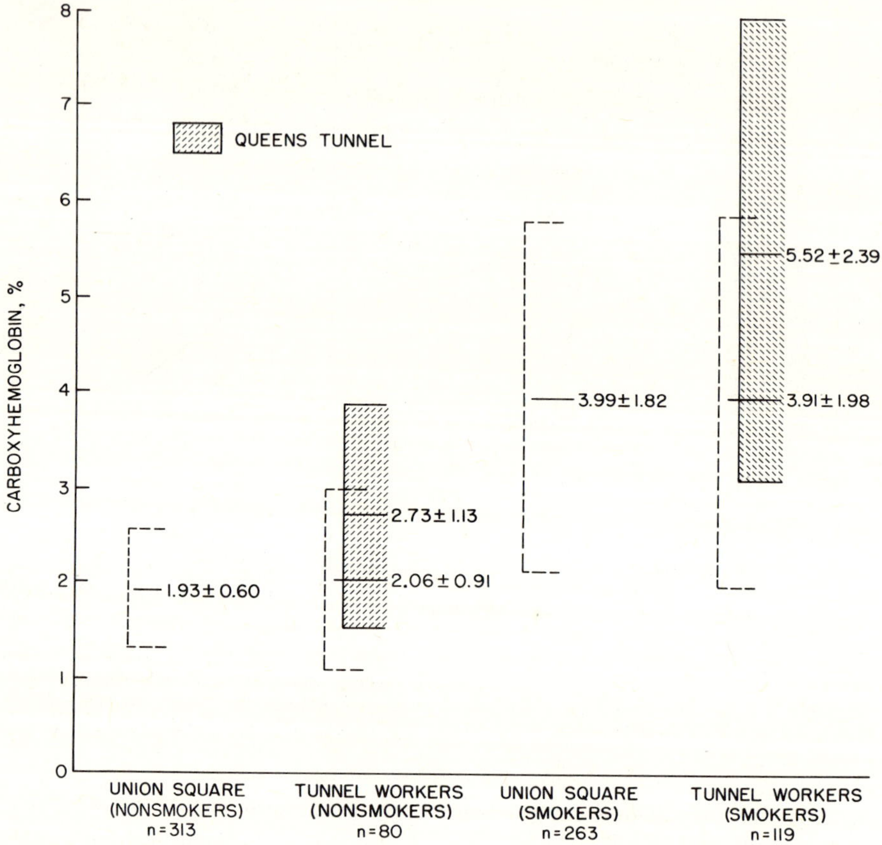

Figure 3.7b Carboxyhemoglobin concentrations calculated from expired air samples in a group of New Yorkers. The group studied in Union Square represents an average urban population. The tunnel workers are members of the Triborough Bridge and Tunnel Authority; some work on well ventilated bridges, others in tunnels. Mean values and brackets indicating one standard deviation are shown. The hatched area shows those workers assigned to the Queens Midtown Tunnel. Note that this latter group has a mean carboxyhemoglobin concentration of 5.52% and a large standard deviation, indicating a significant number of men with concentrations above 8% saturation.

or oxygen is to reduce the quantity of oxygen available to the tissues for energy production. Not only is the amount of hemoglobin available for oxygen transport decreased, but the function of the available hemoglobin is also altered because the uncombined hemoglobin holds on to oxygen with "increased tenacity."[3] Details of the effects of carboxyhemoglobinemia on oxygen-hemoglobin equillibria were presented in 1944.[14]

Toxicity of Carbon Monoxide

The toxicity of carbon monoxide has been the subject of many reviews since John Scott Haldane reported the series of courageous experiments

Table 3.7c
TOXIC EFFECTS OF VARIOUS CARBOXYHEMOGLOBIN CONCENTRATIONS

Carboxyhemoglobin Concentration, %	*Physiological and Subjective Symptoms*
10	No appreciable effects except shortness of breath on vigorous muscular exertion, possible tightness across the forehead and dilation of the cutaneous blood vessels
20	Shortness of breath on moderate exertion, occasionally slight headache with throbbing in the temples
30	Decided headache, irritable, easily fatigued, judgment disturbed, possible dizziness and dimness of vision
40–50	Headache, slight confusion, collapse and fainting on exertion, increased pulse and respiratory rates
60–70	Unconsciousness, intermittent convulsions, respiratory failure leading to death if exposure is long continued
80	Rapidly fatal
Ovcr 80	Immediately fatal

performed upon himself in 1895.[15] Table 3.7c shows a review of toxic effects published by his son, John Burdon Sanderson Haldane, in 1927.[16] Table 3.7d lists the experimental evidence indicating that levels considerably below 10% carboxyhemoglobin saturation may produce subtle abnormalities in cardiac and brain functions.[17]

The central nervous system effects summarized in Table 3.7d[19–26] have occasioned much discussion and some controversy because other studies of different design have failed to elicit positive responses. Much of this apparent disagreement may be explained by failure to select sufficiently sensitive tests to and by the presence of uncontrolled variables such as boredom or shifting motivation. It has been clearly shown that visual acuity is the most sensitive test followed by critical flick fusion, immediate memory, choice reaction time, pursuit motor tests and muscular coordination.[18]

Air Quality Standards

A threshold level for carbon monoxide does not exist but a straight line relationship between dose and effect can be expected. Arthur DuBois has suggested rephrasing the question of carbon monoxide thresholds to, "How much decrement in function are we willing to tolerate under conditions of public exposure for a specified time at a given concentration profile and in specified locations?"[27] (For threshold limit values see Section 5.3.) Exposures should be low enough to prevent individuals from developing 4 to 5% carboxyhemoglobinemia.

Table 3.7d
CARDIOVASCULAR AND NERVOUS SYSTEM EFFECTS OF RELATIVELY LOW CONCENTRATIONS OF CARBOXYHEMOGLOBIN

Investigator	*Lowest Concentration, %*	*Physiologic Effect*
McFarland	2.0	Decreased visual acuity
Halperin et al.	4.5	Decreased visual acuity
Lilienthal and Fugitt	5–10 (at 5,000 feet)	Impaired flicker fusion
Trouton and Eysenck	5	Abnormal coordination
Schulte	3	Increased errors in motor tests
Beard and Wertheim	2–3	Increased errors in judging time intervals
Chevalier et al.	4–5	Increased oxygen debt at exercise
Ayres et al.	5	Abnormal myocardial metabolism, decreased venous oxygen tension

Air Quality Standards for carbon monoxide are generally expressed as 8-hour averages since equilibrium levels for carboxyhemoglobin may be expected after that time period. The Air Quality Document prepared by the National Air Pollution Control Administration suggested 9 ppm for an 8-hour period.[1] Fifty parts per million for the same period is currently used in this country as an industrial hygiene standard, while the Soviet Union has picked eighteen parts per million for the same purpose. Such standards suffer from the difficulties in relating ambient air to blood levels. Ambient concentrations below 10 ppm (for 8 hours) should keep carboxyhemoglobin concentrations below 2 to 3% in nonsmokers and will probably not produce significant abnormalities in most members of the population.

REFERENCES

1. "Air Quality Criteria for Carbon Monoxide," National Air Pollution Control Administration, March 1970.
2. McMichael, W. F. and A. H. Rose, Jr.: "A Comparison of Automotive Emissions in Cities of Low and High Altitudes," Presented at the Annual Meeting, Air Pollution Control Association, Toronto, June 1965.
3. Douglas, C. G., J. Haldane and J. B. S. Haldane: The Laws of Combination of Hemoglobin with Carbon Monoxide and Oxygen, *J. Physiol.* (*London*), vol. 44, pp. 275–304, 1912.

4. Forbes, W. H., F. Sargent and F. J. W. Roughton: The Rate of Carbon Monoxide Uptake by Normal Men, *Amer. J. Physiol.*, vol. 143, pp. 594–608, 1945.
5. Pace, N., W. V. Consolazia, W. A. White, Jr. and A. Behnke: Formulation of the Principal Factors Affecting the Rate of Uptake of Carbon Monoxide by Man, *Amer. J. Physiol.*, vol. 147, pp. 352–359, 1946.
6. Lilienthal, J. L. and M. B. Pine: The Effect of Oxygen Pressure on the Uptake of Carbon Monoxide by Man at Sea Level and at Altitude, *Amer. J. Physiol.*, vol. 145, pp. 346–350, 1946.
7. Goldsmith, J. R., J. Terzaghi and J. D. Hackney: Evaluation of Fluctuating Carbon Monoxide Exposures, *Arch. Environ. Health*, vol. 7, pp. 647–663, 1963.
8. Jones, R. H., M. F. Ellicott, J. B. Cadigan and E. A. Gaensler: The Relationship Between Alveolar and Blood Carbon Monoxide Concentrations During Breathholding, *J. Lab. Clin. Med.*, vol. 51, pp. 553–564, 1958.
9. Peterson, J.: Postexposure Relationship of Carbon Monoxide in Blood and Expired Air, *Arch. Environ. Health*, vol. 21, p. 172, 1970.
10. Sjostrand, T.: Method for the Determination of Carboxyhemoglobin Concentrations by Analysis of Alveolar Air, *Acta Physiol. Scand.* (*Stockholm*), vol. 16, nos. 2 and 3, pp. 201–210, 1948.
11. Kjeldsen, K.: Smoking and Atherosclerosis: Investigations on the Significance of the Carbon Monoxide Content in Tobacco Smoke in Atherogeneses, thesis, Univ. of Copenhagen, 1969.
12. Chance, B., M. Erecinska and M. Wagner: Mitochondrial Responses to Carbon Monoxide Toxicity, *Ann. N.Y. Acad. Sci.*, vol. 174, pp. 193–204, 1970.
13. Cooper, D. Y., H. Schleyer and O. Rosenthal: Some Chemical Properties of Cytochrome P-450 and Its Carbon Monoxide Compound (P-450.CO), *Ann. N.Y. Acad. Sci.*, vol. 174, pp. 205–217, 1970.
14. Roughton, F. J. W. and R. C. Darling: The Effect of Carbon Monoxide on the Oxyhemoglobin Dissociation Curve, *Amer. J. Physiol.*, vol. 141, pp. 17–31.
15. Haldane, J.: The Action of Carbonic Oxide on Man, *J. Physiol.*, vol. 18, pp. 430–462, 1895.
16. Haldane, J. B. S.: Carbon Monoxide as a Tissue Poison, *Biochemical J.*, vol. 21, p. 1068, 1927.
17. Biological Effects of Carbon Monoxide, *Ann. N.Y. Acad. Sci.*, vol. 174, art. 1, October 5, 1970.
18. McFarland, Ross A.: The Effects of Exposure to Small Quantities of Carbon Monoxide on Vision, *Ann. N.Y. Acad. Sci.*, vol. 174, pp. 301–312, 1970.
19. McFarland, R. A.: Experimental Evidence of the Relationship between Ageing and Oxygen Want: In Search of a Theory of Ageing, *Aeronomics*, vol. 6, no. 4, p. 339.
20. Halperin, M. H., R. A. McFarland, J. I. Niven and F. J. W. Roughton: The Time Course of the Effects of Carbon Monoxide on Visual Thresholds, *J. Physiol.*, vol. 146, no. 3, pp. 583–593.
21. Lilienthal, J. L. and C. H. Fugitt: The Effect of Low Concentrations of Carboxyhemoglobin on the "Altitude Tolerance" of Man, *Amer. J. Physiol.*, vol. 145, pp. 359–364, 1946.
22. Trouton, D. and H. J. Eysenck: The Effects of Drugs on Behavior, in H. J. Eysenck (ed.), "Handbook of Abnormal Physiology," pp. 634–696, Basic Books, New York.
23. Schulte, J. H.: Effects of Mild Carbon Monoxide Intoxication, *Arch. Environ. Health*, vol. 7, pp. 524–530, 1963.
24. Beard, R. R. and A. Wertheim: Behavioral Impairment Associated with Small Doses of Carbon Monoxide, *Amer. J. Public Health*, vol. 57, pp. 2012–2022, 1967.
25. Chevalier, R. B., R. A. Krumholz and J. C. Ross: Reaction of Nonsmokers to Carbon Monoxide Inhalation: Cardiopulmonary Responses at Rest and During Exercises, *J. Amer. Med. Ass.*, vol. 198, p. 1061.
26. Ayres, S. M., S. Giannelli, Jr. and Mueller: Myocardial and Systemic Responses to Carboxyhemoglobin, *Ann. N.Y. Acad. Sci.*, vol. 174, pp. 268–293, 1970.
27. Dubois, A. B.: Establishment of "Threshold" CO Exposure Levels, *Ann. N.Y. Acad. Sci.*, vol. 174, pp. 425–428, 1970.

3.8 FLUORIDES

A considerable amount of scientific investigation has been devoted to studying the effects of fluoride emissions on man, animals and vegetation. With the exception of certain organic compounds, most compounds containing fluorine are toxic; and in many cases, the toxicity is related to the solubility of the compounds and to the consequent availability of fluoride ion. Table 3.8a lists the fluorine-containing compounds which are of greatest concern as air pollutants. In general, the highly soluble gases, such as hydrogen fluoride, are the most immediately damaging and are toxic to all forms of life. One of the most toxic and corrosive substances known is fluorine in the elemental state, but it is rarely encountered as an air pollutant and is always handled with extreme care by those required to use it. In general, the effects of exposure to elemental fluorine are related more to its extreme oxidizing power than to its ability to increase fluoride levels. It is improbable that elemental fluorine will be found in ambient air, except perhaps in the vicinity of some installations involved in rocket firing or testing and in specialized chemical industrial plants.

Properties of Fluoride Compounds in the Atmosphere

Fluorides may be found in the air as gases or as particulate matter, and very often industrial emissions will contain fluoride compounds in both states. Some difficulty may be encountered in sampling air (Section 4.17) for fluorides due to uncertainties concerning the form in which the compounds exist, and the method of sampling and analysis will sometimes determine the extent to which gaseous or solid forms are measured. Contamination is also a problem during sampling and analysis, and care must be taken to minimize the addition of fluoride into samples (Section 4.8).

All fluoride-containing compounds are likely to be chemically reactive, so the materials of which filters and bubblers are made must be selected to ensure their suitability for this purpose.[1,2,3]

Sources of Atmospheric Fluorides

If a sufficiently large sample of the particulate matter from ambient air is obtained by filtration, it will always be found to contain a small amount

Table 3.8a
FLUORIDE COMPOUNDS

Compound	*Physical State as Pollutant*	*Formula*	*Solubility*
Hydrogen fluoride	gas-liquid	HF	water soluble
Silicon tetrafluoride	gas-liquid	SiF_4	water soluble
Hydrofluosilicic Acid	gas-liquid	H_2SiF_6	water soluble
Sodium fluoride	solid	NaF	water soluble
Sodium aluminum silicofluoride	solid	$Na_5Al(SiF_6)_4$	water soluble
Sodium silicofluoride	solid	Na_2SiF_6	slightly soluble
Aluminum fluoride	solid	AlF_3	slightly soluble
Fluorapatite	solid	$CaFCa_4(PO_4)_3$	slightly soluble
Cryolite	solid	Na_3AlF_6	water insoluble
Fluorspar, fluorite	solid	CaF_2	water insoluble

of fluoride, providing an analytical method of suitable sensitivity is used. Therefore, there is a fluoride background in ambient air and this background may vary considerably depending upon the geographic location. Background levels derive principally from dust from soils containing small amounts of fluoride. Concentrations are somewhat higher in urban areas, however, and may range from approximately 0.1 to 1 $\mu g/M^3$ depending upon the nature of local industrial emissions. In areas adjacent to fluoride-emitting industries, substantially higher concentrations may be found if samples are taken downwind from uncontrolled sources.

A number of industries utilize fluoride-containing compounds or emit fluorides as a result of manufacturing processes. Such industries include the mining and processing of phosphate rock and other fluoride-containing minerals, the fertilizer-manufacturing industry, the aluminum and steel industry, the glass-making industry and the chemical industry. The steel industry is one of the largest users of calcium or sodium fluoride, which is added to molten steel to assist in the release of gaseous products. The electrolytic process for producing aluminum from alumina relies upon the addition of cryolite in molten form, and substantial fluoride emissions result from this process. The production of fertilizer has also resulted in many air pollution problems caused by high concentrations of fluoride-containing gases and particles in emissions; the starting material for fertilizer is frequently phosphate rock which may contain from 2 to 4% fluoride. When the crushed rock is treated with strong sulfuric or other acids, volatile hydrogen fluoride or silicon tetrafluoride is emitted in copious quantities. Glass usually contains a small amount of fluoride, and certain glasses may contain as much as 5 to 6% fluoride in order to obtain certain desired characteristics. When such glasses are melted by means of gas or oil heat, the resulting flue gases are often rich in fluoride compounds and may create serious air pollution problems.

The Effects of Fluorides

The presence of excessive concentrations of fluoride compounds in the atmosphere can be damaging to all forms of life and to inanimate objects as well. It has been stated that the fluoride ion is a protoplasmic poison, and only a small amount can be tolerated by any living cells. In man and animals, fluoride tends to concentrate in the bones, where it is stored as fluorapatite, a mineral similar to bone itself. Continued intake of excessive quantities of fluorides in man results in bone and teeth abnormalities. In addition, elevated concentrations of hydrogen fluorides and other fluorides are irritating and may cause damage to the respiratory system or exposed tissues. It is possible that living in the vicinity of a fluoride-emitting industry could cause the appearance of the symptoms of fluorosis in man, and it has frequently been so alleged. Ordinarily, however, the principal intake of fluoride results from the daily ingestion of food and water.

As an air contaminant, fluorides are most damaging to livestock and vegetation. Whenever fluoride compounds are emitted over a long period of time, the vegetation in the affected areas tends to contain elevated levels of fluoride when compared to natural background, so that animals grazing on such land will ingest considerably more fluoride than usual. The resulting disease, "fluorosis," in animals is well recognized and is most frequently found in grazing cattle, although all grazing animals may be affected. As in man, the effects are attributable to bone deposition, so that the bones undergo abnormal calcification; the teeth are affected; and ultimately the animals may lose weight, become lame and, in the case of milk-producing cows, exhibit a decline in milk production. The disease is caused by ingestion of contaminated forage and almost never by the direct inhalation of fluoride-containing air.

Vegetation damage, by contrast, usually results from the direct fumigation of plants by hydrogen fluoride or other fluoride-containing gases and only rarely from the uptake of fluoride contained in the soil. All plants are susceptible to damage by fluorides, but some are very susceptible, with damage occurring at extremely low levels. Among the most sensitive plants are gladiolus and pine trees, with atmospheric concentrations of hydrogen fluoride less than 1 part per billion being capable of causing visible damage to these plants.

Many materials of construction are also susceptible to damage by fluoride and the well-known etching of glass is frequently noted in areas where soluble fluoride emissions have occurred. Metallic surfaces are also attacked, and increased rates of corrosion will generally be observed on structural steel or other metallic surfaces exposed to air containing hydrogen fluoride.

No ambient air quality standards or emission standards for fluorides have been issued by the U.S. Environmental Protection Agency, as yet; but several states, and some countries, have ambient air standards which tend to be as low as 1 microgram per cubic meter, for total fluorides, or as low as 1 part per billion, when expressed as gaseous hydrogen fluoride. It is probable that secondary standards will be based on the lowest concentrations known to produce damage to vegetation and will be more stringent than primary standards which are designed to prevent adverse effects on human health.

REFERENCES

1. "Fluoride-Bearing Dusts and Fumes (Inorganic)," Hygienic Guide Series, American Industrial Hygiene Association, July to August 1965.
2. Heyroth, F. F.; Halogens, in Frank A. Patty (ed.), "Industrial Hygiene and Toxicology," Vol. II, 2d ed., Wiley, New York, 1962.
3. Stern, A. C. (ed.): "Air Pollution," Vols. I, II and III, Academic Press, New York, 1968.

3.9 HYDROCARBONS

Methane

In relatively unpolluted areas, methane is by far the most abundant hydrocarbon. In remote polar or oceanic locations, there is a uniform methane concentration of not less than approximately 1.2 ppm in air, whereas other hydrocarbons reach unmeasurably low levels in most such locations. The methane concentration can be an order of magnitude higher in areas with localized natural or man-made emissions.

It was recognized long ago that methane is the major constituent of marsh gas which, as the name implies, is a gaseous emanation of swamps. Methane is a product of biological processes not only in swamps but in the oceans as well. There is recent evidence of a considerable excess of methane evolution over disappearance above the ocean's surface in the locations so far measured. This natural evolution rate, it appears, may be large enough to dominate the input part of the atmospheric methane balance, with man's processes contributing only a minor proportion, to date. The exact sources of methane are not yet identified but are probably biological in nature.

Since the "background" atmospheric level of methane seems to remain relatively constant, the rates of production and destruction must be in balance. Methane seems to be inert in photochemical processes (see Section 2.8) as compared to other hydrocarbons, and its fate is still obscure. A plausible theory recently proposed is that it is converted by hydroxyl (OH) radicals to methyl (CH_3) which is then degraded stepwise to formaldehyde (CH_2O) and eventually to carbon monoxide and carbon dioxide. The estimated half-life of methane in the atmosphere based on these assumptions is on the order of 2 years. This is long compared to most other hydrocarbons, whose half-lives in the atmosphere vary from seconds to hours, and explains the reputation of methane for inertness in photochemical reactions.

Volatile Hydrocarbons

Although the proportions may differ, most of the volatile hydrocarbons observed in natural gas, liquefied petroleum gas, gasoline or other petroleum products have also been observed in ambient air. Some of the more abundant types of volatile hydrocarbons are listed in Table 3.9a. (Methane is usually

Table 3.9a
TYPES OF VOLATILE HYDROCARBONS

Compound	*Average Concentration of Measured Components, Los Angeles 1965, ppm (as Carbon)*	
Methane	3.22	
Ethane	0.20	
Propane	0.15	
Isobutane	0.05	
n-Butane	0.26	
Isopentane	0.21	
n-Pentane	0.18	
Total alkanes (except methane)		1.28
Ethylene	0.12	
Propene	0.05	
Total alkenes		0.29
Acetylene	0.08	
Benzene	0.19	
Toluene	0.37	
Total measured		5.43

more abundant than the rest combined except in highly polluted areas.) There are three major hydrocarbon classes: saturated, unsaturated and aromatic. In the saturated class, after methane, there are its homologs: ethane, propane and higher alkanes, decreasing in concentration more or less as the molecular weight increases. The unsaturated class includes olefins and acetylenes. The first member of the aromatic class is benzene but some substituted derivatives such as toluene and *m*-xylene are usually in larger concentration, after which again the higher homologs are present at lower concentrations.

These patterns of abundance reflect both the patterns of the original emission sources (mainly petroleum and natural gas) and the volatility variation within each class. Benzene, for example, is the most volatile aromatic but is less abundant in most full-range petroleum fractions or products than toluene and *m*-xylene. In automobile exhaust, ethylene and acetylene are the two largest hydrocarbon components.

Terpenes

A particular class of volatile hydrocarbons, the terpenes, arises predominantly not from man's operations but from nature. Terpenes are cyclic nonaromatic hydrocarbons found in pine tar and in other wood sources. They are released in large quantities to the atmosphere from forests, especially coniferous ones. Many members of the class are highly reactive in photochemical smog processes. The haze often observed over forested regions is mainly caused by such atmospheric reactions of terpenes.

Nonvolatile Hydrocarbons*

In the size range between about 10 and 20 carbon atoms per molecule, the distribution of hydrocarbons in air is not well known. Below this range, the components can be determined accurately in the vapor phase (Section 4.10). Above it, certain of the aromatic types can be determined as components of the filtrable airborne particles. The number of individual constituents becomes very large and their concentration levels very small, in the particles. The relatively few nonvolatile hydrocarbons identified so far have been studied mostly in connection with studies of carcinogens (Section 3.5).

Atmospheric Hydrocarbon Reactions

Photochemical smog reactions are discussed in Section 2.8. Although many volatile hydrocarbons participate in these reactions, not all do; and the rate and extent of reaction varies markedly with hydrocarbon structure. This has led to numerous proposals for "reactivity" scales which rate hydrocarbons numerically according to how they behave in some aspect of the photochemical smog system. Some of the bases for reactivity scales are:

1. Rate of disappearance of hydrocarbon.
2. Rate of disappearance of nitric oxide.
3. Amount of ozone formed in a fixed time.
4. Amount of product (e.g., aldehyde) formed in a fixed time.
5. Intensity of eye irritation after a fixed time or, inversely, the time to the onset of detectable eye irritation.

All these scales differ quantitatively and to some degree even qualitatively; but in a broad sense, their conclusions are similar. These can be summarized by describing one widely accepted reactivity scale:

Alkanes below pentane, acetylene and benzene:	0
Pentanes and higher saturated hydrocarbons:	1
Aromatic hydrocarbons above benzene:	3
Olefins:	10

It has also been shown that broad range mixtures act approximately as the sum of the reactivity of their constituents would imply, although some interaction does occur.

* See Glossary for definition.

Toxicity and Plant Damage

The aliphatic hydrocarbons are in general biochemically inert, although at high concentrations some of them have anesthetic properties. No effects occur below 500 ppm, far higher than is ever found in the open atmosphere. Aromatic hydrocarbons are active biochemically. The vapors may irritate mucous membranes. Members of this class can cause systemic injury. Chronic exposure to benzene is associated with blood disorders. None of these effects have been reported for levels below 25 ppm, considerably more than concentrations found in open air.

Plant damage caused by hydrocarbons at observed atmospheric levels has been associated with only one component: ethylene. At concentrations frequently found in polluted air, ethylene will damage a number of sensitive plants. The most serious cases (economically) have been injuries to commercially grown orchids and carnations in California.

Both toxicity and plant damage are very much more severe with the products arising in photochemical reactions of hydrocarbons, which are discussed in Section 2.8.

Air Quality Criteria

Federal air quality criteria for hydrocarbons exclude methane and are based not on the hazards of the hydrocarbons themselves but on their capacity to form photochemical oxidants (as described in Section 2.8). The standard is based on the reference methods described in Section 4.10. Hydrocarbons (exclusive of methane) determined by the reference method are not to exceed 0.24 ppm (6 to 9 A.M.) averaged over 3 hours.

BIBLIOGRAPHY

"Air Quality Criteria for Hydrocarbons," Publication No. AP-64, Environmental Protection Agency, U.S. Department of Health, Education and Welfare, March 1970.

3.10 HYDROGEN SULFIDE AND OTHER REDUCED FORMS OF SULFUR

The sulfur in hydrogen sulfide, mercaptans and dimethyl sulfide is present in a chemically reduced form. Mercaptans are aliphatic compounds that have structures similar to alcohols except that the oxygen is replaced by sulfur, hence they are called thioalcohols. Airborne reduced forms of sulfur originate mainly from industrial operations and, although the total quantities of emissions are small (relative to oxidized forms), the resultant effects may be significant on a localized basis.

Sources

Hydrogen sulfide is principally produced from the refining of petroleum, coking of coal, refining of natural gas, manufacturing of viscose rayon and evaporation of black liquor in the kraft pulping process.[1] Hydrogen sulfide removal and recovery processes have been in existence at many industrial plants for a number of years. The collected hydrogen sulfide is generally converted to elemental sulfur or to sulfur dioxide for subsequent sulfuric acid manufacturing.

Table 3.10a contains a summary of emission factors for several sources of hydrogen sulfide.[2,3] It can be seen from this table that petroleum refining is the major source. Table 3.10b portrays emission factors for hydrogen sulfide, methyl mercaptan and dimethyl sulfide from kraft pulp processing.[4] The methyl mercaptan and dimethyl sulfide emissions exceed the hydrogen sulfide emissions from several operations within the processing line.

Effects

Atmospheric concentrations of hydrogen sulfide and other reduced forms of sulfur can create objectionable odors and cause direct effects on materials, plants, animals and humans. Many complaints against air pollution originate as a result of the occurrence of obnoxious odors. Kraft pulp mills generate odor-causing mercaptans, hydrogen sulfide, dimethyl sulfide and dimethyl disulfide. Overloaded and improperly operated wastewater treatment plants can be a source of hydrogen sulfide. Chemical plants may also be a source of hydrogen sulfide as well as other organic sulfur compounds that cause odors. The concentrations of reduced sulfur compounds which can be detected by

Table 3.10a
HYDROGEN SULFIDE EMISSION FACTORS

Industry	*Emission Factors*
Petroleum products	5.6 tons/plant for 10,000 bbl/day consuming 7×10^9 BTU
Chemical and allied products	0.024 tons/plant consuming 10^9 BTU/day
Stone, clay and glass manufacturing	0.17 ton/cement plant producing 4,830 bbl/day and oil as fuel for 5.9×10^4 BTU/day
Nonferrous industry	0.002 tons/plant producing 50 tons/day of casting and requiring 70×10^6 BTU/day
Gray iron foundries	0.023 tons/plant producing 200 tons/day of casting and requiring 55×10^7 BTU/day
Coal combustion	0.004 (S)[a] tons/ton of coal

[a]Sulfur content of coal as a decimal.

Table 3.10b
REDUCED SULFUR EMISSION FACTORS FOR KRAFT PULP PROCESSING, LB GASEOUS POLLUTANTS PER TON DRY PULP PRODUCED

Source	*Hydrogen Sulfide*	*Methyl Mercaptan*	*Dimethyl Sulfide*
Digester blow system	0.1–0.7 a	0.9–5.3 a	0.9–3.8 a
Smelt tank	1	negligible	negligible
Lime kiln	3.6	5	3
Recovery furnace[b]	3.6–7.0 0.7	a a	a a
Multiple-effect evaporator	1.2 0–0.5	0.004 0.003–0.030	a negligible
Oxidation towers	a	a	0.1

[a]Not available.

[b]Gaseous sulfurous emissions are greatly dependent on the oxygen content of flue gases and on furnace operating conditions.

the sense of smell are lower than the concentrations which cause direct effects, hence the odors will not cause any physical or material damage other than simply being an insult to the senses.[5]

Hydrogen sulfide can cause tarnishing of metals, especially in the presence of moisture.[6] The reactions in Equations 3.10(1) and 3.10(2) are simplifications of the tarnishing effect of hydrogen sulfide upon silver and copper, respectively.

$$2\,Ag^+ + H_2S \xrightarrow[O_2]{H_2O} Ag_2S + 2H^+ \qquad 3.10(1)$$

$$Cu^{++} + H_2S \xrightarrow[O_2]{H_2O} CuS + 2H^+ \qquad 3.10(2)$$

The sulfide coating formed on open silver and copper electrical contacts can greatly increase the resistance across these contacts. Hydrogen sulfide can also cause a discoloration of lead-based house paint in accordance with reaction 3.10(3).

$$H_2S + PbO \longrightarrow PbS + H_2O \qquad 3.10(3)$$

When hydrogen sulfide reacts with lead pigments, a brown or black discoloration of the paint occurs. Hydrogen sulfide in air may be eventually converted to sulfuric acid mist and thus become a potential corrosion agent.

Plants and Animals

Since most ambient air concentrations of hydrogen sulfide are less than 0.1 ppm, this is usually not sufficient to be absorbed from the air by plants to the extent that plant damage results. If excessive amounts of hydrogen sulfide are released by accidental industrial discharges, then the potential for plant damage is present. Some plants have been found to exhibit moderate damage patterns when exposed to 20 to 40 ppm of hydrogen sulfide for a period of 5 hours.[7] This plant group includes cosmos, radishes, clover, tomatoes, poppies, salvia, cucumbers, soy beans, asters and tobacco. Other plants which have been studied do not exhibit damage patterns until subjected to hydrogen sulfide concentrations which are greater than 40 ppm. Hydrogen sulfide differs from many other toxic gases in that it scorches growing leaves instead of attacking mature, older leaves.

Some information on the effects of hydrogen sulfide and organic thiols on animals is available from the Poza Rica, Mexico, episode which occurred in 1952.[8] A cursory investigation disclosed that an undetermined number of canaries, chickens, cattle, pigs, geese, ducks and dogs either died or were taken ill by exposure to reduced forms of sulfur which had been accidently released by a natural gas refining plant.

Several studies have been conducted to ascertain the toxic level of selected organic sulfur compounds that may originate from petroleum processing.[9] Investigators from Europe have described the toxicity of dimethyl sulfide, dimethyl disulfide and methyl mercaptans. Studies to determine the acute toxicity of seven aliphatic and two aromatic thiols to animals were conducted.[10] Of the nine compounds tested by inhalation in rats, rabbits and mice, benzenethiol and methylheptanethiol were the most toxic. The 48-hour lethal concentration for 50% (LC_{50}) of the test animals for the former was 59 ppm; and for the latter, the 24-hour LC_{50} was 51 ppm for a single 4-hour exposure. Although very large differences in acute toxicity were found to exist among the thiols tested, they all had the common toxicological property of being soporific, ranging in degree from producing mild stupor to heavy sedation. Methylhepthanethiol was an exception in that it produced a powerful central nervous system stimulation. Considerably less information is available on the toxicity of organic nonthiol sulfur compounds such as the sulfides and disulfides. Dimethyl sulfide is less acutely toxic to rodents (by inhalation) by a factor of 1.5 to 2 than the corresponding methylthiol; in contrast, dimethyl disulfide is more toxic by a factor of 3 or 4 than the comparable methylthiol.[9]

Effects on Humans

In terms of effects on humans, hydrogen sulfide is a colorless gas that has a characteristic disagreeable odor described as that of rotten eggs. It can be detected by the sense of smell at concentrations as low as 0.1 ppm. Exposures for short times can result in a fatigue of the sense of smell because it becomes desensitized after 2 to 15 minutes of exposure to a concentration of 100 ppm or more.[11] Hydrogen sulfide can also be an irritant to the eyes and respiratory system of humans at concentrations considerably above those that cause detectable odors.[12] Unlike what is known of photochemical eye irritants, hydrogen sulfide causes inflammation and edema of the conjunctiva. At still higher concentrations, it may produce systemic effects and can even result in death. There is evidence to support the association of several deaths in Poza Rica, Mexico, with large quantities of hydrogen sulfide that escaped from units of a natural gas refining plant.[8]

Sulfur compounds from petroleum have a potential for providing protection against certain oxidant gases.[13] Both organic thiols and disulfides are highly effective in antagonizing the acute lethal action of either nitrogen dioxide or ozone when inhaled along with these oxidant gases.

In summary, the air pollution effects resulting from the presence of hydrogen sulfide are mainly associated with nuisances from odors and the discoloration of some paints and the tarnishing of metals at very low concentrations of the gas. Concentrations of hydrogen sulfide sufficiently high to cause vegetation damage, irritation of the eyes and respiratory system, systemic effects or death are unlikely to be encountered in the ambient atmosphere. It is considered that these higher concentrations could only result from

Table 3.10c
AMBIENT AIR QUALITY STANDARDS
FOR HYDROGEN SULFIDE

Country	Averaging Time, hours	Standard, ppm	Remarks
United States			
California	1	0.1	adverse level
Missouri[a]	½	0.05	b
	½	0.03	c
Montana	½	0.05	b
	½	0.03	c
New York	1	0.1	
Pennsylvania	1	0.1	
	24	0.005	
U.S.S.R.	24	0.005	
Poland	⅓	0.04	d
	24	0.013	d
	⅓	0.005	e
	24	0.005	e
Federal Republic	½	0.1	f
of Germany	½	0.2	g
Czechoslovakia	24	0.005	
Canada			
Ontario	½	0.03	h

[a] St. Louis metropolitan area.
[b] Not to be exceeded over twice a year.
[c] Not to be exceeded more than twice in any 5 consecutive days.
[d] For protected areas.
[e] For specially protected areas.
[f] Long-term exposure.
[g] Not to be exceeded more than once in any 8 hours.
[h] Maximum ground level concentration from a single source measured on the centerline downwind from the stack.

accidental discharges from point sources rather than from community-wide emissions.

Standards

Table 3.10c contains a summary of the ambient air quality standards for hydrogen sulfide in several selected countries.[14] The standards are a function of averaging time, with typical values for 30-minute periods being between 0.03 and 0.20 ppm, and for 24-hour periods between 0.005 and 0.013 ppm. Table 3.10d portrays hydrogen sulfide emission standards utilized in several countries, with typical effluent gas concentrations of 5 ppm.[14]

Table 3.10d
HYDROGEN SULFIDE EMISSION STANDARDS

Country	*Emission Standard*	*Remarks*
Britain	5 ppm 7.5 mg/m^3 STP	
Czechoslovakia	0.08 kg/hr	Emission rate above which it is necessary to submit a report to the government; where discharge is for less than 1 hr, there is a proportionate reduction in emission permissible without such reporting
New South Wales and Queensland	5 ppm 7.5 mg/m^3 STP	

REFERENCES

1. Spencer, E. F., Jr.: Pollution Control in the Chemical Industry, Chapter 14, p. 14-9, in H. F. Lund (ed.), "Industrial Pollution Control Handbook," McGraw-Hill, New York, 1971.
2. Anderson, D. M. and F. H. Sussman: "Pure Air for Pennsylvania," Pennsylvania Department of Health, Harrisburg, Pennsylvania, 1960.
3. "Metropolitan Air Pollutant Concentrations—Preparation, Sampling and Assay of Synthetic Atmospheres," Stanford Research Institute, Palo Alto, California, November 1956.
4. Rossano, A. T., Jr.: *"Air Pollution Control Guidebook for Management,"* Chapter 2, p. 19, E.R.A., Inc., Stamford, Connecticut, 1969.
5. Ibid, Chapter 6, p. 87.
6. Yocum, J. E.: Effects of Air Pollution on Materials, Chapter 7A, pp. 210–212, in A. C. Stern (ed.), "Air Pollution," Vol. I, Academic Press, New York, 1962.
7. *Air Pollution Manual, Part I, Evaluation*, pp. 58–59, American Industrial Hygiene Association, Detroit, Michigan, 1960.
8. McCabe, L. C. and G. D. Clayton: Air Pollution by Hydrogen Sulfide in Poza Rica, Mexico, *A.M.A. Archives Industrial Hygiene and Occupational Medicine*, vol. 6, p. 199, 1952.
9. Stokinger, H. E.: Effects of Air Pollution on Animals, Chapter 9, pp. 314–315, in A. C. Stern (ed.), "Air Pollution," Vol. I, Academic Press, New York, 1962.
10. Fairchild, E. J. and H. E. Stokinger: Toxicologic Studies on Organic Sulfur Compounds, Part I, Acute Toxicity of Some Aliphatic and Aromatic Thiols, *J. Amer. Industrial Hygiene Association*, vol. 19, p. 171, 1958.
11. Jacobs, M. B.: "The Analytical Chemistry of Industrial Poisons, Hazards and Solvents," 2d ed., Interscience, New York, 1949.
12. Goldsmith, J. R.: Effects of Air Pollution on Humans, Chapter 10, pp. 364–365, in A. C. Stern (ed.), "Air Pollution," Vol. I, Academic Press, New York, 1962.
13. Fairchild, E. J. II, H. E. Stokinger and S. D. Murphy: Protection Against Ozone and Nitrogen Dioxide Poisoning by Sulfur Compounds, *Science*, vol. 130, p. 861, 1959.
14. Stern, A. C.: Air and Water Pollution Quality Standards, Part 1: Air Pollution Standards, Chapter 4, pp. 4-4 to 4-22, in H. F. Lund (ed.), "Industrial Pollution Control Handbook," McGraw-Hill, New York, 1971.

3.11 MERCURY

Elemental mercury is the only metal which exists in the liquid phase at ambient temperatures. Table 3.11a shows the amounts of mercury consumed in various industrial applications during 1968 and projects its use in 1974–1975.[1] In 1968, in excess of 77% of the mercury consumed in the United States was in the form of elemental mercury. Since fossil fuel combustion mainly gives rise to elemental mercury, it is the major source of mercury pollution in the environment.

Published atmospheric emission data for mercury and its compounds are nonexistent. However, Table 3.11b gives estimates of mercury emissions to the atmosphere for 1968 by United States industry.[2]

The principal mercury compounds discharged into the environment and the primary sources of these compounds are: (1) metallic mercury from mining and refining operations, coal combustion, chlor-alkali and various instrumentation plants; (2) inorganic divalent mercury from chlor-alkali plants; (3) phenyl mercury from paints, pulp and paper manufacturing operations; (4) methyl mercury from agriculture; and (5) methoxyethyl mercury from agriculture. With the exception of dimethyl mercury, the vapor pressures of mercury and its compounds are quite low at ambient conditions. However, at equilibrium, the vapor pressures are sufficient to modify the processes of all living species. Tables 3.11c and 3.11d show some of the air concentrations of mercury achievable above elemental sources and the vapor pressures of some compounds commonly encountered.

Sources of Mercury Pollution

In agriculture, mercury fungicides and bactericides are used to control plant diseases. Until recently, mercurials were heavily used in treatment of the seeds of cereal grains and the compounds contained from 0.01 to 0.04% mercury. These compounds fall into three basic structural categories: alkyl mercury compounds, alkoxyl mercury compounds and aryl mercury compounds.

In industry, amalgams of potassium, sodium and zinc are used as reducing agents. Tetraethyl lead production utilizes sodium amalgams. Zinc recovery from drosses employs an electrolytic process in which mercury serves as the cathode.

Table 3.11a
ESTIMATED CONSUMPTION OF MERCURY IN FLASKS

Industrial Application	Estimated Consumption 1968	Flasks Of Mercury[a] 1974–1975
Chlorine and caustic manufacturing	17,424	22,864
Electrical apparatus	17,200	22,700
Paints	10,566	10,725
Control instrumentation		
Switches and relays	2,500	2,650
Other instruments	6,400	6,600
Dental applications	3,500	3,800
Agriculture	3,430	2,650
Laboratory uses	2,075	2,075
Catalysts		
Urethanes	800	1,560
Vinyl chloride monomers	500	250
Anthraquinone derivatives	175	220
Miscellaneous	236	340
Pharmaceuticals	600	650
Pulp and paper	375	250
Amalgams	259	250
Others	7,815	5,960
Totals	73,855	83,544

[a]Flask is an industrial measure corresponding to 76 lb virgin mercury. Flasks are made of wrought iron and spun steel, have a variety of dimensions and weigh from 7 to 15 lb.

Production of urethane elastomers, vinyl chloride polymers and sulfonated anthraquinones uses organic mercury catalysts.

Standard dental amalgams contain about 50% mercury. The vaporization of mercury from fillings appears to be minimal.

A wide variety of electrical products utilize mercury and its compounds. Batteries, fluorescent lamps, power rectifiers, relays, switches, barometers, thermometers, flowmeters and pressure sensors are typical applications. This product area forms the largest single consumer of mercury.

The second largest consumption of mercury is found in the manufacture of chlorine and caustic by the mercury cell process which utilizes a flowing mercury cathode (Section 7.4). About 25% of chlorine manufactured in the United States employs this method. In Europe, a much higher proportion of chlorine production is in mercury cell facilities.

Mercury is widely used in all research, development and testing laboratories. Consumption is required to replace spilled metal.

Table 3.11b
ESTIMATED ATMOSPHERIC EMISSIONS OF MERCURY, 1968

Industrial Source	*Estimated Emission tons/year*
Mining operations	2.6
Processing of Hg ore	55
Secondary processing	11
Chlor-alkali manufacturing	Unknown[a]
Paint	1
Agriculture	1
Catalysts	Negligible
Pulp and paper	Negligible
Pharmaceuticals	Negligible
Amalgams	Negligible
Electrical equipment	3
Dental preparations	1.2
Coal combustion	257

[a]The chlorine industry used 663 short tons of mercury to offset processing losses. However, a major portion was emitted in aqueous effluents.

Table 3.11c
EQUILIBRIUM AIR CONCENTRATIONS OF MERCURY

Temperature, °C	*Equilibrium Concentrations, mg/M³*	*Temperature, °C*	*Equilibrium Concentrations, mg/M³*
0	2.2	20	14.0
5	3.1	25	20.0
10	5.5	30	32.0
15	8.5	35	49.0

Table 3.11d
VAPOR PRESSURE OF MERCURY AND ITS COMPOUNDS

Compound	*Temperature, °C*	*Vapor Pressure, MM of Hg*
Mercury, elemental	36	4.47×10^{-3}
Dimethyl mercuric mercury	94	7.60×10^{2}
Methyl mercury chloride	35	3.2×10^{-2}
Phenyl mercuric acetate	35	5.0×10^{-6}
Phenyl mercuric dimethyldithiocarbamate	35	8.0×10^{-7}

The paint industry uses organic mercurials of low volatility as antifouling marine coatings, as mildew control agents and as preservatives to avoid agglomeration of latex paints. After painting, low levels of mercury-containing vapor are detectable for 2 to 3 years. Mercury concentrations currently vary from about 0.004 to 2.5% depending on the application involved. Phenyl mercuric acetate, phenyl mercuric dodecyl succinate and phenyl dimethyl dithiocarbonate are compounds frequently employed.

The only data relating fossil fuel consumption to mercury emission are based on coal combustion analyses. Coal samples throughout the United States show an average mercury content of approximately 0.50 ppm. Based on estimated use of 508,990,000 tons of hard and soft coal, 255 tons of mercury were released into the United States atmosphere in 1968 from this source. Recent data indicate that in excess of 90% of the mercury contained in coal is emitted and more than 95% of this is in the form of elemental mercury. Based on similar analyses, mercury emitted into the atmosphere from the combustion of residual fuel oil in the United States was estimated to be 37 tons in 1968.

Ambient Air Concentrations of Mercury

Mercury found in coal, fuel oil and incinerator emissions considerably increases the urban air concentrations of mercury over those arising from natural sources in rural areas. Helicopter surveys conducted in Missouri and Illinois during December 1970, and January 1971 measured mercury concentrations in stack emissions for 27 plants.[3] Levels in excess of 50 $\mu g/M^3$ were detected for one incinerator output. However, 15 power plant emissions were generally in the range of from under 50 to 200 ng/M^3. Incinerator emissions from four plants ranged from 200 to over 10,000 ng/M^3. Emissions from eight industrial plants were in the range of from under 50 to 10,000 ng/M^3.

Ground-level ambient mercury concentrations are strongly dependent upon meteorological conditions. Measurements downwind from typical emitters have shown values as high as 200 to 300 ng/M^3 at distances of 1,000 yards during inversion conditions. Low temperatures apparently condense the mercury vapor into particulate form. Rain also significantly reduces the concentration levels.

Samplings of urban industrial area measurements have shown the following average values: Chicago, 10 ng/M^3; Cincinnati, 100 ng/M^3; Charleston, West Virginia, 170 ng/M^3; New York, 1 to 14 ng/M^3.

Data collected by the U.S. Geological Survey has shown air concentrations between 3 and 9 ng/M^3 in nonmineralized areas. Levels up to 20 ng/M^3 were obtained in the vicinity of copper, precious metal and mercury mines. Ambient mercury concentrations for a variety of meteorological conditions were measured in a location south of San Francisco.[4] The concentration averaged about 0.2 ng/M^3 when the wind blew from the Pacific Ocean. It was somewhat higher when the wind was from the nonindustrial southeast. When the wind direction was from the industrial area to the north and northeast, the average was 8 ng/M^3 with peaks as high as 20 ng/M^3. Mercury is frequently associated with dust particles.

Effects of Mercury Pollution

At levels of 50 to 350 μg/M^3, from 75 to 85% of mercury vapor is absorbed by man via the respiratory route. Elemental mercury produces its toxic effects after oxidation (in the blood and tissues) to mercuric ions. Elimination from the brain and nerve tissue is slow, accounting for the clinical observation that the central nervous system is involved in chronic exposure. Highest levels of mercury are found in the kidneys where damage also may occur. Since divalent mercury is captured by sulfhydryl groups (see Glossary), mercurials even at low concentrations are capable of inactivating the sulfhydryl enzymes.

Mercury poisoning may be either chronic or acute. Chronic poisoning usually follows industrial exposure to compounds such as fungicides. Acute poisoning usually results from oral ingestion but may result from the inhalation of vapors of elemental mercury or organic mercury compounds. The symptoms of acute poisoning by the respiratory route include pneumonitis, lethargy or restlessness, fever, tachypnea, cough, chest pain, cyanosis, diarrhea and vomiting. Atelectasis emphysema, hemorrhage and pneumothorax may follow. The toxic action of inorganic mercury on the kidneys may cause diuresis, but at sufficient dosage levels renal damage may be extensive causing oliguria and anuria. Chronic poisoning symptoms including gingivitis, stomatosis, loosening of the teeth, salivation, metallic taste, colitis, progressive renal damage, loss of appetite, anemia, hypertension and peripheral neuritis. The central nervous system involvement may be evidenced by mental depression, irritability, insomnia, tremors, fatigue and even hallucinations.

Quantitative data are not available in determining the effects on animals of mercury-containing vapors and particulates in air. However, Russian data are available which show that inhalation of air containing mercury levels in the range of 100 to 7,000 μg/M^3 have been fatal for laboratory animals. Animals appear to be more sensitive to renal damage than humans.

Plants differ widely in sensitivity to mercury. Roses have been damaged by as little as 10 μg/M^3 air concentrations and by low concentrations of mercury in fungicides. Damage to plants by mercury is usually evidenced by the development of dark spots or blackening of the leaves and stems followed by eventual plant collapse or premature leaf drop. The severity of the effects are dependent upon the dosage experienced by the plant. Sensitivity is particularly severe in plants such as broad bean, butterfly weed, oxalis, sunflower and a number of varieties of roses.

Air Quality Criteria

In December 1971, the U.S. Environmental Protection Agency proposed emission standards for processors of mercury ores and for facilities using mercury chlor-alkali cells. These regulations fix emissions to the atmosphere at less than 2,300 grams (5 lb) per 24-hour day as measured by specified procedures. For both types of facilities, testing shall be carried out every 3 months, utilizing a minimum sampling time of 2 hours and a minimum air sampling volume of 50 cubic feet. Records will be maintained to allow determination of mercury emissions on a 24-hour basis.

The specified methods for the determination of mercury in particulate and gaseous emissions have been reviewed in Section 4.11.

The American Conference of Government and Industrial Hygienists has recommended, as threshold limit values (Section 5.3), a time-weighted average concentration of 100 $\mu g/M^3$ for metallic mercury vapor and inorganic compounds and 10 $\mu g/M^3$ for organic mercury compounds. However, these values appear to be headed toward revision: 10 $\mu g/M^3$ for alkyl mercurials and 50 $\mu g/M^3$ for all other forms of mercury.

In 1968, an international symposium meeting at the Karolinska Institute, Stockholm, recommended that "maximum allowable concentrations" for mercury and its compounds be established as follows:

Inorganic, phenyl and methoxyethyl mercury salts	100 $\mu g/M^3$
Elemental mercury vapor	50 $\mu g/M^3$
Alkyl mercury compounds	10 $\mu g/M^3$

The committee also advised women of child-bearing age to avoid any exposure to the alkyl mercurials.

The U.S.S.R. has established permissible limits of exposure at 10 $\mu g/M^3$ for inorganic mercury and 5 $\mu g/M^3$ for alkyl mercury compounds. However, these are limit values which should never be exceeded even for brief exposures.

The United Kingdom has recommended a safe level of mercury vapor at 100 $\mu g/M^3$. The value recommended by West Germany is 1 $\mu g/M^3$.

REFERENCES

1. Jones, H. R.: "Mercury Pollution Control," Noyes Data Corp., Park Ridge, New Jersey, 1971.
2. Davis, W. E. and Associates: "National Inventory of Sources and Emissions Arsenic, Beryllium, Manganese, Mercury and Vanadium," report prepared for the Environmental Protection Agency, Office of Air Programs, 1971.
3. Mercury in the Air, an Environment Staff Report; *Environment,* vol. 13, no. 4, pp. 24–33, 1971.
4. Williston, S. H.: *J. Geophys. Res.,* vol. 73, 7,051, 1968.

BIBLIOGRAPHY

"Mercury in the Environment," Geological Survey Professional Paper No. 713, U.S. Department of the Interior, 1970.

National Emission Standards for Hazardous Air Pollutants, *Federal Register*, vol. 36, no. 234, pp. 23, 239-23, 251, December 7, 1971.

Report of an International Committee, *Arch. Environ. Health*, vol. 19, pp. 891–905, 1969.

Stahl, Q. R.: "Air Pollution Aspects of Mercury and Its Compounds," report prepared for the National Air Pollution Control Administration by Litton Systems, Inc., Bethesda, Maryland, 1969.

The Hazards of Mercury, *Environmental Research*, vol. 4, pp. 1–69, 1971.

Wallace, R. A., et al.: "Mercury in the Environment—The Human Element," Oak Ridge National Laboratory Report, ORNL-NSF-EP-1, March 1971.

3.12 NITROGEN OXIDES

From an air pollution standpoint, the most important nitrogen oxides in the atmosphere are nitric oxide (NO) and nitrogen dioxide (NO_2). These two oxides are often referred to jointly in the air pollution literature as NO_x. At high concentrations, these oxides are very toxic and many persons have been killed by accidental emissions. They are also important because of the role they play in the chemistry of photochemical smog. The concentrations of NO and NO_2 in the unpolluted atmosphere are in the vicinity of 2 and 8 ppb by weight. In urban air, they may be as high as 1 part per million (ppm) by weight.

Several other oxides of nitrogen exist in the atmosphere (Table 2.1h). Nitrous oxide (N_2O) is a natural, somewhat variable constituent, which is essentially inert in the troposphere. It is produced by soil bacteria, and its concentration is usually about 1 ppm. N_2O_4, N_2O_3, N_2O_5 and NO_3 (not to be confused with the nitrate ion, NO_3^-) are present in very small amounts. Of these, N_2O_5 may have the highest concentration in smog since it is an oxidation product of NO_2 and is quite stable. NO_3 is very reactive and only exists as a very short-lived intermediate in a number of chemical reactions, such as

$$NO_2 + O_3 \longrightarrow NO_3 + O_2 \qquad 3.12(1)$$

$$NO_3 + NO_2 + M \longrightarrow N_2O_5 + M \qquad 3.12(2)$$

where M is any third molecule.

Worldwide, in the troposphere, NO and NO_2 are mainly produced by soil bacteria (Fig. 2.1i). However, combustion is the main source in polluted atmospheres.

Sources

According to "Air Quality Criteria for Nitrogen Oxides," in 1968, coal, oil, natural gas and motor-vehicle fuel combustion accounted for more than 18×10^6 out of 20.6×10^6 tons of man-made NO_x in the United States.[1] Of this, 10×10^6 tons were from stationary combustion and 8×10^6 tons from transportation, which emphasizes the importance of controlling the emissions from stationary sources as well as from automobiles (Tables 2.1c, 2.1g, 2.10a and 3.1d).

Nitrogen dioxide has been detected in the lower stratosphere, although its concentration has not been well established. Infrared spectra suggests that its concentration is probably of the order of 10 ppb by weight at 20 km altitude. The NO_2 is probably produced by the oxidation of NO by ozone. The NO probably reaches the stratosphere from both the troposphere and mesosphere (Fig. 3.1a) and is produced in situ by the reaction of electronically excited atomic oxygen with N_2O.

Recent studies have shown that NO_x takes part in the chemical reactions responsible for the production and destruction of ozone in the stratosphere. Additional NO_x, emitted in jet engine exhaust gases, would be expected to decrease the concentration of stratospheric ozone. Since ozone serves to prevent much of the ultraviolet radiation, which is harmful to mankind, from reaching the earth's surface, even a small decrease in stratospheric ozone is potentially harmful.

Photochemical Smog

Nitrogen dioxide is almost certainly primarily responsible for triggering the series of chemical reactions that produce photochemical smog. The NO_2 absorbs sunlight in the near ultraviolet region which produces photolysis (Fig. 3.12a):[1]

$$NO_2 + h\nu \longrightarrow NO + O \qquad 3.12(3)$$

Where h is Planck's constant and ν is the frequency of the light absorbed. Thus, $h\nu$ is the energy absorbed per quantum of light.

This is followed by the reactions

$$O + O_2 + M \longrightarrow O_3 + M \qquad 3.12(4)$$

$$NO + O_3 \longrightarrow NO_2 + O_2 \qquad 3.12(5)$$

where M is any third molecule which removes the energy of reaction from the ozone (O_3). In the absence of other reactions, 3.12(3) to 3.12(5) would result in a much lower concentration of ozone than is often observed in smog. Either additional ozone must be produced by some reaction other than 3.12(4) or some reaction which oxidizes NO more rapidly than does ozone must occur.

Organic vapors in smog undergo reactions which include the following very important ones:[2]

$$O + \text{olefins} \longrightarrow R + R'O \text{ of } R\overset{\displaystyle O}{\overset{/\,\backslash}{—}}R' \qquad 3.12(6)$$

where R and R' are organic radicals (see Glossary). The ozone reacts with various organic compounds such as olefins to form a multitude of compounds

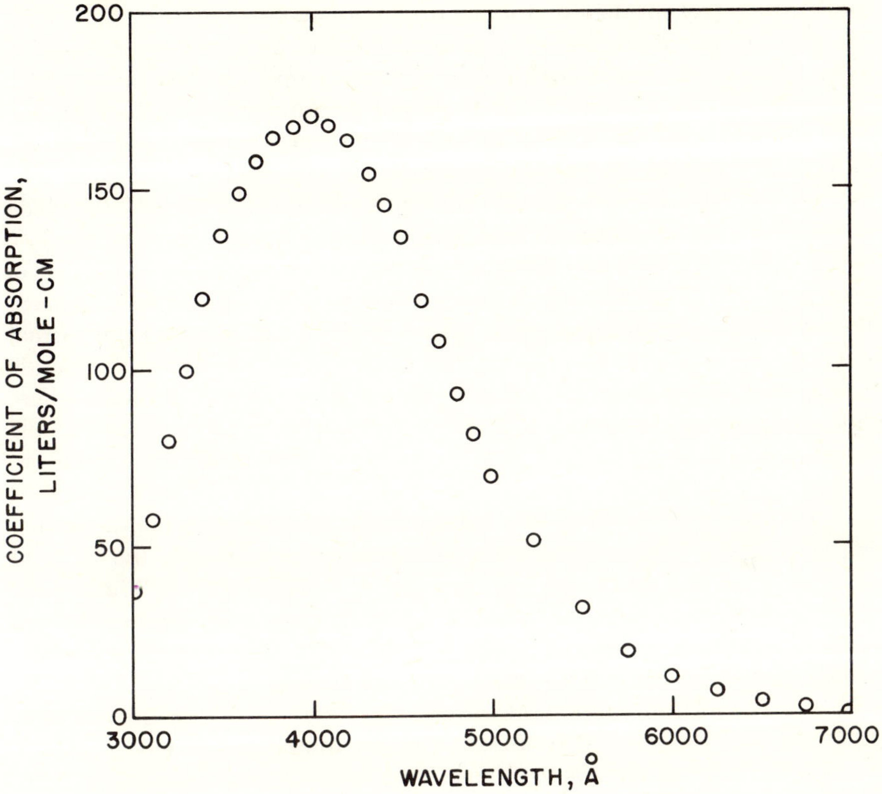

Figure 3.12a NO_2 absorption spectrum.

including free radicals:

$$O_3 + \text{olefins} \longrightarrow \text{products} \tag{3.12(7)}$$

Organic free radicals react rapidly to form peroxy free radicals:

$$R + O_2 \longrightarrow RO_2 \tag{3.12(8)}$$

Possibly these react to form ozone, thus accounting for the large ozone concentration:

$$RO_2 + O_2 \longrightarrow RO + O_3 \tag{3.12(9)}$$

More likely, however, they oxidize NO; and by competing with the reaction in Equation 3.12(5), allow ozone to accumulate:

$$RO_2 + NO \longrightarrow RO + NO_2 \tag{3.12(10)}$$

Another proposed mechanism for NO oxidation involves the formation of the first electronically excited state of oxygen,* $O_2\ (^1\Delta)$, by one of several mechanisms. The $O_2\ (^1\Delta)$ reacts with olefins to form peroxy radicals (RO_2) which oxidize the NO by reaction 3.12(10). Recent determinations of the rate constants for the reactions of $O_2\ (^1\Delta)$ with olefins indicate that such reactions are probably too slow to be of much importance in smog.

NO and NO_2 are free radical scavengers and thus can "break" the chain reactions producing photochemical smog as well as initiate them by reaction 3.12(3). Numerous field and laboratory studies[1,3,4] have shown that a plot of some property of smog, such as ozone concentration, against NO_x concentration, as the independent variable, passes through a maximum value of that property. According to two studies, this occurs at such a low concentration of NO_x that a very large reduction in the emissions of NO_x in cities where photochemical smog is prevalent would be necessary to achieve an appreciable reduction of the smog. In fact, small reductions might make the smog more intense!

This process of breaking free radical chains by NO_x results in the formation of organic nitrogen compounds. One type of compound is the series of peroxyacyl nitrates and especially peroxyacetyl nitrate (PAN). These compounds, which have been identified in smog, damage plants and contribute to the eye irritation produced by smog. Aromatic analogs of PAN, especially peroxybenzoyl nitrate (PBN), have been detected in laboratory "synthetic smog" produced from aromatic hydrocarbons. Eye irritation is greater from PBN than PAN by more than an order of magnitude, but its concentration in smog has not yet been determined.

This finding, however, suggests the use of caution in the widespread substitution of high-aromatic content gasolines for leaded gasoline.

Although PAN is apparently the only type of organic nitrogen compound identified to date in photochemical smog, other organic nitrogen compounds in addition to PBN derived from NO_x are almost certainly produced. For example, NO_2 adds to methyl radicals to form nitromethane:[5]

$$CH_3 + NO_2\ (+M) \longrightarrow CH_3NO_2\ (+M) \qquad 3.12(11)$$

The formation of organic nitro compounds in smog may be important since they are extremely toxic. Much more should be known concerning organic nitrogen compounds in smog and their identity.

Air Quality Standards

Nitrogen dioxide has an orange brown color similar to that of photochemical smog by transmitted light (Fig. 3.12b).[1] The suggestion has often

* When a molecule absorbs light, the molecule can be dissociated, electronically excited or vibrationally excited. In an electronically excited state of a molecule, one or more of the electrons has been promoted to a more energetic quantum state in the molecule.

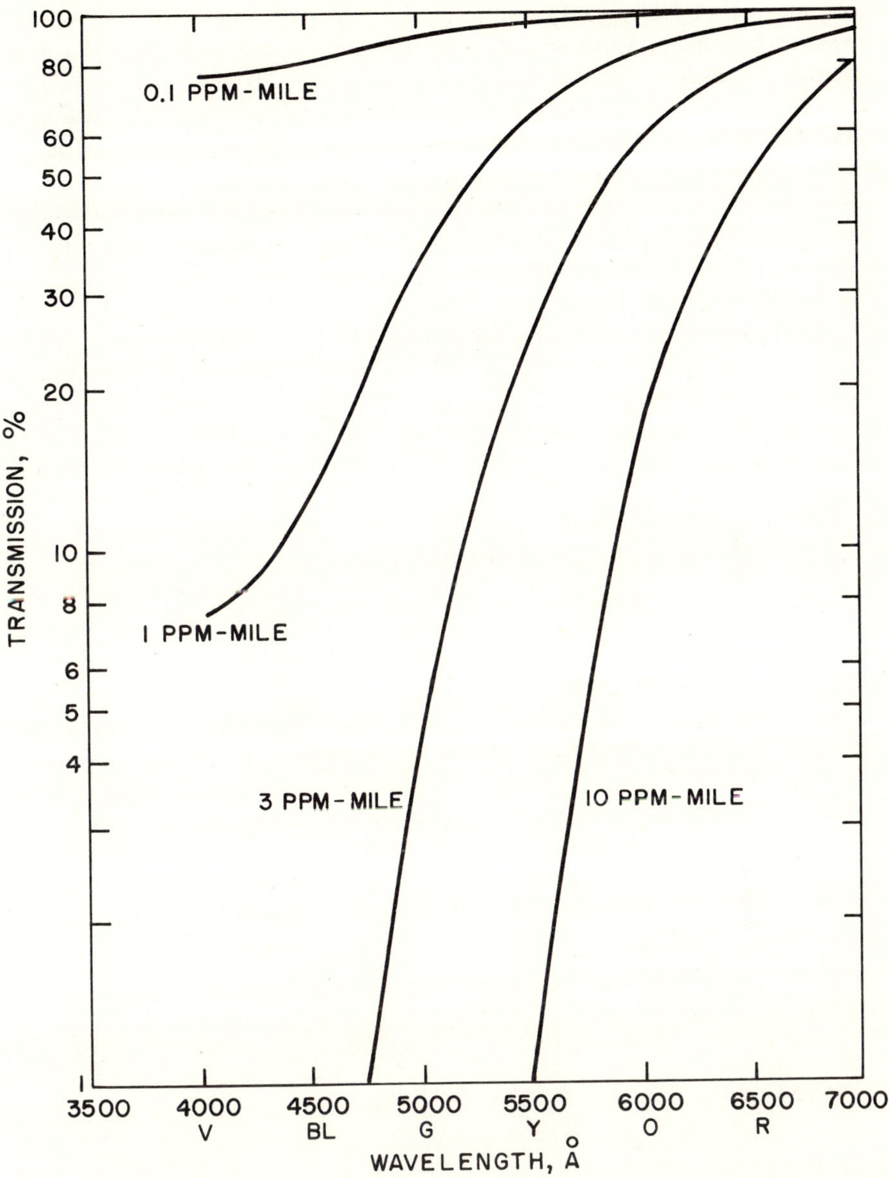

Figure 3.12b Transmittance of visible light at different NO_2 concentrations and viewing distances.

been made that NO_2 is responsible for the color of such smog, but several investigations have demonstrated that the color is due largely to the preferential scattering of blue light by the very small particles.

Nitrogen dioxide is more toxic than NO. Concentrations of the former exceeding 100 ppm are lethal to most animal species, and 90% of the deaths are caused by pulmonary edema.[1] Nitrous oxide is a general anesthetic at concentrations of 8×10^6 ppm and above. Some research has been undertaken on the chronic effects of nitrogen oxides. For example, mice exhibited increased susceptibility to infection after continuous daily exposure to 0.5 ppm NO_2 for 3 months.

Little is known concerning the lethality of peroxyacetyl nitrate. Apparently, it is less lethal to mice than ozone, about the same as NO_2 and more lethal than sulfur dioxide.[6]

A United States national air quality standard has been established for nitrogen oxides.[7] The annual arithmetic mean must not exceed 100 $\mu g/m^3$.

REFERENCES

1. "Air Quality Criteria for Nitrogen Oxides," Air Pollution Control Office Publication No. AP-84, Environmental Protection Agency, Washington, D.C., 1971.
2. Cadle, R. D. and E. R. Allen: *Science*, vol. 167, p. 243, 1970.
3. Schuck, E. A., J. N. Pitts, Jr. and J. K. S. Wan: *Air and Wat. Pollut. Int. J.*, vol. 10, p. 689, 1966.
4. Nicksic, S. W., J. Harkins, and L. J. Painter, *Air and Wat. Pollut. Int. J.*, vol. 10, p. 15, 1966.
5. Allen, E. R. and K. W. Bagley: Ber. Bunsengesellschaft für Physik. Chemie, vol. 72, p. 227, 1968.
6. "Air Quality Criteria for Photochemical Oxidants," National Air Pollution Control Administration Publication No. AP-63, U.S. Public Health Service, Washington, D.C., 1970.
7. Anonymous, *Environ. Sci. Tech.*, vol. 5, p. 503, 1971.

3.13 NOXIOUS METALS

The particulate matter commonly found in city air consists of a great number of chemical substances, including those metals normally present in the earth's crust and a number of metals generally referred to as "trace metals." No rigorous definition of trace metals can be given; but generally, they will constitute less than 1% of the weight of the total particulate matter and frequently will be present in the part per million range. Nearly any metal may be found in particulate matter if a sufficiently large sample is obtained and concentrated. But certain metals are of unique interest inasmuch as they are known to be toxic or in some manner harmful to man or to other forms of life at sufficiently elevated concentrations (Table 3.13a). A complete list of trace metals would include lead, beryllium, cadmium, nickel, vanadium, arsenic, antimony, chromium, mercury, selenium and a number of others. But of necessity, this discussion will be limited to several that are perhaps the most important and are not elsewhere discussed. Lead is undoubtedly the single most important trace metal present in the atmosphere, and extra emphasis will be placed upon the discussion of this metal.

With some exceptions, all the metals exist as part of the particulate matter and are usually present as compounds rather than the free metal itself. Most analytical procedures do not determine the compounds present, however, and instead yield information solely concerning the amount of metal present, expressed as though it were in elemental form. It is of considerable importance to know the forms in which the metals are present; for the chemical nature of these substances determines, to a large extent, their solubility and hence, to some extent, their physiological activity.

The principal physical attribute of airborne particles most influencing their behavior is their size and, to a lesser extent, shape and density. In general, suspended particulate matter consists of particles which are at most a few microns in diameter and range downward to submicron size. When analyses of the total particulates are performed on fractions which have been separated according to their size, however, the results show that not all metals are distributed equally in each size fraction. In Table 3.13b, several metals are seen to have a median size of approximately 1 to 4 μ, while lead is found to be much finer and to appear mainly in the fraction containing particles

Table 3.13a
SELECTED TRACE METALS FOUND IN AIR

Metal	*Approximate Concentration Range, U.S. Cities, $\mu g/M^3$*	
Arsenic	0–0.75	(av. 0.02)
Beryllium	0–0.008	(av. < 0.0005)
Cadmium	0–0.3	(av. 0.002)
Chromium	0.0–0.35	(av. 0.015)
Lead	0.1–5	(higher in traffic)
Manganese	0.01–10.0	(av. 0.10)
Nickel	0.01–0.20	(av. 0.032)
Vanadium	0.0–1.4	

around 0.2 to 0.4 μ in diameter.[1] Data of this kind often yield information concerning the origin of the metals of interest, as well as suggesting something concerning their physiological activity.

The principal reason for concern with trace metals in the air is their possible impact on human health. There is an abundance of information concerning the activity of metals at elevated concentrations, and occupational diseases such as lead poisoning and cadmium poisoning are well known. With occasional exceptions, however, the concentrations of metals found in com-

Table 3.13b
AVERAGE METAL CONCENTRATION
AND SIZE DISTRIBUTION

Metal	*Cincinnati, September 8–23, 1966*		*Fairfax, February 3–16, 1967*	
	Concentration, $\mu gm/M^3$	*MMD, μ*	*Concentration, $\mu gm/M^3$*	*MMD, μ*
Fe	3.12	3.7	1.15	1.4
Cd	0.08	3.1	0.02	10[a]
Mg	7.21	4.5	0.42	7.2[a]
Cr	0.31	1.5	0.28	1.9
Pb	2.78	0.18[a]	0.69	0.42[a]
Cu	0.19	1.2	0.04	Not measured
Total particulate	Not measured	Not measured	44.7	1.2
SO_4	10.0	0.42[a]	7.2	0.42[a]

[a]Estimated

munity air are far below those required to produce disease in this fashion, and it is usually conjectural whether or not they exert any adverse effects on the health of man or animals. Studies are always complicated by the fact that the principal intake for most metals is usually the diet rather than the air, and dietary variations may completely camouflage the effects of inhaled metals on the resulting body burdens. The attitude prevails, nevertheless, that any unnecessary additions of trace metals to the body burden is undesirable; and although proof of damage may be difficult, every effort should be made to minimize levels of exposure.

Lead

Sources of Lead in the Atmosphere

Lead is a useful metal which has served mankind for thousands of years in many ways and has, therefore, found its way into the environment to a much greater extent than most metals. Many industrial uses of lead such as the manufacture of pigments for paints, the manufacture of storage batteries, the melting of brass, the reclamation of metal from scrap and the smelting and refining of ores have contributed and continue to contribute lead to the atmosphere and to the general environment. From the standpoint of air pollution, however, the major source of lead in the atmosphere is the use of leaded fuels by automobiles and other gasoline-powered vehicles. The amount of lead used in the manufacture of alkyl lead compounds such as tetraethyl lead is approximately 250,000 tons per year; and unlike many other lead products, most of the metal is ultimately discharged to the atmosphere from the exhaust pipes of the vehicles consuming leaded gasoline. As a result, lead can be found almost anywhere in the world, and atmospheric levels will generally correlate with the intensity of vehicular traffic. The air of most large cities will contain 1 or 2 $\mu g/M^3$ of lead at most central locations, and a city such as Los Angeles may average in excess of 3 $\mu g/M^3$ (Table 3.13c).[2] If the measurements are made very close to freeways or in vehicles in moving traffic, then substantially higher values such as 5 to 25 $\mu g/M^3$ are commonly found.

Table 3.13c
LEAD CONCENTRATIONS IN "THREE CITIES STUDY"
(Cincinnati, Philadelphia and Los Angeles)

Type of Measurement	*Concentration, $\mu g/M^3$*
Annual average	1–3
Highest monthly concentration	3.1–6.4
Highest single sample concentration	6.4–11.4
Mean concentration, heavy traffic	14–25

When it is realized that total particulate levels may average from 100 to 200 $\mu g/M^3$ at similar sampling locations, it can be seen that lead is sometimes present in high percentages, making it a relatively concentrated trace metal.

Unlike most metals in the air, much of the lead is present as a mixture of lead halides and, to a lesser extent, lead oxide, sulfate or phosphate. The lead halides are formed because organic chloride or bromide compounds are commonly added to gasoline to prevent lead deposits in the engine. Although some of these compounds are known to be rather insoluble, it has been demonstrated that a relatively high proportion of the particles are retained by the lung once inhaled.

Effects on Human Health

Lead at elevated levels produces the occupational disease known as lead poisoning; and when ingested by small children, produces very serious cases of pediatric poisoning. In the case of occupational exposure, the first symptoms tend to be somewhat nonspecific and usually involve disturbances to the gastrointestinal system as made evident by stomach pains and numerous other symptoms. With more severe exposures, the neuromuscular system is affected and very serious consequences can ensue. Children who eat small quantities of paint containing lead may become quite ill and, in severe cases, irreversible brain damage occurs.

None of these effects are known to be associated with low-level exposure by the general population. Instead, there is considerable uncertainty whether any adverse effects result from the constant inhalation of air having a concentration of several micrograms of lead per cubic meter. There is good evidence that the body burden of lead, as indirectly measured by blood or urine levels, is increased slightly in residential areas where lead levels are elevated, but the importance of this finding in regard to health remains controversial (Table 3.13d).[3] One consequence of elevated lead intake is a measurable effect on certain biochemical transformations involved in the body's synthesis of hemoglobin. It has long been known that persons overexposed to lead occupationally may be anemic or may have elevated numbers of certain forms of blood cells. Studies have shown that lead interferes in several ways with the synthesis of heme and that specifically a substance known as delta-aminolevulinic acid, or ALA, is found in elevated levels in urine as a result of the action of lead on the enzyme which normally transforms ALA into other products. Some evidence indicates that even the very low concentrations of lead in the general atmosphere influence the activity of this enzyme to some extent; and on this basis, it has been stated that atmospheric levels of lead in some cities are excessive. Opinions differ on the importance of ALA variations, inasmuch as even the slight reductions in enzyme activity which have been found do not necessarily constitute any health impairment and changes are entirely within what are usually considered normal limits.

Table 3.13d
BLOOD LEAD LEVELS AND ESTIMATED AMBIENT AIR AND OCCUPATIONAL EXPOSURES OF SELECTED POPULATION BY SMOKING HABIT[a]

Type of Population	*Estimated Exposure, μg/M³*	*Male, μg/100 g blood*			*Female, μg/100 g blood*		
		Smoker	*Undifferentiated*	*Nonsmoker*	*Smoker*	*Undifferentiated*	*Nonsmoker*
Community population without occupational exposures							
Remote California mountain residents	0.12	—	12	—	—	9	—
Composite rural U.S.	0.3	—	16	—	—	10	—
Suburban Philadelphia	1.0	15	—	11	16	—	11
Composite urban U.S.	1.4	—	21	—	—	16	—
L.A. aircraft workers	1.9	22	—	18	19	—	14
Pasadena city employees	2.0	22	—	16	—	—	—
Downtown Philadelphia	2.5	25	22	18	—	—	18
Population occupationally exposed[b]							
Cincinnati policemen	2.8	25	—	23			
Philadelphia policemen	4.6	—	30	—			
L.A. traffic policemen	5.3	22	—	20			
Cincinnati auto test lane inspectors	5.3	—	30				
Cincinnati garage workers	7.3	—	38				
Boston Sumner Tunnel employees	15.8	—	30				

[a] U.S. Dept. of Health, Education and Welfare
[b] Estimated exposure levels are based on weighted hourly averages, assuming 8-hr occupational and 16-hr ambient level exposures. No data for estimating occupational exposure of Philadelphia policemen are available.

Lead is toxic to other animals as well as man; and air polluted by excessive quantities of lead has produced sickness in cattle, horses and other domestic animals. The illness does not come from inhalation of lead in the air, however, but rather from ingestion of forage containing excessive amounts of lead deposited on the plants or ground from nearby industrial sources. There is no evidence that lead from vehicular exhaust has produced such effects on animals; but in several locations where lead ores have been refined, such episodes have been reported.

Plant life is not usually affected by lead directly, inasmuch as plants seem to regulate the amount of lead taken up from the soil. It is possible, however, for plants to contain elevated levels of lead on both internal and external surfaces; but the resulting hazard arises from the ingestion of the plants by animals and not from damage to the plant itself.

Air Quality Standards

As a result of the known and suspected action of lead on human health, there has been a movement to remove the lead from gasoline in order to lower atmospheric levels. No general agreement exists as to what atmospheric levels are tolerable, and numerous standards ranging from 10 to 1 $\mu g/M^3$ have been suggested. There is no federal ambient air quality standard for lead at present; but on the basis of actions taken in California and elsewhere, it would appear probable that such a standard might be as low as 1 or 2 $\mu g/M^3$. In the event such a standard is promulgated, it will have the effect of making it necessary to lower the lead concentration in gasoline substantially or discontinue its use entirely. The consequences of such an action are complex and such change will be made only after extensive deliberation and study.

Beryllium

Beryllium compounds are found in the air as a result of a rather limited number of industrial activities and, in more widespread fashion, from the combustion of coal (Table 3.13e) and other nonspecific processes which produce dust. The refining of beryllium ore into beryllium metal, beryllium oxide and other useful compounds can result in the emission of substantial quantities

Table 3.13e
METALS COMMONLY FOUND IN FLYASH FROM COAL OR OIL

Principal constituents, not trace metals		*Trace metals*	
Silicon	Calcium	Nickel	Mercury
Aluminum	Magnesium	Beryllium	Selenium
Iron		Vanadium	Manganese

of beryllium into the air. In the past, such emissions did occur at several locations in the United States; but at present, all such plants are operating under very stringent controls and emissions are minimal. Unlike lead, beryllium is not always found in the particulate matter from cities; and when found, it is present at much lower concentrations. Thus, the average atmospheric concentration for one hundred stations in the United States has been reported at less than 0.0005 $\mu g/M^3$, with a maximum value of 0.008 $\mu g/M^3$. Presumably, these very low levels represent the contribution from coal burning and other industrial emissions which may contain traces of beryllium.

Properties

Beryllium is one of the lightest of metals, and this property makes it very attractive for certain selected industrial usages. But its most important property as an air pollutant is the chemical form in which it is found. The ore beryl and certain other highly insoluble forms of beryllium are thought to be less toxic than the metal and certain other salts. The U.S. Environmental Protection Agency has classified beryllium as a hazardous substance and has published emissions standards designed to prevent ambient air levels in excess of 0.01 $\mu g/M^3$.

Health Effects

Occupational exposure to beryllium dust has produced the disabling lung disease called "berylliosis"; and although similar in many ways to the lung disease silicosis, it has been shown that the amount of beryllium necessary to produce berylliosis in some persons is far less than the quantity of silica necessary to produce silicosis. Evidence has also been reported that non-occupational exposure to beryllium compounds resulting from living near beryllium emitting industries or living in the home of beryllium workers has produced the disease, with the result that beryllium is one of the few toxic metals which has been directly implicated in causing illness at elevated environmental levels.

Although beryllium is toxic to experimental laboratory animals and has been rather extensively studied in this regard, it is not ordinarily a source of damage to farm animals as a result of excessive intake from the air or forage.

Air Quality Standards

The experience with neighborhood cases of beryllium poisoning led to the establishment of an ambient air quality standard by the Atomic Energy Commission long before the modern concept of establishing ambient air quality standards had been established. On the basis of epidemiological studies and other considerations, the Atomic Energy Commission adopted standard has been 0.01 $\mu g/M^3$, averaged over a period of 1 month. Higher concentrations are permissible for shorter periods of time, within certain limitations, but the monthly average is the basic standard.

3.13 Noxious Metals

Cadmium

Relatively little or no cadmium is found in community air. In a recent survey of 58 cities in the United States, particulates from thirty-six cities were found to contain cadmium in concentrations ranging from 0.002 to 0.370 $\mu g/M^3$, while the rest did not contain cadmium. Measurements made by the National Air Sampling Network of the U.S. Public Health Service showed that the average atmospheric concentration of cadmium in 1964 was 0.002 $\mu g/M^3$ with a maximum recorded value of 0.350 $\mu g/M^3$. There are relatively few sources of cadmium and, in general, it tends to be associated with the emissions of zinc. Most cadmium is found naturally as a mineral mixed with zinc, and the processing and subsequent usage of zinc will generally result in cadmium emissions along with the zinc. There are several important uses for cadmium such as the manufacture of alkaline storage batteries, the coating of some metallic parts with a cadmium layer, the production of certain paint pigments and the formulation of solders and other cadmium-containing alloys. The manufacture or use of all these substances may contribute to cadmium emissions, but they are relatively limited in magnitude and do not constitute a large scale air pollution problem.

Health Effects

The presence of any cadmium in the air or in the general environment is of concern, however, for the metal is decidedly toxic at elevated concentrations, and there is evidence that it is injurious in a different way at very low levels. The most serious consequence of occupational exposure to cadmium oxide fumes is a severe lung injury generally considered to be a chemical pneumonia, which can prove fatal after exposure to a high concentration of the fume such as is generated by welding on cadmium plated metal. Chronic exposure to low concentrations of cadmium has been shown to result in damage to the kidney, commonly detected by the presence of a certain kind of protein in the urine of exposed workers. From the standpoint of exposure to cadmium in the general environment, however, neither the pneumonia nor the kidney damage is likely to occur; but investigations in connection with the effects of trace metals on experimental animals have shown a correlation between the cadmium content of animal kidneys and hypertension. Similar observations have been made in respect to humans, and significant correlations have been reported between heart disease and cadmium concentrations in the air. Although cadmium is also toxic to animals and possibly to plants, the amounts found in the air have not been reported to cause effects on either animals or plants. There is no established ambient air quality standard for cadmium; but in all likelihood, when one is established, it will be similar in magnitude to that for lead, or possibly lower.

Nickel

Large quantities of nickel are used annually in the production of steel, particularly stainless steel in which nickel is a major component along with chromium. The emissions from steel-making plants, therefore, may contain nickel; and, in addition, other industries such as the nickel-plating industry also emit nickel compounds to the air. A more important source of nickel in community air, however, is the emissions from coal or oil burning power-generating installations; for certain coal ashes have been shown to contain nickel ranging from 3 to 10,000 ppm. Oil may also contain nickel, and burning heavy residual fuel oils may yield an ash which is rich in nickel. Some oil ashes have been reported to contain in excess of 10% nickel oxide. It is not surprising, therefore, that particulate matter usually contains some nickel. A mean value for a 4-year period, with measurements made in 30 United States cities, was 0.045 $\mu g/M^3$. Little is known about the chemical or physical forms of atmospheric nickel; and in all likelihood, nickel exists as a small component of particles such as flyash which are made up of many other elements (Table 3.13e).

Health Effects

Nickel compounds are toxic at levels much higher than found in community air, causing skin problems and other disorders. But of the greatest interest in respect to air pollution is the ability of some compounds of nickel to cause cancer in experimental animals. One of the most toxic compounds of nickel is actually a gas, nickel carbonyl; this compound is intensely toxic, causing severe lung damage and death at relatively low concentrations. At even lower levels, laboratory animals chronically exposed to nickel carbonyl developed lung cancer. It has also been observed that nickel workers tend to have a higher than expected incidence of pulmonary and nasal cancer.

It is understandable, therefore, that some concern has been expressed about the possibility of increased lung cancer in the general population resulting from exposure to atmospheric levels of nickel. However, at present, there is no evidence to support this contention. In general, although lung cancer is known to be more prevalent in urban populations than in rural populations, it has not been possible to determine which, if any, of many air pollutants are responsible for the phenomenon.

It is probable that air quality standards for nickel will eventually be established, and they can be expected to be quite restrictive, in view of the unknown cancer-producing potential of this metal. The gradual reduction of emissions of flyash to the atmosphere as part of the overall air pollution control effort will, of course, tend to reduce nickel levels along with levels of other substances generally found in flyash.

Vanadium

In common with nickel, vanadium compounds are emitted to the air in substantial quantities as a result of burning fossil fuels for power production. Many coals have been shown to contain vanadium in concentrations ranging from 16 to 176 ppm, while the ash resulting from combustion of such coal can contain as much as 1,000 ppm of vanadium. Some crude oils contain even more vanadium than does coal. In one reported instance, the ash from a residual oil was found to contain 63% vanadium pentoxide. It is not surprising, therefore, that the particulate matter in the air of cities contains vanadium; and concentrations greater than 1 microgram per cubic meter have been frequently found, although on the average only a fraction of a microgram per cubic meter is found.

Health Effects

In common with all metals, excessive concentrations of vanadium can cause adverse health effects; and high levels of vanadium pentoxide, in particular, produce irritation of the lungs, the eyes and the upper respiratory system. Vanadium is also very much involved in some biochemical reactions involving the synthesis of cholesterol and certain other important biological substances. There is no evidence, however, that ambient levels of vanadium exert any adverse effect on human health, and no ambient air quality standards have been proposed for vanadium or its compounds.

Manganese

Manganese is a rather abundant metal for which there are many industrial uses. It is one of the principal alloying elements in the manufacture of steel and most of the industrial emissions of manganese arise from steel-making or from the manufacture of ferromanganese, which is required in making steel. Manganese is also found in coal and oil so that flyash from burning these fuels may contain substantial concentrations. Emissions from coal-fired power plants have been measured to be as high as 60 to 400 $\mu g/M^3$, while some residual fuel oils have produced ashes containing as much as 55 mg/M^3. Samples of particulate matter collected by the National Air Sampling Network usually contain manganese at concentrations ranging from several hundredths of a microgram per cubic meter to more than 10 $\mu g/M^3$, and daily averages of several micrograms per cubic meter are not uncommon (Table 3.13e).

Health Effects

Manganese is considered to be a toxic metal and is known for its neurotoxicity, or ability to damage the nervous system. Occupational exposure to manganese has been described many times in workers engaged in mining and refining manganese. A variety of symptoms resulting from damage to the nervous system are commonly found. These include such general effects as weakness, sleepiness, muscular twitching and emotional disturbances. Inhalation of manganese compounds, commonly manganese dioxide, has also been shown to cause a particular kind of pneumonia; and this disease has actually occurred as a result of excessive air pollution. A number of years ago in Norway, a plant discharged very large quantities of manganese dioxide into the air and studies demonstrated that the incidence of pneumonia in the area was substantially greater than would have been expected. When analyses of lung tissue from deceased residents in the area were conducted, they also contained considerably more manganese than normal.

It appears that these levels were extremely high, and the much lower ambient levels reported as occurring in United States cities are not presently known to result in any adverse health effects. No air quality standard for manganese has been proposed or established to date.

Chromium

Chromium is also a major alloying element in the production of steel, and the steel industry is a major source of chromium emissions to the atmosphere. Another of the most common industrial uses of chromium is plating steel to obtain the "chrome plated" appearance. Chromium plating is performed in a very strong acidic solution containing hexavalent chromium compounds, and the resultant mist which forms is intensely irritating and corrosive (Sec. I. 7.20). Whenever plating shops emit such a mist, there are certain to be complaints from those immediately adjacent concerning damage to property or fear of damage to health. However, such plants do not constitute a major source of chromium in the general atmosphere; and chromium levels in the particulate matter are usually quite low, averaging slightly more than 0.01 $\mu g/M^3$ over a several year period. The highest reported value in that period was 0.35 $\mu g/M^3$.

Health Effects

Hexavalent chromium compounds from plating operations are irritating and toxic and, in general, all compounds of chromium are very likely toxic. Some compounds of chromium have been shown to be carcinogenic to animals; and in addition, chromium workers display a higher incidence of lung cancer than found in the general population. It is not known whether ambient levels of chromium contribute in any way to the incidence of lung cancer in the general population, and there is no general evidence that ambient levels of chromium exert any known effects on health. No air quality standard has been set for chromium; but on the basis of concern for its carcinogenic potential, it is probable that an effort will be made to keep chromium at the lowest levels possible.

REFERENCES

1. Lee, R. E., Jr., R. K. Patterson and J. Wagman: Particle-Size Distribution of Metal Components in Urban Air, *Environmental Science and Technology*, vol. 2, no. 4, p. 289, 1968.
2. The Working Group on Lead Contamination: "Survey of Lead in the Atmosphere of Three Urban Communities," U.S. Department of Health, Education and Welfare, 1965.
3. The Working Group on Lead Contamination: "Survey of Lead in the Atmosphere of Three Urban Communities," U.S. Department of Health, Education and Welfare, 1965.

BIBLIOGRAPHY

Litton Systems, Inc., Reports: "Air Pollution Aspects of Beryllium and Its Compounds," "Air Pollution Aspects of Cadmium and Its Compounds," "Air Pollution Aspects of Chromium and Its Compounds," "Air Pollution Aspects of Manganese and Its Compounds," "Air Pollution Aspects of Vanadium and Its Compounds" and "Air Pollution Aspects of Nickel and Its Compounds," prepared for the National Air Pollution Control Administration, Consumer Protection and Environmental Health Service, DHEW, compiled by N. L. Durocher, Y. C. Athanassiadis and Ralph J. Sullivan, September 1969.

Patty, F. A. (ed.): "Industrial Hygiene and Toxicology," vol. II, Wiley, New York, 1962.

Stern, A. C. (ed.): "Air Pollution," vols. I, II and III, Academic Press, New York, 1968.

3.14 OXIDANTS

Sections dealing with various aspects of oxidants are as follows: Section 2.7, Weather; 2.8, Photochemical Smog; 2.11, Health; 3.2, Aldehydes; 3.5, Carcinogens; 3.9, Hydrocarbons; 3.12, NO_x; 5.3, Threshold Limit Values.

Among the photochemical oxidants, ozone is the most important. Within this group are included:

1. Ozone
2. Free radical oxygen forms
3. Peroxyacetyl nitrate (PAN) homologues
4. Oxides of nitrogen

Photochemical oxidants play a key role in the development and formation of the photochemical ("Los Angeles" type) smog with attendant deleterious health effects, plant damage and reduction of visibility. Hence, the current moves to control these oxidants in the atmosphere.

Types and Sources

Oxidants are formed in the atmosphere by photochemical reactions involving nitrogen oxides, organic vapors and solar radiation. In the Los Angeles basin at 8 A.M. on a typical day, both motor vehicle traffic and emissions of hydrocarbons, NO_x, and other pollutants reach a maximum. Meteorological conditions often prevent quick dispersion and dilution of these pollutants (Fig. 6). As time passes, solar radiation acts on these chemicals to produce ozone and then photochemical aerosols that comprise smog.

Photochemical reactions in the atmosphere can only be initiated by molecules that can absorb solar radiation. The list of absorbing and non-absorbing chemicals is given in Table 3.14a.

Formation of nitric oxide occurs in any air combustion process, automobiles being the major source at the present. Equilibrium concentration in the combustion product gas is a function of temperature (Table 3.14b).

The primary photochemical reaction consists of NO_2 absorbing solar radiation in the 4000 Å wavelength range (Fig. 3.12a) to form highly reactive oxygen atoms [Equation 3.12(3)]. The atomic oxygen produces ozone in accordance with the reaction given in Equation 3.12(4). The secondary reac-

Table 3.14a
ABSORPTION OF SOLAR RADIATION

Absorbing Molecules[a]	*Nonabsorbing Molecules*[a]
O_2	N_2
O_3	H_2O
NO_2	CO
SO_2	CO_2
HNO_2-HNO_3	NO
Particulates	SO_3
*R*CO	*R*CH
*R*COO	*R*COOH

[a] *R* represents any radical.

tions of ozone and oxygen atoms are fairly complex and can produce aldehydes, ketones, alkyl and acyl free radicals, CO, H_2O, epoxides, acids and alcohols.

Peroxyacetyl nitrates (PAN) are presumably formed by the following reactions:

$$O_3 + RH \longrightarrow R'CHO + R'COO \quad 3.14(1)$$

$$R'COO + NO \longrightarrow R'CO + NO_2 \quad 3.14(2)$$

$$R'CO + O_2 \longrightarrow R'COOO+ \quad 3.14(3)$$

$$R'COOO + NO_2 \longrightarrow R'COOONO_2\ (PAN) \quad 3.14(4)$$

where RH is a hydrocarbon and R is an aromatic or aliphatic radical.

The oxidants do not have identical concentrations in each photochemical smog.[1,2]

Table 3.14b
NO CONCENTRATION IN COMBUSTION PRODUCTS,
$N_2 + O_2 \rightleftharpoons 2\,NO$

Temperature, °F	*Equilibrium Concentration of NO,* ppm
4400	25,000
2050	3,700
850	0.30
70	0.009

Effects

Injury to vegetation caused by photochemical smog has been found in many areas.[3] Ozone produces stipple on grape and fleck on tobacco. PAN is considered to be a major phytotoxicant causing undersurface silvering or bronzing of leaves. Field exposures of 0.02 to 0.05 ppm of oxidants cause damage to vegetation in 4 to 6 hours.[4]

Ozone and oxidants cause major damage to the respiratory tract in animals. Oxidants reduce the respiratory function in animals directly after exposure, and chronic injury to lungs is produced by exposure to a 1 ppm concentration. Also, ozone denatures protein, and oxidants affect visual acuity.[4,5]

Figure 6.33f provides detailed data concerning the toxic effects of and human tolerance for ozone. Oxidants and ozone are easily detectable in the 0.02 to 0.05 ppm range by their odor. Even at 0.05 ppm, irritation of nose and throat are reported. Over 0.2 ppm, visual acuity is seriously impaired. At levels of 0.8 ppm onward, pulmonary congestion is reported.[6] PAN has been shown to be a major factor in eye irritation caused by photochemical smog.

Considerable study has been made on the effects of mixtures of ozone and other oxidants. It is reported that acute effects of ozone are slightly reduced by NO_2 (Section 3.12).

Air Quality Standards

National primary and secondary standards have been published for oxidants.[7] These define the permissible level of all photochemical oxidants at a maximum 1 hour concentration of 160 $\mu g/M^3$. These standards do not distinguish between the various individual photochemical oxidants. (For individual threshold limit values see Section 5.3.) Neutral potassium iodide colorimetry can be used to measure all the photochemical oxidants and the results are usually reported as equivalent ozone concentration. Ozone sensors are described in Section 4.14. Peroxyacyl nitrates and nitrites are measured by infrared[8] and gas chromatography[9] techniques.

REFERENCES

1. Leighton, P. A.: "Photochemistry of Air Pollution," Academic Press, New York, 1961.
2. Altshuller, A. P. and J. J. Bufalini: *Photochemistry Photobio.*, vol. 34, p. 245, 1965.
3. Middleton, J. T. and A. J. Haagen-Smit: *J. Air Pollution Cont. Ass.*, vol. 11, p. 125, 1961.
4. Stern, A. C. (ed.): "Air Pollution," Vol. I, Chapter 12, 13, Academic Press, New York, 1968.
5. Scheel, L. D., O. J. Dobrogoski, J. T. Mountain, J. L. Svirbely and H. E. Stokinger: *J. Appl. Physiol.*, vol. 14, p. 67, 1959.
6. Griswold, S. S., L. A. Chambers and H. L. Motley: *Arch. Ind. Health*, vol. 15, p. 108, 1957.
7. "National Primary and Secondary Air Quality Standards," *Federal Register*, April 30, 1971.
8. Lipták, B. G. (ed.): "Instrument Engineers' Handbook," Vol. I, Section 8.3, Chilton, Philadelphia, 1969.
9. Ibid, Section 8.1.

3.15 PARTICULATE MATTER

Particulate matter (also called aerosols) present in the atmosphere may be organic or inorganic in composition and in liquid or solid physical state. Particulate matter in the atmosphere is known to have many adverse effects upon health and welfare, and reduction of emissions of this pollutant is of prime importance. These particles are of very small size. Among the most common are carbon or soot particles, oily or tarry droplets, acid droplets, metallic oxides and salts, silicates and other dust and metallic fumes. The particulate matters in the atmosphere represent a chemically diverse class of substances; however, they do have in common a number of physical properties. Figure 3.15a provides correlation between particle sizes and nomenclature for aerosols.[1]

Definitions[1]

Particulate Matter	Any material, except uncombined water, in the form of solid or liquid in the atmosphere or in a gas stream.
Aerosol	Solid or liquid particles of microscopic size, such as smoke, fog or mist.
Dust	Solid particles larger than colloidal size and capable of suspension in air and other gases.
Flyash	Particles formed as a result of combustion of fuel.
Fog	Visible aerosols where liquid is the dispersed phase.
Fume	Particles formed by condensation, sublimation or chemical reaction, e.g., condensed metal oxides.
Mist	Low concentration dispersion of relatively small liquid droplets.
Particle	Small, discrete mass of solid or liquid matter.
Smoke	Small particles resulting from incomplete combustion of combustible material.
Soot	An agglomeration of tarry carbon particles formed by incomplete combustion of carbonaceous material.

(For more detailed definitions refer to the Glossary.)

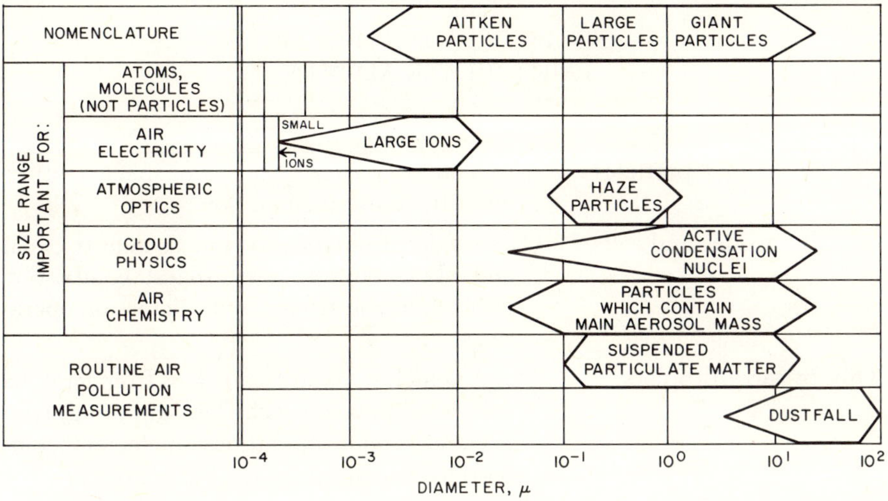

Figure 3.15a Sizes of atmospheric particulate matter and ranges of particle size (diameter) of various types of particulate matter which are found in the earth's atmosphere.

Sources

Approximately 12 million tons per year of particulate matter are emitted into the atmosphere. Table 3.15b summarizes the major sources or particulate matter and the quantities emitted in tons per year.[2] The most important sources of particulate matter emissions by type of particulate matter are summarized in Table 3.15c.

Table 3.15b
SOURCES OF PARTICULATE MATTER AND QUANTITIES EMITTED

Source	*Quantities, millions of tons annually*
Industrial	6.0
Power generation	3.5
Incineration	1.0
Space heating	1.0
Mobile	0.5
Total	12.0

Table 3.15c
SOURCES AND TYPES OF PARTICULATE MATTER

Type	Sources
Carbon or soot particles	Automobiles, power generation, space heating, aircraft, refuse incineration
Metallic oxides and salts	Catalyst dust from refinery; metals industry, coal and oil combustion; auto exhaust, combustion of black liquor, drying and calcination operations
Oily or tarry droplets	Motor vehicles, exhaust emissions from gasoline and diesel powered vehicles; saturating, paving and roofing operations; inefficient combustion; aircraft, ships and locomotives; refuse incineration
Acid droplets	Power plants, metal working and plating operations, storage battery reclamation, waste rendering and incineration, acid plants
Silicates and other inorganic dusts	Quarrying operations, sand and gravel plants
Metallic fumes	Metal industry

Combustion Sources

Combustion of coal and oil results in emission of millions of tons of particulate matter, mostly in the form of flyash. Refuse burning and incineration produces dust, fume, smoke, flyash and like matter.

Power generation using coal and oil as fuel is the major source of emission of particulate matter.[3,4,5] The ash content in coal varies from a low of 7.5% to a high of 22%. Approximately 250 million tons of coal per year is used to produce electric energy, and this requirement is expected to increase to over 600 million tons per year by 2000. In a balanced draft boiler, 85% of the ash is released with the flue gas from the boiler. Assuming average ash content of 15% in coal, the yearly emissions from boilers without any collection equipment is 31,875,000 tons. A 99% efficient dust collection system will reduce the emission to 318,750 tons/year and a 99.9% emission control will result in an emission of 31,875 tons/year. The dust loadings in the flue gas leaving the boiler will range from 2 to 13 gr/SCF of dry gas. Figure 3.15d shows the estimated size distribution for particles emitted from pulverized fuel-fired furnaces.[4]

Fuel oil (Bunker c or grade 6) is used in boilers with steam capacities of 34,500 lb/hr or greater, and the ash content of this oil varies from a low of 0.002% to a high of 0.3% by weight. The particulate loading of stack gases depends primarily upon combustion efficiency and on the rate of build-up of boiler deposits. The dust loading in the flue gas ranges from 0.02 to 0.06 gr/SCF of dry gas. The size distribution of particles emitted depends upon the degree of atomization of oil, mixing efficiency, flame temperature, firebox design and flue gas path. The particulate emission from oil fired boilers is insignificant compared to that from coal fired boilers.

Incineration is one of the most prominent means of solid-waste volume reduction and it ranges from large municipal multiple-chamber incinerators to small domestic contrivances.[6,7] The refuse burning, both in incinerators and in dumps, produces approximately 1 million tons of particulate matter per year. Although refuse burning creates less than 15% of the total particulate matter emissions, such emissions are concentrated in populated areas. The particulate emission factors and particle size distribution for uncontrolled incinerators are presented in Table 3.15e.[1,2]

Space heating contributes approximately 1 million tons per year to the

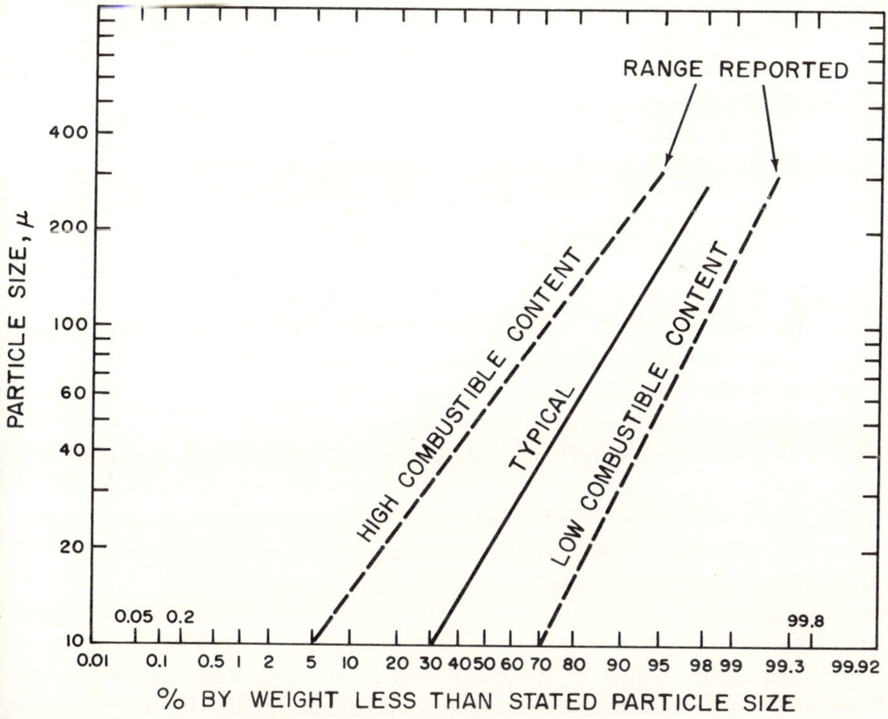

Figure 3.15d Estimated size distribution for particles emitted from pulverized-fuel-fired furnaces (before collectors).

Table 3.15e
PARTICULATE EMISSION FACTORS AND PARTICLE SIZE DISTRIBUTION FOR REFUSE INCINERATORS WITHOUT POLLUTION CONTROLS

Type of Incinerator Unit	Particulate, lb/ton of refuse	Particle Size Distribution, %				
		>44μ	20–44 μ	10–20 μ	5–10 μ	<5 μ
Municipal (multiple chamber)	17	40	20	15	10	15
Commercial (multiple chamber)	3	40	20	15	10	15
Commercial (single chamber)	10	40	20	15	10	15
Flue-fed	28	40	20	15	10	15
Domestic (gas-fired)	15	40	20	15	10	15

total emission of particulate matter. Space heating is a major source of low-level emission of pollutants. In this category, apartment houses and commercial and industrial establishments are included. In most cases, the control of particulate emission is impractical or costly. The practical solution to effectively control emissions is to use fuels such as low sulfur oil and natural gas.

Industrial Sources

Approximately 6 million tons of dust, fume and mist are emitted from industrial sources per year. Table 3.15f lists many of the industries that release large quantities of particulate matter. These include dust, oil and tar mists, acid mists and fumes.

In iron and steel mills, blast furnaces, open hearth, basic oxygen and electric furnaces and sintering plants are major sources of particulate emissions. Table 3.15g summarizes emissions from steel mills, with and without pollution controls being used.[7] Table 3.15h provides a summary of sources, types and quantities of particulate matter emitted from various metallurgical processes, including steel and iron-making processes, secondary brass and bronze-melting processes, secondary aluminum and zinc-melting and lead-refining processes.

In petroleum refineries, the major sources of particulate matter are fired heaters, boilers, catalyst regenerators, asphalt stills and sludge burners. Catalyst dust losses from regenerators equipped with cyclones can be as high as 100 to 350 lb/hr. The loss depends on the size, age and design basis of the unit.

With the use of efficient precipitators, the emissions can be reduced to as low as 40 to 60 lb/hr. The airblowing of asphalts releases oil and tar mists. The sludge (mixture of heavy residues, clay, sand and acids), when incinerated, releases the inorganic matter in the form of dust.

The mineral products industries include the processing of nonmetallic

Table 3.15f
INDUSTRIAL SOURCES OF PARTICULATE EMISSIONS

Industry	*Particulate Emissions*	
	Type	*Major Sources*
Iron and steel	Iron oxide dust and smoke	Blast furnaces, other steel-making furnaces, sintering machines
Gray iron foundries	Iron oxide dust, smoke, oil, grease, metal fumes	Cupolas, shakeout systems, core making
Nonferrous smelters	Smoke, metal fumes, oil and grease	Smelting and melting furnaces
Petroleum refineries	Catalyst dust, ash, acid mist, liquid aerosols	Catalyst regenerator, air-blowing of asphalt and sludge incineration
Cement	Alkali and cement dust	Kiln, dryers, coolers, materials handling
Pulp mills	Chemical dusts, Na_2CO_3, Na_2SO_4, Na_2S, CaO, $CaCO_3$, mists	Recovery furnace, dissolving tank, lime kiln
Asphalt batch plants	Aggregate dusts	Dryers and materials handling
Phosphoric and sulfuric acids	Mist, dust	Grinding, handling, rock acidulating, absorption towers
Coke manufacture	Coal and coke dusts, coal tars	Oven cells, quenching, materials handling
Glass furnace and glass fiber	Acid mist, dust, alkaline oxides, resin aerosols	Materials handling, glass furnaces, fiber forming and curing
Carbon black	Carbon black	Generators
Coal cleaning	Coal dust	Coal dryers
Gypsum processing	Dust	Calciner, dryer, grinding and materials handling
Soap and detergent	Dust	Spray dryers, materials handling
Coffee processing	Oil aerosols, coffee dust, ash	Roasters, dryers, coolers, waste heat boilers, conveying equipment
Cotton ginning	Cotton fiber, dust and smoke	Gins, incineration

Table 3.15g
EMISSIONS FROM STEEL MILLS

Operation	Before control: Stack Loading, gr/scf	Before control: lb/ton of Product	With control: Control Used[a]	With control: Stack Loading, gr/scf	With control: lb/ton of Product	With control: Approx. Efficiency, %	Approximate Volume of Gases Handled
Blast furnace	7–10	200	Preliminary cleaner (settling chamber or dry cyclone)[b]	3–6	—	60	87,000 scfm for 1000-ton/day furnace
			Primary cleaner (wet scrubber)[b]	0.05–0.3–0.7[c]	5.4	90	
			Secondary cleaner (E.S.P. or V.S.)[b]	0.004–0.008	0.1–1.4	90	
Sintering machine	0.5–3.0	5–20–100	Dry cyclone E.S.P. (in series with dry cyclone)	0.2–0.6 0.01–0.05	2 0 1.0	90 95	120,000–160,000 scfm for a 1000-ton/day machine
Sinter machine discharge crusher, screener, and cooler	6.0	22	Dry cyclone	0.4	1.5	93	17,500 scfm for a 1000-ton/day machine
Open hearth (not oxygen lanced)	0.1–0.4–2.0	1.5–7.5–20.0	E.S.P. V.S. Baghouse	0.01–0.05 0.01–0.06 0.01	0.15 0.15–1.1 0.07	98 85–98 99	35,000 scfm for a 175-ton furnace
Open hearth (with oxygen lance)	0.1–0.6–2.5	9.3	E.S.P. V.S.	0.01–0.05 0.01–0.06	0.2 0.2–1.4	98 85–98	35,000 scfm for a 175-ton furnace
Electric arc furnace	0.1–0.4–6.0	4.5–10.6–37.8	High-efficiency scrubber E.S.P. Baghouse	0.01 0.01–0.04 0.01	0.2 0.3–0.8 0.1–0.2	Up to 98 92–97 98–99	Highly variable depending on type of hood May be about 30,000 scfm for a 50-ton furnace
Basic oxygen furnace	5.8	20–40–60	V.S. E.S.P.	0.03–0.12 0.05	0.4 0.4	99 99	Varies with amount of oxygen blown; 20–25 scfm/cfm of oxygen blown
Scarfing machine	0.2–0.8	3 lb/ton of steel	Settling chamber	No data	No data	No data	85,000 scfm for 45-in., 4-side machine
Coke ovens (by-product type)	No data	0.1% of coal processed (rough estimate)	Emissions can be minimized through equipment design and operational techniques	No data	No data	No data	No data

Table 3.15h
METALLURGICAL PROCESSES—
TYPES, SOURCES AND QUANTITIES
OF EMISSIONS

Industry	*Type*	*Quantity*
Steel-manufacturing processes		
open hearth furnace	Fumes or oxides of various metals	0.10–2.0 gr/ft^3 16 lb/ton of material charged 65% below 5 μ
Electric arc furnace	Fumes, dust	5–30 lb/ton of metal charged 0.3–0.5 gr/SCF 45–75% below 5 μ
Electric induction furnace	Fumes, dust	—
Iron casting		
Cupola furnaces	Dust, fumes, smoke and oil vapor	18–25% below 5 μ 1–3 gr/SCF 4–25 lb/ton of charge
Electric arc furnaces	Dust, fumes	5–10 lb/ton of metal processed
Induction and reverberatory furnaces	—	Essentially negligible emission
Secondary brass and bronze melting processes	Dust, metallic fumes, flyash, carbon, zinc, lead and other oxides	ZnO, PbO 0.03–0.3 μ 0.2–0.7 gr/ft^3 3–25 lb/ton of metal charged
Secondary aluminum melting processes	Aluminum chloride fumes, dust, chlorides, fluorides and oxides of metals	100% below 2 μ 90–95% below 1 μ 2–30 lb/ton of metal charged
Secondary zinc melting processes	Smoke or oily mist Zn and ZnO fumes, dust and ammonium chloride	5–90 lb/ton of zinc charges
Lead refining	Smoke, fumes, oxides, sulfides and sulfates of lead, tin, arsenic, copper and antimony	0.07–0.4 μ 1.4–4.5 gr/ft^3 150–300 lb/ton of metal charged
Primary metals industry		
Aluminum ore reduction	Fluorides, alumina and carbon	0.05–0.75 μ 75–85 lb F/ton of aluminum
Copper lead and zinc smelters	Fume, dust	—

Table 3.15i
MINERAL PRODUCT INDUSTRIES—
TYPES, SOURCES AND QUANTITIES
OF EMISSIONS

Industry	*Type*	*Source*	*Quantity*
Asphalt roofing	Dust, oil mist	Asphalt air-blowing saturators	4 lb/ton of asphalt 65 lb/hr
Asphalt concrete batch plants	Dust	Dryer	5.0 lb/ton of product (cyclone outlet 35% less than 5 μ)
Calcium carbide plants	Dust	Electric furnace	1.7 lb/ton of product
		Hood	
		Coke dryer	after cyclone 0.2 lb/ton of product
		Furnace room vents	2.6 lb/ton of product
		Stack	2.0 lb/ton of product (after scrubber)
Cement plant	Dust and fumes	Kiln	15–55 lb/barrel of cement (wet process) 35–75 lb/barrel of cement (dry process) 22% less than 5 μ
Ceramic and clay processes			
ceramic clay	Dust	Dryer	15 lb/ton of charge (after cyclone)
bisque	Dust	—	2 lb/ton of charge (after scrubber)
catalytic material from clay	Dust	—	6 lb/ton of charge (after scrubber)
other processes	—	—	No data
Frit manufacture	Dust, metallic fumes	Frit smelters	16.0 lb/ton of charge 45% less than 5 μ
Glass manufacture	Dust	Melting furnace	2 lb/ton of glass 25% below 5 μ
Lime plant	Dust	Rotary kiln	5–15% by weight of lime produced
		Vertical kiln	1% by weight of lime produced 2% less than 5 μ
Rockwood manufacture	Dust	Cupola	22 lb/ton of charge
		Reverberatory furnace	5 lb/ton of charge (0.5% less than 5 μ)
	Fibers	Blow chamber	22 lb/ton of charge
		Curing oven	4 lb/ton of charge
		Cooler	2 lb/ton of charge
Perlite manufacture	Dust	Furnace	21 lb/ton of charge 32% less than 5 μ

Table 3.15j
PARTICULATE EMISSIONS
FROM KRAFT PULP MILLS

Source	*Particulate Matter*	*Quantity of Emission lb/ton of A.D. pulp*
Recovery furnace	Sodium carbonate Sodium sulfate carbon	150–200
Lime kiln	Lime and soda dust	18–20
Dissolving tank	Sodium carbonate Sodium sulfide	20

substances such as glass, rock, sand, clay and various other materials. Particulate matter is the main air pollutant from these industries. Table 3.15i summarizes the mineral product industries, the types and sources for contaminants and emission quantities.

In the pulp and paper industry, the recovery furnace, lime kiln and dissolving tank are major sources for particulate emissions. (See Table 3.15j.) The dust is mainly sodium carbonate, sodium sulfate, sodium sulfide, lime, soda and carbon. Efficient recovery of these particulate emissions is at present being attained.

In most of the operations in the chemical process industry, the main emissions are gaseous in nature. In the manufacture of paint and varnish, varnish cooking fume losses average 3 to 6% of the feed; in alkyresin production, 4 to 6%; in cooking and blowing of oils, 1 to 3%; and in heat polymerization, 1% of the feed. The fume is mainly comprised of aldehydes, ketones, phenols, terpenes and glycerine. The particle size ranges from 2 to 20 microns.

Most of the particulate matter emitted from acid manufacturing plants is from the production of sulfuric and phosphoric acids. Acid mists are the primary air pollutants. The phosphoric acid mist (particle size range from 0.4 to 2.6 μ) is formed in the thermal process. In the wet process, except for material handling and grinding operations, few particulates are generated. In the sulfuric acid manufacture, without any acid mist collection device, the emissions of acid mist (98% less than 3 μ) range from 0.3 to 7.5 lb/ton of acid produced.

The food and agricultural industries are major contributors of particulate emission because of their volumes of production. Dust is their main emission; and Table 3.15k summarizes the type, source and quantity of emissions.

Table 3.15k
FOOD AND AGRICULTURAL INDUSTRY—
EMISSION TYPE, SOURCE AND QUANTITY

Industry	Type	Source	Quantity
Alfalfa dehydrating	Dust	Primary cyclone air-meal separators	1–3.5% by weight of production Particle size 1.5–10 μ
Coffee roasting plant	Dust	Roaster	
		Direct fired	7.6 lb/ton of green beans
		Indirect fired	4.2 lb/ton of green beans
		Stoner and cooler	1.4 lb/ton of green beans
		Spray drier	1.4 lb/ton of green beans
Cotton ginning process	Trash dust lint	Ginning operation	12 lb/500 lb bale
Feed and grain mills	Grain dust	Wheat air cleaner	0.2 lb/ton of product
		Alfalfa meal mill	4.0 lb/ton of product
		Barley flour mill	3.1 lb/ton product
		Orange pulp dryer	11.3 lb/ton of product
Starch manufacture	Dust	Flash drier	8 lb/ton of starch

Construction and Demolition

The construction and demolition of buildings and roads create particulate air pollution. The handling of dusty materials, truck movements and breaking of masonry walls are a few of the major dust generating operations.

The movement of trucks across dry, unpaved roads is the major source of *road dust.* The *open burning* of wood and other combustibles results in particulate emission. The cleaning of stone and concrete surface by *sand-blasting* also creates air pollution.

The pollution caused by road dust, demolition, open burning and sand-blasting is difficult to control. But, by taking simple precautions such as spraying walls before demolition, black topping or oiling unpaved roads and stopping the practice of open burning, dust emissions will be minimized.

Mobile Sources

The emissions of air pollutants from mobile sources, namely aircraft, automobiles, trucks and buses, are similar to those from other combustion sources.

The conventional four-engined jet aircraft emits about 34 lb of particulate matter per flight (including takeoff and landing) below 3,500 feet. The emissions from other types of aircrafts, namely four-engine fan jet, two-engine and four-engine turboprops and two-engine and four-engine piston are respectively 7.4, 0.6, 2.5, 0.6 and 1.4 pounds per flight.

The particulate emissions from automobiles consist of carbon particles, lead compounds, oil and nonvolatile reaction products. Automobiles contribute significantly to particulate pollution problems, since aerosols are formed in the reaction products from hydrocarbons and nitrogen oxides in the photochemical reaction. For every 1,000 gallons of gas, about 12 pounds of particulates are emitted, and essentially all the particles are less than 5 microns in size. The particulate emission from diesel engine vehicles is 110 lb/1,000 gal of diesel fuel and 62.5% of particles are less than 5 microns in size.

Effects of Particulate Matter[1]

The difficulty in assessing damages resulting from particulate air pollution lies in isolating the effects of particulates from those of oxides of sulfur and nitrogen, ozone and others. The cost of painting steel structures, commercial laundering, cleaning and dyeing, transportation delays, car washing, household maintenance, residential properties, human health and productivity loss are some visible economic effects of air pollution due to particulate matter.

Very little is known about the effects of particulate matter on vegetation and most of the experiments done to date have dealt with specific kinds of dusts rather than the mixture normally encountered in the atmosphere. There is considerable evidence that certain fractions of cement-kiln dusts when deposited on moist leaf surfaces form a hard adherent crust and damage plant tissue and inhibit growth. A marked reduction in the growth of poplar trees 1 mile from a cement plant was observed.

Fluorides in particulate form are less damaging to vegetation than gaseous fluorides. Fluorides absorbed or deposited on plants may be detrimental to animal health (Section 3.8). Magnesium oxide deposits on soils have been shown to reduce plant growth, while iron oxide deposits may be beneficial. Sulfuric acid aerosols cause leaf spotting.

The particulate matter in the atmosphere scatters and absorbs light from the sun and thus reduces the visible radiation available to cities and the solar radiation that reaches the earth. The increased emission of fine particles into the atmosphere may also cause changes in the delicate heat balance of the earth's atmosphere, thus altering worldwide climatic conditions (Sections 2.1 and 2.7). The particles in the atmosphere can affect the weather by serving as condensation nuclei that influence the formation of clouds, rain and snow.

Visibility is reduced by particulate matter in the atmosphere. When the relative humidity exceeds 70%, many particles exhibit deliquescent behavior and grow into fog droplets. Sodium chloride particles act as condensation nuclei and show rapid and large changes in size. In addition to aesthetic degradation of the environment, reduced visibility has consequences on the safe operation of aircraft and motor vehicles.

The particulate matter in the atmosphere, including soot, dust, fumes and mist, can cause a wide range of damage to materials. Through their own corrosiveness and through the action of corrosive chemicals absorbed or adsorbed on them, the particulate matter can cause direct chemical damage to materials. The ability of particulates to damage buildings, sculpture and other structures is particularly great in cities where large quantities of coal and sulfur-bearing fuel oil are burned. Automobile finishes have also been damaged by particulate matter in the atmosphere. The water soluble chlorides and sulfates may cause blistering and enhance the weathering of painted surfaces. The soiling of clothing, curtains and similar textiles is also caused by aerosols.

Particulate emissions represent a major portion of the total quantity of air pollutants and their soiling and nuisance properties are important. There are indications that the toxic effect of sulfur dioxide and trioxide is enhanced by the presence of particulate matter. The reduction in visibility and formation of photochemical smog are other significant effects of particulate matter.

Table 3.15l
ALLOWABLE PARTICULATE EMISSIONS,
San Francisco Bay Area

Process rate		Rate of Emission, lb/hr	Process rate		Rate of Emission, lb/hr
(lb/hr)	(ton/hr)		(lb/hr)	(ton/hr)	
100	0.05	0.551	16,000	8.00	16.5
200	0.10	0.877	18,000	9.00	17.9
400	0.20	1.40	20,000	10.00	19.2
600	0.30	1.83	30,000	15.00	25.2
800	0.40	2.22	40,000	20.00	30.5
1,000	0.50	2.58	50,000	25.00	35.4
1,500	0.75	3.38	60,000	30.00	40.0
2,000	1.00	4.10	70,000	35.00	41.3
2,500	1.25	4.76	80,000	40.00	42.5
3,000	1.50	5.38	90,000	45.00	43.6
3,500	1.75	5.96	100,000	50.00	44.6
4,000	2.00	6.52	120,000	60.00	46.3
5,000	2.50	7.58	140,000	70.00	47.8
6,000	3.00	8.56	160,000	80.00	49.0
7,000	3.50	9.49	200,000	100.00	51.2
8,000	4.00	10.4	1,000,000	500.00	69.0
9,000	4.50	11.2	2,000,000	1,000.00	77.6
10,000	5.00	12.0	6,000,000	3,000.00	92.7
12,000	6.00	13.6			

Air Quality Standards

The Environmental Protection Agency (EPA) standards for ambient air set the annual arithmetic mean limit at 75 $\mu g/M^3$ and the maximum 24-hour mean at 260 $\mu g/M^3$.

Emission standards in the past have usually been based on the total emission rate (total mass of material discharged from the stack), as illustrated by Table 3.15l. But more recently, the EPA issued emission standards for steam generating plants (Table 5), sulfuric and nitric acid plants, cement plants and incinerators. Standards of similar natures are being developed for some forty additional industries.

REFERENCES

1. "Air Quality Criteria for Particulate Matter," National Air Pollution Control Administration Publication No. AP-49.
2. "Control Techniques for Particulate Air Pollutants," National Air Pollution Control Administration Publication No. AP-51.
3. "Atmospheric Emissions from Fuel-Oil Combustion," Public Health Service Publication No. 999-AP-2.
4. "Atmospheric Emissions from Coal Combustion," Public Health Service Publication No. 999-AP-24.
5. "Emissions from Coal-fired Power Plants," Public Health Service Publication No. 999-AP-35.
6. "Air Pollution Engineering Manual," Public Health Service Publication No. 999-AP-40.
7. "Compilation of Air Pollutant Emission Factors," Public Health Service Publication No. 999-AP-42.

3.16 PHENOLS AND ORGANIC ACIDS

Organic acids are carbon compounds that contain a carboxyl group as the functional group (formula R—COOH, where R is group containing C attached to carboxyl group). Phenols are hydroxybenzenes, C_6H_5OH, where one or more ring hydrogens may be substituted by another group.

Sources

Emissions of phenols and organic acids are numerous, but there is relatively little quantitative information available (Table 3.16a). It is probable that these compounds are present in gases and aerosols from most combustion processes. In addition to emissions during the manufacture and use of these compounds as monomers and intermediates, these compounds are part of automobile exhaust,[1] tobacco smoke, plastics incineration[2] and petroleum refining.[3] Acids are formed from fish meal and manure plants, rendering, soap and synthetic detergent manufacture,[4] preparation of synthetic fibers, cattle feed lots,[5] and varnish cookers.[6] Phenols have been reported in emissions from varnish cookers, beehive coke ovens, coke conversion plants, phenol resin plants, plastics and rubber compounding,[6] wire coating ovens[6] and from kraft black liquor pyrolysis.[7] Diesel exhaust contains complex phenols and nitrophenols as well as paraffinic, olefinic and aromatic acids.[8] Emission inventory estimates for acids are generally available, but few have been made for phenols.[1,9,10]

Anaerobic biological degradation of almost all organic matter releases low molecular weight acids. Complex phenol and acid derivatives exist in soils and plants and may become airborne. Fatty acids and epoxy acids have been identified on the surface of airborne urediospores.[11]

Atmospheric Measurements, Interactions, Removal

Phenols have been measured in ambient air and particulate samples from many urban areas have shown several unidentified phenols; 2-fluorenol was positively identified in New York City air.[15] In Russia, 700 samples taken within 0.3 to 3 kilometers of a chemical plant complex in 1964 to 1966 were analyzed for phenol content.[16] Phenol has also been found near a ferrous metallurgical plant. Not surprisingly, low molecular weight acids and fatty

Table 3.16a
EMISSIONS OF ORGANIC ACIDS AND PHENOLS

Process	*Quantity of Emitted Compounds*	*Reference*
Combustion Sources	acids	9
bituminous coal	1.10 (lb/10^6 BTU)	9
residual coal	0.11 (lb/10^6 BTU)	9
distillate oil	0.11 (lb/10^6 BTU)	9
natural gas	0.06 (lb/10^6BTU)	9
Refuse incineration		
single chamber	3 lb/ton (as acetic)	6
multiple chamber	0.22 lb/ton (as acetic)	6
wood waste—multiple chamber	acids, 0.54–1.2 lb/ton	6
pathological waste incinerators	acids, 1.5–2.0 lb/ton	6
pathological waste incinerators equipped with secondary burners	acids, 1 lb/ton	6
Meat Smokehouses	90–125 ppm (formic)	6
	460–500 ppm (acetic and higher)	6
	20–30 ppm (phenol)	6
exhaust	acids, 30–156 ppm	6
using afterburner	acids, 0–76 ppm	6
Hickory wood smoke	guaiacol, eugenol	12
Drying of Douglas fir and Ponderosa pine	abietic, fatty	13
Coffee roasting		
continuous	acids, 223 ppm	6
batch	acids, 175 ppm	6
Debonding of brake shoes	acids, 0.22 lb/hr	6
Reclamation of electric equipment windings	acids, 0.35 lb/hr	6
Phenolic resin manufacturing	phenols, 45 mg/M^3 (1.5 lb/hr)	14

acids are found near oil refineries, manufacturing plants, synthetic fatty acid plants and soap manufacturing installations.

The phenol concentrations indoors of a single dwelling built entirely of plastics were found to be especially high in winter; total organics in indoor air changed from 20–32 to 11–15 mg/M^3 in 4 years.[17]

3.16 Phenols and Organic Acids

Phenol and low molecular weight acids are water soluble and are probably removed by rainfall; they could enter into aerosol formation, and high molecular weight homologues may form settled particles. However, there is no quantitative information on atmospheric reactions and removal mechanisms.

Effects

Organic acids are contributors to odorous emissions; odor thresholds have been published.[6] Workers exposed to phenol in phenol-formaldehyde manufacturing reported chest pains, headaches, and skin rashes; 5 ppm phenol showed no significant pathological damage in rats, mice and monkeys after 90 days. Threshold concentrations in humans for acetic acid (mg/M^3) are: 0.6 (odor), 0.48 (eye irritation), 0.29 (electroencephalogram).[18]

Acids (acetic, formic) in confined spaces may contribute to corrosion of metals (especially Zn, Cd, steel, Mg, Pb).[19]

Plant and animal destruction as a result of 2, 4-D and 2, 4, 5-T spraying has occurred.[20]

Air Quality Standards

No criteria have been published and no standards have been set. For Russian cities, maximum allowable concentrations for phenol of 0.10 mg/M^3 (daily average) have been recommended. The threshold limit value recommended by the ACGIH is: 5 ppm (19 mg/M^3) phenol; 10 ppm (25 mg/M^3) acetic acid; 5 ppm (9 mg/M^3) formic acid[9] (Section 5.3).

REFERENCES

1. Mayer, M.: "A Compilation of Air Pollutant Emission Factors for Combustion Processes, Gasoline Evaporation, and Selected Processes," U.S. Dept. of HEW, May 1965.
2. Ball, G. et al.: *Am. Ind. Hyg. Assn. J.*, vol. 31, no. 5, pp. 572–578, 1970.
3. Steigerwald, B. J. and A. H. Rose: U.S. Public Health Service, Div. of Air Pollution, PHS Publication 763, 1960.
4. Air Pollution Control Association, TI-2 Committee, Technical Manual #3, Report #8, p. 49, 1968.
5. *Air Pollution Abstracts*, vol. 2, APTIC No. 28141, June 1971.
6. "Air Pollution Engineering Manual," 999-AP-40, U.S. Dept. of HEW.
7. Brink, D. L., A. A. Pohlman and J. F. Thomas: *Tappi*, vol. 54, no. 5, pp. 714–720, 1971.
8. Skala, H., F. G. Padrta and P. C. Samson: Paper presented at 1st Nat. Symp. on Heterogeneous Catalysis for Control of Air Pollution, Franklin Institute, Philadelphia, November 21, 1968.
9. Sax, N. Irving: Dangerous Properties of Industrial Materials, 3d ed., 1968.
10. N.Y. State Dept. of Health, Comprehensive Area Survey Report No. 7, Air Pollution, Capital District, March 1968.
11. Laseter, J. L. and R. Valle: *Env. Sci. & Techn.*, vol. 5, p. 631, 1971.
12. Fiddler, W. et al.: *J. Agr. Food Chem.*, vol. 14, no. 6, pp. 659–665, 1966.
13. *Air Pollution Abstracts*, vol. 2, 14129, November 1971.
14. Altwicker, E.: Private Communication.

15. Sawicki, E. et al.: *Microchim. Acta,* no. 5, pp. 1025–1039, 1968.
16. *Chemical Abstracts,* vol. 72, 15514f.
17. *Chemical Abstracts,* vol. 75, 52461u.
18. *Chemical Abstracts,* vol. 71, 24536w.
19. *Air Pollution Abstracts,* vol. 2, APTIC No. 23551 (09), February 1971.
20. *Air Pollution Abstracts,* vol. 2, 10978, June 1971.

3.17 PESTICIDES, HERBICIDES AND OTHER ECONOMIC POISONS

Pesticides have been recognized as environmental pollutants for the past two decades, but relatively little thought has been given to their role as air pollutants. These chemicals are found worldwide in the air, not only in agricultural communities, but also in urban and other nonpesticide consuming communities. It is evident that air is their prime carrier.

Whenever pesticides are used, a certain percentage miss the target and are released to the atmosphere. The portion that is deposited on the target can also contribute through vaporization to secondary pollution of the atmosphere.

Pesticides released into the air are transported by wind, not only locally but also intercontinentally. Little is known of their persistence in the atmosphere, though chemicals which resist degradation on earth surfaces are considered to be persistent in the air also.

The existence of pesticides in the air has been confirmed by air and rainwater analyses. The concentration of pesticides in air varies widely, ranging from 1×10^{-15} to 1×10^{-3} g/M^3, depending on the site and time of sampling. Higher concentrations are obtained in areas where and when chemical spray is applied, while low concentrations are usually atmospheric residues measured in remote areas.

Without having a worldwide monitoring system with complete background information regarding the history of pesticide use previous to and at the time of sampling, interpretation of the implications of pesticides in the air is exceptionally difficult. At present, it is impossible to estimate the total amount of pesticides in the air. The health hazard to man and to other nontarget organisms, particularly at lower concentrations, is also difficult to assess.

Insecticides

Of the many insecticides monitored as air pollutants, organochlorine insecticides, particularly DDT and its homologues (DDE and DDD), and dieldrin are the most common. The organochlorine insecticides, because they were widely used during the past 25 years (though their use at the present time is significantly reduced) and because most are stable under ordinary environmental conditions, can be expected to be found in the atmosphere. In Table 3.17a and Figure 3.17b, summaries of some of the recent analytical data from different sources and locations are illustrated.[15] There is a distinctive difference in the concentrations of these chemicals between three different locations: (1) communities that have no connection with pesticide use; (2) communities that are located near pesticide treated areas; and (3) in and around fields that are being treated.

Organophosphorus and carbamate compounds have been commonly used and will be used more widely in the future because of the enforced reduction in the use of organochlorine insecticides, but their presence has not been extensively investigated. Although generally more volatile than organochlorine insecticides, organophosphorus insecticides are not found as frequently in the air because most are less persistent than organochlorine insecticides in the natural environment. One of the reasons for not finding organophosphorus compounds in general monitoring may be due to methods of sampling. To date, no air pollution data are available for carbamate and inorganic insecticides, in spite of the latters' persistence.

Other Pesticides

Other pesticides include acaricides (miticides), rodenticides, fungicides, nematicides and fumigants. Of these, fungicides are most likely to exist in the atmosphere because of the large amounts used. However, most fungicides now available are less toxic and less persistent than their predecessors, thus very little study has been made of these compounds as general air pollutants. Concentration of the fungicide captan in the air was measured to reduce possible environmental pollution from spray applications, but its persistence in the air is unknown. The rate of evaporation of methyl and ethyl mercury compounds when used as fungicides and seed disinfectants was also studied. Though mercuric compounds are long lasting, the total amounts used as seed disinfectants have been very small compared to those used in industry. In many countries, the use of mercuric compounds as seed disinfectants has been banned.

Fumigants may be found in the air owing to their volatility, but no information regarding this is available. Little is known of the role of many other pesticides as air pollutants.

Herbicides

The use of herbicides has been steadily increasing and this trend will probably continue. In 1967, herbicide sales in the United States exceeded insecticide sales for the first time. Unlike insecticides, the chemical nature of herbicides is so varied that it is difficult to characterize them as a group. "Defoliants" are scientifically classed as herbicides. Major herbicides include 2,4-D in its ester forms, 2,4,5-T and s-Triazines. The vapor pressure of most

Table 3.17a
PESTICIDE CONCENTRATION IN THE AIR AT SEVERAL LOCATIONS

Source of Pesticide	*Type of Community and Sampling Location*	*Time of Sampling*	*Pesticides*	*Concentrations*	*Comments*	*References*
	Remote (Barbados)	Oct '65–Sept '66	DDT homologues and dieldrin	6.5–380 × 10^{-15} g/m^3	Carried by trade wind from European-African land; pesticides were measured on dusts; concentrations of pesticides on dusts: 41 × 10^{-9} g/g of dust (Collecting efficiency: 50%)	1
	Urban (Pittsburgh)	Jan–Dec '64	*p,p'*-DDT	Not detectable—1.14 × 10^{-9} g/m^3	Figures are less than real value owing to loss during sampling; proved the existence of DDT in a heavily industrialized urban area; glass fiber filter used	2
	Urban (Baltimore, Md.)	1968	γ-BHC *p,p'*-DDT	1.1 × 10^{-9} g/m^3 1.0 × 10^{-9} g/m^3	Sampling made with a combination of filter, scrubber (solvent: hexylene glycol) and alumina traps	3
Unknown	Urban (London)	Aug '65	*p,p'*-DDT dieldrin	3.9 × 10^{-9} g/m^3 27.2 × 10^{-9} g/m^3	Bacteriological membrane filter combined with solvent scrubber (solvent: Dimethylformamide); TLC and GLC were combined; a total of 8000 1 of air was sampled	4
	Urban (Cincinnati, Ohio)—Rainwater analysis	1965	DDT DDE	70–340–1300 × 10^{-12} g/g 5–20–20 × 10^{-12} g/g	Minimum, average and maximum value, respectively; 5 other chlorinated insecticides and isooctyl ester of 2.4.5-T were also found	5
	Urban (London)—Rainwater analysis	Feb–July '65	γ-BHC *p,p'*-DDT dieldrin	40–90 × 10^{-12} g/g 70–400 × 10^{-12} g/g 10–95 × 10^{-12} g/g	Limit of detection: 5 × 10^{-12} g/g (γ-BHC) 10 × 10^{-12} g/g (*p.p'*-DDT and dieldrin)	6
	Rural (Central England)—Rainwater analysis	Nov '64–Feb '65	γ-BHC *p,p'*-DDT dieldrin	100 × 10^{-12} g/g 3 × 10^{-12} g/g 20 × 10^{-12} g/g	Toxicity of γ-BHC collected from rain was compared with analytical grade γ-BHC and found to be equally toxic against Drosophila	7
	Agriculture and Industry (Fort Valley, Ga.)	May–June '64 Sept '64	DDT DDT	0.3–9.9 × 10^{-9} g/m^3 0.4–4.0 × 10^{-9} g/m^3	Combined drift from agricultural and industrial use; drift from industry only	8

Known but only that a spray was made in the vicinity	Urban (Eastern city, U.S.A.)	1964	DDT malathion	100–8000 × 10^{-9} g/m^3 and 2–11 × 10^{-9} g/m^3 1–30 × 10^{-9} g/m^3 and 0.1–1 × 10^{-9} g/m^3	During spray application and 1 day after spraying, respectively	8
	Agriculture (Stoneville, Miss.)	Jan–Oct '68	*p,p'*-DDT	7.0–950.0 × 10^{-9} g/m^3	Range of concentrations found the first day of each week; sampling was made with a combination of filter and scrubber (solvent: hexylene glycol); 5 other pesticides found	3
	Agriculture (Pullman, Wash.)	April–Aug '64	2.4-D (esters) 2.4.5-T (Me-ester)	219 × 10^{-9} g/m^3 45 × 10^{-9} g/m^3	Both aerosol and gaseous herbicides were collected; average of 113-day tests; scrubbers solvent: n-decane and aq. $NaHCO_3$	9
	Agriculture (Othello. Wash.)	Aug–Sept '64	ethion	3210 × 10^{-6} g/m^3 and 35.7 × 10^{-6} g/m^3	Applied by airplane, center of corn field; 12 hrs and 14 days after application, respectively	10
	Orchard (St. Hilaire, Quebec)	May '51	parathion	14000 × 10^{-6} g/m^3 and 90 × 10^{-6} g/m^3	During spray application within a few feet of sprayer operator breathing zone and 1 day later; air collected by combination of an impinger and fritted glass scrubber (solvent: alcohol)	11
Known definitely	Orchard (Vineland, Ont.)	May–Aug '70	captan	58 × 10^{-6} g/m^3 and 11 × 10^{-6} g/m^3	40 m and 130 m downwind from source during spraying; airblast sprayer; collected on Florisil® sampler	12
	Forest (New Brunswick)	1968	*p,p'*-DDT	Trace –0.646 × 10^{-6} g/m^3	Airplane application; combination of Florisil sampler and solvent scrubber (solvent: dimethyl formamide); sample sites were 2–9 miles from source of spray	13

DDT: dichlorodiphenyltrichloroethane
p,p'-DDT: 1,1,1-trichloro-2,2-bis(*p*-chlorophenyl)ethane
DDE: (*p,p'*-DDE): 1,1-dichloro-2,2-bis(*p*-chlorophenyl)ethylene
γ-BHC: gamma isomer of 1,2,3,4,5,6-hexachlorocyclohexane
DDD: dichlorodiphenyldichloroethane
TLC: thin-layer chromatography
GLC: gas-liquid chromatography or gas chromatography
2,4-D: 2,4-dichlorophenoxy acetic acid
2,4,5-T: 2,4,5-trichlorophenoxy acetic acid
Me-ester: methyl ester
aq. $NaHCO_3$: $NaHCO_3$ aqueous solution
Florisil® sampler: particulate pesticide collecting device in the air in which Florisil (synthesized magnesium silicate) is used as adsorbent

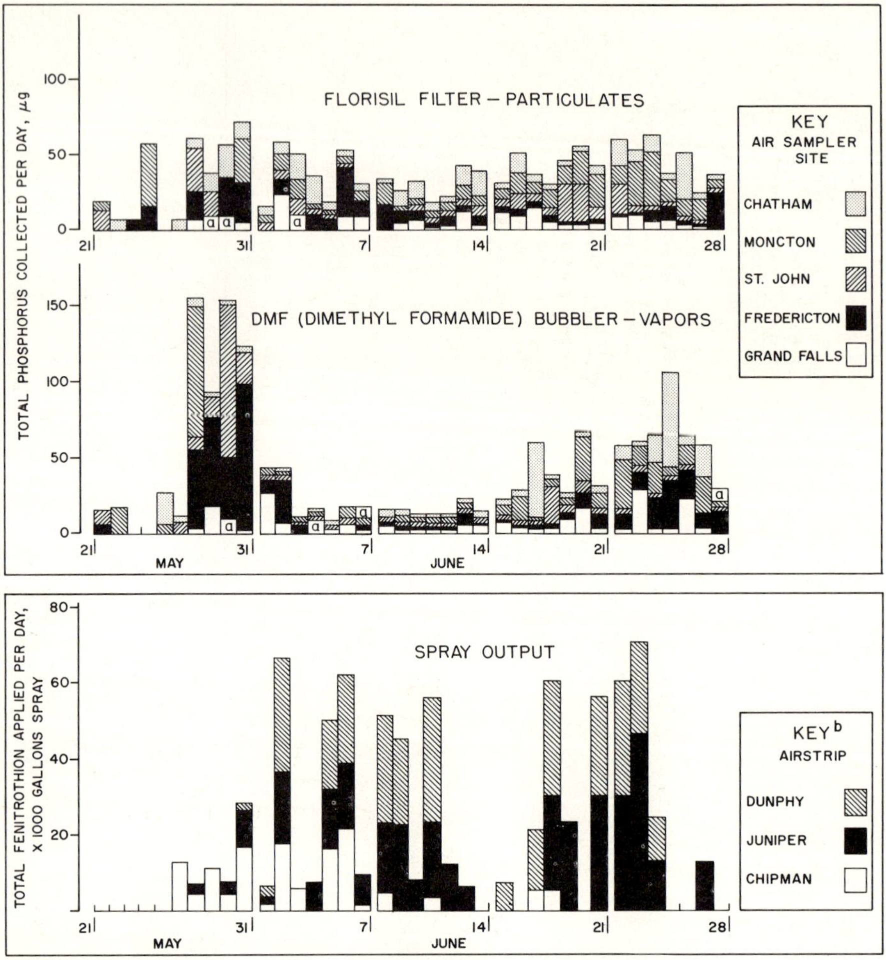

Figure 3.17b Daily output of fenitrothion by airstrip, together with total phosphorus collected as particulates and as vapors by air sampler station.

[a]Sample lost.

[b]The various shadings are associated with names of sampling sites which are explained in the frames. Each day 17.28m^3 of air was taken at each site.

herbicides is not very high; but 2,4-D esters (especially methyl isopropyl and buthyl esters) are exceptions, and their rapid evaporation makes them hazardous to susceptible nontarget crops. However, the greatest source of airborne herbicides is drift from spray applications, the extent of which is considered to be significant. Therefore, extensive studies have been conducted to reduce possible drift from spray applications. However, air sampling for herbicides has not been as extensive as for organochlorine insecticides.

Daily analyses for vaporization and drift of 2,4-D esters from a treated field showed concentrations ranging from nondetectable amounts to 5×10^{-6} g/M^3 and, depending on ambient temperature, both vapor and particulate forms were found in varying ratios. In general monitoring, herbicides, if found, were at low concentrations. Analysis of rainwater in the United States found atrazine and 2,4-D ester at 0.1×10^{-9} g/g each.

Their Sources

The largest use of pesticides is undoubtedly in the agricultural community. However, use in other areas, such as forests, cities and recreation areas, for control of mosquito and other pests is also contributing significantly to the total consumption. In 1962, approximately 27% of the insecticides marketed were used in nonagricultural areas in the USA. Two-thirds of the DDT produced during 1966 and 1967 was exported to other countries, mainly for mosquito control.

Pesticides introduced into the air during application are usually in the form of fine dust or droplets and these are transported for variable distances. In a 3 mph wind, a 2-micron aerosol drop travels 30 kilometers while falling 3 meters, whereas a 100-micron droplet travels only 15 meters. Dust creates a greater drift problem than fine droplets of wettable powder formulations or emulsifying concentrates. More drift results from airplane applications than from ground applications because of turbulence, greater height of release and greater areas covered in a given time period.

The quantity of pesticides released into the air varies significantly depending on many factors which can be classified into four main groups: (1) nature of chemicals including formulation; (2) method of atomization and delivery; (3) climatic factors; and (4) nature of target. Vaporization of deposited pesticides is a secondary route into the atmosphere. Volatile compounds evaporate rapidly, but their stabilities under different climatic conditions decide the real life of these chemicals as air pollutants.

Soil and water also contribute to air contamination. Soil particles eroded by wind can carry pesticides over large distances while trace amounts of pesticides in water (river, lake and sea) are codistilled with water into the air. Figure 3.17c[16] illustrates possible sources of pesticides in the air.

Effects of Atmospheric Pesticides

Studies of the effect of airborne pesticides on man and other nontarget organisms are insufficient. Although acute oral and dermal toxicity data are valuable in making judgments on the possible influence of these chemicals on organisms, the overall effects must be measured differently. The formula in Equation 3.17(1) defines the "percent toxic dose" (PTD) as a function of dermal and respiratory exposures and offers a guideline for estimating the toxicity of airborne chemicals.[14]

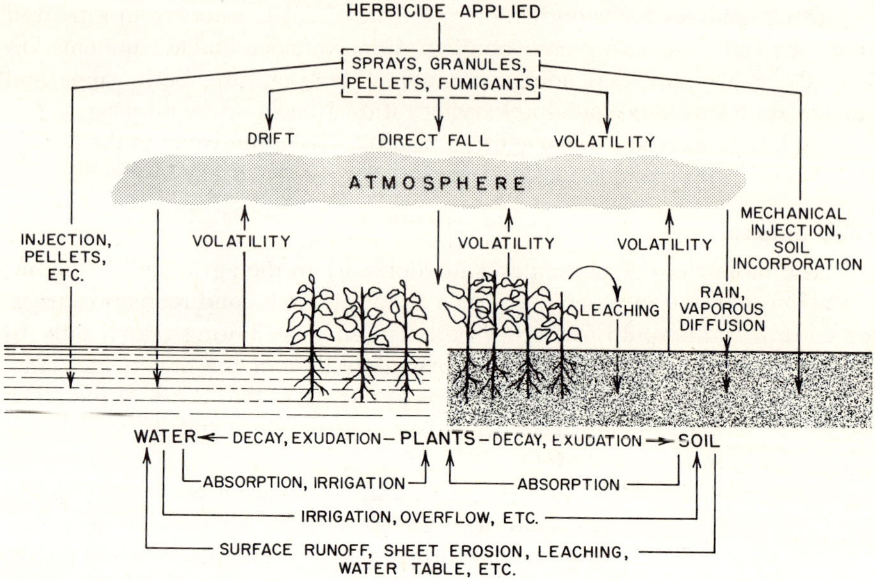

Figure 3.17c The possible distribution and fate of herbicides and their degradation products in the biosphere.

$$\text{PTD} = \frac{\text{dermal exposure (mg/hr)} + \text{respiratory exposure (mg/hr)} \times 10}{\text{dermal LD}_{50}\ \text{(mg/kg)} \times 70} \times 100 \quad 3.17(1)$$

Where LD_{50} is the median lethal dose.

Equation 3.17(1) may be useful to measure the significance of pesticide pollution to factory workers, spray operators and people living close to spraying areas; but its real significance in terms of the effects of airborne pesticides on the general population is difficult to judge.

Air Quality Standards

The American Conference of Governmental Industrial Hygienists suggests "threshold limit values of airborne contaminants," as a means of evaluating exposure (Section 5.3). "Threshold limit values" refer to airborne concentrations of substances to which most workers (based on a 40-hour week with a 7 or 8-hour work day) may be exposed repeatedly without adverse effect (see Glossary). These data do not suggest a relative hazard or toxicity index and should not be used as a basis for community air pollution control. Threshold limit values for pesticides are listed in Table 3.17d to show the levels of concentration tolerable under a specific condition. Another way to assess the possible exposure affect is to calculate the total amount of pesticide taken

Table 3.17d
THRESHOLD LIMIT VALUES OF AIRBORNE CONTAMINANTS[a]

Contaminant	*Classification*[b]	*(mg/M³)*[c]
Aldrin-skin	I	0.25
Azinphos-methyl (Guthion®)-skin	I	0.2
Carbaryl (Sevin®)	I	5.0
Chlordane-skin	I	0.5
2,4-D	H	10.0
DDT-skin	I	1.0
Demeton (Systox®)	I	0.1
Dichlorvos (DDVP)	I	1.0
Dibrom®	I	3.0
Endosulfan (Thiodan®)-skin	I	0.1
Endrin-skin	I	0.1
EPN[b]-skin	I	0.5
Ferbam	F	15.0
Heptachlor-skin	I	0.5
Lead arsenate	I	0.5
Lindane-skin	I	0.5
Malathion®-skin	I	15.0
Methoxychlor	I	15.0
Nicotine-skin	I	0.5
Paraquat-skin	H	0.5
Parathion-skin	I	0.1
Pentachlorophenol-skin	H	0.5
Phosdrin (Mevinphos®)-skin	I	0.1
Pyrethrum	I	5.0
Ronnel	I	10.0
Rotenone (commercial)	I	5.0
2,4,5-T	H	10.0
TEPP[b]-skin	I	0.05
Warfarin	R	0.1

[a]Excerpted from "Threshold limit values of airborne contaminants and intended changes, adopted by ACGIH for 1970," with the permission of American Conference of Governmental Industrial Hygienists.
[b]I: Insecticide, H: Herbicide, F: Fungicide, R: Rodenticide, EPN: O-ethyl O-*p*-nitrophenyl phenylphosphonothionate, TEPP: tetraethyl pyrophosphate.
[c]Approximate milligrams of particulate per cubic meter of air.

in with the air breathed per day. The figures obtained may then be compared with those suggested as the "maximum daily intake of pesticides" (FAO and WHO), although these examples are based on food ingestion.

Milk produced by cows fed on pastures contaminated by drift contained high levels of pesticide residues. This type of side effect is discussed in Section 5.15 in Vol. III. There is no evidence that relatively low levels of airborne pesticides directly affected nontarget organisms.

The effects of herbicides on neighboring plants are direct and severe, though effects on man and animal are judged to be relatively minor.

REFERENCES

1. Risebrough, R. W. et al.: *Science,* vol. 159, p. 1223, 1968.
2. Antommaria, P. et al.: *Science,* vol. 150, 1476, 1965.
3. Stanley, C. W., et al.: *Environ. Sc. Tech.,* vol. 5, p. 430, 1971.
4. Abbott, D. C., et al.: *Nature,* vol. 211, p. 259, 1966.
5. Cohen, J. M. and C. Pinkerton: *ACS Advances in Chemistry,* No. 60, p. 163, 1966.
6. Abbott, D. C., et al.: *Nature,* vol. 208, p. 1317, 1965.
7. Wheatley, G. A. and J. A. Hardman. *Nature,* vol. 207, p. 486, 1965.
8. Tabor, E. C., *Trans. N.Y. Academy of Sciences,* vol. 28, p. 569, 1966.
9. Bamesberger, W. L. and D. F. Adams. *ACS Advances in Chemistry,* No. 60, p. 219, 1966.
10. Hindin, H., et al.: *ACS Advances in Chemistry,* No. 60, p. 132, 1966.
11. Kay, K., et al.: *A.M.A. Arch. Ind. Hyg. Occup. Med.,* vol. 6, p. 252, 1952.
12. Chiba, M., et al.: *Proc. Int. Symp. Ident. Meas. Env. Pollut.,* 1971.
13. Yule, W. N. and A. F. W. Cole. *Proc. 4th Int. Agric. Aviat. Congr.,* 1969.
14. Durham, W. F. and H. R. Wolfe. WHO *Bull.,* vol. 26, p. 75, 1962.
15. Permission to use granted by Springer-Verlag, New York. From Yule, W. N., Cole, A. F. W. and I. Hoffman: A Survey for Atmospheric Contamination Following Forest Spraying with Fenitrothion, *Bulletin of the Environmental Contamination and Toxicology,* vol. 6, p. 289, 1971.
16. Permission to use granted by Springer-Verlag, New York. From Foy, C. L. and S. W. Bingham: Minimizing Herbicides in the Environment, *Residue Reviews,* vol. 29, p. 109, 1969.

3.18 RADIOACTIVE POLLUTION

Natural Radioactivity

Primary and secondary cosmic rays and their products cause a major part of the radiation dose to the gonads, bone lining cells and bone marrow from naturally occurring radiation.[1] Primary cosmic rays (galactic, solar and Van Allen belt) are those rays which reach the earth's surface without interaction with atomic nuclei present in the earth's atmosphere. Secondary cosmic rays are primary rays which have interacted with atomic nuclei, mostly nitrogen and oxygen, in the earth's atmosphere. The major cosmic ray particles in the earth's atmosphere are nucleons, mesons, leptons and photons. The dose at the earth's surface from the ionizing component of cosmic rays is highly latitude- and altitude-dependent and varies with the solar cycle. The dose rate at sea level at mid-latitude, which does not underestimate the dose to the gonads, is 34 mrad/yr.*

The neutron dose reaches a maximum at 20 to 30 kilometers; but at sea level at mid-latitudes, it is only 1.1 mrad/yr to the gonads. Passengers on supersonic transport planes flying at 20 kilometers in polar regions will receive a total dose equivalent of 2.0 mrem/hr with an average quality factor† of 2.7. The average dose from a 3-hour transatlantic flight would be 6 millirem in contrast to the 4.6 millirem for the 8-hour travel by present day jets at 12 kilometers. Dose rates per 24 hours on space flights have been measured from 12 millirads at 300 kilometers through 100 millirads at 700 kilometers to 2,000 millirads at 1,500 kilometers.[2] External dose rates from radon and thoron in the air vary depending on the distribution of radionuclides in geologic materials and have been measured at 1.4 to 11 mrad/yr with a mean rate of 3.[1] The thoron dose is approximately one-half to one-third of the radon dose. Internal doses to the basal epithelial cells of segmental bronchi from inhaled radon daughters range from 50 to 200 mrad/yr.

Fallout

Radioactive materials from explosions of nuclear devices are the main cause of radioactivity in the environment. Nuclear testing has put 1,700 megacuries

* (mrad) (quality factor) (absorbed dose distribution factor) (modifying factors) = mrem.

† The quality factor denotes the modification of the effectiveness of a given absorbed dose by linear energy transfer.

of tritium into the environment in comparison to a steady-state inventory from cosmic ray interactions with atmospheric nitrogen and oxygen of 27 megacuries. If one assumes that all tritium is in the form of tritiated water, then one would ingest 2,500 ml/day of water and inspire 480 ml/day by the lungs. The critical organ (i.e., the one receiving the greatest percentage of the permissible dose) would be the soft tissue from drinking water if the concentration of tritium was the same in air as in water. The dose commitment (the cumulative dose) from the global inventory to the soft tissues is from 0.4 to 1.7 millirads.

The quantity of krypton-85 injected into the atmosphere as a result of testing of nuclear devices is from 2.1 to 3.5 megacuries. By 1970, the amount left in the atmosphere was 1.1 to 1.7 megacuries which if uniformly spread over the world would result in a gonadal dose of 7×10^{-6} mrad/yr.[2]

Mining and Milling

The production of gaseous wastes in the fuel cycle[3] is detailed in Figure 3.18a. To date, the most serious effects of gaseous releases have been in the underground mining of the uranium ore. Despite the example of Schneeburg and Jochimstal (radium mines pitchblend), ventilation in some of the United States mines was insufficient to prevent the incidence of lung cancer among uranium miners due primarily to the radon daughter products. The highest incidence of lung cancer (which has a latent period of 10 to 20 years) is occurring in the group of miners who have worked in mines in which the average concentration of radon daughters was usually higher than 10 working levels (any combination of radon daughters in one liter of air that will result in the emission of 1.3×10^5 MEV of potential alpha energy), whose cumulative exposure is greater than 1,000 WLM (working level months) and who are moderate to heavy smokers.[4] The hazards in the milling operations are primarily due to the silica dust.

Reactors

The major generation of radioactive gaseous wastes occurs in the reactor, though these gases are not necessarily released directly to the environment. The gaseous wastes are generated by the fissioning of U^{235}, primarily H^3, I^{131}, Kr^{85}, Xe^{138}, and those isotopes having a gaseous precursor such as Sr^{90}; the activation of a coolant (air, Ar^{40}; helium, H^3; carbon dioxide, O^{19}, N^{16} and C^{14}); and volatile elements such as Cs^{137}.

Pressurized water reactors release less gaseous wastes than do boiling water reactors, where (due to the shorter holdup period of 20 to 30 minutes) more of the shorter life isotopes are discharged when the air and noncondensable gases (noble gases) are ejected from the condenser. Grab samples from the Yankee pressurized water reactor[5] and the Dresden boiling water reactor,[6] though not necessarily representative, do give some indication of

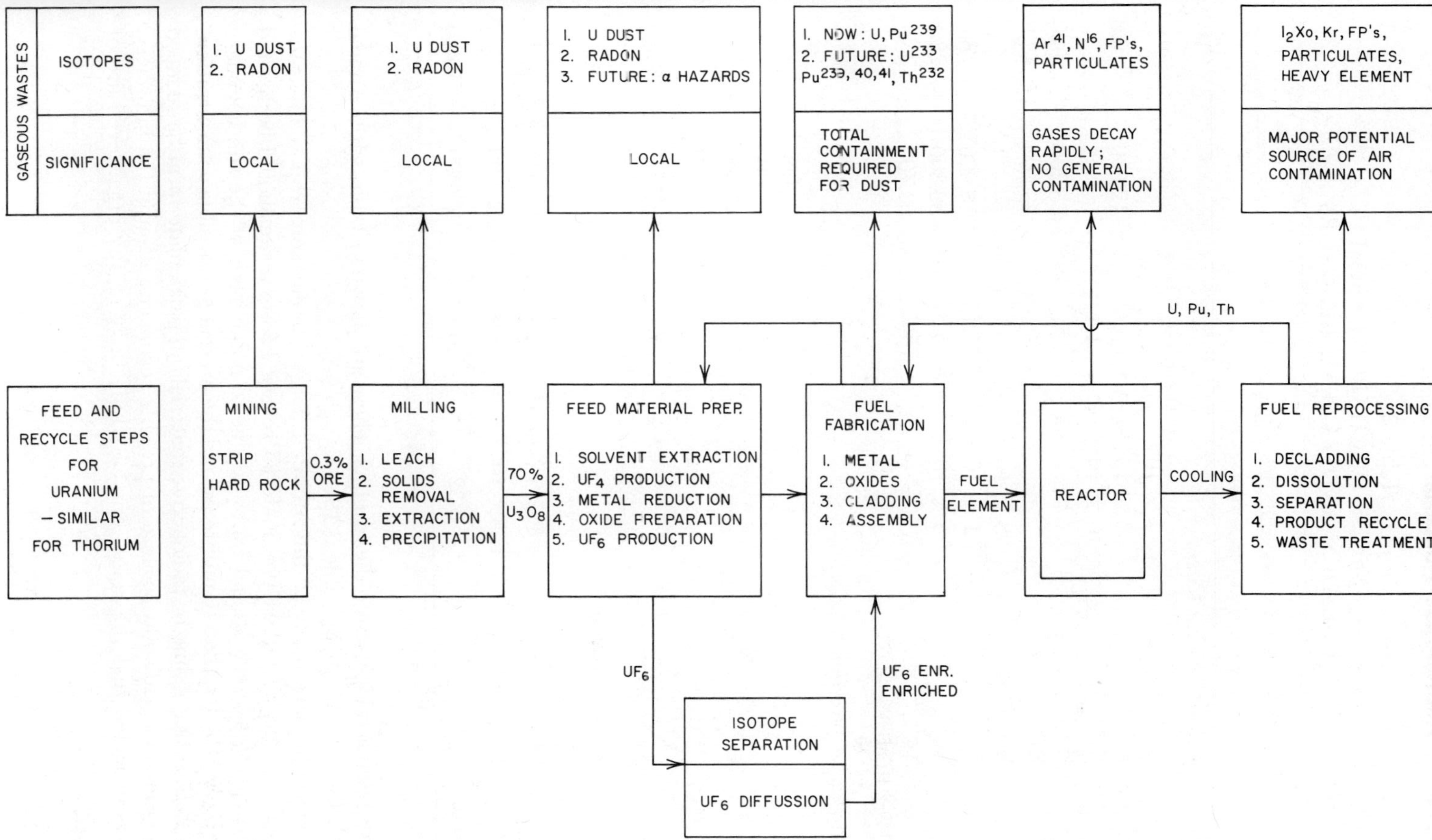

Figure 3.18a Wastes from nuclear fuel cycle; FP, Fission product.

Table 3.18b
GASEOUS DISCHARGES FROM
DRESDEN AND YANKEE REACTORS

Nuclide	*Yankee*		*Dresden*	
	Ci/yr	%	*Ci/yr*	%
Fission Product Noble Gases				
Ar^{41}	4×10^{-3}			
Kr^{85m}	2×10^{-5}		12,600	6.0
Kr^{85}	6×10^{-2}	1.1	33	—
Kr^{87}			25,200	14.0
Kr^{88}			23,200	12.9
Xe^{133m}	8×10^{-7}			
Xe^{133}	1.1×10^{-1}	2.1	15,900	8.8
Xe^{135}	3×10^{-1}	5.6	39,800	22.1
Xe^{138}	—	75.2	63,700	35.4
Short lived Kr + Xe	4			
Particulate Radionuclides			$\times 10^6$	
Co^{58}			660	
Co^{60}			660	
Sr^{89}			30,500	
Sr^{90}			130	
Cs^{137}			1,330	
Ba^{140}			13,300	
Gaseous Fission Products			$\times 10^6$	
I^{131}			.029,200	
C^{14}	6.8×10^{-1}	12.8		
H^3	1.7×10^{-1}	3.2	.266,000	
Total	5.32	100	180,000	100.0

the gaseous releases from the reactors. These are shown in Table 3.18b. Annual releases are shown in Table 3.18c.[9]

On January 18, 1972, the Atomic Energy Commission proposed that the total quantity of radioactive material released from a site should be limited so that the average annual exposure due to noble gases at the site boundary or beyond be less than 10 millirems, and the present permissible concentrations of radioactive iodine or radioactive materials in particulate form be reduced by a factor of 100,000 because of possible reconcentration in the biosphere. It is expected that the actual annual whole body or any organ exposure of

Table 3.18c

GASEOUS RADIOACTIVITY DISCHARGED TO THE ENVIRONMENT IN 1970 FROM COMMERCIAL NUCLEAR POWER STATIONS AND FUEL REPROCESSING PLANTS IN THE UNITED STATES

Nuclear Plant	*Date of Initial Operation*	*Rated Power MWe*[a]	*Type of System*	*Plant Location*	*Airborne Effluent, Ci/yr*[e]		
					Gross[f]	*% of limit*	*Halogens and Particulates*[g]
Dresden I	1959	210	BWR[b]	Morris, Illinois	914,000	4.2	3.3
Yankee	1960	185	PWR[c]	Rowe, Massachusetts	17.2[h]	0.25	1.8×10^{-6}
Indian Point I	1962	265	PWR	Indian Point, New York	1,750	0.033	7.5×10^{-2}
San Onofre	1967	450	PWR	San Clemente, California	1,610[h]	0.28	ND[j]
Connecticut Yankee	1967	600	PWR	Haddam Neck, Connecticut	700	0.002	4.6×10^{-4}[k]
Oyster Creek	1969	515	BWR	Toms River, New Jersey	112,000	1.2	3.2×10^{-1}
9-Mile Point	1969	500	BWR	Scriba, New York	9,500	0.04	ND
Ginna	1969	420	PWR	Ontario, New York	9,500	2.8	ND
Nuclear Fuel Services (NFS)	1966	—	[d]	Ashford, New York	300,000[i]	6.5	1.9×10^{-1}

[a] MWe—megawatt, electric.
[b] BWR—boiling water reactor.
[c] PWR—pressurized water reactor.
[d] Fuel reprocessing plant.
[e] Data from facility operations reports.
[f] Noble gases and gaseous activation products.
[g] All releases were less than 5% of the limit.
[h] H^3 discharge of 8.9 Ci at Yankee and 10.8 Ci at San Onofre were reported.
[i] Reported to be Kr^{85} (1969 value).
[j] ND—not detected.
[k] Does not include halogens.

an individual member of the public will not exceed 5 millirem. These levels can be achieved by the use of larger holdup times (16 to 24 hours for krypton and 10 to 15 days for xenon) and by the addition of high efficiency and charcoal filters.

The major sources of gaseous wastes from the United Kingdom gas-cooled reactors are the shield cooling air, main and auxillary CO_2 pressure circuits, contaminated area ventilation, pressure vessel cooling and maintenance of total pressure circuits. The design estimate of Ar^{41} discharges in the shield cooling air is 2×10^5 curies per year per reactor.

Fuel Reprocessing Plants

The major fraction of I^{131}, Kr^{85} and H^3 is released when the fuel elements are dissolved at the fuel reprocessing plant. The amount of I^{131} can be controlled by larger cooling times because of its short half-life, 8 days. The releases from the only commercially operating plant in the United States, Nuclear Fuel Services, have been analyzed. During 1968, the following isotopes were released into the air: Kr^{85}, I^{129}, I^{131}, Xe^{131m}, Xe^{133}, H^3, Ru^{106}, Ce^{133}, Cs^{137}, Sr^{90}, Ce^{144}, Zr^{95}/Nb, Mn^{54}, Co^{60}, Pu^{238}, Pu^{239}, U^{234} and U^{238}. These resulted in dose commitments from air in 1968 to a typical individual of less than 0.1 millirem[7]. In the long term, however, H^3, Kr^{85} and I^{129}, with half-lives of 12.3, 10.8 and 1.7×10^7 years, respectively, may become restrictive on a worldwide basis.[8] Based on power plant projections, reactor produced tritium will only approach a total of 720 megacuries by the year 2000.[2] Krypton-85 will increase to about 9,300 megacuries by the year 2000, resulting in a gonadal dose of 4×10^{-2} mrad/yr.[2] Iodine-129 in the environment will total 63,000 curies by 2000 and will produce an annual thyroid dose of 0.2 mrem/person.[2] Local concentrations of H^3 and Kr^{85} could be restrictive around reprocessing plants though not likely to be a worldwide problem.

REFERENCES

1. Report of the United Nations Scientific Committee on the Effects of Atomic Radiation, 21st Session, Supplement No. 14 (A/6314), United Nations, New York, 1966.
2. Report of the United Nations Scientific Committee on the Effects of Atomic Radiation, United Nations, New York, 1972.
3. Culler, F. L.: "Nature of Radioactive Wastes," vol. 1, Statement for the Record, Hearings on Industrial Radioactive Waste Disposal, JCAE, Eighty-sixth Congress of the United States, p. 72, August 1959.
4. Federal Radiation Council, Guidance for the Control of Radiation Hazards in Uranium Mining, Report No. 8, revised, September 1967.
5. Environmental Protection Agency, "Radiological Surveillance Studies at a Pressurized Water Nuclear Power Reactor," National Environmental Research Center, Cincinnati, Ohio, August 1971.
6. Department of Health, Education and Welfare, "Radiological Surveillance Studies at a Boiling Water Nuclear Power Reactor," Public Health Service, March 1970.

7. Shleien, B.: "An Estimate of Radiation Doses Received by Individuals Living in the Vicinity of a Nuclear Fuel Reprocessing Plant in 1968," Department of Health, Education and Welfare, Public Health Service, May 1970.
8. World Health Organization, "Public Health Implications of Radioactive Waste Releases," Fourth United Nations International Conference on the Peaceful Uses of Atomic Energy, Geneva, Switzerland, September 6 to 16, 1971.
9. Kahn, B., B. Shleien and C. Weaver: "Environmental Experience with Radioactive Effluents from Operating Nuclear Power Plants," Fourth United Nations International Conference on the Peaceful Uses of Atomic Energy, Geneva, Switzerland, September 6 to 16, 1971.

BIBLIOGRAPHY

International Atomic Energy Agency, "Environmental Contamination by Radioactive Materials," Proceedings Series (STI/PUB/226) Vienna, March 1969.

International Atomic Energy Agency, "Application of Meteorology to Safety at Nuclear Plants," Safety Series No. 29 (STI/PUB/211) Vienna, April 1967.

International Atomic Energy Agency, "Treatment of Airborne Radioactive Wastes," Proceedings Series (STI/PUB/195), New York, August 1968.

First, Melvin W. and James M. Morgan, Jr. (eds.): "Proceedings of the AEC Air Cleaning Conference," Harvard University and U.S. Atomic Energy Commission, annually.

3.19 RADIOACTIVE POLLUTION, NONIONIZING ELECTROMAGNETIC RADIATION

Physical Definitions

Nonionizing electromagnetic radiation is the type of radiation that in impinging upon matter does not produce ionization.

This definition is based on the photon concept of radiation described by Planck's equation:

$$E = h\nu \qquad 3.19(1)$$

where E is the photon energy, ν is frequency and h is Planck's constant.

The fundamental *wave* relationship for electromagnetic radiation is given by Equation 3.19(2):

$$C = \lambda\nu \qquad 3.19(2)$$

where λ is wavelength and C is the velocity of propagation. Combining Equations 3.19(1) and (2), the wavelength and frequency associated with photons having an energy E electron-volts (ev) are given respectively by Equations 3.19(3) and 3.19(4):

$$\lambda = \frac{hC}{E} = \frac{12397 \times 10^{-8}}{E} \text{ [cm]} \qquad 3.19(3)$$

and

$$\nu = 2.41814 \times 10^{14} \text{ E [sec}^{-1}] \qquad 3.19(4)$$

The energy required to produce ionization in atoms and molecules is typically of the order of at least a few electron-volts (1 ev = 23.06 Kcal/mole). Thus, only radiation having an energy level substantially less than about 1 electron-volt should be considered as nonionizing.

From Equation 3.19(3), the corresponding upper limit of frequency for an energy level $E = 1$ ev is 2.418×10^{14} Hz.

This frequency range embraces the following frequency bands: VLF (and all below this range), LF, MF, HF, VHF, UHF, SHF, EHF, infrared (IR), visible light (V) and part of the ultraviolet spectrum UV.*

*The near UV is included also as nonionizing radiation.

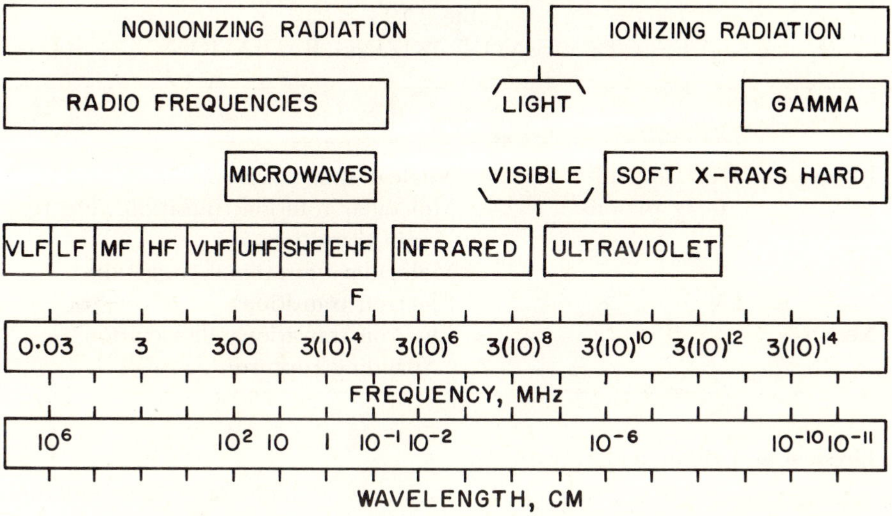

Figure 3.19a Electromagnetic frequency spectrum.

The first eight frequency bands: VLF, LF, MF, HF, VHF, UHF, SHF and EHF are the common bands of the RF spectrum and correspond to those used in radio telegraphy, AM broadcast and television whereas the last three bands comprise the microwave (Mw) spectrum (Fig. 3.19a).

However, this definition of nonionizing radiation is based solely on the energy carried by a single photon according to Equation 3.19(1) and does not take into consideration the density of radiation.

So-called nonionizing radiation may produce heavy ionization in matter if its radiation density is high enough. Examples are lasers (operating in the IR region), Mw discharge tubes (typically operating at 2,450 MHz) and LF discharge tubes. Ionization is produced by the high energy density via thermal processes, or as a result of high electric field strengths acting on the electrons of the outer shells of atoms and molecules.

It is necessary to emphasize the paramount importance of some “elementary particles” that do not produce ionization directly but by induction through secondary processes. As with any other “particle,” these also are considered as electromagnetic radiation on account of the dual nature of radiation expressed by the de Broglie equation:

$$\lambda = \frac{h}{mv} \qquad 3.19(5)$$

where mv is the momentum of the particle.

Among these particles are included neutrons and other elementary electrically neutral particles and their corresponding antiparticles.

Table 3.19b
EFFECTS OF NONIONIZING RADIATION

Type of Radiation	*Effect*
RF region (MHz region)	Nuclear spin transitions
MW region (GHz region)	Molecular rotation transitions, electron spin transitions
IR	Molecular vibrational transitions
Visible band V	Electron transitions
Near UV	Electron transitions, dissociation (banding rupture)

Physical and Biological Effects

Apart from the ionizing effects associated with high nonionizing radiation density, the physical processes taking place in matter in the different frequency bands of interest can be enumerated broadly as in Table 3.19b. In the case of neutrons, no primary ionization is produced; but ionization is caused indirectly by products of neutron collision or capture.

Spectroscopic studies in the frequency regions of RF to UV have shed considerable light on the internal molecular structure and have aided in the identification of the parameters involved in the interaction of electromagnetic radiation with living systems.

In contrast with ionizing radiation, the biological effects of nonionizing radiation in living systems have yet to be systematically investigated.

An enormously rich spectrum of interaction phenomena has been encountered, some potentially beneficial to mankind and some detrimental. Examples are as follows:

Nervous tissue taken from the sciatic nerve of a rat irradiated by a HF (27 MHz) field showed morphological changes which include varying degrees of demyelination, morphological changes in the appearance of the collagen and varying degrees of axonal and Schwann cell damage.[1]

EEG recordings made on domestic fowl revealed interaction between pulsed microwave (SHF) field and the central nervous system.[2]

Increased egg production observed in chickens exposed to a very low intensity microwave field (0.2 $\mu W/cm^2$ to 360 $\mu W/cm^2$) was accompanied by a mortality rate almost double that of the control (unradiated) colony.[3]

Chickens, pigeons, seagulls (ringbill) and parakeets exhibit various escape reactions when exposed to a pulsed microwave field in the intensity range 5 mW/cm^2 to 50 mW/cm^2.[4,5]

Microwave fields have a profound effect on the water metabolism of plants.[6]

Detrimental effects on health have been observed in people living and working near radar installations.[7-10]

The complexity of the problem is apparent if one considers the many ways in which radiation may interact with living systems. While the physical processes taking place are relatively well understood, very little is known of their biological concomitants. This is especially true of microwave radiation where the wavelength of the radiation is comparable to the physical dimensions of the systems under study. A block diagram showing the many ways in which Mw radiation may interact with a living system is given in Figure 3.19c.

Nonionizing Radiation as a Pollutant[11]

Pollutants in general can be divided into two main categories, according to their origin:

1. Man-made pollutants
2. Naturally occurring pollutants

Man-made nonionizing radiation is a pollutant and a potential hazard in all industrial, domestic, research and military applications.

The following devices are nonionizing radiation generators: IR and UV lamps, lasers and masers, radio broadcasting and radar installations, microwave communication systems, microwave ovens, particle accelerators and nuclear reactors.

Atmospheric pollution may arise either directly from ground installations, radio broadcasting and the like or indirectly from reflection of radiation from such sources by ionic plasmas occurring in the ionosphere.

Natural nonionizing radiation as a pollutant is much more difficult to describe and the processes leading to the emission of such radiation are highly complex. The three main sources are:

1. Outer space—including the solar system and earth's ionosphere
2. Earth
3. Earth's atmosphere

Nonionizing radiation from outer space may contain a considerable amount of radiation in one or in a number of the frequency bands listed. The earth's atmosphere and the earth itself also are sources of radiation through the interaction of atoms and molecules (actual atmospheric pollutants) with radiation from outer space. This leads to excitation followed by single or cascade de-excitation or by molecular dissociation and further chemical reaction with other substances (e.g., production of O_3 through UV action on O_2).

Ionic plasmas in the ionosphere also contribute with radiated fields in the nonionizing region of the frequency spectrum.

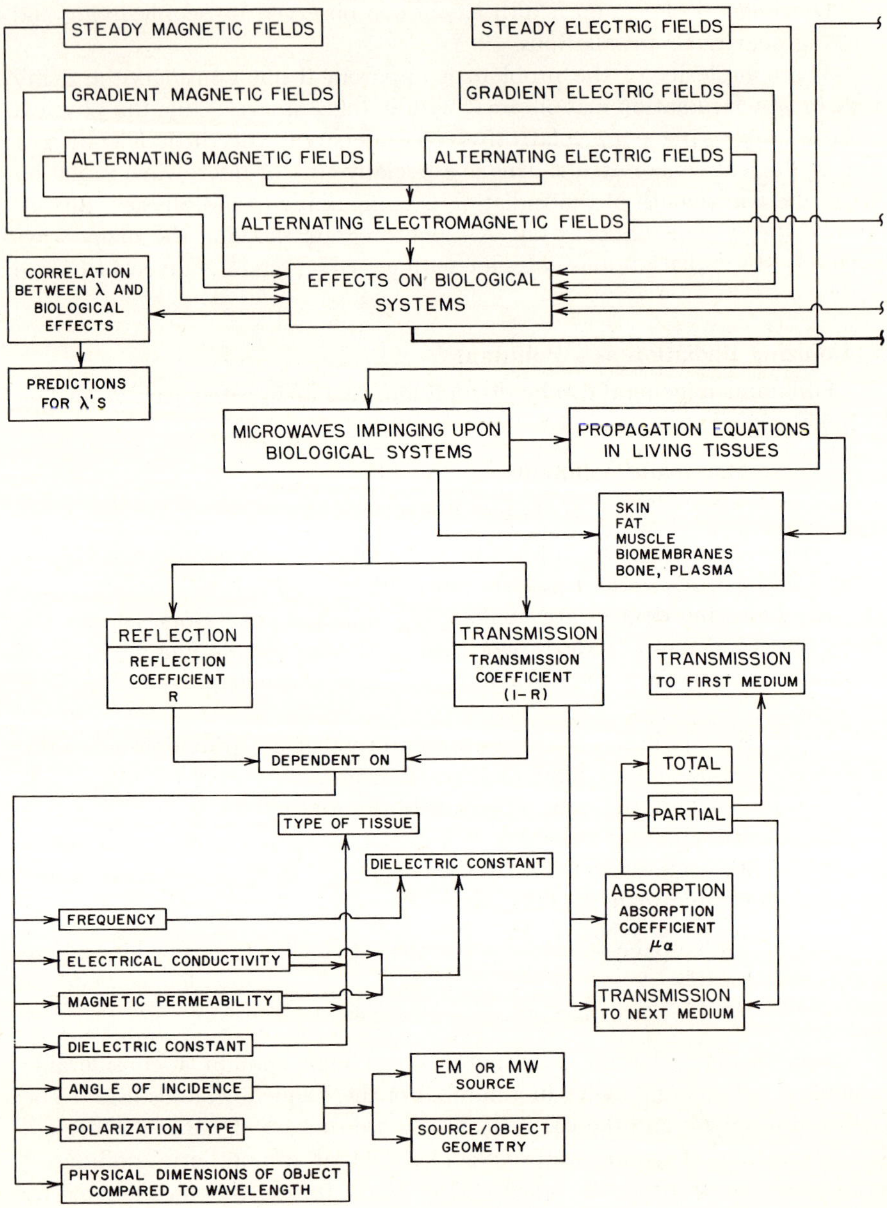

Figure 3.19c Microwaves—physical effects in biological systems.

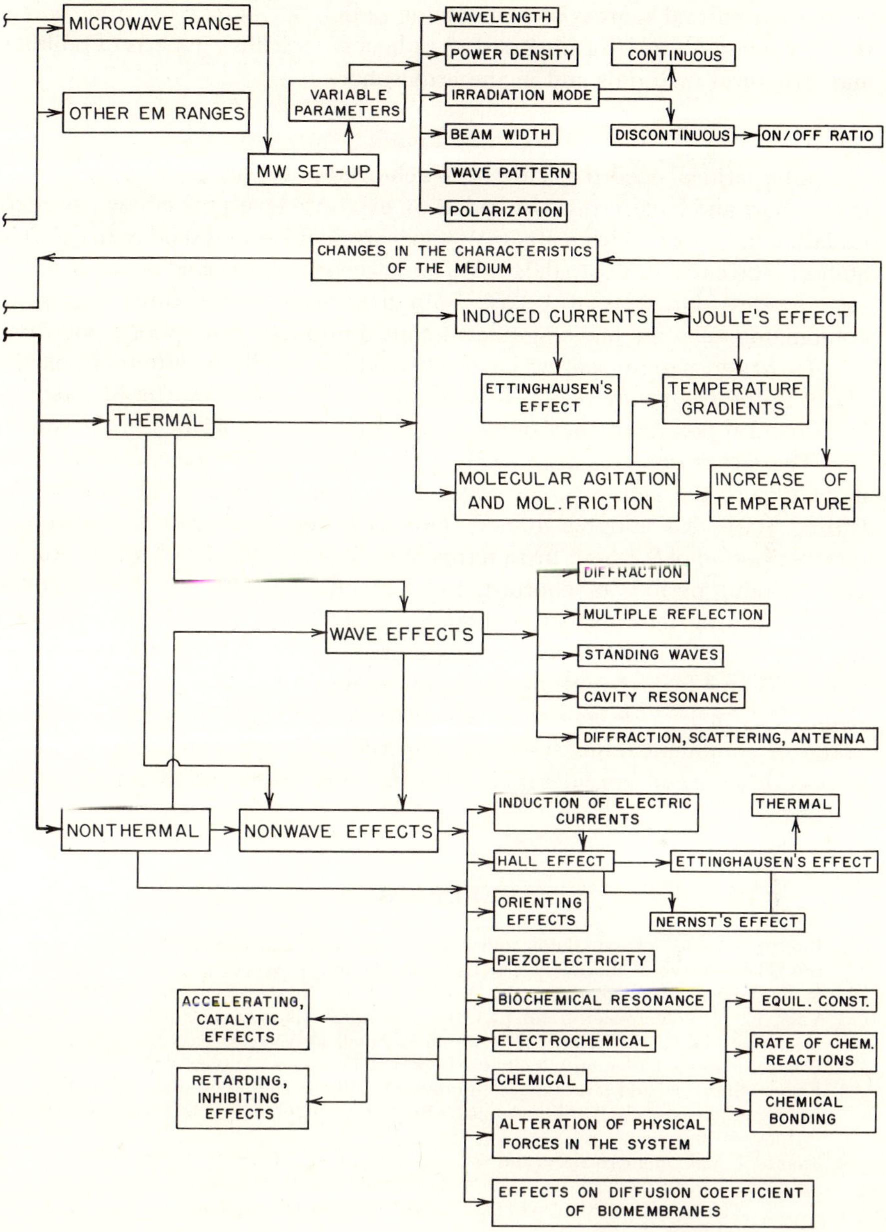
MICROWAVE RANGE
OTHER EM RANGES
VARIABLE PARAMETERS
MW SET-UP
WAVELENGTH
POWER DENSITY
IRRADIATION MODE
BEAM WIDTH
WAVE PATTERN
POLARIZATION
CONTINUOUS
DISCONTINUOUS
ON/OFF RATIO
CHANGES IN THE CHARACTERISTICS OF THE MEDIUM
INDUCED CURRENTS
JOULE'S EFFECT
ETTINGHAUSEN'S EFFECT
TEMPERATURE GRADIENTS
THERMAL
MOLECULAR AGITATION AND MOL. FRICTION
INCREASE OF TEMPERATURE
DIFFRACTION
MULTIPLE REFLECTION
WAVE EFFECTS
STANDING WAVES
CAVITY RESONANCE
DIFFRACTION, SCATTERING, ANTENNA
INDUCTION OF ELECTRIC CURRENTS
THERMAL
NONTHERMAL
NONWAVE EFFECTS
HALL EFFECT
ETTINGHAUSEN'S EFFECT
ORIENTING EFFECTS
NERNST'S EFFECT
PIEZOELECTRICITY
ACCELERATING, CATALYTIC EFFECTS
BIOCHEMICAL RESONANCE
EQUIL. CONST.
ELECTROCHEMICAL
RATE OF CHEM. REACTIONS
RETARDING, INHIBITING EFFECTS
CHEMICAL
CHEMICAL BONDING
ALTERATION OF PHYSICAL FORCES IN THE SYSTEM
EFFECTS ON DIFFUSION COEFFICIENT OF BIOMEMBRANES

Other natural sources of nonionizing pollutants include neutron sources that are found in soils (forming part of radioactive chains), in parts of building and structural materials and in the atmosphere.

Safety Levels[12-14]

International accord has not been achieved as yet on safe exposure levels in the short and long terms. Mw radiation exposure levels for safe whole body radiation have ranged from 100 mW/cm^2, originally established in the United States in the early 1950s, to 0.01 mW/cm^2 recently established by the U.S.S.R. for exposure durations of 1 day. Both man-made and natural sources of nonionizing radiation make significant contributions to atmospheric pollution.

However, natural sources have received practically no attention, particularly in the frequency region from the low RF band up to the Mw region. Such natural pollutants may prove to be of utmost importance in the future.

The status quo of safety levels established by different countries is an indication of the lack of knowledge of the range of biological effects. The United States has adopted 10 mW/cm^2 as a safe level averaged over any possible period of 0.1 hour or in terms of energy density, 1 mWh/cm^2 during any 0.1-hour period. In contrast, the U.S.S.R. radiation standard specifies 1 mW/cm^2 for intermittent exposures of not more than 20 minutes per working day, 0.1 mW/cm^2 for exposures up to 2-hour durations and 0.01 mW/cm^2 for exposure durations of an entire day. The imminent risk to man is from man-made generators which are finding widespread and increasing usage in communications systems, industrial processes and domestic appliances. Very careful consideration must be given to the implications for the future.

REFERENCES

1. Romero-Sierra, C., Susan Halter and J. A. Tanner: Effect of an Electromagnetic Field on the Sciatic Nerve of the Rat, in N. L. Wulfsohn and A. Sances, Jr. (eds.), "The Nervous System and Electric Currents," Vol. 2, Plenum Press, New York, pp. 81–85, 1971.
2. Villa, F., C. Romero-Sierra and J. A. Tanner: "EEG Changes in Chickens Subjected to Microwave Radiation," National Research Council, Division of Mechanical Engineering, Control Systems Laboratory Technical Report, LTR-CS-30, October 1969.
3. Romero-Sierra, C. and J. A. Tanner: Microwave Radiation and Egg Production in Chickens, in "Proceedings of IMPI Fifth Annual Microwave Symposium," Scheveningen, Holland, October 1970.
4. Tanner, J. A., C. Romero-Sierra and S. J. Davie: Nonthermal Effects of Microwave Radiation on Birds, *Nature,* vol. 216, no. 5120, p. 1139, 1967.
5. ______: The Effects of Microwaves on Birds: Preliminary Experiments, *Jour. Microwave Power,* vol. 4, no. 2, 1969.
6. Romero-Sierra, C., J. Bigu del Blanco and J. A. Tanner: "The Effects of Microwaves on Plants—Preliminary Experiments," Control Systems Laboratory Technical Report LTR-CS-38, Division of Mechanical Engineering, National Research Council of Canada. Ottawa, 1970.
7. Presman, A. S., Y. U. Kamenskii and N. A. Levitina: "The Biological Effects of Microwaves," U.S. Joint Publications Research Service 9451, June 1961.

8. Smurova, Ye. I., T. Z. Rogovaya, I. L. Yakub and S. A. Troitskiy: "General Health of Persons Working with HF, UHF and VHF Generators in Physiotherapy Machines," translation of Kazanskiy Meditsinskiy Zhurnal, No. 2, pp. 82–84, 1966.
9. Presman, A. S.: "Electromagnetic Fields and Life," Plenum Press, New York, 1970.
10. Marha, K., J. Musil and H. Tuha: "Electromagnetic Fields and the Life Environment," translated from the Czech, San Francisco Press, San Francisco, 1971.
11. Romero-Sierra, C. and J. A. Tanner: Nonionizing Electromagnetic Radiation and Pollution of the Environment, in "Proceedings of the International Symposium on Identification and Measurement of Environmental Pollutants," Ottawa, June 14–17, 1971.
12. Schwan, H. P.: Radiation Biology, Medical Applications and Radiation Hazards, in E. C. Okress (ed.), "Microwave Power Engineering," Vol. 2, Academic Press, New York, pp. 215–234, 1968.
13. Peyton, M. F. (ed.): "Biological Effects of Microwave Radiation," Plenum Press, New York, 1961.
14. Cleary, S. F. (ed.): "Biological Effects and Health Implications of Microwave Radiation," U.S. Department of Health, Education and Welfare, Bureau of Radiological Health, BRH/DBE 70-2, 1969.

3.20 SULFUR OXIDES

Several sulfur oxides are known: sulfur monoxide (SO), sulfur dioxide (SO_2), sulfur trioxide (SO_3) and disulfur monoxide (S_2O). The existence of a dimeric sulfur monoxide, $(SO)_2$, and the compound sulfur monoperoxide, SOO, have been postulated; but direct proof of their existence is still lacking.[1] Sulfur dioxide is the only one of these that ever exists to an appreciable extent in the open atmosphere. When fossil fuels are burned, any sulfur they contain is oxidized to sulfur dioxide and to a much smaller extent to sulfur trioxide. However, except at the very high temperatures that sometimes exist in stacks, the SO_3 rapidly reacts with water vapor to form droplets of sulfuric acid.[2] When SO_3 concentrations are listed in analyses of polluted atmospheres, they probably always refer to various sulfates and to sulfuric acid calculated as SO_3.

Atmospheric Concentrations

The concentrations of SO_2 in air at locations far distant from sources of pollution vary markedly with time and location (Table 2.1e). For example, the average concentration of SO_2 in the rain forests of Panama has been found to be about 1 ppb by volume. In Antarctic air, the concentrations ranged from < 0.5 ppb to 1.6 ppb. Much higher concentrations are found over most of Europe. The frequency distribution of sulfur dioxide levels in several American cities is shown in Figure 3.20a. Note that West Coast cities, where there is little coal burning and the smog is largely "photochemical," have relatively small concentrations of SO_2. The difference is almost certainly the result of the relatively low concentration of sulfur in most petroleum products.

There are a number of natural sources of atmospheric sulfur compounds, and "natural" atmospheric sulfur probably exceeds that introduced by man.[3] Natural sources are the decomposition and combustion of organic matter, sea salt particles from the oceans and volcanoes (Fig. 2.1f). The "Air Quality Criteria for Sulfur Oxides"[4] lists the primary sources of sulfur dioxide pollution as the combustion of fossil fuels, petroleum refining, the smelting of ores containing sulfur, sulfuric acid manufacture, refuse burning, paper making and the burning or smoldering of coal refuse banks.

In 1966, burning of coal accounted for about 60% of total SO_2 emissions,

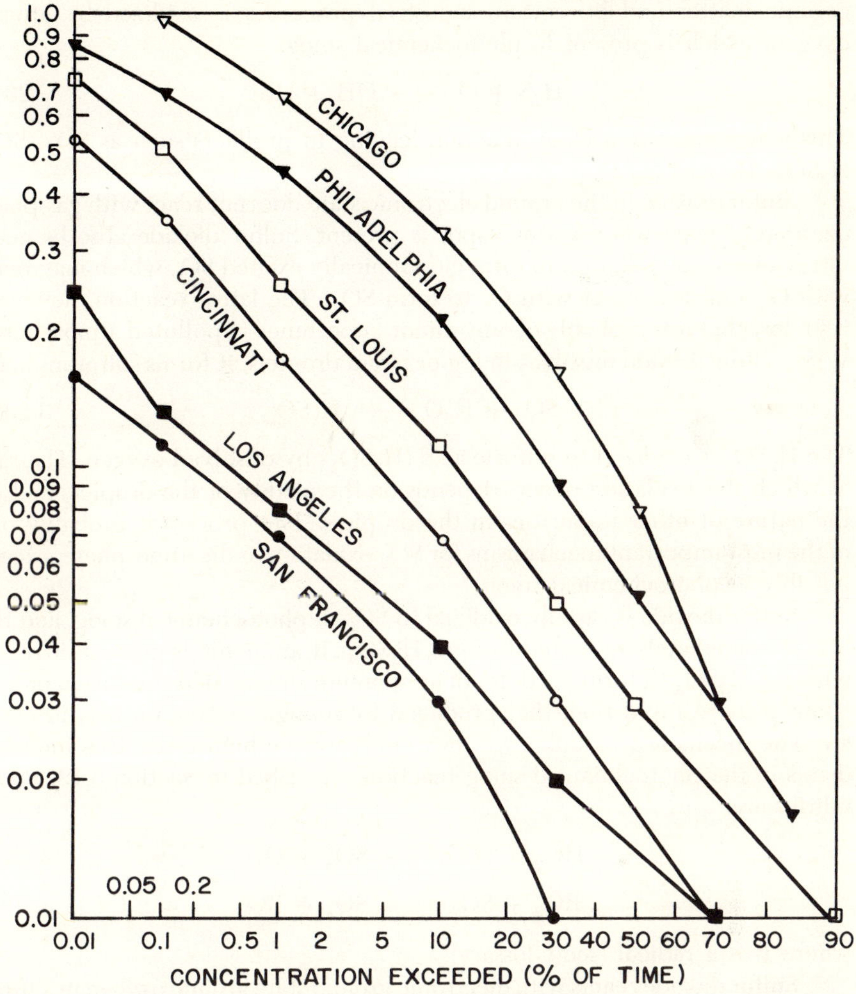

Figure 3.20a Frequency distribution of sulfur dioxide. Levels in selected American cities, 1962 to 1967 (1-hour averaging time).

petroleum combustion about 20%, ore smelting 12% and refinery operations 5%.[4] The sources vary to some extent seasonally, since sulfur-containing fuels (especially coal and fuel oil) will produce a seasonal increase in SO_2 emissions. As of 1971, about one hundred million tons of sulfur dioxide were being emitted worldwide per year[3] (Table 2.1d).

Atmospheric Reactions

Sulfur dioxide is both produced and destroyed in the atmosphere by chemical reactions. Hydrogen sulfide, produced by the decomposition of organic matter and by certain industrial processes, is oxidized by atomic oxygen, which is present in photochemical smog:

$$H_2S + O \longrightarrow OH + HS \qquad 3.20(1)$$

This is followed by a chain reaction leading to products such as SO_2, SO_3, H_2SO_4, H_2 and H_2O.

Sulfur dioxide in the ground electronic state does not react with gas-phase ozone (O_3) even when water vapor is present. Sulfur dioxide absorbs near-ultraviolet solar radiation to form electronically excited SO_2 which may react with O_3 and does react with O_2 to form SO_3. The latter reaction, however, is so slow that it is probably of only minor importance in polluted atmospheres. When sulfur dioxide dissolves in fog or cloud droplets, it forms sulfurous acid:

$$SO_2 + H_2O \longrightarrow H_2SO_3 \qquad 3.20(2)$$

The H_2SO_3 is oxidized to sulfuric acid (H_2SO_4) by dissolved oxygen. The rate at which this oxidation occurs depends on the acidity of the droplets and on the nature of other impurities in the droplets. This process is probably one of the most important mechanisms for SO_2 oxidation in the atmosphere, except possibly in photochemical smog.

Sulfur dioxide is rapidly oxidized to SO_3 in photochemical smog, and the SO_3 is immediately hydrated to form H_2SO_4. If ammonia is present, it reacts with the H_2SO_4 to form particles of ammonium sulfate. This oxidation of SO_2 is much more rapid than that produced by sunlight acting on SO_2 in clean air. The mechanism is unknown, but must involve highly reactive intermediates in the photochemical smog reactions described in Section 3.12. Possibilities are:

$$HO_2 + SO_2 \longrightarrow SO_3 + OH \qquad 3.20(3)$$

$$RO_2 + SO_2 \longrightarrow SO_3 + RO \qquad 3.20(4)$$

where R is a radical (see Glossary).

Sulfur dioxide reacts with electronic ground state atomic oxygen in a three body reaction to form SO_3:

$$SO_2 + O + M \longrightarrow SO_3 + M \qquad 3.20(5)$$

where M is a third molecule or atom. This reaction is probably not important in smog, but is probably an important mechanism for the production of droplets of sulfuric acid in the stratosphere.

Urone et al. discovered that sulfur dioxide in air is oxidized very rapidly even without sunlight in the presence of powdered oxides of aluminum, calcium, chromium, iron, lead and vanadium.[5] Airborne dust particles may have a similar effect, but more needs to be known before this extrapolation is reasonable.

The presence in photochemical smog of highly reactive free radicals and sulfur dioxide strongly suggests that organic sulfur compounds must also be present, formed by reactions between these two species. However, there is little, if any, direct information concerning the existence of such compounds in photochemical smog; emphasizing once again our woeful lack of detailed knowledge of the organic chemistry of air pollution.

Another chemical reaction which is currently receiving considerable attention is that between SO_2 and NH_3, when both are in the gas phase. Scott and Lamb[6] showed that the anhydrous reaction produces a yellowish compound with a stoichiometric ratio 1:1 and a white compound with a 2:1 ratio, provided the temperature is maintained below about 10°C:

$$NH_3 + SO_2 \rightleftharpoons NH_3 \cdot SO_2 \qquad 3.20(6)$$

$$2NH_3 + SO_2 \rightleftharpoons (NH_3)_2SO_2 \qquad 3.20(7)$$

As these equations imply, the compounds are in equilibrium with the gaseous reactants. Recent unpublished results obtained at room temperature at the National Center for Atmospheric Research (NCAR) suggest that when water vapor is also present, ammonium sulfite and ammonium sulfate are produced, the relative amount of the latter proportionately increasing with relative humidity. The kinetics of the reactions are unknown, but the NCAR studies indicate that the reactions are rapid even at concentrations occurring in polluted air. The following mechanism has been suggested for the formation of ammonium sulfate:

$$2NH_3 + SO_2 + \tfrac{1}{2}\, O_2 \longrightarrow (NH_4)_2SO_4 \qquad 3.20(8)$$

Effects

Much of the early concern with regard to SO_2 as a pollutant resulted from the plant damage it can produce. Smelters of sulfide ores were especially harmful to nearby vegetation. Unfortunately, the effects of various dosages are rather hard to predict since they vary with environmental conditions as well as with plant characteristics.[4] Acute injury is usually quite different from chronic damage. The former may consist of sharply defined regions of bleaching; while in the latter case, the major symptom may be a gradual yellowing. Several studies have shown that several-hour exposure of certain plants, such as alfalfa and conifers, to concentrations of SO_2 exceeding 0.3 ppm may cause damage.

Sulfuric acid droplets also injure vegetation, at least when the leaf surfaces are wet. Under the latter conditions, a "spotted injury" develops. Such injury has been reported in the Los Angeles area during periods of air pollution accompanied by fog.

The most obvious acute effect of sulfur dioxide at relatively low concentrations in air on man is its odor, which is similar to that of newly lighted matches.

The odor threshold is about 0.2 ppm by weight, although it varies with the individual.[7] At somewhat higher concentrations, a few ppm, broncho-constriction and general respiratory irritation occurs; in sensitive individuals, such as asthmatics, there may be bronchospasms.

Both laboratory and epidemiological studies indicate that a synergistic irritant effect exists between airborne particles and sulfur dioxide, the irritating effects of the sulfur dioxide being enhanced even by inert particles. This enhancement by some types of particles may result from catalysis by the particles of the oxidation and hydration of the SO_2 to H_2SO_4 droplets. Many of the laboratory studies have involved concentrations much higher than those encountered in urban smog. There is a danger in drawing conclusions from epidemiological studies, by ascribing symptoms to SO_2 and particulate matter, that these readily measured substances may only be indicators of air pollution in general. Thus, whether the indicated synergistic effect in smog is real remains questionable.

Relatively high concentrations of sulfuric acid mist, especially "wet" mists at concentrations exceeding 10 mg/M^3, are very irritating, producing coughing. According to the "Air Quality Criteria for Sulfur Oxides,"[4] sulfuric acid droplets are much more irritating than sulfur dioxide at low concentrations in air, but so little information is available that this is open to question.

Air Quality Standards

National air quality standards have been established for sulfur dioxide in the United States.[8] The annual arithmetic mean is not to exceed 80 $\mu g/M^3$, and the maximum 24-hour concentration which is not to exceed 365 $\mu g/M^3$ more than once a year.

REFERENCES

1. Nickless, G. (ed.): "Inorganic Sulfur Chemistry," Elsevier, New York, 1968.
2. Goodeve, C. F., A. S. Eastman and A. Dooley: *Trans. Faraday Soc.*, vol. 30, p. 1127, 1934.
3. Kellogg, W. W., R. D. Cadle, E. R. Allen, A. L. Lazrus and E. A. Martell: *Science* (in press).
4. "Air Quality Criteria for Sulfur Oxides," U.S. Dept. of Health, Education and Welfare, National Air Pollution Control Administration Publication No. AP-50, Washington, D.C., 1969.
5. Urone, P., H. Lutsep, C. M. Noyes and J. F. Parcher: *Environ. Sci. Technol.*, vol. 2, p. 611, 1968.
6. Scott, W. D. and D. Lamb: *J. Am. Chem. Soc.*, vol. 92, p. 3943, 1970.
7. Stern, A. C. (ed.): "Air Pollution," 2d ed., Academic Press, New York, 1968.
8. Anonymous, *Environ. Sci. Technol.*, vol. 5, p. 503, 1971.

Chapter 4

AIR POLLUTION CONTROL INSTRUMENTATION AND AIR QUALITY DETECTORS

L. J. Bollyky A. Brodgesell
M. Chiba L. Dubois R. J. Gordon
R. A. Herrick J. S. Jacobson
R. C. C. Lao B. G. Lipták D. H. F. Liu
J. L. Monkman D. J. Sibbett
R. G. Smith A. Turk J. I. Yellott

CONTENTS OF CHAPTER 4*

*For a glossary of terms, refer to the back of this volume and for a list of abbreviations used in this chapter refer to the listing given after these contents, preceding Section 4.1.

Contents of Chapter 4

Abbreviations

Å	Angstrom (10^{-10} meters)
AA	Atomic Absorption
AAS	AA Spectrophotometer
ACFM	Actual Cubic Feet per Minute
AEC	Atomic Energy Commission
ALPM	Actual Liter per Minute
AM	Ante Meridiem
bp	Boiling Point
°C	Degree Centigrade
cc	Cubic Centimeter
cm	Centimeter
COH	Coefficient of Haze
DC	Direct Current
DIA	Diameter
ECD	Electron Capture Detector
EPA	Environmental Protection Agency
FID	Flame Ionization Detector
ft	Foot
g	Gram
GC	Gas Chromatography
GLC	Gas Liquid Chromatography
ID	Inside Diameter
IR	Infrared
IUPAC	International Union of Pure and Applied Chemistry

KI	Potassium Iodide
KV	Kilovolt (1,000 volts)
l	Liter
LF	Linear Feet
lpm,LPM	Liters per Minute
ly	Langley
m,M	Meter
mg	Milligram (10^{-3} gram)
ml	Milliliter
mm	Millimeter
MS	Mass Spectrometry
N	Nitrogen or Normal
NBS	National Bureau of Standards
NDIR	Nondispersive Infrared
ng	Nanogram (10^{-9} gram)
nm	Nanometer (10^{-9} meter)
OD	Outside Diameter
OECD	Organization of Economic Cooperation and Development
OSHA	Occupational Safety and Health Administration
PAH	Polynuclear Aromatic Hydrocarbons
PCB	Polychlorinated Biphenyls
pg	Picogram (10^{-12} gram)
ppb	Parts per Billion
ppm	Parts per Million
ppt	Parts per Thousand (also parts per Trillion)
psig	Pounds per Square Inch Gauge
RUDS	Reflectance Units of Dirt Shale
SCFH	Standard Cubic Foot per Hour
SCFM	Standard Cubic Foot per Minute
Sec	Second
UV	Ultra Violet
V	Volt
WHO	World Health Organization
μ	Micron (10^{-6} meter)
μg	Microgram (10^{-6} gram)
$\mu g/m^3$	ppm (24,500/molecular weight)

4.1 SUMMARY OF AVAILABLE INSTRUMENTS

This Section provides a condensed summary of the instrumentation hardware available for air pollution control. These application and selection guides are organized in tabular format and are grouped by the function which the instrument serves. In many of the tables, a reference is given to the corresponding section of the "Instrument Engineers' Handbook,"[1] where the noted device is discussed in greater detail. Also included is information on the instrument symbols that are recommended for use by the Instrument Society of America.

In 1973, the Environmental Instrumentation Group of the Lawrence Berkeley Laboratory under the sponsorship of the National Science Foundation published the first volume of a five volume survey titled: "Instrumentation for Environmental Monitoring". This is the most informative and comprehensive survey yet published on the topic.

[1]Lipták, B. G. (ed.): Instrument Engineers' Handbook, Volumes I, II and Supplement One, Philadelphia, Chilton, 1969, 1970 and 1972.

Table 4.1a
ORIENTATION TABLE FOR LEVEL DETECTORS

Section[a]	Service / Instrument Type	Liquids: Switches		Liquids: Local Indicators		Liquids: Self-Contained Local Controller		Liquids: Transmitters		Solids	
		Clean Fluid	Foam	Accounting	Standard	Clean	Hard to Handle	Clean	Hard to Handle	Switches	Transmitters
1.1	Float Level Switches	G	—	—	—	—	—	—	—	—	—
1.2	Displacer Level Switches	E	—	—	—	—	—	—	—	—	—
1.3	Capacitance Probes	G	F-G	P-F	F	—	—	F	P-F	G	P-F
1.4	Vibrating Reed Level Switches	G	P-F	—	—	—	—	—	—	G	—
1.5	Ultrasonic Level Detectors	G	P-F	P-F	F	—	—	G	P-F	G	E
1.6	Conductivity Probes	F	P	—	—	—	—	—	—	P-F	—
1.7	Radiation Gauges	G	G	E	G	—	—	G	E	G-E	F
1.8	Air Bubblers	P-F	—	P-F	P-F	P-F	P-F	P-F	P-F	—	—
1.9	Diaphragm Level Sensors	P-F	—	P-F	P-F	P-F	P-F	P-F	P-F	G-E	—
1.10	Thermal Level Sensors	G	P-F	P-F	P-F	—	—	P-F	P-F	P-F	P-F
1.11	Tape Level Gauges	G	—	E	G	—	—	G	P-F	G	G-E
1.12	Std. & Magnetic Gauge Glasses	F	—	F	E	—	—	—	—	—	—
1.13	Slip Tubes	—	—	P-F	P-F	—	—	—	—	—	—
1.14	Differential Pressure Level Detectors	G	—	P-F	G	G	P-F	G-E	G-E	—	—
1.15	Displacement Level Sensors	F	—	P-F	F-G	G-E	P-F	F	P-F	—	—
1.16	Float Level Indicators	—	—	P-F	F	—	—	P-F	P-F	—	—
1.17	Rotating Paddle Level Switches	—	—	—	—	—	—	—	—	G	—
1.18	Metric Tape Level Devices	—	—	P-F	F	—	—	P-F	P-F	—	P-F
1.19	Level Control Valves	P-F	—	—	—	F	P-F	—	—	—	—

E—Excellent G—Good F—Fair P—Poor

[a] The noted section numbers are those of the Instrument Engineers' Handbook, Volume I and the reader may consult this reference for further details concerning the selected sensor.

Table 4.1b
ORIENTATION TABLE FOR PRESSURE DETECTORS

Category	Section[a]		Type of Design	Inline Device	Laboratory or Pilot Plant Device	Local Readout (Gauge)	Remote Readout Trans.
High Vacuum	2.1	Ionization	Hot Cathode		√	√	√
			Cold Cathode		√	√	√
			Radiation Alphatron		√	√	√
	2.2	Thermal	Thermocouple	√	√	√	√
			Thermopile	√	√	√	√
			Resistance Wire	√	√	√	√
	2.3	Mechanical	McLeod	√	√	√	
			Molecular Momentum	√	√	√	
			Capacitance	√	√	√	√
Moderate Positive & Vacuum Press.	2.4	Diaphragm	Abs. Press. Motion Balance	√	√	√	√
			Abs. Press. Force Balance	√	√		√
			Atm. Press. Ref. Motion Bal.	√	√	√	√
			Atm. Press. Ref. Force Bal.	√	√		√
			Press. Repeaters	√	√		√
	2.5	Bellows	Abs. Press. Motion Balance	√	√	√	√
			Abs. Press. Force Balance	√	√		√
			Atm. Press. Ref. Motion Bal.	√	√	√	√
			Atm. Press. Ref. Force Bal.	√	√		√
			Aneroid Manostats	√	√		

Applicable Pressure Ranges

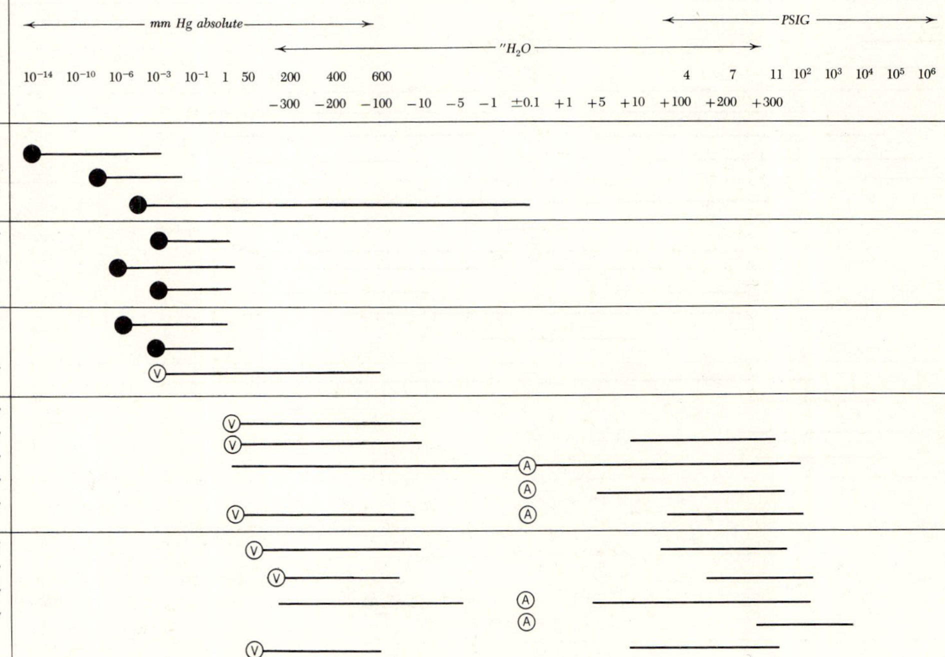

Table 4.1b (Continued)

Category	Section[a]		Type of Design	Inline Device	Laboratory or Pilot Plant Device	Local Readout (Gauge)	Remote Readout Trans.
Moderate Positive & Vacuum Press.	2.6	Bourdon	C. Bourdon	√	√	√	√
			Spiral Bourdon	√	√	√	√
			Helical Bourdon	√	√	√	√
			Quartz Helix		√	√	√
	2.7	Manometers	Inverted Bell	√	√	√	√
			Ring Balance	√	√	√	√
			Float Manometer	√	√	√	√
			Barometers	√	√	√	
			Visual Manometers	√	√	√	
			Micro Manometers		√	√	
			Cartesian Divers	√	√		
	2.8	Elec-tronic	Strain Gauge	√	√	√	√
			Electronic Transmitters	√	√		√
			Capacitive Sensors	√	√		√
High Press. Sensors	2.9	Special	Dead Weight Piston Gauge		√	√	
			Bulk Modulus Cell	√	√		√
			Manganin Cell		√		√

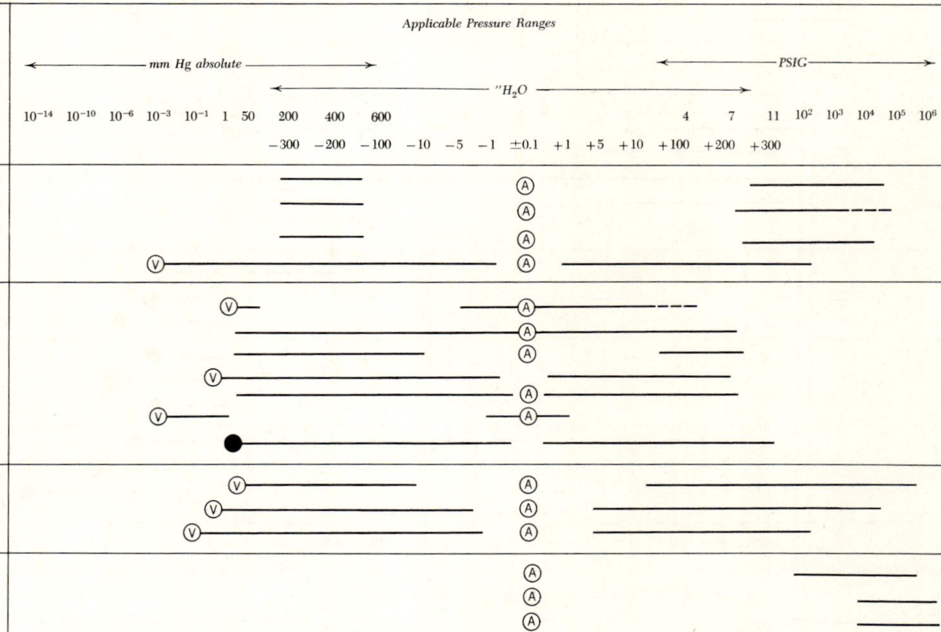

Ⓥ Indicates that the device uses full vacuum reference in its operation.
Ⓐ Indicates that the instrument uses atmospheric pressure reference.
● Indicates that the operating principle used does not involve the use of reference pressures.

[a]The noted section numbers are those of the Instrument Engineers' Handbook, Volume I and the reader may consult this reference for further details concerning the selected sensor.

Table 4.1c
ORIENTATION TABLE FOR DENSITY SENSORS

Section Number[a]	LIQUID Density Sensor Design	Applicable To: Clean Process Streams	Applicable To: Slurry Service	Applicable To: Viscous or Polymer Streams	Minimum Span Based on Water SG = 1.0	Accuracy (% span)	Design Pressure and Temperature Limitations (PSIG/°F)	Temperature Compensation Available	Direct Local Indicator	Transmitter
3.1	Hydrometers	✓			0.1	1%	100/200	✓	✓	✓
3.2	Chain Balanced Float	✓			0.005	1–3%	125/400	✓	✓	✓
3.2	Electro-Magnetic Suspension	✓			0.01	1%	200/350	✓		✓
3.2	Buoyant Force Displacer	✓			0.005	1%	600/400	N.S.		✓
3.3	Hydraulic Head Type	✓	✓	✓	0.05	½–1%	5,000/350	N.S.	✓	✓
3.4	Radiation Gauges	✓	✓	✓	0.05	1%	Unlimited	✓		✓
3.5	Weight of Fixed Volume Bulb	✓			0.05	1%	2,400/500	✓	✓	✓
3.5	Weighing of U or Straight Tube	✓	✓	L	0.02	1–2%	200/300	N.S.		✓
3.6	Vibrating U-Tube	✓			0.05	1–3%	1,000/300	✓		✓
3.7	Angular Position of Displacers	✓			0.1	½%	1000/500	N.S.		✓

N.S. = Non-Standard L = Limited

Section Number[a]	GAS Density Sensor Design	Minimum Span Based on Air SG = 1.0 or in lbm/ft³ Units	Accuracy (% span)	Design Pressure and Temperature Limitations (PSIG/°F)	Manually Operated Indicator	Continuous Indicator	Transmitter
3.8	Manual Displacement Elements	0.02 SG	1%	Near Ambient	✓		
3.8	Electro-Magnetic Suspension	0.01 SG	1%	Near Ambient			✓
3.8	Gas Column Balance	0.25 SG	0.002 SG	Near Ambient		✓	✓
3.8	Thermal Type	0.5 SG	0.01 SG	Near Ambient			✓
3.8	Viscous Drag Type	0.1 SG	1–2%	Near Ambient		✓	✓
3.9	Buoyant Force Displacer	1.0 lbm/ft^3	0.25%	1,500/200	✓	✓	✓
3.9	Centrifugal Type	1.0 lbm/ft^3	0.1–0.5%	2,000/300		✓	✓

[a]The noted section numbers are those of the Instrument Engineers' Handbook, Volume I and the reader may consult this reference for further details concerning the selected sensor.

Table 4.1d
ORIENTATION TABLE FOR TEMPERATURE SENSORS

Section[a]	Type
4.1	Glass Stem Thermometer
4.2	Bi-Metallic Elements
4.3	Filled Elements—Liquid
	Vapor
	Gas
	Mercury
4.4	Color Indicators
4.5	Pyrometric Indicators
4.6	Resistance Bulbs—Nickel
	Platinum
4.7	Thermistors
4.8	Quartz Crystals
4.9	Thermocouples—Type T
	Type J
	Type K
	Type R & S
4.10	Radiation Pyrometers
4.11	Manual Optical Pyrometers
4.12	Infrared Pyrometers
4.13	Misc.—Spectroscopy
	—Paramagnetic Salts
	—Diodes
	—Transistors
	—Liquid Crystals
	—Carbon Resistors
	—Pneumatic Probe
	—Fluidic Sensor
	—Ultrasonic
	—Thermography

▬▬ Recommended ■■■■ Available but not recommend

[a]The noted section numbers are those of the Instrument Engineers' Handbook, Volume I and the read may consult this reference for further details concerning the selected sensor.

Available Span			Accuracy		Cost ($)			Sensor Size			Available With										
Under 100°F	Between 100–1000°F	Above 1000°F	% Full Scale	Best Attainable in °F	Under $50	Between $50–500	Above $500	Small	Medium	Large	Recorder	Indicator	Controller	Stability	Repeatability	Response Time	Sensitivity	① Interchangeable	Linear	② Complete System	Maximum Distance to Readout (ft) ③
✓	✓		0.1–2	0.01	✓				✓			✓		E	E	G	G	✓	✓	✓	N
✓	✓		1–2	0.2	✓			✓	✓		✓	✓	✓	E	F	G	G		N	✓	N
✓	✓		0.5–2	0.1	✓	✓		✓	✓		✓	✓	✓	E	F	F	G		✓	✓	40
✓	✓		0.5–2	0.6	✓	✓			✓		✓	✓	✓	E	F	G	F		N	✓	200
	✓	✓	0.5–2	1.2	✓	✓				✓	✓	✓	✓	E	F	F	F		✓	✓	150
✓	✓	✓	0.5–2	0.25	✓	✓		✓	✓		✓	✓	✓	E	F	F	G		✓	✓	50
✓			—	8	✓							✓					F			✓	N
✓			—	8	✓			✓	✓			✓					F			✓	N
✓	✓		0.1	0.2	✓	✓	✓	✓	✓		✓	✓	✓	G,E	E	G,E	G,E	✓	N	N	1000
✓	✓	✓	0.1	0.05		✓	✓	✓	✓		✓	✓	✓	G,E	E	G,E	G,E	✓	✓	N	3000
✓	✓		0.1	0.2		✓		✓			✓	✓	✓	G,E	G	G,E	E	✓	N	N	3000
✓	✓		0.1	0.2			✓		✓	✓	✓	✓		E	G	G	E	✓	✓	N	1000
	✓		0.3–1	1	✓	✓		✓	✓		✓	✓	✓	G	G,E	E	G,E	✓	N	N	3000
	✓	✓	0.3–1	2	✓	✓		✓	✓		✓	✓	✓	G	G,E	E	G,E	✓	N	N	3000
	✓	✓	0.3–1	2		✓		✓	✓		✓	✓	✓	G	G,E	E	G,E	✓	✓	N	3000
	✓	✓	0.3–1	3		✓		✓	✓		✓	✓	✓	G	G,E	E	G,E	✓	N	N	3000
		✓	0.5–1	5			✓		✓	✓	✓	✓	✓	G	G,E	G	F		N	N	2000
		✓	1–2	10		✓	✓		✓	✓		✓				F,G			N	✓	N
	✓	✓	1–2	5			✓		✓	✓	✓	✓	✓			E			N	✓	1000
		✓	1				✓			✓		✓		G	G,E	G			N	✓	
✓			1	0.005			✓			✓		✓		G	G,E	E	F,G		N	✓	
✓			2			✓		✓	✓			✓		G	G	G	G		✓	N	
✓			2	1		✓		✓	✓			✓		G	G	G	G		✓	N	
✓				1	✓					✓		✓		E	G,E	G	E,G			✓	
✓				0.5		✓		✓	✓			✓		F,G	G	G	G		N	N	
	✓		2			✓				✓		✓		F	F	F	F		N	N	
	✓		2			✓				✓		✓		F	F	F	F		N	N	
✓	✓	✓	5				✓			✓		✓		G	F,G	E	E,G		N	N	
	✓			1	✓	✓						✓					F			✓	

Terminology

N—No or None
E—Excellent
G—Good
F—Fair

① Interchangeable sensor, without recalibration of entire system.

② System is complete when sensor and readout is sold as single unit. When several readouts can be used with the same sensor, system is not considered to be complete.

③ Without special compensation.

Table 4.1e
ORIENTATION TABLE FOR FLOW SENSORS

Section[a]	Type of Design	Applicable to Detect the Flow of: Clean Liquid	Viscous Liquid	Slurry	Gas	Solids	Direct Mass—Flow Sensor	Volumetric Flow Detector	Flow Rate Sensor	Inherent Totalizer	Direct Indicator	Transmitter Available	Linear Output	Rangeability	Pressure Loss Thru Sensor	① Approx. Straight Pipe-Run Requirement (Upstream Diam/Downstream Diam.)	Accuracy ° ±% Full Scale °° ±% Rate
5.1	Orifice (Plate or Integral Cell) (V)	✓	L		✓			✓	✓			✓	SR	3:1②	H	20/5	¼–2°
5.2	Venturi Tubes and Flow Nozzles	✓	L	L	✓			✓	✓			✓	SR	3:1②	M	20/5	¼–3°
5.3	Pitot Tubes	✓			✓			✓	✓			✓	SR	3:1②	M	40/10	2–5°
5.4	Elbow Taps	✓	L	L	✓			✓	✓			✓	SR	3:1②	N	25/10	5–10°
5.5	Target Meters (V)	✓	L	L	✓			✓	✓		SD	✓	SR	4:1	H	15/5	½–2°
5.6	Weirs, Flumes	✓	L	L				✓	✓			✓	SD	100:1	M	See Text	2–5°
5.7	Rotameters (V)	✓	L	L	✓			✓	✓		✓	✓	✓	10:1	A	N	½–2°
5.8	Magnetic Flowmeters (V)	✓	✓	✓				✓	✓			✓	✓	20:1	N	N	½–1°
5.9	Turbine Flowmeters (V)	✓	L	SD	L			✓	✓			✓	✓	20:1	H	15/4	¼°°
5.10	Swirl Flowmeter	L	L	L	✓			✓	✓			✓	✓	100:1	H	10/0	½–1°°
5.11	Laminar Flowmeters	✓	L	L	✓			✓	✓			✓	✓	10:1	H	15/5	1–2°
5.12	Thermal Flowmeters	✓	L	L	✓		✓		✓			✓	L	20:1	A	5/3	1–2°
5.13	Ultrasonic Flowmeters (V)	✓	L	L				✓	✓			✓	✓	10:1	N	5/3	2°
5.14	Angular Momentum Types (V)	✓	L	L	✓		✓		✓	SD	SD	✓	✓	12:1	H	5/3	½–1°
5.15	Liquid Displacement Meters (V)	✓	L					✓		✓	✓	SD	✓	20:1	A	N	¼–1°°
5.16	Gas Displacement Meters (V)				✓			✓		✓	✓	SD	✓	20:1	A	N	½–1°
5.17	Metering Pumps (V)	✓	✓	✓				✓		✓		SD	✓	20:1	—	N	⅒–1°°
5.18	Solids Flowmeters (V)		SD	SD		✓	SD	SD	✓	✓	SD	✓	✓	20:1	—	5/3	½–3°
5.19	Self Contained Regulators	✓						✓	✓		SD			20:1	H	N	2–10°

L = Limited
SD = Some designs
H = High
A = Average
M = Minimal
N = None

(V)—Design so noted is immune to viscosity variations. (Either up to a maximum limit or completely.)
SR—Square Root
①—The data in this column is for general guidance only. The reader shall consult Figure A-36 in th appendix for precise information.

[a] The noted section numbers are those of the Instrument Engineers' Handbook, Volume I and the reade may consult this reference for further details concerning the selected sensor.

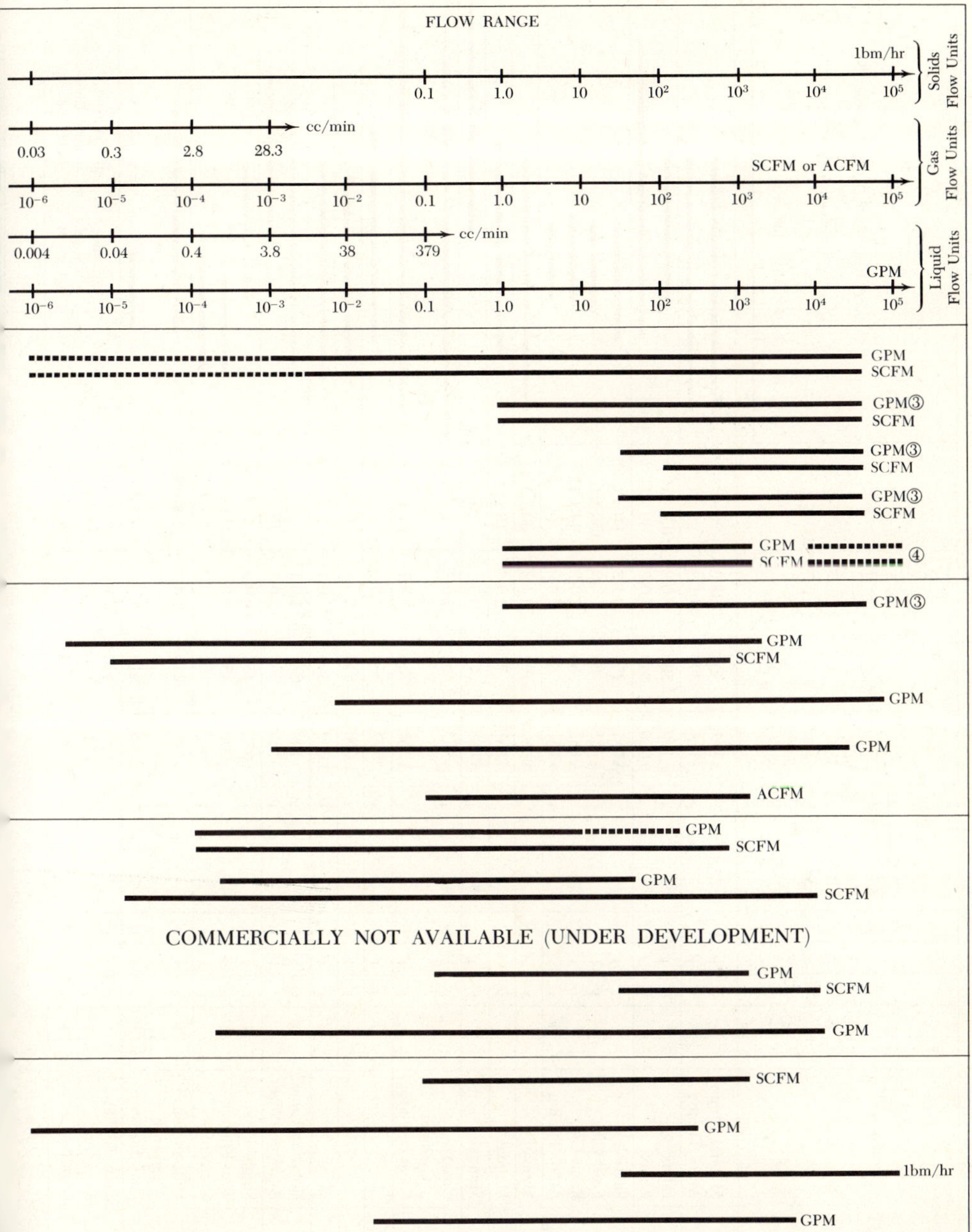

■■■■■■ = Non standard range

)—Inherent rangeability of primary device is substantially greater than shown. Value used reflects limitation of differential pressure sensing device, when maximum readout accuracy is desired.

)—Pipe size establishes the upper limit.

)—Practically unlimited with the probe type design.

Table 4.1f

ORIENTATION TABLE FOR VISCOMETERS

Section[a]	Type of Design	Features	Provides Continuous Signal	In-Line Device	Laboratory Device	Local Readout	Remote Readout Trans.	Temp. Compensation	Fluids: Gas	Fluids: Newtonian	Fluids: Non-Newtonian	Maximum Design Pressure, PSIG.	Maximum Design Temperature, °F	Accuracy, % (1) Based on Full Scale (2) Based on Measurement	Minimum Sample Size or Flow Rate	APPLICABLE VISCOSITY RANGES° Centipoises
6.1	Capillary Tube	Manual Timing			√	√				√		ATM.	300	0.35(2)	20 CC	
		Auto. Timing		√	√	√	√			√		ATM.	300	0.01(2)	20 CC	
6.2	Capillary Extrusion	Influx			√	√				√	√	100	300	2.0(2)	0.7 CC	
		Efflux			√	√	√			√	√	5,000	640	2.0(2)	30 CC	
6.3	Efflux Cup	Saybolt			√	√				√		ATM.	250	0.1(2)	60 CC	
		Ford Cup			√	√				√		ATM.	80	2.0(2)	150 CC	
		Zahn Cup			√	√				√		ATM.	80	2.0(2)	44 CC	
		Auto. Timing		√	√	√	√			√	√	ATM.	80	5.0(2)	—	
6.4	Falling Ball	Manual			√	√			√	√		ATM.	300	0.1(2)	30 CC	
6.5	Bubble Time	Manual			√	√				√		ATM.	77	2.0(2)	13 CC	
6.6	Sliding Plate	Manual			√	√				√	√	ATM.	77	0.5(2)	2 CC	
6.7	Rotational Laboratory	Coaxial-Cylinder			√	√	√			√	√	ATM.	2,500	1.0(2)	1 CC	
		Cone & Plate			√	√	√			√	√	ATM.	750	0.5(2)	0.1 CC	
6.8	Falling Piston			√	√	√	√	√		√	√	500	650	1.0(1)	—	
6.9	Ultrasonic		√	√	√	√	√	√		√	√	1,000	650	2.0(1)	0.6 GPH	
6.10	Vibrating Reed		√	√	√	√	√	√		√	√	3,000	300	1.0(1)	—	
6.11	Rotational Industrial	Rotating Cone	√	√	√	√	√	√		√	√	100	650	1.0(1)	—	
		Agitator Power	√	√		√	√			√	√	125	200	~5.0(1)	15 GPM	
6.12	Float	Single Float	√	√	√	√	√			√	√	650	450	4.0(2)	0.75 GPM	
		Two-Float	√	√	√	√				√		300	450	2.0(2)	0.25 GPM	
		Concentric	√	√	√	√	√	√		√	√	650	450	2.0(2)	2 GPM	
6.13	Continuous Capillary	Differential Pressure	√	√	√	√	√			√		115	240	1.0(1)	1 GPH	
		Back Pressure	√	√	√	√	√			√		65	210	0.5(1)	1 GPH	
6.14	Plastometer	Cone & Plate			√	√	√				√	100	400	0.5(1)	25 CC	Arbitrary Units Are Used
		Kneader	√	√	√	√	√				√	ATM.	570	1.0(1)	80 CC	Mooney % Scale MI & CIL
		Capillary	√	√	√	√	√				√	5,000	570	2.0(1)	0.6 #/HR	0–200 Points 0–1,000 Division 0–200 & 0–100

° ———— Normal Range – – – – – With Special Modifications.

Table 4.1g
ORIENTATION TABLE FOR WEIGHING SYSTEMS

Type of Scale / Features	Range	Installed Cost	Accuracy	Sensitivity	Temperature Sensitivity	Ease of Installation	Amount of Site Preparation Required	Amount of Maintenance Required	Applicable to Outdoor Installation	Applicable to Dynamic Loads	Remote Readout is Conveniently Available	Inherently Explosion Proof	Applicable to Corrosive Atmospheres	Primary Measurement Signal Generated
Mechanical Lever Scales	L–M	M–H	**E**	**E**	**L**	F	H	H	**Y**	F	N	**Y**	N	Displacement
Hydraulic Load Cells	M–H	**L**	G	G	M	**E**	**L**	**L**	**Y**	**E**	**Y**	**Y**	**Y**	Hydraulic Pressure
Pneumatic Load Cells	L	L–M	G	G	**L**	**E**	**L**	**L**	N	F	**Y**	**Y**	**Y**	Pneumatic Pressure
Strain Gauge Load Cells	**Un-limited**	M–H	G–E	**E**	H	**E**	M	**L**	**Y**	**E**	**Y**	N	**Y**	Electronic

L—Low
M—Medium
H—High

E—Excellent
G—Good
F—Fair

Y—Yes
N—No

▬ Compares Favorably with Others

Table 4.1h
ORIENTATION TABLE FOR ANALYZERS

Type	Principle of Operation	Sample Phase: Liquid	Gas	Solid	Accuracy	Continuous Semi-Continuous or Discontinuous	Specific or Non-Specific	Maintenance	Special Features
Chromatograph (Sect. 8.1)[a]	Chromatographic Separation of Components in Sample	√	√		E	S.C	S	Moderate	Multi-components
Ultra-Violet (Sect. 8.2)[a]	UV Energy absorbed by Measured Compound	√	√		E-G	C	S	Low	
Infra-Red (Sect. 8.3)[a]	IR Energy absorbed by Measured Compound	√	√		E-G	C	S	Low	
Colorimeters (Sect. 8.4)[a]	Visible Energy absorbed by Measured Sample	√		√	G	C	S	Low	
Thermal Conductivity (Sect. 8.5)[a]	T.C. of Sample Compared to Standard		√		G-P	C	N.S.	Low	
Refractometer (Sect. 8.6)[a]	Refractive Index of Sample Measured	√			E-G	C	N.S.	Low	In-Line Possible
Oxygen Detectors (Sect. 8.7)[a]	Deflection of test body due to paramagnetism of oxygen indicates concentration		√		E-G	C	S	Low	H_2 in sample interferes with measurement
	Thermal effect of oxygen's paramagnetism is used as a measure of its concentration		√		G	C	S	Low	Rugged unit. H_2 in sample interferes with measurement

	Fuel is burned with oxygen in sample and generated heat of combustion indicates concentration		√	E-G	C	S	Moderate	Requires continuous supply of fuel
	Oxygen in sample and reference gases is ionized by heat and relative ion concentrations are detected		√	E-G	C	S	Low-Moderate	Sample cannot contain combustibles. Fast response
	Electrode reduces oxygen into hydroxide and the resulting ion current is measured as concentration. (Polarographic or Galvanic cells.)	√	√	E-G	C	S	Low-Moderate	Probe design may eliminate sampling system
pH (Sect. 8.8)[a]	Glass membrane develops emf. prop. to H_2 ion concentration	√		E-G	C	S	Moderate	In-Line Possible
Moisture Analyzers (Sect. 8.9, 10 & 11)[a]	Latent Heat of Water		√	F	C & D.C.	S	Low for D.C. Moderate for C.	
	Dimension change of Fiber		√	P	C	S	Moderate	
	Solution Conductivity		√	F	C	S	Moderate	
	Surface Conductivity		√	E-G	C	S	Low	
	Color Change		√	P	D.C.	S	Low	Detects Moisture in excess of a limit

Table 4.1h (Continued)

Type	Principle of Operation	Sample Phase			Accuracy	Continuous Semi-Continuous or Discontinuous	Specific or Non-Specific	Maintenance	Special Features
		Liquid	Gas	Solid					
Moisture Analyzers (Sect. 8.9, 10 & 11)[a] *Continued*	Electrolysis of Water	√	√		E-G	C	S	Moderate	Special Sampling for liquids
	Change of Sample Dielectric Constant	√	√	√	E-G Fair for Solids	C	S	Low	
	Change of Probe Impedance	√	√		E-G	C	S	Low	
	Oscillating Frequency of a crystal changes with moisture content		√		E-G	C	S	Moderate	
	Heat of adsorption	√	√		G	C	S	Moderate	Special Sampling for liquids
	Infra-red energy absorbed by sample	√		√	G	C	S	Moderate	
	Micro-wave energy absorbed by sample	√		√	G	C	S	Low	
	Neutron Moderation	√		√	E	C	N.S.	Low	Density Measurement included
	Electrical Resistance of sample	√		√	P	C	S	Low	
Turbidity (Sect. 8.12)[a]	Light Attenuation	√			G	C	N.S.	Moderate	
	Light Scattering	√			G	C	N.S.	Low	

(Sect. 8.13)[a]	[illegible] shaped sensor	√		F	C	N.S.	Moderate	
	Differential pressure in a flow bridge	√		G	C	N.S.	Moderate	
Molecular Weight (Sect. 8.14)[a]	Osmotic Pressure	√		F	D.C.	N.S.	Moderate	
	Vapor Pressure Lowering	√		F	D.C.	N.S.	Moderate	
	Viscosity	√		F	D.C.	N.S.	Low	
	Light Scattering	√		E-G	D.C.	N.S.	Low	
	Chromatographic sorting of molecules by size (Gel Permeation)	√		E-G	S.C. & D.C.	N.S.	Moderate	On Stream Measurement possible
	Molecules magnified to visible size by an electron microscope	√	√	E	D.C.	N.S.	Moderate	
	Sorting of molecules by centrifugal forces	√		F	D.C.	N.S.	Low	Independent viscosity measurement required
	Count number of distinct atom groups in a sample of known molecular structure	√		E-G	D.C.	S	—	
Conductivity (Sect. 8.15)[a]	Measure Resistance of sample between electrodes of known area & spacing	√		E-G	C	N.S.	Low	
	Electrical coupling by the sample of two coils	√		G	C	N.S.	Low	

E-excellent
G-good
F-fair
P-poor

C-continuous
SC-semi-continuous
DC-discontinuous
S-specific
NS-non-specific

[a] The noted section numbers are those of the Instrument Engineers' Handbook, Volume I and the reader may consult this reference for further details concerning the selected sensor.

Table 4.1i
ANALYZER SELECTION TABLE

Grouped by the method of analysis applied

Code	Sensor Type / *Manufacturers of Online Analyzers:*	Agricultural Control Systems, Inc.	American Instrument Co.	Anacon, Inc.	Applied Automation, Inc.	Automated Environmental Systems	Automation Mfg. & Service Co.	Bacharach Instrument Co.	Bailey Meter Co.	Barton ITT	Beckman Instruments, Inc.	Bendix Corp.	Bristol Div., ACCO	Brookfield Engineering Labs., Inc.	Cambridge Systems, Inc. (EC&G)	CEC (Bell & Howell)	Corning Scientific Instruments Co.	Customline Control Panels, Inc.	Dasie, Corp.	Davis Instrument Co.	Delta Scientific Corp.	DeZurik Corp.	E. I. du Pont de Nemours & Co.	Electron Machine Corp.	Electronic Associates, Inc.	Farrand Optical Co.	Fischer & Porter Co.	The Foxboro Co.	GAM-RAD, Inc.	General Electric Co.	Hach Chemical Co.	Hallikainen Instruments	Hartman & Braun	Hays Corp.	Honeywell, Inc.	Houston Atlas, Inc.
	ELECTROMAGNETIC RADIANT ENERGY																																			
1	Ultrasonic																																			
2	Microwave																																			
3	Infrared			✓							✓	✓											✓							✓			✓			
	Visible																																			
4	Colorimeter		✓	✓							✓										✓		✓			✓	✓				✓	✓	✓			✓
5	Flame photometer																				✓		✓													✓
6	Refractive index			✓																				✓												
7	Turbidity	✓	✓						✓		✓										✓		✓				✓	✓	✓		✓	✓	✓		✓	
8	Ultraviolet		✓	✓				✓	✓		✓												✓			✓							✓			
9	Radiation (gamma, neutron)																													✓			✓			
	ELECTROCHEMICAL																																			
11	Electrolytic cells															✓																				
12	Galvanic cells			✓					✓												✓													✓		
13	pH										✓						✓				✓							✓					✓		✓	
14	Ion-selective electrodes										✓						✓				✓							✓						✓	✓	
15	ORP (Red-Ox)																				✓							✓							✓	
16	Polarographic cells																										✓								✓	
17	Conductivity										✓										✓							✓					✓		✓	

Grouped by the method of analysis applied

Manufacturers of Online Analyzers:

Code	*Sensor Type*	Agricultural Control Systems, Inc.	American Instrument Co.	Anacon, Inc.	Applied Automation, Inc.	Automated Environmental Systems	Automation Mfg. & Service Co.	Bacharach Instrument Co.	Bailey Meter Co.	Barton ITT	Beckman Instruments, Inc.	Bendix Corp.	Bristol Div., ACCO	Brookfield Engineering Labs., Inc.	Cambridge Systems, Inc. (EG&G)	CEC (Bell & Howell)	Corning Scientific Instruments Co.	Customline Control Panels, Inc.	Dasie, Corp.
	INFERENTIAL PROPERTY																		
17	Conductivity-resistivity										✓								
20	Consistency													✓					
21	Density, thickness								✓			✓							
22	Dielectric constant																		
23	Flash, melt, pour point																		
24	Flow index																		
25	Molecular weight										✓	✓							
26	Plastometry																		
27	Thermal conductivity		✓						✓		✓	✓							
28	Vapor pressure																		
29	Viscosity											✓		✓					
	COMBINATION, MISCELLANEOUS																		
	Chromatography																		
31	Adsorption				✓		✓				✓	✓				✓			
32	Partition				✓		✓				✓	✓				✓			
33	Gel-permeation																		
34	Total carbon, hydrocarbon																		
	Ionization (mass spectrometry)																		
35	Thermal & flame										✓	✓				✓			
36	Electrostatic															✓			
37	Radiation											✓							
38	Paramagnetic											✓							
39	Catalytic combustion							✓											
40	Piezoelectric																		
41	Wet chemistry (auto-titrators)																		

Code	*Sensor Type*	Davis Instrument Co.	Delta Scientific Corp.	DeZurik Corp.	E. I. du Pont de Nemours & Co.	Electron Machine Corp.	Electronic Associates, Inc.	Farrand Optical Co.	Fischer & Porter Co.	The Foxboro Co.	GAM-RAD, Inc.	General Electric Co.	Hach Chemical Co.	Hallikainen Instruments	Hartman & Braun	Hays Corp.	Honeywell, Inc.	Houston Atlas, Inc.
	INFERENTIAL PROPERTY																	
17	Conductivity-resistivity		✓							✓					✓		✓	
20	Consistency			✓					✓	✓								
21	Density, thickness									✓				✓				
22	Dielectric constant													✓				
23	Flash, melt, pour point													✓				
24	Flow index																	
25	Molecular weight																	
26	Plastometry																	
27	Thermal conductivity														✓	✓		
28	Vapor pressure													✓				
29	Viscosity								✓	✓				✓				
	COMBINATION, MISCELLANEOUS																	
	Chromatography																	
31	Adsorption																	
32	Partition																	
33	Gel-permeation																	
34	Total carbon, hydrocarbon		✓															
	Ionization (mass spectrometry)																	
35	Thermal & flame																	
36	Electrostatic						✓											
37	Radiation																	
38	Paramagnetic															✓		
39	Catalytic combustion	✓														✓		
40	Piezoelectric				✓													
41	Wet chemistry (auto-titrators)		✓										✓					

Table 4.1i (Continued)

Grouped by component for which analysis is being made

Sensor Type \ Manufacturers of Online Analyzers:	Agricultural Control Systems, Inc.	American Instrument Co.	Anacon, Inc.	Applied Automation, Inc.	Automated Environmental Systems	Automation Mfg. & Service Co.	Bacharach Instrument Co.	Bailey Meter Co.	Barton ITT	Beckman Instruments, Inc.	Bendix Corp.	Bristol Div., ACCO	Brookfield Engineering Labs., Inc.	Cambridge Systems, Inc. (EG&G)	CEC (Bell & Howell)	Corning Scientific Instruments Co.	Customline Control Panels, Inc.	Dasie, Corp.	Davis Instrument Co.	Delta Scientific Corp.	DeZurik Corp.	E. I. du Pont de Nemours & Co.	Electron Machine Corp.	Electronic Associates, Inc.	Farrand Optical Co.	Fischer & Porter Co.	The Foxboro Co.	GAM-RAD, Inc.	General Electric Co.	Hach Chemical Co.	Hallikainen Instruments	Hartman & Braun	Hays Corp.	Honeywell, Inc.	Houston Atlas, Inc.
Chlorine																				√						√									
Combustibles							√	√			√								√														√		
Dissolved oxygen																				√													√	√	
Flame								√																											
Humidity		√										√		√																					
H_2S									√		√																								√
Moisture in gas or liquid samples			√							√	√	√			√							√									√	√			
Moisture in solid samples			√																													√			
Oxygen			√				√	√		√	√									√												√	√		
Ozone																		√		√						√									
Sampling systems						√	√			√	√						√														√		√		
Smoke and dust density					√			√														√													
SO_2 and SO_3					√				√							√			√																
Water quality monitoring packages					√															√														√	

Grouped by the method of analysis applied

Code	Sensor Type / *Manufacturers of Online Analyzers:*	Ikor, Inc.	Industrial Nucleonics Corp.	Infrared Industries, Inc.	Jacoby-Tarbox Corp.	Kay-Ray, Inc.	Kollmorgen Corp.	Leeds & Northrup Co.	Lockwood & McLorie, Inc.	Microwave Instruments Co.	Mine Safety Appliances, Inc.	Monsanto Research Corp.	Norcross Corp.	Nuclear Chicago Corp.	NUS Corp.	Ohmart Corp.	Olkon Corp.	Panametrics, Inc.	Permutit Co.	Precision Scientific Development Co.	Princo Instruments, Inc.	Process Analyzers, Inc.	Research Appliance Co.	Seiscor Div., Seismograph Service Corp.	Taylor Instrument Process Control Div.	Technicon Corp.	Teledyne Analytical Instruments, Inc.	Thermo-Lab Instruments, Inc.	UGC Industries, Inc.	Union Carbide Corp.	Vap-Air Div., Vapor Corp.	Waters Associates, Inc.	Westinghouse Electric Corp.
	Electromagnetic Radiant Energy																																
1	Ultrasonic														✓																		
2	Microwave									✓																							
3	Infrared			✓				✓			✓														✓		✓						
	Visible																																
4	Colorimeter						✓																				✓						
5	Flame photometer																																
6	Refractive index								✓																								
7	Turbidity				✓																						✓			✓			
8	Ultraviolet																										✓						
9	Radiation (gamma, neutron)		✓			✓								✓		✓		✓					✓		✓								
	Electrochemical																																
11	Electrolytic cells								✓													✓								✓			✓
12	Galvanic cells								✓													✓					✓	✓					
13	pH							✓																						✓			
14	Ion-selective electrodes							✓																									
15	ORP (Red-Ox)							✓																						✓			
16	Polarographic cells																																
17	Conductivity							✓																						✓			

Table 4.1i (Continued)

Manufacturers of Online Analyzers:

Grouped by the method of analysis applied

Code	*Sensor Type*	Ikor, Inc.	Industrial Nucleonics Corp.	Infrared Industries, Inc.	Jacoby-Tarbox Corp.	Kay-Ray, Inc.	Kollmorgen Corp.	Leeds & Northrup Co.	Lockwood & McLorie, Inc.	Microwave Instruments Co.	Mine Safety Appliances, Inc.	Monsanto Research Corp.	Norcross Corp.	Nuclear Chicago Corp.	NUS Corp.	Ohmart Corp.	Olkon Corp.	Panametrics, Inc.	Permutit Co.	Precision Scientific Development Co.	Princo Instruments, Inc.	Process Analyzers, Inc.	Research Appliance Co.	Seiscor Div., Seismograph Service Corp.	Taylor Instrument Process Control Div.	Technicon Corp.	Teledyne Analytical Instruments, Inc.	Thermo-Lab Instruments, Inc.	UGC Industries, Inc.	Union Carbide Corp.	Vap-Air Div., Vapor Corp.	Waters Associates, Inc.	Westinghouse Electric Corp.
	INFERENTIAL PROPERTY																																
17	Conductivity-resistivity							√																						√			
20	Consistency							√					√																				
21	Density, thickness		√			√								√	√	√			√		√	√			√				√				
22	Dielectric constant																	√															
23	Flash, melt, pour point																			√													
24	Flow index																																
25	Molecular weight											√												√									
26	Plastometry											√												√									
27	Thermal conductivity							√			√											√					√						
28	Vapor pressure																			√													
29	Viscosity											√	√				√							√									
	COMBINATION, MISCELLANEOUS																																
	Chromatography																																
31	Adsorption										√											√											
32	Partition										√											√											
33	Gel-permeation																															√	
34	Total carbon, hydrocarbon																					√								√			
	Ionization (mass spectrometry)																																
35	Thermal & flame										√																√						
36	Electrostatic										√																						
37	Radiation										√			√																			
38	Paramagnetic										√																						
39	Catalytic combustion										√																						
40	Piezoelectric																																
41	Wet chemistry (auto-titrators)																									√				√			

Grouped by component for which analysis is being made

Sensor Type / Manufacturers of Online Analyzers:	Ikor, Inc.	Industrial Nucleonics Corp.	Infrared Industries, Inc.	Jacoby-Tarbox Corp.	Kay-Ray, Inc.	Kollmorgen Corp.	Leeds & Northrup Co.	Lockwood & McLorie, Inc.	Microwave Instruments Co.	Mine Safety Appliances, Inc.	Monsanto Research Corp.	Norcross Corp.	Nuclear Chicago Corp.	NUS Corp.	Ohmart Corp.	Olkon Corp.	Panametrics, Inc.	Permutit Co.	Precision Scientific Development Co.	Princo Instruments, Inc.	Process Analyzers, Inc.	Research Appliance Co.	Seiscor Div., Seismograph Service Corp.	Taylor Instrument Process Control Div.	Technicon Corp.	Teledyne Analytical Instruments, Inc.	Thermo-Lab Instruments, Inc.	UGC Industries, Inc.	Union Carbide Corp.	Vap-Air Div., Vapor Corp.	Waters Associates, Inc.	Westinghouse Electric Corp.
Chlorine																																
Combustibles			√							√																						
Dissolved oxygen																													√			
Flame																																
Humidity																	√													√		
H_2S																					√											
Moisture in gas or liquid samples							√			√							√				√		√			√				√		
Moisture in solid samples					√				√	√		√			√									√								
Oxygen							√	√													√					√	√					√
Ozone																																
Sampling systems								√		√													√									
Smoke and dust density	√			√																		√							√			
SO_2 and SO_3							√			√											√				√							
Water quality monitoring packages																									√				√			

Table 4.1j
ORIENTATION TABLE FOR CONTROL VALVES

Features / Type of Control Valve Body	*Split Body, Single Seated*	*Angle Valve, Single Seated*	*Globe Valve, Top-Bottom Guided, Double Seated, V-port*	*Cage Valve, Top Entry, Single Seated*	*Butterfly Valve 60° Open*	*Butterfly Valve 90° Open*	*Saunders Diaphragm Valve*
Applicable Section[a]	1.3	1.3	1.3	1.3	1.9	1.9	1.10
Capacity [$C_v = (Cd)(d^2)$] Critical	$10d^2$	$14d^2$	$10d^2$	$12d^2$	$12d^2$	$20d^2$	$14d^2$
Non-critical	$12d^2$	$28d^2$	$12d^2$	$14d^2$	$17d^2$	$35d^2$	$22d^2$
Flow Characteristics	Good	Good	Good	Good	Fair–Good	Fair	Poor[4]
Cost	Medium	Medium–High	Medium	Medium	Low	Low	Low–Medium
Available Size Range (in.)	½–12	1–6	1–24	1–4	1–150	1–150	¼–20
Max./Min. Operating Pressure (PSIG)	6,000/vacuum	6,000/vacuum	6,000/vacuum	6,000/vacuum	2,000/vacuum	2,000/vacuum	200/<1 mm Hg[9]
Max. Pressure Drop (PSID)	3,000	3,000	2,000	1,000	1,000	1,000	100[5,9]
Max./Min. Operating Temp. (°F)	1,200/−40	1,200/−40	1,200/−40	1,200/−40	2,000[2]	2,000[2]	350/−80[7]
Available in Jacketed Construction	No	Yes	Yes	Yes	No	No	Yes
Applicable to Flashing Service	Yes	Yes	Limited	Yes	No	No	No
Applicable to Cavitating Service	Limited	Limited	Yes	Yes	No	No	No
Applicable to High Viscosity ($\mu > 10^4$ cps) Service	Limited	Limited	No	No	Limited	Limited	Yes—Limited
Applicable to Slurry Service	Yes	Yes–Limited	No	Limited	Limited	Limited	Yes
Tight Shut-off Available[1]	Yes[3]	Yes[3]	No (½–2% leakage)	Yes[3]	Yes[3]	Yes[3]	Yes
Available Rangeability[9]	Up to 50:1	Up to 50:1	30:1 to 50:1	30:1	20:1	30:1	15:1[6]
Has "Fail-safe" (Open or Closed) Action	Yes	Yes	Yes	Yes	Yes	Yes	Yes
Available in Variety of Materials	Yes	Yes	Yes	Yes	Yes	Yes	Limited[7]
Flow in Either Direction	Yes	Yes	Possible	No	Some designs	Some designs	Yes
Unobstructed Streamline Flow	Not fully	Some designs	No	No	Not fully	Not fully	Almost[8]
Has Tendency to Cavitate Due to High Recovery	Some	Yes	No	No	Yes	Yes	Yes
Stem or Shaft Sealed from Process by	Packing	Packing	Packing	Packing	Packing and O-ring	Packing and O-ring	Diaphragm
Valve Trim Easily Removed	Yes	No	Yes	Yes	No	No	No
Low Maintenance	Yes	No	No	Yes	Yes	Yes	No

[1] See Section 1.21[a] for a discussion of leakage.
[2] Metal-to-metal seated designs only.
[3] Soft-seated designs only.
[4] Fair with dual range design.
[5] Diaphragm life short at high ΔP.
[6] Better with dual range design.
[7] Limited by diaphragm material.
[8] Yes, with straight-through design.
[9] Refer to section 1.21[a] for the definition of rangeability.

Applicable Section[a]	1.11	1.12	1.13	1.14	1.14	1.14
Capacity [$C_v = (Cd)(d^2)$] Critical	$14d^2$	$14d^2$	$25d^2$	$5d^2$–$10d^2$	$18d^2$	$8d^2$
Non-critical	$24d^2$	$22d^2$	$60d^2$	$6d^2$–$12d^2$	$22d^2$	$9d^2$
Flow Characteristics	Good	Good	Poor[12]	Fair	Fair	Fair
Cost	Medium	Medium	Low	Medium	Low	Low
Available Size Range (in.)	⅛–42	2–16	⅛–24	½–6	2–36	1–3
Max./Min. Operating Pressure (PSIG)	10,000/<1 mm Hg	350/<1 mm Hg	150/vacuum	300/vacuum	10^{14}	10,000/vacuum
Max. Pressure Drop (PSID)	1,000	200	60	300	10^{14}	5,000
Max./Min. Operating Temperature (°F)	1,000/−300	600	350	450	1,000	750
Available in Jacketed Construction	Yes	No	Yes	No	No	No
Applicable to Flashing Service	No	No	No	Yes	No	Yes
Applicable to Cavitating Service	No	No	No	Yes	No	Yes
Applicable to High Viscosity ($\mu > 10^4$ cps) Service	Yes—Limited	Limited	Yes	No	Limited	No
Applicable to Slurry Service	Yes	Yes	Yes	No	Limited	No
Tight Shut-off Available[1]	Yes	Yes	Yes	Yes[3]	Yes[3]	Yes[3]
Available Rangeability[9]	Over 50:1	Over 50:1	15:1	40:1	15:1	40:1
Has "Fail-safe" (Open or Closed) Action	Yes	Yes	No	Yes	No	Yes
Available in Variety of Materials	Yes	Limited	Limited[13]	No	No	No
Flow in Either Direction	Yes	No	Yes	No	Yes	Yes
Unobstructed Streamline Flow	Not fully	No	Yes	No	No	No
Has Tendency to Cavitate Due to High Recovery	Yes	Yes	Yes	No	No	No
Stem or Shaft Sealed from Process by	O-ring	O-ring/packing	Sleeve	Packing	Packing	Delrin seals
Valve Trim Easily Removed	Yes[11]	No	No	No	Yes	No
Low Maintenance	Yes	No	No	No	Yes	No

[11] Only in top-entry design.
[12] Better in second half of stroke and with the "shutter closure" design.
[13] Limited by sleeve material.
[14] More in sizes under 10 in.

Table 4.1j (Continued)

Features / Type of Control Valve Body	Plug Valve V-Ported Type	Plug Valve Adjustable Cylinder Type	Plug Valve Expanding Seat Plate Type	Plug Valve Eccentric Shaft Type	Plug Valve Retractable Seat Type	Plug Valve Eccentric Rotating Type (Camflex)	Ball and Cage Type Valves
Applicable Section[a]	1.15	1.15	1.15	1.15	1.15	1.15	1.16
Capacity [$C_v = (Cd)(d^2)$] Critical	$12d^2$	$12d^2$	$10d^2$	$14d^2$	$14d^2$	$12d^2$	$14d^2$
Non-critical	$17d^2$	$14d^2$	$16d^2$	$24d^2$	$24d^2$	$14d^2$	$20d^2$
Flow Characteristics	Good	Fair	Fair	Fair	Fair	Good	Good
Cost	Low	Medium	Low	Medium	Low	Medium	Medium–High
Available Size Range (in.)	½–24	⅜–6	2–16	4–60	4–12	1–12	¼–14
Max./Min. Operating Pressure (PSIG)	700/vacuum	125	700	600/vacuum	10,000	1,200/vacuum	2,500/vacuum
Max. Pressure Drop (PSID)	400	40[15]	500	200	5,000	300[16]	2,500
Max./Min. Operating Temperature (°F)	450	400	400	450	450/−100	750/−320	1,800/−425
Available in Jacketed Construction	Yes	No	No	No	No	No	No
Applicable to Flashing Service	No	No	No	No	No	Yes	Yes
Applicable to Cavitating Service	Limited	No	Limited	No	No	Yes	Yes
Applicable to High Viscosity ($\mu > 10^4$ cps) Service	Limited	Limited	Limited	No	No	Yes—Limited	Yes—Limited
Applicable to Slurry Service	Yes	No	No	No	No	Yes	Yes
Tight Shut-off Available[1]	Yes	No	Yes	No	Yes	Yes	Yes
Available Rangeability[9]	20:1	10:1	20:1	Over 50:1	Over 50:1	Over 50:1	Over 50:1
Has "Fail-safe" (Open or Closed) Action	Yes	Yes	Yes	No	No	Yes	Yes
Available in a Great Variety of Materials	Yes	No	No	Yes	Yes	No	Average
Flow in Either Direction	No	No	Yes	No	No	Yes	No
Unobstructed Streamline Flow	No	Not Fully	No	Not fully	Not fully	Not fully	Not fully
Has Tendency to Cavitate Due to High Recovery	Some	No	Some	Yes	Yes	Some	No
Stem or Shaft Sealed from Process by	Packing	Precision fit	Packing	Packing	Packing	Packing	Packing
Valve Trim Easily Removed	Yes	No	Yes	No	No	No	Yes
Low Maintenance	Yes	Average	Yes	Average	Average	Average	Average

[15] ½ PSID on gas service. [16] More with positioner.

Table 4.1j (Continued)

Features / Type of Control Valve Body		Expansible Tube or Diaphragm Type Valve	Fluid Interaction Valve
Applicable Section[a]		1.17	1.18
Capacity [$C_v = (Cd)(d^2)$]	Critical	$10d^2$	$10d^2$
	Non-critical	$12d^2$	$14d^2$
Flow Characteristics		Function of pilot	Diverting
Cost		Low–Medium	Low–Medium
Available Size Range (in.)		1–12	½–4
Max./Min. Operating Pressure (PSIG)		1,500	100[17]
Max. Pressure Drop (PSID)		1,200	50[17]
Max./Min. Operating Temperature (°F)		150/0	Above 1,000[18]
Available in Jacketed Construction		No	No
Applicable to Flashing Service		Limited	—
Applicable to Cavitating Service		Limited	—
Applicable to High Viscosity ($\mu > 10^4$ cps) Service		No	No[19]
Applicable to Slurry Service		No	Limited[20]
Tight Shut-off Available[1]		Yes	No
Available Rangeability[9]		20:1	10:1
Has "Fail-safe" (Open or Closed) Action		No	No
Available in a Great Variety of Materials		No	Yes
Flow in Either Direction		No	No
Unobstructed Streamline Flow		No	Yes
Has Tendency to Cavitate Due to High Recovery		No	Yes
Stem or Shaft Sealed from Process by		Sleeve	No moving parts
Valve Trim Easily Removed		No	Has no trim
Low Maintenance		Average	Yes

[17] Operating principle places no limit; values given are those of existing installations.
[18] Can handle molten metals.
[19] Operation requires a minimum Reynolds number value.
[20] Pilot line needs to be kept from plugging.

[a] The noted section numbers are those of the Instrument Engineers' Handbook, Volume II and the reader may consult this reference for further details concerning the selected valve.

Table 4.1k
SUMMARY OF VALVE SIZING FORMULAS

Basis for Selecting the Sizing Formula	*State of the Fluid*	*Liquid*	*Gas*	*Vapor (for steam, K = 2.1)*
Basic Equations For Cold, Non-viscous, Non-cavitating or Non-flashing Liquids or for Vapors and Gases Where the Expansion at the Valve Seat Can Be Neglected. ($\Delta P \ll P_1/5$	Volumetric Flow (GPM or SCFH)	$C_v = \dfrac{Q_f\sqrt{G_f}}{\sqrt{\Delta P}}$	$C_v = \dfrac{Q}{963}\sqrt{\dfrac{(G_f)(T_1)}{\Delta P(P_1 + P_2)}}$	
	Mass Flow (lbm/hr)	$C_v = \dfrac{W}{500\sqrt{(\Delta P)(G_f)}}$	$C_v = \dfrac{W}{3.22\sqrt{(\Delta P)(G_f)(P_1 + P_2)}}$	$C_v = \dfrac{W}{K}\sqrt{\dfrac{1}{\Delta P(P_1 + P_2)}}$
Critical Flow Due to Cavitation $\Delta P > \Delta P_{allowed}$ $P_v < 0.7\ P_1$	$P_v < \dfrac{P_c}{10}$	$C_v = \dfrac{Q_f\sqrt{G_f}}{C_f\sqrt{P_1 - P_v}}$		
	$P_v > \dfrac{P_c}{10}$	$C_v = \dfrac{Q_f\sqrt{G_f}}{C_f\sqrt{P_1 - R_cP_v}}$ (1)		
Pipe Reducer Effect $D/d > 1.0$	Non-critical $P_v \simeq 0$	$C_v = \dfrac{F_1Q_f\sqrt{G_f}}{\sqrt{\Delta P_t}}$	$C_v = \dfrac{F_1Q\sqrt{(G_f)(T_1)}}{825\ C_f\ P_1\ X}$	$C_v = \dfrac{1.16F_1W}{(K)(C_f)(P_1)(X)}$
	Critical $\Delta P_t > \Delta P_{t\,allowed}$ $P_v < 0.7\ P_1 <$ $< 0.1\ P_c$	$C_v = \dfrac{Q_f\sqrt{G_f}}{C_{fr}\sqrt{P_1 - P_v}}$	$C_v = \dfrac{Q\sqrt{(G_f)(T_1)}}{825(C_{fr})(P_1)}$	$C_v = \dfrac{1.16W}{(K)(C_{fr})(P_1)}$

High Viscosity Resulting in $Re_{valve} \leq 500$	$C_v = \frac{F_2 Q_f \sqrt{G_f}}{\sqrt{\Delta P}}$		
Superheated Non-critical Vapor Flow			$C_v = \frac{W\left(1 + \frac{1,400}{T_s}\right)}{K} \sqrt{\frac{1}{\Delta P(P_1 + P_2)}}$
Critical Vapor or Gas Flow $\Delta P > \Delta P_{allowed}$		$C_v = \frac{Q\sqrt{(G_f)(T_1)}}{825(C_f)(P_1)}$	$C_v = \frac{1.16W}{(K)(C_f)(P_1)}$
Generalized Sizing Formula for Compressible Flow		$C_v = \frac{Q\sqrt{(G_f)(T_1)}}{825(C_f)(P_1)(X)}$ (2)	$C_v = \frac{1.16W}{(K)(C_f)(P_1)(X)}$ (3)
Flashing Liquids	If percent flash is known, use the sum of equations (1) and (2) or (3). If $(C_m)(P_v - P_2)$ product curves are available, use mixture density technique. If neither of above is known, use (1) as qualified in text.		
Gas-liquid Mixtures	Calculate gas and liquid C_v's separately and base valve selection on their sum.		

C_d	Valve discharge coefficient (C_v/d^2)
C_{dr}	Valve reducer discharge coefficient (C_{vr}/d^2)
C_{dt}	Discharge coefficient for valve-reducer combination
C_f	Pressure recovery factor ($C_1/36$)
C_{fr}	Pressure recovery factor for valve-reducer combination
C_g	Supercritical gas flow coefficient (C_1C_v)
C_v	Valve capacity coefficient
C_{vr}	Equivalent valve coefficient of the valve reducers
C_1	Pressure recovery coefficient (36 C_f or C_g/C_v)
d	Nominal valve size (in.)
D	Nominal pipe diameter (in.)
D_c	Flow characteristics distortion coefficient
F_1	Pipe reducer correction factor
F_2	Viscosity correction factor
g	Gravitational acceleration (32.17 lbm-ft/lbf-sec^2)
Gf	Specific gravity at inlet temperature, relative to air or water
G_m	Specific gravity of liquid-gas or of liquid-vapor mixture, relative to water
G_0	Specific gravity of flowing stream at valve outlet, relative to water
K	Vapor constant (Table 1.2q)
L	Equivalent or actual pipe length
$P_{allowed}$	Valve outlet pressure that produces maximum flow at given inlet conditions (PSIA)
P_c	Pressure corresponding to thermodynamic critical point, where fluid state is undefined (PSIA)
P_v	Liquid vapor pressure at inlet temperature (PSIA)
P_{vc}	Pressure at vena contracta (PSIA)
P_1	Valve inlet pressure (PSIA)
P_2	Valve outlet pressure (PSIA)
ΔP	Valve pressure drop ($P_1 - P_2$)
$\Delta P_{allowed}$	Apparent pressure drop at which choked flow first occurs (PSID)
Q	Volumetric flow (SCFH)
Q_t	Liquid flow (GPM)
R_c	Critical pressure ratio [$f(P_v/P_c)$]*
T_s	Degrees superheat (°F)
T_1	Valve inlet temperature [°R = °F + 460]
$\overline{V}$	Velocity (ft/sec)
W	Mass flow (lbm/hr)
X	Correction factor for expansion at valve port
ρ	Density (lbm/ft^3)
μ	Viscosity (centipoises or lbf − sec/ft^2)

Table 4.11
ORIENTATION TABLE FOR NON-VALVE TYPE FINAL CONTROL ELEMENTS

Service \ Hardware	*Throttling Electrical Energy*					*Adjustment of Drive Speeds*				*Pumps and Feeders as Final Control Elements*								*Dampers*	
	SCR	*Ignitron*	*Saturable Core-Reactor*	*Amplifier*	*Contactor*	*Electric Motor*	*Magnetic Coupling*	*Mechanical Variable-Speed Drive*	*Hydraulic Drive*	*Plunger Pump*	*Diaphragm Pump*	*Opposed Centrifugal Pump*	*Ball Check Feeder*	*Belt Feeder*	*Self-Powered Feeder*	*Roll Feeder*	*Screw Feeder*	*Vane Dampers*	*Variable-Orifice Valve*
DC Output Required	E	E	G	P	P														
High Operating Voltage	P	E	G	P	P														
Reliability at Frequent Overloads and Transients	G	E	E	P	G	G	E	G	E										
Zero Output Required with Zero Input Signal	E	E	P	P	E														
Continuous Speed Adjustment Not Required						E	G	E	G										
Very Wide Speed Regulation Range						E	G	P	G										
High Output Power						G	E	P	G										
Vibration at Driver or Load						P	E	P	G										
Shock Loads						P	G	P	E										
High Output Pressure										E	G	G	G						
Low Output Pressure, Low Flow										E	E	G	E						
Coarse Slurry Service										P	P	E	P						

Table 4.11
Continued

Hardware / Service	Throttling Electrical Energy					Adjustment of Drive Speeds				Pumps and Feeders as Final Control Elements								Dampers	
	SCR	*Ignitron*	*Saturable Core-Reactor*	*Amplifier*	*Contactor*	*Electric Motor*	*Magnetic Coupling*	*Mechanical Variable-Speed Drive*	*Hydraulic Drive*	*Plunger Pump*	*Diaphragm Pump*	*Opposed Centrifugal Pump*	*Ball Check Feeder*	*Belt Feeder*	*Self-Powered Feeder*	*Roll Feeder*	*Screw Feeder*	*Vane Dampers*	*Variable-Orifice Valve*
Very Low Slurry Flows										P	G	E	G						
Very Thick Slurry with Fine Solids										P	P	P	E						
Viscous Fluids										P	G	P	P	E	G	G	G		
Power Source Not Available														P	E	P	P		
Low Flow of Fine Powders														E	G	E	G		
Solids Blending														E	P	G	G		
Low-Pressure Gas																		E	E
Throttling Solids Flow																		P	G
Large Flow Rates																		E	E
Low Pressure Loss																		E	E
Accuracy	E	E	E	E	P	E	E	G-P	E	G	G	G	P-G	E	P	G	G	P	E

E—Excellent, G—Good, P—Poor

Table 4.1m
MEANINGS OF FUNCTIONAL INSTRUMENT IDENTIFICATION LETTERS

	First Letter		*Succeeding Letters*		
	Measured or Initiating Variable	*Modifier*	*Readout or Passive Function*	*Output Function*	*Modifier*
A	Analysis		Alarm		
B	Burner Flame		User's Choice	User's Choice	User's Choice
C	Conductivity (Electrical)			Control	
D	Density (Mass) or Specific Gravity	Differential			
E	Voltage (EMF)		Primary Element		
F	Flow Rate	Ratio (Fraction)			
G	Gauging (Dimensional)		Glass		
H	Hand (Manually Initiated)				High
I	Current (Electrical)		Indicate		
J	Power	Scan			
K	Time or Time-Schedule			Control Station	
L	Level		Light (Pilot)		Low
M	Moisture or Humidity				Middle or Intermediate
N	User's Choice		User's Choice	User's Choice	User's Choice
O	User's Choice		Orifice (Restriction)		
P	Pressure or Vacuum		Point (Test Connection)		
Q	Quantity or Event	Integrate or Totalize			
R	Radioactivity		Record or Print		
S	Speed or Frequency	Safety		Switch	
T	Temperature			Transmit	
U	Multivariable		Multifunction	Multifunction	Multifunction
V	Viscosity			Valve, Damper, or Louver	
W	Weight or Force		Well		
X	Unclassified		Unclassified	Unclassified	Unclassified
Y	User's Choice			Relay or Compute	
Z	Position			Drive, Actuate or Unclassified Final Control Element	

Table 4.1n
RELAY FUNCTION SYMBOLS

Symbol	*Function*
1-0 or ON-OFF	Automatically connect, disconnect, or transfer one or more circuits, provided that this is not the first such device in a loop.
Σ or ADD	Add or totalize (add and subtract), with two or more inputs.
Δ or DIFF.	Subtract (with two or more inputs)
± + [−]	Bias (single input)
AVG.	Average
% or 1:3 or 2:1 (typical)	Gain or attenuate (input:output), with single input
[×]	Multiply (two or more inputs)
÷	Divide (two or more inputs)
[√] or SQ. RT.	Extract square root
x^n or $x^{1/n}$	Raise to power
f(x)	Characterize
1:1	Boost
[>] or HIGHEST (Measured Variable)	High-select. Select highest (higher) measured variable (not signal, unless so noted).
[<] or LOWEST (Measured Variable)	Low-select. Select lowest (lower) measured variable (not signal, unless so noted).
REV.	Reverse
	Convert
a. E/P or P/I (typical)	For input/output sequences of the following: Designation — Signal E — Voltage H — Hydraulic I — Current (electrical) O — Electromagnetic or sonic P — Pneumatic R — Resistance (electrical)
b. A/D or D/A	For input/output sequences of the following: A — Analog D — Digital
∫	Integrate (time integral)
D or d/dt	Derivative or rate
1/D	Inverse derivative

Note: The use of a box enclosing a symbol is optional. The box is intended to avoid confusion by setting off the symbol from other markings on a diagram.

Table 4.1o
MISCELLANEOUS SYMBOLS

Instrument Line Symbols

All lines should be fine in relation to process piping lines.

Connection to process, or mechanical link, or instrument supply

Pneumatic signal 

The pneumatic signal symbol applies to a signal using any gas as the signal medium. If a gas other than air is used, the gas is identified by a note on the signal symbol or otherwise.

Electric signal

Capillary tubing (filled system) 

Hydraulic signal 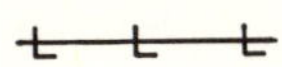

Electromagnetic or sonic signal (without wiring or tubing) 

Electromagnetic phenomena include heat, radio waves, nuclear radiation, and light.

Undefined signal 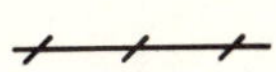

Power Supply Abbreviations

The following abbreviations are suggested to denote the types of power supply. These designations may also be applied for purge fluid supplies.

AS	Air supply
ES	Electric supply
GS	Gas supply
HS	Hydraulic supply
NS	Nitrogen supply
SS	Steam supply
WS	Water supply

The power supply level may be added to the instrument supply line, e.g., AS 100 for a 100-PSIG air supply; ES 24DC for a 24-volt direct current supply, etc.

Instrument Symbol Balloons

APPROXIMATELY 7/16" DIAMETER

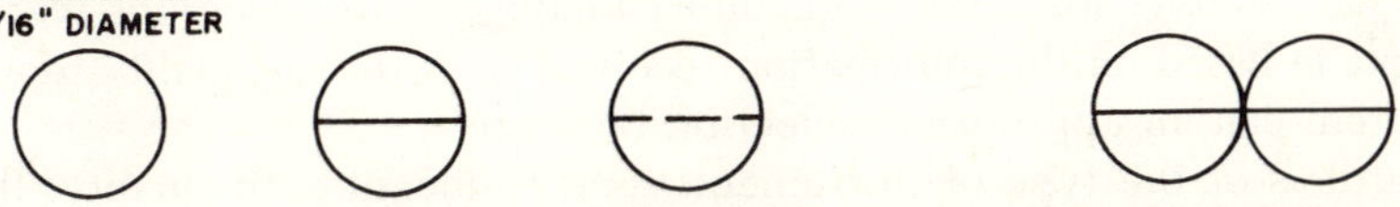

LOCALLY MOUNTED

MOUNTED ON MAIN BOARD

MOUNTED BEHIND THE BOARD

Instrument for two measured variables or single variable instrument with more than one function. Additional tangent balloons may be added as required.

4.2 AIR QUALITY MONITORING SYSTEMS

Sensor Types:	(*a*) *Static sensors,* (a1) Dustfall jars, (a2) Lead peroxide candles, (a3) Sodium carbonate on filter, (a4) Steel exposure plates; (*b*) *Manual analyses,* (b1) High-volume sampler, (b2) Wet sampling train; (*c*) *Instrumental methods.*
Air Samplers:	See Section 4.15.
Cost of Large Systems:	In the range of $1 million.
Partial List of Installed Large System Suppliers:	International Business Machines, Litton Industries, Inc.

There are many options in the type of air quality information which can be collected, and the cost of air quality monitoring systems can vary by orders of magnitude. Only with a thorough understanding of the decisions which must be made based on the information received from the air quality monitoring system can an appropriate selection be made.

Regardless of the type of instruments used to measure air quality, the data are only as good as they are representative of the sampling site selected.

The simplest air quality monitors are static sensors which are exposed for a given length of time and are later analyzed in the laboratory. In some cases, these devices provide all the information which is required. More commonly, a system of automatic instruments measuring several different air quality parameters will be used. With more than a few instruments, the signals from these instruments can be retained on magnetic tape rather than on recorder charts.

The most common errors in the design of air quality monitoring systems are poor site location and the acquisition of more data than necessary to accomplish the purpose of the installation.

Purpose of Monitoring

The principal purpose of air quality monitoring is frequently the acquisition of data for comparison to regulated standards. In the United States, standards have been promulgated by the federal government[1] and by many of the states. Where possible, these standards have been based on the physiological effect of the air pollutant in question, so that human health is protected. The averaging time over which various concentrations standards must be maintained is different for each pollutant. Tables 4.2a and 4.2b are tabulations of the ambient air quality standards promulgated by the federal government.

Some air quality monitoring systems are operated to determine the impact of a single source or a concentrated group of sources of emission on the surrounding area. In this case, it is important to determine the background level, the maximum ground level concentration in the area and the geographical extent of the air pollutant impact of the source. When the source is isolated, e.g., a single industrial plant in a rural area, the system design is straightforward. Utilizing the meteorological records which are normally

Table 4.2a
PRIMARY AIR QUALITY STANDARDS
(levels of air quality which are deemed necessary, with an adequate margin of safety, to protect the public health)

Pollutant	*Long-term Concentration Level* ($\mu g/m^3$)	*Short-term Concentration Level*[a] ($\mu g/m^3$)
Particulate matter	75—annual geometric mean	260—maximum 24-hour concentration
Sulfur oxides (SO_2)	80 (0.03 ppm)—annual arithmetic mean	365 (0.14 ppm)—maximum 24-hour concentration
Carbon monoxide	—	10,000 (9 ppm)—maximum 8-hour concentration or 40,000 (35 ppm)—maximum 1-hour concentration
Photochemical oxidants	—	160 (0.08 ppm)—maximum 1-hour concentration
Hydrocarbons	—	160 (0.24 ppm)—maximum 3-hour concentration (6–9 A.M.)
Nitrogen dioxide	100 (0.05 ppm)—annual arithmetic mean	

[a] Stated concentrations not to be exceeded more than once per year.

Table 4.2b
SECONDARY AIR QUALITY STANDARDS
(levels of air quality which are deemed adequate to protect the public welfare from any known or anticipated adverse effects)

Pollutant	Long-Term Concentration Level ($\mu g/m^3$)	Short-term Concentration Level[a] ($\mu g/m^3$)
Particulate	60—annual geometric mean	150—maximum 24-hour concentration
Sulfur dioxide	60 (0.02 ppm)—annual arithmetic mean	260 (0.1 ppm)—maximum 24-hour concentration
Carbon monoxide	—	10,000 (9 ppm)—maximum 8-hour concentration or 40,000 (35 ppm)—maximum 1-hour concentration
Photochemical oxidants	—	160 (0.08 ppm)—maximum 1-hour concentration
Hydrocarbons	—	160 (0.24 ppm)—maximum 3-hour concentration (6–9 A.M.)
Nitrogen dioxide	100 (0.05 ppm)—annual arithmetic mean	—

[a] Stated concentrations not to be exceeded more than once per year.

available from nearby airports or from government meteorological reporting stations, a wind rose can be prepared to estimate the direction of principal drift of the air pollutant from the source. Dispersion calculations (Sections 2.6 and 5.45) can be performed to estimate the location of the expected point of maximum ground level concentration. As a rule of thumb, with stacks between about 50 and 350 feet tall, this point of maximum concentration will be approximately ten stack heights downwind. The air quality monitoring system should include at least one sensor at the point of expected maximum ground level concentration. Additional sensors should be placed not less than 100 stack heights upwind (prevailing) to provide a background reading, and at least two or three sensors should be placed between 100 and 200 stack heights downwind to determine the extent of the travel of the pollutants from the source in question. If adequate resources are available, sampling at the intersection points of a rectilinear grid with its center at the source in question is advisable.

With such a system for an isolated source, adequate data can be obtained in a period of 1 year to determine the impact of the source on the air quality of the area. There are very few instances where less than 1 year of data collection will provide adequate information because of the variability in climatic conditions on an annual basis. If a study of this type is performed to determine the effect on air quality of process changes, it may be necessary to continue the study for 2 to 5 years to develop information which is statistically reliable.

Some air quality monitoring is designed for the specific purpose of investigating complaints concerning an unidentified source. This usually happens in urban situations for odor complaints. In these cases, a triangulation technique is used by human observers to correlate the location of the observed odor with wind direction over a number of days. Plotting on a map can pinpoint the offending source in most cases. While this is not an air pollution monitoring system in an instrumental sense, it is a useful tool in certain situations.

The research needs of air pollution call for a completely different approach to air quality monitoring. Here, the purpose is to define some unknown variable or combination of variables. This can be either a new atmospheric phenomenon or the evaluation of a new air pollution sensor. In the former case, the most important consideration is the proper operation of the instrument used. In the latter case, the most important factor is the availability of a reference determination against which the results of the new instrument can be compared.

Monitoring in Urban Areas

Urban situations are the area of major interest in air pollution in the United States, since most of the population lives in urban areas. The most sophisticated and expensive air quality monitoring systems in the United States are those of large cities (and one or two of the largest states) where data collection and analysis are centralized at a single location through the use of telemetry. On-line computer facilities provide data reduction.

The three philosophies which can be utilized in the design of an urban air quality monitoring system are the location of sensors on a uniform area basis (rectilinear grid); the location of sensors at locations where pollutant concentrations will be high; and the location of sensors in proportion to population distribution. The operation of these systems is nearly identical but the interpretation of the results can be radically different.

The most easily designed systems are those where the sensors are located uniformly on a geographical basis according to a rectilinear grid. Adequate coverage of an urban area frequently requires at least 100 sensors, so this concept is usually applied only with static or manual methods of air quality monitoring.

The location of air quality sensors at points of maximum concentration indicates the highest levels of air pollutants which are encountered throughout the area. Typically, this will include the central business district and the industrial areas on the periphery of the community. Data of this type are extremely useful when interpreted in the context of total system design. One or two sensors are usually placed in "clean" or "background" locations, so the average concentration of air pollutants over the entire area can be estimated approximately. The basis of this philosophy is that if the concentrations in the dirtiest areas are below the air quality standards, there will certainly be no problem in the cleaner areas.

The design of air quality monitoring systems based on population distribution calls for placement of air quality sensors in the most populous areas. While this may not include all the high pollutant concentration areas in the urban region, it generally encompasses the central business district and is a measure of the levels of air pollutants to which the bulk of the population is exposed. The "average" concentrations from a sampling network of this type are an adequate description from a public health standpoint. This system design may, however, miss some localized high concentration areas.

Agreement on the purpose of air quality monitoring must be reached between the system designer and those responsible for interpreting the data before the system is designed.

Sampling Site Selection

Once the initial layout has been developed for an air quality monitoring system, specific sampling sites must be located as close as practical to the ideal locations. The major considerations are the lack of obstruction from local interferences and the adequacy of the site to represent the air mass of interest, accessibility and security.

Local interferences can cause major disruptions to air quality sensor sites. A sampler inlet placed at a sheltered interior corner of a building is most undesirable because of poor air motion. Tall buildings or trees immediately adjacent to the sampling site can also invalidate most readings.

The selection of sampling sites in urban areas is complicated by the canyon effect of streets and by the high density of pollutants, both gaseous and particulate, immediately at street level. In order to be representative of the air mass to be sampled, the purposes of the study must again be reviewed. If the data are to be collected for the determination of area-wide pollutant averages, it may be best to locate the sampler inlet in a city park, vacant lot or other completely open area. If this is not possible, the sampler inlet could be at the roof level of a one or two story building so that street level effects could be minimized. On the other hand, if the physiological impact of the air pollutants is a prime consideration, the samplers should be at or near the breathing level of the people exposed. As a rule of thumb, an elevation of 3 to 6 meters above the ground is suggested as an optimum elevation.

Sampling site location can be different for the same pollutant depending upon the purpose of the sampling. Carbon monoxide sampling is an example. The federally promulgated air quality standards for carbon monoxide include both an 8-hour and a 1-hour concentration limit. Maximum 1-hour concentrations are likely to be found in a high-traffic density center-city location. It is unlikely that people would ordinarily be exposed to these concentrations for 8-hour periods. When sampling for comparison with the 1-hour standard, the sensor should be located within about 20 feet of a major traffic intersection. When sampling for comparison with the 8-hour standard, the sensor should be located near a major thoroughfare in either the center-city area or in the suburban area with the sampler less than about 50 feet from the intersection. The purpose for two different locations of sampling sites is to be consistent with the physiological effects noted from carbon monoxide exposure and to be consistent with the living pattern of the bulk of the population. If only one site can be selected, the location described for the 8-hour averaging time is preferred.

When the sampling instruments are located inside a building and an air sample is to be drawn in from the outside, it is frequently advantageous to utilize a sampling pipe with a small blower to bring outside air to the instrument inlets. This improves sampler line response time. An air velocity of approximately 700 feet per minute in the pipe is a good choice to balance the problems of gravitational and inertial deposition of particulate matter, where particulates are to be sampled.

The sampling site should be accessible to the operation and maintenance personnel. Since most air pollution monitoring sites are unattended much of the time, sample site security is a very real consideration; the risk of vandalism is high in many areas.

Static Methods of Air Monitoring

Minimum capital cost is attained when static sensors are used to monitor air quality. While averaging times are in terms of weeks and sensitivity is frequently low, there are many cases where static monitors provide the most information for the amount of investment. They should not be rejected out of hand but should be considered as useful adjuncts to more sophisticated systems.

Dustfall Jars

The simplest of all air quality monitoring devices is the dustfall jar (Fig. 4.2c). This device measures the fallout rate of coarse particulate matter, generally above about 10 microns in size. Dustfall and odor are two of the major reasons for citizen complaints concerning air pollution. Dustfall is offensive because it builds up on porches, automobiles and such; and it is highly visible and gritty to walk upon.

Dustfall seldom carries for distances in excess of $1/2$ mile because these

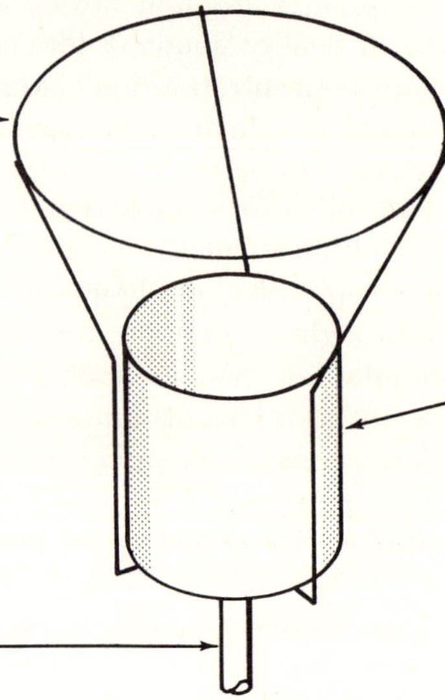

Figure 4.2c Dustfall Jar.

large particles are subject to strong gravitational effects. For this reason, dustfall stations must be more closely spaced than other air pollution sensors, if a detailed study of an area is desired.

Dustfall measurements conducted in large cities in the United States in the 1920s and 1930s commonly indicated dustfall rates in hundreds of tons per square mile per month. These levels would be considered excessive today, as evidenced by the dustfall standards of 15 to 30 tons per square mile per month promulgated by many of the states. While the measurement of low or moderate values of dustfall does not indicate freedom from air pollution problems, measured dustfall values in excess of 50 to 100 tons per square mile per month are a sure indication of the existence of excessive air pollution.

The large size of the particulate matter found in dustfall jars makes it amenable to chemical analysis or to physical analysis by such techniques as microscopy. These analyses can be useful to identify specific sources.

Lead Peroxide Candles

For many years, an indication of sulfur dioxide levels has been gained through the use of lead peroxide candles. These devices, known as candles because they are a mixture of lead peroxide paste spread on a porcelain cylinder about the size and shape of a candle, are normally exposed for periods of 1 month. Sulfur gases in the air react with lead peroxide to form lead sulfate. Sulfate is analyzed according to standard laboratory procedures to give an indication of the atmospheric levels of sulfur gases during the period of exposure.

A modification of this technique which simplifies the laboratory procedure is the use of a fiber filter cemented to the inside of a plastic petri dish.* The filter is saturated with an aqueous mixture of lead peroxide and a gel, commonly gum tragacanth, and is allowed to dry. These dishes or "plates" are exposed in an inverted position for periods of 1 to 4 weeks.

Lead peroxide estimation of sulfur dioxide suffers from inherent weaknesses. All sulfur gases, including reduced sulfur, react with lead peroxide to form lead sulfate. More importantly, the reactivity of lead peroxide is dependent upon its particle size distribution. For this reason, the results from different investigators are not directly comparable. Nevertheless, a network of lead peroxide plates over an area provides a good indication of the relative exposure to sulfur gases during the particular exposure period. This is useful for determining the geographical extent of sulfur pollution.

Other Static Methods

The use of a fiber filter cemented to a petri dish has been modified by the use of sodium carbonate rather than lead peroxide for the measurement of sulfur gases. This method also shows some success in indicating the relative concentration of other gases, including nitrogen oxides and chlorides. Relative levels of gaseous fluoride air pollution have been measured using larger filters, e.g., 3-inch diameter, dipped in sodium carbonate and placed in shelters to protect them from the rain. With all these static methods, the accuracy is extremely low and the data cannot be converted directly into ambient air concentrations. They do, however, provide a low cost indicator of relative levels of pollution in a given area.

The corrosive nature of the atmosphere has been evaluated, using standardized steel exposure plates for extended periods to measure the corrosion rate. This provides a gross indication of the corrosive nature of the atmosphere. As with other static samplers, the results are not directly related to the concentration of air pollutants.

Manual Analyses

Manual analyses for air quality measurements are those which require that the sample be first collected and then analyzed in the laboratory. Manual instruments provide no automatic indication of pollution levels.

The manual air sampling instrument in widest use is the high-volume sampler, discussed in Section 4.15. With this method, ambient air is drawn through a pre-weighed filter at a rate of approximately 50 ACFM for a period

* A flat bottom dish with shallow walls used for biological cultures.

of 24 hours. The filter is then removed from the sampler, returned to the laboratory and weighed. The gain in weight, in combination with the measured air volume through the sampler, allows the calculation of particulate mass concentration expressed in micrograms per cubic meter.

Reference methods for nearly all gaseous air pollutants involve the use of a wet sampling train in which air is drawn through a collecting medium for some period of time. The exposed collecting medium is then returned to the laboratory for chemical analysis. Sampling trains have been developed which allow the sampling of five or more gases simultaneously into separate bubblers. There are also sequential samplers which automatically divert the airflow from one bubbler to another at preset time intervals.

These methods of sampling can be accomplished with a modest initial investment; but the manpower required to set out and to pick up the samples, in combination with the laboratory analysis, raises the total cost to a point where automated systems may be more economical for long-term studies.

Instrumental Methods

As the need for accurate data which can be statistically reduced in a convenient manner becomes more the rule than the exception, automated sampling systems become more and more desirable. The elements of an automated system include the airflow handling system, the sensors, the data transmission storage and display apparatus and the data processing facility. The overall system is no more valuable than the weakest link of this chain.

Sensors

The variety of sensors for individual air pollutants is discussed in detail in the other sections of this chapter.

The reliability of the output from an air quality sensor is dependent upon its inherent accuracy, sensitivity, zero drift and calibration. The inherent accuracy and sensitivity are a function of the design of the instrument and of the principle upon which it operates. Zero drift can be either an electronic phenomenon or an indication of difficulties with the instrument. In those instruments using an optical path, lenses can become dirty. In wet chemical analyzers, the flowrates of reagents can vary, causing a change in both the zero and the span (range) of the instrument. Because of these potential problems, every instrument should have routine field calibration at an interval determined in field practice to be reasonable for the sensor.

The calibration of an air quality sensor is frequently accomplished using either a standard gas mixture or a prepared, diluted gas mixture using permeation tubes. In some cases, the airstream entering the sensor is concurrently sampled by a reference wet chemical technique.

The operator of air quality sensors should always have an adequate supply of spare parts and tools so that downtime can be held to a minimum. Operator training should, at a minimum, include instruction to recognize the symptoms of equipment malfunction and vocabulary to describe the symptoms to the individual responsible for instrument repair. Ideally, each operator would receive a short training session from the instrument manufacturer or someone trained in the use and maintenance of the instrument, so that he can make repairs on site. Since this is seldom possible in actual practice, the recognition of symptoms of malfunction becomes increasingly important.

Data Transmission

The output signal from a continuous monitor used in an air quality monitoring system is typically the input to a strip chart recorder, magnetic tape data storage or to an on-line data transmission system.[2] The output of most air quality sensors is in analog form.* This is suitable for direct input to a strip chart recorder; but in the case of automated systems, the analog signal is commonly converted to a digital signal.† For those sensors which have a linear output, the signal can go directly to the recorder or transmission system. In the case of sensors with a logarithmic output, it may be desirable to first convert this signal to a linear form.[2]

Many of the early automated air quality monitoring systems in the United States had their major difficulty with the data transmission step in the system. In some cases, this was caused by attempts to overextend the lower range of the sensors so that the signal-to-noise ratio‡ was unfavorable. In other cases, the matching of sensor signal output to the data transmission system was poor. With developments and improvements in systems over the past 5 years or so, these early difficulties have been nearly completely overcome.

On-line systems, i.e., those which can provide an instantaneous readout of the dynamic situation monitored by the system, add an additional step in the complexity to air quality monitoring systems. In almost all cases, the data from continuous monitors will be stored on magnetic tape for later processing and statistical reduction.[2] In an on-line system, this can be done instantaneously. The added expense of this sophistication must be evaluated in terms of the purposes of the air quality monitoring system. When decisions with substantial community impact must be made within a very short time, this real-time capability may be necessary.

Data Processing and Interpretation of Results

The concentration of many air pollutants has been found to follow a log normal rather than a normal distribution. In a log normal distribution, a plot

* An uninterrupted signal such as is displayed by the speedometer of an automobile.

† A signal where the pulse rate is an indication of magnitude. The odometer of an automobile represents the display of a digital totalizer.

‡ Noise can be caused by an extraneous disturbance and it is that portion of the signal which does not convey useful information.

of the logarithm of the measured value more closely approximates the bell-shaped Gaussian distribution curve than does a plot of the numerical data. Suspended particulate concentrations are a prime example of this type of distribution. When this is the case, the geometric mean is the statistical parameter which best describes the population of data. The arithmetic mean is of limited value because it is dominated by a few occurrences of high values. The geometric mean, in combination with the geometric standard deviation, is a complete description of a frequency distribution for a log normally distributed pollutant.

The averaging time over which the sample results are reported is a consideration in the processing and interpretation of air quality data. For sulfur dioxide, air quality standards have been promulgated by various agencies based on annual arithmetic average, monthly arithmetic average, weekly arithmetic average, 24-hour arithmetic average, 3-hour arithmetic average and 1-hour arithmetic average concentrations. The output of a continuous analyzer can be averaged over nearly any discrete time interval. In order to reduce the computation time, the time interval over which the continuous analyzer output is averaged to obtain a discrete input for the calculations must be considered. If a 1-hour average concentration is the shortest time interval of value in interpreting the study results, it is not economical to use a 1 or 2-minute averaging time for inputs to the computation program.

Caution must be exercised in using strip chart recorders for the acquisition of air quality data. The experience of many organizations, both governmental and industrial, is that the reduction of data from strip charts is a tedious chore at best.[2] Many organizations have decided that they "did not really need all that data in the first place," once they find a large backlog of unreduced strip charts. Two cautions are suggested by this experience. First, if you do not plan to use the data, do not collect it. Second, be aware of the advantages of magnetic tape data storage followed by computer processing.[2]

The visual display of air quality data has considerable appeal to many nontechnical personnel.[2] Long columns of numbers can be deceptive if only one or two important trends are to be shown. The use of bar charts or graphs is frequently advantageous, even though they do not show the complete history of the air quality over the time span of interest.

REFERENCES

1. *Federal Register*, Part II, vol. 36, p. 84, April 30, 1971.
2. Lipták, B. G. (ed.): "Instrument Engineers' Handbook," Vols. I, II and Supplement One, Chilton, Philadelphia, 1969, 1970 and 1972.

4.3 ANALYTICAL TECHNIQUES IN AIR POLLUTANT DETECTION

Analysis Types:	(*a*) *Particulate matter,* (a1) Sampling techniques, (a1.1) Settling, (a1.2) Filtration, (a1.3) Impingement, (a1.4) Precipitation, (a1.5) Centrifugal, (a1.6) Hybrid, (a2) Particle analysis, (a2.1) Optical microscope, (a2.2) Electron microscope, (a2.3) Photometric methods, (a2.4) Crystal frequency methods, (a3) Inorganic particulate composition, (a3.1) Transmittance and reflectance measurements, (a3.2) Ring oven technique, (a3.3) Atomic absorption, (a3.4) Specific ion electrode, (a4) Organic particulate composition; (*b*) *Gaseous air pollutants,* (b1) Inorganic gases, (b2) Organic gases.
Methods of Inorganic Gas Analysis:	See Table 4.3c.
Methods of Organic Gas Analysis:	See Table 4.3e.
Cost of Particulate Matter Analyzers:	See Table 4.3a.
Suppliers of Commercially Available Instruments:	See references 48 to 53, in addition to the other sections of this chapter.

Particulate Air Pollutants

Analysis of Number and Size

SAMPLE COLLECTION METHODS

For details see Sections 4.15 to 4.18.

Settling. This is the simplest sampling technique but it is limited to particle diameters of about 10 microns (μ). It is widely used for determining particle fallout or dust fall in urban environments.[2] (Section 4.2). The samples collected may be analyzed by sieving or by microscopic evaluation, and the water insoluble fractions may be estimated by extraction and weighing. General data on dust fall concentrations are expressed in terms of weight per unit area per 30 days involved.

Filtration. Samples are generally collected on filters using an airflow rate of 1 to 2 M^3/minute. There are a number of media available to provide high filtration efficiencies. The controlled porosity membrane filters are the most popular for low-volume samples. Membranes with selective pore sizes from 0.005 to 10 μ and collection efficiencies in excess of 99.7% are available. The structure, application and properties of membrane media can be found in the literature.[24]

Impingement. A number of instruments have been designed to sample particulate matter by impingement. The most widely used devices are the Greenburg-Smith (standard) and midget impingers, Cascade impactor and the Andersen sampler. Impingers employ a wet collecting method. The Cascade and Andersen samplers are dry collectors.[42] Samples collected on plates of glass, stainless steel or filter may be directly examined microscopically or washed off the plates for further chemical analysis. Impingement collection efficiency is high for particles with diameters of 1μ or greater.[31]

Precipitation. Two kinds of precipitators are available as particulate samplers, electrical and thermal. The electrical one uses a potential of 12 to 30 KV; the thermal technique relies on a temperature gradient of 750 to 3,000°C/cm. Collection efficiencies are nearly 100% for particles in the range 0.01 to 10 μ while tolerating flowrates as low as 10 to 20 cm^3/minute.[42]

Centrifugal. Based on the cyclone principle, these devices have a limited efficiency range for small particles but are successful in collecting particles larger than 5 μ, such as fly ash.[2,4]

Hybrids. A number of new particulate samplers are being developed which combine older collection techniques with current ideas for concentration estimation. For example, instruments have been developed to precipitate particulates onto the surfaces of crystal oscillators whose frequency changes may be co-related to particulate mass concentrations.[40]

Analysis of Particles

Optical microscope. A portion of the collected dust sample is first uniformly suspended in a liquid and then placed in a standard cell for counting. Particle sizing (see Glossary) usually is prefaced by sieving and the dry sample is placed directly in the light field for size analysis.[24,42]

Electron microscope. Smaller particles in the submicron region require increased magnification to permit size analysis. Electron microscopy in particulate analysis is generally used as a qualitative tool and it has seen little use in particle counting.[42]

Photometric methods. Light-scattering aerosol counters (see Glossary) and photometers have been used for the continuous estimation of number, concentration and size of particles[53] (Section 4.15). Automatic filter-paper samplers have been available for estimating particles.[24] The design of these samplers is based on the measurement of light transmission or reflectance of particles deposited on the sampling media.[53]

Crystal frequency. As precipitated particles coat on oscillating quartz crystal, the mass change of the crystal causes a frequency change which may be related to the total mass of the particulates sampled. The method is semicontinuous for mass concentration but offers no sizing capability.[40]

Composition of Inorganic Particulates

Dustfall measurements. The amount of particulate matter collected may be determined gravimetrically. Data are usually presented in units of $mg/cm^2/month$ or $tons/mile^2/month$. The acidity or basicity of the sample, its heavy-metal content, anions and the free carbon can also be determined.[34]

Airborne particulates and aerosols. These consist of either liquid or solid particles with diameters ranging from 0.01 to 100 μ. The chemical analysis of this class of air contaminant is dependent on the sampling technique and medium. General chemical and instrumental methods are summarized in Table 4.3a.

Dust or smoke stain indexes. These stain indexes are derived values representing the transmittance and reflectance measurements of light. The results are reported as coefficients of haze (COH) or reflectance units of dirt shade (RUDS).[17,18] Smoke concentrations may often be expressed as the number of COH units per 1,000 linear feet of air sampled through the filter. The soiling index of zero RUDS is defined as absolutely black. Further information as to definitions, standard procedures for smoke-stain estimation and proposed calibration curves are available.[17,18,23,28]

Ring oven technique. The ring oven (see Glossary) is a simple microchemical technique for separating and measuring inorganic ions. This method finds particular use in the analysis of trace amounts of airborne contaminants trapped on filter media. Determinations in the micro to nanogram region are possible with accuracies in excess of 90%. The instrumentation is easily operated, inexpensive and commercially available.[44,46]

Atomic absorption (AA) spectrometry. Since the last decade, AA spectrometry (see Glossary) has become the most widely used technique for the analysis of the inorganic moiety (fraction) of airborne metallic particulates. Simply operated and amenable to data processing, it provides specificity, high sensitivity and high sample throughput. The development of flameless AA techniques has lowered the minimum detection limits from submicrogram to subnanogram quantities. The advances have also permitted direct analysis of both solids as well as the usual solution extracts.

Table 4.3a

SUMMARY OF METHODS FOR ANALYSIS OF PARTICULATE MATTER[42]

Method	Application	Sensitivity	Reliability and Speed	Approximate Cost, $
Atomic absorption spectrometry	Metals	μg–ng (10^{-9} g)	Specific and reliable, simple and fast	5,000–15,000
Colorimetry	Ions	mg–μg	Reliable, simple and fast	300–8,000 if automated
Coulometry (titrimetric)	Ions	mg–μg	Nonspecific, precise if automated, fast	300–5,000
Electron microscopy	Molecules	Qualitative	Slow, training required	50,000–100,000
Emission spectrometry	Metals	mg–μg$^+$	Specific, fast, training required	10,000–70,000 if automated
Flame photometry	Metals	mg–μg	Specific, fast	1,000–5,000
Fluorimetry	Ions and metals	μg–ng	Selective, fast	8,000–20,000
Gravimetry	Ions and total mass	mg	Selective, slow	500–2,000
Microprobe	Elements	<pg (10^{-12} g)	Reliable, slow, training required	40,000–100,000
Neutron activation	Element	ng–pg	Specific, fast, training required	50,000 and up
Optical microscope	Molecules	Qualitative	Reliable, slow, skill required	300–4,000
Polarography	Metals	mg–μg	Reliable, slow, skill required	500–4,000
Ring oven	Ions	μg–ng	Selective and reliable, simple and fast	300–700
Spectrophotometry	Ions	mg–ng	Selective and reliable, simple and fast	5,000–40,000 if automated
Spark source mass spectrometry	Metals	mg–ng	Specific, slow, training required	80,000 and up
Titrimetry	Ions	mg	Nonspecific, skill required	100 and up

Specific ion electrode. This new technique (see Section I.4.11) has already found wide application for the measurements of fluorides and it permits the analysis of a growing number of cations and anions.[15]

The sampling technique and sample medium limits the quality of the analysis results. Membranes, while containing few impurities, can limit particulate sample weight because of sample volumes. Glass fibers, while permitting higher volumes of air to be sampled, have a number of undesirable contaminants whose concentrations have reduced this medium's usefulness for certain analytical investigations. The importance of sample blanks (see Glossary) has also been emphasized.[14]

Composition of Organic Particulates

The major contributors of organic particulates to the environment are incomplete combustion of carbonaceous fuels and, to a lesser degree, secondary chemical reactions in the air. Many of the compounds found in the organic moiety extractable from airborne particulates are biologically active.[19,20] Because of the low concentrations of organic particulate matter in air, high-volume sampling is the most generally applied method. Aliquot portions of the filter medium are extracted with organic solvent, fractionated and analyzed.[10,12,35]

There are a number of sample preparation and separation schemes developed for the organic particulates (Fig. 4.3b).[33]

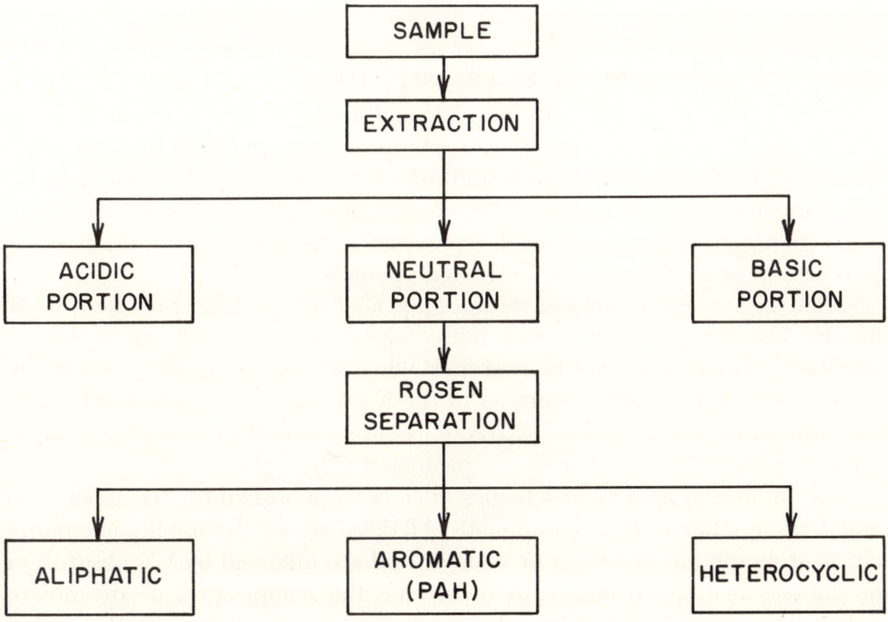

Figure 4.3b Organic particulate preparation and separation methods.

Acidic portion. This portion contains predominantly nonvolatile fatty acids and on occasion nonvolatile polyphenols. Numerous quantitative analytical procedures for the fatty acids have been published.[42] The polyphenols are analyzed by gas chromatography and by flame ionization detection.[42,53]

Basic portion. This portion contains only a small amount of the organic matter. Its chemical composition is virtually unknown but traces of N-heterocyclic aromatics have been found. Analytical techniques applicable to these compounds have also been developed.[42]

Neutral portion. The neutral portion of the extracts consists of three basic classes of organic compounds: aliphatic, aromatic and heterocyclic. Separation on silica gel successfully fractionates the neutral portion. The aliphatic fraction contains mainly saturated and some unsaturated hydrocarbons whose analyses by gas chromatography or by a combined approach with gas chromatography and mass spectrometry have been well documented.[42,53]

The heterocyclic fraction consists of oxygenated sulfur (S) and N-heterocyclic compounds. These compounds have been analyzed by gas chromatography.[37,43] In addition, a number of methods using liquid-solid and thin-layer chromatography in combination with UV absorption and fluorescence spectroscopy have also been reported.[27,35,42,43,53]

The aromatic fraction contains a wide spectrum of benzene-derivative compounds. Separation and determination of the constituent components have been accomplished with variable success by using chromatography and spectroscopy.

Polynuclear Aromatic Hydrocarbons (PAH)

The PAH compounds with four, five and six rings in the aromatic fraction are the most extensively studied trace components in polluted air. The great interest in PAH materials stems from the fact that several of these compounds are known to be carcinogenic[5,19,20,30] (Section 3.5). A number of urban atmospheric studies have reported concentrations of benzo(a)pyrene, benzo(e)-pyrene and other biologically active compounds.[6,8,9,38,42]

The high-volume sampler with a glass fiber filter is used to collect PAH in air. The sample is extracted with organic solvents such as cyclohexane.[10,35,42] Great care must be exercised when the extract is concentrated by solvent evaporation since a number of the low molecular weight PAH materials are appreciably volatile. Extracted samples stored in solvent are stable for a considerable time if they are protected from light.

A number of analytical schemes have been published for the separation and determination of PAH compounds. All these procedures use liquid column chromatography in one form or another and are followed by UV absorption or fluorescence spectroscopy to determine the composition of the eluate fraction.[8,9,10,20,35,36,42,53] The advantages and disadvantages of these two optical methods are well documented.[8,9,35,36,53] The common difficulty for the accurate determination of PAH by these methods is that reference spectra of primary standard PAH with known purity are needed. However, many available PAH compounds have been found to be insufficiently pure for their use as reference standards. The result in quantitative PAH measurement can have an error of orders of magnitude. Most recently, mass spectrometry has been successfully applied to the analysis of PAH materials.[26] The technique is much less susceptible to interference by impurities. If mass spectrometry is coupled with gas liquid chromatography (GLC-MS), the system may prove to be the most discriminating technique available for PAH analysis.

PESTICIDES

Pesticides collected on the filter from an air sample are found in the neutral portion of Fig. 4.3b. The common chlorinated-hydrocarbon pesticides such as DDT are in the aromatic fraction, and the thiophosphor ester malathion in the heterocyclic fraction.[42] A number of analytical methods for pesticide measurement have been developed since the last decade (Section 4.16). However, the available methods for the analysis of pesticides in air are inadequate, and much additional effort is required.[42] Gas liquid chromatography (GLC) with an electron capture detector (ECD) (see Glossary) appears to be the most popular technique.[53] The system is relatively inexpensive and easily operated. However, the technique is time-consuming and, in many cases, it cannot separate and identify pesticides found in the animal-body tissues.[3] In addition, the GLC-ECD system fails in determining polychlorinated biphenyls (PCBs).[29] As in other cases, the combination of GLC-MS will appear to be one of the best choices for pesticide and PCB analyses.[3,41]

Gaseous Air Pollutants

The gaseous air pollutants are, in general, highly toxic to the health of man, animals and vegetation (Section 2.11). Their concentrations are usually low and are reported in units of $\mu g/m^3$ of the air at 25°C and 760 mm of mercury pressure or in parts per million (ppm) and parts per billion (ppb). The conversion for a gas is: μg per $m^3 = 24{,}500 \times$ ppm/molecular weight and ppb $=$ ppm $\times 10^3$.

A great many batch, yet specific, chemical methods have been developed in different countries for the analysis of gases. However, only a limited number of these methods have been accepted and recommended by such international organizations as IUPAC, OECD and WHO.[24] These agencies are making an effort to standardize techniques or reach agreement on recommended procedures.[28] In selecting and evaluating methods for gas analysis, the following factors must be considered: the efficiency of sample collection; the stability of reagents and products; the ease of calibration procedure; the degree of specificity, simplicity, sensitivity, accuracy, precision and reliability of the procedure.[1,2,4,22,24,25,28,34,42,53]

Table 4.3c
ANALYTICAL METHODS FOR THE DETERMINATION OF INORGANIC GASEOUS POLLUTANTS

Pollutant and Method	*Detection Principle*	*Sensitivity*	*Interference*
Sulfur Dioxide (Section 4.20)			
Nonautomatic			
West and Gaeke[45a]	Colorimetry	0.015–0.6 mg/m^3 of air sample	Sulfides, nitrites, salts of iron and other heavy metals
Acidimetric H_2O_2[22,24]	Titrimetry	0.03 mg/m^3	Acidic gases and NH_3
H_2O_2[24,42]	Conductrimetry	0.015–3 mg/m^3	Same as above
Iodimetric[22,24]	Spectrophotometry	7 μg/m^3	N_2S, NO_2, O_3 and oxidants
$BaSO_4$[1,24]	Turbidimetry	2 μg/3 ml solution	H_2S, H_2SO_4 mist
Polarographic[2,24]	Polarogram	0.06 mg/m^3	Nitrogen oxides
Others[1,24,42]			
Automatic[53]			
p-rosaniline[21,24,42]	Colorimetry	0.002 μg/ml solution	NO_2 and O_3
Electrolytic[24]	Conductrimetry	28 μg/m^3	Not specific
I_2 generation[39]	Coulometry	0–0.3 mg/m^3	O_3
Others[1,24,25,32,42]			
Hydrogen Sulfide (Section 3.10)			
Nonautomatic			
Methylene blue[22,24,42]	Spectrophotometry	0.05 μg/ml solution	NO_2, Cl_2, and SO_2
AgS_2[24,42]	Spectrophotometry	0.2 μg/ml solution	Specific, less sensitive
Automatic			
Thomas[24]	Conductrimetry	15 μg/m^3	Not specific
Potentiometric[24,42]	Potentiometry	Not sensitive	Not specific
Paper tape[34]	Colorimetry	1.5 μg/m^3	HF, SO_2

Sulfuric Acid Mist			
Nonautomatic			
NaOH[7,24]	Titrimetry	16 μg/m³	NH_3, soluble electrolytes
$Na_2B_4O_7$[7,24]	Titrimetry	1 μg/m³	NH_3, soluble bases
Specific[13]	Titrimetry	0.4 μg/m³	Specific
Others[22,24,28]			
Carbon Monoxide and Carbon Dioxide (Section 4.5)			
Nonautomatic			
I_2O_5[22,34]	Coulometry	1.4–3.5 μg of CO	None, highly specific
NBS indicator-tube[22,24]	Colorimetry	1.25–12.5 mg/m³	Organic substances
Conversion to CH_4[11]	Chromatography	2.5 mg/m³	Specific
Conversion to CO_2[24] (for CO and CO_2)	Titrimetry	1.25 mg/m³	Specific
Automatic[53]			
Infrared absorption[22,24] (for CO and CO_2)	Spectrometry	0–25 mg/m³ or greater	None, highly specific
Nitrogen Oxides and Nitrates (Section 4.12)			
Nonautomatic			
NO_2: Griess-Ilosvay[24]	Spectrophotometry	1 μg/ml solution	None, sensitive
NO: oxidation to NO_2[1,2]	Spectrophotometry	1 μg/ml solution	None, low conversion efficiency (60–70%)
Nitrates: 2,4-xylenol[24,42]	Spectrophotometry	0.1–10 μg/m³	Chlorides, nitrites
Automatic[53]			
NO_2: sulfanilamide diazotization[22,24]	Spectrophotometry	0.08 mg/m³	SO_2, but can be eliminated
NO: Griess-Ilosvay[24]	Spectrophotometry	0.014 mg/m³	O_3
NO_2 and NO combined[34]	Colorimetry	0–2 mg/m³	O_3, SO_2
Ozone and Oxidants (Section 4.14)			
Nonautomatic			
Neutral buffered KI[24,34]	Spectrophotometry	0.02 mg/m³ in terms of Ozone (O_3)	SO_2, reducing gases, but simple and accurate
Alkaline KI[24,34]	Spectrophotometry	0.04 mg/m³ in O_3	Same as above

Table 4.3c (Continued)

Pollutant and Method	Detection Principle	Sensitivity	Interference
Automatic[53]			
Phenolphthalein[24]	Spectrophotometry	0–2 mg/m^3 (O_3)	H_2S, acidic gases
I_2 generation	Coulometry	0–0.2 mg/m^3 (O_3)	SO_2
Ammonia (*Ammonium Salts*) *and Hydrogen Cyanide* (*Cyanides*)			
Nonautomatic			
NH_3: Nesslerization[22]	Colorimetry	1 $\mu g/ml$ solution	H_2S, aliphatic amines
HCN: palladium chelate[1,24,34]	Fluorimetry	0.02 $\mu g/ml$ solution	HF, fluorides
Automatic[53]			
Potentiometric	Potentiometry	1.3 $\mu g/m^3$	Acidic gases
Halogen Compounds (Section 4.8)			
Nonautomatic			
F (fluorides):			
alizarin lake[2,28]	Spectrophotometry	0.6 $\mu g/m^3$	None, specific
microdiffusion[28]	Spectrophotometry	0.2–5.0 $\mu g/ml$ solution	Specific
Ion electrode[39]	Potentiometry	0.3 $\mu g/m^3$	Specific
Free chlorine:			
o-tolidine[22]	Colorimetry	0.01 $\mu g/ml$ solution	ClO_2, not specific
HCl (chlorides):			
$AgNO_3$[1,22]	Titrimetry	2 μg of Cl^-/ml solution	HCN, halogen acids
nephelometric or turbidimetric[1,22]	Spectrophotometry	1 μg of Cl^-/ml solution	HCN, halogen acids

[a]This method is more specific than the others; it is applicable to determine SO_2 in ambient air with the concentration range of 0.015–15 mg/m^3.

Inorganic Gases

The sensitivity of the analytical procedure is of primary concern. The analytical methods listed in Table 4.3c have been used for the determination of inorganic gaseous pollutants, but not all of them have been adequately evaluated.[24,28,53]

Rapid advances in GLC, MS, optical spectroscopy and detection methods such as flame ionization and electron capture have made available a great many new instruments as manufactured items for inorganic gaseous analyses. (See other Sections of this chapter.) For routine measurement and research in air pollution, a large number of samples have to be analyzed and rapid services are needed. As one approach, auto-analyzers have been developed and are commercially available for batch and continuous measurement.[47,53] The reagents used in such an analyzer and the sensitivities for some typical substances are listed in Table 4.3d.[47]

Organic Gases

The importance of organic gases in air lies in the fact that they are the reactants and products in photochemical air pollution (Section 2.8). Almost all of these gases are toxic though their concentrations are normally lower than those of inorganic gases in air.

Table 4.3d
SENSITIVITY OF AN AUTO-ANALYZER FOR TYPICAL SUBSTANCES AND REAGENTS USED

Substance	*Range*	*Reproducibility or Deviation*	*Reagent*
Nitrates as NO_3	0.05 mg N/liter	0.003	Griess-Ilosvay method
Nitrites as NO_2	0.02–5.0 mg N/liter	±1.5%	Sulfanilic acid, naphthylamine hydrochloride
Fluorides	0–2.0 ppm	±1.0%	Alizarin complex, zirconium lake
SO_2	0–5.0 μg/ml	±0.5%	Modified West-Gaeke method
NO_2	0.1–5.0 μg/ml	±0.5%	Azo dye
Phosphates	0–100 ppm	±1.0%	Molybdovanadic acid
Chlorides	0.1–100 μg/ml	±1.0%	Mercuric thiocyanate
Chlorine	0–1.0 ppm	±1.0%	0-tolidine dihydrochloride
Cyanides	0–0.1 ppm	±1.0%	Pyridine, benzidine, bromine
Silica (water analysis)	0–200 ppb	±0.5%	Ammonium molybdate, 1-amino-2-naphthol-4-sulfonic acid
Tetraethyllead	0–1.0 μg/ml Pb	—	Dithizone reaction

Table 4.3e
ANALYTICAL METHODS FOR THE DETERMINATION OF ORGANIC GASEOUS POLLUTANTS

Compound	Method	Sensitivity	Interference
Acrolein	4-Hexylresorcinol spetrophotometry[34]	0.05–25 mg/m^3	Dines will cause minor interference
Formaldehyde	Colorimetry:		
	Chromotropic acid method[1,24]	0.004 μg/ml solution	Phenols, olefins, aromatics
	Fuchsin sulfate[1,24]	1 μg/3 ml solution	Aldehydes
Aliphatic aldehydes	MBTH[a] Spetrophotometry[34]	2.68 μg/m^3 in terms of formaldehyde	Aromatic amines, imino compounds
Methanol	Determination as formaldehyde[22]	0.08 μg/ml solution	Same as of formaldehyde
Phenols	Spetrophotometry:		
	p-Aminodimethylaniline sulfate[22]	1 μg/m^3	Other organics with same color reactions
	4-Aminoantipyrine[1,22]	0.1 μg/ml solution	Relatively specific
	Colorimetry:		
	p-Nitroaniline[1,22]	0.2 μg/ml solution	Cresols
Acetylene	Colorimetry:		
	Cuprous chloride[42]	1 μg/ml solution	H_2S, mercatans, acetylenic compounds

[a] 3-methyl-2-benzothiazolone hydrazone.

The procedures for organic gaseous pollutant analysis are based on the use of instrumentation.[4,24,42,53] Unlike the case of inorganic gaseous analysis, most of these are available only as laboratory techniques; and very few instruments have as yet been developed for the continuous analysis of organic gases, except for infrared and flame ionization detector hydrocarbon analyzers.[4,42,53]

For details of the application, sensitivity, accuracy and limitation of each procedure, a fair number of publications can be consulted.[1,2,4,22,24,25,32,34,42,53] Table 4.3e outlines chemical methods available for the determination of some organic gaseous pollutants.

REFERENCES

1. Alekseeva, M. V.: "The Determination of Atmospheric Pollutants," Medgiz, Moscow, 1963.
2. "ASTM Standards on Methods of Atmospheric Sampling and Analysis," Philadelphia, 1962.
3. Bagley, G. E.: *J. Assoc. Official Analyt. Chem.*, vol. 53, p. 251, 1970.

4. Chambers, L. A.: "Encyclopedia of Instrumentation for Industrial Hygiene," University of Michigan, 1956.
5. Clar, E.: "Polycyclic Hydrocarbons," vol. I and II, Academic Press, New York, 1964.
6. Commins, B. T.: *National Cancer Institute Monograph*, vol. 9, p. 201, 1962.
7. Commins, B. T.: *Analyst*, vol. 88, p. 364, 1963.
8. Commins, B. T. et al.: *British J. Indust. Med.*, vol. 22, p. 128, 1965.
9. Dubois, L. and J. L. Monkman et al.: *J. Air Pollut. Control Assoc.*, vol. 14, p. 295, 1964.
10. Dubois, L. and J. L. Monkman: *Inter'l J. Air Pollut.*, vol. 9, p. 833, 1965.
11. Ibid, vol. 16, p. 135, 1966.
12. Dubois, L. and J. L. Monkman: *J. Chromatography*, vol. 28, p. 317, 1967.
13. ______: *Mikrochimica Acta*, p. 269, 1969.
14. ______: *Inter'l J. Environ. Analyt. Chem.*, in press, 1972.
15. Durst, R. A.: *American Scientist*, vol. 59, p. 353, 1971.
16. Gambrill, C. M.: *Analytical Chemistry*, vol. 37, p. 143R, 1965.
17. Gruber, C. W. et al.: *J. Air Pollu. Contr. Ass.*, vol. 4, p. 143, 1954.
18. Hemeon, W. L. C. et al.: *J. Air Pollu. Contr. Asso.*, vol. 3, p. 22, 1953.
19. Hoffmann, D. J.: *J. Air Pollu. Contr. Ass.*, vol. 15, p. 162, 1965.
20. Hoffmann, D. J. et al.: *Proc. Am. Ass. Cancer Res.* vol. 6, p. 29, 1965.
21. Huitt, A. A. and J. P. Lodge: *Analyt. Chem.*, vol. 36, p. 1305, 1964.
22. Jacobs, M. B.: "The Chemical Analysis of Air Pollutants," Interscience, New York, 1960.
23. Katz, M. and H. P. Sanderson: *Spec. Tech. Publ. ASTM*, no. 250, 1958.
24. Katz, M.: "Measurement of Air Pollutants," World Health Organization, Geneva, 1969.
25. Katz, M. (ed.): Abstract of Tentative Methods on Intersociety Committee, *J. Assoc. Offi. Analy. Chem.*, vol. 54, p. 224, 1971.
26. Lao, R. C. et al.: *Proc. Inter'l Symp. Identification and Measurement of Environ. Pollut.*, Ottawa, Canada, June 1971.
27. Monkman, J. L. et al.: *"National Cancer Institute Monograph,"* vol. 3, p. 153, 1961.
28. Organization for Economic Co-op. and Develop (OECD), *Working Party on Methods of Measuring Air Pollut.*, Paris, 1964.
29. Papageorge, W. B.: *Proc. of PCB8s in Environ.*, Nat'l Water Quality Lab. Fed. Water Quality Admin., Duluth, Minnesota, March 1970.
30. Patter, A. M. et al.: "The Ring Index," 2d. ed., Am. Chemical Soc., Washington, D.C., 1960. (Suppl. I, 1963; Suppl. II, 1964.)
31. Ranz, W. E. et al.: *Indus. and Eng. Chemistry*, vol. 44, p. 1371, 1952.
32. Ronicke, G.: "Working Instructions for the Measurement Stations," Commission for Research of the Effects of Air Pollutants, German Research Society, Schallstadt, German, 1969.
33. Rosen, A. A. et al.: *Analytical Chemistry*, vol. 27, p. 790, 1955.
34. Saltzman, B. E.: "Selected Methods for the Measurements of Air Pollutants," No. 999-AP-11, Interbranch Chem. Advisory Committee, R. A. Taft Sanitary Eng. Center, Cincinnati, Ohio, 1965.
35. Sawicki, E. and K. Cassel (eds.): Analysis of Carcinogenic Air Pollutants, *Nat'l Cancer Institute Monograph*, no. 9, U.S. Public Health Service, Washington, D.C., 1962.
36. Sawicki, E. et al.: *Inter'l J. Air Pollut.*, vol. 2, p. 273, 1960.
37. ______: *Inter'l J. Air and Water Pollut.*, vol. 9, p. 515, 1965.
38. Sawicki, E.: *Proc. Arch. Environ. Health*, vol. 14, p. 46, 1967.
39. Schulze, F.: *Analytical Chemistry*, vol. 38, p. 748, 1966.
40. Sem, et al.: *Atmospheric Environment*, vol. 5, p. 653, 1971.
41. Stalling, D. L. et al.: *J. Ass. Off. Analy. Chem.*, vol. 54, p. 801, 1971.
42. Stern, A. C. (ed.): "Air Pollution," vol. 2, pp. 1–280, Academic Press, New York, 1968.
43. Van Duuren, B. L. et al.: *Nat'l Cancer Institute*, vol. 31, p. 41, 1963.
44. Weisz, H.: "Microanalysis by the Ring Oven Technique," 2d. ed., Pergamon Press, Oxford, 1970.
45. West, P. W. and G. C. Gaeke: *Anal. Chem.*, vol. 28, p. 1816, 1956.
46. West, P. W. et al.: *Anal. Chim. Acta*, vol. 37, p. 374, 1967.
47. Zaleiko, N. S.: *J. Air Pollu. Contr. Ass.*, vol. 13, p. 531, 1963.
48. Laboratory Guide, published annually by *Analytical Chemistry* in the July issue.
49. Annual Review, published by *Analytical Chemistry* in the April issue.

50. Annual Review of Instruments, in the December issue of the *J. Air Pollu. Contr. Ass.*
51. Guide to Scientific Instruments, published annually in *Science*.
52. Reports on Instrumental Development and Evaluation, the Environmental Protection Agency, U.S. Public Service.
53. Lipták, B. G. (ed.): "Instrument Engineers' Handbook," (Vols. I, II and Supplement One), Philadelphia, Chilton, 1969, 1970 and 1972.

4.4 ANALYZERS, EXISTING AND NEW ONES

Analyzer and Sampler Types:	(*a*) *Particulate matter detection,* (a1) Piezoelectric particle mass monitor, (a2) Andersen Hi Vol® sampler, (a3) Alternating quadrupole electric field, (a4) Light scattering, (a5) Condensation, (a6) Electric counting; (*b*) *Metal and cation detection,* (b1) Atomic absorption, (b2) X-ray fluorescence, (b3) Neutron activation, (b4) Polarography, anodic stripping voltammetry; (*c*) *Anion detection,* (c1) Colorimetric methods, (c2) Polarography, (c3) Specific ion electrodes; (*d*) *Gaseous air pollutants,* (d1) Spectroscopic methods, (d1.1) Infrared spectroscopy, (d1.2) Laser and lidar, (d2) Chromatographic methods, (d3) Colorimetric analyzers, (d4) Electrochemistry, (d4.1) Electroconductivity, (d4.2) Coulometric methods, (d4.3) Polarography, (d5) Flame ionization detector, (d6) Chemiluminescent monitor, (d7) Modulated beam quadrupole mass spectrometer.
Cost:	The cost of glassware and of conventional chemicals used in these manual methods is inconsequential compared to the cost of the skilled labor required.

Particulate Air Pollutants

Particulate air pollutants are of major health concern, particularly in the micron and submicron sizes. Such very small particulates are able to penetrate the human respiratory tract. Consequently, it becomes important to measure the size and number of particles and/or their mass concentration.[1,2] The particulates also contain metallic cations, anions and organic substances as such or adsorbed on the particles. All these compounds are of interest from a health point of view.

Size and Number of Particles

Most of the particulate collection methods are based on filtration, impaction or electrostatic precipitation.[3,4] Depending upon the collection method used, the size, number and concentration of the particles are determined by photometry, optical and electron microscopy.[5-11] In all these methods, some of the particles after collection are different from their original state before entering the collecting apparatus. Indeed, they can agglomerate or fragment during the collection. A study on respirable dust sampling and the related problems was made by Lippmann.[2]

Two new instruments are of some interest, namely the piezoelectric particle mass monitor and the Hi Vol Andersen sampler.[12] Although able to measure total mass concentration over short periods of time, the piezoelectric particle mass monitor does not measure either the size or the number of particles, and consequently its use is limited. However, with an appropriate pre-filter, or a large size particle cutoff device, it can measure the total mass concentration of the respirable particles.

The Andersen Hi Vol sampler has the great advantage of a larger sample over the usual Andersen sampler. The particles, being separated in five different ranges, can be weighed so that a range of mass concentrations can be obtained. Theoretically, it should be possible with this instrument to evaluate the number of particles by a simple calculation assuming a uniform density for the collected particles. This instrument being relatively new, it will take some time before enough data have accumulated to permit a full assessment of the apparatus. Losses may occur on the walls between the collection stages.

The apparatus which makes use of an alternating quadrupole electric field (see Glossary) shows great promise in particle sizing and mass concentration measurement.[13,14] Particulate matter has been measured continuously in the range 0.001 to 10 μ, using three different principles: light scattering, condensation and electric counting.[15,36] A continuous field instrument is still needed for the measurement of particle size and mass concentration.

The substances of which the dusts are composed, or the substances adsorbed on dusts, are of great interest. These substances may be either organic or inorganic.

Organic Compounds

The number and complexity of organic substances present in particulate matter are great. Only a few research workers have been active in the field. Carcinogenic compounds, such as polycyclic aromatic hydrocarbons, have been measured in daily, weekly or monthly samples after a soxhlet extraction (see Glossary) with an organic solvent.[16-18] All the other groups, aliphatics, oxygenated organic compounds and sulfur and nitrogen compounds, have received little attention. Any improvement in the automation of separation techniques (chromatography, sublimation) and in the measurement techniques (ultraviolet spectroscopy fluorescence and mass spectrometry) would be welcome.

Metals and Cations

The inorganic compounds, collected in general on glass fiber sheets, are dissolved and digested according to different procedures and are subsequently measured by atomic absorption spectroscopy (see Glossary), which is at the moment the most widely used method.[19-21] There is a need for this technique to be adapted so that continuous measurement of metals could be achieved. X-ray fluorescence, dispersive and nondispersive, is also used for direct measurement of metals on filters. Although it is a nondestructive method, its lack of sensitivity, matrix interference and the expensive equipment needed make its use somewhat difficult.[22] In the last few years, neutron activation, atomic fluorescence and anodic stripping voltammetry have been gaining more interest.[23,24] Tables 4.4a and 4.4b list the advantages and disadvantages of the different techniques and the limits of detection for some metals measured in particulate matter.

Anions

Anions are usually measured by colorimetric methods after extraction from the filters.[25] Since these methods are likely to be long and tedious, analyzers should be developed which can measure them faster. Polarography and specific electrodes (Section I.4.11), if their sensitivity were increased, could

Table 4.4a
FEATURES OF METAL AND CATION DETECTORS

Technique	*Cost*	*Sample Preparation*	*Interference*	*Continuous Measurement*
Atomic absorption	Low	Extraction, digestion	Low	Possibly continuous
X-ray fluorescence	Expensive	No preparation	Matrix	No
Neutron activation	Expensive	Expensive	Low	No
Anodic stripping voltammetry	Low	Extraction	Low	Possibly semicontinuous

Table 4.4b
LIMITS OF DETECTION OF METAL AND CATION SENSORS

Metal Analyzed for	Atomic Absorption, μg/ml		Polarography, μg/ml		Neutron Activation, Neutron flux 10^{12}
	Normal	Graphite Furnace	Normal	Anodic Stripping Voltammetry	Total μg
As	5.00	0.020000	0.10	....	0.04
Bi	1.00	0.000500	2.00	0.05000	
Cd	0.03	0.000002	0.10	0.00005	
Cr	0.05	0.000050	1.00	....	0.02
Cu	0.10	0.000020	0.20	0.05000	0.01
Fe	0.10	0.000060	0.50	....	1.50
Ni	0.20	0.000200	0.20	....	1.50
Pb	0.30	0.000150	0.50	0.00010	
Se	2.00	0.010000	0.50	....	0.01
Zn	0.03	0.000002	0.10	0.00020	0.01

be used to measure anions. A semicontinuous method could also be developed if the sample was not to be taken on filters.[26]

Gaseous Air Pollutants

The same principles apply to gaseous air pollutants as for particulate matter; i.e., it is better to measure the concentrations of gases as present in air rather than by using adsorption in solution or on adsorbants.

Spectroscopic Methods

Spectroscopic methods seem to be most generally applicable. However, since the concentration of air pollutants is low, long light path spectroscopy has to be used. Consequently, the investigation of such a technique has been made in the infrared region.[27] However, some difficulties are encountered due to water vapor adsorption and light scattering by particles. Only a few gases can be measured because of serious spectral interferences. The technique is of interest for SO_2, CO, CH_4 and O_3. Nondispersive infrared analyzers[36] are also used for the measurement of CO and hydrocarbons; but, they are too bulky, and the response is affected by CO_2 and water vapor. These instruments are inaccurate below 1 ppm. Instruments to measure CO below these concentrations are badly needed. Theoretically, multireflection cells could be used; but unless interferences due to CO_2 and water can be eliminated, this improvement will be only theoretical.

In the last few years, lidar (laser-radar) has been used to measure the transport and diffusion of plumes and clouds of particulates.[28] It should be possible to use it in the detection of certain gases (CO, CH_4, SO_2). However, it would seem that lasers are easier to apply in the measurement of gases than lidars. The differences in the absorption of two lasers operating at the same wavelength but located at different distances from a central station should permit the measurement of a given pollutant over that given distance. These laser and lidar techniques should be ideal since there is no physical contact with the sample. However, probably only a few gases could be measured due to water vapor interferences.

Chromatography

Gas chromatography[36] is a very useful technique in the laboratory, but its application in the field is limited for reasons of bulk and weight. Some instruments are being used for the semicontinuous measurement of methane and carbon monoxide in urban air.[29] Other developments are being made to allow measurement of SO_2, H_2S and mercaptans. After separation on a teflon-polyphenyl ether column, the sulfur-containing compounds are directed to a flame ionization detector and the intensity of the sulfur line at 394 nanometers is measured.[30,31,36] The great advantage of these chromatographic methods is their specificity. However, difficulties are encountered in finding the appropriate column and the concentration of the pollutants can only be measured on a semicontinuous basis.

Recently, an instrument which permits the concentration of the sample by frontal analysis* has been developed and used for the measurement of carbon monoxide in air (Fig. 4.4c). This instrument could be used for any pollutant since the columns and the detectors can be changed. Such an instrument is of definite interest in the measurement of low levels of air pollutants.[32] All the disadvantages mentioned in connection with the gas chromatographic techniques are also inherent in this method. Because of their specificity, however, the chromatographic methods do enjoy great popularity.

Colorimetry

Colorimetric methods for gaseous pollutants and even continuous colorimetric analyzers are equipped with absorbers to remove interfering substances.[36] Unfortunately, some of the gas of interest is also removed.

Electrochemistry

There are two main types of electrochemical instruments: the electroconductivity and the coulometric. The electroconductivity instrument is not specific and it measures the total conductivity of the substances absorbed in

* A component of a gas sample can be concentrated by frontal analysis if the gas is passed over an adsorbent for a period of time long enough so that the adsorbent is saturated with the component of interest. An equilibrium between the concentration of the component in the gas sample and the concentration of the component on the adsorbent is then reached.

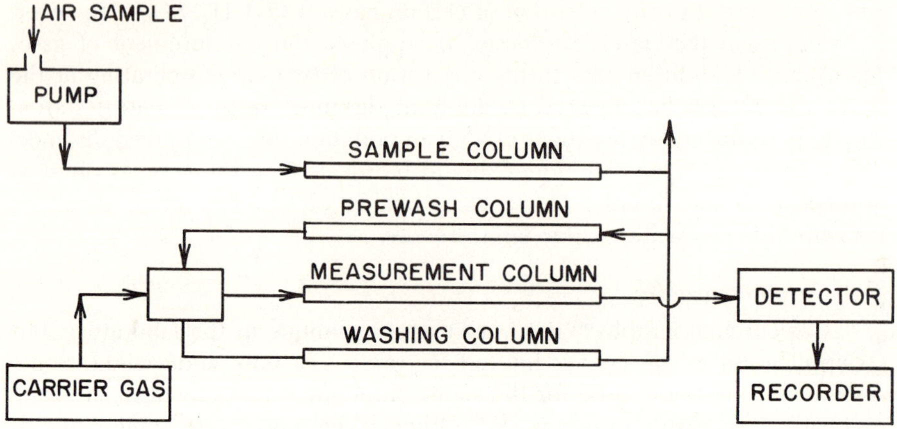

Figure 4.4c Chromatographic gas analyzer used for CO detection.

the solution (Section I.4.8). Since scrubbers are incorporated in the coulometric instruments, it follows that some of the pollutant being measured can be removed in this sample preparation step.[36]

Two recent papers have suggested that polarographic techniques could be used for the monitoring of gaseous pollutants.[33,34,36] This is an interesting breakthrough in the measurement of gaseous pollutants since specific response could be obtained. It is also expected that in the next few years some efforts will be made to apply specific ion electrodes (Section I.4.11) to the measurement of these pollutants. The sensitivity of these electrodes, however, has first to be increased.

Others

The flame ionization detector for the measurement of volatile hydrocarbons is close to being an ideal instrument. It is specific for volatile carbonaceous material and the sample is introduced into the instrument without any alteration.

It is easier to design such instruments since a whole class of compounds (hydrocarbons) are measured rather than one specific component. However, there is a need for an apparatus which could monitor classes or groups of compounds on a continuous basis, and specific detectors such as a sulfur detector should be designed. Detectors or instruments based on specific reactions can and could be used with great advantage. Such an instrument is the chemiluminescent monitor which uses the reaction of nitric oxide with ozone.[35]

Recently, the modulated beam quadrupole mass spectrometer has been developed, in which the gas sample to be analyzed is presented to the ionizer as a molecular beam running through high vacuum rather than as gas flowing into the background. Its background-free characteristics and high sensitivity (it can detect a sample in the 10^{-9} to 10^{-12} gm range) make it extremely valuable for the accurate determination of traces of gaseous air pollutants and their isotopic ratios which are now used to investigate the sources of pollution emission. The commercial availability of this mass spectrometer is desirable.

Conclusion

A lidar instrument for the monitoring of air pollutants is a most promising development since this can be used to measure the pollutants where they are and without any alteration due to sampling. These instruments are in the development stages and the preliminary results obtained are encouraging. Particulate concentration has been measured; and with variable wavelength lasers, some specific pollutants could also be determined.

Their great advantage is the fact that air pollution levels over large areas can be measured, and this could be done from a centrally located station where measurements could be made without the need of expensive networks. Their use, however, will probably be limited to a few pollutants only. Any new specific air pollution instrument is welcome in this challenging field.

REFERENCES

1. Dautrebande, L.: "Microaerosols," Academic Press, New York, 1962.
2. Lippman, M.: *Amer. Ind. Hyd. Ass. J.*, vol. 31, p. 138, 1970.
3. Orr, C. and J. M. Dalla Valle: "Fine Particle Measurement Size, Surface and Pore Volume," Macmillan, New York, 1960.
4. Yaffe, C. D., D. H. Byers and A. D. Hosey (eds.): "Encyclopedia of Instrumentation for Industrial Hygiene," University of Michigan Press, Ann Arbor, Michigan, 1956.
5. Jacobs, M. B.: "The Analytical Toxicology of Industrial Inorganic Poisons," Interscience, New York, 1967.
6. Wells, B. J.: *Health Physics*, vol. 13, p. 1001, 1967.
7. Benarie, M. and J. P. Quetier: *Aerosol Sci.*, vol. I, p. 77, 1970.
8. Moll, S. H. and R. E. Ogilvie: *Anal. Chem.*, vol. 39, p. 867, 1967.
9. Ferguson, J. S. and E. G. Sheridan: *J. Air Pollu. Contr. Ass.*, vol. 16, p. 669, 1966.
10. Blanco, A. J. and G. B. Hoidale: *Atmos. Environ.*, vol. 2, p. 327, 1968.
11. Goetz, A.: *Environ. Sci. Technol.*, vol. 3, p. 154, 1969.
12. Olin, J. G. and G. J. Sem: *Atmospheric Environment*, vol. 5, p. 653, 1971.
13. Aoyama, M. and S. Masuda: *Electrical Engineering in Japan*, vol. 90, no. 3, p. 10, 1970.
14. Masuda, S. and K. Fujibayashi: *Electrical Engineering in Japan*, vol. 90, no. 3, p. 1, 1970.
15. Peterson, C. O. and H. J. Paulus: *Amer. Ind. Hyg. Assoc. J.*, vol. 29, p. 111, 1968.
16. Dubois, L., A. Zdrojewski, J. Baker and J. L. Monkman: *J. of Air Pollu. Contr. Ass.*, vol. 17, no. 12, p. 818, 1967.
17. Dubois, L., A. Zdrojewski, P. Jènnawar and J. L. Monkman: *Atmospheric Environment*, vol. 4, p. 199, 1970.
18. Sawicki, E., T. W. Stanley, W. C. Elbert, J. Meeker and S. McPherson: *Atmospheric Environment*, vol. 1, p. 131, 1967.
19. Zdrojewski, A., H. Chung, L. Dubois and J. L. Monkman: The Accurate Measurement of Lead in Airborne Particulates, 18th Spectroscopy Symposium, Spectroscopy Society of Canada, Toronto, October 1971.
20. Thompson, R. J., G. B. Morgan and L. J. Purdue: *Atomic Absorption Newsletter*, vol. 9, p. 53, 1970.

21. Robinson, J. W.: "Atomic Absorption Spectroscopy," Marcel Dekker, New York, 1966.
22. Leroux, J. and M. Mahmud: *J. Air Pollu. Contr. Ass.*, vol. 20, no. 6, p. 402, 1970.
23. Dams, R., J. A. Robbins, K. A. Rahn and J. W. Winchester: *Anal. Chem.*, vol. 42, no. 8, p. 861, July 1970.
24. Matson, W. R.: Trace Substances in Environmental Health IV, Proceedings of University of Missouri's 4th Annual Conference on Trace Substances in Environmental Health, June 1970.
25. Jacobs, M. B.: "The Chemical Analysis of Air Pollutants," Interscience, New York, 1960.
26. Ion Selective Electrodes, Proceedings of a Symposium Held at the National Bureau of Standards, Gaithersburg, Maryland, January 30–31, 1969.
27. Beard, J. T. and W. R. Teele: The Quantitative Determination of Gaseous Air Pollutants by Long Path Infrared Transmission Spectroscopy, 61st Annual Meeting, The Air Pollution Control Association, 1968.
28. Johnson, W. B.: *J. of the Air Pollu. Contr. Ass.*, vol. 19, no. 3, p. 176, 1969.
29. Stevens, R. K., A. E. O'Keefe and G. C. Ortman: A Gas Chromatographic Approach to the Semi-Continuous Monitoring of Atmospheric Carbon Monoxide and Methane, Proceedings of the 11th Conference on Methods in Air Pollution on Industrial Hygiene Studies, Berkeley, California, 1970.
30. Stevens, R. K., A. E. O'Keefe, J. D. Mullk and K. J. Krost: Gas Chromatography of Reactive Sulfur Gases in Air at the Part Per Billion Level, 1: Direct Chromatographic Analysis, 157th ACS Meeting, April 1969.
31. Stevens, R. K., A. E. O'Keefe and G. C. Ortman: *ISA Transactions*, vol. 9, no. 1, p. 1, 1970.
32. Dubois, L. and J. L. Monkman: Continuous Determination of Carbon Monoxide by Frontal Analysis, in A. Zlatkis (ed.), "Advances in Chromatography," 1971.
33. Chand, R. and P. R. Cunningham: *IEEE Transactions on Geoscience Electronics*, vol. GE-8, no. 3, p. 158, July 1970.
34. Kuempel, J. R. and W. D. Shults: *Analytical Letters*, vol. 4, no. 2, p. 107, 1971.
35. Fontijn, A., A. J. Sabadell and R. J. Ronco: *Anal. Chem.*, vol. 42, no. 6, p. 575, 1970.
36. Lipták, B. G. (ed.): "Instrument Engineers' Handbook," (Vol. I, II and Supplement One) Philadelphia, Chilton, 1969, 1970 and 1972.

4.5 ALDEHYDE ANALYSIS

Aldehyde Analysis Methods:	(a) Formaldehyde detection by chromotropic acid reagent; (b) Detection by bisulfite reagent; (c) Analysis with MBTH reagent; (d) Schiff's reagent; (e) Acrolein detection by 4-hexylresorcinol.
Air Sample Size:	25 liters (a,c), 28 liters (d).
Detectable Formaldehyde Concentration, ppb:	10 (d), 20 (e), 30 (c), 100 (a).
Automated Aldehyde Analyzers:	Have been reported but are not yet commercially available.

There are numerous methods for aldehyde determination which are based on condensation reactions, sometimes followed by dehydration (Table 4.5a). Although useful in some circumstances, it is difficult to achieve the high sensitivity needed for the low concentrations of aldehydes normally found in the atmosphere with most reagents. The reactions in general form are as follows:

$$\mathrm{RCHO} + \mathrm{HX} \longrightarrow \underset{\displaystyle \mathrm{OH}}{\mathrm{RCH}} - \mathrm{X} \qquad 4.5(1)$$

$$\underset{\displaystyle \mathrm{OH}}{\mathrm{RCH}} - \mathrm{X} \longrightarrow \mathrm{RCH} = \mathrm{X}' + \mathrm{H_2O} \qquad 4.5(2)$$

Table 4.5b lists a number of such condensation reagents.

Formaldehyde Detection by Chromotropic Acid[1]

The first and smallest aldehyde, formaldehyde, is ordinarily the major aldehyde component found in air. Although it sometimes undergoes the same reactions as higher homologues, this is not always the case. Formaldehyde is very water-soluble, whereas most higher aldehydes are not. It is often desirable to use a method specific for formaldehyde and determine total

Table 4.5a
ALDEHYDE ANALYSES

Reagent	*Reactions Used*	*Method of Analysis*	*Limitations*
Bisulfite ($NaHSO_3$, aqueous)	$RCHO + NaHSO_3 \longrightarrow RCH(OH)—SO_3Na$	Destroy excess reagent, add buffer, determine HSO_3^- by iodimetry	Lower molecular weight aldehydes only, sensitivity moderate (useful for collecting for other methods)
Schiff's reagent 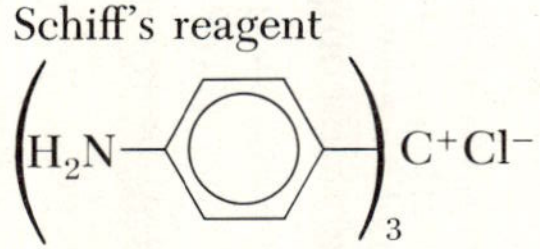 (saturated with SO_2)	Not well established	Colorimetric	Responds more to formaldehyde than other aldehydes, high purity dye essential

3-Methyl-2-benzothiazolone hydrazone (MBTH) 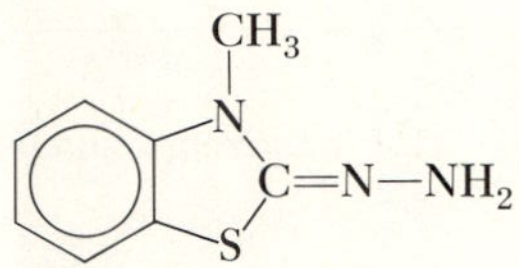	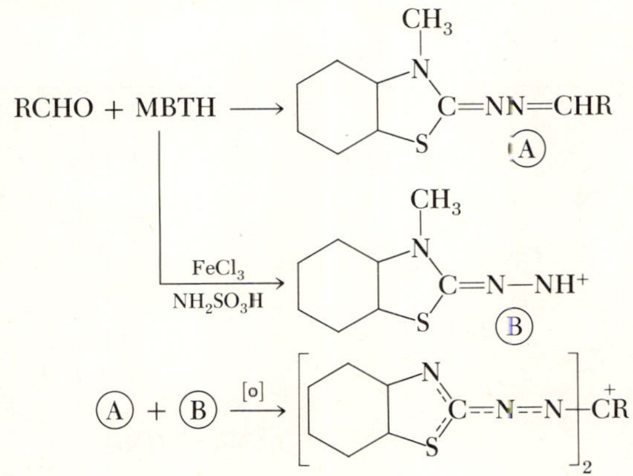	Colorimetric	Aliphatic aldehydes
Chromotropic acid 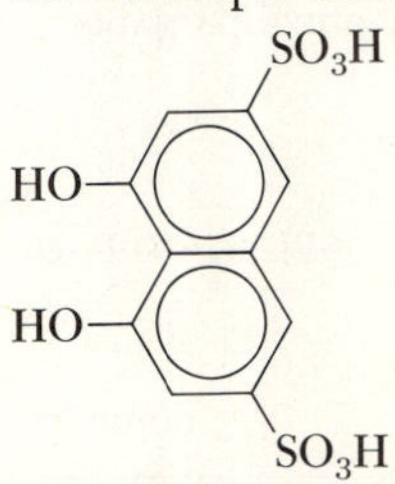	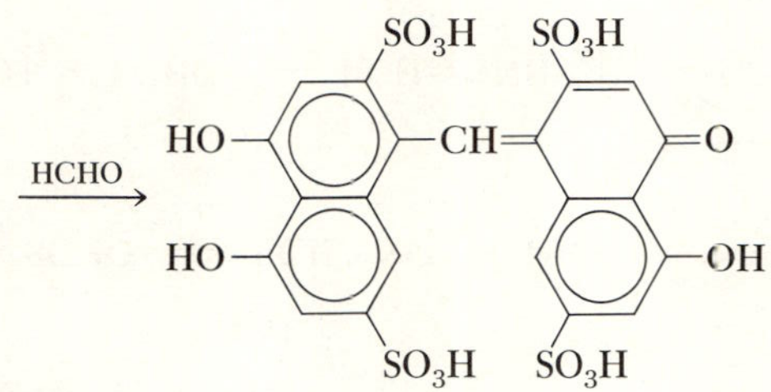	Colorimetric	Formaldehyde only; small negative interferences from phenols, alcohols and certain olefins and aromatic hydrocarbons

Table 4.5b

CONDENSATION REAGENTS

Reagent	*Reactions*	*Remarks*
Cyanide	$HCN + RCHO \longrightarrow RCH(OH)-CN$	Hazardous reagent
Hydroxylamine	$NH_2OH + RCHO \longrightarrow RCH{=}NOH + H_2O$	Excess hydroxylamine hydrochloride is titrable, allows quantitative analysis
Phenylhydrazine	$C_6H_5NHNH_2 + RCHO \longrightarrow RCH{=}NNHC_6H_5 + H_2O$	Used with sugars
2,4-Dinitrophenylhydrazine	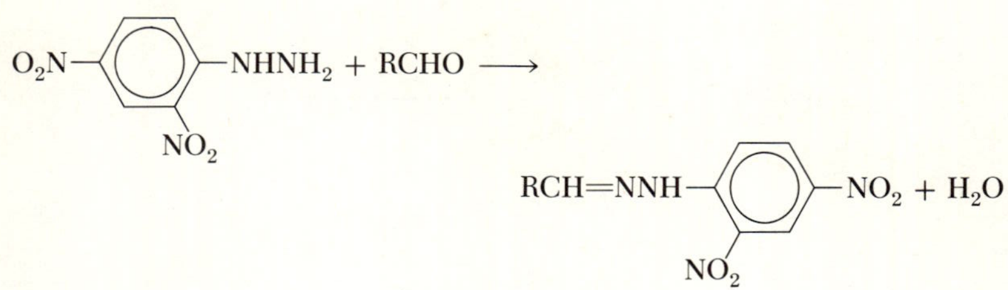	Good reagent to collect total carbonyls, gravimetric procedure lengthy and not sensitive enough for atmospheric work
2-Hydrazinobenzothiazole	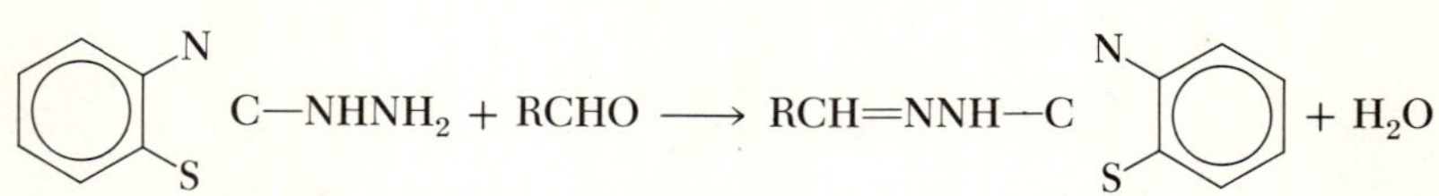	

aldehydes separately. Chromotropic acid is the reagent most commonly used for formaldehyde alone.

In the chromotropic acid method, the air sample is passed through two bubblers in series containing water. Chromotropic acid (4,5-dihydroxy-2,7,-naphthalenedisulfonic acid disodium salt) reagent solution is added, then acidified with concentrated sulfuric acid and is cooled. Color intensity is measured at 580 nanometers. Sodium formaldehyde bisulfite solutions, standardized iodimetrically, are used to prepare reference color standards.

The reactions occurring in this determination are not known with certainty. The following sequence is proposed:

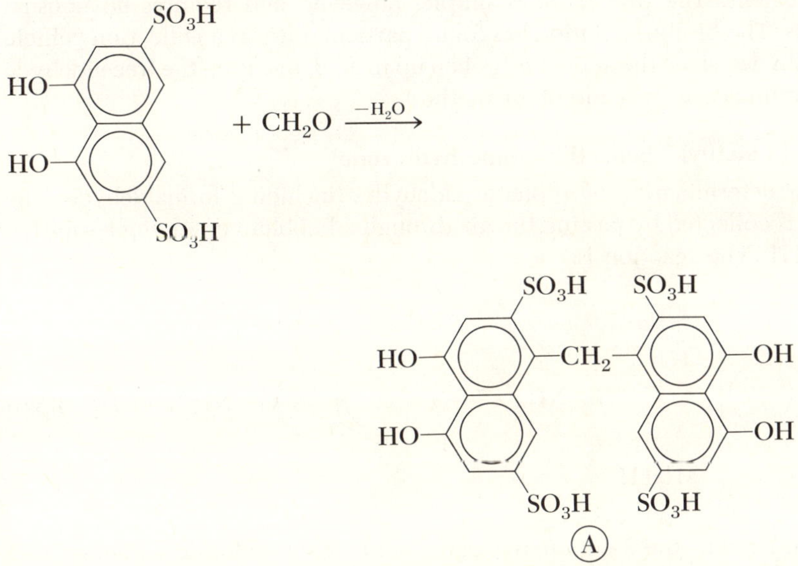

4.5(3)

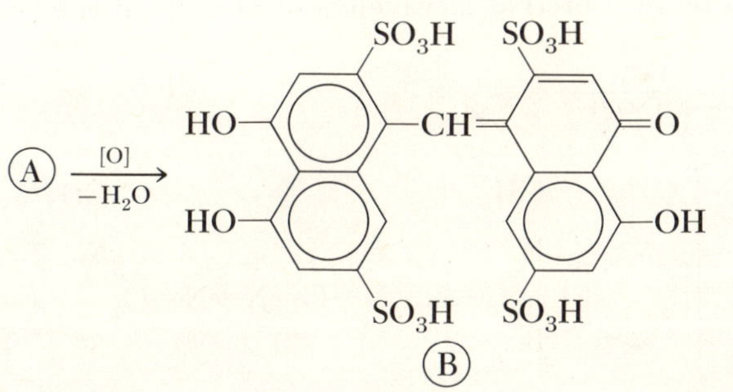

4.5(4)

The product B is purple in color. A concentration of 0.1 ppm formaldehyde can be determined in a 25 liter air sample. This method is recommended by the U.S. Environmental Protection Agency.

4.5 *Aldehyde Analysis*

Bisulfite[2]

Lower molecular weight aldehydes (including formaldehyde) will condense with aqueous sodium bisulfite:

$$RCHO + HSO_3Na \longrightarrow \underset{\displaystyle |\atop OH}{RCHSO_3Na} \qquad 4.5(5)$$

The excess reagent is destroyed with iodine, after which the aldehyde adduct is broken up by pH adjustment and determined iodimetrically. Since the sensitivity is only moderate, it is of limited use for atmospheric aldehyde measurement. The procedure is simple, however, and requires no unusual reagents. The bisulfite solution has found particular use as a collection vehicle for aldehydes, since the adduct is broken up in acid, releasing the free aldehyde for determination by some other method.

MBTH (3-methyl-2-benzothiazolone hydrazone)[3]

For determination of aliphatic aldehydes (including formaldehyde), the sample is collected by passing the air through a bubbler containing a solution of MBTH. The reaction is:

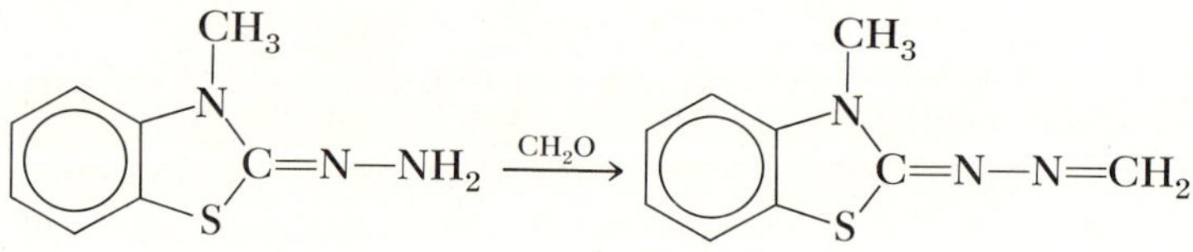

4.5(6)

MBTH (A)

After collection, the solution is treated with ferric chloride–sulfamic acid solution to convert the azine A to a blue cationic dye C, the color intensity of which can be measured at a wavelength of 628 nanometers.

$$\text{MBTH} + \text{(A)} \xrightarrow[NH_2SO_3H]{FeCl_3}$$

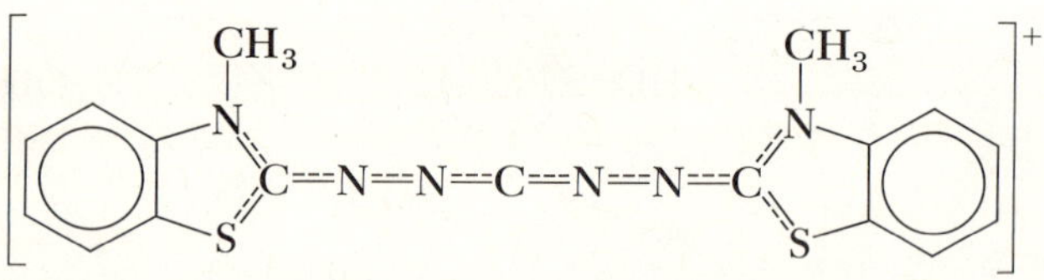

4.5(7)

There are no known interferences of consequence in airborne samples. A concentration of 0.03 ppm aldehyde can be measured in a 25 liter air sample. This procedure is recommended by the U.S. Environmental Protection Agency.

Schiff's Reagent[4]

As originally developed by Schiff over a century ago, the reagent was fuchsin bleached by sulfur dioxide. There have been numerous improvements since then. A recent variation is suitable either for manual or continuous analyzer use. The reagents are sodium sulfitodichloromercurate (a rather unstable product of sodium tetrachloromercurate and sodium sulfite) and purified *p*-rosaniline in aqueous hydrochloric acid. For manual samples, the air is sampled through water which is then mixed in succession with the reagents. The color intensity is read at a wavelength of 560 nanometers. An air sample of 28 liters will give a sensitivity of 0.01 ppm formaldehyde. The instability and toxicity of the reagent are drawbacks to this method.

Acrolein Detection by 4-Hexylresorcinol[5]

Acrolein is an unsaturated aldehyde with powerful lachrymatory properties. A specific method has been developed for its determination which involves 4-hexylresorcinol. The air sample is passed through a fritted tube bubbler with an ethanolic solution of the hexylresorcinol with mercuric chloride and trichloroacetic acid. Time is allowed (15 minutes at 60°C, 2 hours at room temperature) for color development, and the blue purple color intensity is read at a wavelength of 605 nanometers 15 minutes later (within 2 hours). The method is calibrated against freshly distilled acrolein. The range is about 20 ppb to at least 10 ppm. The reference acrolein and the reagents mercuric chloride and trichloroacetic acid are all highly toxic or corrosive and must be handled with great care.

Assessment

Only one automated aldehyde analyzer has been reported[4] (for formaldehyde) and none are commercially available. All other methods involve grab samples (brief measured airflow through bubblers) and subsequent analysis by colorimetry or titrimetry. It is probably preferable to determine formaldehyde separately by a specific procedure (chromotropic acid) and total aldehydes by another method such as the MBTH procedure.

REFERENCES

1. Altshuller, A. P., D. L. Miller and S. F. Selva: *Anal. Chem.*, vol. 33, pp. 621–625, 1961.
2. Wilson, K. W.: *Anal. Chem.*, vol. 30, pp. 1127–1129, 1958.
3. Altshuller, A. P. and T. J. Leng: *Anal. Chem.*, vol. 35, pp. 1541–1542, 1963.
4. Lyles, G. R., F. B. Dowling and V. J. Blachard: *J. Air Pollu. Contr. Ass.*, vol. 15, pp. 106–108, 1965.
5. Cohen, I. R. and A. P. Altshuller: *Anal. Chem.*, vol. 33, pp. 726–733, 1961.

4.6 CARBON DIOXIDE SENSORS

Method of Detection:	Infrared.
Sample Pressure:	Up to 15 psig, but atmospheric or near atmospheric is normal.
Sample Temperature:	Up to approximately 120°F. Not a consideration when freeze-out trap is used.
Sample Flowrate:	Generally less than 0.5 ACFM; typically 1 to 2 LPM.
Accuracy:	Can be as high as ±0.2 ppm; typically ±1% of full-scale range.
Ranges:	From ppm to 100%.
Response:	Determined by cell volume and sampling rate. Typically less than 30 seconds.
Cost:	$3,000 to $5,000 without recorder, depending on sample conditioning accessories.
Partial List of Suppliers:	Beckman Instruments, Inc.; Mine Safety Appliances Co.; Peerless Instrument Co., Inc.; Intertech Corp.; Infrared Industries, Inc.; Wilks Scientific Corp.

The measurement of carbon dioxide in the ambient air is a primary concern of the geophysicist rather than of the air pollution control engineer. The precise measurement of the carbon dioxide content of the atmosphere is of significant concern in determining long-term changes in the composition of the atmosphere (Section 2.1). The measurement techniques used by geophysicists are highly precise compared to the techniques for air pollution evaluations.

In air pollution, carbon dioxide is hardly ever measured in the ambient air. It is measured at emission points since some combustion equipment regulations are stated in terms of allowable pollutant discharges corrected to 50% excess air.

Ambient Air Measurement

The precise knowledge of ambient air concentrations is necessary. An increase in global carbon dioxide concentration of only 1%, or about 3 ppm, has significant consequences (Fig. 2.1b). The instruments used to measure atmospheric carbon dioxide concentrations are, of necessity, highly precise.

The universal instrument for this purpose is the nondispersive infrared analyzer.[1] Since the absorption bands of water and carbon dioxide overlap somewhat, a freeze-out trap ($-80°C$) is used in the sample preparation system. When using prepared calibration standard mixtures of carbon dioxide in nitrogen, an accuracy of ± 0.2 ppm is attainable. At normal atmospheric levels of approximately 314 ppm, this is the equivalent of $\pm 0.06\%$ accuracy.

Source Measurement

Measurement of the carbon dioxide, carbon monoxide and oxygen concentration of flue gases from boilers has been done for many years.[2] These measurements allow precise setting of boiler operating variables for maximum fuel economy.

Before the use of the infrared analyzer became accepted practice, mechanical instruments were used to continuously determine the carbon dioxide content of flue gases. Their operation was based on the reduction in gas volume resulting from the absorption of carbon dioxide in a strong alkaline solution. This is the principle of the Orsat analyzer,[3] still used as the standard method for manual determination of combustion gas composition.

For air pollution testing purposes, the carbon dioxide content of the flue gas is determined only during the few hours of the test. The usual procedure is to slowly withdraw a low-volume integrated sample into a plastic bag over the duration of the test. The bag sample is then analyzed manually using an Orsat analyzer or instrumentally using a nondispersive infrared analyzer.

REFERENCES

1. Lipták, B. G. (ed.): "Instrument Engineers' Handbook," Vol. I, Section 8.3, Chilton, Philadelphia, 1969.
2. Ibid, Vol. II, Section 10.6.
3. Ibid, Vol. I, Section 8.7.

4.7 CARBON MONOXIDE ANALYZERS

Type of Detection Method:	(a) Nondispersive infrared; (b) Mercury vapor; (c) Gas chromatography; (d) Electrochemical analyzers; (e) Catalytic analysis.
Reference Method:	(a).
Range, ppm:	0 to 25 (a); 0 to 50 (a,b); 0 to 100 (a); 0 to 200 (c).
Sensitivity, ppm:	0.05 (b); 0.1 (c); 0.5 to 1.0 (a); 1.0 (d).
Response Time, minutes:	1 to 5 (a).
Cost:	$2,500 to $5,000 (a); $3,000 (b); $8,000 (c).
Partial List of Suppliers:	Bacharach Instrument Company (b); Beckman Instruments, Inc. (a,c); Bendix Environmental Science Div., Bendix Corp. (a); Calibrated Instrument Co. (a); Intertech Corp. (a); Mine Safety Appliances Co. (a,c); Morgan Co. (a); Tracor, Inc. (c); Union Carbide Corp. (c).

Calibration Techniques

Methods used for analysis of carbon monoxide at atmospheric levels (Table 4.7a) require calibration with gas mixtures of known concentration. Such mixtures may be prepared by volumetric dilution of pure carbon monoxide with carbon monoxide–free nitrogen or helium. If the volumes used in dilution are accurately known, the concentration may be calculated. A known small volume of pure carbon monoxide is placed in an evacuated tank of known volume and the tank is then filled with the diluent gas. The pressures of the carbon monoxide volume and the final mixture must be known (or else known to be equal). Smaller samples may be prepared in plastic bags by injecting pure carbon monoxide from a gas syringe into a stream of diluent gas metered accurately into the bag with a flow device such as a wet test meter.

Table 4.7a
CARBON MONOXIDE ANALYZERS

Detection Method	*Range, ppm*	*Sensitivity, ppm*	*Advantages*	*Disadvantages*
Nondispensive infrared (NDIR)	0–25, 50, 100	0.5–1	U.S. reference method, accurate, stable, dry gases	Sensitive, water and CO_2 interferences (correctable), zero gas problems
Mercury vapor (Hot HgO + CO releasing Hg vapor)	0–50	0.05	Sturdy, accurate, dry gases, high sensitivity	Interferences by water and other gases
Gas chromatography (reduction of CO to CH_4, flame ionization detection)	0–200	0.1	Accurate, high sensitivity, also read CH_4, dry gases	Complex and expensive

4.7 Carbon Monoxide Analyzers

Cylinder nitrogen commonly has small amounts of carbon monoxide present and must not be assumed to be pure without verification. Pure helium is reliably free of carbon monoxide. A useful procedure is to zero the detector on helium, then read the content of the available cylinder nitrogen and make the corresponding correction. This permits using the less expensive nitrogen for most calibrations.

A reference gas mixture whose carbon monoxide content is not accurately known may be analyzed gravimetrically or volumetrically by various methods, but these techniques are usually less convenient and often require relatively large amounts of gas.

Nondispersive Infrared Analyzers

NDIR (Nondispersive infrared) analysis[1] is the reference method for the U.S. National Air Quality Standard for carbon monoxide. It allows continuous analysis based on the capacity of carbon monoxide to absorb infrared radiation. A schematic diagram of a typical NDIR analyzer is shown in Figure 4.7b. Infrared radiation from a hot filament is chopped to pass alternately through sample and reference cells to be absorbed in the detector cell divided by a pressure-sensitive diaphragm. If the sample contains carbon monoxide, it will absorb part of the radiation, causing that half of the detector to exert less pressure on the diaphragm, whose distortion is converted to an electrical signal for rectification and amplification.

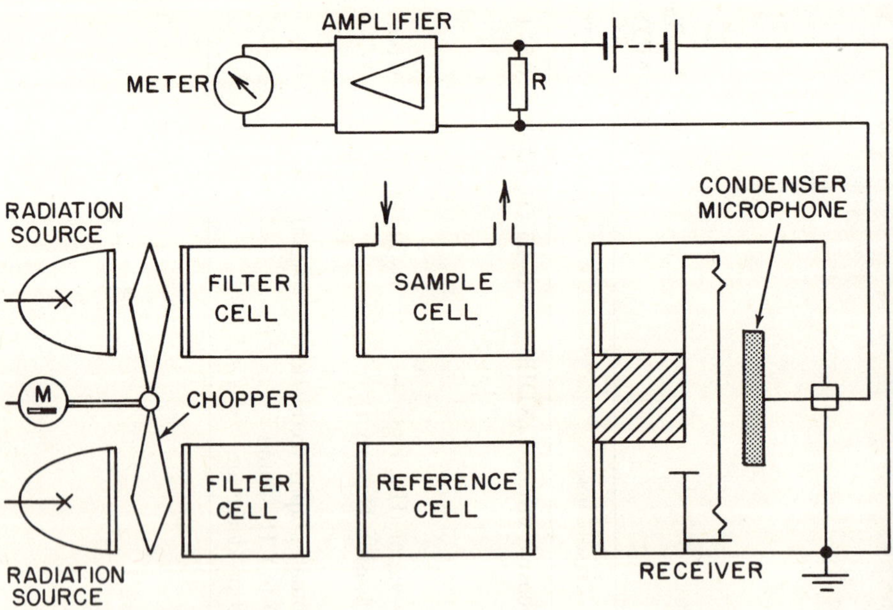

Figure 4.7b Nondispersive infrared carbon monoxide analyzer.

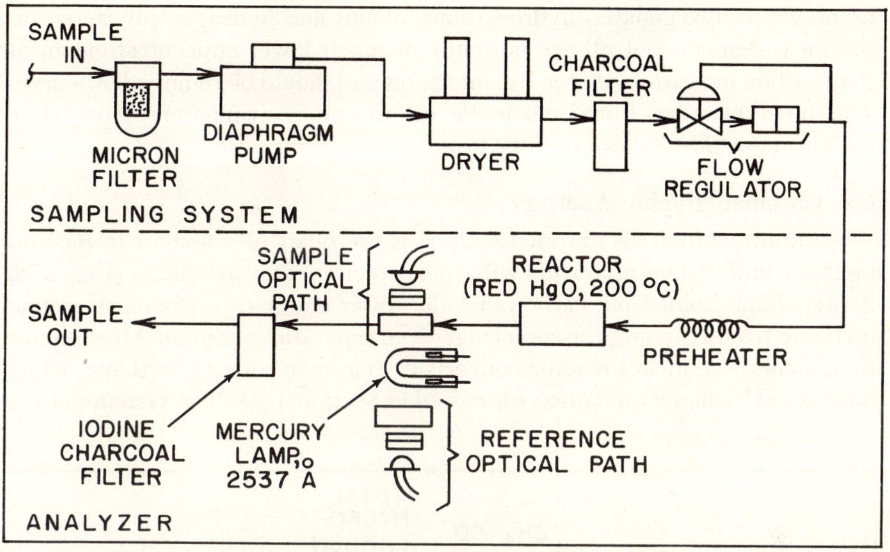

Figure 4.7c Mercury vapor carbon monoxide analyzer.

Sample airflow is continuous at or sometimes above atmospheric pressure. The cell is commonly 0.5 meter long. Measuring ranges usually extend from a minimum of 0.5 to 1 ppm up to a full-scale range of 25 to 100 ppm. Response times are in the range of less than 1 to 5 minutes. Although NDIR response is nonlinear, it is assumed to be linear over the limited calibration range in use. Some instruments correct for nonlinearity in the output amplifier. Instruments can be operated by nontechnical personnel.

Carbon dioxide and water vapor, in the sample, interfere with the measurement. Filter cells filled with these gases, or optical filters, when placed in front of the cells can minimize effects of normal atmospheric levels of these interfering gases. The control of water vapor interference by its removal with desiccants (e.g., silica gel) or a refrigerator condenser is preferable in many cases, even though at the cost of some increase in response time.

Mercury Vapor Analyzer

Carbon monoxide is oxidized by hot mercuric oxide[2] as follows:

$$CO + HgO(s) \xrightarrow{(210°C)} CO_2 + Hg(g) \qquad 4.7(1)$$

The mercury vapor released may be measured photometrically. An analyzer based on this principle (Fig. 4.7c) can be operated continuously. It has higher sensitivity than does the NDIR type, but it also suffers from some interferences. Detection levels go down to 0.025 ppm and changes of a tenth of that can be observed. Oxygenated hydrocarbons, olefins and hydrogen interfere with the measurement, but all are normally at much lower concentrations in air than carbon monoxide. Water also interferes and should be removed by a dryer. This instrument has found particular use in nonurban measurements where carbon monoxide levels are very low.

Gas Chromatographic Analyzer

An automated gas chromatograph[3] is the heart of a system to measure methane and carbon monoxide with high precision and specificity (Fig. 4.7d). A precolumn holds back carbon dioxide, water and hydrocarbons other than methane from reaching the molecular sieve separation column. After separation, a catalytic nickel reactor converts carbon monoxide to methane, which is detected by flame ionization (discussed in Section 4.10). The system permits

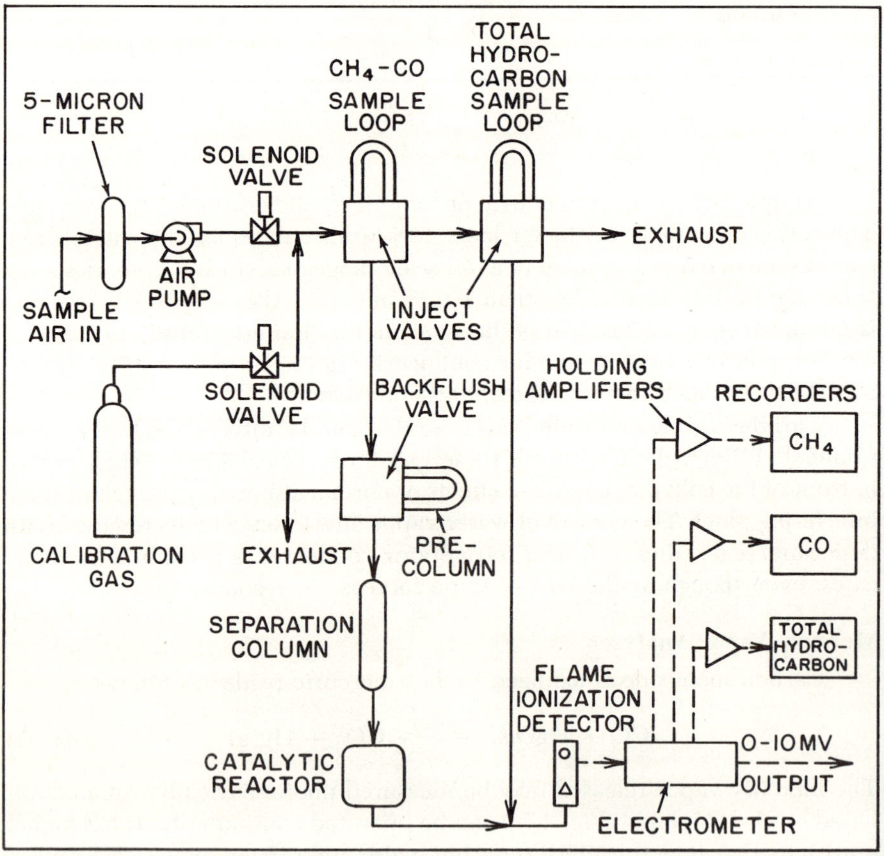

Figure 4.7d Methane–carbon monoxide analyzer.

determination of both methane and carbon monoxide about once every 5 minutes. The output is linear for both components and can be read from 0.1 to 200 ppm. The instrument is relatively complex and expensive, however, and requires technically trained operators.

Electrochemical Analyzers

A galvanic cell for continuous carbon monoxide analysis[1] is based on its reaction with iodine pentoxide:

$$5\ CO + I_2O_5 \xrightarrow{(150°C)} 5\ CO_2 + I_2 \qquad 4.7(2)$$

The iodine liberated is absorbed by an electrolyte and reaches the cathode of a galvanic cell where it is reduced. The resulting current is measured by a galvanometer. Interferences by mercaptans, hydrogens sulfide, hydrogen, olefins and acetylenes may be minimized by sampling through an absorption tube of mercuric sulfate on silica gel. Water vapor interference is removed by a drying column.

The same reaction is used in a coulometric method with a modified Hersch-type cell. The iodine is passed into the cell and the current flow is measured by an electrometer. The interferences are the same as for the galvanic analyzer.

The minimum detectable concentrations are 1 ppm with good precision if flowrates and temperature are controlled. Careful column preparation is required and the response time is relatively slow.

Catalytic Analysis

The catalyst Hopcalite will oxidize carbon monoxide to carbon dioxide.[1] The resultant temperature rise may be recorded continuously as a measure of carbon monoxide concentration. The catalyst temperature and residence times must be controlled to avoid interference by hydrocarbons. The method is not suitable for most air monitoring applications because of low sensitivity.

Spot Sampling Methods

When only intermittent analyses are required, it is convenient to collect samples in the field for later analysis in the laboratory. Rigid glass bulbs or stainless steel tanks may be evacuated, then simply opened briefly to collect the air sample. Plastic bags may be filled by means of a small air pump. The samples may be analyzed later by various means, including use of a continuous analyzer at some other location. The samples may be analyzed in a central laboratory by an infrared spectrophotometer with a long-path gas cell or by suitable gas chromatographic apparatus.

Some colorimetric methods are available for carbon monoxide analysis; although, in general, the sensitivity and precision are low for atmospheric work. An NBS colorimetric indicating gel, if freshly prepared, will give accuracies of 5 to 10%, with detectability down to 0.1 ppm. The technique is simple but lengthy and tedious, with interferences by oxidizing and reducing gases.

Assessment

For simple, not too precise measurement of short-term carbon monoxide levels, there are gel tube methods. For continuous and precision methods, more sophisticated equipment is necessary. The method in most widespread use currently is the nondispersive infrared analysis; but, in addition, there are now available two new types of analyzers. If it is necessary to also measure methane, the combination gas chromatographic methane–carbon monoxide analyzer, though expensive, is worthy of consideration. The other new type, the mercury vapor analyzer, is not yet thoroughly tested but seems particularly suitable where the carbon monoxide levels are low.

REFERENCES

1. "Air Quality Criteria for Carbon Monoxide," U.S. Dept. of Health, Education and Welfare, National Air Pollution Control Administration Publication No. AP-62, Ch. 5, March 1970.
2. Robbins, R. C., K. M. Borg and E. Robinson: *J. Air Pollu. Contr. Ass.*, vol. 18, pp. 106–110, 1968.
3. Stevens, R. K., A. E. O'Keefe and G. C. Ortman: "A Gas Chromatographic Approach to the Semicontinuous Monitoring of Carbon Monoxide and Methane," 156th National Meeting of the American Chemical Society, Atlantic City, September 1968.

4.8 FLUORIDE SENSORS

Method of Sampling:	(a) Automatic sequential collection with impingers or bubblers; (b) Automatic sequential collection with dual tape samplers; (c) Automatic sequential collection with coated glass tubes.
Type of Analysis:	(d) Automatic sequential microdistillation and photometric analysis; (e) Potentiometry.
Sampling and Analysis:	(f) Automatic sequential collection on a coated glass tube and photometric analysis; (g) Automatic sequential collection on coated glass tubes and potentiometric analysis.
Airflow Rate:	Varies with device. Generally between 7 and 30 lpm.
Sample Duration:	Varies with device, concentration of fluoride in air, sensitivity of analytical method and purpose of monitoring program. Usual range for different methods is from 0.5 to 24 hours.
Materials of Construction:	Surfaces of equipment that contact air or liquid containing fluoride are usually composed of stainless steel, Teflon,® epoxy, polyethylene or polypropylene.
Concentrations Measured:	Usual range of interest for ambient air monitoring is less than 1 $\mu g/m^3$ to more than 50 $\mu g/m^3$. Lower concentrations usually require longer sample durations; higher concentrations may require slower airflow rates or sample dilution.

4.8 Fluoride Sensors

Sensitivity:	Depends on magnitude of blank values, sensitivity of analytical method, collection rate and length of sampling period. Limit of detection should be three to five times lower than lowest concentration of interest.
Calibration:	Procedure varies with method and equipment. Frequently, the sampling step is not included in the calibration procedure and errors due to incomplete sampling or contamination may go undetected.
Accuracy:	In most cases, information is insufficient for reliable estimates of accuracy. Probable range of absolute error: ±10 to ±50%; with percentages higher when the concentration is below 1 $\mu g/m^3$. Accuracy highly dependent on type and frequency of calibration.
Precision:	Range of relative errors for repetitive measurements of selected methods under laboratory conditions: ±5 to ±21%. May be less precise at low concentrations and under outdoor conditions.
Partial List of Suppliers:	Analytic Systems Associates (f); Beckman Instruments, Inc. (e); Bendix Corp. (a, b); Coleman Instruments (e); Corning Glass Works (e); Denki Kagaku Keiki Co. (g); Hemeon Associates (c); Leigh Instruments Ltd. (c, f); Orion Research Co. (e); Precision Scientific (a); Research Appliance Co. (a, b, c); Technicon Corp. (d).
Costs:	\$240 to \$2,200 (a, b, c); \$1,000 (manual) to \$10,000 (automated) (d, e); \$4,500 to \$9,000 (f, g).

There are two major operations in every atmospheric fluoride monitoring system: collection and analysis. Each monitoring system must possess fluoride sensing devices which are compatible with the available methods of collection.[1,2]

Selection of Analysis Methods

The circumstances and requirements of the monitoring program and the purposes for which the data are obtained dictate the choice of detection methods.[3] Air quality criteria for fluorides should also be examined[4] and the need for separating gaseous and particulate atmospheric components determined.[5] When selecting the methods of analysis, it should be kept in mind that the experience and capability of the supervisor, engineer, technician or analyst are extremely important because monitoring methods are very susceptible to both positive and negative biases. Comparative testing and measurement of accuracy, precision and reliability of methods of monitoring atmospheric fluorides lag far behind the rate of development and change in techniques and equipment.

Equipment

Details about individual sampling, analytical and sample treatment methods are provided in Table 4.8a.[20,21] Included in the table are the references which should be consulted for more detailed information.

Types of Monitoring Systems

There are a great many combinations of sampling and analysis methods currently in use. Basic equipment can be purchased and assembled into monitoring systems to satisfy particular requirements.[22] The fluorometric analyzer developed some years ago[23] is not commercially available. Wet chemical methods of analysis (Section 4.22), such as spectrophotometry, colorimetry and titrimetry, are used extensively.[7] The proliferation of methods, without thorough comparative testing, is highly undesirable because it leads to wide variations in results among laboratories. Reference should be made to the published standardized methods[6,20,24] and approved procedures should be followed exclusively.

Source Monitoring

Sampling fluoride near the source of emissions, such as in stack gases, usually requires special techniques. Higher concentrations of fluoride, particulate matter and water vapor are encountered and higher temperatures and reactivity of the effluent gases must be overcome. An automatic dosimeter instrument for measurement of gaseous fluorides has been developed for stack monitoring.[25] In a recently developed method, atmospheric fluorides react to form silicon tetrafluoride before collection of particulate and gaseous fluoride and analysis is performed by manual methods.[26] This method has been modified to provide potentiometric analysis and calibration with hydrogen fluoride permeation tubes.[27]

Table 4.8a ORIENTATION TABLE FOR ATMOSPHERIC

Sampling Methods — *Device*	*References*	*Material Collected for Analysis*				*Sampling Duration, hr*	*Automatic Sequential Collection Equipment Available*
		Wet	*Dry*	*Gas*	*Particles*		
A. Impinger	6	✓		✓	✓	U 6–24	✓
B. Bubbler	7	✓		✓	P	U 6–24	✓
C. Scrubber	8	✓		✓	P	0.5–2.5[f]	✓
D. Aluminum tube, impingers	9	✓	✓	✓[a]	✓	U 6–24	
E. Glass fiber filter	9, 10		✓	✓	✓	U 24–72	
F. Acid-treated or heated membrane or cellulose filter	11–13		✓	[b]	P[d]	0.5–4	✓
G. Alkali-treated membrane or cellulose filter	9, 12, 13		✓	✓	P[d]	0.5–4	✓
H. Coated glass tube	9, 12, 13		✓	✓		U 6–24	✓
I. Coated silver beads	14, 15, 28		✓	✓	[e]		
J. Limed filter paper	16–19		✓	✓[c]	✓[c]	U 1 month	

[a] By difference.
[b] Significant quantities of gaseous fluoride may be collected depending on the type and treatment of the filter and amount of dust accumulated.
[c] Fluoride collected statically by absorption and deposition.
[d] Collection efficiency for particles dependent on flowrate, type of filter, porosity and amount of dust accumulated.
[e] Larger particles will be collected unless excluded by using a Herpertz cap (Ixfeld, 1971).
[f] Functions as dosimeter.

Additional Research and Development

Published data on atmospheric concentrations of fluorides and factors affecting fluoride distribution in the atmosphere are scanty. The development of new instrumentation has taken place with inadequate knowledge of concentrations and fluctuations in concentration with time and distance from sources and with insufficient comparative testing of methods on known fluoride atmospheres.

REFERENCES

1. Hochheiser, S., F. J. Burmann and G. B. Morgan: Atmospheric Surveillance: The Current State of Air Monitoring Technology, *Environmental Science & Technology*, vol. 5, pp. 678–684, 1971.

FLUORIDE MEASUREMENT METHODS

Appropriate Sample Treatment Methods[h]					*Preferred Analytical Methods*			*Separate Measurement of Gaseous and Particulate Fluorides*
Ashing or Combustion	*Acid Distillation*[g]	*Micro-diffusion*	*Ion Exchange*	*Extraction or Elution*	*Titrimetry*	*Colorimetry*[g]	*Potentiometry*	
	✓	✓	✓		✓	✓	✓	A. After F
	✓	✓	✓		✓	✓	✓	B. After F
			✓			✓	✓	C. After F
	✓	✓	✓		✓	✓	✓	D. Impinger in parallel
✓	✓	✓			✓	✓	✓	E. After F
✓	✓	✓	✓	✓		✓	✓	F. Before A, B, C, E, G, H or I
✓	✓	✓	✓	✓		✓	✓	G. After F
	✓	✓	✓	✓		✓	✓	H. Before F
	✓	✓	✓	✓		✓	✓	I. Before F
✓	✓	✓	✓	✓	✓	✓	✓	

[g]Equipment for automatic sequential analysis is available.
[h]Sample treatment methods may be needed to concentrate fluoride or to avoid interference with analytical measurements. In many cases, fluoride analyses are performed without concentrating or purifying the samples.
NOTE: The information in this table is for general guidance only. The reader should consult references for specific information.
P = partial; U = usually

2. Hendrickson, E. R.: Air Sampling and Quantity Measurement, in A. C. Stern, (ed.), "Air Pollution," vol. II, 2d ed., Academic Press, New York, 1968.
3. Jacobson, J. S., L. H. Weinstein and G. H. Farrah: A Review of Methods for Monitoring the Fluoride Content of Air, Proceedings of the Environmental Control Symposium, 1972 Annual AIME Meeting, Metallurgical Society of AIME, in press.
4. McCune, D. C.: On the Establishment of Air Quality Criteria, with Reference to the Effects of Atmospheric Fluorine on Vegetation, *Air Quality Monograph*, no. 69-3, American Petroleum Institute, New York, 1969.
5. National Academy of Sciences, Committee on Biologic Effects of Atmospheric Pollutants, "Fluorides," pp. 51–65, Washington, D.C., 1971.
6. ASTM, "Standard Method of Test for Inorganic Fluoride in the Atmosphere," D 1606-60, Book of ASTM Standards, part 23, pp. 443–453, 1971.
7. Farrah, G. H.: Manual Procedures for the Estimation of Atmospheric Fluorides, *J. Air Pollu. Contr. Ass.*, vol. 17, pp. 738–741, 1967.
8. Adams, D. F. and R. K. Koppe: Automatic Atmospheric Fluoride Pollutant Analyzer, *Analytical Chemistry*, vol. 31, pp. 1249–1254, 1959.
9. Pack, M. R., A. C. Hill, M. D. Thomas and L. G. Transtrum: Determination of Gaseous

and Particulate Inorganic Fluorides in the Atmosphere, ASTM, Special Technical Publication No. 281, pp. 27–44, 1959.

10. Pack, M. R. and A. C. Hill: Further Evaluation of Glass Fiber Filters for Sampling Hydrogen Fluoride, *J. Air Pollu. Contr. Ass.*, vol. 15, pp. 166–167, 1965.
11. Habel, K.: The Separation of Gaseous and Solid Fluorine Compounds During Air-Quality Measurements, *Staub* (English translation), vol. 28, no. 7, pp. 26–31, 1968.
12. Mandl, R. H., L. H. Weinstein, G. J. Weiskopf and J. L. Major: Separation and Collection of Gaseous and Particulate Fluorides, Paper CD-25A, Second International Clean Air Congress, Washington, D.C., 1970.
13. Weinstein, L. H. and R. H. Mandl: "The Separation and Collection of Gaseous and Particulate Fluorides," VDI Berichte, Nr. 164, pp. 53–63, 1971.
14. M. Buck and H. Stratmann, "Ein Verfahren zur Bestimmun sehr geringer Konzentrationen von Fluor-Ionen in der Atmosphäre," *Brennstoff-Chemie Bd.*, vol. 46, pp. 231–235, 1965.
15. Svoboda, K. and H. Ixfeld: "Probenahme und automatische Analyse gasförmiger Fluor-Immissionen," *Staub*, Bd. 31, pp. 1–8, 1971.
16. Robinson, E.: Determining Fluoride Air Concentrations by Exposing Limed Filter Paper, *American Industrial Hygiene Association Quarterly*, vol. 18, pp. 145–148, 1957.
17. Adams, D. F.: A Quantitative Study of the Limed Filter Paper Technique for Fluorine Air Pollution Studies, *Inter. J. Air and Water Pollu.*, vol. 4, pp. 247–255, 1961.
18. Wilson, W. L., M. W. Campbell, L. D. Eddy and W. H. Poppe: Calibration of Limed Filter Paper for Measuring Short-Term Hydrogen Fluoride Dosages: The Effect of Temperature, Humidity, Wind Speed, and Dose, *American Industrial Hygiene Association Journal*, vol. 28, pp. 254–259, 1967.
19. Mukai, K. and H. Ishida: The Alkaline Filter Paper Method for Surveying Fluorides in the Atmosphere, Paper No. A70-10, TMS-AIME Annual Meeting, Denver, Colorado, 1970.
20. Intersociety Committee on Methods for Ambient Air Sampling and Analysis, "Tentative Method of Analysis for Fluoride Content of the Atmosphere and Plant Tissues (Manual and Semiautomated Methods)," 12204-01-68T and 12204-02-68T, Health Laboratory Science, vol. 6, pp. 64–91 and 94–101, 1969.
21. ASTM, "Analysis of Fluoride in Ambient Air Samples with the Selective Ion Electrode," Committee D-22, in press.
22. West, P. W., G. R. Lyles and J. L. Miller: Spectrophotometric Determination of Atmospheric Fluorides, *Environmental Science and Technology*, vol. 4, pp. 487–491, 1970.
23. Thompson, C. R., L. F. Zielenski and J. O. Ivie: A Simplified Fluorometric Fluoride Analyzer, *Atmospheric Environment*, vol. 1, pp. 253–259, 1967.
24. Association of Official Analytical Chemists, "Official Methods of Analysis," 11th ed., "Fluorine," Sections 25.029–25.035, 1970.
25. Adams, D. F.: An Automatic Hydrogen Fluoride Recorder Proposed for Industrial Hygiene and Stack Monitoring, *Analytical Chemistry*, vol. 32, pp. 1312–1316, 1960.
26. Dorsey, J. A. and D. A. Kemnitz: "A Source Sampling Technique for Particulate and Gaseous Fluorides," *J. Air Pollu. Contr. Ass.*, vol. 18, pp. 12–14, 1968.
27. Elfers, L. A. and C. E. Decker: Determination of Fluoride in Air and Stack Gas Samples by Use of an Ion Specific Electrode, *Analytical Chemistry*, vol. 40, pp. 1658–1661, 1968.
28. Ixfeld, H.: "Fluor-Immissionsmessung in Lande Nordrhein-Westfalen," VDI Berichte, Nr. 164, pp. 71–75, 1971.

4.9 HUMIDITY AND DEW POINT SENSORS

Design Pressure:	To 600 PSIG.
Design Temperature:	−25°F to 170°F.
Element Material:	Hair, glass, stainless steel.
Cost:	\$50–\$1000.
Accuracy:	±5% relative humidity.
Range:	5% 100% relative humidity.
Partial List of Suppliers:	Foxboro Co., Honeywell, Industrial Div., Palmer Instruments, Inc., Taylor Instrument Companies, Vap-Air Div., Vapor Corp.

Wet and Dry Bulb Hygrometers

When water changes phase from liquid to vapor, an amount of heat equal to the latent heat of vaporization must be supplied from the environment for each unit mass of water evaporated. The vaporization, in addition to its dependence on heat available, is also a function of the degree of saturation or relative humidity of the atmosphere surrounding the water. The wet and dry bulb hygrometer takes advantage of this fact to effect a relative humidity measurement.

The instrument consists of two temperature sensitive elements exposed to the atmosphere whose moisture level is to be measured. One of the two elements, the wet bulb, is wrapped with a wick soaked in water; the other element, the dry bulb, is left bare. Water evaporating from the wick lowers its temperature and that of the temperature element. The temperature values indicated by the two thermometers are related to the relative humidity of the sample atmosphere. Relative humidity can be calculated as shown on the following pages using the wet and dry bulb temperature readings, or relative humidity can be read from a psychometric chart such as the one shown in Figure 4.9a. For a dependable measurement, the sample velocity should be in excess of 10 feet per second. The sensing elements should, therefore, be

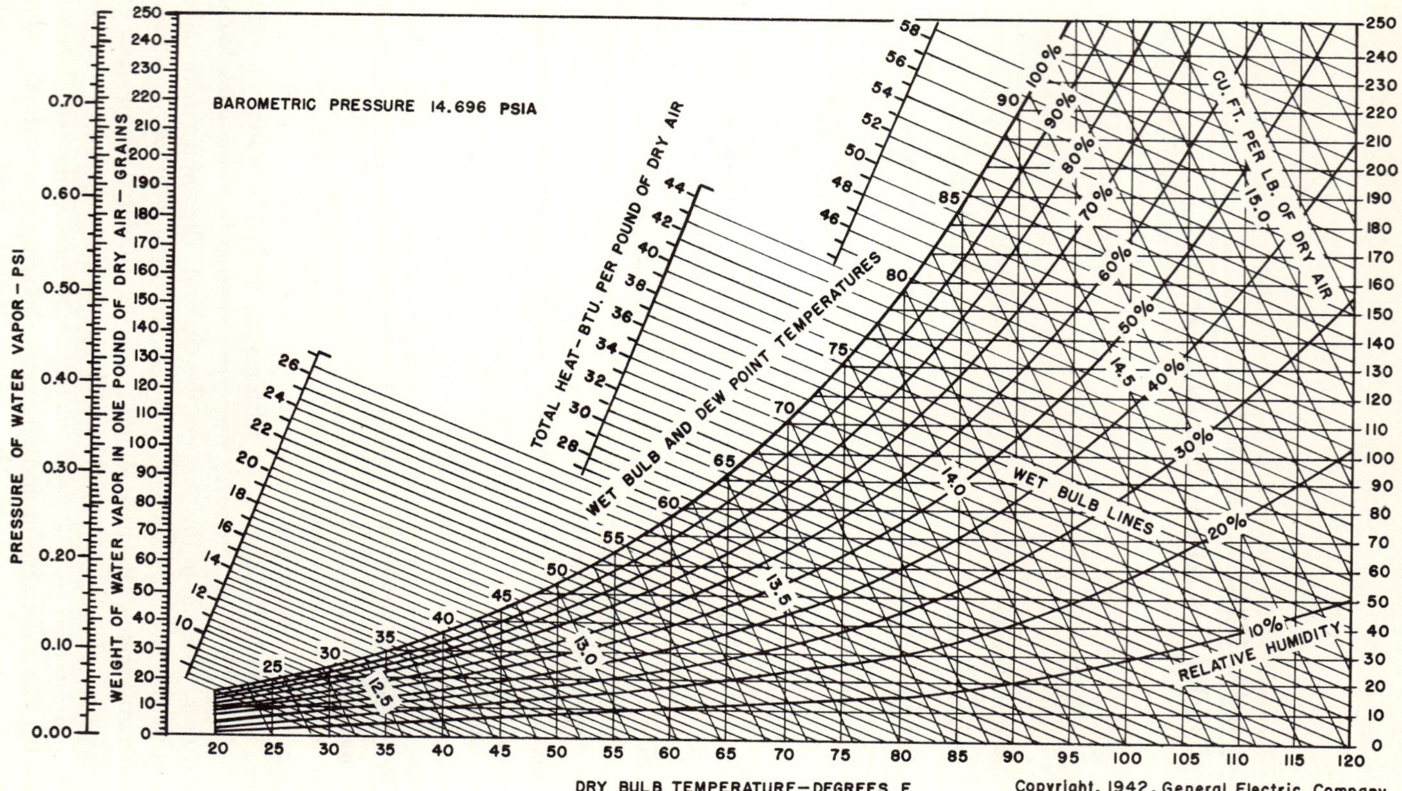

Figure 4.9a Psychrometric chart.

mounted where there is adequate circulation, or forced circulation must be provided by means of a fan.

Figure 4.9b shows a sling psychrometer consisting of wet and dry bulb thermometers and mounting arrangement such that the unit can be whirled manually when a reading is to be made.

Figure 4.9c shows a wet and dry bulb assembly for mounting in a duct. The wick is kept wet by a water reservoir externally mounted, or a continuous water supply can be piped to the unit. A similar device is available for mounting in a room or compartment. Generally, these units are connected to a two-pen recorder providing continuous, automatic monitoring of moisture level in the atmosphere.

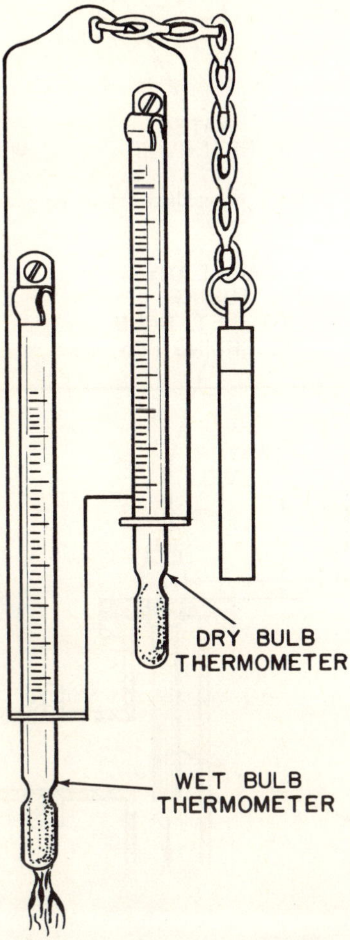

Figure 4.9b Sling psychrometer.

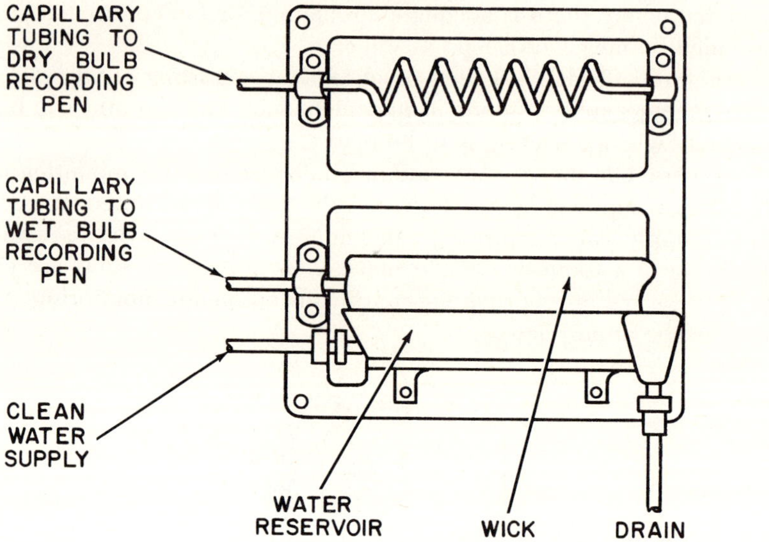

Figure 4.9c Wet-dry bulb hygrometer using filled system temperature sensors arranged for duct mounting.

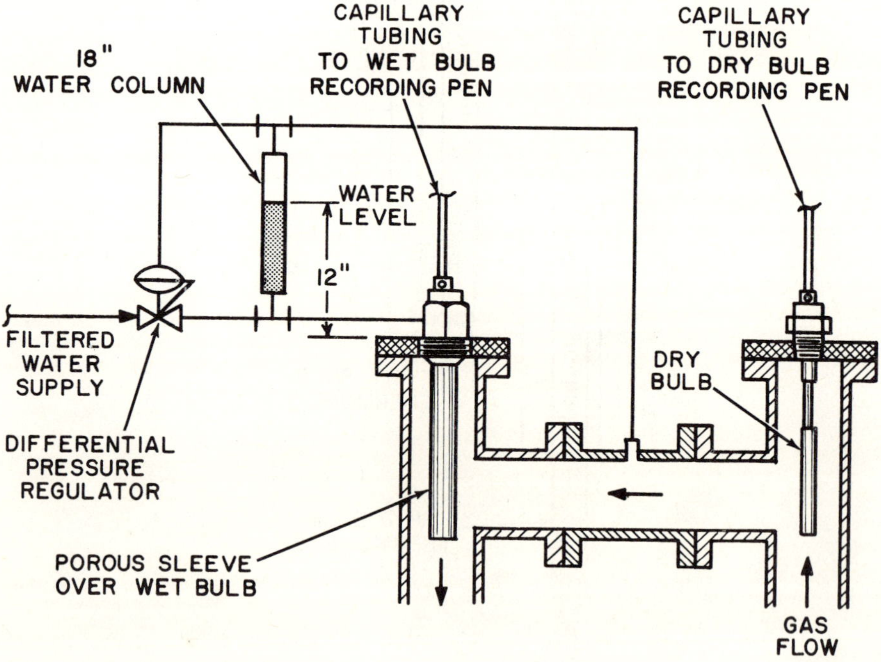

Figure 4.9d Wet-dry bulb hygrometer using filled system temperature sensors arranged for pressure pipe-mounting.

Figure 4.9d shows a wet bulb element for installation in a duct or pipe line operating at elevated pressures. The wick in this case is replaced by a porous sheath through which water can evaporate. The water supply pressure to the porous sleeve must be slightly higher than the process pressure. For process lines at constant low pressure, a standpipe at the wet bulb element can be used to provide the required water pressure. Usually, however, a differential regulator as shown in the diagram, is used to maintain the differential pressure across the porous sheath.

Calculation of Relative Humidity

Relative humidity can be calculated directly from wet and dry bulb thermometer readings in conjunction with a steam table; however, the calculation is valid for air at ambient conditions only and should not be applied to other gases or where conditions of temperature and pressure differ significantly from ambient.*

S_1, S_2 = specific humidity
t_1, t_2 = temperature (°F)
he_2 = enthalpy of evaporation (BTU/pound mole)
hv_1 = enthalpy of saturated vapor (BTU/pound mole)
hw_2 = enthalpy of saturated water (BTU/pound mole)
Pv_1, v_2 = partial pressure of water vapor (PSIA)
Ps_1 = saturation pressure (PSIA)
P = total gas pressure (PSIA)

Subscripts 1 and 2 refer to dry and wet bulb conditions respectively. Values of Ps_1, Pv_2, hv_1, hw_2, and he_2 are obtained from a steam table.

$$S_2 = \frac{0.622\ Pv_2}{P - Pv_2} \qquad 4.9(1)$$

$$S_1 = \frac{0.622\ Pv_1}{P - Pv_1} \qquad 4.9(2)$$

$$S_1 = \frac{0.24(t_2 - t_1) + S_2 he_2}{hv_1 - hw_2} \qquad 4.9(3)$$

$$\text{Relative Humidity (\%)} = \frac{Pv_1}{Ps_1} \times 100 \qquad 4.9(4)$$

Equations 4.9(1) and 4.9(2) are derived from the definition of specific humidity assuming water vapor in an ideal gas. Equation 4.9(3) is derived from the first law of thermodynamics. Equation 4.9(4) is the definition of relative humidity for an ideal gas.

* Thermodynamics—G. J. Van Wylen, J. Wiley & Sons, Inc. 1965.

Example:

Determine the relative humidity of air at 14.7 PSIA when dry and wet bulb thermometer readings of 80°F and 60°F respectively were obtained.

From the steam table:

$$Ps_1 = 0.5069 \text{ PSIA} \qquad Pv_2 = 0.256 \text{ PSIA}$$
$$hv_1 = 1096.6 \text{ BTU/Lbm} \qquad hw_2 = 28.06 \text{ BTU/Lbm}$$
$$he_2 = 1059.9 \text{ BTU/Lbm}$$

$$S_2 = 0.622 \frac{0.256}{14.7 - 0.256} = 0.011 \qquad 4.9(1)$$

$$S_1 = \frac{0.24(60 - 80) + 0.011(1059.9)}{1096.6 - 28.06} = 0.00642 \qquad 4.9(3)$$

$$.00642 = 0.622 \frac{Pv_1}{14.7 - Pv_1} \qquad 4.9(2)$$

$$Pv_1 = 0.150 \text{ PSIA}$$

$$\text{Relative Humidity} = \frac{0.150}{0.5069} \times 100 \qquad 4.9(4)$$

$$\text{Relative Humidity} = 28\%$$

From the psychrometric chart, a value of 30% relative humidity can be read for the given data.

Hair Hygrometers

The principle of measurement utilized by this device is the change of length of certain organic and synthetic fibers when these are exposed to a moist atmosphere. A mechanical linkage is used to amplify the element movement for readout. Figure 4.9e shows a hair hygrometer consisting of several fibers mounted together to form a band; one end of the band is fixed while the other end is free to move, driving a pen arm through a linkage. Hair hygrometers, similar to wet and dry bulb hygrometers, require good circulation of the measured gas. In ducts there is generally adequate gas velocity to insure a dependable measurement; however, if the instrument is to be mounted in a room, the location must be carefully chosen. The instrument should not be mounted near doors or other openings where it will be exposed to spurious drafts; flush mounting on a panel should be avoided because the atmosphere in the back of panel is stagnant. The hair element can be mounted on top or on the back of the instrument case depending on the installation. The element can also be mounted on an extension in the back of the instrument such that the sensing portion is in the room or compartment where relative humidity is to be measured, while the readout device is surface

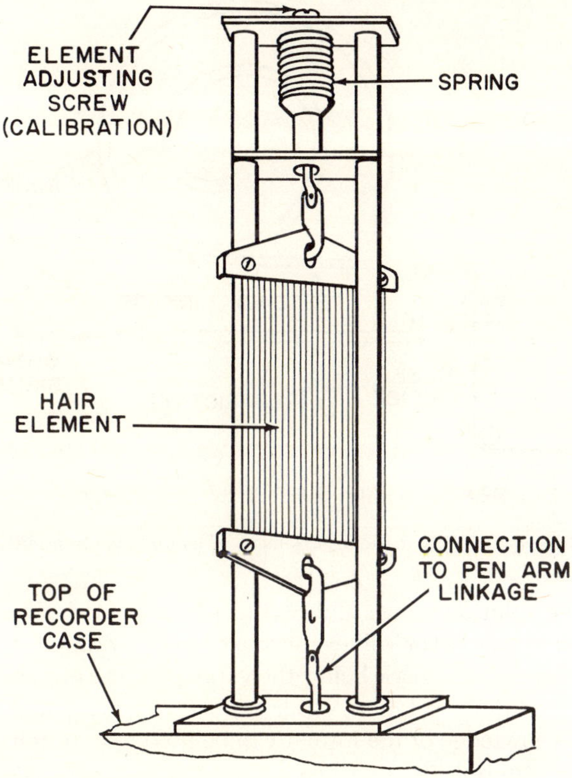

Figure 4.9e Hair hygrometer element.

mounted on the wall outside. Recorders are generally available as two-pen instruments with the second pen recording temperature.

Figure 4.9f shows a pneumatic transmitter with the fiber element connected to a nozzle-baffle bleed to vary air output pressure as a function of relative humidity. The transmitting units can be installed in a room or directly in a duct.

Dew Point Hygrometers

Solution Conductivity Type

All solids condense and absorb minute quantities of water on their surfaces. If the solid is soluble, a layer of saturated solution is formed whose vapor pressure is lower than that of pure water at the same temperature. Some solids form saturated solutions in water whose vapor pressure is less than the partial pressure of water vapor in the atmosphere contacting the solution. When that is the case, water vapor will continue to condense until the vapor

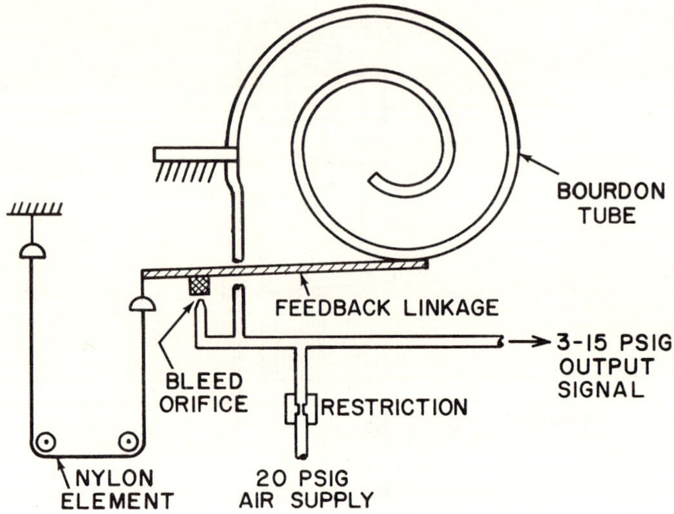

Figure 4.9f Schematic arrangement of hygrometer transmitter.

pressure of the solution equals the partial pressure of the water vapor. On the other hand, water vapor will evaporate from the solution should the water vapor partial pressure decrease below the vapor pressure of the solution. Under steady-state conditions, solution vapor pressure equals water vapor partial pressure, and knowledge of the former can be used to determine the moisture content of the atmosphere.

The sensing element consists of a thin-walled, hollow-metal socket wrapped with tape impregnated with salt crystals (lithium chloride). Two wires are wrapped over the tape and connected to a regulated, alternating current voltage source; the electric circuit between the wires is completed by the salt crystals. When the sensing element is exposed to the sample atmosphere, water condensing on the crystals forms an ionic solution permitting an electric current to flow between the wires. This current in turn heats the solution and raises its vapor pressure. As more water condenses, more current flows resulting in a further increase of the solution vapor pressure until equilibrium is reached. On decreasing moisture content of the measured gas, water evaporates from the element, decreasing the current flow and resulting in a new equilibrium at a lower solution vapor pressure. A temperature sensor inside the hollow socket is used to detect sensor temperature and to provide a signal for readout.

The output calibration can be in terms of dew point temperature water vapor pressure, or specific humidity. Values of specific humidity must be corrected for pressure variations of the sample. The correction factor is (Pc − Ps)/(P − Ps), where Pc is the sample pressure the instrument is cali-

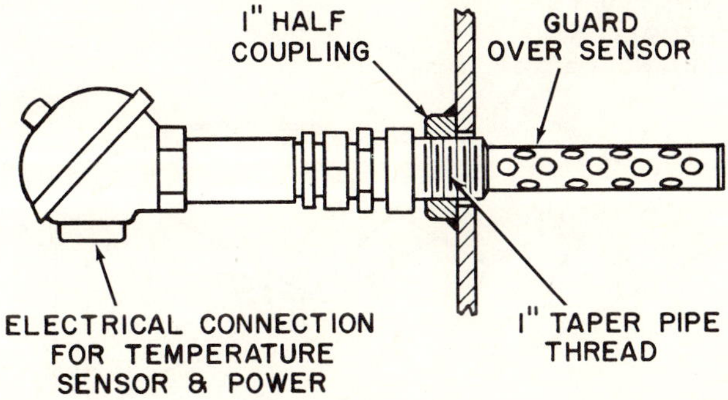

Figure 4.9g Solution conductivity probe with resistance temperature element.

brated for, Ps is the water saturation pressure at the sample dew point, and P is the actual process pressure. Relative humidity is obtained by the use of a chart converting sample operating and dew point temperature readings to percent relative humidity. Therefore, a sample temperature measurement is required if moisture level in terms of relative humidity is desired (Fig. 4.9g).

Installation

The instrument does not require a high velocity sample to perform satisfactorily; in fact, sample velocity in excess of one foot per second will result in a poor measurement and shortened element life. The reason is evident when one considers the measurement principle involved; namely, the change of solution vapor pressure and temperature on the sensing element by heating to establish equilibrium with the sample. At high sample velocities, convective heat losses swamp the measurement signal and operation at higher power causes a reduction of element life. The sensing element must, therefore, be located in a relatively quiescent zone or must be protected from direct impingement of the sample. In ducts, a sheet metal hood installed over the element and open on the downstream side is adequate. In pipe lines, the element can be installed through the side outlet of a tee or can be mounted separately in a sampling chamber piped to the process line. The latter installation is preferred for samples under pressure since the element can be serviced without shutting down the pipe line (Fig. 4.9h).

Sampling System

A sampling system is required when:

a. The sample temperature is above the operating limit of the sensing element. (Operating limit approximately 200°F.)
b. The process line cannot be shut-down for maintenance of the element.

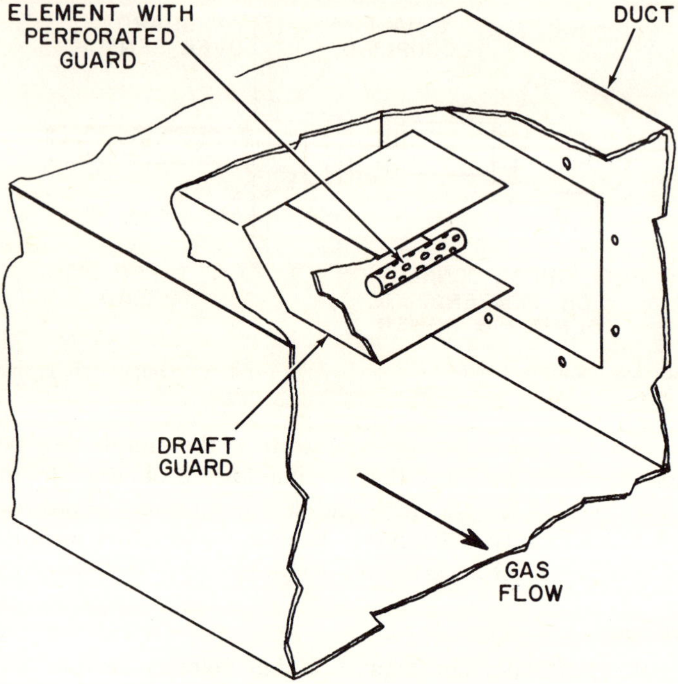

Figure 4.9h Duct installation with draft guard.

When the process temperature is above the operating temperature limit of the sensor, a sample cooler must be installed upstream of the measuring element. The sampling system must be so designed that the sample will never be cooled to its dew point under any possible process or ambient conditions. Sample tubing should be small in diameter while the sensing element is mounted in a larger diameter sampling chamber, as shown in Figure 4.9i, to reduce sample velocity at the element.

The sensor does require periodic maintenance and must be removable from the line. Whenever such removal would result in a process shutdown, the sensor should be installed in a separate sampling chamber and shutoff valves provided in the sample piping.

Limitations

The instrument can be used on all non-corrosive gases that do not react with the salt crystals or water. Specifically such gases as ammonia, sulfur dioxide, sulfur trioxide, chlorine react with water and this instrument is not applicable for such service. The instrument is not suitable for samples containing diethylene glycol or triethanolamine.

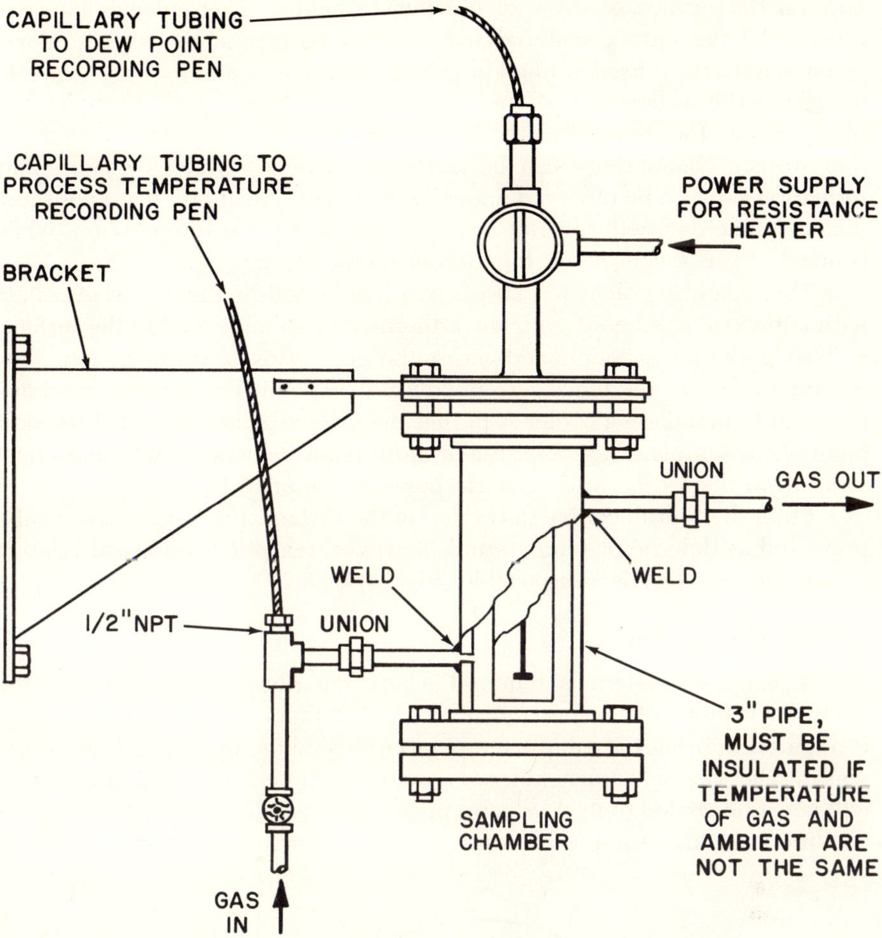

Figure 4.9i Probe installed in sampling chamber.

Surface Conductivity Type

Every object in a moist atmosphere has water molecules on its surface; the concentration of these molecules is related to the temperature of the object and the dew point of the atmosphere. If the temperature of the surface is above the dew point of the atmosphere, the thin layer of molecules is invisible; however, as the surface is cooled to the dew point, the density of water molecules at the surface becomes so great that water condenses on the surface and dew can be seen. At surface temperatures above the dew point, the moisture density at the surface can be detected electrically although the water vapor is not visible to the eye. This water vapor will permit a current to flow on the surface of even an excellent insulator. This current flow is a function of the surface material and moisture density at the surface. For a given material and fixed applied potential, the current will increase logarithmically as the surface temperature decreases to the dew point of the atmosphere. Below the dew point, the logarithmic relationship does not hold and thus surface conductivity can be related to a dew point. Of course, dew formation can also be observed optically; however, visual observation is more difficult since dew will form at or below the dew point temperature, while conductivity is a continuous function of temperature.

The measuring element consists of a highly polished inert surface inlaid with an intermeshed gold grid, and a thermocouple imbedded in the surface. A fixed potential is maintained across the gold grids, and the current flow is compared to the reference current flow at dew point. This signal is amplified and used to modulate a cooler such that the surface is maintained at the dew point of the sample. The cooler is a bismuth-telluride crystal that pumps heat away from the sensor when electric power is supplied to it.

From the thermocouple imbedded in the surface, the output is normally presented as dew point temperature; however, readout in terms of relative or specific humidity is also possible (Figs. 4.9j and k).

Sampling System

A sampling system is required where the sample gas is outside the operational limits of the instrument. At high temperatures, a sample cooler is required to bring the sample temperature below the operating limit of the sensor. The sample cooler, however, must be chosen such that the sample will never be cooled to its dew point under any possible operating or ambient condition. At high sample pressures, a pressure reducing valve is required.

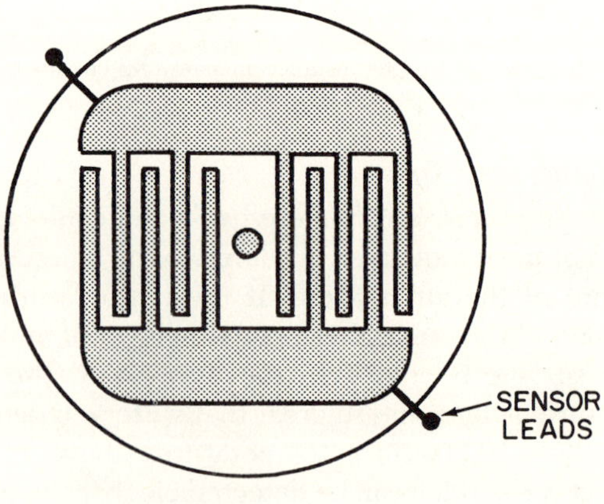

Figure 4.9j Sensing surface.

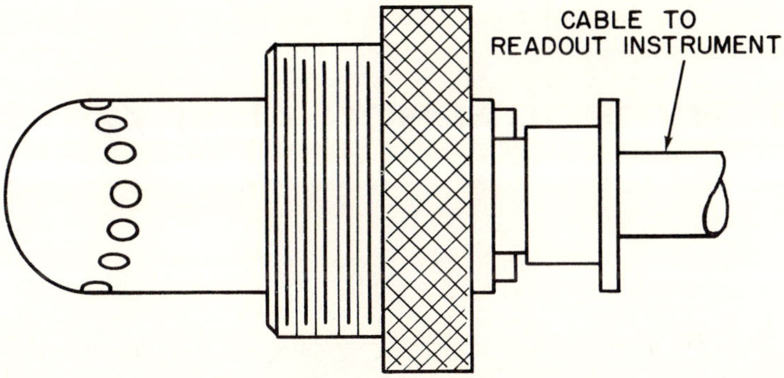

Figure 4.9k Sensor

For measurement of stagnant atmospheres, low pressure samples or closed systems, a pump may be required to draw the sample over the sensor.

Limitations

The instrument can be used with all the common atmospheric gases such as nitrogen, oxygen, carbon dioxide, etc. The instrument is not suitable for samples containing components that react with water such as chlorine or sulfur dioxide, and components with a dew point higher than that of water.

Moisture Indicators

Sometimes it is desirable to detect only a change of the amount of moisture in the sample, while the exact concentration is of only secondary importance. Such may be the case on the outlet of drying beds where moisture breakthrough will indicate the need for regeneration. Certain salt crystals indicate the presence of moisture by a change in color and this is the most economical method for detecting changes of moisture level. The crystals are packed into a small chamber with glass window for observation of the color. A sample side stream is piped through the chamber, and increase of moisture in the sample is indicated to the operator by a color change.

Conclusions

The instruments discussed above are generally applicable to such easy-to-handle fluids as the atmospheric gases. Operating temperature and pressure are usually limited to atmospheric conditions although with some designs the operating limits can be extended. The principal feature common to all the above instruments is the fact that the standard output form is not in terms of moisture content directly, but in terms of a parameter related to moisture content. Accuracy varies widely depending on design; best accuracy can be obtained with the conductivity type hygrometer and poorest accuracy with the moisture indicator. Generally, a sampling system is required only for special applications.

4.10 HYDROCARBON ANALYZERS

Type of Measurement:	(al) Total hydrocarbon by flame ionization; (b1) Methane by chromatography, (b2) Methane by flame ionization, (b3) Methane by mass spectrometry; (c1) Hydrocarbon classes by flame ionization, (c2) Hydrocarbon classes by mass spectrometry, (c3) Hydrocarbon classes by infrared; (d1) Individual hydrocarbons by chromatography.
Reference Method:	GC with flame ionization for nonmethanes.
Sensitivity:	0.1 ppm (a1, b2, c1).
Costs:	\$2,000 to \$5,000 (a1); \$3,000 to \$5,000 (b2, c1), \$5,000 and up (b3); \$10,000 and up (b1, c3, d1), \$20,000 and up (c2).
Partial List of Suppliers:	Beckman Instrument, Inc. (a1,b1, b2,c1,d1); Bendix Environmental Science Div. Bendix Corp. (a1,b2, c1); Davis Instrument Co. (a1,b2,c1); Delphi Industries (a1,b2,c1); Mine Safety Appliances Co. (a1,b1,b2, c1,d1); Power Design, Inc. (a1,b2,c1); Tracor, Inc. (b1,d1); Union Carbide Corp. (b1,d1); Wilks Scientific (c3).

Unlike most gaseous air pollutants, the hydrocarbons comprise a class containing scores of individual components (Table 4.10a).[1] The most abundant hydrocarbon, methane, is unreactive in photochemical reactions and otherwise not a hazardous air pollutant (see Section 3.9). The remaining hydrocarbons vary widely in their reactivity. For this reason, the type of hydrocarbon analysis to be used will depend on the purpose for which data are needed.

Table 4.10a
ATMOSPHERIC HYDROCARBON ANALYZERS

Hydrocarbon Type	*Method*	*Limitations and Interferences*
Total (as carbon)	Flame Ionization Detection (FID)	Some response to carbon-containing nonhydrocarbons
Methane	Gas chromatography	Expensive equipment (can also be used for carbon monoxide)
	"Methane-only" subtractive column and FID	Column preparation fussy
	Mass spectrometry	Freeze-out required, expensive
Aromatics, olefins, paraffins	Subtractive columns and FID	Column preparation fussy
	Mass spectrometry	Freeze-out required, expensive, data reduction requirements large
	Infrared spectrometry	Freeze-out required, expensive, not total class coverage
Individuals	Gas chromatography	Expensive, data reduction requirements large

To obtain highly detailed analyses for numerous hydrocarbons requires expensive and sophisticated equipment, whereas a single measurement for total hydrocarbons is simpler and somewhat less expensive. There are also methods of intermediate complexity which permit determination of methane separately from other hydrocarbons.

Calibration Methods

Only dynamic calibration is completely satisfactory. In this procedure, known, realistic concentrations of the pollutant in air are passed into the analytical system, preferably in exactly the same way as unknown air samples are collected. This may be difficult to accomplish because of the low concentrations and possible reactivity or condensability of some pollutants.

Standard calibration mixtures of one or several of the common hydrocarbons in air are available commercially in high-pressure gas cylinders. These should always be checked against a reference standard, or at least against a previously calibrated analyzer. If much calibration is necessary, it may be useful to set up a gas-handling and dilution system to prepare standard mixtures. One fairly simple technique for hydrocarbons, such as butane (which can be liquefied at moderate pressure), uses permeation tubes. These are sealed tubes of a specific plastic, partially filled with liquid hydrocarbon. The hydrocarbon will effuse through the walls of a given tube at a rate depending only on temperature, as long as some liquid remains. To generate a known concentration, air is passed at a known flowrate over the tube in a temperature controlled vessel. The tube may be calibrated gravimetrically.

In another technique, a known amount of hydrocarbon is added by syringe or by crushing an ampoule, either in a large rigid vessel of known air volume, or into the metered flow of air passing into a plastic bag. The bag must be of inert plastic; and it has the advantage of collapsing as the sample is withdrawn, whereas a rigid vessel must be large relative to sample size to avoid substantial pressure differentials.

Flame Ionization Detectors

Flame ionization detection (FID) is the most suitable means of analysis for most hydrocarbons[2,5] at the levels found in polluted air. It may be used alone for the measurement of total hydrocarbons, or as a detector after separation by a column device such as a gas chromatograph. In FID analyzers, a sensitive electrometer detects the increase in ion intensity in a hydrogen flame when air containing organic compounds is introduced. The response is approximately in proportion to the number of organically bound carbon atoms in the sample, so the detector is basically a carbon atom counter. Carbon atoms bound to oxygen, nitrogen, or halogens, however, give reduced response. There is no response to nitrogen, carbon monoxide, carbon dioxide or water vapor. The results are usually expressed in terms of the calibration gas used, e.g., ppm carbon as methane. Response is rapid and can be as sensitive as 0.1 ppm. The response to various hydrocarbons is not perfectly uniform and such variations should be taken into account in interpreting FID data. The FID analyzer is a reference method for total hydrocarbons in U.S. Air Quality Standards. Basically, it is an air metering device attached to a flame ionization detector (Fig. 4.10b).

Gas Chromatography

The only practical method for general analysis of specific hydrocarbons is gas chromatography (GC).[6] In GC, the basic apparatus (Fig. 4.10c) is a carefully prepared tubular column of a finely divided solid with provisions for passing a steady flow of a carrier gas (often helium) through it. The column is temperature controlled and the packing usually supports or is coated by a nonvolatile liquid phase. Preceding the column there is means for sample introduction (and sometimes sample splitting) and following the column is a detector. A high sensitivity column useful for atmospheric work usually employs a flame ionization detector. There are a great many optional varia-

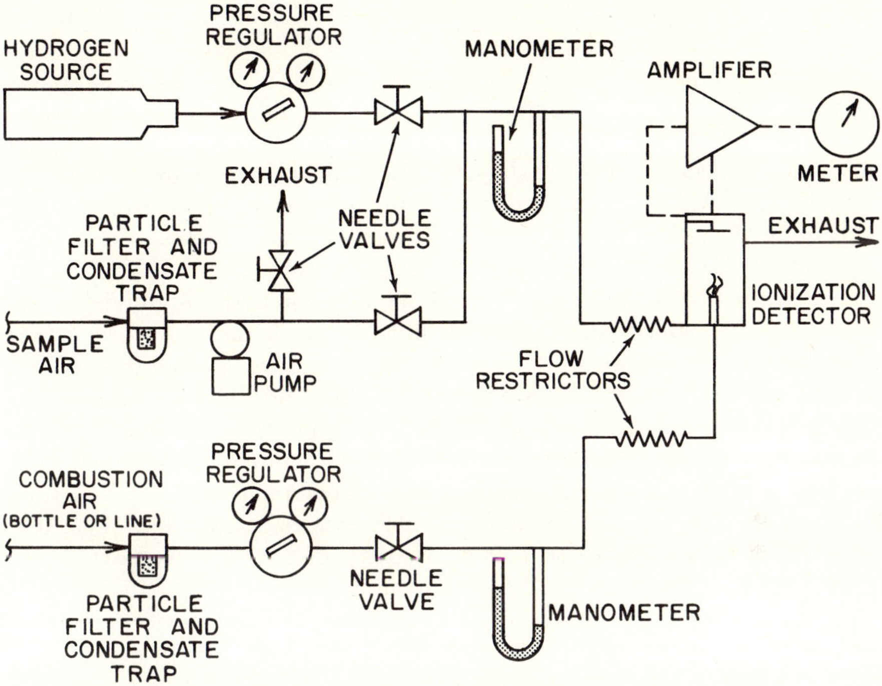

Figure 4.10b Hydrocarbon analyzer, hydrogen flame ionization detector.

tions in the makeup of a GC instrument involving sample valves, precolumns, columns, programmed temperature changes, stream splitters and detectors. The output may be recorded on a strip-chart (possibly with a peak integrator) or converted to digital form for computer handling. In full-scale analyses for scores of hydrocarbons, data reduction from analog strip-chart recordings

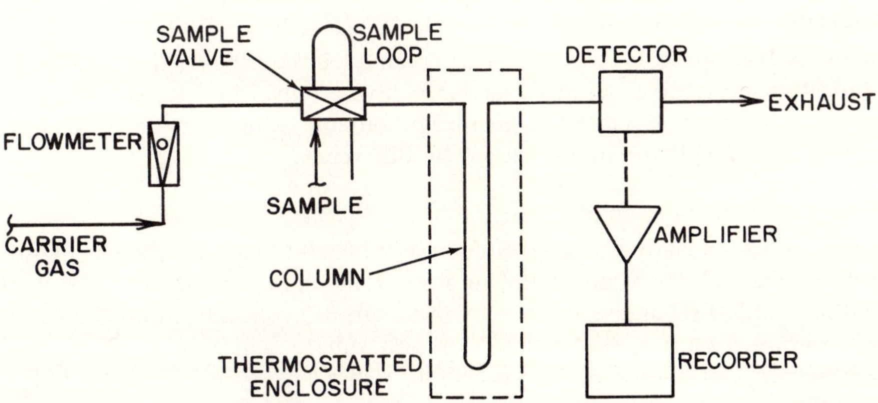

Figure 4.10c Gas chromatograph.

becomes very tedious and time-consuming. In that event, the addition of computerized readout may be worth the increase in cost.[6]

Supplemental techniques increase the utility of GC for atmospheric hydrocarbons. A common method is to pass a sample of air through a small freeze-out trap, sweep out the air with helium, then warm the trap and introduce the condensables to the GC column in one concentrated slug. This extends the lower limits of sensitivity to the ppb range. Subtractive columns may be used in parallel or in series with conventional columns to help relate the various recorded peaks to specific hydrocarbon classes.

Complete GC calibration would require hundreds of different pure hydrocarbons and is never achieved fully, although there are usually few unknowns left in careful work except in the more complex higher molecular weight ranges. Calibration is based on the fact that under carefully reproduced conditions, a given hydrocarbon will always require the same length of time to pass through the column to the detector. Ambiguities arising where components overlap may be resolved by change in column packing or sometimes by the use of subtractive columns. An elaborate but useful technique is to attach another analytical instrument such as an infrared or mass spectrometer to the outlet of the detector for peak identification.

In Section 4.7 on carbon monoxide sensors, an instrument is described for combined analysis of methane and carbon monoxide. This is a simple GC, set up to determine methane, with a programmed valving arrangement and a catalytic carbon monoxide–to–methane converter. The resultant is a readout of two peaks, one for methane directly and the other for methane formed by reduction of carbon monoxide.

Subtractive Columns

Nonmethane Hydrocarbons

A column of adsorptive charcoal may be treated with methane just until breakthrough occurs; that is, no more methane will be adsorbed, but other hydrocarbons are still retained.[3] This column can then be used with a flame ionization detector as an analyzer for methane only. Used in parallel or in switched alternation with the detector without any column, it allows determination of nonmethane hydrocarbons by difference.

Reactive Hydrocarbons

Certain specially prepared columns can be used to adsorb specific hydrocarbon classes.[4] These are useful in gas chromatography and could be used with a simple FID analyzer alone to give analysis by broad classes. A column of crushed fire-brick supporting mercuric sulfate will adsorb olefins and acetylenes, and one of palladium sulfate will then adsorb aromatic hydrocarbons (except benzene). A combination of these subtractive columns has been applied successfully to auto-exhaust analysis and shows promise for use in atmospheric work.

Spectrometric Methods

Either infrared or mass spectrometry may be used with considerable advantage for individual hydrocarbon determination, although the sensitivity of both methods would require concentration such as by a freeze-out trap. Both types of instruments are complex and expensive. Since the calibration requirements are about equal to those needed for gas chromatography, the data interpretation is generally more complicated and the extent of coverage is less; these methods have been widely supplanted by gas chromatography for atmospheric hydrocarbon detection work. These instruments may be useful for analysis of a particular type or class of hydrocarbon and might make it unnecessary to set up a gas chromatographic capability in some cases.

Assessment

For total atmospheric hydrocarbon measurement, the flame ionization analyzer is the most generally accepted detector. It requires technical attention and compressed gases but is reliable and accurate. The methods for specific classes add complexity and reduce precision and reliability but are not much more expensive than the flame ionization analyzer itself. (The spectrometric methods, unless already in use for other reasons, are not competitive.) If individual hydrocarbon determination is required, there is no good alternative* to gas chromatography (frequently augmented with a freeze-out step), even though the data reduction is time-consuming (if manual) or expensive (if computerized), and calibration requirements are demanding.[6]

*NDIR sensors can only identify hydrocarbon classes, while two-color IR detectors are able to identify individual hydrocarbon species. In this design sensitivity is traded for specificity.

REFERENCES

1. Altshuller, A. P.: in A. C. Stern (ed.), "Air Pollution," Ch. 18, 2d ed., Academic Press, New York, 1968.
2. Andreatch, A. J. and R. Feinland: *Anal. Chem.*, vol. 32, pp. 1021–1024, 1960.
3. Ortman, G. C.: *Anal. Chem.*, vol. 38, pp. 644–646, 1966.
4. Klosterman, D. L. and J. E. Sigsby, Jr.: *Environ. Sci. Techn.*, vol. 1, pp. 309–314, 1967.
5. Lipták, B. G. (ed.): "Instrument Engineers' Handbook," Supplement One, Section 1.15, Chilton, Philadelphia, 1972.
6. Ibid, Volume I, Section 8.1.

4.11 MERCURY DETECTION IN AIR

Methods of Detection:	(a) Ultraviolet, flameless atomic absorption; (b) Atomic absorption spectrophotometer.
Typical Ambient Concentrations:	1 to 200 ng/m^3.
Costs:	\$500 to \$3,000 (a-sample analyzers); \$3,200 to \$30,000 (a-integrated sampler-detector); \$6,000 to \$18,000 (b).
Partial List of Suppliers:	Anti-Pollution Tech. Corp. (a); Bacharach Instrument Company (a); Bausch and Lomb, Inc. (b); Beckman Instruments, Inc. (a,b); Bendix Corp. (b); Carle Instruments (b); Fisher Scientific Co. (b); Geomet Inc. (a); Jarrel-Ash Div., Fisher Scientific Co. (b); Laboratory Data Control (a); Olin Corp. (a); Perkin-Elmer Corp. (a,b); Scintrex Co. (a); Varian Co. (b).

Data concerning the levels of mercury in the atmosphere are given in Section 3.11.

Air saturated with mercury will contain between 4 and 100 mg/M^3 in the temperature range between 45 and 110°F. Since saturation is not approached under normal circumstances with moving air, sensitive analytical and sampling methods are needed when data on actual ambient levels of mercury are required. For practical use, the methods employed must be relatively insensitive to interferences and be otherwise appropriate for the collection of data under a wide range of operating conditions. For regulatory purposes, related to emissions, three or four levels of sensitivity are of importance: 100, 50, 10 and perhaps 5 μg/M^3 of mercury in the air. Accuracy within $\pm$10% at these levels is required. However, ambient air monitoring requires the ability to determine levels in the 1 to 200 ng/M^3 range.

Sample Collection and Concentration

Collection of mercury samples for accurate determination of concentrations found at ambient conditions or in industrial plants requires that: (1) the size of the sample taken must be adequate to satisfy the sensitivity threshold requirements of the analytical technique utilized; (2) the rate of sampling must be controlled so that efficient collection is achieved; (3) the sampling interval selected must be appropriate for the required monitoring function; (4) the collection device or procedure must be compatible in capacity and chemical reactivity of material components with the levels and properties of mercury and mercury compounds to be sampled; (5) effects of potential interferences must be avoided; (6) elapsed time between collection and analysis must be minimized in order to avoid losses; and (7) it must be possible to evaluate all unknown factors in collection and analysis by calibration procedures.

In the sampling of mercury and its compounds from the air, two basic procedures are available. These procedures utilize either the chemical reactivity of the sampled components in various aqueous media, or the extraordinarily efficient extraction (amalgamation) of mercury from air by the noble metals. In the latter case, reduction of mercury compounds to elemental mercury is required before the sampling process is carried out. Currently, a wet chemical procedure for sampling is recommended by the Environmental Protection Agency; it requires considerable labor and upkeep to perform effectively.

Impinger Collection Methods

The Environmental Protection Agency has published two methods for sampling and analyzing mercury in gaseous and particulate emissions in conjunction with publication of proposed National Emission Standards for Hazardous Air Pollutants.[1] The methods are similar with the exception that particulates must be sampled isokinetically, while gaseous samples are passed through filters to remove particles. This latter procedure requires equilibration of the filter prior to generation of accurate data at low mercury levels.

In principle, samples are drawn through acidic aqueous iodine monochloride solutions utilizing stringent controls over flowrates, acidity and other process variables. This procedure gives excellent recovery of organic mercury compounds and of mercury vapor. Samples obtained with this procedure may be stored.

Figure 4.11a shows the sampling train utilized for particulate and vapor samples. It consists of: (1) a pyrex-lined, heated probe section containing a thermocouple for temperature measurement and a pair of pitot tubes and a manometer for gas flow measurement; (2) five impingers of the Smith-Greenburg type placed in an ice bath; (3) a filter for particle removal prior

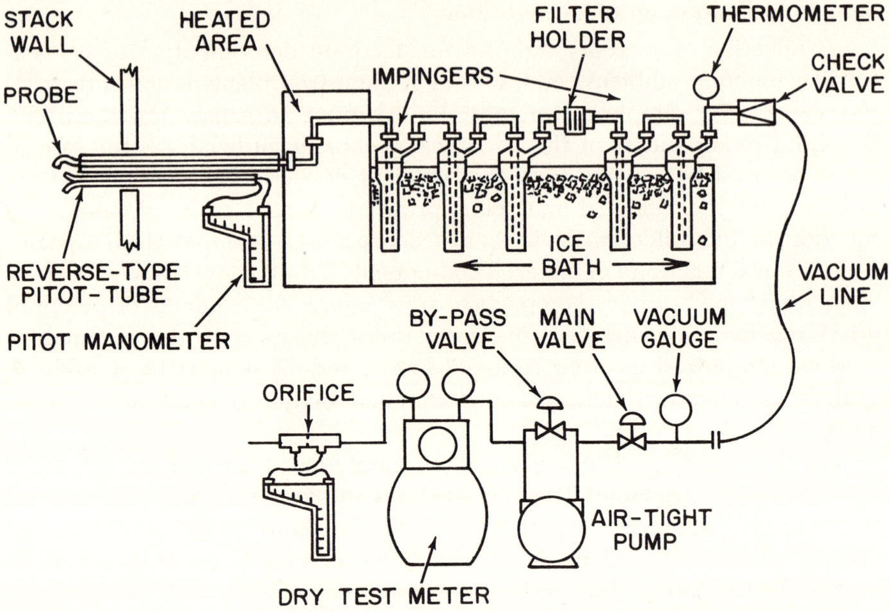

Figure 4.11a Particulate-sampling train.

to passage of the gases through the two final absorber stations; and (4) miscellaneous components for control and calibration of the sample airflow. Figure 4.11b shows a simpler technique used for sampling vapors without particulates. In this method, particulates are removed by a filter at the inlet into the absorption train. For sampling particulates, the initial rate of gas flow is 2 SCFH. Flow should be maintained at a rate proportional to the flow through the stack sampled.

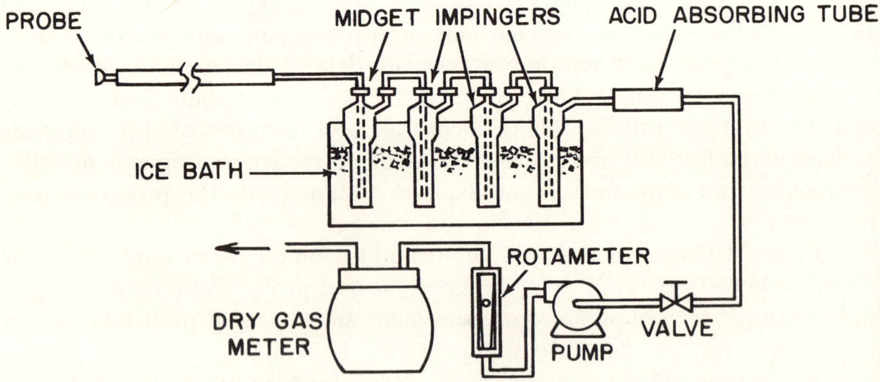

Figure 4.11b Vapor sampling train.

Samples obtained by this procedure may be analyzed by a procedure utilizing a flameless atomic absorption spectrophotometer which has been accepted by the Environmental Protection Agency for investigations.

A number of other collection procedures employing impingers have been examined. These utilized water, ethyl alcohol, isopropyl alcohol, potassium permanganate-sulfuric acid, potassium permanganate-nitric acid or iodine-potassium iodide. However, for effective collection of alkyl mercury compounds, acidic iodine monochloride solutions are preferable.

Amalgamation on Wettable Metals

A number of less laborious methods of sampling mercury vapor, compounds and particulates have been developed which utilize absorption (amalgamation) of vapor on silver and gold and on alloys of these materials. In order to sample particulates and compounds of mercury, the trapping procedure must be preceded by a decomposition step. A catalytic process which achieves this objective is described later. Examples of procedures for collecting elemental mercury vapor include Figure 4.11c, which shows a simple absorption tube filled with gold or silver wool. The tube may be sized to satisfy any specific parameters of the sampled gas stream. Fine wire may also be used as packing for flowrates under 50 liters per minute, and mercury vapor removal can be made quantitative. Gold materials are required for gases containing hydrogen sulfide or sulfur dioxide. When a simple absorption tube of this type is employed, the mercury vapor collected may be released for analysis in a flameless atomic absorption spectrophotometer by utilizing an induction furnace. In this procedure, extensively used by the U.S. Geological Survey, a continuously flowing airstream is used to pass the mercury vapor pulse which results during heating by the furnace into the AAS cell.

A versatile method of sampling gas streams, employed by the Canadian Department of the Environment, uses a packing which consists of finely woven silver mesh (Fig. 4.11d). Other packings may be used such as those reviewed in Section 6.27. One version of the sampling tube is 5 inches high and utilizes a $\frac{1}{4}$ inch I.D. pyrex tubing. Separation between the inner and outer tube is $\frac{3}{64}$ inches. In order for this collection procedure to function effectively, relatively low airflow rates through the wire mesh must be used. Sampling rates of 1 to 5 liters per minute have been employed, but long sampling periods are required to make measurements at low air concentrations.

Figure 4.11e shows schematically, a collection device contained in an integrated mercury analyzer. In the collection section of this device, air at

Figure 4.11c Absorption tube type mercury collection device.

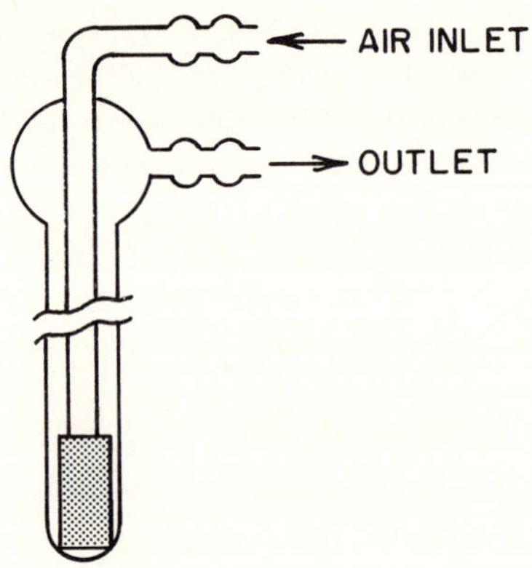

Figure 4.11d Mercury gas sampling tube.

rates from 1 to 200 liters per minute is passed through a grid containing 160 feet of 22 gauge wire arranged in a carefully spaced annulus. Mercury vapor is desorbed by direct passage of electrical current through the wire. Analysis is achieved by the flameless atomic absorption method in a photometer.

Activated Absorption

Use of a silver on charcoal absorbent is reviewed in Section 6.27. This absorbent is highly efficient and may be used to remove mercury vapor from many gas streams. Extraction of the mercury vapor as a concentrated pulse for analysis, however, is relatively difficult. Induction heating is required to maximize the vapor concentration. Trapping is also possible with iodine-activated charcoal and mineral wool. Regeneration of these materials for repeated use presents difficulties not yet resolved.

Conversion of Mercury Vapors

In order to analyze for total mercury including particulates and vapor, reduction of compounds such as mercuric chloride, mercuric sulfide, dimethyl and diethyl mercury, pesticides and fungicides to the elemental form must be achieved. Three techniques are in use: (1) thermal decomposition at 1,500

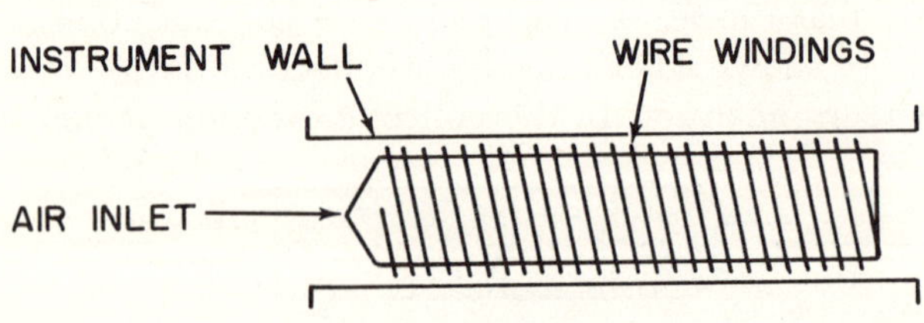

Figure 4.11e Mercury collection device.

to 2,000°F; (2) catalytic reduction or disproportionation, usually at 1,200 to 1,300°F over materials such as cupric oxide; and (3) chemical conversion in solutions such as acidic iodine monochloride. Commercial equipment is available which may be used in conjunction with sampling and/or analysis instrumentation.

Methods of Detection

No direct analytical method is capable of attaining accurate measurements in the range required for ambient air without the use of a concentrator discussed earlier.

Ultraviolet Light Absorption

A wide variety of methods have been developed to measure airborne mercury-containing materials in the atmosphere based on the strong absorption of ultraviolet light at 253.7 nm by elemental mercury vapor. Since many other compounds absorb light in this range, although much higher concentrations are required than with mercury, some method of separating interferences is needed when low levels of mercury are measured.

Atomic Absorption Spectrophotometry

Mercury is conveniently analyzed by atomic absorption techniques because the elemental vapor exists in atomic form under normal ambient conditions. Thus, when a cloud of mercury vapor is irradiated at an appropriate wavelength, mercury atoms absorb a portion of the energy of the beam. In the standard atomic absorption technique, liquid samples are aspirated into an acetylene air flame where the atomic mercury is released and the absorption of light at 253.7 nm is measured by a spectrophotomultiplier (Fig. 4.11f). The sensitivity of 0.01 μg of mercury per ml of sample is readily attained by this

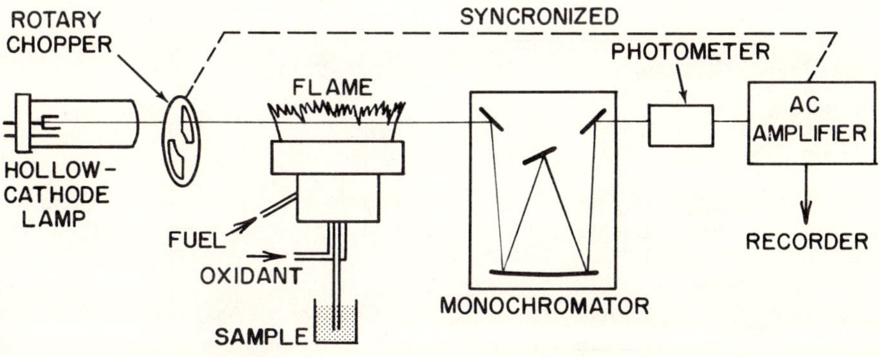

Figure 4.11f Atomic absorption spectrophotometer.

procedure. A continuous series of calibration checks are required for maximal precision and accuracy.

Flameless Atomic Absorption Spectroscopy

The burner-aspirator and flame indicated in the schematic of the conventional AAS (Fig. 4.11f) may be replaced by a quartz windowed glass cell and the sample may be generated by reducing an aqueous sample of mercury compounds to elemental mercury. Figure 4.11g shows the apparatus which may be used in conjunction with an AAS. In this procedure, a sample collected by utilizing a separate collection train is pipetted into the gas washing bottle. To this is added a solution of a reducing agent such as hydroxylamine sulfate or stannous chloride and the bottle is closed. With clean air supplied at a rate of 1.3 liters per minute, aeration is commenced and continued until a maximum peak height readout is obtained on the recorder. By calibration against acidic mercuric chloride standard solutions, the peak heights obtained with unknown samples can be compared with the calibration curve for direct determination of the mercuric ion concentration. This value may be converted to the gas phase concentration of mercury.

At least seven types of commercial units are available which use the principle of flameless atomic absorption spectroscopy for determination of mercury. Properties of these instruments are shown in Table 4.11h. Affirmative designations in column 4 indicate that samples must be taken by use of a liquid sampling train and transported to the instrument.

Other devices based on the use of absorption of radiation at 253.7 nm have been developed by Barringer Research and by Williston and co-workers (Cordero Mining Company).[2] The Barringer method takes advantage of the pressure-broadening of the mercury emission lines to minimize interferences due to other compounds.

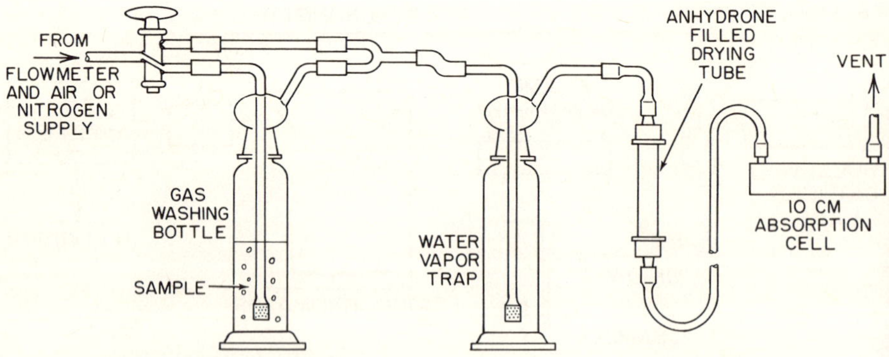

Figure 4.11g Apparatus for flameless atomic absorption.

Table 4.11h

ESTIMATED PERFORMANCE PARAMETERS OF COMMERCIALLY AVAILABLE MERCURY MONITORS

(utilizing flameless atomic absorption photometers)

Manufacturer	*Integral Collector*	*Collection Process*	*Discrete Samples Required*	*Sensitivity Range*	*Total Hg*	*Installation*	*Price Range, $*
Antipollution Technology Corporation (Thermatron)	No	—	Yes—liquid	10 μg/M³	With extra train	Portable	1,000
Bacharach Instrument Company	No	—	No	0.1 mg/M³	No	Portable	500
Beckman Instruments	No	—	No	0.005–0.1 and 0.03–3.0 mg/M³	No	Portable	800–1,200
GEOMET, Incorporated	Yes	Amalgamation	No	1–1,000,000 ng/M³	Yes	Portable	7,000–10,000
Laboratory Data Control	No	—	Yes—liquid	1–5,000 ng/M³	With extra train	Portable	1,800
Olin Corporation	Yes	Liquid Impinger	No	1 ng/M³	Yes	Fixed	20,000–30,000
Perkin-Elmer	No	—	Yes—liquid	10 ng/M³	With extra train	Fixed & portable	3,800–14,500
Scrintrex	Yes	Palladium Fluoride Filter	Continuous	0.5 ng/M³	No	Portable	3,250–18,000

4.11 Mercury Detection in Air

Colorimetric Methods

DITHIZONE

In this method, air is drawn through an acidic solution (10% H_2SO_4) of potassium permanganate (4 g/100 ml) where the elemental mercury, inorganic mercury compounds and the readily decomposable mercury-organic compounds are converted to mercuric ions. Methyl mercury compounds require the use of permanganate at 100°C. The determination of mercury may be concluded colorimetrically whence the sample is titrated with dithizone and compared with the colors developed by standards. A spectrophotometer may also be used for quantitative comparison between the color developed by test samples and those of controlled standard quantities of mercury.

In the method developed by the American Conference of Government Industrial Hygienists, the sample is collected from the air by passage through an aqueous solution of iodine (0.25%) and potassium iodide (3%). Mercury is extracted from a buffered solution (ammonium citrate, hydroxylamine hydrochloride and phenol red at pH 8.5) with dithizone in chloroform (20 mg dithizone/5 ml) until all color is removed. After washing the chloroform phase, the mercury is extracted into acidic (0.1 N HCl) potassium bromide (8%), shaken with a chloroform solution of 10 mg/liter dithizone, and filtered through cotton into a colorimeter cell. The optical density at 485 nm is used to determine the mercury. Calibration is carried out with mercuric chloride.

The dithizone method is relatively insensitive, requiring the use of large samples (about 10 μg are required) when the levels of mercury are low. It also requires considerable care on the part of the chemist to avoid the loss of organic mercury compounds.

SELENIUM SULFIDE AND OTHERS

Active selenium sulfide applied to paper as a coating may be used to detect the presence of elemental mercury vapor. The paper develops a black color on exposure to air containing mercury. The degree of blackening is a function of the concentration of mercury and the time of exposure. Quantitation may be made with a simple densitometer.

Mercury can also be detected with copper, iodide paper, selenium paper and gold chloride on silica gel. These methods appear to have a limiting sensitivity in the 500 to 1,000 $\mu g/M^3$ range with an accuracy of $\pm 5\%$.

Gas Chromatographic Methods for Mercury Organics

Organic mercurials can be determined at levels of approximately 1 ppb by gas chromatographic methods. Samples must be extracted into benzene and injected into a gas chromatographic column (Carbowax® 20M on Chromasorb® W) operating at 180°C which separates methyl mercury, ethyl mercury, methoxyethyl mercury, phenyl mercury and dimethyl mercury. Quanti-

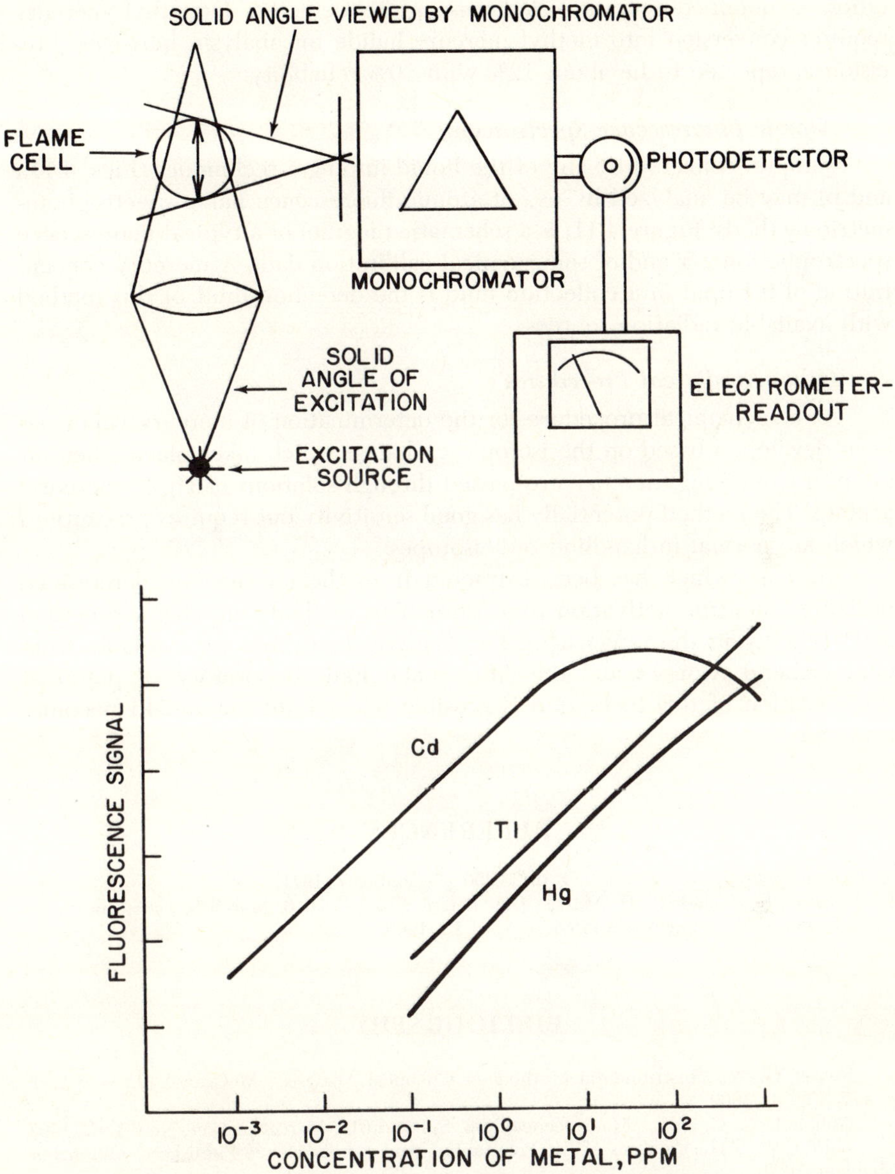

Figure 4.11i Atomic fluorescence spectrophotometer.

tation is obtained with an electron-capture detector. Dimethyl mercury requires conversion into methyl mercury halide for analysis, however. Precision is reported to be about 12% with 20% reliability.

Atomic Fluorescence Spectroscopy

Samples collected by any of the liquid impinger techniques (Figs. 4.11a and b) may be analyzed by use of atomic fluorescence flame spectrophotometric methods. Figure 4.11i is a schematic diagram of a typical fluorescence spectrophotometer and of some typical calibration data. A mercury concentration of 0.1 ppm in a collection fluid is the detection limit of this method with available radiation sources.

Other Analytical Procedures

A radiochemical procedure for the determination of mercury vapor has been developed based on the isotopic exchange which takes place when air samples containing mercury are passed through solutions of Hg^{203} mercuric acetate. The method potentially has good sensitivity but requires precautions which are normal in handling radioisotopes.

Mercury which has been extracted from the air may be determined utilizing a neutron activation procedure. This method can have a precision and accuracy in the range of ±2%. Analytical services are available from Gulf General Atomics and the AEC Idaho Falls Laboratory. A potential simplification of this technique is possible when Californium-246 becomes available.

REFERENCES

1. *Federal Register,* vol. 36, no. 234, p. 23,250, December 7, 1971.
2. Williston, S. H. and M. H. Morris: U.S. Patent No. 3,173,016, March 9, 1965; and S. H. Williston, U.S. Patent No. 3,178,572, April 13, 1965.

BIBLIOGRAPHY

Ewing, G. W.: "Instrumental Methods of Chemical Analysis," McGraw-Hill, New York, p. 177 (Fig. 3), 1969.

Guilbault, G. G. (ed.): "Fluorescence," pp. 581 and 618, Marcel Dekker, New York, 1967.

Leithe, W.: "The Analysis of Air Pollutants," Ann Arbor Science Publishers, Ann Arbor, Michigan, 1970.

Vaughn, W. W.: "A Simple Mercury Vapor Detection for Geochemical Prospecting," Geological Survey Circular No. 540, U.S. Department of the Interior, 1967.

4.12 NITROGEN OXIDE ANALYZERS

Method of Analysis:	(a) Colorimetric; (b) Coulometric; (c) Chemiluminescent; (d) Gas Chromatography; (e) Electrochemical.
Reference Method:	(a), applied to integrated samples collected in alkaline solution.
Costs:	\$2,000 to \$5,000 (a,b,e); over \$10,000 (c,d).
Partial List of Suppliers:	Aerochem Res. Labs. Inc. (c); Atlas Electric Device Co. (a); Beckman Instruments, Inc. (a,b); Dynasciences Corp. (e); Environmetrics, Inc. (e); Intertech Corp. (e); Mast Development, Inc. (b); Monitor Labs, Inc. (a); Pollution Monitors, Inc. (a); Precision Scientific (a); Raditron, Inc. (a); Scientific Industries (a); Technicon Corp. (a); Wilkens-Anderson Corp. (a).

Calibration Methods

Dynamic calibration of a gas analyzer is always a better check on the reliability of the system than is static calibration, but it is also much more difficult. Nitric oxide (NO) is oxidized to nitrogen dioxide (NO_2) by air, especially rapidly at high concentrations, and NO_2 condenses and dimerizes at high concentrations. For these reasons, dynamic calibration for NO and NO_2 requires great care. In colorimetric or coulometric analysis, the NO is in fact not measured directly at all but is measured after oxidation to NO_2, so that calibration is made only for NO_2.

Dynamic calibration for NO_2 requires preparation of a sample of inert gas containing a known concentration of NO_2. This may be accomplished by gas-dilution techniques (with special precautions in handling NO_2), gravimetrically, electrolytically or by use of a permeation tube. The permeation tube (as used with hydrocarbons) is described in Section 4.10. NO_2 permeation tubes are moisture-sensitive, however. They should be protected from moisture in storage and only dry diluent gases should be used. If these precautions are observed, the permeation tube is probably the most convenient method of dynamic calibration for NO_2.

Static calibration is carried out with standard solutions of nitrite. Since this procedure does not involve the components of the gas sampling line, it is not as complete a test as is dynamic calibration, but is much simpler. The stoichiometric ratio of nitrite to NO_2 is usually not unity, but is a constant value under controlled conditions. In the Griess-Saltzman method (although there has been controversy), the consensus is that 0.72 mole of nitrite gives the color of 1 mole of NO_2; and in the Jacobs-Hochheiser method, this factor has been found to be 0.63.

NO-NO_2 Combination Analysis

In continuous analysis, it is customary to determine both NO and NO_2 as NO_2 (Table 4.12a). The NO is oxidized to NO_2 by means of potassium permanganate or dichromate or by chromium trioxide, in various formulations. The efficiencies of conversion seem to depend on the length of service and, for the chromium oxidizers, on humidity. Aqueous permanganate seems to be the best choice even though it may not be completely efficient (hence NO may be underestimated).

In series analysis for NO and NO_2 (Fig. 4.12b), the air passes through an NO_2 analyzer for measurement and removal of NO_2, then through an oxidizer to convert NO to NO_2, and finally through a second NO_2 analyzer. The second analyzer gives a measure of NO concentration. In one type of parallel analysis, two equal airstreams are analyzed for NO_2, one of them after passage through an oxidizer. The latter gives total oxides of nitrogen (NO + NO_2) from which NO is found by difference from the other parallel analyzer. In a second type of parallel analysis, one stream is analyzed directly for NO_2; the second is scrubbed free of NO_2 by passage through Ascarite, followed by oxidation of NO to NO_2 and NO_2 analysis.

Colorimetric Determination

There are two important colorimetric methods[1] for NO_2 determination, the Griess-Saltzman and the Jacobs-Hochheiser. The first is used in most continuous colorimetric NO_2 analyzers. It is based on the reaction of NO_2 with sulfanilic acid to form a diazonium salt which couples with N-(1-napthyl)-ethylenediamine dihydrochloride to form a deeply colored azo dye. Air is passed into the reagent solution for not over 30 minutes, time is allowed for development and the color is measured at 550 nm. The range of concentrations measurable is 0.02 to 0.75 ppm. In manual use, the color is developed for 15 minutes and should be read within 1 hour (on a colorimeter or spectrophotometer). In a continuous analyzer (Fig. 4.12c), the gas and liquid flowrates

Table 4.12a
NITROGEN OXIDE ANALYZERS

General Method	*Type*	*Advantages*	*Disadvantages*
Colorimetric	Griess-Saltzman	Precise, thoroughly tested, widely used, continuous analysis	Short life of collected sample, sensitive reagents, NO oxidation required
	Jacobs-Hochheiser	Precise, stable after collection	Not adapted to continuous analysis, sensitive reagents, NO oxidation required
Coulometric		Simple apparatus, continuous analysis	Sensitive to other oxidants, NO oxidation required
Chemiluminescent		Dry gases only, sensitive photometry, continuous analysis	Requires ozone generator NO_2 catalytic reduction
Gas chromatography		Specific, frequent analysis	Not a developed instrument, expensive and complex
Electrochemical		Simple apparatus, continuous analysis	Sensitivity not high, NO oxidation required

are adjusted for optimum response and the developed color is read in a flow cell photometrically, using a 550 nm filter. Response times are usually 5 to 15 minutes.

The Jacobs-Hochheiser method is the standard reference method for U.S. National Air Quality Standards. The reason for this is that the Air Quality Standard is an annual average; and this method allows collection of up to 24-hour integrated samples and delays in analysis of at least 2 weeks, whereas Griess-Saltzman samples must be quickly analyzed. In the Jacobs-Hochheiser method, the air is passed through aqueous sodium hydroxide, converting NO_2 to nitrite ion. The solution is freed of possible sulfur dioxide by treatment with hydrogen peroxide and acidified. The rest of the procedure is the same as for the Griess-Saltzman method except that sulfanilamide is used rather than sulfanilic acid. Efficiencies found with this procedure in the U.S. National Air Surveillance Network are approximately 35%.

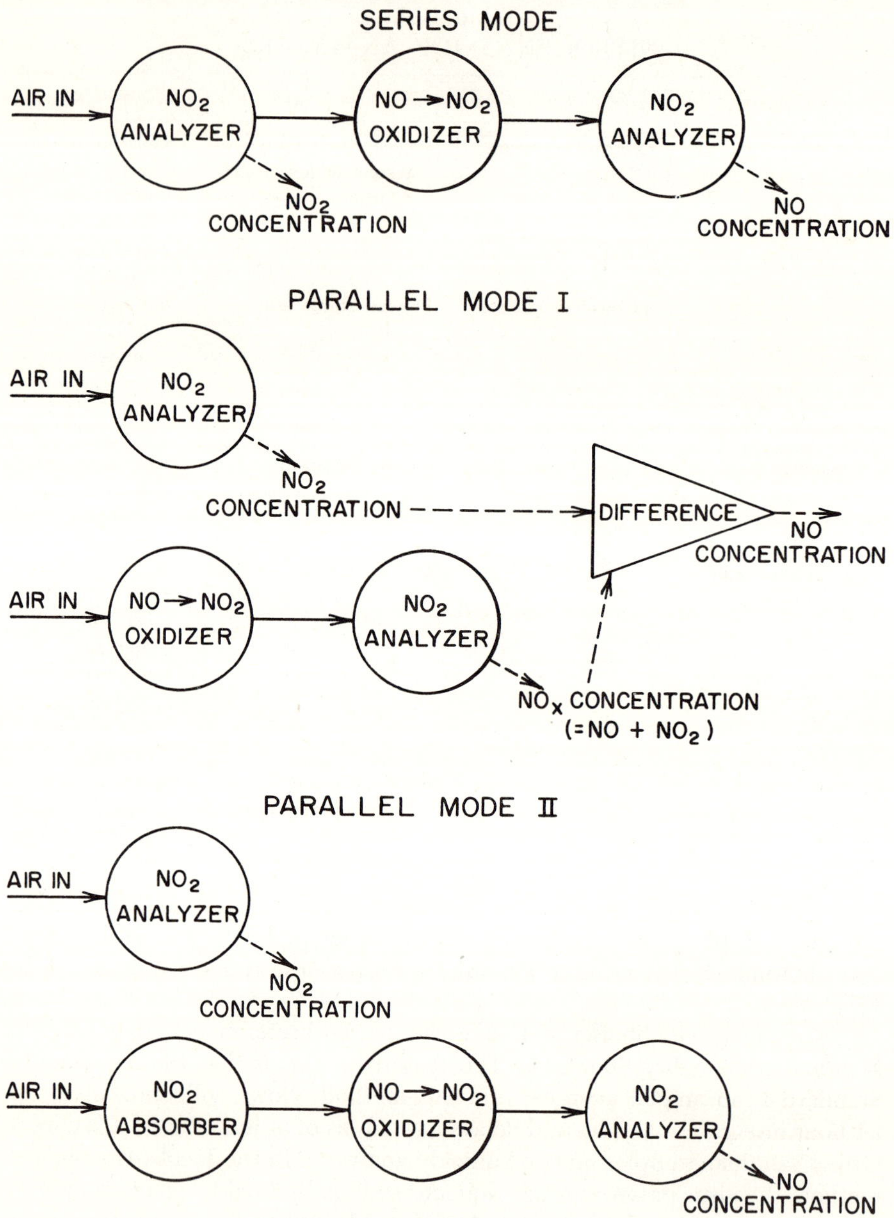

Figure 4.12b Nitrogen oxides analyzer modes.

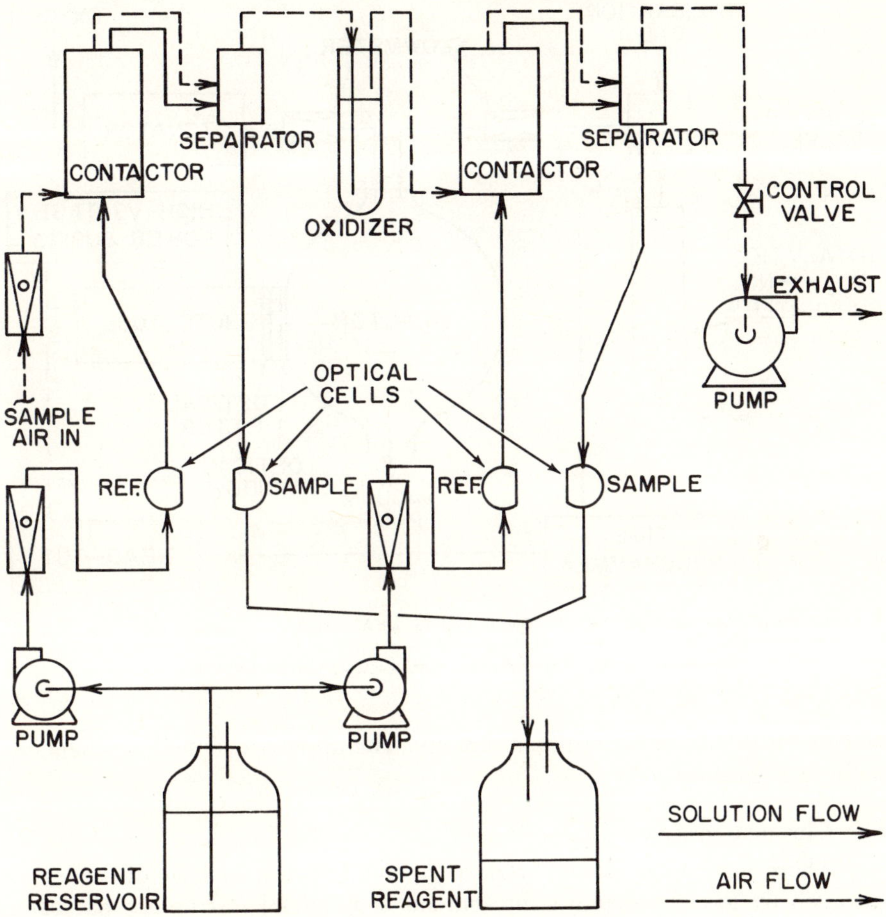

Figure 4.12c Colorimetric nitrogen oxides analyzer.

Chemiluminescent Determination

A recent development in nitrogen oxides[2] determination makes use of the fact that nitric oxide (NO) reacts with ozone, forming nitrogen dioxide (NO_2) with chemiluminescence:

$$NO + O_3 \longrightarrow NO_2 + hv \text{ (light } 0.6 - 3\,\mu) \qquad 4.12(1)$$

The light emitted can be measured photometrically as an indication of the extent of reaction. With excess ozone, the emission is proportional to the amount of nitric oxide. Instruments based on this operating principle are available commercially. For NO_2 determination (which reacts with ozone rather slowly), some analyzers include a catalytic converter to reduce NO_2

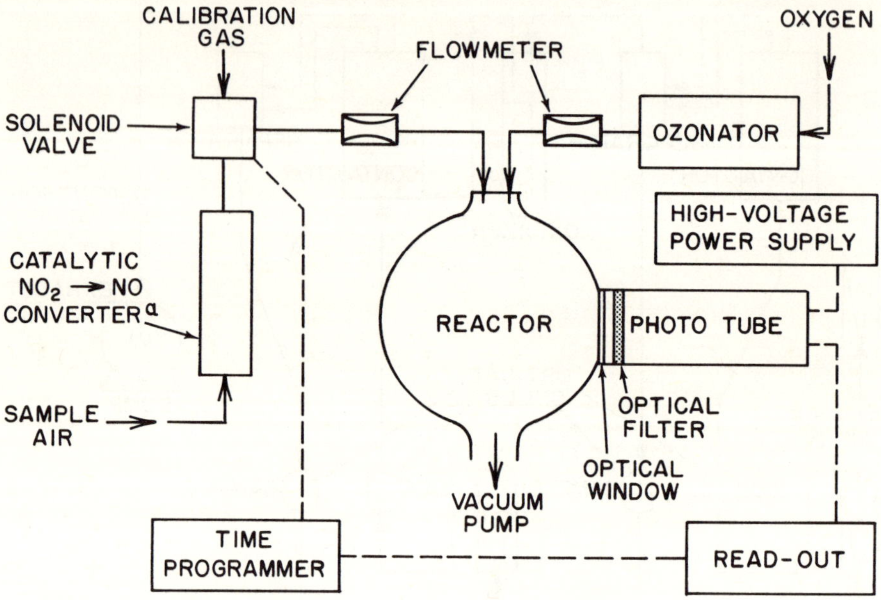

Figure 4.12d Chemiluminescence nitric oxide analyzer.

to NO which can be determined directly. The apparatus is shown schematically in Figure 4.12d.

Gas Chromatography

There have been limited reports of methods for analysis of oxides of nitrogen by gas chromatography, but no commercial instrument has been marketed. Such methods involve fairly short columns, as a rule, with electron capture detectors. These are the same types of instruments used for analysis of peroxyacyl nitrates (discussed in Section 3.14).

Assessment

The conventional method in widest use is colorimetric, based usually on the Griess-Saltzman reagent (a diazotization method). This is a fairly precise and dependable method but requires a great deal of attention. The colorimetric method is specific for nitrogen dioxide. In order to analyze for nitric oxide, an oxidation step is required. Of various oxidizer columns, the most commonly used is permanganate, but none are completely satisfactory. It is important to calibrate over a range of nitric oxide concentrations and at various humidity levels. A newer technique which is probably less demanding of attention is the chemiluminescent method utilizing the reaction between ozone and nitric oxide. This is a dry gas method requiring an ozone generator and compressed gas. In the chemiluminescent method, a catalytic reduction of nitrogen dioxide to nitric oxide is required.

REFERENCES

1. "Air Quality for Nitrogen Oxides," U.S. Environmental Protection Agency, Air Pollution Control Office Publication No. AP-84, Ch. 5, 1971.
2. Fontijn, A., A. J. Sabadell and R. J. Ronco: *Anal. Chem.*, vol. 42, pp. 575–579, 1970.

4.13 ODOR MEASUREMENT

Methods of Detection:	(a) Organoleptic; (b) Chemical/instrumental.
Sensitivity of Detection:	0.01 ppm by instrumental methods, 0.0002 ppm by the human olfactory system.
Partial List of Odor Detecting Chromatograph Suppliers:	Analytical Instrument Development Inc.; Beckman Instruments, Inc.; Carle Instruments, Inc.; Gow-Mac Instrument Co.; Hewlett-Packard Co., Hewlett-Packard, Avondale Div.; Leeds & Northrup Co.; Mine Safety Appliances Company; Perkin-Elmer Corp.; Tracor Inc.; Varian Aerograph; WEMCO Instrumentation Co.

This section compares organoleptic and chemical/instrumental methods for odor measurement and indicates the general shortcomings of the chemical/instrumental methods when compared to the reference organoleptic method.

The organoleptic methods, which utilize the human olfactory system, are completely subjective. However, there are techniques available that can convert subjective measurements into useful objective results. Chemical/instrumental methods generally suffer from two major shortcomings:

1. Sensitivity. The human olfactory system is generally three orders of magnitude more sensitive than currently available chemical/instrumental methods. It can detect and identify odors present in quantities to which commercially available instrumentation and chemical methods are completely insensitive.
2. Flexibility. The human olfactory system is capable of detecting and identifying a wide variety of chemical structures and giving different responses to different materials. Commercially available instrumentation and chemical methods are generally restricted

to particular chemical structures and give a similar response to all compounds with that structure.

The most successful instrumental methods for the measurement of odor are those using a gas chromatograph. A gas chromatography method has been developed using a flame photometric detector which can determine the concentrations of sulfur dioxide, hydrogen sulfide and other odorous gases produced by kraft paper mills down to about 0.01 ppm.[1] Success in detecting these sulfur-containing compounds using a coulometric cell with platinum electrodes as the detector has also been reported.[2] Levels as low as 0.1 ppm were detected by this method. The threshold levels detected by the human nose[3] are as low as 0.00021 ppm (trimethylamine). Thus, it is apparent that except for special situations, the human nose, with its attendant sociological, psychological and physiological complications, is the only odor sensor available for general odor measurement.

The Human Olfactory System

The human olfactory system actually involves more than the nose. As a stream of air is drawn in through the nostrils, it is warmed and filtered by passing over the three baffle-shaped turbinate bones in the upper part of the nose (Fig. 4.13a). Some of the air swirls past the olfactory receptors located

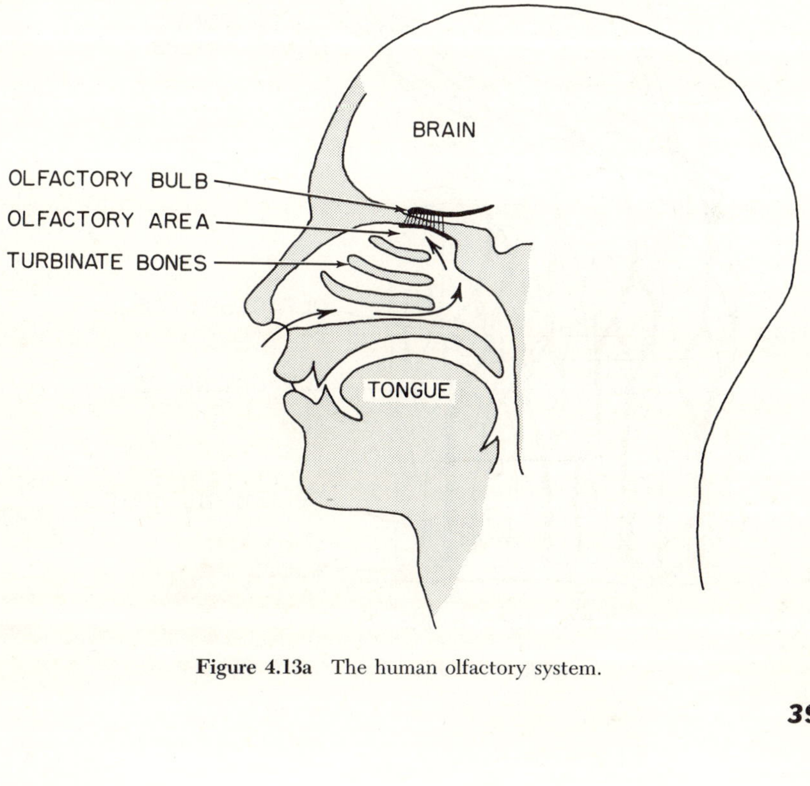

Figure 4.13a The human olfactory system.

high up in the nasal passages just below the brain. These odor receptors consist of hair-like filaments attached to the end of the fibers of the olfactory nerve and the trigeminal nerve endings (Fig. 4.13b). On being stimulated by odorous materials, these receptors send signals to the olfactory bulb, where they are relayed to higher centers of the brain. In these higher centers, the signals are integrated and interpreted in terms of the character and intensity of the odor.

Because of the extreme sensitivity of the human nose, the concentrations of odorants in air can be extremely low, yet give a strong response. These low concentrations present a problem in the manner by which samples are presented to the human sensor. Improper handling of an ambient air sample or improper preparation of an odorant standard can result in erroneous results because of adsorption of the odorants on the walls, incomplete mixing of an odorant with dilution air or impure dilution air. Errors can also result from the type of system used to bring the sample in contact with the olfactory sensors. The use of syringes or other means in which the nose is not immersed in the sample yields lower apparent intensities than are obtained by fully

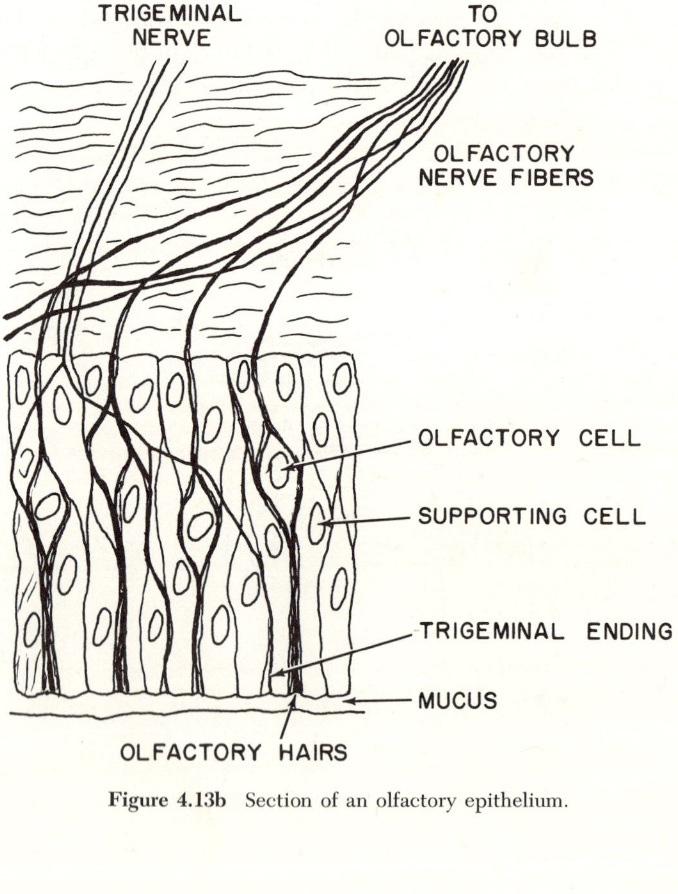

Figure 4.13b Section of an olfactory epithelium.

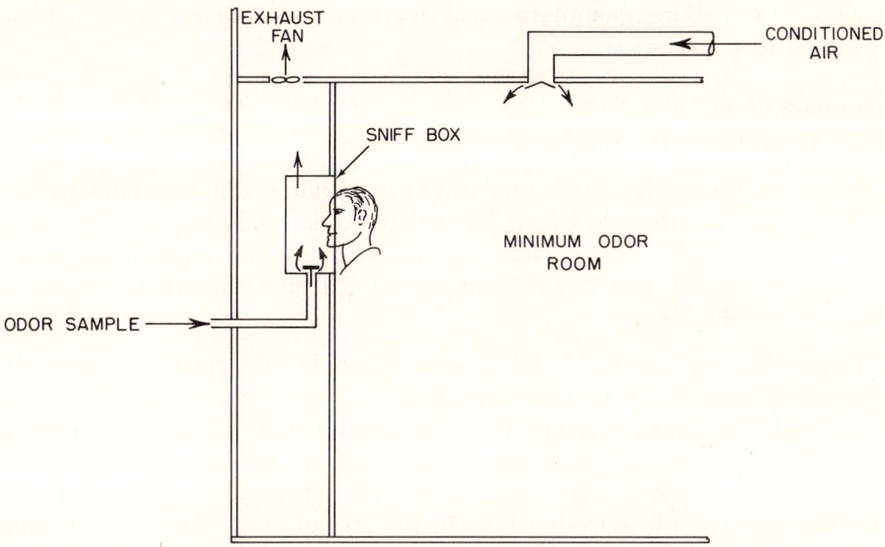

Figure 4.13c Diagram of sniff box odor test system.

exposing the nose.[4] The use of an odor room can also yield misleading results because of natural odors generated by the body and adsorption by clothing. The best approach appears to be the "sniff-box" shown in Figure 4.13c. The human sensor is kept in an air conditioned room into which charcoal filtered air is blown to maintain a slight positive pressure. This prevents the sample in the sniff-box and other contaminated air from entering the room. For measurement, the whole face is placed in the sample stream within the sniff-box.

Odor Panels

The use of humans as measurement instruments introduces a variability that is difficult to control. This variability is due to moods, biases and other vagaries associated with this complicated "sensor." Each human being is a unique creation, with differences in intelligence, persistance, sensitivity, experience and interests. These differences make the response of an individual an unreliable indicator in itself. The use of a panel of individuals for the measurement of odor levels statistically eliminates the unreliability of the individual.

An odor panel generally consists of 5 to 10 people. The people who compose the odor panel need not have unusual olfactory abilities. However, they should be able to distinguish among odors of different intensities, discriminate among different odor qualities and communicate the perceived sensations in terms of reference standards. They should also be emotionally receptive to making quantitative and discriminatory judgments without expressing their preference.

Training of an Odor Panel

The primary tasks of an odor panel are to:

1. Judge the relative intensity of an odor at different dilutions.
2. Discriminate among the different odor qualities.
3. Combine the intensity and quality of an odor to give a composite profile that can be communicated to the scientist in charge of the test.

The panelists are expected to follow instructions, but also render independent judgments reflecting their own sensations.

With these tasks in mind, three tests can be used to train the panelists, in addition to exposing them to various levels of odors similar to those to be measured. These are the triangle test, the intensity rating test and the multicomponent identification test. In the triangle test, three samples are presented to the panelist at the same time. Two are identical, the third is different in either intensity or quality. The panelist must select the odd sample.

The intensity rating test consists of a series of perhaps 20 dilutions of an odorant in an odorless medium. One sample is removed from the series and the panelist is asked to determine the sample's proper place in the series according to its odor intensity.

The last test is the multicomponent odor identification test in which three mixtures are presented to the panelist. These mixtures contain, in sequence, two, three and four odors out of a possible total of eight known standards. The panelist is told how many components to look for and is asked to identify each of them.

This group of three tests develops a panelist's ability to distinguish between different odor intensities, discriminate among odor qualities, and communicate their sensations in terms of predetermined standards. After some experience working as a group under a competent test director in actual measurement situations, highly accurate measurements of odor levels can be made.

REFERENCES

1. Stevens, R. K., A. E. O'Keefe, J. D. Mulick and K. J. Krost: "Gas Chromatography of Reactive Sulfur Gases in Air at the Parts Per Billion Level, 1. Direct Chromatographic Analysis," National Air Pollution Control Administration, Cincinnati, Ohio, 1969.
2. Applebury, T. E. and M. J. Schaur: "Analysis of Kraft Pulp Mill Gases by Process Gas Chromatography," Dept. of Chem. Eng., Montana State Univ., Bozeman, Montana, 1968.
3. Leonardos, G.: "The Profile Approach to Odor Measurement, Proceedings of Mid-Atlantic States Section, Air Pollution Control Association Semiannual Technical Conference on Odors:

Their Detection, Measurement and Control," Rutgers Univ., New Brunswick, New Jersey, May 13, 1970.
4. Reckner, L. R. and R. E. Squires: "Diesel Exhaust Odor Measurement Using Human Panels," SAE Paper 680444, 1968.

BIBLIOGRAPHY

Amoore, J. E., J. W. Johnston, Jr. and M. Rubin: The Stereochemical Theory of Odor, *Scientific American*, vol. 210, no. 2, p. 42, 1964.

Sullivan, R. J.: "Preliminary Air Pollution Survey of Odorous Compounds," A literature review prepared under contract no. PH 22-68-25, National Air Pollution Control Administration Publication No. APTD 69-42, 1969.

4.14 OZONE SENSORS

Methods of Detection:	(a) Chemiluminescence; (b) Ultraviolet Spectroscopy; (c) Coulometric; (d) Thermal Conductivity; (e) Colorimetric.
	NOTE: In the feature summary below, the letters "a" through "e" refer to the listed detection methods.
Reference Method:	(a), ethylene reaction of gas phase O_3, calibrated against neutral KI method.
Sample Flowrate lpm:	0.14 (c), 0.33 (d), 1.0 (a,b), 7.0 (b).
Interferences:	Oxidizing agents interfere with "c."
Ranges:	0.9 to 30 ppt (d), 1 ppb to 10 ppm (a), 0.01 to 1 ppm (c), 0.01 to 10 ppm (b).
Sensitivity:	1 ppb (a), 0.01 ppm (c), 0.1 ppm (b).
Reproducibility or Accuracy:	2% (a,d), 2 to 3% (b), 4 to 5% (c).
Cost of Sensor Without Recorder:	\$800 to \$3,000 (c,d); \$2,000 to \$4,200 (b): \$3,500 to \$4,500 (a).
Partial List of Suppliers:	Atlas Electric Devices Company (c); Beckman Instruments, Inc. (c,e); Bendix Corp. (a); Dasibi Corp. (b); Mast Development Co. (c); McMillan Electronics Corp. (a); Meloy Laboratories, Inc. (a); Monitor Labs Inc. (e); Ozone Research (c); Pollution Monitors Inc. (a); Raditron, Inc. (e); Rem Inc. (a); Spectrometrics of Florida Inc. (b); Technicon Corp. (e); Triangle Instrument Co. (a); Welsbach Corp. (d).

Ozone is an unstable triatomic allotrope of oxygen which naturally occurs in the atmosphere below 0.1 ppm concentration at sea level and at substantially higher concentrations at high altitudes.[1] It is produced by electric storms and irradiation of atmospheric oxygen by sunlight. Pollutants present in the atmosphere can increase ozone production leading to dangerously high ozone concentrations. Ozone concentrations close to 1 ppm have been recorded in Los Angeles on muggy days. Ozone is a powerful oxidizing agent if present in high concentrations, and attacks the pulmonary tracts.

Ozone is also used to control pollutants in both air and water. Its most desirable features are that: (1) it is capable of destroying most pollutants by oxidation; (2) the oxidation products are usually nontoxic and biodegradable; (3) the unreacted excess ozone decomposes to oxygen leaving no undesirable residue.

This section reviews those detectors which measure the ozone concentration in air or gas. For the measurement of ozone in water, see Section I.4.18. Gas phase ozone detectors are used to monitor atmospheric pollution and to detect leaks in ozone generator installations.

The highly specific and very sensitive chemiluminescence detectors were introduced in the last few years. The operating principles of the other sensors have been used for other measurements and the detectors have been on the market for many years.

Chemiluminescence Ozone Detectors

This measurement is based on the chemiluminescent reaction between ozone and ethylene. The air sample and ethylene are continuously drawn into a reaction chamber by a pump. In this chamber, a photomultiplier is mounted on a flat window. The chemiluminescent light emission measured by the photomultiplier tube is proportional to the ozone concentration in the sample. The signal from the photomultiplier is amplified and displayed on a meter or recorded.

The photomultiplier tube assembly and reaction chamber are enclosed in a thermoelectrically cooled housing. A low (about 10°C) constant temperature is maintained to avoid long-term drift.

These detectors are designed for long-term, unattended operation in weather and pollution monitoring stations. The ethylene consumption is low: 15 to 20 cc per minute. The ozone measurement is specific. Oxidizing gases or other usual components of the atmosphere do not interfere with it. Chemiluminescence sensors are the most sensitive ozone detectors available.

Ultraviolet Ozone Detectors

These instruments take advantage of the strong ultraviolet absorption of ozone at the wavelength of 254 nm.

The drawn air sample is divided into two portions. From one portion, the ozone is removed by a manganese dioxide (MnO_2) catalyst filter. Then the two portions are alternately introduced into an air absorption chamber (Fig.

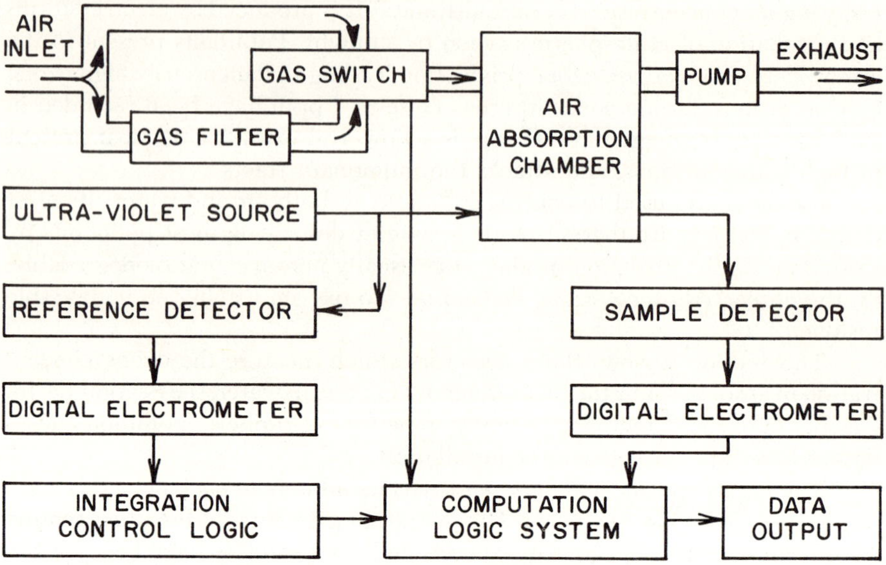

Figure 4.14a Ultraviolet ozone detector.

4.14a). A UV light beam is directed through the absorption chamber into a detector. The difference between the signals from the ozone-free filtered and unfiltered air samples is displayed. The unit is calibrated for ozone on the basis of Beer's law.[2]

These detectors are somewhat less sensitive than the chemiluminescent type. They measure ozone specifically and the normally occurring components in the atmosphere do not interfere with the measurement.

Coulometric Ozone Detectors

Ozone can generate iodine from KI according to the reaction in Equation 4.14(1) and the iodine can be measured coulometrically.

$$O_3 + 2\,KI + H_2O \longrightarrow O_2 + 2\,KOH + 2\,I \qquad 4.14(1)$$

The air sample is passed through a sensing cell over a thin film of potassium iodide solution which is continuously pumped through the cell (Fig. 4.14b). The solution film containing the freshly produced iodine is passed over some electrodes. The current produced at the electrode is proportional to the iodine content and ultimately to the ozone concentration.

These detectors are not specific to ozone because they measure total oxidants. There are filters available to decrease interference by other oxidizing agents. Their most satisfactory application is to measure total oxidants, or ozone when other oxidizing agents are absent.

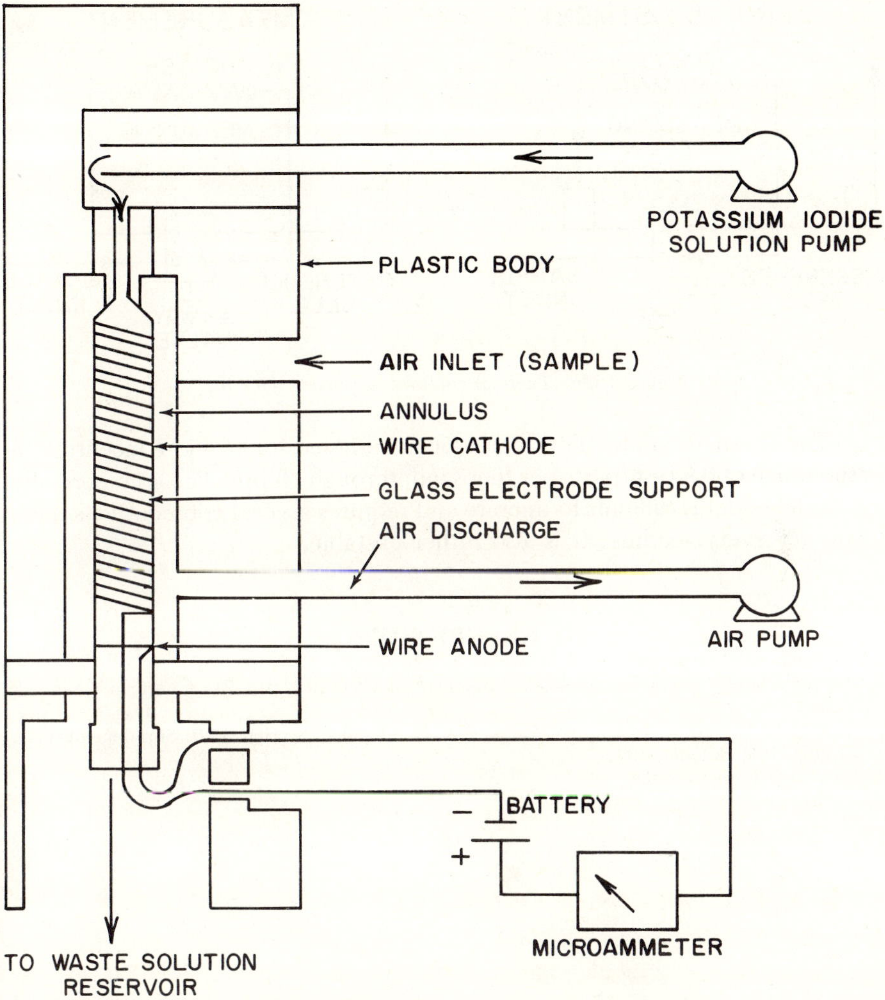

Figure 4.14b Coulometric oxidant (ozone) analyzer.

Thermal Conductivity Ozone Detector

The measurements of thermal conductivity can be correlated with the type and concentration of the gas sample.[3]

This ozone detector functions as indicated in Figure 4.14c. The gas sample enters the meter and passes through the first thermal conductivity cell and then through a heater that decomposes the ozone. The ozone-free gas is then introduced into the second thermal conductivity cell. The resistances of the two cells are balanced against each other by the Wheatstone bridge and the differential is measured and calibrated in terms of ozone concentration.

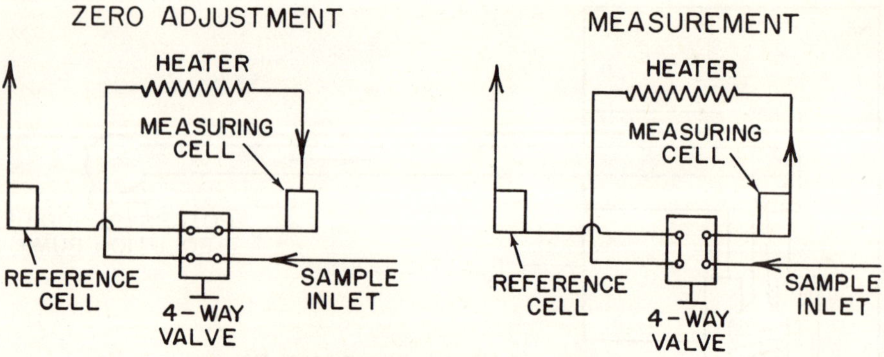

Figure 4.14c Thermal conductivity ozone detector.

The thermal conductivity detector is designed for ozone concentration in the range of 0.9 ppt (parts per thousand) through 30 ppt. The commercially available model is difficult to operate and requires several consecutive adjustments for every reading. It is also rather unstable.

REFERENCES

1. "Ozone Chemistry and Technology," *Advances in Chemistry,* no. 21, American Chemical Society.
2. Lipták, B. G. (ed.): "Instrument Engineers' Handbook," Section 1.10, Supplement One, Chilton, Philadelphia, 1972.
3. Ibid., Vol. I., Section 8.5.

4.15 PARTICULATE MONITORS AND SMOKE OR DUST DENSITY SENSORS

Methods of Detection:	(a) High-volume sampler; (b) Light scatter; (c) Piezoelectric crystal; (d) Tape sampler; (e) Light attenuation; (f) Impaction devices.
Reference Method:	(a).
Sample Flowrate:	Variable with instrument; from 1 ALPM to 50 ACFM.
Accuracy:	For high-volume sampler ±3%; others less accurate.
Ranges:	From 1 $\mu g/M^3$ to 1 g/M^3.
Responses:	Instantaneous for methods "b," "c" and "e." One to twenty-four hours for designs "a," "d" and "f."
Cost:	Designs "a" and "f" \$100 to \$400. Designs "d" and "e" \$400 to \$1,500. Designs "b" and "c" \$2,000 to \$5,000.
Partial List of Suppliers:	Bendix Environmental Science Div., Bendix Corp. (a, d, f); Curtin Scientific Co. (a); General Metal Works, Inc. (a); Mine Safety Appliances Company (a, f); Misco Scientific (a); Precision Scientific, Subsidiary of GCA Corp. (a, d); Billings and Gussman, Inc. (a); Radeco, Inc. (a); The Staplex Co. (a); Weather Measure, Div. of KMS Industries (a); Research Appliance Co. (a, d); Environmental Research Corp. (d); Leigh Systems, Inc. (d); Meteorology Research, Inc. (b); Thermo-Systems, Inc. (c); Atlantic Research Co., A Div. of the

Susquehanna Corp. (c); 2000, Inc. (f); Monsanto Enviro-Chem Systems (f); Intertech Corp. (e); Bailey Meter Co., A Subsidiary of Babcock & Wilcox Co. (e).

Particulate matter is the world's most ubiquitous air pollutant. It is generated naturally by fires, volcanic action, vegetation and wind (Section 3.15). Man-made activity contributes the major portion of the particulate burden in the atmosphere.

Gravitational effects are a major factor in removing dust above a particle size of about 10 μ. For these larger particles, settling to the ground occurs very close to their point of origin. Smaller particles, particularly those below about 2 μ, are predominantly subject to wind forces and not to gravity. They remain suspended in the air for long periods and are carried to and fro by the wind.

Any measurement technique for particulates must be selected with an understanding of the purpose of the measurement and the particle size selectivity of the sampler because of the difference in physical travel of different sizes of particles.

High-volume Sampler

The basic instrument for determining the airborne concentration of particulate matter in the United States is the high-volume sampler (Fig. 4.15a). A high-speed, multistage blower is used as the suction source. An adapter section allows the use of a flat 8 $\times$ 10-inch glass fiber filter. This filter has an efficiency of well over 99% for 0.3 μ particles. An orifice plate on the back of the blower allows the use of either manual or recording type static pressure gauges to monitor the airflow rate through the sampler. Twenty-four hours is a common sampling time, during which period the buildup of particulates increases the filter resistance and decreases the airflow rate. When the decrease is small, e.g., less than 10%, the initial and final readings are averaged to determine the sampling rate. If the decrease is significant, as might be expected in locations where the particulate concentration is high, it is desirable to use a pressure recorder on the instrument and determine the flowrate from the chart record.

The high-volume sampler provides an actual mass concentration of particulate matter in the air. The roof of the standardized sampling shelter has an annular opening sized so that particles larger than a nominal size of 100 μ are not drawn into the sampler. Varying airflow rate and the effect of winds make this method an approximation of a particle size cut at best; but it can furnish a large body of ambient air quality data, and air quality standards have been promulgated based on the use of this standardized instrument.

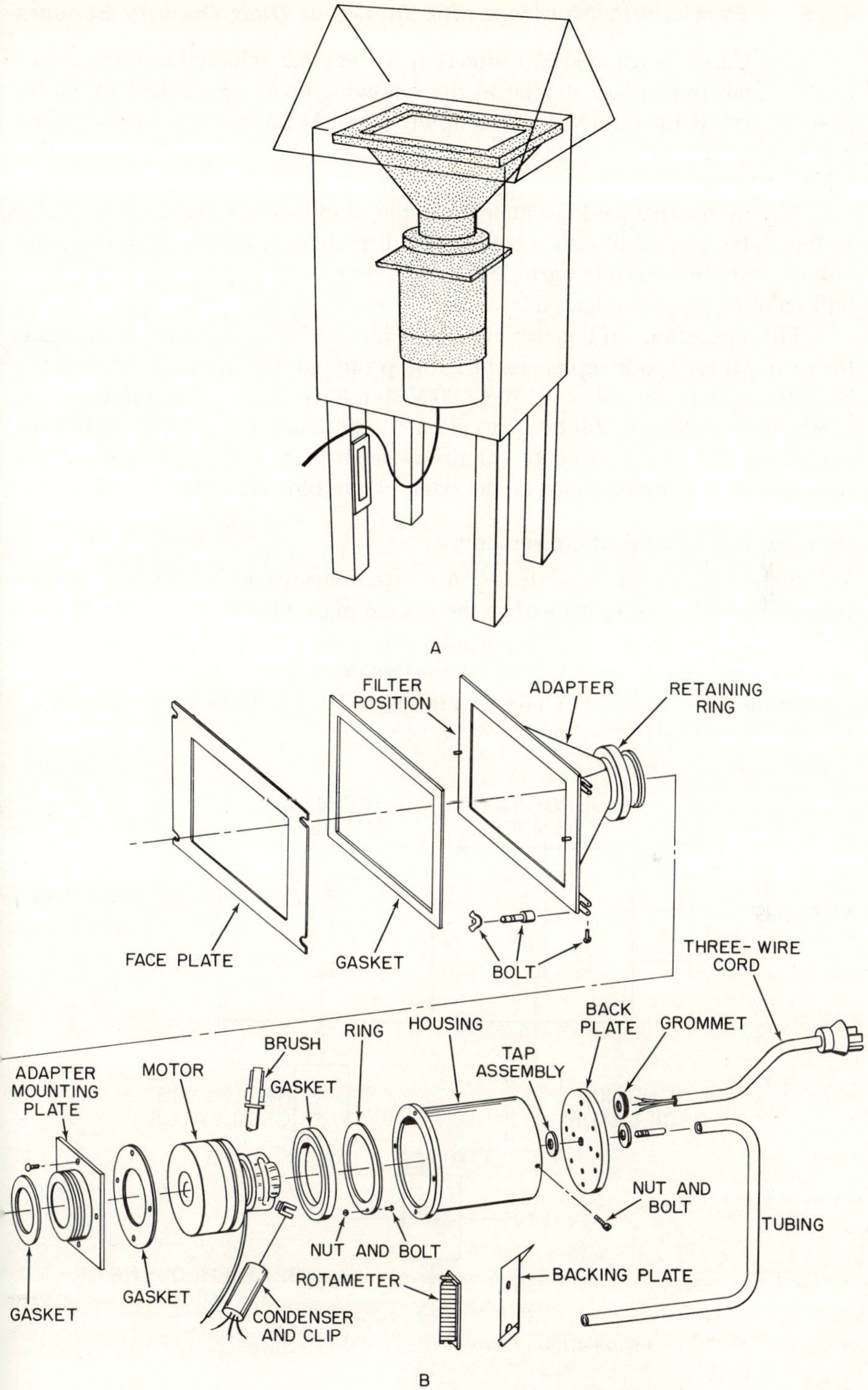

Figure 4.15a High-volume sampler. (A) Assembled sampler and shelter; (B) Exploded view of typical high-volume air sampler.

The U.S. Environmental Protection Agency has selected as ultimate air quality goals (secondary standards) the following levels of particulate matter as measured by the high-volume sampler: 60 $\mu g/M^3$, annual geometric mean; and 150 $\mu g/M^3$, maximum 24-hour concentration not to be exceeded more than once per year.[1]

Any more advanced instrumentation used to measure particulate matter in the ambient air will either have to be directly comparable with the high-volume sampler's unique particle size selectivity or new numerical standards will have to be promulgated.

The operation and maintenance of high-volume samplers is straightforward. Motor brush replacement is the principal maintenance item. It has been found that the use of a 110 to 90 V stepdown transformer increases the brush life from about 500 hours to about 1,500 hours. Instrument calibration is performed by calibrating actual airflow against the static pressure reading obtained at the orifice plate at the back of the blower.

Piezoelectric Crystal Measurements

Some crystals, when excited by an alternating current, exhibit a frequency response which is proportional to the crystal mass. Quartz is commonly used and is readily available for this purpose. This principle is used to continuously monitor the mass concentration of particulate matter in air, using instrumentation shown schematically in Figure 4.15b. Particles are deposited on the sensing crystal but not on the reference crystal. A beat frequency develops

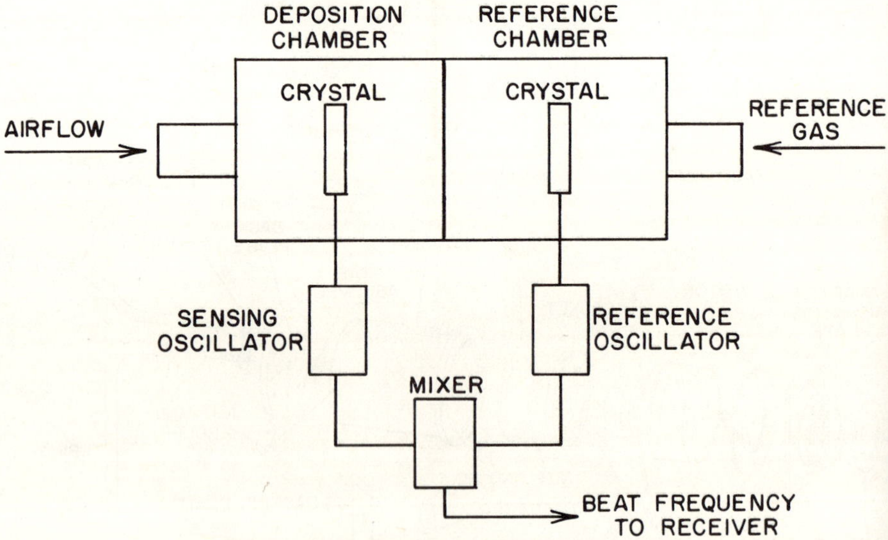

Figure 4.15b Piezoelectric crystal mass monitor.

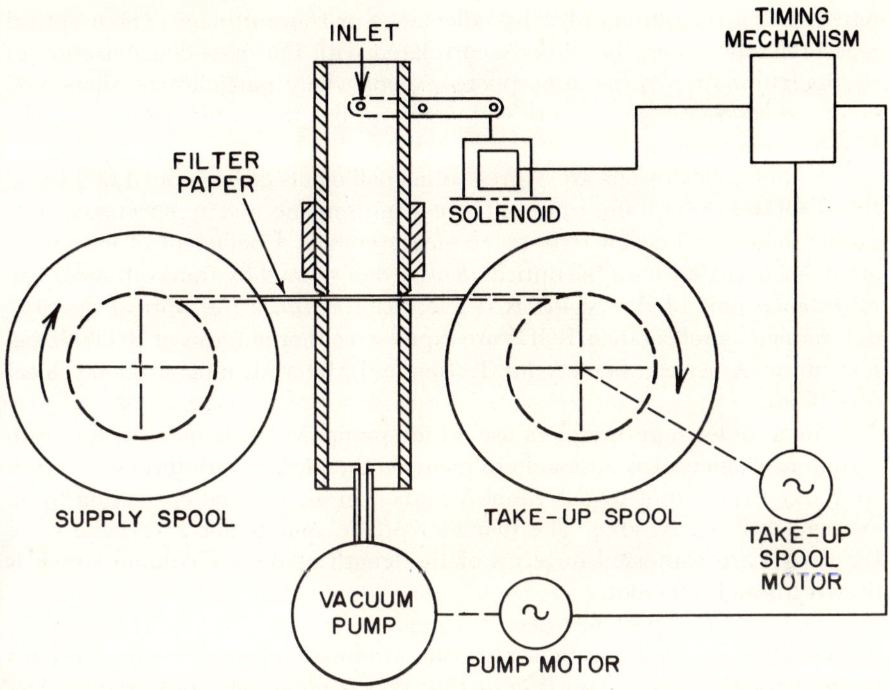

Figure 4.15c Tape sampler.

and the particle mass is determined based on the mass-frequency response of the crystal. Commercially available quartz crystals show a frequency shift of 2,000 Hz/μg of material deposited.

The major problems to be overcome before this instrument gains wide acceptance involve the deposition and retention of particles on the sensing crystal. Two techniques, impaction and electrostatic precipitation, have been used commercially. Both approaches have weaknesses in collecting and retaining particles over the wide-size range encountered in atmospheric aerosols.

The piezoelectric crystal technique could be applicable to both ambient and source concentrations of particulate matter, although the low flowrate (on the order of 0.5 ALPM) would be awkward when dealing with high-source concentrations because of deposition and resulting plugging of the sample handling system leading to the sensor.

Tape Sampler

The tape sampler (Fig. 4.15c) provides an indication of the soiling properties of the sampled air. Air is drawn through a paper tape at a rate of about 0.25 ACFM for periods of 1, 2 or more hours. Filter paper and membrane tapes have been used. The soiling index (optical properties) of the filtered particulate is measured either by reflectance or transmittance. These optical measurements cannot be directly correlated with the mass concentration of particulate matter in the atmosphere, except where particle size, shape and color are constant and a series of comparative tests are made using the high-volume sampler.

Soiling index results are expressed in coefficients of haze per 1,000 lineal feet (COHs/1,000 LF) for transmittance measurements and in reflectance units of dirt shade (RUDS) for reflectance measurements. Coefficient of haze units are defined as 100 times the optical density determined by transmittance. The reflectance unit of dirt shade is defined as 100 times the optical density* determined by reflectance. RUDS are expressed either in terms of 10,000 lineal feet of air (American Society for Testing and Materials unit) or 1,000 lineal feet of air.

Some older tape samplers use an air pump which is quite sensitive to system resistance. It is advisable to measure the airflow rate through a clean spot and a dirty spot to determine whether an air volume correction to an average rate is advisable. The diameter of the spot is not a variable since the results are expressed in terms of the length of the air column which is drawn through the spot.

The tape sampler has been adapted to quantitative measurement of particulate mass by using beta radiation attenuation to determine the mass of the deposit. This system (Fig. 4.15d) uses a membrane tape rather than paper tape because of the higher particulate collection efficiency and consistent mass of the tape. This concept shows promise for both ambient and source measurement.

Light Attenuation

The visible appearance of stack plumes has been used for many years by regulatory agencies to provide a quick indication of particulate emissions from sources. The earliest quantification of the technique was the use of the Ringelmann scale, a series of four visual comparison charts with 20, 40, 60 and 80% black shading on a white background. These are designated as Ringelmann values of 1, 2, 3 and 4, respectively (Section 1.4). This was originally prepared to judge the appearance of black plumes from coal-fired boilers. The concept has been extended to equivalent opacity, and observers have been trained to evaluate plumes of all colors. By the addition of pure white and solid black to the four shaded charts, a scale of 0 to 100% opacity (or Ringelmann values 0 to 5) is used.

Various refinements of the Ringelmann chart are based mainly on the use of photographic film or on the use of semitransparent devices to aid the

*Optical density = $\log_{10} (Io/I)$

where Io = light intensity transmitted through clean paper
I = light intensity transmitted through the sample

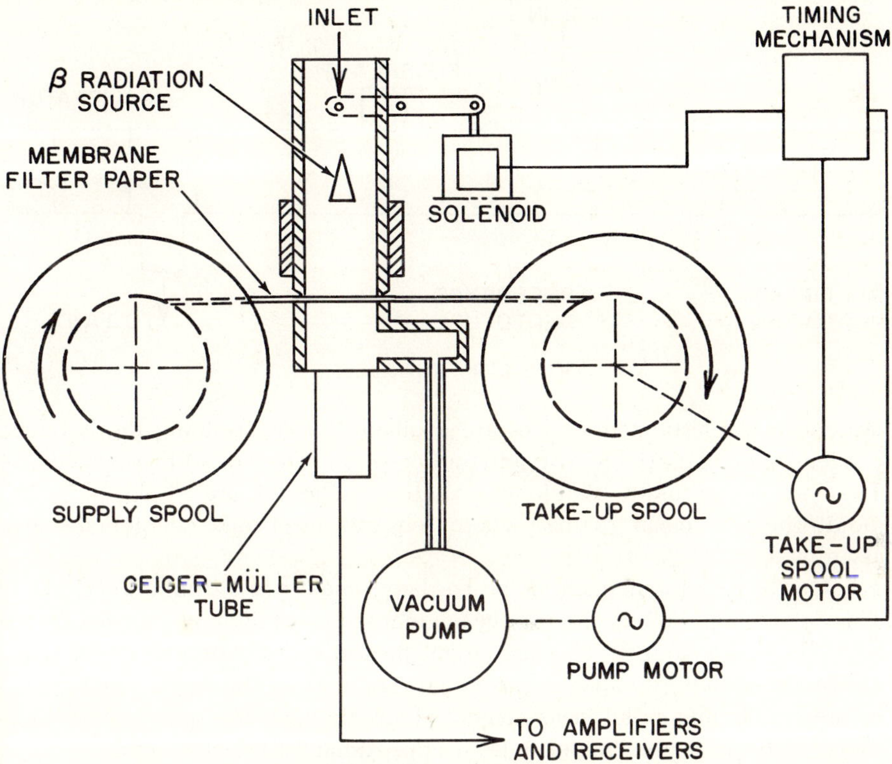

Figure 4.15d Beta gauge tape sampler.

observer, but there is no substitute for training. An experienced observer can consistently read plumes to ±0.5 Ringelmann units.

Instrumental methods for determining the optical properties of suspensions of particulates in air are simple in design but complex in principle.[2]

In its simplest form, a stack opacity meter consists of a light source on one side of the stack and a sensing cell on the other. Light attenuation is the principal measured parameter, although light scatter also has some effects. These devices are called transmissometers. Some commercial instruments based on this principle measure light energy in specific wavelengths.

Most stack measurements are made with transmissometers.

Light Scatter

Upon striking a particle, light scatters based on the size, shape and color of that particle. Light scatter can be measured as forward scatter, angular scatter or back scatter.

Suspensions of particulate matter in the air, scatter the light in all

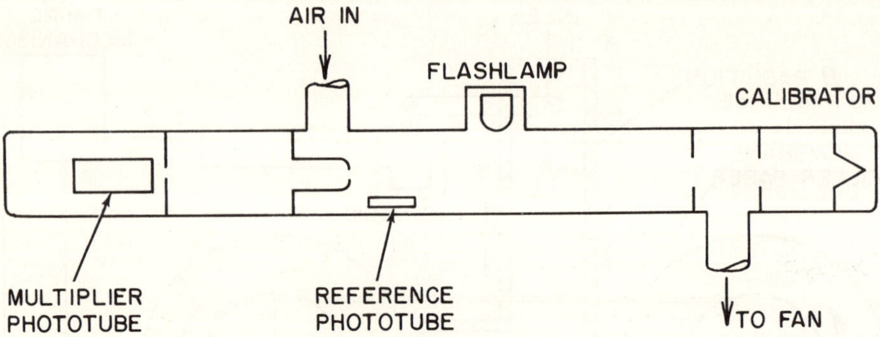

Figure 4.15e Nephelometer.

directions. Sophisticated devices are available for determining the number concentration of particles in air and for determining the size of these particles. These are not normally used in air pollution studies but are employed in the monitoring of "clean rooms" where sensitive electronic components are assembled.

One device, the nephelometer, has had some use in air pollution studies. This device (Fig. 4.15e) measures the angular scatter of light over a wide range in a flowing air sample. The instrument measures the scattering coefficient, which is inversely proportional to visual range. Thus, the nephelometer can be used to obtain a point measurement of visual range. The span of measured range with the nephelometer is from approximately 1 to 100 miles.

Impaction Devices

Determination of the particle size distribution in aerosols is frequently desired. While light scattering devices can provide an indication of particle size distribution, most atmospheric measurements are made using impaction devices. In principle, more inertial energy is required to impact small particles than large ones when they are impinged on a plane surface. This characteristic has been utilized through a series of impaction stages with succeeding stages operated at higher velocities (Fig. 4.15f).

The upper practical size limit for impaction devices is about 10 μ. They are particularly useful in the size range of 0.5 to 5 μ, which corresponds to the approximate size range of particulate matter in the ambient air (Fig. 4.17e).

Impaction devices can be used for evaluating stack emissions if the particulate concentration is low enough, or if adequate dilution is added to the sample stream, to allow sampling for a long enough time to obtain a representative sample.

In most cases, the impaction plate is removed from the sampling device and is weighed to determine the amount of sample. High plate tare weight is a potential source of inaccuracy.

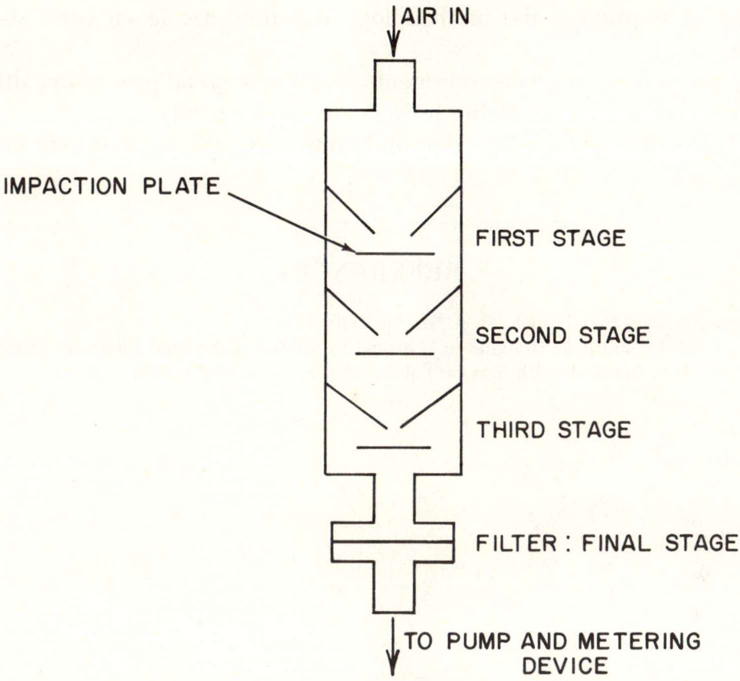

Figure 4.15f Schematic impaction device.

Calibration of impaction devices is tedious but necessary. The normal practice involves the generation of a mono-disperse aerosol of known particle size. Polystyrene spheres, as manufactured for latex paint, have been used. Assuming that the dispersion technique generates a mono-disperse aerosol (often the most difficult part of the procedure), the calibration is performed using a series of sizes. The calibration is in terms of aerodynamic size, i.e., the inertial properties which are a combination of particle size and specific gravity. The results from impaction sampling must state the specific gravity, whether measured or assumed, with the stated size distribution.

Summary

The measurement of particulate matter, either in the ambient air or in stacks, is in a much earlier stage of instrumental development than are the measurements of most gaseous air pollutants.

The high-volume sampler, the standard instrument for measuring the particulate burden in the ambient air, has an ill-defined upper limit on the particle size drawn into the sampler. It is basically a manual rather than an instrumental method of measurement. More sophisticated devices such as the beta gauge tape sampler and the piezoelectric crystal mass monitor hold the promise of improving the methodology but they are in an early stage of development.

Quantitative source measurements are still a manual procedure, although some instruments are available to indicate stack opacity.

There is a major need for new and improved instrumentation to measure particulates.

REFERENCES

1. *Federal Register*, Part II, vol. 36, p. 84, April 30, 1971.
2. Conner, W. D. and J. R. Hodkinson: "Optical Properties and Visual Effects of Smoke-Stack Plumes," U.S. Public Health Service Publication No. 999-AP-30, 1967.

4.16 PESTICIDE AND HERBICIDE SENSORS

Sampling System and Detector Components:	(a) Quantitative pesticide sampler; (b) Bacteriological filters for down to 0.2 μ; (c) Impactors for separating 1 to 40 μ size particles into four to six stages; (d) Scrubber with 250 ml volume and 100 ml solvent; (e) Vacuum pump; (f) Flowmeter; (g) Gas chromatograph with electron capture, flame photometric or thermoionic detector.
Filter Types:	Glass fiber for over 1 μ particles at 1,000 LPM sampling rates or bacteriological filters for particles down to 0.2 μ at 10 LPM sampling rates.
Costs:	For sampling, \$300 to \$3,000; for analysis (gas chromatograph and recorder), \$3,500 to \$15,000.
Partial List of Suppliers:	A. A. Andersen (c); Chemical Project Ltd. (a); Fisher Scientific Company (d); Gelman Instrument Co. (e); Kontes Glass Co. (f); Millipore Corp. (b); Misco Scientific (a); Varian Aerograph (g).

Any kind of air sampling equipment may be used (Sections 4.15 and 4.17) to sample airborne pesticides. Care must be taken, however, to ensure that nonpreferential samples are obtained, because pesticides exist in the air both as vapors and as particulates of varying sizes.

Air Sampling

A filter and solvent-scrubber combination is used to quantitatively collect pesticides from the air. A flow diagram of a commercially available sampler

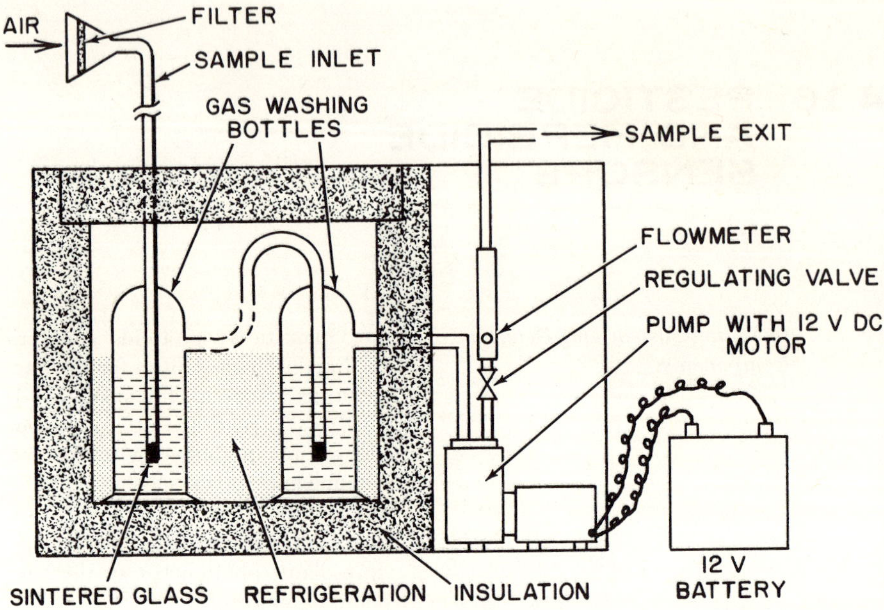

Figure 4.16a Airborne pesticides sampler.

is illustrated in Figure 4.16a. Bacteriological membrane filters are commonly used because they are capable of retaining particles as small as 0.2 μ, whereas glass fiber filters are useful only for particles greater than 1 μ. Other filter materials available are Florisil, silica gel and alumina.

A solvent-scrubber equipped with a sintered glass inlet and containing 100 ml of solvent, can handle a 12 LPM airflow rate. The sintered glass inlet ensures maximum exposure of the sample to the solvent. In some conditions, extra scrubbers or other traps may be required to ensure retention of all the vapor fractions.

At ambient temperatures, the solvent choice is somewhat limited: ethylene glycol (bp 198°C), hexylene glycol (bp 190 to 194°C) and dimethylformamide (bp 153°C) have been used. Common solvents, such as n-hexane and benzene are useful if an adequate cooling system is available. Aqueous $NaHCO_3$ solution is suitable for free acid forms of herbicides.

Until recently, attention was focused on organochlorine insecticides almost exclusively. Because of the ease of operation, large volume (up to 1,200 liters/min) sampling, utilizing a glass fiber filter, has been used to collect particulate pesticides only. DDT was collected by the filter as fine dust or adsorbed to dust particles in almost all air samples investigated. The efficiency of collection is largely dependent on the fineness of the filter. Air volume must be adjusted when a scrubber is employed to collect pesticides as vapors.

When the source of pesticides is known, the concentrations found are usually higher than when the source is unknown. Accordingly, the use of impactors (or impingers) capable of separating particles of 1 to 40 μ into four to six stages, when combined with a scrubber, are exceptionally useful to study the distribution and drift potential of pesticides. In some applications, paper or treated surfaces are used to collect large airborne particulates and are convenient because of their simplicity.

Limitations and Difficulties

Sampling sites must be chosen at different topographical locations to obtain a realistic picture of the pesticide distribution in the air, because concentrations of pesticides in the air vary widely. When the source of the pesticide is known, local meterological conditions, such as wind velocity and direction, temperature and humidity, become critical in evaluating pesticide movement and distribution. Sampling sites must be chosen at different heights and varying directions and distances from the source. In practice, it is difficult to satisfy all the requirements because resources are usually limited.

Pesticides in the particulate form are usually collected at each sampling site using a filter or impingement surface. However, particulates remaining on the filter will vaporize to some degree and over an extended collection period could give a misleading indication of particulate concentrations. The vaporized pesticides will be trapped by the solvent scrubber giving a higher than actual vapor fraction. The orientation and size of the air intake can also alter the particulate to vapor ratio owing to particulate trajectories at the intake. Air sampling devices using air ingestion may misrepresent particulate concentrations.

Analysis

For cleanup, the routine methods used in pesticide residue analyses are adequate. Analysis of collected samples is usually made by gas–liquid chromatography[1] because of the high sensitivity and wide availability of this equipment. However, it is recommended that confirmatory tests be carried out (when the source of pesticides is unknown) by additional means, such as thin-layer chromatography, mass spectrometry and chemical analysis or by a combination of these methods.

REFERENCE

1. Lipták, B. G. (ed.): "Instrument Engineers' Handbook," Section 8.1, Vol. I, Chilton, Philadelphia, 1969.

4.17 SAMPLING OF AMBIENT AIR

Sampling Methods:	(a) *Gas and vapor sampling,* (a1) Collection in bags or containers, (a2) Absorption devices and bubblers, (a3) Impingers, (a4) Adsorption devices, (a5) Freeze-out sampling; (b) *Particulate matter sampling,* (b1) Filtration, (b1.1) Membrane filters, (b1.2) Fiberglass filters, (b2) Impingers, (b3) Impactors, (b3.1) Cascade impactors, (b4) Electrostatic precipitation, (b5) Thermal precipitators.
Suitable Sampling Bag Materials:	Mylar,® Saran, Scotchpak,® Teflon.
Partial List of Suppliers:	Anderson Samplers (b3.1); Mine Safety Appliances Company (b4).

All substances in the ambient air exist as either particulate matter or gases or vapors. In general, the distinction is easily made; gases and vapors consist of substances dispersed as molecules in the atmosphere, while particulate matter consists of aggregates of molecules sufficiently large that they are said to behave like particles. Particulate matter, or particulates, are filterable, may be precipitated and, in still air, may be expected to settle out. By contrast, gases and vapors do not behave in this fashion and are homogeneously mixed with the air molecules. A substance, such as carbon monoxide, may exist only as a gas; or an inorganic compound like iron oxide may exist only as a particle. Many substances may exist as either particles or vapors, however; and additionally, substances which are gases may be attached by some means to the particulate matter in the air. If sampling is to be intelligently conducted, prior knowledge of the physical state in which a substance exists must be available or else a judgment must be made. Devices which are suitable for collecting particulate matter (Section 4.15) do not usually collect gases or vapors and, hence, an incorrect selection of sampling method may lead to erroneous results. Fortunately, a considerable body of experience has evolved concerning the more common pollutants, and it is not difficult in most cases to select a sampling method reasonably suitable for the substances of interest.

General Air Sampling Problems

Certain general observations relative to sampling ambient air must be recognized. For example, the quantity of a given substance contained in a volume of air is likely to be extremely small, and it will necessitate a sample of sufficient size for the analytical method subsequently employed to be adequate. Even air which is heavily polluted is not likely to contain more than a few milligrams per cubic meter of most contaminants; and more frequently, the amount present will best be measured in micrograms, or even nanograms, per cubic meter. Consider, for example, the current air quality standard for particulates: 75 $\mu g/M^3$. A cubic meter of air, or 35.3 cubic feet, is a relatively large volume for many sampling devices, and a considerable sampling period may be required to draw such a quantity of air through the sampler. If atmospheric mercury analyses (Section 4.11) are to be attempted, then it must be realized that background levels are likely to be as low as several nanograms per cubic meter and, in general, most substances tend to be of concern at quite low levels in the ambient air.

In addition to problems arising from the low concentrations of substances being sampled, other problems are caused by the reactivity of the substances, resulting in changes after collection and necessitating special measures to minimize such changes.

Whenever something is removed from a volume of air by sampling procedures, some alteration of the substance of interest may take place and analysis may be less informative than desired, or may even be misleading. It would be ideal, therefore, to perform analyses on the unchanged atmosphere if possible, using direct-reading devices which could give accurate information concerning the chemical and physical state of contaminants as well as concentration information. Such instruments do exist for some substances and many more are being developed, but conventional air sampling methods are still used in many instances and doubtlessly will continue to be required for some time to come.

Gas and Vapor Sampling

Collection in Containers or Bags

The simplest method of collecting a sample of air for subsequent analysis is to fill a bottle or other rigid container with it, or more conveniently to use a bag made of a suitable material. Although it may be relatively easy to sample by this method, the sample size is distinctly limited and it may not be possible to collect a sufficiently large sample to permit subsequent analysis. Bottles larger than several liters in capacity are awkward to transport; and while bags of any size are conveniently transported when empty, they may be difficult to deal with when inflated. Nevertheless, it may prove more convenient to collect a number of samples in small bags than to take more complex sampling apparatus to a number of sampling sites. If it is possible to analyze for the contaminant of interest by means of gas chromatographic procedures, or by gas phase infrared spectroscopy for example, then samples as small as a liter or less may be adequate and can be quickly and easily collected by means of bags.

There are several methods of filling a rigid container such as a bottle; one of them is to evacuate the bottle beforehand, then fill it at the sampling site by drawing air into the bottle and seal it again (Fig. 4.17a). Alternatively, a bottle may be filled with water, which is then allowed to drain and fill with the air. A third method consists of passing a sufficient amount of air through the bottle by means of a pumping device until the original air is completely displaced by the air being sampled. Plastic bags are frequently filled by means of a simple hand-operated squeeze bulb with valves on each end (Fig. 4.17a) and then connected to a piece of tubing attached to the sampling inlet of the bag. In most cases, this procedure is satisfactory; but care should be taken to avoid contamination of the sampled air by the sampling bulb or possible losses of the constituent on the walls of the sampling bulb. To avoid problems of this kind, it is possible to place the bag in a rigid container such as a box and then withdraw air from the box so that a negative pressure is created, which results in air being drawn into the bag.

Bag materials must be selected with care, for some will permit losses of contaminants by diffusion through the walls, and others may contribute contaminants to the air being sampled. A number of polymers have been studied, and several have been found to be suitable for many air sampling purposes. Materials suitable for use as sampling bags include Mylar, saran, Scotchpak (a laminate of polyethylene, aluminum foil and Teflon) and Teflon.

Even though the bag may be made of relatively inert materials, it is always possible that gas phase chemical reactions will occur, so that after a period of time the contents of the bag may not be identical in composition with the air originally sampled. Thus, a reactive gas like sulfur dioxide or nitric oxide can be expected to gradually oxidize, depending upon the storage temperature. It is generally advisable to perform analyses as soon after collecting the samples as possible. Losses by adsorption or diffusion will also tend to be greater with the passage of time and will occur to some extent even though the best available bag materials have been used.

The use of small bags may permit the collection of samples to be analyzed for a relatively stable gas such as carbon monoxide at a number of locations throughout a community of interest, thus permitting routine air quality measurements which might otherwise be inordinately expensive.

Absorption

Most air sampling for gases and vapors is performed by absorbing the contaminant of interest in a suitable sampling medium. Ordinarily, this me-

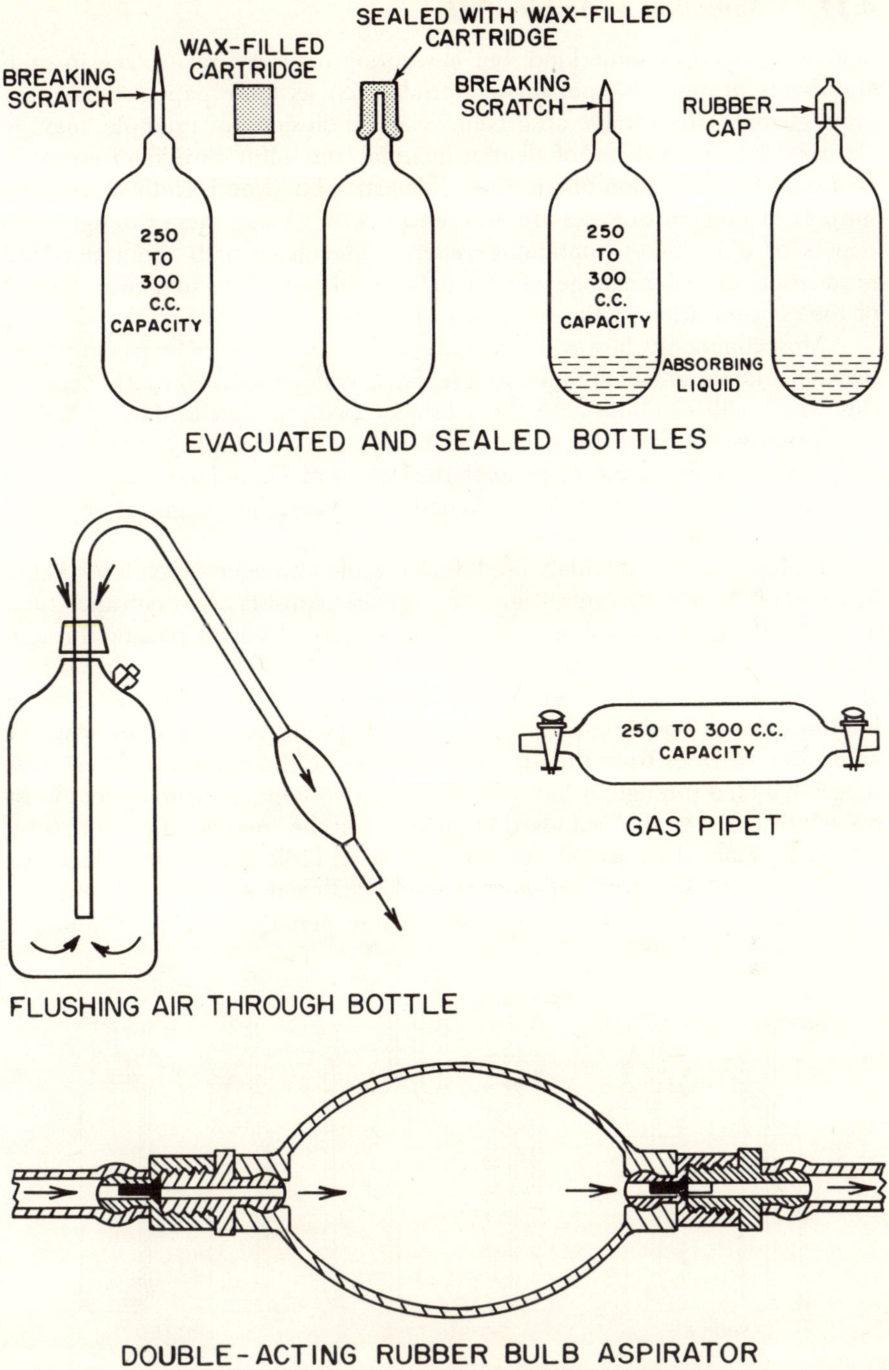

Figure 4.17a Devices for obtaining grab samples.

dium is a liquid of some kind, but absorption may also take place in solid absorbents or upon supporting materials such as filter papers which are impregnated with suitable absorbents. Carbon dioxide, for example, may be absorbed in a granular bed of alkaline material, and sulfur dioxide is frequently measured by absorption of reactive chemicals placed on a cloth or ceramic support. A number of gases are also detected by passing them through filter papers or glass tubes containing reactive chemicals with the immediate production of a color change which can be evaluated by eye to give a measure of the concentration of the substance of interest.

Most commonly, however, gases and vapors are absorbed by passing them through a liquid in which they are soluble, or which contains reactive chemicals which will combine with the substance being sampled. Many different absorption vessels have been designed, ranging from simple bubblers made by inserting a piece of tubing beneath the surface of a liquid to rather complex gas-washing devices which are designed to increase the length of time the air and liquid are in contact with each other (Fig. 4.17b).[1]

Probably the most widely used device is the impinger which is available in several sizes and configurations. An impinger consists of an entrance tube terminating in a small orifice, causing the velocity of the air passing through the orifice to greatly increase. When this jet of air strikes a plate or the bottom of the sampling vessel, at an optimal distance from the orifice, an intense impingement or bubbling action occurs which results in much more efficient absorption of gases from the airstream than would take place if the air was simply bubbled through at low velocity. The two impingement devices most frequently used are the "standard impinger" and the "midget impinger" (Fig. 4.17c); they are designed to operate best at an airflow of 1 cubic foot per minute and 0.1 cubic foot per minute, or 28.3 and 2.8 liters per minute respectively.[2] Using such devices for sampling periods of 10 or 30 minutes

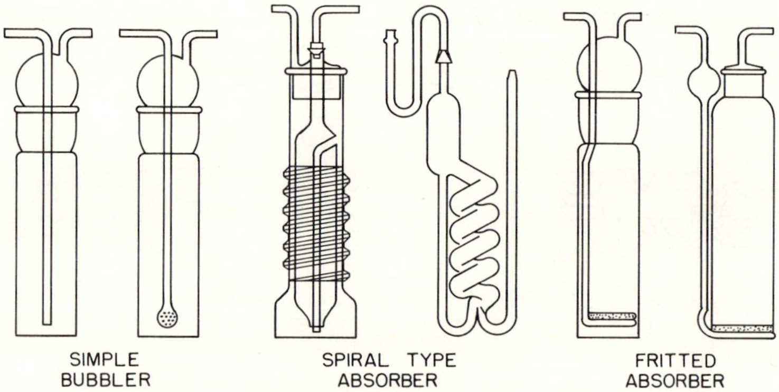

Figure 4.17b Gas absorbing vessels.

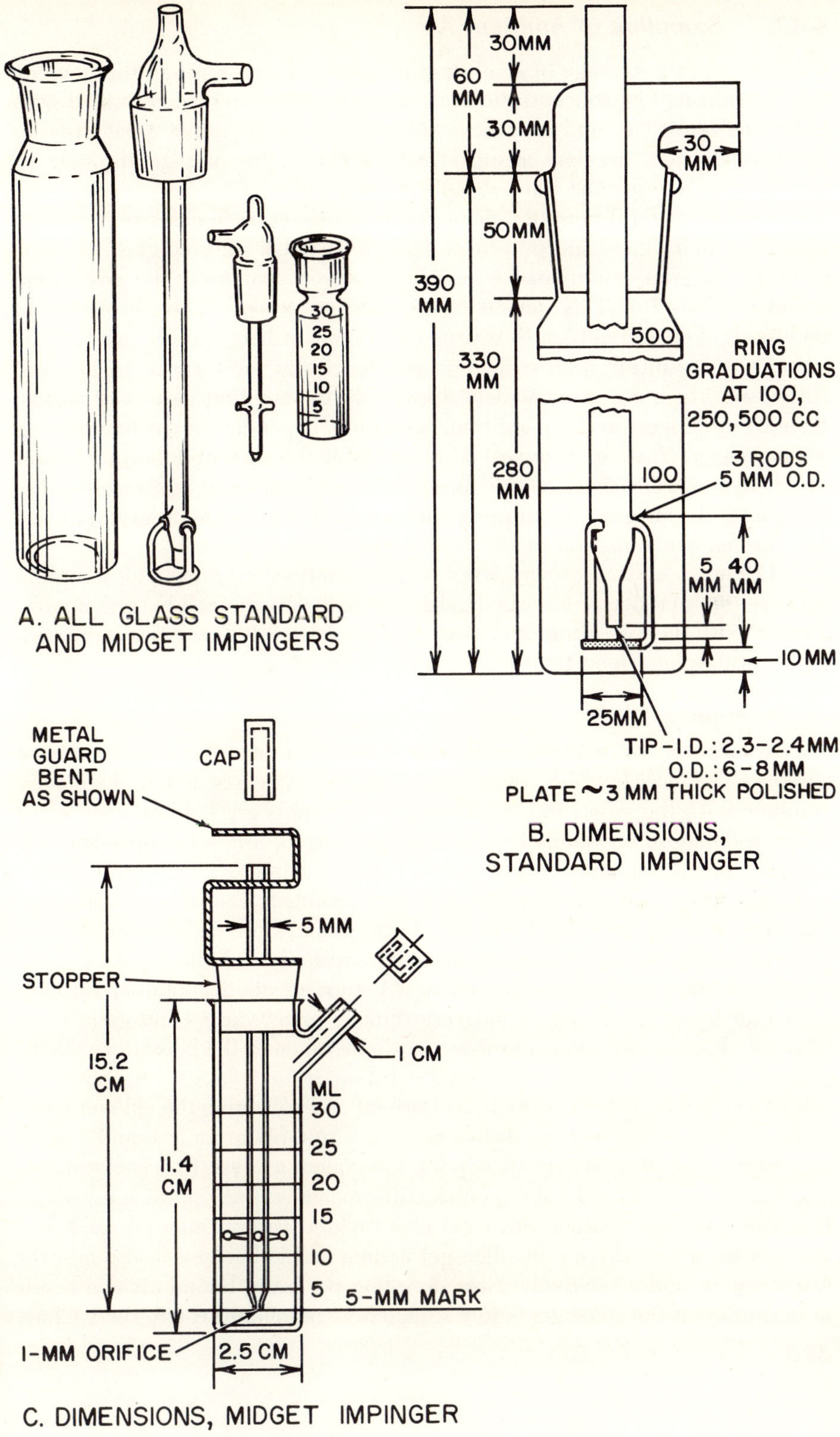

Figure 4.17c Standard and midget impingers.

will result in the passage of a substantial amount of air through the devices, thus permitting low concentrations of trace substances to be determined with improved sensitivity and accuracy. Many relatively insoluble gases, such as nitrogen dioxide, are not quantitatively removed by passage through an impinger containing the usual sampling solutions.

The most useful sampling devices for absorbing trace gases from air are those in which a gas dispersion tube made of fritted or sintered glass, ceramic or other material is immersed in a suitable vessel containing the absorption liquid (Fig. 4.17b). This device causes the gas stream to be broken into thousands of small bubbles, thus promoting contact between the gas and the liquid with resulting high collection efficiencies in most cases. In general, fritted absorbers are more widely applicable to sampling gases and vapors than are impingers and, in addition, are not as dependent upon flowrates as are impingers. They are standard items available from scientific supply houses and come in various sizes suitable for many sampling tasks. It is often advisable to prefilter the air prior to sampling with a fritted absorber in order to prevent the gradual accumulation of dirt within the pores of the frit.

The use of solid absorbents is not widely practiced in ambient air sampling because the quantity of absorbed gases is usually determined by gravimetric means. With the exception of carbon dioxide, relatively few gases of interest in the atmosphere lend themselves to this type of analysis.

Adsorption

Adsorption, by contrast with absorption, consists of the retention of gaseous substances by solid adsorbents which in most cases do not chemically combine with the gases. Instead, the gases or vapors are held by adsorptive forces and may subsequently be removed unchanged. Any solid substance will adsorb a small amount of most gases; but to be useful as an adsorbent, a substance must have a large surface area and be able to concentrate a substantial amount of gas in a small volume of adsorbent. Most widely used for this purpose are activated carbon or charcoal and activated silica gel. A small quantity of either adsorbent placed in a U-tube or other container through which air is passed will quantitatively remove many vapors and gases from a large volume of air and may subsequently be taken to the laboratory where desorption will remove the collected substances for analysis. Desorption is commonly achieved by heating the adsorbent and collecting the effluent gases or by eluting the collected substances with a suitable organic liquid.

In the case of most organic vapors, subsequent analysis by gas chromatographic, infrared or ultraviolet spectroscopic means is usually most convenient. For some purposes, either silica gel or activated carbon may be used; but in other instances, the use of silica gel is undesirable because it also adsorbs water vapor, and a relatively short sampling period in humid air can result in saturation of the silica gel before sufficient contaminant is adsorbed. Charcoal does not adsorb water and hence may be used in humid environments and may be sampled for days or even weeks if the concentration of contaminant is low.

The ease of sampling by using adsorbents is offset somewhat by the difficulty of quantitatively desorbing the samples for analysis. When literature data are not available to assist in predicting the behavior of a new substance, it is advisable to perform tests in the laboratory to determine both the collection efficiency and the success of desorption procedures after sample collection.

Freeze-out Sampling

Vapors or gases which are condensable at low temperature may be removed from the sampled airstream by passage through a vessel which is immersed in a refrigerating liquid. Table 4.17d lists liquids which are commonly used for this purpose and, usually, it is desirable to form a sampling train in which two or three coolant liquids progressively lower the air temperature in its passage through the system. All freeze-out systems are hampered somewhat by the accumulation of ice resulting from water vapor and may eventually become plugged with ice. Flowrates through a freeze-out train are necessarily limited also; for in order that the air will be cooled to the required degree, a sufficient residence time in the system must be provided. For these reasons, and because of the general inconvenience of assembling freeze-out sampling trains, they are not generally used for routine sampling purposes unless no other approach is feasible. However, freeze-out sampling is an excellent means of collecting substances for research studies, inasmuch as the low temperatures tend to arrest further chemical changes and ensure that the material being analyzed will remain in the sampling container ready for analysis after warming. Analysis is most frequently conducted by means of gas chromatographic, infrared or ultraviolet spectrophotometry, or by mass spectrometric means.

Table 4.17d
COOLANT SOLUTIONS FOR FREEZE OUT SAMPLING

Coolant	*Temperature (°C)*
Ice water	0
Ice and salt (NaCl)	−21
Dry ice and acetone	−78.5
Liquid air	−147
Liquid oxygen	−183
Liquid nitrogen	−196

Particulate Matter Sampling

Filtration

Particulate matter is most conveniently removed from air by passage through a filter (Fig. 4.15a). Before filtration is used to obtain a sample, however, consideration should be given to the purpose for which the sample is being taken. Many filters collect particulates efficiently, but thereafter it may be impossible to remove the collected matter except by chemical treatment. If the samples were initially collected for the purpose of examining the particles and measuring their size, or noting morphological characteristics, then many filters are not suitable because the particles are imbedded in the fibrous web of the filter and cannot readily be viewed nor removed. If the sample is collected for the purpose of performing a chemical analysis, then it is important that the filter itself does not contain significant quantities of the substance for which analysis is required. If the purpose of sampling is to collect a sufficient amount of particulate matter to permit weighing, then it is necessary to select a filter which can be weighed with the required precision; for many filtration materials are hygroscopic and change weight appreciably in response to changes in the relative humidity. Filters may be made of many substances and, in fact, almost any solid substance could probably be made into a filter. In practice, however, fibrous substances, such as cellulose or paper, fabrics, asbestos, and a number of plastics or polymerized materials are generally used. The most readily available filters are those made of cellulose, or paper, and used in chemical laboratories for filtering liquids. Such filter papers come in a variety of sizes and range in efficiency from rather loose filters which remove only the larger particles to papers which will remove very fine particles with high efficiency. All filters display similar behavior, and ordinarily a high collection efficiency is accompanied by increased resistance to airflow.

Certain kinds of filtration media are more suited to air sampling than most paper or fibrous filters. Of these, the class of filters known as membrane filters are of greatest utility, and commercially available membrane filters combine extremely high collection efficiencies with relatively low resistance to flow. Such filters are not made up of a fibrous mat, but instead are usually composed of gels of cellulose esters or other polymeric substances in such fashion that a smooth surface of predictable characteristics is formed. Such filters contain many small holes, or pores, and may be made to exacting specifications in this regard so that their performance characteristics may be predicted. In addition, the filters are usually of high chemical purity and are well suited to trace metal analyses. Some of the membrane filters can also be rendered transparent and thus permit direct observation of collected particles with a microscope. Alternatively, the filters can be dissolved in an organic solvent and the particles may be isolated and studied. Most membrane filters are not greatly affected by relative humidity changes and may be weighed before and after use to obtain reliable gravimetric data.

Another kind of filter that is widely used in sampling ambient air is the "fiberglass filter." These filters are originally made of glass fibers in an organic binder; and subsequently, the organic binder may be removed by firing, leaving a web of glass which is very efficient in collecting fine particles from the air. The principal advantage of using this type of filter is its relatively low resistance to airflow and its virtually unchanging weight regardless of relative humidity. The filters are not well suited to particle size studies, however; and in addition, they are not chemically pure and care must be taken to be certain that a substance for which analysis is to be made will not be contributed in unknown quantities by the filter itself. In the United States, most of the data relating to suspended particulate matter in our cities have been obtained on filters made of fiberglass and used in conjunction with a sampling device referred to as a high-volume sampler (Fig. 4.15a).

Many other kinds of filters are available but most sampling needs are well met by membrane or fiberglass filters. Extensive data concerning the types of filters available and their performance characteristics will be found in the bibliography cited at the end of this section.

Impingement and Impaction

The impingers previously described in relation to sampling gases and vapors (Fig. 4.17c) may also be used for the collection of particles and, in fact, were originally developed for that purpose. However, in ambient air sampling they are not used, inasmuch as the efficiency of collection tends to be low and unpredictable for the fine particles which may be present in ambient air. The relatively low sampling rates also make them less attractive than filters for general air sampling, but instances do arise when impingers can be satisfactorily used. When used, it is important that the correct sampling rates be maintained, for the collection efficiency of impingers for particles may vary drastically when flowrates are other than optimal.

Of more widespread use in ambient air sampling are devices known as impactors, in which air is passed through small holes or orifices and made to impinge or impact against a solid surface. When such devices are constructed so that the air passing through one stage is subsequently directed onto another stage containing smaller holes, the resulting device is known as a cascade impactor and has the capability of separating particles according to their size. Various commercial devices are available. Figure 4.17e portrays one that is widely used and consists of several layers of perforated plates through which the air must pass. Each plate contains a constant number of holes, but the hole size is progressively decreased so that the same volume of air passing through each stage will impinge at an increased velocity. The result is that coarse particles are deposited on the first stage and successively

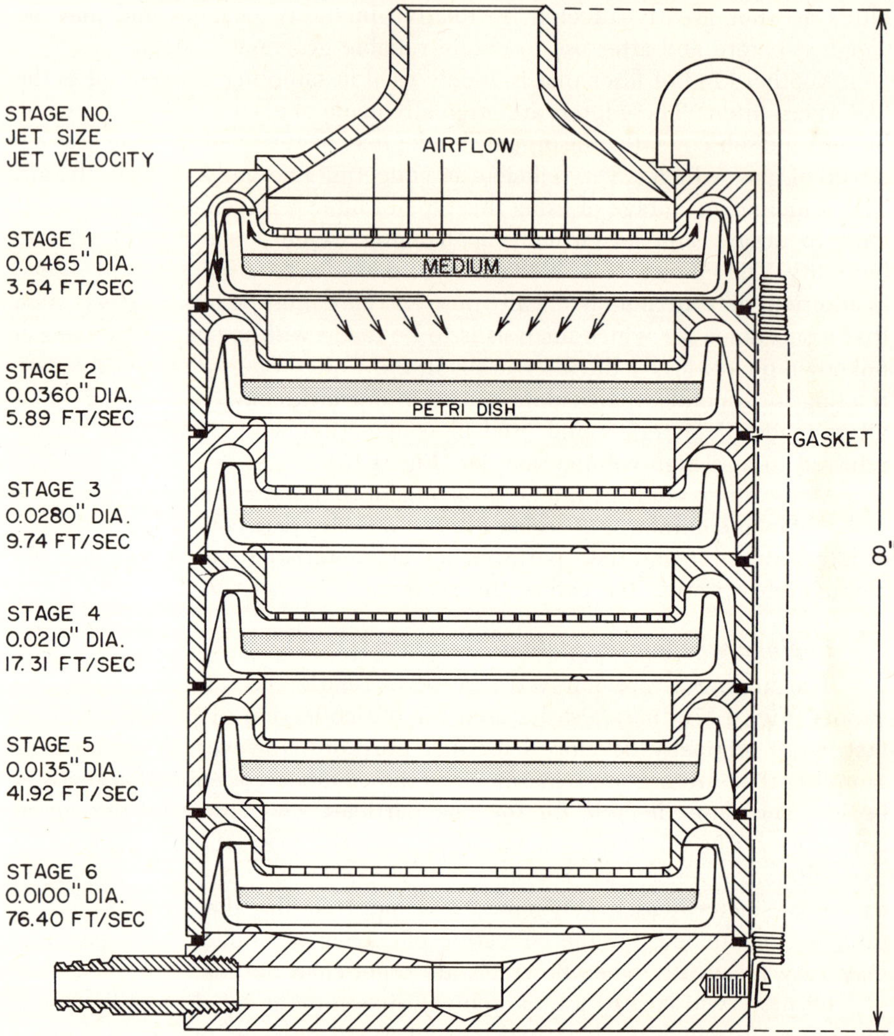

Figure 4.17e Cascade impactor (Andersen sampler).

finer particles are removed at each subsequent stage. Although exact particle size fractions are not achieved by such instruments, they do perform predictably when the characteristics of the aerosol being sampled are known. In use, a cascade impactor is assembled after scrupulously cleaning each stage and applying, if necessary, a sticky substance or a removable surface on which the particles are to be deposited. After a suitably long period of sampling, during which time the volume of air is metered, the stages may be removed and the total weight of each fraction determined, as well as their chemical composition. Such information may be much more useful than a single weight or chemical analysis of the total suspended particulate matter without regard to its particle size.

Electrostatic Precipitation

Particulate matter can be quantitatively removed from air by means of instruments known as electrostatic precipitators. Devices which operate on the same principle but are very much larger are frequently used to remove particulate matter from stack gases prior to discharging into the atmosphere (Sections 5.22 to 5.24). There are several commercially available electrostatic precipitators which may be used for air sampling; and all of them operate on the same general principle, that of passing the air between charged surfaces, imparting a charge to particles in the air and collecting the particles on an oppositely charged surface or plate. In one of the more widely used commercial devices (Fig. 4.17f), a high voltage discharge is made to occur along a central wire; the collecting electrode is a metallic cylinder which is placed around the central wire, while the air to be sampled is passed through the tube. An intense corona discharge takes place on the central wire, the particles entering the tube are charged and are promptly swept to the walls of the tube where they remain firmly attached. By this means, it is possible to collect a sample for subsequent weighing or chemical analysis; and it is also possible to examine the particles and study their particle size and shape, but the intense electrical forces may produce aggregates of particles which are different than those which existed in the sampled air.

Electrostatic precipitators are not as widely used as filters for ambient air sampling, because they are generally less convenient and tend to be heavy due to the power pack necessary to generate the required high voltage. Nevertheless, when available, they are excellent instruments for obtaining samples for subsequent analysis, and sample at relatively high flow rates and very low resistance.

Thermal Precipitators

Whenever a strong temperature gradient exists between two adjacent surfaces, there will be a tendency for particles to be deposited on the colder of these surfaces. Collection of aerosols by this means is termed "thermal precipitation" and, in practice, several such commercial devices are available. Because thermal forces are so weak, it is necessary to have a rather large temperature difference maintained in a small area; and in addition, the rate of airflow between the two surfaces must be low in order not to destroy the temperature gradient and to permit particles to be deposited before moving out of the collection area. As a result of these requirements, most devices use a heated wire as the source of the temperature differential and deposit a narrow ribbon of particles on the cold surface. Airflows are very small, being

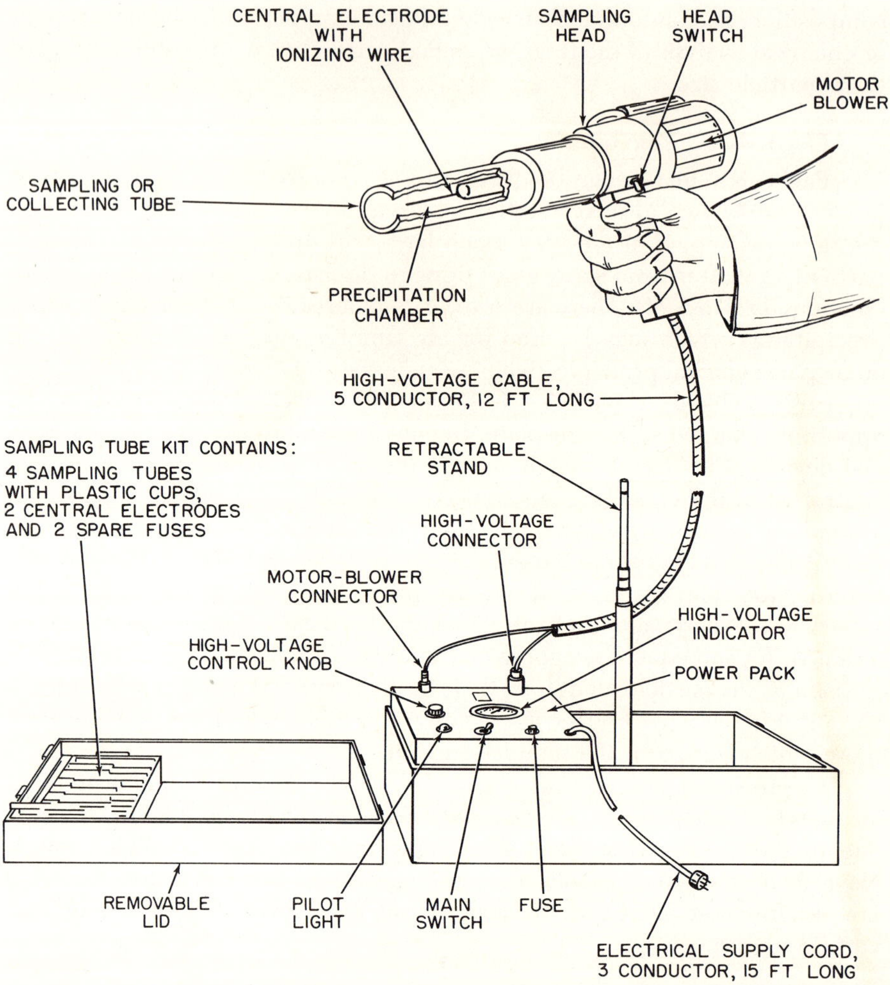

Figure 4.17f Electrostatic precipitator. (*Mine Safety Appliances Company*)

on the order of 10 to 25 ml per minute. At such low rates, the amount of material collected will normally be insufficient for chemical analysis or weight determinations, but will be ample for examination by optical or electron microscopy.

Collection for particle size studies is, in fact, the principal use for thermal precipitation units, and they are well suited to collecting samples for such investigations. Because the collecting forces are gentle, it is generally believed that the particles are deposited unchanged. The microscopic examination gives information which can be translated into data concerning the number of particles and their morphological characteristics. It is convenient also to use a small grid suitable for insertion into an electron microscope as the collecting surface, thus making it unnecessary to perform additional manipulations of the sample prior to examination by electron microscopy.

Summary

Particulate matter may be collected in many ways, and the selection of a method will ultimately be determined on the basis of the use to be made of the data obtained. If simple gravimetric determinations are required, then almost any device can be made to yield satisfactory results; whereas if chemical analysis or particle size studies are needed, then the nature of the aerosol to be studied may favor the use of some methods and not others. Other means of sampling in addition to those described are available, including the placing of sticky papers in the air and the collection of particles which fall into open buckets. Although such methods do find use in specialized instances, they are not, in general, applicable to sampling suspended particulate matter.

REFERENCES

1. "1971 Annual Book of ASTM Standards," part 23, ASTM, Philadelphia, Pa.
2. Jacobs, M. B.: "The Analytical Toxicology of Industrial Inorganic Poisons," vol. 22., Interscience, New York, 1967.

BIBLIOGRAPHY

"Air Sampling Instruments," American Conference of Governmental Industrial Hygienists, 1967, Cincinnati, Ohio.

"1971 Annual Book of ASTM Standards," part 23, Water: Atmospheric Analysis, American Society for Testing and Materials," Philadelphia, Pennsylvania.

Jacobs, Morris B.: "The Analytical Toxicology of Industrial Inorganic Poisons," vol. 22, Chemical Analysis, Interscience, New York, 1967.

Stern, A. C. (ed.): "Air Pollution," 2d ed., Vol. II, Analysis, Monitoring, and Surveying, Chapter 16, Academic Press, New York, 1968.

4.18 SAMPLERS FOR STACKS

Type Sample:	Gas containing particulates.
Design Pressure:	Generally atmospheric or near atmospheric.
Design Temperature:	−25°F to 1,200°F.
Sample Flowrate:	0.1 to 1 SCFM.
Materials of Construction:	Stainless steel (316 or 304) for pitot tubes, stainless 316 or 304, quartz or Inconel® for sampling probes.
Cost:	$2,500 to $5,000.
Partial List of Suppliers:	Bendix Environmental Science Div., Bendix Corp.; Joy Manufacturing Company, Western Precipitation Div.; Research Appliance Co.; Scientific Glass Apparatus Co., Inc.

Stack sampling[1-3] involves a number of problems which are not usually encountered in open-air or gas sampling described in the preceding section. Since the gases are confined in a stack or flue, some sort of opening must be provided for sampling. Considerations should also be given to flowrates and concentrations of the pollutant which usually vary considerably across a stack. Thus, when determining mass rate of emission both volumetric flowrates and concentrations should be obtained at a number of sampling points.

Measurement of Gas Flow

The first step in stack sampling is the measurement of gas or air flow[7] through the stack at which the pollutant concentration is to be determined. Proper sampling for particulates also rests upon knowing the velocity and direction of the ambient atmosphere being sampled.

Velocity Measurement

Stack-gas velocity can be determined from the gas density and from measurement of the velocity head using a pitot tube.[7] The velocity head is

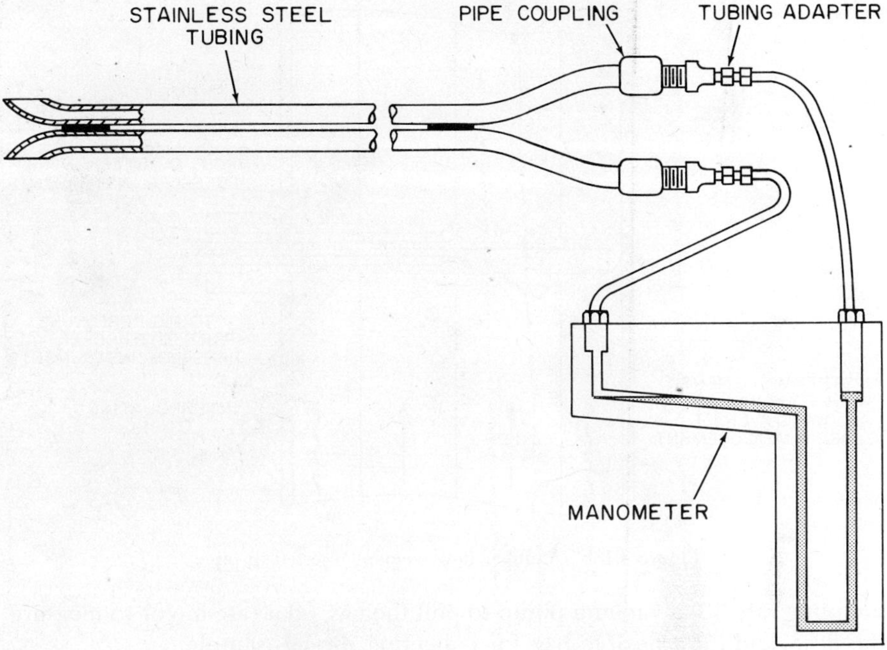

Figure 4.18a Type-S pitot tube–manometer assembly.

the difference between the impact pressure (measured against the gas flow) and the static pressure (measured perpendicular to the gas flow).

Figure 4.18a illustrates the pitot tube–manometer assembly. The Type-S (Stauscheibe or reverse type) pitot tube consists of two opposing openings, one made to face upstream and the other facing downstream during the measurement.

The Type-S pitot tube is designed for easy entry into small holes in the stack wall and because of its relatively large openings does not readily plug in presence of high concentrations of particulate matter. Calibration is achieved by measuring the velocity head at some point in the flowing gas stream with both a Type-S pitot tube and a standard type pitot tube with a known coefficient.[1,3,7] The annubar flow element results in improved measurement accuracy (Fig. 4.18b).

Molecular Weight Determination

Figure 4.18c shows the diagram of an integrated gas-sampling train. A stainless steel or Pyrex® probe fitted with an integral filter (e.g., glass or Pyrex wool) is placed in the stack at the sampling point. The filter removes particulates at stack temperature. The sampling train also includes (1) an air-cooled condenser to remove any excess moisture, (2) a needle valve to control the

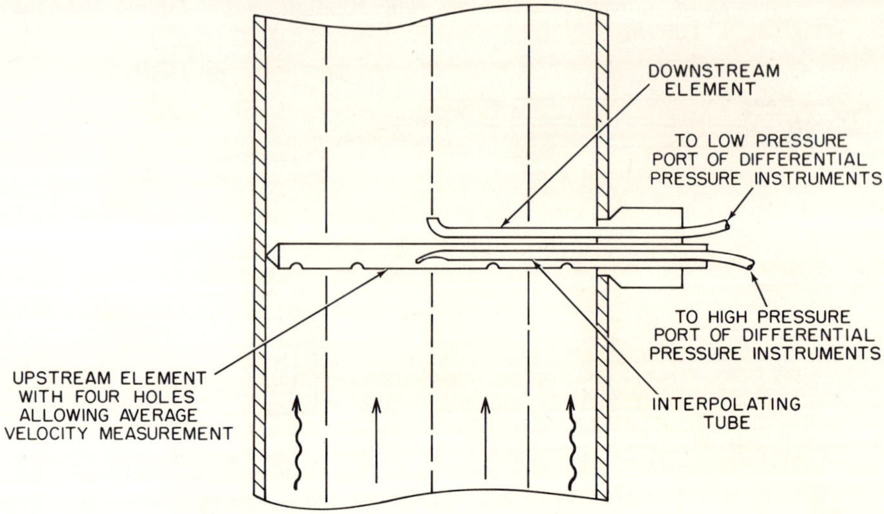

Figure 4.18b Annubar flow element inserted in pipe.

sampling rate, (3) a vacuum pump to pull the gas, (4) a rate meter to measure the flow, and (5) a flexible bag for collecting the gas sample.

The fixed gases (N_2, CO_2, O_2) in the sample are readily measured by the Orsat analyzer.[8] In the Orsat analyzer, a measured gas sample is passed through several reagents which absorb the various gases. The amount of gas contained

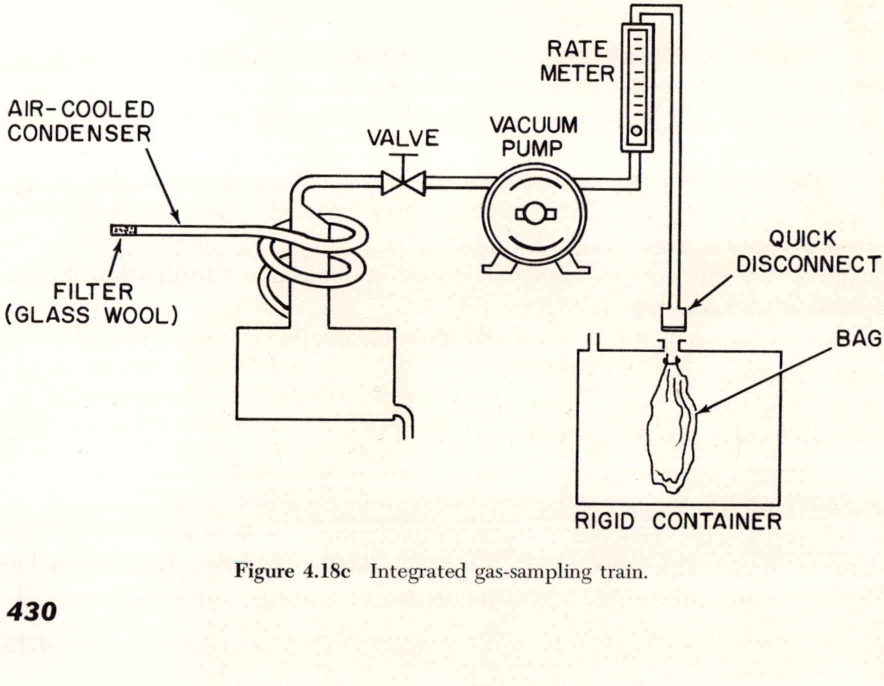

Figure 4.18c Integrated gas-sampling train.

in the sample is measured by the decrease in volume after passage through each reagent. Reagents that are selective absorbers include 20 to 40% aqueous KOH for CO_2 (and other acidic gases), alkaline pyrogallol for O_2 and acidic cuprous chloride for CO. Nitrogen is determined by difference.

Sampling and Velocity Traverses

The sampling point and the number of sampling or traverse points should be so selected that representative samples will be obtained. The sampling point should be located at a distance equal to at least eight stack or duct diameters downstream and two diameters upstream from any source of flow disturbance, such as a bend, expansion, contraction, valve, fitting or visible flame. For rectangular stacks, the equivalent diameter can be calculated from the following equation:

$$\text{Equivalent diameter} = 2(\text{length} \times \text{width})/(\text{length} + \text{width}) \qquad 4.18(1)$$

After determining the sampling location(s), provisions must be made to "traverse" the stack. The number of traverse points is 12, if the earlier noted sampling point criteria can be met. If this criteria is not met, the required number of traverse points are a function of sampling point distances from the nearest upstream and downstream disturbances, which can be estimated by using Figure 4.18d. The higher of the two numbers of traverse points in Figure 4.18d is selected, such that for circular stacks the number is a multiple of four.

The cross-sectional layout and location of traverse points are as follows:

1. For circular stacks, the traverse points should be located on two perpendicular diameters according to Figure 4.18e and Table 4.18f.[3]
2. For rectangular stacks, the cross section is divided into as many equal rectangular areas as traverse points, such that the ratio of the length to the width of the elemental areas is between one and two. The traverse points are to be located at the centroid of each equal area according to Figure 4.18e.

The velocity heads at various traverse points are measured using the pitot tube–manometer assembly shown in Figure 4.18a. Gas samples are collected at a rate proportional to the stack gas velocity and analyzed for CO, O_2 and CO_2. Samples for molecular weight determination are collected by using the apparatus shown in Figure 4.18c. Other data needed for calculating the volumetric flow are stack temperature at each traverse, stack and barometric pressures and wet bulb and dry temperatures of the gas sample.

Table 4.18g presents the equations for converting pitot tube readings into velocity and mass flow, and a typical data sheet for stack-flow measurements.

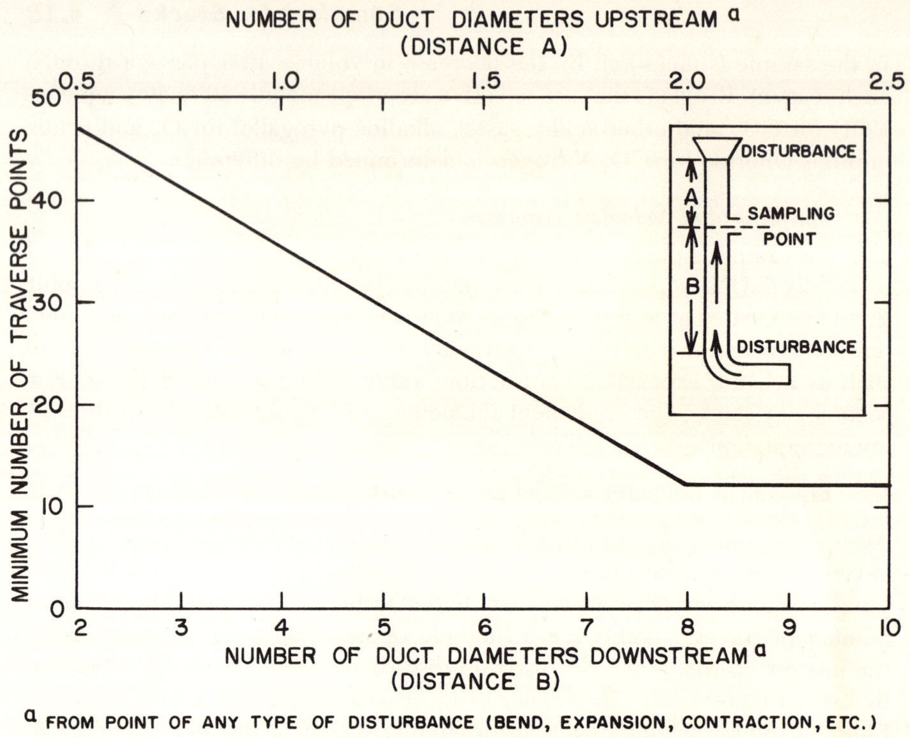

Figure 4.18d Minimum number of traverse points.

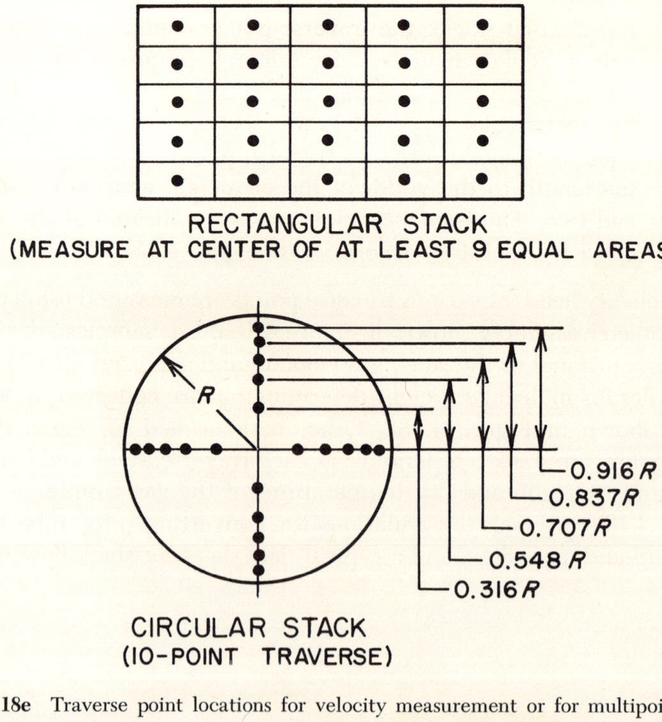

Figure 4.18e Traverse point locations for velocity measurement or for multipoint sampling.

Table 4.18f
LOCATION OF TRAVERSE POINTS IN CIRCULAR STACKS

Traverse Point Number On a Diameter	*Number of Traverse Points on a Diameter*[a]									
	6	*8*	*10*	*12*	*14*	*16*	*18*	*20*	*22*	*24*
1	4.4	3.3	2.5	2.1	1.8	1.6	1.4	1.3	1.1	1.1
2	14.7	10.5	8.2	6.7	5.7	4.9	4.4	3.9	3.5	3.2
3	29.5	19.4	14.6	11.8	9.9	8.5	7.5	6.7	6.0	5.5
4	70.5	32.3	22.6	17.7	14.6	12.5	10.9	9.7	8.7	7.9
5	85.3	67.7	34.2	25.0	20.1	16.9	14.6	12.9	11.6	10.5
6	95.6	80.6	65.8	35.5	26.9	22.0	18.8	16.5	14.6	13.2
7		89.5	77.4	64.5	36.6	28.3	23.6	20.4	18.0	16.1
8		96.7	85.4	75.0	63.4	37.5	29.6	25.0	21.8	19.4
9			91.8	82.3	73.1	62.5	38.2	30.6	26.1	23.0
10			97.5	88.2	79.9	71.7	61.8	38.8	31.5	27.2
11				93.3	85.4	78.0	70.4	61.2	39.3	32.3
12				97.9	90.1	83.1	76.4	69.4	60.7	39.8
13					94.3	87.5	81.2	75.0	68.5	60.2
14					98.2	91.5	85.4	79.6	73.9	67.7
15						95.1	89.1	83.5	78.2	72.8
16						98.4	92.5	87.1	82.0	77.0
17							95.6	90.3	85.4	80.6
18							98.6	93.3	88.4	83.9
19								96.1	91.3	86.8
20								98.7	94.0	89.5
21									96.5	92.1
22									98.9	94.5
23										96.8
24										98.9

[a] Figures in body of table are % of stack diameter from inside wall to traverse point.

Sampling for Particulate Emissions

As illustrated in Figure 4.18h, representative sampling is obtained only if the velocity of the stack-gas stream entering the probe nozzle is the same as the velocity of the stream passing the nozzle. If the sampling velocity is too high, a convergent stream will develop at the nozzle face, with an excessive amount of lighter particles entering the probe. Because of the inertia effect, the heavier particles, especially those in the range of 3 μ or greater, travel past the edge of the nozzle and are not collected. The result is a sample indicating an excessively high concentration of lighter particles and the sample weight is in error on the low side. Conversely, if the sampling velocity is below that of the flowing gas stream, portions of the gas stream approaching at a

Table 4.18g
PITOT TUBE CALCULATION SHEET

STACK VOLUME DATA

STACK NO.________ STATION________ DATE________ PAGE________

NAME OF FIRM________

Point	*Position, in.*	*Reading, H, in. of H_2O*	$\sqrt{H}$	*Temp., t_s, °F.*	*Velocity, V_s, ft./sec.*
1					
2					
3					
4					
5					
6					
7					
8					
9					
10					
11					
12					
13					
14					
15					
16					
	Totals				
	Average				
	Abs. temp., $T_s = t_s + 460 =$ °R.				

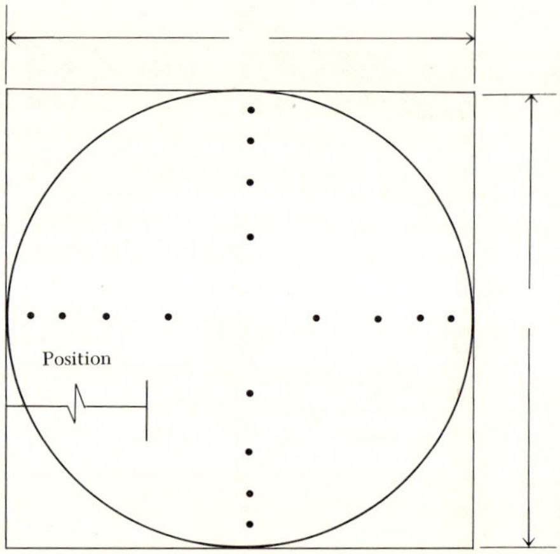

Dry bulb temp., $t_d =$ ________ °F.

Wet bulb temp., $t_w =$ ________ °F.

Absolute humidity, $W =$ ____ lb. H_2O/lb. dry gas

Stack area, $A_s =$ ________ sq. ft.

Barometer, $P_b =$ ________ in., Hg

Stack gage pressure = ________ in., H_2O

Stack abs. pressure, $P_s = \_\_\_\_ \frac{\text{in., } H_2O}{13.6} \pm P_b = \_\_\_\_$ in., Hg

Pitot correction factor, $F_s =$ ________

Component	*Vol. fraction, dry basis*	× *mol. wgt.*	=	*wgt. fraction, dry basis*
Carbon dioxide		44	=	
Carbon monoxide		28	=	
Oxygen		32	=	
Nitrogen		28	=	

Average dry gas molecular weight, $M =$ ____

Specific gravity of stack gas, $G_s = \frac{0.62\, M\, (W + 1)}{18 + MW} = \frac{0.62 \times \_\_\_ \times \_\_\_}{18 + \_\_\_} = \_\_\_\_\_\_$
(Ref. dry air at same conditions)

Velocity, $V_s = 2.9\, F_s \sqrt{\frac{29.92 \times T_s}{P_s \times G_s}} \sqrt{H} = 2.9 \times \_\_\_ \sqrt{\frac{29.92 \times \_\_\_}{\_\_\_ \times \_\_\_}} \sqrt{H} = \_\_\_\_\_\_$ ft./sec.

Volume = ______ ft./sec. × ______ sq. ft. × 60 ______ = ______ cfm.

Standard volume = cfm. $\times \frac{530}{T_s} \times \frac{P_s}{29.92} = \_\_\_ \times \frac{530}{\_\_\_} \times \frac{\_\_\_}{29.92} = \_\_\_\_\_\_$ scfm.

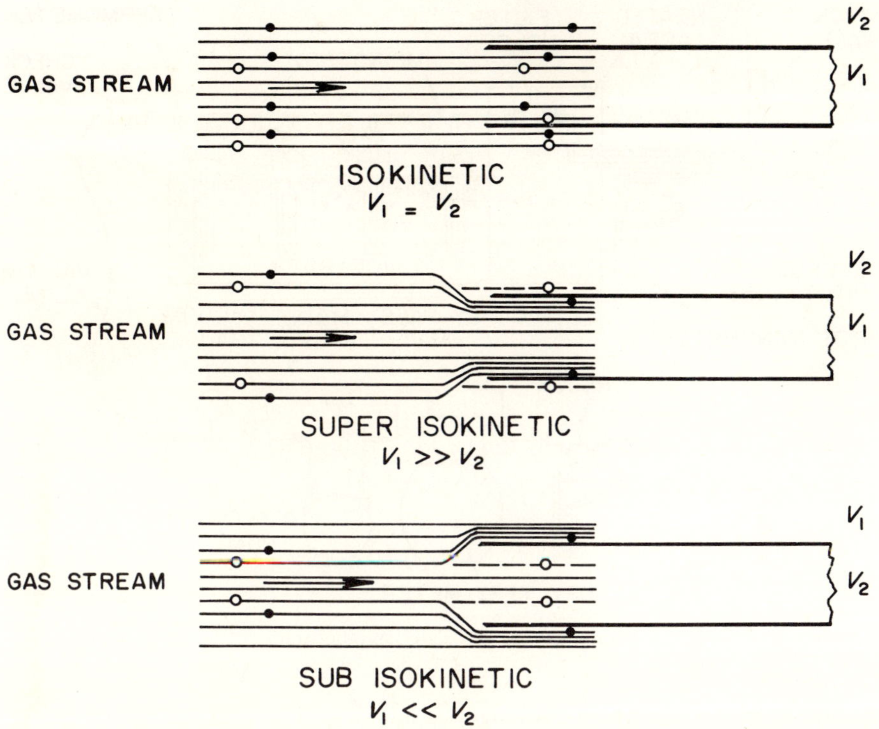

Figure 4.18h Particle collection and sampling velocity

higher velocity are deflected. The lighter particles follow the deflected stream and are not collected, while the heavier particles because of their inertia continue into the probe. The result is a sample indicating high concentration of heavier particles and the sample weight is in error on the high side.

Apparatus

The sampling train for determination of particulate emission is shown in Figure 4.18i.[3] The probe is made of either Pyrex or 316 stainless with a heating system capable of maintaining a gas temperature at 250°F. The probe is connected to a Pyrex glass filter. The filter recommended by EPA is glass fiber, MSA 1106 BH filter or its equivalent. Following the filter are four impingers connected in series with glass ball joint fittings. The first, third and fourth impingers are of the Greenburg-Smith design. To reduce pressure drop, the tips are removed and replaced with a ½-inch I.D. glass tube extending to ½ inch from the bottom of the flask. The second impinger is of the Greenburg-Smith design with the standard tip. During sampling for particulates, the first and second impingers are filled with 100 milliliters of distilled and

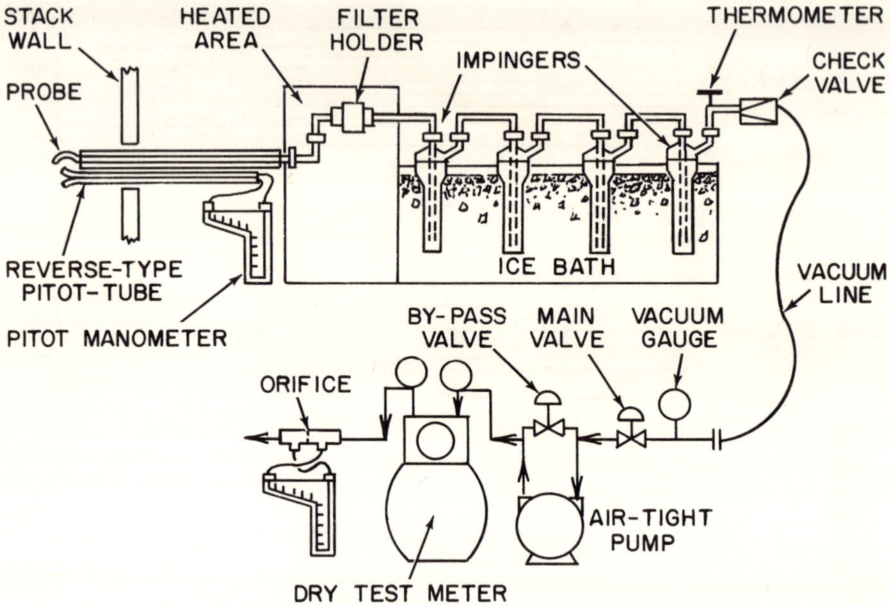

Figure 4.18i Particulate-sampling train.

deionized water. The third impinger is left dry to remove entrained water. The last impinger contains 200 grams of precisely weighed silica gel (6 to 16 mesh, dried at 350°F for 2 hours) for complete removal of any remaining water. Crushed ice should be added to the ice bath during the run so that the temperature of the gas leaving the last impinger is at 70°F or less.

The metering system is connected by vacuum line with the last Greenburg-Smith impinger. The pump intake vacuum is monitored with a vacuum gauge just after the quick disconnect. A bypass valve in parallel with the vacuum pump provides fine flow control and permits recirculation of gases at low sampling rate so that the pump motor is not overloaded. Downstream of the pump and bypass valve are thermometers, dry gas meter and a calibrated orifice with an inclined/vertical manometer. The calibrated orifice and inclined manometer indicate the instantaneous sampling rate. The dry gas meter gives an integrated gas volume. The average of the two temperatures on each side of the dry gas meter gives the temperature at which the sample is collected. The addition of the atmospheric pressure and orifice pressure drop gives the meter pressure.

Procedure

Isokinetic sampling starts at the first traverse point after completion of gas flow measurement and is maintained throughout the sampling period

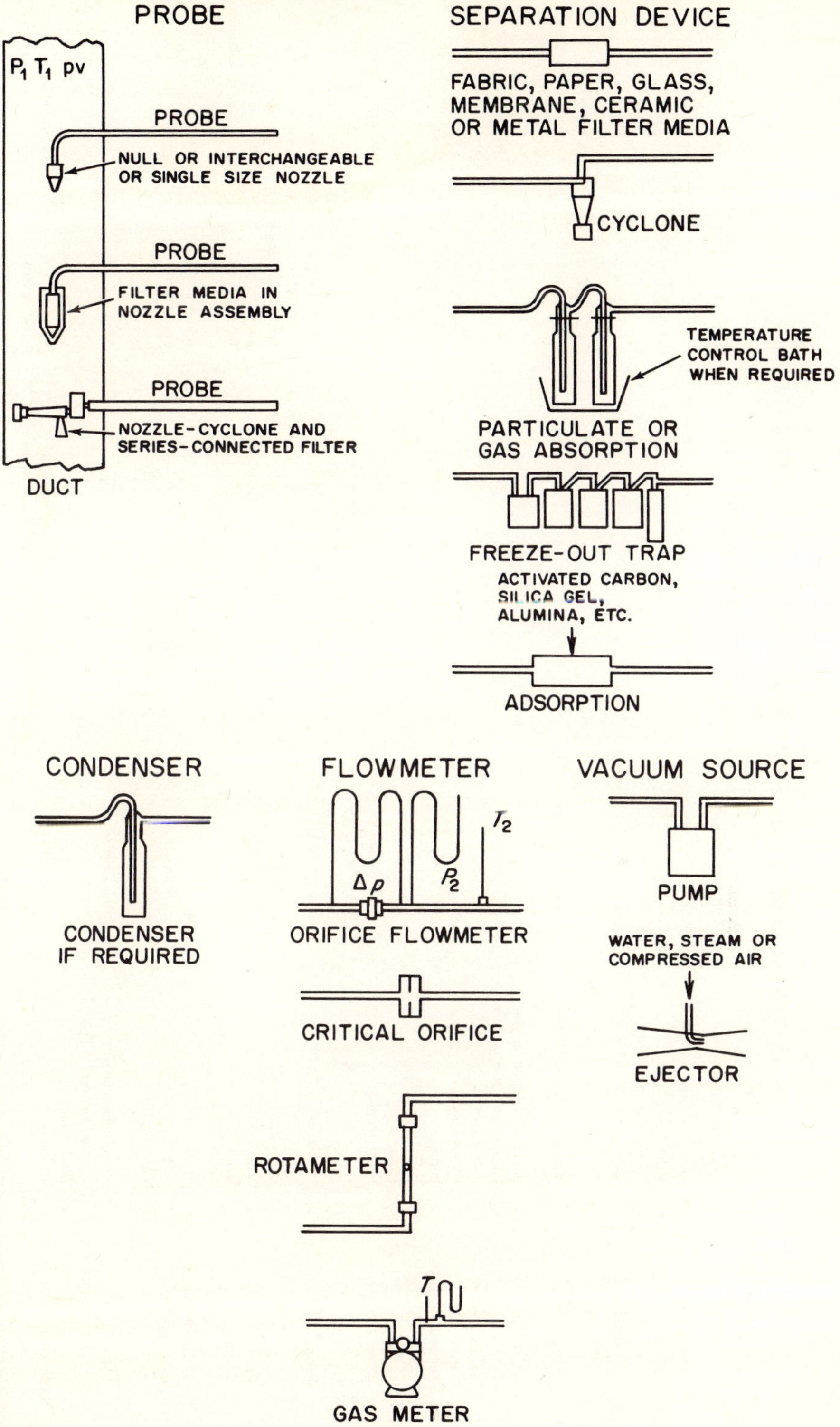

Figure 4.18j Components of common sampling systems.

Table 4.18k
SAMPLING OUTLINE FOR SOME SPECIFIC POLLUTANTS[a]

Pollutant	*Collecting Device*	*Collecting Medium*	*Sampling Rate* (ft^3/min)	*Minimum Sampling vol* (ft^3)	*Analytical Method*	*Threshold Limit Value*[b]
Acetone	Mylar bag	None			Gas chromatograph	1000 ppm
	Combustible gas indicator	None				
Acid vapors and mists	Impinger or fritted bubbler	Aqueous alkali or water	Up to 1	Sulfuric acid-60 ft^3 Nitric acid-0.1 ft^3 Hydrochloric acid-5 ft^3		1.0 mg/m^3 5 mg/m^3 7.0 mg/m^3
Aldehydes	Impinger or fritted bubbler	1% $NaHSO_3$ solution	0.1	1	Titration	Formaldehyde-5 ppm Acetaldehyde-200 ppm
Alkali mists	Impinger or fritted bubbler	Water	Up to 1	20		
	Impinger or fritted bubbler	Dilute H_2SO_4	Up to 1	20	Titration	Hydroxide-2 mg/m^3
Ammonia	Fritted bubbler	0.1 *N* H_2SO_4	0.1	1	Colorimetric	50 ppm
Aniline	Fritted bubbler	Dilute H_2SO_4	0.05	1	Colorimetric	5 ppm
Antimony	Impinger	Water	1	10	Colorimetric or Spectrograph	
Arsenic	Impinger	Water	1	3	Colorimetric	0.5 mg/m^3
Arsine	Midget impinger	Silver diethyldithiocarbonate in pyridine	0.1	3 or Color change from yellow to red	Colorimetric	0.05 ppm
Benzene (Benzol)	Bubbler	Isooctane	0.25	3	UV Spectrophotometer	
	Mylar bag	None			Gas chromatograph	25 ppm
Beryllium	Millipore filter	None	3	10	Spectrograph	0.002 mg/m^3
Cadmium	Impinger	Water	1	10	Polarograph	0.1 mg/m^3 as cadmium oxide
Carbon dioxide	Mylar bag	None			Infrared	5000 ppm
Carbon disulfide	Fritted bubbler	Alcoholic KOH	0.05	0.1	Colorimetric	20 ppm
	Fritted bubbler	Diethylamine-Cu acetate	0.05	0.1		
Chlorinated hydrocarbons	Combustion tube with fritted bubbler	Na_2CO_3-arsenite reagent	0.1	1	Titration	
	Mylar bag	None			Gas chromatograph	
Chromic acid	Impinger	Water or aqueous alkali	Up to 1	Sample until yellow color is detected	Titration or colorimetric	0.1 mg/m^3
Cobalt	Millipore filter	None	3	15	Spectrograph	0.1 mg/m^3[c]

Copper	Millipore filter	None	3	15	Spectrograph	Fume 0.1 mg/m^3 Dusts & mists
Cresol	Fritted bubbler	Aqueous alkali	0.05	1	Colorimetric	1.0 $mg./m^3$
Cyanides	Fritted bubbler	Aqueous alkali	0.1	2.5	Colorimetric	5 ppm
Dimethylaniline	Fritted bubbler	Alcohol	0.1	1	Colorimetric	5 mg/m^3 as CN
Dinitrotoluene	Fritted bubbler	Alcohol	0.1	5	Colorimetric	5 ppm
Ethane	Mylar bag	None			Gas chromatograph	1.5 mg/m^3
	Ethane detector	None				
Ethyl alcohol	Fritted bubbler	H_2SO_4-K_2CrO_7	0.05	0.5	Titration	1000 ppm
	Mylar bag	None			Gas chromatograph	
Fluorides (inorganic salts)	Electrostatic precipitator	None	3	15		2.5 mg/m^3 as F
Halogenated compounds	See chlorinated hydrocarbons					
Hydrogen cyanide	Fritted bubbler	Aqueous alkali	0.1	2.5	Colorimetric	10 ppm
Hydrogen fluoride	Fritted bubbler	Aqueous alkali	Up to 1	15	Colorimetric	3 ppm
Hydrogen sulfide	Bubbler column	1% $CdCl_2$	0.05	3 or Until yellow precipitate appears	Gravimetric	10 ppm
Lead-in-air	Impinger	Dilute HNO_3	1	10	Polarograph	0.2 mg/m^3
	Millipore filter	None	3	10		
Manganese	Impinger	Water	1	10	Colorimetric	5.0 mg/m^3
	Millipore filter		3	10		
Mercury	Impinger	Dilute HNO_3	1	10	Colorimetric	0.1 mg/m^3
	Mercury vapor detector	None				
Methane	Mylar bag	None			Gas chromatograph	
	Ethane detector	None				
	Combustible gas indicator	None				
Methanol	Combustible gas indicator					200 ppm
Nickel carbonyl	Mylar bag	None			Gas chromatograph	
Nitrobenzene	Grab sample flask	None				
				500 ml	Colorimetric	1 ppm
Nitrogen dioxide	Grab sample flask	None		500 ml	Colorimetric	5 ppm
Nitrogen oxides	Grab sample flask	None		500 ml	Colorimetric	5 ppm
Oil mist	Impinger	Perchlorethylene	1	20	Infrared	5 mg/m^3
Ozone	Fritted bubbler	0.1 *N* NaOH + 1% KI solution	0.5	10	Colorimetric	0.1 ppm
Phenols	Fritted bubbler or impinger	Aqueous alkali	0.1	10	Colorimetric	5 ppm
Phosgene	Fritted bubbler	Sat. aniline + sat. diphenylurea	0.1	1	Gravimetric	0.1 ppm

Table 4.18k (Continued)

Pollutant	*Collecting Device*	*Collecting Medium*	*Sampling Rate* (ft^3/min)	*Minimum Sampling vol* (ft^3)	*Analytical Method*	*Threshold Limit Value*[b]
Phosphoric acid	Impinger or fritted bubbler	Aqueous alkali	Up to 1	10	Colorimetric	1 mg/m^3
Phosphine	Midget impinger	Silver diethyldithio-carbonate in pyridine	1	3 or color change from yellow to red	Colorimetric	0.3 ppm
Propane	Mylar bag	None			Gas chromatograph	
	Combustible gas indicator	None				
Selenium	Impinger	Water	1	10	Gravimetric	0.2 mg/m^3
Stibine	Bubbler	6% $HgCl_2$ + 6 *N* HCl	0.1	10	Colorimetric	0.1 ppm
Sulfur dioxide	Fritted bubbler	2% glycerol in 0.05 *N* NaOH	Up to 1	10	Colorimetric	5 ppm
Tellurium	Impinger	Water	1	20	Spectrophotometer	0.1 mg/m^3
Thallium	Fritted bubbler	10% HNO_3	0.1	3	Spectrograph or polarograph	0.1 mg/m^3
Toluene	Mylar bag	None			Gas chromatograph	200 ppm
Tolylene diisocyanate	Fritted bubbler	HCL + acetic acid	0.1	3	Colorimetric	0.02 ppm
Xylene	Mylar bag	None			Gas chromatograph	100 ppm

[a] Courtesy of the Michigan Department of Public Health.

[b] American Conference of Governmental Industrial Hygienists threshold limit value for 8 hours per day—5 days per week adult worker exposure (1967) (Section 5.3).

[c] Proposed value listed under notice of intended change.

(generally, 5 to 10 minutes per traverse point). Isokinetic sampling involves observing the pitot tube–manometer reading and using nomographs (such as APTD-0576) to aid in rapid adjustment of the sampling rate as required.[5] Temperature, manometer pressure and dry gas volume should be recorded at the start and at the end of each traverse point and whenever a change occurs.

Detailed procedures for recovery of particulates collected and methods for calculating the concentration of particulate matter in the stack gas from the experimental data are specified in reference 3.

Sampling for Gases and Vapors

Some commonly used components in stack sampling systems are pictured in Figure 4.18j. Using ball-and-socket joints and compression fittings, any arrangement of components can be readily set up for field use. The stack sampling components are selected depending upon the source to be sampled, the pollutants involved, and the data desired. All systems incorporate a combination of probes to go into the stack and collection or separating devices for removing the pollutants being measured from the gas that is being sampled through the probe.

The sampling train and procedure for determining sulfur dioxide, nitrogen oxide and sulfuric acid mist emissions are described by the Environmental Protection Agency.[3] (See Sections 4.12 and 4.20.)

Table 4.18k is a summary of sampling procedure outlines developed by industrial hygienists for some specific pollutants. The procedure outlines can serve as a starting point in assembling a stack sampling system, after consideration has been given to the complications that might arise because of the presence of interfering substance in the gas samples. Other recommended sampling procedures for gases and vapors are given in reference 6.

REFERENCES

1. Stern, A. C. (ed.): "Air Pollution," 2d ed., vol. II, chapter 28, Academic Press, New York, 1968.
2. Air Pollution, *Anal. Chem.*, vol. 43, IR, 1971; vol. 41, IR, 1969; vol. 39, 10R, 1967.
3. Environmental Protection Agency, Standards of Performance for New Stationary Sources, Notice of Proposed Rule Making, *Federal Register*, vol. 36, no. 159, pp. 15,704–15,722, August 17, 1971.
4. Morrow, N. L., R. S. Brief and R. R. Bertrand: Sampling and Analyzing Air Pollution Sources, *Chem. Eng.*, vol. 79, no. 2, pp. 84–98, January 24, 1972.
5. Rom, Jerome J.: "Maintenance, Calibration and Operation of Isokinetic Source Sampling Equipment," Environmental Protection Agency, APTD-0576.
6. American Society for Testing and Materials, "Standards of Methods for the Sampling and Analysis of Atmospheres," part 23, 1971.
7. Lipták, B. G. (ed.): "Instrument Engineers' Handbook," vol. I, chap. 5, Chilton, Philadelphia, 1969.
8. Ibid, Section 8.7.

4.19 SOLAR RADIATION SENSORS

Types of Solar Radiation Detectors:	(a) Thermal radiometers; (b) Photovoltaic sensors.
Cost:	\$150 to \$500 (b); \$500 to \$1,500 (a).
Partial List of Suppliers:	The Eppley Laboratory (a); Matrix, Inc. (b).

The intensity of solar radiation can be measured by (1) absorbing the incoming radiation and determining the rate of heat absorption and (2) by making use of photovoltaic or photoresistive transducers. Among the first accurate measurements of solar radiation were those made with miniature waterflow calorimeters in which the flowrate was kept constant and the temperature rise due to absorption of solar radiation was determined by high-accuracy mercury-in-glass thermometers and later with thermopiles.

Thermal Radiometers

The term *pyranometer* is applied to a solar radiometer which measures the shortwave radiation coming directly from the sun and diffusely from all parts of the sky. *Pyrheliometer* is the word used to identify an instrument which measures only the direct radiation from the solar disk itself.

Pouillet in France (1830) and Ericsson in the United States (1870) devised pyrheliometers which measured the rate of temperature rise of blackened metal masses located at the bottom of tubes that excluded all but the direct rays of the sun. C. G. Abbott of the Smithsonian Institution perfected this technique with his *silver-disk* pyrheliometers which are still in use as secondary standards. He had earlier built a waterflow calorimeter in which a precisely controlled flow of distilled water carried away the heat absorbed by a blackened cone located at the bottom of a collimating tube. A platinum-resistance thermometer was used to measure the temperature rise and then an exactly equal rise was created by passing the water through an electrical heating coil. Precise measurement of the heating current and the coil resistance enabled the user of the instrument to make first-order determinations of the rate at which the incoming solar energy was being absorbed. This instrument was long used as the primary standard for measuring solar radiation.

Today's primary standard radiometer was originated in 1899 by K. J. Ångström of Sweden. His instrument uses two blackened strips of manganin, each of which can be heated either electrically or by the rays of the sun. The measurement is made by exposing one strip to the sun, measuring its temperature precisely and then measuring the current required to heat the adjacent strip to exactly the same temperature. By reversing the operation of the strips several times, an average value of the radiation intensity within 1% of the absolute value can be obtained. The results may be expressed in langleys (1 ly = 1.0 g cal/cm^2/min = 221.2 Btu/hr/ft^2) or in electrical units (1 ly = 69.7 mw/cm^2).

The solar radiometer which is in widest use today employs a multijunction thermopile to detect the difference in temperature between whitened and blackened segments of divided or concentric circles (Figs. 4.19a and 4.19b) which are generally protected by hemispherical glass domes. These instru-

Figure 4.19a Thermopile-type pyranometer for measuring total radiation from sun and sky. (*The Eppley Laboratory, Inc.*)

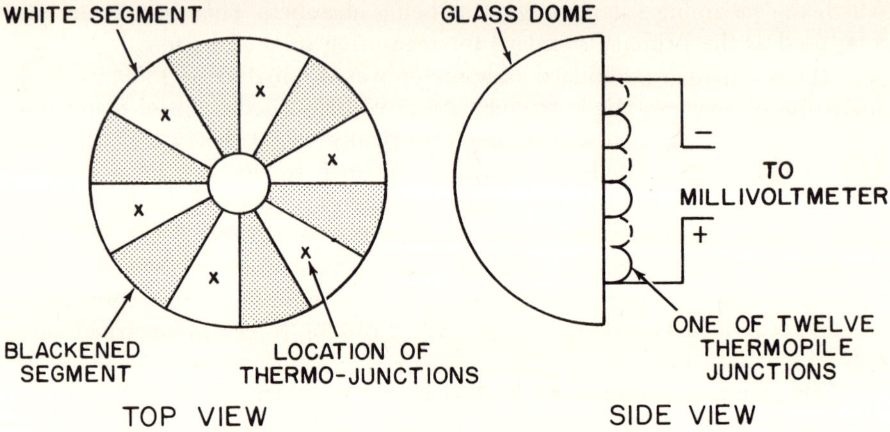

Figure 4.19b Thermopile-type pyranometer.

ments characteristically produce from 2.0 to 10.0 mv/ly of incoming radiation and their output can be readily measured with indicating or recording millivoltmeters. Mechanical or electrical integrators can be used to total daily or monthly insolation. Since the response of the thermopile instruments is independent of wavelength, they can be used with filters to determine radiation intensity in selected portions of the solar or longwave spectrums.

Photovoltaic Sensors

The newest form of solar radiometer uses silicon photovoltaic cells which produce a short-circuit current that is directly proportional to the intensity of the solar radiation falling on them. Widely used in the space program to provide power for space probes and satellites, silicon cells are rugged, sensitive and capable of producing currents large enough to operate self-powered integrators.

BIBLIOGRAPHY

Yellott, J. I.: Solar Radiation Measurement, Low Temperature Applications of Solar Energy, pp. 19–25, ASHRAE, New York, 1967.

MacDonald, T. H.: Some Characteristics of the Eppley Pyrheliometer, *Monthly Weather Review,* vol. 79, no. 8, pp. 153–159, August 1951.

Yellott, J. I. and A. L. Pittinger: Development of an Indicating and Integrating Solar Radiometer, ASME Paper 67-WA/Sol-3, Winter Annual ASME Meeting, 1967.

4.20 SULFUR OXIDE SENSORS

Methods of Detection:	(a) Colorimetric; (b) Conductimetric; (c) Correlation spectrometry; (d) Coulometric; (e) Electrochemical; (f) Flame photometric; (g) Infrared; (h) Ultraviolet.
Reference Method:	(a), Pararosaniline.*
Accuracy:	On the order of ±2% of full-scale range, limited by an inherent precision of ±0.005 ppm for ambient measurement.
Ranges:	Generally 0.01 to 5 ppm for ambient air sensors "a", "b", "d" and "f". With adequate dilution systems these sensors can measure up to 5,000 ppm. Generally 500 to 10,000 ppm for units "c", "g" and "h" as commercially available stack gas sensors.
Sample Pressure:	Generally atmospheric or near atmospheric.
Sample Temperatures:	At sensor, limited to about 120°F for designs "a", "b", "d", "e" and "f" for ambient air analysis. In stack monitoring applications, designs "c", "g" and "h" are not limited since they are not in direct contact with the sample.
Sample Flowrate:	Range of 0.1 to 1.0 ACFM is typical for ambient air analysis.
Materials of Construction:	Glass, Teflon and stainless steel preferred in all applications due to reactivity of sulfur oxide gases. Es-

*The EPA study (CPA 70-101) found coulometry to be equivalent to this reference method. Of the coulometers evaluated, the Philips PW-9700 gave the best correlation.

pecially avoid polyvinyl chloride for this reason.

Response: Nearly instantaneous for designs "c", "e" and "f". Designs "a", "b" and "d" rely on automated wet chemistry and the response time is determined almost entirely by the liquid capacitance of the system.

Cost: Generally $3,000 to $8,000 not including strip chart recorder. Some portable models of designs "b", "d" and "e" may cost slightly less than this, while models of designs "a", "c" and "f" can be more expensive if specially fitted for multiple gas and/or multiple point sampling duty.

Partial List of Suppliers: Atlas Electric Devices Co. (a, d); Barringer Research Ltd. (c); Barton Instrument Corp. (d); Beckman Instruments, Inc. (b, d); Bendix Environmental Science Div., Bendix Corp. (f); Calibrated Instruments, Inc. (b); Combustion Equipment Associates (b); Davis Instrument Co. (b); Devco Engineering (b); Dynasciences (e); E. I. Dupont de Nemours & Co., Inc. (g, h); Envirometrics, Inc. (e); Erickson Instruments (e); Instrument Devel. Co. (b, e); Intertech Corp. (g); ITT Environmental Products Div. (d); Kimoto (e); Leeds & Northrup, Inc. (b, g); Litton Industries, Inc. (a); Meloy Laboratories, Inc. (f); Monitor Labs, Inc. (a, b); Philips Electronic Instruments (d); Pollution Monitors, Inc. (a); Precision Scientific (a); Process Analyzers, Inc. (d); Scientific Industries (b); Scott-Davis Instruments (b); Technicon Industrial Systems (a); Theta Sensors, Inc. (e); Tracor Analytical Instruments Division (f); Varian-Aerograph, Inc. (f); Wilkens-Anderson Company (a).

The predominant sulfur oxide in the atmosphere is sulfur dioxide, SO_2. Some sulfur trioxide, SO_3, is formed in combustion processes but this rapidly hydrolyzes to sulfuric acid which is considered to be a particulate matter. This section will deal with sulfur dioxide.

The measurement of sulfur dioxide in air has been performed for many years because of the simplicity of performing manual analysis based on the ability of sulfur dioxide to reduce a starch-iodine solution. The earliest instruments to detect sulfur dioxide were automated wet chemical devices which measured the increase in the conductivity of a solution after air had been intimately mixed in by some contacting procedure.

The U.S. Environmental Protection Agency has deleted conductivity type analyses from their approved list of ambient air sampling instruments because this technique is nonspecific for SO_2, but measures all acid gases. EPA has designated the colorimetric technique using pararosaniline as the referee procedure, with flame photometric and coulometric detection as acceptable alternate procedures.[1]

Ambient Air Sampling

In the United States, the ultimate air quality goals (secondary standards) for sulfur dioxide are 60 $\mu g/M^3$ (0.02 ppm) annual arithmetic average, 260 $\mu g/M^3$ (0.1 ppm) maximum 24-hour concentration not to be exceeded more than once per year and 1,300 $\mu g/M^3$ (0.5 ppm) maximum 3-hour concentration not to be exceeded more than once per year.[2] Instrument range is generally selected for 0 to 1 ppm full-scale range except where higher levels are anticipated. There is a sensitivity problem, with few instruments capable of accurately reading at levels of 0.01 ppm and below. Statistical procedures are usually required to determine arithmetic average concentrations.

Colorimetric Analyzers

The reference technique for determining the sulfur dioxide content of atmospheric air involves the absorption of SO_2 from the air sample in a solution of sodium tetrachloromercurate. Upon addition of formaldehyde and pararosaniline dye, a strong purple dye complex is formed. This procedure has been automated by numerous manufacturers (Fig. 4.20a).

This family of analyzers requires frequent maintenance because of the complexity of the plumbing and the tendency of the dye complex to plate out on the cell windows, reducing its sensitivity. Properly maintained, they provide an excellent record of sulfur dioxide concentrations in air since they are based on a chemical reaction which is specific for sulfur dioxide.

Conductimetric Analyzers

There is more experience with the operation of this family of analyzers than with any other for sulfur dioxide detection. The instruments are fairly

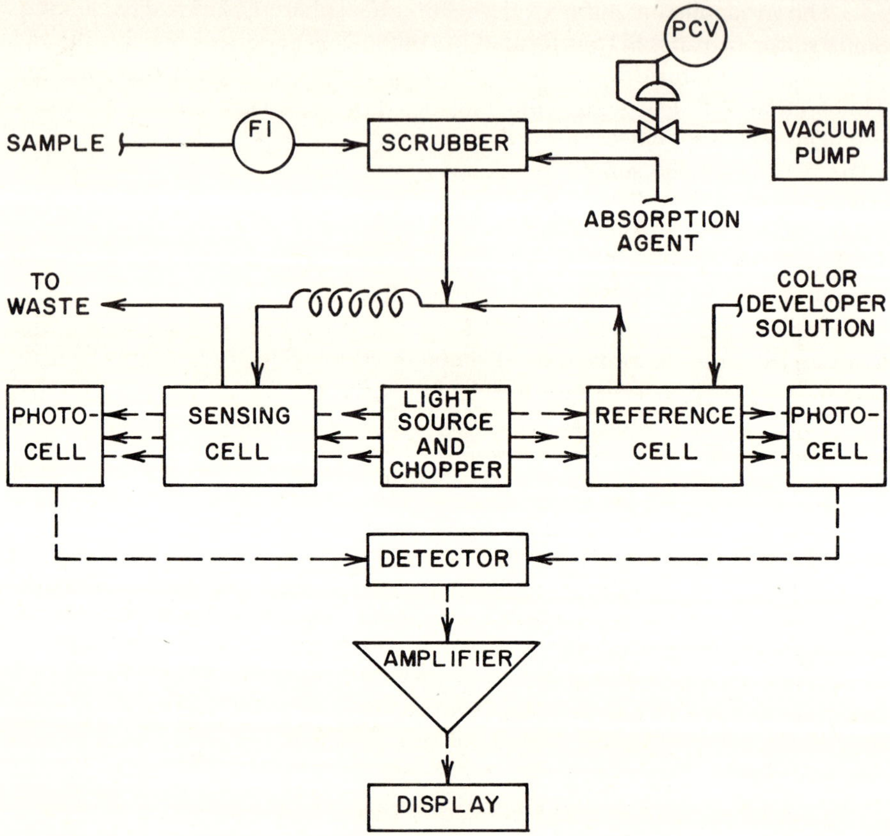

Figure 4.20a Colorimetric analyzer for oxides of sulfur.

simple electronically, while the plumbing creates some maintenance problems. A basic circuit diagram is shown in Figure 4.20b. There are various techniques for contacting the air with the reagent as discussed in Sections 4.17 and 4.18. Some of the available instruments use a weak solution of sulfuric acid and hydrogen peroxide as an absorbent, while others use distilled water.

It is important to utilize a temperature correcting element in the conductimetric analyzer circuitry since the change in conductance of the reagent over the normal ambient temperature range is nearly as large as the conductance changes at the sulfur dioxide levels of interest.[3]

Conductimetric analysis is not specific for sulfur dioxide. Other acid gases such as hydrogen chloride are a positive interference, while basic gases such as ammonia are a negative interference (Section I.4.8).

Coulometric Analyzers

The reducing action of sulfur dioxide will electrogenerate free bromine or iodine from a solution. These active elements can be detected to give a measure of the SO_2 content of the airstream (Fig. 4.20c).

The reagent supply (Fig. 4.14b) for coulometric analyzers is frequently much smaller than for either colorimetric or conductimetric analyzers. At least one of the commercially available coulometric analyzers is capable of measuring ozone concurrently with sulfur dioxide.

Flame Photometric Analyzers

When an airstream containing sulfur is burned in a hydrogen-rich flame, radiation is generated in a wavelength band centered at 394 mμ. This phenomenon is utilized in air sampling instruments to detect sulfur gases down to levels less than 0.01 ppm. The electrical output signal (Fig. 4.20d) is logarithmic. Typical instruments cover two decades of concentration, e.g., 0.01 to 1.0 ppm, with the capability to switch to a higher range.

This instrument measures total sulfur in the sample stream. In order to provide discrimination between various sulfur compounds, additional hardware is available from some manufacturers. An all Teflon gas chromatographic column preceding the flame photometric analyzer separates sulfur compounds quantitatively. This has proved extremely useful for the measurement of

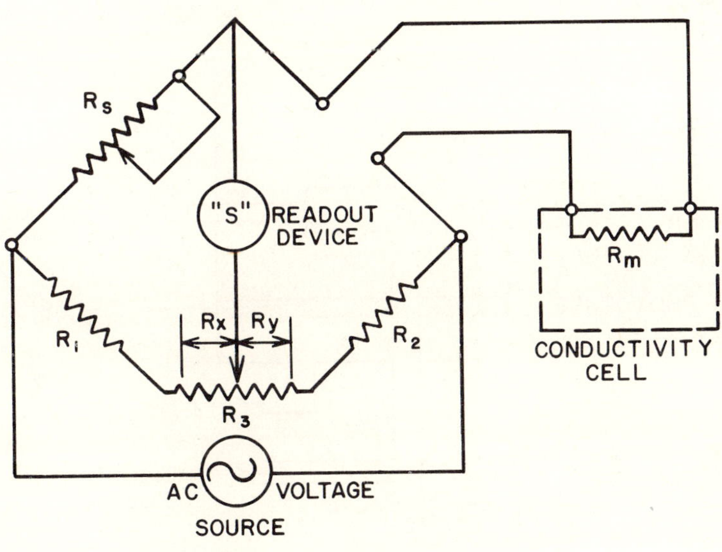

Figure 4.20b Conductivity measuring circuit.

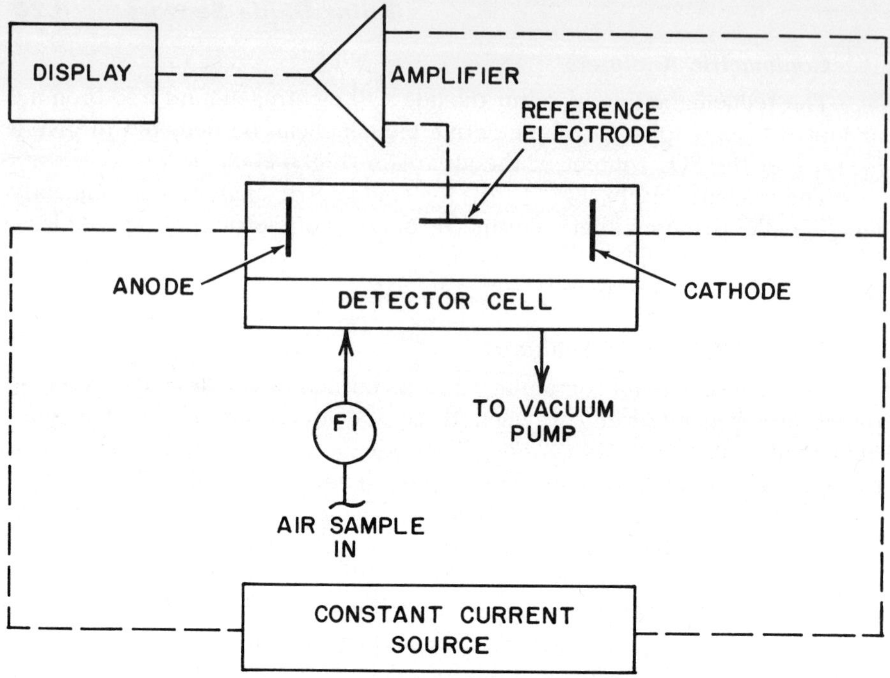

Figure 4.20c Coulometric analyzer for oxides of sulfur.

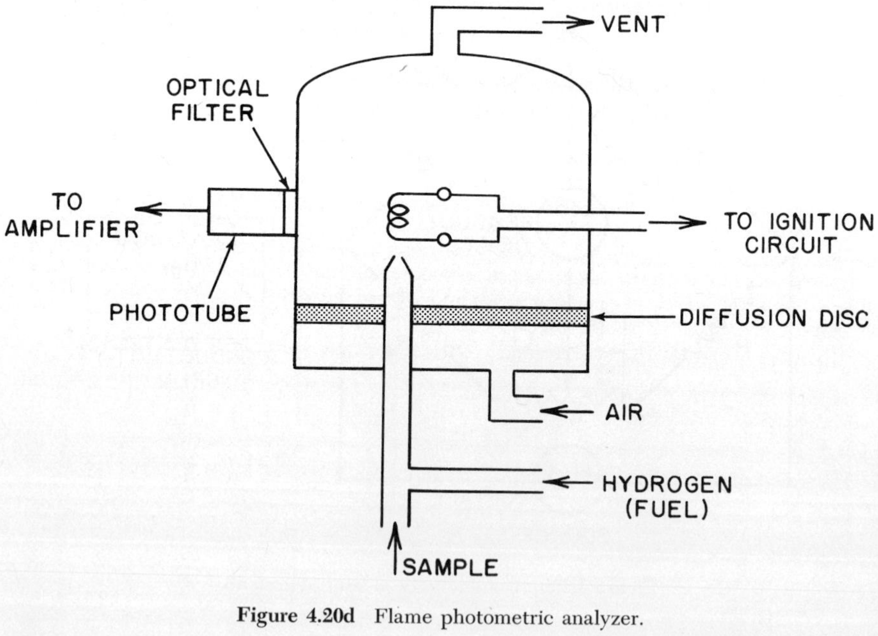

Figure 4.20d Flame photometric analyzer.

reduced sulfur compounds, e.g., hydrogen sulfide and dimethyl disulfide, in the air.

Electrochemical Analyzers

When air is passed over an element consisting of a semipermeable membrane, an electrolyte and a voltage-sensitive pickup, certain gases may migrate selectively across the membrane and generate a signal in the electrolyte (Fig. 4.20e). This principle has been used to measure many gases including sulfur dioxide.

The electrochemical principle for the measurement of air pollutants has not had extensive field use. Operational problems can develop in maintaining the proper moisture content of the membrane and the proper electrolyte strength because of the migration of water vapor.

Source Sampling

Stack effluents contain orders of magnitude higher concentrations of sulfur dioxide than does ambient air. Few of the principles used to detect the trace amounts of sulfur dioxide in ambient air are applicable to source monitoring except by massive dilution.

The colorimetric procedure is not applicable to source monitoring at all, mainly because of the interference of nitrogen oxides. The electrochemical principle may be directly applicable to source monitoring. Instruments which are generally applicable to source monitoring (but not to ambient monitoring) are IR, UV and correlation spectrometry.

Infrared and Ultraviolet Absorption

Sulfur dioxide absorbs light energy over a broad band which includes both the infrared and ultraviolet regions of the spectrum. Some instruments utilize a positive filtering scheme (shown in Figure 4.20f and described in detail elsewhere).[3] Others utilize photometric analysis (Fig. 4.20g). The lowest

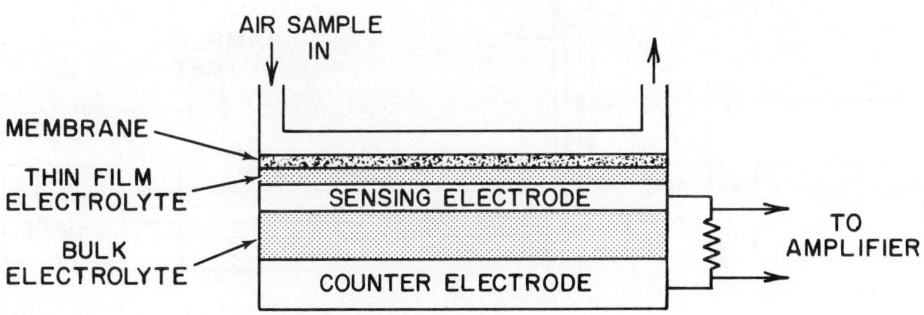

Figure 4.20e Electrochemical gas analyzer.

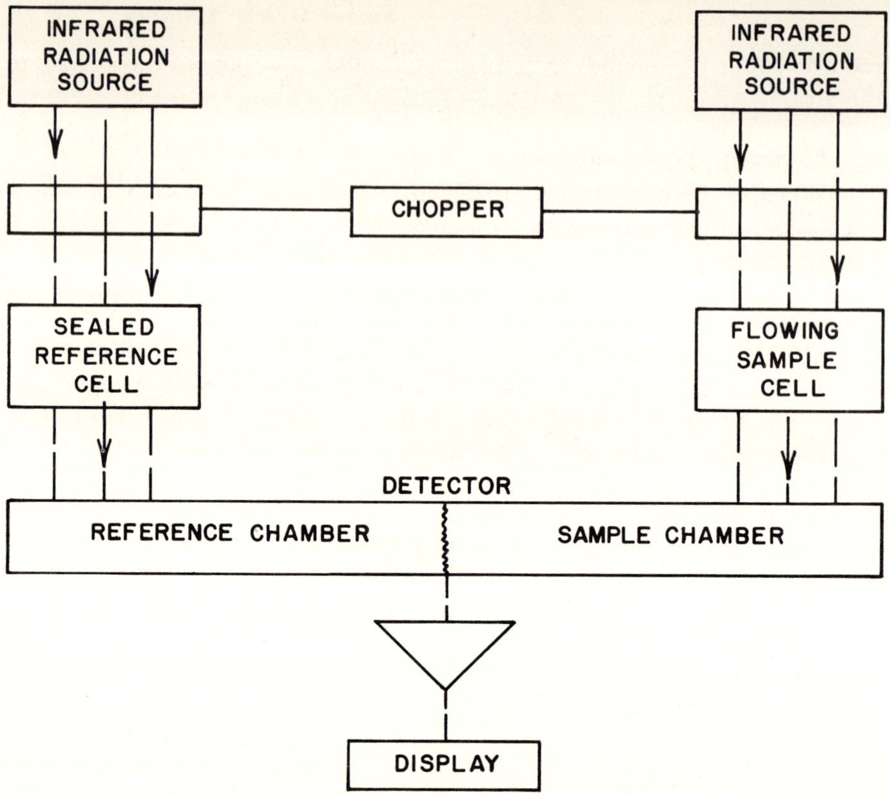

Figure 4.20f Nondispersive infrared analyzer.

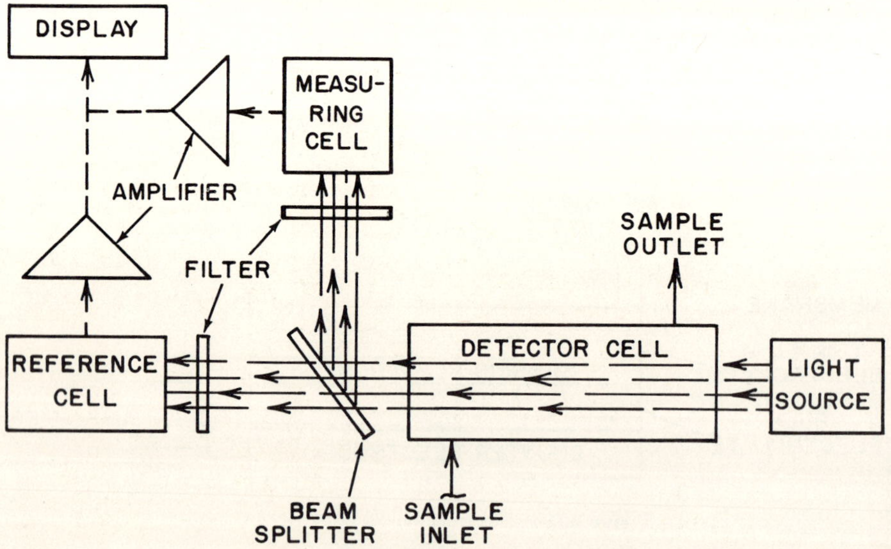

Figure 4.20g Photometric analysis for oxides of sulfur.

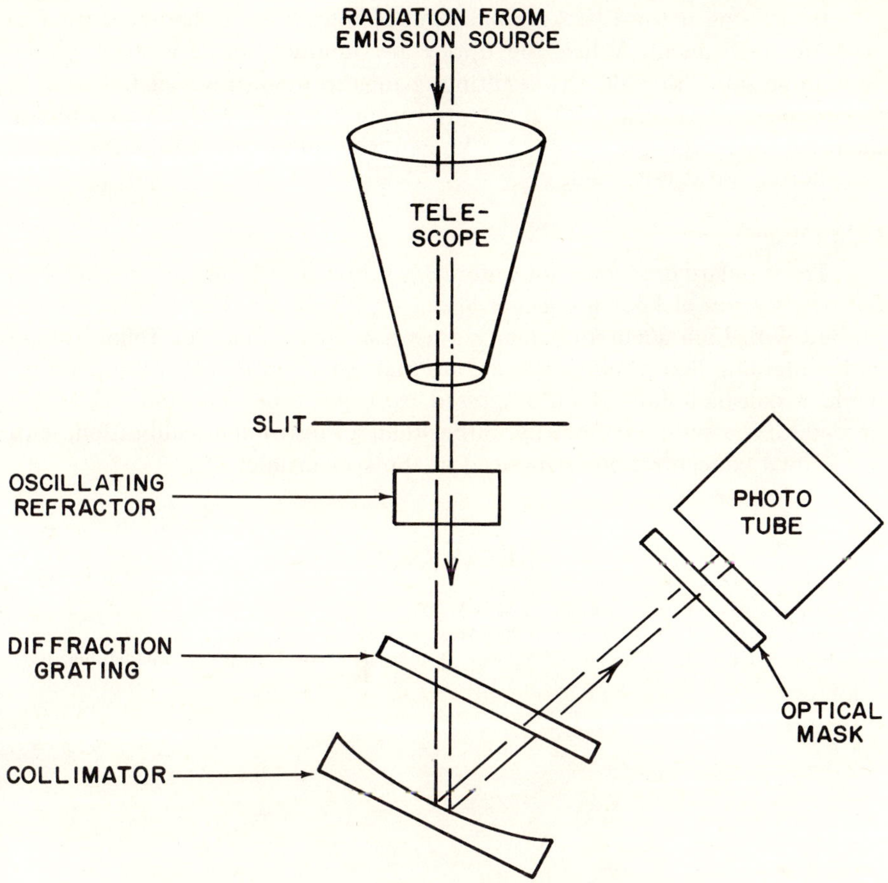

Figure 4.20h Remote sensing spectrometer.

practical full-scale range for these instruments is 0 to 200 ppm, with the upper limit in the range of whole percentages. This adequately covers the range of concentrations encountered in practice.

The gases must be filtered before reaching the instrument to prevent a buildup of particulate matter in the cell. Likewise, the moisture content of the sample gas must be controlled to prevent condensation.

Correlation Spectrometry

The correlation spectrometer can be used at a location remote from the source to determine the sulfur dioxide content of stack gases. The instrument in Figure 4.20h is specific for SO_2 because of the photographic optical mask of the SO_2 spectrum.

In use, the natural background level of ultraviolet radiation is used to zero the instrument. When the optics are pointed across a stack plume containing sulfur dioxide, the resulting change in the ultraviolet radiation in the wavelength pattern of SO_2 is detected at the instrument. If the stack plume diameter and stack gas density are known, the sulfur dioxide concentration can thereby be determined.

Calibration

The standard procedure for calibrating sulfur dioxide instruments involves the preparation of known concentrations of SO_2 in air. This can be accomplished with high accuracy when permeation tubes, made of Teflon tubing and containing liquid sulfur dioxide, are maintained at constant temperature while a constant flow of air is passed over the tube. The most desirable procedure is a system calibration rather than an instrument calibration, with the known concentration introduced at the system inlet.

REFERENCES

1. *Federal Register,* Part II, vol. 36, p. 158, August 14, 1971.
2. *Federal Register,* Part II, vol. 36, p. 84, April 30, 1971.
3. Lipták, B. G. (ed.): "Instrument Engineers' Handbook," Vol. I, Chilton, Philadelphia, 1969.

4.21 TOTAL OXIDANT ANALYZERS

Methods of Total Oxidant Detection:	(a) KI colorimetry; (b) KI coulometry; (c) KI galvanometry; (d) Ferrous ammonium sulfate; (e) Phenolphthalin.
Reference Method:	Chemiluminescence, gas phase O_3-ethylene reaction.
Costs:	\$1,000 to \$4,000 (b, c); \$2,500 to \$7,000 (a).
Partial List of Suppliers:	Atlas Electric Devices Co. (b, c); Beckman Instruments, Inc. (a, b, c); Mast Development Co. (b, c); Monitor Labs, Inc. (a); Ozone Research (b, c); Radiation, Inc. (a); Technicon Corp. (a).

Reference and Calibration

The use of total oxidant analyzers[1] is widespread and is principally a matter of convenience and simplicity. The total oxidant in air is taken to be the measure of those compounds other than atmospheric oxygen which will oxidize some reference material (Table 4.21a). In actual fact, the air may contain both oxidizing and reducing compounds (sometimes in considerable variety) which may react with a reference material in varying and even opposite fashion, although ozone is usually the predominant oxidant.

Almost universally, the reference compound is some type of potassium iodide solution (KI). The iodide may be oxidized to iodine which can be estimated analytically, usually by its ultraviolet light absorption. Although most atmospheric oxidants will react with KI to some degree, the rates and extent of reaction differ. Moreover, some reducing constituents interfere in the reverse direction, so that the KI analysis measures the *net* oxidant in the atmosphere in a very nonspecific way. It is obviously desirable and necessary to measure the specific individual oxidants themselves; for example, ozone, nitrogen dioxide, peroxyacyl nitrates and peroxides. (See Sections 3.11, 3.14,

Table 4.21a
TOTAL OXIDANT ANALYZERS

General Method	Type	Advantages	Disadvantages
Potassium iodide (KI) colorimetric	Neutral buffered	Precise, well-tested, widely accepted, continuous analysis, reference	Interferences by NO_2 and SO_2; collected sample not stable
	Alkaline	Stable after collection	More complex and not widely used; not continuous
KI coulometric and KI galvanometric		Simple, inexpensive, continuous	Interferences by NO_2 and SO_2; sometimes not in accord with neutral KI colorimetric method
Ferrous ammonium sulfate	Manual	High sensitivity	Reducing interferences; not in accord with neutral KI colorimetric method; not continuous
Phenolphthalein	Manual	Precise (once widely used)	Lengthy and tedious; sensitive to pH and temperature; results twice those obtained by neutral KI method

4.12 and 4.14.) Nevertheless, the KI determination remains in use as a primary standard because of its precision against specific reference oxidants and its ease of application.

The general reference method (manual) for oxidant, specified for National Air Quality Standards, is the neutral phosphate-buffered KI colorimetric technique of Saltzman[2], in which the air is passed through a neutral buffered 1% aqueous KI. The liberated tri-iodide (KI_3) is measured colorimetrically at a wavelength of 325 nm. One mole of ozone releases one mole of iodine with apparently only minor dependence on bubbler design and on phosphate concentration. Since ozone is usually the principal oxidant in air, this equivalence is advantageous. (For the same reason, the manual KI method is often used to calibrate ozone sources used for specific ozone sensors; see Section 4.14.)

This manual procedure for atmospheric oxidants is useful if analysis is completed immediately after sampling, because the iodine color fades with time. With care the method has a useful range of 0.01 to 10 ppm.

Continuous Oxidant Analysis by Colorimetric KI

When KI determination of oxidant is made continuously, the neutral buffered KI absorbing medium is normally of 10 to 20% concentration (to minimize effects of reagent depletion by reaction). Gas–liquid contact is fairly efficient and color development is very rapid. The solution is passed through a flow cell for a photometric measure of absorption at a filtered optical wavelength. Figure 4.21b is a schematic of a colorimetric KI oxidant analyzer.

The negative interference by sulfur dioxide (SO_2), a reducing gas, may be of consequence during periods of high SO_2 concentrations. It can be reduced by passing the air over chromium trioxide, although this also converts nitric oxide (NO) to nitrogen dioxide (NO_2) with variable efficiency depending on humidity. NO_2 interference (positive) may be rather serious. The results

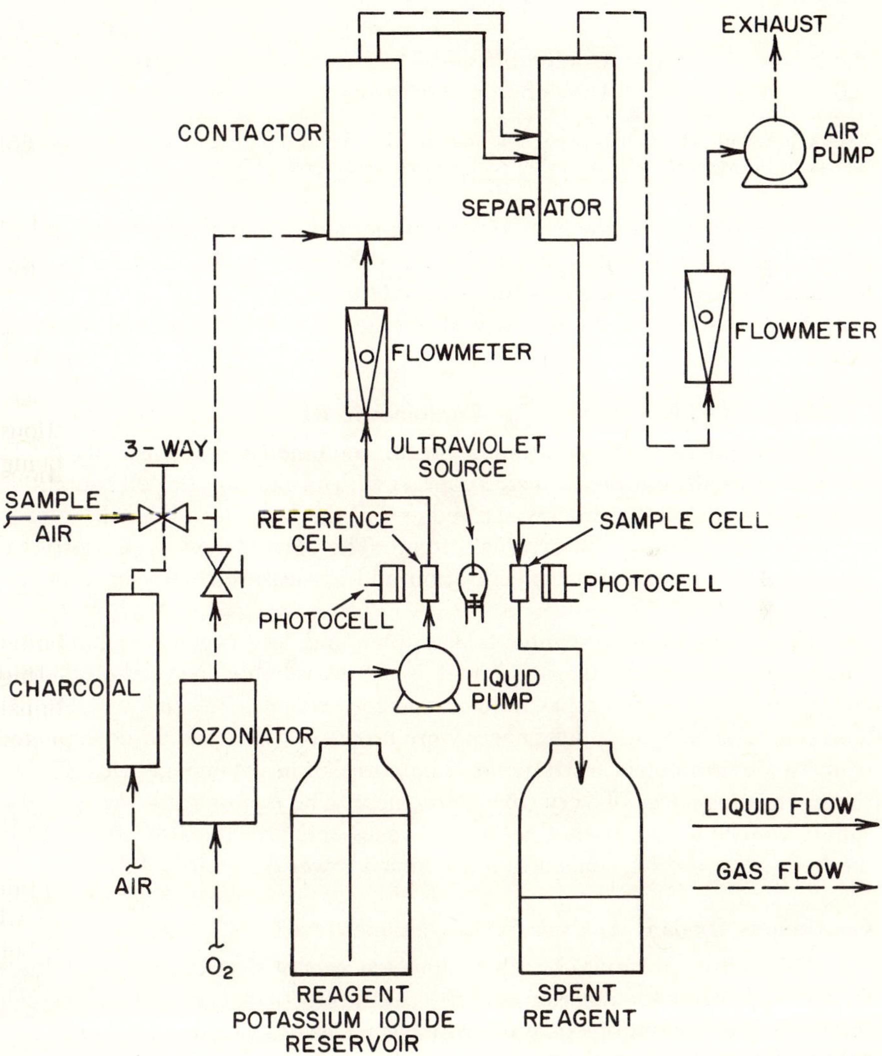

Figure 4.21b Colorimetric oxidant analyzer.

Table 4.21c
NO_2 INTERFERENCE IN TOTAL OXIDANT DETECTION

NO_2 in Air, ppm	*Method*	*NO_2 Concentration, % as Ozone*[a]
<36	1% KI, pH7 (manual)	8–11
1.5–5.5	2% KI, pH7 (manual)	6.4
2.0	10% KI, pH7 (continuous)	21
<1	20% KI, pH7 (continuous)	30
0.24–0.43	20% KI, pH7 (continuous)	12–47 (average 25)
2.0	Coulombic cell	10
2.0	U.V. photometer (ozone)	2

[a] NO_2 reacts with KI just like ozone, but more slowly, so that it is read as ozone to the extent shown for the various methods and for different concentrations of NO_2.

of a study of NO_2 interferences in various methods are shown in Table 4.21c.[1] NO_2 calculated as ozone in neutral buffered KI may reach 47% interference. It is useful to calibrate a given analyzer system at two or three concentrations of NO_2 for their ozone equivalence, if parallel NO_2 determinations are being made.

Continuous Oxidant Analysis by Coulometric KI

Mast-analysis is the common term for coulometric oxidant analysis.[3] In Mast-analysis, the sample air makes contact with an electrolytic cell containing KI. The free iodine thus liberated is reduced at the cathode of the cell, causing a current flow through an external circuit. The current flow is proportional to the iodine (or oxidant) concentration and is measured with an ammeter (Fig. 4.14b).

The coulometric instrument is simpler and less expensive than most colorimetric oxidant detectors, but it is not as reliable. Nevertheless, field comparisons of the two types showed that for oxidant levels below 0.20 ppm (corrected for NO_2), the differences were no greater than would be expected from two colorimetric instruments. The correlation coefficient was .87, but there were instances of very poor agreement. The coulometric method also suffers from NO_2 interference, although to a lesser degree than the colorimetric method, since the KI concentration is much lower (see Table 4.21c).

Continuous Oxidant Analysis by Galvanometric KI

This method is similar to the coulometric except that no external voltage is applied. Instead, the reaction occurring at the electrodes generates a current, related to the amount of oxidation, which is measured galvanometrically (Fig. 4.21d). The interferences are similar to those affecting coulometry.

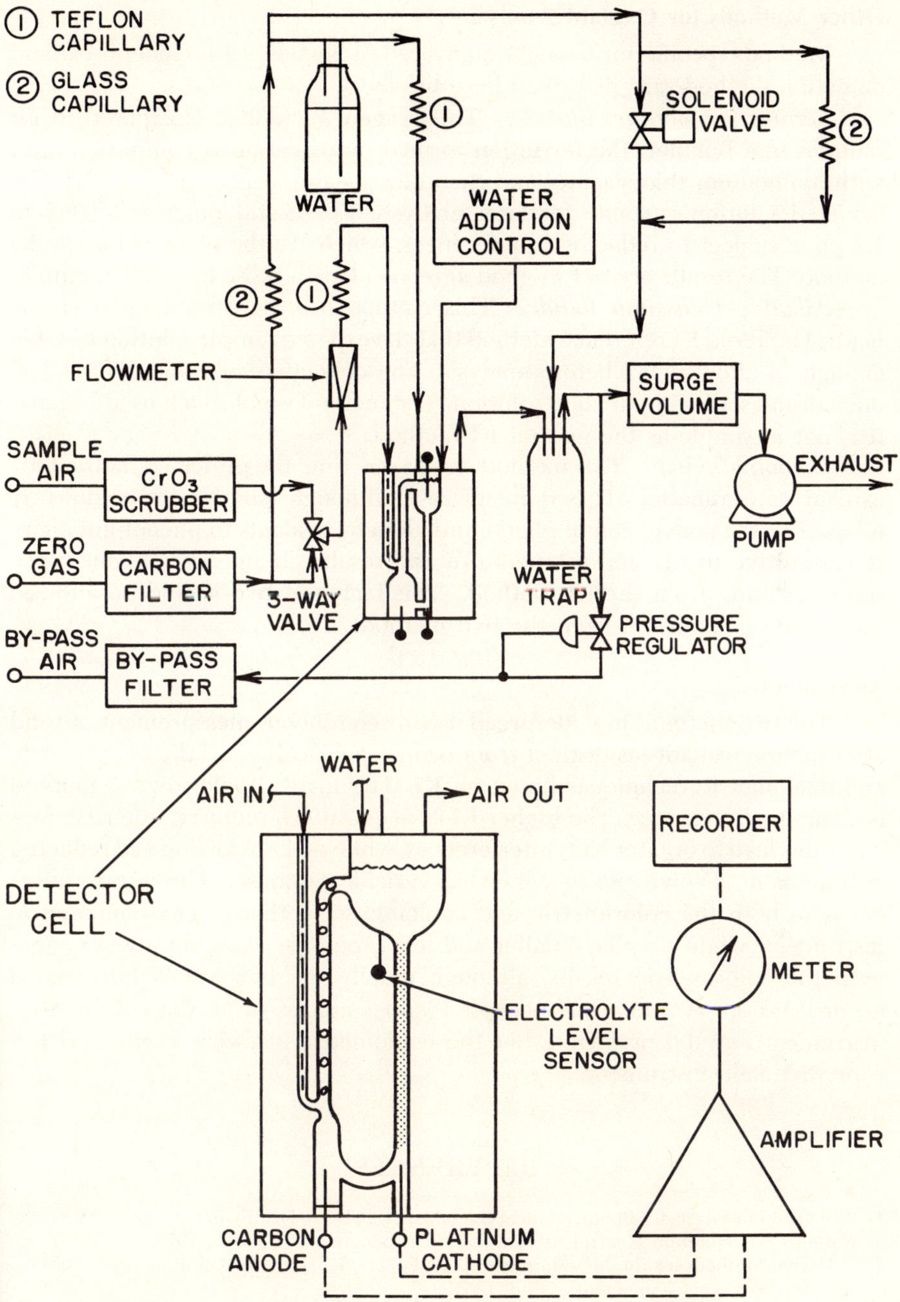

Figure 4.21d Galvanic oxidant analyzer.

Other Methods for Oxidant Analysis

Methods specific for ozone are covered in Section 4.14. General oxidant detection methods are described here briefly.

Ferrous Ammonium Sulfate. This reagent (acidified) is exposed to air samples in a bubbler, the ferric ion formed by oxidation is complexed later with ammonium thiocyanate, and the color measured photometrically. The oxidized solution is stable for later analysis. The useful range is 0.0005 to 1.2 ppm, subject to reducing interferences, which are the same as for the KI method. The results are not in good agreement with those by other methods.

Alkaline Potassium Iodide. This method has the advantage over the neutral buffered KI reference method that the exposed sample solution is stable enough to allow delay before analysis. The collection solution is 1% KI in one normal sodium hydroxide solution. The method is not much used because it is not as simple as the neutral KI method.

Phenolphthalein. This method was at one time the preferred method for oxidant determination. It is quite precise, although lengthy and tedious. It is based on the conversion of phenolphthalein by oxidants to phenolphthalein. It is sensitive to pH and temperature and results obtained are about twice those resulting from the KI method. (This factor of two should be allowed for in data comparison from the two methods.)

Assessment

The two methods in widespread use for continuous measurement of total atmospheric oxidant (as distinct from ozone specifically) are the colorimetric and coulometric techniques based on KI. The manual colorimetric method is a standard reference. The higher KI concentrations required in continuous analyzers lead to greater NO_2 interferences, while other oxidizing and reducing pollutants in air give rise to somewhat variable response. (The interferences occur in both the colorimetric and coulometric methods.) The coulometric instrument, while simpler, smaller and less expensive, does not always agree with the colorimetric results, although usually the two are in fairly good accord. Where accurate and accepted values are essential, the colorimetric instrument may be preferred, but the coulometric is a very useful and less expensive field instrument.

REFERENCES

1. "Air Quality Criteria for Photochemical Oxidants," U.S. Dept. Health, Education and Welfare, National Air Pollution Control Administration Publication No. AP-63, 1970.
2. "Selected Methods for the Measurement of Air Pollutants," Div. Air Pollution, PHS Publication No. 999-AP-11, Cincinnati, Ohio, 1965.
3. Mast, G. M. and H. E. Saunders: *Instr. Soc. Amer. Trans.*, vol. 1, pp. 325–328, 1962.

4.22 WET CHEMISTRY AND COLORIMETRY METHODS

Wet chemistry and colorimetry methods are becoming obsolete nowadays. However, they are still used as reference methods and often are the only specific methods available to measure some pollutants. The accuracy of any measurement depends on the sampling as well as the actual measurement itself. Improper sampling leads to worthless data.

Gas Sampling

For the chemical analysis of air and gases, two basic methods of sampling are of general interest (Sections 4.17 and 4.18). In the first method, a definite volume of air is trapped in a container which is usually evacuated before sampling; while in the second method, a measured volume of air is passed through a solution, an absorbing medium or a solid absorbant (Figs. 4.17a, b and c). For colorimetric methods, this last sampling technique is the most appropriate. The pollutant can be concentrated to suit the working range of the method by sampling for a sufficient length of time. Some problems, such as the evaporation of the solvent, when collecting in aqueous medium, could limit the sampling time. In order to get accurate results, the measurement of the volume of air sampled and the efficiency of collection of the absorbing medium, under specific conditions, have to be precisely known.

Solid Sampling

Solids are, in most cases, collected by filtration on membrane, glass fiber, polystyrene and paper filters (Sections 4.15, 4.17 and 4.18). Different types and sizes of pumps are available, but the Hi-Vol sampler is the most useful for the collection of the particulates from large volumes of air (Fig. 4.15a). An area aliquot of the exposed filter is then used for the measurement of the pollutant of interest. The aliquot size to be taken depends upon the concentration of the compound present in the air sampled and on the mass present upon the area aliquot taken. Preliminary experiments are often required to decide upon optimum sampling conditions which will match the sensitivity available from the analytical finish.

For gases, no special sample preparation is needed since they are generally collected in solution. Similarly, for solids or particulate matter, it is sometimes possible to recover the compounds by a simple extraction. However, in most cases, the samples have to be digested by a procedure such as wet ashing with oxidizing acids to produce a prepared sample which is acceptable to the analytical method or instrument used.

Extraction

The solvent and the type of extraction used will depend on the nature of the pollutant. For the measurement of anions such as SO_4^{--}, NO_3^-, Cl^- and F^-, distilled water is the appropriate solvent. For metals, dilute acids such as hydrochloric may replace water. These extractions can be accomplished in two different ways. In some cases, an aliquot of the sample is extracted in a Soxhlet, while in others extraction may be done by repetitive washings with small volumes of warm solvent.[1,2] The extracts are then ready for measurement or for separation by conventional techniques such as ion exchange.

Digestion

In order to obtain a quantitative extraction of some compounds, it may be necessary to remove organic and carbonaceous material with oxidizing acids. Depending on the compound to be measured and the filter used, the samples are digested with acids such as nitric acid; mixtures of nitric and sulphuric; or mixtures of nitric, sulphuric and perchloric acids. A proper choice of the acid has to be made so as not to interfere with further color development. All these details are described in the methods mentioned in Tables 4.22a and 4.22b.

Once the digestion is completed, the measurement can be carried out directly on the solution or by extraction of the metal with an organic solvent following formation of a complex. Since much air sampling is based on the use of glass fiber sheets, it is worthwhile to mention one elegant method of sample preparation in such cases, namely the dissolution of the filter with hydrofluoric acid.[3] In this case, complete disintegration of the filter takes place, and the digestion can proceed with other acids as needed. The metals can again be extracted for colorimeter analysis. Figure 4.22c is a schematic diagram of sample preparation for the complete analysis of air samples taken on glass fiber filters.

Measurement

Tables 4.22a and b list reference methods for the measurement of air pollutants. These are tentative methods published by the Intersociety Committee on Recommended Analytical Methods.[4-7] Since all these methods are colorimetric, they make use of the same principle of the absorption of light by solutions at specific wavelengths.

Table 4.22a

REFERENCE COLORIMETRIC METHODS FOR THE MEASUREMENT OF PARTICULATE POLLUTANTS

Compounds	*References*	*Reagent*	*Separation*	*Range (μg/ml)*	*Collection*
Arsenic	9, 10	Silver diethyl dithiocarbamate	Reduction to arsenic	0–1.5	Glass fiber sheets, membrane or filter paper
Fluoride	11–15	Lanthanum-Alizarin Complexone,	Distillation	0–0.5	
		Zirconium-Eriochrome Cyanine R,	or ion exchange	0–1.40	
		Zirconium-Spadns	or diffusion	0–1.40	Paper filter
Antimony	24, 25	Rhodamine B	Extraction	0–10	Membrane, cellulose or glass fiber filters
Selenium	26, 27	2,3 Diaminonaphthalene	Ion exchange	0–0.1 fluorometric 0–4 spectrophotometric	Membrane, cellulose or glass fiber filters
Manganese	24, 28, 29	Potassium periodate	No separation	0.01–0.6	Membrane, cellulose or glass fiber filters
Molybdenum	30, 31	Potassium thiocyanate	Extraction	0–0.6	Membrane, cellulose or glass fiber filters
Phenolic compounds	42–45	4 aminoantipyrine	Distillation	0.05–2.0	Glass fiber filters
Iron	51, 52	Bathophenanthroline	Extraction	0.1–2.0	Membrane, cellulose filters. Electrostatic precipitation
Nitratea	48, 49	2,4 Xylenol	Extraction	0–10	Glass fiber filters
Beryllium	54, 55	Morin	No separation	0.001–0.1 solution	Paper, membrane or glass fiber filters. Fluted pad

Table 4.22b
REFERENCE COLORIMETRIC METHODS FOR THE MEASUREMENT OF GASEOUS POLLUTANTS

Compounds	References	Reagent	Separation	Range, μg/ml	Collection
Fluoride	11–15	Lanthanum-Alizarin Complexone, Zirconium-Eriochrome Cyanine R, Zirconium-Spadns	Distillation or ion exchange	0–0.5 0–1.40 0–1.40	Filter paper impregnated with calcium based fixative agents *or* water or alkaline solution
Nitrogen dioxide	16	Diazotization of sulfanilic acid then coupling with N-(1-naphthyl) ethylenediamine dihydrochloride	No separation	0–12	In a fritted bubbler containing an azo dye forming reagent
Sulfur dioxide	17–19	Pararosaniline	No separation	0–1.4	Midget impingers, standard fritted bubbler, multiple jet bubbler containing sodium tetrachloromercurate
Carbon monoxide	20, 21	Silver salt of p-sulfaminobenzoic acid	No separation	0–0.05	Evacuated erlenmeyer flask
Formaldehyde	22, 23	Chromotropic acid-sulfuric acid solution	No separation	0.1–2.0	Fritted bubbler containing distilled water
Formaldehyde	38, 39	3-methyl-2-benzothiazolone hydrazone hydrochloride	No separation	0–0.7	Fritted bubbler with 3-methyl-2-benzothiazolone hydrazone hydrochloride
Oxidizing substances	32, 33	Potassium iodate	No separation	0.02–0.21 solution	Midget impingers with potassium iodate
Hydrogen sulfide	34–37	N,N-dimethyl-p-phenel-enediamine and ferric chloride	No separation	0–0.2 solution	Midget impingers with alkaline suspension of cadmium hydroxide
Acrolein	40, 41	4-hexylresorcinol	No separation	0.1–3.0 solution	2 fritted bubblers containing 4-hexylresorcinol in trichloroacetic acid
Phenolic compounds	42–45	4-aminoantipyrine	Distillation and extraction	0.05–2.0 solution	Impingers containing 0.1N NaOH
Total nitrogen oxides	46, 47	Phenoldisulfonic acid	No separation	0–0.0188	Evacuated glass sampling flask
Mercaptans	50	N,N-dimethyl-p-phenel-enediamine and ferric chloride	No separation	0–1.0	Midget impingers containing mercuric acetate
Free chlorine	53	Methyl orange	No separation	0–1.0 solution	Fritted bubbler containing an acid solution of methyl orange

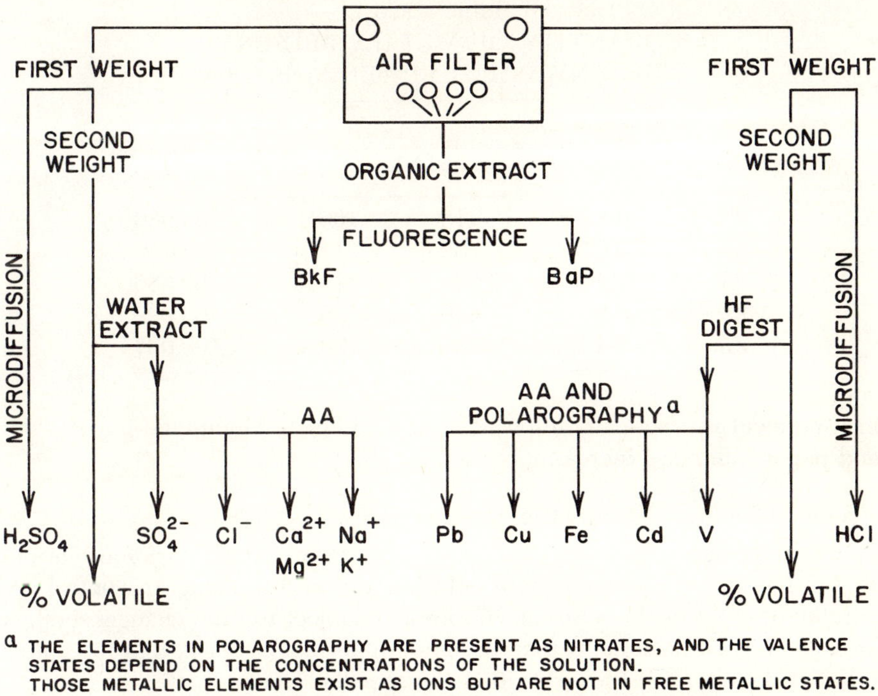

Figure 4.22c Sample preparation for air sample analysis.

Blank

In any analytical method, a blank* has to be run at the same time the sample is being studied. By so doing, the operator can detect any possible contamination which could occur during the measurement. The blank value should be as low as possible. Consequently, all reagents used should be purified. Water should be double distilled. The blank should be obtained by following as closely as possible the same procedure followed for the sample.

Since glass fiber sheets are used on a large scale for the measurement of metals in air, a special note of caution has to be added. These sheets may contain a large number of chemical elements in high concentrations. Table 4.22d gives the concentration of some of the metals present in an 8 × 10-inch filter sheet as used for high-volume sampling. The acid digested blanks were obtained after destruction of the sheets with hydrofluoric acid. It can be observed that the concentrations of the elements listed are so high that their

*Blank is a determination made as close as possible under the same conditions as a true determination, but with the omission of the substance to be tested for the purpose of ascertaining the effect due to associated factors (such as impurities in the reagents).

Table 4.22d
QUANTITY OF METALS PRESENT IN UNWASHED FILTER SHEETS

Metal	Blanks Aqueous Extracts		Blanks Acid Digests	
	μg/cm²	*μg/sheet*	*μg/cm²*	*μg/sheet*
Ca	0.8	418	166.9	84,980
Mg	0.07	37	11.3	5,750
Ba	7.2	3,669	313.8	159,686
K	0.8	426	125.7	63,992
Zn	0.4	203	196.6	100,097

measurement is practically impossible on such filters. Membrane polystyrene and paper filters are increasingly used for this reason.

Factors Affecting Colorimetric Measurements

As a change in hydrogen ion activity can change the color of a solution, it is important in the majority of colorimetric measurements to control accurately the pH of the solution. In solutions subject to such changes, buffers are added when possible. It then becomes important to check that the buffer solution does control the pH within the specified limits.

The shape of the absorption curve of a colored solution sometimes varies with temperature. These changes can be reversible or irreversible, depending on the colored complex. One should avoid variations in temperature higher than 5°C during color development.

The range of concentrations measured is, in general, in the linear portion of the curve of concentration against optical density or absorbance. With methods where the fluorescence of a complex is measured rather than the absorbance, the curve of concentration against fluorescence intensity does go through a maximum. Consequently, two concentrations correspond to a given fluorescence reading; one on each side of the maximum. A given reading, obtained on an unknown solution, may possibly be on either side of the maximum. In such a situation, it is recommended to dilute the solution and make another measurement. The second measured concentration will differ from the first value if the first reading was on the wrong side of the maximum.

Under the action of light, the colored compound can be permanently destroyed. Consequently, the samples should not be left too long in the beam of the spectrophotometer. It is good practice to measure the color of the sample as soon as the color development is made so that it is not exposed to light for long periods of time. Direct sunlight on the sample before measurement should never be allowed.

Interferences

The presence of interfering compounds may contribute to the nonlinearity of the curve representing the relationship between concentration and absorbance. Although analytical methods, in general, deal with specific interfering substances, it is possible that such interfering compounds can be introduced during the measurement by impurities present in distilled water or in reagents. To avoid such possibilities, extreme purity of solvent and reagents is a necessity.

Samples collected on glass fiber sheets, even after filtration through analytical grade filter paper, still contain tiny glass fibers which scatter the light. The extracts from these sheets have to be filtered through a membrane filter if color development is to be made directly on the extracts. Neither filter paper nor glass wool should be used to remove water droplets or turbidity in the colored solution before measurement, since some colored complexes are retained on paper and small glass fibers contribute an artificial absorbance. Lastly, cuvettes should be carefully cleaned before a reading is taken.

Conclusion

Colorimetric methods, as a general rule, are lengthy, tedious and complicated. Efforts designed to make them more rapid usually contribute to a decrease in their specificity. Their chief advantage is that no expensive equipment is required. The West-Gaeke method for sulfur dioxide may be used as a reference method. In practical applications, they have to be inferior to methods which utilize an instantaneous reaction in the gas phase such as the flame ionization detector for hydrocarbons. There is bound to be an increase in the use of such direct-reading instruments when they can be shown to be reliable. It may soon be established, for example, that instrumental methods for measuring oxides of nitrogen and ozone in air may be more reliable than the colorimetric iodide method which is now used for the calibrating reference.

Wet chemistry, on the other hand, has been of great utility in the measurement of metals in air. The value of a measurement of an air sample is of course, affected by the previous treatment given to the sample. It is important in the preparation of a sample to avoid volatilization losses, precipitation or co-precipitation or any treatment which may lead to inaccurate results. Wet chemical methods involving aqueous extraction or acid digestion lead logically to the use of atomic absorption (AA) spectroscopy. The blank due to the use of digesting acid must be kept low by using best quality material in minimal amount. With the advent of flameless AA (Fig. 4.11g), a breakthrough may be expected where wet digestion may be simplified or eliminated in some cases. The increase in sensitivity which flameless AA brings will undoubtedly lead to simplification and improvement of the wet chemistry.

REFERENCES

1. "Air Pollution Measurements of the National Air Sampling Network." U.S. Department of Health, Education and Welfare, Cincinnati, Ohio, 1965.
2. Dubois, L., R. S. Thomas, T. Teichman and J. L. Monkman: *Mikrochimica Acta,* pp. 1268–1275, 1969.
3. Zdrojewski, A., H. Chung, L. Dubois and J. L. Monkman: The Accurate Measurement of Lead in Airborne Particulates, in press, *Journal of Environmental Analytical Chemistry.*
4. Intersociety Committee, "Manual of Methods of Ambient Air Sampling and Analysis," April 1969.
5. Ibid, January 1970.
6. *Health Laboratory Science,* vol. 7, no. 3, 1970.
7. Ibid, vol. 8, no. 1, p. 53, 1971.
8. Zdrojewski, op. cit., vol. 1, no. 2, p. 113, November 1971.
9. Determination of Arsenic in Air, *American Conference of Governmental Industrial Hygienists Manual,* 1955.
10. Vasak, W. and V. Sedivec: *Chem Listy,* vol. 46, pp. 341–344, 1952.
11. Belcher, R. and T. S. West: *Talanta,* vol. 8, p. 853, 1961.
12. Hall, R. J.: *Analyst,* vol. 88, p. 76, 1963.
13. Megregian, S.: *Anal. Chem.,* vol. 26, p. 1161, 1954.
14. Nielsen, J. P. and A. D. Dangerfield: *A.M.A. Arch. Ind. Health,* vol. 11, p. 61, 1955.
15. Rowley, R. J. and G. H. Farrah: *Amer. Ind. Hyg. Ass. J.,* vol. 23, p. 314, 1962.
16. Saltzman, B. E.: *Anal. Chem.,* vol. 26, p. 1949, 1954.
17. West, P. W. and G. C. Gaeke: *Anal. Chem.,* vol. 28, p. 1816, 1956.
18. Zurlo, N. and A. M. Griffini, *Med. Lasaro,* vol. 53, p. 330, 1962.
19. Scaringelli, F. P., B. E. Saltzman and S. A. Frey: *Anal. Chem.,* vol. 39, p. 1709, 1967.
20. Ciuhandu, G. Z.: *Anal. Chem.,* vol. 155, p. 321, 1957.
21. Levaggi, D. A. and M. Feldstein: *Amer. Ind. Hyg. Ass. J.,* vol. 25, p. 64, 1964.
22. Altshuller, A. P., L. J. Leng and A. F. Wartburg: *Int. J. Air Wat. Pollu.,* vol. 6, p. 381, 1962.
23. MacDonald, W. E.: *Amer. Ind. Hyg. Ass. Quarterly,* vol. 15, p. 217, 1954.
24. American Conference of Governmental Industrial Hygienists: "Manual of Recommended Analytical Methods," 1963.
25. Maren, T. H.: *Anal. Chem.,* vol. 19, p. 487, 1947.
26. Lott, P. F., P. Cukor, G. Moriber and J. Solga: *Anal. Chem.,* vol. 35, p. 1159, 1963.
27. Parker, C. A. and L. G. Harvey: *Analyst,* vol. 87, p. 558, 1962.
28. Richards, M. B.: *Analyst,* vol. 55, p. 554, 1930.
29. Willard, H. H. and T. H. Greathouse: *J. Am. Chem. Soc.,* vol. 39, p. 2366, 1917.
30. Hibbits, J. O. and R. T. Williams: *Anal. Chem. Acta,* vol. 26, p. 363, 1962.
31. Furman, N. H. and W. M. Murray, Jr.: *J. Am. Chem. Soc.,* vol. 58, p. 1689, 1936.
32. Byers, D. H. and B. E. Saltzman: *J. Am. Ind. Hyg. Ass.,* vol. 19, p. 251, 1958.
33. Saltzman, B. E. and N. Gilbert: *Anal. Chem.,* vol. 31, p. 1914, 1959.
34. Jacobs, M. B., M. M. Braverman and S. Hockheiser: *Anal. Chem.,* vol. 29, p. 1349, 1957.
35. Bamesberger, W. L. and D. F. Adams: *Environ. Sci. and Tech.,* vol. 3, p. 258, 1969.
36. Mecklenburg, W. and R. Rozenkranzer: *A. Anorg. Chem.,* vol. 86, p. 143, 1914.
37. Boström, C. E.: *Air and Water Pollu. Int. J.,* vol. 10, p. 435, 1966.
38. Sawicki, E., T. R. Hauser, T. W. Stanley and W. Elbert: *Anal. Chem.,* vol. 33, p. 93, 1961.
39. Hauser, T. R. and R. L. Cummins: *Anal. Chem.,* vol. 37, p. 679, 1964.
40. Cohen, I. R. and A. P. Altshuller: *Anal. Chem.,* vol. 33, p. 726, 1961.
41. Altshuller, A. P. and S. P. McPherson: *J. Air Pollu. Contr. Ass.,* vol. 13, p. 109, 1963.
42. Emerson, E. J.: *J. Organic Chem.,* vol. 8, p. 417, 1943.
43. Etlinger, M. D., C. C. Ruchhoft and R. J. Lishka: *Anal. Chem.,* vol. 23, p. 1783, 1951.
44. Mohler, E. F. and L. N. Jacob: *Anal. Chem.,* vol. 29, p. 1369, 1957.
45. Smith, R. G., J. D. MacEwen and R. E. Barrow: *Amer. Ind. Hyg. Ass. J.,* April 1959.
46. Chamot, E. M., D. S. Pratt and H. W. Redfield: *J. Amer. Chem. Soc.,* vol. 31, no. 1909, p. 922; vol. 32, no. 1910, p. 630; vol. 33, no. 1911, p. 336.
47. Saltzman, B. E.: *Anal. Chem.,* vol. 26, p. 1949, 1954.

48. Huey, N. A.: Public Health Service Publication No. 999-AP-11, pp. J-1 to J-4, May 1965.
49. Barnes, H.: *Analyst,* vol. 75, p. 388, 1950.
50. Moore, H., H. L. Helwig and R. J. Graul: *Ind. Hyg. J.,* vol. 21, p. 466, 1960.
51. Smith, C. F., W. H. McCurdy and H. Diehl: *Analyst,* vol. 77, p. 418, 1952.
52. Reynolds, R. G. and J. L. Monkman: *A.I.H.A. Journal,* vol. 23, p. 415, 1962.
53. Taras, M.: *Anal. Chem.,* vol. 19, p. 342, 1947.
54. Sill, C. W. and C. P. Willis: *Anal. Chem.,* vol. 31, p. 598, 1959.
55. Walkley, J.: *A.I.H.A. Journal,* vol. 20, p. 241, 1959.

Chapter 5

GENERAL AIR POLLUTION CONTROL TECHNIQUES

R. A. Armer R. L. Bennett J. H. Black
D. E. Cole B. Dimitriades
R. H. Essenhigh J. W. T. Ferretti
R. G. Gantz W. J. Hatcher, Jr. D. A. Hirscher
D. H. F. Liu C. J. Santhanam I. S. Shah

CONTENTS OF CHAPTER 5*

*For a glossary of terms used, refer to the back portion of the volume and for a list of abbreviations used in this chapter, refer to the listing given after these contents, preceding the first section.

Contents of Chapter 5

Abbreviations

Å	Angstrom (10^{-10} meter or 10^{-4} micron)
acfm, ACFM	Actual Cubic Feet per Minute (2.12 liter/sec)
A/F	Air to Fuel Ratio

Contents of Chapter 5

ACGIH	American Conference of Government Industrial Hygienists
atm, ATM	Atmosphere (14.7 PSI)
BTU	British Thermal Unit
cal	Calories (gram calorie)
cfm, CFM	Cubic Feet per Minute (2.12 liter/sec)
cgs	Centimeter-gram-second
C/H	Carbon/Hydrogen Ratio
cm/sec	Centimeter per second
CO	Carbon Monoxide
DIF	Dry-Inert-Free
ERG	Exhaust Gas Recirculation
fpm, FPM	Foot per Minute
fps, FPS	Foot per Second (30.5 cm/sec)
FRP	Fiber Glass Reenforced Plastic
GPH, gph	Gallon per Hour
GPM, gpm	Gallon per Minute
g, gm	Gram (15.4 grains)
gr, grain	15.4 grains in 1 Gram
HC	Hydrocarbon
H-E	High Efficiency
$''H_2O$	Inches of water column
hp	High Pressure
HP	Horsepower
HV	High Voltage
ID	Inside Diameter
ISA	Instrument Society of America
lbf	Pound Force
lbm	Pound Mass
liter/sec	Is equivalent of 0.47 CFM
lp	Low Pressure
LP	Liquified Petroleum
m, M	Meter
M	Million
ma	Milliampere
MAC	Maximum Allowable Concentration
mg, mgm	Milligram (gram/1000 or 0.015 grains)
mppcf	Million Particles per Cubic Foot
Mw	Megawatt (10^6 watts)
NH_3	Ammonia
NDIR	Nondispersive Infrared
NO_x	Nitrogen Oxides
N_s	Separation Number
N_t	Number of Transfer Units

P-A	Pease-Anthony
pphm	Parts per hundred Million
ppm	Parts per Million
psia, PSIA	Pounds per square inch absolute
psig, PSIG	Pounds per square inch gauge
P_t	Total Contacting Power
RPM, rpm	Revolution per Minute
SCFM	Standard cubic feet per Minute (2.12 liter/sec)
S-F	Svenska-Flaktfabriken
sq	Square
SSF	Saybolt Furol Seconds
SSU	Saybolt Universal Seconds
STP	Standard Temperature Pressure
TCA	Turbulent Contact Absorber
TLV	Threshold Limit Values
TVA	Tennessee Valley Authority
USPHS	U.S. Public Health Service
W-A	Wet Approach
μ	Micron (10^{-6} meter)

5.1 TYPES OF AIR POLLUTION CONTROL EQUIPMENT

This section serves as a general introduction and orientation to the types of equipment that are available for use in air pollution control work.

The various types of air pollution control equipment can be broken down into the following categories:

1. Thermal combustion
2. Electrostatic precipitators
3. Dry collectors
4. Wet collectors
5. Other, newer designs

Thermal Combustion Equipment

This class of equipment is applied to effect complete combustion of all chemical components into water, carbon dioxide and the ash of the incombustibles. This category of pollution abatement hardware includes:

1. Direct flame incinerators
2. Afterburners
3. Fume destructors
4. Catalytic afterburners
5. Thermal oxidizers

All of these are thermal combustion equipment and are used to control the gaseous (nonparticulate) emissions. Thermal combustion equipment is most suitable for the abatement of the following pollutants and pollution sources:

1. Aldehydes
2. Lithographic processes
3. Mercaptans
4. Oil mists
5. Organic solvents
6. Paint drying
7. Phenols
8. Phthalic anhydrides
9. Rendering odors

10. Sewage gases
11. Sludge incinerators
12. Smokehouses
13. Sponge rubber curing processes

The thermal combustion system can be of the direct flame or the catalytic combustion (flameless) type.

The sole difference between these two processes is the temperature at which they operate. The drawback to direct flame combustion is its high cost of operation. The catalytic combustion process is similar to the catalytic cracking methods used in oil refineries.

Direct Flame Combustion

Afterburners and incinerators are discussed in some detail in the later sections of this chapter. When used for air pollution control, they require that the total BTU input and the dwell time of the system (0.2 to 1 second) be within acceptable limits in order to guarantee complete combustion. Operating costs associated with this type of equipment are usually high and can be reduced by the installation of heat recovery equipment.

Catalytic Combustion (Flameless)

Catalytic combustion is a flameless, low temperature (500° to 700°F) combustion process. Most catalysts used belong to the platinum family and can operate at vapor velocities of about 300 feet per minute. The important design criteria include the temperature and homogenization of the exhaust mixture. Where the exhaust temperature is high, a heat recovery section is usually included.

A catalyst that is applied to a horizontal gas stream usually appears in square form similarly to a filter in a filter bank. The vertical gas stream catalyst is usually cylindrical to reduce the pollutant stream leakage and to ensure greater contact area.

For the reliable performance of catalytic oxidation systems, it is important to maintain constant both the flowrate and the composition of the gas stream being treated.

Once a catalyst is designed for a specific exhaust, it should not vary in composition because contamination of the catalyst bed can render it ineffective.

Satisfactory applications of the catalytic combustion hardware include:

1. Asphalt oxidizing
2. Petroleum refining
3. Oil quenching
4. Paint and varnish cooking
5. Pharmaceutical manufacturing

6. Sewage treatment off gases
7. Food processing
8. Oven processing
9. Metal decorating
10. Paint baking
11. Wire enameling
12. Brake shoe bonding and burn off
13. Fabric curing
14. Paper printing and impregnating
15. Fish and vegetable oil processing

Electrostatic Precipitators

The electrostatic precipitator design is based on passing a dust-laden gas stream through a highly charged electrostatic field which then charges the dust or smoke and causes agglomeration and precipitation of the accumulated particles.

The precipitators first found use in such industries where the material reclaimed had a monetary value. Later improvements and economies have broadened its applications to the air conditioning field; the chemical industry; flyash, smoke and dust abatement; gas cleaning; and reclamation.

The units operate by producing a unidirectional electrostatic field between two electrodes and by "sweeping" the dust from the air stream as it passes through the field. The dust is deposited on the inside surfaces of the chamber where it is collected and later removed by periodic shaking. The apparatus usually consists of a bundle of vertical metallic tubes through which the dusty air flows. In each pipe, an insulated wire electrode is mounted. When a potential difference of 15,000 to 75,000 volts is applied between the pipe and wire electrodes, the dust particles are charged and move toward the wall. The electric potential as well as the number, size and length of the tubes and the air velocity required are dependent upon the amount and size of dust particles.

A somewhat different method of collection has been used in air conditioning applications. Here, the airstream first passes through an ionizing field at a potential of 10,000 to 12,000 volts. As the air passes through this field, the particles and molecules are electrically charged. The charge that a particle of material can carry is a function of its area. After being charged, the particles and molecules in the air pass between collector plates. These plates are at a potential of 5,000 volts. The ionized particles are attracted to the plates and are deposited there and lose their charge. The rate of recombination of the ions which charge the molecules of air is extremely rapid.

The electrostatic precipitators are frequently used in conjunction with air filters which take out the larger dust particles and prevent rapid loading up of the precipitator plates. The nonautomatic units can be cleaned with a water spray. In the automatic units, the precipitator plates are periodically renewed by being dipped into an oil bath to restore the viscous film.

A newer development is the wet electrostatic precipitator. This has not, at this writing, been practically applied nor come into its own by acceptance but it must be noted to have several advantages. Whereas true gas molecules would have difficulty in being collected by a dry precipitator, the wetting in this new development does effect a more complete treatment of the variety of pollutants within a single stream.

Electrostatic precipitator capacities used to be limited to the high flow-rate (over 150,000 SCFM) applications; but in recent years, several smaller units became available. This trend has brought the electrostatic precipitator into applications outside the traditional uses, which are the standard coal-burning power plants, steel-making furnaces and cement plants.

Dry Collectors

The following types of equipment can be found within this category:

1. Cyclones
2. Filters
3. Baghouses
4. Moving filters

Cyclones

The cyclone or centrifugal dust separator is primarily used to collect the heavier dust particles and materials such as sawdust, shavings and dust from grinders. This does not preclude its use in the separation of fine particles.

The operation of the centrifugal dust separator utilizes the vortex motion to obtain the centrifugal force which causes the dust particles to move outward and to concentrate in the outer layer of the moving air. The air and dust enter the cylinder tangentially at substantially the same velocity and they move in long sweeping spirals down the collector. The air motion forms a free vortex which is whirling faster at the central part of the collector than near its periphery. As the air turns and moves upward through a central discharge pipe, some of the dust still carried by the airstream will also be discharged. One important design criterion is to maintain the residence time required for the dust particle to move through the airstream; and another is to maintain the airflow with a minimum of turbulence.

Where the finest separation is desired, large collectors will not function as well as several smaller size units. These multicyclones give better cleaning and provide a more compact arrangement but also require more pressure drop.

Several styles are available, including the cone shapes and the cylindrical bodies. The collector with the air entrance at a greater distance from its center line will have a higher pressure loss and will do finer cleaning. The multicyclone has a high length-to-diameter ratio and the static pressure loss through such units is around 2 to 3″ H_2O.

There are various modifications of the dry collector. If the problem is one of cleaning the air, a small part of the air current together with the concentrated dust may be discharged to the atmosphere. If the task is one of collecting the dust, the smaller air current with its dust may be passed through another and smaller collector.

Filters

Few filters depend upon the lodgment of particles in the crevices of porous material. This is because the pores are seldom as small as the dust particles which are intended to be filtered out; and if they were, they would be very difficult to clean. Generally, the principles of operation are impingement, inertia or centrifugal action.

Dry filters are frequently of the throw-away type or are arranged to be cleaned by shaking or by water or steam. The filter pack may be cloth, paper, hair, glass, wire or metal plates in the cleanable designs. Throw-away filters are usually packed with some form of cellulose material.

The dry filter has also been used for ultra-filtration at higher pressure drops in "clean-room" applications.

Baghouses

A baghouse is a flexible media fixed in a position to act as a stationary filter. Usually the media of the baghouse is a textile product woven to specific requirements of dust loading and temperature. Baghouses are a grouping of fabric filters, usually cylindrical in shape, having the ability to collect the agglomerated dust which is trapped within the system.

To remove accumulated dust, several methods are used, including: (1) "Puff-back" which is the reversing of the air pressure to provide a partial or complete collapse of the bag to release the accumulated dust or (2) the physical touching of the bag by a ring running up and down its length to loosen the dust trapped in the bag.

Baghouse applications include:

1. Building materials dust removal
2. Grain processing
3. Soap powders
4. Dry chemical recovery
5. Talc dust recovery
6. Dry food processing
7. Pneumatic conveying
8. Metals dust recovery
9. Fertilizer dust removal

Moving Filters (Disposable)

A recent development is the moving filter made of a disposable media. The air carrying oil mist or other pollutants is directed toward and impinged upon a filter media that travels at a slow speed as it collects the contaminant. This technique can also be applied to stack emission control.

Wet Collectors

Wet collectors are designed to develop an interface between the scrubbing liquid and the gas being cleaned. The intimacy of interface developed is directly proportional to the efficiency of cleansing the pollutant from the gas (vehicle) stream.

One of the first scrubber types is the baffle design which extracts a variety of dust-particle sizes by impinging the high-velocity particle upon a water barrier through a change of direction technique. The particle-laden gas stream enters the water through an orifice plate which breaks up the gas stream into bubbles. Thereby, the wetted impingement orifice plate brings about an intimacy of contact between liquid and gas.

Another early scrubber design, the "venturi" type takes a dirty gas stream and subjects it to a powerful (high-pressure) water stream with a high wetting capability of the particulates, thereby effecting a high efficiency of scrubbing.

Scrubbers have been successfully applied to the following industries:

Blast furnaces
Basic oxygen furnaces
Open hearth furnaces
Sintering plants
Cevient kilns
Aluminum furnaces
Ore roasting
Sulfuric acid plants
Triple superphosphate plants
Electric furnaces
Pulp mill lime kilns
Fertilizer dryers
Coal dryers
Asphalt plants
Foundry shake-out
Pickle liquor concentrators
Hydrogen plants

Integrated Systems

In many applications, several air contaminants are to be removed and no single piece of equipment is capable of fulfilling all the requirements. In case of coal burning plants, for example, the flyash removal problem can be solved by the use of electrostatic precipitators; but the removal of sulfur dioxide requires some other piece of equipment. As a result, the total air pollution control system usually contains several pieces of equipment such as a dry cyclone in front of an electrostatic precipitator followed by a scrubber or a chemical injection system followed by a scrubber-cooler and then a baghouse.

In this chapter, the reader will find the description of the various pieces of equipment that are available for air pollution control work. Chapter 6 is devoted to the coverage of the removal methods of specific air contaminants while Chapter 7 discusses the treatment of industrial and municipal air pollution sources. The cross referencing between these three chapters should provide the reader with the tools needed to apply the most advanced methods of air pollution abatement.

5.2 AIR POLLUTION CONTROL COST ESTIMATING

There are two types of costs. The first type is called capital cost, capital investment, investment cost or first cost. The second type is known as operating cost, manufacturing cost or repeated cost.

The capital investment is the amount of money which is invested in plant and equipment and risked against failure, in order to get a project started. Such an investment is made in the hope of a future return. The capital investment is composed of two parts: (1) the fixed capital investment to provide the physical facilities and (2) the working capital investment to provide a revolving fund to keep the facility in operation.

The operating cost is different from the capital cost because the operating cost is a repeated cost while the capital cost is spent only once. When the plant is bought and paid for, the buyer owns it and does not "rebuy" it every year. When a plant is running, however, there are certain items for which the owner must pay repeatedly, every day, every month and every year. These are repeated costs and must be paid to keep the plant operating.

Antipollution devices, by and large, are addenda to a plant, albeit expensive ones. For this reason, certain capital and operating cost components will be small, or even nonexistent, compared with those same items in an integrated process plant.

This section will deal with the components of both capital and operating costs and how these components can be either obtained or estimated.*

Capital Investment Costs

The total amount of money required to put a project in operation is known as the *capital investment.* This is composed of the fixed capital investment to buy such items as land, buildings, utilities, process units, storage facilities, auxiliary facilities and spare parts. The capital investment also includes working capital for such items as raw materials in stock or storage, finished product in storage, semifinished product in process, accounts receivable, accounts payable, taxes payable and cash for the payment of operating expenses for a month or two. This capital investment figure is important because it represents the amount of money risked by the owner against the possibility of failure.

*A good collection of estimating data can be found in EPA publication EPA—R5-73-023, "Capital and Operating Costs of Pollution Control Equipment Modules." (July 1973.)

Because most antipollution measures are either an addition to an existing plant or generally an adjunct to a proposed plant, certain capital items should be quite small. For instance, the working capital component for a wet scrubber used to remove particulate matter from a stack gas should be very small because there very likely would be no charge for product inventory nor accounts receivable for the particulate matter. There would be a small component of working capital, however, for such items as water, electricity and any inventory of scrubbing agents. Likewise, one would expect the addition of a particulate wet scrubber to the stack gas train to have very little effect on the cost of the guardhouse, the steam plant or the main process units. For this reason, any estimates of capital costs derived by applying percentages, based on industry averages, to other capital items should be approached with caution and tempered with engineering judgment. One example is using the chemical industry average of 15% of total capital investment for working capital. If one blindly applies this factor to the particulate wet scrubber, the working capital estimate is bound to be high.

Estimates of capital costs are made for any one of several reasons. Some of the more important ones include:

1. Feasibility studies
2. Selection among alternative designs
3. Selection among alternative investments
4. Appropriation of funds
5. Presentation of bids

It is obvious that a capital cost estimate intended for bidding purposes needs to be much more accurate than one intended for a feasibility study.

To provide for these varying needs and because the accuracy of an estimate varies directly with the time and trouble expended on it, different types of estimates have been developed. The American Association of Cost Engineers[4] has recommended the classification of estimates into five basic types, according to use and accuracy. These are summarized in Table 5.2a, which also shows the uses for the various types of estimates and their probable accuracies.

The amount of information needed to prepare the different types of estimates varies directly with the desired or required accuracy. As a project develops, more definite information becomes available and, hence, a more accurate estimate can be made. The items required for each type of estimate are shown in Table 5.2b. It is obvious that the estimates can be more accurate when more information is available. It follows also that the estimates become more expensive and take longer to prepare as they become more detailed and more accurate.

Typical of the order-of-magnitude type of estimate is the kind based on unit costs. For instance, a unit capital cost for a wet scrubber might be given

Table 5.2a
ESTIMATE TYPES

Name(s)	*Description and Uses*	*Probable Error or Accuracy, %*
Order-of-magnitude estimate; ratio estimate	Usually based on previous similar cost information.	over ±30
Study estimate; factored estimate	Better than the first but accuracy insufficient for budgeting; a guide to further interest.	up to ±30
Preliminary estimate; budget authorization estimate; scope estimate	Data ample for estimate to be budgeted.	±20
Definitive estimate; project control estimate	Considerable data available but completed drawings and specifications are unavailable.	±10
Detailed estimate; firm estimate; contractor's estimate	Prepared from complete engineering specifications, drawings and site surveys.	±5

as $1.225/ACFM (actual cubic feet per minute). This means that for a stack gas discharge of 8,150 ACFM, the wet scrubber would cost $9,980.

Typical of the study estimate type are any of the factored estimates. Of these, the most famous is the so-called "six-tenths" factor rule. By this rule, the cost of an installation of one capacity can be estimated from the cost of a similar installation of different capacity by multiplying the known cost by the ratio of capacities raised to the 0.6 power. For example, if a wet scrubber to clean 8,150 ACFM of stack gas costs $9,980, we can estimate the cost of a wet scrubber to clean 12,200 ACFM of gas by applying the "six-tenths" rule. Thus, the cost of a 12,200 ACFM scrubber is estimated to be $\$9{,}980\left(\frac{12{,}200}{8{,}150}\right)^{0.6}$, or $12,750.

Typical of the preliminary estimate type is the percent of equipment cost estimate. In this method, the capital cost is estimated by applying factors to the delivered cost of all major pieces of equipment, down to and including pumps, in order to find the total capital investment. This method is preferred by many estimators because it represents the best compromise between speed and accuracy. There are several sources of suggested factors.[2,3,5] Table 5.2c[1] gives one suggested set of factors for the process industries which should be typical of pollution control equipment. This serves as a checklist, but it should be borne in mind that the factors must be tempered to fit the circumstances.

Table 5.2b
REQUIRED INFORMATION FOR VARIOUS ESTIMATE TYPES

Type of Information	*Type of Estimate*				
	Order-of-Magnitude	*Study*	*Preliminary*	*Definitive*	*Detailed*
Site					
location		X	X	X	X
general description			X	X	X
soil bearing			X	X	X
location and dimensions of RR, roads and fences			X	X	X
well-developed site plan				X	X
well-developed site facilities					X
Process Flowsheet					
rough sketches		X			
preliminary drawings			X		
engineering drawings				X	X
Equipment List					
preliminary sizing		X	X		
engineering specifications				X	X
vessel sheets				X	X
preliminary arrangement			X	X	
engineered arrangement					X
Building and Structures					
approximate sizes and types		X	X		
foundation sketches			X	X	
architectural and construction			X	X	X
preliminary structural design			X		
general arrangements				X	X
detailed drawings					X
Utility Requirements					
rough quantities		X			
preliminary heat balance			X		
preliminary flowsheets			X		
engineered heat balance				X	X
engineered flowsheets				X	X
well-developed drawings					X

Table 5.2b (Continued)

Type of Information	*Type of Estimate*				
	Order-of-Magnitude	*Study*	*Preliminary*	*Definitive*	*Detailed*
Piping					
preliminary flowsheet		X	X		
engineered flowsheets				X	X
piping layouts					X
Insulation					
rough specifications			X		
preliminary list of equipment and piping to be insulated				X	
insulation specifications and schedules				X	X
well-developed drawings					X
Instrumentation					
preliminary instrument list			X		
engineered list and flowsheet				X	X
well-developed drawings					X
Electrical					
preliminary motor list		X	X		
engineered list				X	X
numbers and sizes—substations			X	X	X
distribution specifications					X
preliminary lighting specifications			X		
preliminary control wiring				X	
power and light diagrams				X	X
well-developed drawings					X
Man Hours					
engineering and drafting		X	X	X	X
labor by craft					X
supervision					X
Project Scope					
product capacity, location, and site requirements; utility and service requirements; building and auxiliary handling and storage requirements.	X				

Table 5.2c
FACTORS FOR ESTIMATING CAPITAL INVESTMENTS, BASED ON DELIVERED EQUIPMENT COSTS[a]

Types of Expenses	*Types of Processes*		
	Solid-Processing Plant[b]	*Solid-Fluid-Processing Plant*[c]	*Fluid-Processing Plant*[d]
Direct Costs			
purchased equipment—delivered	100	100	100
equipment installation	45	39	47
instrumentation and controls	9	13	18
piping (installed)	16	31	66
electrical (installed)	10	10	11
buildings	25	29	18
yard improvements	13	10	10
service facilities	40	55	70
land purchase (if required)	6	6	6
Total Direct Plant Cost	264	293	346
Indirect Costs			
engineering and supervision	33	32	33
construction expenses	39	34	41
Total Direct and Indirect Plant Costs	336	359	420
contractor's fee (about 5% of direct and indirect)	17	18	21
contingency (about 10% of direct and indirect)	34	36	42
fixed-capital investment	387	413	483
working capital (about 15% of total)	68	74	86
Total Capital Investment	455	487	569

[a] In % of delivered equipment cost.
[b] A coal briquetting plant would be typical of a solid-processing plant.
[c] A shale oil plant with crushing, grinding, retorting and extraction would be typical of a solid-fluid processing plant.
[d] A distillation unit would be typical of a fluid-processing plant.

Thus, there would be no cost for land if a wet scrubber installation is being added to an existing plant. The advantage to the checklist, however, is that no items will be overlooked; and therefore, each potential cost item can be considered and estimated individually.

For example, if a wet scrubber is to be added to an existing plant, it is necessary to install the scrubber, a pump, a settling tank and a booster fan. The delivered cost for this equipment is estimated to be $35,000. Because wet scrubbing of particulate matter is a solid-fluid processing system, factors from the middle column in Table 5.2c should be used. Certain items can be eliminated immediately such as buildings, yard improvements, service facilities, land and working capital. The remaining factors should be modified if necessary. Thus, the full 39% for installation and the full 10% for electricals are justified. Because this is a relatively simple plant addition, the factors for instrumentation, piping and engineering might be cut in half. Because it is a plant addition, however, it might be wise to double the factor for construction expense since the contractor must shoe-horn the equipment among existing pieces of equipment. The contractor's fee and the contingency factors should probably remain at 5 and 10%, respectively, of the direct and indirect cost. A total capital investment factor of 260% results from these considerations. Therefore, the addition of a wet scrubber should cost 2.60 × $35,000, or $91,000.

The other two estimate types require detailed information. In such estimates, each item which goes into the finished plant is priced individually. This is the type of estimate prepared by engineering constructors for their clients. Table 5.2b shows the type of information needed for such estimates; and it is clear that the estimates are exact down to the number of feet of various sizes of pipe, wire and lumber. If this information is available, it is a simple but tedious procedure[6,7] to price each item.

Sources and Representation of Cost Data

The most reliable source of capital cost data is the vendor of the equipment or engineering contractors. For preliminary evaluations, however, the engineer often needs data for many alternatives and, thus, the vendors' patience is soon exhausted. The next best source is company records of previous, similar installations. Such data must, of course, be updated to allow for the effect of inflation.

If the data are unavailable from vendors or from company files, the next most reliable source is published information. Many of the sections in this handbook have some cost information, but there are also many other fruitful sources.[2,3,5,7–10] One must be particularly careful in using undefined cost data. Thus, some published data, in an article for instance, may include some disassociated items installed at the same time; or alternatively, perhaps the quoted figure represents only the processing unit itself and not any of the necessary auxiliaries.

The most common method of representing cost data is by the so-called cost-capacity relationship based on the "six-tenths" rule. Thus according to Equation 5.2(1):

$$C_2 = C_1\left(\frac{Q_2}{Q_1}\right)^x \qquad 5.2(1)$$

in which C_2 is the desired cost at capacity Q_2, and C_1 is a known cost at known capacity Q_1. By taking logs of both sides of Equation 5.2(1),

$$\log C_2 = \log C_1 + X \log \frac{Q_2}{Q_1} \qquad 5.2(2)$$

According to Equation 5.2(2), the logarithm of the cost can be plotted against the logarithm of the size or capacity to give a straight line relationship. Plotting of the data on log-log graph paper is the most convenient method of presenting it. A plot of this type is shown in Figure 5.2d for wet scrubber installations for brass/bronze reverberatory furnaces.[10]

The most common value for the cost-capacity factor, X, is 0.6, hence the name "six-tenths" factor rule. The value for X, or slope of the cost versus capacity plot, is 0.75 in Figure 5.2d; but values of cost-capacity factors can vary from as low as 0.2 to values over 1.0. Table 5.2e gives some typical values of cost-capacity factors. This "six-tenths" rule, or cost-capacity factor rule, also applies to entire plants.[5]

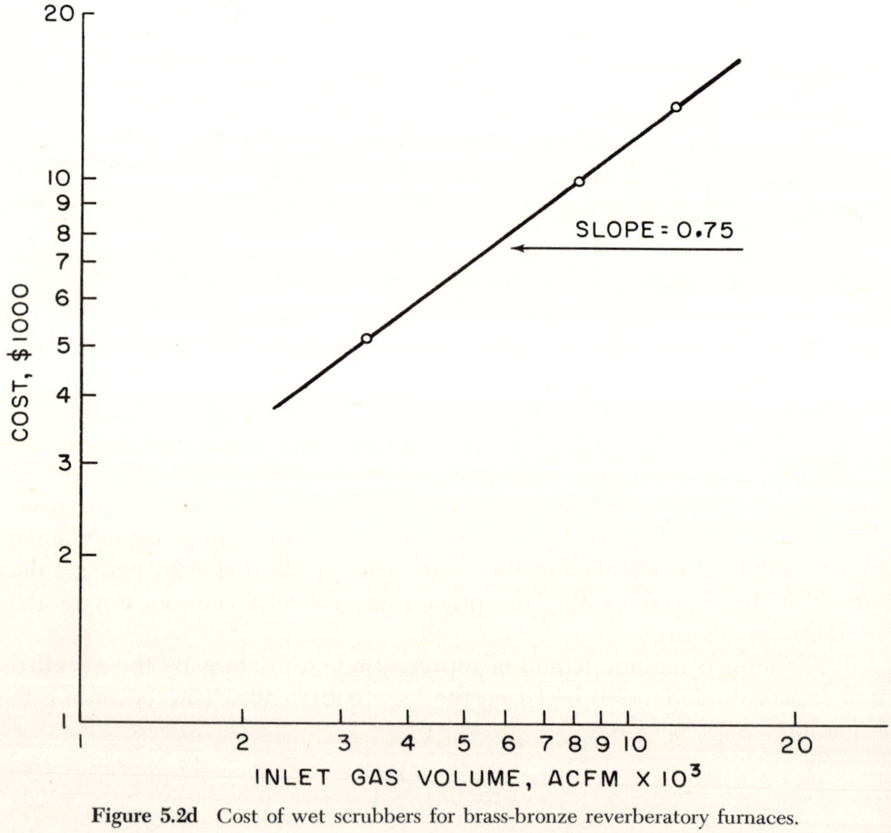

Figure 5.2d Cost of wet scrubbers for brass-bronze reverberatory furnaces.

Table 5.2e
COST-CAPACITY FACTORS FOR PLANTS AND EQUIPMENT

Type of Plant or Equipment	*Capacity or Size*	*Units*	*Cost Capacity Factor*
Plants			
carbon black	1	tons/day	0.53
ethyl alcohol	10 MM	gal/yr	0.60
coal power plants	100	Mw(elec)	0.88
Equipment			
continuous drum filters	30	sq ft	0.40
horizontal steel storage tanks	5,000	gal	0.70
electrostatic precipitators	1,000	Cfm	0.90

The cost data from these various sources apply to some particular point in time, either when the original equipment was installed or when the data were published or accumulated. Unfortunately, costs are measured in dollars which are not a fixed unit of measure; the dollar depreciates in value and seems to fall continuously. This means that all published cost data are rapidly out of date. A means for converting past to present costs is usually accomplished by means of a cost index which gives the relative cost of an item in terms of the cost at some particular base period. There are many cost indexes for many different purposes. Some of the more common ones are shown in Table 5.2f. For example, if we know that a wet scrubber cost \$9,980 in 1970, by the use of the Marshall and Stevens equipment cost index, the scrubber should cost $\$9{,}980 \times \frac{324}{303}$, or \$10,680 in 1971.

Pitfalls in Capital Cost Estimating

The cost estimator is under continuous pressure to provide detailed, accurate estimates, while having time allotments and information suitable for order-of-magnitude estimates. In spite of any qualifying statements about accuracy or availability of information, the first estimates have a habit of sticking and all the qualifying statements are overlooked or ignored.

The real pitfalls in capital cost estimating are an indefinite scope or a changing scope. The indefinite scope can be recognized when off-the-cuff figures are requested for a project or plant to make or treat a definite product, but other information such as size or quantity is vague. The changing scope can be recognized after the project has been estimated and decided upon by such changes as capacity, location or product mix. Often, extraneous projects will be combined with the one under consideration, which certainly represents a change in scope.

Table 5.2f
MAJOR COST INDEXES

Year	Engineering News-Record: Construction	Engineering News-Record: Building	Marshall and Stevens Installed-Equipment	Chemical Engineering Plant Construction
	1913 = 100		1926 = 100	1957–1959 = 100
1950	510	375	168	74
1951	543	401	180	80
1952	569	416	181	81
1953	600	431	183	85
1954	628	446	185	86
1955	660	469	191	88
1956	692	491	209	94
1957	724	509	225	99
1958	759	525	229	100
1959	797	548	235	102
1960	824	559	238	102
1961	847	568	237	102
1962	872	580	239	102
1963	901	594	239	102
1964	936	612	242	103
1965	971	627	245	104
1966	1,019	650	253	107
1967	1,070	671	263	110
1968	1,154	721	273	114
1969	1,270	790	285	119
1970	1,380	835	303	126
1971	1,581	948	324	135
1972	1,753	1,048	336	139
1973	1,895	1,138	353	149

Operating or Repeated Costs

Operating costs, or manufacturing costs or repeated costs, are expenses which recur each year, as distinct from capital costs or first costs which occur only once. Operating costs most commonly are divided into direct or variable costs and into indirect or fixed costs.

The direct or variable costs apply directly to the specific process or product and, in general, vary linearly with production rates. These direct cost items include raw-materials costs, by-product credits, direct labor, direct supervision, direct payroll charges, utilities, maintenance, factory supplies, laboratory charges, packaging costs and running royalties.

The indirect or fixed costs are those cost items which must be paid regardless of production rate and are, therefore, fixed. These costs include depreciation, property taxes, interest on any borrowed capital, insurance, plant overhead and general headquarters expense.

Some authors[11,12] have suggested that operating cost elements are really semivariable in nature. It means that there is a fixed cost floor and a variable cost element on top of the fixed cost floor. Maintenance, for instance, varies in this fashion, for there is a maintenance cost even at zero production. Operating labor affords another example because it takes as much labor to operate equipment at half capacity as it does at full capacity. Packaging and shipping labor, on the other hand, varies directly with production.

In estimating operating costs for projects of any type, a tabular form is very useful and constitutes a valuable checklist. A sample form or checklist is shown in Figure 5.2g. This form is designed to calculate costs on an annual basis rather than a daily or unit-of-product basis. The annual basis is preferred because the effect of seasonal variations is evened out. The plant onstream time or operating factor is considered, it permits more rapid calculation of the cost of operation at less than capacity; and it provides an expeditious way of considering infrequently occurring but large expenses, such as annual turn-around costs or replacement costs of catalyst. Daily costs or unit-of-product costs can be easily obtained from annual costs by merely dividing the annual cost by the operating days per year or by the annual units of production, respectively.

The accuracies of operating cost estimates vary, like the capital cost estimates, with the accuracy and quantity of available information. If all the quantities and unit costs of the elements shown in Figure 5.2g are known, it is a simple matter to fill out the form, extend the quantities and add up the figures to get the annual production cost. The following paragraphs will discuss what is included in each cost item and how to estimate it if the information is not available from company records.

The first item among the direct costs is raw-material cost. Most anti-pollution installations would not have raw material costs, but some others would. Information on the quantities of raw materials could come from published sources, vendors, laboratory experiments or from theoretical calculations to which reasonable yields are applied.

Direct labor can be estimated in a number of ways, depending upon what information is available. In the later project stages, the labor can be estimated from flowsheets, keeping in mind such items as the type and arrangement of equipment, company policy and multiplicity of units. Care must be exercised so that no worker is in an area by himself. In the very early stages of a project, the quantity of direct labor can often be estimated from company

Figure 5.2g
PRELIMINARY MANUFACTURING COST ESTIMATE

Date ______ By ______

Location: ______ Product: ______

Manufacturing Capital: ______ Process: ______

______ Cost Index: ______ Operating Days: ______

Total ______ Annual Production: ______

Raw Materials	*Annual Quantity*	*Unit Cost*	*$/Year*	*$/*
		Gross Raw-Material Cost:		
By-product (credit)				
		Total Credit:		
		Net Raw-Material Cost:		

Direct Cost	*Unit*	*Quantity*	*$/Unit*		
Steam, hp	M lb				
Steam, lp	M lb				
Water, Plant	M gal				
Water, Process	M gal				
Electricity					
Fuel ( )					
Inert gas	M cu ft				
Compressed Air	M cu ft				
Refrigeration	tons				
Containers ( )					
Labor					
Supervision					
Payroll Charges					
Repairs					
Factory Supplies					
Laboratory					
Royalty					
			Total Direct Conversion Cost:		

Indirect Cost		
Depreciation		
Real Estate Taxes		
General Plant Overhead		
General Head-quarters Expense		
Total Indirect Conversion Cost:		
Total Conversion Cost:		
Total Manufacturing Cost:		
Packing and Shipping Expense		
Total cost F.O.B. Plant:		

experience with similar projects or from literature information on similar processes. A correlation for labor in the process industries, to which most antipollution projects would be similar, is available in the literature.[2,5] Very likely, most antipollution equipment could be added to proposed projects, or to existing plants, without increasing the labor requirements significantly. Labor rates are obtained from company records or from Department of Labor publications.

Direct supervision covers the expense for foremen and for other direct operating supervision. In the absence of more exact figures from company records, the cost of direct supervision may be taken as 10 to 25% of the cost of direct labor.

Payroll charges comprehend payments for such costs as pensions, social security, vacation pay, group insurance, disability pay and unemployment compensation. In the absence of records, payroll charges can be estimated as 12 to 20% of the cost of direct labor and direct supervision.

Utilities include such items as steam, process and plant water, fuel, refrigeration, inert gas and compressed air. Utility requirements can often be estimated from available information about similar installations if exact information is lacking. Alternatively, utilities can be estimated from preliminary process designs, flowsheets and energy balances. A factor should be applied to estimated quantities of utilities to allow for equipment efficiencies, line losses, lighting, heating, contingencies and sanitary facilities. Appropriate factors are suggested in the literature.[3]

The repair and maintenance item is commonly estimated as a percentage of the plant investment cost. If another method has not been developed from company experience, the maintenance costs can be estimated as 3 to 6% of the plant investment, the costs being dependent upon the severity of operating conditions and the corrosiveness of the processing materials involved. Under very severe conditions, maintenance costs can go as high as 9 or 10% of investment.

Factory supplies include such items as wiping cloths, lubricating oil, instrument charts, janitorial supplies, test chemicals and gaskets. In the absence of specific information, a usual allowance for this cost is 0.5 to 1% of the plant investment.

Laboratory charges cover the cost of laboratory analyses for process control and for product control tests. It is usually calculated by estimating the man-hours involved and using $10 to $20 per analytical man-hour as the rate. For very quick estimates, this cost can be taken as 20% of the direct labor cost.

Most antipollution projects would probably not have a packaging or shipping expense, except for such installations as the manufacture of sulfuric acid from stack gas sulfur dioxide; however, this item is included in the checklist for completeness.

Royalties are payments for the use of any patented process or equipment. Paid royalties become part of the capital investment, but running royalties are part of the operating costs. Information about the cost of royalties must eventually come from the company holding the patent. For preliminary estimates, the royalties can be estimated as 1 to 5% of sales.

The indirect, or fixed, costs in Figure 5.2g include depreciation, property taxes, interest, insurance, plant overhead and general headquarters expense.

Depreciation is an allowance, as a cost, for exhaustion and wear and tear on property used in a business. The depreciation rate, or the expected equipment life, is usually fixed by Publication 456 (7-62) of the Internal Revenue Service. Depreciation can be estimated from company experience with similar equipment. The three widely used methods of calculating depreciation are the straight-line, the double-declining balance and the sum-of-the-years' digits methods.[2] The choice among these methods is a matter of company policy.

Property taxes are assessed by local governments and are based on the value of the plant and equipment. Tax rates at existing plant locations can be obtained from company records. For new locations, effective tax rates on similar property might be obtained from local sources. A figure of 2% of plant cost is a normal figure.

Interest is the cost of capital borrowed to pay for the plant and working capital. In preliminary estimates, a reasonable interest rate might be 5 to 6%.

The cost of insurance may be taken as 1% of the capital investment.

The plant overhead may be approximated either as 40 to 60% of the direct labor or as 15 to 30% of the direct conversion cost. Very likely, the addition of antipollution equipment to a plant or project would not increase this item.

The general headquarters expense includes the cost of selling, research and administrative functions in the company's general office. From 5 to 15% of the net annual sales will ordinarily cover these expenses. It is doubtful that this item would be a factor in most antipollution projects.

The sum of the costs shown in Figure 5.2g represents the total manufacturing cost or the plant operating cost. A summary of the suggested factors and sources of information is presented in Table 5.2h.

Pitfalls in Operating Cost Estimating

The estimates of operating costs are also sensitive to the same pitfalls of indefinite and changing scope as are the capital cost estimates, since some operating cost items, such as depreciation and property taxes, depend on the size of the capital investment. In addition, any changes in operating cost components would directly affect the operating cost itself.

Another pitfall in operating cost estimating involves the blind application of factors. For example, the factor for plant general office overhead calls for

Table 5.2h
SOURCES AND FACTORS FOR OPERATING COST ESTIMATING

Direct Costs

Raw materials: quantities from literature, laboratory experiments or stoichiometric calculations. Prices from *Oil, Drug and Paint Reporter* or from company records.

By-product credits: quantities and prices as above.

Direct labor: estimated from literature, company records or flowsheets. Labor rates from company records or Department of Labor information.

Direct supervision: 10 to 25% of direct labor cost.

Payroll charges: company records or 12 to 20% of the cost of labor plus supervision.

Utilities: estimated from similar processes in company records or literature, or calculated from flowsheets. Rates from company records or utility companies.

Repair and maintenance: 3 to 10% of plant investment.

Factory supplies: 0.5 to 1% of plant investment.

Laboratory charges: $10 to $20 per man-hour or 20% of direct labor.

Packaging and shipping expense: estimated from cost of packages, plus labor and freight.

Royalties: estimated from information from licensor, or 1 to 5% of sales.

Indirect Costs

Depreciation: estimated from company records or Publication 456(7–62). Probably 5 to 10% is a good range.

Property taxes: company records or 2% of plant investment.

Interest: 5 to 10% of borrowed capital.

Insurance: 1% of capital investment.

Plant general office overhead: from plant records or 40 to 60% of direct labor or 15 to 30% of the direct conversion cost.

General headquarters expense: company records or 5 to 15% of sales.

a figure of 15 to 30% of the direct conversion cost. It is extremely doubtful that the addition of antipollution equipment to a plant or project would increase the plant general office overhead expenses; however, the direct conversion cost of the main product will be increased by the addition of this equipment. For this reason, each item in the preliminary manufacturing cost estimate must be considered carefully and individually.

The whole problem of allocated costs constitutes another pitfall in operating cost estimation. Many of the cost items in the estimate are based on percentages applied to capital investment, labor or direct conversion cost. Before these percentages are used, the estimator should convince himself that the cost item will really vary with changes in the capital investment, labor or the direct conversion cost. It may be that the cost item should stay constant in spite of such changes in the basic costs.

REFERENCES

1. Osburn, J. O. and Karl Kammermeyer: "Money and the Chemical Engineer," Prentice-Hall, Englewood Cliffs, N.J., 1958.
2. Jelen, F. C.: "Cost and Optimization Engineering," McGraw-Hill, New York, 1970.
3. Peters, M. S. and K. D. Timmerhaus: "Plant Design and Economics for Chemical Engineers," 2d ed., McGraw-Hill, New York, 1968.
4. *AACE Bulletin,* vol. 1, no. 1, p. 12, November 1958.
5. Chilton, C. H.: "Cost Engineering in the Process Industries," p. 58, McGraw-Hill, New York, 1960.
6. Godfrey, R. S.: "Building Construction Cost Data," R. S. Means Company, Duxbury, Massachusetts, published annually.
7. Bauman, H. C.: "Fundamentals of Cost Engineering in the Chemical Industry," Reinhold Publishing, New York, 1964.
8. Popper, Herbert: "Modern Cost Engineering Techniques," McGraw-Hill, New York, 1970.
9. Alonso, J. R. F.: Estimating the Costs of Gas Cleaning Plants, *Chemical Engineering,* vol. 78, no. 28, pp. 86–96, December 13, 1971.
10. Hardison, L. C. and H. R. Herington: "Study of Technical and Cost Information for Gas Cleaning Equipment in the Lime and Secondary Nonferrous Metallurgical Industries—Final Report," Industrial Gas Cleaning Institute, Inc., Rye, New York, 1970.
11. Hackney, J. W.: Estimate Production Costs Quickly, *Chemical Engineering,* vol. 68, no. 8, pp. 179–184, 1961.
12. Black, J. H.: "Estimating Operating Costs for New Ventures," *AACE Bulletin,* vol. 4, no. 2, p. 42, 1962.

5.3 AIR CONTAMINANT SIZES AND THRESHOLD LIMIT VALUES

In sizing cleanup equipment, it is necessary to know the quantity, size and density of the polluting particles. Emission factors* for the various pollutants discharged by industrial and municipal sources are given in Chapters 1, 2, 3 and 6. The purpose of this section is to provide information on the sizes and threshold limit values of air contaminants.

Sizes of Airborne Contaminants

Figure 5.3a provides information on the sizes of airborne contaminants. The numbers in this illustration indicate the reference sources for the data presented.

References for Figure 5.3a

1. Alexander, P., Kitchner, J. A., and Briscoe, H. V. A.: *Inert Dust Insecticide II. Nature of Effective Dust,* Ann. Applied Biol., 31:150–156 (1944).
2. Avy, P.: *The Filtration of Industrial Dusts and Fumes, Mem. services chem. etat.,* 30:257–271 (1943).
3. Breed, R. S., Murray, E. G. D., and Hitchens, A. P.: *Bergey's Manual of Determinative Bacteriology,* 6th ed., Baltimore, Williams and Wilkins (1948).
4. Frampton, V. L.: *On the Size and Shape of the Tobacco Mosaic Virus Protein Particle,* Science, 95:232–233 (1942).
5. Greeman, Nelson, Jr.: (O. D.). Private Communication.
6. Heverly, J. R.: *Similitude of Artificial and Natural Fogs,* Trans. Amer. Geophys. Un., 30:15–18 (1949).
7. Holden, F. R., Hemeon, W. C. L., and Hyatt, E. C.: *Appraising Exposure to Silica Dust,* J. Ind. Hyg. Toxicol., 29:265–268 (1947).
8. Houghton, H. G., and Radford, W. H.: *On the Measurement of Drop Size and Liquid Water Content in Fogs and Clouds,* Pap. Phys. Ocean. Meteor. Mass. Inst. Tech. Woods Holes Ocean. Instn., 6:(4)1–31 (1938).
9. Jennison, M. W.: *Atomizing of Mouth and Nose Secretions into the Air as Revealed by High Speed Photography,* Aerobiology, Washington, Am. Assoc. for the Advancement of Science, Publication No. 17, 1942, p. 106.
10. Lapple, C. E.: *Dust and Mist Collection,* Chemical Engineer's Handbook, 3rd ed., New York, McGraw-Hill, 1950, p. 1016.
11. Leonard, B. R., Auderegg, J. W., Kaeberg, P., Schulman, S., and Beeman, W. W.: *The Size, Shape, and Hydration of Southern Bean Mosaic Virus and Tobacco Necrosis Virus in Solution,* J. Chem. Phys., 19:793–794 (1951).
12. Lipscomb, W. N., Rubin, T. R., and Sturdivant, J. A.: *Investigation of a Method for the Analysis of Smokes According to Particle Size.,* J. Applied Phys., 18:72–79 (1947).
13. Magill, P. L.: *Sampling Procedures and Measuring Equipment,* Chapter 6, Air Pollution Abatement Manual, Washington, D.C., Manufacturing Chemist Association, 1952, p. 7.

*For a complete listing see EPA publication AP-42, "Compilation of air pollutant emission factors." (April, 1973.)

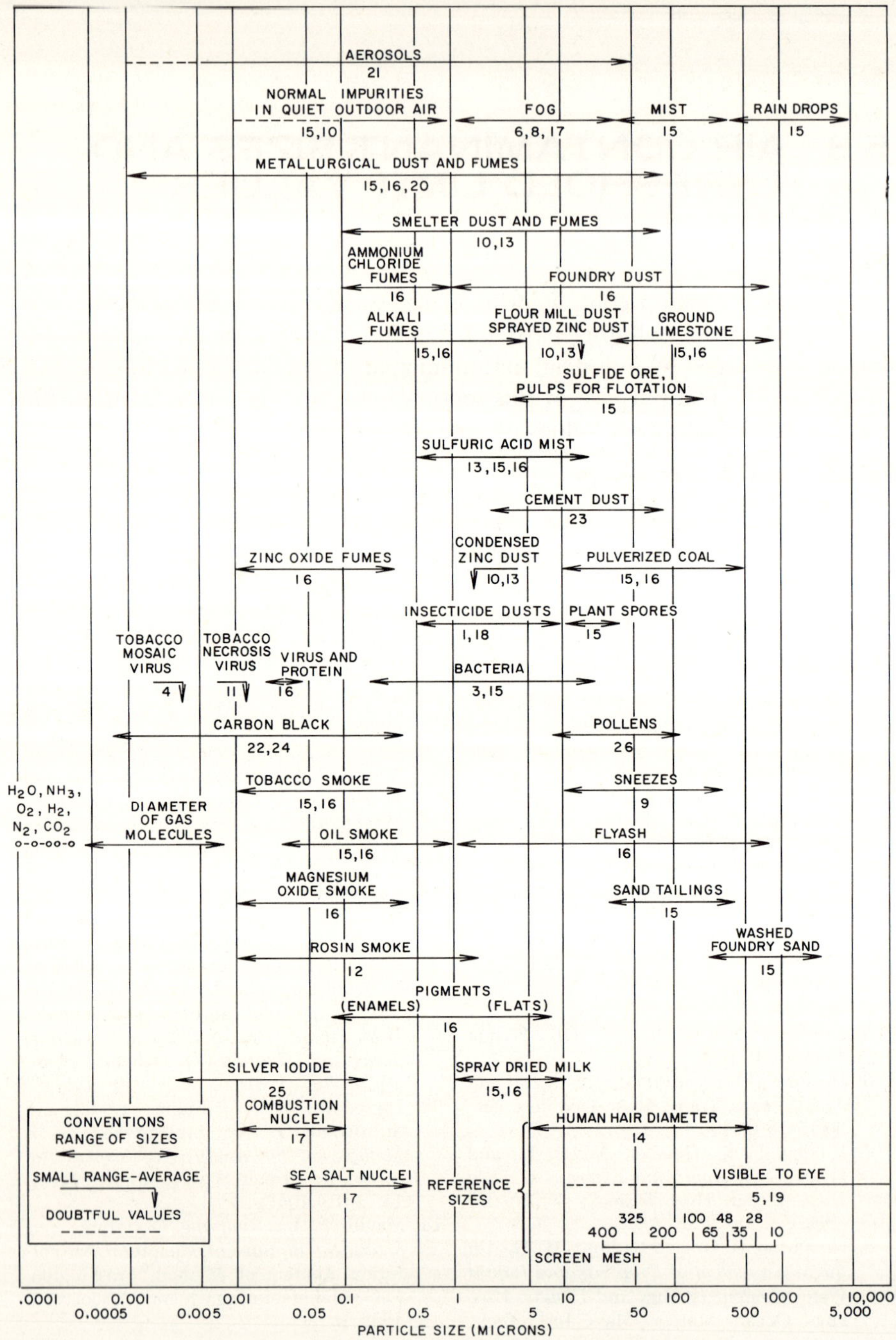

Figure 5.3a Sizes of airborne contaminants.

14. Maximow, A. A., and Bloom, W.: *A Textbook of Histology*, 5th ed., Philadelphia, Saunders, 1948, p. 336.
15. Miller, C. E.: *Pointers on Selecting Equipment for Industrial Gas Cleaning*, Chem. Met. Eng., 45:132–135 (1938).
16. Munger, H. P.: *Present Status of Air-Pollution Research*, Mech. Eng., 73:405–411 (1951).
17. Neiburger, M. and Wurtele, M. G.: *On the Nature and Size of Particles in Haze, Fog, and Stratus of the Los Angeles Region*, Chem. Rev., 44:321–335 (1949).
18. Parkin, E. A.: *Control of the Granary Weevil With Finely Ground Mineral Dusts*, Ann. Applied Biol., 31:84–88 (1944).
19. Robertson, J. K.: *Introduction to Physical Optics*, 3rd. ed., New York, van Nostrand, 1941, p. 284.
20. Silverman, Dr. Leslie: Private Communication, December 28, 1954.
21. Sinclair, D.: *Physical Properties of Aerosols*, Air Pollution, Proceedings of the United States Technical Conference on Air Pollution, New York, McGraw-Hill, 1952, pp. 169–181.
22. Sittenfield, M.: *The Control of Air Pollution in the Chemical Industry.*, Chem. Eng. Progress, 47:481–483 (1951).
23. Sperberg, L. R., and Barton, H. M.: *Chemical and Physical Properties of Philblack Carbons*, Philblack Bull. 12, Akron, Phillips Petroleum Company.
24. Steinherz, A. R.: *Fineness of Portland Cement*, Rock Products, 52:100–103, 122–123 (1949).
25. Sweitzer, C. W., and Goodrich, W. C.: *The Carbon Spectrum for the Rubber Compounder*, Rubber Age, 55:469–478 (1944).
26. Vonnegut, Bernard: *Nucleation of Supercooled Water Clouds by Silver Iodide Smokes*, Chem. Rev., 44:277–289 (1949).
27. Wodehouse, R. P.: *Atmospheric Pollen*, Aerobiology, Washington, Am. Assoc. for the Advancement of Science, Publication No. 17, (1942).

"Threshold Limit Values of Airborne Contaminants and Physical Agents for 1971"*

Threshold limit values refer to airborne concentrations of substances and represent conditions under which it is believed that nearly all workers may be repeatedly exposed day after day without adverse effect. Because of wide variation in individual susceptibility, however, a small percentage of workers may experience discomfort from some substances at concentrations at or below the threshold limit, a smaller percentage may be affected more seriously by aggravation of a pre-existing condition or by development of an occupational illness.

Simple tests are now available (J. Occup. Med. 9, 537, 1967; Ann. N.Y. Acad. Sci., 151 Art. 2, p. 968, 1968) that may be used to detect those individuals hypersusceptible to a variety of industrial chemicals (respiratory irritants, hemolytic chemicals, organic isocyanates, carbon disulfide). These tests may be used to screen out by appropriate job placement the hyperreactive worker and thus in effect improve this "coverage" of the TLVs.

Threshold limit values refer to time-weighted concentrations for a 7 or 8-hour workday and 40-hour workweek. They should be used as guides in the control of health hazards and should not be used as fine lines between safe and dangerous concentrations. (Exceptions are the substances listed in Appendices A and E and those substances designated with a "C" or Ceiling value, Appendix C.)

Time-weighted averages permit excursions above the limit provided they are compensated by equivalent excursions below the limit during the workday. In some instances it may be permissible to calculate the average concentration for a workweek rather than for a workday. The degree of permissible excursion is related to the magnitude of the threshold limit value of a particular substance as given in Appendix C. The relationship between threshold limit and permissible excursion is a rule of thumb and in certain cases may not apply. The amount by which threshold limits may be exceeded for short periods without injury to health depends upon a number of factors such as the nature of the contaminant, whether very high concentrations—even for short periods—produce acute poisoning, whether the effects are cumulative,

*Reprinted with permission from the American Conference of Governmental Industrial Hygienists.

the frequency with which high concentrations occur, and the duration of such periods. All factors must be taken into consideration in arriving at decision as to whether a hazardous condition exists.

Threshold limits are based on the best available information from industrial experience, from experimental human and animal studies, and, when possible, from a combination of the three. The basis on which the values are established may differ from substance to substance; protection against impairment of health may be a guiding factor for some, whereas reasonable freedom from irritation, narcosis, nuisance or other forms of stress may form the basis for others.

The committee holds to the opinion that limits based on physical irritation should be considered no less binding than those based on physical impairment. There is increasing evidence that physical irritation may initiate, promote or accelerate physical impairment through interaction with other chemical or biologic agents.

In spite of the fact that serious injury is not believed likely as a result of exposure to the threshold limit concentrations, the best practice is to maintain concentrations of all atmospheric contaminants as low as is practical.

These limits are intended for use in the practice of industrial hygiene and should be interpreted and applied only by a person trained in this discipline. They are not intended for use, or for modification for use, (1) as a relative index of hazard or toxicity, (2) in the evaluation or control of community air pollution or air pollution nuisances, (3) in estimating the toxic potential of continuous uninterrupted exposures, (4) as proof or disproof of an existing disease or physical conditions, or (5) for adoption by countries whose working conditions differ from those in the United States of America and where substances and processes differ.

Ceiling vs Time-Weighted Average Limits. Although the time-weighted average concentration provides the most satisfactory, practical way of monitoring airborne agents for compliance with the limits, there are certain substances for which it is inappropriate. In the latter group are substances which are predominantly fast acting and whose threshold limit is more appropriately based on this particular response. Substances with this type of response are best controlled by a ceiling "C" limit that should not be exceeded. It is implicit in these definitions that the manner of sampling to determine compliance with the limits for each group must differ; a single brief sample, that is applicable to a "C" limit, is not appropriate to the time-weighted limit; here, a sufficient number of samples are needed to permit a time-weighted average concentration throughout a complete cycle of operations or throughout the work shift.

Whereas the ceiling limit places a definite boundary which concentrations should not be permitted to exceed, the time-weighted average limit requires an explicit limit to the excursions that are permissible above the listed values. The magnitude of these excursions may be pegged to the magnitude of the threshold limit by an appropriate factor shown in Appendix C. It should be noted that the same factors are used by the Committee in making a judgment whether to include or exclude a substance for a "C" listing.

"Skin" Notation. Listed substances followed by the designation "Skin" refer to the potential contribution to the over-all exposure by the cutaneous route including mucous membranes and eye, either by airborne, or more particularly, by direct contact with the substance. Vehicles can alter skin absorption. This attention-calling designation is intended to suggest appropriate measures for the prevention of cutaneous absorption so that the threshold limit is not invalidated.

Mixtures. Special consideration should be given also to the application of the TLVs in assessing the health hazards which may be associated with exposure to mixtures of two or more substances. A brief discussion of basic considerations involved in developing threshold limit values for mixtures, and methods for their development, amplified by specific examples are given in Appendix B.

Nuisance Dusts. In contrast to fibrogenic dusts which cause scar tissue to be formed in lungs when inhaled in excessive amounts, so-called "nuisance" dusts have a long history of little adverse effect on lungs and do not produce significant organic disease or toxic effect when exposures are kept under reasonable control. The nuisance dusts have also been called (biologically) "inert" dusts, but the latter term is inappropriate to the extent that there is no dust which does not evoke *some* cellular response in the lung when inhaled in sufficient amount. However, the lung-tissue reaction caused by inhalation of nuisance dusts has the following characteristics: 1) The architecture of the air spaces remains intact. 2) Collagen (scar tissue) is not formed to a significant extent. 3) The tissue reaction is potentially reversible.

Excessive concentrations of nuisance dusts in the workroom air may seriously reduce visibility (iron oxide), may cause unpleasant deposits in the eyes, ears and nasal passages (Portland Cement dust), or cause injury to the skin or mucous membranes by chemical or mechanical action per se or by the rigorous skin cleansing procedures necessary for their removal.

A threshold limit of 10 mg/M^3, or 30 mppcf, of total dust $<1\%$ SiO_2, whichever is less, is recommended for substances in these categories and for which no specific threshold limits have been assigned. This limit, for a normal workday, does not apply to brief exposures at higher concentrations. Neither does it apply to those substances which may cause physiologic impairment at lower concentrations but for which a threshold limit has not yet been adopted. Some "inert" particulates are given in Appendix D.

Simple Asphyxiants—"Inert" Gases or Vapors. A number of gases and vapors, when present in high concentrations in air, act primarily as simple asphyxiants without other significant physiologic effects. A TLV may not be recommended for each simple asphyxiant because the limiting factor is the available oxygen. The minimal oxygen content should be 18 percent by volume under normal atmospheric pressure (equivalent to a partial pressure, pO_2 of 135 mm Hg). Atmospheres deficient in O_2 do not provide adequate warning and most simple asphyxiants are odorless. Several simple asphyxiants present an explosion hazard. Account should be taken of this factor in limiting the concentration of the asphyxiant. Specific examples are listed in Appendix E.

Physical Factors. It is recognized that such physical factors as heat, ultraviolet and ionizing radiation, humidity, abnormal pressure (altitude) and the like may place added stress on the body so that the effects from exposure at a threshold limit may be altered. Most of these stresses act adversely to increase the toxic response of a substance. Although most threshold limits have built-in safety factors to guard against adverse effects to moderate deviations from normal environments, the safety factors of most substances are not of such a magnitude as to take care of gross deviations. For example, continuous work at temperatures above 90°F or over-time extending the workweek more than 25%, might be considered gross deviations. In such instances judgment must be exercised in the proper adjustments of the threshold limit values.

"Notice of Intent." At the beginning of each year, proposed actions of the Committee for the forthcoming year are issued in the form of a "Notice of Intent." This Notice provides not only an opportunity for comment, but solicits suggestions of substances to be added to the list. The suggestions should be accompanied by substantiating evidence. The list of Intended Changes follows the Adopted Values in the TLV booklet.

As Legislative Code. The Conference recognizes that the TLVs may be adopted in legislative codes and regulations. If so used, the intent of the concepts contained in the Preface should be maintained and provisions should be made to keep the list current. These values are reviewed annually by the Committee on Threshold Limits for revision or additions, as further information becomes available.

Reprint Permission. This publication may be reprinted provided that written permission is obtained from the Secretary-Treasurer of the Conference and that it be published in its entirety.

Adopted Threshold Limit Values

Substance	*ppm*[a)]	*mg/M*^{3} [b)]
Abate	—	15
Acetaldehyde	200	360
Acetic acid	10	25
Acetic anhydride	5	20
Acetone	1,000	2,400
Acetonitrile	40	70
Acetylene	E	—
Acetylene dichloride, see 1, 2-Dichloroethylene	—	—

a) Parts of vapor or gas per million parts of contaminated air by volume at 25°C and 760 mm. Hg pressure.

b) Approximate milligrams of particulate per cubic meter of air.

5.3 Air Contaminant Sizes and Threshold Limit Values

	Substance	ppm[a)]	mg/M³ [b)]
	Acetylene tetrabromide	1	14
	Acrolein	0.1	0.25
	Acrylamide-Skin	—	0.3
	Acrylonitrile-Skin	20	45
	Aldrin-Skin	—	0.25
	Allyl alcohol-Skin	2	5
	Allyl chloride	1	3
°°C	Allyl glycidyl ether (AGE)	10	45
	Allyl propyl disulfide	2	12
	Alundum (Al_2O_3)	—	D
	2-Aminoethanol, see Ethanolamine	—	—
	2-Aminopyridine	0.5	2
°°	Ammonia	50	35
	Ammonium sulfamate (Ammate)	—	15
	n-Amyl acetate	100	525
	sec-Amyl acetate	125	650
	Aniline-Skin	5	19
	Anisidine (o, p-isomers)-Skin	—	0.5
	Antimony & compounds (as Sb)	—	0.5
	ANTU (alpha naphthyl thiourea)	—	0.3
	Argon	E	—
	Arsenic & Compounds (as As)	—	0.5
	Arsine	0.05	0.2
	Azinphos-methyl-Skin	—	0.2
	Barium (soluble compounds)	—	0.5
C	Benzene (benzol)-Skin	25	80
	Benzidine-Skin	—	A[1]
	p-Benzoquinone, see Quinone	—	—
	Benzoyl peroxide	—	5
	Benzyl chloride	1	5
	Beryllium	—	0.002
	Biphenyl, see Diphenyl	—	—
	Bisphenol A, see Diglycidyl ether	—	—
	Boron oxide	—	15
	Boron tribromide	1	10
C	Boron trifluoride	1	3
	Bromine	0.1	0.7
°	Bromine pentafluoride	0.1	0.7
	Bromoform-Skin	0.5	5
	Butadiene (1, 3-butadiene)	1,000	2,200
	Butanethiol, see Butyl mercaptan	—	—
	2-Butanone	200	590
	2-Butoxy ethanol (Butyl Cellosolve)-Skin	50	240
	Butyl acetate (n-butyl acetate)	150	710
	sec-Butyl acetate	200	950
	tert-Butyl acetate	200	950
	Butyl alcohol	100	300
	sec-Butyl alcohol	150	450
	tert-Butyl alcohol	100	300
C	Butylamine-Skin	5	15

° 1970 Addition (throughout)
°° See Notice of Intended Changes
Capital letters refer to Appendices.

	Substance	*ppm*[a)]	*mg/M^3*[b)]
C	tert-Butyl chromate (as CrO_3)-Skin	—	0.1
	n-Butyl glycidyl ether (BGE)	50	270
*	Butyl mercaptan	0.5	1.5
	p-tert-Butyltoluene	10	60
	Cadmium (Metal dust and soluble salts)	—	0.2
*C	Cadmium oxide fume (as Cd)	—	0.1
	Calcium carbonate	—	D
	Calcium arsenate	—	1
	Calcium oxide	—	5
**	Camphor (Synthetic)	2	—
	Carbaryl (Sevin®)	—	5
	Carbon black	—	3.5
	Carbon dioxide	5,000	9,000
	Carbon disulfide-Skin	20	60
	Carbon monoxide	50	55
	Carbon tetrachloride-Skin	10	65
	Cellulose (paper fiber)	—	D
	Chlordane-Skin	—	0.5
	Chlorinated camphene-Skin	—	0.5
	Chlorinated diphenyl oxide	—	0.5
*	Chlorine	1	3
	Chlorine dioxide	0.1	0.3
C	Chlorine trifluoride	0.1	0.4
C	Chloroacetaldehyde	1	3
	α-Chloroacetophenone (phenacylchloride)	0.05	0.3
	Chlorobenzene (monochlorobenzene)	75	350
	o-Chlorobenzylidene malononitrile (OCBM)	0.05	0.4
	Chlorobromomethane	200	1,050
	2-Chloro-1, 3-butadiene, see Chloroprene	—	—
	Chlorodiphenyl (42% Chlorine)-Skin	—	1
	Chlorodiphenyl (54% Chlorine)-Skin		0.5
	1-Chloro, 2, 3-epoxypropane, see Epichlorhydrin	—	—
	2-Chloroethanol, see Ethylene chlorohydrin	—	—
	Chloroethylene, see Vinyl choride	—	—
C	Chloroform (trichloromethane)	50	240
	1-Chloro-1-nitropropane	20	100
	Chloropicrin	0.1	0.7
	Chloroprene (2-chloro-1, 3-butadiene)-Skin	25	90
	Chromic acid and chromates (as CrO_3)	—	0.1
	Chromium, sol. chromic, chromous salts as Cr	—	0.5
	Metal & insol. salts	—	1
	Coal tar pitch volatiles (benzene soluble fraction) anthracene, BaP, phenanthrene, acridine, chrysene, pyrene)	—	0.2
	Cobalt, metal fume & dust	—	0.1
	Copper fume	—	0.1
	Dusts and Mists	—	1
	Corundum (Al_2O_3)	—	D
	Cotton dust (raw)	—	1
	Crag® herbicide	—	15
	Cresol (all isomers)-Skin	5	22
	Crotonaldehyde	2	6
	Cumene-Skin	50	245
	Cyanide (as CN)-Skin	—	5

	Substance	ppm[a]	mg/M³ [b]
°	Cyanogen	10	—
	Cyclohexane	300	1,050
	Cyclohexanol	50	200
	Cyclohexanone	50	200
	Cyclohexene	300	1,015
	Cyclopentadiene	75	200
	2, 4-D	—	10
	DDT-Skin	—	1
	DDVP, see Dichlorvos	—	—
	Decaborane-Skin	0.05	0.3
	Demeton®-Skin	—	0.1
	Diacetone alcohol (4-hydroxy-4-methyl-2-pentanone)	50	240
	1, 2-Diaminoethane, see Ethylenediamine	—	—
	Diazomethane	0.2	0.4
	Diborane	0.1	0.1
C	1, 2-Dibromoethane (ethylene dibromide)-Skin	25	190
	Dibutyl phosphate	1	5
	Dibutylphthalate	—	5
°C	Dichloroacetylene	0.1	0.4
C	o-Dichlorobenzene	50	300
	p-Dichlorobenzene	75	450
	Dichlorodifluoromethane	1,000	4,950
	1, 3-Dichloro-5, 5-dimethyl hydantoin	—	0.2
	1, 1-Dichloroethane	100	400
	1, 2-Dichloroethane	50	200
	1, 2-Dichloroethylene	200	790
C	Dichloroethyl ether-Skin	15	90
	Dichloromethane, see Methylenechloride	—	—
	Dichloromonofluoromethane	1,000	4,200
C	1, 1-Dichloro-1-nitroethane	10	60
	1, 2-Dichloropropane, see Propylenedichloride	—	—
	Dichlorotetrafluoroethane	1,000	7,000
	Dichlorvos (DDVP)-Skin	—	1
	Dieldrin-Skin	—	0.25
	Diethylamine	25	75
	Diethylamino ethanol-Skin	10	50
°°C	Diethylene triamine-Skin	10	42
	Diethylether, see Ethyl ether	—	—
	Difluorodibromomethane	100	860
C	Diglycidyl ether (DGE)	0.5	2.8
	Dihydroxybenzene, see Hydroquinone	—	—
	Diisobutyl ketone	50	290
	Diisopropylamine-Skin	5	20
	Dimethoxymethane, see Methylal	—	—
	Dimethyl acetamide-Skin	10	35
	Dimethylamine	10	18
	Dimethylaminobenzene, see Xylidene	—	—
	Dimethylaniline (N-dimethylaniline)-Skin	5	25
	Dimethylbenzene, see Xylene	—	—
	Dimethyl 1, 2-dibromo-2, 2-dichloroethyl phosphate (Dibrom)	—	3
	Dimethylformamide-Skin	10	30
	2, 6-Dimethylheptanone, see Diisobutyl ketone	—	—
	1, 1-Dimethylhydrazine-Skin	0.5	1
	Dimethylphthalate	—	5

	Substance	ppm[a)]	mg/M³ [b)]
	Dimethylsulfate-Skin	1	5
	Dinitrobenzene (all isomers)-Skin	—	1
	Dinitro-o-cresol-Skin	—	0.2
	Dinitrotoluene-Skin	—	1.5
	Dioxane (Diethylene dioxide)-Skin	100	360
	Diphenyl	0.2	1
	Diphenyl amine	—	10
	Diphenylmethane diisocyanate (see Methylene bisphenyl isocyanate [MDI])	—	—
	Dipropylene glycol methyl ether-Skin	100	600
	Di-sec, octyl phthalate (Di-2-ethylhexylphthalate)	—	5
	Emery	—	D
°	Endosulfan (Thiodan®)-Skin	—	0.1
	Endrin-Skin	—	0.1
	Epichlorhydrin-Skin	5	19
	EPN-Skin	—	0.5
	1, 2-Epoxypropane, see Propylene oxide	—	—
	2, 3-Epoxy-1-propanol, see Glycidol	—	—
	Ethane	E	—
	Ethanethiol, see Ethylmercaptan	—	—
	Ethanolamine	3	6
	2-Ethoxyethanol-Skin	200	740
	2-Ethoxyethylacetate (Cellosolve acetate)-Skin	100	540
	Ethyl acetate	400	1,400
	Ethyl acrylate-Skin	25	100
	Ethyl alcohol (ethanol)	1,000	1,900
	Ethylamine	10	18
	Ethyl sec-amyl ketone (5-methyl-3-heptanone)	25	130
	Ethyl benzene	100	435
	Ethyl bromide	200	890
	Ethyl butyl ketone (3-Heptanone)	50	230
	Ethyl chloride	1,000	2,600
	Ethyl ether	400	1,200
	Ethyl formate	100	300
	Ethyl mercaptan	0.5	1
	Ethyl silicate	100	850
	Ethylene	E	—
	Ethylene chlorohydrin-Skin	5	16
	Ethylenediamine	10	25
	Ethylene dibromide, see 1, 2-Dibromoethane	—	—
	Ethylene dichloride, see 1, 2-Dichloroethane	—	—
C	Ethylene glycol dinitrate and/or Nitroglycerin-Skin	0.2[d)]	—
	Ethylene glycol monomethyl ether acetate, see Methyl cellosolve acetate	—	—
	Ethylene imine-Skin	0.5	1
	Ethylene oxide	50	90
	Ethylidine chloride, see 1, 1-Dichloroethane	—	—
	N-Ethylmorpholine-Skin	20	94
	Ferbam	—	15
	Ferrovanadium dust	—	1
	Fibrous glass	—	D
	Fluoride (as F)	—	2.5

[d)] An atmospheric concentration of not more than 0.02 ppm, or personal protection may be necessary to avoid headache.

	Substance	*ppm*[a)]	*mg/M*^{3} [b)]
	Fluorine	0.1	0.2
	Fluorotrichloromethane	1,000	5,600
°°C	Formaldehyde	5	6
	Formic acid	5	9
	Furfural-Skin	5	20
	Furfuryl alcohol	50	200
	Gasoline	—	A[3]
	Glycerine mist	—	D
	Glycidol (2, 3-Epoxy-1-propanol)	50	150
	Glycol monoethyl ether, see 2-Ethoxyethanol	—	—
	Graphite, (Synthetic)	—	D
	Guthion,® see Azinphosmethyl	—	—
	Gypsum	—	D
	Hafnium	—	0.5
	Helium	E	—
	Heptachlor-Skin	—	0.5
	Heptane (n-heptane)	500	2,000
	Hexachloroethane-Skin	1	10
	Hexachloronaphthalene-Skin	—	0.2
	Hexane (n-hexane)	500	1,800
	2-Hexanone	100	410
	Hexone (Methyl isobutyl ketone)	100	410
	sec-Hexyl acetate	50	300
	Hydrazine-Skin	1	1.3
	Hydrogen	E	—
	Hydrogen bromide	3	10
C	Hydrogen chloride	5	7
	Hydrogen cyanide-Skin	10	11
	Hydrogen fluoride	3	2
	Hydrogen peroxide	1	1.4
	Hydrogen selenide	0.05	0.2
	Hydrogen sulfide	10	15
	Hydroquinone	—	2
°	Indene	10	45
	Indium and compounds, as In	—	0.1
C	Iodine	0.1	1
	Iron oxide fume	—	10
	Iron salts, soluble, as Fe	—	1
	Isoamyl acetate	100	525
	Isoamyl alcohol	100	360
	Isobutyl acetate	150	700
	Isobutyl alcohol	100	300
	Isophorone	25	140
	Isopropyl acetate	250	950
	Isopropyl alcohol	400	980
	Isopropylamine	5	12
	Isopropylether	500	2,100
	Isopropyl glycidyl ether (IGE)	50	240
	Kaolin	—	D
	Ketene	0.5	0.9
	Lead	—	0.2
	Lead arsenate	—	0.15
	Limestone	—	D
	Lindane-Skin	—	0.5
	Lithium hydride	—	0.025

	Substance	ppm[a)]	mg/M[3 b)]
	L. P. G. (Liquefied petroleum gas)	1,000	1,800
	Magnesite	—	D
	Magnesium oxide fume	—	15
	Malathion-Skin	—	15
	Maleic anhydride	0.25	1
C	Manganese and compounds, as M_n	—	5
	Marble	—	D
°°	Mercury-Skin	—	0.1
°°	Mercury (organic compounds)-Skin	—	0.01
	Mesityl oxide	25	100
	Methane	E	—
	Methanethiol, see Methyl mercaptan	—	—
	Methoxychlor	—	15
	2-Methoxyethanol, see Methyl cellosolve	—	—
	Methyl acetate	200	610
	Methyl acetylene (propyne)	1,000	1,650
	Methyl acetylene-propadiene mixture (MAPP)	1,000	1,800
	Methyl acrylate-Skin	10	35
	Methylal (dimethoxymethane)	1,000	3,100
	Methyl alcohol (methanol)	200	260
	Methylamine	10	12
	Methyl amyl alcohol, see Methyl isobutyl carbinol	—	—
°	Methyl isoamyl ketone	100	475
	Methyl (n-amyl) ketone (2-Heptanone)	100	465
C	Methyl bromide-Skin	20	80
	Methyl butyl ketone, see 2-Hexanone	—	—
	Methyl cellosolve-Skin	25	80
	Methyl cellosolve acetate-Skin	25	120
°°C	Methyl chloride	100	210
	Methyl chloroform	350	1,900
	Methylcyclohexane	500	2,000
	Methylcyclohexanol	100	470
	o-Methylcyclohexanone-Skin	100	460
	Methyl ethyl ketone (MEK), see 2-Butanone	—	—
	Methyl formate	100	250
	Methyl iodide-Skin	5	28
	Methyl isobutyl carbinol-Skin	25	100
	Methyl isobutyl ketone, see Hexone	—	—
	Methyl isocyanate-Skin	0.02	0.05
°	Methyl mercaptan	0.5	1
	Methyl methacrylate	100	410
	Methyl propyl ketone, see 2-Pentanone	—	—
C	Methyl silicate	5	30
C	α Methyl styrene	100	480
C	Methylene bisphenyl isocyanate (MDI)	0.02	0.2
	Methylene chloride (dichloromethane)	500	1,740
	Molybdenum (soluble compounds)	—	5
	(insoluble compounds)	—	15
	Monomethyl aniline-Skin	2	9
C	Monomethyl hydrazine-Skin	0.2	0.35
	Morpholine-Skin	20	70
	Naphtha (coal tar)	100	400
	Naphthalene	10	50
	β-Naphthylamine	—	A[1]
	Neon	E	—

Substance	ppm[a)]	mg/M³ [b)]
Nickel carbonyl	0.001	0.007
Nickel, metal and soluble cmpds, as Ni	—	1
Nicotine-Skin	—	0.5
Nitric acid	2	5
Nitric oxide	25	30
p-Nitroaniline-Skin	1	6
Nitrobenzene-Skin	1	5
p-Nitrochlorobenzene-Skin	—	1
Nitroethane	100	310
Nitrogen	E	—
Nitrogen dioxide	5	9
Nitrogen trifluoride	10	29
Nitroglycerin-Skin	0.2	2
Nitromethane	100	250
1-Nitropropane	25	90
2-Nitropropane	25	90
N-Nitrosodimethylamine (dimethylnitrosoamine)-Skin	—	A[1]
Nitrotoluene-Skin	5	30
Nitrotrichloromethane, see Chloropicrin	—	—
Nitrous oxide	E	—
Octachloronaphthalene-Skin	—	0.1
° Octane	400	1,900
° Oil mist, particulate	—	5[h)]
° Oil mist, vapor	[i)] A[3]	—
Osmium tetroxide	—	0.002
Oxalic acid	—	1
Oxygen difluoride	0.05	0.1
Ozone	0.1	0.2
Paraquat-Skin	—	0.5
Parathion-Skin	—	0.1
Pentaborane	0.005	0.01
Pentachloronaphthalene-Skin	—	0.5
Pentachlorophenol-Skin	—	0.5
Pentaerythritol	—	D
° Pentane	500	1,500
2-Pentanone	200	700
Perchloroethylene	100	670
Perchloromethyl mercaptan	0.1	0.8
Perchloryl fluoride	3	13.5
° Petroleum Distillates (naphtha)	[i)] A[3]	—
Phenol-Skin	5	19
p-Phenylene diamine-Skin	—	0.1
Phenyl ether (vapor)	1	7
Phenyl ether-Biphenyl mixture (vapor)	1	7
Phenylethylene, see Styrene	—	—
Phenyl glycidyl ether (PGE)	10	60
Phenylhydrazine-Skin	5	22
Phosdrin (Mevinphos®)-Skin	—	0.1
Phosgene (carbonyl chloride)	0.1	0.4
Phosphine	0.3	0.4
Phosphoric acid	—	1

[h)] As sampled by method that does not collect vapor.
[i)] According to analytically determined composition.

	Substance	ppm[a)]	mg/M[3 b)]
	Phosphorus (yellow)	—	0.1
	Phosphorus pentachloride	—	1
	Phosphorus pentasulfide	—	1
	Phosphorus trichloride	0.5	3
	Phthalic anhydride	2	12
	Picric acid-Skin	—	0.1
	Pival® (2-Pivalyl-1, 3-indandione)	—	0.1
	Plaster of Paris	—	D
	Platinum (Soluble Salts) as Pt	—	0.002
	Polytetrafluoroethylene decomposition products	—	A[2]
	Propane	E	—
	β Propiolactone	—	A[1]
	Propargyl alcohol-Skin	1	—
	n-Propyl acetate	200	840
	Propyl alcohol	200	500
	n-Propyl nitrate	25	110
	Propylene dichloride	75	350
	Propylene imine-Skin	2	5
	Propylene oxide	100	240
	Propyne, see Methylacetylene	—	—
	Pyrethrum	—	5
	Pyridine	5	15
	Quinone	0.1	0.4
	RDX-Skin	—	1.5
	Rhodium, Metal fume and dusts, as Rh	—	0.1
	Soluble salts	—	0.001
	Ronnel	—	10
	Rotenone (commercial)	—	5
	Rouge	—	D
	Selenium compounds (as Se)	—	0.2
	Selenium hexafluoride	0.05	0.4
	Silicon carbide	—	D
	Silver, metal and soluble compounds	—	0.01
	Sodium fluoroacetate (1080)-Skin	—	0.05
	Sodium hydroxide	—	2
	Stibine	0.1	0.5
	Starch	—	D
°	Stoddard solvent	200	1,150
	Strychnine	—	0.15
°°C	Styrene monomer (phenylethylene)	100	420
	Sucrose	—	D
	Sulfur dioxide	5	13
	Sulfur hexafluoride	1,000	6,000
	Sulfuric acid	—	1
	Sulfur monochloride	1	6
	Sulfur pentafluoride	0.025	0.25
	Sulfuryl fluoride	5	20
	Systox, see Demeton®	—	—
	2, 4, 5 T	—	10
	Tantalum	—	5
	TEDP-Skin	—	0.2
	Teflon® decomposition products	—	A[2]
	Tellurium	—	0.1
	Tellurium hexafluoride	0.02	0.2

	Substance	ppm[a)]	mg/M^3 [b)]
	TEPP-Skin	—	0.05
C	Terphenyls	1	9
	1, 1, 1, 2-Tetrachloro-2, 2-difluoroethane	500	4,170
	1, 1, 2, 2-Tetrachloro-1, 2-difluoroethane	500	4,170
	1, 1, 2, 2-Tetrachloroethane-Skin	5	35
	Tetrachloroethylene, see Perchloroethylene	—	—
	Tetrachloromethane, see Carbon tetrachloride	—	—
	Tetrachloronaphthalene-Skin	—	2
	Tetraethyl lead (as Pb)-Skin	—	0.100[j)]
	Tetrahydrofuran	200	590
	Tetramethyl lead (as Pb)-Skin	—	0.150[j)]
	Tetramethyl succinonitrile-Skin	0.5	3
	Tetranitromethane	1	8
	Tetryl (2, 4, 6-trinitrophenyl-methylnitramine)-Skin	—	1.5
	Thallium (soluble compounds)-Skin as Tl	—	0.1
	Thiram	—	5
	Tin (inorganic cmpds, except SnH_4 and SnO_2)	—	2
	Tin (organic cmpds)	—	0.1
	Tin oxide	—	D
	Titanium dioxide	—	D
	Toluene (toluol)	200	750
C	Toluene-2, 4-diisocyanate	0.02	0.14
	o-Toluidine-Skin	5	22
	Toxaphene, see Chlorinated camphene	—	—
	Tributyl phosphate	—	5
	1, 1, 1-Trichloroethane, see Methyl chloroform	—	—
	1, 1, 2-Trichloroethane-Skin	10	45
	Trichloroethylene	100	535
	Trichloromethane, see Chloroform	—	—
	Trichloronaphthalene-Skin	—	5
	1, 2, 3-Trichloropropane	50	300
	1, 1, 2-Trichloro 1, 2, 2-trifluoroethane	1,000	7,600
	Triethylamine	25	100
	Trifluoromonobromomethane	1,000	6,100
°	Trimethyl benzene	25	120
	2, 4, 6-Trinitrophenol, see Picric acid	—	—
	2, 4, 6-Trinitrophenylmethylnitramine, see Tetryl	—	—
	Trinitrotoluene-Skin	—	1.5
	Triorthocresyl phosphate	—	0.1
	Triphenyl phosphate	—	3
	Tungsten & compounds, as W		
	Soluble	—	1
	Insoluble	—	5
	Turpentine	100	560
	Uranium (natural) sol. & insol. compounds as U	—	0.2
C	Vanadium (V_2O_5 dust)	—	0.5
	(V_2O_5 fume)	—	0.1
	Vinyl benzene, see Styrene	—	—
°°C	Vinyl chloride	500[k]	1,300
	Vinylcyanide, see Acrylonitrile	—	—
	Vinyl toluene	100	480
	Warfarin	—	0.1

[j)] For control of general room air, biologic monitoring is essential for personnel control.
[k)] In 1974 this limit was reduced to 50 ppm.

Substance	*ppm*[a)]	*mg/M*^{3} [b)]
Xylene (xylol)	100	435
Xylidine-Skin	5	25
Yttrium	—	1
Zinc chloride fume	—	1
Zinc oxide fume	—	5
Zirconium compounds (as Zr)	—	5

Radioactivity: For permissible concentrations of radioisotopes in air, see U.S. Department of Commerce, National Bureau of Standards, Handbook 69, "Maximum Permissible Body Burdens and Maximum Permissible Concentrations of Radionuclides in Air and in Water for Occupational Exposure," June 5, 1959. Also, see U.S. Department of Commerce National Bureau of Standards, Handbook 59, "Permissible Dose from External Sources of Ionizing Radiation," September 24, 1954, and addendum of April 15, 1958.

Mineral Dusts

Substance	*m.p.p.c.f.* [e)]
Silica	
Crystalline	
°° Quartz, Threshold Limit calculated from the formula	$\frac{250^{f)}}{\%SiO_2 + 5}$
°° Cristobalite, Threshold Limit calculated from the formula	$\frac{250^{f)}}{\%SiO_2 + 5}$
Amorphous, including natural diatomaceous earth	20
Silicates (less than 1% crystalline silica)	
°° Asbestos, all types	5
Mica	20
Portland Cement	50
Soapstone	20
Talc (non-asbestiform)	20
Talc (fibrous), use asbestos limit	—
Tremolite, see asbestos	—
Graphite (natural)	15
°° "Inert" or Nuisance Particulates see Appendix D	50 (or 15 mg/M^3 whichever is the smaller) of total dust $<1\%$ SiO_2

Conversion factors

mppcf × 35.3 = million particles per cubic meter
= particles per c. c.

Notice of Intended Changes (for 1970)

These substances, with their corresponding values, comprise those for which either a limit has been proposed for the first time, or for which a change in the "Adopted" listing has been proposed. In both cases, the proposed limits should be considered trial limits that will remain in the listing for a period of at least two years. During this time, the previously Adopted Limit

e) Millions of particles per cubic foot of air, based on impinger samples counted by light-field technics.
f) The percentage of crystalline silica in the formula is the amount determined from airborne samples, except in those instances in which other methods have been shown to be applicable.
°° See Notice of Intended Changes for Mineral Dusts.

will remain in effect. If, after two years no evidence comes to light that questions the appropriateness of the values herein, the values will be placed in the "Adopted" list. Documentation is available for each of these substances.

	Substance	*ppm*	*mg/M*^{3}
	2-Acetylaminofluorene-Skin	—	A^1
+	Allyl glycidyl ether	5	22
	4-Aminodiphenyl-Skin	—	A^1
+	Ammonia	25	18
+	Ammonium chloride fume	—	10
	Asphalt (petroleum) fumes	—	5
+	Butyl lactate	1	5
+	Camphor (synthetic)	2	12
+	Diazinon-Skin	—	0.1
+	2-N Dibutylamino ethanol-Skin	2	14
	Dichlorobenzidine-Skin	—	A^1
+	Diethylene triamine-Skin	1	4
	4-Dimethylaminoazobenzene	—	A^1
	Fibrous glass	[g)]—	D
+C	Formaldehyde	2	3
+	Iron pentacarbonyl	0.01	0.08
+	Mercury (Alkyl compounds)-Skin	—	0.01
	Mercury (all forms except alkyl)	—	0.05
+	Methyl chloride	100	210
	Methyl 2-cyanoacrylate	2	8
+	Methylcylopentadienyl manganese tricarbonyl (as Mn)	0.1	0.2
	Methyl demeton-Skin	—	0.5
	Methyl parathion-Skin	—	0.2
	Phenothiazine-Skin	—	5
+	Rosin Core Solder, pyrolysis products	—	0.1 (as aldehyde)
	Styrene	100	420
+C	Subtilisins (Proteolytic enzymes)	—	0.0003 (as 100% pure crystalline enzyme)
+	Vanadium (V_2O_5 Fume) as V	—	0.05
	Vinyl acetate	10	30
+	Vinyl chloride	200°	770
+	Wood dust (non allergenic)	—	5

+1970 Revision or Addition
Capital letters refer to Appendices
[g)] $<$5–7 μ Diameter. No TLV for coarse fibrous glass has yet been set.
° Reduced to 50 ppm in 1974.

	Substance	*TLV*
+	Asbestos (All types)	5 fibers/ml $>$ 5μ in length[k)]
+	Coal dust (bituminous)	2 mg/m^3 (respirable dust)[m)]
	Cristobalite	Use one-half the value calculated from the count or mass formulae for quartz.
+	'Inert' or Nuisance Particulates	10 mg/M^3 or 30 mppcf (whichever is the smaller) of total dust $<$1% SiO_2[n)]

Substance	*TLV*
+ Quartz	TLV in mppcf: $\frac{300}{\%SiO_2 + 10}$
	TLV for respirable dust in mg/m3: $\frac{10 \text{ mg/M}^{3p)}}{\% \text{ Respirable quartz} + 2}$
	TLV for "total dust," respirable and nonrespirable: $\frac{30 \text{ mg/M}^3}{\% \text{ quartz} + 3}$
Silica, fused	Use quartz formulae
Tridymite	Use one-half the value calculated from formulae for quartz.

k) As determined by the membrane filter method at 430× magnification phase contrast illumination. Concentrations >5 fibers/ml, but not to exceed 10, may be permitted for 15-minute periods each hour up to five times daily.

m) "Respirable" dust as defined by the British Medical Research Council Criteria (1) and as sampled by a device producing equivalent results (2).

(1) Hatch, T. E. and Gross, P., Pulmonary Deposition and Retention of Inhaled Aerosols, p. 149. Academic Press, New York, New York, 1964.

(2) Interim Guide for Respirable Mass Sampling, AIHA Aerosol Technology Committee, AIHA J. *31,* 2, 1970, p. 133.

n) This automatically reduces all particulate substances in Adopted list with TLV of 15 mg/M^3 to 10 mg/M^3.

p) Both concentration and per cent quartz for the application of this limit are to be determined from the fraction passing a size-selector with the following characteristics:

Aerodynamic Diameter (μ) (unit density sphere)	*% passing selector*
$\leq$ 2	90
2.5	75
3.5	50
5.0	25
10	0

Appendix A

A[1] Because of the high incidence of cancer, either in man or in animals, no exposure or contact by any route, respiratory, oral or skin should be permitted for the compounds:

2-Acetylaminofluorene	beta-Naphthylamine
4-Aminodiphenyl	4-Nitrodiphenyl
Benzidine & its salts	N-Nitrosodimethylamine
Dichlorobenzidine	beta-Propiolactone
4-Dimethylaminoazobenzene	

Because of the extremely high incidence of bladder tumors in workers handling beta-naphthylamine and the potential carcinogenic activity of the other compounds, the State of Pennsylvania prohibits the manufacture, use and other activities that involve human exposure without express approval by the Department of Health.

A[2] *Polytetrafluoroethylene* decomposition products.* Thermal decomposition of the fluorocarbon chain in air leads to the formation of oxidized products containing carbon, fluorine and oxygen. Because these products decompose in part by hydrolysis in alkaline solution, they can be quantitatively determined in air as fluoride to provide an index of exposure. No TLV is recommended pending determination of the toxicity of the products, but air concentrations should be minimal.

A[3] *Gasoline and/or Petroleum Distillates.* The composition of these materials varies greatly and thus a single TLV for all types of these materials is no longer applicable. In general, the aromatic hydrocarbon content will determine what TLV applies. Consequently the content of benzene, other aromatics and additives should be determined to arrive at the appropriate TLV (Elkins, et al. A.I.H.A.J. *24*, 99, 1963).

Appendix B
B.1 Threshold Limit Values for Mixtures

When two or more hazardous substances are present, their combined effect, rather than that of either individually, should be given primary consideration. In the absence of information to the contrary, the effects of the different hazards should be considered as additive. That is, if the sum of the following fractions,

$$\frac{C_1}{T_1} + \frac{C_2}{T_2} + \cdots \frac{C_n}{T_n}$$

exceeds unity, then the threshold limit of the mixture should be considered as being exceeded. C_1 indicates the observed atmospheric concentration, and T_1 the corresponding threshold limit (See Example 1A.a.).

Exceptions to the above rule may be made when there is good reason to believe that the chief effects of the different harmful substances are not in fact additive, but independent as when purely local effects on different organs of the body are produced by the various components of the mixture. In such cases the threshold limit ordinarily is exceeded only when at least one member of the series (C_1/T_1+ or $+C_2/T_2$ etc.) itself has a value exceeding unity (See Example 1A.b.).

Antagonistic action or potentiation may occur with some combinations of atmospheric contaminants. Such cases at present must be determined individually. Potentiating or antagonistic agents are not necessarily harmful by themselves. Potentiating effects of exposure to such agents by routes other than that of inhalation is also possible, e.g. imbibed alcohol and inhaled narcotic (trichloroethylene). Potentiation is characteristically exhibited at high concentrations, less probably at low.

When a given operation or process characteristically emits a number of harmful dusts, fumes, vapors or gases it will frequently be only feasible to attempt to evaluate the hazard by measurement of a single substance. In such cases, the threshold limit used for this substance should be reduced by a suitable factor, the magnitude of which will depend on the number, toxicity and relative quantity of the other contaminants ordinarily present.

Examples of processes which are typically associated with two or more harmful atmospheric contaminants are welding, automobile repair, blasting, painting, lacquering, certain foundry operations, diesel exhausts, etc., (Example 2.).

Threshold Limit Values for Mixtures (Examples)

1A. General case, where *air* is analyzed for each component

a. *Additive effects.* (*note: It is essential that the atmosphere be analyzed both qualitatively and quantitatively for each component present, in order to evaluate compliance or noncompliance with this calculated TLV.*)

$$\frac{C_1}{T_1} + \frac{C_2}{T_2} + \frac{C_3}{T_3} + \cdots = 1$$

*Trade Names: Algoflon, Fluon, Halon, Teflon, Tetran

Example No. 1: Air contains 5 ppm of carbon tetrachloride (TLV = 10 ppm) 20 ppm of ethylene dichloride (TLV = 50 ppm) and 10 ppm of ethylene dibromide (TLV = 25 ppm)

Atmospheric concentration of mixture =

$$5 + 20 + 10 = 35 \text{ ppm of mixture}$$

$$\frac{5}{10} + \frac{20}{50} + \frac{10}{25} = \frac{25 + 20 + 20}{50} = 1.3$$

Threshold Limit is exceeded. Furthermore, the TLV of this mixture may be calculated by reducing the total fraction to 1.0; i.e.

$$\text{TLV of mixture} = \frac{35}{1.3} = 27 \text{ ppm}$$

Example No. 2: Air contains 200 ppm of hexane (TLV = 500 ppm) 100 ppm of methylene chloride (TLV = 500 ppm) and 20 ppm of perchlorethylene (TLV = 100 ppm)

Atmospheric concentration of mixture =

$$200 + 100 + 20 = 320 \text{ ppm of mixture}$$

$$\frac{200}{500} + \frac{100}{500} + \frac{20}{100} = \frac{200 + 100 + 100}{500} = \frac{400}{500} = 0.8$$

Threshold Limit is *not* exceeded. The TLV of this mixture $= \frac{320}{0.8} = 400$ ppm

1B. Special case when the source of contaminant is a liquid mixture and the atmospheric composition is *assumed* to be similar to that of the original material; e.g. on a time weighted average exposure basis, all of the liquid (solvent) mixture eventually evaporates.

a. *Additive effects* (approximate solution)

1. The percent composition (by weight) of the liquid mixture is known, the TLVs of the constituents must be listed in mg/M^3.

(*NOTE: In order to evaluate compliance with this TLV, field sampling instruments should be calibrated, in the laboratory, for response to this specific quantitative and qualitative air-vapor mixture, and also to fractional concentrations of this mixture; e.g., ½ the TLV; ¹⁄₁₀ the TLV; 2 × the TLV; 10 × the TLV; etc.*)

$$\text{TLV of mixture} = \frac{1}{\frac{f_a}{TLV_a} + \frac{f_b}{TLV_b} + \frac{f_c}{TLV_c} + \cdots \frac{f_n}{TLV_n}}$$

Example No. 1: Liquid solvent contains (by weight) 50% heptane (TLV = 2000 mg/M^3) 30% methylene chloride (TLV = 1740 mg/M^3) 20% perchlorethylene (TLV = 670 mg/M^3)

$$\text{TLV of mixture} = \frac{1}{\frac{0.5}{2000} + \frac{0.3}{1740} + \frac{0.2}{670}} = \frac{1}{.00025 + .00017 + .0003}$$

$$= \frac{1}{.00072} = 1390 \text{ mg/M}^3$$

Of this mixture: 50% or 695 mg/M^3 is heptane, 30% or 417 mg/M^3 is methylene chloride and 20% or 278 mg/M^3 is perchlorethylene

These values can be converted to ppm as follows:

heptane: 2000 mg/M^3 = 500 ppm
1 mg/M^3 = 0.25 ppm
695 mg/M^3 = 174 ppm

methylene chloride: 1740 mg/M^3 = 500 ppm
1 mg/M^3 = 0.287 ppm
417 mg/M^3 = 119 ppm

perchlorethylene: 670 mg/M^3 = 100 ppm
1 mg/M^3 = 0.15 ppm
278 mg/M^3 = 42 ppm

The TLV of this mixture = 174 + 119 + 42 = 335 ppm.

1B.b. General Exact Solution for Mixtures of N Components with Additive Effects and Different Vapor Pressures.

(1) $$\frac{C_1}{T_1} + \frac{C_2}{T_2} + \cdots + \frac{C_n}{T_n} = 1;$$

(2) $$C_1 + C_2 + \cdots + C_n = T;$$

(2.1) $$\frac{C_1}{T} + \frac{C_2}{T} + \cdots + \frac{C_n}{T} = 1.$$

By the Law of Partial Pressures,

(3) $$C_1 = ap_1,$$

and by Raoult's Law,

(4) $$p_1 = F_1 p_1^{\circ}.$$

Combine (3) and (4) to obtain

(5) $$C_1 = aF_1 p_1^{\circ}.$$

Combining (1), (2, 1) and (5), we obtain

(6) $$\frac{F_1 p_1^{\circ}}{T} + \frac{F_2 p_2^{\circ}}{T} + \cdots + \frac{F_n p_n^{\circ}}{T} = \frac{F_1 p_1^{\circ}}{T_1} + \frac{F_2 p_2^{\circ}}{T_2} + \cdots + \frac{F_n p_n^{\circ}}{T_n}$$

and solving for T,

(6.1) $$T = \frac{F_1 p_1^{\circ} + F_2 p_2^{\circ} + \cdots + F_n p_n^{\circ}}{\frac{F_1 p_1^{\circ}}{T_1} + \frac{F_2 p_2^{\circ}}{T_2} + \cdots + \frac{F_n p_n^{\circ}}{T_n}}$$

or

(6.2) $$T = \frac{\sum^{i} = n\, F_1 p_1^{\circ} \quad i = 1}{i = n \quad \sum \quad i = 1 \frac{F_1 p_1^{\circ}}{T_1}}$$

T = Threshold Limit Value in ppm.
C = Vapor concentration in ppm.
p = Vapor pressure of component in solution.
p° = Vapor pressure of pure component.
F = Mol fraction of component in solution.
a = A constant of proportionality.

Subscripts 1, 2, . . . n relate the above quantities to components 1, 2, . . . n, respectively.
Subscript i refers to an arbitrary component from 1 to n.
Absence of subscript relates the quantity to the mixture.

1B.c. Solution to be applied when there is a reservoir of the solvent mixture whose composition does not change appreciably by evaporation.

Exact Arithmetic Solution of Specific Mixture

	Mol. wt.	*Density*	*TLV*	*p°at 25°C*	*Mol fraction in half-and-half solution by volume*
Trichloroethylene	(1) 131.4	1.46 g/ml	100	73 mm Hg	0.527
Methylchloroform	(2) 133.42	1.33 g/ml	350	125 mm Hg	0.473

$$F_1p_1^\circ = (0.527)\ (73) = 38.2$$
$$F_2p_2^\circ = (0.473)\ (125) = 59.2$$

$$\text{TLV} = \frac{38.2 + 59.2}{\dfrac{38.2}{100} + \dfrac{59.2}{350}} = \frac{(97.4)\ (350)}{133.8 + 59.2} = \frac{(97.4)\ (350)}{193.0} = 177$$

$$\text{TLV} = 177\ \text{ppm}$$

(Note difference in TLV when account is taken of vapor pressure and mol fraction in comparison with above sample where such account is not taken.)

2. A mixture of one part of (1) parathion (TLV, 0.1) and two parts of (2) EPN (TLV, 0.5).

$$\frac{C_1}{0.1} + \frac{C_2}{0.5} = \frac{C_m}{T_m}$$

$$C_m = 3C_1 \qquad\qquad C_2 = 2C_1$$

$$\frac{C_1}{0.1} + \frac{2C_1}{0.5} = \frac{3C_1}{T_m}$$

$$\frac{7C_1}{0.5} = \frac{3C_1}{T_m}$$

$$T_m = \frac{1.5}{7} = 0.21\ \text{mg/M}^3$$

1C. TLV for Mixtures of Mineral Dusts.

For mixtures of biologically active mineral dusts the general formula for mixtures may be used.

For a mixture containing 80% talc and 20% quartz, the TLV for 100% of the mixture is given by:

$$\text{TLV} = \frac{1}{\dfrac{0.8}{20} + \dfrac{0.2}{2.5}} = 8.4\ \text{mppcf}$$

Essentially the same result will be obtained if the limit of the more (most) toxic component is used provided the effects are additive. In the above example the limit for 20% quartz is 10 mppcf.

For another mixture of 25% quartz, 25% amorphous silica and 50% talc:

$$\text{TLV} = \frac{1}{\dfrac{0.25}{2.5} + \dfrac{0.25}{20} + \dfrac{0.5}{20}} = 7.3\ \text{mppcf}$$

The limit for 25% quartz approximates 8 mppcf.

Appendix C
Permissible Excursions for Time-Weighted Average (TWA) Limits

The Excursion TLV Factor in the Table automatically defines the magnitude of the permissible excursion above the limit for those substances not given a "C" designation; i.e., the TWA limits. Examples in the Table show that nitrobenzene, the TLV for which is 1 ppm, should never be allowed to exceed 3 ppm. Similarly, carbon tetrachloride, TLV 10 ppm, should never be allowed to exceed 20 ppm. By contrast, those substances with a "C" designation are not subject to the excursion factor and must be kept below the TLV.

These limiting excursions are to be considered to provide a 'rule-of-thumb' guidance for listed substances generally, and may not provide the most appropriate excursion for a particular substance. Efforts are being made to develop such specific excursions, when indicated to be significantly different from that recommended by the present excursion factors.

Substance	*TLV*	*Excursion Factor*	*Max. Conc. Permitted for short time*
Nitrobenzene	1 ppm	3	3 ppm
Carbon tetrachloride	10 ppm	2	20 ppm
Carbon monoxide	50 ppm	1.5	75 ppm
Acetone	1000 ppm	1.25	1250 ppm
Boron trifluoride	C 1 ppm	—	1 ppm
Butylamine	C 5 ppm	—	5 ppm
Styrene monomer	C 100 ppm	—	100 ppm

For all substances:

TLV = 0 – 1 (ppm or mg/m^3), Excursion Factor = 3
TLV = 1 – 10 (ppm or mg/m^3), Excursion Factor = 2
TLV = 10 – 100 (ppm or mg/m^3), Excursion Factor = 1.5
TLV = 100 – 1000 (ppm or mg/m^3), Excursion Factor = 1.25

Basis for Assigning Limiting "C" Values

By definition in the Preface, a listed value bearing a "C" designation refers to a 'ceiling' value that should not be exceeded; all values should fluctuate below the listed value. This, in effect, makes the "C" designation a maximal allowable concentration (MAC). In general, the bases for assigning or not assigning a "C" value rest on whether excursions of concentration above a proposed limit for periods up to 15 minutes may result in a) intolerable irritation, b) chronic, or irreversible tissue change, or c) narcosis of sufficient degree to increase accident proneness, impair self-rescue or materially reduce work efficiency.

Appendix D
Some "Inert" or Nuisance Particulates[q)]

Alundum (Al_2O_3)
Calcium carbonate
Cellulose (paper fiber)
Portland Cement
Corundum (Al_2O_3)
Emery
Glycerine Mist
Graphite (synthetic)
Gypsum
Vegetable oil mists (except castor, cashew nut, or similar irritant oils)
Kaolin
Limestone
Magnesite
Marble
Pentaerythritol
Plaster of Paris
Rouge
Silicon Carbide
Starch
Sucrose
Tin Oxide
Titanium Dioxide

[q)] When toxic impurities are not present, e.g. quartz $<1\%$.

Appendix E
Some Simple Asphyxiants—
"Inert" Gases and Vapors

Acetylene	Hydrogen
Argon	Methane
Ethane	Neon
Ethylene	Nitrogen
Helium	Nitrous Oxide
	Propane

5.4 REDUCING POLLUTION BY SOURCE CORRECTION METHODS

A successful industrial air conservation program begins at the conceptual phase of a process or product development program. At this early stage, the research and development team should ask themselves questions concerning the proposed process or product: "Will the process or product have an effluent?" "Will it create an effluent during manufacturing operations, during use or on disposal?" "Will this effluent create potential air pollution problems?" If the answers to these questions are affirmative, then the best time to solve a potential air pollution problem could well be at that point. The exact cause of the potential problem should be pinpointed and viable alternatives investigated before the development proceeds. By evaluating the alternatives, methods or products that do not contribute to air pollution or have a minimum air pollution potential can be selected. Once the manufacturing process has become commercialized, reducing the air pollution contribution by source correction methods becomes more difficult. Yet, often there are changes that can still be made without upsetting the economy of the operation. These changes can be in raw materials substitution, modification of process, alteration of particular equipment components and removal of pollutants at the source. Some methods may use various combinations of these alterations.

Raw Materials Substitution

If one raw material results in a pollution problem while a substitute material does not, obviously the substitute would be more desirable. In some instances, the raw material may contain a nonessential ingredient that is a pollution source. If the nonessential ingredient can be removed prior to processing the raw material, the pollution potential is minimized.

One example of raw material substitution is to use nonreactive aqueous solution solvent systems to replace reactive organic solvents. Organic solvents when emitted to the atmosphere can eventually become photochemical smog. Another example is to use low-sulfur fuels to replace high-sulfur ones; however, low-sulfur fuels are in short supply.

The combustion of high-sulfur fuels results in gaseous sulfur oxides, the largest gaseous pollutant by quantity. Methods of removing sulfur from heavy fuel oils are being developed (see Section 6.1). These methods involve high-

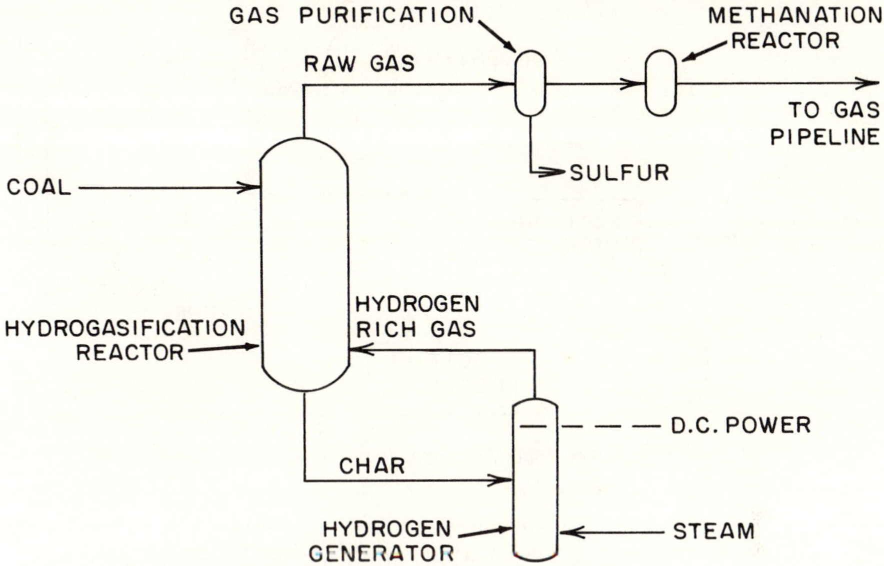

Figure 5.4a Coal gasification plant.

pressure catalytic reactors in which hydrogen combines with the sulfur to form hydrogen sulfide which is then recovered. Some sulfur can be washed from coal, although much of coal's sulfur is combined chemically with the coal so that it cannot be removed by physical treatment. Large-scale research is being carried out on methods of gasifying the coal to produce a synthetic fuel gas containing mainly methane. In coal gasification, the sulfur is recovered by passing the sulfur-bearing gases through a liquid purification solution. A schematic of a coal gasification plant is shown in Figure 5.4a.

Process Modifications

Pollution abatement can frequently be achieved by new or modified processing methods. The use of exhaust hoods and ducts over several types of industrial ovens has allowed the recovery of valuable solvents that could have become air pollutants. In the petroleum refining industry, hydrogen sulfide was once passed to the flare system and burned. This resulted in sulfur oxide emissions to the atmosphere. Currently, the industry uses the Claus process or a combination Claus-Stretford process that will yield elemental sulfur as a refinery by-product, in addition to preventing an air pollution problem. Figure 5.4b illustrates sulfur recovery by the Claus process.

Another example of a process modification concerns the combustion of fuels. During combustion of fuels, nitrogen oxides are formed by the reaction of nitrogen and oxygen in the air used for combustion. Most of the nitrogen

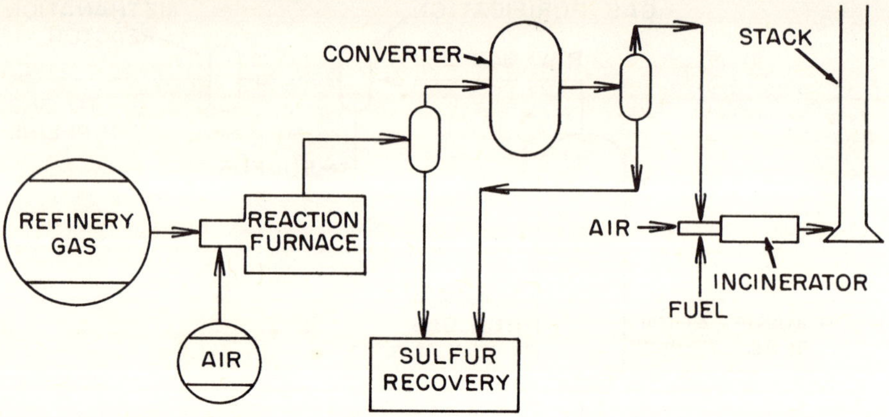

Figure 5.4b Sulfur recovery from refinery gases.

oxides pollution is caused by this nitrogen fixation reaction occurring during the combustion of fossil fuels for various applications. Potential combustion control techniques for reducing nitrogen oxides emissions consist of modifying the operating techniques and equipment design features that affect combustion parameters. Some of the possible operating modifications include low excess air firing, two stage combustion, flue gas recirculation, water injection and combinations of these techniques.

Two-stage combustion can be particularly effective. This process consists of firing all the fuel with substoichiometric amounts of primary air in the first stage and in the second stage injecting secondary air to complete the burnout of the fuel. In the first stage, nitrogen oxide formation is limited by the unavailability of oxygen. Removal of heat between stages reduces the temperatures achieved when the secondary air is added and, thereby, limits the reaction rate of nitrogen oxide formation. Nitrogen oxides emissions have been lowered as much as 90% by using this technique.

Another process modification that may reduce nitrogen oxides emissions is to use a fluidized bed for the combustion boiler. The lower flame temperature in the fluidized bed minimizes nitrogen oxide formation. This technique is under active research in several parts of the world. Further discussion of the fluidized bed boiler is given in Chapter 6.

Equipment Alteration

Equipment alterations such as the use of floating roof tanks rather than vented tanks can cut down on evaporation losses. In many cases, newer types of equipment are less pollution-prone. For example, the basic oxygen furnaces that are replacing the open-hearth furnaces in the steel industry are less of a pollution problem. Newer types of equipment in the pulp and paper industry also cut down on the quantity of pollutants emitted.

Development of a suitable replacement for the internal combustion engine as the automobile's power source is another example of equipment alteration. It may be that an electric car drawing its power from a nuclear power plant or a pollution-free conventional plant could replace the internal combustion engine driven car in urban areas. Another possible power source for the automobile is the fuel cell.

Other modifications to automobile equipment have already been made. The positive crankcase ventilation valve prevents blow-by gases containing hydrocarbons from being emitted onto the roadway. Evaporative losses from the fuel tank and carburetor have been decreased by the use of an activated charcoal canister that adsorbs hydrocarbon vapors and then releases them through special valves under controlled conditions to be burned in the engine. Various emissions sources on an uncontrolled car are illustrated by Figure 5.4c.

Removal of Pollutants at Source

Another concept in emissions control should be to remove emissions near their source. This concept applies to those cases where emissions cannot be prevented and pollution control equipment is necessary to remove them from the main gas stream. It is normally more economical to remove these pollutants while they are present in a relatively high concentration rather than at some point away from the source where the pollutants are diluted by other process gases or air.

Procedural changes can be used to ease the effects of air pollution. Automatic operating changes can be made during levels of high-ambient pollution. For example, the TVA switches to low-sulfur fuels for power generation during adverse meteorological conditions.

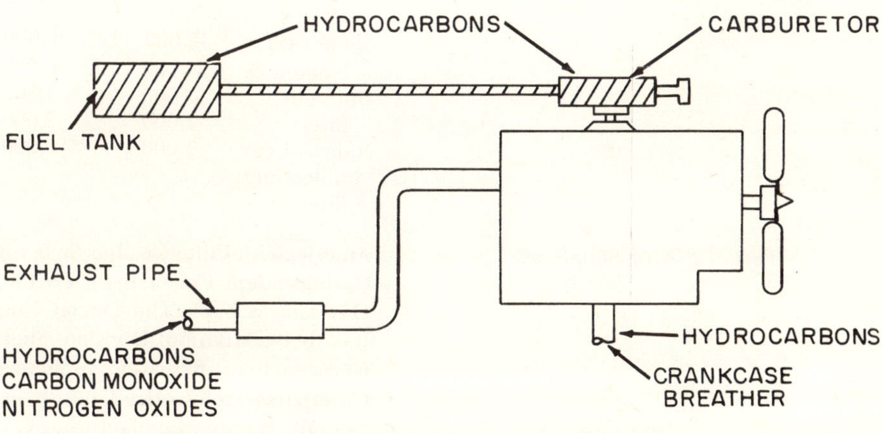

Figure 5.4c Emissions from uncontrolled car.

5.5 BAG AND FABRIC FILTERS—TYPES, FEATURES AND OPERATION

Type of Filter Media: (a) Paper (for special, no reuse applications); (b) Woven fabric (with low air/cloth ratio); (c) Felted fabric (with high air/cloth ratio).

Cleaning Methods: By various methods of shaking, including manual, mechanical, pneumatic, bubble cleaning, ejector, reverse air flexing and sonic techniques or by reverse flow methods including repressurizing, atmospheric and reverse jet techniques.

Operation: It can be intermittent, automatically cycled or fully continuous.

Uninstalled Equipment Costs:

Filter Type	*Capacity, ACFM*	*Cost, $*
Reverse jet filter with feeder and airlock	10,000	15–18,000
	30,000	45–55,000
	60,000	85–105,000
Shaker type baghouses	12,000	20,000
	50,000	60,000
Bin vent filters	500	1,800
	1,000	2,000
Railroad car unloading filter	1,000	5–6,000

Partial List of Suppliers: American Air Filter Co., Inc. (a, b, c); Carborundum Co. (a, b, c); Carter-Day Co. (a, b, c); The Ducon Co., Inc. (b, c); Mikropul Division, Slick Corp. (a, b, c); Research Appliance Co. (a); Research-Cottrell, Inc. (b, c); The W. W. Sly Manufacturing Co. (b, c); Wheelabrator Corp. (b, c).

Baghouses and fabric filters are used for efficient removal of particulates from exhaust air. Fabric filters and inertial dry collectors (see Sections 5.8 and 5.9) share the following functional characteristics:

1. Particulates are collected dry and in usable condition.
2. Exhaust gas is not cooled or saturated with moisture.
3. Subsequent solid handling units must be carefully designed, to avoid secondary dust problems.
4. Unlike scrubbers, filters and dry collectors do not add moisture to the cleaned exhaust and, hence, do not create plume.

In a bag filter, the dust-laden gas flows through a porous medium (paper, woven or felted fabric) and the particulates are deposited in the voids. As voids fill up, pressure drop increases until a point when the fabric must be cleaned.

The deposition of dust and the consequent separation in a bag filter is achieved by a combination of the following mechanisms: direct deposition, impingement, diffusion and electrostatic action.

Filter Types

Filters can be classified according to their (1) media form, (2) type of media, (3) method of cloth cleaning and (4) cloth housing.

Filter Forms

Available filter media forms include envelope type bags and tubular type bags.

Envelope-Type Bags. (See Fig. 5.5a.) Dust collects on the outside of the

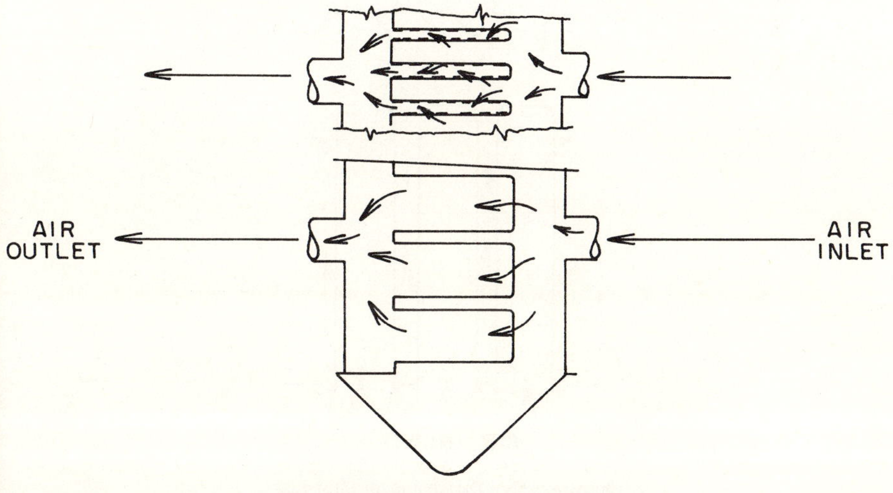

Figure 5.5a Envelope type filter bag.

bags, which are prevented from collapsing by internal screens. Envelope-type filter bags provide maximum square footage of filtering area per unit volume but do sometimes suffer from dust bridging problems, and the bag changing can also be laborious. Filters using paper as filter media usually employ envelope-type bags.

Tubular-Type Bags. (See Fig. 5.5b.) Tubular bags are open at the inlet and closed at the other end. Direction of filtering may be inside-out or outside-in. If the direction is outside-in, a frame is necessary to prevent the collapse of the bag. Outside-in design increases the difficulty of bag replacement and frequently leads to shorter bag life because of bag-frame contact. Tubular bags may be sewn together in groups when they are called *Multibags* (Fig. 5.5c). This design requires the loop-top suspension of the bag, which usually does not permit adequate adjustment for dimensional instability of the fabric.

Other multitubular bag configurations include the *unibag* shown in Figure 5.5d. In this configuration, several bags are retained on frames with a common cell-plate ceiling and a cell-plate floor. The gas enters at the top and flows into the bags which are open at the top and bottom. This design may result in dead pockets that can cause problems due to condensation, but it presents a minimum of reentrainment problems.

Type of Media

Filtration media in bag and fabric filters may be made of (1) paper, (2) woven fabric and (3) felted fabric.

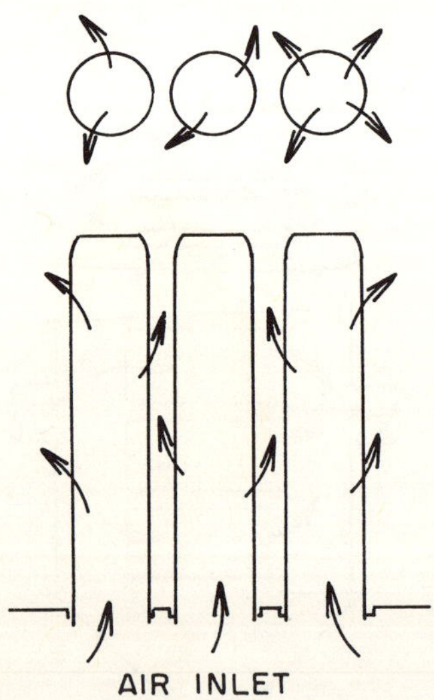

Figure 5.5b Tubular type filter bag.

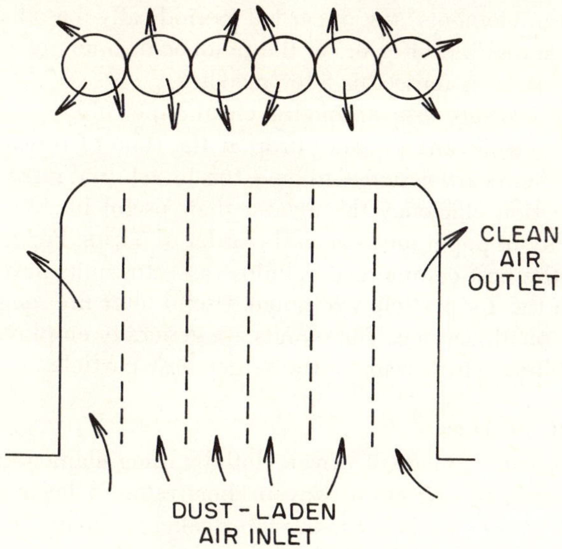

Figure 5.5c Multifilter bags.

Paper Filters

Filters employing paper as collecting substrate are employed in sampling and analysis of air, collection of radioactive or bacterial dust, clean rooms in pharmaceutical and electronic plants and the like. Paper filters are not employed in large commercial units.

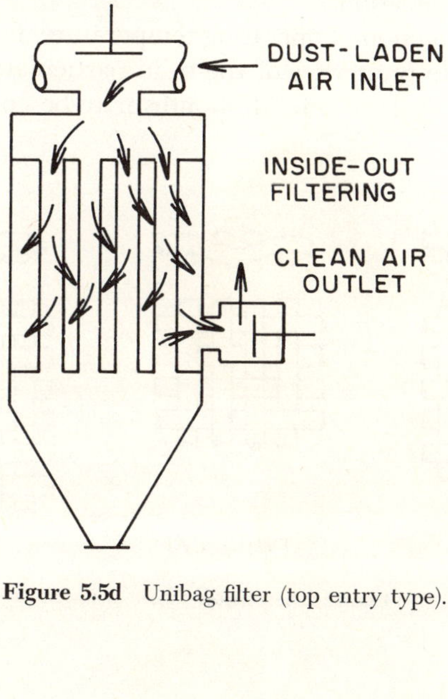

Figure 5.5d Unibag filter (top entry type).

Paper filter elements are discarded periodically based on either the pressure drop across the filter or on the number of hours of use. Generally, no attempt is made to reuse this filter media.

The initial pressure drop through a clean paper filter is relatively large, (0.5 to 1″ H_2O), while the pressure drop at the time of replacement is 1 to 2″ H_2O. Paper filters are designed to have this high initial resistance to ensure adequate collection efficiency throughout their useful life.

Gas velocity in paper filters is on the order of 5 fpm. Paper filter efficiencies are high; the best commercial cellulose-asbestos units have an efficiency of over 99% in the 1 μ particle size range. Useful filter life ranges from 1,000 to 5,000 hours on air service. These units are generally employed in conjunction with prefilters which remove the larger dust particles.

Woven Fabric Filters

Filter bags can be made of woven cloth by using filament or staple filters in various weaves. In operation, woven fibers retain a layer of dust on the surface which serves as a grid for the filter cake. Normally, woven bags are employed with low air/cloth ratios. The average values range from 1.5:1 to 6:1 cfm/ft^2.

The woven fabric may be plain, twill or sateen (Fig. 5.5e). Twill and sateen are the most common. Performance of woven fabric is influenced by (1) the thread density, (2) the fiber composition and (3) the nap, i.e., texture of the cloth surface.

The characteristics of some of the common fabrics are listed in Table 5.5f. Cotton is the least expensive, while glass fiber cloth is often used for high-temperature applications.

Fabric life is a function of operating temperature, frequency and method of cleaning and of the properties of the solid particulates and the gas being cleaned. Fabric filter lives of 6 to 18 months may be considered satisfactory averages.

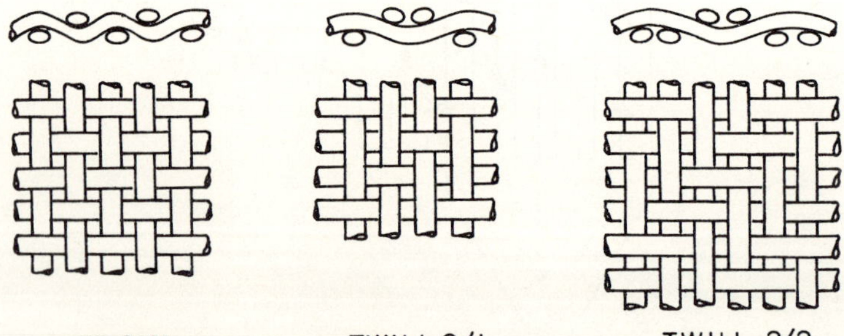

Figure 5.5e Filter fabric weaves.

Table 5.5f
CHARACTERISTICS OF FILTER FABRICS

Fabric	Maximum Operating Temperature °F	Specific Gravity	Physical Resistance					Chemical Resistance					Relative Cost
			Dry Heat	Moist Heat	Abrasion	Shaking	Flexing	Mineral Acids	Organic Acids	Alkalies	Oxidizing Agents	Solvents	
Cotton	180	1.6	G	G	F	G	G	P	G	F	F	E	Low
Dacron	275	1.4	G	F	G	E	E	G	G	F	G	E	Moderate
Orlon	250	1.2	G	G	G	G	E	G	G	F	G	E	Moderate
Nylon	250	1.1	G	G	E	E	E	P	F	G	F	E	Moderate
Dynel	180	1.3	F	F	F	P–F	G	G	G	G	G	G	Moderate
Polypropylene	225	0.9	G	F	E	E	G	E	E	E	G	G	Low
Creslan	275	1.2	G	G	G	G	E	G	G	F	G	E	Moderate
Vycron	300	1.4	G	F	G	E	E	G	G	G	G	E	Moderate
Nomex	450	1.4	E	E	E	E	E	P–F	E	G	G	E	High
Teflon	500	2.3	E	E	P–F	G	G	E	E	E	E	E	High
Wool	200	1.3	F	F	G	F	G	F	F	P	P	F	Moderate
Glass	550	2.5	E	E	P	P	F	E	E	G	E	E	High

E-Excellent G-Good F-Fair P-Poor

Dimensional stability is an important factor for all fabrics. Cotton and wool have to be preshrunk. All synthetic materials have to be heat treated. Glass requires a treatment with silicones.

Felted Fabric Filters

If the material can be felted, then a felted bag may serve as a mat for filtration. Felted bags are more expensive than woven bags, but their use permits employment of high air/cloth ratios. Usual values range from 4:1 to 12:1 cfm/ft^2.

Wool is the only material producing a true felt. Most synthetic fibers can be needled to function as a felted fabric, but cotton and mineral fibers cannot be satisfactorily felted for such use.

Felted materials are not very suitable for extremely fine fumes since the particulates get imbedded in the felt and cleaning becomes difficult. Felted fabrics, in general, require thorough cleaning.

Cleaning Methods

Manual. Small baghouses with up to 400 or 500 square feet of filtering area may be cleaned by manual shaking, using hand levers. The framework from which the filter is suspended, can be manually rapped to shake the dust loose. Thorough cleaning requires vigorous shaking for several minutes.

Mechanical. Mechanical shaking is probably the most widely used method for woven fabrics. A cam or an eccentric in the shaker assembly translates the rotary motion of a motor into an oscillatory motion. The common procedure as shown in Figure 5.5g is to isolate a section of the baghouse and mechanically shake the dust loose. Compared to reverse flow methods, this technique reduces the flexing of the bags. Sometimes, reverse flow and mechanical shaking are used together.

It is essential that there be no pressure inside the bag during the shaking cycle, because even a minute pressure ($<0.1''$ H_2O) may still interfere with proper cleaning. Butterfly-dampers (unless they are the possitive seating type) cannot completely seal off a section while the blower is operating. A small amount of reverse airflow can be used to ensure complete bag collapse during cleaning. Another possibility is to shut off the blower during cleaning.

Air Shaking. Bags can be wind-whipped by blowing air between rows of bags as shown in Figure 5.5h. The dancing action created removes the dust from the bag.

Bubble Cleaning. In this method, as shown in Figure 5.5i, the top of the bag is caused to take up the shape of a bubble at the beginning of the cleaning cycle. As the bubble travels downward, the rippling action cleans the bags. This method requires high amounts of compressed air.

Ejector. This method is usually employed with felted fabrics and with the airflow direction being outside-in, thus requiring a frame to hold the bag.

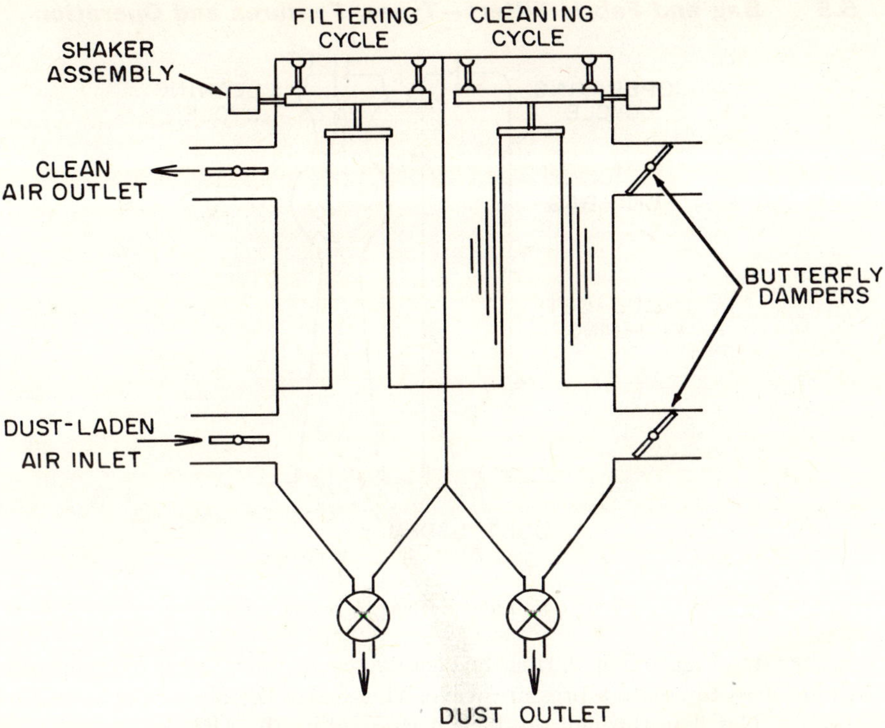

Figure 5.5g Filter cleaning by mechanical shaker.

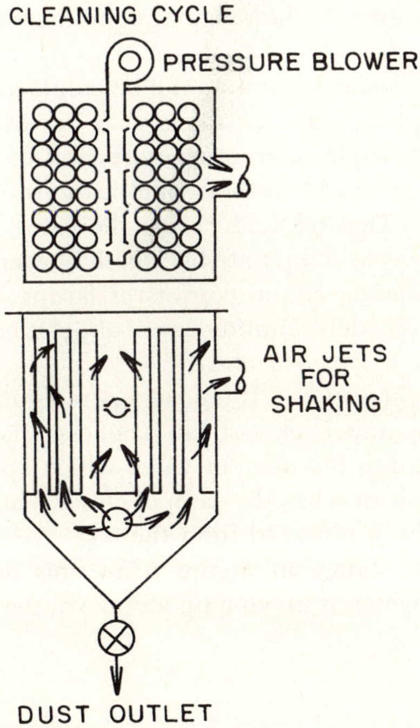

Figure 5.5h Filter cleaning by air shaking.

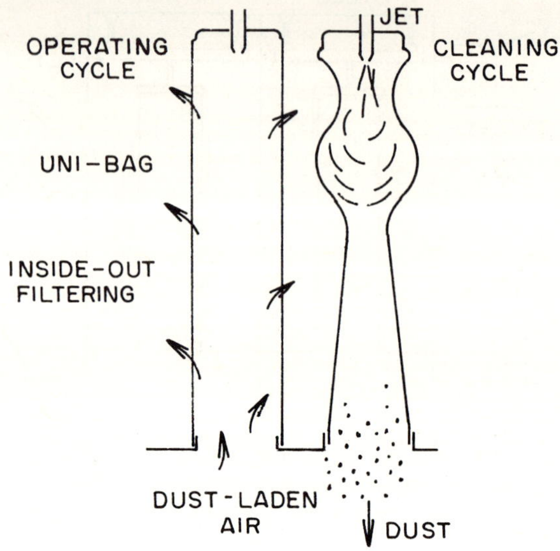

Figure 5.5i Filter cleaning by bubble action.

As shown in Figure 5.5j, during the cleaning cycle, short bursts of compressed air are used to create a pressure wave. This shakes the filter cake loose from the bag. Gas flow through the bags is stopped by this cleaning action, even though the bag is not isolated.

Reverse Air Flexing. This is an improved ejector design where the bag is sealed off as it is being cleaned (Fig. 5.5k), thus causing a positive back-washing action. This extends the range of operation of the ejector-type filter bag cleaner.

Sonic Methods. Sonic generators can be employed to clean the bags but usually require supplementary cleaning with reverse flow or shaking. Its cleaning action is not superior to other types and the recent regulations on noise (Section III.4) may add further complications on its use.

Repressurizing. This method cleans the filter bags by backwashing. During the cleaning cycle, a separate backwash blower (Figure 5.51) provides the cleaning air. Cleaning action consists of flexing and creasing the cloth which contributes to bag deterioration. Usual air/cloth ratio in backwash filters is low, ranging from 1:1 to 3:1 cfm/ft^2.

Atmospheric Reverse Flow. If the bag operates under suction (subatmospheric pressure), a separate backwash fan is not essential. An atmospheric vent can be incorporated into the damper valve which opens to the atmosphere to obtain the backwash air when the damper closes. This method is satisfactory only if the dust is easily removed from the bag.

Reverse Jet. As shown in Figure 5.5m, this design uses a traveling compressed air ring which is moving up and down the bag and removing the

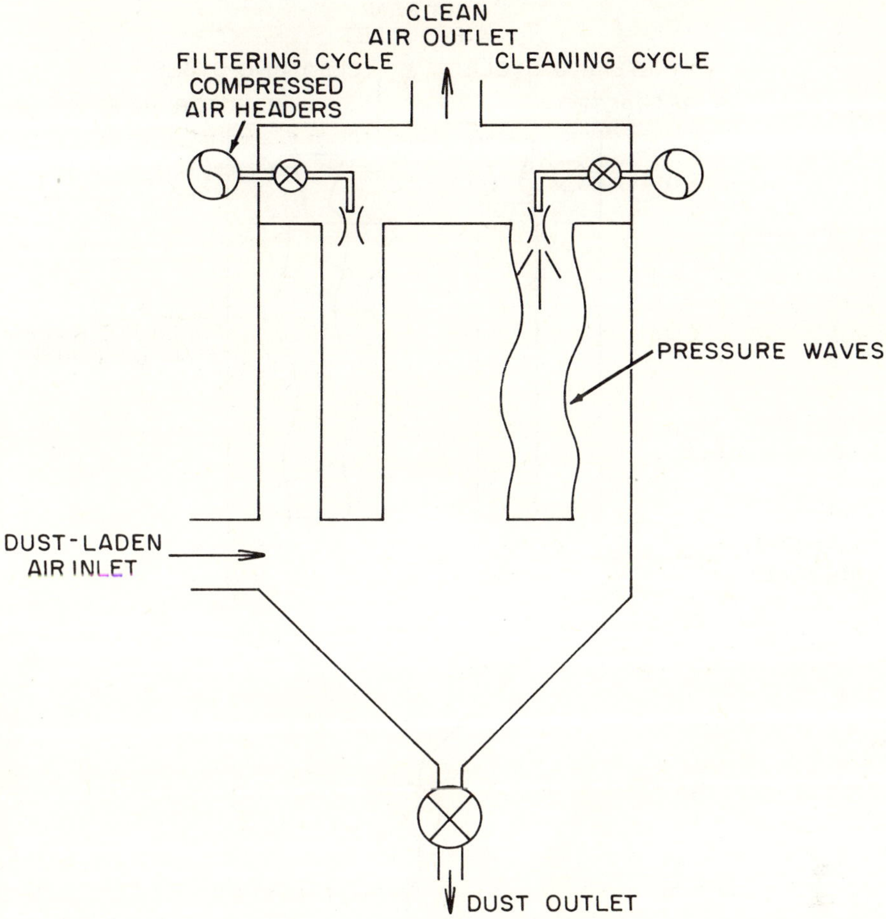

Figure 5.5j Filter cleaning by ejector.

dust by its reverse jet action. For fine dust removal, this method requires substantial pressure drops due to the reentrainment of the dust. This design is successful in the removal of coarse dust at high air/cloth ratios. The reverse jet method is employed only with felted bags.

General Design Considerations

Pressure Drop

Total pressure drop is the result of the combined resistance of the clean cloth and the dust mat[1]. At average airflow rates, resistance of the clean cloth rarely exceeds 0.1″ H_2O. Pressure drop evaluations which are required for the selection of the proper baghouse are discussed in Section 5.6.

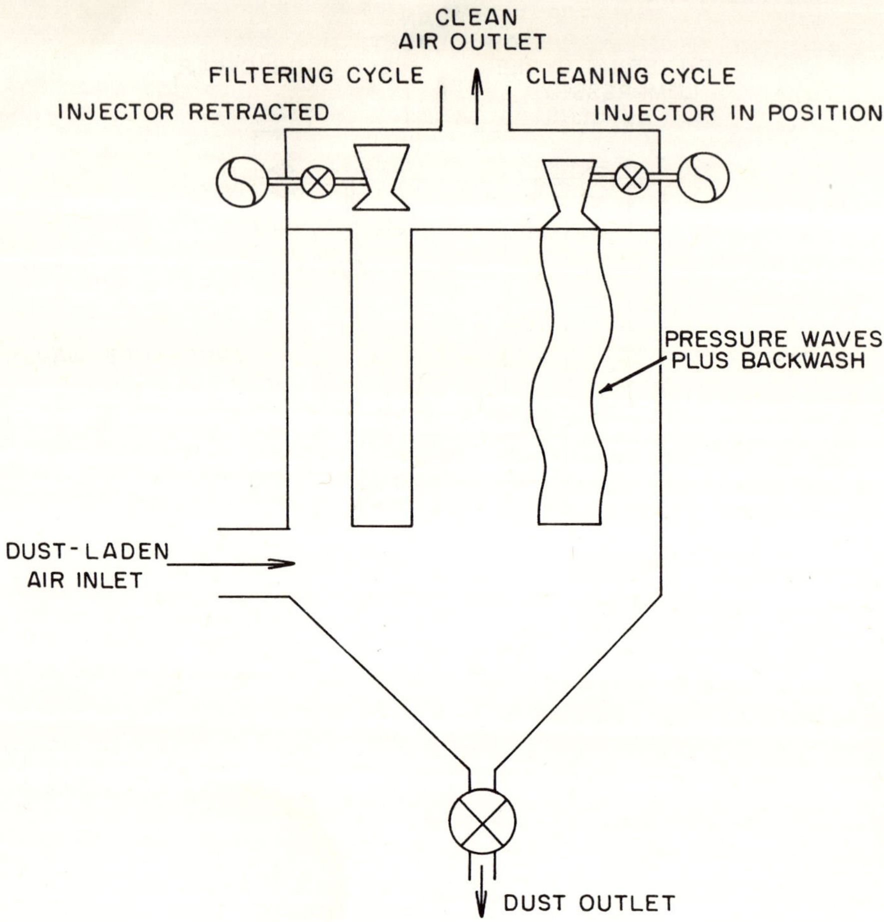

Figure 5.5k Cleaning by reverse air flex.

Air/Cloth Ratio

Air-to-cloth ratio or filtering velocity is a critical factor in system design. The value of this ratio depends on the (1) type of fabric, (2) method of cleaning, (3) properties of the particulates and (4) gas properties.

Typical air/cloth ratio values for some processes are given in Table 5.5n.

Bag Size and Shape

Tubular bag diameters can be 6 or 12 inches and their length can vary from 5 to 25 feet. Length-to-diameter ratios are not yet standardized.

Precooling

Gases from furnaces or other high-temperature sources must be precooled before entering the baghouse. Available cooling methods include (1) cooling

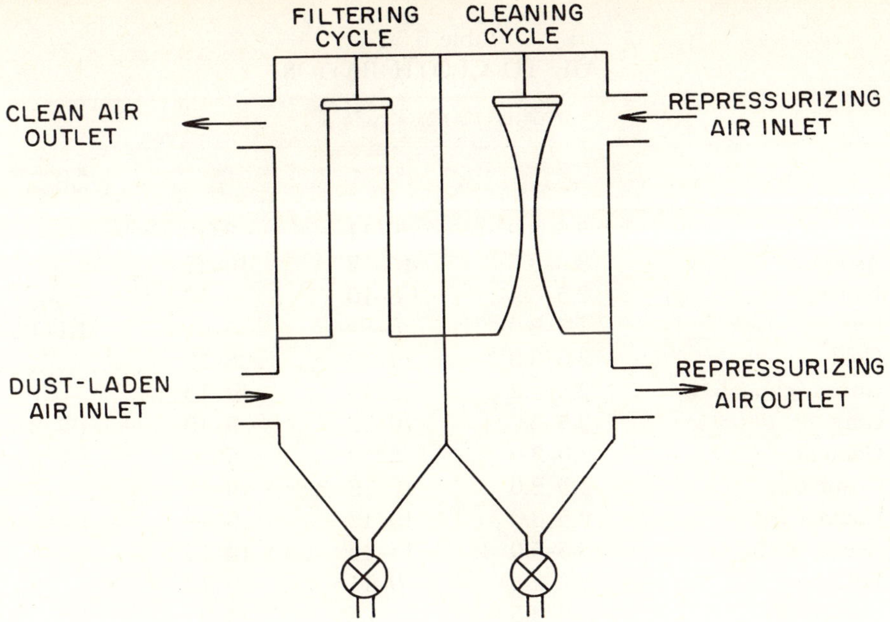

Figure 5.5l Filter cleaning by repressurizing.

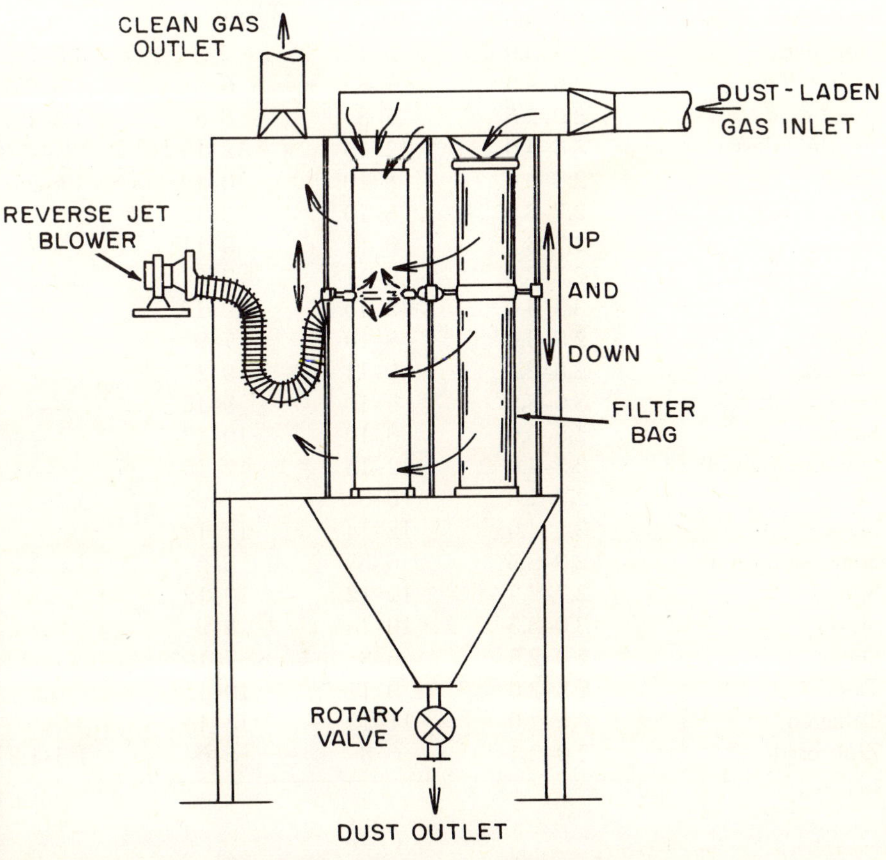

Figure 5.5m Reverse jet type filter cleaning.

Table 5.5n
AIR TO CLOTH RATIOS

Dust	Usual Air-Cloth Ratio, CFM/sq. ft. Shaker Collector	Reverse Jet	Pulse Jet	Reverse Air Collapse
Alumina	2.5–3.0	9–11	8–10	—
Asbestos	3.0–3.5	10–12	10–12	—
Bauxite	2.5–3.2	8–10	8–10	—
Carbon black	1.5–2.0	6–8	5–6	1.1–1.5
Coal	2.5–3.0	10–12	8–10	—
Cocoa, chocolate	2.8–3.2	12–15	12–15	—
Clay	2.5–3.2	10–12	9–10	1.5–2.0
Cement	2.0–3.0	8–10	8–10	1.2–1.5
Cosmetics	1.5–2.0	10–12	10–12	—
Enamel frit	2.5–3.0	10–12	9–10	1.5–2.0
Feeds, grain	3.5–5.0	14–16	14–15	—
Feldspar	2.2–2.8	9–11	9–10	—
Fertilizer	3.0–3.5	—	8–9	1.8–2.0
Flour	3.0–3.5	12–14	12–15	—
Graphite	2.0–2.5	7–9	5–6	1.5–2.0
Gypsum	2.0–2.5	8–10	10–12	1.8–2.0
Iron ore	3.0–3.5	10–12	11–12	—
Iron oxide	2.5–3.0	9–11	7–8	1.5–2.0
Iron sulfate	2.0–2.5	7–8	6–8	1.5–2.0
Lead oxide	2.0–2.5	7–8	6–8	1.5–1.8
Leather dust	3.5–4.0	12–15	12–15	—
Lime	2.5–3.0	8–10	10–12	1.6–2.0
Limestone	2.7–3.3	8–10	8–10	—
Mica	2.7–3.3	9–11	9–11	1.8–2.0
Paint pigments	2.5–3.0	9–11	7–8	2.0–2.2
Paper	3.5–4.0	12–15	10–12	—
Plastics	2.5–3.0	8–10	7–9	—
Quartz	2.8–3.2	9–11	9–11	—
Rock dust	3.0–3.5	9–10	9–10	—
Sand	2.5–3.0	8–10	10–12	—
Sawdust (wood)	3.5–4.0	12–15	12–15	—
Silica	2.3–2.8	6–8	7–9	1.2–1.5
Slate	3.5–4.0	12–14	12–14	—
Soap, detergents	2.0–2.5	6–8	5–6	1.2–1.5
Spices	2.7–3.3	10–12	10–12	—
Starch	3.0–3.5	10–14	8–9	—
Sugar	2.0–2.5	6–8	7–10	—
Talc	2.5–3.0	9–11	10–12	—
Tobacco	3.5–4.0	11–12	13–15	—
Zinc oxide	2.0–2.5	6–8	5–6	1.5–1.8

by convection, radiation or by the use of waste heat boilers, (2) tempering with ambient air and (3) cooling by water sprays. A combination of these methods is often employed.

Fire Hazards

Separation of oxidizable particulates can be a fire hazard. Fires in baghouses can quickly attain high temperatures unless airflow is shut off and sprinklers started to control the fire at once. All large installations handling such particulates must have automatic fire protection devices.

REFERENCE

1. Danielson, J. (ed.): "Air Pollution Engineering Manual," Chap. 4, Public Health Service, Cincinnati, Ohio, 1967.

BIBLIOGRAPHY

Adam, H. L.: High Temperature Cloth Collectors, *Chemical Engineering Progress*, vol. 62, April 1966.

Borgwardt, R. H., R. E. Harrington and P. W. Spaite: Filtration Characteristics of Flyash from a Pulverized Coal-fired Power Plant, *J. Air Pollu. Cont. Ass.*, vol. 18, no. 6, pp. 387–390, June 1968.

Culhane, F. R.: Production Baghouses, *Chemical Engineering Progress*, vol. 64, p. 65, January 1968.

Filtration by Fibrous Filters, pp. 244–296, in W. Strauss, "Industrial Gas Cleaning," Pergamon Press, New York, 1968.

Gaden, E. L. and A. E. Humphrey: Fibrous Filters for Air Sterilization, *Industrial Engineering Chemistry*, vol. 48, pp. 2172–2176, 1956.

Iinoya, K. and C. Orr, Jr.: Source Control by Filtration, pp. 409–435, in A. C. Stern (ed.), "Air Pollution," Academic Press, New York, 1968.

Loffler, F.: Collection of Particles by Fiber Filters, pp. 337–376, in W. Strauss (ed.), "Air Pollution Control, Part I," Interscience, New York, 1971.

Morash, N., M. Krouse and W. P. Vosseller: Removing Solid Mist Particles from Exhaust Gases, *Chemical Engineering Progress*, vol. 63, p. 70, 1967.

Sargent, D. D.: Dust Collection Equipment, *Chem. Eng.*, pp. 130–150, January 27, 1969.

York, O. H. and E. W. Poppele: Two Stage Mist Eliminators for Sulfuric Acid Plants, *Chemical Engineering Progress*, vol. 66, p. 67, November 1970.

5.6 BAG AND FABRIC FILTERS—ACCESSORIES AND INSTRUMENTS

Types of Accessories: (a) Blowers; (b) Hoppers; (c) Discharge Valves; (d) Conveyors; (e) Instruments.

Costs: (a) For blowers developing 12 inch H_2O pressure differential, \$8,000 for 30,000 ACFM capacity, \$10,000 for 65,000 ACFM and \$13,000 for 80,000 ACFM; (b) Hoppers are usually part of bag filters and cost under \$1/ft^3; (c) For discharge valve costs, see Section 5.10; (d) Conveyor costs start at around \$1,500; (e) Instrumentation costs are included in the figures given in Section 5.5.

Partial List of Suppliers: See Section 5.10.

The bag or fabric filter auxiliaries and instrumentation must be matched to the filter in order to give a smoothly operating system.

Blowers

The prime mover that supplies the pressure drop for the whole system is the blower. It may be located on either side of the baghouse. If it is located ahead of the baghouse, the system is called the push-through type. If the blower is located after the baghouse, it is called the pull-through type.

The push-through system has the advantages of (1) not requiring an airtight construction for the baghouse, (2) operating with open sides as long as weather protection is provided and (3) permitting a greater degree of cooling when handling hot gases. Its disadvantages are that (1) the blower handles the entire dust load which causes substantial wear to the blower and (2) the dust deposits require frequent dynamic balancing of the blower.

The merits of the pull-through system include the facts that (1) the blower handles clean air and is subject to less wear and maintenance and (2) it is sometimes convenient to place the blower on the roof. Its main disadvantage is that it requires airtight construction in the baghouse and in the associated ductwork.

Blower Types

The centrifugal blowers generally employed in baghouse systems include the following types.

Forward Curved Blade Type (See Fig. 5.6a.) In this design, the leading

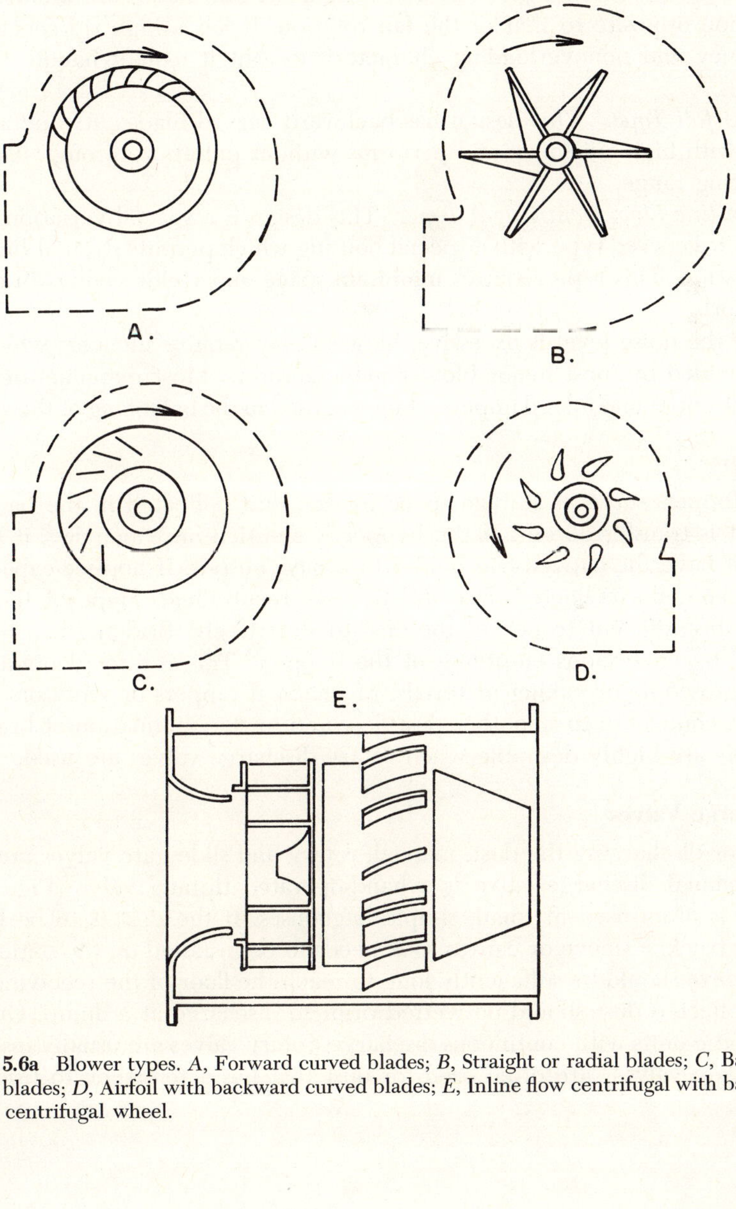

Figure 5.6a Blower types. *A*, Forward curved blades; *B*, Straight or radial blades; *C*, Backward curved blades; *D*, Airfoil with backward curved blades; *E*, Inline flow centrifugal with backward curved centrifugal wheel.

edge of the fan blades curves toward the direction of rotation. These blowers have low space requirements and low tip speeds and are relatively quiet in operation. They are used in the low to moderate static pressure range and are not recommended for dust or fume handling.

Radial Blade Type. This is the most common design. It can handle materials which are likely to clog the fan blades. Such fans have medium tip speeds and a medium noise factor.

Backward Blade Type. In this design, the fan blades are inclined in a direction opposite to that of the fan rotation. It has a high tip speed, high efficiency and nonoverloading characteristics but it cannot handle dust or fumes.

Airfoil Type. This design has backward curved blades, its operation is quiet with high efficiency and performs without pulsation through its entire operating range.

In-line Flow Centrifugal Type. This design is a special variation of the backward curved type with a special housing which permits straight line duct installation. This type requires minimum space and yields savings in cost of ductwork.

If the noise level is excessive, blowers may require silencers which can be provided by most major blower manufacturers. Most manufacturers can provide outlet and inlet dampers which are used in the balancing of the system.

Hoppers

Hoppers serve as storage space for the dust collected in the baghouse until it is transferred out. If the hopper is emptied once per day, it should at least have the capacity to hold 2 to 3 days output. If hopper capacity is insufficient, the collected dust will become reentrained. Slope of the sides should be sufficient to permit the dust to slide freely. Bridging can be prevented by continuous emptying of the hoppers. This is also recommended for hygroscopic or caking materials. Mechanical rappers or vibrators at the bottom contribute to smoother dust flow and to the elimination of bridging. Rappers are highly desirable when rotary discharge valves are used.

Discharge Valves

For discharging the dust, manual, rotary and slide gate valves are used. The manual discharge valve is a hand-operated tipping valve (Fig. 5.10c) which is often used in small, simple baghouses. If the dust is to be loaded onto a truck, a sleeve or canvas sock needs to be installed on the outlet side. The sleeve should be sufficiently long to reach the floor of the receiving unit. The collected dust should be wetted prior to discharge at a dump. On fully automatic units with continuous discharge, rotary valves are usually installed. Slide gate valves[2] are often used for dust that has a tendency to bridge or contains fibrous materials. In locations where the available headroom is limited, this type is preferable to rotary valves.

Conveyors and Accessories

In large installations, the dust collected is continuously removed from the hopper. This is conveniently done by a conveyor, the screw type being the most common. For added details, refer to Sections 5.10 and 5.34. Discharge from several hoppers can be continuously handled by a single screw to a single point of discharge.

The miscellaneous accessories include: the support for the baghouse including base, walkways, angle legs and braces; the dampers used for the balancing of the system in case more than one airstream is fed into one baghouse; the hooding and ductwork;[1] and, if the gas temperature in the baghouse is significantly different from that of ambient air, insulation, which is usually recommended to avoid condensation.

Filter Cleaning Controls

Instrumentation and control systems are used to automate the cleaning cycle.

Manual

In small bag filters, or if once a shift cleaning is adequate, the electrical or pneumatic shakers are manually activated. Interlocks can be provided to ensure that the fan is shut off before the shaker can be activated.

Semiautomatic

In intermittent processes, where this mode of control is frequently employed as the blower is turned off, a timer is activated. The shaking cycle is automatically started after a delay to permit the blower to come to rest. An interlock prevents the fan from being started until the cleaning cycle is over.

Automatic

Compartmented baghouses can be automated on a programmed cycle. The operation is illustrated in Figure 5.6b for a two-compartment unit. At a predetermined pressure drop across the units detected by dPIS-1 or dPIS-2, the cleaning cycle is initiated. The section with the high-pressure drop is isolated from the process and the shaker drive is started. This cleaning action may be assisted by reverse airflow through UV-5 or UV-6. At the end of the cleaning cycle, the section is automatically returned to the operating cycle. The time period required for cleaning one section varies from 3 to 6 minutes while the operating time may be 15 minutes to several hours.

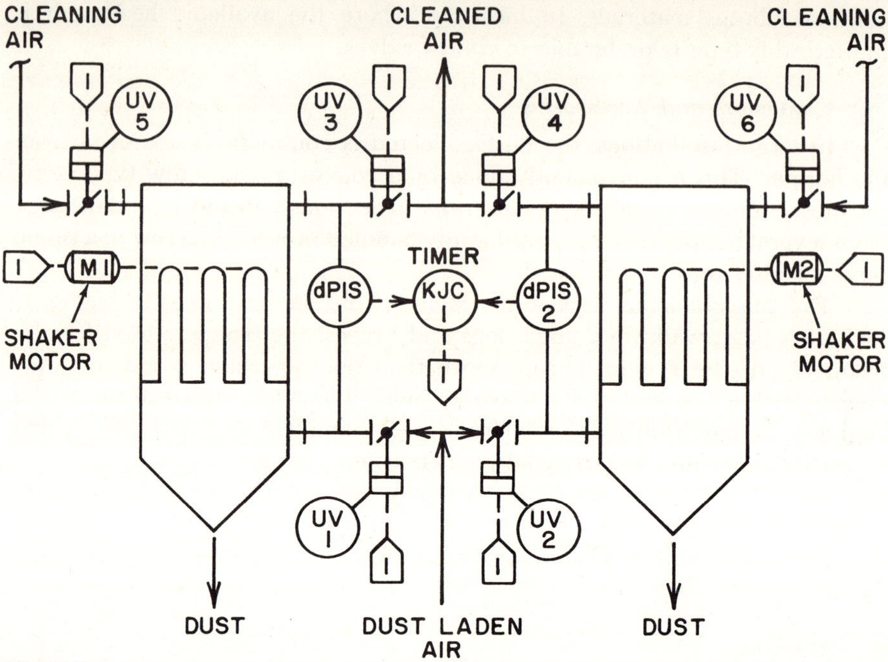

NOTE:
INTERLOCK FLAG [I] REFERS TO THE FOLLOWING OPERATING LOGIC:

OPERATING STATES \ CONTROLLED DEVICES	dPIS-1	dPIS-2	KJC-1	M-1	M-2	UV-1	UV-2	UV-3	UV-4	UV-5	UV-6
dPIS-1 AND-2 ARE NOT HIGH, KJC-1 IS OFF	LOW	LOW	OFF	OFF	OFF	O	O	O	O	C	C
dPIS-1 STARTED KJC-1 CYCLE BECAUSE HIGH DIFFERENTIAL	HIGH	LOW	ON	ON	OFF	C	O	C	O	O	C
dPIS-2 STARTED KJC-1 CYCLE BECAUSE HIGH DIFFERENTIAL	LOW	HIGH	ON	OFF	ON	O	C	O	C	C	O

O-OPEN, C-CLOSED

Figure 5.6b Automatic baghouse cleaning cycle.

Compartmented baghouses should be so designed as to provide adequate filtering area during all phases of operation.

For efficient operation without excessive over-design, six or more compartment sections are frequently needed.

Continuous

The reverse jet type units are compatible to fully continuous cleaning. In this design, compressed air is blown into the fabric by a moving ring, as shown in Figure 5.5m.

REFERENCES

1. "Industrial Ventilation—A Manual of Recommended Practice," American Conference of Governmental Industrial Hygienists, Lansing, Michigan, 1970.
2. Liptak, B. G. (ed.): "Instrument Engineers' Handbook," Volume II, Section 1.14, Chilton, Philadelphia, 1969.

NOTE: For Bibliography, see Section 5.5.

5.7 BAG AND FABRIC FILTERS—SELECTION AND APPLICATION

Table 5.7a provides guidance for the selection and application of bag filters, considering the plant requirements together with dust and gas properties.

Relative to the dust being collected, the primary considerations include:

1. Dust loading (usually expressed in gr/scf)
2. Particle size (usually expressed in microns)
3. Properties including hygroscopicity, bridging, caking characteristics and angle of slide
4. The permissible levels of dust in the final exhaust

The gas flow is usually expressed in ACFM. The expected variations in flowrate, temperature, pressure, humidity and corrosiveness are all important design considerations. In addition, the application engineer must know if the collected dust is reused or dumped and should also be aware of the space available for the mounting of the equipment.

In Table 5.7a, the recommended designs are listed in the order of preference and are identified by the following numbers:

1. Small package units with paper or woven fabric filters and with shaker type cleaning
2. Woven or felted fabric filter with shaker type cleaning
3. Felted fabric filter with pulse jet type cleaning
4. Felted fabric filter with reverse jet type cleaning
5. Same as (4) but for high-temperature service

Baghouse Design and Selection

The determining factor in sizing a baghouse is the pressure drop across the filter fabric. Pressure drop[1] varies directly with the time period between cloth cleanings and exponentially with filtering velocity (or air/cloth ratio).

In high-temperature applications, the cooling of the gas is vital. The trend toward higher operating temperatures tends to reduce cooling requirements but increases the gas volume to be filtered.

If the gas is at or below its dew point, baghouses are not suitable unless

Table 5.7a
FILTER SELECTION GUIDE

Type of Application	*Dust Loading in grains/scf*	*50% of the Dust Particles is in the Size Range Given* μ	*Recommended Selections Listed in Order of Preference*[o]
Ceramics			
Materials handling[a]	0.5–2	2–7	2–4
Fettling and grinding[b]	2–5 and up	2–15	2, 4, 3
Spraying	0.5–3	7–15	2–4
Chemicals			
Materials handling[c,d]	0.5–5 and up	2–15 and up	2–4
Crushing and grinding[d]	2–5 and up	2–15 and up	2–4
Weighing and screening[d]	0.5–5	2–15	4, 2, 3
Roasters, kilns and dryers[d–f]	over 5	7–15	5, 2, 3
Bin ventilation	0.5–2	2–15	1, 3, 4
Foundry			
Abrasive cleaning[f]	3–5	7–15	2–4
Swing frame grinders[g]	3–5	7–15	2–4
Shakeout[f]	0.5–5	2–15	2, 3
Sand handling[f]	3–5	2–15	2, 3
Tumbling mills[f]	over 5	7–15	2–4
Cupola[e,h–j]	3–5	2–15 and up	5, 2
Nonferrous melting[c,i,j]	0.5–5 and up	0.5 2	5, 2, 3
Grain Operations			
Materials handling[k]	0.5–2	7–15	4, 3, 2
Feed mills[l]	3–5	7–15	4, 3, 1, 2
Coolers	3–5	7–15	4, 3, 2
Bin ventilation	0.5–2	7–15	3, 4, 1
Metal Working			
Production grinding[g]	0.5–2	over 15	2, 3
Toolroom[g]	0.5–2	7–15	1
Polishing and buffing[f]	0.5–2	2–15 and up	2, 1, 3
Cast iron machining[m]	3–5	2–15 and up	2, 1, 3
Food and Pharmaceutical			
Mixing, grinding, blending[d]	0.5–2	7–15	1, 3, 4, 2
Coating pans[f]	0.5–5 and up	2–15	3, 2, 4
Sugar handling[d]	0.5–2	2–15	2–4
Sugar granulation[f]	3–5	2–15	

Table 5.7a (Continued)

Type of Application	*Dust Loading in grains/scf*	*50% of the Dust Particles is in the Size Range Given μ*	*Recommended Selections Listed in Order of Preference*[o]
Rock Products and Mining			
Materials handling[f]	3–5 and up	2–15	2, 3
Crushing and screening[f]	over 5	7–15	2, 3
Dryers and kilns[e,f]	3–5 and up	2–15	5, 2, 3
Coolers[f]	3–5	over 15	2, 3
Rubber and Polymers			
Mixers[f]	3–5	2–7	2–4
Batchout rolls[f]	0.5–2	2–7	2–4
Talc dusting	3–5	7–15	2–4
Grinding and buffing[g]	3–5	over 15	2, 3
Polymeric material handling[c,d]	3–5	7–15	2, 3
Polymer finishing	0.5–2	2–15	1, 3, 4
Steel Mills			
Basic oxygen furnace[j]	2–5 and up	0.5–2	—
Electric arc furnace[e,i,j,n]	0.5–2	0.5–2	5, 2
Open hearth[j]	2–5 and up	2–7	—
Blast furnace[j]	over 5	2–15 and up	—
Kish[o]	2–3	7–15	—
Coal and coke handling[f]	3–5	7–15	2, 3
Sintering machines[e,f]	2–5 and up	2–15	5, 2
Coke screening[f]	2–5 and up	7–15	2, 3
Materials handling[f]	2–5 and up	2–15	2, 3
Woodworking			
Saws[g]	3–5	2–15 and up	1, 2
Sanding[g]	3–5	2–7	1, 2
Hogging[g]	over 5	2–15 and up	1, 2
Other Operations			
Bakeries[g]	3–5	7–15	1, 3
Brake lining operations[f]	over 5	7–15	2–4
Bronzing machines[g]	3–5	7–15	1, 2
Coal			
Materials handling[f]	3–5	7–15	2, 3
Bunker ventilation[g]	0.5–2	2–7	3
Cosmetics	0.5–5 and up	2–15	1–4

Table 5.7a (Continued)

Type of Application	*Dust Loading in grains/scf*	*50% of the Dust Particles is in the Size Range Given* μ	*Recommended Selections Listed in Order of Preference*[o]
Cotton flocking[g]	0.5–2	7–15	1
Lead battery plants[f]	0.5–2	2–15	2–4
Leather buffing[f]	3–5	7–15	1, 4
Leather sanding[f]	3–5	2–15	4, 1, 3
Metal powders	0.5–5 and up	2–15	2, 3
Metal reclaiming[e,j]	0.5–5 and up	0.5–2	2, 5
Paper cutting	3–5	7–15	1, 2
Paper grinding	3–5 and up	7–15	1, 2
School	3–5	2–15	1

[a]Dust from bin filling, weighing, mixing, pressing and forming.
[b]Vitreous enamelling, pottery and tile operations.
[c]Conveying, elevating, mixing and packaging.
[d]Seek manufacturers' recommendation.
[e]Gas cooling may be needed for fabric filter use.
[f]Wet collectors are preferred.
[g]Dry centrifugal collectors are preferred.
[h]Dry centrifugal collectors can be used when code permits 0.20 grains per SCF in discharge.
[i]Custom-built units are commonly used due to the large exhaust volumes involved.
[j]Venturi scrubbers are preferred.
[k]Car unloading, conveyors, weighing, bulk storage.
[l]Grain handling, bins, hammermills, mixers, feeders, conveyors and bagging operations.
[m]Oil bath air cleaning is preferred.
[n]Direct shell evacuation method is often used.
[o]See text for the description of the numerical identifications given in this column.

external heating is applied ahead of the baghouse. If the dust is hygroscopic or if the gas stream can generate sparks, baghouses should not be used. In special cases, use of spark arresters and/or inert mineral deposits on the filter fabric can be considered.

The following steps are involved in selecting a baghouse system:

1. Determine the dust loading and the minimum volume to be vented.
2. Estimate the maximum desirable baghouse pressure drop (6 to 12″ H_2O).
3. Select blower type and operating point to determine minimum operating volume at maximum resistance.
4. Estimate the minimum baghouse resistance after bags are cleaned and determine the second operating point on the blower curve.

5. Determine the required filtering area from maximum air/cloth ratio (Table 5.5n) and from choice of filter fabric.
6. Size blower and determine best configuration and layout for plant.

REFERENCE

1. Danielson, J. A. (ed.): "Air Pollution Engineering Manual," U.S. Public Health Service, Cincinnati, Ohio, 1967.

NOTE: For Bibliography, see Section 5.5.

5.8 DRY COLLECTORS—SETTLING CHAMBERS

Standard Designs: These are large, rectangular expansion chambers with gas inlet at one end and outlet at the opposite end or top. Settling occurs due to the reduction in gas velocity. (Fig. 5.8a). These chambers are usually field fabricated.

Cost: See Figure 5.8b for the cost of settling chambers which remove 95% of the 50 μ size particles.

The Process of Settling

The performance of a settling chamber is described by Equation 5.8(1), if the velocity in the chamber is low enough to minimize turbulence and if the gas is uniformly distributed at the chamber inlet and outlet.

$$\zeta = \frac{100\ U_t L}{HV} \qquad 5.8(1)$$

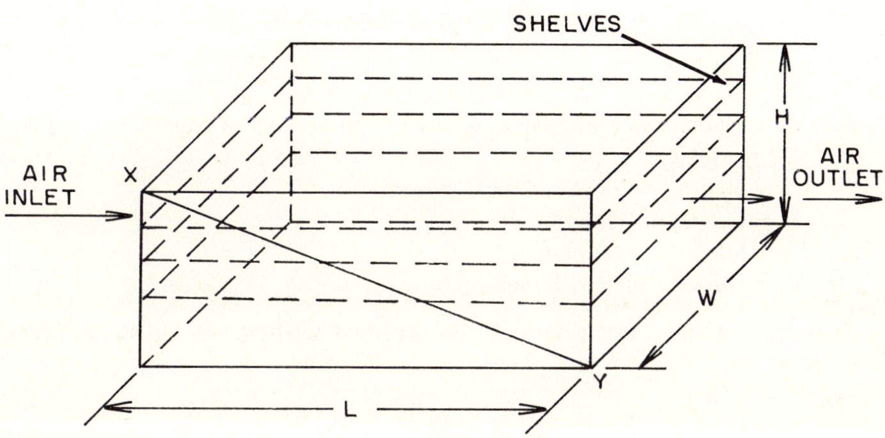

Figure 5.8a Settling chamber.

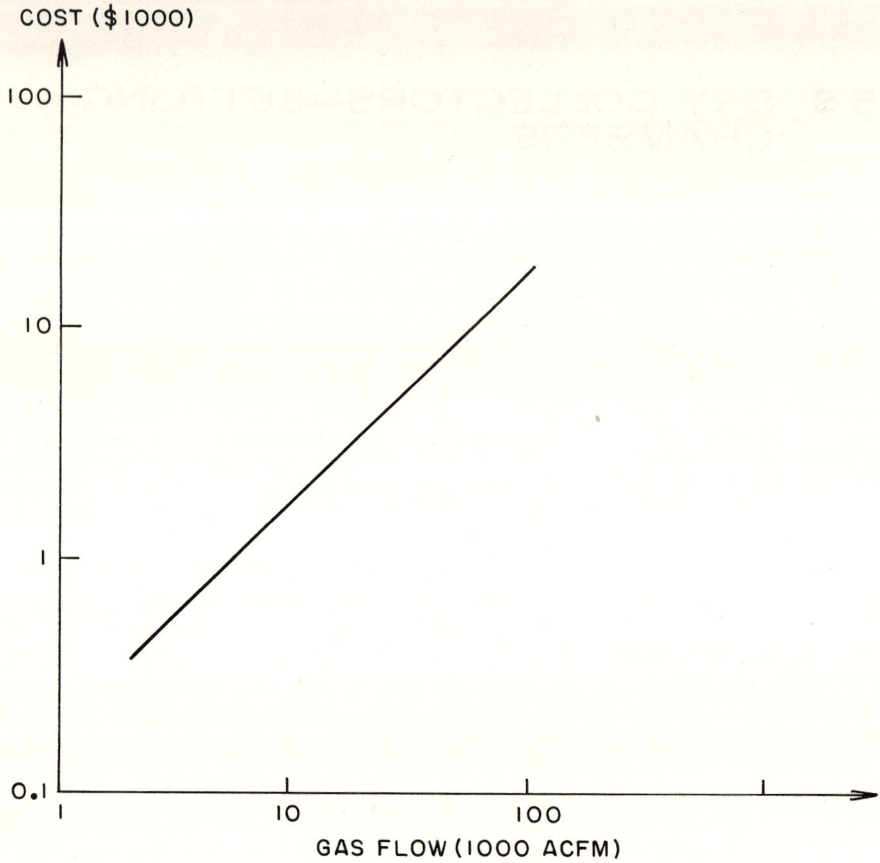

Figure 5.8b Settling chamber costs.

where ζ = efficiency of unit expressed as weight percent of particles of settling velocity U_t

U_t = settling velocity of dust, ft/sec

L = length of chamber, ft

H = height of chamber, ft

V = velocity of gas, ft/sec

If the particles are spherical, the terminal settling velocity U_t is given by Stokes' law:

$$U_t = g_c D_p^2 \frac{(\rho_s - \rho)}{18\,\mu} \qquad 5.8(2)$$

where D_p = particle size, ft
μ = gas viscosity, lbm/(ft)(sec)
g_c = gravitational acceleration 32.2 ft/sec^2
ρ_s = particle density lb/ft^3
ρ = gas density lb/ft^3

Combining Equations 5.8(1) and 5.8(2):

$$D_P = \sqrt{\frac{18\,\mu HV}{g_c L(\rho_s - \rho)}} \qquad 5.8(3)$$

where D_P = minimum size particle collected at 100% efficiency, ft

Horizontal plates may also be inserted into the chamber to form shelves as shown by the dotted lines in Figure 5.8a. These shelves increase the efficiency of separation by reducing the vertical distance between settling surfaces.

The performance of such a chamber is given by Equation 5.8(4):

$$\eta = \frac{NU_tWL}{q} \qquad 5.8(4)$$

$$D_P = \sqrt{\frac{18\mu Nq}{g_c WL(\rho_s - \rho)}} \qquad 5.8(5)$$

where N = number of plates
W = width of chamber, ft
Q = gas flowrate ft^3/sec

Application of Settling Chambers

The gravity settling chamber is the simplest dust collection device. It lends itself to almost any type of construction, if due consideration is given to the temperature and corrosive nature of the gas. Industrial applications of settling chambers are limited because they are able to remove only the larger particles (25 to 50 μ in diameter). Below this size, the settling velocity required becomes very small and the volume requirements become unmanageably large. Particles that can be removed in settling chambers are seldom important pollutants.

Careful design to provide good gas distribution at inlet and outlet is essential. Gradual transitions, splitters or perforated distribution plates are employed. The gas velocity in the chamber should be kept below 10 ft/sec.

Settling chambers with horizontal plates offer greater efficiency and require less space but present cleaning problems; and the plates tend to warp at high temperatures. One cleaning method is to flush the plates with water sprays.

5.8 Dry Collectors—Settling Chambers

In the future, settling chambers are likely to serve only as precleaners ahead of more effective devices for separation of particulates.

BIBLIOGRAPHY

Gottschlich, C. F.: "Gravity, Inertial, Sonic and Thermal Collectors," American Petroleum Institute, New York, 1961.

Perry, John H. et al.: "Chemical Engineers' Handbook," 4th ed., McGraw-Hill, New York, 1963.

Stern, A. C. (ed.): "Air Pollution," Vol. III, Academic Press, New York, 1968.

5.9 DRY COLLECTORS—CENTRIFUGAL COLLECTORS

Types of Designs:	(a) Cyclones; (b) Rotary stream dust separators; (c) Impingement collectors; (d) Dynamic precipitators.
Cost:	See Figure 5.9a.
Characteristics:	See Table 5.11a.
Relative Merits of Cyclones:	See Table 5.9h.
Partial List of Suppliers:	Aerodyne Development Corp. (b); American Air Filter Co., Inc. (a, c, d); Belco Pollution Control Corporation (a); Buell Engineering Co. (a, c); Carborundum Co. (a, c, d); Carter-Day Co. (a); Fisher-Klosterman, Inc. (a, c); Joy Manufacturing Company (a, c); Zurn Industries, Inc. (a).

The dust particles suspended in air do possess inertia and are subject to the forces of gravity. Hence, if the gas stream is forced to change direction, centrifugal forces will act on the particles. This mechanism is the basis for separation in centrifugal collectors.

Cyclones

A cyclone separates particulate matter from the air by transforming the velocity of the inlet stream into a double vortex confined within the structure of the cyclone. The main vortex is formed by the inlet gas and spirals downward while the vortex core spirals upward and terminates at the outlet (Fig. 5.9b).

Types of Cyclones

The cyclone inlet produces the vortex, the axial outlet discharges the clean air and the dust drops out at the bottom. Figure 5.9c illustrates the basic cyclone arrangements:

A. Tangential inlet and axial dust discharge

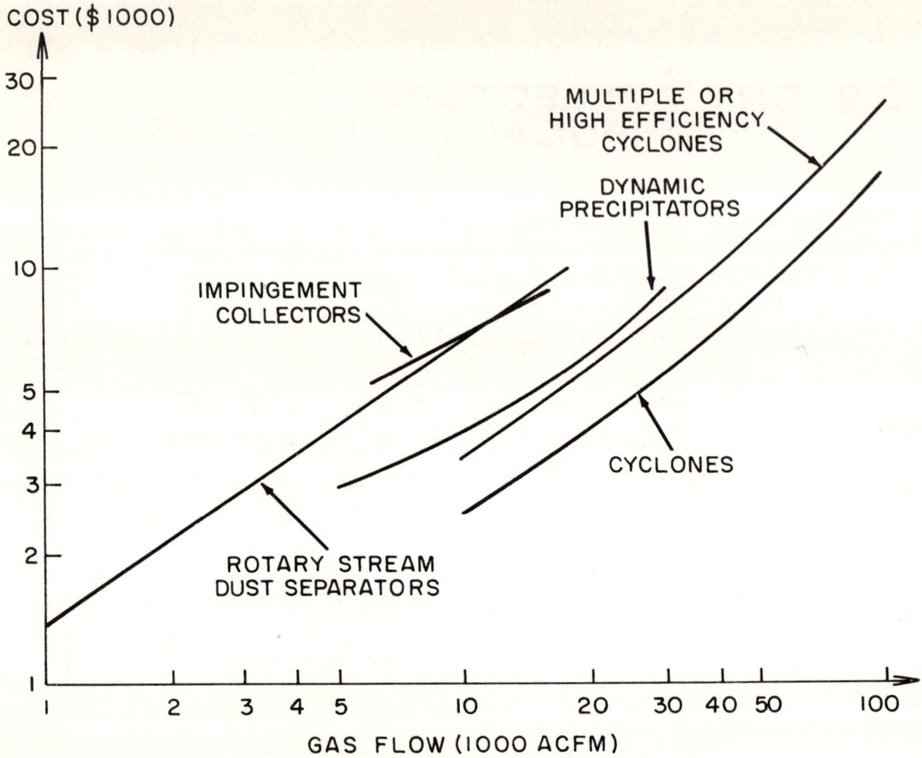

Figure 5.9a Centrifugal collector costs.

B. Tangential inlet with peripheral dust discharge
C. Axial inlet through swirl vanes with axial dust discharge
D. Axial inlet through swirl vanes with peripheral dust discharge

The cyclone inlet is usually rectangular because a circular inlet would bring the inlet stream close to the center of the unit and hence would lower the efficiency. Inside surfaces of cyclones must be smooth to facilitate easy movement of collected dust down the wall.

Typical geometry and dimensional ratios for a standard cyclone are given in Figure 5.9d.

Separation Mechanism

The gas entering the cyclone creates the main vortex which continues down to the bottom of the cone where the vortex reverses its axial direction but retains its direction of rotation. In the annular region between the cyclone body and the gas outlet near the top, the tangential gas velocity is lowest

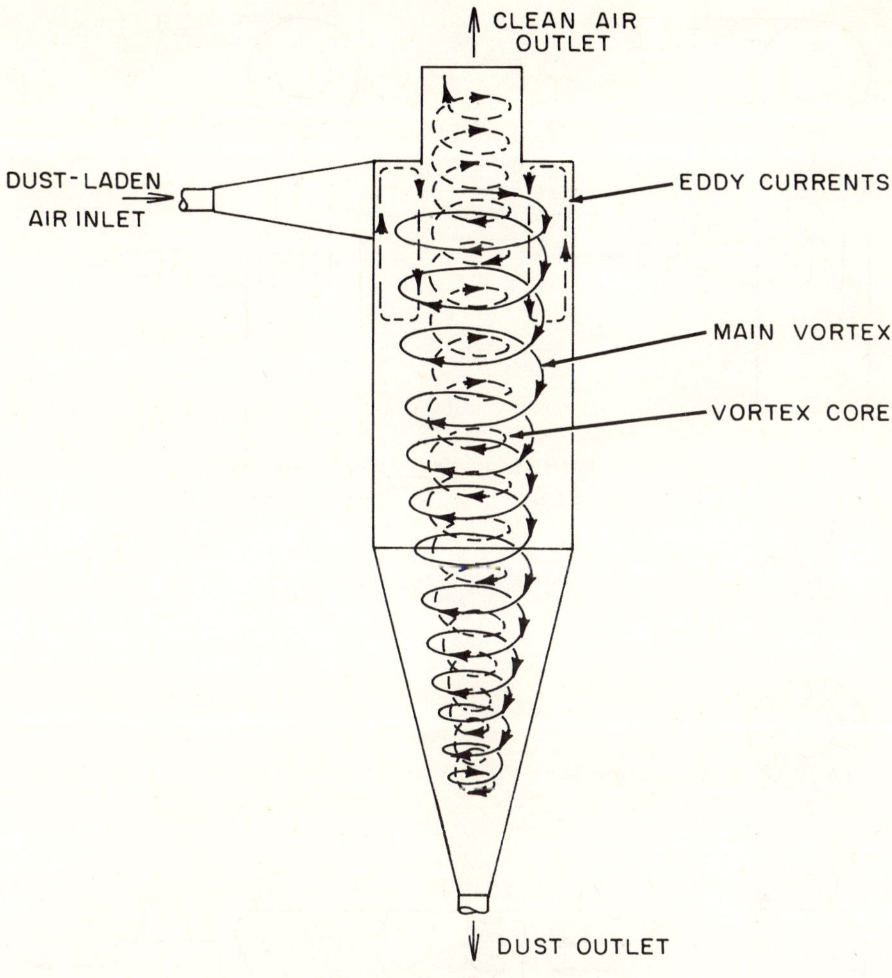

Figure 5.9b Flow patterns in cyclones.

at the wall and increases near the outlet, which causes some eddy currents in this region. The longer the gas outlet projection into the body, the more pronounced these currents will be. They contribute to the loss of efficiency because they bring the dust collected near the wall back to the outlet where it is carried out by the airstream. Axial inlet cyclones inherently overcome this problem.

The centrifugal force applied to the particles varies with the square of inlet velocity and inversely with the radius of the cyclone. These comprise a dimensionless factor called "separation factor" (S).

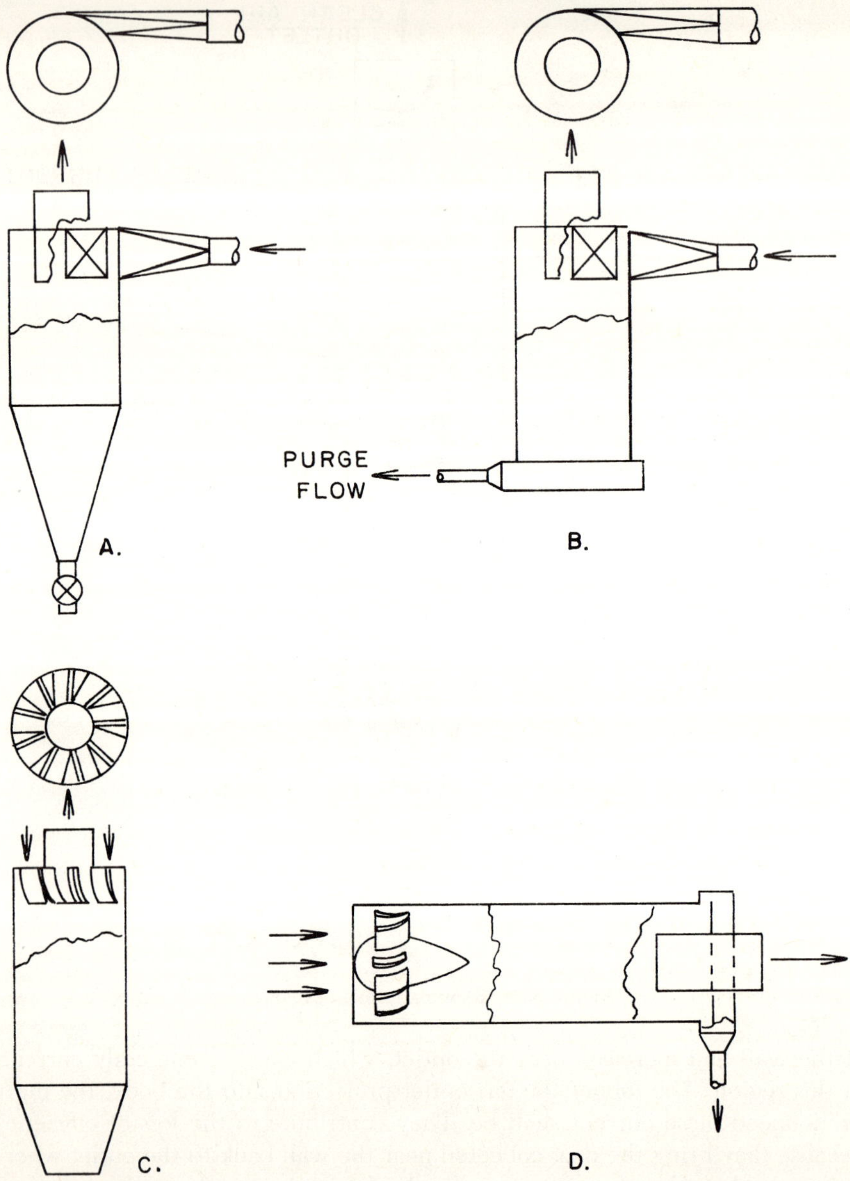

Figure 5.9c Types of cyclones. *A*, Tangential inlet and discharge; *B*, Tangential inlet and peripheral discharge; *C*, Axial inlet and discharge; *D*, Axial inlet and peripheral discharge.

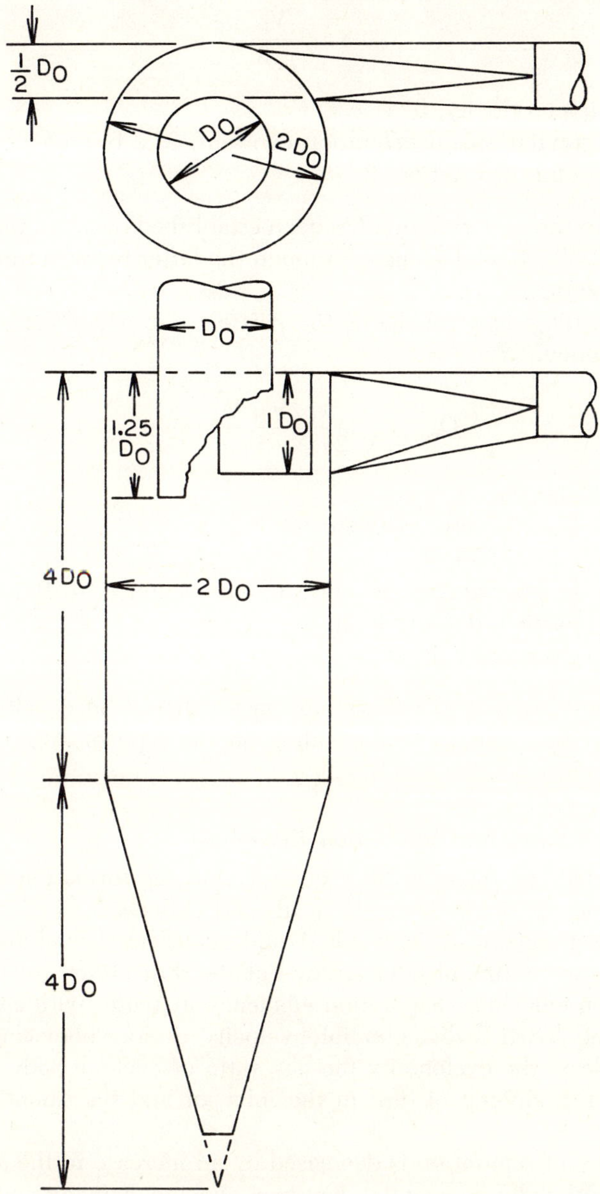

Figure 5.9d Typical dimension ratios of a cyclone.

$$S = \frac{V_i^2}{g_c R} \qquad 5.9(1)$$

where V_i = inlet velocity, ft/sec
g_c = gravitational acceleration constant, 32.2 ft/sec^2
R = radius of rotation, ft

At present, no direct correlation has been established between the separation factor S and collection efficiency, although the latter is also a function of the separation factor.

Cut size (D_{cp}) in a cyclone is the particle size which can be collected at 50% efficiency:

$$D_{cp} = \sqrt{\frac{9 \mu W_i}{2\pi \, Ne \, V_i(\rho_p - \rho)}} \qquad 5.9(2)$$

where D_{cp} = cut size, ft
μ = gas viscosity, lb/ft sec
W_i = inlet width, ft
Ne = number of effective turns in the cyclone (5 to 10 in most cases)
ρ_p = particle density, lb/ft^3
ρ = gas density, lb/ft^3

Whenever possible, actual calibration curves should be developed for the particular cyclone, instead of depending on the type of approximations as given by Equations 5.9(1) and 5.9(2).

Dust Discharge and Separation Efficiency

The centrifugal force in the cyclone causes the formation of a layer of dust which slowly swirls down the walls and collects in the bottom of the hopper. Reentrainment of the solids should not occur during discharging. A small purge (up to 10% of inlet airflow) at the dust outlet greatly increases the separation efficiency. Separation efficiency increases with an increase in particle density, particle diameter, inlet velocity, cyclone body length, number of turns made in the cyclone by the gas, ratio of cyclone body diameter to outlet diameter, amount of dust in the inlet air and the smoothness of the inner wall.

Efficiency of separation is decreased by an increase in the air viscosity, inlet duct width, inlet area, outlet diameter, cyclone diameter and the density of the air.

Collection efficiency in a cyclone is decreased by the installation of inlet vanes or guide vanes. If the inlet dust concentration is very high (above 50 grains/SCF) some agglomeration of collected dust is likely to occur and this increases collection efficiency.

High-efficiency cyclones have smaller inlet thickness and body diameters

Table 5.9e
EFFICIENCY OF CYCLONES

	Efficiency range % collected	
Particle Size Range μ	*Conventional*	*"High-efficiency"*
Less than 5	Less than 50	50–80
5–20	50–80	80–95
15–40	80–95	95–99
Greater than 40	95–99	95–99

and, hence, produce greater separating forces. Cyclones with up to 9-inch body diameters are often considered to be "high efficiency" cyclones.

Cyclones, in general, provide low collection efficiencies (Table 5.9e)[1] and are the least expensive among air pollution control devices.

Pressure Drop

A satisfactory method for calculating the pressure drop through a cyclone is not available and empirical methods are used for pressure drop predictions.

The pressure drop may be roughly approximated by Equation 5.9(3).

$$\Delta P = \frac{k_c q P \rho}{T} \qquad 5.9(3)$$

where ΔP = pressure drop, $''H_2O$
k_c = cyclone constant (Fig. 5.9f)
q = airflow rate, cfm
P = absolute pressure of air, atmospheres
T = temperature of gas, °R

Usually the pressure drop in cyclones varies from 1 to 6″ H_2O.

Cyclone Performance

Figure 5.9g shows a typical relationship between velocity and efficiency. As inlet velocity is increased, a rapid linear rise in efficiency results. The slope of the line is greater for larger particles and for smaller cyclones. At about 70% efficiency, the slope decreases rapidly; and, therefore, cyclones are not used in this region. The inlet velocity, at which the flat portion of the curve is reached, is lower for small diameter cyclones and for larger particle sizes. In the absence of test data for a particular application, the variations in efficiency with flowrate may be estimated by Equation 5.9(4).

$$\eta_1/\eta_2 = \sqrt{q_1/q_2} \qquad 5.9(4)$$

where η_1 and η_2 are the weight percent collection efficiencies at flows q_1 and q_2.

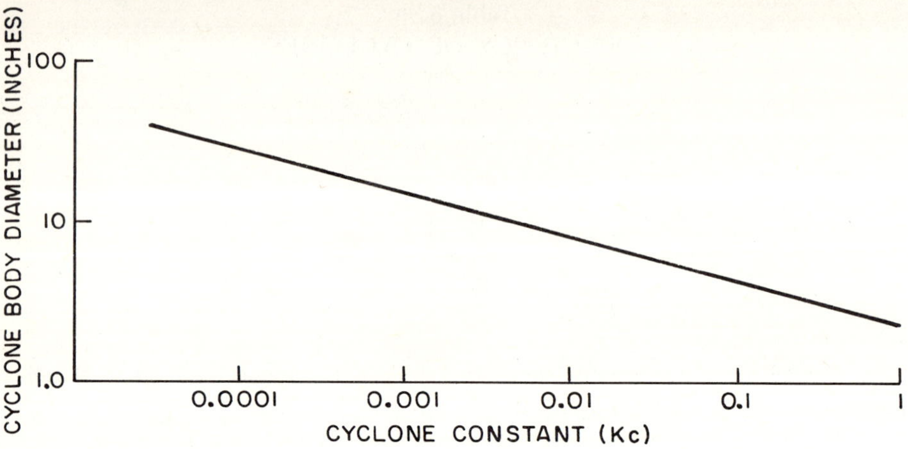

Figure 5.9f Cyclone constants for pressure drop approximation.

Cyclone efficiency is also affected by particle size distribution. Fractional efficiency curves are obtained by actual tests; and if the size distribution of the dust to be cleaned is known, the overall efficiency can be calculated.

Pressure drop decreases and collection efficiency increases as the inlet dust loading increases. This is due to the higher momentum and the presence of agglomerates in highly dust-laden gases.

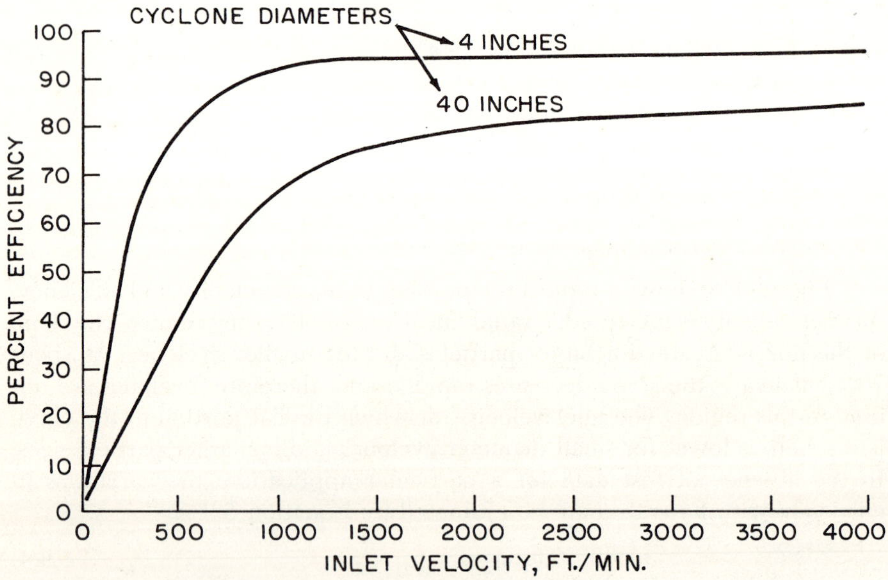

Figure 5.9g Influence of inlet velocity and cyclone diameter on cyclone efficiency.

The viscosity of the gas through which the dust particle must travel also has an effect on collection efficiency. The relationship between efficiency and gas viscosity at constant flowrate can be estimated by Equation 5.9(5).

$$\frac{100 - \eta_1}{100 - \eta_2} = \sqrt{\mu_1/\mu_2} \qquad 5.9(5)$$

The effect of gas density variations on efficiency can usually be neglected.

Multiple Cyclones

In order to handle large gas volumes using small diameter cyclones, they may be manifolded in parallel. If the number of parallel cyclones is large, a common inlet plenum chamber, dust storage bin and outlet plenum chamber are used. In order to obtain efficiencies near those of a single tube of the same size, it is necessary to equalize the gas and dust load distribution between the cyclones to prevent backflow, plugging or reentrainment from the dust bin.

Cyclones are also operated in series. If dust agglomerates are subject to break-up due to high gas velocities, it may be advantageous to operate two cyclones in series at low velocities. In some cases, a large diameter cyclone may be used as a precleaner to remove large or abrasive particles that would otherwise foul the smaller secondary cyclone. To guard against unintentional emissions in case of equipment failure, a backup cyclone is sometimes provided in a series arrangement.

Plugging

A cyclone may become fouled by plugging of the dust outlet or by the accumulation of dust on the cyclone wall. Plugging of the outlet may result from overfilling of the dust bin, by large pieces of caked material breaking off the walls of the cyclone or by foreign materials contained in the inlet gas. The buildup of materials on the wall is mainly a function of the physical and chemical properties of the dust. Moisture in the dust may lead to condensation on the cyclone wall and contribute to the accumulation of material there.

Liquid Droplet Removal Services

Cyclones are frequently employed to remove liquid droplets from gas streams or to prevent entrainment from boiling liquids.

Cyclones can also be operated in a wet state for the collection of dry dust. In such units, water is sprayed countercurrent to the air at a rate of 3 to 10 gpm per 1,000 ACFM of airflow. The efficiency of such wet units is high because the dust particles are entrapped in a liquid film and are not easily reentrained. In such units, spray nozzles are used to wet all surfaces inside the cyclone to prevent caking and plugging of unwetted surfaces. One

Table 5.9h
COMPARISON OF VARIOUS CYCLONES

Collector Type	Space Requirements	Volume Range	Efficiency by Weight	Pressure Loss In Inches H_2O	Temperature Limitations (°F)	Power HP per 1,000 CFM Gas	Costs per cu ft Gas ¢	Application Areas
Conventional cyclone	Large	Normal range up to 50,000 cfm	50% on 10μ	1–3	700–1000° limited only by materials of construction	0.24–0.73	.03–.10	Woodworking, paper and buffing fibers, well suited for dry dust particles 20μ and above in diameter
High-efficiency cyclone	Medium	Normal range up to 12,000 cfm	80% on 10μ	3–6	700–1000° limited only by materials of construction	0.73–1.2	.07–.15	Woodworking, material conveying and product recovery; well suited for dry dust particles 10μ and above in diameter
Multitube cyclones	Small	Normal range up to 100,000 cfm	90% on 5μ	3–6	700–1000°	1.1	.09–.15	Precollector for electrostatic precipitator on flyash, product recovery; well suited for dry dust particles 5μ and above in diameter

disadvantage of the wet operation is the production of a dust slurry which requires further treatment prior to disposal.

For a summary of the features of the various cyclone designs, refer to Table 5.9h.

Rotary Stream Dust Separators

In rotary dust separators (Figure 5.9i), the main gas stream forms the inner vortex. A secondary airstream forms an outer vortex flowing in the same tangential but opposite axial direction. The combined centrifugal forces and radial gas flows collect the dust in the annular ring. A baffle ring at the top stops a portion of the uppermost jet flow and greatly increases efficiency.

In this design, two pressure drops are involved: the drop in main gas stream and the drop in secondary air jets. Efficiency values for a typical case are given in Table 5.9j.

Rotary dust separators are superior to common cyclones at comparable power costs. As in cyclones, best overall efficiency is obtained by using multiple bank installations of small diameter tubes. In such units, it is difficult to maintain uniform dust load distribution.

One important advantage of rotary dust separators is the elimination of erosion problems and of the need for smooth walls, because the dust collection takes place in an annular ring and not at the wall. On the other hand, these units require a source of high-pressure secondary air and about 30 to 40% of the total airflow through the unit is secondary air. In most cases, the clean exhaust air is recycled by a secondary blower for this duty.

Inertial Collectors

Inertial separation devices cause sudden changes in the direction of gas stream and, thereby, separate the particles by inertia, by impingement on a target or by centrifugal forces.

When dust-laden air impinges on a body, the air will be deflected while the dust due to its greater inertia will tend to collect at the surface.

Theoretical methods have been developed,[1] correlating separation efficiency for simple-shaped bodies. However, these are of little use to evaluate commercial inertial separators on the market; and, therefore, performance information is best developed based on vendors' data and tests.

Inertial separator efficiencies are comparable to those of high-efficiency cyclones. Those designs which have no moving parts can be constructed of any material. The pressure drop requirements in commercial units range up to 6″ H_2O.

Dynamic Precipitators

Dynamic precipitators utilize the centrifugal forces generated by rotating impellers to affect separation. The dust particle in the rotating wheel is subject

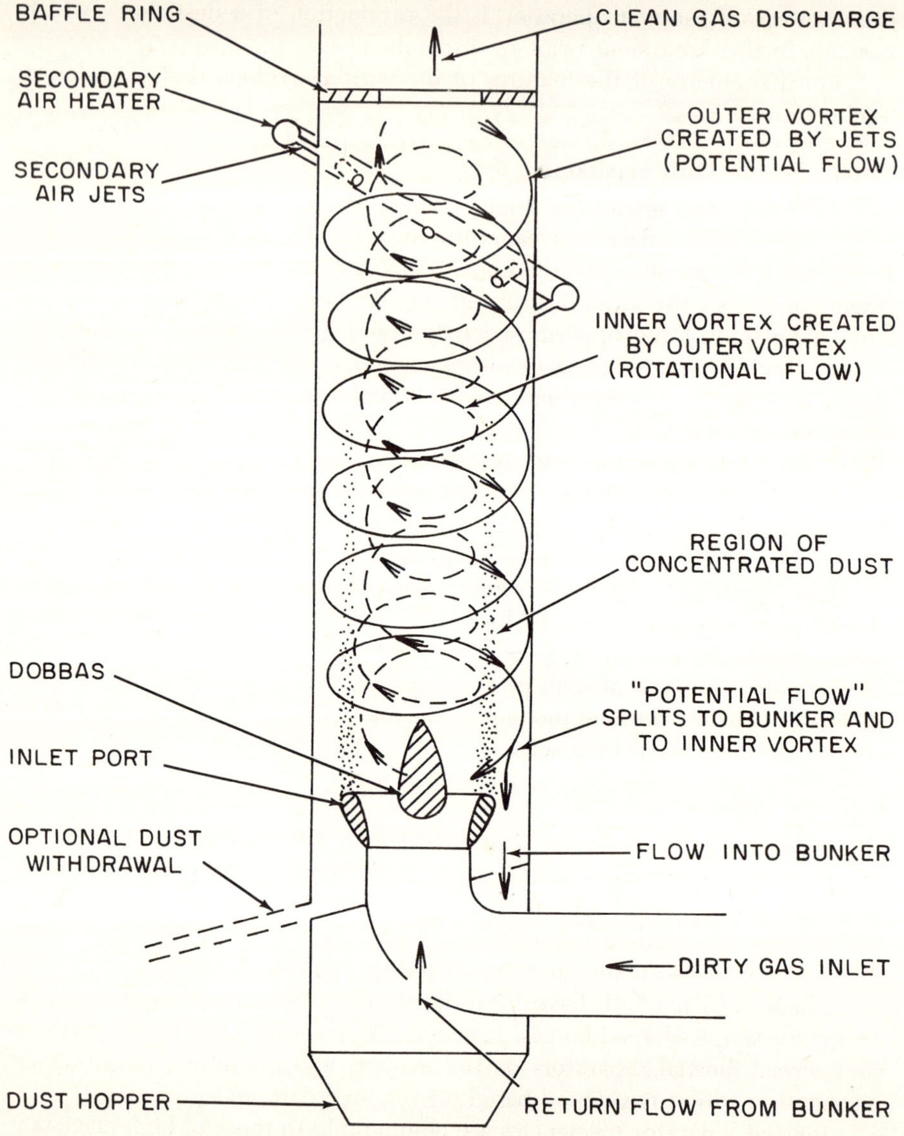

Figure 5.9i Rotary dust separator.

to a force many times that of gravity. These units (Fig. 5.9k) use large numbers of hyperbolically contoured impeller blades to intercept the dust particles and force them into a converging path toward the blade tip. The concentrated dust particles are then removed from the blade peripheral regions by a secondary airstream and are pneumatically conveyed to a secondary hopper.

Table 5.9j
COLLECTION EFFICIENCY OF ROTARY STREAM DUST SEPARATORS

Particle Size Frequency Maximum, μ	*% Collection Efficiency*
2.5	92.0
5.0	95.0
10.0	98.5

General correlations on the performance of these units are not available. Evaluations should be based on the experience and data gathered by manufacturers.

Dynamic precipitators function as prime movers (fans or blowers) as well as dust collectors. These units require more power input than does a centrifugal fan operating at the same volume and system resistance; but total power requirement is less than what is needed by a separate centrifugal collector and fan.

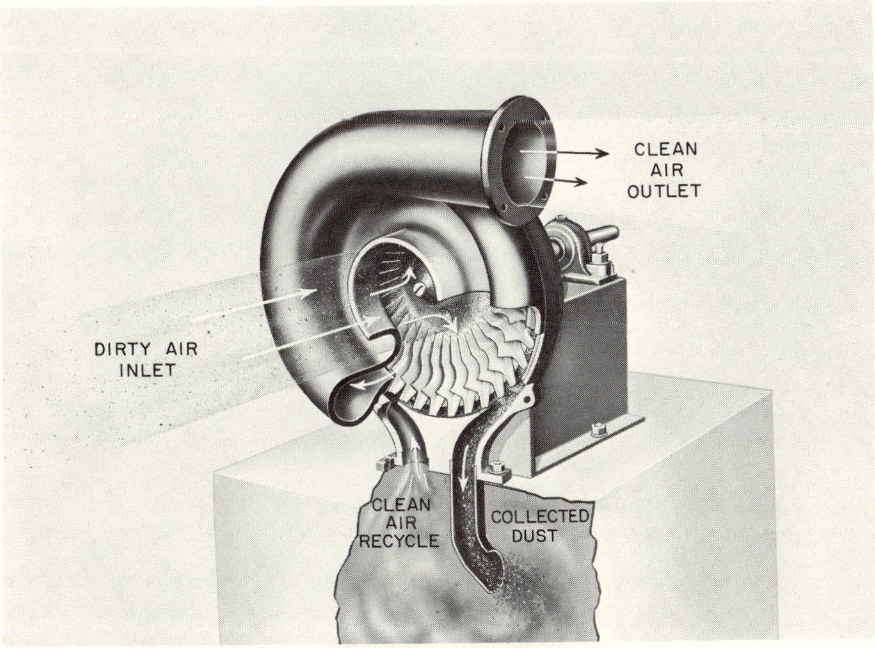

Figure 5.9k Dynamic precipitator. (*Courtesy of American Air Filter Co., Inc.*)

REFERENCE

1. Caplan, K. C.: Source Control by Centrifugal Forces and Gravity, in A. C. Stern (ed.), "Air Pollution," Vol. III, Chap. 43, Academic Press, New York, 1968.

BIBLIOGRAPHY

"Air Pollution Manual—Part II," American Industrial Hygiene Association, Detroit, 1968.

Danielson, J. A. (ed.): "Air Pollution Engineering Manual," Chap. 4, U.S. Public Health Service, Cincinnati, Ohio, 1967.

Gas Cleaning Devices, Chap. 4, in "Control Techniques for Particulate Air Pollutants," National Air Pollution Control Administration, Washington, D.C., 1969.

Munson, J. A.: Dry Mechanical Collectors, *Chem. Eng.*, pp. 147–152, October 14, 1968.

Sittig, M.: "Air Pollution Control Processes and Equipment," Noyes Development Corp., Park Ridge, New Jersey, 1968.

5.10 DRY COLLECTORS—DUST DISCHARGE DEVICES

Types of Discharge Devices:	(a) *Hoppers;* (b) *Valves,* (b1) Slide Gate Type, (b1.1) Manual, (b1.2) Air Cylinder Operated, (b2) Double Tipping Type, (b3) Rotary Type, (b4) Diverter Type; (c) *Conveyors,* (c1) Screw Type, (c2) Belt Type; (d) *Blowers.* NOTE: In the feature summary below, the letters "a" to "d" refer to the above listed devices.
Costs:	$1/ft^3 (a); $300 (6 inches, b1.1); $500 (6 inches, b1.2); $600 (4 inches, b3); $800 (6 inches, b2 or 4 inches, b4); $1,000 (8 inches, b3); $1,200 (6 inches, b4); $1,500 (6 inches, SS, b2); over $1,500 (c1); over $2,000 (c2). For blower costs see Section 5.6.
Partial List of Suppliers:	Acrison, Inc. (a), B I F, a unit of General Signal Corporation (a, b); Black, Sivalls & Bryson, Inc. (a, b); Buffalo Forge Co. (d); DeZurik Corp. (a, b); Jeffrey Manufacturing Co. (b, c); Joy Manufacturing Company (b, c); New York Blower Co. (d); Nooter Corporation (a); Northern Blower Div., Buell Engineering Co. (d); Rex Chainbelt, Inc. (b); Sprout-Waldron & Co., Inc. (a, c); Vibra Screw, Inc. (a, c); Wallace & Tiernan Division Pennwalt Corporation (c).

This section concentrates on the discharging of the dust. Feeding of this dust to some other equipment at a controlled rate requires a feeder. Feeders are discussed in Chapter I.6.

Hoppers

Hoppers may be made of wood, steel or concrete, although steel is the most common because dust slides more easily on steel. It is also easier to provide probe holes in a steel hopper to help the flow of particulates by prodding.

Hoppers are usually of the same cross-sectional shape as the bin above (they are the lower part of the bin) and usually taper to one or more discharge points either under the bin or offset to one side. A hopper may have multiple outlets which improve the flow characteristics of the bin because they increase the ratio of outlet area to bin area. An offset outlet improves the hopper flow characteristics. The inside surface of the hopper should be smooth with all welding splatter and surface imperfections removed.

An important factor in hopper design is the *angle of slide.* This is the minimum angle at which a material will flow on an inclined plane. Most hoppers are designed with the inclined surfaces at 30 to 60 degrees from vertical. If the dust has a tendency to bridge or cake, one of the following flow aids may be used: vibrators, antipacking devices, agitating devices and fluidizing devices.

Figure 5.10a illustrates the importance of spout design. When a spout feeds an enclosed space, venting of the air displaced by the flow of solids is essential. Similarly, warm materials flowing through a cold spout can cause condensation problems.

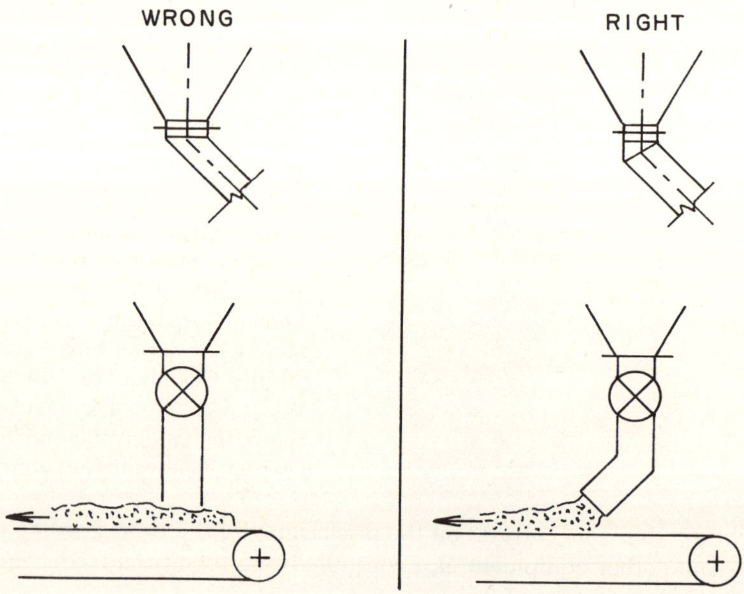

Figure 5.10a Spout designs.

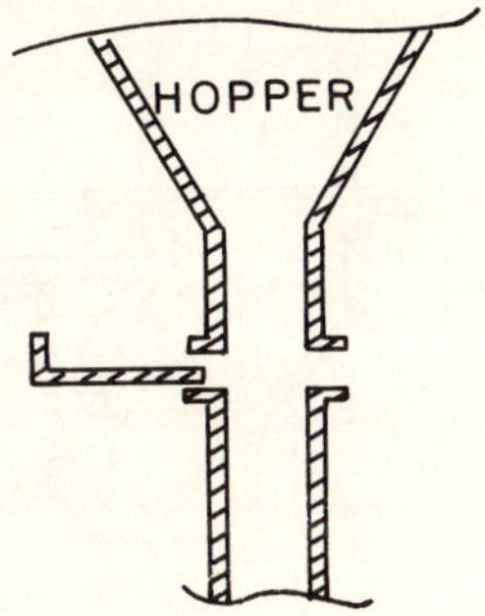

Figure 5.10b Slide gate valve.

For a discussion of hopper design, see the Section on Feeders in Chapter I.6.

Discharge Valves

The simplest in this category is the slide gate valve shown in Figure 5.10b. This is a movable plate inserted below the hopper which can be mechanically or manually moved, thus permitting downward flow of collected dust from the hopper.

A modification of this design is the flap or tipping type valve (Fig. 5.10c).

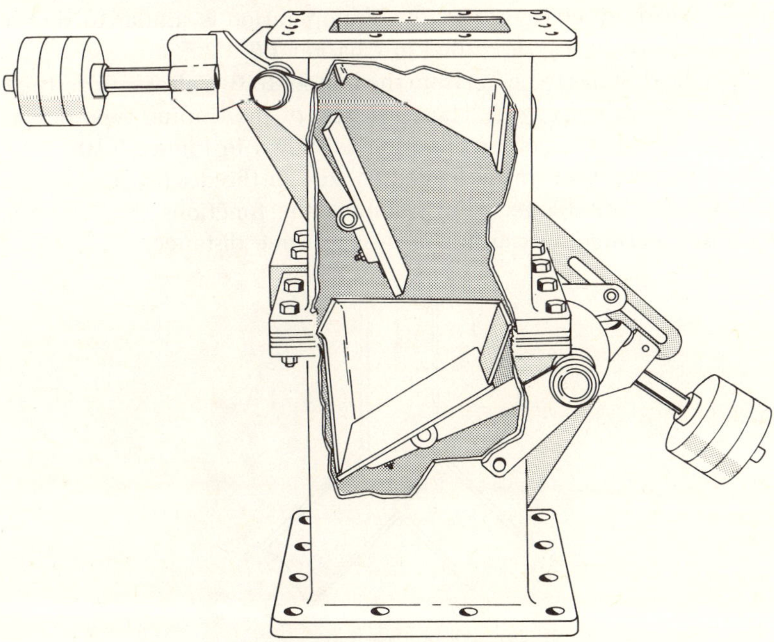

Figure 5.10c Tipping valve.

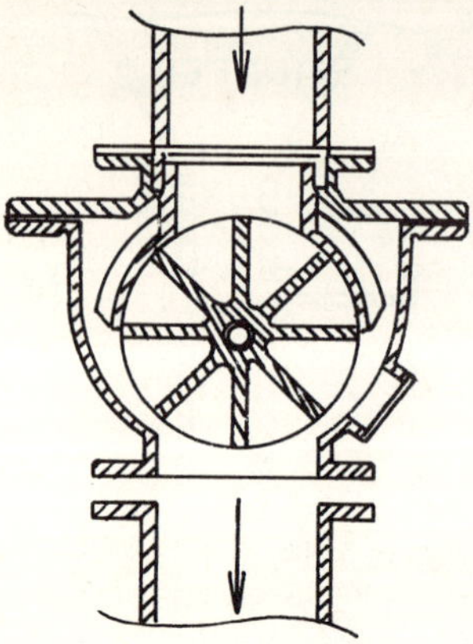

Figure 5.10d Rotary valve.

These employ a counterweight on the flaps to permit easier operation. Tipping valves can be manually operated or motorized and can be made to operate when the weight of material on the flaps is greater than the counterweights. In rigorous applications where free flow of solids is required and the hopper is under vacuum or pressure, spherical segment double tipping valves are used to ensure a positive seal.

Rotary valves (Fig. 5.10d) are employed in large or continuous units and permit the steady discharge of solids. The operation is similar to that of the rotary feeders which are described in Chapter I.6.

When the dust has to be fed from the hopper to two alternate destinations, diverter valves are employed. These are also of the tipping type design and permit flow in one or another direction as shown in Figure 5.10e.

Screw feeders with spring loaded chokes on the discharge side can also be used to effect discharge. They combine the functions of a valve and a conveyor and permit the transfer of dust to some distance.

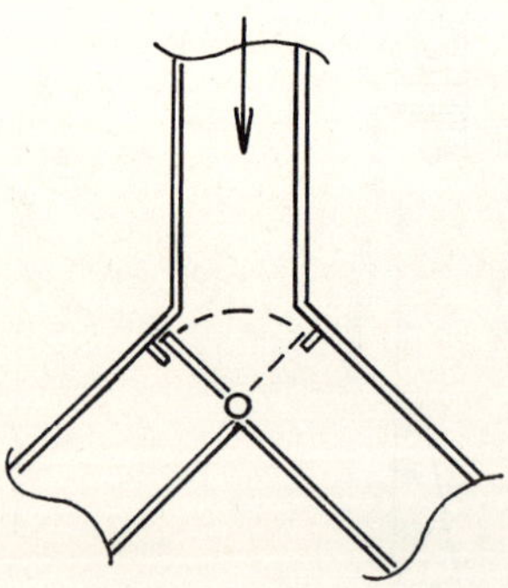

Figure 5.10e Diverter valve.

Conveyors

In addition to valves, the larger dust collection systems also use conveyors to transfer the dust for further handling. The conveyors are located directly below the discharge valves. Screw and belt conveyors are both employed. The spout feeding a belt conveyor should be inclined toward the direction of flow. For additional information on conveyors[1] also refer to Sections 5.6 and 5.34.

REFERENCE

1. Lipták, B. G. (ed.): "Instrument Engineers' Handbook," Vol. I, Section 7.6. Chilton, Philadelphia, 1969.

BIBLIOGRAPHY

Johanson, J. R.: Feeding, *Chem. Eng.*, pp. 75–83, October 13, 1969.

Perry, John H. et al.: "Chemical Engineers' Handbook," 4th ed., McGraw-Hill, New York, 1963.

Stepanoff, A. J.: "Gravity Flow of Bulk Solids and Transportation of Solids in Suspension," Wiley, New York, 1969.

5.11 DRY COLLECTORS—SELECTION AND APPLICATION

Table 5.11a provides a summary of dry collector features and characteristics. Table 5.11b is a selection guide listing some of the more common dry collector applications. Similarly to Table 5.7a, the recommended choices of dry collectors are listed in their order of preference and are identified by the following numbers:

1. Standard cyclone
2. Multiple cyclone
3. Inertial collector
4. Dynamic precipitators

Gravity chambers and rotary dust separators are not listed because the first is only used as a precollector and the second has only recently become commercially available in the United States.

In selecting the dry collector best suited for a particular application, one must consider the airflow rate, temperature, pressure, dust loading (grains/scf), dust particle size, distribution and characteristics, legal requirements on emission concentrations, the end use of the collected dust, layout, space, maintenance and economic factors. On proprietary equipment, as are most impingement collectors and dynamic precipitators, the best source of information is the supplier.

Table 5.11a
DRY COLLECTOR CHARACTERISTICS

Features

Types	*Space Required*	*Maximum Capacity ACFM*	*Efficiency and Micron Size to which it Applies*	*Pressure Drop, inches H_2O*	*Temperature Limit, °F*	*Cost*	*Typical Applications*
Gravity chamber	large	available space is the only limit	over 50 μ	0.2–0.5	700–1000	low	Precollectors for other devices
Cyclone	large to medium	50,000	50–80% on 10 μ	1–6	700–1000	low	Woodworking, buffing fibers, pneumatic conveyors
Rotary dust separator	medium	18,000	90% on 2 μ	to 8	to 500	moderate	Tar separation, cement dust handling
Multiple cyclone	small	100,000	90% on 5 μ	3–6	700–1000	moderate	Precollectors for electrostatic precipitation
Inertial collectors	small	75,000	90% on 2 μ	3–6	to 700	moderate	Pulverized coal handling
Dynamic precipitators	small	25,000	90% on 2 μ	none; it is a fan	to 500	moderate	Metalworking; coal bunkers

Table 5.11b
DRY COLLECTOR SELECTION GUIDE

Type of Application	*Dust Loading in grains/scf*	*50% of the Dust Particles is in the Size Range Given Below in Microns*	*Recommended Choices Listed in the Order of Preference*[k]
Chemicals			
Materials handling[a–c]	0.5–5 and up	2–15 and up	1–3
Crushing and grinding[b,c]	2–5 and up	7–15	1–3
Weighing and screening[b,c]	0.5–5	2–15	1, 2
Roasters, kilns and dryers[b,c]	over 5	7–15	1–3
Pneumatic conveyors[c]	over 5	over 15	1, 3, 2
Fertilizers			
Materials handling[a,b]	2–5 and up	2–15	—
Dryer, cooler[b,d]	over 5	2–15	1, 2
Flyash			
Coal, stoker fired[e]	3–5	2–15	3
Coal, pulverized[e]	over 5	2–15	3
Wood burning[e]	0.5–5 and up	over 15	3
Food Products			
Materials handling[a,c]	0.5–3	7–15	1, 2
Roasters, dryers and coolers[b,c]	2–3	7–15	1, 2
Mixers, grinders and blenders[c]	0.5–2	7–15	1, 2
Pneumatic conveying[c]	over 5	over 15	1, 2
Grain Operations			
Handling and storage[c,f]	0.5–2	7–15	1, 2, 4
Feed mill[c,g]	3–5	7–15	1, 2
Metalworking			
Production grinding	0.5–2	over 15	4, 3
Toolroom	0.5–2	2–7	4
Cast iron machining[h]	3–5	2–15 and up	4, 3
Woodworking			
Saws[c]	3–5	7–15	1–3
Sanding[b,c]	3–5	2–7	4
Hogs[c]	over 5	2–15	1–3
Miscellaneous			
Asphalt plants[b,d,i]	over 5	2–15	1, 2
Bakeries[c]	3–5	7–15	4
Bronzing machines	3–5	7–15	4

Table 5.11b (Continued)

Type of Application	*Dust Loading in grains/scf*	*50% of the Dust Particles is in the Size Range Given Below in Microns*	*Recommended Choices Listed in the Order of Preference*[k]
Coal bunkers	0.5–2	2–7	4, 3
Cotton flocking	0.5–2	7–15	4
Foundry cupola[c,d,j]	3–5	0.5–15	3
Leather sanding and buffing[c]	3–5	7–15	3
Plastics finishing[c]	0.5–2	2–15	1, 2
Rubber grinding[b,c]	3–5	over 15	3
Tobacco leaf separation	3–5	over 15	3

[a] Includes conveying, mixing, elevating and packaging.
[b] Wet collectors preferred.
[c] Fabric collectors preferred.
[d] High-energy wet collector (Venturi) preferred.
[e] Electrostatic precipitator often required as final collector.
[f] Includes car unloading, conveyors and weighing.
[g] Includes hammer mills, mixers, feeders and bagging operation.
[h] Oil bath or spray cleaner preferred.
[i] Inertial and centrifugal collectors preferred to recover solids in dry state.
[j] Inertial collectors used if code permits 0.20 GR/SCF discharge.
[k] In this column (1) stands for the standard cyclone, (2) for the multiple cyclone, (3) for the inertial collector and (4) for the dynamic precipitator.

NOTE: See Sections 5.9 and 5.10 for bibliography.

Scrubbers

(Sections 5.12–5.21 are devoted to this discussion.)

5.12 MECHANISM OF WET COLLECTION

Types of Scrubbers:	(a) Spray Towers and Chambers; (b) Packed Towers; (c) Cyclonic and Centrifugal Scrubbers; (d) Impingement Scrubbers; (e) Venturi Scrubbers; (f) Mechanically Induced Spray Scrubbers; (g) Mist Eliminators and Irrigated Filters; (h) Spray Nozzles and Liquid Distributors. NOTE: In the feature summary below, the letters "a" to "h" refer to the above listed scrubber designs.
Scrubber Selection and Application:	See Section 5.21.
Performance Data:	See Sections 5.13 to 5.20.
Materials of Construction:	Materials of construction normally used are carbon steel, carbon steel with protective lining (from corrosion and erosion point of view), stainless steel, 304, 316 and 316L, fiber glass reinforced plastics (FRP) and other plastics, wood, etc.
Costs:	See Table 5.21f.
Partial List of Suppliers:	Air Correction Div. of UOP (a, b, c, e, g); Airetron Engr. Corp. (a, b, c, e); American Air Filter Co., Inc. (a, c, d, e, f); Automotive Rubber Co. (a, e); Bete Fog Nozzle, Inc. (h); Buell Engineering Co. (b, c, e); Buffalo Forge Co. (a, c); The Ceilcote Co. (a, b, c, d); Centrispray Co. (a, c, d, e, f); Chemical Construction Corp. (a, b, c, e, g); Croll-Reynolds Engineering Co., Inc. (e); The Ducon Co., Inc. (a, e); Fly Ash Arrestor Corp. (e); Heil Process Equipment Corporation

(a, b, c, e, g); Koch Engineering Co., Inc. (b, h); Krebs Engineers (a); Monsanto Company (g); National Dust Collector Corp. (b, d); Pangborn Corp. (e); Peabody Engineering Corp. (a, c, d, e); Research-Cottrell, Inc. (e); Schmieg Industries (a, c, f); Claude B. Schneible Co. (c, d, f); Schutte and Koerting Co. (e, h); Spray Engineering Co. (h); Spraying Systems Co. (h); U.S. Stoneware Co. (a, b, e, g, h); The W. W. Sly Manufacturing Co. (c, d, e); Western Precipitation Division of Joy Manufacturing (a, c, d, e, f); Otto H. York Co., Inc. (g).

One of the principal methods for removing fine particulate matter from flue gas streams is by contact with a liquid in wet collectors.

The wet scrubbers remove the dust from a flue gas stream by contacting it with a scrubbing liquor; and by recovering the dust as a slurry or solution that requires further processing, it can be used as a product or disposed of as waste.

In a wet scrubber:

Gases as well as particulates can be removed.
The flue gas is both cooled and cleaned.
The stack effluent usually will be well cleaned but will contain some fines, mists and steam plume.
The temperature and moisture content of the inlet gas is essentially unlimited.
The corrosive gases can be neutralized by proper choice of scrubbing medium.
The scrubbing liquor can be a clear solution or a slurry.
Consideration of freezing conditions is important.
Hazards of explosive dust-air mixtures are reduced.
Equipment is simple and occupies only a moderate amount of space.

Wet scrubbers are flexible and versatile dust collection devices. Their advantages include:

High dust collection efficiency which can be varied by changing the power input (in terms of pressure drop) to the scrubber.
The ability to collect from submicron to coarse-size particulate matter.
Compact size and generally lower capital cost than that of dry collection equipment.

No limitations on dust loadings, temperature and humidity of inlet flue gas.

Simultaneous removal of gaseous pollutants such as sulfur dioxide, hydrogen sulfide and nitrogen oxides.

Lower maintanence cost as the mechanical construction is simple with few or no moving parts.

Theory and Mechanism of Wet Collection

The dust-scrubbing process is extremely complex and there is no simple theoretical analysis of it, similar to the processes for electrostatic precipitation, filtration or dry inertial separation. In the wet collectors, the fine particles are "conditioned" to increase their effective size, thus making them easier to collect. The reentrainment of the collected particles is minimized by trapping them in a liquid film. The effective size of the particle may be increased by promoting condensation on fine particles, which act as nuclei. The trapping or forced conditioning of dust particles on liquid droplets is attained by impact due to inertial forces.

The principal mechanisms by which aerosols are collected in liquids include:

1. Wetting of the dust particles by intimate contact with a liquid droplet. The liquid droplets can be formed either by high-velocity impact of flue gas with the scrubbing liquor or by spraying the scrubbing liquor into the gas stream.
2. Impingement of wetted or unwetted dust particles on collecting surfaces which are then flushed by a liquid.

The wetting of the particles occurs in accordance with the following mechanisms:

1. *Spray droplet impingement.* A spray of scrubbing liquor across the flow path of the dust particles impinges upon the particles. The efficiency is dependent upon the number and size of droplets and upon the energy imparted to them.
2. *Brownian movement or diffusion.* The liquid droplets are dispersed among the dust particles, which deposit on the droplets and thus get collected. This is the principal mechanism for the collection of submicron particles.
3. *Condensation.* Cooling the flue gas below its dew point causes the moisture to condense onto the dust particles which act as condensation nuclei. This results in an increase in particle size and thus makes collection easier. The condensation alone can help to remove only relatively small amounts of dust, because the amount of condensation required to remove large concentrations is greater than can be achieved.

4. *Electrostatic Forces.* The electrostatic force helps agglomeration of smaller particles and thus makes collection easier. Experimental data have shown that to attain a particular outlet dust loading, less pressure drop is required in the scrubber when it handles the flue gas leaving the electrostatic precipitator than when it handles the flue gas before the precipitator. Electrostatic forces result when dust particles and liquid droplets are electrically charged, which may be induced by friction or by the presence of electrically charged matter.

The principal mechanism involved in wet collection of particulate matter is the impingement or impaction of the individual dust particles upon the scrubbing liquor droplets. The dust particles, due to their inertia flow in a nearly straight path, tend to collide with the liquid droplet and get collected by it. The droplets, being substantially larger in size and mass than the dust particles, are readily collected by gravitational or inertial forces. The contribution of impaction or impingement to particle collection efficiency is related to the "separation number" (N_s), a dimensionless number. The "separation number" for particles smaller than 100 μ diameter is given by Equation 5.12(1).

$$N_s = Dp^2(dp\text{-}dg)\ Vo/18\,\mu D_b \qquad 5.12(1)$$

where Dp = dust particle diameter, μ
Db = liquid droplet diameter, μ
dp = dust particle density, g/cm^3
dg = flue gas density, g/cm^3
μ = flue gas viscosity, poise
Vo = the relative velocity of dust particles with respect to liquid droplets, cm/sec

Dust Scrubber Efficiency

The collection mechanisms mentioned are of theoretical significance, and they do not offer a practical basis for wet scrubber design. In most cases, design or selection of a scrubber is based on actual performance tests.

In recent years, various investigators have developed an empirical method for correlation of scrubber efficiency data, applicable to the principal classes of scrubbers.[1-4] The efficiency of wet scrubbers is related to the total energy expended in the gas-liquid contacting process. For a given aerosol, the scrubbing efficiency is a function of the pressure drop across the scrubber; and the scrubber design, geometry or size has very little influence on it.

This concept of power consumption in scrubbing can be stated as follows: In the gas-liquid contacting process, power is dissipated in fluid turbulence and, ultimately, as heat. This power, expressed as power per unit of volumetric gas flowrate, is the criterion of scrubber efficiency and is called "contacting power." The contacting power represents only that power which is usefully consumed in contacting the gas with the liquid and not the gross power input which includes electrical and mechanical losses in motors, bearings, pumps, fans and, possibly, net additions to the energy of the gas stream. It does not include the friction losses through those pieces of equipment that are operating dry. With the same total power consumption, all wet scrubbers designed for *optimum* utilization of energy provide essentially the same dust collection efficiency for a specific dust, regardless of their collection mechanisms or their design features.

The relationship between scrubber collection efficiency η and number of transfer units N_t is given by Equation 5.12(2).

$$N_t = \ln\left(\frac{1}{1-\eta}\right) \qquad 5.12(2)$$

Table 5.12a shows the numerical relationship of efficiency to number of transfer units. The relationship between the number of transfer units and the total contacting power P_t is given by Equation 5.12(3).

$$N_t = a(P_t)^b \qquad 5.12(3)$$

On a logarithmic plot, this equation represents a straight line with *b* being the slope and *a* the value of the intercept with the ordinate where P_t equals unity. (See Table 5.12b and Figure 5.12c.) The contacting power derived from the gas stream P_g is approximated by Equation 5.12(4):

$$P_g = 0.157\, F_s \qquad 5.12(4)$$

Across the scrubber, changes in the kinetic energy of the gas stream are usually small, and the friction loss F_s is essentially the same as the pressure drop.

The contacting power derived from the liquid stream P_l may be calculated from Equation 5.12(5):

$$P_l = 0.583\, p_f\left(\frac{q_l}{q_g}\right) \qquad 5.12(5)$$

Table 5.12a
RELATIONSHIP BETWEEN SCRUBBER EFFICIENCY AND NUMBER OF TRANSFER UNITS

Number of Transfer Units, N_t	*Collection Efficiency %, η*	*Number of Transfer Units, N_t*	*Collection Efficiency %, η*
0.10	9.52	5.00	99.33
0.50	39.35	6.00	99.752
1.00	63.21	7.00	99.909
2.00	86.47	8.00	99.967
3.00	95.02	10.00	99.9955
4.00	98.17		

Table 5.12b
SUMMARY OF SCRUBBER DATA

Curve Number in Fig. 5.12c	Aerosol	Scrubber	Correlation Parameters a	Correlation Parameters b	Remarks
	Lime Kiln Dust and Fume (Kraft Mud Kiln)				
1	Raw gas	Venturi and cyclonic spray	1.47	1.05	Lime dust and soda fume
2	Prewashed gas	Venturi, pipeline, and cyclonic spray	0.915	1.05	Mostly soda fume
3A	Talc dust	Venturi	2.97	0.362	
3B		Orifice and pipeline	2.70	0.362	
4	Black liquor Recovery furnace Fume	Venturi and cyclonic spray	1.75	0.620	Cold scrubbing water Furnace gases humid
5		Venturi, pipeline and cyclonic spray	0.740	0.861	Hot fume solution for scrubbing; furnace gases humid
6		Venturi evaporator	0.522	0.861	Hot black liquor for scrubbing. Furnace gases dry
7	Phosphoric acid mist	Venturi	1.33	0.647	
8	Foundry cupola dust	Venturi	1.35	0.621	
9	Open hearth steel furnace fume	Venturi	1.26	0.569	
10	Talc dust	Cyclone	1.16	0.655	
11A	Copper sulfate	Solivore Mechanical spray generators	0.390	1.14	Power estimated
11B		Hydraulic nozzles	0.562	1.06	
12	Ferrosilicon furnace fume	Venturi and cyclonic spray	0.870	0.459	
13	Odorous mist	Venturi	0.363	1.41	
	Blast furnace dust and fume	Venturi, orifice, cyclonic spray, tower and disintegrator			

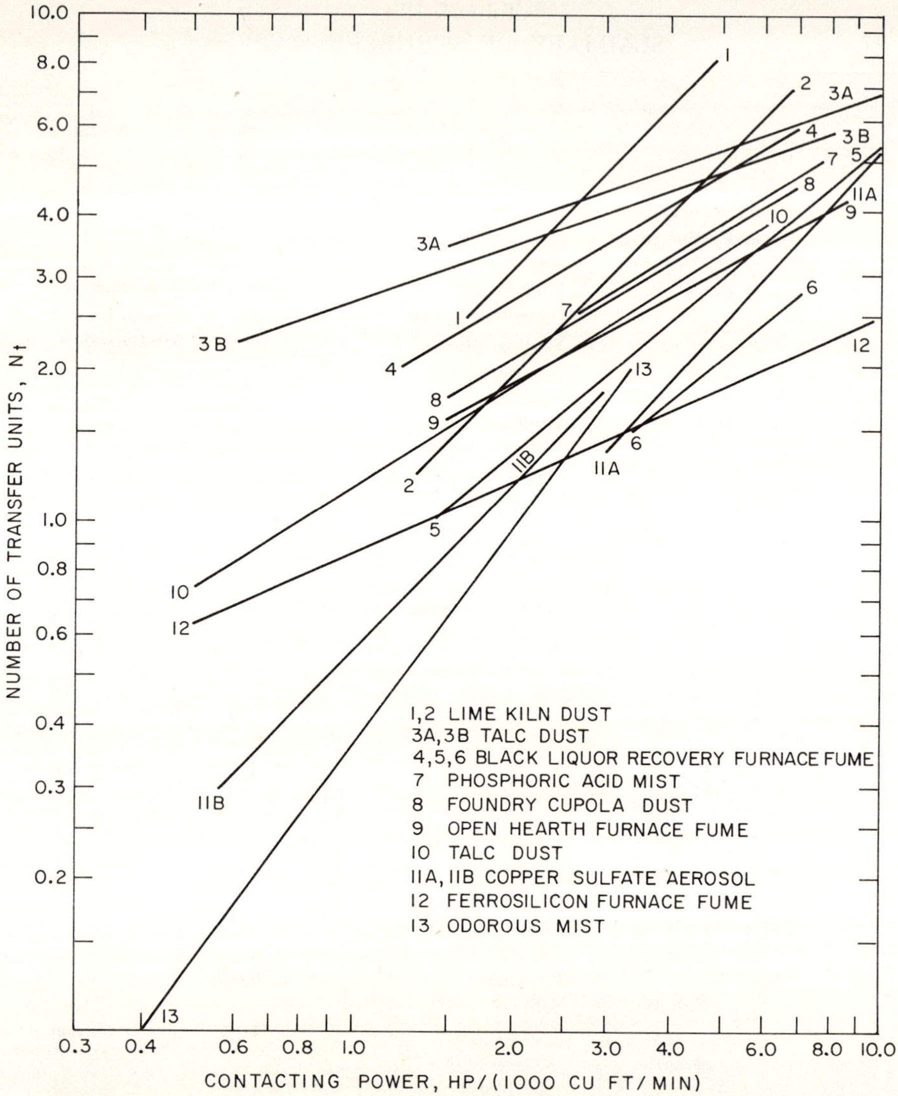

Figure 5.12c Performance curves for aerosol scrubbing systems described in Table 5.12b.

where p_f = liquid feed pressure, lb/in.^{2} gauge

q_l = liquid feed rate, gal/min

q_g = gas flowrate, cu ft/min

P_l = contacting power derived from the liquid stream, HP/(1,000 cu ft/min)

$P_t = P_g + P_l$

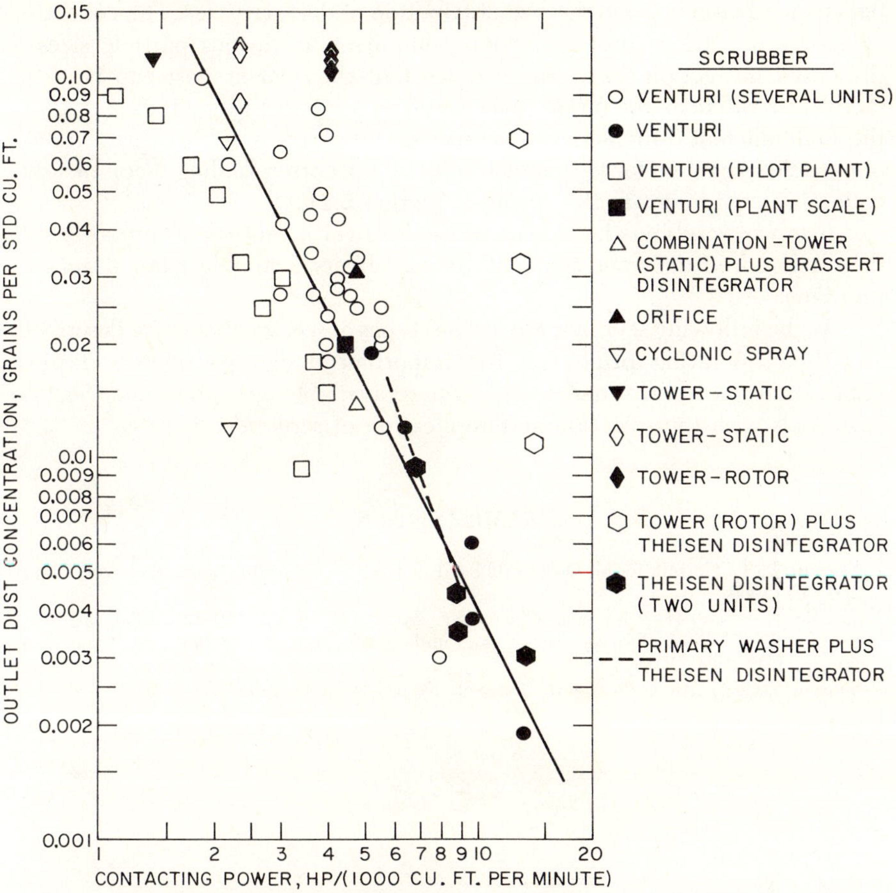

Figure 5.12d Performance of various scrubbers on blast furnace dust and fume.

Semrau[1] has used this method of correlation for a variety of aerosols and scrubbing equipment. Table 5.12b and Figure 5.12c summarize some basic scrubber data.[1] In Figure 5.12d, the outlet dust loading of blast furnace flue gas is plotted as a function of the contacting power, for a variety of scrubbers. This substantiates the "contacting power rule" which states that the efficiency of a scrubber on a given aerosol is essentially dependent only upon the power per unit of volumetric gas flowrate that is dissipated in the gas-liquid contacting process, and not upon the geometry or size of the unit or on the manner in which the power is applied to the contacting process.

Collection Efficiency and Particle Size

To attain the desired collection efficiency, pressure drop requirements increase with decreasing particle sizes. The efficiencies of various wet scrubbers depend largely upon the size distribution of dust particles. The collection efficiency of various dust collection equipment at various particle sizes is presented in Section 5.21. One can see that the coarser dust can be satisfactorily treated in the simple wet scrubbers, while the finer dusts require the high-efficiency–high-energy venturi scrubbers. A graphical presentation of particle size range for various aerosols and the corresponding recommended collecting equipment is also given in Section 5.21.

The wet scrubbers have been in use for over a century. There are more than a hundred different types of wet scrubbers available from dozens of suppliers.

In the following sections, the various types of wet scrubbers are described; and their operating principles, the important design parameters, typical applications and performance data, where available, are presented. Section 5.21 deals with the selection and application of scrubbers.

REFERENCES

1. Semrau, K. T.: Correlation of Dust Scrubber Efficiency, *J. Air Pollu. Contr. Ass.*, vol. 10, pp. 200–207, 1960.
2. Lapple, C. E. and H. J. Kamack: *Chem. Eng. Progr.*, vol. 51, pp. 110–121, 1955.
3. Semrau, K. T., C. W. Marynowski, K. E. Lunde and C. E. Lapple: *Ind. Eng. Chem.*, vol. 50, pp. 1615–1620, 1958.
4. Semrau, K. T.: *J. Air Pollu. Contr. Ass.*, vol. 13, pp. 587–594, 1963.

5.13 SPRAY CHAMBERS AND TOWERS

Types of Designs:	(a) *Spray chambers* (Fig. 5.13a), Cocurrent, countercurrent and crosscurrent flow configurations; (b) *Spray towers* (Fig. 5.13b); (c) *Elbair scrubber* (Fig. 5.13c).
Principle of Operation:	Interception and collision of dust particles with spray droplets, due to difference in their velocities.
Pressure Drop:	0.5 to 1″ H_2O.
Liquor Flow:	5 to 10 gal/1,000 ACFM.
Collection Efficiency:	Coarse Particles, 10 μ or larger are collected at 80% efficiency. Elbair scrubber is more efficient on both coarse and fine dust.
Applications:	Soot blowing operation, preconditioner in fertilizer industry, boiler stacks, cooler and humidifier.
Cost:	\$0.25 to \$1 per cfm (installed cost). No cost figures available for Elbair scrubber.
Partial List of Suppliers:	See Section 5.12. Elbair is supplied by Krebs Engineers.

Spray Chambers

The spray chamber is the simplest wet scrubber. It can be round or rectangular in shape. The scrubbing liquor is introduced by means of spray nozzles. Figure 5.13a shows a typical spray chamber with countercurrent flow configuration. The flue gas velocity decreases as it enters the spray chamber; the dust particles, which are wetted by sprayed liquor, settle and are collected at the bottom of the chamber. The scrubbing liquor rate is normally in the range of 5 to 10 gallons per 1,000 ACFM.

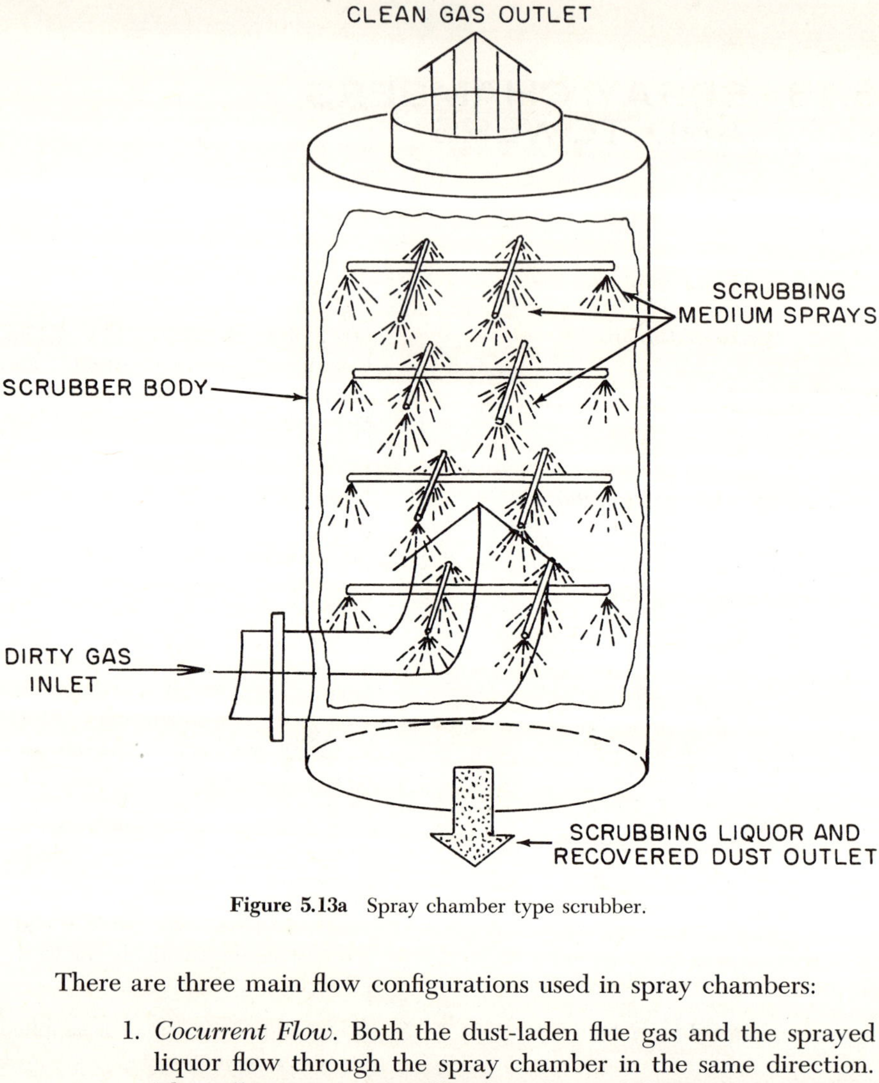

Figure 5.13a Spray chamber type scrubber.

There are three main flow configurations used in spray chambers:

1. *Cocurrent Flow.* Both the dust-laden flue gas and the sprayed liquor flow through the spray chamber in the same direction. The collisions are caused by the velocity difference between the spray droplets and flue gas stream containing dust. The collection efficiency is at a minimum.
2. *Countercurrent Flow.* The scrubbing liquor and flue gas flow in opposite directions; normally the gas flows upward and the liquor flows downward. As a result of countercurrent flow, the relative velocity of the particulate matter and liquor droplets is at a maximum, thus providing maximum collection efficiency.

3. *Cross Flow.* In this configuration, the liquor is introduced at 90 degrees to the flue gas direction. The gas flow can be upward or downward in the chamber or it can flow in a horizontal direction. The relative velocity is somewhat higher than that for cocurrent flow but lower than that for countercurrent flow.

Normally, a mixed flow configuration exists in a spray chamber. This is due to turbulence, liquid droplet inertia and gravitational forces.

In addition to removing coarse particles at low-pressure drops from the flue gas, the spray chamber acts as a cooler and humidifier. It can also provide preconditioning of the particulate matter by water vapor condensation.

The flue gas velocity and the proper pattern of sprays are the important design parameters.

Spray chambers can be used as gas coolers, humidifiers and preconditioners (in phosphate fertilizer), as dust collectors during soot blowing operations and as final cleanup devices for exhaust gases from recovery furnaces in pulp mills. This type of scrubber is also used for scrubbing boiler stacks by installing the sprays in the stack.

The spray chamber has a low dust *collection efficiency* for all except coarse dust particles (over 10 μ). Compared to packed scrubbers, the spray chambers are inefficient. Where collection efficiency below 80% is satisfactory, spray chambers can be an economical solution.

Spray Towers

In spray towers (see Fig. 5.13b), the liquid droplets are produced by either spray nozzles or atomizers, and the liquor and gas flow countercurrent to each other. The flue gas enters at the base of the tower and passes through a distribution plate, which can be a perforated plate. The rising gas flow contacts the liquid droplets created by a bank of spray nozzles. The terminal settling velocity of the liquor droplets must be greater than the velocity of the flue gas (from 2 to 5 feet per second) to avoid entrainment and carry over of the droplets.

The collection efficiency is affected by the difference between the free falling velocity of droplets and the flue gas velocity. The spray towers are effective on particles of 10 μ or larger.

Elbair Scrubber

In this scrubber, liquor is sprayed at high pressure (100 to 200 psig) against a membrane in the direction of gas flow. The membrane consists of vertically mounted, closely spaced, parallel rods. The type and pattern of sprays and the distance of the spray nozzles from the membrane are set so that a rebound zone of fast moving atomized liquid droplets is established near the membrane

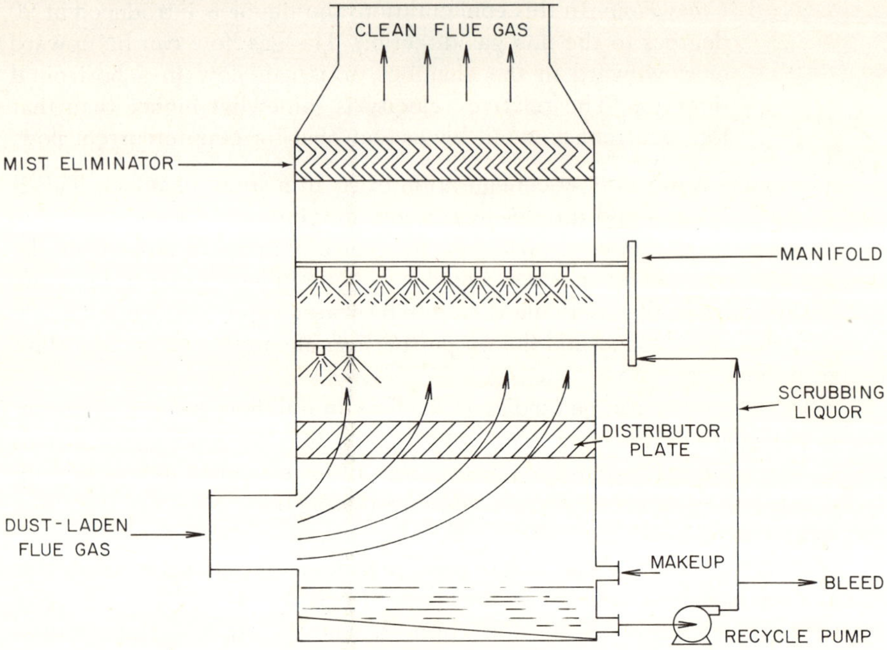

Figure 5.13b Spray tower.

surface. The flue gas is subjected to a highly intense scrubbing action, as it passes through the rebound zone (Fig. 5.13c). The atomized droplets impact on the dust and fume particles at high velocity, thus causing a thorough wetting of particles that are normally difficult to wet. Further scrubbing takes place as the flue gas is accelerated through the narrow spaces between the rods. The major portion of the dust is removed immediately upstream and downstream of the membrane.

The scrubbing water requirement is 3 to 6 gal/1,000 ACFM per stage at 100 to 200 psig pressure. The pressure drop on the gas side is only about 1″ H_2O per stage.

The Elbair scrubber is a high efficiency, simple and proven design. The collection efficiency of one, two and three stage units for a variety of dust and fume applications is presented in Table 5.13d.[1] Detailed dust loading data for various applications are given in Table 5.13e. Commercial units with capacities as high as 100,000 CFM have been installed.

REFERENCE

1. "Elbair," Bulletin E 869, Krebs Engineers, Gas Kinetic Division.

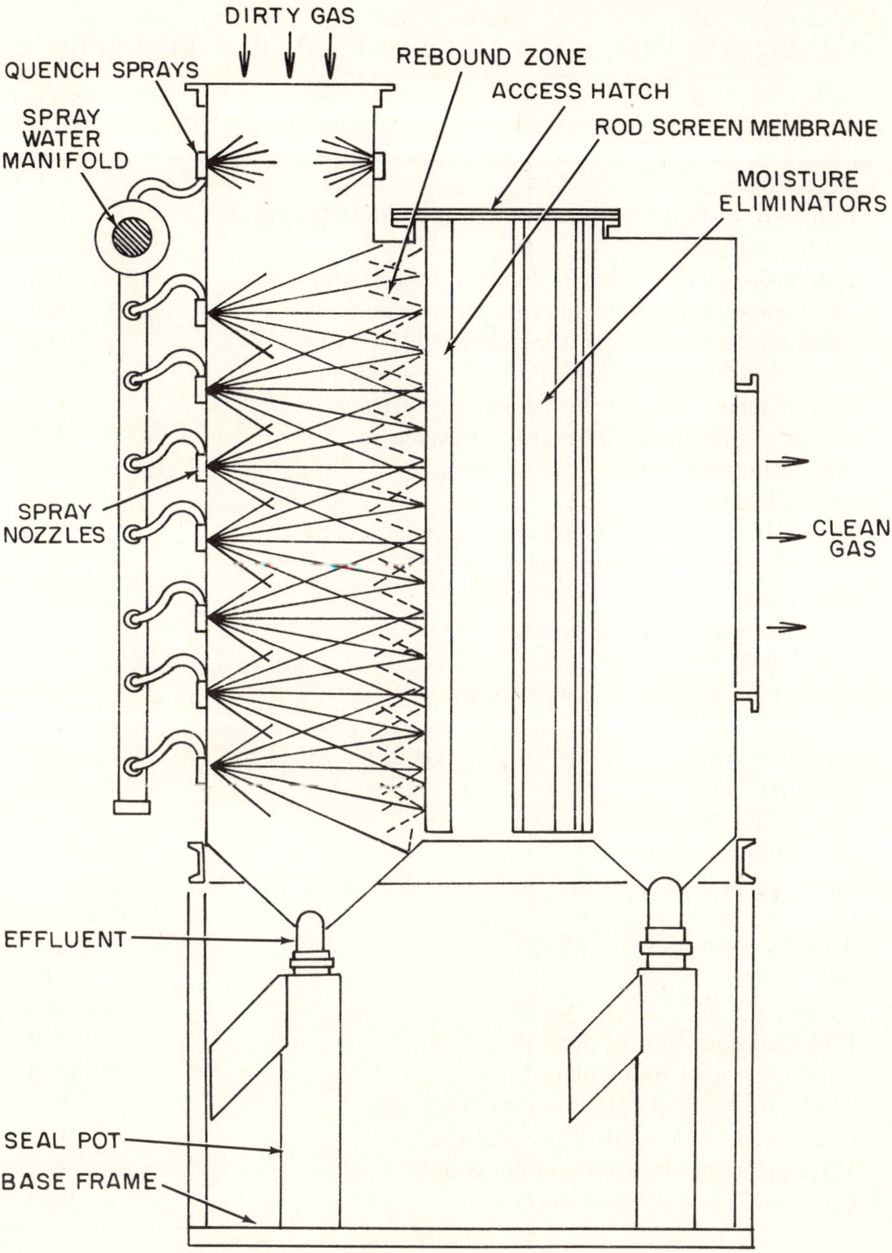

Figure 5.13c Elbair scrubber.

Table 5.13d
COLLECTION EFFICIENCY DATA FOR THE ELBAIR SCRUBBER

Application	*Collection Efficiency by Weight, %*
3-stage Elbair	
Fumes from basic oxygen steel furnace (during oxygen blowing periods)	99.80
Calcine from precycloned fluidized bed roaster gases	99.9
Lead fumes from Newnham hearth furnace	99.3
Lead fumes from rotary furnace and blast furnace	98.5
Lead chloride fumes from de-leading furnace	99.4
Zinc chloride fumes from de-leading furnace	99.9
Copper fumes from arc melting furnace	97.2
SO_3 from fluidized bed roaster gas	91.5
2-stage Elbair	
Fumes from oxygen lanced steel furnace (averaged over entire operating cycle)	98.5
Zinc oxide furnace fumes	+97.0
Lead fumes from blast furnace	98.0
Paper dust from pulverizers	99.0
Copper fumes from arc melting furnace	97.0
Lead fumes from cooling drum in sinter plant	99.0
Flotation concentrate dust from rotary drier exhaust gas	99.0
Fluorine	96.0
Lead chloride fumes	+97.0
Zinc chloride fumes	+97.0
Single Stage Elbair	
Crusher dust	99.0
Asbestos dust	98.3
Sulfur dust	99.7
Flyash	99.0
Flotation concentrate drier dust	97.0
Dust from lead sinter plant	98.0
Blast furnace dust (platinum recovery plant)	98.7
Calcine from Edwards type gold roaster	98.0
Ultrafine silica dust (66% minus 5 μ)	95.7
Coal dust from thermal dryer	+99.0

Table 5.13e
PERFORMANCE DATA FOR THE ELBAIR SCRUBBER

Application	*Number of Stages*	*Inlet Dust Loading, grains/std cu ft*[a]	*Outlet Dust Loading, grains/std cu ft*[b]	*Efficiency, %*
Lead fumes and dust from rotary furnace and lead blast furnace	3	4.0 6.0	0.06 0.09	98.5
Lead and zinc fumes from zinc de-leading furnace	3	0.366 lead 17.30 zinc	0.003 lead 0.019 zinc	99.4 99.9
Lead fumes from Newnham hearth furnace	3	3.5	0.033	99.3
Copper fumes from arc melting furnace	3	3.03	0.081	97.2
Gases from fluidized-bed gold/copper roaster	3 3	10.7 calcine 24.3 SO_3	0.016 calcine 2.03 SO_3	99.9 91.5
Gases from fluidized-bed copper/cobalt roaster	3 3 2	8.04 2.05 2.40	0.00181 0.00043 0.00059	99.9 99.9 99.9
Fumes from basic oxygen steel furnace—sampled during oxygen blowing periods	3	20.0	0.033	99.8
Nickel/copper/platinum blast furnace gas	1	3.03	0.04	98.7
Taconite pelletizer plant exhaust gas	1	3.69	0.074	98.0
Flue gases, Edwards type gold roaster	1	3.7	0.075	98.0
Asbestos dust from finishing mill	1	1.96	0.0337	98.3
Silica dust (ultrafine, 66% minus 5 μ)	1	0.675	0.0288	95.7
Sulfur dust	1	1.065	0.0034	99.7

[a]Cubic feet based on standard conditions of 70°F and 14.7 psia.
[b]Grains per dry standard cu ft.

5.14 PACKED, FLOATING AND FLOODED BED SCRUBBERS

Types of Designs:	(a) *Fixed bed packed scrubbers,* Flow configurations: cocurrent (Fig. 5.14e), countercurrent (Fig. 5.14c), cross flow (Fig. 5.14d); (b) *Floating bed scrubbers,* floating bed scrubber (Fig. 5.14f), turbulent contact absorber (Fig. 5.14g), polysphere gas scrubbers, Hydro-filter®; (c) *Flooded bed scrubbers* (Fig. 5.14h). NOTE: In the feature summary below, the letters "a" to "c" refer to the above scrubber designs.
Principle of Operation:	The wetted packing provides the impingement surface for the deposit of dust (Fig. 5.14a).
Pressure Drop Requirements:	3 to 4 inch H_2O per stage. For fixed packed beds, 0.5 to 1.0 inch per foot of packing. For cross flow, 0.2 to 0.3 inch per foot of packing.
Liquor Flow Required per 1,000 ACFM:	Fixed bed, countercurrent flow: 10 to 20 GPM, cocurrent flow: 7 to 15 GPM, cross flow: 5 to 10 GPM; Floating bed: 10 to 20 GPM; Flooded bed: 2 to 3 GPM.
Types of Packing:	Raschig rings, Saddles, Pall rings, spheres, Tellerettes® (Fig. 5.14b).
Collection Efficiency on 2 μ or Larger Particles:	80 to 95% (a); 95 to 99% (b); 99% (c).
Applications:	Fixed Bed: Acid mist, phosphate and fluoride dusts, and other soluble dusts; Floating Bed: Flyash, viscid material and materials which form

gummy precipitates and gels and phosphate dust; Flooded Bed: Carbon black, flyash, cupola dust.

Cost: Between that of spray chambers and electrostatic precipitators. (See Table 5.21f.) Depends on application, types of packing and materials of construction.

Partial List of Suppliers: Air Correction Div. of UOP(b); Buell Engineering Co. (b); The Ceilcote Co. (a); Koch Engineering Co., Inc. (a); National Dust Collector Corp. (a, b, c); U.S. Stoneware (a). Also refer to Section 5.12.

In packed scrubbers, the dust-laden flue gas is passed through a bed of packing which is fixed or moving with the scrubbing liquor and evenly distributed over the surface of the packing to keep it clean and to prevent reentrainment of deposited particles. The wetted packing provides an impingement surface for the deposit of dust, and the scrubbing liquor washes the dust off and transfers it into a slurry or solution form. Figure 5.14a[1] shows the dust collection phenomena and the impingement principle when using Tellerette, ring and saddle-type packings.

Packed scrubbers are generally used for absorption of gaseous pollutants but have also been successfully used for dust and mist collection. In addition to dust collection, direct-contact cooling also occurs.

All the packed scrubbers utilize some kind of packing, such as Raschig rings, pall rings, saddles, tiles, marbles or polypropylene plastic spheres. The packing may be held in place, as is the case in a fixed bed scrubber; or it may be free to move, as is the case in a floating bed scrubber; or it may be covered with liquor, as is the case in flooded bed scrubbers.

The liquor and gas flow configurations can be countercurrent, cocurrent or of the cross-flow type. The countercurrent packed scrubber has practically no solids-handling capacity, since solids tend to plug the packing and support plates, which can be cleaned only after removal. The cross-flow type scrubbers can handle dusts, as the face of the packing can be washed with spray nozzles, while the body of the packing is irrigated from the top. In floating bed type scrubbers, the movement of packing keeps it clean.

Types of Packed Scrubbers

The different types of packed scrubbers and their flow configurations for liquor and gas are:

Type	*Flow Configuration*
Fixed bed packed scrubber	Countercurrent Cocurrent Counter–Cocurrent combination and Cross flow
Floating Bed Scrubber	Countercurrent
Turbulent Contact Absorber (TCA)	Countercurrent
Flooded Bed Scrubber	Countercurrent

In the cross-flow configuration, the flue gas moves horizontally through the packing, whereas the scrubbing liquor flows downward by gravity, irri-

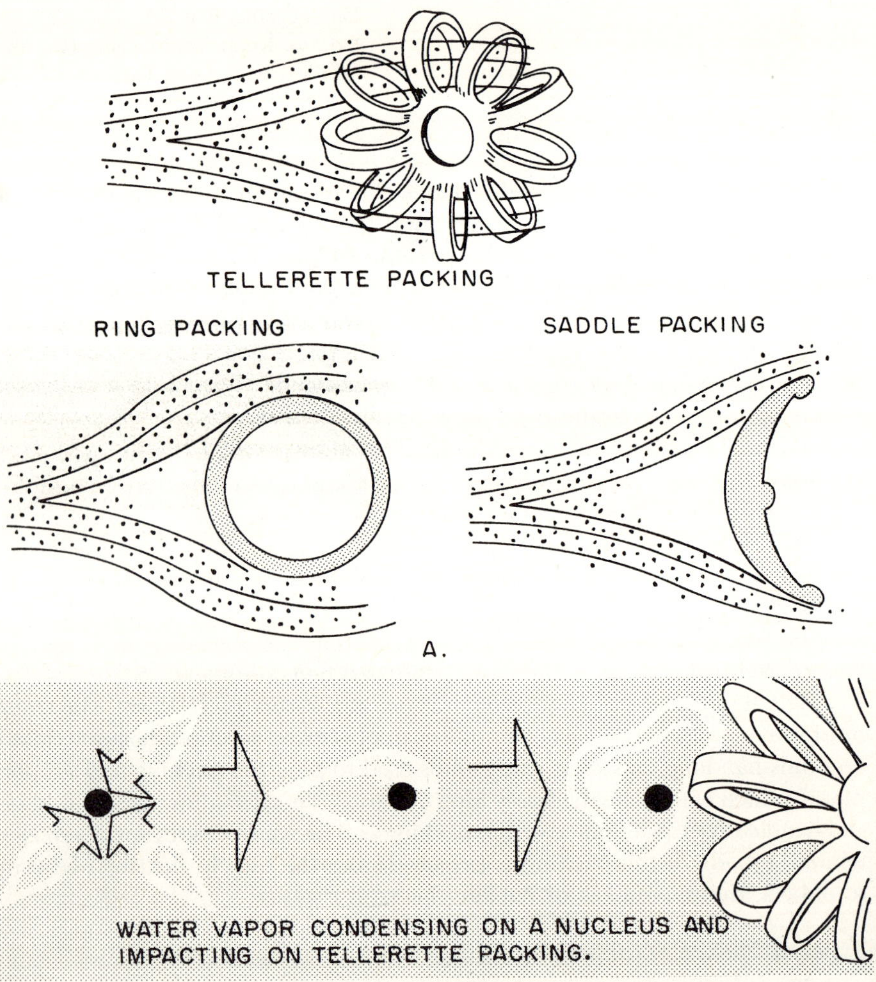

B.

Figure 5.14a Impingement (A) and nucleation (B) in dust collection.

gating the packing. The countercurrent flow is the most common configuration in packed beds, where the flue gas flows upward through the packing against the countercurrent gravity flow of the liquid. In case of cocurrent flow, both the flue gas and scrubbing liquor flow through the packing in the same direction. In this configuration, the scrubbing liquor that irrigates the packing also keeps it from being plugged; and both the flue gas and the scrubbing liquor help to wash the solids through the bed.

In another configuration of cocurrent flow, both the flue gas and scrubbing liquor flow upward through the packing with some of the liquor, before being collected, flowing downward and in opposite direction to the gas flow. This provides a combination of cocurrent and countercurrent flow configurations.

Types of Packing

A variety of packings, available in various shapes, are used in packed scrubbers. The main types of packings used are Raschig rings, Berl® and Intalox saddles, pall rings, Tellerettes, spheres and ping pong balls. Packings have been made of ceramic, polypropylene and metal, depending on their applications. Figure 5.14b shows the various types of packings used in wet scrubbers.

The packings should provide a relatively large surface area, low weight per unit volume and a large, free cross section. Plastic packings are lightweight, essentially unbreakable and can be used in almost all the corrosive conditions. Due to its unique shape, the Tellerette type packing has been used effectively as an impingement device at low pressure drops. The plastic spheres (light in weight) have been used as packing for floating bed and turbulent contact scrubbers.

Collection Efficiency

For particles 2 μ or larger in diameter, the collection efficiencies of the fixed bed, the floating bed and the flooded bed scrubbers are 80 to 95%, 95 to 99% and 99%, respectively. The collection efficiency of packed scrubbers is mainly dependent on the time of contact between the flue gas and scrubbing liquor on the packed bed surface. For the collection of dust particles, 10 μ or larger, a coarsely packed bed is used, and the flue gas velocities are limited to 5 to 7 feet per second. The finely packed beds have a tendency to plug and are used mainly to clean flue gases with low dust loadings or having liquid entrainment. The fine packed beds are used for removing dust in the size range of 1 to 5 μ and the velocity through the bed is kept below 1 foot per second.

Applications of Packed Scrubbers

The packed scrubbers are generally used for the absorption of gaseous pollutants but have also been successfully used for dust and mist collection and for the simultaneous removal of both particulate matter and gaseous pollutants. The fixed bed packed scrubbers are used to remove acid mists,

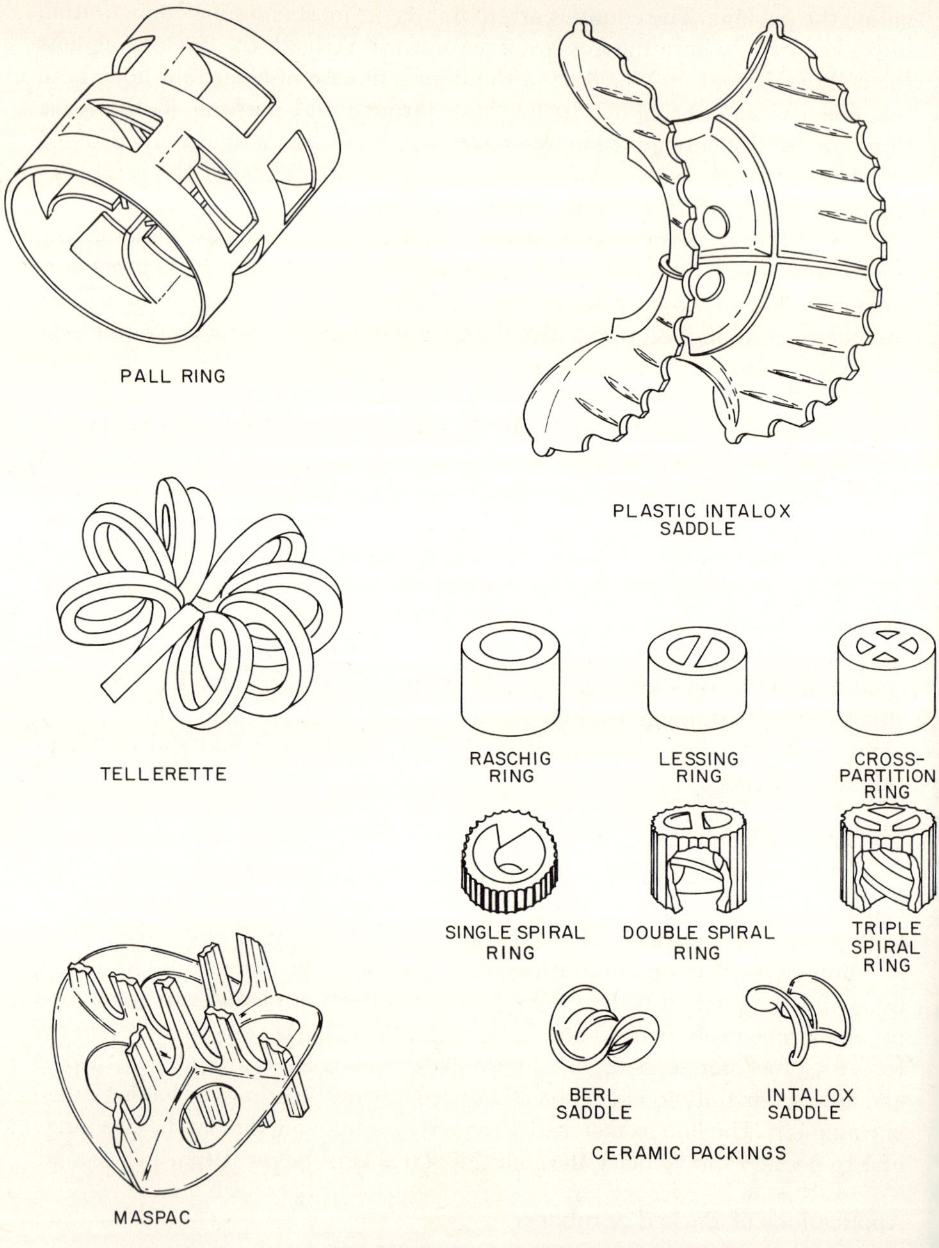

Figure 5.14b Packing types.

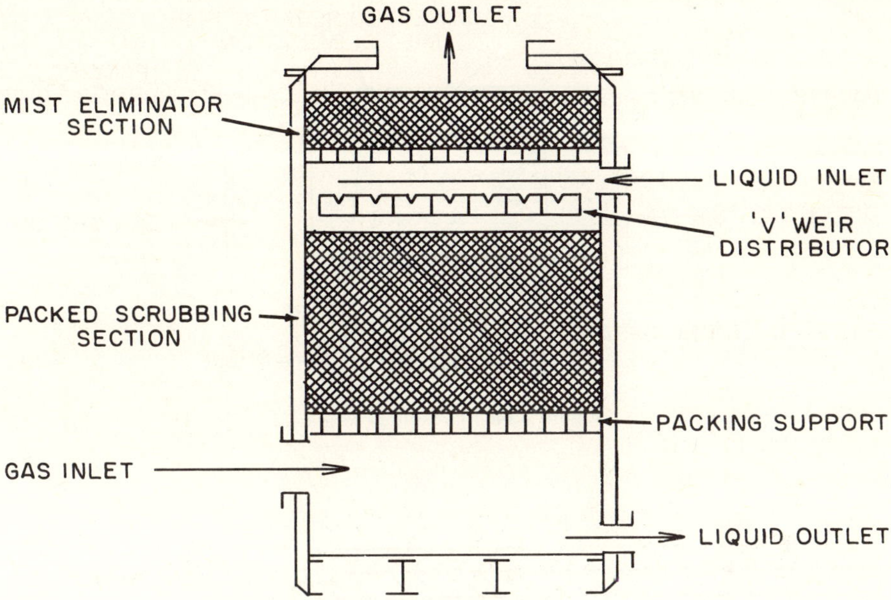

Figure 5.14c Countercurrent flow scrubber.

phosphate dust, fluoride dust and other soluble dusts. The floating bed and turbulent contact type scrubbers are also used for simultaneous removal of flyash and sulfur dioxide from coal-fired boilers. The packed scrubbers can also serve as coolers, preconditioners or as entrainment devices.

Cost of Packed Scrubbers

The cost of packed scrubbers varies widely with the types of packing, materials of construction and size of the equipment. For comparatively coarse and thin beds, the costs are comparable to those for a spray chamber. The costs of deep beds with fine packing are about the same as those for electrostatic precipitators.

Fixed Bed Packed Scrubbers

In this type of scrubber, the packing is held in a fixed position with the flow configurations being countercurrent, cocurrent or cross flow. Figure 5.14c shows a typical countercurrent flow packed scrubber, Figure 5.14d illustrates a typical cross-flow scrubber, and Figure 5.14e represents a cocurrent flow scrubber.

The most common flow configuration is the countercurrent flow. This is most effective when the pressure drop and flue gas velocity are such that the liquor is held up in and on top of the packed bed. The use of entrainment

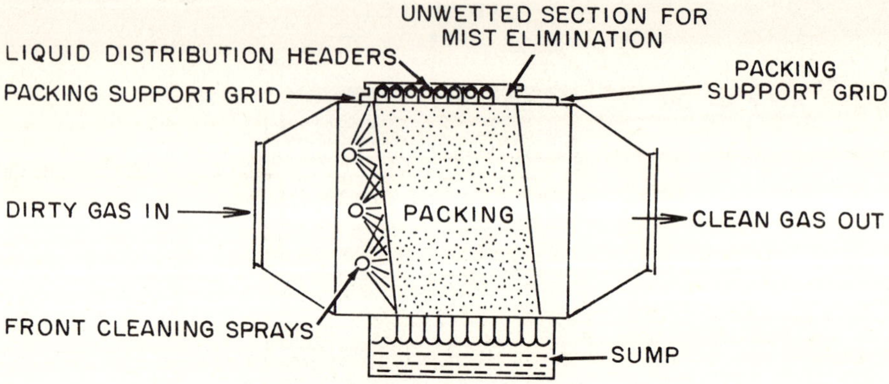

Figure 5.14d Cross-flow scrubber.

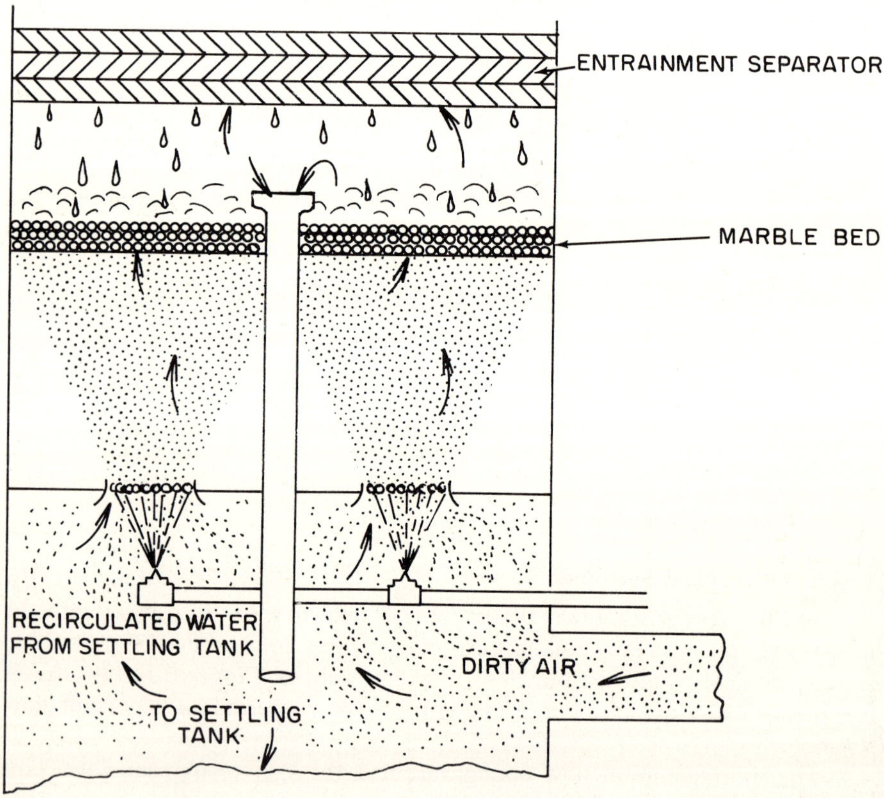

Figure 5.14e Cocurrent flow scrubber.

separators is essential. Flue gas velocities are normally 5 to 7 feet per second and the liquor flowrate ranges from 10 to 20 GPM/1,000 ACFM of gas. The pressure drop across the packed bed is in the range of 0.5 to 1.0″ H_2O per foot of packing. The type of packing, the gas velocity and the liquor flow affect the pressure drop.

In the cocurrent flow configuration, both gas and liquor flow downward. The liquor flow requirement is only 7 to 15 GPM/1,000 ACFM of gas. The pressure drop on the gas side ranges from 1 to 4 inches of water per foot of packing and is dependent upon the type of packing and flue gas velocity. No entrainment separators are required.

In the Hydro-filter type cocurrent flow scrubber (see Fig. 5.14e), the scrubbing liquor is sprayed upward and toward the bottom of a bed of spheres. The dirty flue gas flows in the same upward direction and through the packed bed along with the liquor. In passing through the packing, the high-velocity gas atomizes the liquor, causing the dust particles and liquor droplets to collide. The entrained agglomerated dust particles and liquor droplets pass through the bed of marbles, where the droplets on which the dust particles are entrapped are collected and drained to a sump. The cleaned flue gas then enters the entrainment separator before being discharged to atmosphere.

The cross-flow type scrubbers (Fig. 5.14d) are commonly used for removing dust particles as well as gaseous pollutants. In the cross-flow scrubber, the flue gas moves in the horizontal direction through the packing and is contacted by scrubbing liquor flowing down through the packing.

The liquor flow requirement is almost half of that required for countercurrent flow configuration, and the gas side pressure drop is about one-third of that in a countercurrent flow configuration.

The cross-flow packed scrubber can handle dust more easily than the countercurrent flow scrubber, as the collected dust deposits near the front of the bed where it can be washed off by cleaning sprays. The first few inches of the bed are normally more heavily irrigated than the rest of the bed to reduce buildup of solids.

A separate mist eliminator is not necessary, because the last 6 to 12-inch section of the bed is maintained dry and, thus, acts as a demisting section. The packed bed, facing the gas flow, is usually inclined in the direction of the gas flow, so that the face of the bed can be completely wetted and washed.

The packed scrubbers are normally used for absorption of gaseous pollutants and for mist removal but are also effective in removing solid particulates by impingement or interception, although they usually require extensive maintenance in these services due to plugging problems. The cross-flow type unit has a better ability to handle dust because it is easily cleaned.

The packed scrubbers are not used to handle flue gas with dust loadings higher than 5 grains per cubic foot and they are effective in collecting larger than 2 to 3 μ particles.

Floating Bed Scrubbers

In the floating bed scrubbers, the packing is kept in motion; and the flow configuration is always countercurrent with the gas flowing upward and the liquor downward. This category includes the floating bed scrubbers,[2,3] the turbulent contact absorbers (TCA)[2,3] and the polysphere gas scrubbers.[4]

These scrubbers were originally developed to answer a need of the aluminum industry to absorb fluorides from flue gas containing tarry substances.

In the floating bed scrubbers, the packing is low density spheres of polyethylene or polypropylene and is continually in motion between the upper and lower retaining grids (Fig. 5.14f).[2,3] The flue gas flowing upward at a critical velocity keeps the lightweight spheres in turbulent motion. The random, turbulent motion of the spheres creates intimate mixing between gas and liquor not only on the wetted surfaces of the spheres, but throughout the entire spray zone. The tumbling actions of the spheres in motion provide continuous cleaning, thus preventing solids buildup and plugging of the bed.

The flue gas velocity is very critical and is in the range of 7 to 13 feet per second. If the velocity falls below the minimum limit, the spheres will settle down on the bottom supporting grid, as they no longer can remain in suspension, whereas if the velocity exceeds the maximum, the spheres will adhere to the top retaining grid.

The floating bed scrubber can be provided with one or more stages, each with a supporting grid and retaining grid where the latter becomes the supporting grid for the next stage. In the floating bed scrubbers, the grids are spaced about 18 inches apart, and the static bed of spheres in each stage is about 12 inches deep. The pressure drop per stage is about 3 to 4″ H_2O, column, when the flue gas velocity and liquor flowrates are in the range of 7 to 10 feet per second and 10 to 20 gallons per minute per square foot.

The TCA is essentially a modified floating bed scrubber, in that the grids are about 4 feet apart while the static bed of spheres is only 8 to 12 inches deep (Fig. 5.14g). By increasing the distance between the supporting and retaining grids, increased liquor and gas flowrates can be attained at pressure drops equivalent to those obtained in the floating bed scrubber.

The advantages of the mobile packing concept are: (1) no plugging, fouling or channeling, (2) high gas and liquor throughputs, (3) wider operating range and (4) higher transfer rates.

The floating bed scrubbers are most suitable to collect viscid material, or materials which form gummy precipitates and gel after being scrubbed from the flue gas.

Successful applications include:

1. Removal of dust from flue gases leaving the reduction furnaces in the metal refining operations. This dust becomes sticky when contacted with scrubbing water.

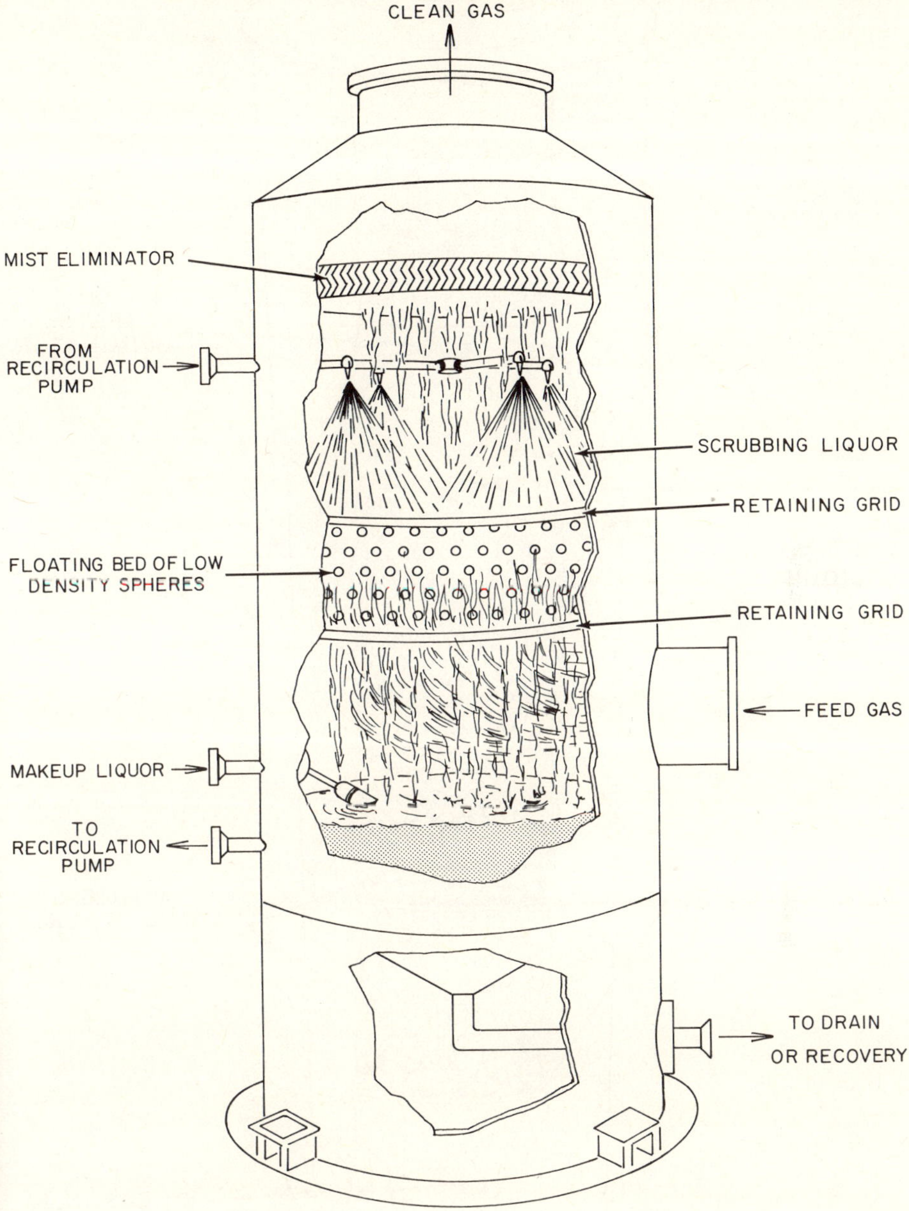

Figure 5.14f Floating-ball (fluid-bed) packed scrubber.

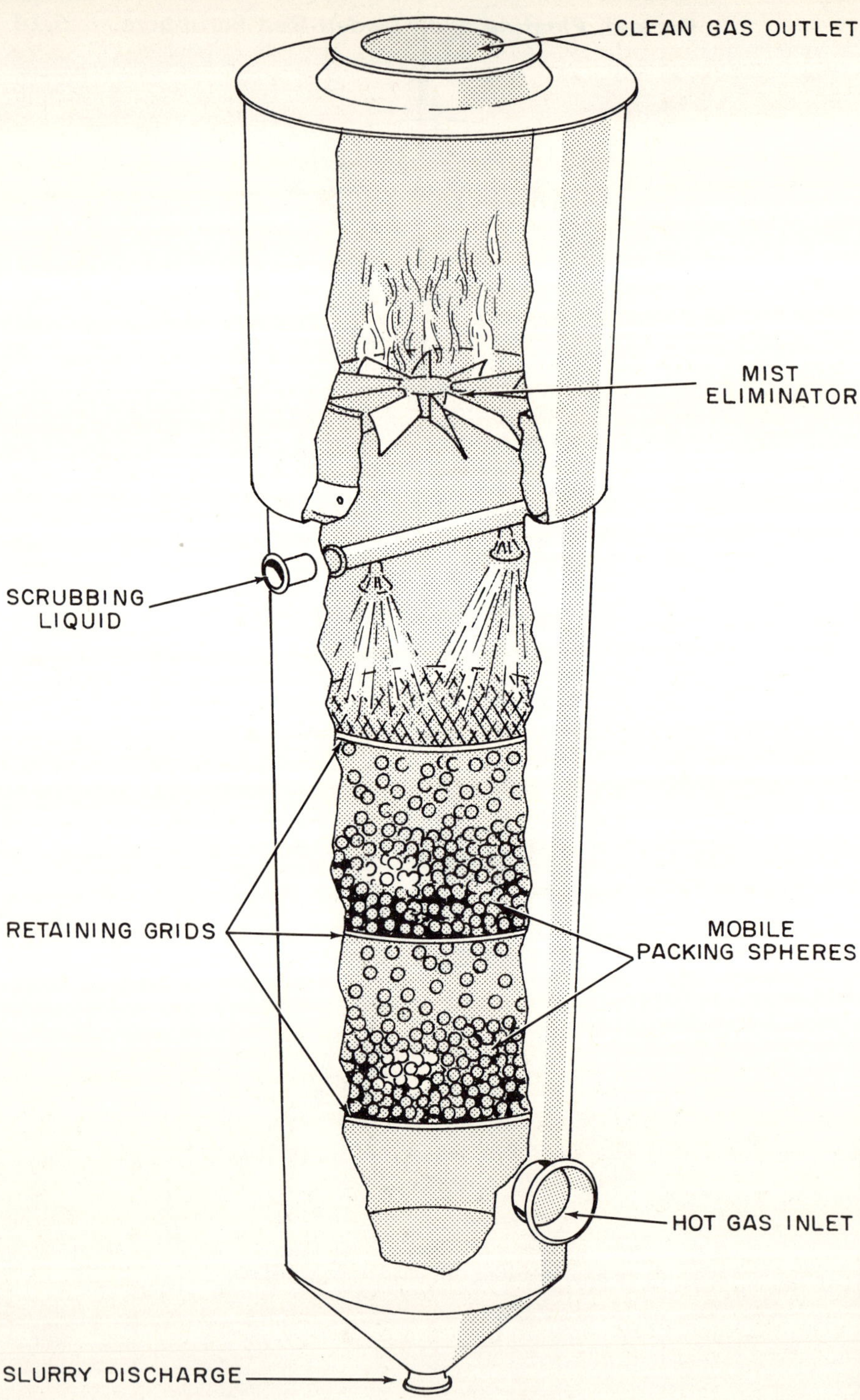

Figure 5.14g Turbulent contact absorber wet scrubber.

2. Removal of free chlorine, aluminum chloride and aluminum oxide dust, from off gases evolved during chlorination of molten aluminum. The dust particles, in the form of extremely dense submicron fume, precipitate into a thick slurry when scrubbed with caustic solution.
3. Collection of hydrogen fluoride, fluoride dust and a tar mist produced during electrolytic reduction aluminum.

The floating bed scrubber is being used to simultaneously remove both flyash and sulfur dioxide from flue gas leaving coal-fired boilers, but the breakage of plastic spheres due to flyash erosion is a serious problem.

Other applications include the removal of gaseous and particulate pol-

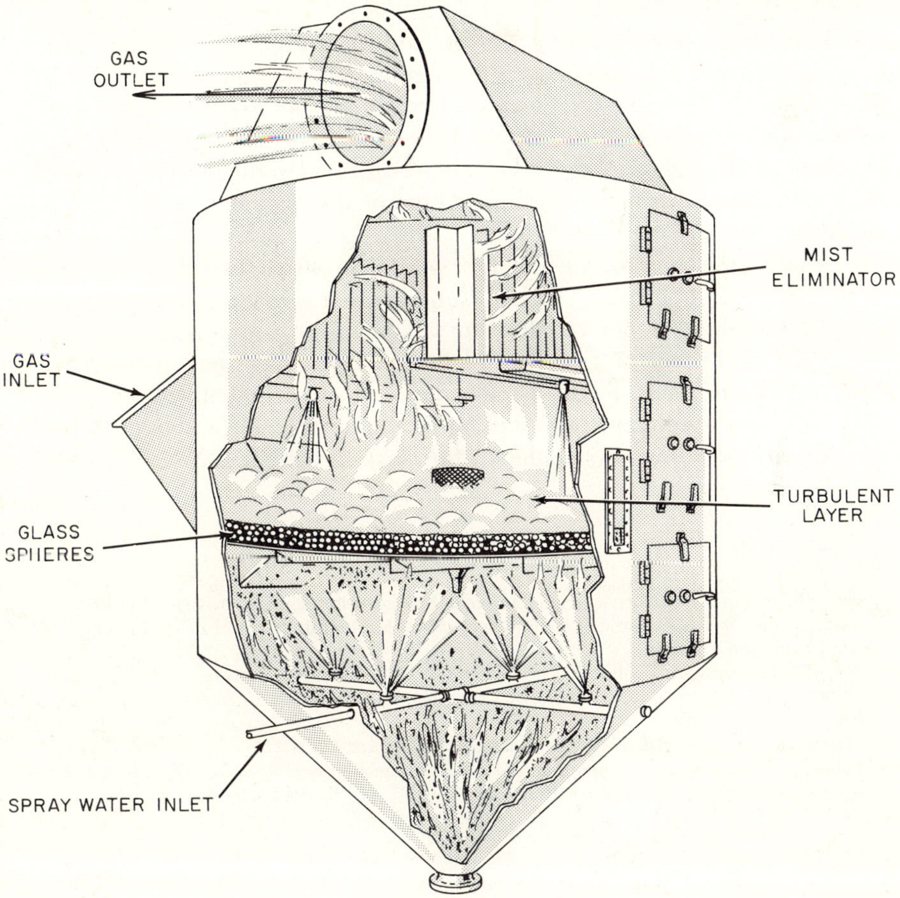

Figure 5.14h Flood-bed scrubber.

lutants in the phosphate fertilizer industry, and the removal of other gaseous pollutants such as hydrogen sulfide, chlorine and gas cooling.

Collection Efficiency

The floating bed scrubbers are found to be effective on 1 μ and larger dust particles. Hydro-filter[5] is reported to provide collection efficiencies in excess of 99% for 2 μ and larger particulate matter. Norblo® Poly-Sphere[4] provides 99% collection efficiency, on a flue gas stream containing 56% minus 10 μ particles. With the use of single stage and two stage TCA scrubbers[3], efficiency of 95% and 98%, respectively, can be attained on flyash from coal-fired boilers. Pilot plant work[6] carried out at a power plant showed flyash collection efficiency of 95 to 99%, with a two-stage TCA scrubber.

Flooded Bed Scrubbers

In the flooded bed packed scrubbers (see Fig. 5.14h), the flow configuration is countercurrent with the flue gas flowing up and the liquor down. The dirty flue gas enters below the bed of spherical glass marbles, passes through a spray section and up through the flooded bed of spheres. The packing usually consists of a 6-inch bed of glass marbles. The bubbles formed in the bed create a 6-inch deep turbulent layer. The constant motion of the packing provides the self-cleaning action.

A 99% efficiency for 2 μ or larger particles, with a 4 to 6″ H_2O pressure drop, has been reported.[5] The high solid content slurry can be recirculated; and the scrubbing liquor requirement is low, approximately 2 to 3 gallons per 1,000 cubic feet of gas. The flooded bed scrubbers have been used to remove carbon black, flyash, cupola dust, other particulate matter and to remove gaseous pollutants and acid mists. In performance and operation, the flooded bed scrubbers are similar to the floating bed types.

REFERENCES

1. Ceilcote Company Advertisement, *Chemical Engineering*, p. 95, February 10, 1969.
2. Tomany, James P.: "A Guide to the Selection of Air Pollution Control Equipment," UOP Air Correction Division.
3. "UOP Wet Scrubbers," Bulletin No. 608, UOP Air Correction Division.
4. "Norblo Poly-Sphere Gas Scrubbers," Catalog No. PS-120, Northern Blower Division of Buell Engineering Co., Inc.
5. "Hydro Filter," Bulletin N-20, National Dust Collector Corp.
6. Pollock, W. A., James P. Tomany and Garry Frieling: Removal of Sulfur Dioxide and Flyash from Coal Burning Power Plant Flue Gases, ASME publication, 66-WA/CD-4.

5.15 CYCLONIC AND CENTRIFUGAL SCRUBBERS

Types of Designs: (a) Pease-Anthony® cyclonic scrubber (Fig. 5.15b); (b) Multiwash scrubber (Fig. 5.15d); (c) Cyclonic spray scrubber (Fig. 5.15a); (d) Centrifugal scrubber with vaned baffles (Fig. 5.15c); (e) Irrigated cyclone (Fig. 5.15f); (f) Cyclone scrubber with helical baffles (Fig. 5.15e).
NOTE: In the feature summary below, the letters "a" to "f" refer to the above listed scrubber designs.

Principle of Operation: Centrifugal motion of spinning flue gas increases the velocity difference between liquor droplets and the flue gas. Centrifugal motion is initiated by tangential entry, vanes, impellers and helical baffles.

Pressure Drop Requirement: 1 to 8″ H_2O.

Liquor Flow Requirement: 4 to 10 GPM/1,000 ACFM.

Spray Nozzle Pressure: 50 to 400 psig.

Collection Efficiency: 90% + on particles 2 μ or larger. See Table 5.15g for type "a" and Table 5.15h for type "c."

Applications: See Tables 5.15g and 5.15h for types "a" and "c." Other typical applications include aluminum and magnesium dusts, kilns, dryers, pigment dusts and fumes.

Cost: \$0.50 to \$1.50 per cubic foot of gas handled if mild steel construction is used.

5.15 Cyclonic and Centrifugal Scrubbers

Partial List of Suppliers: Buffalo Forge Co. (c); Chemical Construction Corp. (a); Claude B. Schneible Co. (b); also refer to Section 5.12 for additional suppliers.

The cyclonic and centrifugal scrubbers are more efficient than the spray towers and spray chambers. The flue gas is introduced tangentially into all cyclonic and centrifugal scrubbers. The centrifugal force of a spinning gas stream increases the velocity difference between that of the liquor droplets and the flue gas stream, which results in higher collection efficiency.

By the tangential entry into the scrubber, a rising vortex of gas is formed, into which the scrubbing liquor is sprayed. The dust particles get collected on liquor droplets as a result of impaction and interception. The centrifugal force accelerates both dust particles and liquor droplets to the vessel wall, where collection of dust particles takes place. The wetted wall also enhances the collection.

The scrubbing liquor is introduced into the gas path by spraying through nozzles mounted on the walls of the scrubber (Fig. 5.15a)[5] or from nozzles mounted on a pipe in the center of the vessel (Fig. 5.15b).[1] The spinning motion is initiated by introducing the flue gas tangentially (as shown in Fig.

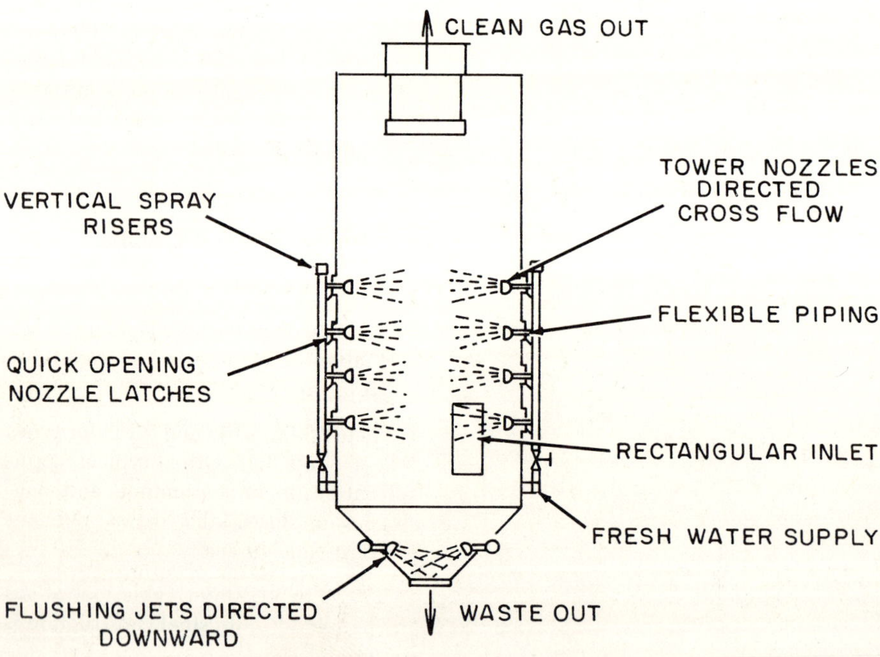

Figure 5.15a Cyclonic spray scrubber.

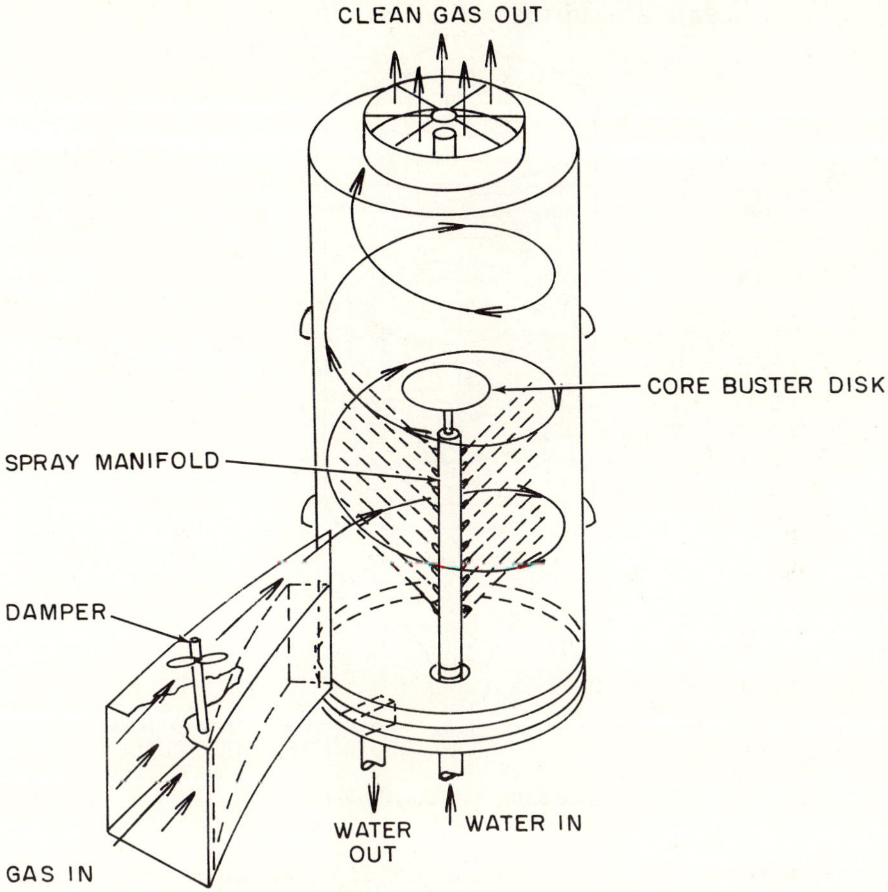

Figure 5.15b Pease-Anthony cyclonic scrubber.

5.15b) or by having the scrubbing liquor introduced (or sprayed) tangentially inward from the wall (Fig. 5.15a). This imparts a centrifugal motion to the flue gas. In a third method, the spinning motion is given by fixed vaned baffles (Fig. 5.15c) or by fixed vanes and impellers (Fig. 5.15d).[3] Another method is to use helical baffles to prolong the centrifugal action of the flue gas (Fig. 5.15e).

In the cyclonic and centrifugal scrubbers, the fine spray droplets are trapped in the spinning gas stream and are thrown upward toward the walls of the scrubber by centrifugal force. These droplets collide with and collect the dust particles. The scrubbing liquor with the collected particles runs down the walls to the scrubber bottom and out, whereas the cleaned gas leaves through the top.

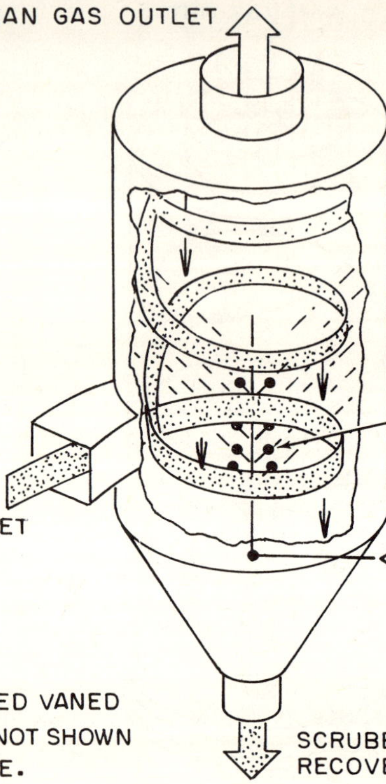

Figure 5.15c Centrifugal scrubber.

The pressure drop ranges from 1 to 8″ H_2O, and the scrubbing liquor rates vary from 4 to 10 GPM per 1,000 ACFM of gas handled. The tangential velocities at the entrance are in the range of 100 to 120 ft/sec. The scrubbing liquor is introduced at pressures of 50 to 400 psig. The smaller the particles (down to 1 μ size), the higher the pressure (up to 400 psig) used. These scrubbers are inadequate for the collection of submicron range particles.

Low pressure drops on the gas side, low liquor flowrate requirements and ease of fabrication and maintenance are some of the advantages of cyclonic scrubbers; whereas inability to handle scrubbing liquors containing solids (due to nozzle plugging), inadequate removal of submicron particles and the requirement for higher liquor supply pressures are the disadvantages.

The main types of cyclonic and centrifugal scrubbers are illustrated in Figures 5.14a, b, c, d, e and f.

Pease-Anthony Cyclonic Scrubber

The Pease-Anthony cyclonic scrubber (Fig. 5.14b) can remove dusts and liquid aerosols which are 1 μ or larger in size.[1,2] It is sometimes used in series

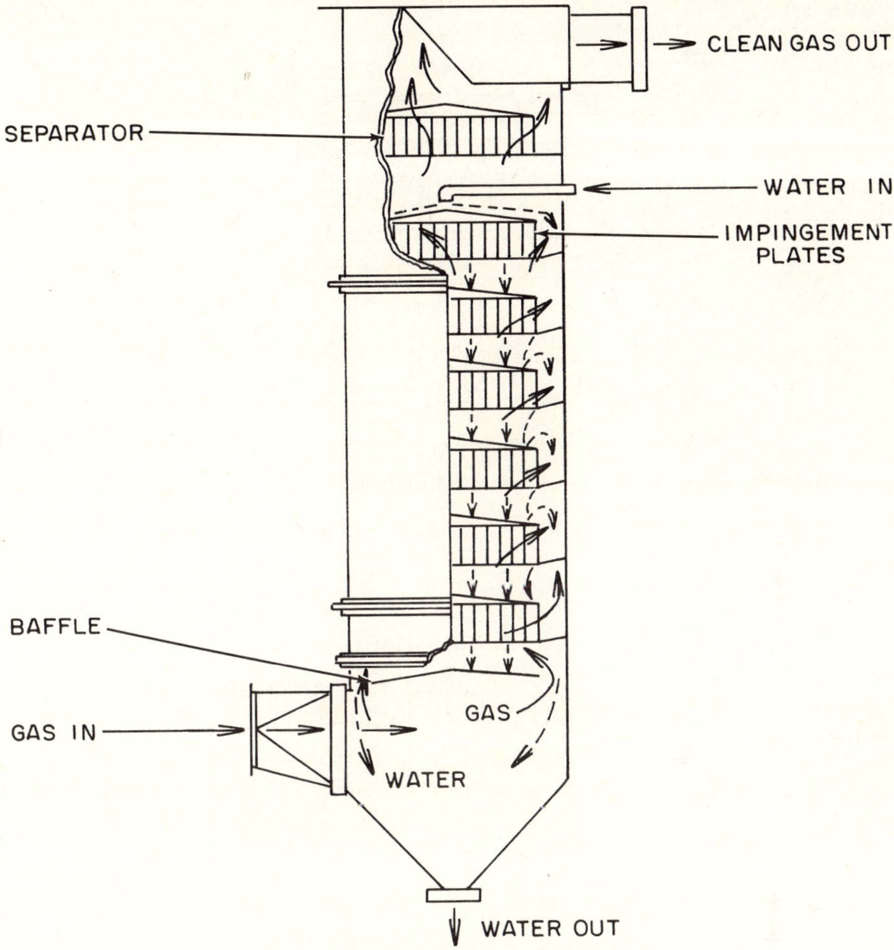

Figure 5.15d Multiwash scrubber.

with the venturi scrubber on applications where secondary scrubbing action is required.

In the Pease-Anthony scrubber, a finely atomized liquor spray is passing through a spinning gas stream. The flue gas enters tangentially at the bottom of the scrubber and the liquor spray is introduced from a manifold in the lower part of the unit. The spray nozzles are in the center of the scrubber but, if necessary, can also be located near the walls. The tangential velocity at the entrance is very critical to maintain the spinning action of the flue gas. A damper can adjust the velocity if the gas flow to the scrubber varies. The upper half of the scrubber vessel acts as a cyclonic separator which makes a supplementary entrainment separator unnecessary.

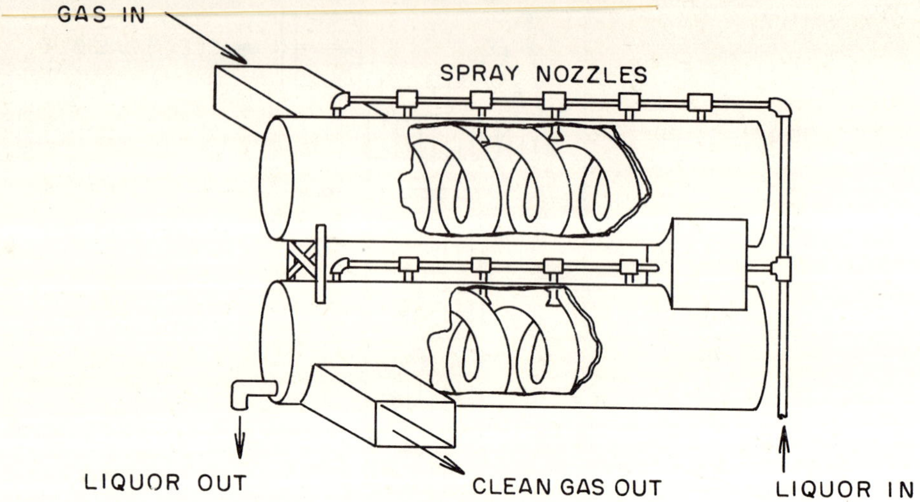

Figure 5.15e Cyclone scrubber with helical baffles.

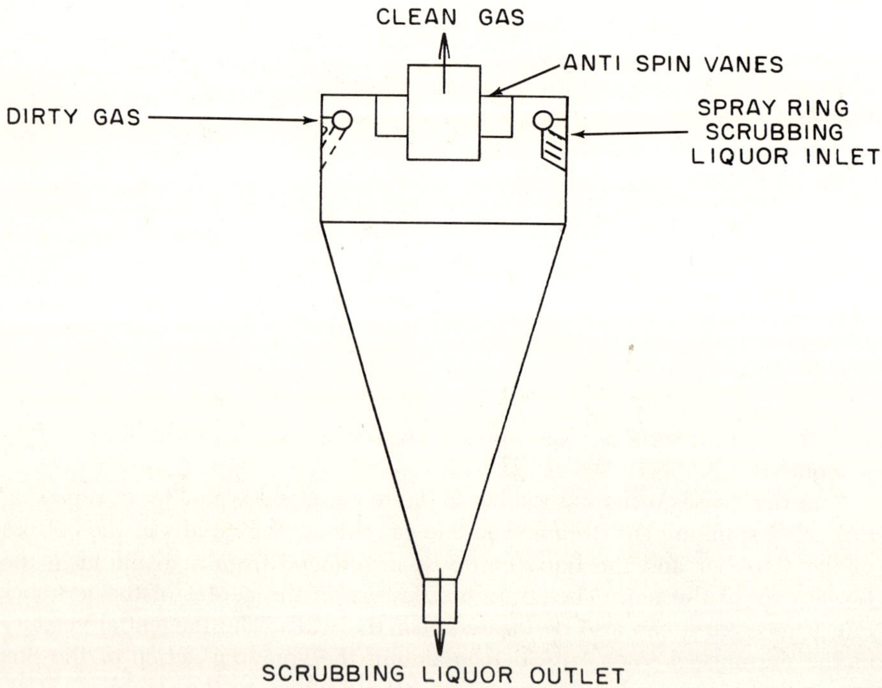

Figure 5.15f Irrigated cyclone.

The P-A (Pease-Anthony) cyclonic scrubber has been successfully applied to blast furnace gas, kraft recovery furnace flue gas and to lime kiln flue gas services. In addition to being an efficient remover of 1 μ and larger dusts, it has a low gas side pressure drop (2 to 4 inches H_2O) and a low liquor flow requirement (5 to 10 GPM/1,000 ACFM). The maintenance and initial costs are also reasonably low. Table 5.15g shows some of the performance data for the P-A cyclonic scrubber.

Multiwash Scrubber

The Multiwash scrubber[3,4] is shown in Figure 5.15d. This scrubber has been successfully applied in all types of foundry operations and in industrial and chemical processes, including alumina and magnesium dusts, exhaust gases

Table 5.15g
PERFORMANCE DATA OF THE PEASE-ANTHONY CYCLONIC SCRUBBER

Source of Gas	*Type of Dust (Particle Size Range, μ)*	*Dust Loading grains/cu ft*		*Removal Efficiency, %*
		Inlet	*Outlet*	
Lime kiln (raw stone)	Lime (2–40)	9.2	0.08	99
Lime kiln (Kraft mill)	Lime and soda (1–25)	7.7	0.25	97
Reverberatory (lead furnace)	Lead compounds (0.5–2)	—	—	97
Blast furnace (iron)	Iron ore and coke (0.5–20)	3–24	0.03–0.08	99
Rotary dryer	Ammonium nitrate	—	—	99+

Source of Gas	*Type of Mist*	*Mist Loading, mg/cu ft*		*Removal Efficiency, %*
		Inlet	*Outlet*	
Superphosphate den and mixer	Fluorine compounds	312.0	7.2	97.8
Air bodying of castor oil	Castor oil	6.3	1.4	78.0

from various kilns, dryers, pigment dusts and fumes, nitrates and phosphate dusts.

In the multiwash scrubber, the dust-laden flue gas enters at the bottom, just above the cone. The cone serves as a wet cyclone, collecting the larger and heavier dusts and creating a vortex which sends the flue gas upward in a spiraling motion. The fixed vanes and impellers also enhance the rotating motion given to the flue gas. The scrubbing liquor is introduced just below the entrainment separator stage and cascades evenly over the impingement stage elements, carrying the collected material to the slurry outlet. The velocity of the flue gas induces a turbulent action into the scrubbing liquor which breaks down into finely divided spray patterns. This intimate contact and impingement causes the collection of the dust particles.

The minimum liquor requirement is 3 GPM per 1,000 ACFM; but in case of the higher dust loadings, 4 to 5 GPM per 1,000 ACFM are recommended. The scrubbing liquor thoroughly cleans the scrubber during operation and it can be recycled. The pressure drops vary from 2 to 12″ H_2O.

In this scrubber, the arrangements of vanes, deflectors and shelves are such that they provide not only the maximum impingement area but also a longer time of contact between the flue gas and scrubbing liquor by providing a lengthy, tortuous path for the flue gas. This greater impingement area and the longer contact time result in higher collection efficiency of dust particles.

The multiwash scrubber is 99% efficient in collecting particles 2 μ and larger in size. For particles 1 μ and larger, its efficiency is over 90%. As the multiwash unit is constructed in stages, more stages can be added to increase the contact time and, hence, the collection efficiency.

Cyclonic Spray Scrubber

In this scrubber,[5] shown in Figure 5.15a, the scrubbing liquor is tangentially sprayed inward from the scrubber wall and, thus, imparts the centrifugal motion to the flue gas.

This scrubber is also called a "hydraulic scrubbing tower." It is a high efficiency, wet centrifugal collection device which utilizes a finely atomized spray for particle conditioning and an inertial separation of the contaminant from the flue gas stream. The particle conditioning occurs when spray droplets move radially through the spinning gas and agglomerate with dust particles. The upper portion of the vessel acts as the cyclonic entrainment separator.

The scrubbing liquor requirements are approximately 5 gallons per 1,000 cubic feet of gas, and the liquor is introduced through nozzles at a pressure of 100 to 400 psig. Approximately 40% of the spray nozzles in the scrubber are located in such a manner that they cover the gas inlet section with a dense concentration of spray droplets. The remainder of the spray nozzles are distributed uniformly throughout the lower half of the cylindrical section of the scrubber. The high-pressure spray nozzles have small orifice openings and

Table 5.15h

PERFORMANCE DATA FOR THE CYCLONIC SPRAY SCRUBBER[5]

Source of Gas	Dust	Particle Size Range, μ	Dust loading		Efficiency, grains/cu ft	Pressure Drop, inches H_2O	Spray Pressure psig
			Inlet	Outlet			
Open galvanizing operation	Ammonium and zinc ammonium chloride	0.3–2.0; mostly below 1.0	0.1	⩽0.01	90%+	2.0	400
Aluminum chloride production	Metallic fume, aluminum chloride and hydroxide	0.3–0.5	1.0–3.5		88–93.5%	1.0	
Forge press smoke exhaust fumes	Dense smoke, oil mist and graphite particles				90%+	1.5	230 GPM at 100 psig
Acetylene production lime hydrate plant	Calcium hydroxide, coke, calcium carbonate, calcium oxide and calcium carbide	89% less than 10μ and over 5% less than 1 μ	4.5–5.0		95%+		360
Coke drying	Coke dust	9% less than 10 μ	2.0–3.4		98%+		400
					96%+		160
Composite efficiency based on particle size distribution		0–0.5			78.6		400
		0.5–1.0			83.4		
		1.0–2.0			89.7		
		2.0–3.0			94.0		
		3–4			96.4		
		4–6			97.5		
		6–8			98.6		
		8–10			99.2		

recirculation of the scrubbing liquor is not possible unless the solids content of the liquor is first removed.

Collection efficiencies as high as 96% for particles 2 μ or larger, at operating pressure drops of 1.3 to 2.3″ H_2O have been reported. At approximately 2″ H_2O pressure drop and using 400 psig water spray pressure, a collection efficiency of 90%+ has been attained in the galvanizing operations where ammonium and zinc ammonium chloride fumes are emitted, with a particle size range of 0.3 to 2.0 μ and most of the material below 1 μ in size. Performance data are presented in Table 5.15h, which also includes composite efficiencies based on particle size distribution. Applications include phosphate rock drying, pulverized coal handling and thermal catalytic reactors.

Centrifugal Scrubber with Vaned Baffles

This scrubber is shown in Figure 5.15c. Its operation is similar to that of the cyclonic spray scrubber, except that the liquor is sprayed from a central spray arrangement and the fixed vanes enhance the spinning and spiraling motion of the flue gas.

Irrigated Cyclone

This is essentially a dry cyclone which is converted to a wet scrubber (see Fig. 5.15f) by the introduction of a scrubbing liquor. Both gas and liquor enter tangentially at the top of the cyclone. The wetted walls primarily collect the dust particles from the flue gas. Best collection is obtained by spraying the liquor countercurrent to the gas flow. The liquor flow requirements are in the range of 5 to 15 gallons per 1,000 cubic feet of gas, and the pressure drop is in excess of 3″ H_2O.

Cyclone Scrubber with Helical Baffles

In this scrubber (Fig. 5.15e), the prolonged centrifugal action is imparted to the flue gas by using helical baffles. The spray contact time is increased by the use of multiple spray nozzles.

REFERENCES

1. "Chemico P-A (Pease-Anthony) Gas Scrubbers," Chemico Bulletin M-102, Chemical Construction Corporation.
2. "Chemico Gas Scrubbers For Industry," Bulletin M-104, Chemical Construction Corporation.
3. "Multiwash Collectors and Scrubbers," Technical Bulletin No. 266, Claude B. Schneible Co.
4. Bulletin Nos. 462 and 861, Claude B. Schneible Co.
5. "Hydraulic Scrubbing Towers," Bulletin AP525A, Buffalo Forge Company.

5.16 IMPINGEMENT OR ORIFICE TYPE SCRUBBERS

Types of Designs:	*Impingement Scrubbers:* W-D Tuyere scrubber (Fig. 5.16a); Impinjet® scrubber (Fig. 5.16b); Peabody direct-contact scrubber (Fig. 5.16d). *Orifice Type Scrubbers:* Turbulaire® scrubbers (Fig. 5.16e); Circular-wedge scrubbers (Fig. 5.16f); Rotoclones (Fig. 5.16g). *Self-Induced Spray Scrubbers* (Fig. 5.16h): Doyle® scrubber; Schmieg® swirl-orifice scrubber; Blaw-Knox® liquid vortex scrubber.
Principles of Operation:	*Impingement Scrubbers:* High-velocity gas while passing through orifices impinges with a layer of scrubbing liquor on a baffle. *Orifice Scrubbers:* High-velocity gas impinges on a pool of scrubbing liquor. *Self-induced spray scrubbers:* Same as orifice scrubbers, except the orifice is partially submerged.
Pressure Drop Required:	2 to 30″ H_2O.
Liquor Flow Requirement:	2 to 5 GPM/1,000 ACFM.
Collection Efficiency:	90 to 99%+ for 1 μ or larger, and 90%+ for below 1 μ, for impingement type scrubbers. Orifice type and self-induced type are efficient on coarser dust.
Applications:	Flyash, oil mists, ammonium nitrate dust, dust from cupola and coal dryers, pyrites roaster, blast furnace, sinter machines, crushers and roller mills, dryers and calciners and lime kiln.

Costs: See Table 5.21f.

Partial List of Suppliers: See Section 5.12.

In the impingement or orifice type scrubbers, the high-velocity flue gas, impinging on the liquor surface, results in an intimate contact between the two phases. The flow of flue gases through a restricted passage partially filled with scrubbing liquor, results in the dispersion of the liquor. The centrifugal forces, impingement and the turbulence caused, accomplish the wetting of the particles and their collection.

In the impingement plate scrubbers, a preagglomerating stage is sometimes provided to enhance the collection efficiency of fine particles. In the self-induced spray scrubbers, a spray curtain is induced by gas flow through a partially submerged orifice or through streamlined baffles.

These are low-pressure drop scrubbers and are efficient on dust particles larger than 1 μ.

Impingement Scrubbers

W-D Tuyere Scrubber

This scrubber[1] is schematically shown in Figure 5.16a and employs a washing tuyere at the bottom and a drying tuyere at the top. The scrubbing liquor is introduced into the lower section of the scrubber in the form of sprays. The flue gas and the sprayed liquor are thoroughly mixed, while passing through the blades of the scrubbing tuyere. The dust particles and the spray droplets impinge on the blades. The presence of spray droplets (about 10 μ in size) enhances the collection of fine dust. The upper tuyere separates the liquid from the flue gas and, in addition, performs a certain amount of secondary scrubbing.

This scrubber is primarily used to remove acid fumes from chlorine and sulfuric acid manufacturing plants and sulfides from refinery gases. It is also used to remove flyash from boiler flue gases and to remove oil mists from food processing plants.

Impinjet Scrubber

The Impinjet scrubber[2] (Fig. 5.16b) utilizes an impingement baffle plate as the contacting device between flue gas and scrubbing liquor.

The Impinjet first cools and humidifies the entering flue gas with the use of a spray section. Here, in addition to the collection of the larger dust particles, condensation and agglomeration of fine particles also occur. The under-plate spray action also keeps the plate and walls clean and prevents buildup. The cooled, humidified flue gas then enters the impingement baffle plate on which a uniform liquid level is maintained by the use of an adjustable weir. The impingement causes a jet-swirl interaction between the flue gas

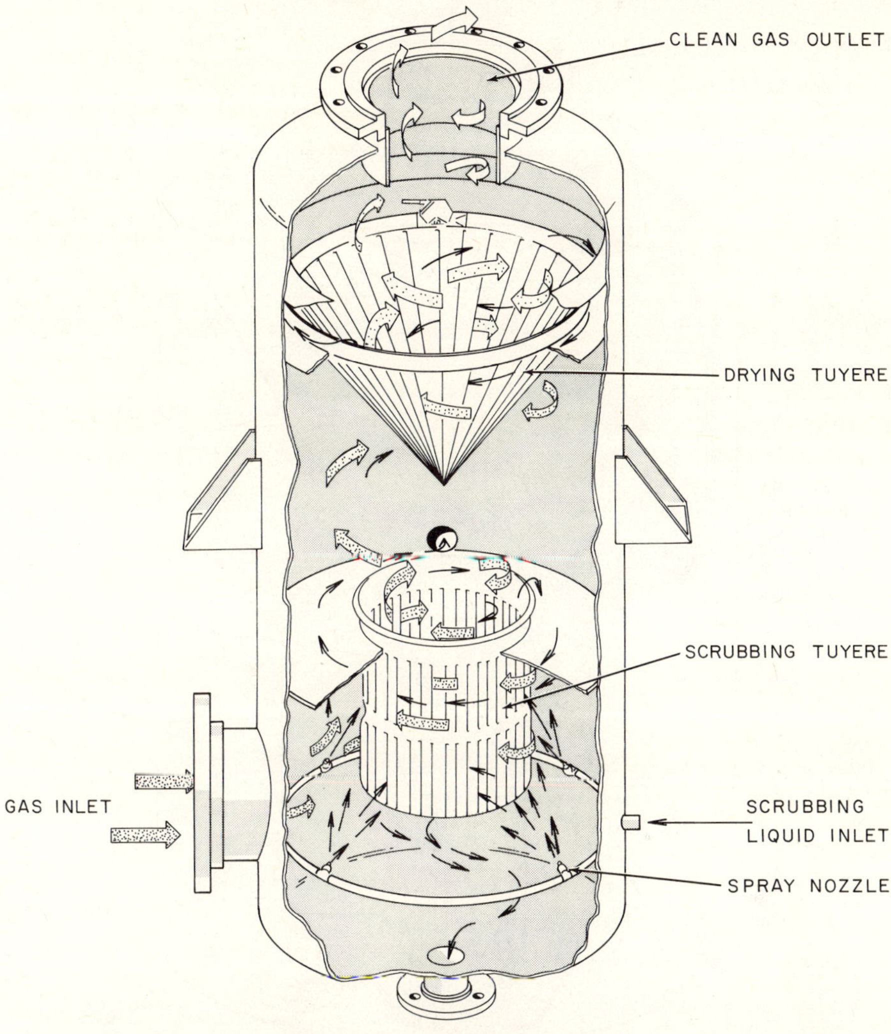

Figure 5.16a W-D tuyere scrubber.

and scrubbing liquor. This turbulent effect provides for the thorough wetting of dust particles and, hence, for efficient collection.

The formation of minute liquor droplets at the plate orifices, and the higher difference in velocities of droplets and dust particles, results in higher collection efficiency.

The liquor requirement for the spray section is usually 0.5 to 1.0 gallon at 20 psig per 1,000 cubic feet of gas; and for the impingement plate, it is approximately 1 to 3 gallons per 1,000 cubic feet of flue gas. The quantity

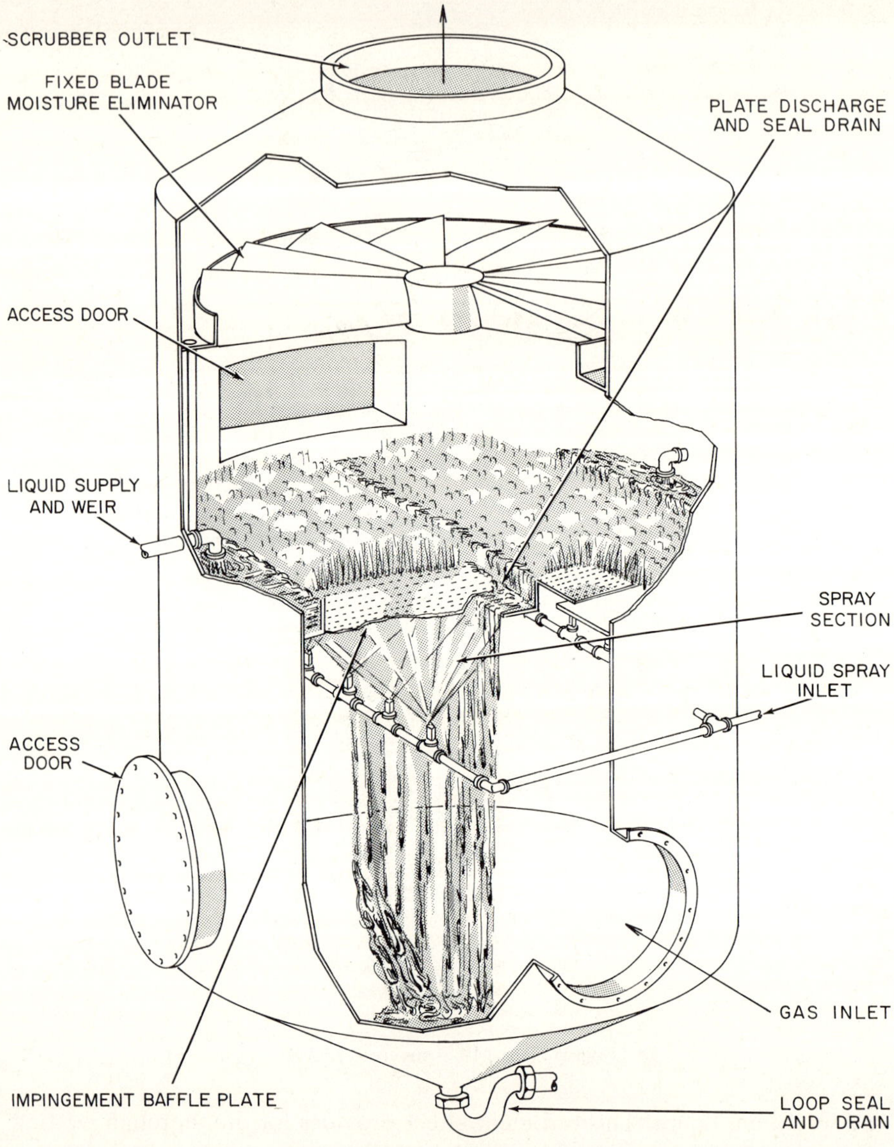

Figure 5.16b Impinjet scrubber.

of water on the plate is dependent on the inlet dust loading. The liquor can be recycled provided that the solids concentration does not exceed 10% by weight.

The Impinjet scrubber removes dust particles ranging from submicron fumes to micron size. Figure 5.16c[2] shows the relationship of collection effi-

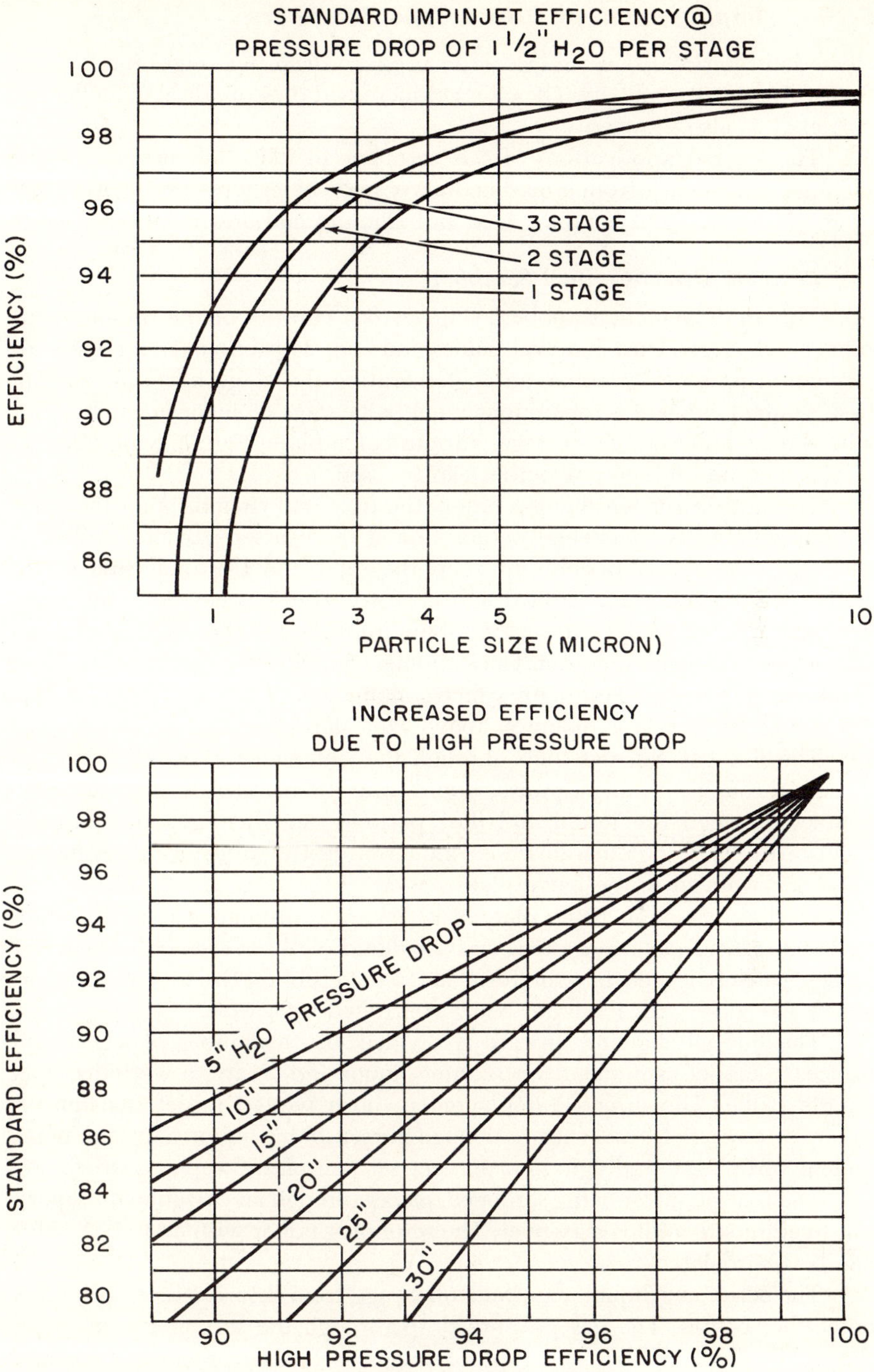

Figure 5.16c Particle size and pressure drop effect on Impinjet scrubber efficiency.

ciency and particle size, at $1\frac{1}{2}''$ H_2O pressure drop per stage, for 1, 2 and 3 stage Impinjet scrubbers. The effect of increased pressure drop on collection efficiency is also shown.

The typical applications of this scrubber include the removal of ammonium nitrate dust from a fluid bed dryer and cooler, dust from cupola flue gases, dust from kilns handling lead and zinc ore and dust from coal dryers.

Peabody Direct-Contact Scrubber

This impingement scrubber[3] (Fig. 5.16d) consists of one or more impingement stages, mist removal baffles and a spray section. To agglomerate submicron particles or fume, an agglomerating slot stage is employed. The basic contact element is the impingement baffle stage alone or in combination with the agglomerating slot stage. Various baffle plate stages having different orifice sizes and flue gas velocities can be used.

The hot dust-laden flue gas enters the base and rises through the spray section, where it is quenched, saturated and the coarser particles collected. The spray section also helps to keep the shell and the underside of the agglomerating slot stage flushed and cleaned. The flue gas then enters the agglomerating slot stage where the submicron particles, fumes, certain dusts, tars or resins are agglomerated. The agglomerating slot stage comprises a liquid blanketed horizontal diaphragm, which contains one or more rectangular slots. A rectangular nozzle is mounted above each slot.

The flue gas can pass only through the slots and the nozzles. The gas passing through the slots aspirates liquid from the blanket, forcing turbulent contact between the liquid and dust particles and between the particles themselves. This agglomerates the submicron dust or fume particles, before they are removed in the baffle plate stage.

The impingement baffle plate stage comprises a plate having from 600 to 3,000 orifices per square foot and a baffle directly over each orifice. The baffles are submerged in a uniform layer of scrubbing liquor, which flows across the stage with its depth being controlled by a weir.

The flue gas flowing upward is divided into a number of jets by the orifices. Each jet aspirates the scrubbing liquor and creates a wetted surface on the baffle. The directed impingement on a wetted target dynamically precipitates particles and entraps them in the scrubbing liquor. The continuous violent turbulence and agitation prevent settling of entrapped particles.

The flue gas, after being scrubbed, passes through an entrainment separator, to eliminate liquid carry-over. The scrubbing liquor with up to 65% solids can be recirculated.

The scrubbing liquor requirements range from 3 to 5 gallons per 1,000 cubic feet of gas. The gas pressure drop through the scrubber depends on the application, number and type of stages used. The pressure drop for a

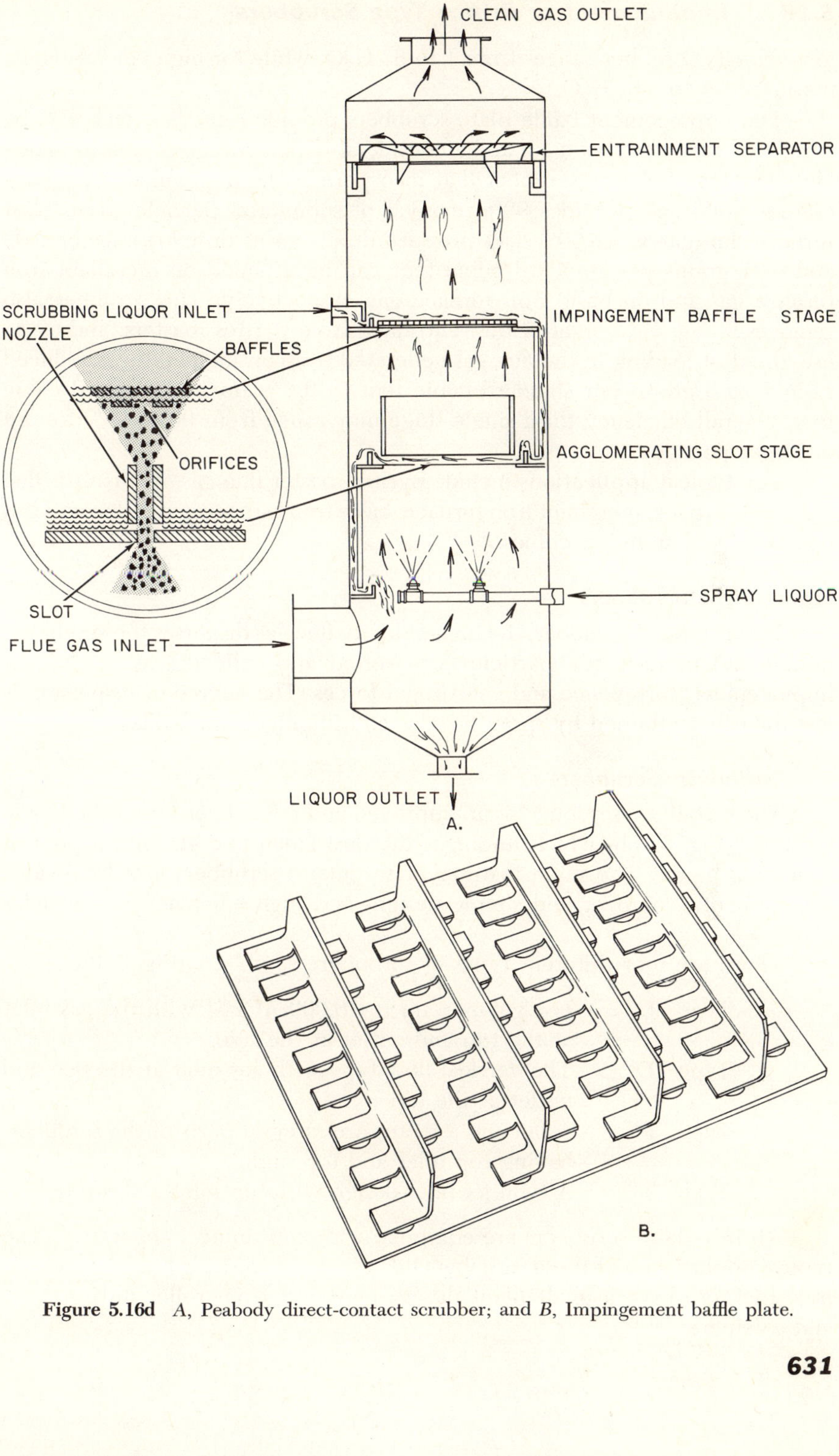

Figure 5.16d *A*, Peabody direct-contact scrubber; and *B*, Impingement baffle plate.

low-velocity scrubber ranges from 1 to 8″ H_2O, while for high velocity units it can be up to 40″ H_2O.

The impingement baffle plate scrubbers provide removals over 97% by weight of particles above 1 μ and a considerable percentage of smaller particles. The high-velocity scrubber that incorporates the high-velocity agglomerating slot stage provides 99% removal of submicron particles from blast furnace flue gases. At 30″ H_2O pressure drop, outlet dust loadings of 0.02 and 0.01 grains per standard cubic foot can be attained on merchant iron furnace gas and on basic iron furnace gas, respectively. This scrubber also removes 99%+ by weight of dust entrained from pyrites roasters; and by its use, the dust loading in the flue gas generated from an incinerator is reduced from 2 to 5 grains per standard cubic foot to 0.2 grain per standard cubic foot. Overall efficiency for a single stage may range from 90 to 98% for 1 μ or larger particles.

The typical applications include pyrites roaster flue gas, incinerator flue gas, blast furnace, merchant iron furnace, basic iron furnace gases, dryers, kilns flue gases and sinter machines.

Orifice Type Scrubbers

In this type of scrubber, the high-velocity flue gas disperses the scrubbing liquor; and in turn, the particles are wetted and collected as a result of impingement, turbulence and centrifugal forces. The degree of dispersion is less than that attained by spray nozzles and impingement baffles.

Turbulaire Scrubbers

The turbulaire scrubber[4] is an improved and refined version of the Doyle scrubber (Fig. 5.16h) which impinges the dust-laden flue gas into a pool of scrubbing liquor. The main features of turbulaire scrubbers are: low water requirements, compact and simple designs and high efficiency on particles above 1 μ.

There are four different types of turbulaire scrubbers (Fig. 5.16e):

- Type "C" — For volumes up to 18,000 ACFM with the gas inlet at the top and outlet at the side.
- Type "D" — The vertical flow type with gas inlet at the side and outlet at the top.
- Type "F" — A fan and a motor are integral parts of the scrubber, having top inlet and top outlet.
- Type "T" — A twin jet inlet design with top inlet and top outlet.

All turbulaire scrubbers are equipped with entrainment separators. The typical applications of these scrubbers include stone crushing plants, incinerators, rock crushers, asphalt plant dryers, clay dust from roller mills, dryers and calciners.

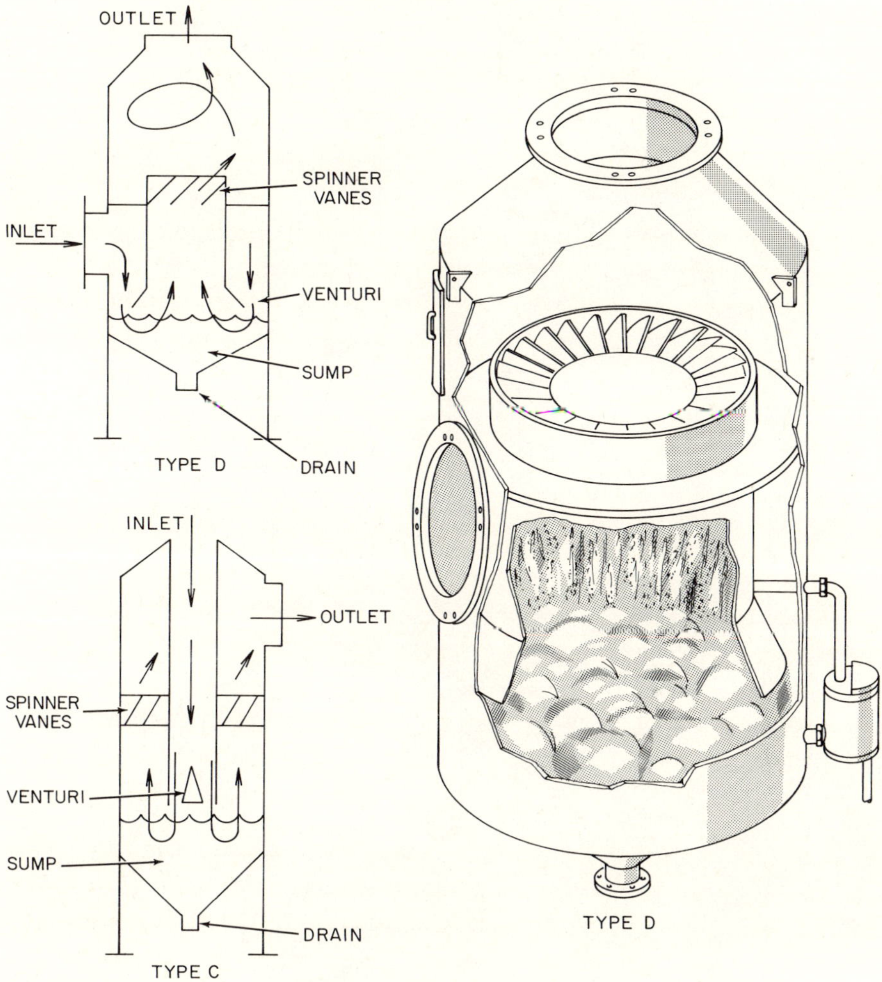

Figure 5.16e Turbulaire scrubbers.

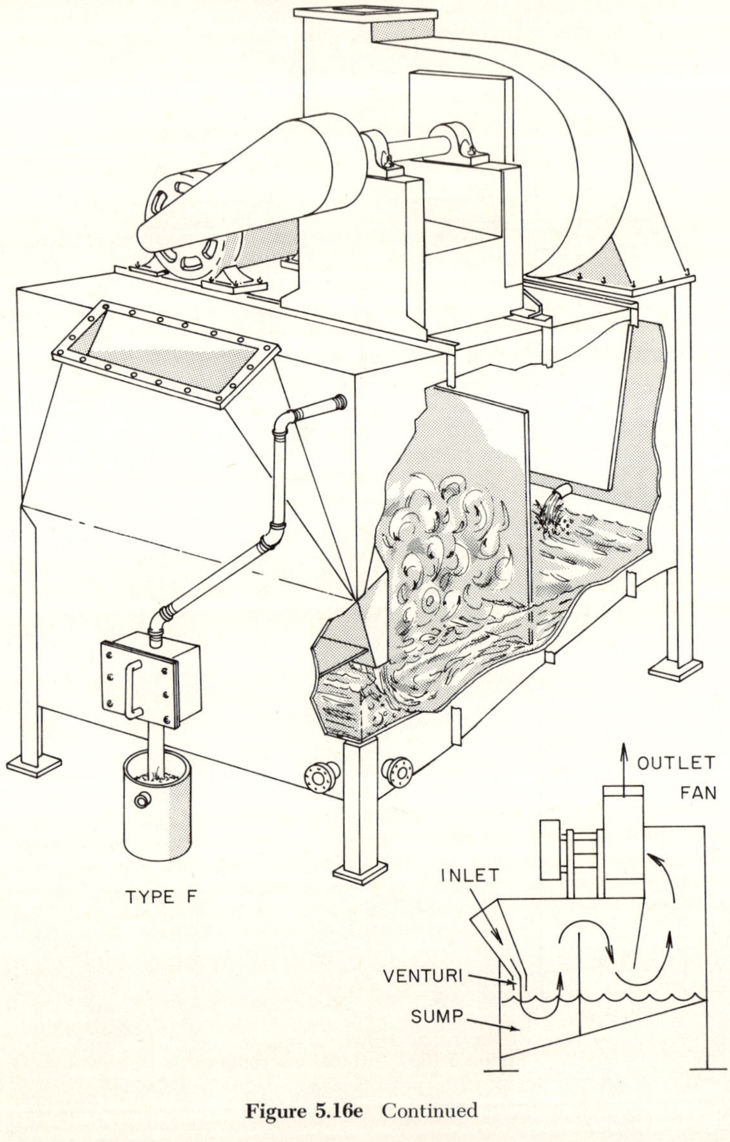

Figure 5.16e Continued

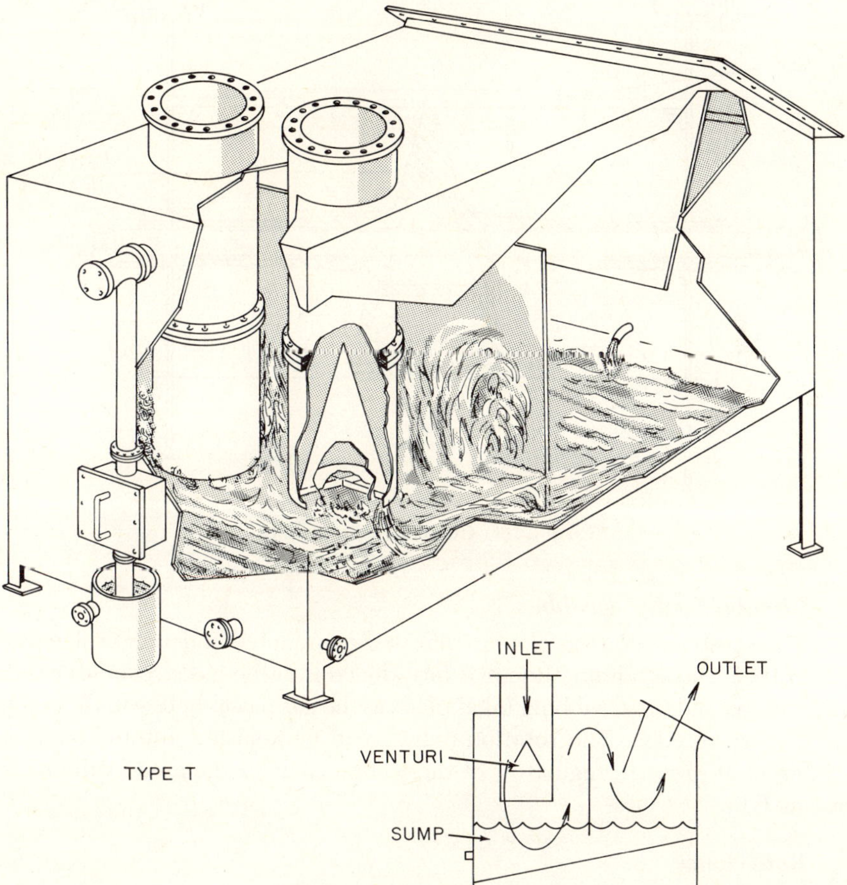

Figure 5.16e Continued

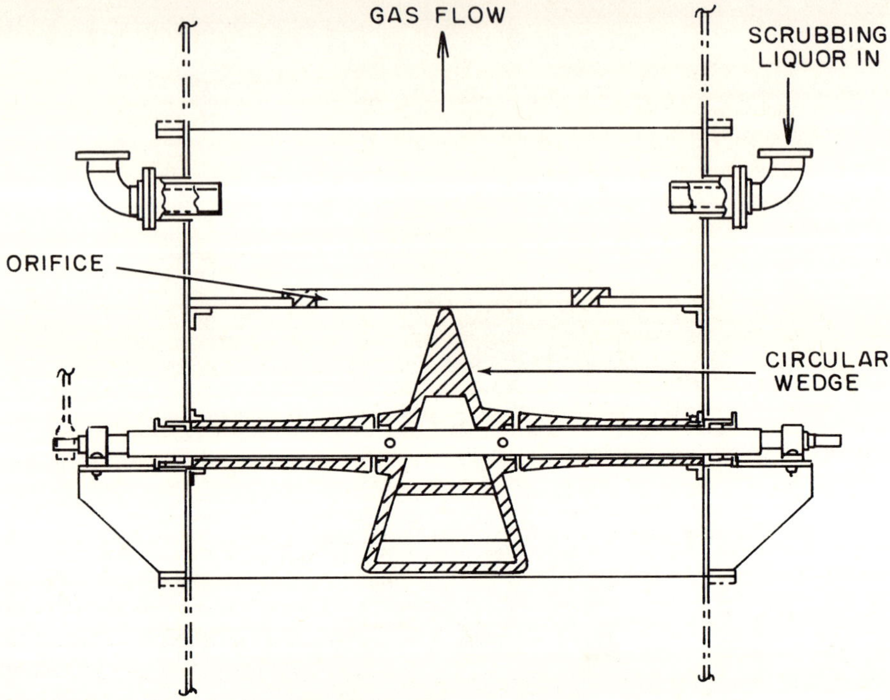

Figure 5.16f Circular-wedge scrubber.

Circular-Wedge Scrubber

This scrubber[1] consists of a circular wedge mounted under an orifice plate (Fig. 5.16f). The scrubbing liquor is introduced into the gas stream above the orifice plate, and the scrubbing takes place as the gas passes between the wedge and the orifice edge. The rotation (which can be adjusted automatically by a differential pressure regulator) of the wedge changes the area available for the gas flow.

Roto-clones

The Roto-clones[5] clean the flue gas by the combined action of centrifugal force and a thorough intermixing of scrubbing liquor and dust-laden flue gas. The flue gas flows through the stationary impeller at high velocity and carries the scrubbing liquor in a heavy turbulent sheet (Fig. 5.16g). The centrifugal force exerted by the rapid changes in the direction of flow causes the dust particles to penetrate the water film and get collected. The scrubbing liquor is continually reused; and since the water curtain is produced by the flue gas flow, no pumps or nozzles are required. The scrubbing liquor level is main-

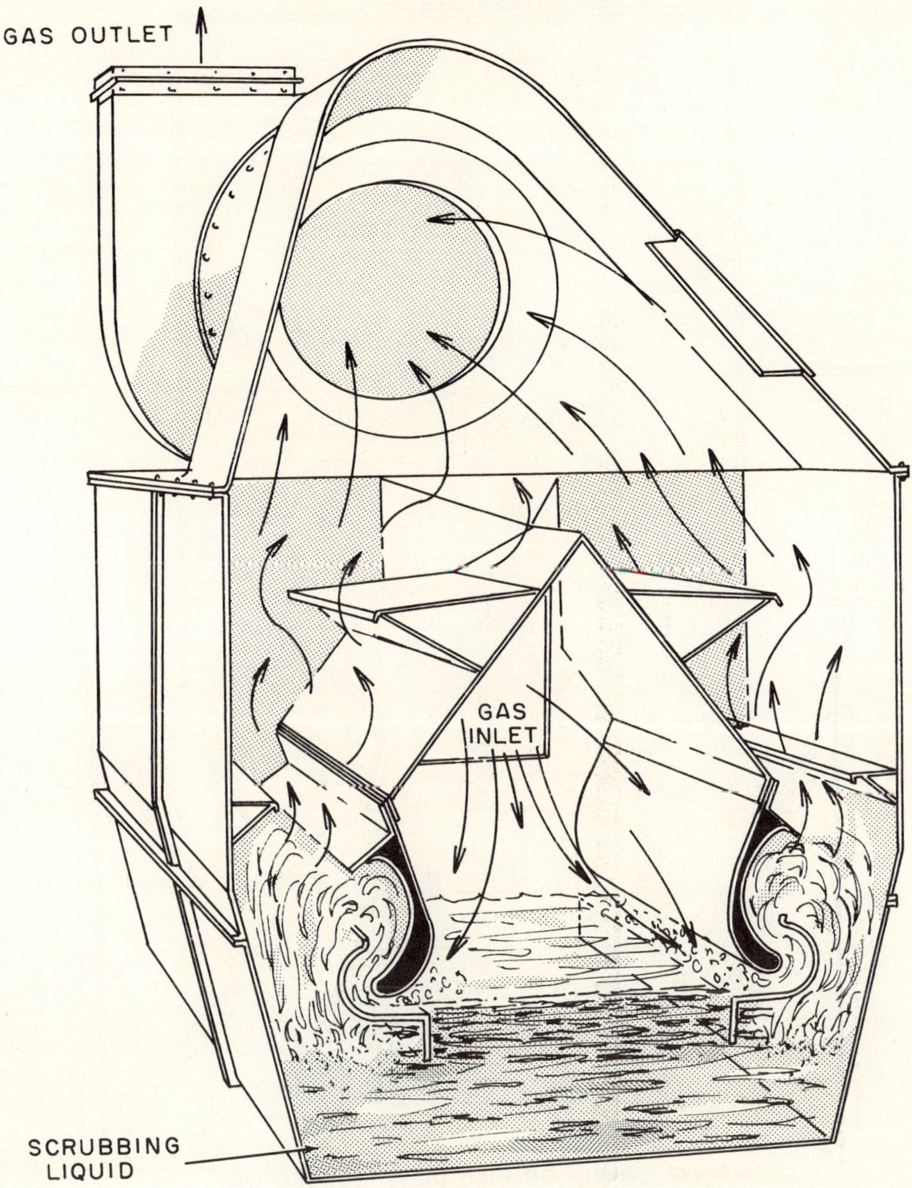

Figure 5.16g Type N roto-clone.

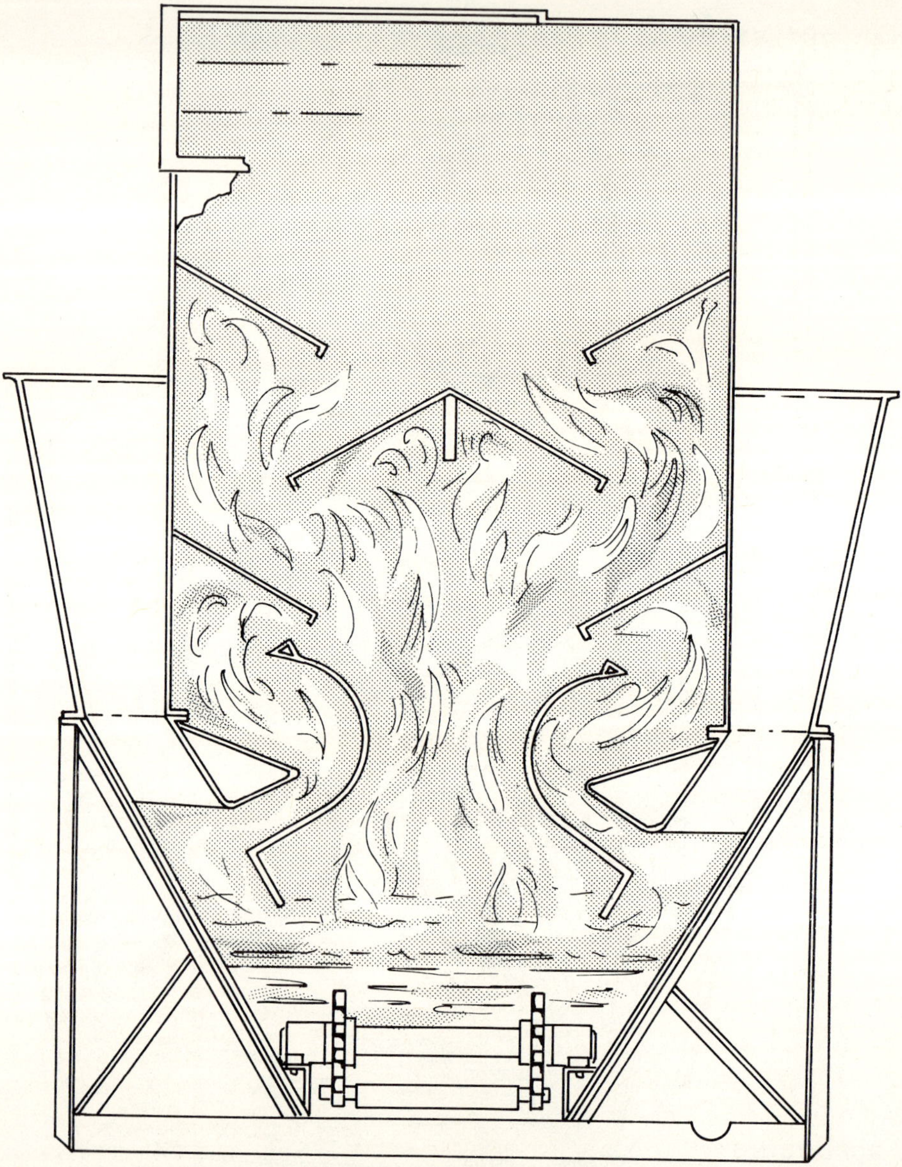

SCHMIEG SWIRL–ORIFICE DUST COLLECTOR

Figure 5.16h Self-induced spray scrubbers.

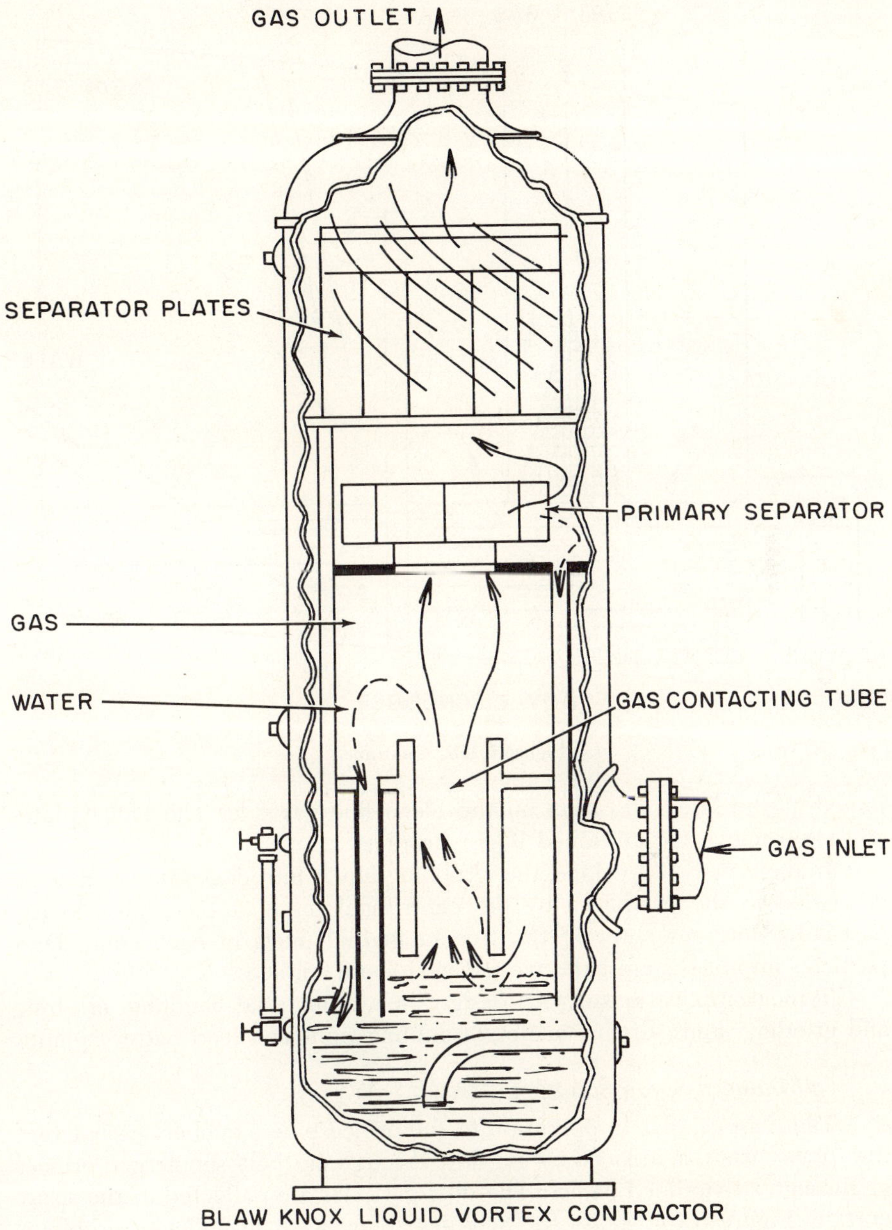

Figure 5.16h Continued

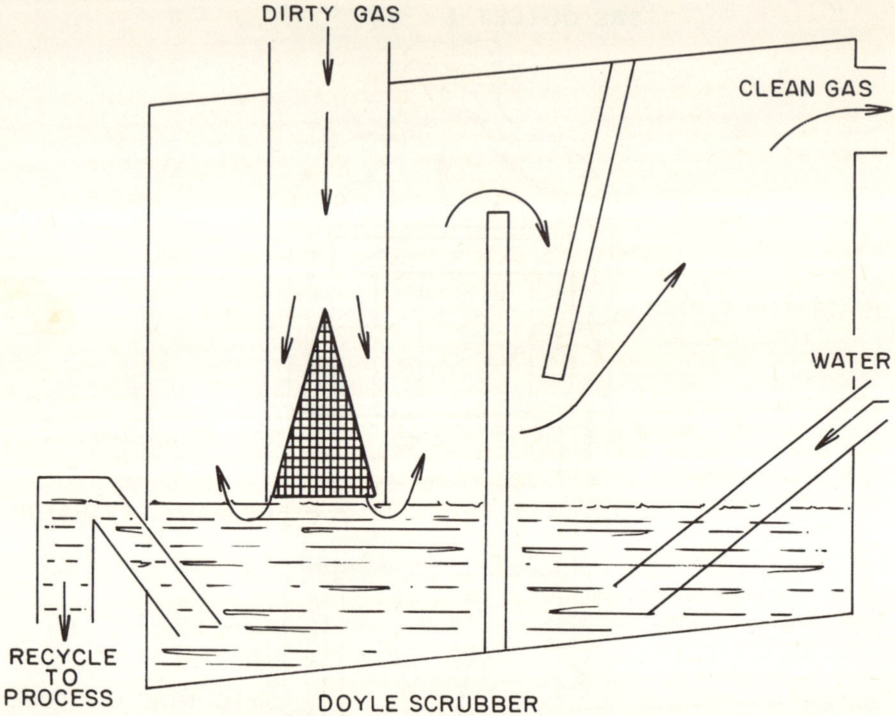

Figure 5.16h Continued

tained at the same elevation on the clean flue gas side. The baffle plates (zig-zig) remove the entrained liquor droplets.

In the type R Roto-clone, the scrubbing liquor introduced into each cone is carried to the periphery by the centrifugal force of the high-velocity, dust-laden flue gas entering the two tangential inlets of each cone. Dust particles impinge against the wetted peripheral surfaces.

Typical applications of this design include material handling, crushing and grinding, kilns, dryers, coolers, tumbling mills and lead battery plants.

Self-induced Spray Scrubbers

These are similar in operation to the orifice type scrubbers except that the spray curtain is induced by gas flow through partially submerged orifices or through streamlined baffles. The dust particles are collected in the spray curtain. Figure 5.16h shows the Schmieg, Blaw Knox and Doyle scrubbers. The impingement gas velocity is about 50 feet per second in the submerged orifice Schmieg scrubber,[6] whereas in the Blaw Knox and Doyle scrubbers,[7] the impingement velocities range from 120 to 180 feet per second.

The pressure drop range is 2 to 15″ H_2O and the scrubbing liquor requirements are about one gallon per 1,000 cubic feet of gas. The ability to handle high dust loadings and concentrated slurries are their main advantages.

REFERENCES

1. Hughson, Roy V.: *Chemical Engineering*, Reprint no. 008, August 29, 1966.
2. "Impinjet Gas Scrubbers," Catalog 151, The W. W. Sly Manufacturing Company.
3. "Peabody High-Efficiency Gas Scrubbers," Bulletin 203-C, Peabody Engineering Corp.
4. "Turbulaire Gas Scrubbers for Dust and Fume Control," Bulletins T-105 and TD101R, Western Precipitation Division, Joy Manufacturing Company.
5. "Roto-clone," Bulletins 277E and 293, American Air Filter.
6. "Schmieg-Swirl-orifice Dust Collectors," Catalog 100A, Schmieg Industries.
7. Doyle, H. and A. F. Brooks: *Ind. Eng. Chem.*, vol. 49, no. 12, pp. 57A–62A, December 1957.

5.17 VENTURI SCRUBBERS

Types of Designs:	Pease-Anthony venturi scrubber (Fig. 5.17b); Kinpactor® venturi scrubber (Fig. 5.17d); S-F venturi scrubber (Fig. 5.17e); Flexiventuri® scrubber (Fig. 5.17f); Airetron® venturi scrubber (Fig. 5.17g); W-A venturi scrubber (Fig. 5.17i); Flood-disc scrubber (Fig. 5.17j); Aeromix® scrubber (Fig. 5.17l); Venturi-sphere, Venturi-slot, Oriclone® venturi and Heil® venturi (all on Fig. 5.17m); Multiventuri scrubber (Fig. 5.17o); Ejector venturi scrubber (Fig. 5.17p).
Principles of Operation:	High-velocity gas impinges (or impacts) upon the liquid stream in the throat of a venturi-shaped section, atomizing the liquor into fine droplets, on which the dust particles are then collected. Sometimes, spray nozzles are used to create droplets. In case of the Ejector type venturi, the high-velocity energy of the liquid jet is used to form the droplets.
Pressure Drop Required:	6 to 80″ H_2O.
Liquor Flow Requirement:	10 to 20 GPM/1,000 ACFM.
Introduction of Liquor:	By distribution box, weir or tangential nozzles.
Collection Efficiency:	99%+ for 1 μ or larger sized particles; 90 to 99% for below 1 μ size.
Applications:	Almost all industries including steel, pulp and paper, power plants, chemical and metallurgical processing plants.

Costs: See Table 5.21f.

Partial List of Suppliers: See Section 5.12.

Venturi Scrubbers

Venturi type wet collectors have come into prominence since the postwar period because of their ability to provide high-collection efficiency for submicron dust particles and their simplicity of installation and maintenance.

The dust-laden flue gas enters the scrubber and is accelerated to a high velocity while passing through the converging section and approaching the throat section. The velocity of the flue gas is the maximum at and in the throat section of the scrubber. The high-velocity gas impinges (impacts) upon the liquid stream in the throat, thereby atomizing the liquor into large numbers of fine droplets. The high-differential velocity between the gas and the atomized droplets causes liquor droplets and dust particles to collide, impact and agglomerate. As the flue gas decelerates, further impaction and agglomeration of the droplets take place. After the particles have been trapped within the liquor droplets, the resulting agglomerates are readily removed from the gas stream in the separator[1] (Fig. 5.17a).

Energy in the form of pressure drop is expended to perform the scrubbing action. For particles larger than 1 μ, low-pressure drops are required; whereas for submicron particles, higher pressure drops are needed. The higher the pressure drop, the higher the removal efficiency of submicron particles. The pressure drop can be increased in a venturi scrubber by increasing the gas velocity and/or the scrubbing liquor flowrate, but the first is more effective.

The size of the dust particles, and of the liquor droplets formed by atomization, plays a very important role in determining the collection efficiency and the collection mechanism. If the particles are 0.1 μ or smaller in diameter, the collection mechanism by which they may be collected is the Brownian diffusion, whereas if the particles are larger than 0.1 μ, the mechanism of impaction becomes predominant. The average diameter of the droplets (D), formed by atomization of the scrubbing liquor in a venturi throat, may be estimated[2] from Equation 5.17(1):

$$D = \frac{16{,}000}{U} + 1.4(L)^{1.5} \qquad 5.17(1)$$

where D = volume-surface mean diameter of spray droplets, μ
L = liquor-to-gas ratio, gal/1,000 ft^3
U = gas velocity at venturi throat, ft/sec

Types of Venturi Scrubbers

The only variation between the several venturi scrubber designs is in the method of liquor introduction in the throat and in the types and shapes of

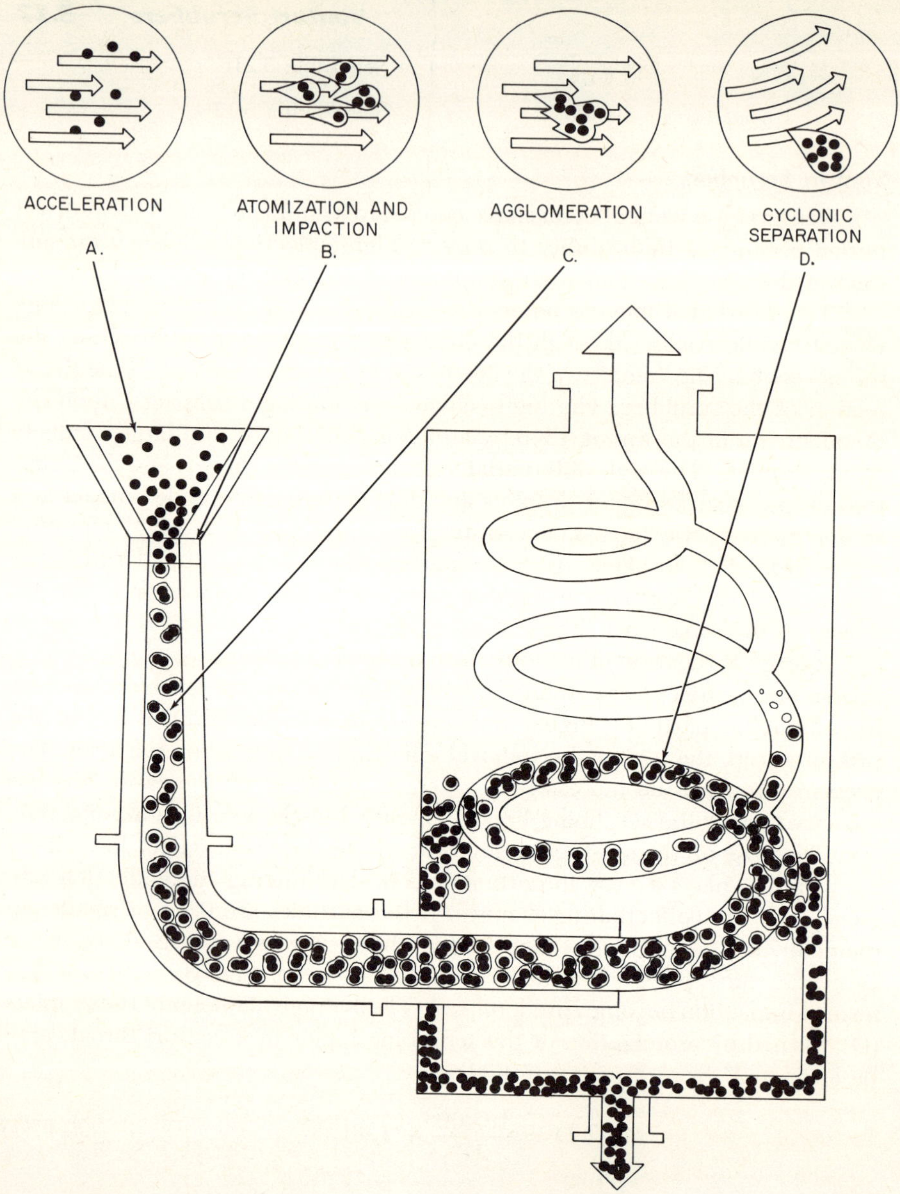

Figure 5.17a The venturi scrubber. *A*, *Acceleration*. The contaminated gas enters the venturi and is accelerated to high velocity; *B*, *Atomization and impaction*. Scrubbing liquid introduced into the gas stream is atomized into fine droplets by the high velocity of the gas, simultaneously trapping the particles to be removed; *C*, *Agglomeration*. The liquid droplets collide with each other, tending to combine and intercept additional particles from the gas stream; *D*, *Cyclonic separation*. The gas stream continues into a cyclonic separator, where the particle-laden scrubbing liquid and the gas are thoroughly separated by centrifugal force to complete the cleaning operation.

throats. The performance of all types of venturi scrubbers on same application will be essentially same, provided the throat pressure drop is same.

The most common types and shapes of throats are rectangular, circular, oval and annular. The width and length (or diameter in the case of circular throat) of the throat are very important to attain full coverage of the throat with scrubbing liquor. The circular throat could be of any diameter when the scrubbing liquor is introduced by spray nozzles located above the throat. This type is normally used for low-pressure drop applications. In the case of high-pressure drop applications and when the liquor is introduced in the throat by horizontal sprays or by other methods, the maximum opening size used is 18 to 20 inches. In case of rectangular or oval throat cross sections, the maximum width is about 18 to 20 inches; and the lengths are approximately 8 to 10 feet. The requirement for complete coverage of the throat by scrubbing liquor limits the width of the throat (a blow hole can otherwise be developed by the high-velocity gas), whereas the limitation on throat length is due to the limitation on available scrubbing liquor flow.

To handle larger gas volumes, a more commonly used throat type is an annulus formed by two circles. Here again the throat gap is maintained at about 18 to 20 inches. Examples of circular throats are given in Figures 5.17 l, m and o. The rectangular and oval throats are offered in designs shown in Figures 5.17b, e and i, whereas the scrubber shown in Figure 5.17j has an annular throat. The scrubbing liquor is introduced by spray nozzles located either in or above the throat by weir arrangement or by tangential pipe nozzles.

The collection efficiency, dust particle size and pressure drop are interrelated, with the pressure drop as the main criteria for collection efficiency. Dust collection efficiency can be maintained constant by maintaining the pressure drop across the throat essentially constant. For varying gas flowrates, an adjustable throat is used to adjust the gas velocity and, hence, the pressure drop across the throat. Following each of the venturi scrubbers, an efficient entrainment separator must be installed to remove the entrained liquor from the flue gas.

Pease-Anthony (P-A) Venturi Scrubber

The P-A venturi scrubber (Fig. 5.17b) has been proven effective and economical for difficult scrubbing applications such as the removal of submicron dusts, fumes and mists.[1,3-5] In the P-A venturi, scrubbing liquor is introduced through jets which form a curtain of liquor across the throat. The velocity force of the accelerated gas atomizes the liquor into fine droplets, which trap the particles from the flue gas stream. The P-A venturi scrubber can be mounted horizontally or vertically, depending on the plant layout. The throat can be circular or rectangular in cross section and it can be adjustable to maintain high dust collection efficiency at varying gas flowrates.

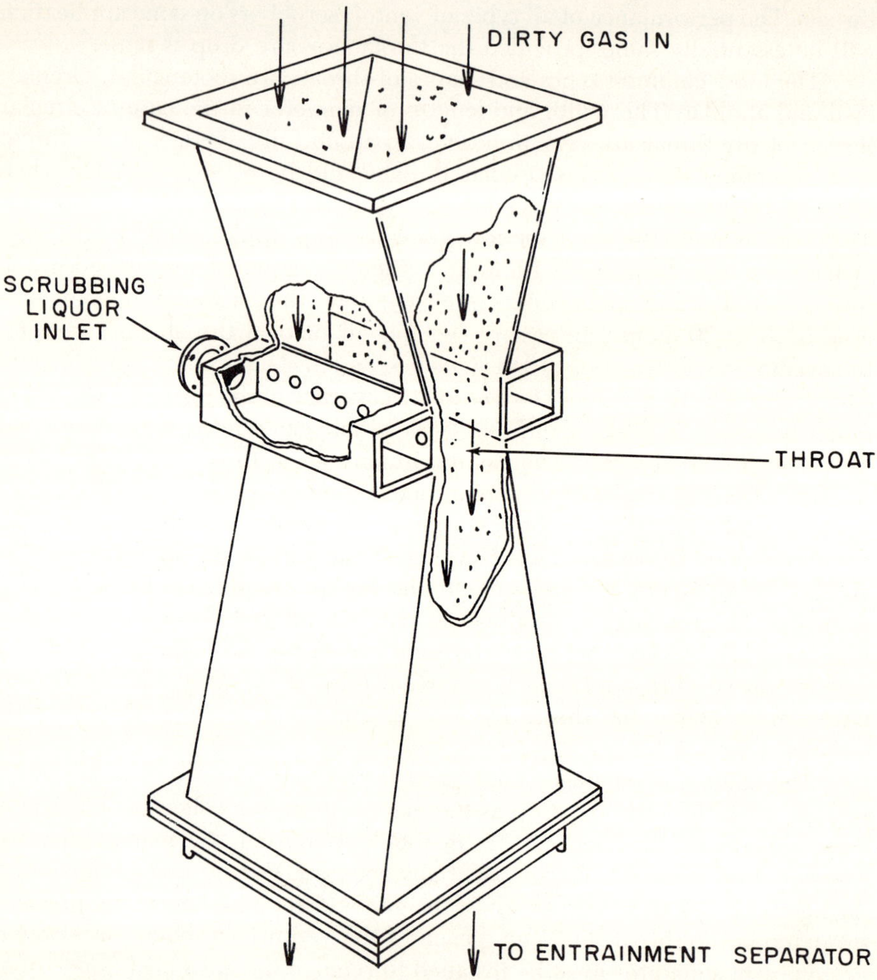

Figure 5.17b Type P-A venturi scrubber.

The liquor flowrates range from 3 to 12 gallons per 1,000 cubic feet of saturated gas, and velocities range from 150 to 400 feet per second. The pressure drop requirements depend on the type of application and on the required collection efficiency. The pressure drop requirements range from 10 to 80″ H_2O. The scrubbing liquor can be recycled; but there is a limitation of slurry concentration that can be used without plugging the jets. Each jet is individually mounted and is provided with a reamer for manual or automatic cleaning.

The P-A venturi scrubber has been used in the iron, steel, chemical, nonferrous metals and pulp and paper industries. Its applications and performance are summarized[1] in Table 5.17c. The P-A venturi scrubber provides 90 to 99%+ collection efficiency on dust particles ranging in size from 0.05 to 100 μ.

The converging sides of the scrubber can be kept wetted to prevent wet-dry line buildup on the converging sides.

Kinpactor Venturi Scrubber

This scrubber[6,7] is shown in Figure 5.17d. Its operation is very similar to that of the P-A venturi scrubber except that the water is injected in the form of a sheet rather than as jets or atomized sprays. The water jets are impacted on the splash plate to convert them into sheets.

The kinpactor is generally designed to use 8 gallons of scrubbing liquor per 1,000 cubic feet of saturated gas at the throat, and the scrubbing liquor is usually recycled. For dusts larger than 1 μ size, it requires pressure drop of 10″ H_2O; while for submicron particles, the pressure drop requirements may be as high as 100″ H_2O. The performance and applications of this scrubber are similar to those of the P-A venturi scrubber.

S-F Venturi Scrubber

The S-F (Svenska-Flaktfabriken) venturi scrubber[1,8] is particulary recommended for the removal of "sticky" solids from gases, for the recycling of heavy slurries and for the recovery of process materials in concentrated form. In the S-F venturi scrubber, the scrubbing liquor is introduced by two different methods.

In one method, the scrubbing liquor is introduced through troughs (or weirs) at the top of the unit (Fig. 5.17e). The liquid flows downward in a continuous film along the sloping walls to the deflecting lips, which direct it across the throat of the venturi to form a curtain of liquor. The accelerating high-velocity gas impinges on the curtain and shatters it into fine droplets on which the dust particles are collected.

In the second method, the liquor is introduced in a tangential manner on the shelf (Fig. 5.17e). The liquor flows through open pipe nozzles where it then swirls down the converging sides (like a dentist bowl effect) to the deflecting lip, which throws the liquor across the throat to form a curtain.

As there are no nozzles being used to introduce the scrubbing liquor and as the scrubbing liquor is always in motion, heavy slurries and concentrated solutions can be used as scrubbing liquor. The continuous swirling motion of the liquor on the converging side helps keep the surface clean.

In the S-F venturi scrubbers, the throat can be oval, rectangular, circular or annular in shape. The lip size and location are the most critical design parameters. Normally, the lip is located at the entrance of the throat. The

Table 5.17c

PERFORMANCE DATA AND APPLICATIONS OF THE P-A VENTURI SCRUBBER

Source of Gas	Contaminants	Approximate Size Range, μ	Loading, grains/cu ft		Average Removal Efficiency, %
			Inlet	Exit	
Iron and Steel Industry					
Gray iron cupola	Iron, coke, silica dust	0.1–10	1–2	0.05–0.15	95
Oxygen steel converter	Iron oxide	0.5–2	8–10	0.05–0.08	98.5
Steel open hearth furnace (scrap)	Iron and zinc oxide	0.08–1	0.5–1.5	0.03–0.06	95
Steel open hearth furnace (oxygen lanced)	Iron oxide	0.5–2	1–6	0.01–0.07	99
Blast furnace (iron)	Iron ore and coke dust	0.5–20	3–24	0.008–0.05	99
Electric furnace	Ferro manganese fume	0.1–1	10–12	0.04–0.08	99
Electric furnace	Ferro silicon dust	0.1–1	1–5	0.1–0.3	92
Rotary kiln—iron reduction	Iron, carbon	0.5–50	3–10	0.1–0.3	99
Crushing and screening	Taconite iron ore dust	0.5–100	5–25	0.005–0.01	99.9
Chemical Industry					
Acid—humidified SO_3	H_2SO_4 mist				
(a) Scrub with water		—	303[a]	1.7[a]	99.4
(b) Scrub with 40% acid		—	406[a]	2.8[a]	99.3
Acid concentrator	H_2SO_4 mist	—	136[a]	3.3[a]	97.5
Copperas roasting kiln	H_2SO_4 mist	—	198[a]	2.0[a]	99
Chlorosulfonic acid plant	H_2SO_4 mist	—	756[a]	7.8[a]	98.9
Dry ice plant	Amine fog	—	25[a]	2.0[a]	90+
Wood distillation plant	Tar and acetic acid	—	1080[a]	58.0[a]	95
$TiCl_4$ plant, TiO_2 dryer	TiO_2-HCl fumes	0.5–1	1–5	0.05–0.1	95
Spray dryers	Detergents, fume & odor	—	—	—	95
Flash dryer	Furfural dust	0.1–1	1–1.5	0.05–0.08	95+
Phosphoric acid plant	H_3PO_4 mist	—	192[a]	3.8[a]	98+

Nonferrous Metals Industry					
Blast furnace (sec. lead)	Lead compounds	0.1–1	2–6	0.05–0.15	99
Reverberatory lead furnace	Lead and tin compounds	0.1–0.8	1–2	0.12	91
Ajax furnace—aluminum alloy	Aluminum chloride	0.1–0.9	3–5	0.02–0.05	95
Zinc sintering	Zinc and lead oxide dusts	0.1–1	1–5	0.05–0.1	98
Reverberatory brass furnace	Zinc oxide fume	0.05–0.5	1–8	0.1–0.5	95
Mineral Products Industry					
Lime kiln	Lime dust	1–50	5–10	0.05–0.15	99+
Lime kiln	Soda fume	0.3–1	0.2–5	0.01–0.05	99
Asphalt stone dryer	Limestone and rock dust	1–50	5–15	0.05–0.15	98+
Cement kiln	Cement dust	0.5–55	1–2	0.05–0.1	97+
Petroleum industry					
Catalytic reformer	Catalyst dust	0.5–50	0.09	0.005	95+
Acid concentrator	H_2SO_4 mist	—	136[a]	3.3[a]	97.5
TCC catalyst regenerator	Oil fumes	—	756[a]	8.0[a]	98+
Fertilizer Industry					
Fertilizer dryer	Ammonium chloride fumes	0.05–1	0.1–0.5	0.05	85+
Superphosphate den and mixer	Fluorine compounds	—	309[a]	5.5[a]	98+
Pulp and Paper Industry					
Lime kiln	Lime dust	0.1–50	5–10	0.05–0.15	99+
Lime kiln	Soda fume	0.1–2	2–5	0.01–0.05	99
Black liquor recovery boiler	Salt cake	—	4–6	0.4–0.6	90

[a] Milligrams per cubic ft

NOTE: *The efficiencies shown above are average values for a particular plant or group of installations operating under a specific set of conditions.*

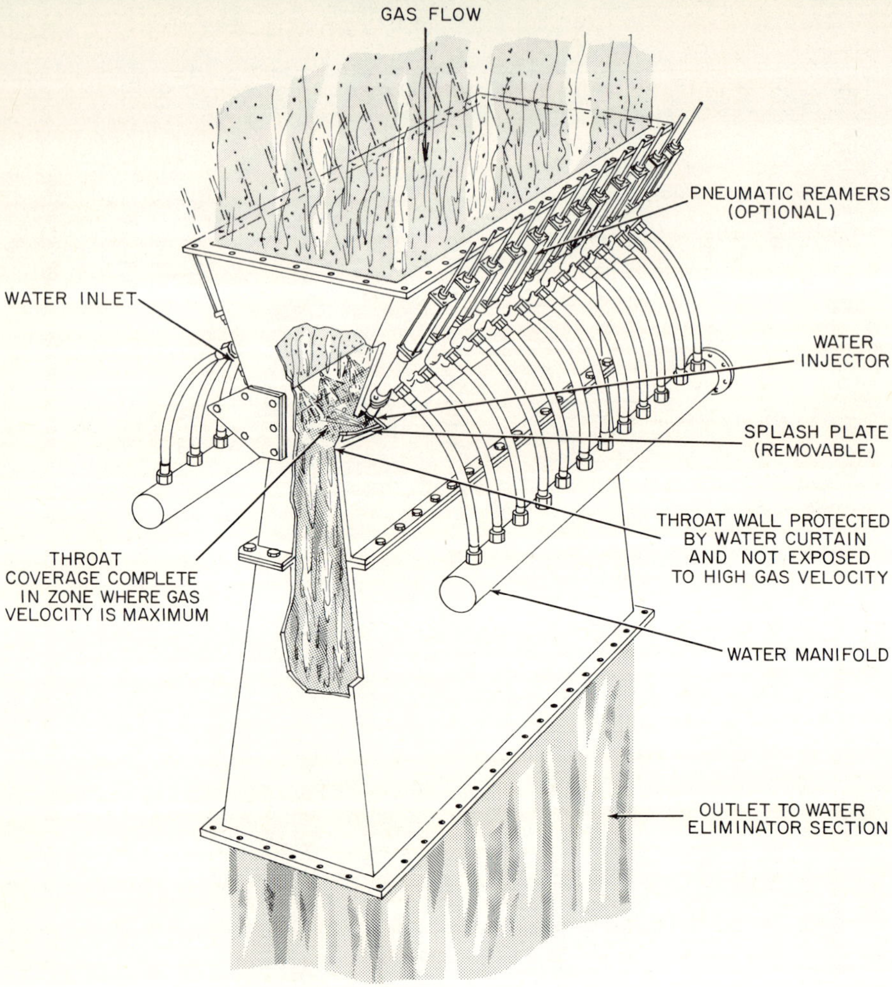

Figure 5.17d Kinpactor venturi scrubber.

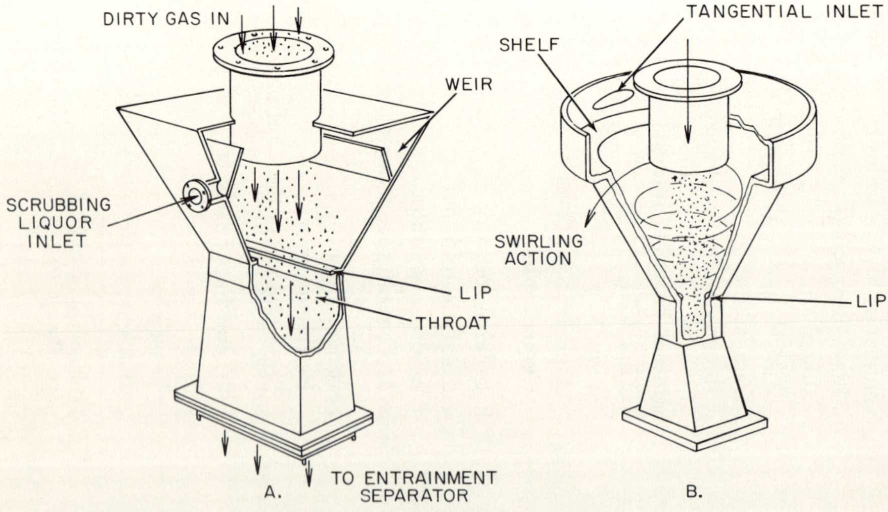

Figure 5.17e S-F venturi scrubber. *A*, Weir type liquor inlet; *B*, Tangential liquor inlets.

throat velocities range from 75 ft/sec to as high as 350 ft/sec (normally 250 ft/sec), and the scrubbing liquor flow ranges from 5 to 50 gallons per 1,000 cubic feet of saturated gas. The design pressure drops range from 5 to 40″ H_2O.

The S-F venturi scrubber has been widely used to remove dust from lime kilns and on kraft recovery furnace or in phosphate fertilizer operations. It has been successfully used to collect sulfuric acid and phosphoric acid mists. On equal-pressure drop basis, the performance of this scrubber is equivalent to that of the P-A venturi scrubber. On dust particles ranging in size from submicron to micron and larger, the collection efficiency is in the range of 90 to 99%+.

Flexiventuri Scrubber

This scrubber[9] is very similar to the S-F venturi scrubber except that instead of a deflecting lip at the throat, it utilizes a weir arrangement, as shown in Figure 5.17f, to throw the scrubbing liquor across the throat and thereby form a curtain of liquor.

Airetron Venturi Scrubber

This scrubber produces droplets pneumatically, when the unscrubbed gas is caused to accelerate through the throat region, and shatters the distributed liquid into minute droplets.[10] The liquor rate is generally 5 gallons per 1,000

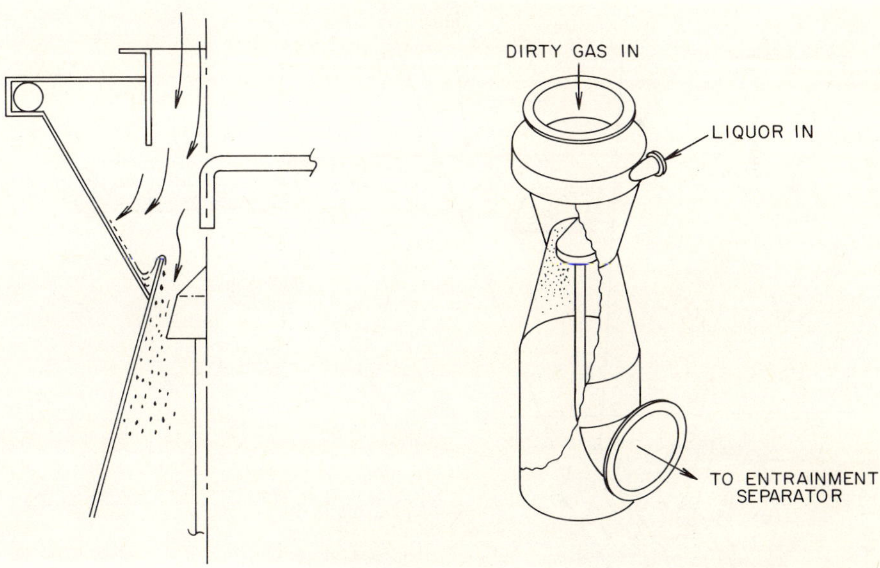

Figure 5.17f Flexiventuri scrubber.

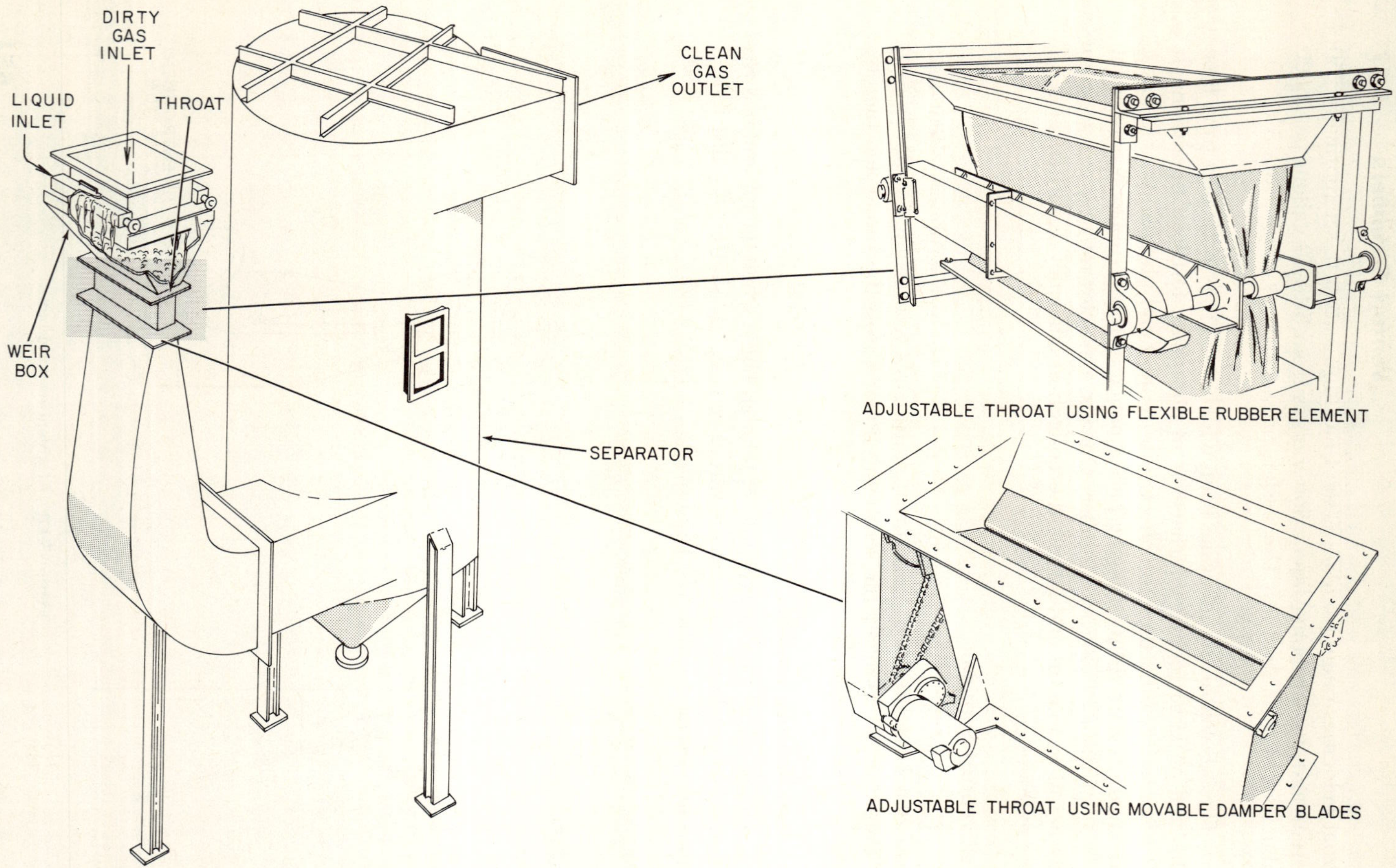

Figure 5.17g Airetron venturi scrubber with adjustable throat.

Table 5.17h
PERFORMANCE DATA AND APPLICATIONS OF THE AIRETRON VENTURI SCRUBBER

Application	*Gas Discharge Loading at Various Pressure Drops, inches H_2O*					*Units of Discharge*
	5	*10*	*15*	*20*	*30*	
Ore roasting H_2SO_4 mist	—	15	10	4	1½–2	mgm/Scf
Chlorosulfonic acid H_2SO_4 + mist	—	20	—	5	2	mgm/Scf
Reverberatory Furnace ZnO	—	—	0.3	0.1	0.01–0.02	gr/Scf
Blast furnace PbO	—	—	0.2	—	0.005–0.01	gr/Scf
Phosphoric acid concentrator P_2O_5	—	30	—	4	2	mgm/Scf
Mix fertilizer drier NH_4Cl	0.2	0.11	—	0.05	0.02–0.03	gr/Scf
Induction furnace strip galvanizer NH_4CL	—	—	—	0.06	0.02	gr/Scf
TCC unit oil mist	—	32	24	16	7	mgm/Scf
Asphalt plant rock dust	0.16	0.008	—	—	—	gr/Scf
Dry ice plant MEA mist	—	—	1.5	—		mgm/Scf

cubic feet of saturated gas, and the scrubbing liquor can be recycled. Table 5.17h lists various applications and outlet dust loadings at various pressure drops across the scrubber.

Two types of adjustable throats are available (Fig. 5.17g).[10] The standard adjustable throat utilizes moveable damper blades; whereas for special applications, a flexible rubber throat is available.

Wet Approach (W-A) Venturi Scrubbers

In the W-A scrubber,[11] the liquor is introduced tangentially through pipe nozzles. The liquor swirls down the converging walls and forms a liquor curtain in the throat. Figure 5.17i shows a typical W-A scrubber for handling large gas volumes (200,000 to 600,000 ACFM). The scrubber and entrainment separator are both in one common vessel.

Normally, an annular throat cross section is used to handle gas volumes

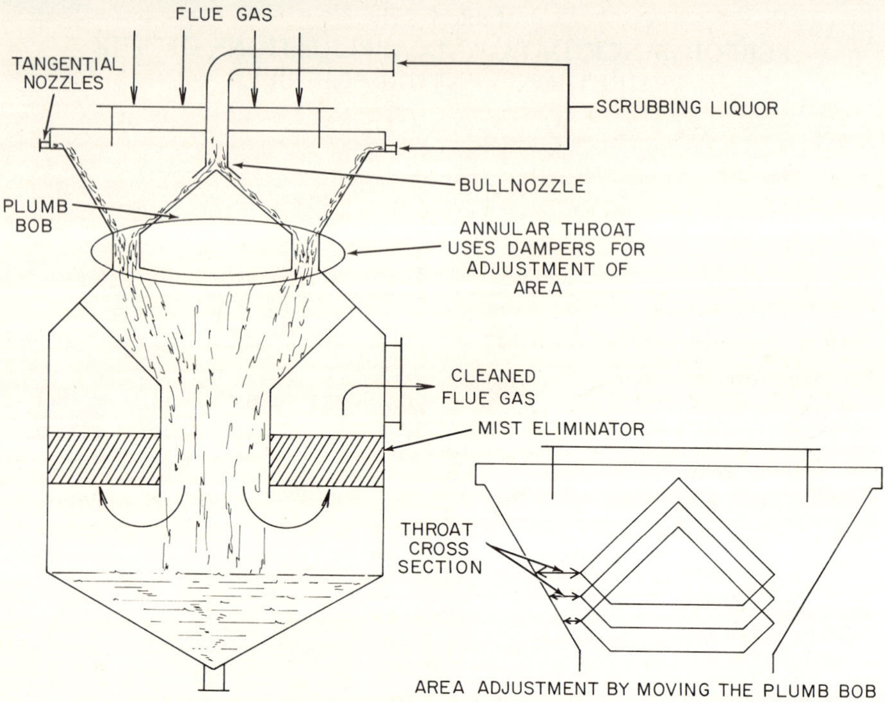

Figure 5.17i W-A venturi scrubber.

larger than 100,000 ACFM; but in the case of smaller gas volumes, the throat could be oval or rectangular in shape. The width of the throat gap is a critical design parameter and, depending on gas velocity, it could vary from 2 to 20 inches.

In case of an annular scrubber, the recycled scrubbing liquor is divided into two portions. One portion is introduced through tangential nozzles to cover the converging section, and the other portion is introduced on the central cone (also called plumb bob) through open bull nozzles.

The adjustment of the throat area is achieved either by utilizing damper blades or by moving the plumb bob up or down. The liquor flow requirements range from 10 to 20 gallons per 1,000 cubic feet of saturated gas, the gas velocities are in the range of 75 to 250 ft/sec, and the pressure drops range from 5 to 40″ H_2O.

The W-A scrubber has been successfully used in applications such as pulp and paper, fertilizer, phosphoric and sulfuric acid manufacture, coal and oil-fired generating stations. The performance of the W-A scrubber, on an equal pressure drop basis, is similar to that of the P-A venturi scrubber.

Flooded-disc Venturi Scrubber

The flooded-disc venturi scrubber (Fig. 5.17j) consists of a flooded disc contactor and a demister.[12,13] It is essentially a variable throat orifice scrubber, in which the scrubbing liquor is introduced at the center of the orifice plate (disc) through a pipe. The liquor flows upward through the center disc support column and impinges on the liquor distribution cone; and it is distributed radially across the face of the disc. The high-velocity gas, while passing through the annular orifice, atomizes (by the shearing action) the scrubbing liquor as it passes off the periphery of the disc. Collection of the dust particles by the liquor droplets is by inertial impaction.

The vertical movement of the disc in the converging throat varies the annulus area, thus providing a means to adjust the pressure drop (or maintain it constant) as the gas flow varies. The vertical motion is accomplished by hydraulic drives which are either manually or automatically controlled.

The performance of this scrubber (according to the testing program) is equivalent to that of other types of inertial impaction scrubbers. Table 5.17k summarizes performance data.[13]

Aeromix Wet Scrubber

This is a low-energy venturi scrubber (Fig. 5.17l)[14] which differs from the conventional venturi scrubbers in that the gas flow is upward (more turbulent mixing) and the scrubbing liquor recirculation is by gravity (no pumps).

The flue gas, flowing upward at high velocity through the throat, mixes with the scrubbing liquor introduced from the feed tank. In the diffuser section, intimate liquor-gas contact results, due to continuous internal refluxing. The liquor-gas stream, with its collected dust particles, passes upward to the mist eliminator where fixed vanes impart a centrifugal motion, resulting in liquor impingement on the wall and the collection of liquor into a trough.

The absence of pumps and nozzles permits recirculation of heavy slurries (concentrations as high as 25%). Collection efficiencies of 98% or more for particles 1 μ or larger in size can be attained at pressure drops of 4 to 8″ H_2O. The typical applications of this scrubber include the removal of sand from casting operations; the removal of dusts from fertilizer operations; and the removal of dusts from crushing, drying and conveying operations in mineral product industries.

Other Venturi Scrubbers

Other venturi scrubbers are shown in Figure 5.17m.

The *Ventri-Sphere*® is a combination of a venturi scrubber and a mobile bed scrubber.[14] It has a circular throat with liquor being introduced in the venturi by overflow weir and in the mobile bed scrubber by spray ring header. The throat is adjustable.

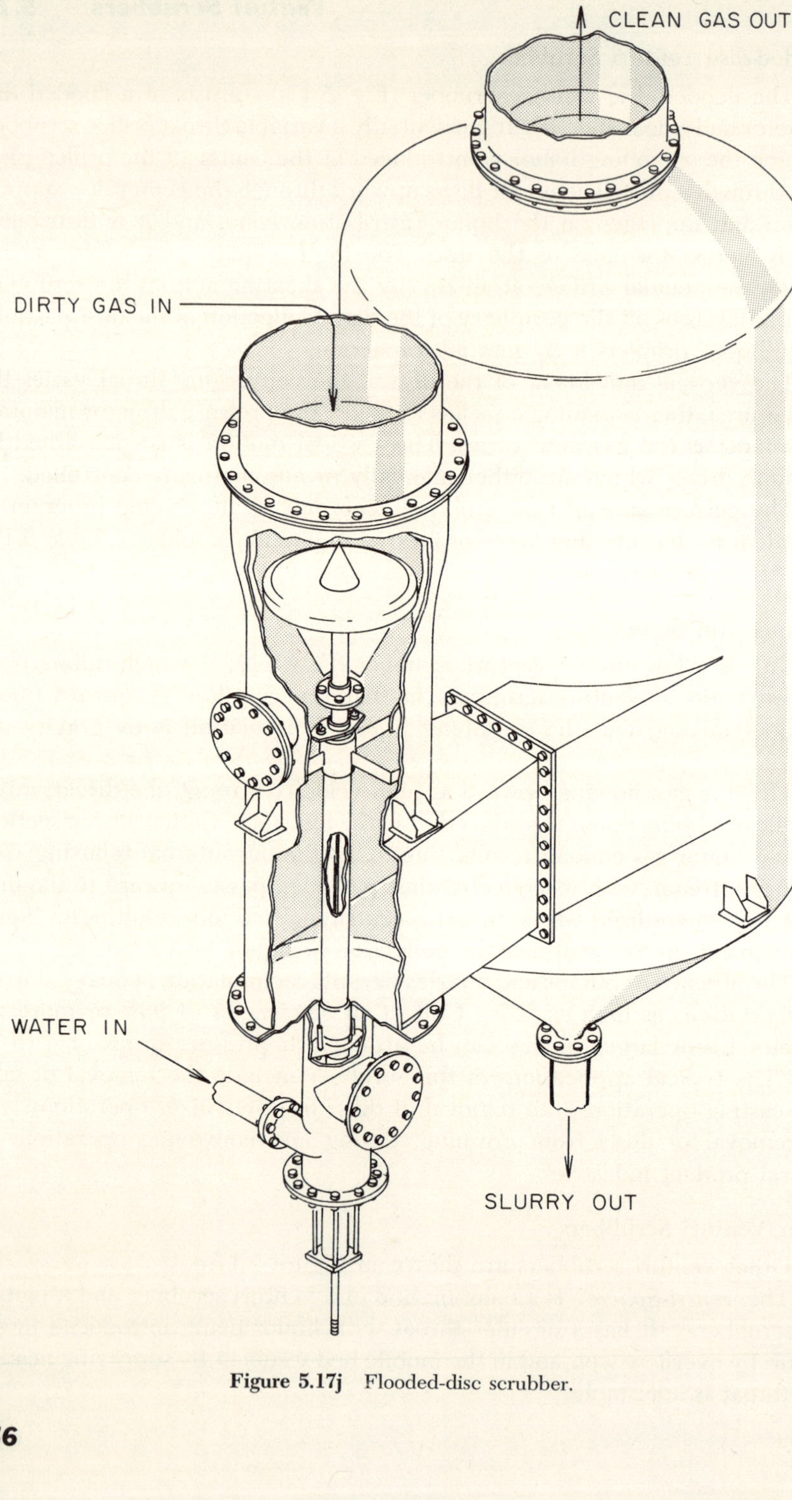

Figure 5.17j Flooded-disc scrubber.

Table 5.17k
PERFORMANCE DATA FOR FLOODED DISC SCRUBBER (PILOT AND FULL-SCALE INSTALLATIONS)

Application	*Liquor Flow, GPM/1,000 ACFM*	*Pressure Drop, inches*	*Dust loading Grains/SCF*	
			Inlet	*Outlet*
Lime kiln	3.5–5	8–15		0.03–0.1
Electric furnace				
Plant A				
oxygen blow	10	45	10.5	0.013–0.035
total heat		45		0.019
Plant B	10	40		0.003–0.009
Open hearth (pilot plant)		15–40	3.75–9.9	93–98.5%
Blast furnace				
pilot plant A	9.85 avg (2.1–30)	10–45	0.51–3.88	0.1–0.01
pilot plant B	8.0 avg (4.5–14.9)	40–70	1.98–7.40	0.01–0.07
commercial		45		0.03
Basic oxygen furnace (pilot plant)	7.8 avg (3.4–12.2)	15–40	1.5–2.3	0.04–0.34

The *ventri-slot* design has a rectangular throat with liquor being introduced by a distribution box and through pressure sprays.[15] It also has a damper-type adjustable throat.

The *Oriclone®* is a horizontal venturi wrapped around a cyclonic separator. Liquor is introduced by spray nozzles and the throat is of the adjustable orifice type. See Table 5.17n for performance data.[16]

The *Heil® venturi scrubber* is a vertical venturi with a horizontal separator, or horizontal venturi with vertical separator[17] into which liquor is introduced by spray nozzles.

Multiventuri Scrubber

In the multiventuri scrubber (Fig. 5.17o), the scrubbing liquor is sprayed by nozzles into the belled venturi entrance so that the spray strikes the throat wall and bounces at right angles to the flue gas stream. The dust is removed in the spray and the gas is further scrubbed by the liquor droplets generated by the bouncing. The dust collection efficiencies are high, and the main design features are the low liquor flow and low-pressure drop requirements.

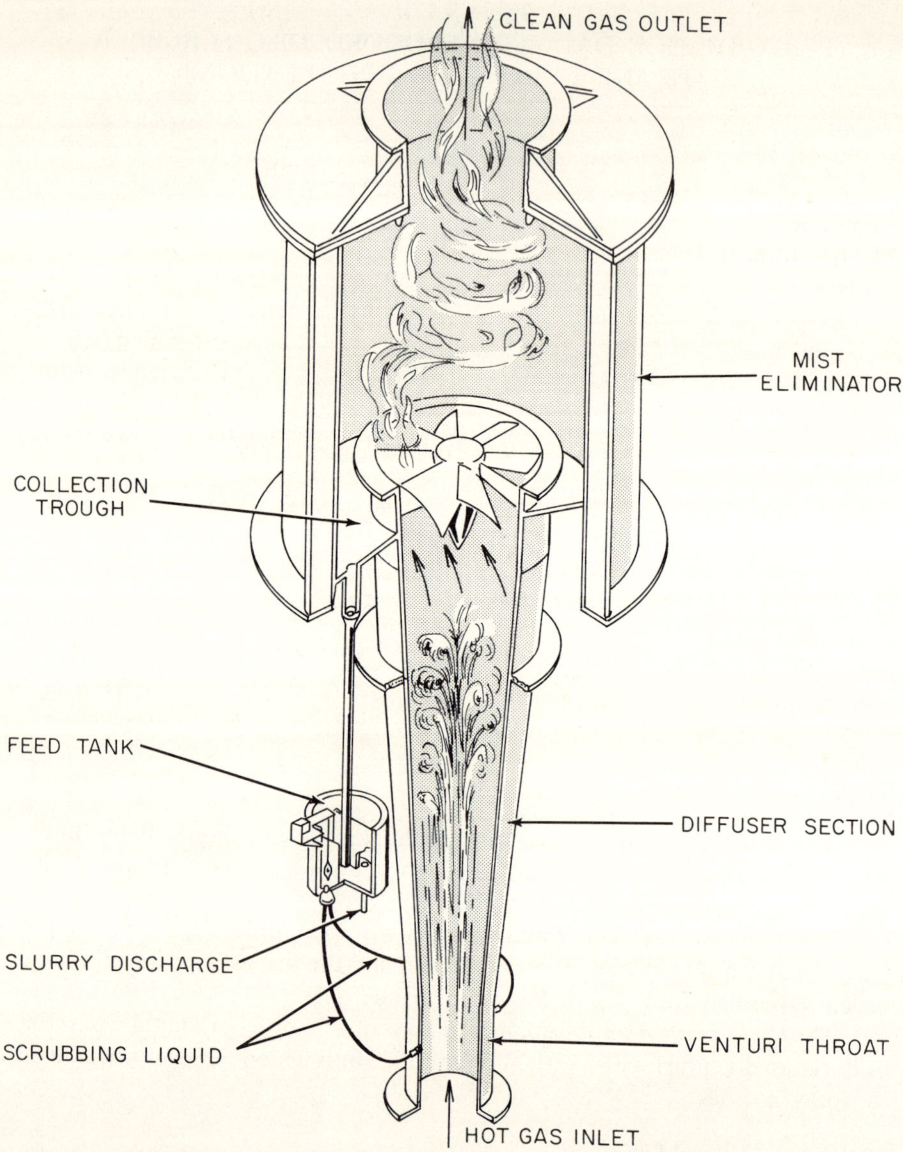

Figure 5.17l Aeromix wet scrubber.

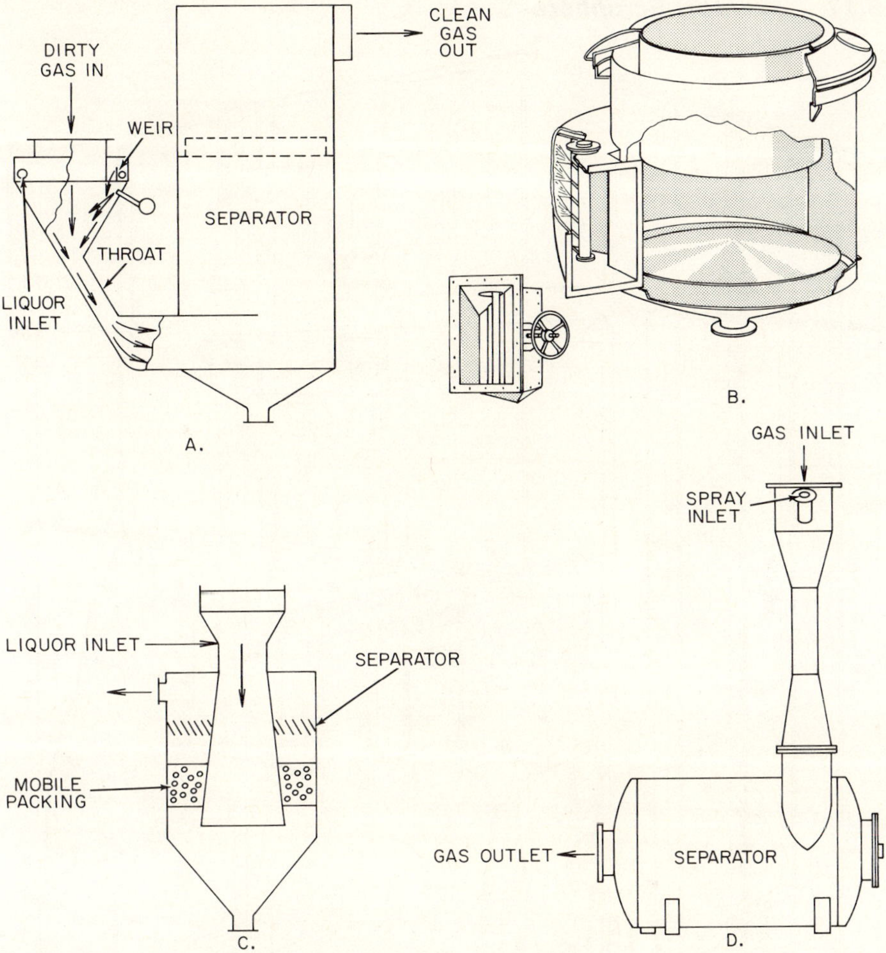

Figure 5.17m Other venturi scrubber designs. *A*, Venturi-slot scrubber; *B*, Oriclone venturi scrubber; *C*, Venturi-Sphere scrubber; *D*, Heil venturi scrubber.

Table 5.17n
PERFORMANCE DATA OF ORICLONE SCRUBBER

Type of Dust and Source	*Particle Size Range, μ*	*Pressure Drop, inches H_2O*	*Dust loading, grains/SCF*	
			Inlet	*Outlet*
Silica, flyash and metal oxides from cupola	0–5	25	10	0.048
		10	10	0.3
Petroleum coke dust	—	30	0.1–0.3	0.012
		10	0.1–0.3	0.016
Catalyst dust	0–5	23	1–3.4	0.0055
		10	1–3.4	0.024
Pharmaceutical dust	—	27.2	1.3	0.013
		13.0	1.3	0.062

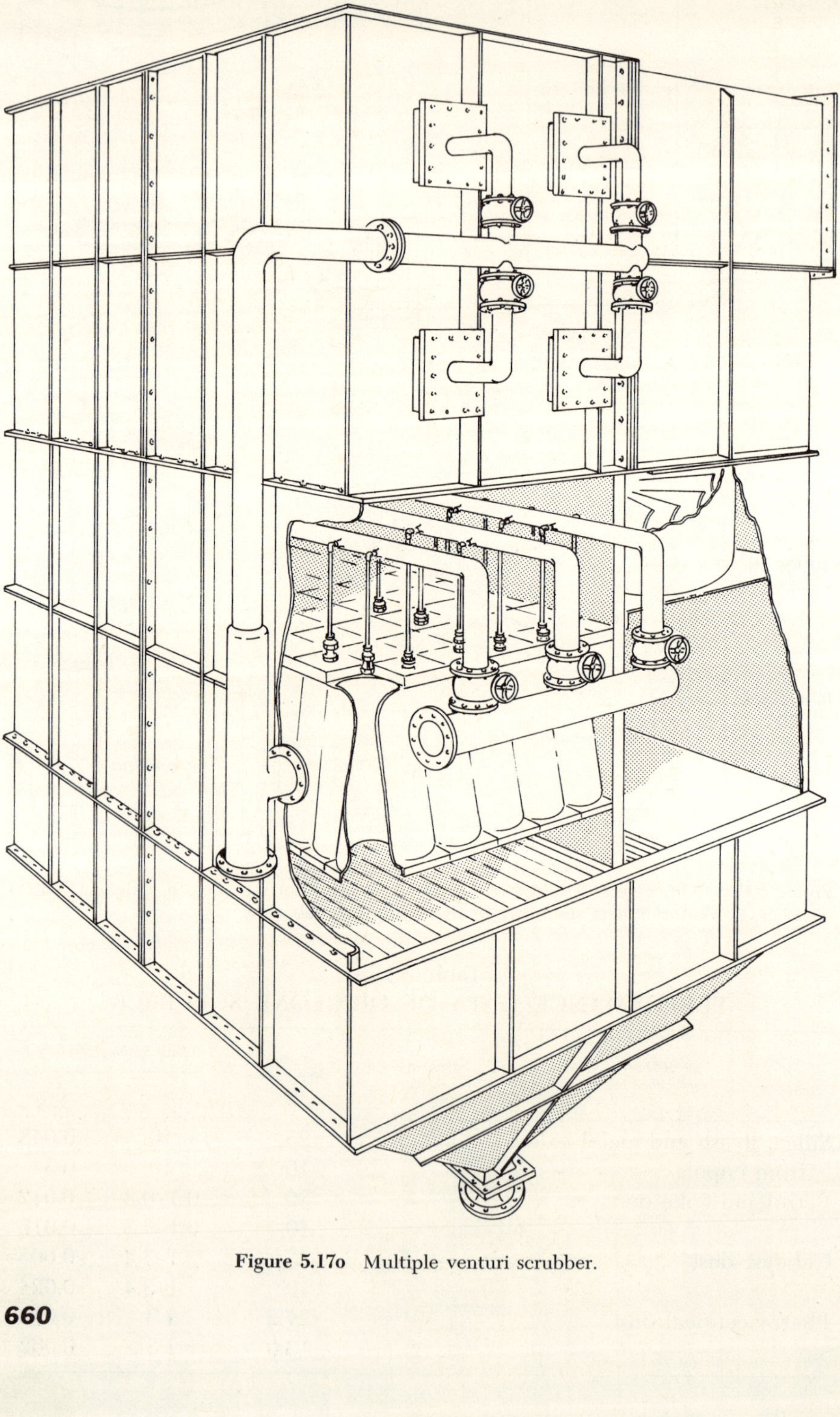

Figure 5.17o Multiple venturi scrubber.

Ejector Venturi Scrubbers

The ejector venturi scrubber uses the high-velocity energy of the liquid jet stream to break up and distribute the liquor into numerous small droplets and to pump the flue gas through the scrubbing system and connecting ductwork (Fig. 5.17p). In ejectors, the atomization of liquor is highly efficient but the pumping of gas is not. The design differs from the conventional venturi scrubber in that no separate blower is required. The energy input for pumping is the total energy required for the scrubber.

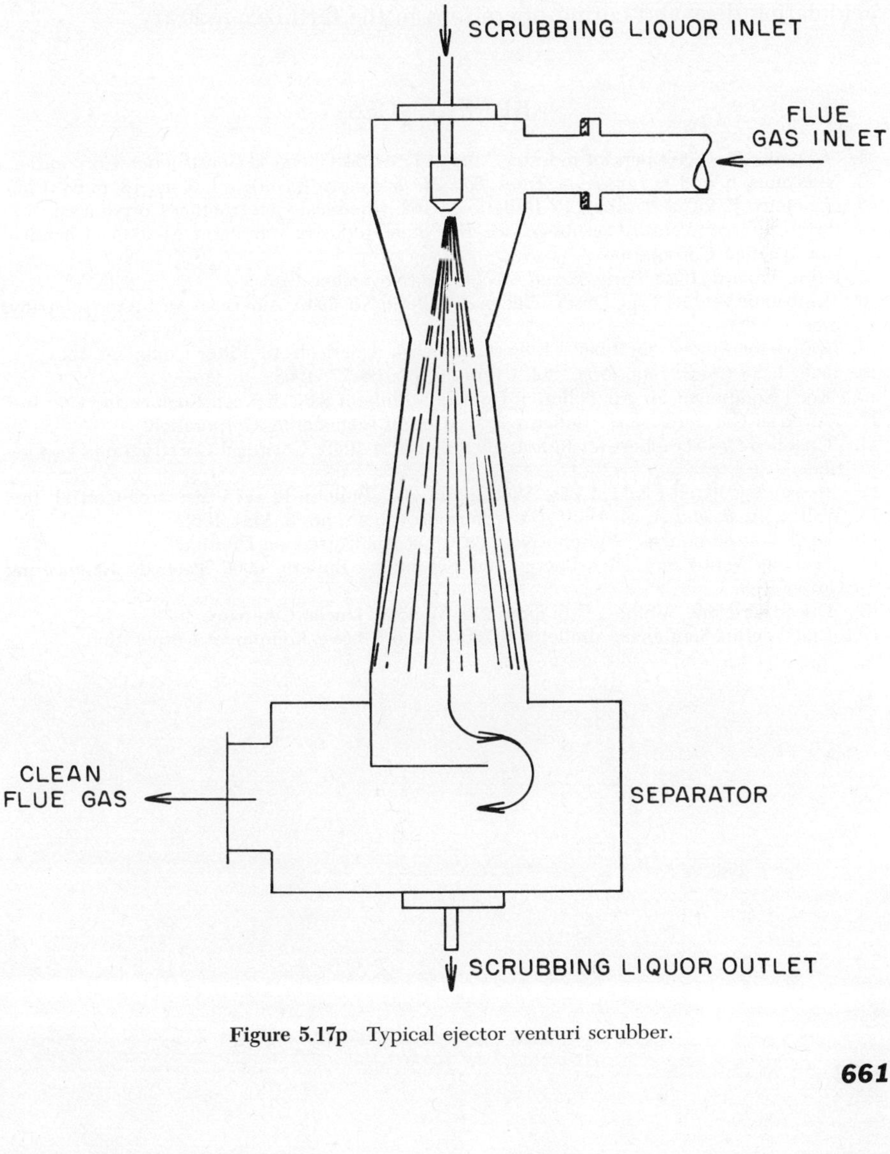

Figure 5.17p Typical ejector venturi scrubber.

The scrubbing liquor enters the ejector, under pressure, through a nozzle located at the top and is forced at high velocity through a venturi type throat. The velocity of the entering liquid produces a pumping action which pulls the flue gases into the scrubber. The extreme turbulence created by the high velocity of the liquid provides efficient contact between the two phases.

The ejector type scrubbers are designed for removal of fumes, dust, gases and vapors. It provides satisfactory removal efficiency for dust particles larger than 1 μ. The efficiency depends on the pressure of the scrubbing liquor (minimum of 15 psig) and on the dust loading of the flue gas.

The ejector type scrubbers are used to remove fumes and dust from acidulation dens and curing operations in the fertilizer industry.

REFERENCES

1. "Chemico Gas Scrubbers for Industry," Bulletin M-104, Chemical Construction Corporation.
2. Nukiyama, S. and Y. Tanasawa: *Trans. Soc. Mech. Engrs. (Japan)*, vol. 5, no. 18, p. 68, 1939.
3. "Chemico P-A Gas Scrubbers," Bulletin M-102, Chemical Construction Corporation.
4. Pease-Anthony Venturi Scrubbers for the Steel Industry," Bulletin M-103A, Chemical Construction Corporation.
5. Basse Bernard, *Blast Furnace and Steel Plant,* November 1956.
6. "Kinpactor Venturi Type Dust Collector," Bulletin No. 294A, American Air Filter Company, Inc.
7. "Kinpactor Kinetic Scrubber," Bulletin No. 294, American Air Filter Company, Inc.
8. Shah, I. S.: *Chem. Eng. Prog.*, vol. 64, no. 9, pp. 66–77, 1968.
9. "Koch Equipment for Air Pollution Control," Bulletin KPC-1, Koch Engineering Co., Inc.
10. "Airetron Gas Scrubbers," Bulletin 767, Airetron Engineering Corporation.
11. "Chemico Gas Scrubbers for Industry," Bulletin M-104R, Chemical Construction Corporation.
12. "Research-Cottrell Flooded Disc Wet Scrubbers," Bulletin RC-975, Research-Cottrel, Inc.
13. Walker, A. B. and R. M. Hall: *APCA Journal,* vol. 18, no. 5, May 1968.
14. "UOP Wet Scrubbers," Bulletin No. 608, UOP Air Correction Division.
15. "Peabody Ventri Slot High-Energy Gas Scrubber," Bulletin 6006, Peabody Engineering Corporation.
16. "Ducon Oriclone Scrubber," Bulletin No. W-8560, Ducon Company, Inc.
17. "Heil Venturi Scrubbers," Bulletin B-720-1, Heil Process Equipment Corporation.

5.18 MECHANICALLY INDUCED SPRAY SCRUBBERS

Types of Designs:	(a) *Mechanical Scrubber* designs include the Schmieg vertical-rotor dust collector (Fig. 5.18a) and the center spray high-velocity scrubber (Fig. 5.18b); (b) *Mechanical centrifugal collector with water sprays* are the disintegrator scrubbers, the dynamic scrubber (Centrifugal fan wet scrubber) (Fig. 5.18c) and the inline wet scrubbers (Fig. 5.18d).
Principles of Operation:	Sprays are generated by mechanical means such as a rotary paddle, cage, drum or disk, or are generated by spray nozzles. Liquid is sheared mechanically to break into droplets for collection by inertial impaction between droplets and dust particles.
Pressure Drop Required:	3 to 10″ H_2O.
Power Requirement:	5 to 20 HP/1,000 cfm flue gas.
Liquor Flowrate Required:	0.5 to 5 GPM/1,000 ACFM.
Collection Efficiency:	99%+ on 5 to 10 μ dust particles; 90 to 95% on 1 μ or larger.
Typical Applications:	Blast furnace flue gas, steel industry foundries, coal processing and metal mining industries.
Costs:	See Table 5.21f.
Partial List of Suppliers:	See Section 5.12.

In *mechanical type scrubbers,* the sprays are generated by mechanical means, such as a rotary paddle wheel, cage, drum or disk, which are rotating elements partially submerged in a pool of liquor. The sprays are normally generated at right angles to the direction of the flue gas flow. The scrubbing liquor is usually recirculated, and the rates of recirculation and the degree of dispersion will vary with the different types of rotating elements. Dust collection takes place in a restricted passage where the sprays generated cause extreme turbulence and collision between dust particles and spray droplets.

In the *mechanical-centrifugal collector* (also called dynamic wet scrubber), liquor is sprayed at the inlet to enhance the collection efficiency. The collection mechanism is impingement of dust particles on the rotating blades.

In case of *high-pressure spray scrubbers*, the scrubbing liquor is sprayed at high pressure to increase the number and size of droplets and, thereby, to increase the collection efficiency.

Mechanical Scrubbers

The Schmieg vertical-rotor dust collector (Fig. 5.18a)[1] and center spray high-velocity scrubber (Fig. 5.18b)[2] are examples of mechanical scrubbers. In this type of scrubber, the rotor, which is partially submerged in a pool of scrubbing liquor, produces the high-velocity sprays. The atomization of the liquid is done by the rotor and, in addition, by the impaction of the liquor against the outer wall of the scrubber. The flue gas flowing vertically upward contacts with the horizontally formed high-velocity sprays. The dust particles are scrubbed due to the impaction of the high-radial velocity spray droplets with the particles in the flue gas. The mist eliminator section is required to remove liquid carryover by the flue gas.

This scrubber is very compact, and the scrubbing liquor requirements are relatively low, about 3 to 5 gallons per 1,000 cubic feet of flue gas. The pressure drop on the gas side, and the power required to generate the spray droplets, depends upon the dust particle size and on the desired scrubbing efficiency. The erosion of the rotor due to large and abrasive dust particles is the major defficiency in this type of scrubber. High-dust loadings can be handled if the units are designed for continuous removal of sludge.

Mechanical, Centrifugal Collector with Water Sprays

Disintegrator Scrubbers

The disintegrator type scrubber utilizes both rotating and stationary bars to break up the scrubbing liquor into fine droplets. The disintegrators are squirrel-cage type devices in which the scrubbing liquor is churned violently in one direction, while the dust-laden flue gas passes through the unit in the opposite direction. The barred rotor is located within the barred stator, and the scrubbing liquor is injected axially through the rotor shaft. The high-velocity difference between the rotor and stator bars mechanically shears the injected liquor and breaks it up into fine droplets. The inertial impaction

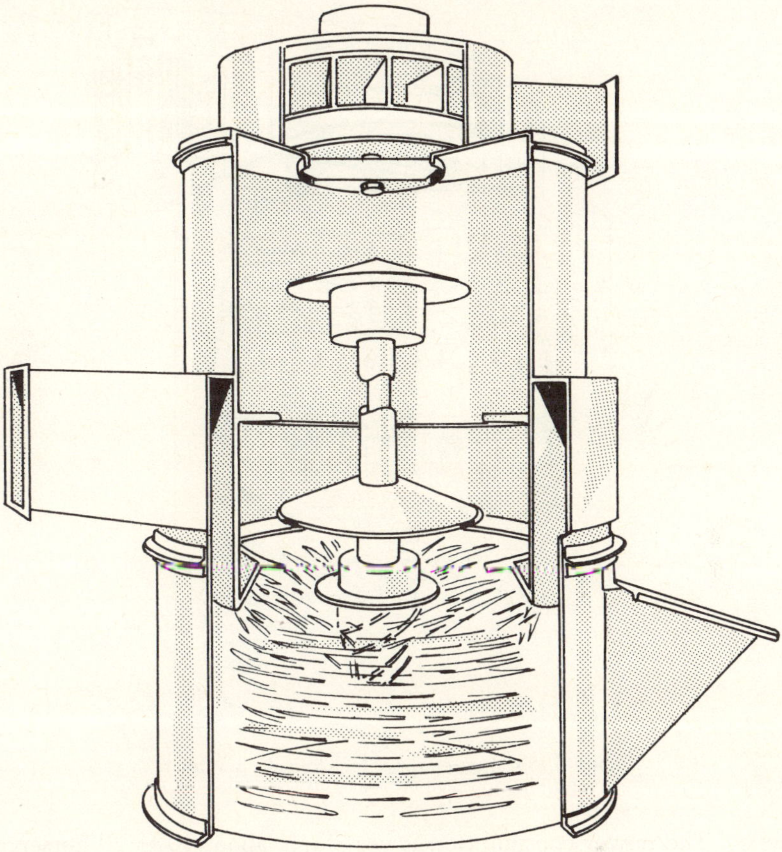

Figure 5.18a Schmieg vertical-rotor dust collector.

between the high-velocity droplets and dust particles causes the dust particles to be collected.

Disintegrators have been used for many years in the steel industry for gas cleaning purposes. The venturi scrubbers have replaced these devices. An outlet dust loading of 0.004 grains per standard cubic foot can be accomplished with disintegrator type scrubbers used for the cleaning of blast furnace flue gas.[3]

The scrubbing efficiency is essentially independent of the inlet dust loading, and it can be improved by increasing either the scrubbing water flowrate or the number of stator and rotor bars or both. For dust particles, $1\ \mu$ or larger, this type of scrubber can provide 95% or better collection

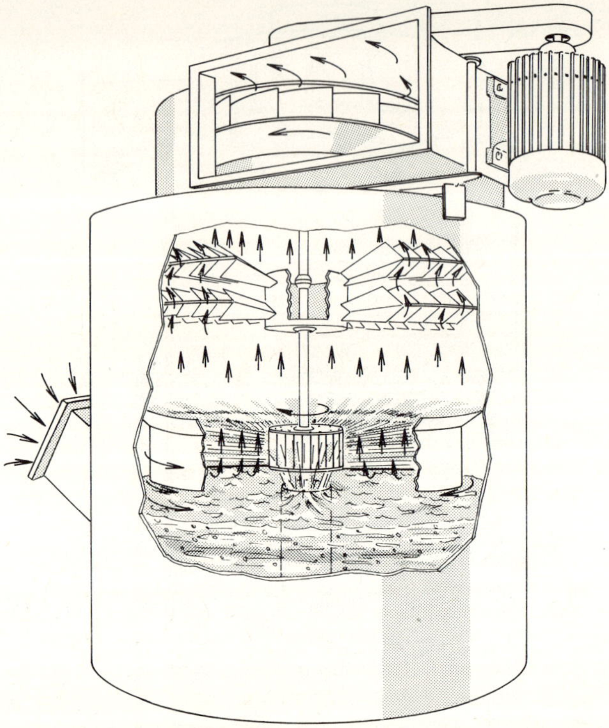

Figure 5.18b Center spray high-velocity scrubber.

efficiency. The power consumption is very high, about 10 to 20 horsepower per 1,000 cfm of gas.

The buildup of dust particles on the rotor must be avoided to prevent rotor unbalance. (The rotor speed is 350 to 750 rpm.) This can be achieved by lowering the temperature (prequenching the flue gas to attain saturation) and by reducing the inlet dust loading (precleaning of the flue gas using a cyclone or centrifugal collector).

Its compact size and high-collection efficiency (even for submicron dust particles) are the two main advantages, whereas the large power requirements and buildup problems are the principal disadvantages.

Dynamic Scrubber

The dynamic scrubber[4] (also called centrifugal fan wet scrubber) is designed to combine the scrubbing effect of water with the effect of dynamic precipitation. This type of scrubber (Fig. 5.18c) consists of a multiple blade centrifugal blower. In its simplest form, the scrubbing liquor is sprayed into the suction of the fan with the wetted impeller and fan housing preventing

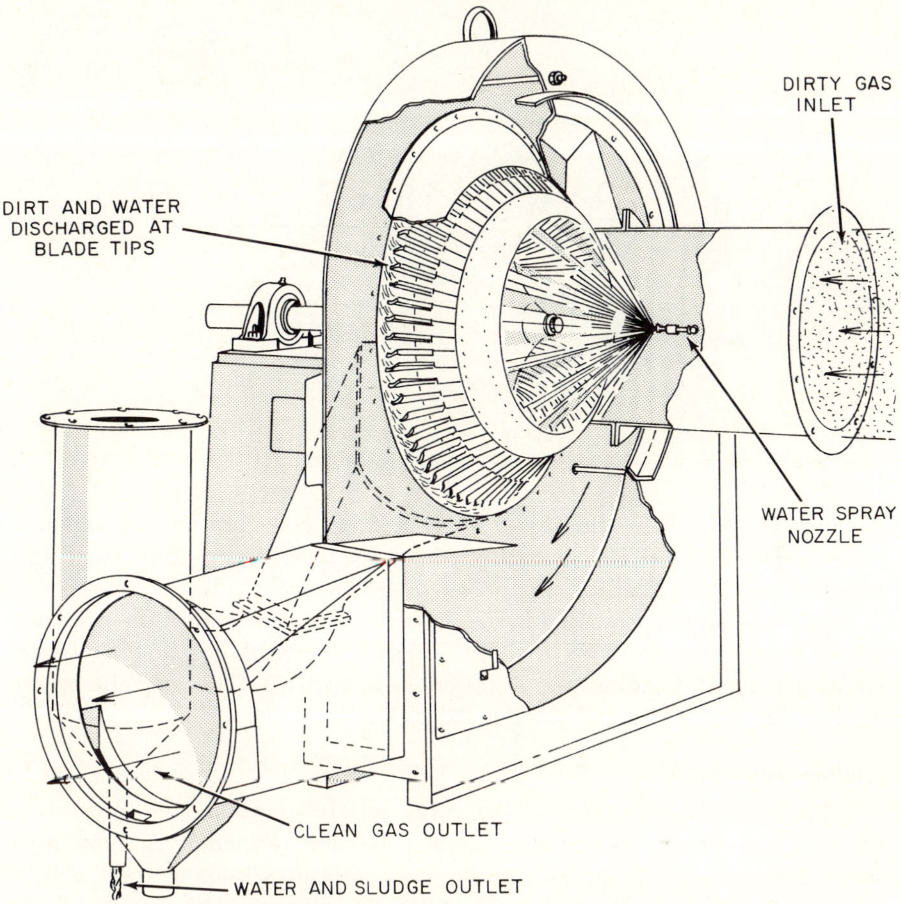

Figure 5.18c Centrifugal fan wet scrubber.

dust particles from reentrainment. The dust particles impinge on the rotating blades, and the spray flushes away the collected dust.

The relative efficiency for fine particles is high. The pressure drop ranges from 6.5 to 9″ H_2O, whereas the scrubbing liquor requirements are 0.5 to 1.5 gallons per 1,000 cubic feet of gas.

The main features of this scrubber are high dust collection efficiency at moderate power requirements and low water and space requirements. The typical applications include foundries, coal processing and metal mining.

Inline Wet Scrubbers

This scrubber (Fig. 5.18d)[5] consists of a spray chamber followed by impingement screens, which are followed by a centrifugal section. The dust

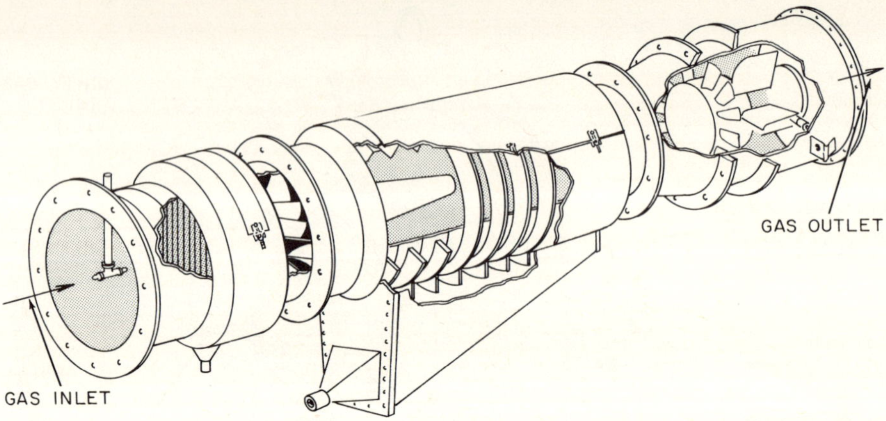

Figure 5.18d Inline wet scrubber.

particles are wetted by the liquor spray and are collected by impingement on the baffle screens. The wetted particles are separated by the centrifugal action.

For dust particles ranging in size from 5 to 10 μ, the collection efficiency is 99%+ at a pressure drop of about 5″ H_2O. The scrubbing liquor requirements are about 1 gallon per 1,000 cubic feet of gas. These scrubbers are used in the coal and metal mining industries.

High-pressure Sprays

Increasing the pressure at the spray nozzles increases the number of droplets and, hence, increases the dust collection efficiency. A number of scrubbers operate with spray nozzle water pressures ranging from 300 to 600 psig (normally most scrubbers use 100 to 150 psig). The nozzle orifices are very small and, therefore, the recirculation of scrubbing liquor is not practical. The location and design of nozzles is important to minimize collision between liquor droplets while ensuring maximum collision between dust particles and spray droplets.

REFERENCES

1. "Schmieg Vertical-Rotor Dust Collector," Catalog 101A, Schmieg Industries.
2. Jamison, R. M., V. W. Hanson and O. M. Arnold, *Air Engineering Magazine.*
3. Rice, O. R. and C. G. Bigelow: Amer. Inst. of Mining and Metallurgical Eng., February 1950.
4. "Roto-clone Dynamic Precipitator Type W," Bulletin 274-F, American Air Filter.
5. "The Joy Microdyne Dust Collector," Bulletin J110, Western Precipitator Company.

5.19 MIST ELIMINATORS AND IRRIGATED FILTERS

Types of Designs:	(a) *Irrigated Filters* (Fig. 5.19a); (b) *Mist Eliminators,* which can be of the Wet fiber type, Brink type (Figs. 5.19b and c), Wire mesh type (Fig. 5.19d), Impingement baffle type (Fig. 5.19f), Vane type (Fig. 5.19f) and the Packed bed type.
Principles of Operation:	Inertial impaction or impingement. In case of the Brink mist eliminator, the predominant mechanism is Brownian diffusion.
Pressure Drop Required:	0.5 to 15″ H_2O.
Liquor Flow:	8 to 10 gallons per 1,000 cubic feet of gas, for irrigated wet filters.
Collection Efficiency:	Irrigated wet filters, 90 to 99%+ on TiO_2 dust and on acid mist. Brink mist eliminator, Low velocity, 99%+ on acid mists 3 μ or smaller; High velocity, 90 to 96% on mists 3 μ or smaller. Wire mesh mist eliminator, 90 to 99%+ on acid mist. Baffle types, 85 to 90% on 16 to 200 mesh size particles; 95%+ on 40 μ or larger; Vane types, 99% on particles 11 μ or larger.
Applications:	Mist removal from sulfuric, phosphoric and nitric acid plants, coke quenching operations or oil mists.
Costs:	See Table 5.21f.
Partial List of Suppliers:	See Section 5.12.

The main mechanism for separation and collection of mist and dust particles from flue gas is the inertial impaction or impingement. In case of wet fiber mist eliminators, in addition to inertial impaction, the mechanism of Brownian diffusion is also involved. In case of a Brownian diffusion mechanism, the collection efficiency increases with the decrease in gas velocity as more retention time is attained in the filter bed. This mechanism assists in the collection of particles, 3 μ and smaller, by direct interception. The Brownian diffusion is the primary mechanism in the collection of mists (largely submicron size) in fiber beds having large specific surface areas and operating at flue gas velocities of 0.08 to 0.5 feet per second. For particles larger than 3 μ and at gas velocities higher than 30 feet per second in coarse filter beds, the dominating collection mechanism is inertial impaction. The collection efficiency of the inertial impaction process increases with an increase in flue gas velocity.

The mechanism by which dust and mist are collected in irrigated wet filters is the interception by individual filaments in the filter. Simultaneous removal of gaseous pollutant and particulate matter can be accomplished in these wet filters.

Irrigated Wet Filters

A wet filter consists of a spray chamber with filter pads of glass fibers, knitted wire mesh or other fibrous materials. Two or more filter stages can be arranged in series with the number, depending on the characteristics of the flue gas and on the collection efficiency required.

The dust is collected on the filter pads, and sprays are used to wash it off (Fig. 5.19a). The pads are about 3 to 8 inches thick and have about 20 inches square surface area. The gas velocities range from 3 to 5 feet per second, and the liquor flow requirements range from 8 to 10 gallons per 1,000 cubic feet of gas which is equivalent to approximately 3 gallons per minute per square foot of filter area. The pressure drop (0.2 to 3″ H_2O per 4-inch bed depth) is a function of the gas flow, liquor flow and type of filter.

The liquor can be recycled, but the use of spray nozzles restricts the allowable solids concentration. The final stage can be a dry bed containing small diameter fibers used as a final entrainment separator stage. Sometimes a wetted impingement plate is used preceding the first filter stage to reduce the dust loading.

The filter pads commonly used are coarse fibers, fine glass wool fibers or knitted wire mesh. The coarse fibers are not efficient for the collection of fine dust, whereas the fine glass wool fibers are efficient; but because of the matting and sagging when wetted, their usefulness is limited. Pilot plant data for the removal of titanium dioxide dust and sulfuric acid mist from calciner exhaust gases have been reported, using an irrigated, thin, felted fiber filter made of polyester fibers and polypropylene.[1] The flue gas containing

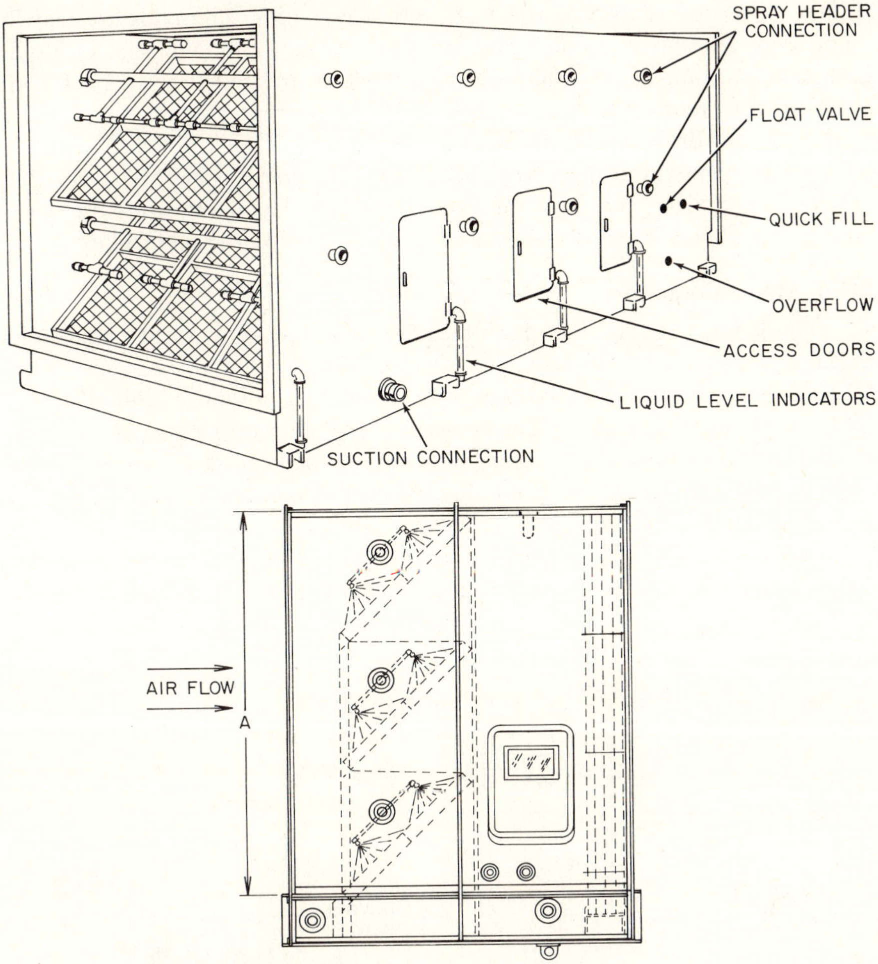

Figure 5.19a Irrigated wet filter.

0.044 grain of TiO_2 per standard cubic foot and 1.56 grains of H_2SO_4 acid mist per standard cubic foot were treated. The outlet loadings were essentially zero for TiO_2, and 0.02 to 0.09 grains per standard cubic foot of acid mist. A minimum liquor flowrate of 8 gallons per 1,000 cubic feet of gas is required to obtain high-mist removal efficiencies, and the pressure drops range from 5.5 to 10″ H_2O at gas flowrates of 1 to 5 feet per second.

Wet Fiber Mist Eliminators

This type of mist eliminator consists of fiber pads through which a mist-laden gas is forced to travel. While passing through the pads, the mist coalesces into droplets that run back into a collector. The pads are usually made from glass fiber or teflon and sometimes from thin metal wire of highly corrosion resistant alloys. These mist eliminators are widely used in sulfuric acid plants to remove acid and oleum mists from effluent gases. Two mechanisms, namely Brownian diffusion and inertial impaction, are involved in the separation of mist and dust particles in the wet fiber mist eliminators. The collection efficiency and pressure drop are largely dependent upon the diameter of the fiber and on the distance between the two adjacent fibers.

Brink Mist Eliminators

The Brink mist eliminator utilizes the Brownian movement of particles in the low and submicron ranges to effect high collection efficiencies. Figure 5.19b shows a typical low-velocity (5 to 30 feet per minute), high-efficiency (H-E) Brink mist eliminator.[2] The construction of fiber mist eliminators is simple since they are only a packed bed of fibers retained between two concentric screens. Large mist particles are collected on the fibers by inertial impaction and direct interception, and smaller particles are collected by Brownian diffusion. Mist particles coalesce to form a liquid film that wets the fibers. The liquid on the fibers moves horizontally through the fiber bed due to the

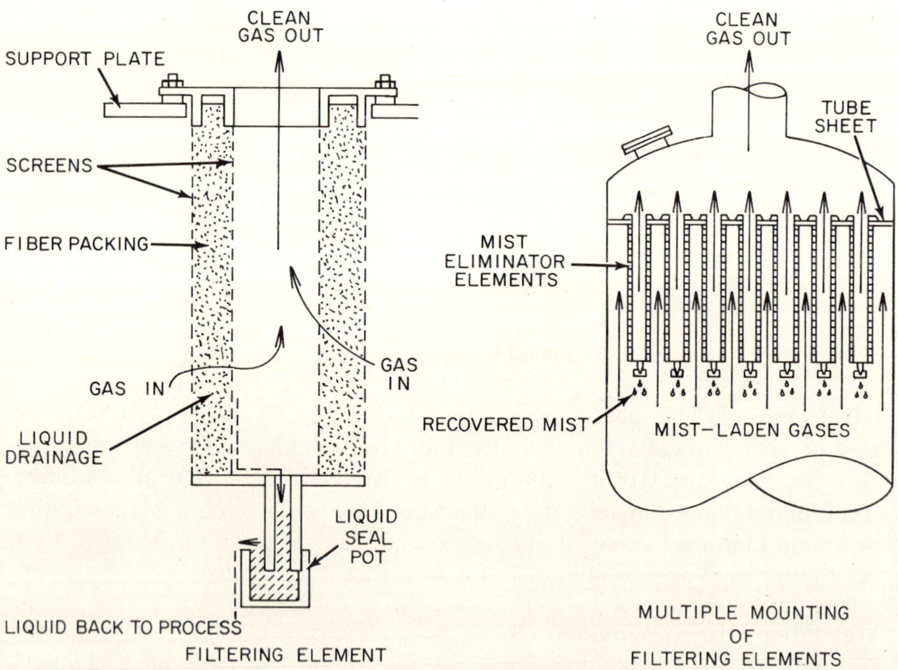

Figure 5.19b Low-velocity Brink mist eliminator.

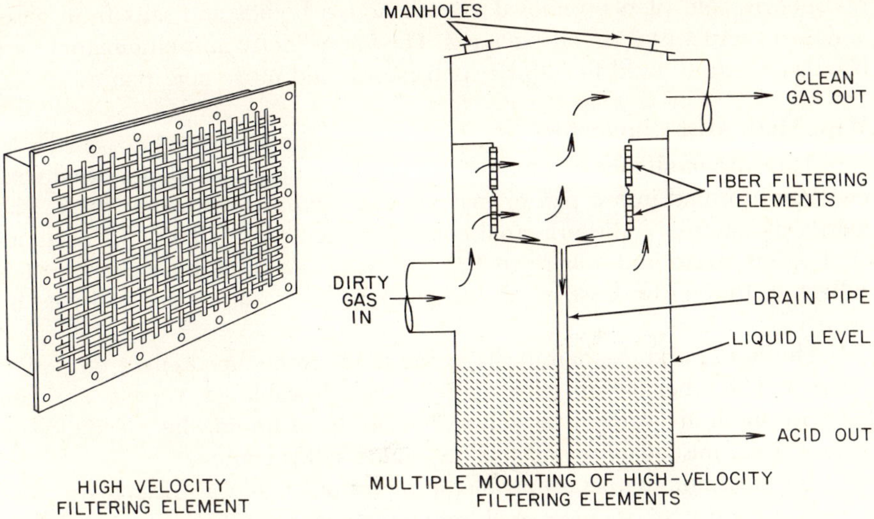

Figure 5.19c High-velocity Brink mist eliminator.

drag of the gases. Then it moves downward (due to gravity) around the inner screens to the bottom of the element and through a liquid seal into a collection reservoir.

Extremely high collection efficiencies, as high as 99.98% on acid mists 3 μ and smaller, have been reported.[2] Since the Brownian diffusion is a predominant mechanism for mist collection, the efficiency increases slightly with a decrease in gas velocity through the fiber bed. The pressure drops range from 5 to 15″ H_2O. The low velocity filtering elements can be mounted in multiplicity, as shown in Figure 5.19b. These units range in capacity from 3 to 20,000 cfm. Installed costs are considerably less than that of other equipment with equivalent capacity.

Figure 5.19c shows the high-velocity (30 to 90 feet per minute) filtering element and a multiple mounting installation of it. The liquid flow pattern is similar to that of the low-velocity filter element. The high-velocity filter element is a flat bed element rather than a cylindrical design. The thickness of the fiber bed can be varied by changing the frame which holds down the screen on the fiber bed. The flue gas passes through the fiber bed into a sealed semicircular chamber behind the bed and then out at the top.

A collection efficiency of 92% on particles 3 μ and smaller at a pressure drop of 3 to 3.4″ H_2O has been reported[3] when no oleum was being produced. The collection efficiency dropped to 39% when oleum production was at maximum. At a pressure drop of 7.5 to 8.0″ H_2O, collection efficiencies as high as 96% were obtained on particles 3 μ and smaller.

The typical applications of high-velocity mist eliminators are sulfuric and phosphoric acid plants, removal of radioactive liquids and salts from gases and nitric acid mist from nitrogen gas. The low velocity mist eliminators have been successfully used in sulfuric, phosphoric and nitric acid plants.

Wire Mesh Mist Eliminators

The wire mesh mist eliminator (Fig. 5.19d) is constructed of knitted wire mesh and formed into a pad or mat, usually from 4 to 6 inches thick. The inertial impaction is the principal collection mechanism. The mist particles or droplets strike and adhere to the fine wires. As additional particles are collected, they grow in size and weight until they run off and fall free by gravity.

The liquid and gas density have the most pronounced effect on design velocity. The other factors which govern the allowable gas velocity through the wire mesh are surface tension and viscosity of liquid, the specific wire-surface area, mist loading and the suspended solids content.

The mist eliminators are available in various materials of construction, namely carbon steel, stainless steels, polyethylene, polypropylene and Teflon. In the composite mesh construction, multifilament yarns, namely glass fiber and synthetic and natural fibers, can be incorporated into the basic wire mesh structure.

The allowable velocity of the composite mesh mist eliminators is approximately one-half of that for the wire mesh styles. The pressure drop for wire mesh mist eliminators ranges from 1 to 3″ H_2O; whereas for composite mesh mist eliminators, the range is from 3 to 10″ H_2O.

This type of mist eliminator is widely used in phosphoric and sulfuric

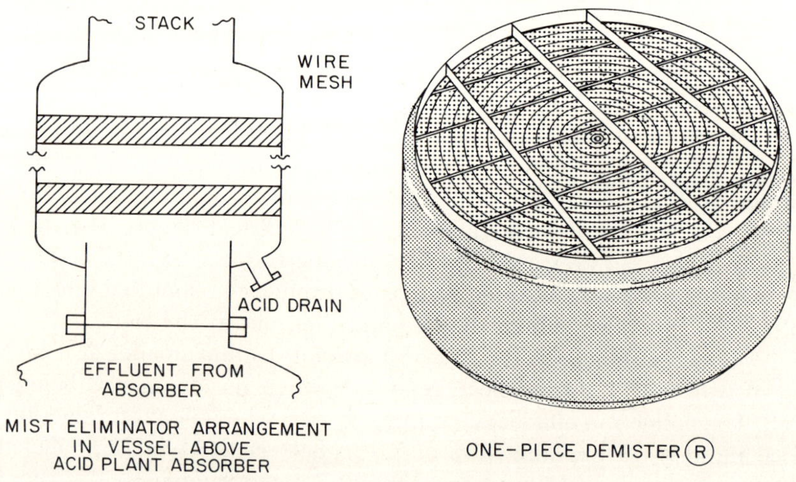

Figure 5.19d Wire mesh mist eliminator.

Table 5.19e
PERFORMANCE DATA OF WIRE MESH MIST ELIMINATORS FOR REMOVAL OF SULFURIC ACID MIST

Mesh Type 2-stage	*Pressure Drop, inches H_2O*	*Gas Velocity, ft/sec*	*mg H_2SO_4/cu ft*		*Efficiency, %*
			Inlet	*Outlet*	
Wire mesh	1.0	12–15	8.8	0.70	92.5
	0.75		4.1	0.74	82.0
	1.25		13.8	0.92	93.4
	1.25		66.9	0.77	98.9
	1.50		32.5	0.98	97.0
	1.50		13.0	0.76	94.1
Wire mesh	1.5–2.0	15–18	38–150	2.1–2.3	94–99+
Composite	2.5–9.0	9–18	25–200	0.7–1.7	97–99+

acid plants to remove acid mists. The performance data[4] for wire mesh and composite mesh filters are given in Table 5.19e. Outlet acid mist loading of less than 1 mg per cubic foot can be attained. The composite mesh provides better collection efficiency than does the wire mesh filter, but only at the expense of a higher pressure drop. The typical applications are sulfuric and phosphoric acid plants, oil mist removal and recovery of precious metals.

Impingement Baffle and Vane Type Mist Eliminators

These are simple devices normally used for the removal of large particles and liquid carryover. *Baffles* are usually employed as entrainment separators, which are so arranged that the flue gas is forced to take a zigzag path. The dust particles impinge on the baffle and are collected. Washing sprays are usually provided to wash off the collected dust and to keep the baffles clean. The sprays can be continuous or intermittent.

Figure 5.19f shows some commonly used mist eliminators. The flue gas velocity through the mist eliminator, when used as dust collector, may vary from 15 to 40 feet per second; whereas when used as an entrainment separator, the velocity may not exceed 11 feet per second.

Mist eliminators are successfully used in the reduction of emissions from the coke quenching operation in steel mills. An 85 to 90% collection efficiency has been reported[5] on particulate matter ranging from 16 to 200 mesh. At a pressure drop of 1 to 2″ H_2O, removal efficiency as high as 95% can be achieved on mist particles 40 μ or larger.

The *vane type* mist eliminators operate at gas velocities ranging from 10 to 50 feet per second and at a pressure drop range of 0.1 to 2″ H_2O. Collection efficiencies of 99% for particles 11 μ or larger have been reported.[6]

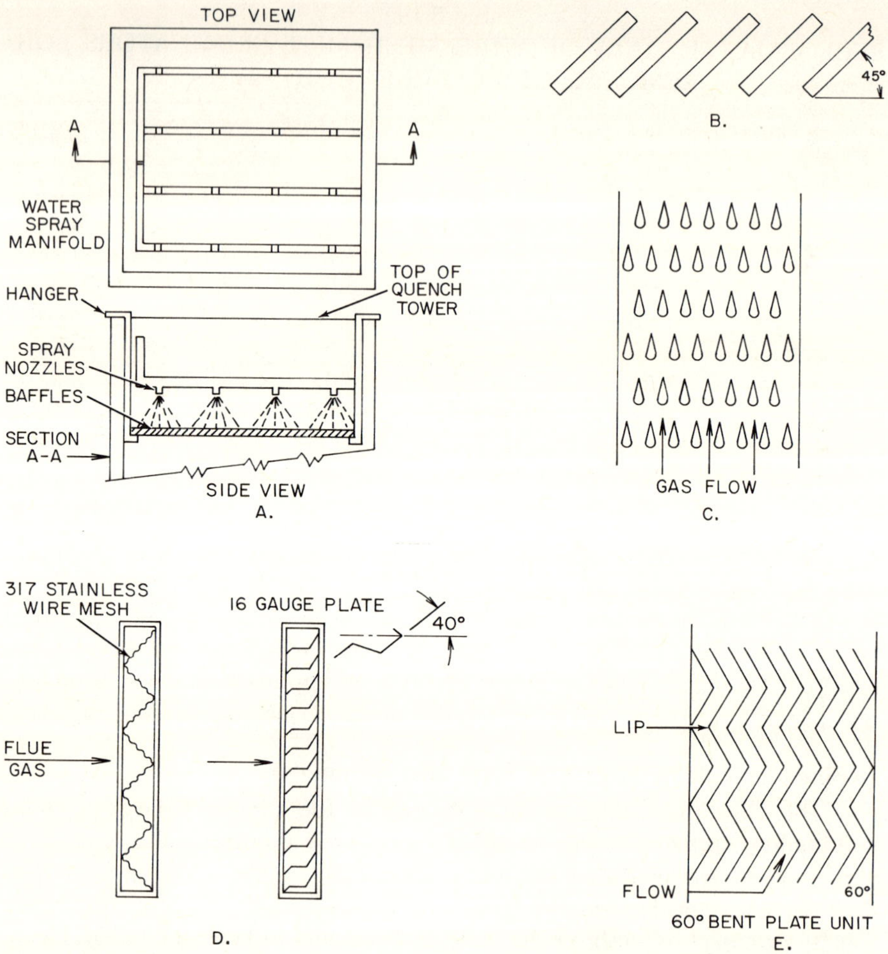

Figure 5.19f Mist eliminators. *A*, Spray and baffle arrangement; *B*, Mist eliminator baffles; *C*, Streamlined mist eliminator baffles; *D*, Screen and mist eliminator; *E*, Baffle design.

Packed Bed Mist Eliminators

Packed beds are used for tail gas cleanup in sulfuric and phosphoric acid plants. The operation of packed bed mist eliminators is similar to that of packed bed scrubbers (Section 5.14). Normally fixed bed types are used as mist eliminators with a variety of packings, including saddles, pall rings or marbles. The flue gas velocities are limited to a maximum of 10 feet per second. These mist eliminator designs are not suitable for dust collection, due to plugging problems encountered.

REFERENCES

1. Morash, N. et al.: *Chem. Engr. Prog.* vol. 63, no. 3, pp. 70–74, 1967.
2. Brink, J. A., Jr.: *The Canadian Journal of Chem. Eng.*, vol. 41, pp. 134–138, June 1963.
3. Brink, J. A., Jr. et al.: *Chem. Engr. Prog.*, vol. 60, no. 11, pp. 68–73, 1964.
4. York, Otto H. and E. W. Poppele: *Chem. Engr. Prog.*, vol. 59, no. 6, pp. 45–50, 1963.
5. Fullerton, R. W.: *J. Air Pollu. Contr. Ass.*, vol. 17, no. 12, pp. 807–809, 1967.
6. "Chemical Engineers' Handbook," 4th ed., pp. 18–85, McGraw-Hill, New York, 1963.

5.20 SPRAY NOZZLES AND LIQUID DISTRIBUTORS

Types of Designs: (a) *Spray nozzle* designs include the hollow cone type (Fig. 5.20a), the full cone type (Fig. 5.20b), the flat and atomizing types or the impingement and rotating types (Fig. 5.20c); (b) *Liquid distributor* designs (Fig. 5.20d) include the weir box V-notch type, the weir risers and the submerged orifice plate.

Applications: Dust collection and gas absorption using packed scrubbers, spray towers and chambers, cyclonic and centrifugal scrubbers.

Partial List of Suppliers: Bette Fog Nozzle, Inc. (a); Koch Engineering Co., Inc. (b); Schutte & Koerting Co. (a); Spray Engineering Co. (a); Spraying Systems Co. (a); U.S. Stoneware Co. (b).

Spray Nozzles

Almost all the wet scrubbers use spray nozzles[1-3] for the introduction of scrubbing liquor into the scrubber. The spray nozzles are used to create small droplets to provide large surface areas for particulate collection and gas absorption, to distribute the scrubbing liquid in a specific pattern, to provide proper coverage in a scrubber. The spray nozzle should be capable of handling slurries because in most cases, the scrubbing liquor is recycled.

The spray nozzles normally used in wet scrubbers can be classified in accordance to their spray patterns, namely hollow cone, full cone, flat or atomizing. They can also be classified as pressure nozzles (hollow and solid cone, impingement and impact), rotating nozzles (spinning atomizers) and miscellaneous types.

Hollow Cone Spray Nozzles

Figure 5.20a shows some of the hollow cone spray nozzles.[1-3] Here, the liquor is fed to a swirl chamber either through a tangential inlet or through

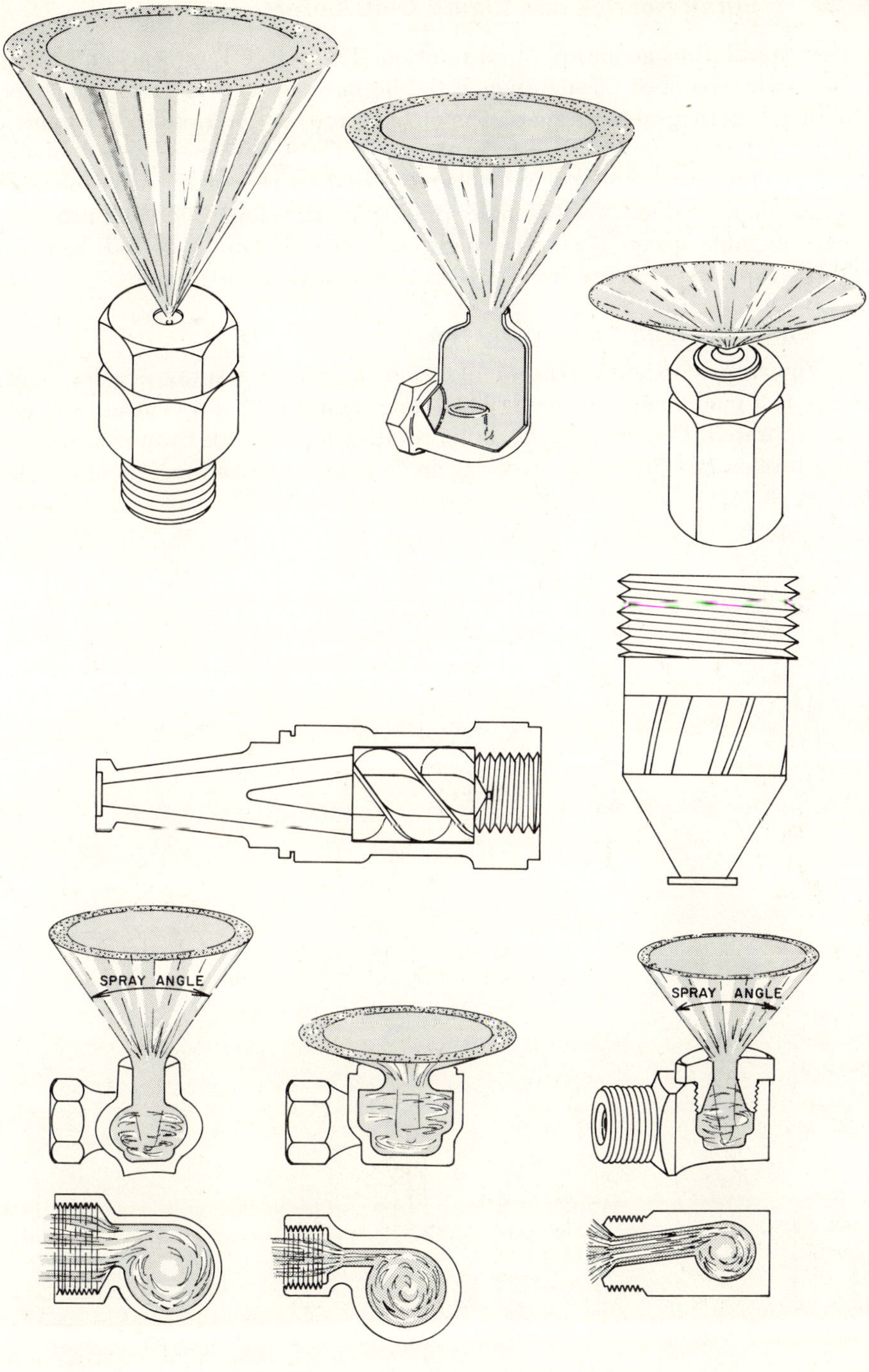

Figure 5.20a Hollow cone spray nozzle designs.

a fixed spiral, thus acquiring rapid rotation. The orifice is on the center axis of the swirl chamber through which the liquor exits as a hollow cone sheet and then breaks up into droplets. The nozzle pressure determines the number and size of droplets and the angle of spray. The spray angle also depends upon the dimensions of the swirl chamber and on the pitch of the nozzle. A short pitch produces a wide-angle spray, whereas a long pitch will produce a narrow-angle spray. The spray angle may range from 15 to 165 degrees, and the liquor flow range from 0.04 to 2,360 gallons per minute.

Solid Cone (Full Cone) Spray Nozzles

Figure 5.20b shows some of the solid or full cone spray nozzles. This nozzle is a modification of the hollow cone design in that a central axial jet is incorporated. The jet strikes the outer rotating liquid inside the nozzle orifice and is broken into droplets. The spray angle is essentially a function of nozzle

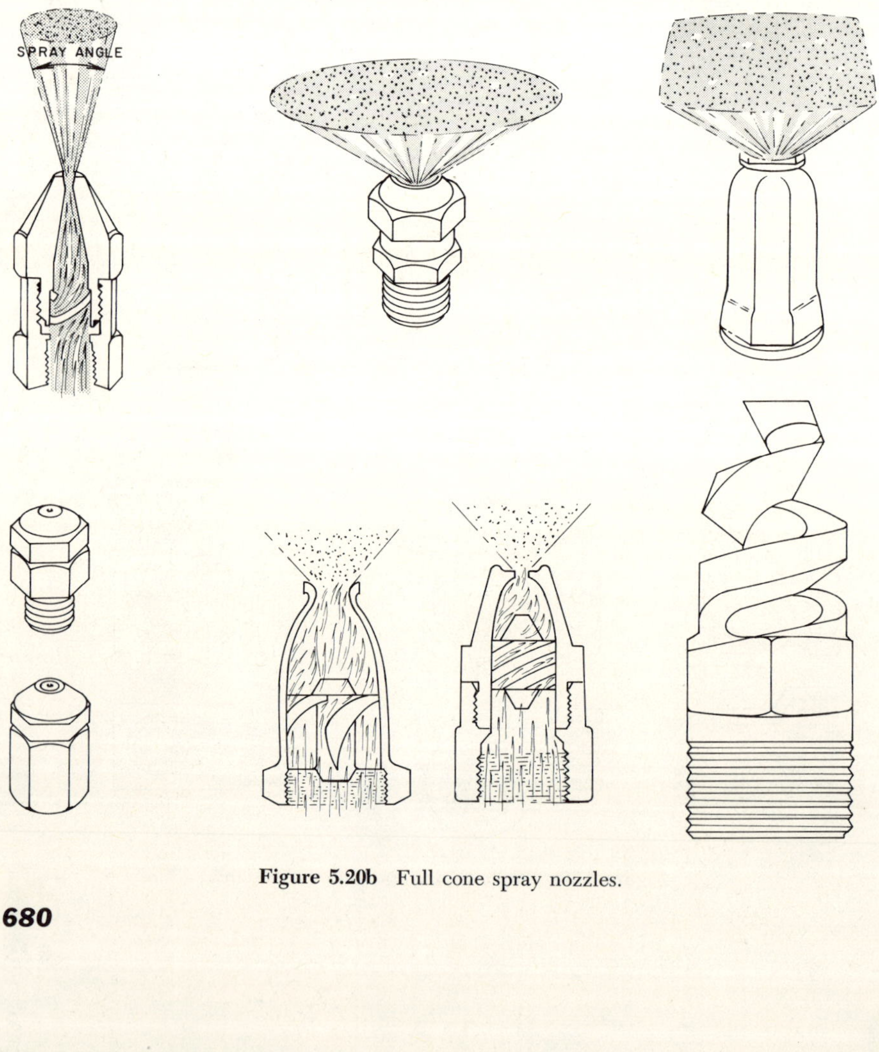

Figure 5.20b Full cone spray nozzles.

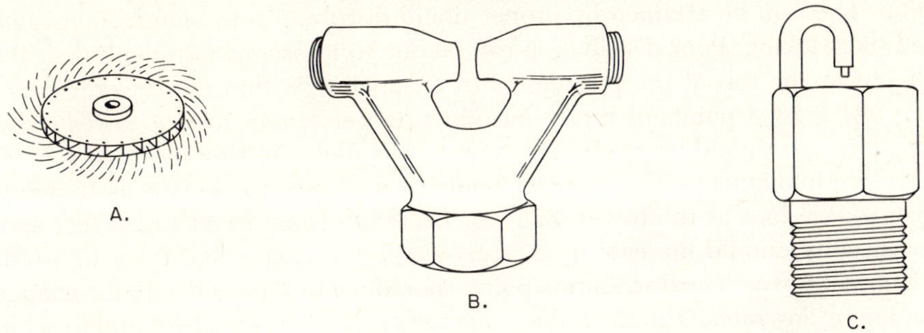

Figure 5.20c Impingement and rotating nozzles. *A*, Spinning disc spray nozzle; *B*, Opposed jet spray nozzle; *C*, Pin jet impingement spray nozzle.

design and does not depend upon the liquor supply pressure. The solid cone nozzle, consisting of an orifice and of an external helical spiral, is essentially nonclogging. The spray angles range from 30 to 100 degrees. The liquid distribution can be improved by using several small sprays instead of one larger spray of the same capacity, as is the case in a cluster type spray nozzle.

Impingement and Rotating Nozzles

In the impingement type nozzle,[1-3] a high-velocity liquid jet is directed at a solid target or at a liquid stream. By proper orientation and shape of the target, or by controlling the size and shape of two streams, one can produce a hollow cone, fan or dish-shaped spray pattern.

The rotating nozzles (also called spinning atomizers) produce droplets which are uniform in size and can be controlled by changing the speed of the disc. These are less frequently used than either the hollow or solid cone nozzles. The spinning atomizers can easily handle slurries which might plug conventional nozzles. Figure 5.20c shows impingement and rotating type nozzles. Other, miscellaneous nozzles include the fog jet and fan jet designs.

The installation of nozzles is very important, as they frequently wear or clog and require replacement and maintenance. They should be installed such that they can be easily removed and replaced.

Full cone and hollow cone spray nozzles are widely used in wet scrubbers. In jet ejectors, the hollow cone spray nozzles with spray angles of from 15 to 20 degrees are used. The rotating nozzles are utilized as liquid distributors.

Liquid Distributors

The liquid distribution[4,5] is of utmost importance in wet scrubbers. In spray chambers and towers, by suitable location of nozzles, a proper distribution of liquor can be attained for efficient contact with flue gas. The effectiveness of a packed scrubber depends on the availability of a large, exposed liquid film. This can be attained by proper liquid distribution to uniformly irrigate all the packing. Poor distribution can be due to improper introduction of the liquor at the top of the packing or to channeling within the tower.

At least 4 points of liquor introduction per square foot of tower cross-section area should be provided to ensure complete wetting for towers 4 feet or more in diameter. This number should be increased to as many as 16 points per square foot as the tower diameter decreases from 4 feet to 2.5 feet and, eventually, should increase to as many as 32 per square foot for a 1.25 foot diameter tower. The distribution points should not be changed with the change in liquor flowrates. The small diameter towers have shorter beds and thus are more sensitive to good liquor distribution.

The liquid rate must be sufficient to wet the packing but not to flood the tower. Solid cone spray nozzles are excellent distributors but tend to plug with solid particles. In randomly packed towers, the liquor tends to channel toward the walls, as the density of the packing is lower near the walls. By use of proper liquid distributor design, the channeling problem and uneven distribution of liquor can be avoided.

Most of the liquid distributors (Fig. 5.20d) are of the weir box V-notch, weir riser and submerged orifice plate type.[4,5] The spray nozzles, because of wear or clog characteristics, are not dependable liquid distributors; whereas the weir box distribution is dependable and requires little maintenance after initial leveling. The pumping power requirements are lower in case of distributors (as compared to spray nozzles). For vertical packed beds depending on the height of packing, liquid redistribution may be required to avoid

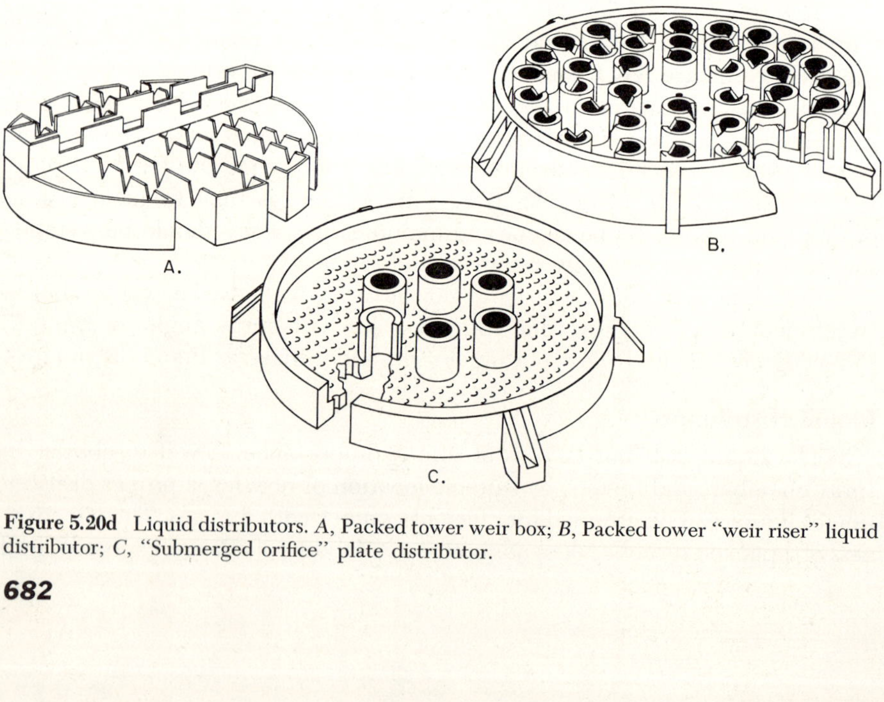

Figure 5.20d Liquid distributors. *A*, Packed tower weir box; *B*, Packed tower "weir riser" liquid distributor; *C*, "Submerged orifice" plate distributor.

channeling. Normal drip point requirements for weir box distribution range from 8 to 10 points per square foot for vertical packed beds and from 15 to 30 points for cross-flow scrubbers.

REFERENCES

1. "Spraco Spray Nozzles," Catalog 68, Spray Engineering Company.
2. "Industrial Spray Nozzles," Catalog 25A, Spraying Systems Company.
3. "Bette Fog Nozzles," Catalog of Bette Fog Nozzle Inc.
4. Eckert, John S.: *Chem. Engr. Prog.* vol. 57, no. 9, pp. 54–58, 1961.
5. "Design Information for Packed Towers," Bulletin DC-10, U.S. Stoneware.

5.21 SELECTION AND APPLICATION OF WET COLLECTORS

For a given application, the selection of equipment is primarily based on performance and reliability. The equipment and operating cost is the secondary criteria because the cost to attain acceptable performance and reliability of operation far exceeds the initial savings resulting from low equipment and operating costs.

To determine the performance of the equipment under consideration, one can obtain performance data for similar applications from the manufacturer. For a new application, a pilot plant test is the best solution.

To make a preliminary selection of a scrubber, the basic data required are inlet dust loading, particle size distribution, gas flowrate and the required collection efficiency. Table 5.21a summarizes the data required for equipment selection.[1] This is also the information which equipment suppliers need to select and offer the best equipment for the application.

Selection Procedure

The steps one should follow in the selection of suitable equipment are as follows:

1. *Definition of the problem.* This should include the study of present emission problems at the plant and of pollution codes, present as well as future.
2. *Gathering of basic information.* This should include items listed in Table 5.21a.
3. *Preparing the specifications* for vendors to bid on.
4. *Evaluation and screening.* The vendors' quotes must be carefully evaluated. This includes comparison of guarantees, operating and equipment costs, reliability of materials of construction. In addition, vendors' reputation, capabilities and past performance are also evaluated.
5. *A visit to operating installations* on similar applications. Visits can usually be arranged by the selected vendors or through other contacts.
6. *Pilot plant test.* This is desirable if the application is new, such

Table 5.21a
PROCESS DATA FOR SCRUBBER SELECTION

Flue Gas Characteristics	*Particulate Matter Description*	*Effluent Considerations*
Flowrate (minimum, normal and maximum)	Source	Collection efficiency desired or code requirement
Temperature and static pressure	Particle size distribution	Method of disposal
Humidity or moisture content	Concentration (range if necessary)	
Dust loading (range if necessary)	Composition	
Other components of flue gas	Particle density and bulk density	
Corrosive nature of flue gas		

as a wet scrubber application for flyash removal from coal-fired flue gas.

7. *Final evaluation and selection.*

Both the dry and wet methods of collection should be evaluated. If a dry collector (Section 5.11) can provide performance to meet the codes, it is preferable to collect the particulate matter in a dry state, provided the pollution problem is caused by particulate matter alone. The most important factor favoring the selection of a wet scrubber is its ability to simultaneously remove both gaseous and particulate pollutants. The simultaneous removal of flyash and sulfur dioxide from the flue gases of coal-fired boilers is the best example.

In the evaluation of wet scrubber designs, one should consider the following operating parameters:

1. Pressure drop, on gas side.
2. Liquor flow, gallons per 1,000 cubic feet of gas.
3. Ability to handle suspended solids.
4. Protection against corrosion and errosion, materials of construction.
5. Overall operating and equipment costs.
6. Performance data on similar applications.
7. Maintenance requirements.
8. Reliability of operation.
9. The effect of varying gas flowrates on efficiency.
10. Capability to simultaneously remove gaseous pollutants.
11. Entrainment separator design and its method of cleaning. The

entrainment separation is very important as efficient work by the scrubber can be canceled if the entrainment separator is not highly efficient.

12. Advantages and disadvantages of hot fan (located before the wet scrubber) versus cold wet fan (located after the wet scrubber).
13. What happens if the liquor flow to the scrubber is stopped? If this happens, can the scrubber handle hot gases? Are there provisions to cool the gas so that it can pass through the scrubber? Does the scrubber have to be bypassed? Will the failure of the scrubber require the shut down of the process unit that

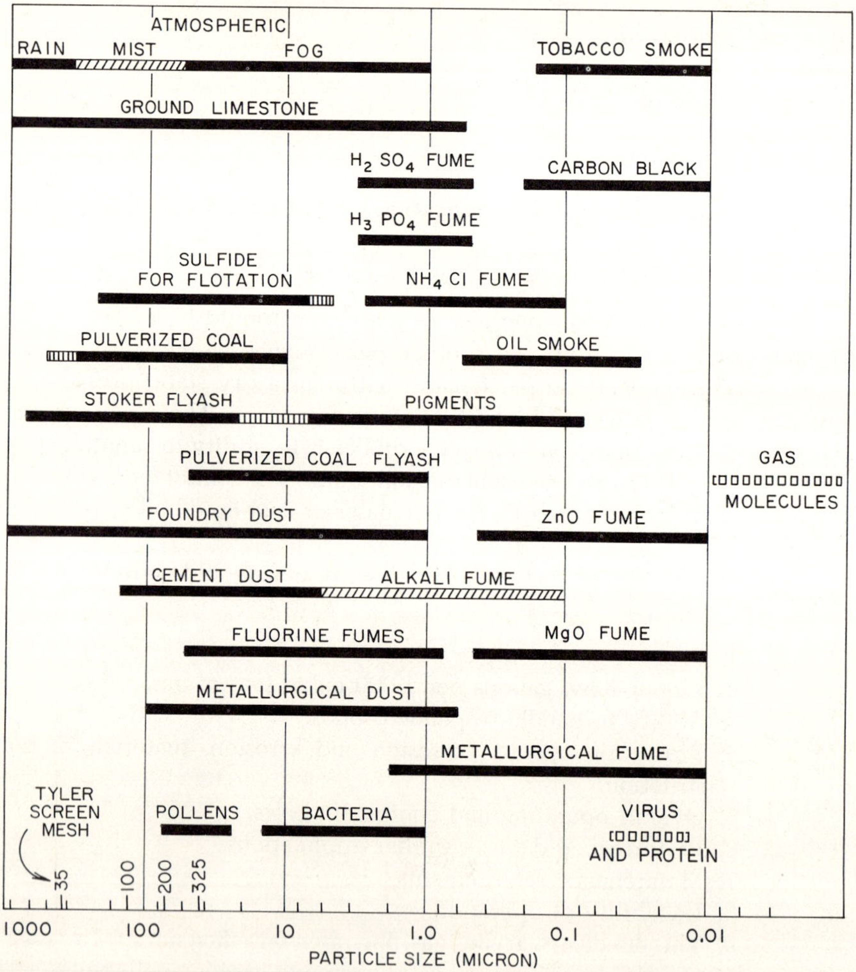

Figure 5.21b Particle size range of various aerosols.

Table 5.21c
INDUSTRIAL APPLICATIONS OF VARIOUS WET SCRUBBERS

Type of Scrubber	*Applicable Section*	*Industrial Application*
Spray chambers	5.13	Kraft mill recovery furnace, phosphate fertilizer, soot blowing operations and electroplating
Spray towers	5.13	Same as spray chambers
Elbair scrubber	5.13	Coal-fired boilers, basic oxygen steel furnace, blast furnace, roasters and sinter plant
Fixed packed bed	5.14	Acid mist, fertilizer industry and other soluble dusts
Floating bed	5.14	Flyash, aluminum ore reduction, aluminum foundries and viscid, gummy and gel type materials in the phosphate industry
Flooded bed	5.14	Carbon black, flyash and cupola dust
Pease-Anthony cyclonic	5.15	Kilns and dryers, fertilizer industry, Kraft recovery furnace, flyash, blast furnace and lime kilns
Multiwash scrubber	5.15	Foundry, alumina and magnesium dusts, kilns and dryers, pigment dusts and fumes, nitrate and phosphate dusts
Cyclonic spray scrubber and centrifugal scrubber	5.15	Galvanizing operation, phosphate rock drying, materials handling, dryers and kilns, crushers and classifiers
Impingement	5.16	Acid fumes, flyash, oil mists, fluid bed dryers and coolers, cupola dust, calciners, blast furnace, merchant iron furnace, roasters, incinerators, fertilizer and sinter plant
Orifice type	5.16	Crushing and grinding, incinerator, dryer, calciner, roller mills, materials handling and lead battery plants

Table 5.21c (Continued)

Type of Scrubber	*Applicable Section*	*Industrial Application*
Self-induced spray scrubbers	5.16	Mining (coal and ore) incinerators
Venturi scrubbers	5.17	Almost all industries, and all applications, namely iron and steel, pulp and paper, chemical, nonferrous metals, power plants, kilns, dryers, roasters, boilers, steel making furnaces and acid manufacturing, fertilizer manufacture
Mechanically induced spray	5.18	Foundries, cupolas, paints, blast furnace, mining, coal processing, pharmaceuticals and paints
Irrigated wet filters	5.19	Acid mists; titanium dioxide
Wet fiber mist eliminators	5.19	Sulfuric, phosphoric and nitric acid mists, oil mists, radioactive liquids and dusts
Wire mesh mist eliminators	5.19	Acid mists
Baffles and vane types	5.19	Coke quenching, plating and kraft pulp mill
Packed beds	5.19	Acid mists

generates the flue gas? This is a very important consideration; because in power plant flue gas applications, the failure of the scrubber might necessitate the shut down of the power plant and, hence, the generation of electricity.

14. Guarantees. Will the equipment supplier stand behind his equipment and make a reasonable effort to meet the performance guarantees?
15. Future code requirements. Can the equipment be modified to meet anticipated codes? Is the fan static pressure adequate to operate scrubbers requiring higher pressure drops to attain higher efficiency?

The selection of suitable dust collection equipment is a challenge because of the heterogeneous nature of particulates in flue gas streams and the wide variety of wet scrubber designs. Frequently, the operating principles are yet undefined (or not understood), which makes it necessary to rely upon the

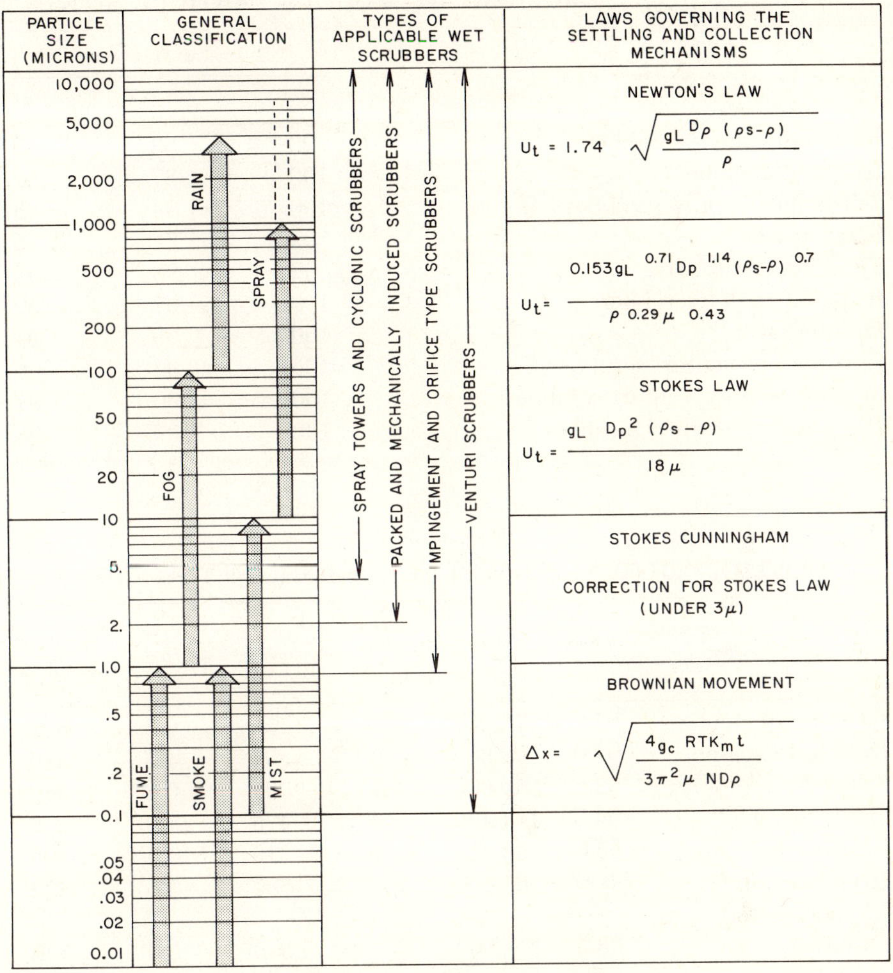

Figure 5.21d Scrubber comparison chart.

equipment manufacturers for proprietary design data and for performance guarantees.

Scrubber Costs and Applications

In Figure 5.21b, the particle size range for various aerosols is shown. The finer the particle size, the higher the pressure drop required to attain the same collection efficiency. Table 5.21c summarizes some typical industrial applications of wet scrubbers, which can be used as a guide for selection of wet collectors. In Figure 5.21d, several scrubber designs are compared as to their applicability for the removal of various dust particle sizes.[2] The laws

Table 5.21e
EFFICIENCY OF SCRUBBERS AT VARIOUS PARTICLE SIZES

Type of Scrubber	*Percentage Efficiency at*		
	50 μ	*5 μ*	*1 μ*
Jet-impingement scrubber	98	83	40
Irrigated cyclone	100	87	42
Self-induced spray scrubber	100	94	48
Spray tower	99	94	55
Fluid bed scrubber	99+	98	58
Irrigated target scrubber	100	97	80
Disintegrator	100	98	91
Low energy venturi scrubber	100	99+	96
Medium energy venturi scrubber	100	99+	97
High energy venturi scrubber	100	99+	98

Table 5.21f
COST OF 60,000 ACFM WET SCRUBBERS, 1965 FIGURES

Type of Scrubber	*Efficiency, %*	*Capital Cost,[a] $*	*Average Pressure Drop, Inches H_2O*	*Yearly Power Requirement,[b] 1,000 Kwh*	*Liquor Flow, gal/1,000 ft^3 of gas*	*Maintenance Cost Annually, $*
Jet impingement	87.9	78,000	8.0	750	0.8	832
Irrigated cyclone	91.0	48,880	3.9	406	4.0	410
Self-induced spray	93.6	54,600	6.1	564	0.6	546
Spray tower	94.5	114,400	1.4	476	18.0	832
Floating bed	95.0	43,420	2.4	310	14.5	1,300
Irrigated target (flooded bed)	97.9	67,600	6.4	580	3.0	832
Disintegrator (wet dynamic)	98.5	111,800	—	4,540	5.0	546
Low-energy annular throat venturi	99.5	83,200	12.5	1,174	7.0	832
Medium-energy venturi	99.7	88,400	20.0	1,882	7.0	832
Medium-energy annular throat venturi	99.8	88,400	20.5	1,934	7.0	832
High-energy venturi	99.9	96,200	31.5	2,974	7.0	832

[a]Capital cost (based on mild steel construction only) includes auxiliaries such as fans, pumps and motors (but not solids disposal equipment) and includes an allowance for site clearance and erection of plant.

[b]Annual operating time is taken as 8,000 hours.

that govern the settling and collection mechanism of various size particles are also presented. Table 5.21e summarizes the efficiency of various wet scrubbers at 50, 5 and 1 μ dust particle sizes.

The cost of the scrubber alone, compared to the total cost of ductwork, piping, fans and pumps and waste disposal systems, is relatively small. Table 5.21f summarizes typical costs[3] for wet scrubbers of various types, handling 60,000 ACFM of flue gas at 68°F and containing fine silica dust particles 30% of which are smaller than 10 μ and 12% below 2.5 μ. This table should be used only as a guide or for preliminary comparison purpose.

REFERENCES

1. Sargent, G. D.: *Chem. Eng.*, pp. 130–150, January 27, 1969.
2. "Koch Equipment for Air Pollution Control," Bulletin KPC-1, Koch Engineering Co.
3. Stairmand, C. J.: *The Chemical Engineer*, pp. 310–326, December 1965.

5.22 ELECTROSTATIC PRECIPITATORS—TYPES, FEATURES, OPERATION

Types of Designs:	Single or two stage for dry or wet operation.
Applications:	Single stage dry units remove solid particulates from industrial gases while the wet design can also handle liquid mists. The two stage design is for air conditioning applications.
Flue Gas Limitations:	Flowrate up to 4×10^6 ACFM, temperature up to 800°C, pressure up to 700 PSIG, velocity up to 25 ft/sec., pressure drop is under 1″ H_2O.
Power Requirement:	50 to 300 watts per 1,000 ACFM.
Yearly Power Requirement:	1 to 5×10^6 watt-hours per 1,000 ACFM.
Dust Particle Size Range:	From under 0.1 μ to over 200 μ.
Particle Density Range:	10^{-4} to 10^2 mg/liter (mg/l = 0.436 gr/ft^3).
Collection Efficiency:	90 to 99.5%.
Capital Cost:	$100,000 for 50,000 ACFM and high efficiency or 100,000 ACFM and medium efficiency. $200,000 for 100,000 ACFM at high efficiency and $600,000 for 500,000 ACFM at high efficiency.
Yearly Maintenance Cost:	$10 to $30/1,000 ACFM.
Partial List of Suppliers:	American Air Filter Co., Inc.; Buell Engineering Co.; Buffalo Forge Co.; Chemical Construction Corp.; General Electric Company; Heil Process

Equipment Corp.; Honeywell, Inc.; Koppers Co., Inc.; Lawler Company; Lodge-Cottrell, Inc.; Monsanto Enviro-Chem. Systems, Inc.; Precipitair Pollution Control, Inc.; Research-Cottrell, Inc.; Seversky Electronatom Corp.; The W. W. Sly Manufacturing Co.; Western Precipitation Div., Joy Manufacturing Company; Westinghouse Electric Corporation; Wheelabrator Corp.; John Zink Company.

Operating Principle

Electrostatic, or electrical, precipitators are a variety of equipment utilizing electric forces to separate particulate matter and aerosols from gases. Some types are truly electrostatic while others are not. The fundamental steps of operation involved are (1) electrical charging of suspended particles, (2) collection of the charged particles in an electric field and (3) removal of the collected material to an external receptacle.

The high-voltage direct-current corona method is used for charging the particles. Usually the corona is established between a fine wire (the active electrode) and a smooth cylindrical or plate electrode at ground potential. In industrial precipitators, the wire has negative polarity and the positive ions formed in the corona glow are attracted to the wire. The negative ions (equal in number to the positive ones) are also formed in the corona, and these are attracted to the ground electrode. Therefore, most of the gas space between the two electrodes contains negative ions. Particles in the gas are bombarded by these negative ions and quickly become charged. Typical particle charges achieved are 300 electron charges for 1 μ size particles and 30,000 electron charges for 10 μ particles.

The charged particles are collected by subjecting them to an electric field. This collecting field may be a continuation of the corona, as in the Cottrell or single-stage type design, or it may be an electrostatic field between non-discharging electrodes (the two-stage design). Single-stage designs are used in cleaning industrial gases (Fig. 5.22a and b),[1] and the two-stage type is used principally for cleaning air when low-ozone generation is necessary (Fig. 5.22c). In either case, the coulomb force, which is proportional to the product of the particle charge and the electric field intensity, is the separating force on the particle. The motion of the particle is resisted by viscous forces governed at least approximately by Stokes' law. Thus, a particle attains a velocity determined by the equilibrium between the coulomb and Stokes forces. This velocity is called the migration velocity and is about 3 cm/sec for 0.5 μ particles at normal precipitator conditions.

The last phase is the removal of the collected material to external receivers. Some solid particles are very susceptible to reentrainment in the (turbulent) gas stream and, therefore, are difficult to remove from the precipitator. Reentrainment normally does not occur with liquid droplets or with solid particles that are cohesive and form larger agglomerates.

The size and type of precipitator are determined by the physical properties of the particles and of the carrier gas, the gas flowrate and the required collection efficiency. For collection of liquid droplets and for smaller gas flowrates, the pipe type electrode is used. Plate type systems are used for dry solids and large gas flowrates.

Electrical Field Characteristics

In order to ionize the gas, the electrical field strength must exceed, at least locally, the electrical breakdown strength of the gas. A local discharge that does not propagate is called a corona. Sparking, then, is an advanced stage of corona in which the breakdown of gas is complete along a given path. Most industrial precipitators are operated with a negative corona because higher voltages may be used without sparking. Positive coronas cause less ozone formation and, therefore, are generally used in air conditioning applications.

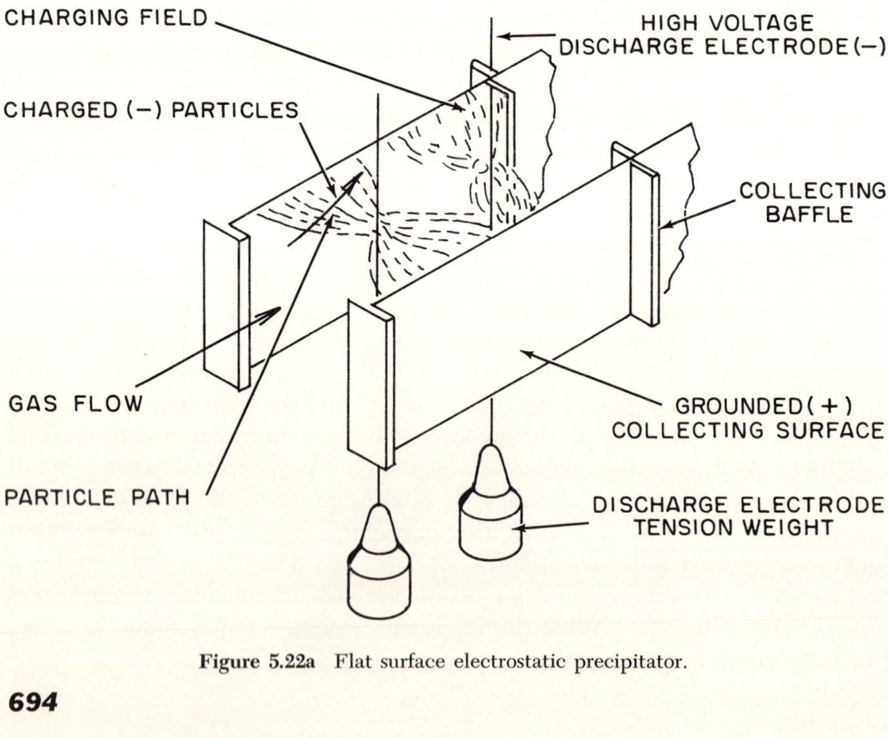

Figure 5.22a Flat surface electrostatic precipitator.

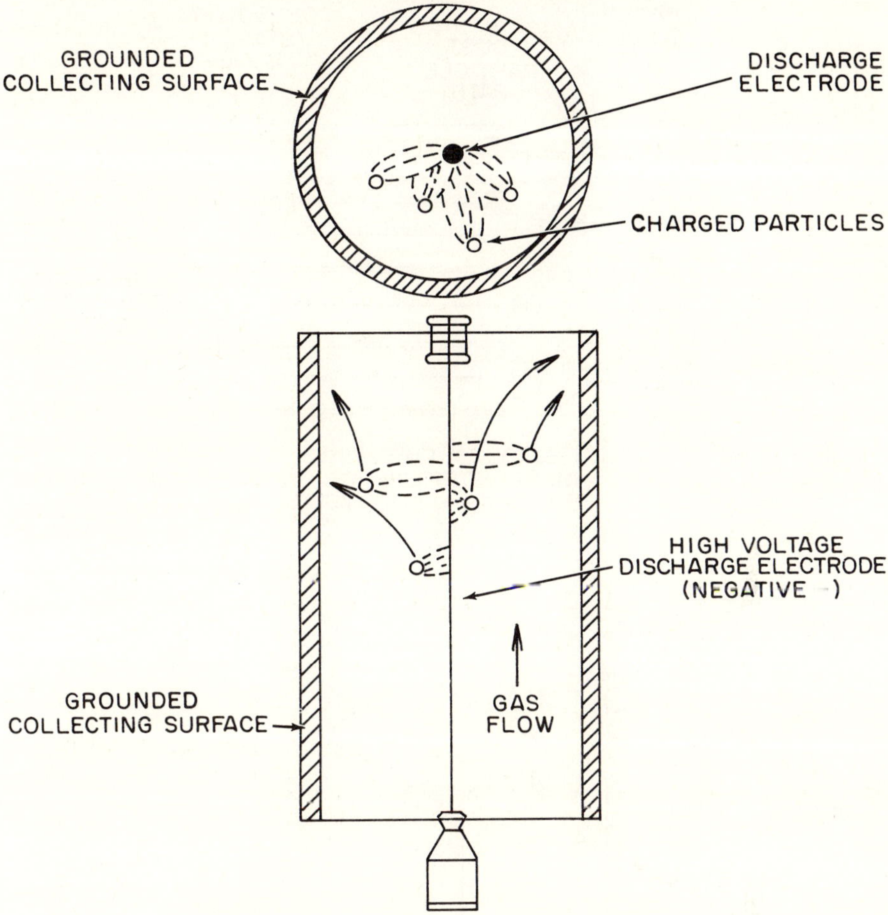

Figure 5.22b Tubular surface electrostatic precipitator.

The electric field for concentric cylinder electrodes is given by Equation 5.22(1).

$$E = [(2\,I/\lambda) + (E_o r_o/r)^2]^{1/2} \qquad 5.22(1)$$

where E is the potential gradient, I is the corona current per unit length of wire, λ is the ionic mobility, E_o is the critical corona field for ionization, r_o is the radius of the corona-glow around the wire and r is the radial coordinate. For values of voltage less than the corona-onset voltage, the current is zero and

$$E = \frac{V}{r \ln D_t/D_w} \text{ for } V \leq V_o \qquad 5.22(2)$$

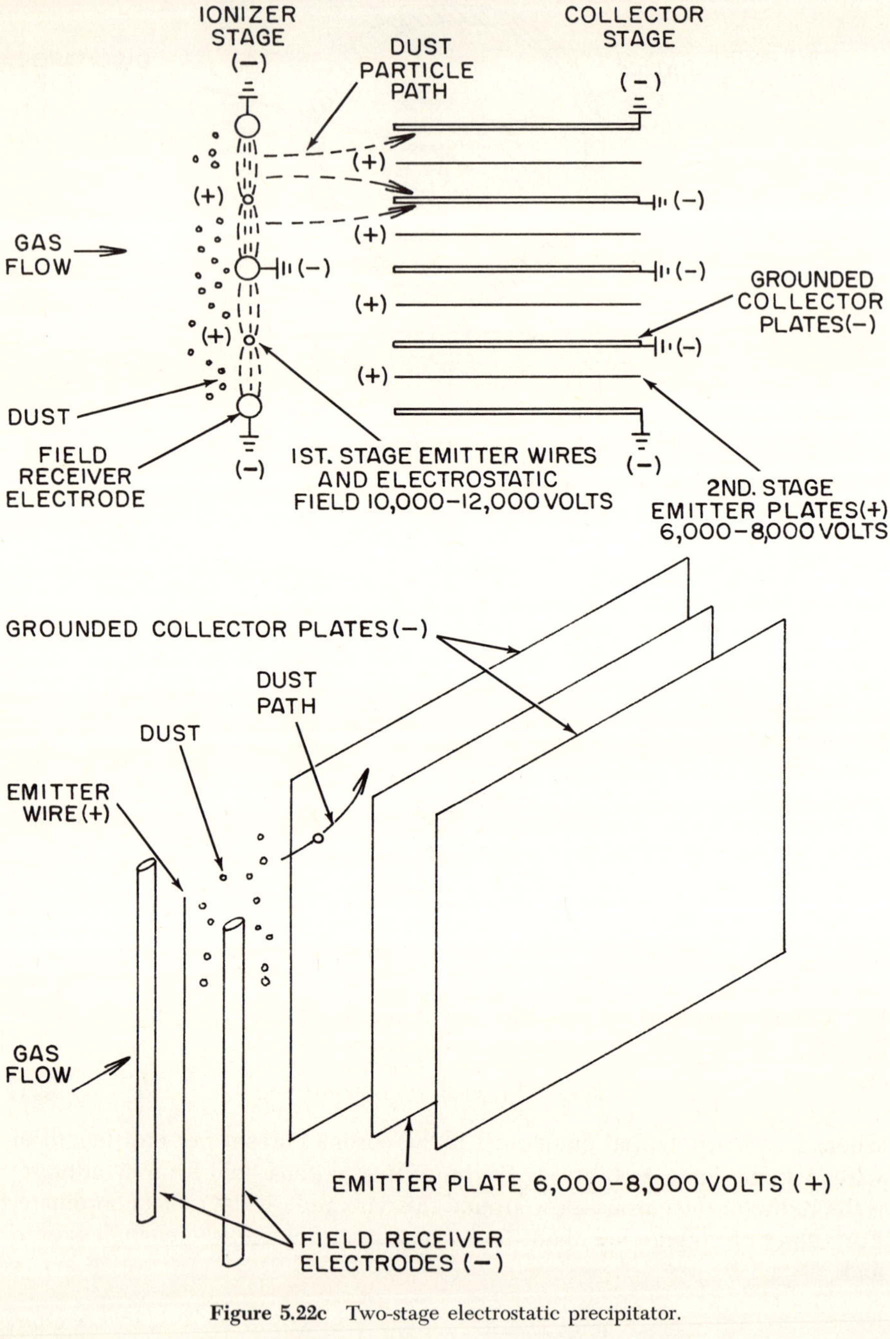

Figure 5.22c Two-stage electrostatic precipitator.

Table 5.22d
SPARKING POTENTIALS IN MOIST AIR

Pipe Diameter, cm	*Sparking Potential, volts*	
	Peak	*Root Mean Square*
10	59,000	45,000
15	76,000	58,000
23	90,000	69,000
30	100,000	77,000

where D_t is the inside diameter of the cylinder and D_w is the outside diameter of the wire. For large corona currents and for distances well away from the wire, the second term in Equation 5.22(1) becomes negligible and

$$E = \sqrt{2\ I/\lambda} \qquad 5.22(3)$$

Table 5.22d lists some sparking potentials for moist air at 38°C, and Figure 5.22e illustrates the effect of moisture on the sparking potential at temperatures between 38 and 370°C.

Particle Charging

Almost all particulates and aerosols carry some electrical charge; however, these charges are a small fraction of that required for efficient precipitation. Since doubling the charge doubles the coulomb force, with other factors being equal, doubling the charge will cut the required precipitator size in half. Therefore, the particle charge should be made as high as possible. This is best done by the unipolar high-voltage corona discharge. In the corona field of the precipitator, particle charging occurs by two mechanisms. The first, field or impact charging, occurs by ion attachment in the electric field and is the predominant mechanism for particles larger than 0.5 μ. The second mechanism, diffusion-charging, is due to ion diffusion and depends only on the thermal energy of the ions. Diffusion charging predominates for particles less than 0.2 μ, and both mechanisms are important for intermediate size particles.

For field charging, it can be shown that the fractional approach to the saturation charge is related to the time in the electrical field by:

$$\frac{n}{n_s} = \frac{t}{t + t_o} \qquad 5.22(4)$$

where n/n_s is the fraction of the saturation charge, t is time and t_o is a time constant.

$$t_o = \frac{1}{\pi\sigma\epsilon\lambda} \qquad 5.22(5)$$

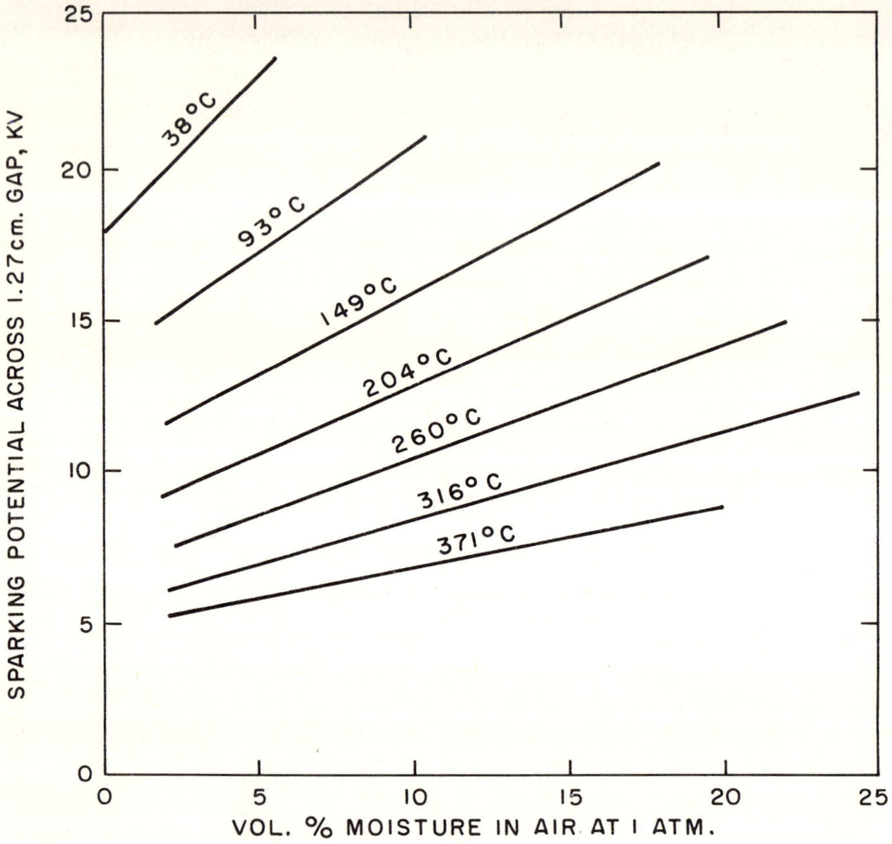

Figure 5.22e Sparking potential for negative point-to-plane. *(From Sproul, W. T. and Y. Nakada: Operation of Cottrell Precipitators—Effects of Moisture and Temperature,* Ind. Eng. Chem., *vol. 43, p. 1356, 1951)*

σ is the ion density, no./cc; ϵ the elementary electrical charge; λ the ionic mobility, cm/sec/volt/cm. The number of elementary electrical charges acquired by a particle at saturation is given by Equation 5.22(6).

$$n_s = \left(1 + 2\frac{k-1}{k+2}\right)\frac{E_o D_p{}^2}{4\epsilon} \qquad 5.22(6)$$

where E_o is the corona field strength, D_p the particle diameter and k the dielectric constant.

The diffusion charging process occurs because ions in the gas diffuse through the gas according to the laws of kinetic theory. The equation for diffusion charge n based on kinetic theory of the ions is

$$n = \frac{D_p kT}{2\epsilon^2} \ln\left(1 + \frac{\pi D_p C \sigma \epsilon^2 t}{2kT}\right) \qquad 5.22(7)$$

where k is the Boltzmann's constant, T is the temperature °K, C is the root mean square velocity of the ions and t is the time. At σ of 5×10^8 ions/cm and 300°K and particle diameter of 0.2 μ, for a time of 0.1 sec, n is 7 electrons and for 1 sec n is 11 electrons.

Particle Collection

The second step in the precipitation process is the collection of the charged particles. The theory for undisturbed collection is well established; however, several disturbances to collection occur in actual practice such as the upset in gas flow, particle reentrainment and disturbances of the corona. Nevertheless, the theory of undisturbed collection is useful in design guidance; and a more complete theory, taking the disturbances into consideration, could become too complex for practical use.

The movement of a particle suspended in gas and acted on by an electric field can be calculated by equating the coulomb and viscous forces. The resulting equation is

$$W = \left(1 + 1.72\frac{L}{D_p}\right)\frac{D_p E_o E_p}{4\pi\mu} \qquad 5.22(8)$$

where W is drift velocity, E_p is precipitation field density, L is mean free path of gas and μ is the absolute viscosity of the gas. The second term in the parenthesis of Equation 5.22(8) arises from the Cunningham correction factor to Stokes' law.

The theoretical collection efficiency for uniform size spherical particles has been developed from probability concepts. The expression is

$$\eta = 1 - e^{-(A/Q)W} \qquad 5.22(9)$$

where η is the fractional collection efficiency, A is the collection surface and Q is the volumetric gas flowrate.

Since most dusts and dispersoids contain a range of particle sizes, other expressions for efficiency are necessary. Many of these dusts follow the log-normal particle size distribution where the size frequency is

$$\gamma(t) = \frac{1}{\sqrt{2\pi}} \exp\,(-t^2/2) \qquad 5.22(10)$$

$$\text{where } t = \frac{\ln X - \ln \bar{X}_g}{\ln \sigma_g} \qquad 5.22(11)$$

where $\bar{X}_g$ = geometric mean

σ_g = geometric standard deviation

Then, the overall fractional efficiency for the total dust can be obtained from

$$\eta = 1 - \frac{1}{\sqrt{2\pi}} \int_{-\infty}^{\infty} \exp(-t^2/2) \cdot \exp(-k\bar{X}_g e^{t \ln \sigma_g})dt \qquad 5.22(12)$$

where $k = \frac{A}{Q}\frac{E_oE_p}{4\pi\mu}$ 5.22(13)

All quantities in Equations 5.22(12) and (13) should be in the cgs system. The above integral can be evaluated numerically or graphically for specific applications where $\bar{X}_g$ and σ_g are known.

Particle Resistivity

The particle resistivity can affect precipitator performance in two ways. First, if the resistivity is high (greater than 2×10^{10} ohm-cm), the voltage and current at which the precipitator can operate is reduced, thus lowering performance. If the resistivity is too low, reentrainment of the particles can occur.

In precipitator operation, high-particle resistivity usually results in excessive sparking at moderately lowered voltages or excessive current at greatly lowered voltages. These effects cause a loss in efficiency; the loss increases with increasing resistivity. The resistivity of the particle not only depends on the chemical composition and structure of the particle, but it also depends on the surroundings. The solids adsorb gases such as water vapor, and these adsorbed materials provide electrical conductivity in excess of that provided of the dry particles. The operating temperature affects the resistivity in part because it affects the adsorption. Low moisture, temperatures in the range of 120 to 220°C and absence of certain chemical impurities are likely to cause high-particle resistivity. The effects of temperature and moisture on particle resistivity are shown in Figure 5.22f for a typical flyash.

Conditioning agents are used to modify the particle resistivity. These agents are injected into the gas to modify the resistivity of the particle due to their presence on the particle surface. Sulfur trioxide at the 10 ppm level has been effective in reducing flyash resistivity in several commercial operations. Ammonia has been used to lower the resistivity of powdered catalyst dust and has also been used for flyash.

Other methods for overcoming high resistivity have been studied on a small scale. These include maintaining a thin film of water on the collecting electrodes or temperature controlling the collecting electrodes.

Energization

The electrical energy requirement to separate particles from a gas stream can be estimated from the drag resistance according to Stokes' law.

$$D = 3\pi\mu D_p W \qquad 5.22(14)$$

where D is the frictional resistance or drag. The work required to move one particle a distance s is

$$W_o = Ds = 3\pi\mu D_p Ws \qquad 5.22(15)$$

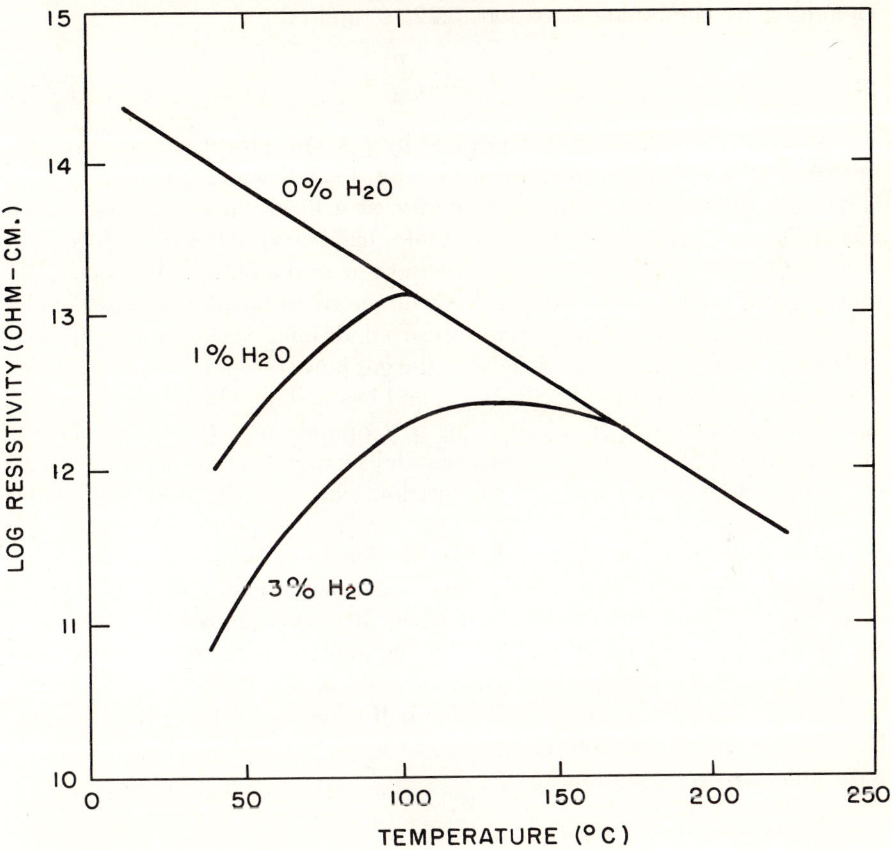

Figure 5.22f Effects of temperature and humidity on particle resistivity of typical flyash.

The drift rate W has been shown to be proportional to the total corona current in accordance with Equation 5.22(16).

$$W = \frac{k_1}{A}\left(\frac{V_p + V_m}{2}\right)I \qquad 5.22(16)$$

where V_p is the peak voltage; V_m is the minimum voltage; A is the collecting area; and k_1 is a proportionality constant depending on the gas, particulates and precipitator design. The average corona power P_c may be approximated by

$$P_c = \left(\frac{V_p + V_m}{2}\right)I \qquad 5.22(17)$$

Therefore, by combining Equations 5.22(16) and (17)

$$W = k_1 \frac{P_c}{A} \qquad 5.22(18)$$

If a large precipitator is energized by a single power supply, the performance will be only a fraction of that for a small scale precipitator. This is because the sparking rate and intensity, at a fixed voltage, increase with increasing precipitator size. This scale factor can be overcome by high-tension sectionalization, dividing the large precipitator into a number of electrically independent sections. An additional advantage of sectionalization is that the voltage can be adjusted to suit the local conditions. Series or longitudinal sectionalization is important because as the gas flows through the precipitator, the particle concentration becomes less and less and may be nil at the outlet. Each section then can be adjusted for best operation at that particular set of conditions. In large precipitators, parallel sectionalization provides a way of coping with uneven dust and gas distribution across the inlet face of the precipitator.

Half-wave operation allows double the degree of sectionalization for the same number of rectifier sets. Full-wave versus half-wave operation requires an economic comparison of the alternatives. It is sometimes advantageous to use both in large precipitators. Figure 5.22g depicts such a system. Here the inlet sections with high-particle concentration and collection operate with high voltage and low currents and have half-wave sets. The outlet has relatively clean gas and, therefore, operates at lower voltage and higher current with full-wave energization.

Gas Distribution

Optimum precipitator performance is associated with uniform, low-turbulence gas flow. Disturbing effects such as eddy formation and flow separation decrease collection efficiency, sometimes very seriously. Nonuniform flow in the precipitator can affect performance in two fundamental ways: (1) the uneven gas treatment lowers efficiency in high-velocity zones to a degree not compensated for in the low-velocity sections, and (2) the particles may be blown off in high-velocity zones and this can cause reentrainment. In severe circumstances, the second effect is the predominant one and can result in overall collection efficiencies in the 60% range.

Other effects from unbalanced flow may be (1) the loss of particles from hoppers by reentrainment, (2) gas bypassing the precipitation zone, (3) excessive dust collection on corona electrodes in zones of low gas velocity and (4) excessive sparking in some regions due to unbalanced electrical characteristics of gas and dust.

Scale models of the fluid flow system are quite useful in the evaluation of precipitator design. The general principles of fluid mechanics may be

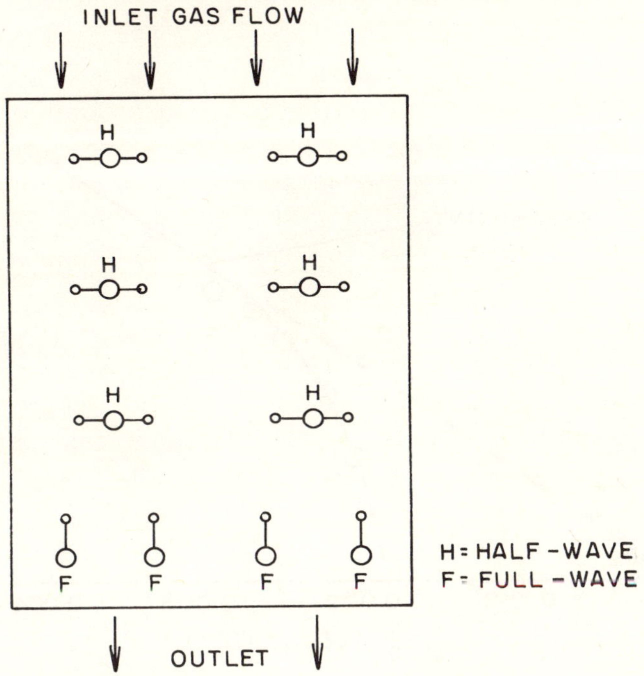

Figure 5.22g Sectionalization, using half-wave and full-wave energization.

applied to yield useful results from scaled down models. Flow patterns can be determined by electrical anemometers and pitot tubes. Use of smoke or other visual gas tracer injections in transparent plastic mock-ups provide qualitative data on turbulence and general flow characteristics.

Particle Reentrainment

There are several methods for determining whether or not particle reentrainment is occurring. Two of the methods require measurement of the collection efficiency at varying volumetric gas rates. The basic efficiency equation can be rearranged as follows:

$$\frac{1}{Q} = -\frac{1}{AW}\ln(1-\eta) \qquad 5.22(19)$$

Equation 5.22(19) indicates that a plot of the reciprocal of the volumetric gas rate Q versus the logarithm of $1 - \eta$ should yield a straight line. If the experimental data indicate a curve rather than a straight line, particle reentrainment is occurring. This is illustrated by Figure 5.22h. Another method would be to plot the precipitation rate W versus the volumetric flowrate Q. In the theoretical equation, W should be constant with increasing Q. Onset

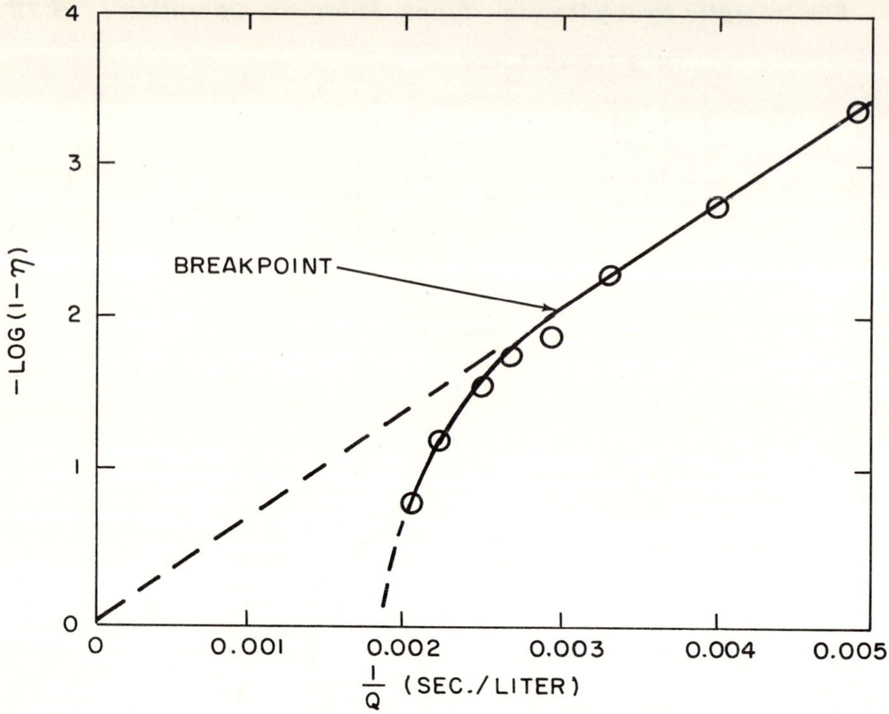

Figure 5.22h Onset of reentrainment.

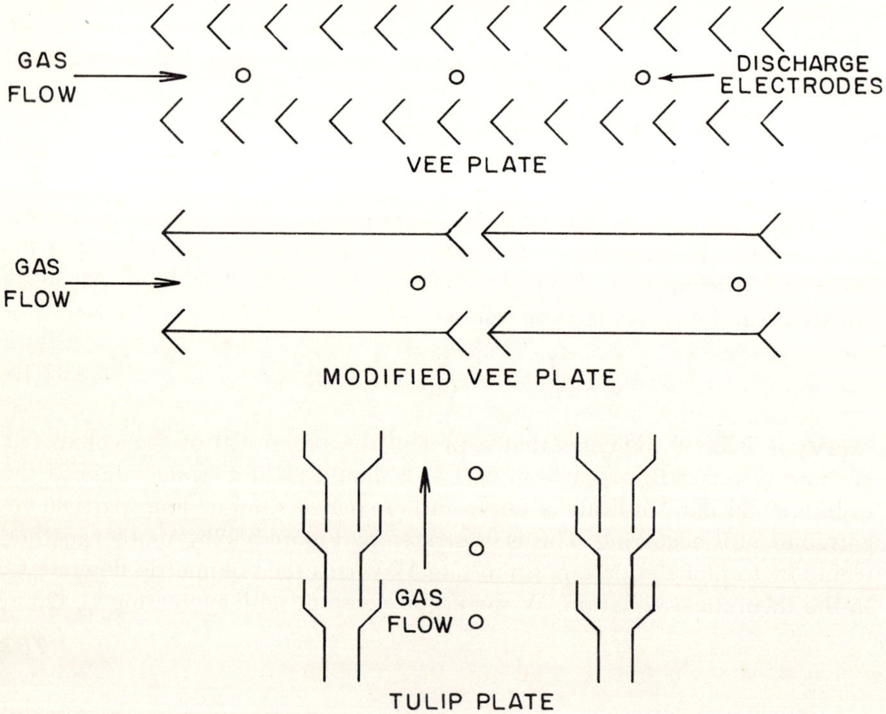

Figure 5.22i Shielded collecting electrodes.

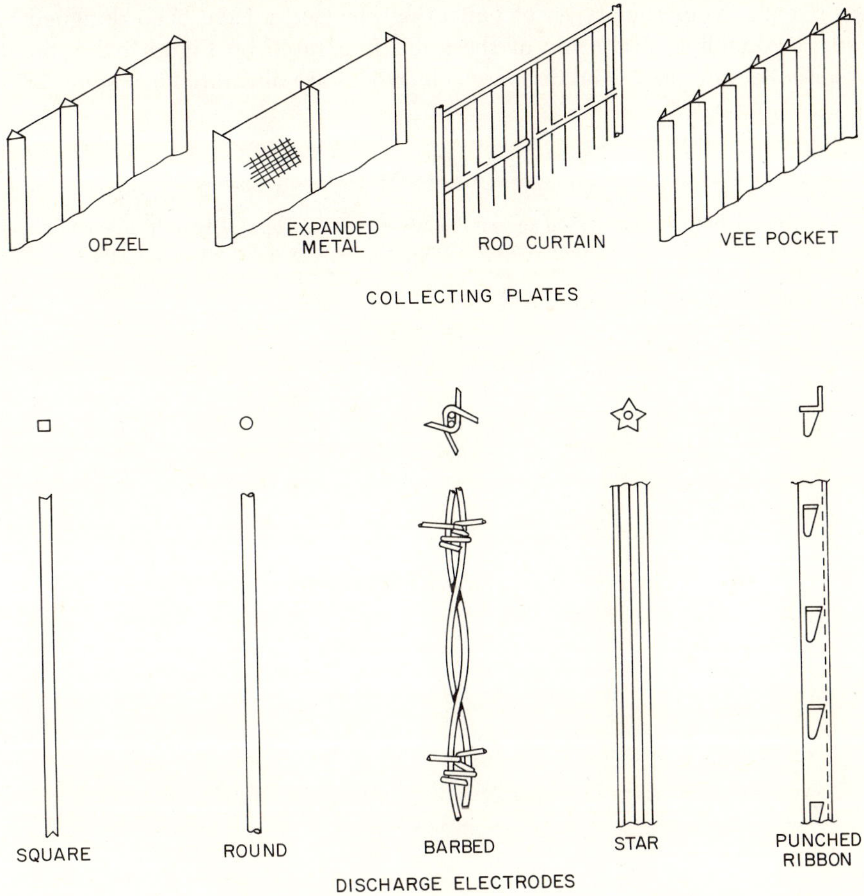

Figure 5.22j Electrostatic precipitator collecting plates and discharge electrodes.

of reentrainment causes a loss of efficiency in excess of that predicted by the theoretical formula and causes a sharp drop in W.

A third method of detecting reentrainment is useful when data at varying gas rates are not available. This method is carried out by examination of the particle size of the particles in the precipitator outlet gas. The presence of relatively large particles or of agglomerated particles is a strong indicator of reentrainment. Comparative particle-size analysis and microscopic examination of the particulates entering and leaving the precipitator provide a practical and sensitive method for detection of reentrainment.

One of the primary methods of minimizing the effect of reentrainment is to provide shielded collection zones protected from the direct blast of the gas stream. Numerous types of collecting electrodes have been designed to provide shielding, and some of these are illustrated by Figure 5.22i. Some other collecting plates and discharge electrodes are illustrated in Figure 5.22j.

REFERENCE

1. "Control Techniques for Particulate Air Pollutants," U.S. Dept. of Health, Education and Welfare, National Air Pollution Control Administration, Washington, D.C., January 1969.

NOTE: For Bibliography see Section 5.24.

5.23 ELECTROSTATIC PRECIPITATORS—ACCESSORIES AND INSTRUMENTS

Rectifiers

Best precipitator operation can be obtained with pulsating waveforms primarily due to the fact that higher voltages can be maintained by intermittent voltages under sparking conditions commonly occurring in precipitators. Therefore, most precipitators are powered by high-voltage unfiltered rectifier sets. Synchronous mechanical rectifiers were first used but are now obsolete. Electronic tube rectifiers came into wide-scale use in about 1950. Present day tubes have operating lives of 20,000 to 30,000 hours or more. Filament voltage adjustment is very critical in tube life because, for example, a 5% reduction in filament voltage can double the tube life.

Solid-state rectifiers of the selenium type have been in use as long as the electronic tube rectifiers. The silicon rectifier,[1] however, is technically superior to the selenium type and is in wide-scale use in precipitator applications.

Providing a high level of corona power is the purpose of high-voltage equipment; and fulfilling this requirement means proper current and voltage output ratings, proper voltage waveform, and stability against sparking. Automatic voltage control is necessary in many applications. Half-wave voltage is preferred over full wave in cases where sparking is an important consideration.

Electrode Rappers

When the particulates have been precipitated on the collecting electrodes, they must be removed to a collection hopper. This is necessary not only to remove the collected material, but also to maintain optimum electrical conditions. Mechanical jarring, vibrating or rapping is used to dislodge the collected dust. This step is critical to the overall performance, because improperly operated or adjusted rappers can cause reentrainment of the particulates. In some cases, particles can build up on the discharge electrodes and affect the corona discharge.

The types of rappers include mechanical (electric motor driven), pneumatic and magnetic-solenoid operated design (Fig. 5.23a). Most new installa-

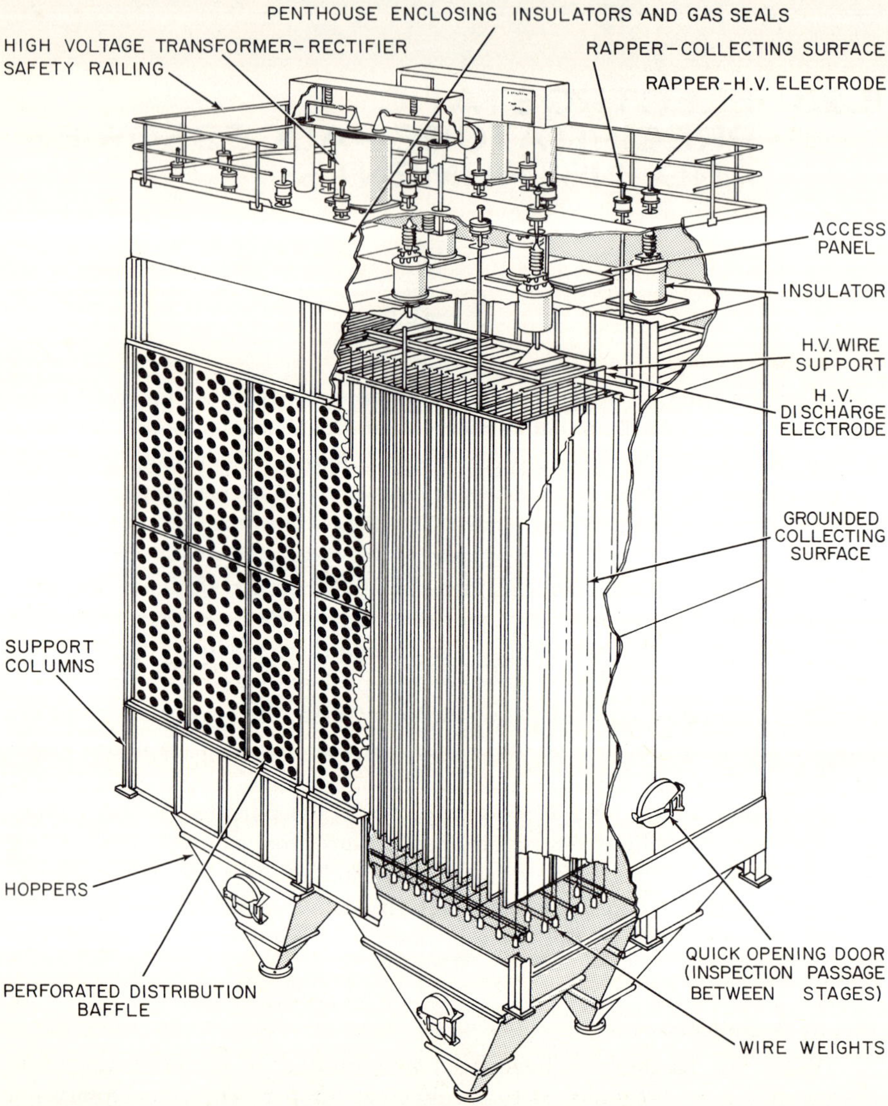

Figure 5.23a Flat surface electrostatic precipitator with accessories. *(Courtesy of UOP Air Correction Division)*

tions have the solenoid type rappers that can be accurately adjusted to set the intensity and frequency of the raps. Usually impact or knocking designs are used on collecting electrodes and vibrating types are used on discharge electrodes.

Normally, the dust buildup is allowed to reach 6 to 25 mm before it is rapped off. This way, the collected material falls off in large agglomerates and will not be reentrained. Optimum adjustment of rapper intensity and frequency is best made by direct visual observation through sight ports (Fig. 5.23a) with high-intensity illumination. Factors that influence the rapper settings include the agglomeration characteristics of the particulates, the rate of collection of the particulates, the effect of the collected material on the electrical operation and reentrainment tendencies.

In some cases where the reentrainment problem is severe, sections of the precipitator are blocked off by dampers during rapping; while normally, rapping is carried out during continuous operations. If the collected material is liquid, or if wet type operation (irrigated precipitator) is used, rapping is unnecessary.

Dust Hoppers

Hoppers are normally located directly under the collecting electrodes to catch and store the collected material. Vertical baffles are commonly installed in the hoppers in order to minimize gas by-passing the precipitator. This scheme is shown in Figure 5.23b. The angle between the hopper walls is normally 60 degrees. This angle sets the slope of the hopper walls greater than the angle of repose of most particulate solids so that they slide down the hopper without bridging.

The collected dust can be removed from the hopper continuously or periodically (see Section 5.10). If the dust level is too high in the hopper, reentrainment can occur; and also there is danger of shorting the discharge electrodes. Therefore, adequate storage volume must be provided.

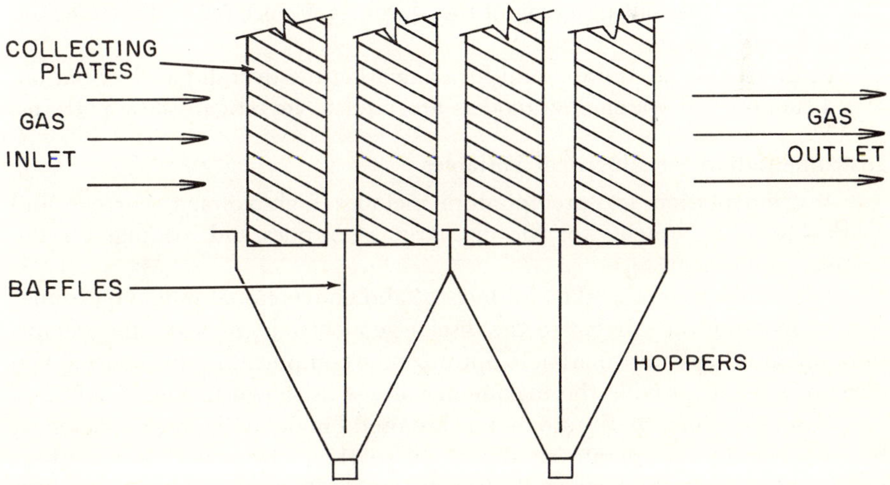

Figure 5.23b Dust hoppers with baffles.

The solids can be removed from the bottom of the hopper through knife valves, slide gates, star feeders, rotary valves or screw conveyors. The system can be designed for continuous operation including automatic solids level control in the hoppers (see Section 5.6). Vibrators may be required if the dust does not move freely.

Inlet Gas Distribution

Gas flow quality, as measured by uniform gas velocities, is mandatory for proper precipitator operation. Acceptable gas distribution, as defined by the Industrial Gas Cleaning Institute, is when 85% of all local or point velocities are within ±25% of their average and no single reading differs from the average by more than ±40%.

Specific means of accomplishing the good gas flow patterns include guide vanes, screens or diffuser elements, flue transitions; and plenum chambers. Guide vanes are used to maintain flow patterns through sharp turns; they do not improve flow patterns, but they prevent them from becoming worse. The vanes should be long enough to cover the projection of the opening through which the gas is approaching. Normally three to five vanes should be used in a 100 to 200 cm wide duct; and more vanes should be used with larger ducts, keeping the spacing to a maximum of 60 cm.

Diffuser elements are used to break up large eddies or swirls. They are used at the inlet of most precipitators and at the outlet of many. Diffusers are either a perforated plate with holes of 1 to 2 inches in diameter or a grid of large rectangular openings in a flat sheet. The open area of the plates normally is 25 to 50% of the total area. The diffusers are very likely to become covered with dust deposits, and good engineering design requires rappers on the diffusers to prevent plugging of the openings. Figure 5.23c illustrates the use of turning vanes.

Scale models are quite useful for evaluating the overall gas distribution. Wind tunnels and water flow models are used to investigate flow patterns.

Instrumentation for Electrical Variables

Instrumentation for precipitators includes high voltage, current and spark-rate measurements for the discharge electrodes and readings for the rectifier equipment.

High-voltage meters (kilovoltmeters) and conventional milliampere meters, to measure the current to the discharge electrode, provide the precipitator operator with information to optimize the equipment performance.[2] The current is set to provide the maximum voltage as shown in Figure 5.23d.

Direct-reading, spark-rate meters are useful in obtaining good efficiency. Some precipitators operate at almost no sparking. However, many others appear to perform best when the spark rate is 50 to 100 per minute. The best rate depends on several factors, including characteristics of the particulate

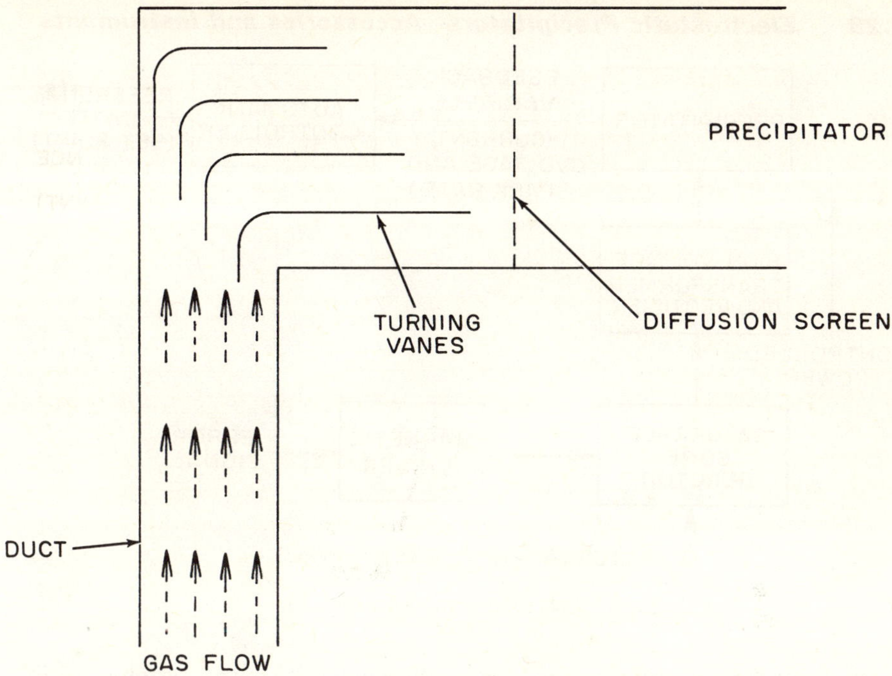

Figure 5.23c Gas distribution system.

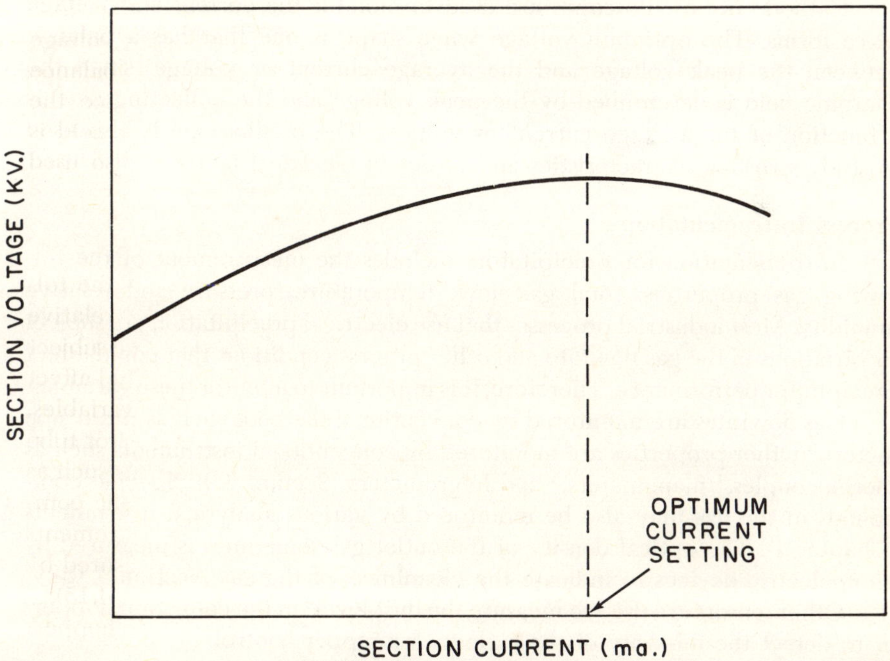

Figure 5.23d Optimum electrical set adjustment.

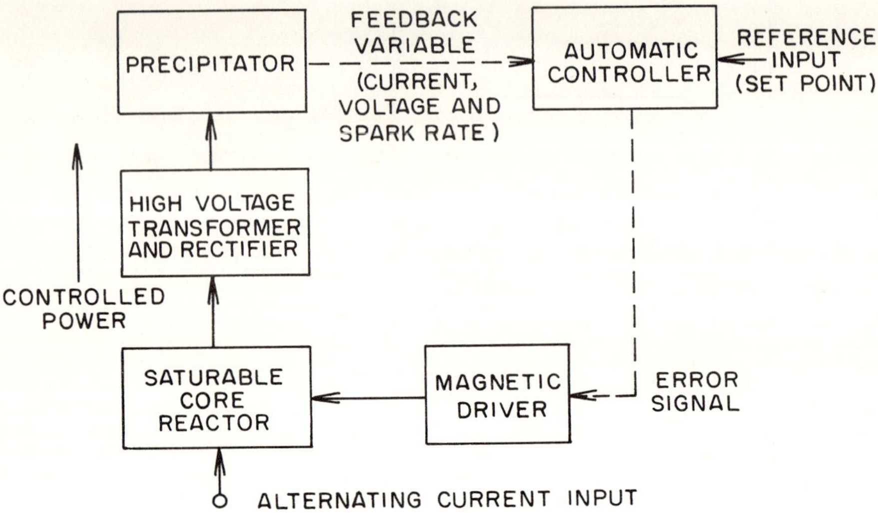

Figure 5.23e Automatic control system.

matter and the precipitator size. Automatic feedback control systems are in wide use to control one of the following electrical variables: current, voltage or spark rate. Figure 5.23e depicts a typical control loop.

Cathode ray oscilloscopes are used to monitor the current and voltage wave forms. The optimum voltage wave shape is one that has a balance between the peak voltage and the average current or voltage, since the charging field is determined by the peak voltage and the collecting field is a function of the average current or voltage. The oscilloscope is also used to study sparking characteristics and to detect electrical faults.

Process Instrumentation

Instrumentation for precipitators includes the measurement of the following gas properties: total gas flow, temperature, pressure and relative humidity. Most industrial processes that use electrical precipitation are subject to variations in the gas flowrate and other process conditions that could affect precipitator performance. Therefore, it is important to monitor these variables.

Gas flowrates are monitored by conventional methods such as pitot tube meters.[3] Other properties are monitored by conventional instruments such as thermocouples,[4] manometers,[5] and hygrometers (Section 4.9). Specific compounds in the gas may also be monitored by various analytical instruments (Chapter 4). The optical density of the outlet gas sometimes is measured by photoelectric devices to indicate the cleanliness of the gas (Section 4.15).

Other sensors are used to measure the dust level[6] in the collection hoppers or to detect the intensity of the rappers for rapper control.

REFERENCES

1. Lipták, B. G. (ed.): "Instrument Engineers' Handbook," Vol. II, Section 2.1, Chilton, Philadelphia, 1970.
2. Ibid, Supplement 1, Section 1.17.
3. Ibid, Vol. I, Section 5.3.
4. Ibid, Vol. I, Section 4.9.
5. Ibid, Vol. I, Section 2.7.
6. Ibid, Vol. I, Chapter I.

NOTE: For Bibliography see Section 5.24.

5.24 ELECTROSTATIC PRECIPITATORS—SELECTION AND APPLICATION

Electrical precipitators have many advantages over other types of particulate collection devices. Precipitators are unique in that the collecting force is applied only to the material being collected and not to the total gas stream. This results in relatively low operating costs, since the pressure drop through the system is small and the power requirements are low. Precipitators have the ability to handle very large gas rates and to remove very small particles at high efficiency. They can operate over a wide range of conditions, temperatures up to 800°C and pressures of 50 atmospheres, and can be used to remove tar and acid mists which are difficult to remove by any other method.

Their largest disadvantage is the high initial cost; the investment for the precipitator is normally greater than for any other equipment for air pollution control. Another disadvantage is that precipitators are not easily adapted to changing process conditions. Automatic voltage control helps this situation, yet precipitators are most efficient at constant operating conditions. Also, the high-voltage equipment in electrical precipitation necessitates special safety precautions.

Selection Guidelines

Specifications for electrostatic precipitators describe the recovery efficiencies at operating conditions of the process. The precipitator performance depends largely on the properties of the particulates and of the gas. Particulate properties include particle size distribution, particle composition and structure, concentration of particulates and the resistivity of the particles at operating conditions. The gas properties include total volumetric flow, static pressure, temperature, composition and corrosive properties. These gas and particulate properties (including expected variations) should be specified as accurately as possible for good precipitator design. If there is uncertainty or unknown variables, the design will probably be on the conservative side and may cost more than is actually necessary.

Some basic design parameters are the corona power and power density, the duct width, the gas velocity, migration velocity and collection surface.

The classic Deutsch equation can be rearranged to yield a relationship between the collection surface and efficiency,

$$\frac{A}{Q} = -\frac{1}{W}\ln(1 - \eta) \qquad 5.24(1)$$

where A/Q is the collection surface per volumetric gas rate; W, the migration velocity of the particles; and η, the collection efficiency. The duct width, or pipe diameter in pipe type precipitators, forms the cross-sectional flow area; and the gas velocity is the volumetric flowrate divided by the flow area.

The corona power can be related to the collection efficiency by

$$\frac{P_c}{Q} = k\ln(1 - \eta) \qquad 5.24(2)$$

where P_c is the useful corona power required to achieve a desired collection efficiency, and k is a proportionality constant best determined from field data. A comprehensive checklist for precipitator specifications is given in Table 5.24a.

Construction Details

Precipitator shells and hoppers may be made of a variety of materials. These are selected based on the operating conditions and the corrosive nature of the process gas and particulate matter. Materials of construction that have been used include mild steel, alloy steel, lead-coated steel, plastics or other types of linings on steel, acid-resisting brick, cement, concrete, aluminum and wood.

The discharge electrodes have been made in a variety of shapes such as round wires, ribbons, barbed wire and twisted rods of square cross section. (Fig. 5.22j) The material of construction is normally selected based on corrosion resistance. Steel alloys are the most common materials; but copper, aluminum, nichrome, silver, hastelloy and monel have also been used. The ribbon and twisted rod designs are useful for heavy dust concentrations and for sticky particles where keeping the electrodes clean is a severe problem.

The collecting electrodes have been constructed in a great variety of designs. These include perforated or expanded metal plates, rod curtains and various hollow electrode shapes (Fig. 5.22j). There are several fundamental requirements for the collecting electrodes, including high sparkover voltage characteristics, aerodynamic shielding of surface to minimize reentrainment, good rapping characteristics and high mechanical strength.

Performance Testing

Performance tests on precipitator equipment are made to establish the actual collection efficiency on a mass basis. Standard testing techniques have

Table 5.24a
ELECTROSTATIC PRECIPITATION SPECIFICATION CHECKLIST

Particle Properties:
- Source; composition
- Particle concentration; particle structure
- Particle-size distribution; in situ resistivity

Gas Properties:
- Total gas flowrate; composition
- Pressure; corrosive properties
- Temperature

Minimum Collection Efficiency Required

Mechanical Details:
- Types of access facilities for maintenance and inspection
- Structural details (dimensions and weights)
- Number of dust hoppers
- Maximum pressure drop through system
- Effective collection plate area
- Maximum precipitation zone gas velocity
- Type of rapping
- Type of wire for discharge electrode

Electrical Details:
- Type of rectifiers
- Type of energization, half-wave or full-wave
- Type of stabilizing circuit; resistance, monocyclic, saturable reactor
- Rectifier power input
- Corona power, maximum and average expected
- Number of high-voltage bus sections

Control System: Current, Voltage or Spark Rate Control

been established and should be followed. Particle sampling determinations are usually made by filter-thimble techniques for dry dusts. Systematic traverses must be made for representative samples; isokinetic sampling rates must be maintained; and a large number of sample points must be used. Volumetric gas rates also are measured using traverses and standard techniques. Photoelectric methods can be used with empirical correlations to monitor the cleanliness of the outlet gas. The intensity of the scattered light is proportional to the mass concentration. Additional details on sampling and measurements are provided in Chapter 4.

Precipitator Troubleshooting

Troubleshooting of precipitator operations can be divided into three areas of concern. These are operational problems, mechanical troubles and basic

5.25 AFTERBURNERS—TYPES, FEATURES AND OPERATION

Types of Afterburners:	(a) Direct-fired; (b) Catalytic.
Nature of Pollutants Handled:	Combustible aerosols, vapors and smoke able to be oxidized at 1600°F or less, but which have insufficient fuel value to burn without assistance.
Removal Efficiency:	Over 98% destruction of pollutant.
Capacity:	No theoretical limit. Units with capacities over 30,000 scfm of process fumes are relatively common.
Operating Temperature:	Direct-fired units: 1200 to 1600°F; Catalytic units: 700 to 1000°F.
Inlet Stream Temperature:	20 to 800°F typical. Equipment can be designed to accommodate temperatures outside this range.
Installation Cost:	Typically $4 to 15 per scfm of contaminated fume stream. (For a 30,000 scfm unit about $5/scfm.) Cost will vary widely with nature of contaminant, afterburner size and accessory equipment. (Also see Section 5.33.)
*Partial List of Suppliers:**	C-E Inc., Air Preheater Div.; Dorr-Oliver, Inc.; Eclipse Fuel Engineering Co.; Hirt Combustion Engineers; Maxon Corp.; North American Manufacturing Co.; Oxy-Catalyst, Inc.; Surface Combustion Div., Midland-Ross Corp.; Thermal Research & Engineering Corp.; Thermal Transfer Corp.; UOP Air Correction Div.

* Some of the listed manufacturers supply only afterburner components but not complete units.

5.25 Afterburners—Types, Features and Operation

An afterburner is an incineration chamber in which objectionable aerosols or gases issuing from some process are oxidized to less offensive forms, ideally to carbon dioxide and water vapor. Supplemental heat, usually required to sustain the oxidation reactions, is supplied by burning auxiliary fuel in either a preheat chamber or in the reaction chamber itself, or by passing the incoming pollutant stream through a heat exchanger. Specialized applications of afterburners for the removal of gaseous contaminants are discussed in Sections 5.32 and 5.33.

More often than not, the afterburner is a separate structure through which the exhaust gases are ducted. Afterburners have been successfully applied to paint-baking ovens, solid waste incinerators, metal chip dryers, coffee roasters and blood dryers. Well-designed afterburners, correctly applied and properly maintained, can achieve over 98% destruction of pollutants. Any source generating aerosols or fumes that can be oxidized at temperatures of 1600°F or less is a good potential application for an afterburner.

Afterburners can be direct-fired or catalytic. Gaseous pollutants capable of being burned in air, with little or no auxiliary fuel, are generally treated in waste gas incinerators, which are described in Sections 5.35 to 5.38.

Direct-fired Afterburners

Direct-fired (or thermal) afterburners are generally refractory-lined steel enclosures, cylindrical or rectangular in shape, in which the pollutant stream is mingled with the hot combustion gases of a gas or oil burner (Fig. 5.25a). The objective is to elevate the pollutant stream to the desired incineration temperature (1200 to 1600°F) and to keep it at that temperature until oxidation of the pollutant is complete.

The important design features are as follows.

1. The operating temperature must be high enough for substantially complete oxidation of the pollutant. In most cases, 1200 to 1600°F is sufficient.
2. The turbulence must be sufficient for thorough mixing of the pollutant with the afterburner's hot combustion gases.
3. The pollutant must be held at the operating temperature long enough to ensure complete incineration. This dictates an afterburner chamber of sufficient volume to contain all heated gases for a *residence time* or *retention time* of about ½ to 1 second. Field experience has determined the residence time requirements for many common pollutants. However, local codes may require longer retention times than theoretically necessary.
4. Enough oxygen must be provided for the complete oxidation of both the pollutant and the fuel required to burn it. Good

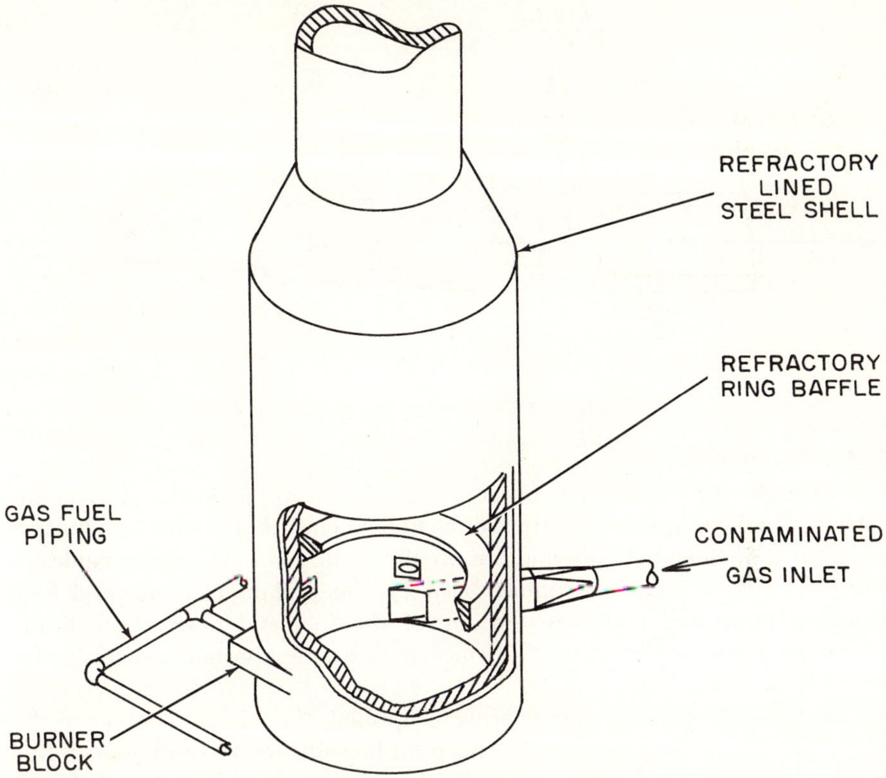

Figure 5.25a Direct-fired afterburner.

engineering practice is to provide excess oxygen to ensure complete combustion. The oxygen source for afterburners is air which can enter with the pollutant stream and/or through the afterburner combustion system.

Combustion Systems

Virtually all afterburners use gaseous fuel as a heat source because it burns quickly and cleanly and, thus, does not complicate the main task of incinerating the pollutant. The majority of commercially available afterburners will operate on natural, LP or clean manufactured gas. Suppliers should be consulted about the suitability of gases not specifically recommended for their equipment.

Gas burners for afterburners can be divided into two categories: straight ("raw") gas or gas-air burners.

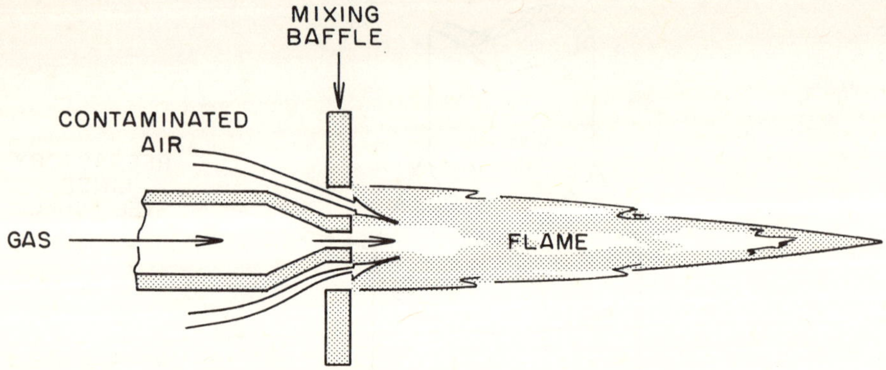

Figure 5.25b Straight gas burner.

Straight Burners

Straight gas or raw gas burners (Fig. 5.25b) are basically devices for injecting fuel gas into the afterburner chamber. They derive their combustion air from the effluent stream and are usually equipped with mixing baffles to foster turbulent, intimate contact between the pollutant stream and fuel, ensuring rapid and complete burning of the fuel and destruction of the pollutant. These burners also usually incorporate a pilot burner or spark igniter to initiate combustion.

Before a straight gas afterburner is applied, the oxygen content of the pollutant stream must be known. This must be sufficient to burn enough fuel gas to raise the entire pollutant stream to incineration temperature, with enough excess for complete oxidation of the pollutant itself. These requirements must be calculated for each potential application. Experience has shown that field conditions rarely follow predictions, therefore an appropriate safety factor should be used. As a general rule, the effluent stream should contain at least 16% oxygen before a straight gas burner system is considered.

Straight gas burners evolved from a single pipe or lance inserted through the wall of the duct carrying the contaminated airstream into the modern grid or plate-type fume burner. Section 5.32 describes this design evolution. Grid burners are either built to size or are assembled from component sections including a perforated plate that completely fills the duct cross section, forcing all contaminated air through the perforations. A drilled manifold, to which the perforated grid is attached, discharges jets of gas adjacent to the perforations. These jets mix with the contaminated airstream and burn. All contaminants in the air must pass through the resulting flame envelopes which raise them to high temperatures and thereby initiate the oxidation process. Oxidation is completed as the contaminant is sustained at temperature in its passage through the afterburner combustion chamber. Major components of an afterburner with a grid-type raw gas burner are shown in Figure 5.25c.

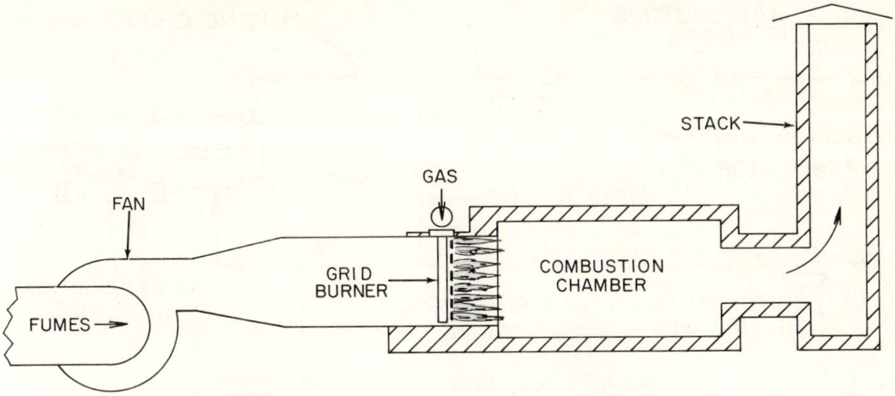

Figure 5.25c Afterburner with grid burner.

Gas-Air Burners

There are numerous instances where conditions prohibit the use of straight gas burners. Oxygen content of the contaminated airstream may be insufficient, or other constituents in the airstream may adversely affect burner operation, even though the oxygen supply is adequate. The presence of excessive water vapor is a good example. It may quench the flames of a straight gas burner, causing frequent nuisance shutdowns or preventing complete combustion of fuel and pollutant. Coffee roasters and blood dryers usually present such conditions. The simplest solution to this problem is a gas-air burner, which has an independent air supply. Standard, general purpose industrial or commercial units are normally used.

Air-gas burner systems fall into two groups: premixing (Fig. 5.25d) and nozzle mixing (Fig. 5.25e). In premixing systems, air and gas are proportioned and mixed in a device (usually a venturi) ahead of the burner. The burner

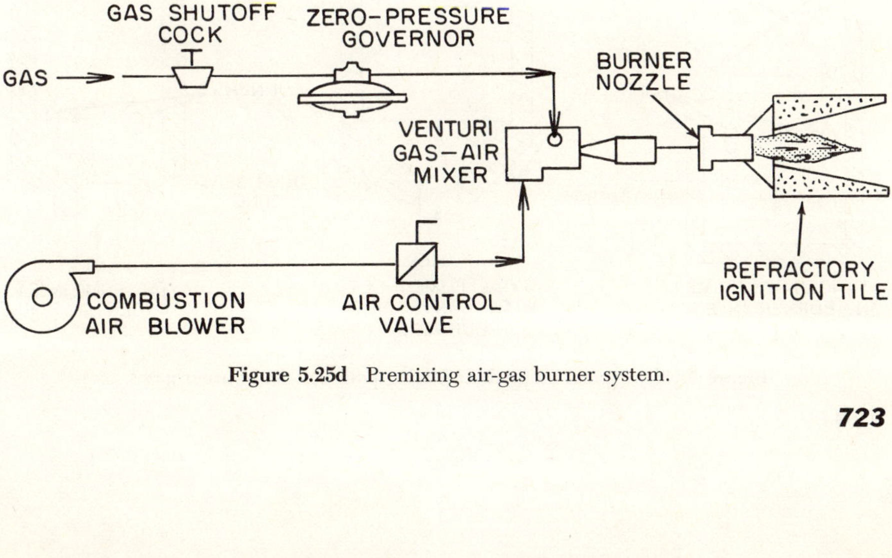

Figure 5.25d Premixing air-gas burner system.

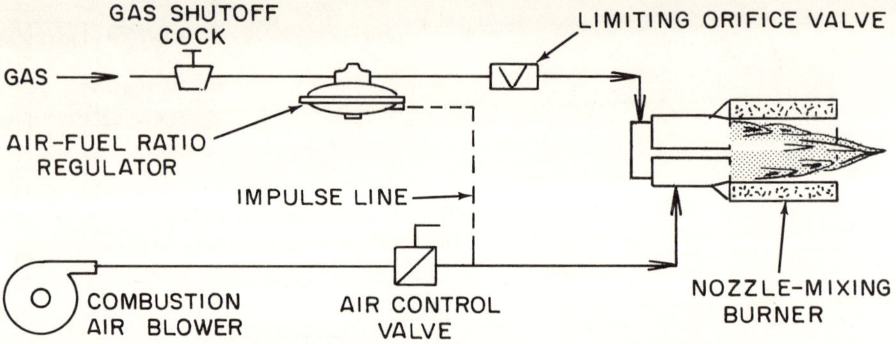

Figure 5.25e Nozzle mixing air-gas burner system.

itself injects this mixture into the afterburner chamber and ignites it. In nozzle mixing systems, proportioned quantities of air and gas are fed to the burner through separate lines. They mix at the burner discharge nozzles, just prior to ignition. Nozzle mixing burners have gained wider acceptance because of their greater operating flexibility.

In contrast to grid burners with their multiple small flames, most gas-air burners are relatively compact and produce a single, relatively large flame. This, plus the fact that the contaminated airstream does not pass through the burner itself, makes good contact between the pollutant and hot burner gases more difficult to accomplish. Some successful mixing schemes are shown in Figure 5.25f. Whatever the technique used, it must ensure rapid, complete mixing of the gas streams. However, it must also delay the onset of mixing long enough to first permit nearly complete combustion of the burner fuel. Failure to do so can result in incomplete combustion of both pollutant and burner fuel.

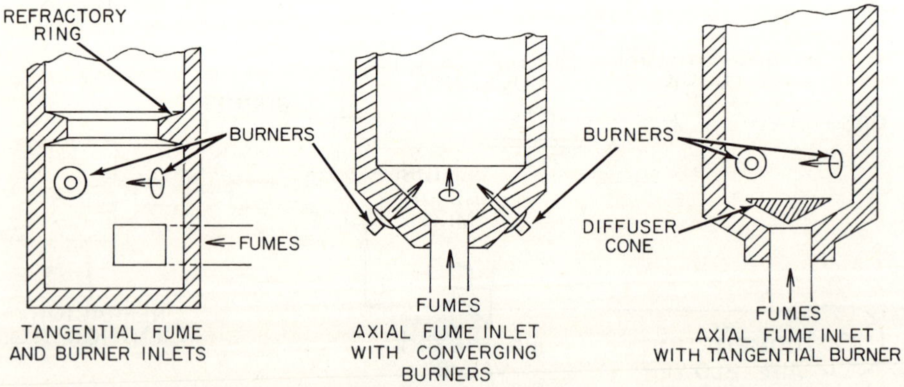

Figure 5.25f Typical methods of mixing pollutant and burner gases.

The afterburner must burn enough fuel to raise the mixture of its own combustion products, the contaminated airstream and any additional induced (secondary) air to the required incineration temperature. If the auxiliary fuel input is modulated to accommodate swinging effluent loads, the burner must also operate at peak efficiency when the firing rate is reduced.

Combustion Chambers

Combustion chambers for direct-fired afterburners are usually cylindrical or rectangular sheet steel enclosures suitably reinforced and lined with refractory. The cylindrical shape is favored because its cross-sectional symmetry makes gas flow patterns simpler to predict and control.

Internal volume of the chamber is dictated by residence time requirements. It must be great enough to retain the hot mixture of contaminated gas, burner combustion products and secondary air (if any) for the required retention time period. Chamber height will vary according to burner flame length, the presence of internal baffles or refractory checkers, dimensional limitations imposed by adjacent structures and the need to avoid resonant reinforcement of combustion noise. A minimum length-to-diameter (or diagonal, in rectangular chambers) ratio of two to one is strongly advised. Afterburner chambers are normally installed with their long axis vertical and burner and fume inlet ports near the bottom.

The refractory material lining the chamber has the purposes of keeping the steel shell below its safe temperature limit (about 850°F) and fostering fuel economy by retarding heat loss from the chamber. Chemical action by certain contaminants in the pollutant stream may lead to rapid deterioration of refractories at temperatures well below their service limits.

Heat Recovery

One of the chief complaints voiced about direct-fired afterburners is that they produce no saleable product, yet their fuel consumption is often greater than that of the production equipment whose emissions they control. This high operating cost can be partly offset by heat recovery. The 1500°F exhaust of a direct-fired afterburner contains considerable latent and sensible heat. A suitable heat exchange device can transfer much of this heat from the exhaust gases to plant makeup air heating systems, ovens or process heaters or to the waste gas stream entering the afterburner, preheating it and thereby lowering the afterburner's fuel requirements.

Catalytic Afterburners

In the presence of a catalyst, many chemical reactions will proceed to completion at temperatures well below those normally required. This principle has been used with great success in the design of catalytic afterburners.

Here, the pollutant and air are passed over an oxidizing catalyst at 700 to 1000°F. Molecules of the reactants are adsorbed on the catalyst surface and react chemically. The oxidation products are then released and exhausted from the afterburner.

Comparison of the operating temperature range of the catalytic afterburner with the 1200 to 1600°F range of thermal units indicates the economic advantages of the first. In the catalytic afterburner, less auxiliary fuel is needed to elevate the pollutant stream to incineration temperature. In fact, many installations require no auxiliary heat at all, because the process exhaust stream enters the afterburner at or above the required incineration temperature. Other installations achieve the desired level of preheat by passing the incoming fumes through a heat exchanger, where the afterburner exhaust is used as the heat source.

The lower operating temperatures of catalytic units also eliminate the need for the extensive use of refractories and heat-resisting alloys in the chamber construction. Usually, a steel enclosure, lined with insulation to retard heat losses, is adequate.

Catalytic units are best suited to the treatment of those waste gas streams in which the pollutant is a relatively unstable chemical and is entirely in the vapor or gaseous phase at incineration temperatures. The waste gas stream must also be free of substances that poison the catalyst or form impervious deposits on its surface. Complete incineration of pollutants with high chemical stability often requires catalytic incinerator temperatures over 1000°F. This negates much of the economic advantage of the catalytic units. Pollutants which remain in the liquid or solid phase at incineration temperatures will not oxidize fast enough in the presence of the catalyst and, therefore, residual pollutant levels in the afterburner exhaust are apt to be excessive. The presence of heavy metal vapors and other known catalyst "poisons" will quickly lower the efficiency of the catalyst, necessitating frequent replacement. The same phenomenon occurs with contaminants that leave solid residues on the catalyst surface, masking it from the gases to be incinerated.

Catalytic Afterburner Components

Figure 5.25g is a schematic representation of a typical catalytic afterburner. All the components shown may not be required on a given unit.

Because combustion takes place in the catalyst bed itself (and to a minor degree, in the preheat zone), combustion chamber residence requirements, as discussed under direct-fired afterburners, do not apply. Destruction of the pollutant is expected to be substantially complete at the exit of the catalyst bed.

Preheat Section

If a contaminated stream enters a catalytic afterburner at temperatures below those required for effective incineration, it must be preheated.

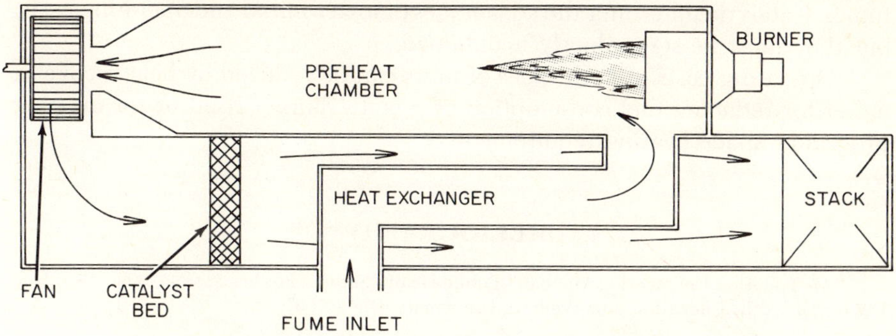

Figure 5.25g Typical catalytic afterburner (plan view).

Preheat burners can be used—similar in construction to the burner in Figure 5.25e—to burn the required auxiliary fuel in the presence of the incoming contaminant stream. Vanes and baffles are often used for improved mixing.

Heat exchanges provide an alternate method by extracting sensible heat from the afterburner exhaust gases and transferring it to the incoming fume stream. Some afterburners use a preheat burner in conjunction with the economizer heat exchanger, either as a supplementary heat source or as a start-up device. In the latter case, the burner is shut off once the afterburner exhaust comes up to temperature. The heat exchanger then takes over the entire task of preheating.

Catalyst

The most common catalysts in commercial use are metals of the platinum family deposited on mats of metal ribbon or on ceramic shapes for maximum exposure of surface area. The ribbons or ceramic shapes are installed in frames or panels for ease of handling and replacement.

Each type of catalyst element is most efficient at a certain gas velocity through the element. "Sizing" the catalyst, therefore, involves providing sufficient element area facing the fume stream to keep gas velocities at the required level.

Because the static pressure of the exhaust fan feeding the afterburner may be insufficient to overcome the flow resistance of the catalyst bed, most catalytic afterburners are equipped with a fan to provide the necessary pressure boost.

Conclusions

Afterburners, either direct-fired or catalytic, are effective control devices for a wide variety of combustible fumes and aerosols. Their function may be the elimination of visible smoke or essentially complete destruction of noxious fumes. Catalytic units offer the advantages of lower initial and operating costs, but they must be scrupulously maintained.

Operating costs of both types of units can be reduced by heat recovery, either for reducing fuel consumption of the afterburner itself or for meeting process or space heating requirements.

BIBLIOGRAPHY

"Air Pollution Engineering Manual," Public Health Service Publication No. 999-AP-40 U.S. Dept. of Health, Education and Welfare, Cincinnati, Ohio, 1967.

5.26 AFTERBURNERS—ACCESSORIES AND INSTRUMENTS

Afterburners require numerous control accessories to ensure safe operation, good fuel economy and thorough incineration. These controls can be divided into three groups: temperature controls, fuel flow controls and safety controls.

Temperature Controls

Two types of temperature controls are typically used on afterburners: throttling operating controls and limit controls.

Operating temperature controls regulate the fuel input to the afterburner to maintain the desired combustion chamber temperature. If chamber temperature is allowed to fall below the optimum level, oxidation of the pollutant will become incomplete. Operating temperatures higher than necessary will increase fuel consumption without a corresponding increase in afterburner efficiency.

Operating temperature controllers require a sensing element in the afterburner chamber. This element must be located where temperatures are representative of the entire mass of heated gases and where fluctuations in gas temperature can be quickly detected. The output signal of the sensing element is transmitted to the temperature controller, which compares it to a set point corresponding to the desired combustion chamber temperature. If the controller detects a deviation from the set point, it adjusts the firing rate of the burner.

Direct-fired afterburner temperatures are usually monitored with a thermocouple[1] located near the exhaust end of the combustion chamber. In this location, mixing and combustion of contaminants and fuel should be complete. Fluid-filled bulbs or thermocouple sensors are used on catalytic afterburners. They are located on the discharge side of the catalyst bed.

With a two-position controller governing fuel input, the afterburner combustion system operates either at full firing rate or at a fixed, low firing rate. This system cannot accommodate swings in effluent loading without furnishing either too much or too little auxiliary fuel, and its response to these swings is usually unacceptably slow. Accordingly, two-position controllers are rarely used on afterburners.

Modulating, or proportioning, temperature controllers[2] generate an output signal proportional to the required firing rate. The proportioning control system is desirable because it responds rapidly to temperature fluctuations and it permits the burner system to operate at a firing rate consistent with the flowrate of flue gases to the afterburner. This achieves the optimum combination of incineration efficiency and fuel economy.

Temperature limit controls are safety devices. If combustion chamber temperatures rise to the point where the afterburner may be damaged and may not be corrected by the operating temperature controls, the temperature limit control will shut down the afterburner.

This high-temperature switch is usually located in or near the afterburner exhaust stack. It is a normally closed switch in the power circuit to an electrically operated fuel shutoff valve. On high temperature, the switch opens, interrupting power to the fuel valve and closing it.

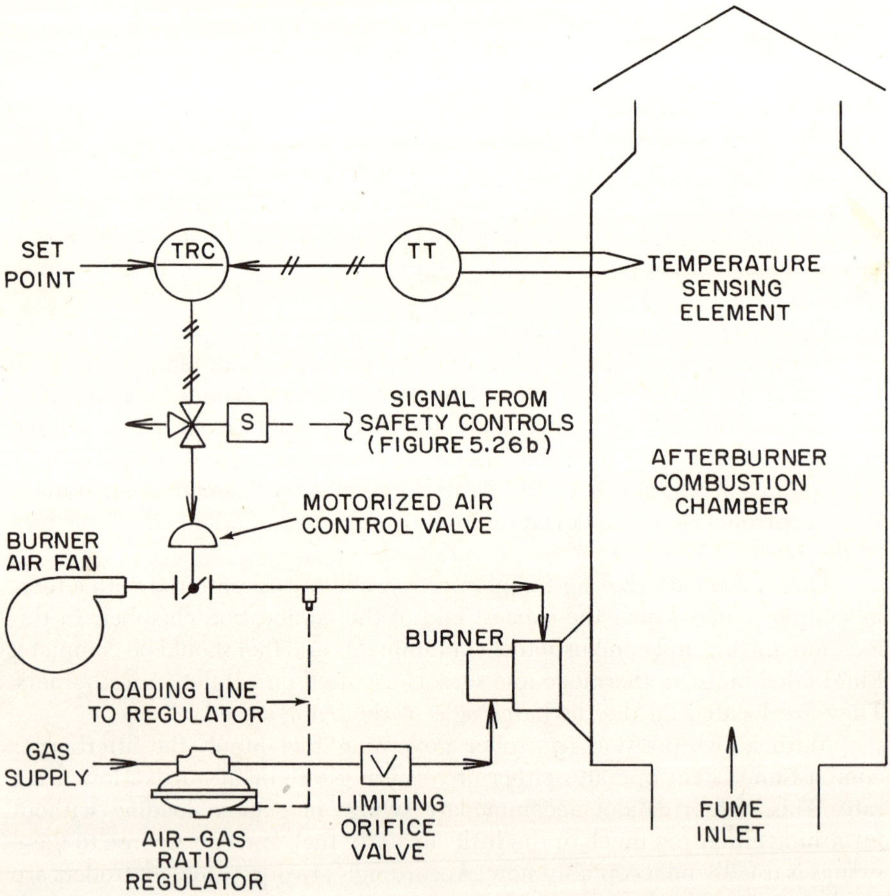

Figure 5.26a Afterburner temperature-fuel control system.

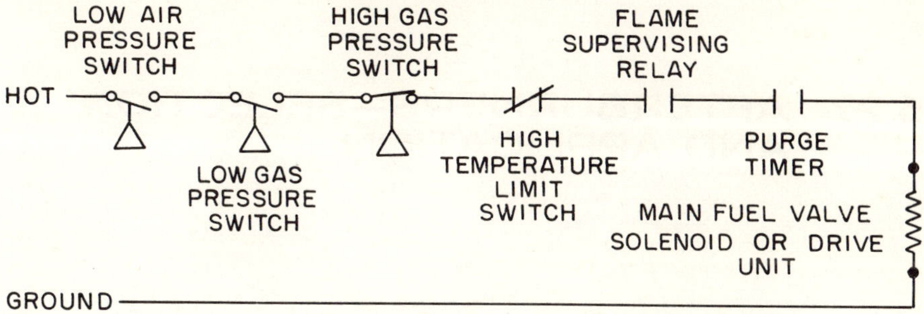

Figure 5.26b Typical afterburner combustion safety circuit.

Fuel Flow Controls

Fuel flow controls[3] include those devices which regulate fuel input to the afterburner and proportion it to combustion air (in burners which are consuming both).

Fuel input is customarily controlled by a motor-driven butterfly valve which is located in the gas line of a straight gas burner and in the air line of a gas-air burner. The valve is throttled by the temperature controller. Fuel flow is usually proportioned to the air (in nozzle-mixing gas-air burner systems) by a diaphragm-type pressure ratio regulator which is loaded by the pressure of the air supply line. In premixing gas-air burner systems, fuel and airflows are held in fixed proportion by the mixing device. Figure 5.26a shows the principal elements of a typical temperature/fuel control system.

Safety Controls

Safety controls include various pressure switches to prove safe pressure levels in the fuel and air lines and flame safety devices to prove the presence of sustained, stable combustion at the burner. These limit controls are all connected in series with the high-temperature limit switch and, possibly, with a timer to ensure adequate purging of the afterburner chamber on start-up. Current passes through this circuit to permit the opening of the main fuel valve (Fig. 5.26b). Unless all limits are satisfied, the fuel valve cannot be opened. If, at any time during the operation of the afterburner, one of the limit devices senses an unsafe condition, it breaks the circuit to the main fuel valve thereby closing it and shutting down the combustion system until corrective action can be taken.

REFERENCES

1. Lipták, B. G. (ed.): "Instrument Engineers' Handbook," Vol. I, Section 4.9, Chilton, Philadelphia, 1969.
2. Ibid, Vol. II, Chap. IV.
3. Ibid, Vol. II, Sections 10.6 and 10.7.

5.27 AFTERBURNERS—SELECTION AND APPLICATION

An afterburner installation must be sized to handle the maximum anticipated throughput and must also be capable of handling reduced flows with no loss of efficiency. Since field conditions often vary from the original design basis, it is desirable to include an overload factor in the maximum design capacity of the unit.

In many cases, afterburners are supplied as packaged units. They may be custom-built to satisfy the requirements of a particular application or they may be standard, preengineered units.

Plants with capable engineering staffs often undertake the design of their own afterburner facilities, utilizing proprietary components. Unless they have had extensive past experience in this field, they should consult burner, refractory and control equipment suppliers on the features and limitations of their products.

Table 5.27a lists the information required for proper selection or design of an afterburner.

Having a complete laboratory analysis of the fume stream is always desirable and it is vital where mixed wastes are being handled.

Because of differences in afterburner chamber designs, combustion accessories, catalyst elements and heat exchangers and since incineration parameters vary for different pollutants, equipment selection is best done in consultation with qualified manufacturers or suppliers. The design of an afterburner installation involves the following steps:

1. Determination of combustion system and/or heat exchanger size necessary to elevate the entire fume stream to incineration or preheat temperature.
2. Determination of the minimum combustion chamber volume consistent with residence time requirements (for direct-fired afterburners) or the optimum preheat chamber size for uniform preheating of the fume stream (catalytic incinerators).
3. For catalytic incinerators, selection of the correct number of catalyst elements for complete incineration.
4. Design of the afterburner structure itself, including selection of

Table 5.27a
DESIGN DATA FOR AFTERBURNERS

Fume Stream Characteristics
- Maximum and minimum volumetric flowrates
- Minimum temperature at point of entry into the afterburner
- Static pressure of fume stream at entry to afterburner
- Chemical composition of contaminant(s) to be incinerated
- Range of concentrations, by weight or volume, of contaminant(s) in stream
- Nature and concentrations of any noncombustible materials in stream (dirt, ash, water vapor); include potential catalyst poisons
- Minimum percent volume of oxygen in stream (if sufficient, this may permit cost savings through use of raw gas burners.).
- Any other unusual conditions that may affect selection and operation of the afterburner

Afterburner Fuel Available
- Type of fuel available, e.g., natural gas, no. 2 fuel oil
- Calorific value of fuel
- Fuel pressure available at point of use

Physical Requirements
- Space limitations, if any
- Proximity of workers or equipment that might be affected by heat or noise emitted from afterburner

Code Requirements
- Applicable electrical, insurance or safety codes
- Air quality codes applying to physical construction of the afterburner, combustion chamber retention times or residual contaminant levels permitted in the afterburner exhaust

Miscellaneous
- Electrical specifications for fan motors, temperature controls and other accessories

suitable metals and refractories and design and location of fume inlet ports and burners.

5. Selection of burners, fans, control accessories and auxiliaries.

Removal Of Gaseous Pollutants

Sections 5.28 through 5.34 describe various abatement techniques for gaseous pollutant emissions. Economic or other reasons frequently necessitate the use of a combination of these methods. For example, if pathogenic substances are present in the emission, it is likely that they will have to be destroyed by direct incineration. Having this incinerator as a component in the pollution control system will most likely reduce the load on the other control component(s) in the system.

In other applications, it might be desirable to combine scrubbing and condensation techniques, where the latter assists in the coalescing of the contaminants which in turn makes the scrubbing operation more effective. Similarly, the operating cost of an additive injection system can be reduced by the upstream introduction of a carbon absorption bed; or the load on an absorption bed can be reduced by the use of upstream condensation or scrubbing. In smaller installations, the high operating costs of direct incinerators and the long regeneration periods of carbon absorption systems are serious limitations.

The removal of some specific gaseous pollutants is discussed in Chapter 6, while the industrial and municipal pollution sources are covered in Chapter 7.

5.28 CONDENSATION METHODS

Partial List of Condenser Suppliers: AB Rosenblads Patenter, Stockholm; Acme Industries, Inc.; Artisan Industries Inc.; Buflovak Div., Blaw-Knox Food & Chemical Eqpt., Inc.; CVI Corp., Sub. of Pennwalt Corp.; Centri-Spray Corp.; Chem-Pro Equipment Corp.; Clermont Engineering Co. Inc.; Combustion Equipment Assoc., Inc.; Dean Products Inc.; The Ducon Co., Inc.; Dunham-Bush, Inc.; Economy Mfg. Co.; Edwards Engineering Corp.; Englehard Minerals & Chemicals Corp.; First Machinery Corp.; Goslin-Birmingham, Inc.; Graham Mfg. Co., Inc., H & P Equipment Co.; Hankison Corp.; Haveg Industries Inc.; Heil Process Equipment Corp.; Industrial Process Engineers; Ingersoll-Rand; Jet-Vac Corp.; Lawler Co.; Loeb Equipment Supply Co.; Manning & Lewis Engrg. Co.; McKenzie-Ris Mfg. Corp.; Monsanto Enviro-Chem Systems Inc.; The Patterson-Kelley Co.; Penberthy-Div. of Houdaille Ind. Inc.; Perfex Corp.; Perry Products Co.; Puregas Equipment Corp. Div. of General Cable Corp.; Renneburg & Sons Co.; Rotodyne Mfg. Corp.; Schutte & Koerting Co.; Swenson Div. of Whiting Corp.; Tailor & Co., Inc.; The Trane Co.; Union Carbide Corp.; Vulcan Mfg. Co.; Welding & Steel Fabrication Co., Inc.; Whitlock Mfg. Co.; Worthington Corp.

5.28 Condensation Methods

Vapor phase pollutants can be removed by condensation initiated by the cooling or compression of the total gas stream. In air pollution abatement work, temperature reduction is usually the more economical method.

When pollutants are present in high concentrations and with dewpoints above 30°C, condensation should be considered as a method of removal. Condensation is commonly used in series with other control equipment; the condenser reduces the load on a more expensive control device such as an adsorber, absorber or catalytic incinerator.

Condenser designs can be surface or contact types. In surface condensers, heat is removed through a heat exchange surface and the coolant does not come in contact with the gas. In contact condensers, the coolant and the polluted gas are mixed together.

Types of Condensers

Most surface condensers are heat exchangers of the multitube and shell type. Normally, the coolant is in the tubes if the condenser is installed horizontally, and the vapors condense on the outside of the tubes. In vertical installations, vapors can be condensed inside or outside of the tubes. Chilled water is the most common coolant. When the mixture of the condensable vapor and noncondensable gas is exposed to a surface colder than the dewpoint of the mixture, some condensation occurs. A liquid film of condensate forms on the cooling surface, and a film of gas collects next to the liquid film. The concentration of vapor in the gas film is lower than is its concentration in the main gas stream. Because of differences in the partial pressures of the vapor in the bulk gas mixture and in the liquid-gas film interface, the vapor diffuses from the bulk gas through the gas film to liquefy on the cold surface. The rate of condensation is thus governed by the diffusion of the vapor through a film of noncondensable gas.

Direct contact condensers can be the spray scrubber and the jet condenser types. The spray scrubber type is simply a large vessel in which the gas-condensable vapor mixture is contacted by the liquid coolant as the coolant is injected into the vessel through spray nozzles (Sections 5.13 and 5.18). Baffles are provided to ensure intimate contact between the liquid spray and the gaseous mixture.

In the jet condenser type unit, the gas-vapor mixture is entrained into a high-speed water jet (Fig. 5.17p). This operation involves heat transfer and mass transfer between the two phases, and the condensed pollutant leaves with the liquid stream. The direct contact condenser requires less cooling water but may result in an unacceptable water contamination. Figure 5.28a illustrates both types of direct contact condensers.

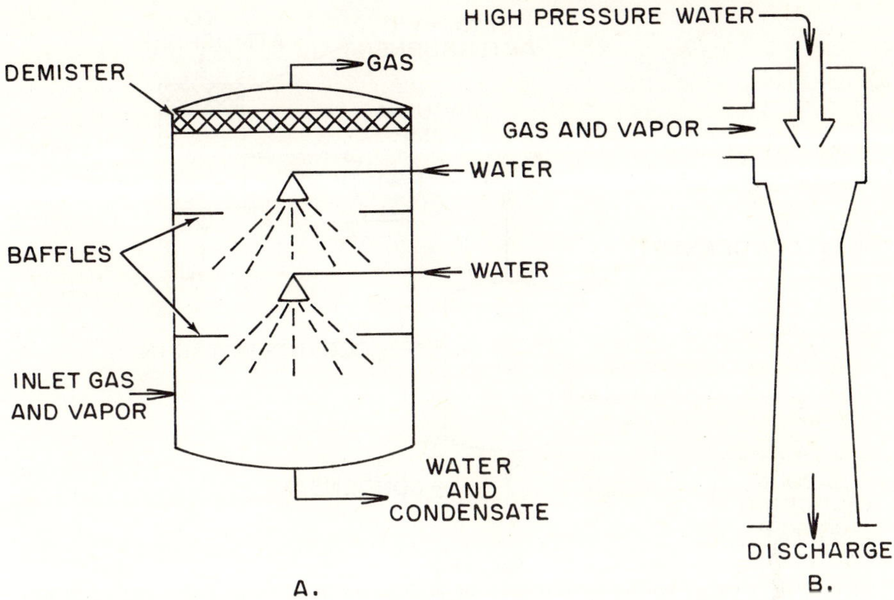

Figure 5.28a Condensers. *A*, Spray contact; *B*, Jet contact.

Typical Installations

Condensers are normally used in conjunction with other control equipment. They are dependent on the concentration of the pollutant vapor in the gas stream, and, therefore, this method becomes very inefficient as the pollutant concentration is lowered by the removal of condensate. It is normally economically attractive to remove the bulk of the pollutant with a relatively inexpensive condenser and then follow it up with a more efficient control device for finishing the job. Thereby, the condenser reduces the load on the more expensive equipment. A typical scheme for a fat rendering operation is depicted in Figure 5.28b.

Condensers are applicable when the vapor is present in relatively high concentrations and the dewpoint is above 30°C. Applications in the petroleum refining and petrochemical industries include the removal of solvents and organic vapors from air.

Design of Condensers

The design of a surface condenser requires the knowledge of the amount of heat to be transferred and of the surface area required for heat transfer. The amount of heat to be transferred can be calculated from an enthalpy

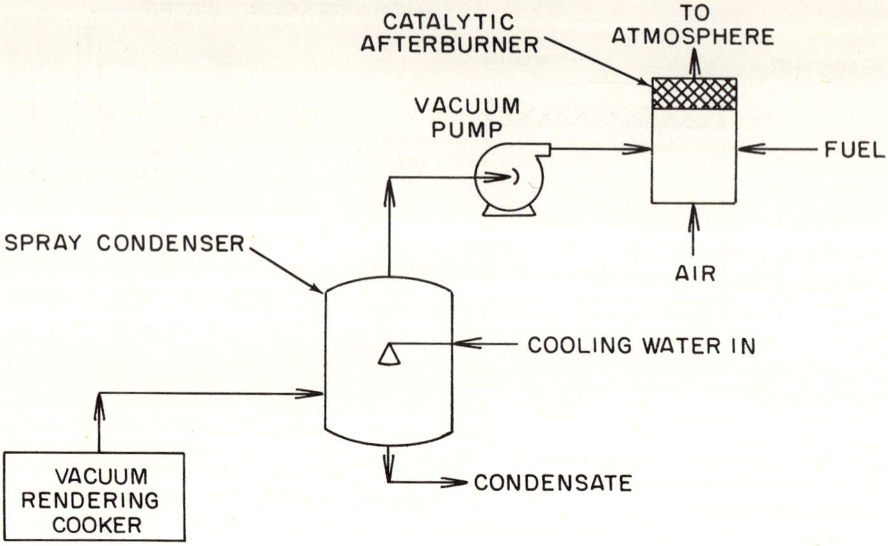

Figure 5.28b Contact condenser and catalytic afterburner in an air pollution control system.

balance on the gas-vapor stream. The amount of heat that must be removed from the stream is the difference in the total enthalpy of all components of the gas that enter the exchanger and the total enthalpy of the stream at the exit. This heat will flow from the vapor-gas stream through a solid retaining wall into the cooling water. Equation 5.28(1) relates the area required to transfer this amount of heat:

$$q = UA(\Delta T) \qquad 5.28(1)$$

where q is the heat transfer rate, U is the overall heat transfer coefficient, A is the area required and ΔT is the mean temperature difference between the gas and the cooling water. The overall heat transfer coefficient varies greatly during the condensation of a vapor from a noncondensable gas which is initially at its dewpoint, since the potential for diffusion varies greatly as the vapor is removed from the bulk gas stream leaving a higher percentage of noncondensable gas. The numerical value of U can decrease as much as by a factor of 100 from inlet to outlet. Therefore, the design should be based on an elemental area as

$$dA = \frac{dq}{U(\Delta T)} \qquad 5.28(2)$$

Then the total area is solved by graphical methods.[1]

Sizing of the direct contact condenser involves the calculation of the amount of heat to be removed from the gas mixture and the time of contact required between the gas and liquid phases. Psychrometric charts are useful in obtaining the heat and mass transfer requirements in direct contact systems.[1]

REFERENCE

1. Kern, D. Q.: "Process Heat Transfer," New York; McGraw-Hill, 1950.

5.29 ADSORPTION TECHNIQUES

Types of Designs:	Regenerative or nonregenerative and fixed or fluidized bed designs.
Applications:	Regenerative fixed bed units are utilized in solvent recovery, nonregenerative fixed beds in odor removal and fluidized beds in specialized applications.
Design Temperature:	Normally below 40°C, maximum up to 350°C.
Design Pressure:	Normally low, but up to 1,000 PSIG or more in special cases.
Gas Velocities:	0.5 to 3 fps, 20 to 100 cm/sec.
Contact Time:	0.02 to 0.05 sec.
Bed Depth:	1 inch or less for thin beds, 10 to 30 inches for thick beds.
Costs:	Vary with pollutant concentration and with recovery equipment complexity. $10 to $15 per ACFM for units under 6,000 ACFM size and $3 to $8 per ACFM for units in the 30,000 ACFM size range.
Partial List of Suppliers:	American Air Filter Co., Inc.; C&I/Girdler, Inc.; Dravo Corporation; Heil Process Equipment Corporation; Norton Company; Pall Corp.; Research-Cottrell, Inc.; Vulcan Mfg. Co.; Witco Chemical Corp.

Adsorption Theory

Adsorption is a process whereby molecules of a liquid or gas contact and adhere to the surface of a solid. In air pollution control, adsorption is used to separate pollutants from industrial gas streams or air. For example, valuable solvent vapors which would become pollutants if discharged to ambient air can be so recovered from dilute mixtures with air. These operations exploit the ability of certain solids to preferentially adsorb specific substances onto their surfaces.

There are two major categories of adsorption phenomena, physical and chemical. Physical, or van der Waals, adsorption results from attraction forces between molecules of the substance being adsorbed and the solid. A gas will condense on the solid surface, even though its partial pressure is less than the vapor pressure, if the attractive forces between the gas and solid are greater than those between the molecules of the gas. The adsorbed substance does not penetrate the lattice structure of the solid; however, the substance will penetrate the capillaries of a porous solid if it wets the solid. An equilibrium sorption is reached when the partial pressure of the adsorbed substance equals that of the gas phase. The adsorbed gas is removed or desorbed by raising the temperature or lowering the pressure of the gas phase.

Chemical adsorption, or chemisorption, involves a chemical bonding between the adsorbed substance and the solid surface. This occurs at high temperatures, usually above 200°C. Chemisorption is of importance in catalysis but is not a primary mechanism in air pollution adsorption operations.

Adsorbent Materials

There is a wide variety of solids possessing adsorptive ability. The most common of these adsorbents for air pollution control is activated charcoal, or active carbon as it is sometimes called. Other commercial adsorbents include alumina, silica gel, other metallic oxides, synthetic zeolites or "molecular sieves" and resins as outlined in Table 5.29a.

Activated charcoal is made from hard woods, coconut and other nut shells, bituminous coal and petroleum residues. The nature of the raw material and the treatment conditions in the activating process determine the ultimate properties of the activated charcoal. Among these properties are the hardness, the size and quantity of the micropores in the charcoal. The internal surface area created by the activation process can be 1,000 m^2/gm (10,760 ft^2/gm) or more. The surface of the charcoal (being uncharged carbon atoms) has no significant electrical potential gradients that would selectively bind polar molecules. Therefore, activated charcoal is effective in adsorbing organic molecules even in the presence of highly polar water molecules. The larger, less polar organic molecules are selectively adsorbed because the water molecules attract each other more than does the nonpolar carbon surface. It is this property of activated charcoal that makes it particularly useful for air pollution control. Many metallic oxides will adsorb hydrocarbons from dry gas streams; but because of the electrical characteristics of the oxide surface, water vapor will be selectively adsorbed from wet gas streams.

Table 5.29a
ADSORBENT CHARACTERISTICS

Material	*Size*	*Internal Porosity*	*External Void Fraction*	*Dry Bulk Specific Gravity*	*Average Pore Diam, Å*[a]	*Surface Area, m^2/gm*
Active alumina	3–8 mesh[b]	0.25	0.49	0.80	34	250
Silica gel	0.3 cm	0.34	0.36	0.80	21	650
Zeolites	0.3 cm	0.50	0.36	0.71	3–12	770
Magnesia-silica gel	3–8 mesh	0.33	0.25	0.48	—	300
Carbon, shell-based	4–6 mesh	0.50	0.37	0.48	20	1000
Carbon, wood-based	4–6 mesh	0.60	0.40	0.37	30	1200
Carbon, peat-based	0.3 cm	0.55	0.40	0.41	35	1100
Carbon, coal-based	10–30 mesh	0.45	0.40	0.51	—	1100
Carbon, petroleum-based	—	0.70	0.30	0.32	20	1000
Anhydrous calcium sulfate	—	0.38	0.45	0.95	—	—
Iron oxide	4–6 mesh	0.22	0.37	1.45	—	20
Magnesia	—	0.75	0.45	0.40	—	2
Phenolic resin	10–50 mesh	—	0.35	0.35	—	3

[a]Å = 10^{-10} meter or 10^{-4} micron
[b]Mesh = number of lines per inch in a screen

The solid adsorbent should be strong and hard enough to support its own weight in beds without fracturing or crumbling apart. It also should have a large surface area per unit weight and it must concentrate the pollutant substance preferentially to all other substances in a gas mixture. Some aspects of the preferential adsorption phenomena are not well understood; and, therefore, empirical observation is the basis for most adsorbent selections.

Design Principles

In many adsorbent systems, the gas mixture is passed through a stationary bed of the solid adsorbent. When the gas is first introduced into the solid bed, the front layer of solids will adsorb the pollutant (adsorbate or solute) very rapidly and the little solute left will be taken up by the deeper layers of solids. After a time, the layers of solid at the entrance will become saturated and will not adsorb any new solute. The adsorption zone then moves deeper into the bed. At a still later time, any further operations will result in a "breakthrough" of pollutant. Therefore, some provision must be made for renewing the adsorbent; in the meantime, operations are switched to a second bed. Figure 5.29b depicts the adsorption wave as it passes through the bed.

The design of a fixed-bed adsorber in air pollution control applications[1] and the determination of the adsorption cycle time require a knowledge of the breakthrough curve.

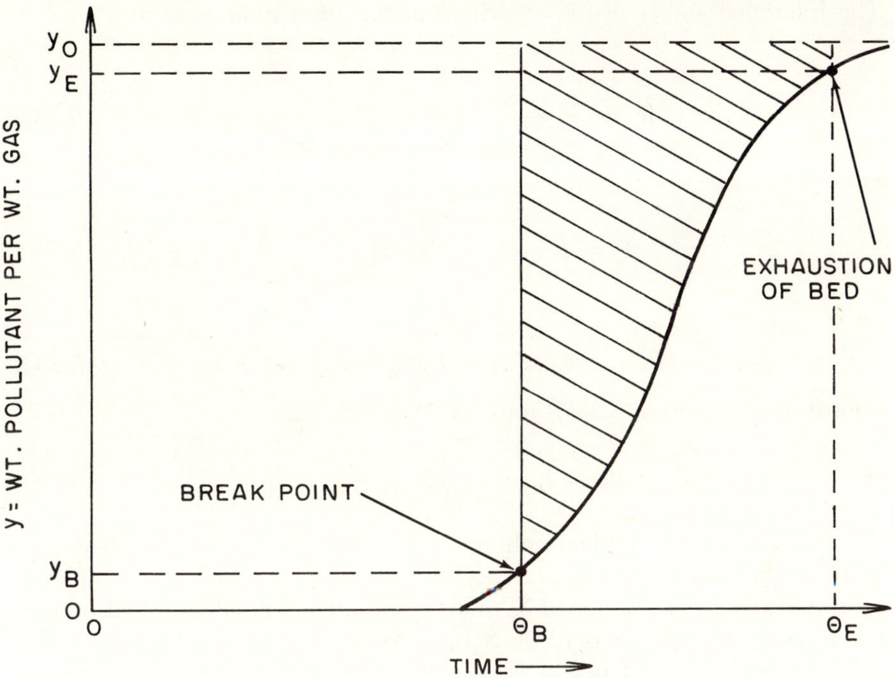

Figure 5.29b Adsorption breakthrough curve.

The mass velocity of the total gas stream is G_s, mass rate per cross-sectional area of bed. The initial mass fraction of pollutant in the inlet gas is y_o, and the breakpoint mass fraction is y_B. The adsorbent has essentially reached its capacity at some arbitrarily chosen effluent concentration y_E. The primary concern is the time elapsed from the start of the adsorption cycle to the moment when breakpoint occurs θ_B and the shape of the curve from θ_B to θ_E. The adsorption zone L_A in the bed, of constant height, is that part of the bed in which the concentration is in the process of being changed from y_B to y_E. If θ_F is the time required from startup until the adsorption zone is completely formed and L is the overall height of the bed,

$$L_A = L\frac{\theta_E - \theta_B}{\theta_E - \theta_F} \qquad 5.29(1)$$

The quantity of solute removed from the gas S from θ_B to θ_E can be determined by the shaded area in Figure 5.29b, having the units of mass solute per bed cross section. This is given by Equation 5.29(2):

$$S = G_s \int_{\theta_B}^{\theta_E} (y_o - y)d\theta \qquad 5.29(2)$$

The fractional ability of the adsorbent in the adsorption zone is

$$F = \frac{S}{y_o G_s(\theta_E - \theta_B)} = \frac{\int_{\theta_B}^{\theta_E} (y_o - y)d\theta}{y_o(\theta_E - \theta_B)} \qquad 5.29(3)$$

or

$$F = \int_o^{1.0} (1 - y/y_o)d\left(\frac{\theta - \theta_B}{\theta_E - \theta_B}\right) \qquad 5.29(4)$$

and

$$\theta_F = (1 - F)(\theta_E - \theta_B) \qquad 5.29(5)$$

Substituting Equation 5.29(5) into 5.29(1) yields

$$L_A = L\frac{\theta_E - \theta_B}{\theta_E - (1 - F)(\theta_E - \theta_B)} \qquad 5.29(6)$$

The adsorption column contains $L\rho_B$ mass of adsorbent per bed cross section, when ρ_B is the bulk density of the bed. If the bed was completely saturated at a concentration of X_s (mass adsorbate per unit mass of adsorbent), the mass of adsorbate would be $X_sL\rho_B$. At breakpoint, the bottom part of the adsorption zone of height L_A is still in the column, but the remainder of the column $L - L_A$ is saturated. The mass of adsorbate at breakpoint is

$$(L - L_A)\rho_B X_s + L_A\rho_B(1 - F)X_s \qquad 5.29(7)$$

The fractional approach to saturation of the entire sorbent bed at breakpoint is

$$\eta = \frac{(L - L_A)\rho_B X_s + L\rho_B(1 - F)X_s}{L\rho_B X_s} = \frac{L - FL_A}{L} \qquad 5.29(8)$$

Over the differential height dL, the transfer from gas to solid phase is

$$G_s dY = K_Y a_P(y - y^*)dL \qquad 5.29(9)$$

where $K_Y a_P$ is the mass transfer coefficient K_Y times the transfer area a_p and y^* is the equilibrium concentration. Figure 5.29c depicts an equilibrium adsorption curve.

For the adsorption zone,

$$N_{OG} = \int_{y_B}^{y_E} \frac{dy}{y - y^*} = \frac{L_A}{H_{OG}} = \frac{L_A}{G_s/K_Y a_P} \qquad 5.29(10)$$

where N_{OG} is the overall number of gas transfer units in the adsorption zone and H_{OG} is the height of one gas transfer unit in the adsorption zone.

For any value of L less than L_A,

$$\frac{\text{L at y}}{L_A} = \frac{\theta - \theta_B}{\theta_E - \theta_B} = \frac{\displaystyle\int_{y_B}^{y} \frac{dy}{y - y^*}}{\displaystyle\int_{y_B}^{y_E} \frac{dy}{y - y^*}} \qquad 5.29(11)$$

In using this method, experimental data of the adsorption equilibrium values must be available and the mass transfer coefficient must be estimated from correlations or obtained from experimental measurements. First, $(\theta - \theta_B)/(\theta_E - \theta_B)$ values must be obtained by graphical integration of the right hand side of Equation 5.29(11). The ratio of $(\theta - \theta_B)/(\theta_E - \theta_B)$ should be plotted against $1 - y/y_o$ and F determined by the area under the curve, Equation 5.29(4). Then the number of gas transfer units N_{OG} can be calculated by graphical integration of Equation 5.29(10), and the height of the adsorption

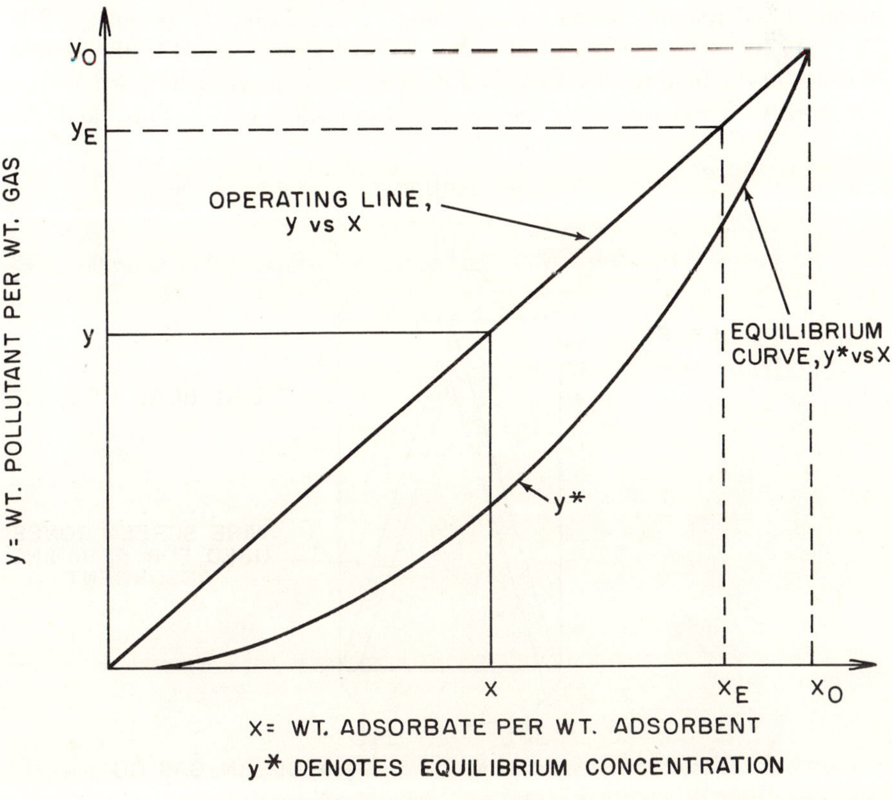

Figure 5.29c Equilibrium adsorption curve.

zone L_A calculated as the product of N_{OG} times H_{OG}. For a given bed size L, the fractional approach to saturation can be calculated by Equation 5.29(8), and the mass of adsorbate in the bed at breakpoint is $\eta L \rho_B X_s$.

Fixed-bed Systems

The fixed-bed adsorber can be designed as a thin bed of granular solids. The main advantage of this type is the low resistance to gas flow. Flat, cylindrical, conical and pleated shapes of solids have been used. The granular bed is sandwiched between perforated sheets of metal or wire screens (Fig. 5.29d).

Thick-bed adsorbers use bed depths of 30 to 100 cm (10 to 30 inches). They provide larger adsorbing capacity and sometimes better efficiency. The inlet gas is made to flow down to prevent bed lifting. The bed typically is installed in horizontal, cylindrical vessels as shown in Figure 5.29e.

Fluidized-bed Systems

Fluidized beds have also been used in adsorption operations. In the fluidized-bed system, the gas passes upflow through a bed of granular solids. The velocity of the gas is such that the pressure drop is equal to the weight of solids in the bed. In this dense-bed fluidized system, the solid particles are continually in motion; thus, they are at a constant temperature, and the

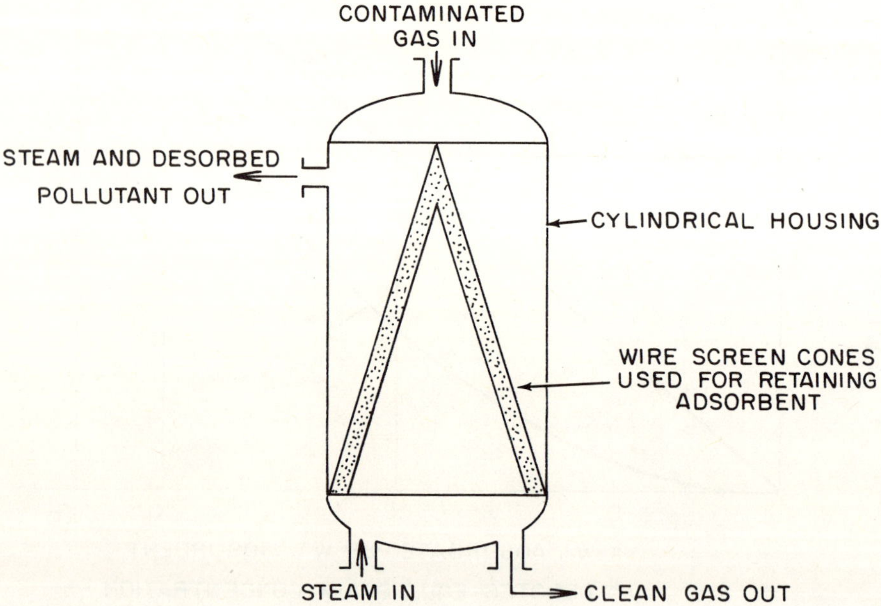

Figure 5.29d Thin-bed adsorber.

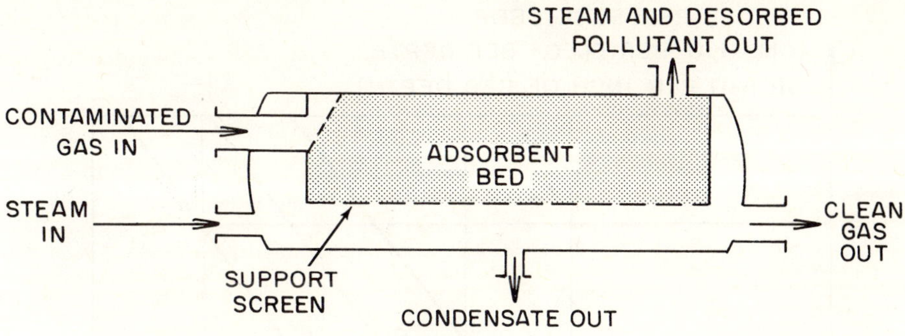

Figure 5.29e Thick-bed adsorber.

amount of sorbed material on the solids is also uniform throughout the bed. An advantage of this type of bed is that it is relatively easy to achieve continuous solids circulation between the adsorber vessel and a regenerator. On the other hand, its operation is more complex than a fixed-bed system and the operating costs are higher.

Types of fluidized-bed adsorber systems include pneumatic transport systems where the gas velocity is high enough so that the adsorbent particles are carried along with the gas and raining solids systems where the adsorbent particles fall countercurrent to a stream of rising gas.

Regeneration

Nonregenerative systems are those where the adsorbent is replaced after its performance has dropped off. They are economical when the quantity of pollutant is small, such as in odor control. The removed adsorbent may be discarded or it may be regenerated in external equipment.

Regenerative systems reactivate the adsorbent on site (Section 5.30). They also can allow for the recovery of the adsorbate. If the pollutant concentration is above 0.1%, the regenerative system generally is the more economical. In regenerative systems, the adsorbate (pollutant) can be either destroyed or recovered; and this choice is a function of the economic value of the adsorbate. Normally, in solvent recovery applications, the value of the solvent is such that recovery is justified.

Effect of Process Variables

The bed design should provide enough pressure drop to guarantee uniformity of gas distribution through the solid bed while avoiding excessive pressure drops that would cause high operating costs. The rate of adsorption is rapid, and adsorbent bed volumes are sized to give contact times of 0.02 to 0.05 seconds. This means that a thin granular bed of 1 to 2 cm thickness

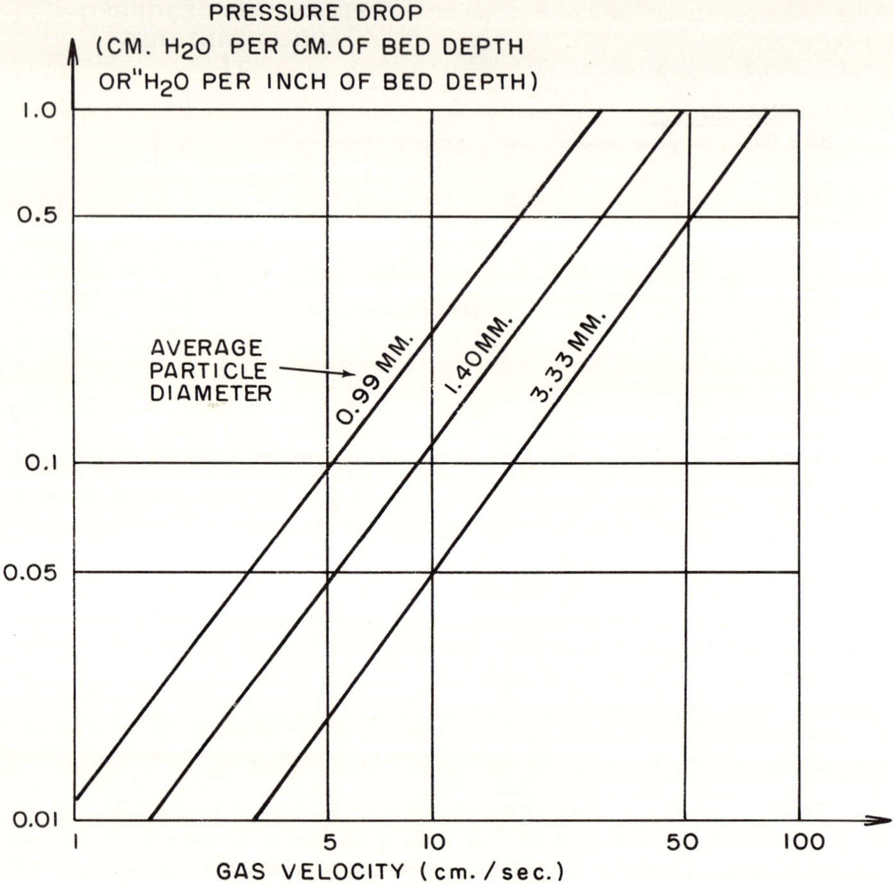

Figure 5.29f Effect of gas flowrate on pressure drop with varying particle size.

(0.5 to 1 inch) could operate at linear velocities of 20 to 100 cm/sec (0.5 to 3 fps). The resulting pressure drops are shown in Figure 5.29f.

The capacity of a solid adsorbent decreases with increasing temperature. Therefore, operating temperatures above 40°C are usually avoided. The adsorption mechanism is exothermic, and the heat liberated tends to increase the adsorber operating temperature. For relatively dilute organic solvent-air mixtures, the temperature rise due to adsorption is about 10°C which has no substantial effect on the operation. However, concentrated hydrocarbon streams can cause adsorber temperatures to rise to dangerously high levels.

The maximum time period for which an adsorber can be expected to operate can be calculated from the ratio of the weight of solute that can be held in the bed to the flowrate of the solute to the bed or

$$t = \frac{SW}{MQC_v/RT} \qquad 5.29(12)$$

where t = duration of adsorber, sec
S = fractional retentivity of adsorbent, mass solute/mass adsorbent
W = mass of adsorbent in the bed, kg
M = molecular weight of solute, kg/gmol
Q = volumetric flowrate of total gas, liters/sec
R = gas constant, 0.082 l-atm/gmol°K
T = temperature, °K
C_v = volume fraction of vapor in total gas

Table 5.29g gives some typical retentivities of activated charcoal.

Since the capacity of adsorbents decreases with temperature, operating temperatures are normally kept as close to ambient as possible. Excessive temperature fluctuations should be avoided since periods of temperature rise can cause massive desorption. Some temperature rise can be expected during the adsorption cycle since adsorption is exothermic; this effect can be minimized by bed design and by dilution of the feed gas if necessary.

Operational Problems

Many of the operational problems associated with adsorbers concern the availability of the adsorbent surface. For example, in surface coating opera-

Table 5.29g
RETENTIVITY BY ACTIVATED CARBON

Adsorbate	*Saturation Weight Fraction*[a]	*Approximate Retentivity, Weight Fraction*[b]	*Retentivity after Regeneration*[c]
Benzene	0.45–0.55	0.25	0.06
Carbon tetrachloride	0.8–1.1	0.45	0.20
Gasoline	0.1–0.2	0.07	0.02
Methyl alcohol	0.5	—	0.01
Isopropyl alcohol	0.5	0.18	0.01
Ethyl acetate	0.58	0.20	0.05
Acetone	0.51	0.10	0.03
Acetic acid	0.70	0.30	0.03

[a] Weight of adsorbate per weight of dry carbon.
[b] Weight adsorbate per weight carbon retained in dry air stream at 20°C.
[c] Regeneration with steam at 150°C for 1 hour.

tions, some of the solvent vapors adsorbed are high-molecular weight compounds that are very difficult to desorb. Under normal regeneration conditions, these high-molecular weight compounds are not removed and they accumulate on the adsorbent surface over a period of time. Therefore, the effective area for adsorption decreases with time. In order to remove these materials, high-temperature stripping with steam (at 300°C or above) is used periodically.

Particulate matter also can impair the ability to adsorb by coating the surface of the adsorbent. Particulates can adhere to the surface and become almost impossible to remove. In addition, particulates can plug up the interparticle spaces and cause excessive pressure drops. In some operations, it is necessary to place a filter at the inlet to the adsorber in order to protect against the entry of particulate matter.

Recovery of polar compounds by steam regeneration presents additional operating complexity because additional processing steps are necessary to separate the polar compound from the aqueous solution. The separation is usually by fractional distillation.

In some operations, clean air is used to cool down the adsorbent bed following regeneration and prior to the next adsorption cycle. However, if high-temperature stripping has been used to regenerate the bed, air must not be applied because of the possibility of reaching the lower explosive limit at the elevated temperature.

Corrosion can be a problem in adsorbers with steam regeneration. The severity of the problem increases with increasing steam temperature. A phenolic resin coating can be applied to the interior vessel parts or stainless steel can be used to eliminate this problem.

Applications in Emissions Control

Activated charcoal is the most common adsorbent for air pollution control because of its suitability for removing organic vapors. It can be used to recover almost any organic solvent or combination of solvents, from air at ambient conditions even at low solvent concentrations and high humidity. Since organic solvents have very low vapor pressures at ambient temperatures, the adsorption operation is efficient at low solvent concentrations. Therefore, adsorption can be economically attractive because of the recovery of the valuable solvent.

Solid adsorption is an effective method of removal of odorous emissions even at such low odorant concentrations as 100 parts per billion or less. Many types of industries have malodorous emissions that can be controlled by adsorption. These include the food processing industry, the chemical manufacturing industry, pulp and paper manufacturing, foundries and several other industries.

In food processing, odors arise from cooking operations such as baking, broiling, frying or boiling. They also occur in meat, fish and poultry processing, dehydration operations, waste rendering, coffee roasting and in fermentation processes. In the chemical process industries, odors can arise from the loss of solvents, products or by-products into the atmosphere. Some of these materials are highly odorous, especially those in certain pesticides, fertilizers, glues, cements and pharmaceuticals.

Adsorption has been considered as a method for removing other gaseous pollutants such as sulfur and nitrogen oxides. The Reinluft process for removing sulfur oxides from power plant stack gases is an example. Other examples of adsorption applications are discussed in Chapter 7.

REFERENCE

1. Michaels, A. S.: *Ind. Eng. Chem.*, vol. 44, p. 1922, 1952.

BIBLIOGRAPHY

American Industrial Hygiene Association, "Air Pollution Manual," Part II: Control Equipment, Detroit, 1968.

Brunauer, S.: "The Adsorption of Gases and Vapors," Princeton Univ. Press, Princeton, 1945.

Danielson, J. A. (ed.): "Air Pollution Engineering Manual," PHS Publn. 999-AP-40, HEW, Cincinnati, 1967.

Hassler, J. W.: "Activated Carbon," Chem. Pub. Co., New York, 1963.

Mantell, C. L.: "Adsorption," 2d ed., McGraw-Hill, New York, 1951.

Perry, J. H. (ed.): "Chemical Engineers' Handbook," 4th ed., McGraw-Hill, New York, 1963.

Treybal, R. E.: "Mass Transfer Operations," 2d ed., McGraw-Hill, New York, 1968.

5.30 REGENERATION OF ADSORPTION BEDS

Regeneration involves removing the adsorbate from the adsorbent. Normally, when the breakpoint in an adsorber bed is reached, the adsorption operation is halted and regeneration is started. Meanwhile, adsorption operations are switched to a second, standby bed. Regeneration is accomplished by raising the temperature of the adsorbent or by lowering the gas phase pressure. Either action causes the equilibrium between the adsorbate on the surface and in the gas phase to be shifted in favor of the gas. Thus, the sorbed molecules are detached or desorbed from the adsorbent surface and enter the gas stream.

The gas flow through the adsorbent bed during regeneration is normally countercurrent to gas flow during adsorption. That is, if the gas flow in the adsorption step was downflow, then it will be upflow during regeneration. The advantage of the countercurrent flow is due to the adsorbate concentration gradient across the bed. The adsorption cycle inlet side of the bed can be saturated before the breakpoint is reached at the outlet. The countercurrent flow will cause the desorbed material to flow back through the saturated adsorbent and, thereby, minimize readsorbing-desorbing occurrences. Therefore, the countercurrent flow is more efficient.

Carbon Regeneration

Regeneration of activated carbon is normally accomplished by using steam to heat up the adsorbent bed and also to act as a carrier gas to remove the desorbed vapors from the vessel. Saturated steam at 110°C is sufficient to remove most organic solvents; however, temperatures as high as 300°C may be necessary to reactivate the carbon to its original condition. The steam flow during regeneration is 10 to 20% of the gas flow (on a volumetric basis) during the adsorption cycle. The total steam requirement is 5 to 15 times the weight of the solvent recovered.

Design Considerations

Regeneration in fixed beds requires a minimum of two beds for continuous operations. One of the beds is on the adsorption cycle while the other is being regenerated. (For the description of a similar system, refer to Figure 5.6b.)

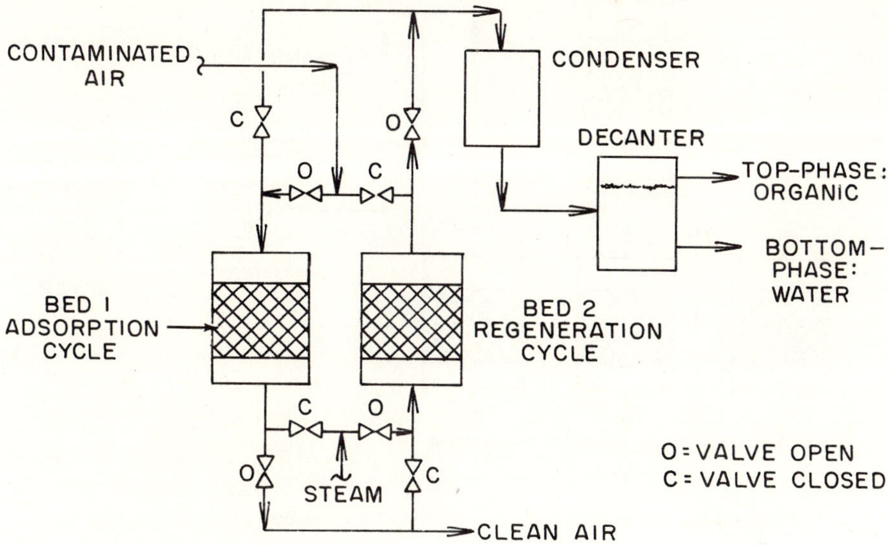

Figure 5.30a Operation and regeneration of a two-bed adsorption system.

There must be enough time to cool the regenerated bed to ambient temperature if elevated temperatures are used during regeneration and to remove water if steam is used.

Most of the adsorbate is released quickly during regeneration. The adsorbate concentration in the regeneration gas reaches a maximum early in the regeneration cycle and then tails off gradually. In order to save time, not all the adsorbate is stripped from the adsorbent. If the regeneration and cooling can be completed before the breakpoint is reached in the operating adsorption bed, then two beds are sufficient. A schematic diagram of such a system for recovering organic solvents is depicted in Figure 5.30a. The switching of beds from regeneration to cooling to adsorption can be accomplished automatically by a cycle timer and control valves (Fig. 5.6b).

If the regeneration and cooling time is greater than the breakpoint time, three fixed-bed adsorbers have to be used as shown in Figure 5.30b. In this system, at any one time, two beds are used in series. The clean effluent from the first bed is passed through a second bed which is still hot and wet from regeneration. Evaporation of the water in the second bed cools the adsorbent while the water is also being removed. Therefore, the second bed is on a cooling cycle while the first is adsorbing and the third is regenerating. At the end of this cycle, the first bed is switched to regeneration, the second to adsorption and the third to cooling. During the third phase, the first bed is cooling, the second regenerating and the third is adsorbing. In many cases, the cooling cycle is really a cooling plus adsorbing cycle. In such cases, the

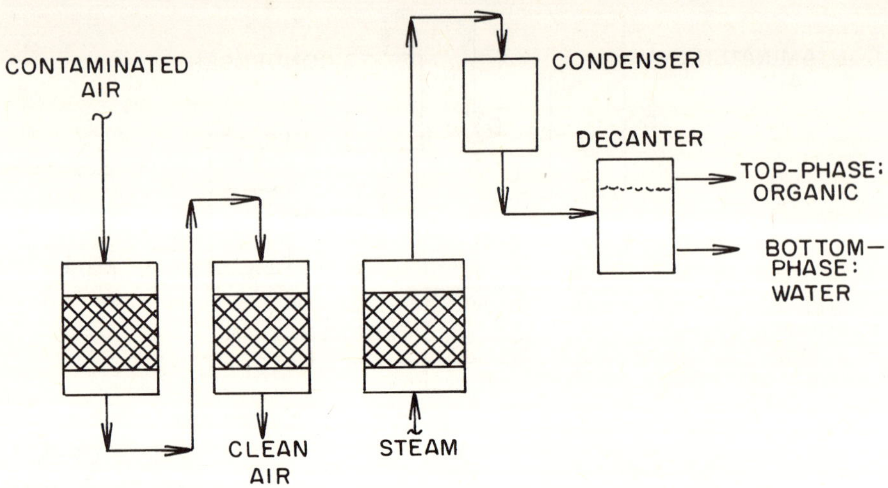

Figure 5.30b Operation and regeneration of a three-bed adsorption system.

cooling wave travels through the bed quite rapidly and then the cooled bed acts as an adsorber. Since the cooled bed is in series with the bed on adsorption cycle, the first bed can be operated beyond its breakpoint, thereby increasing the capacity of the system.

Recovery of Adsorbate

If the value of the adsorbate can justify recovery, as is the case in organic solvents, separation and recovery equipment is installed to follow the adsorber beds. A recovery scheme for organic solvents is illustrated by Figure 5.30a. Here, the effluent from the regenerating bed is a mixture of steam and desorbate that has been removed from the bed. This mixture is condensed by indirect heat exchange; and since the liquid organic solvent is immiscible in water, the organic phase can be separated in a decanter.

The adsorbate may be miscible with water and, if this is the case, fractional distillation can be used to separate the mixture.

Destruction of Adsorbate

When the concentration of the pollutant in the total gas stream is less than 2 ppm, a nonregenerative system is normally used. For concentrations greater than 1,000 ppm, regeneration is necessary because of the high cost of frequent adsorbent replacement, and the adsorbate is recovered if it has economic value. In the intermediate range between 2 ppm and 1,000 ppm, the cost of the recovery equipment usually outweighs the worth of the adsorbate, and its destruction by oxidation is an acceptable alternative.

For this type of operation, the adsorbent is impregnated or coated with a small amount of oxidation catalyst. This catalyst is normally a metallic oxide. During the adsorption cycle, the adsorbent functions normally, without catalytic action. During regeneration, the heating of the bed to a higher temperature initiates the catalytic combustion of the adsorbate. The operating temperature in the bed during regeneration is 350 to 400°C. The hot airstream continues to be injected into the bed until all the adsorbate has been oxidized to innocuous products such as carbon dioxide and water vapor.

Care must be taken to select an adsorbent that will not be burned during the catalytic oxidation step. Some activated carbons have kindling temperatures above 400°C, while others do not. Carbons with high kindling temperatures can undergo many regenerations by oxidation without altering their surface structure and adsorptive ability.

Early in regeneration cycle, the effluent should be passed back to the bed which is on its adsorption cycle. This is needed because during this early period the bed being regenerated is warming up and some vapors are desorbed before the catalytic combustion temperature is reached. These vapors should not be released to the atmosphere. Once the combustion temperature is reached, the effluent from the regenerative bed can be released to atmosphere.

For References and Bibliography see Section 5.29.

5.31 SCRUBBING TECHNIQUES

For costs, equipment suppliers and additional information see Section 5.12.

The gaseous pollutants from flue gas can be efficiently removed using a suitable scrubbing liquor in a mass transfer device. The liquor and gas can contact each other while both are flowing in the same direction (cocurrent flow), in opposite directions (countercurrent flow) or while one flows perpendicular to the other (crossflow). The scrubbing liquor used for the removal of gaseous pollutants can be a slurry or a solution. In chemical engineering terminology, the alternate term for scrubbing is absorption.

The Absorption Process

Absorption is probably the most important gas-purification technique and is common to a great number of processes. It involves the transfer of a pollutant (mass-transfer) from the gaseous phase to the liquid phase through the phase boundary. The absorbed material may dissolve physically in the liquid or react chemically with it. Desorption or stripping represents the reverse case of this operation in which the material moves from the liquid to the gaseous phase.

Many of the mass-transfer operations are carried out under constant flow conditions which tend to result in constant concentration profiles in the equipment used. Gas absorptions often use solvents that are reactive with the gas or which contain a solute that is reactive with it. When a chemical reaction takes place in the liquid with the absorbed molecules as they are diffusing, the concentration profiles are altered and, hence, the rate of absorption is affected. It is convenient to interpret the results in terms of the two-film theory,[1] although there is no proof that stagnant fluid films actually exist at the gas-liquid interface.

The Two-film Theory

Since the solute is diffusing from the gas phase into the liquid, there must be concentration gradients in the direction of diffusion within each phase. These concentration gradients are shown graphically in terms of the distance through the phases in Figure 5.31a, which represents a section through the

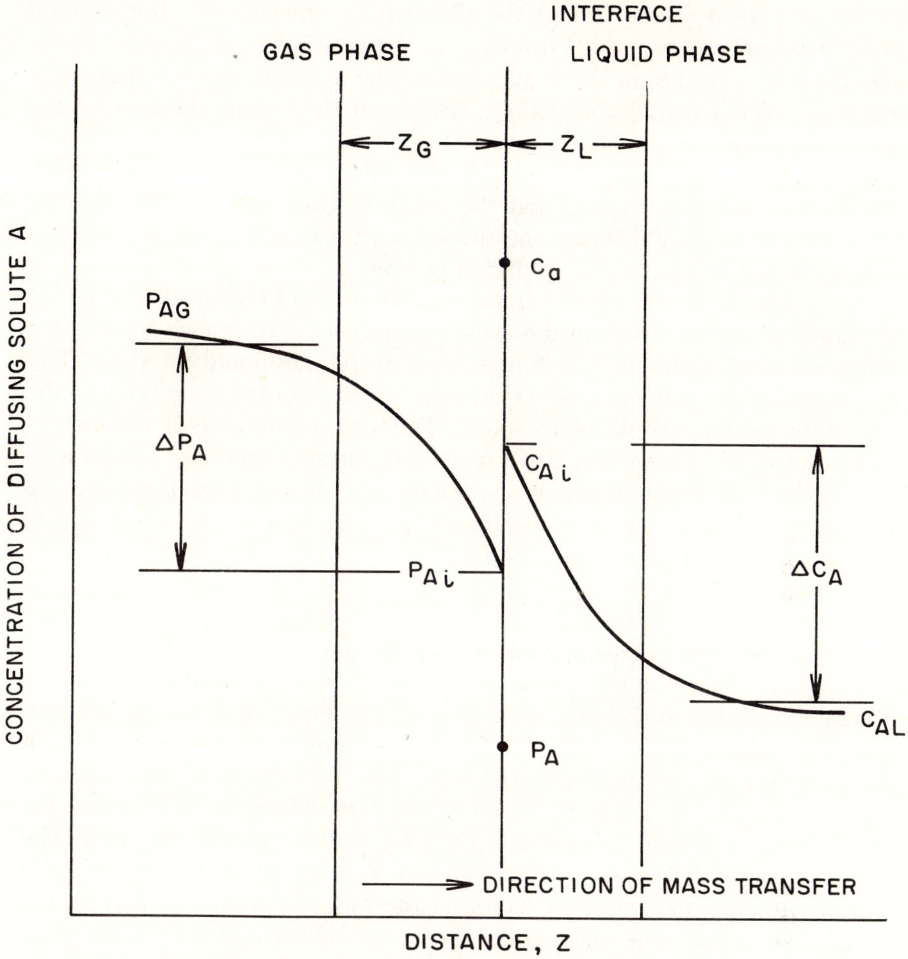

Figure 5.31a Two-film theory of interphase diffusion.

two phases in contact. In the gas phase, the diffusional resistance which the concentration gradient overcomes lies largely within a thin laminar film at the interface between the two phases, but partly also within the turbulent main core of the gas. The entire diffusional resistance may be described in terms of a fictitious or effective film whose resistance to molecular diffusion is the same as that of the true molecular and eddy diffusional resistances actually present. A similar situation exists within the liquid phase.

The average concentration of component A in the gas phase in terms of its partial pressure is P_{AG}. This concentration falls to P_{Ai} at the interface as the component moves through the effective film thickness Z_G. In the liquid phase, the concentration falls from C_{Ai} at the interface to C_{AL} which is the average concentration in the liquid phase. The average concentrations P_{AG} and C_{AL} are not equilibrium values; because if they were, diffusion of the solute would not occur. Whitman and Lewis[1,2] have suggested that there is no resistance to the transfer of the diffusing solute across the interface itself. This is the same as assuming that the concentrations within each phase at the interface are equilibrium concentrations, such as if the surfaces were in contact for an indefinitely long period of time.

The kinetic theory of simultaneous diffusion and chemical reaction in the liquid phase[3] is based on the assumption that the resistance to diffusion is concentrated within a thin film adjacent to the gas-liquid interface. This film is assumed to have negligible capacity (relative to the main body of the liquid) for holding the dissolved solute. The main liquid body is assumed to be so thoroughly mixed that there is no concentration gradient within it.

In the case of simultaneous absorption and chemical reaction, one can assume that a solute gas A is absorbed from a gas mixture by a solution of a substance B, which combines into AB.

$$A + B \longrightarrow AB \qquad 5.31(1)$$

As the solution first comes in contact with the gas, A will dissolve and react immediately with B at the phase boundary. The product AB will begin to diffuse toward the main body of the liquid. The liquid near the surface will soon be depleted of B, which will begin to diffuse from the main body of the liquid toward the interface. The rapid removal of B from the zone near the surface makes it necessary for incoming A to diffuse through part of the liquid film to meet the substance B diffusing in the opposite direction. The zone of reaction between A and B will thus move away from the gas-liquid interface, taking up a position such that the rate of diffusion of A from the gas phase is equal, mole for mole, to the rate of diffusion of B from the main body of the liquid. This process requires only a very short time during which the absorption rate falls rapidly. Within a few seconds at the most, the reaction zone reaches this equilibrium. If the reaction is not very rapid compared to the rate of diffusion, the reaction zone will not be concentrated in a thin region but will be spread throughout the liquid film.

Scrubbing Techniques

Scrubbing techniques can be divided into physical and chemical methods. In all scrubbing techniques, the flue gas containing the gaseous pollutant is intimately contacted with the scrubbing liquor.

In *physical scrubbing*, the gaseous pollutant diffuses into the liquid phase and dissolves in the scrubbing liquor. The flue gas and scrubbing liquor temperatures, the concentrations of the pollutant in both gaseous and liquid phases and the solubility of pollutant in the scrubbing liquor are the major factors that determine the effectiveness of this method. The scrubbing liquor (usually water) is the solvent and the gaseous pollutant is the solute. The following are some of the gaseous pollutants which can be removed by physical scrubbing techniques:

1. Absorption of hydrogen chloride gas in water
2. Absorption of SO_3 with sulfuric acid
3. Absorption of fluoride with water

This technique is not effective for the removal of low concentration of gaseous pollutants from flue gases.

In the *chemical scrubbing technique*, the gaseous pollutant is chemically reacted with a component (a solvent) in the scrubbing liquor. This is the most efficient and widely used method to remove gaseous pollutants from flue gases. Factors, such as reaction rates, temperature, pH of liquor and concentrations, that affect chemical reactions are also important in determining the effectiveness of this method. Some of the examples of the chemical scrubbing technique are the absorption of sulfur dioxide using calcium carbonate, calcium oxide, magnesium oxide, sodium hydroxide and other alkali; the absorption of nitrogen oxides using urea solution; and the absorption of hydrogen sulfide using green liquor and sodium hydroxide.

This technique is successfully utilized to produce or recover saleable products. For example, gypsum can be produced by scrubbing sulfur dioxide pollutants from flue gases with calcium-based alkali. Another example is the production of liquid sulfur dioxide, elemental sulfur and sulfuric acid which can be produced by scrubbing sulfur dioxide with magnesium oxide slurry or with a sodium sulfite–sodium bisulfite solution. (See Chapter 6 for details.)

As another example, hydrogen sulfide, removed from process gas streams in a petroleum refinery to meet air pollution regulations with further processing, can be converted to elemental sulfur which is a valuable product.

Scrubbing of Gaseous Pollutants

The gaseous air pollutants most commonly controlled by absorption include sulfur dioxide, nitrogen oxides, hydrogen sulfide, hydrogen chloride, chlorine, ammonia and light hydrocarbons. Table 5.31b summarizes the more commonly used scrubbing media used for the removal of various gaseous pollutants. Chapter 6 describes in detail the scrubbing of sulfur dioxide, hydrocarbons and nitrogen oxides. Section 7.4 deals with chlorine scrubbing.

In selecting the scrubbing media, one should remember that removal of gaseous pollutants should not result in water pollution or waste disposal problems. The use of calcium-based scrubbing liquor for the removal of sulfur dioxide results in a disposal problem; because calcium sulfite, calcium sulfate and the soluble magnesium, potassium, and sodium compounds (present as

Table 5.31b
GASEOUS POLLUTANTS AND COMMONLY USED SCRUBBING MEDIA

Gaseous Pollutants	Scrubbing Media
Sulfur dioxide	$NaOH$, Na_2CO_3, Na_2SO_3, KOH, K_2CO_3, K_2SO_3, MgO or $Mg(OH)_2$ ZnO, CaO, $Ca(OH)_2$, $CaCO_3$, NH_3 and MgO-MnO_2.
Nitrogen oxides	$Ca(OH)_2$, $Mg(OH)_2$, Na_2SO_3-$NaHSO_3$ and urea.
Hydrogen sulfide	$NaOH$, Na_2CO_3, KOH, K_2CO_3, and green liquor.
Hydrogen chloride	Water, $NaOH$, NH_3, $Ca(OH)_2$ and $Mg(OH)_2$.
Chlorine	Water, $NaOH$ and NH_3.

impurities with lime or limestone) are potential water pollution sources. The regeneration of the scrubbing alkali and recovery of the gaseous pollutant in the form of a saleable product is the most practical solution. (See Chapter 6 for SO_2 recovery processes.)

One can easily and efficiently remove the gaseous pollutant from flue gas using available scrubbing technology, but it is also important to pay attention to the costs and the resulting water pollution problems. The SO_2 regeneration-recovery processes do essentially eliminate water pollution problems and reduce the alkali cost, but they increase regeneration expenses in order to produce a saleable product. The selection of scrubbing liquor is very important, as the alkali that is most effective for the removal of the gaseous pollutant from flue gas may also be the most difficult to regenerate and may cause water pollution problems. The sodium-based alkali, namely caustic soda (NaOH), soda ash ($NaCO_3$) and sodium sulfite (Na_2SO_3), are very effective in absorbing sulfur dioxide from flue gases; but because of its very high solubility, the regeneration of the alkali is more difficult. The calcium-based alkali, namely calcium oxide (CaO) and calcium carbonate ($CaCO_3$), are least effective for removal of sulfur dioxide and are most difficult to regenerate. The magnesium-based alkali, namely magnesium oxide (MgO) and magnesium carbonate ($MgCO_3$) can effectively remove sulfur dioxide and can be effectively regenerated, although it is not compatible with all scrubber designs because of plugging problems.

Scrubbing Equipment

A variety of equipment, namely packed, plate and spray towers and venturi scrubbers, are used to remove gaseous pollutants from flue gases. (See Sections 5.12 to 5.19.)

Packed towers are the most efficient mass transfer devices, using saddles, pall rings and tellerettes as packing to provide mass transfer surface. Normally, the liquor and gas flow countercurrent to each other, the gas flowing upward and liquor flowing downward. The use of packed towers is limited to clean gases, as any precipitate or slurry will cause plugging of the packing. Thus, packed towers are not suitable for use with calcium and magnesium-based slurry scrubbing liquors.

The *plate towers* also have the same limitations as packed towers. In the *spray towers*, the flow arrangements are normally countercurrent and cross-flow. The mass transfer surface is provided by droplets created by spray nozzles. When slurry is used as scrubbing liquor, spray nozzles should be carefully designed and selected to prevent plugging.

Venturi scrubbers are successful in absorbing gaseous pollutants, using both clear liquor and slurry as scrubbing liquors. In this design, the gas and liquor flow in the same direction. The liquor is introduced either by spray nozzles or tangentially through pipe nozzles, and the gas is accelerated through the converging section reaching the maximum velocity at the throat.

REFERENCES

1. Whitman, W. G.: *Chem. Met. Eng.*, vol. 29, p. 147, 1923.
2. Lewis, W. K. and W. G. Whitman: *Ind. Eng. Chem.*, vol. 16, p. 1215, 1924.
3. Hatta, S. (ed.): "Absorption and Extraction," 2d ed., McGraw-Hill, New York, 1952.
4. "Chemical Engineers Handbook," 4th ed., McGraw-Hill, New York.

5.32 DIRECT INCINERATION

The scope of this section is limited to the discussion of direct incineration as it applies to the removal of gaseous pollutants. For costs, suppliers and, in general, a detailed treatment of incineration and afterburner hardware, the reader is referred respectively to Sections 5.35 to 5.39 and to Sections 5.25 to 5.27.

All combustion processes require a combustible fuel (which can be the pollutant), oxygen (which can be carried by the flue gas), a high enough temperature, sufficient turbulence and residence time. It is also important that the products of combustion should not be pollutants in themselves. These requirements cannot be met by some gaseous pollutants and, therefore, direct incineration is not applicable for the removal of the following types of gases.

1. Oxides, e.g., sulfur dioxide.
2. Compounds too chemically stable to be broken down under normal combustion conditions, such as hydrogen chloride.
3. Compounds which can be chemically altered by combustion, but whose combustion products are harmful or offensive. An example is hydrogen sulfide, which produces water vapor and sulfur dioxide when burned. (Nevertheless, incinerators are sometimes used to convert these pollutants to their combustion products if these products can be collected more easily in additional stages of control equipment.)

There are two basic methods of gaseous incineration: direct flame and thermal. Catalytic incineration is a special form of thermal incineration and is discussed in Section 5.33.

Direct Flame Incineration

If the waste gas has sufficient heating value to fuel its own destruction and if it is generated at a steady enough rate to sustain continuous operation of an incinerator, direct flame incineration is the most logical choice. Direct flame incinerators (Fig. 5.32a) burn the waste gas without any auxiliary heat source. Components of these incinerators include:

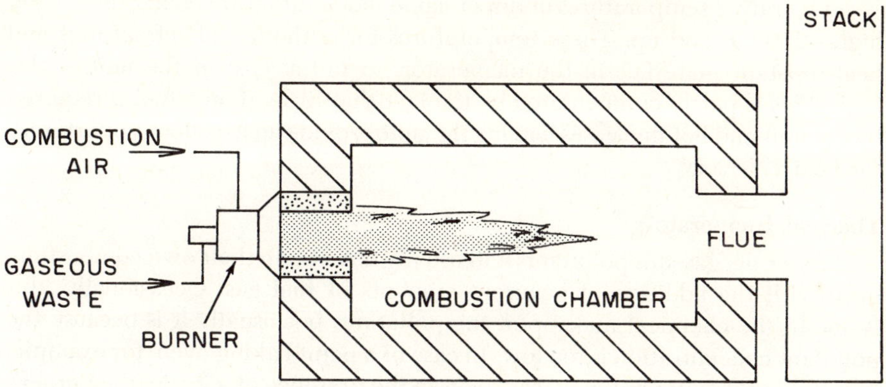

Figure 5.32a Direct flame incinerator.

1. *Combustion chamber* of sufficient size to ensure that all combustion reactions are carried to completion before the waste gases are exhausted.
2. *Combustion system,* which includes one or more burners and all the accessories necessary to ensure that correct proportions of gas and air will reach the burner. Customary practice is to furnish slightly more air than is stoichiometrically required for complete oxidation of the gas. This safety factor ensures that no waste gas will escape unburned. One of the burner types used is the atmospheric burner, which draws in atmospheric air without the help of a blower. The pressure burner relies on a blower or compressor as an air source. Mixing of the air and gas takes place either in a mixing tee ahead of the burner or at the discharge of the burner nozzle. Atmospheric burners are less expensive to install and operate, but also have inferior operating characteristics; because the mixing of air and gas is less thorough, the range of firing rates is limited and their gas-air ratio is affected by atmospheric conditions.
3. *Control accessories* include temperature controls, which regulate gas and air inputs for maximum operating efficiency, and safety controls, which shut down operation if an unsafe condition (loss of flame, air and gas or excessive combustion chamber temperature) arises.

Frequently, the gaseous pollutant is not quite able to sustain itself in combustion or its composition fluctuates enough to render it a marginal fuel. Enrichment with a fuel such as natural gas permits it to be burned in direct flame incinerators.

5.32 Direct Incineration

Operating temperatures of direct flame incinerators are characteristically high—2500°F and up. These temperatures force the use of refractories and heat-resistant materials in the incinerator, so initial cost of the unit is also high. This is partly compensated by the fact that little, if any, fuel is required by the unit and hot flue gases leaving the unit provide an excellent opportunity for heat recovery.

Thermal Incinerators

As a rule, gaseous pollutants will usually not sustain their own combustion, even with the addition of generous amounts of fuel gas. Occasionally, this is due to the chemical makeup of the pollutant; but usually it is because the pollutant concentration is too low. In case of a paint baking oven, for example, flammable solvents are being driven off in the presence of a flame (the burner). Fire codes require solvent fumes to be diluted to one-fourth of the concentration corresponding to their lower limit of flammability, which effectively prevents oven explosions but results in too low solvent concentration for direct flame incineration of the oven exhaust.

Thermal incineration involves the burning of a fuel, usually natural gas or propane, and mixing the waste gas stream with the hot combustion products. If the pollutant can be held at temperatures of 1200 to 1500°F for up to one second, it will nearly be completely oxidized.

Thermal incinerators (Fig. 5.32b) differ from direct flame incinerators in the following aspects:

1. Sufficient fuel is required to raise the entire waste gas stream to 1200 to 1500°F.
2. The burner may not need an air source, since the waste gas stream often contains enough oxygen for combustion of both fuel and pollutant (16 to 21% oxygen is usually required).

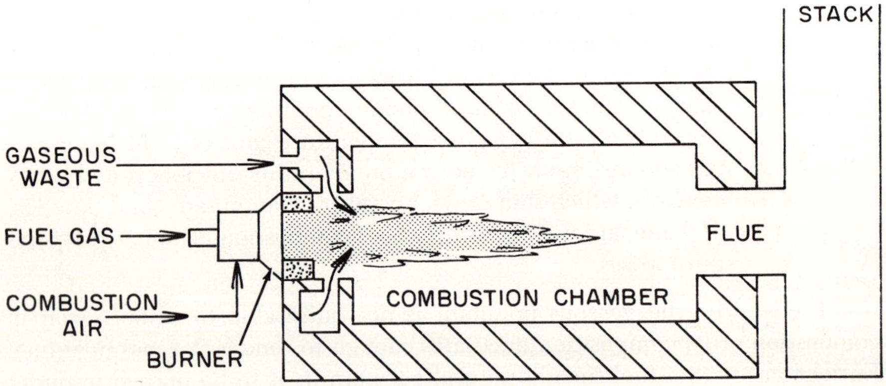

Figure 5.32b Thermal incinerator.

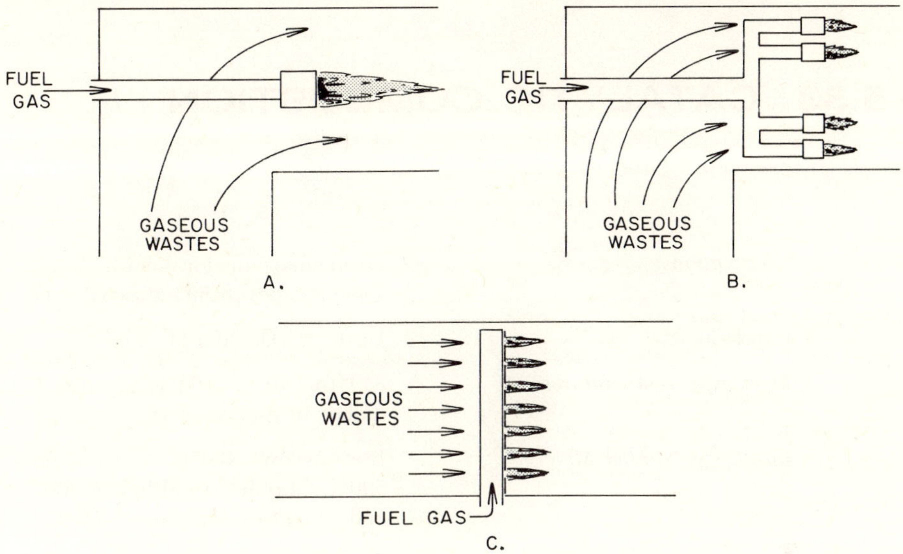

Figure 5.32c Evolution of grid-type burner. *A*, Single nozzle; *B*, Multiple nozzle; *C*, Grid.

3. Less expensive materials of construction can be used, since operating temperatures are lower. Some refractory lining of the combustion chamber is still usually required, and heat-resistant alloys are necessary at critical locations.

Much of the success of thermal incinerators depends on the thorough mixing of the pollutant with the burning fuel. Slow or incomplete mixing can result in incomplete combustion of the pollutant or in excessive residence time requirements. To assure thorough mixing, burner designs have undergone intensive development. Experience shows that multiple small nozzles usually give better results because of increased surface contact between the pollutant stream and the flame. The multiple nozzle burners evolved into grid or perforated plate burners (Fig. 5.32c) which fill the entire duct carrying the waste gas. The waste gas stream is forced to divide into many smaller streams to get through the burner grid. Gas jets adjacent to the grid openings assure intimate contact between pollutant and flame. Normally, these burners are raw gas units and draw their combustion air from the waste gas streams.

Where the waste gas stream contains insufficient oxygen or excessive water vapor, raw gas burners cannot be used and an independent air supply must be provided.

5.33 CATALYTIC COMBUSTION

Applications:	Oxidizable emissions such as organic solvent vapors from industrial ovens.
Capacity:	Up to 40,000 ACFM.
Operating Temperatures:	340 to 540°C (600°F to 1000°F), higher in special cases.
Catalyst Chamber Materials:	Iron below 400°C, steel below 590°C, stainless steel below 700°C and refractory lined over 700°C.
Catalyst Support Materials:	Brick-sized porcelain blocks, cylindrical porcelain rods, metallic-ribbon or wire screens and honeycomb-shaped elements of metal, alumina or porcelain.
Catalyst Materials:	Platinum, nickel, copper chromite, oxides of copper, cobalt, chromium and manganese.
Capital Cost ($/*ACFM*):	$1 to $2 at 30,000 ACFM capacity, $2 to $5 for 10,000 ACFM sizes and over $5 for units at or below the 2,000 ACFM capacity.
Yearly Operating Cost:	$100 to $400/1,000 ACFM with an average of about $200/1,000 ACFM.
Partial List of Suppliers:	BSP Division of Envirotech Systems, Inc.; C&I/Girdler, Inc.; Gas Drying, Inc.; Intercontek; Lawler Company, Machinery & Equip. Co., Inc.; Matthey Bishop, Inc.; Mid-South Mfg. Corp.; Monsanto Enviro-Chem. Systems, Inc.; Oxy-Catalyst, Inc.; Puregas Equip. Corp.; Edward Renneburg & Sons Co.; Robbins Aviation, Inc.; Ross Engr., Machinery Division, Midland-Ross Corporation; UOP Air Correction Div.

This section is devoted to the discussion of catalytic combustion applications in gaseous pollution control. For a general discussion of catalytic afterburners, the reader is referred to Sections 5.25 to 5.27.

If there is economic incentive to recover the gaseous emissions, adsorption on activated charcoal can be used; otherwise, combustion is a suitable alternative.

Advantages of catalytic oxidation stem from the moderate temperatures required for the combustion reaction on the catalyst surface and because quite dilute mixtures can also be reacted. The capital costs are significantly lower since the lower temperatures allow the use of less expensive materials of construction and require less heat exchange equipment. Fuel economies also result from the lower operating temperatures and from the elimination of the requirement to maintain a combustible mixture. The catalytic system also provides a safer installation because combustible mixtures need not be maintained. Disadvantages of the catalytic units are that the catalyst can be poisoned by atmospheric contaminants and inadvertent overheating can cause loss of catalytic activity.

Platinum and nickel have been the most widely used catalysts; however, their high costs provide incentive for developing alternate choices. Other possible catalysts include copper chromite and the oxides of copper, cobalt, chromium and manganese, all of which have exhibited catalytic properties for the combustion reaction.

The solid catalyst has a highly porous, very adsorptive surface. The catalytic oxidation mechanism proceeds by (1) diffusion of the combustibles and oxygen into the pores of the catalyst, (2) adsorption of the gases on the active surface of the catalyst, (3) chemical reaction between the adsorbed oxygen and combustible to yield carbon dioxide and water as the primary products, (4) desorption of these products from the surface and (5) diffusion of the combustion gases out of the catalyst pores and into the main gas stream.

Applications

Catalytic combustion has been primarily used to control organic vapor emissions. Examples of these vapors are solvents and other organic compounds exhausted from industrial ovens used in metal decorating and metal coating operations. Emissions from foundry core ovens, fabric-coating and fabric-backing ovens or from dry cleaning plants can also be controlled by catalytic combustion. When lean mixtures of a combustible vapor and air are present in any industry, catalytic combustion is a possible emissions control method.

Operation

The catalytic afterburner consists of a catalyst section preceded by a preheat section if necessary (Fig. 5.33a). In the preheat zone, the gases are heated by a natural gas burner to the temperature required to support catalytic

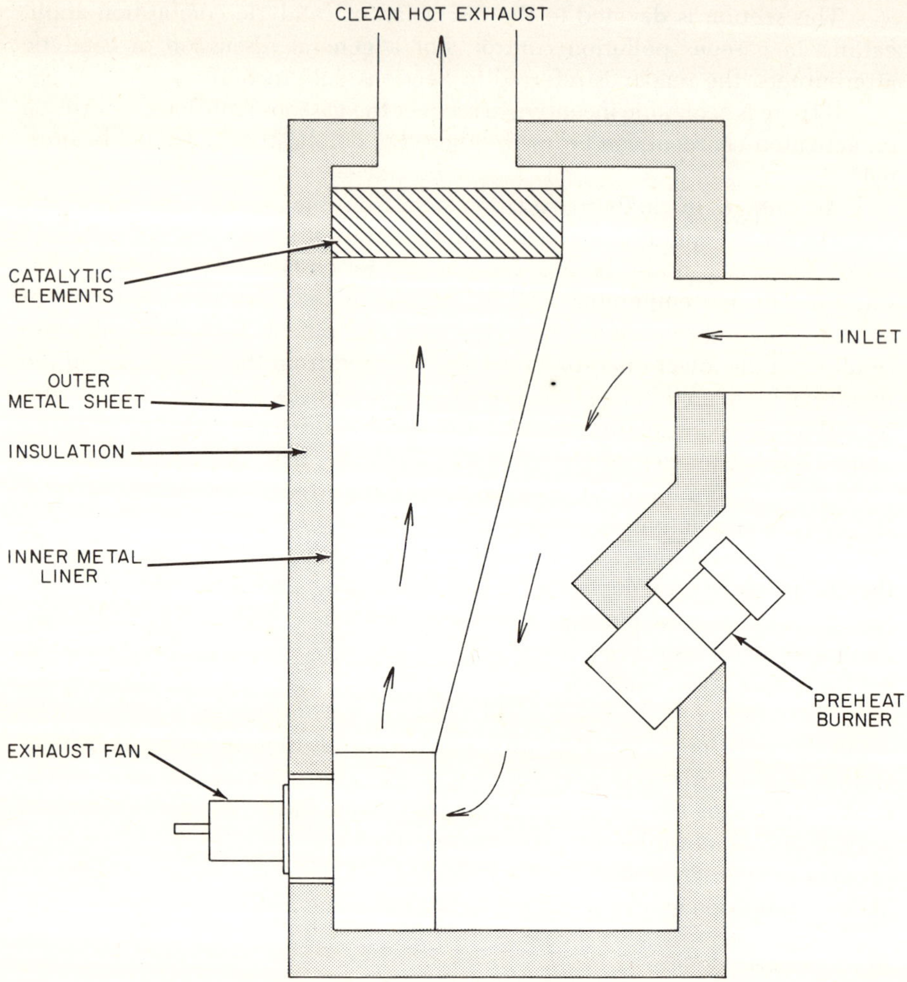

Figure 5.33a Catalytic afterburner chamber.

combustion. The preheated gas is pulled through the catalyst elements by the exhaust fan and the combustibles are burned by catalysis.

If the contaminated gas contains a high concentration of combustibles, it is diluted with air drawn in by a suction fan prior to entering the preheat zone. When the preheat burner is on the discharge side of a fan, the pressure in the afterburner chamber is positive, and a premix type of burner is used. On the other hand, when the fan is between the burner and the catalyst bed, the preheat section is under a negative pressure and an atmospheric burner can be used.

At startup, a maximum natural gas fuel consumption is experienced; but as normal operating temperature is reached, the fuel consumption is reduced. This normally is accomplished automatically by temperature control of the discharge gas.

After the gases pass through the catalyst system, they may be sent through heat recovery equipment prior to being discharged (Fig. 5.33b).

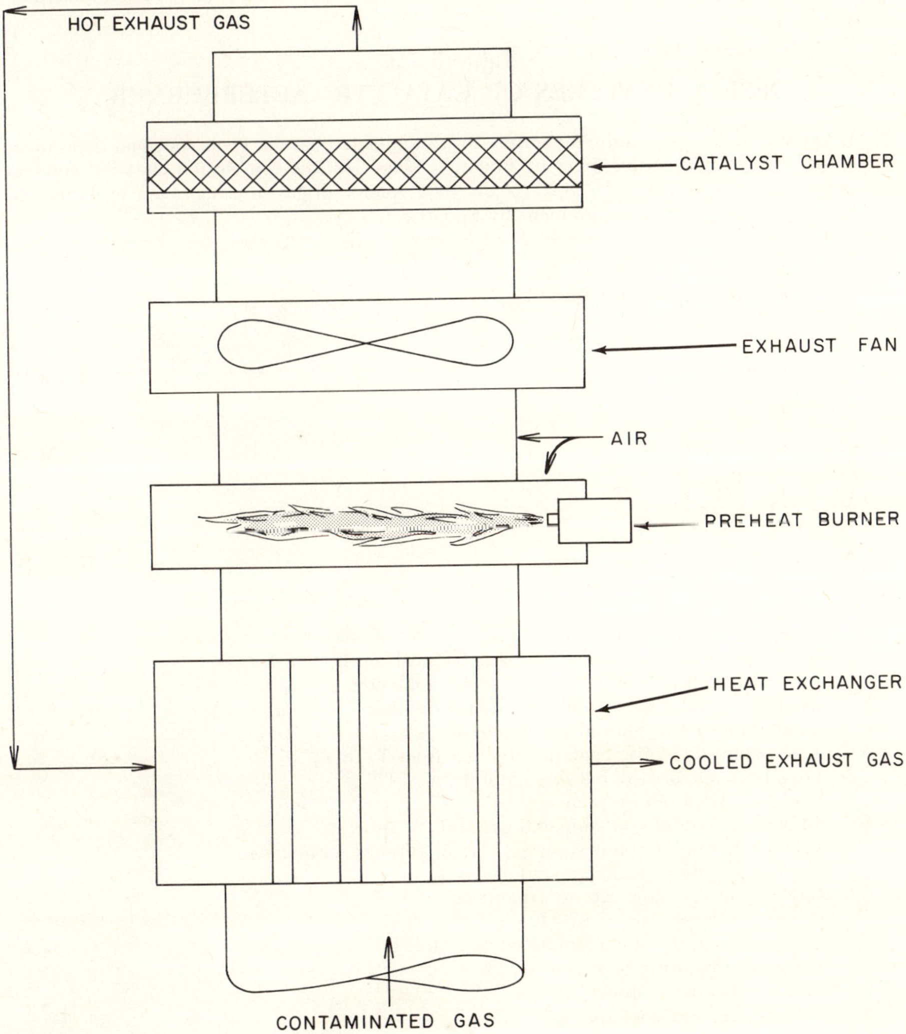

Figure 5.33b Heat recovery with catalytic afterburners.

Design Considerations

Several factors influence the efficiency of the catalytic combustor. These include the contact time of the gases with the catalyst, the operating temperature and the concentration of oxygen and combustibles in the gas.

The catalyst system is designed to expose a large catalyst surface to the gas, while keeping the pressure drop through the system at a reasonable level. One method of achieving this is to electrolytically deposit platinum catalyst on thin nickel-chrome ribbons which are formed into mats similar to the air filter mats. Catalysts are supported by ceramic bricks, rods, honeycomb shapes

Table 5.33c
DESIGN FEATURES OF CATALYTIC AFTERBURNER

If the volumetric flowrate and temperature of contaminated gas are given, one can determine the (1) natural gas fuel requirement by Equation 5.33(1), (2) the combustion chamber volume for preheat zone by Equation 5.33(2), (3) the total volume of gas to exhaust fan by Equation 5.33(3) and (4) the catalyst requirement by Equation 5.33(4).

1. $q = w\,C_p(T_2 - T_1) = (Q/\rho)(C_p)(T_2 - T_1)$ 5.33(1)
   $q_t = q + q_{\text{losses}} \cong 1.1\,q$
   $n = q_t/h_n$

2. $q_g = n \cdot h_g$ 5.33(2)
   $V_p = q_g/\tau$

3. $Q_c = n(11.45)T_2/T_0$ 5.33(3)
   $Q_g = Q \cdot T_2/T_1$
   $Q_t = Q_c + Q_g$

4. $A = Q_t/3$ meters per sec 5.33(4)

Nomenclature

- A, cross-sectional area of catalyst
- C_p, heat capacity of contaminated gas $\cong$ 0.25 cal/gm-K
- h_g, gross heating value of natural gas $\cong$ 9.8 Kcal/liter
- h_n, net heating value of natural gas $\cong$ 7.0 Kcal/liter
- n, volumetric rate of natural gas required
- q, heat required to raise contaminated gas from T_1 to T_2
- q_g, gross heat released by burning natural gas
- q_t, total heat required by preheater
- Q, volumetric rate of contaminated gas at T_1
- Q_c, volumetric rate of combustion gas (11.45 l/liter of natural gas)
- Q_g, volumetric rate of contaminated gas at T_2
- Q_t, total volumetric flow rate to exhaust fan
- T_0, standard temperature = 273°K
- T_1, inlet temperature of contaminated gas, °K
- T_2, preheater operating temperature, °K
- V_p, preheat chamber volume, liters
- w, mass rate of contaminated gas
- ρ, density of contaminated gas at T_1
- τ, preheat sizing parameter = 124 cal/liter-sec

and other forms by coating the support with a liquid solution containing the catalyst, which upon heating will decompose and leave a catalyst deposit on the support surface. The catalyst supports are designed to create turbulence which provides good contact between the gas and catalyst.

The efficiency of a given system must be determined experimentally for a particular combustible and catalyst type. Manufacturers' guidelines are to design for about a 3 meters per second (10 fps) superficial velocity for the metallic ribbon type catalyst system, about 2.5 to 5 standard liters per second (5 to 10 SCFM) per catalyst element for porcelain bricks or rods and about 30 standard liters per second (60 SCFM) per liter (0.035 ft^3) of catalyst support for the honeycomb type. Table 5.33c lists other design features that are needed in specifying an afterburner system.

Operational Problems

Operational problems with catalytic combustion systems normally center around the catalytic activity. Obstruction of the catalyst surface causes loss of activity; the obstruction can be deposits of carbonaceous materials resulting from incomplete combustion, particulate matter in the gas stream sticking to the catalyst surface or poisons adsorbing on the surface. Gradual erosion of the surface due to abrasion also causes a loss in activity.

Vapors of metals such as mercury, lead, zinc and organometallic compounds together with fluorine and chlorine are principal poisons. These are adsorbed irreversibly on the catalyst surface, thus decreasing the amount of surface available for catalyzing the combustion reaction.

Extreme temperatures can destroy the catalyst surface by causing physical changes. Maximum temperatures of about 1000°C (1800°F) can sometimes be reached, but this varies with the particular catalyst.

If the loss in activity is due to simple fouling, occasional washing can restore the catalyst. Insurance regulations require that air be added to dilute the rich mixtures to less than one-fourth of the lower explosive limit.

BIBLIOGRAPHY

American Industrial Hygiene Association, "Air Pollution Manual," Part II: Control Equipment, Detroit, 1968.

Danielson, J. A. (ed.): "Air Pollution Engineering Manual," PHS Publn. 999-AP-40, HEW, Cincinnati, 1967.

Smith, J. M.: "Chemical Engineering Kinetics," 2d ed., McGraw-Hill, New York, 1970.

5.34 ADDITIVE INJECTION TECHNIQUES

Partial List of Additive Proportioning Equipment Suppliers: Alexander Chemicals, Div., of North American Car Corp.; Alken-Murray Corp.; American Meter Controls; American Water Softener Section—Rex Chainbelt Inc.; Automation Products, Inc.; Brooks Instrument Div., Emerson Electric Co.; Broughton Corp.; California Filter Co., Inc.; Carborundum Co.; Carman Industries, Inc.; Chemcon, Inc.; Chempump Div., Crane Co.; Clark-Cooper Corp.; Cleveland Mixer Co., Sub. of Clark Cable Corp.; Clayton Mark & Co.; Clermont Engineering Co., Inc.; Cole-Parmer Instrument Co.; Cornell Machine Co.; Crane Co.; Environmental Services, Inc.; First Machinery Corp.; Fisher & Porter Co.; Fluid Metering, Inc.; Foster Pump Works, Inc.; Fuller Co.; Gelber & Sons Inc.; Grimes Instruments & Controls, Inc.; H & P Equipment Co.; Hardman Inc.; Hills-McCanna Co.; Honeywell, Industrial Div.; Howe Richardson Scale Co.; Hull Corp.; Hytek Int'l Corp.; Kemlon Products; Komline-Sanderson Engrg. Corp.; Lawler Co.; Leslie Co.; Lester Labs., Inc.; Loeb Equipment Supply Co.; Luft Instruments Inc.; Machinery & Equipment Co., Inc.; Martin-Decker Corp.; McDowell-Wellman Engrg. Co.; Mec-O-Matic Inc.; Merrick Scale Mfg. Co.; Metritape Inc.; Meyer & Sons, Inc.; Milton Roy Co.; Monsanto Biodize Systems, Inc.; Neptune

Chemical Pump Co.; New York Testing Laboratories, Inc.; Norcross Cos.; Orr & Sembower, Inc.; Pall Trinity Micro Corp.; Precision Control Products Corp.; Rex Chainbelt, Inc.; Roy, Milton, Co.; The W. J. Sackett & Sons Co.; Salina Mfg. Co.; Schore Automations; Scientific Industries, Inc.; Sirco Controls Co.; Sonic Engrg. Corp.; Stevens, Inc., C. W.; Turfgrass Service Co.; Vibra Screw, Inc.; Vulcan Laboratories Inc.; Warren Automatic Controls Corp.; Waugh Controls Corp.; Wide, Inc.

The hardware aspects of feeders are covered in Chapter I.6 and other solids-handling equipment is discussed in Sections 5.6 and 5.10. The purpose of this section is to describe the techniques used in injecting additives.

A general method for removing gaseous pollutants is to convert them into innocuous substances by chemical reaction with an additive. This technique is called additive injection and can be represented by the following reaction:

$$aA + P(g) \longrightarrow \text{harmless products} \qquad 5.34(1)$$

where a moles of A, the additive (solid, gaseous or vapor), react with one mole of P, the gas phase pollutant. Products of the reaction may be solids or gas or both. In systems using a gaseous additive, the chemical reaction can occur noncatalytically in the gas phase or catalytically on a catalyst surface. Reactions with solid additives normally are noncatalytic.

In the evaluation of any potential system using additives, the first step involves the investigation of the extent of chemical reaction. Consider the reaction

$$aA + bB \rightleftharpoons rR + sS \qquad 5.34(2)$$

Thermodynamics allows the calculation of the equilibrium constant K from the standard free energies F of the reacting materials.

$$\Delta F = rF_R + sF_S - aF_A - bF_B = -RT \ln K \qquad 5.34(3)$$

The standard free energy for the reaction ΔF at temperature T together with the gas law constant R permits values of K to be calculated. With a known equilibrium constant, the maximum attainable yield of reaction products can be estimated.[1]

The second step in evaluating a potential system is to determine the rate at which the reaction products will be formed. This step can be complicated since several rate processes such as chemical reaction kinetics, heat transfer and mass transfer can be simultaneously involved.

The following paragraphs on the solids additives will consider kinetic models of the gas-solid reaction, various methods of contacting the solids and gas and other design and operating factors. Paragraphs on the gaseous additives will discuss mechanisms for noncatalytic and catalytic gaseous reactions and other design considerations.

Injection of Solid Reactants

Once a particular system is selected, there is a wide variety of methods to contact the gas with the solid additive. Some of these methods are illustrated by Figure 5.34a. In the fluidized bed, the rising gas stream provides the energy required to fluidize the solids. The solid particles become highly mobile and are in a perfectly mixed state within the bed. Because of this rapid mixing of the solids, there is good contact between the gas and solid particles and excellent temperature control. The pressure drop in the dense phase fluidized bed can be high.

A second method for gas-solid contacting is by use of a pneumatic conveying system. In the transfer line, the mass flow ratio of solids-to-gas is low, normally between 1 and 20 (pounds of solids/pounds of gas = 0.05).

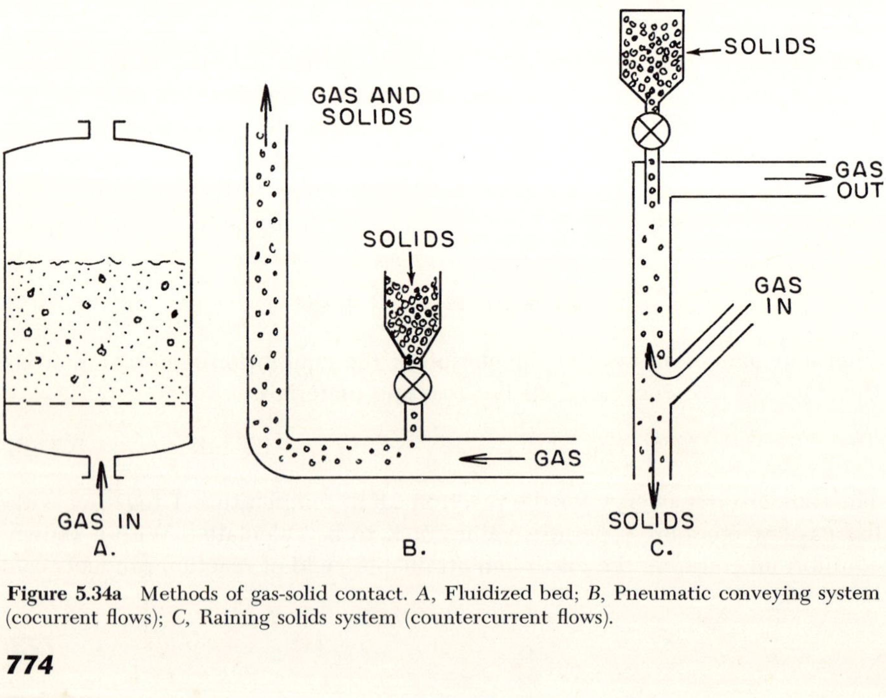

Figure 5.34a Methods of gas-solid contact. *A*, Fluidized bed; *B*, Pneumatic conveying system (cocurrent flows); *C*, Raining solids system (countercurrent flows).

The linear velocity of the gas must exceed the terminal velocity of the largest solid particles in order to prevent their settling out. At these high-gas velocities, only rapid reactions can be accommodated, since the slower reactions would require longer contact times and, therefore, excessive transfer-line lengths.

A third method of contacting the solids and gas is the use of the raining solids system. Here, if velocity of the upflowing gas does not exceed the terminal velocity of the smallest particles, the force of gravity causes the solids to fall through the gas.

If innocuous gaseous products are formed in the pollutant-additive reaction, these may be emitted to the atmosphere. On the other hand, if the products are particulate solids, they are potential pollutants themselves and must be removed from the gas stream. The solid products can be removed by conventional methods such as cyclones, bag filters, electrostatic precipitators, wet scrubbers and packed beds. (See earlier Sections in this Chapter.)

Particles used in solids injection systems are normally small (10 to 2,000 μ) in order to provide a large surface area for chemical reaction.

Process Variables

The chemical conversion of the solid additive can follow one of two extremes. At one extreme, the gaseous pollutant diffuses into the solid particle very rapidly compared to the chemical reaction rate so that the solid reactant is converted more or less uniformly throughout the particle. This is called the continuous reaction model and is depicted in Figure 5.34b. The key assumption is that the gaseous pollutant concentration is the same anywhere within the particle. In terms of the fractional solids conversion X and at a uniform pollutant concentration Cp, the rate expression is

$$\frac{dX}{dt} = kCp\ (1 - X) \qquad 5.34(4)$$

and the integrated form is

$$-\ln\ (1 - X) = kCpt \qquad 5.34(5)$$

where k is the reaction rate coefficient and t is the time required to reach conversion X.

The second extreme case is when the pollutant diffuses into the particle so slowly that the reaction front is restricted to a narrow zone which advances from the outer surface toward the center of the particle. Figure 5.34c illustrates the unreacted core model for solids of unchanging size.

For a reaction described by Equation 5.34(1), the following steps can occur in series:

1. Gaseous pollutant P diffuses through the gas film surrounding the particle to the particle's exterior surface.

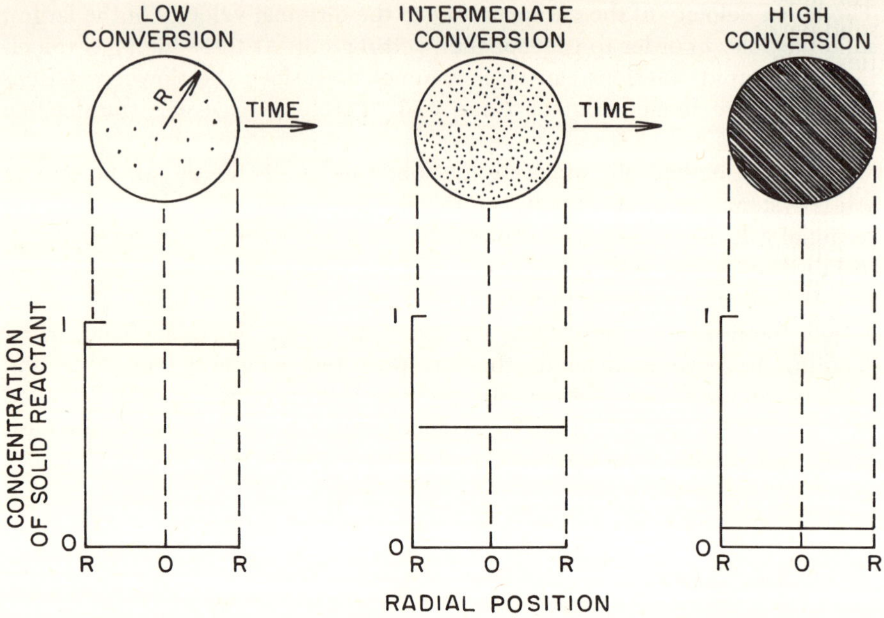

Figure 5.34b Continuous reaction model.

2. P diffuses through the layer of solid product to reach the reaction zone.
3. P reacts with solid A at the reaction zone.
4. Gaseous reaction products diffuse back through the product layer to the exterior surface.
5. Gaseous products diffuse into the main gas stream.

One or more of these steps may not exist for a particular system. For example, if no gaseous products are formed, steps 4 and 5 do not exist. The time required

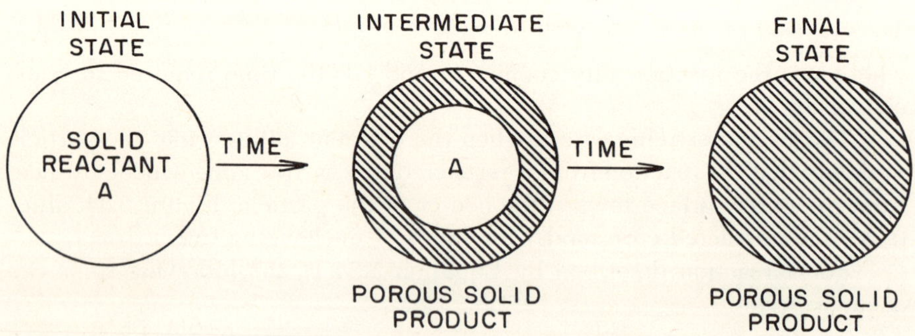

Figure 5.34c Unreacted core model.

for complete conversion of the solid particle at a constant gas concentration is[2]

$$\tau = \frac{R^2 \rho_A}{aDCp\alpha} \qquad 5.34(6)$$

where τ is the time for complete conversion, R is the radius of the solid particle, ρ_A is the molar density of the solid particle, a is the stoichiometric coefficient from Equation 5.34(1), D is the gaseous diffusivity, Cp is the concentration of pollutant in the gas stream and α is a factor described by Equation 5.34(7):

$$\alpha = \left(\frac{1}{6} + \frac{1}{k} + \frac{1}{3h}\right)^{-1} \qquad 5.34(7)$$

where h is a dimensionless mass transfer coefficient and k is a dimensionless reaction rate constant.

$$h = \frac{k_g R}{D} \qquad 5.34(8)$$

$$k = \frac{k_s R}{D} \qquad 5.34(9)$$

The mass transfer coefficient is k_g, and the surface reaction rate constant is k_s. The time required for the reaction zone to reach some radius r_c can be described by Equation 5.34(10):

$$t = \tau\left\{1 - \alpha\left[\frac{1}{k}\left(\frac{r_c}{R}\right) + \frac{1}{2}\left(\frac{r_c}{R}\right)^2 + \frac{1}{3}\left(\frac{1}{h} - 1\right)\left(\frac{r_c}{R}\right)^3\right]\right\} \qquad 5.34(10)$$

If one of the steps, such as mass transfer from the main gas to the solid surface, diffusion through the product layer or surface reaction, is the controlling resistance, Equation 5.34(10) can be simplified considerably. An experimental study of the gas-solids reaction in laboratory scale equipment can provide data to determine values for reaction rate constants and effective diffusivities that will allow scale-up of the system.

Design of Solid Injection Systems

In continuous solids injection systems, the conversion of the solids depends on the rate of reaction and the residence time or contact time of the solids with the gas. A knowledge of the effects of solids particle size, gas pollutant concentration and temperature on the reaction rate allows one to estimate the contact time necessary to achieve the desired degree of pollutant removal. Knowing the volumetric flowrate determines the reactor volume needed for the contact time.

Several types of devices can be used to feed the fine solids into the system.

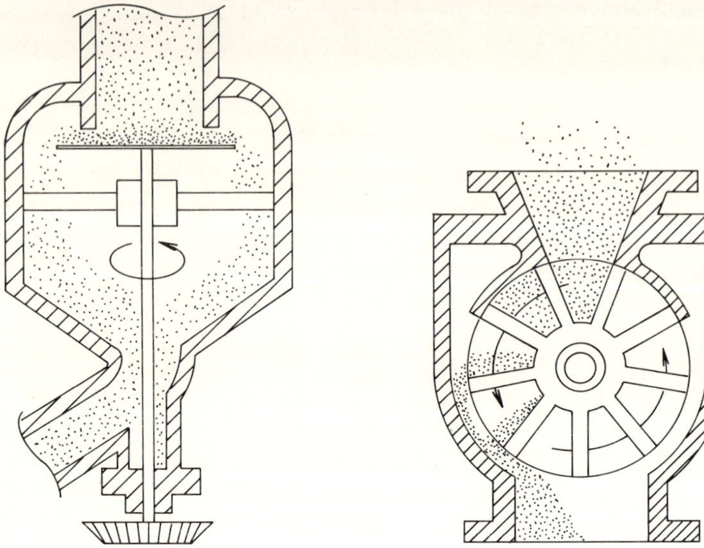

Figure 5.34d Solids feeder equipment.

(For hardware description, see Sections 5.6, 5.10 and I.6.33.) Table feeders are used to feed solids into low pressure systems. If the back pressure is high, a rotary feeder can be used to minimize gas leakage. Both of these devices are depicted in Figure 5.34d. Other controlled feeding devices include cone valves, slide valves, lock hoppers and screw feeders (Figs. 5.10b, c, d and e).

Methods of solid recovery depend largely on the type of contacting system employed. For example, the raining solids system can use a hopper at the bottom to collect the solids. In the transfer-line system, the solids can be removed by dry collection methods (Sections 5.5 to 5.11). In fluidized beds, an overflow line can be installed in the bed for solids withdrawal. Automatic level controllers can also be used to control the fluidized-bed height.

The solids can be discarded or can be regenerated and reused in the system.

Operational Considerations

It is essential to maintain the solids in dynamic suspension because settling of particles can cause the lines to become clogged and threaten the complete shutdown of operations. Special care is necessary to avoid sharp bends in piping and duct work carrying the solids and to the development of reliable start-up and shutdown procedures.

Static electricity will be generated when solids are transported in pipe lines, and there is a potential hazard of the explosion of combustible powders. This hazard is particularly serious with solids of low electrical conductivity and with dry gases. Proper grounding can minimize this hazard.

In dense phase fluidized-bed operations, the solids should be completely removed during shutdown and introduced into the system in small increments during start-up. This will avoid the possibility of having to fluidize a large bed of solids all at once. Start-up with a large bed can be difficult because of the excess pressure needed to lift the solids. Another problem is that the solids may cause erosion.

Injection of Gaseous or Vapor Reactants

The chemical reaction can take place in the gas phase or on a catalytic surface. The general form of the reaction rate expression for the gas phase reaction is

$$-r_p = -\frac{1}{V}\frac{d\,N_p}{dt} = k\,C_A{}^a C_p{}^P \qquad 5.34(11)$$

where $-r_p$ is the rate of disappearance of the gaseous pollutant P, V is the volume of the system, $d\,N_P/dt$ is the rate of change of the moles of P with time, k is the reaction rate constant which is a function of temperature, C_A is the concentration of the gaseous additive and C_p is the concentration of pollutant. The values of the exponents a and p and values for the reaction rate constant must be determined from experimental data using laboratory scale equipment. Knowledge of the reaction rate equation and the residence time distribution of the gas in the reactor system will allow the residence time required for a given level of pollutant removal to be calculated.

The catalytic system involves the injection of a gaseous additive and then contacting the total gas stream, including the pollutant and additive with a catalyst. The chemical reaction between the pollutant and additive takes place on the catalyst surface in several steps:

1. Reactants must diffuse from the main body of gas to the exterior surface of the catalyst.
2. Since the interior of the catalyst contains more surface than the exterior, most of the reactants must diffuse into the interior through the catalyst pores.
3. Reactant molecules close to the catalyst surface are adsorbed on the surface.
4. Chemical reaction takes place between adsorbed molecules.
5. Reaction products are desorbed from the surface.
6. Gaseous products diffuse through the pores to the exterior of the catalyst.
7. Products then diffuse from the mouth of the catalyst pores into the main gas stream.

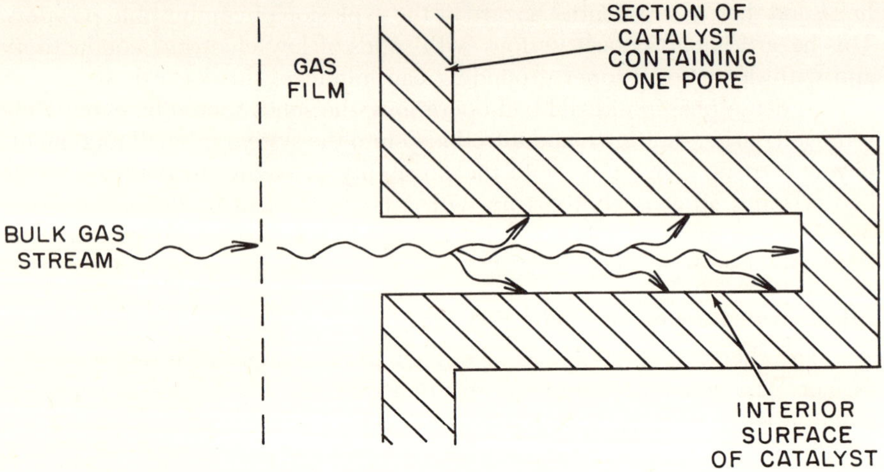

Figure 5.34e Catalyst pore (idealized).

Figure 5.34e shows an idealized sketch of a catalyst pore. The form of the reaction rate expression depends upon the controlling resistance among the listed steps. If the controlling resistance is the reaction between adjacent adsorbed molecules of pollutant and additive, the rate expression for the disappearance of pollutant is

$$-r_p = \frac{k\,(P_P\,P_A - P_G/K)}{(1 + K_A\,P_A + K_P\,P_P + K_G\,P_G + K_I\,P_I)^2} \qquad 5.34(12)$$

where P_A, P_P, P_G and P_I are the gas phase partial pressures of the additive, pollutant, gas product and inerts, respectively; K is the equilibrium constant; K_A, K_P, K_G, K_I the adsorption coefficients; and k the reaction rate constant. The values of the rate constant and adsorption coefficients must be determined from experimental data. Rate expressions for other controlling steps are available.[3,4] For the description of gas feeder hardware, refer to Sections I.4.10 and I.5.23.

Process Variables

The major variables in determining the reaction rate are temperature and additive concentration. Most chemical reactions are very temperature sensitive. The reaction rate constant k is a function of temperature, and the relationship can be described by the Arrhenius equation

$$k = k_o\,e^{-E/RT} \qquad 5.34(13)$$

where k_o is the frequency factor, E is the activation energy for the reaction, R is the gas law constant and T is the absolute temperature.

The concentration of the additive affects the reaction rate as illustrated by Equations 5.34(11) and (12).

Design Considerations

Methods of contacting the additive and the pollutant are depicted in Figure 5.34f. If the only consideration is the reaction of the additive and the pollutant, then it is desirable to have a maximum additive concentration at some point in the system so that all the additive can be injected at one point. If high concentrations of the additive result in undesirable side reactions, then it should be injected at several points along the length of the system.

In catalytic reaction systems, the gas velocity must be high enough to ensure good contact between the gas and the catalyst; yet the velocity must not result in excessive pressure drop. Temperature control may be necessary in some systems. Particulate matter may stick to the catalyst surface and render the catalyst ineffective. Potential corrosion problems should be considered in designing all injection systems.

Applications of Injection Systems

Sulfur oxides are the largest gaseous pollutants in terms of quantity, and many methods of controlling them have been investigated. Some of the methods involve the injection of solid metal oxides. Most of the sulfur oxides are formed during the combustion of high sulfur fuels, and some of the injection systems involve blowing the solid oxides into the combustion chamber. Oxides of calcium, manganese and magnesium have been used, and these compounds form sulfites and sulfates when reacted with the sulfur oxides. The solid sulfites and sulfates must then be recovered from the main gas stream (see Chapter 6).

An application of gaseous injection is the catalytic reduction of nitrogen

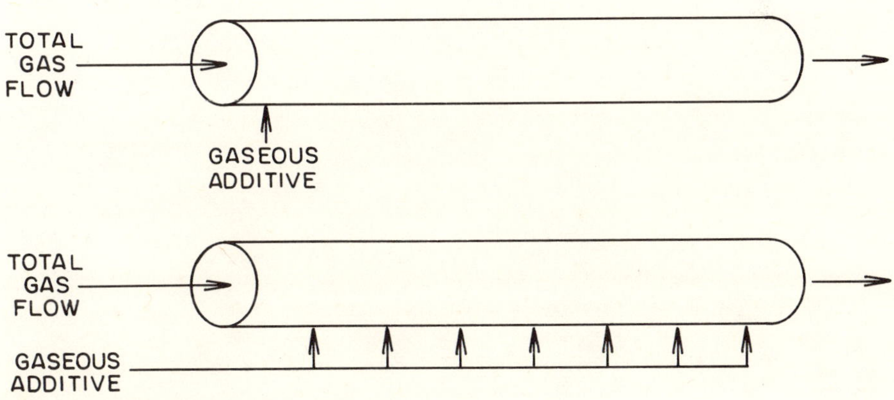

Figure 5.34f Contacting patterns for gaseous additive injection.

oxides by ammonia. Ammonia has been found to selectively reduce nitrogen oxides to form elemental nitrogen and water even in the presence of oxygen. Such reduction of nitrogen oxides with NH_3 has been practiced to a limited extent in nitric acid plants, and it is potentially very attractive for reducing nitrogen oxides in flue gases. Further discussion of the reduction of nitrogen oxides is found in Chapter 6.

REFERENCES

1. Levenspiel, O.: "Chemical Reaction Engineering," McGraw-Hill, New York, 1962.
2. Bischoff, K. B.: *Chem. Engr. Sci.*, vol. 20, p. 783, 1965.
3. Hougen, O. A. and K. M. Watson: "Chemical Process Principles, Part III," Wiley, New York, 1947.
4. Smith, J. M.: "Chemical Engineering Kinetics," 2d ed., McGraw-Hill, New York, 1970.

5.35 INCINERATORS—THE INCINERATION PROCESS

Sections 5.36 to 5.39, Chapters I.5 and III.2 provide further information on the many aspects of incinerator application and technology.

All wastes can be thought of as having only three primary components: combustible, inert and moisture. The inert is generally inorganic ash, i.e., stones, dirt, glass; but it also includes either metals or their oxides whenever the temperatures and residence times are sufficient for the metals to burn.

Most wastes burn with such sufficient release of heat that the combustion is self-sustaining. If the calorific value is low or negative (endothermic reaction), supplementary fuel is required, generally gas or oil.

The broad qualitative outlines of the fundamental processes are well understood. Heat transfer determines whether the waste is hot enough to react with sufficient speed, and fluid flow patterns determine whether the necessary oxygen is available at a given location and time. Some wastes burn as solids; other wastes will pyrolyze, either partially or completely, so that the solid residue (if any) burns by direct oxidation, and the pyrolysis products burn by mechanisms similar to those found during oil or natural gas combustion.

For sizing an incinerator, the quantitative information of how fast the waste will burn is necessary. In the industry of incineration, the level of research has been all too low, mainly because the commercial incentive for improving efficiency has been largely lacking and the necessary research would not pay for itself.

Design Requirements

An incinerator is a steel or refractory box fed with waste, combustion air and auxilliary fuel, if necessary.

Whatever the details of design, the total combustion space can always be divided into a solid bed reaction zone and a gas phase reaction zone. The ratio of the sizes of the two zones is an important design parameter. For the waste to burn, air must be supplied to each zone; and the air proportioning between zones is another important design parameter. The gas phase zone is always the larger in volume since this must accommodate most of the reaction. It may be physically contained in one or more separate but connecting chambers; but what matters most in design is not the number of chambers, but the resultant fluid flow pattern.

The fluid flow pattern is best described with reference to the flame axis, and it is found that the flow pattern is well approximated in most instances by a sequence of subzones in which the flow can be characterized as either "plug flow" or "backmix flow." Backmix flow can be identified experimentally by introducing some convenient foreign gas into the flow and showing that part of it is carried back upstream. This does not happen with plug flow. The number and proportions of such plug flow and backmix subzones can affect the reaction process, the combustion intensity and chamber dimensions. Although evidence in support is limited, the majority of existing incinerators would seem to have only two such subzones, namely, a backmix subzone followed by a plug flow subzone. A backmix region is also referred to as a "stirred reactor" region. The optimum mixing configuration should have a stirred reactor region followed by plug flow. Although incinerators do, in fact, appear to have this desirable mixing configuration, it would seem that the proportions of the two subzones is far from optimum. In particular, the backmix subzone in all probability is generally too small and/or has a very weak level of backmix. An optimized backmix region, with a sufficiently high level of backmix, can be affected by the number, size and location of overfire air ports, for example.

Figure 5.35a illustrates a continuous vertical tube with a grate (or hearth) at the bottom divided into zones and subzones with the flame axis as the vertical. Incinerators whose flame axes depart from the vertical (being often horizontal, hairpin or S-shaped) will still behave similarly if the subzone sequence is the same.

Table 5.35b is a summary of the principal incinerator design parameters.

Quantitative Design

Lacking any other quantitative data on burning rates of wastes, the best existing basis for design is the use of combustion intensities. If an incinerator has a total combustion chamber volume V (ft^3) that is filled with flame when fired with waste having a calorific value B (Btu/lb) at a maximum capacity F (lb/hr), the maximum combustion intensity of that unit I (Btu/ft^3 hr) and the mean volumetric burning rate $\bar{R}_v$ (lb/hr ft^3) are defined by Equation 5.35(1):

$$\bar{R}_v = I/B = F/V \qquad 5.35(1)$$

The values of $\bar{R}_v$ and I calculated from this equation for a given waste are values belonging to a particular unit or design configuration (i.e., number of chambers and air supply locations). Tables of values of $\bar{R}_v$ and I can be used in two ways.

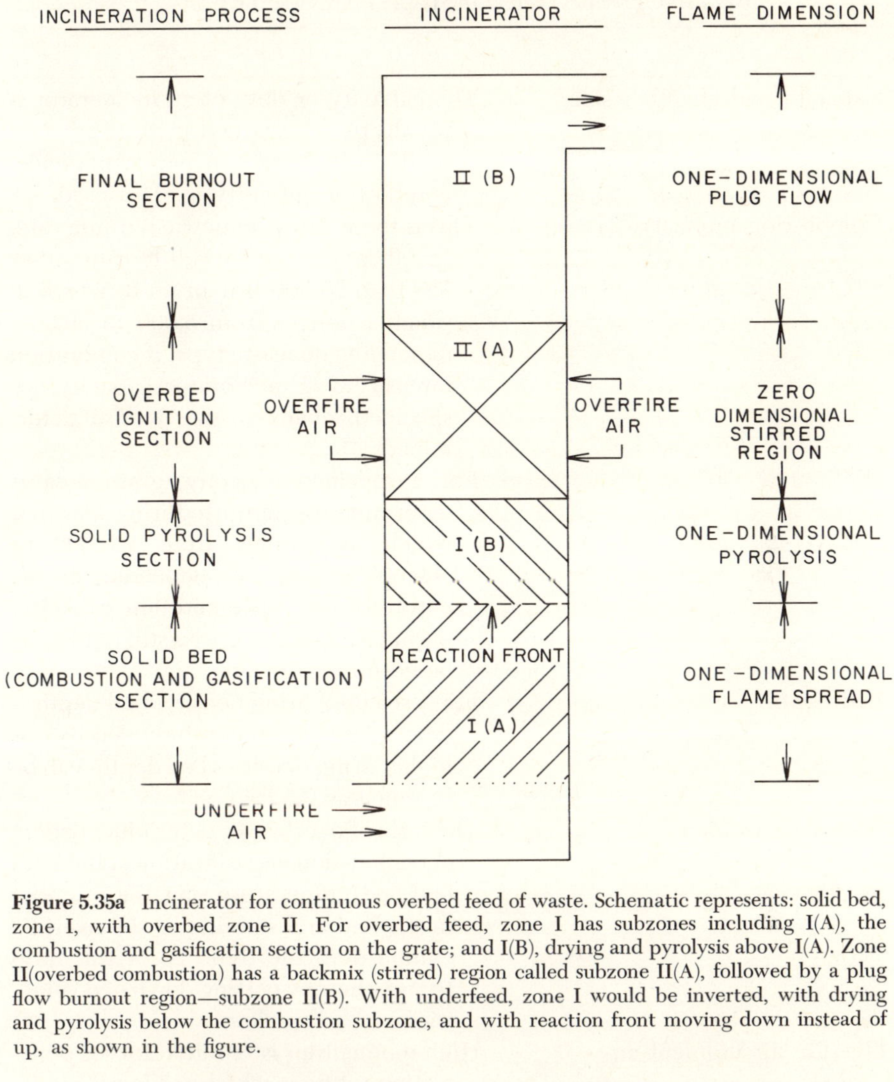

Figure 5.35a Incinerator for continuous overbed feed of waste. Schematic represents: solid bed, zone I, with overbed zone II. For overbed feed, zone I has subzones including I(A), the combustion and gasification section on the grate; and I(B), drying and pyrolysis above I(A). Zone II(overbed combustion) has a backmix (stirred) region called subzone II(A), followed by a plug flow burnout region—subzone II(B). With underfeed, zone I would be inverted, with drying and pyrolysis below the combustion subzone, and with reaction front moving down instead of up, as shown in the figure.

1. Since poor unit designs can be expected to generate lower combustion intensities, tabulated data for a range of comparably sized units can be used to compare incinerator designs.
2. As a basis for new design, it is reasonable to suppose that combustion intensities achieved in the past can be achieved again. If F is specified for a waste of known B, assuming a previous value of I, the required chamber volume V can be calculated from Equation 5.35(1).

Table 5.35b
INCINERATOR DESIGN PARAMETERS

Parameter	*Significance*
Capacity and chamber size	The capacity or duty of an incinerator is usually predetermined; existing practice tends to base size on past experience; combustion intensity is often used.
Combustion intensity	This is the mean volumetric burning rate; specifications mostly call for intensities less than 25,000 Btu/hr cu ft, which in practice ranges from 5,000 to 50,000, depending on waste type; if combustion intensity and incinerator capacity are specified, chamber size is also determined.
Grate area	For a specified capacity, grate area is determined by permitted firing densities, which may range from 100,000 to 400,000 Btu/sq ft hr, depending on the nature of the waste and unit capacity; permitted firing densities still tend to be based on experience.
Bed depth	For a selected firing density, bed depth is determined by the refuse density; at higher firing densities bed depths can be as much as 3.5 ft.
Air proportioning	The ratio of overfire air determines degree of combustion or gasification of the solid bed, bed temperature, slagging potential in the bed and stability of the overbed combustion; the trend is for this ratio to rise, thereby cutting down bed velocities and ash pickup.
Overfire air momentum	High momentum is required for jet penetration without which mixing is inadequate; to obtain penetration, air pressures in excess of 6 inches H_2O are increasingly common.
Overfire air jet configuration	The jet locations, spacing, number and directions determine degree of backmix and dimensions of backmix (stirred reactor) region; this region may have to

Table 5.35b (Continued)

Parameter	Significance
	be 50% or more of the total combustion chamber volume; optimum configuration is probably the biggest unknown at present; current mixing practice is poor.
Excess air	High excess air, ranging from 50 to 400% is common for temperature control to prevent slagging; resulting low temperatures can cause terminal flame quenching, resulting in pollutant emissions.
Gas velocity	High velocity favors pickup and retention of flyash; specifications often call for 35 fps overall and less than 9 fps in any one settling chamber; such specifications will set chamber and gas pass dimensions but should not void overfire air momentum requirements.
Chamber configuration	Current practice favors two or more chambers (exclusive of settling chambers); the practice could be an alternative to purely aerodynamic means of generating a backmix region in the flow pattern.
Grate type	Small units generally have fixed grates and are fired as batch or semi-batch; movable grates can determine supply rate of waste, evenness of combustion and completeness of burnout; gentle stirring is advantageous to break up accumulations tending to clog bed; too active stirring can increase flyash pickup.
Wall type and refractory selection	Refractories must be resistant to slag attack and abrasion; resistance is improved by air cooling; if waste is to be used as fuel for steam raising, water walls can be used, allowing reduction of excess air.

The next step in quantitative design is to size the grate. If a firing density F_A (lb/hr ft^2) can be assumed, again from past experience, the grate area required is given by Equation 5.35(2):

$$A = F/F_A \qquad 5.35(2)$$

Another approach to this problem has been to combine Equations 5.35(1) and (2) by forming, in effect, a dimensionless ratio between the chamber volume V and the grate area A. The resulting equation has the form:

$$F_A = [V^{2/3}/A]\,[I/B]^{2/3} \cdot F^{1/3} \qquad 5.35(3)$$

Since $F^{1/3}$ is approximately proportional to log F, Equation 5.35(3) can be written in the form

$$F_A \simeq K_L \cdot \log F \qquad 5.35(4)$$

where K_L is a so-called "waste" factor given approximately by

$$K_L \simeq 3[V^{2/3}/A]\,[I/B]^{2/3} \qquad 5.35(5)$$

Equation 5.35(4) was originally obtained empirically and has been quite widely proposed as a design equation. There is no *a priori* reason for the area firing rate to increase with capacity as a first order effect. (For example, consider two identical units side by side with a common wall. If the common wall is removed, F is doubled, but F_A should be unchanged.) Equation 5.35(4) implies, when coupled with Equation 5.35(5), that $(V^{2/3}/A)$ remains unchanged (assuming $\bar{R}_v$ also remains unchanged). But this ratio only remains constant as A increases if the effective height above the grate is increased; i.e., Equation 5.35(4) holds only when the available combustion volume per unit area of grate is allowed to increase in a specified relationship to grate area. Use of Equation 5.35(4) is, therefore, not encouraged.

Past experience must, therefore, be used to select values of I and F_A, without any guarantee that the same values will hold in any new device. Any advances in quantitative design must be aimed at developing predictive bases for I and F_A.

Numerical Values

Numerical values of some of the parameters involved in incineration will help to set the process in perspective. Much of the behavior is dominated by the air requirements. It is convenient to define a stoichiometric ratio S as the cubic feet of cold air required per Btu of heat released. If the calorific value of the waste as fired is B (Btu/lb), the air supply is SB (1 + e), if e is the excess air fraction. More generally, if a and M are the ash (inert) and moisture fractions, respectively, and B_o is the dry-inert-free (DIF) calorific value, the air requirement is given by

$$\text{cold air volume (ft}^3\text{/lb)} = B_o(1 - a - M)S(1 + e) \qquad 5.35(6)$$

In the cases where metals are not burned, or not appreciably, and the combustible component is predominantly organic, the value of S is approximately 0.01 (i.e., 1 ft^3 of cold air per 100 Btu). This approximation is valid, generally to within 10 or 20%, for a remarkably wide range of organic fuels whether solid, liquid or gas (including coal, wood and fuel oils). Consequently, variations between wastes depend very largely on their incombustible content (A + M), particularly as the DIF calorific values lie within the relatively narrow range of 8000 to 10,000 Btu/lb. The stoichiometric air requirements of most DIF wastes are therefore 80 to 100 ft^3/lb or 6.4 to 8 lb air/lb waste. If the waste contains 50% combined ash and moisture, the calorific value drops to 4000 to 5000 Btu/lb. If fired at 150% excess air, the air requirements are 8 to 10 lb air/lb waste as fired. These figures indicate that there is much less variation between wastes than is generally realized. Variations in behavior are dominated more often than not by the moisture content. The figures also show the extent to which the mass flow through an incinerator is dominated by the air.

The incombustibles, and particularly the moisture, also have a significant influence on both the volumetric reaction rate and (even more) on the combustion intensity. Data are very scarce. Average volumetric reaction rates of waste as fired range from 4.5 to 5 lb/hr ft^3, for DIF or low ash waste, down to 1.5 to 2 lb/hr ft^3 when combustible contents are down to 10 or 20% of the total mass. With combustible levels as low as that, auxilliary fuel is required. The limit at which self-sustaining combustion is possible without auxilliary fuel would appear to be about one-third combustible and two-thirds incombustibles.

Combustion intensities also drop with increasing moisture, because of the effect mainly on temperature which, in dropping, reduces the reaction rate significantly. Values of combustion intensity may be as high as 40,000 Btu/ft^3 hr for DIF material, dropping to about 15,000 at 50% combustibles. When auxilliary fuel is required, the combustion intensity would appear to have to be maintained at a level between 5000 and 10,000 Btu/hr ft^3. These are values that can be used in Equation 5.35(1) for estimating combustion volumes.

The highest combustion intensities match those in coal firing on grates. Firing densities on the grate, likewise, now match those found in coal firing. The coal rates are 30 to 40 lb/ft^2 hr, or 400 to 500,000 Btu/ft^2 hr with about 6-inch bed depths. Uncompacted waste is one-half to one-third as dense and one-half to one-third of the calorific value, so comparable firing densities on a Btu basis require supply rates of about 60 lb/ft^2 hr with bed depths of 3 to 4 feet.

The Solid Bed Zone

The Basic Firing Density Equation

The area firing rate, or firing density F_A, is determined by behavior in the solid bed. At the very simplest, we can relate the firing density to the rate of air (oxygen) supply and to the fractional combustion. Since we are considering unit area of bed, we write $\bar{G}_A$ as the rate of supply of air per unit area. The oxygen in this air supply will burn all or part of the combustible fraction in the waste. The rate of supply of combustibles is F_A (1 − a − M); and of this quantity, let us also assume that some fraction V_m is lost from the solid bed as unburned pyrolysis products, so that the combustible fraction is further reduced by the fraction (1 − V_m). If the stoichiometric ratio is written as S ft^3 of cold air/lb waste (partially or totally pyrolyzed), then multiplication by this factor yields the cold air supply rate, if all the oxygen is utilized. To allow for incomplete utilization of the oxygen, we introduce the "utilization factor" f as a multiplier for $\bar{G}_A$. Equating yields

$$f \cdot \bar{G}_A = (1 - a - M)(1 - V_m)\, S \cdot F_A \qquad 5.35(7)$$

The utilization factor f can also exceed unity, thus representing a degree of incomplete reaction (e.g., gasification with formation of CO). Equation 5.35(7) otherwise shows that, other things being equal, the firing density can theoretically be increased more or less indefinitely so long as there is a proportional increase in the air supply rate. The theoretical limitation to the firing density is not to be found necessarily on the grate; it depends rather on the size of the overbed combustion volume and the capacity of this overbed volume to burn up the pyrolysis and gasification products streaming out of the solid bed. If total burn up is not attempted, there is not otherwise any theoretical limitation to F_A. There will be a practical limitation, however, since as $\bar{G}_A$ rises to match F_A, the increasing velocities will first be carrying too much flyash into suspension and, ultimately (in a practically absurd limit), would start to lift the bed itself.

Construction of Equation 5.35(7) yields an expression for predicting F_A, if we can first evaluate the parameters f and V_m. To do this requires more detailed consideration of the reaction processes, which in turn depend critically on the method of feed onto the grate. In essence, we have only three idealized types of grate behavior to consider: (1) overbed feed, with fresh fuel supplied on top of a burning bed so that a horizontal ignition plane exists in the bed traveling up into the unburned fuel and with the reaction proceeding below; (2) underbed feed, which can be regarded as an inversion of the overfed bed; and (3) the tumbled bed. In cross-feed, as on a chain grate stoker, each location in space on the grate can be considered as location in time with an underfeed bed, but with the ignition plane traveling down into the fresh fuel till it reaches the grate instead of being maintained stationary in space by continuous underfeed.

Propagation Processes

The three systems differ in their behavior as follows. In top and bottom feed, a definite ignition plane exists, traveling either up or down. In the tumbled bed, it is to be presumed that no ignition plane can exist: the nearest equivalent system would seem to be the fluid bed with reaction somewhat uniform throughout the whole reaction zone. In top and bottom feed, the method of propagation of the ignition plane is by radiation, but respectively aided and hindered by convection. Conduction mostly plays a negligible part. In top feed, the convective flow is of hot combustion products from the main combustion/gasification zone, and this provides substantial thermal flux for pyrolyzing and drying the green fuel ahead of the relatively short-range radiation flux that is still the main source of heat for propagation of the ignition plane. Drying, and perhaps even some pyrolysis, is also aided by radiation to the top of the bed from the overbed flame and/or the containing refractories.

In contrast to top feed, drying with bottom feed has to depend (perhaps entirely) on the cold or barely warm underfire air, perhaps aided by conduction down through the bed. It is a general assumption that solid reaction will not start until the fuel has dried and pyrolyzed. Consequently, an endothermic pyrolysis zone can be expected to be mostly interposed between the hot, radiating bed of burning, pyrolyzed solid and the drying zone. The whole system is further complicated by the flush of moisture and volatiles through the bed of burning, pyrolyzed solid. Both moisture and volatiles can interfere with the bed reaction, pyrolysis products by competing for oxygen and moisture by imposing the endothermic water gas reaction (water vapor reacting with carbon to produce CO and hydrogen). High moisture can therefore lead to erratic burning, probably more so than in the case of overfeed supply. Any heat for pyrolysis and drying must come ultimately from the burning bed. Combustion becomes more stable when burning at such a higher rate that the competing water gas and pyrolysis reactions are more easily accommodated. Reference to Equation 5.35(7) indicates that this should be accomplished by increasing $\bar{G}_A$ at constant F_A whereby the factor f, if greater than unity, should drop, and V_m may also drop.

A greater fraction of the total combustibles will burn in the bed, to a greater degree of completion, thus increasing the heat release rate and hence the bed temperatures. However, of the increased heat released, only a relatively small fraction will be returned to the pyrolysis and drying zones. Consequently, with very wet refuse, the main burning bed may have to be fired too hard, to the point of slagging, to be able to dry the waste at the required rate. This consideration also applies, but at a greatly reduced level, to the overfeed bed which has the advantage that virtually all the extra heat developed from increased underfire air is available to convect up through the green fuel for drying and pyrolysis.

The tumbled bed provides some compromise between top and bottom feed, in addition to the advantage it has in breaking up any accumulations tending to clog the bed. Other arrangements such as inclined grates (particularly with reverse acting bars) or a sequence of traveling grates, tend to accomplish the same purpose. The effect of tumbling is periodically to propel green, wet fuel into the middle of the burning bed where it will dry much faster.

Numerical Values

We may now consider some numerical values of the pyrolysis factor V_m and the oxygen utilization factor f. For the overfeed bed, the position is fairly straight forward. The drying and pyrolysis zone exists between the burning section of the bed and the overbed flame; and in this subzone, all moisture and the total volatile matter are driven off into the overbed flame. In the case of most domestic and municipal wastes where the predominant combustible components are approximates to cellulose, the value of V_m is 0.7 and 0.8 (of the DIF material). The combustible char residue is only 10 or 20% of the original combustible fraction, which partly burns and partly gasifies. If the char burns entirely to CO_2 in the bed ($f = 1$), a dry, low ash waste will burn best with as little as 15 to 25% of stoichiometric air, with the rest supplied as overfire air. If the bed is deep enough (requiring about 1 foot of char or more), gasification will take place with f rising to a possible theoretical maximum of 2. It is then to be expected that the underfire air would have to be increased to offset the endothermic gasification reactions. For most calculations, however, a value of f lying between 1.0 and 1.5 can be adopted until more complete information is available.

For the underfeed bed and for the cross-feed bed, the situation is much less clear. In cross-feed, as the waste first encounters the intense radiation of the primary chamber, the top layers will dry and pyrolyze. This can sometimes occur almost explosively. However, beyond that, if pyrolysis products within the bed are more successful in competing for oxygen than is the char residue, then these will burn preferentially. Consequently, the value of V_m (the fraction of released volatiles breaking the bed surface to burn overbed) must be close to zero. With an allowance for the initial surge of volatiles on entry, an overall V_m factor of 0.05 to 0.1 may be realistic. For the oxygen utilization in the later stages of char gasification and burnout, again a figure of 1.0 to 1.5 for f may be appropriate, at least for a homogeneous bed. However, if blow holes develop so that there is appreciable short circuiting, this can have the overall effect of *reducing* the average value of f below unity. It is not, therefore, surprising to find that in existing practice, when operating with about 150% excess air, about two-thirds of the total air is supplied as underfire air, equivalent to roughly 60% excess of stoichiometric. The range of values is from one-third (i.e., 0.8 of stoichiometric) to all air as underfire air.

The Reaction Process

Values of the pyrolysis factor and oxygen utilization factor can, in principle, be predicted from theory; however, the necessary theoretical developments are still far from complete. In the case of top feed resulting in only char combustion in the bed, the reactions are between carbon and oxygen alone. The bed can be regarded as a porous carbon block containing tortuous channels, with oxygen diffusing to the channel walls and also deeper into the porous char itself to produce a mixture of CO and CO_2. This is the result of a direct attack of the gaseous oxygen on the solid carbon by way of a more or less understood sequence of subreactions (involving the chemisorption of oxygen on the carbon surface) and followed by desorption (with the oxygen carrying away the underlying carbon atom). A competing reaction that tends to dominate, as CO_2 builds up and oxygen runs out, is the gasification process of CO_2, reacting with the carbon to yield CO. At the same time, in the gas phase of the pores, CO reacts with O_2 back to CO_2. Theories to predict burning rates based on these chemical reaction concepts have been developed, but the most successful theories to date have been based on such (physical) assumptions as that the controlling factors in the bed reaction are the rates of diffusion of oxygen and CO_2 from the pore channel to the char wall. Some success has now been achieved, using this model, in predicting burning rates, oxygen utilization factors, char bed depths and temperature profiles.

In the underfeed system, the processes described above take place with further complications from the presence of moisture reactions and pyrolysis combustion. The pyrolysis reactions can be assumed to be essentially gas phase (like the CO to CO_2 conversion), but the moisture will also react with the solid carbon by a series of subprocesses essentially paralleling the oxidation and gasification reactions. A further complication will be the water gas shift reaction, also taking place in the pores. Theories to predict burning rates and bed depths based on these complex considerations are much less advanced than for the char combustion alone, although work is in progress. In the meantime, there has been some success on a phenomenological level by assuming that all pyrolysis products are totally burned and that the exit gas from the top of the bed has reached thermodynamic equilibrium between CO, CO_2, H_2 and H_2O. This is a promising approach that may yield reasonably accurate values of both f and V_m for substitution in Equation 5.35(7) for underfeed and cross-feed systems.

The Overbed Combustion Zone

The Predictive Requirements

The unburned or part burned pyrolysis and gasification products from the solid bed are burned out in the overbed combustion zone. The quantities of materials are determined by the parameters V_m and f in Equation 5.35(7) (as multipliers for F_A), since these represent, respectively, the fraction of volatiles not burned in the solid bed and the degree of incomplete reaction (when $f > 1$). Here we assume that these are known. The requirement for the overbed zone is to predict the burnout distance from the rate of overbed burning since this determines the overbed height required to accommodate the reacting materials. The square of the overbed height is effectively the ratio $(V^{2/3}/A)$ in Equation 5.35(3).

Two possible approaches are open. One is again the use of combustion intensity data, applied in this instance just to the overbed combustion region; the other is again an approach through fundamentals by prediction from rate equations.

Prediction by Combustion Intensity

The use of combustion intensity depends on the relation between overall combustion intensity and the average combustion intensities for the zones I and II in Fig. 5.35a (I_I and I_{II}). If V is the total combustion volume, and V_I and V_{II} are the volumes of the two zones, respectively, then,

$$I = I_I (V_I/V) + I_{II} (V_{II}/V) \qquad 5.35(8)$$

Since V_I is likely to be 5% or less of the total combustion volume, the overall combustion intensity is dominated by I_{II}, if there is any appreciable degree of reaction in the overbed zone. Consequently, $I \simeq I_{II}$; and the average values calculated from Equation 5.35(1) for existing units can be used to back calculate V_{II} since $V_{II} \simeq V$. The procedure is somewhat arbitrary since it is averaged data for a volume where the variation of combustion intensity from point to point can be extreme. Nevertheless, it is a valid procedure in the absence of any better basis.

However, the real validity of the procedure depends on the similarity of design configurations. This includes, in particular, the configuration of mixing subzones (backmix and plug flow), since the relative sizes of the subzones can affect the burning rates profoundly. Here again data are sparse; but in one study, it was estimated that a burnout height of 100 meters would be required if the overbed subzone was plug flow alone, but this could be reduced to 1 or 2 meters by utilizing an optimally sized backmix reactor in which there was 80% fuel conversion, with the final 20% burnout in a plug flow subzone following. This reemphasizes the importance of correctly optimizing the size of the backmix region. Most studies seem to agree that conversion in the backmix subzone should lie between about 60 and 90%.

If the design configuration is significantly different from that where the combustion intensity data were derived, deviations can be expected. One semiempirical method of dealing with the deviations is to use the following relation between combustion intensity and relative pressure drop $(\Delta p/p)$ across the air registers or supply pipes for secondary air:

$$I = K (\Delta p/p)^{1/2} \qquad 5.35(9)$$

where K lies between 1×10^6 and 3×10^6, for I in Btu/ft^3 hr.

The range in values of K apparently depends on some characteristic cross-sectional dimension d of the unit, with K inversely proportional to d according to the best evidence. In the values quoted, 3×10^6 would be for a cross-sectional dimension of about 1 foot, and 1×10^6 for about 10 feet. It must be emphasized, however, that the reliability of the numerical values of K is not too high. Equation 5.35(9) has some basis in theory, but values of K are still empirical and sparse. Nevertheless, the expression has some design value. If a new unit is found not to be able to accommodate the overbed flame at the rated capacity, Equation 5.35(9) can be used to estimate the increase in pressure drop required to bring about the needed increase in combustion intensity (or flame shortening).

The effect of increasing the pressure drop is attributed to improved mixing which, by increasing either the rate of local mixing and/or the level of backmix, results in increased fuel conversion in the backmix region of the overbed volume.

The Reaction Process

Development of rate equations for predicting overbed volumes is still in such an embryonic stage that it will be some time yet before they can be used for design purposes. Nevertheless, there is some point in outlining the reaction processes and procedures being used. The basis of any calculations must depend on the equations used to represent the rate of reaction in some local volume, if the temperatures and fuel and oxygen concentrations are known. As a first approach to this kinetic reaction problem, equations based on phenomenological kinetics have been used:

$$\text{Rate of fuel disappearance} = k_o c_f c_{ox} \exp(-E/RT) \qquad 5.35(10)$$

where c_f and c_{ox} are mass concentrations of fuel and oxygen respectively, T is the local temperature, k_o is known as a characteristic "frequency factor" for the process and E is the activation energy for the reaction. Expressions of this form have now been used quite successfully by computer modeling to predict combustion and burnout in special research incinerators. In particular, present evidence supports values of E lying somewhere between 15,000 and 20,000 cal/mole, which is an informative figure in chemical kinetics.

Conclusions

Although the incineration process is extremely complex, it is at last yielding to investigation. Of all possible systems, the top feed and bottom feed types represent extremes of design with some curiously significant differences between them.

Bottom feed (or cross-feed) presents greater problems on the grate when the waste is very wet than does top feed because of the *more direct and efficient drying process* in the latter. On the other hand, the products from the bottom feed bed are likely to leave the bed at a much *higher temperature* than is the case for the top feed unit, which should make final burnout easier. In one set of experiments, temperatures in the bed (with top feed of dry waste) were found to fall from nearly 1000°C in the center to about 600°C at the bed top due to endothermic pyrolysis, even with an overbed flame in excess of 1000°C. With wet waste, this could provide a significant decoupling effect between the bed reaction and the overbed flame. The bed could not be relied on as a flame holder. Significantly, the experimental unit with top feed was found to emit smoke, even with substantial excess air when most of the air was supplied as underfire. At the same air level, the unit was smokeless when most was supplied overfire and in a configuration designed to promote backmix and flameholding.

Good overbed mixing is important for smoke elimination, particularly where there can be sudden intermittent surges of volatiles (as sometimes happens with traveling grate stokers). Good overbed mixing also has significance for odor removal if the mixing is accompanied by a well-stabilized flame. However, an unexpected result of overbed mixing by backmix is the potential for coupling the overbed and bed zones, as a result of overbed gases backmixing right into the bed, as found in at least one instance experimentally.

A final difference between top and bottom feed is the need for very different levels of underfire air, which thereby affects the different possible degrees of flyash pick-up at a given firing density. The much lower levels of underfire air that are required by top feed systems favor low grit emission.

Overall, patterns of behavior of different systems with a developing analytical basis are now emerging. It is a reasonable expectation that such information will have increasing impact on design and operation of incinerators in the future.

NOMENCLATURE

a ash fraction in waste
A grate area (ft^2)
B calorific value of waste (Btu/lb)
B_o calorific value of dry-inert-free (DIF) waste (Btu/lb)
c_f mass concentration of fuel in gas stream (lb/ft^3)
c_{ox} mass concentration of oxygen in gas stream (lb/ft^3)
e excess air fraction
E activation energy of gas phase reactions (cal/mole)
f oxygen utilization factor
F incinerator capacity (lb/hr)
F_A firing density on grate (lb/ft^2 hr)
$\overline{G}_A$ air supply rate per unit grate area (ft^3/ft^2 hr)
I combustion intensity (av.) for total incinerator (Btu/ft^3 hr)
I_I combustion intensity (av.) for zone I (solid bed) (Btu/ft^3 hr)

I_{II} combustion intensity (av.) for zone II (overbed) (Btu/ft^3 hr)
k_o "frequency factor" for gas-phase reactions (ft^3/lb hr)
K proportionality constant in Equation 5.35(9)
K_L waste constant, see Equation 5.35(4)
M moisture fraction
p pressure in overfire air registers (psi)
Δ_p pressure drop across overfire air registers (psi)
R gas constant
S stoichiometric ratio (ft^3 air (cold)/Btu of heat released)
T absolute temperature
V total combustion chamber volume (ft^3)
V_I volume of zone I (solid bed) (ft^3)
V_{II} volume of zone II (overbed) (ft^3)
V_m unburned volatile fraction leaving zone I (solid bed) and entering zone II (overbed)

BIBLIOGRAPHY

The bibliography on incineration is substantial, but the fundamental research component describing incineration processes is still meager. The following short list has value as much for further references cited therein as for their information content.

American Society of Mechanical Engineers, Biennial National Incinerator Conferences, nos. 1 to 5, 1964, 1966, 1968, 1970, 1972.

"Cleaning Our Environment: The Chemical Basis for Action," Sec. 3: Solid Wastes, American Chemical Society, Washington, D.C., 1969.

Corey, R. C. (ed.): "Principles and Practices of Incineration," Wiley, New York, 1969.

Essenhigh, R. H. and T. Kuo: "Development of Fundamental Basis for Incinerator Design Equations and Standards," Third Mid-Atlantic Industrial Waste Conference Proceedings, University of Maryland, 1970.

Niessen, W. R.: "Systems Study of Air Pollution from Municipal Incineration," (see particularly Vol. II Secs. H. C. and I. C.) Arthur D. Little, Inc. Cambridge, Massachusetts, 1970.

5.36 INCINERATORS—TYPES, FEATURES AND OPERATION

Types of Designs:	(a) *Residential,* (a1) Single dwelling, (a2) Multiple dwellings; (b) *Municipal;* (c) *Industrial,* (c1) Multiple chamber, (c2) Rotary, (c3) Fluid bed. NOTE: The letters "a" to "c3" in the feature summary below refer to the above listed designs.
Residue Density (lb/yd^3):	1,200 for conventional, 2,700 for slagging type incinerator, which operates at 3000°F.
Capital Costs (including scrubbers and controls on larger units):	*Solid Wastes:* \$200 (a) 1½ bushel unit; \$6000 (c1) 100 lb/hr; \$10,000 (b) per ton per day; \$11,000 (c1) 500 lb/hr; \$12,000 (a2) 500 lb/hr; \$20,000 (c1) 1,000 lb/hr; \$36,000 (c2) 100 lb/hr; \$42,000 (c2) 300 lb/hr. *Liquid Wastes:* \$250 to \$450/GPH. *Gaseous Wastes:* \$10,000 for 5,000 SCFM. (Also see Sections 5.25, 5.31 and 5.32.)
Federal Regulations on Particulate Emissions:	0.08 gr/SCF corrected to 12% CO_2. Starved air systems meet this without gas cleaning requirement, conventional incinerators require wet scrubbers or electrostatic precipitators with 99% efficiency for compliance.
Steam Production Potential:	1 lb of steam per lb of refuse in conventional and 3 lbs of steam per lb of solid waste in water-walled or slagging type (high temperature, 3000°F) incinerators.

Partial List of Suppliers: *Small field-erected or packaged units:* Bangor Punta Operations, Inc., Brule Incinerator Corp., C-E Air Preheater, CE-Raymond/Bartlett-Snow, Copeland Systems, Inc., Joseph Goder, Inc., Midland Ross Corp., Surface Combustion Div., Peabody Engineering Corp., Sanitas Technology & Development Corp., The Silent Glow Corp., Zurn Industries, Inc. *Large field-erected complete systems:* Combustion Engineering Inc., Dorr-Oliver, Inc., Envirotech Corp., Morse Boulger, Inc., MSI Industries, Inc., Nichols Engineering & Research Corp., Pyro Industries, Inc., Thermal Research & Engineering Corp., Zimpro, Inc., John Zink Co., Zurn Industries, Inc. *Components for field-erected systems:* Bartlett-Snow, Beaumont Birch Co., Bigelow-Liptak Corp., Coen Co., M.H. Detrick Co., Detroit Stoker Co., Dorr-Oliver, Inc., Envirotech Corp., Klenz-Aire, Inc., Maxon Corp., Morse Boulger, Inc., MSI Industries, Inc., Nichols Engineering & Research Corp., Pyro Industries, Inc., Thermal Engineering & Research Corp., Zimpro, Inc., John Zink Co., Zurn Industries, Inc.

In addition to this section, various aspects of incineration applications and hardware are discussed in other sections of this chapter and sections in Chapters I.5 and III.2.

The advantages of incineration include wide-range ability for handling varying loads and small-space requirement for ultimate disposal; its limitations include fairly high maintenance and operating costs and problems associated with being a potential source of air pollution.

Principles of Combustion

For a detailed discussion of the incineration process, refer to Section 5.35. It was explained there that in terms of fuel and air movement, three basic incineration processes are possible:

1. Underfeed beds in which fuel and air flow in the same direction.
2. Overfeed beds in which fuel and air flow in opposite directions.
3. Crossfeed beds in which fuel and air flow in directions at some angle to each other.

To obtain good combustion in deep fuel beds, it is desirable to have:

1. Underfeed or crossfeed burning.
2. Gas flow from raw fuel to ignited fuel.
3. Overfire air mixed with hot gases flowing out of the fuel bed.
4. Delaying additional mixing with large amounts of excess air (for quenching) so as not to hinder proper combustion.

If combustion is to be complete, time, temperature and turbulence in the entire combustion zone must be within satisfactory limit. For incinerators handling aqueous wastes, the flammability zones are very critical.

Wastes are categorized into seven types as shown in Table 5.36a. Type 5 for industrial wastes requires further study of processes involved. Approximate heating values of wastes are given in Table 5.36b.

Table 5.36a
CLASSIFICATION OF WASTES, DENSITY AND HEATING VALUES

Type	*Description*	*Lb/ft³ BTU's/lb*
0	Trash, highly combustible paper, cardboard, wood boxes, sweepings; up to 10% plastics and rubber	8 to 10 8500 BTU
1	Rubbish; combustible paper, cardboard, wood, foliage, sweepings; up to 20% food waste, but no plastics or rubber	8 to 10 6500 BTU
2	Refuse; approximately even mixture of rubbish and garbage	15 to 20 4300 BTU
3	Garbage; animal and vegetable food wastes	30 to 35 2500 BTU
4	Pathological; human and animal remains	45 to 55 1000 BTU
5	Gaseous, liquid or semiliquid by-product waste	a
6	Solid, by-product waste. Mixtures containing over 10% plastics by weight	a
	Compact, solid waste (Security Documents, I.B.M. Cards, files, bound materials)	35 to 50

[a] Varies with the process, consequently a survey of the waste is required.

Table 5.36b
APPROXIMATE CALORIFIC VALUES OF WASTES

Material	*BTU/lb*	*lbs/ft³*
Animal fats	17,000	60
Brown paper	7,250	7
Citrus rinds	1,700	40
Coated milk cartons	11,330	5
Corn cobs	8,000	15
Corrugated boxes	7,040	7
Coffee grounds	10,000	30
Cotton seed hulls	8,600	30
Latex	10,000	45
Leather	7,250	20
Linoleum	11,000	90
Magazines	5,250	35
Newspapers	7,975	7
Polyethylene	20,000	50
Polyurethane (foamed)	13,000	2
Rags (silk and wool)	8,500	15
Rags (cotton & linen)	7,200	15
Rubber waste	10,000	100
Wood sawdust	8,500	12

Theoretical air requirement is best estimated from a chemical composition analysis of the wastes.[1] In the absence of this data, theoretical air requirement may be taken as 0.7 pounds or 10 cubic feet per 1000 BTU heat released. In developing thermal balances, an excess air of at least 20% is employed.[1,2]

Air Pollution from Incinerators

Emission of air pollutants from incinerators includes particulates such as flyash, unburnt fuel and others. Permissible level[3] of particulate emissions from large incinerators is 0.23 grams per standard cubic meter of exhaust (0.1 gr/SCF) corrected to 12% CO_2. The specification of particulate loading is tied to a given amount of CO_2 to prevent evasion of the law by dilution using air or steam. Smoke includes all liquid and solid matter in the exhaust that hinders visibility. True smoke which may be white, brown or black is distinguished from a steam plume by its persistence. Smoke can be eliminated by mixing the exhaust with hot air to complete combustion, but black smoke requires fairly high air temperatures.

Techniques for particulates removal from combustion products are discussed in Sections 7.2 and 7.10.

To estimate the sulfur oxides in incinerator emissions, one can assume that the entire sulfur content of the refuse is converted to gaseous sulfur dioxide. Sulfur trioxide occurs in small amounts (1 to 5%); hence, the extra oxygen of the trioxide is often neglected in estimates on air requirements.

At present, permissible emissions of sulfur oxides should be determined from EPA standards[4] on ambient air quality and applicable state and local regulations. (See Chapters 1 and 3.)

All combustion processes produce nitric oxide, which is slowly converted to nitrogen dioxide in the atmosphere. It is estimated that every ton of refuse incinerated produces about 2 pounds of nitrogen dioxide.

Residential Waste Incineration

On-site incineration is both simple and convenient for disposal of wastes from houses, apartment buildings and offices. It eliminates some of the problems associated with the collection and transportation of wastes.

Incinerators for Single Dwellings

The presently commercially available units are modifications of the smokeless, odorless prototypes developed under the sponsorship of the American Gas Association.[5] Generation of solid wastes in domestic dwellings varies substantially as shown in Tables 5.36c and d. Most domestic incinerators are designed to handle about 1.2 to 1.5 lb/person/day of wastes.

Domestic wastes require supplementary fuel with an approximate heat requirement of 2600 to 8000 BTU/lb. Typical prototype units for single dwellings are rated for about 40,000 BTU/hr.

A domestic incinerator (Fig. 5.36e) has two stages of combustion, the second one serving the purpose of oxidizing the smoke and odorous material produced. The burner configuration possibilities include:

1. Horizontal firing through a tube into the primary zone with provision to continue in the secondary zone.

Table 5.36c
SOLID WASTES COLLECTED PER CAPITA IN THE UNITED STATES (1969)

Type	*Quantity, lb/Capita-Calendar Day*
Residential (domestic)	1.5–5.0
Commercial (restaurants, businesses or stores)	1.0–3.0
Bulky solid wastes (furniture, fixtures and construction wastes)	0.3–2.5

Table 5.36d
TYPICAL DAILY SOLID WASTE GENERATION

Building Types	*Estimated Waste*
Private homes	5 lb basic + 1 lb per bedroom/day
Apartments	4 lb per sleeping room/day
Warehouses	2 lb per 100 sq ft/day
Office buildings	$1\frac{1}{2}$ lb per 100 sq ft/day
Department stores	4 lb per 100 sq ft/day
Restaurants	2 lb per meal/day
Grade schools	10 lb per room + $\frac{1}{4}$ lb per pupil/day
High schools	8 lb per room + $\frac{1}{4}$ lb per pupil/day
Hospitals	15 to 18 lb per bed/day
Nursing homes	3 lb per person/day
Hotels, class I	3 lb per room + 2 lb per meal/day
Hotels, medium	$1\frac{1}{2}$ lb per room + 1 lb per meal/day
Motels	2 lb per room/day
Trailer camps	6 to 10 lb per trailer/day

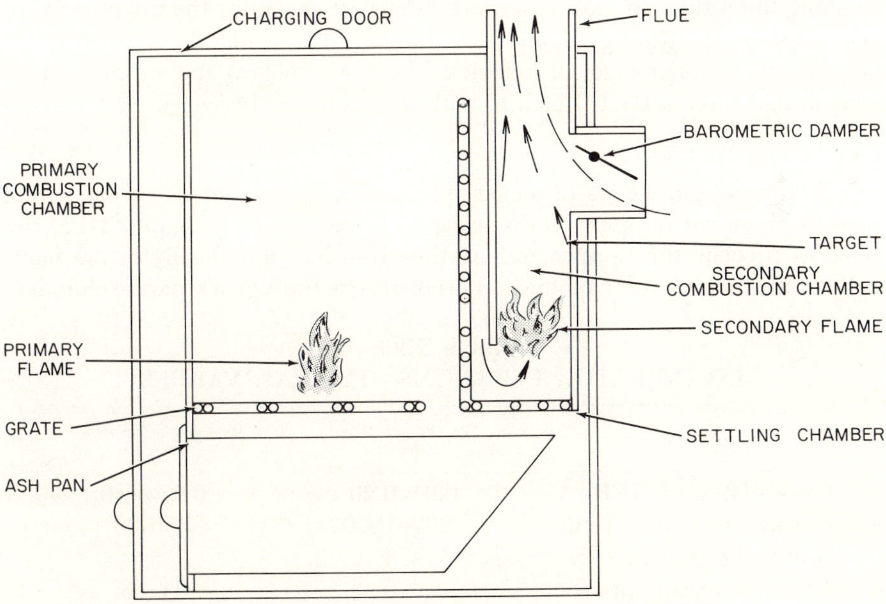

Figure 5.36e Features of a domestic incinerator.

2. Burner across the base of the back of the unit with flames partly directed to the primary zone and mainly to secondary zone.
3. Power burner directing flame to the top of the waste pile.

Air pollution emissions of typical domestic units are given in Table 5.36f, together with comparable emissions from municipal incinerators.

Domestic incinerators are less efficient and more expensive than central municipal incinerators. In addition, large central facilities can permit recovery of materials from wastes.

Incinerator for Multiple Dwellings

A typical dwelling, with a population of 500, produces about 850 lb/day of wastes with an average bulk density of 4 lb/ft.[3] Just storing this refuse requires 52 standard four cubic foot trash cans per day.

On-site incineration has the advantages of producing a sterile, odorless residue that can be stored without health hazard and requiring only about 8% of the volume of the original waste in transportation and landfill.

Flue Fed Incinerators

In flue fed incinerators (Fig. 5.36g), the flue for the combustion products also serves as the charging chute for the wastes.

The single flue incinerator has a box-like combustion chamber with grates above the ashpit. A charging door above the grates is used for igniting or smoking the refuse. In most cases, provisions are made for the use of overfire and underfire air through doors with spinners. The walls are constructed of two layers: an inner layer of firebrick (about $4\frac{1}{2}$ inches) and an outer layer of common brick (about 9 inches) with airspacing in between.

Chute Fed Incinerator

In this design, the waste is charged through hopper doors to a flue which feeds a basement hopper. As discussed in more detail in Chapter III.2, the wastes collect in this hopper and are then transferred manually or mechanically to the incinerator. Combustion products rise through a separate chimney.

Table 5.36f
INCINERATOR EMISSIONS—TYPICAL VALUES

Pollutant	*New Domestic Wastes Incinerator*	*Municipal Incinerator with Scrubber*
Particulates, gr/SCF	0.01–0.20	0.03–0.40
Carbon monoxide, ppm	200–1000	<1000
Ammonia, ppm	<5	—
Nitrogen oxides, ppm	2–5	24–58
Aldehydes, ppm	25–40	1–9

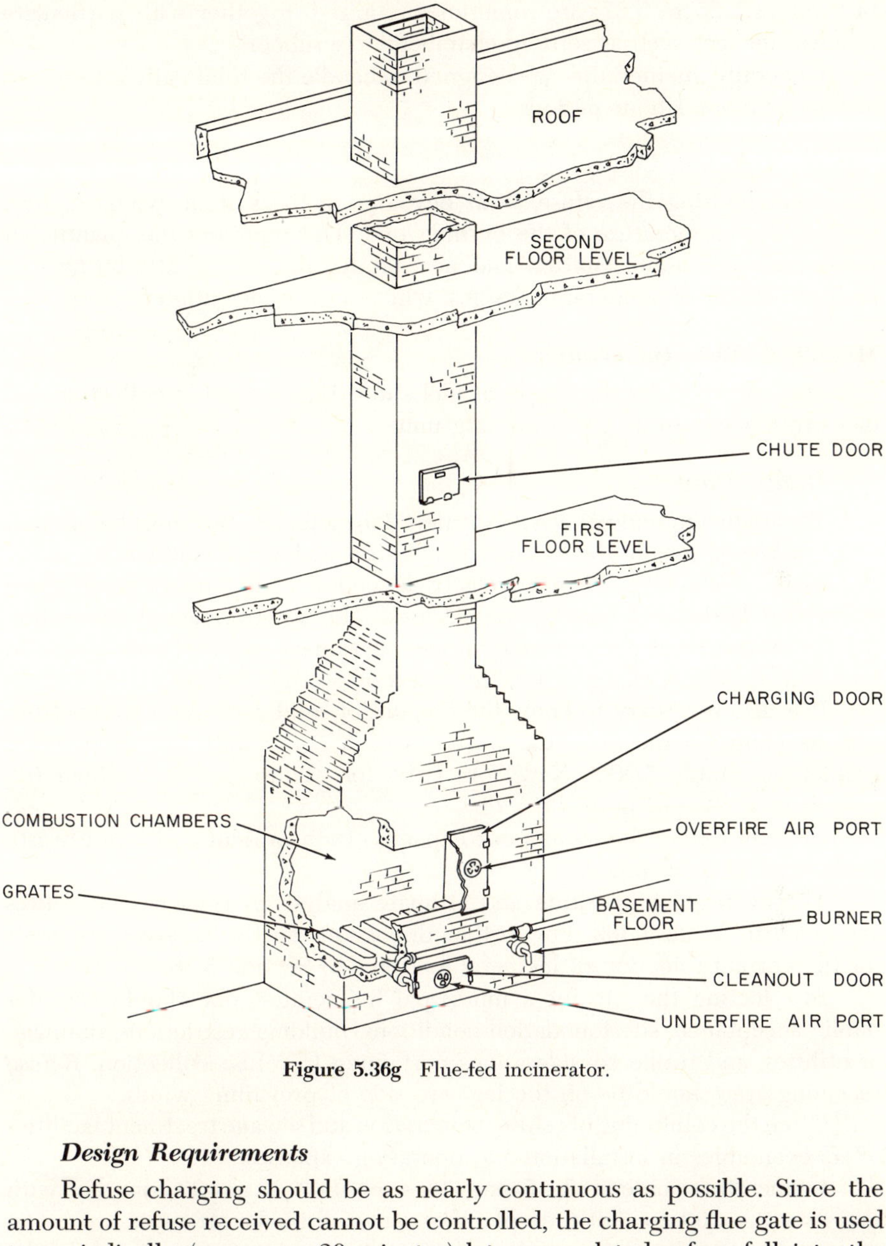

Figure 5.36g Flue-fed incinerator.

Design Requirements

Refuse charging should be as nearly continuous as possible. Since the amount of refuse received cannot be controlled, the charging flue gate is used to periodically (say every 30 minutes) let accumulated refuse fall into the furnace. (See Section 5.37 for a discussion of alternate methods of charging.)

Auxiliary burners are usually installed above the refuse to ignite it and to establish draft at the beginning of the cycle. Afterburners (as discussed in Sections 5.25 to 5.27) are sometimes required, together with particulate removal devices such as settling chambers or scrubbers.

Generally, incinerators are designed to handle the total daily refuse load in four or more 1-hour periods.

Air Pollution Control

After igniting the refuse, draft develops in the system, which in turn reduces the temperature of the burning pile. This increases the quantity of smoke and pollutants emitted and necessitates the use of afterburners or multiple chamber incinerator designs, which are discussed later.

Municipal Waste Incineration

One can consider the wastes as fuel and design the entire collection and incineration system as a fuel burning unit.

Design Data

To design a municipal wastes incineration system,[1,6] one must be familiar with the regulations in the locality, which may include air and water quality standards; zoning, building and electrical codes; occupational, health and safety standards; and sanitary regulations. The Environmental Protection Agency standards of performance for new stationary sources[3] also applies to incinerators with a charging rate in excess of 50 tons per day.

It is also necessary to know the amount of solid and liquid wastes from process plants in the area; the wastes from a community are much more predictable (Table 5.36c). New York City, for example, collects about $6\frac{1}{2}$ pounds of waste per person per day.

In determining the incinerator capacity, both present and future waste loads need to be estimated.

The average composition and ultimate analysis of typical solid wastes are given in Table 5.36h. For further discussions on the data requirements for the correct selection of incinerators, refer to Section 5.38.

In selecting the site for a municipal incinerator, one should consider public acceptance, site foundation conditions, building restrictions, drainage of utilities, and traffic considerations pertaining to refuse collection. Refuse receiving area should be on the leeward side of prevailing winds.

Often the combining of refuse incineration and sewage treatment facilities offers economies in installation and operating expenses.

Incinerators range in size from 60 to over 750 tons per 24 hour day, with many in the 200 to 300 tons per day range. Still, larger units are being proposed to match the increase in urban waste. New York City, for example, produces 26,000 tons per day of municipal wastes.

Incinerators work best when operated at steady, continuous loads. They are usually designed to operate two to three shifts per day and five to six days per week.

Table 5.36h
COMPOSITION OF TYPICAL URBAN SOLID WASTES, OVERALL CALORIFIC VALUE AS RECEIVED IS 3000 TO 6000 BTU PER LBM

Component	Average Weight, %
Food Wastes	18.2
Garden Wastes	7.9
Paper Products	43.8
Metals	9.1
Glass and Ceramics	9.0
Plastics, Leather and Rubber	3.0
Textiles	2.7
Wood	2.5
Rock and Ash	3.7

Ultimate Residue Analysis	
Moisture	15–35
Carbon	15–30
Oxygen	12–24
Hydrogen	2–5
Nitrogen	0.2–1.0
Sulfur	0.02–0.1
Others	15–25

Solid waste is delivered to the plant during the day shift, in several types of vehicles. These are weighed and discharged to the storage pits (Fig. 5.36i). The storage pits provide holding space and are usually sized for the 36-hour capacity of the incinerator. Storage volume requirements are determined on the basis of 350 lb/cubic yard waste density. The pit is so designed that a monorail crane can be used without being obstructed by trucks as they are dumping.

Refuse may be charged into the incinerator by front-end loaders in small installations provided that the storage area and charging hoppers are at the same elevation. Vibrating hoppers or conveyors may also be employed.

Cranes are used at larger installations, utilizing closed scoop buckets or grapples to transfer the wastes. (See Section 5.37.)

Furnaces can be of the batch feed design used for plants with up to 100 tons per day capacity.

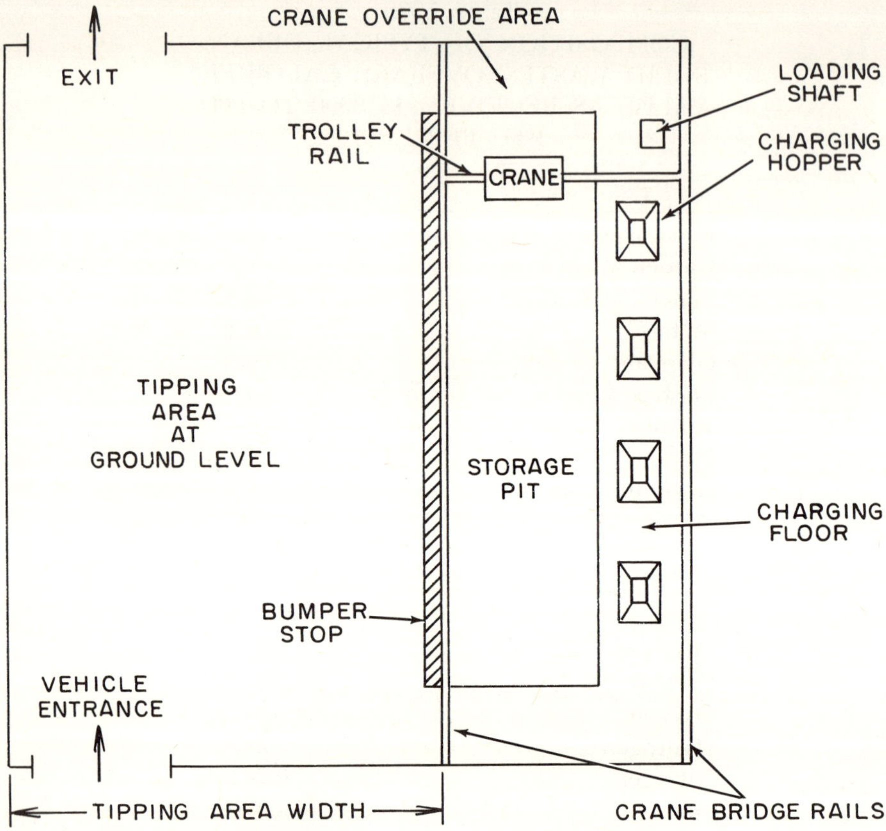

Figure 5.36i Municipal incinerator layout.

In continuous feed type designs used for larger units, the refuse is continuously charged from a large hopper by moving grates.

Furnace Configurations

Furnaces can be vertical circular, rotary, multicell and rectangular in their design. They can be sized on the basis of 18,000 BTU per cubic foot per hour.

Vertical Circular Furnace

In this design, the refuse is charged through a door in the ceiling and drops onto a central cone grate surrounded by a circular grate (Fig. 5.36j). Primary combustion is supported by underfire air. As the cone and arms rotate, the fuel bed is agitated and residue is moved to the discharge gate on the periphery. Stoking doors are provided for manual stoking or for residue

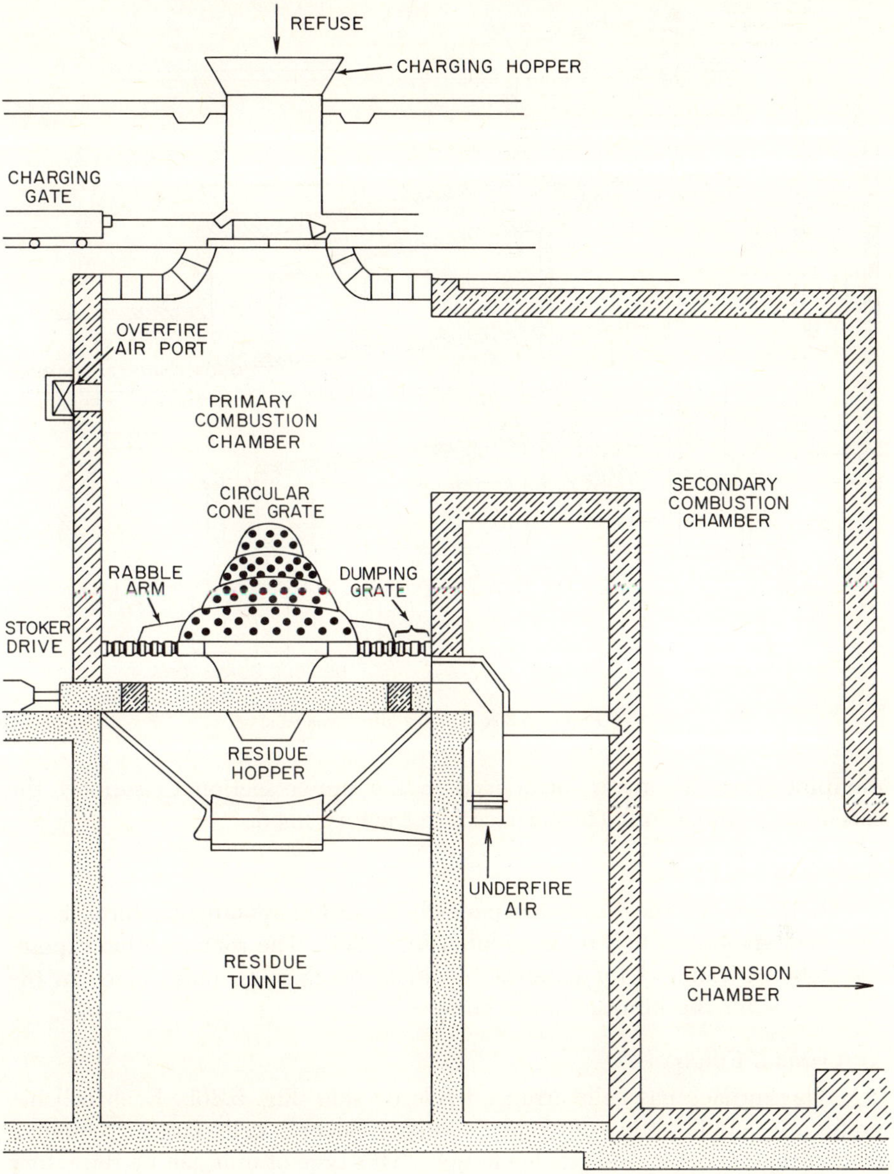

Figure 5.36j Vertical circular furnace.

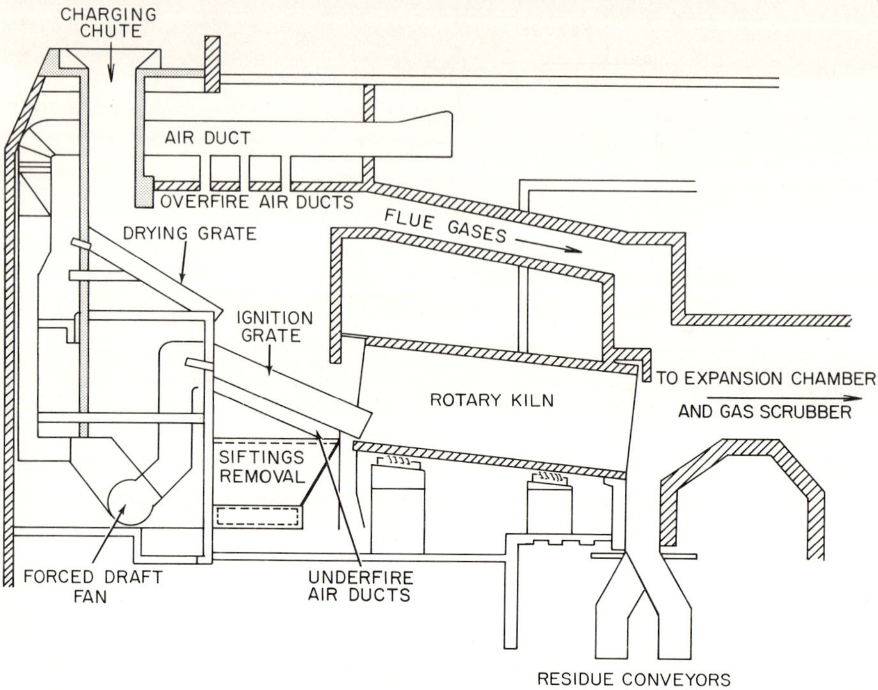

Figure 5.36k Rotary kiln furnace.

dumping. Overfire air is introduced in the upper section to support the secondary combustion. The furnace is refractory lined.

Rotary Kiln Furnace

In this design the wastes are partially burnt in a rectangular furnace and then fed, via grates, to a rotating kiln (Fig. 5.36k). The rotary action exposes the unburnt material for combustion. Final combustion takes place in the chamber after the kiln discharge point.

Multicell Furnace

This furnace has cells arranged side by side (Fig. 5.36l). Each cell has grates for moving the refuse across them. Several cells have a common combustion chamber and residue hopper. This type of unit can be refractory lined or water cooled.

Rectangular Furnace

In this unit, two or more grates are arranged in tiers so that the refuse is agitated as it falls from one level to the next (Fig. 5.36m). Secondary combustion is usually employed.

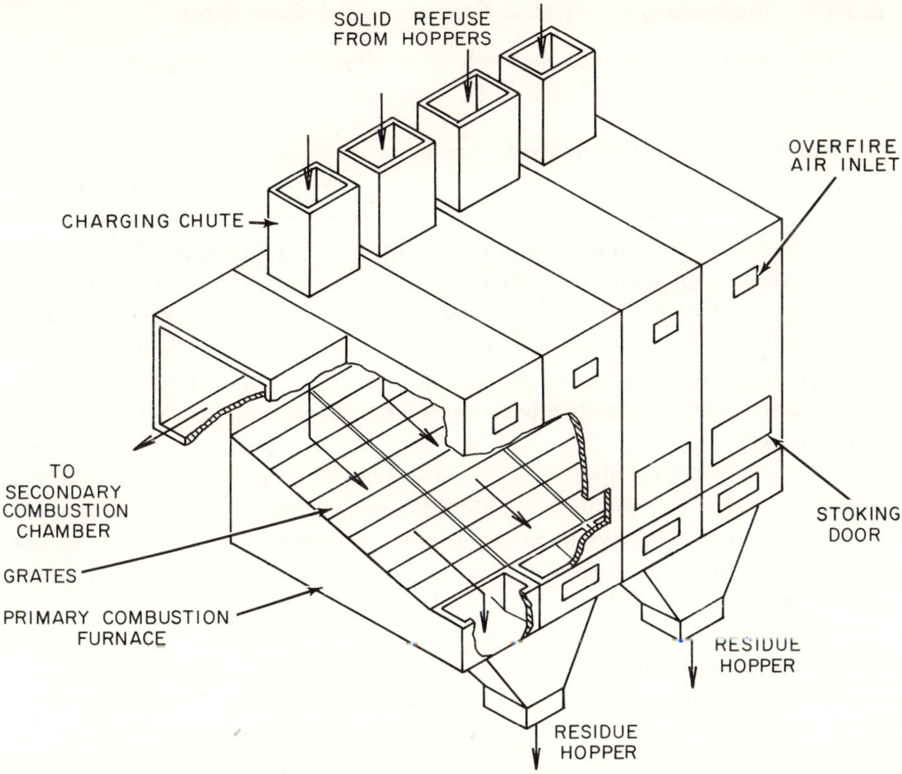

Figure 5.36l Multicell furnace.

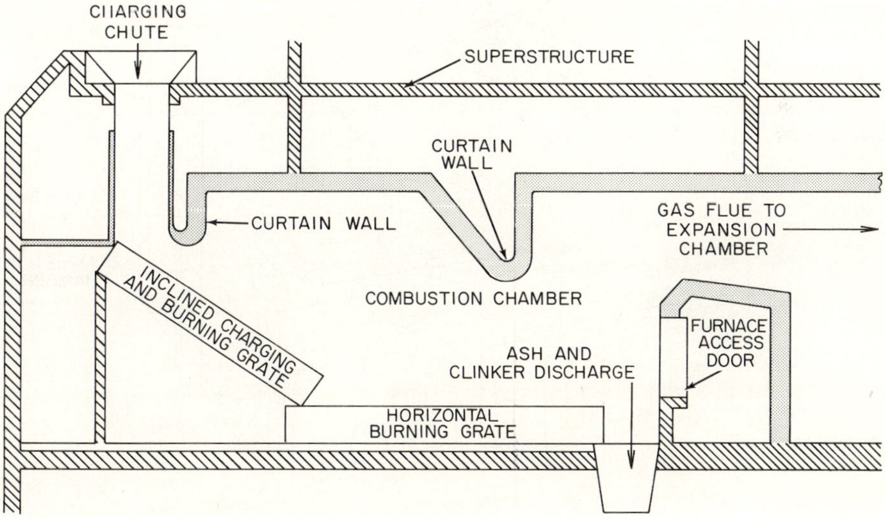

Figure 5.36m Rectangular furnace.

For details on accessories such as grates, dampers, blowers and ash hoppers, refer to Section 5.37.

Heat Recovery (*Water Wall Furnace*)

Incinerators with 250 or 300 tons/day capacity are frequently the straight refractory type. With capacities exceeding 400 tons/day, heat recovery is usually justified. This subject is discussed in detail in Chapter III.3.

In water wall furnaces, part or most of the refractory in the furnace chamber is replaced by water tube walls made of closely spaced special steel tubes welded together to form a continuous wall. Water or steam circulates through these tubes. This design has the advantages of reduced refractory maintenance, higher permissible flame temperatures without excess slag and heat recovery.

The water in the tubes should be at a temperature over 300°F to avoid fireside pitting and it should be deaerated and softened as in boiler practice. Preferably, when water walls are employed, a steam or high-pressure boiler should be included as part of the circulation system and the treated water should be recirculated.

Industrial Waste Incineration

The incineration of solid and liquid industrial wastes is discussed here. For gas and fume incineration, refer to Sections 5.32 and 5.33.

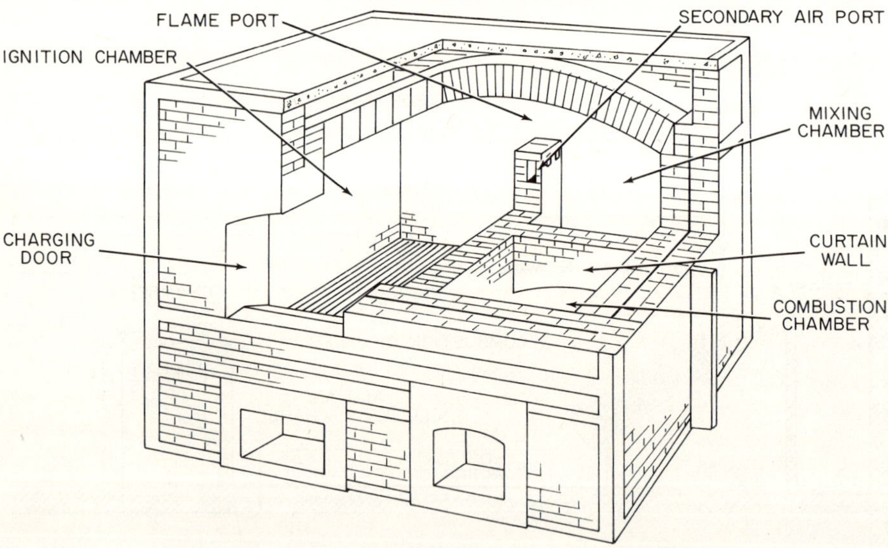

Figure 5.36n Retort type multiple-chamber incinerator.

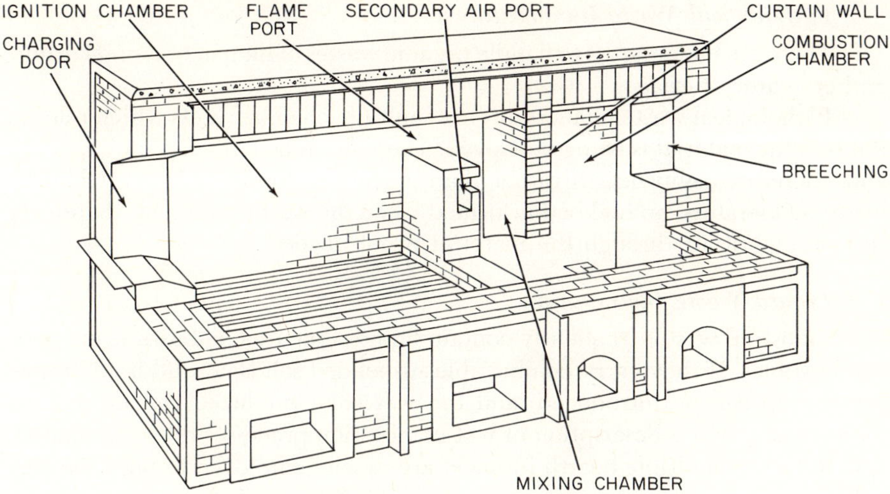

Figure 5.36o Inline multiple-chamber incinerator.

Multiple Chamber Incinerators

Multiple chamber incinerators include the retort type and the line type. In the retort type (Fig. 5.36n), the chambers are so arranged that combustion gases go through U-turns in both lateral and vertical directions. Mixing chambers, flame ports and curtain wall ports have a length-to-width ratio in the range of 1 to 2.4. (See also Chapter III.2.)

In the line type design (Fig. 5.36o), combustion gases go through 90 degree turns only in the vertical direction. Ports and chambers extend across the full width of the unit having length-to-width ratios of up to 5.

Retort type units are employed in the capacity range of 25 to 800 lb/hr, while at higher loads, the line type has better combustion characteristics and is more common.[1,2]

Mobile multiple chamber units are also available for on-site disposal of commercial wastes associated with land clearance, housing construction and so on.

Rotary Incinerators

In this type, the combustion process is conducted in a rotating kiln (Fig. 5.36k). Charging of wastes can be done by a pneumatic ram on an automatic cycle or on a batch basis. These units are available in 100 lb/hr to over 4,000 lb/hr capacity range.

Fluid Bed Incinerators

These are more recent designs for the incineration of wet filter cakes and other chemical wastes, discussed in detail in Chapter III.2.

Pathological Waste Incinerators

These are incinerators to handle organic wastes of human or animal origin and crematory furnaces.

Pathological waste incinerators are multiple chamber units which release fluids as the material is being destroyed. These fluids do not evaporate quickly and, therefore, solid hearth rather than grating is required. In pathological waste incinerators, no fuel bed is formed when the waste is burned; therefore, the passage of air through the material is not needed.

Liquid Waste Incineration

Industrial wastes frequently contain liquids that are combustible or carry combustibles. If they carry combustible suspended solids, a solid-liquid separation step can be introduced, and the wet cake produced can be sent to incineration. (For a description of wet combustion processes refer to Chapter I.5.) Rotary or multiple hearth furnaces are usually employed to burn the wet cake.

If the liquid is at least partially combustible, then direct incineration can be considered. The waste liquid should be pumpable at ambient temperatures or on being heated. Wastes that do not flow at 300 to 400°F are difficult to incinerate as a liquid and are usually handled as sludge or as solids.

Materials with viscosities below 1,000 SSU at handling temperatures can usually be incinerated.

If the liquid is relatively free of suspended solids and has a heating value in excess of 8000 BTU/lb, it can be directly piped to an oil burner. Such materials as light solvents and still bottoms belong in this category. They are usually atomized prior to entering the combustion chamber. When such a system is properly designed, no secondary combustion chamber is required. Forced draft burners with short flame and stable burning characteristics are used. Residence time varies from 0.2 to 1 second.

For wastes that form acidic gases (such as chlorinated organics), the incinerator chamber must be made of acid resistant brick and absorption towers are required to remove the acidic gases.

REFERENCES

1. Corey, R. C. (ed.): "Principles and Practices of Incineration," Wiley, New York, 1969.
2. Danielson, J. A.: "Air Pollution Engineering Manual," Public Health Source, Cincinnati, Ohio, 1967.
3. *Federal Register*, pp. 15704–15722, August 17, 1971.
4. *Federal Register*, p. 1, April 7, 1971.
5. Skipworth, D. W., G. M. Hein and H. W. Nelson: "Design of Domestic Gas-Fired Incinerators for Elimination of Smoke, Odors and Flyash," AGA Research Bull., p. 78, 1958.
6. Marco, J., et al.: "Incinerator Guidelines," Bureau of Solid Wastes Management, Cincinnati, Ohio, 1969.

5.37 INCINERATORS—ACCESSORIES AND INSTRUMENTS

Partial List of Suppliers:

Conveyors: Bartlett-Snow; Beaumont Birch Co.; The Bucket Elevator Co.; Gifford-Wood; S. Howes Co., Inc.; Link-Belt Div., FMC Corp.; Edward Renneburg & Sons Co.; S. G. Sargent's Sons Corp.; Sprout-Waldron & Co., Inc.

Gates: Beaumont Birch Co.; The Bucket Elevator Co.; S. Howes Co., Inc.; Sprout-Waldron & Co., Inc.

Hoppers and Bins: Acrison, Inc.; Caldwell Tanks, Inc.; Carter-Day Co.; Chattanooga Boiler and Tank Co., Inc.; Dean Products, Inc.; Dustex Div., Amer. Precision Ind., Inc.; Kennedy Tank & Mfg. Co., Inc.; Nooter Corp.; Edward Renneburg & Sons Co.; Sprout-Waldron & Co., Inc.; Vibra Screw, Inc.; Wyssmont Co., Inc.

Screens: Ludlow Saylor Wire Cloth, Div. GSI; Multi-Metal Wire Cloth Inc.; The W. S. Tyler Co.; Wemco Div. Envirotech.

Control Panels: AETRON, Div. of Aerojet-General Corp.; Customline Control Products Inc.; Custom Engineering Co.; Honeywell Inc., Special Systems Div.; Lake Erie Electric Mfg. Co.; Mercury Company Inc.; Monitor Panel Co.; Scam Instr. Corp., Panellit Div.; Swanson Engineering Co.; Sycamore Engineering & Mfg. Co.

Fans: Buffalo Forge Co.; The Buhler

Corp.; Ceilcote Co.; Coppus Engineering Corp.; Garden City Fan and Blower Co.; Heil Process Equipment Corporation; Sprout-Waldron & Co., Inc.; Westinghouse Electric Corporation.

Ducts and Stacks: Ceilcote Co.; Chicago Bridge & Iron, Co.; Continental Boiler & Sheet Iron Works; Heil Process Equipment Corporation; Jones & Hunt Inc.; Justin Enterprises, Inc.; Kennedy Tank & Mfg. Co., Inc.; Edward Renneburg & Sons Co.; The Torit Corp.

Dampers: Honeywell, Inc.; Lundy Electronics & Systems Inc.; Penn Vent Co.; Ruskin Manufacturing Co.; Syntron, Inc.

Draft Gauges: Foxboro Co.; Heise Bourdon Tube Co.; Helicoid Gauge Div. of American Chain & Cable Co.; Heraeus-Engelhard Vacuum, Inc.; Manning, Maxwell & Moore, Div. of Dresser Industries; Marsh Instrument Co.; Martin-Decker Corp.; Robertshaw Controls Co.; Fulton Sylphon Div., Seegers Instrument Co.; U.S. Gauge, A Div. of Ametek, Inc.; Wallace & Tiernan Inc., Weksler Instruments Corp.

Temperature and Draft Controllers: Bailey Meter Co.; Black-Sivalls and Bryson, Inc.; Bristol Co.; Fisher Controls; Foxboro Co.; Hagan Controls Corp.; Kieley and Mueller, Inc.; Masoneilan International, Inc.; Moore Products Co.; U.S. Gauge Div. of Ametek.

Grates: Airtrol Co.; Combustion Engineering; Flynn-Emirich; Mourse Boulger Destructor Co.; Nichols Engr. and Research Corp.

Charging Equipment

The type of incinerator (batch or continuous) dictates to a large extent the type of charging devices used.

For municipal incinerators, charging gates (sliding or rolling types) underneath feed hoppers are frequently used. When they are closed, the gates seal the opening at the top of the furnace. Charging gates are made of a refractory lined, inverted steel box. In some incinerator designs, the feed hopper discharges continuously through a chute and no gate is used. In these designs, the chute must be kept full to maintain an air seal. Skip loaders, conveyor belts, oscillating conveyors and rams are also utilized.

In refuse incinerators for apartment houses, the charging should be as nearly continuous as possible or at least in small increments so that momentary overloading is avoided. This may be accomplished by use of a charging gate that opens at 15 to 30 minute intervals or by use of a rotating star feeder. Another type of charging device is the hydraulic plunger or ram which forces the refuse from the bottom of a hopper into the furnace with an alternate forward and rearward movement.

Incinerators for burning sawdust normally use screw or pneumatic conveying systems for charging. The screw conveyor carries the wood waste from the bottom of a feed bin to the incinerator. The screw should be at least 15 cm (6 inches) in diameter to prevent larger pieces of wood from blocking flow. Also, the screw should have a variable pitch to permit more even loading along the length and minimize compression of the shavings at the discharge. Pneumatic conveying is also used to carry sawdust to incinerators (Fig. 5.37a).

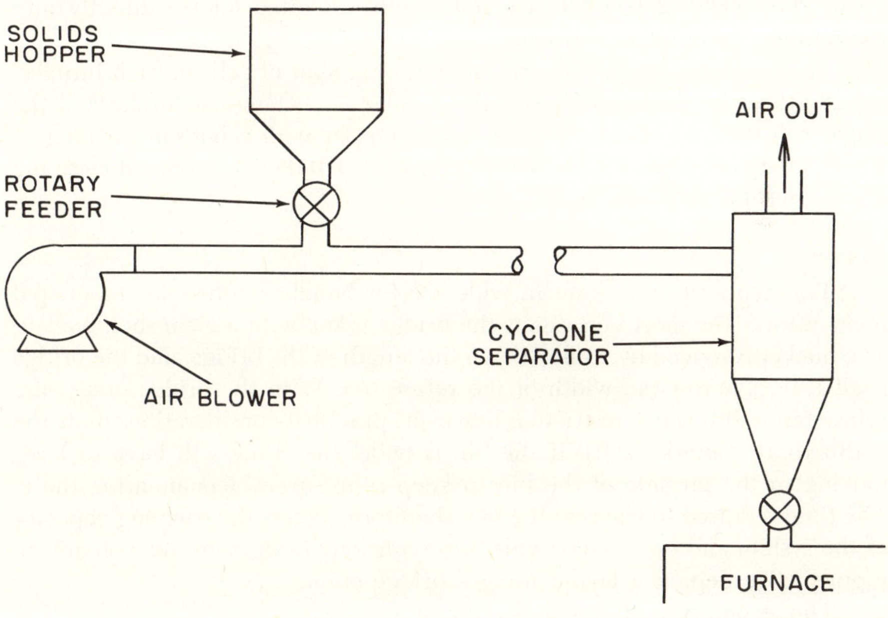

Figure 5.37a Pneumatic conveying system.

These systems require about 0.15 liters/sec of air per kg/hr of sawdust to be burned (0.15ACFM per lbm/hr). Air velocities in the duct should be at least 15 meters/sec (50 fps) to prevent sawdust from blocking the duct work. About 10% of the conveying air should be admitted to the incinerator to spread the sawdust particles and to produce even burning.

Hoppers and Bins

In municipal incinerators, storage space is provided by a storage bin or pit. These pits are at a low level so that refuse can be dumped into them. Construction normally is of reinforced concrete with steel plates or rails along the sides to protect against damage by the crane bucket. The floor should be pitched to facilitate drainage. An average capacity might be 12 to 36 hours of the incinerator burning capacity, depending on the hours of operation and collection schedule. Municipal refuse is quite bulky with specific gravity of 0.15 to 0.20 (relative to water) and requires large storage volumes.

Other types of incinerators, wood waste disposal, for example, can use storage bins constructed of sheet steel. These bins are built with vee bottoms to facilitate solids flow.

Charging hoppers are used to keep an even flow of rubbish into the furnaces. They are located over the furnace, and the refuse drops from the hopper into the furnace. In larger municipal incinerators, cranes are used to move refuse from the storage pit to the charging hoppers. In small furnaces, up to 100,000 kg (220,000 lb) capacity, refuse is often loaded directly into the hoppers from collection trucks.

The number of hoppers varies with the number of cells in each furnace. Usually there is one hopper per cell. Circular or rectangular furnaces of the single cell design typically have a single hopper with a bottom opening of 2.5 to 3.5 square meters (25 to 40 ft^2). Figure 5.37b depicts a typical charging hopper installation.

Cranes

Two types of cranes are in wide use for handling refuse for municipal incinerators. The most versatile is the bridge crane with a clam shell bucket. The bucket is moved by trolley down the length of the bridge, and the bridge itself travels across the width of the refuse bin. With the bridge crane, the refuse bin width is not restricted; however, practical considerations limit the width to 10 meters (30 ft). If the bin is wide, the crane will have to keep moving to the far side of the bin to keep refuse from accumulating there. The time required to traverse the bin, therefore, affects the carrying capacity of the system and excessively wide bins with long bridges are uneconomical. Figure 5.37c depicts a layout using a bridge crane.

The second type of design, the monorail crane, can move in one direction only, along the rail at the center line of the bin. The range of the monorail

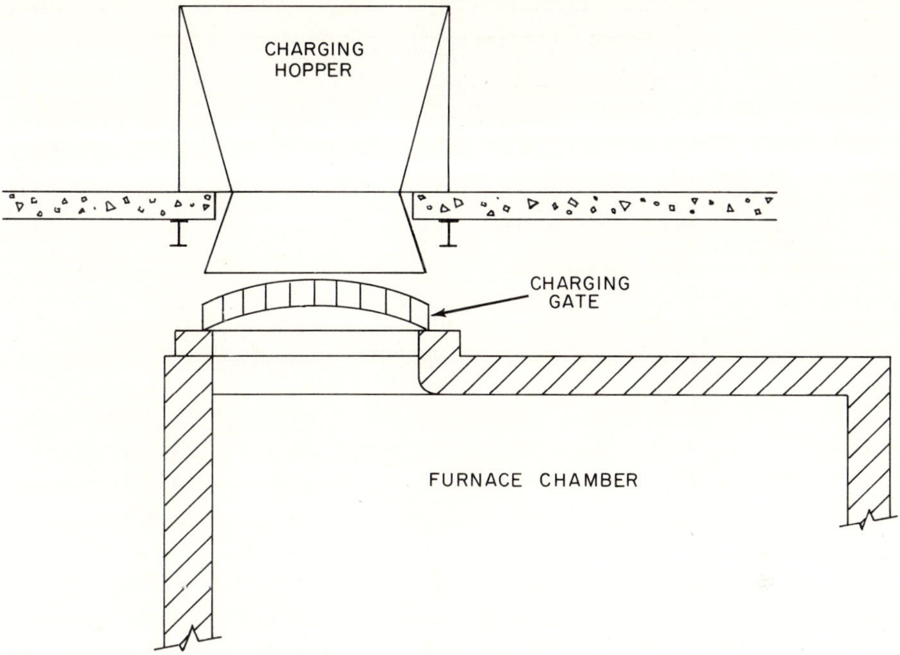

Figure 5.37b Charging hopper for rectangular batchfeed incinerator.

crane is limited in regards to the width of the bin. The bin width is limited to about 1 meter wider than the overall width of the opened bucket. If the bin is too wide, the bucket cannot move the rubbish along the sides of the bin, and the material will accumulate because of its tendency to cling together. The monorail is normally designed to follow a straight path with the bin at one end and several charging hoppers at the other.

The monorail's initial cost is lower than that of the bridge crane since it has no bridge. However, the choice depends on several factors such as the furnace capacity and the plant layout. In large furnaces of more than 300,000 kg capacity, a bridge crane is used. In medium-sized incinerators of 100,000 to 300,000 kg capacity, the monorail is often used.

The maximum operating speeds for cranes are 110 meters per minute (6 fps) for bridge travel and 100 meters per minute for hoisting and trolley travel. Cab control for bridge cranes frequently is used because of better visibility and safer operation. Air conditioning of the cab is essential. Stationary control is standard with monorail-type cranes.

The clam shell grapple or bucket for either type of crane is built of a steel that will withstand abrasion. The normal bucket capacity ranges from 1 to 4 cubic meters (35 to 140 ft^3). An automatic lubrication system for the

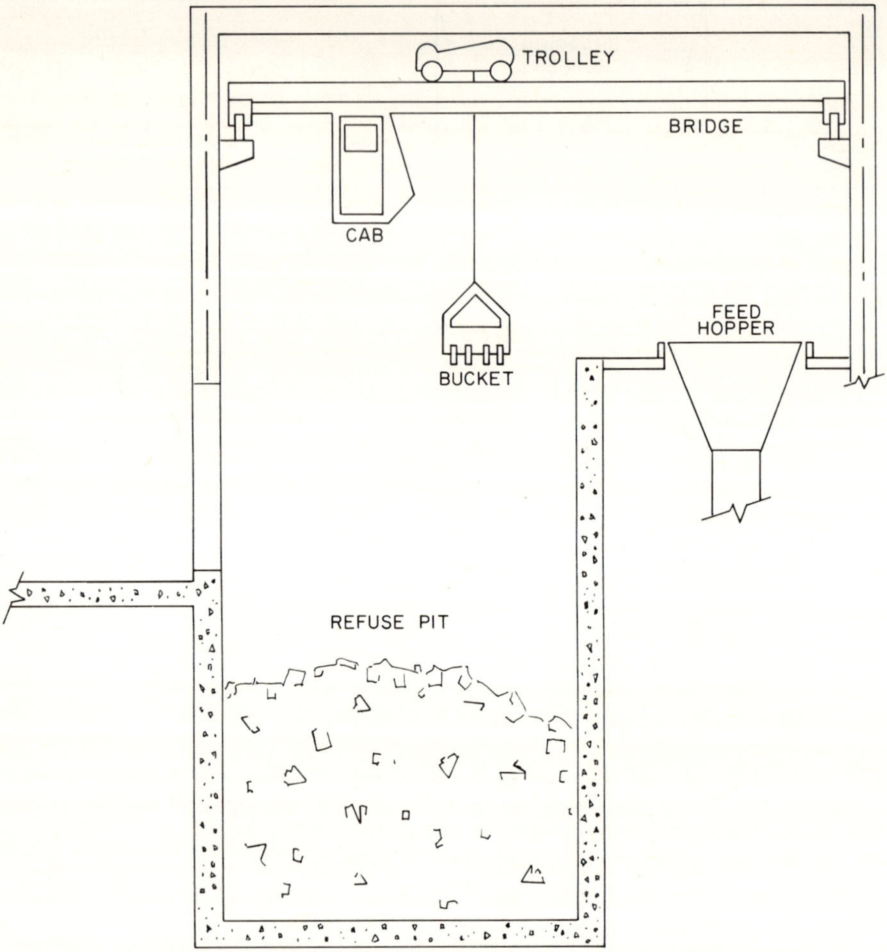

Figure 5.37c Bridge crane installation.

crane is highly desirable, and a good preventive maintenance program is essential. The need for a spare crane is debatable; however, there is general agreement that a spare bucket is needed.

Screens

Wire mesh screens of 4 to 10 mesh size (4 to 10 lines per inch) are commonly used to trap large particulate matter from the incinerator. These flyash screens are made from stainless steel to protect against corrosion and to prevent rapid burnout. They may be located at the top of the stack, inside the stack or in the breeching between the furnace and the stack. In municipal incinerators, a preferred design is one that has a double hung, counterbalanced construction. This moveable type system provides good accessibility for cleaning and maintenance. One screen is in the flue gas stream while the other is available for repair and cleaning.

Control Panels

The control panel is usually a large board of high dielectric strength on which all instruments that record or control draft, temperatures and other operational and safety parameters are mounted. The number and complexity of the instruments vary with the size of the incinerator. At large municipal incinerators, the control panel is usually housed in an air conditioned control room.

Careful planning should go into the panel layout. The most critical instruments should be located near the center of the board and installed at eye level for easy readability. Indications and records must be coordinated and grouped with controls, switches and alarms. That is, if a temperature recorder reading dictates that the operator should flip a certain switch, then that particular switch should be located adjacent to the recorder.

Fans

Fans are used to supply induced-drafts in incinerator systems not using full chimneys, and they are used to supply air in forced-draft systems to be used in both burning and cooling operations. These fans may be divided into two classifications: (1) axial flow or propeller type and (2) radial flow or centrifugal type.

In the induced-draft system, the fan pulls the combustion gases from the furnace. Cooling air is also pulled in through a barometric damper. The duct in which the cooling air mixes with the combustion gases should be sized so that the combined velocity is about 12 meters per second (40 fps). The temperature of the mixed gas depends, of course, on the relative quantities of air and combustion gas. The most economical operating temperature is roughly 300°C. This is a compromise between the use of very expensive, high-temperature fans of small capacity versus cheaper, low-temperature fans that must have much larger capacities due to the excess air pulled into the system. Fans that can operate at 600°C or higher are available, but these must be of stainless steel construction and be equipped with water-cooled bearings. Low-temperature fans with mild steel blades are capable of operating at up to 150°C. The maximum operating temperature of these fans can be increased to about 400°C by the addition of inexpensive bearing coolers. Axial fans have the advantage of being low cost, lightweight and compact and are normally selected for this application.

Forced-draft fans supply air at a positive pressure to the air ports located above and below the burning grate. Their application is normally limited to large incinerators. Air for both burning and cooling can be supplied by the same fan with dampers used to distribute the airflow. In automatic systems, the dampers are modulated by temperature controllers. If precise air control is necessary, several fans can be used to provide air in different locations and at various pressures.

Operating efficiencies of fans are in the range of 40 to 70%. The air power of a fan is given by the following dimensional equation:

$$A = 0.977 \times 10^{-4} Q \cdot h \tag{5.37(1)}$$

where A is the power, kilowatts; Q gas rate, liters/sec; and h, developed head, cm of water. The shaft power is the air power divided by the efficiency. Figure 5.37d illustrates some axial and radial fan designs.

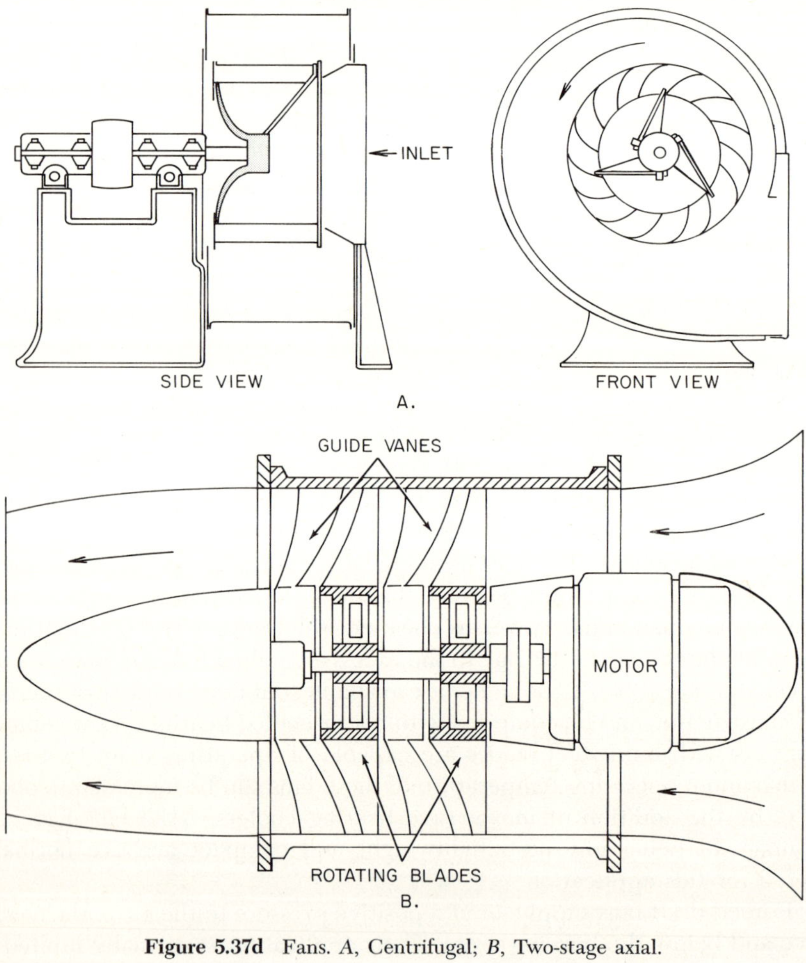

Figure 5.37d Fans. *A*, Centrifugal; *B*, Two-stage axial.

Ducts

Ducts are used to direct gas flow in incinerators. Breechings are high-temperature ducts lined with refractory material that convey combustion gas from the furnace to the chimney. In large incinerators, the breechings are refractory-lined masonry construction. In smaller furnaces, the breeching may be refractory-lined steel duct. Normally, the cross section is larger than that of the chimney. Linear gas velocities are 6 to 12 meters per second (20 to 40 fps) in breechings.

Ducts are used to supply air for burning and cooling in forced draft systems. These ducts may be galvanized sheet steel riveted and soldered. The frictional head losses can be calculated from Fanning's equation:

$$\Delta P = \frac{2\,f\rho v^2 L}{g_c D} \qquad 5.37(2)$$

where the friction factor has been found experimentally to be

$$f = 0.0791/Re^{0.25} \text{ for } 3 \times 10^3 < Re < 10^5 \qquad 5.37(3)$$

and $f = 0.0008 + 0.0552/Re^{0.237}$ for $10^5 < Re < 10^8$ 5.37(4)

ΔP = frictional head loss

Re = Reynolds number, dimensionless $= \dfrac{Dv\rho}{\mu}$

D = duct diameter or equivalent diameter

v = linear gas velocity

ρ = gas density

μ = gas absolute viscosity

f = friction factor, dimensionless

L = length of duct

g_c = conversion factor $\dfrac{980 \text{ gm-cm}}{(\text{gm-wt}) \sec^2}$

Dampers

The amount of draft in a furnace produced by natural draft stacks can be controlled by dampers. Depending on the location of the damper, the materials of construction may be high-temperature resistant steel, refractory-covered steel or water-cooled steel.

Some furnaces operate on an induced-draft system. In these designs, the hot combustion gases are cooled by mixing with ambient air to a temperature that can be handled by an induced draft fan. The cooling air is introduced through manually adjustable and barometric dampers. Manually adjustable dampers are sized to provide enough air to cool the maximum amount of flue gases to the design temperature of the fan. Barometric dampers sustain an adequate and uniform draft in the incinerator.

Many multistory apartment buildings have incinerator stacks that are considerably higher than necessary for providing natural draft. The reason for this usually is to discharge the flue gases above the roof. The draft produced by the tall stack can be reduced by several methods. One way is to use a guillotine, or sliding knife, type of damper in the breeching between the furnace and the stack. The disadvantage of this system is that the damper needs constant adjustment to provide the proper draft, especially during incinerator startup. Another solution is to use a barometric damper in the stack. Once the barometric damper is adjusted, it can automatically maintain the draft at the proper level by letting ambient air enter the stack and mix with the exhaust gases as required. The barometric damper is constructed as a balanced, pivoted plate which changes its opening with the draft.

In flue fed apartment incinerators, a draft control damper is often installed in the stack just above the furnace. This damper receives direct flame impingement and must be made of stainless steel. Dampers of type 302 stainless with a 20-gauge thickness have been used satisfactorily. A swinging, counterweighted design that is pivoted from a rod attached to one wall will allow the damper to be swung flush with the wall. This allows refuse charging through the flue.

All combustion-air inlets to incinerators should have positive control. Circular-spinners with rotating shutters are used in retort-type incinerators for both underfire and overfire air openings. Top-hinged dampers or butterfly, center-hinged-plate dampers are used with rectangular ports for secondary air openings and overhead air openings for in-line incinerators. These air inlet controls can be cast iron. Rectangular sliding dampers tend to warp and are not recommended for these applications.

Grates

Grates are surfaces located within the furnace that support the fuel bed and allow air to pass through the fuel. They are designed to permit removal of unburned residue and may be horizontal or inclined, manual or automatic and stationary or moveable. Since grates are subjected to elevated temperatures, they may be constructed of cast iron or of a castable refractory.

In cylindrical, batch-feed incinerators, dumping grates are located around the periphery. These grates are raised and lowered manually or by hydraulic power cylinders. When they are lowered, the accumulated residue is raked into the ashpit. In rectangular, batch-feed municipal incinerators, the grates are normally of a rocking or reciprocating design. These moving grates carry the burning refuse from the charging end to the dumping grate. Smaller incinerators of this type may have a fixed grate installed horizontally or inclined.

Continuous-feed municipal incinerators always use traveling type grates. These may be traveling chain, rocking or reciprocating grates. Figure 5.37e depicts a continuous-feed incinerator with traveling grates. Another type is

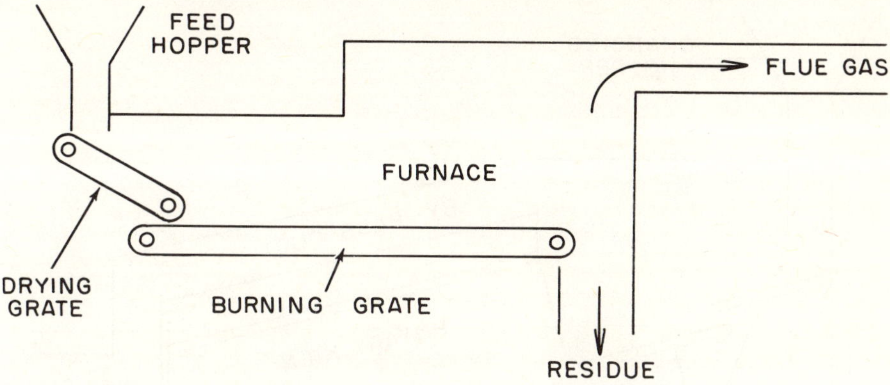

Figure 5.37e Traveling grate system.

the rotating drum grate shown in Figure 5.37f. Sometimes a combination of these types of grates are used with other devices. For example, in a rotating kiln incinerator, mechanically reciprocating, inclined grates can be used to dry and ignite the refuse. Then the burning refuse is discharged into an inclined, rotating kiln where combustion is completed. This design is illustrated in Figure 5.37g.

Small general-refuse incinerators usually use cast iron stationary grates

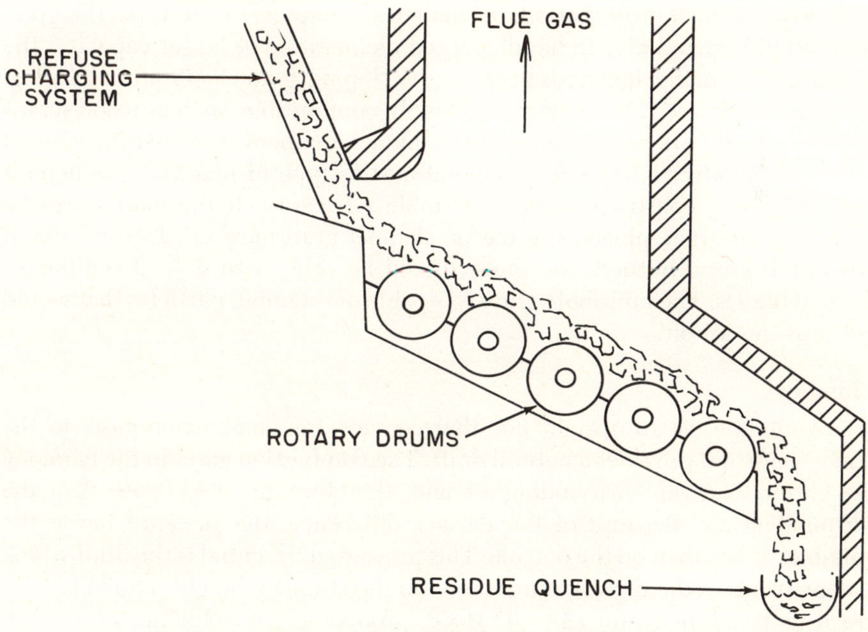

Figure 5.37f Rotating drum grate.

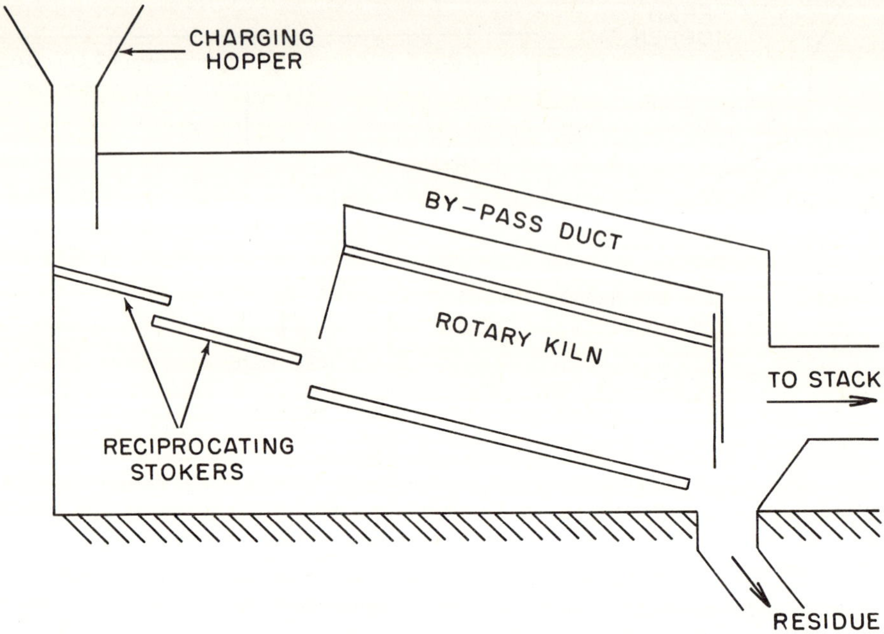

Figure 5.37g Rotary kiln incinerator.

of a tee or channel cross section. In smaller incinerators of this type, the grate is installed horizontally. In hand-charged incinerators of larger capacity, the grates are normally inclined, because the sloping grate facilitates charging.

When the material to be burned is highly combustible, such as wood scraps or sawdust, the grates are subjected to higher temperatures than in general refuse incinerators. Grates for incinerating this type of material can be cast iron of heavier construction or of castable refractory. If the material to be burned is in larger pieces, the tee or channel grates are satisfactory; but if sawdust is being burned, the grate should be solid with 1 to 2-centimeter diameter holes. The multihole grate is capable of retaining particles that would fall into the ashpit.

Chimneys

A chimney is an upright flue that conveys the combustion gases to the atmosphere and provides a natural draft. The combustion gases in the chimney are hotter than the surrounding air and, therefore, are less dense than the surrounding air. Because of the density difference, the pressure inside the chimmey is less than on the outside. This pressure differential is the draft which is dependent on the density difference and the chimney height. Drafts usually are 5 to 10 cm of water (2 to 4″ H_2O), producing gas velocities of 8 to 16 meters per second (25 to 50 fps). If low chimneys are used, draft must be produced by induced draft fans. Materials of construction may be reinforced concrete, masonry or steel. (Also refer to Sections 5.42 to 5.45.)

Reinforced concrete chimneys are strong; they can withstand higher winds than other types. The major disadvantage is that the surface may crack due to temperature and plastic stress conditions. Large chimneys are more economical when they are constructed with concrete rather than other building materials.

Masonry chimneys may be constructed from straight or radial bricks or blocks. They are built with a lining shell of fireclay brick that will resist temperatures of 500°C and an outer shell of ordinary brick. An annular airspace of 5 to 15 cm (2 to 6 in) normally separates the two shells. Linings are also used in concrete chimneys.

Steel chimneys have low initial cost, are easily constructed and are lightweight in comparison to other designs. Disadvantages are high maintenance costs and unsightly appearance. It is more economical to support steel chimneys by guy-wires than to make them self-supporting. A thin lining of bonded fused-glass is normally used on the inside of the steel stack.

Certain accessories are necessary for large chimneys. Some of these are caps, clean-out doors, test openings, ladders and catwalks and lightning arrestors. Cast iron caps prevent chimney deterioration from moisture. Clean-out doors at the bottom of the chimney are necessary for cleaning, inspection and maintenance work. Openings for test samples should be at a point where gas flow is smooth. This point normally should be at least six diameters above the bottom and nine diameters below the top. Two openings for gas samples located 90 degrees apart on the same horizontal plane are sufficient. Temperature reading points are also desirable. All test openings should be capped when not in use. Lightning arrestors in the form of lead-coated copper rods projecting above the chimney cap should be used. These rods must be well grounded. The height of the chimney and warning lights must comply with Federal Aviation Agency regulations. An inspection check-list for chimneys is given in Table 5.37h.

Residue Handling Equipment

At large incinerator plants, particularly those with continuous-feed furnaces, residue from the furnaces is discharged continuously into a water trough. A conveyor system pulls the settled ash from the bottom of the water trough up an incline to an ash hopper, truck or container. Most of the quenching water runs off the ash before it is dumped from the conveyor. This system is shown in Figure 5.37i. The wear on the conveyor system is severe; therefore, the conveyor needs wearing shoes, replaceable bottoms and rails and chains made from hard alloy steels.

Batch-feed furnaces are equipped with ash hoppers or chambers under-

Table 5.37h
CHIMNEY INSPECTION CHECKLIST

1. Is lightning rod system correctly grounded?
2. Are there excessive soot deposits in chimney bottom?
3. Can cracked lining be seen through cleanout door?
4. Is the cleanout door broken, or has it been left open?
5. Is there loose packing where breeching enters chimney?
6. Are there cracks or spalled areas on chimney surface?
7. Are outside metal bands (if any) corroded?
8. Are all clamps on lightning rod cable properly tightened?
9. Is the outside ladder corroded?
10. Are any of the lightning rod points loose or missing?
11. Is the cap (if any) loose or defective?

neath the grates for collection of residue. Normally, the ash hoppers are large enough to store the ash from several hours of burning. This may amount to 5 to 25 weight % of the refuse depending on its nature. The ash should be quenched by a water spray or water bath to reduce fire hazards. A typical installation of an ash hopper is depicted in Figure 5.37j. This system, typical of larger installations, allows dump trucks to drive under the hoppers. A discharge gate on the bottom of the hopper is lowered, and the residue falls into the truck by gravity.

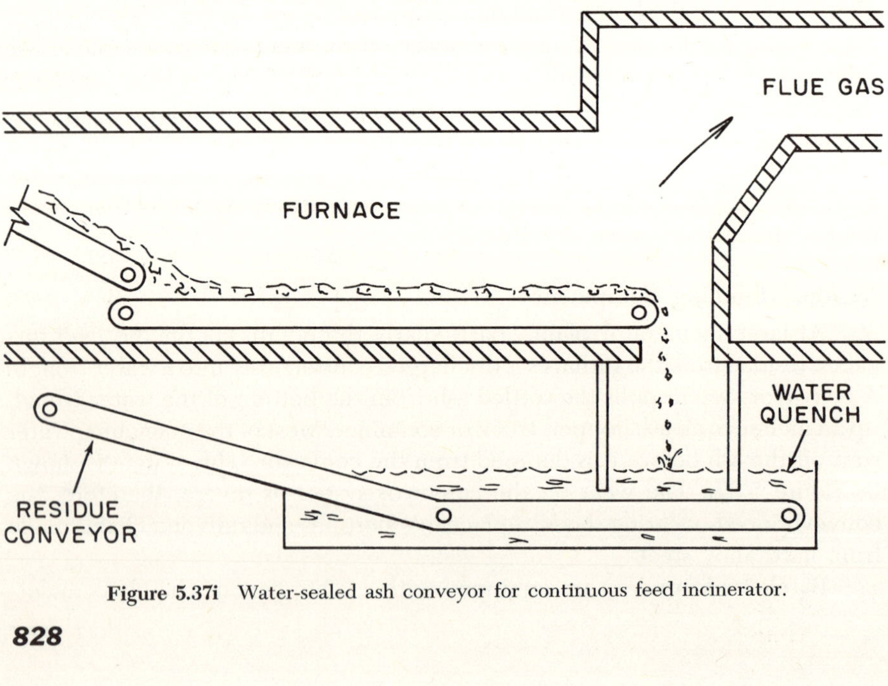

Figure 5.37i Water-sealed ash conveyor for continuous feed incinerator.

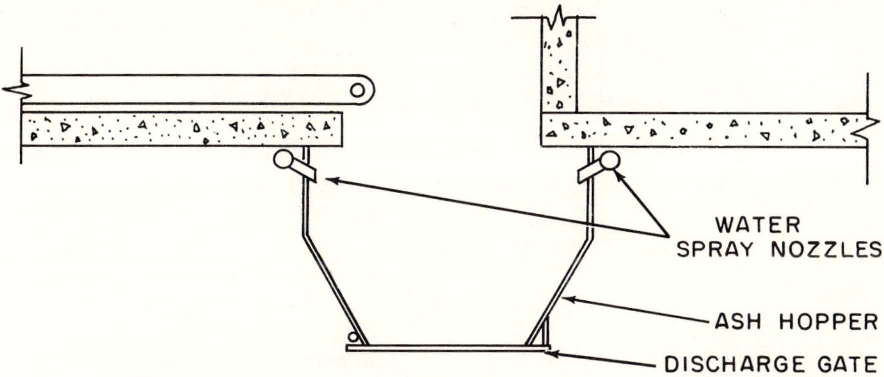

Figure 5.37j Ash hopper layout.

Dust Collection Equipment (Flyash)

The furnace volume should be large enough so that the gas velocity in the furnace is low and only very fine ash particles will escape from the furnace. Various types of dust collection equipment are available to remove particulate matter in order to meet air pollution emission standards. These types will be examined briefly with regard to their applicability to incinerator effluent gases. More detailed discussion of these devices is in the preceding sections of this chapter.

Packed towers containing shaped ceramic pieces are efficient in removing fine flyash. The dust-laden gas comes in intimate contact with wash water so that the particles are scrubbed out with the water. Wet scrubbers with baffles and sprays or venturi tubes are also effective. With any wet collector system, however, the problem of separating solids from the water slurry must be considered in order to prevent sewage disposal problems and permit the reuse of the water. Settling basins are commonly used, with the overflow water being recirculated. After scrubbing, the gas is reheated and discharged through the stack.

Dry collection devices can also be used to collect the flyash. Electrostatic precipitators of proper design can be one of the most efficient of the dry collection devices. Bag filters of silicone-coated glass fibers can be used at temperatures less than 300°C. Both of these devices can have collection efficiencies of over 99%.

Other dry collection devices such as cyclone separators, impingement baffles and gravity settling chambers have relatively low collection efficiencies due to the small particle size material involved. They can be used ahead of and in series with more efficient devices to collect most of the particles prior to the more efficient equipment used as a follow-up. Using two or more types in series is frequently the most economical method for meeting a given overall

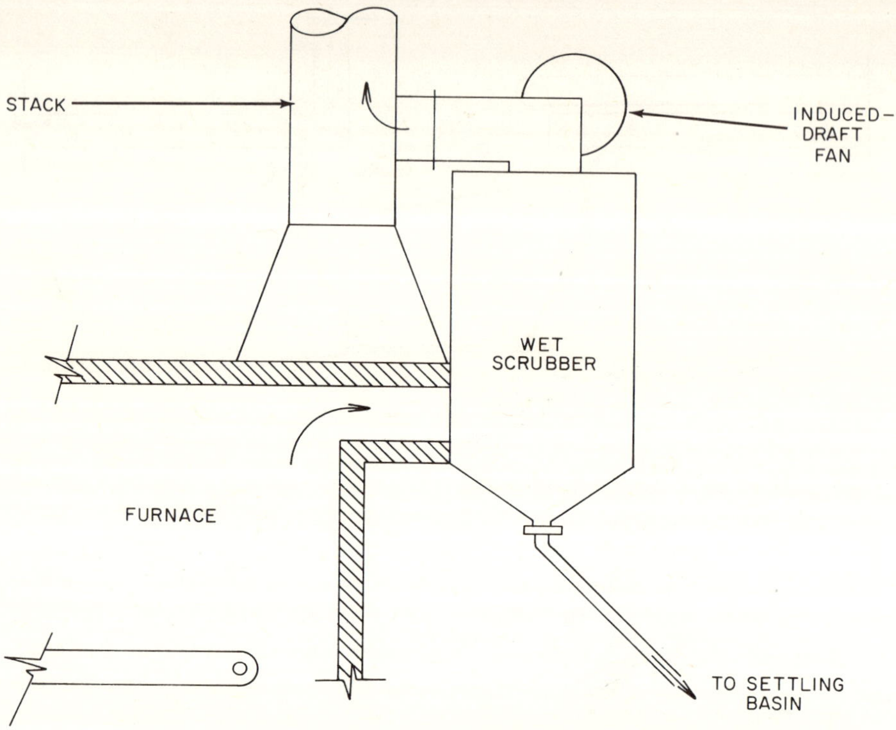

Figure 5.37k Dust collection equipment.

collection efficiency. A typical layout with dust collection equipment is shown in Figure 5.37k.

Temperature Recorders and Controllers

Temperatures at various points in the incinerator should be measured and indicated or recorded. The most common types are the thermocouple potentiometers, millivoltmeter pyrometers and resistance thermometers. Radiation pyrometers and optical pyrometers also may be used.

Platinum-platinum-rhodium thermocouples (ISA types S and R) can be used up to a maximum temperature of about 1700°C; therefore, they give the most flexibility among thermocouple types. Chromel-alumel (ISA type K) can be used up to 1200°C. Both of these types have good resistance to the influence of high-temperature oxidizing atmospheres. Thermocouple protecting tubes or thermowells should be installed at all high-temperature sensor locations. Ceramic tubes such as refractory porcelain (Sillimanite, Mullite) and fused silica are essentially impervious to gases and provide the best protection. The refractory porcelain tubes can be used up to about 1500°C and fused silica to 1050°C.

Temperature records provide the operator with information on trends as an operating guide and allow the supervisors to observe the overall performance and any irregularities which may have occurred. Recorders may be classified as (1) circular-chart or (2) strip-chart recorders. They also may be classified as (1) continuous recorders, using pen and fluid ink on single or multiple records; or (2) multipoint recorders, using a stylus or print wheel and semifluid ink to produce intermittent records. In terms of cost per point recorded, multipoint recorders are the least expensive and represent the simplest method of data gathering. Alarm contacts can be included in most types. The multipoint temperature recorder is, in general, quite rugged and requires little maintenance, although considerable mechanism is involved in the synchronized switching and printing.

Controllers can be used with continuous recorders to automatically adjust temperatures. For example, the furnace temperature in continuous-feed incinerators can be held quite uniform by regulating a forced draft fan. Another typical application is in the control of the inlet temperature to the dust collection system by regulating control valves on a water spray system. Table 5.371 suggests a minimum of instrumentation for a continuous-feed incinerator.

Table 5.371
INSTRUMENT LIST FOR CONTINUOUS FEED INCINERATOR

Temperature Recorders
- Furnace temperature detected at furnace sidewall near outlet, range 38° to 1250°C.
- Stoker compartment temperatures, range 38° to 1250°C.
- Dust collector inlet temperature, range 38° to 500°C.

Temperature Controllers
- Furnace outlet temperature controlled by regulating total air from forced draft fan; set point in 800° to 1000°C range.
- Dust collector inlet temperature controlled by regulating water spray into flue gas; set point in 300° to 400°C range.

Draft Gauges
- Forced-draft-fan outlet duct
- Induced-draft-fan inlet duct
- Furnace outlet
- Stoker compartments
- Differential gauge across dust collector

Draft Controller
- Furnace draft control by regulating damper opening

Oxygen Analyzer
- Furnace outlet

Draft Gauges

Abnormal conditions such as burned out or overloaded fuel beds, fouled gas passages and abnormal furnace draft can be detected by draft gauges at appropriate points. The draft gauge can be a local sensing and indicating device with the indicator being an integral part of the instrument or it can have remote indicating or recording features. The local indicating instrument can be an inclined-tube manometer or it can be a mechanical sensing element. In draft measurements, the static pressure causes movement of a slack diaphragm that is counterbalanced by a restraining spring. The diaphragm then moves the indicator pointer. The draft gauge selected for incinerator application should be fully enclosed for cleanliness, with large, easily read scales. The scale range should be so selected that the pointer will be at midscale during normal operations. In continuous-feed incinerators, the furnace draft can be controlled automatically by a pressure recorder-controller that adjusts the damper openings.

Other Instrumentation

At municipal incinerators, it is desirable to weigh the refuse that is brought in. Scales for weighing refuse trucks may be the manually operated mechanical type or the load cell type with an electromechanical recording device. The latter type is more expensive but is also more accurate, mechanically dependable and tamper proof.

In municipal or industrial incinerator operation, continuous oxygen analyzers and recorders are useful to monitor the oxygen content of the furnace effluent. This measurement indicates the excess air to the furnace and aids the operator in regulating the introduction of air to the furnace.

Flowrate measurements are not made routinely at incinerator installations; however, for special operating performance tests, it is desirable to measure the gas flowrates. Point velocity measurements are made with pitot tube meters or other types of velocity meters. Air pollution control regulations also require monitoring of stack emissions. Instrumentation for these purposes is described in Chapter 4.

BIBLIOGRAPHY

American Public Works Association, "Municipal Refuse Disposal," Public Administration Service, Chicago, 1966.

Corey, R. C. (ed.): "Principles and Practices of Incineration," Wiley Interscience, New York, 1969.

Danielson, J. A. (ed.): "Air Pollution Engineering Manual," PHS Publication 999-AP-40, HEW, Cincinnati, 1967.

Fryling, G. R. (ed.): "Combustion Engineering," rev. ed., Combustion Eng., Inc., New York, 1966.

"Incinerator Standards," Incinerator Inst. of America, New York, 1966.

Lipták, B. G. (ed.): "Instrument Engineers' Handbook," Chilton, Philadelphia, 1969.

Perry, J. H. (ed.): "Chemical Engineers' Handbook," 4th ed., McGraw-Hill, New York, 1963.

5.38 INCINERATORS—SELECTION AND APPLICATION

The first step in selecting an incinerator is to determine the nature and amount of wastes it is expected to handle. The Incinerator Institute of America, in its publication "Incinerator Standards," has classified all solid, liquid and gaseous wastes into seven types, according to their physical nature, moisture content, noncombustible content and heating value when burned (Table 5.36a). The Institute has further divided incinerators into nine classes, according to type of waste handled, burning capacity and method of feeding. It has also set minimum construction and performance standards for each class. National Fire Protection Association Standard 82, "Incinerators and Rubbish Handling," using basically the same classification systems, specifies minimum standards of construction for incinerators and their auxiliaries.

These standards complement each other and, along with state and local codes, represent an excellent starting point for selection.

Most incinerator manufacturers offer standard, preengineered units for types 0, 1, 2 and 3 waste (trash, rubbish, refuse, garbage and mixtures thereof) and type 4 (pathological) waste; so selection becomes primarily a matter of specifying the proper size incinerator. This is not to imply that an incinerator should be simply picked from a sizing chart; many competing designs exist and some of them are obviously better suited than others to a given situation.

Type 5 (waste industrial by-product liquids, semiliquids or gases) and type 6 (waste industrial by-product solids) nearly always require specially designed incinerators. The physical, chemical and combustion characteristics of these wastes vary so widely that designing standard incinerators for them is not practical.

The destruction of waste gases can usually be carried out in a waste gas incinerator or afterburner; these are described fully in Sections 5.25, 26, 27, 31 and 32. Therefore, the balance of this section will deal with liquid and solid waste incinerators.

Liquid Waste Incinerators

Many liquid wastes have sufficient heating value to be excellent fuels, and their destruction is carried out by substituting them for oil in standard or slightly modified industrial oil burners and firing those burners into refractory-lined chambers (Fig. 5.38a). The liquid waste can be incinerated if it is combustible and if its viscosity is low enough (either naturally or as the result of heating) to permit it to be pumped and atomized.

Incineration should not be ruled out for liquid wastes with lower heating value than needed to sustain continuous combustion. Enrichment of the waste with auxiliary fuel, usually gas, permits incineration of these wastes also. Enrichment is usually carried out by simultaneously burning the waste and gas in dual-fuel burners. Where the waste may be intermittently combustible (as with used cutting oils containing varying amounts of water), auxiliary fuel is required; but its consumption can be kept to a minimum by shutting it off when it is not needed or by reducing its flow when the fuel value of the waste rises.

Regardless of its heating value, the waste liquid is treated as a burner fuel. Therefore, many of the tests used to evaluate fuel oils can be applied to it. Other characteristics of interest include corrosiveness and toxicity. Figure 5.38b reproduces part of a typical form on which liquid waste properties can be specified. If the exact properties of the waste are not known, a representative sample should be tested in a commercial laboratory. If the composition or properties of the liquid fluctuate, several samples, representing extreme conditions, should be tested. Cost of a battery of tests may be $150 to $200 per sample.

Equipment sizing is a function of waste flowrate; therefore, maximum and minimum flows must be known. If liquid wastes are generated at an appreciable, steady rate, a continuously operating incinerator is the best solution. If waste is supplied intermittently or at widely fluctuating rates, economics may dictate their accumulation in a storage tank and burning for only a few hours a day.

The problem of corrosion, particularly by liquid wastes from the chemical and plastics industries, can be substantial. Waste-handling equipment (pumps,

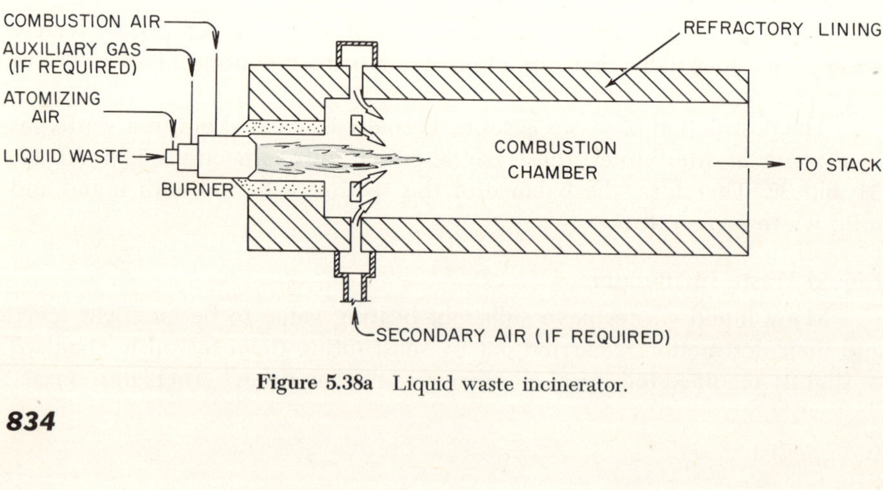

Figure 5.38a Liquid waste incinerator.

%Ash ________ % Sediment ________ % Water ________

Waste material soluble in water? ________ Water content well mixed, emulsified? ________

If there are solids in the liquid, what is their size range? ________

Conradson carbon ________ Corrosion (copper strip) ________

Is the material corrosive to carbon steel? ________ Corrosive to brass? ________

What alloy is recommended for carrying the fluid? ________

Distillation data (if applicable) 10% at ________ °F; 90% at ________ °F; end point ________ °F.

Flash point ________ °F; fire point ________ °F; pour point ________ °F.

Viscosity ________ SSF at 122°F or ________ SSU at 100°F.

pH ________; acid number ________; base number ________

Heating value ________ Btu/gallon. Specific gravity (H_2O = 1.0) ________

Will the material burn readily? ________

Toxic? (explain) ________

Figure 5.38b Liquid waste analysis sheet.

valves, piping and tanks) may have to be built of corrosion-resistant materials. The problem is even more acute with burner components and refractories, because high temperatures in the combustion chamber accelerate corrosion rates.

Many liquid wastes carry fine solids in suspension. If these solids are not filtered or settled out beforehand, they can cause premature failure of pumps and clogging of valves and burner orifices.

Special Purpose Solid Waste Incinerators

Type 6 solid wastes present the greatest challenge to the incinerator designer. Many of these materials have extremely low heating values and, thus, are unable to contribute any significant amount of heat to the incineration process. Consequently, higher than normal auxiliary inputs are required. Other materials are capable of such rapid, intense combustion that they must be fed to the incinerator and burned under very close supervision to avoid severe thermal damage to the incinerator or explosive combustion reactions.

Some materials, typified by many types of plastics and synthetic rubbers, decompose at elevated temperatures, forming generous amounts of free carbon and complex organic molecules. Destruction of these products in secondary combustion chambers is difficult without extremely high auxiliary heat input or turbulence, or both. In some cases, final burning stages are carried out in a third combustion chamber or by recycling the gases through the first incineration chamber.

Many solids, by virtue of their size, shape or compactability, cannot be evenly distributed on the floor or grates of the primary incineration chamber and nonuniform burning results. Grading or size reduction may be necessary prior to incineration. (See Section III.2.11.)

Incinerator selection for type 6 wastes must be the result of a thorough study of the characteristics of the waste and of a complete understanding, by both the purchaser and supplier, of the performance requirements.

Operation of the Incinerator

Even the best of incinerators will malfunction if neglected or operated carelessly. Below are a few basic rules for obtaining optimum performance:

1. Do not exceed design load limits. The temptation to overload hand-charged incinerators is particularly strong.
2. Keep all blowers, burners, grates and refractory surfaces clean.
3. Repair or replace refractories as needed.
4. Be sure all operating and safety controls are in good working order.
5. Consult the designer or manufacturer of the incinerator before attempting to use it to burn materials for which it was not specifically designed.

BIBLIOGRAPHY

"Incinerator Standards," Incinerator Institute of America, New York, New York.

"NFPA Standard 82, Incinerators and Rubbish Handling," National Fire Protection Association, Boston, Massachusetts.

Dispersion of Final Exhaust

(Sections 5.39 to 5.45 are devoted to this discussion.)

5.39 FLARE SYSTEM TYPES, FEATURES AND OPERATION

Types of Flare Systems:	Elevated combustion flares can be inspirated by steam using multiple jets, single nozzle or shroud or can also utilize mechanical air blowing. Ground combustion flares can use water injection or air inspirating burners. Burning pits are used in emergencies.
Design Pressures:	3 PSIG in the header for all types except the air inspirating ground burners, which require 7 PSIG in the header.
Materials of Construction:	Elevated flares have alloy tips (burners) and carbon steel stacks. Ground flares have alloy steel burners and carbon-steel refractory-lined stacks. Burning pits have alloy steel burners and refractory or concrete-lined earthen walls.
Gases Handled:	All types will handle saturated and unsaturated gaseous hydrocarbons, plus hydrogen. Gases must be free of liquid droplets and excessive quantities of noxious inorganic contaminants.
Flow Range:	Elevated flares usually handle up to 1,200,000 lbm/hr of vent gas. Ground flares generally are smaller than 100,000 lbm/hr but could also be larger. Burning pits are unlimited in size.
Costs:	See Section 5.42 for first approximation only.

5.39 Flare System Types, Features and Operation

Partial List of Suppliers: Flaregas Corporation; John Zink Company; National Airoil Burner Co.; Smoke Ban Manufacturing, Inc.

Most plants in the petroleum refining and chemical processing industries must dispose of small continuous hydrocarbon waste gas streams from the process units. These streams could be burned in process furnaces; however, provision must also be made to handle huge quantities of process gases released by pressure relief valves during emergency operating conditions. These emergency conditions occur due to compressor failures, loss of cooling water, vessel overpressure, power failures, fires and the like. It is common practice to tie all the continuous waste gas streams and emergency relief valve outlets into a common header system which vents to a combustion flare.

Environmental Requirements

Most air pollution problems can be avoided if the waste gases are completely burned in the flare stack before release. However, flaring is not a satisfactory solution for continuous waste gas streams which contain excessive quantities of inorganic contaminants such as sulfur, chlorine and fluorine. In these instances, pretreatment is required to remove the objectionable materials before flaring. As discussed in Chapter 1, the pertinent federal, state and local air pollution laws must be studied to determine the maximum quantity of contaminants that may be emitted in each case. Even when the waste gases are free of toxic contaminants, careful consideration must be given to minimize the formation of smoke, caused by incomplete combustion, and noise, due to rapidly escaping gases and steam at the flare tip. Many states now require flares to operate below Ringelmann no. 1 (see Section 1.5 for definition), except during extreme emergency conditions of short duration. Noise especially can be a problem if the plant is located close to a residential area.

Types of Flares

Flares can be classified as follows:

1. Elevated combustion flares
2. Ground combustion flares
3. Burning pits
4. Elevated vent stacks

The fourth type is a nonburning blowdown stack similar to those discussed in Section 5.42 and, therefore, will not be covered here.

Elevated Combustion Flares

Elevated flares are the most commonly used. The combustion tip is usually 100 to 300 feet above grade, which drastically reduces the effects of heat radiation. Consequently, the flare can be located close to the process units.

This minimizes the amount of vent piping and reduces land requirements. The extra height gives the added advantage of better dispersion of the combustion products than with ground flares. Elevated flares can be sized to handle up to 1,200,000 lbm/hr waste gases with smokeless combustion over most of the operating range.

The height of elevated flares can be determined by detailed calculations.[1,2] The basic objective is to prevent excessive radiation from harming employees or from damaging equipment. Once the minimum height has been determined for radiation protection, dispersion calculations (Section 5.45) should be made to establish that ground level contaminant concentrations will meet the ambient air standards. If they do not, the height is then adjusted upward accordingly.

The elevated flares, depending upon the method of achieving smokeless combustion, utilize (1) air inspiration with steam or (2) mechanical air blowing.

Both can be built as self-supporting units, derrick supported or guyed stacks.

When discharging hydrocarbon vapors from a flare, it is difficult to achieve the mixing required with the surrounding air to obtain the proper fuel-to-oxygen ratio for complete combustion. Due to the high temperature, cracking and polymerization reactions occur in the unburned portions of the gas and soot is formed. This condition is especially serious if the waste gas contains unsaturated hydrocarbons such as olefins, diolefins and aromatics. These type compounds contain double-bonded molecules which are very susceptible to cracking. The tendency to smoke increases as the hydrogen-to-carbon ratio diminishes and as the molecular weight increases, although not in a directly proportional relationship.

Air Inspiration with Steam

By injecting steam into the flare tip, the tendency to smoke can be greatly reduced or even eliminated, since the steam has an inspirating effect and draws large quantities of air into the combustion zone. This supplies the necessary oxygen for burning, provides intense mixing and has a diluent cooling effect which reduces cracking and polymerization. Steam also has the ability to react with carbon particles in accordance with Equation 5.39(1).

$$C + H_2O \rightleftharpoons CO + H_2 \qquad 5.39(1)$$

The amount of steam required for a specific application depends on the type of hydrocarbon being burned, the efficiency of mixing with the waste gas and air, plus the degree of smokeless combustion desired.

The principal methods for injecting steam into flares involve the use of multiple jets, single nozzle or a shroud.

A flare tip with multiple jets is shown in Figure 5.39a. As illustrated, the waste gases are exhausted from the open end of the flare tip. A large header located around the periphery of the tip distributes steam to several jets. The jets are so oriented that their discharge covers the tip and creates turbulence and mixing of the waste gases with the surrounding air. It is reported[3] that steam consumption at peak loads is 0.2 to 0.5 pounds of steam per pound of waste gas, depending upon the hydrocarbon being burned. This relatively low steam consumption is usually balanced against the maintenance costs which are slightly higher than for the single nozzle design. A small bleed of steam is required at all times to provide cooling and to prevent plugging. The flare tip is constructed of corrosion-resistant alloy steel and can be equipped with auxiliary fuel gas burners for applications requiring endothermic combustion.

Figure 5.39b shows the single steam nozzle type flare tip, where the steam line enters the flare and continues upward in the center until it terminates several inches below the top of the tip. As the steam exits the supply line, it expands to fill the inside of the flare tip and, in so doing, mixes with the waste gas. The turbulence created is not as great as with multiple jets; consequently, the steam utilization efficiency is lower. However, the system requires less maintenance due to its simple design.

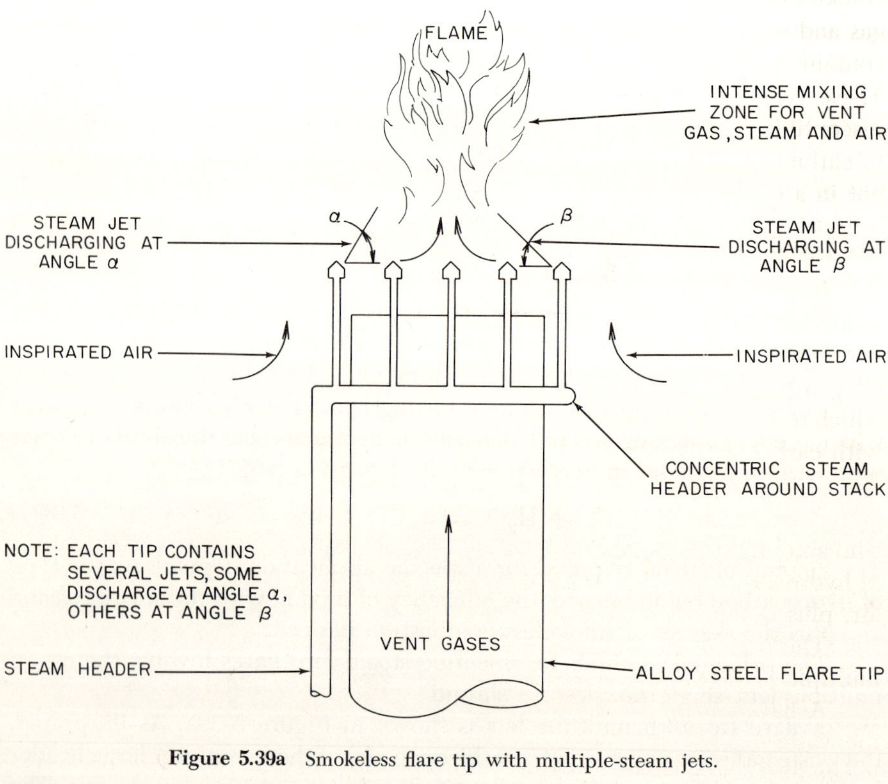

Figure 5.39a Smokeless flare tip with multiple-steam jets.

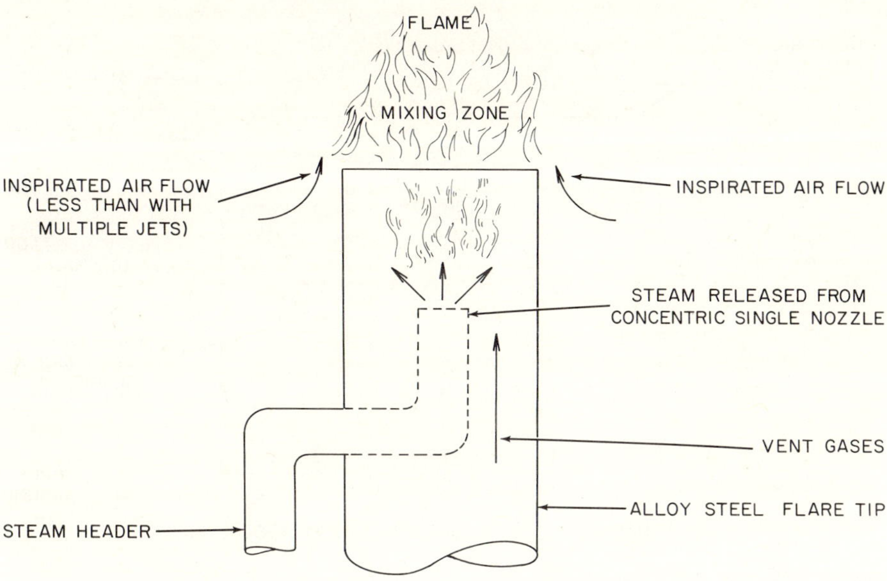

Figure 5.39b Smokeless flare tip with single-steam nozzle.

In the shroud type design (Fig. 5.39c), the flare tip is surrounded by a metal skirt or shroud. This reduces some of the cross-wind effects and forms a turbulent zone for premixing of the air and steam. Waste gas exits radially from the center portion of the tip and travels toward the shroud. Simultaneously, steam and inspirated air move vertically through the shroud, causing intense mixing with the vent gas. The steam utilization efficiency of this design is comparable with that of the multiple steam jet type.

MECHANICAL AIR BLOWING

Blowers can be utilized to provide air for smokeless combustion of small waste gas streams in elevated flares. However, for gas rates over 100,000 lbm/hr, the amount of air required makes the equipment and line sizes very large. The capital investment is usually not competitive with steam inspiration systems, if steam is available. In some instances, mechanical air blowing flares have been built to handle large emergency vent rates with only the first portion of the operating range having been designed for smokeless combustion.

Ground Combustion Flares

As the name implies, ground flares are built near grade level and seldom exceed 60 feet in height. Consequently, heat radiation effects require them to be limited in size and located remotely from the process areas. This raises

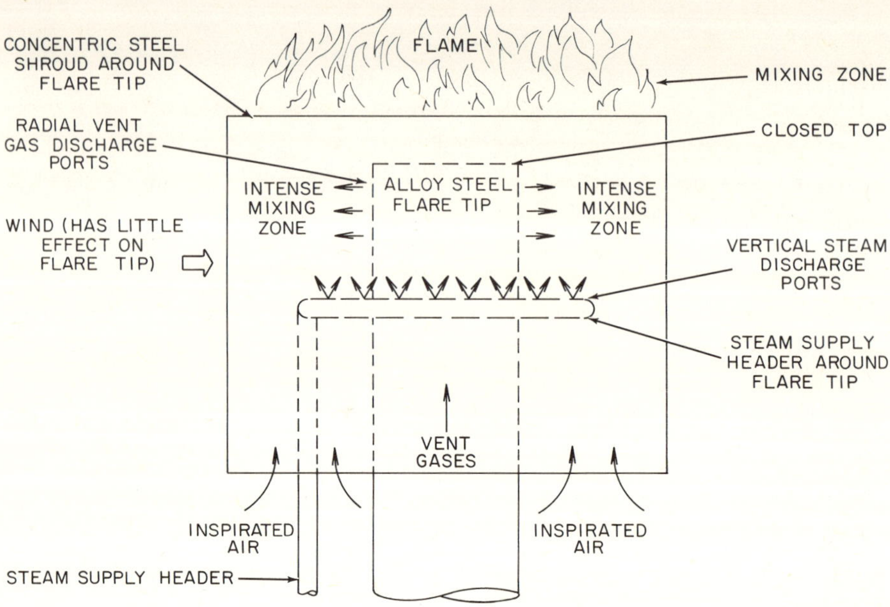

Figure 5.39c Smokeless flare tip with steam shroud.

the piping costs and eliminates them from consideration in plants with little available space and high-vent gas rates. Perhaps their greatest application is for locations where elevated flares would be unsightly and complete smokeless operation is not required.

Ground flares offer an important advantage in that water can be substituted for steam in many cases. This greatly reduces operating costs; and for plants with a shortage of excess steam capacity, it can be a decisive factor. However, small orifice sizes and high water pressures are required for the spray nozzles. As the water requirement increases at high vent gas rates, it becomes increasingly difficult to obtain satisfactory combustion. Therefore, smokeless operation is limited to a maximum of 100,000 lbm/hr gas flowrate.

The use of water sprays is not practical with elevated flares because a very high pump discharge pressure would be required for the water line. Plugging and freezing would also be difficult to control with the nozzles inaccessible for routine inspection. In addition, high winds would be more prone to sweep atomized water away from the burner in an elevated tip.

A typical water injection flare is composed of three concentric stacks as shown in Figure 5.39d. The innermost stack contains the burner and water atomization nozzles. The second stack is slightly larger and serves to confine the tiny water droplets for effective mixing with the incoming air and the vent gases. The outer stack merely directs the flame upward and protects

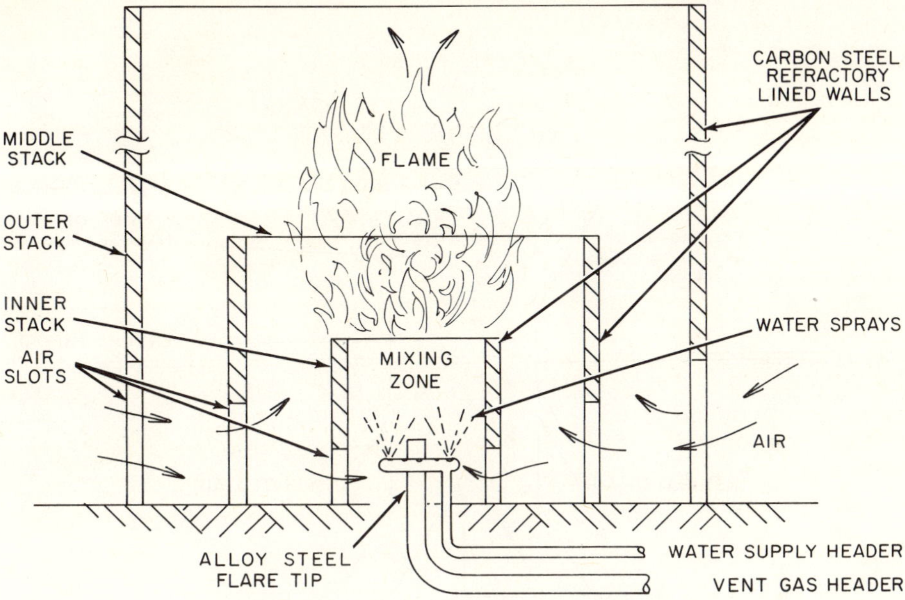

Figure 5.39d Typical water spray ground flare.

against cross winds. Slots are provided near the base of all three stacks to allow the entrance of air by natural draft.

Ground flares can be designed to handle higher vent gas rates by using air inspirating venturi burners. In this design, the back pressure is built up in the vent gas header and then is released through the burners. The energy of the vent gas serves as the motivating force to draw air into the system, similar to a conventional ejector as shown in Figure 5.39e. However, this technique cannot be used in many cases because it requires 1 to 4 PSIG pressure at the burner[4] and even greater back pressures on the relief valve discharge headers (about 7 PSIG) in the process units.

To handle a wide range of vent gas rates, several burners are required. At low rates, only a small venturi is operational. As the waste load increases and the pressure builds, other burners can automatically be placed into service by pressure controlled snap-action valves. This auxiliary control equipment is a significant cost item. A major drawback of the system is that it cannot handle vent rates which substantially differ from the design basis.

Burning Pits

Burning pits are excavated units with alloy steel burners along one or more sides. The walls are usually concrete or refractory-lined. Dense clouds of smoke are evident and the combustion products are not dispersed suffi-

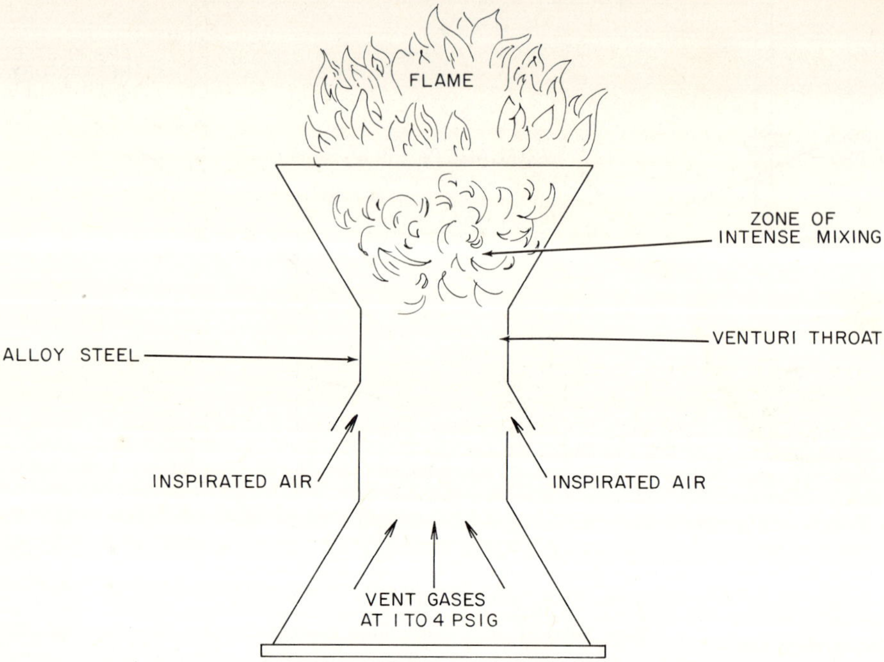

Figure 5.39e Typical air inspirating venturi burner.

ciently. Such pits are generally unacceptable except as a device to handle catastrophic emergency situations.

REFERENCES

1. Kent, George R.: Find Radiation Effect of Flares, *Hydrocarbon Processing,* vol. 47, no. 6, pp. 119–130, 1968.
2. Tan, Soen H.: Flare System Design Simplified, *Hydrocarbon Processing,* vol. 46, no. 1, pp. 174–176, 1967.
3. "Air Pollution Engineering Manual," Air Pollution Control District—County of Los Angeles, U.S. Public Health Service Publication No. 999-AP-40, p. 587, 1967.
4. "Manual on Disposal of Refinery Wastes," vol. II, "Waste Gases and Particulate Matter," American Petroleum Institute, p. 44, 1957.

5.40 FLARE ACCESSORIES AND INSTRUMENTS

The successful operation of a flare depends as much upon the accessories and instrumentation provided as upon the proper selection and sizing of the stack and combustion tip. Generally, the accessories are similar for both elevated and ground flares.

Vent Header and Blowdown Drum

Before reaching the flare, all vent gases should pass through a liquid disengaging drum. Otherwise, flammable liquids will spew from the flare tip and set fire to the surrounding area. The drum should be located at such a distance from the flare that the heat radiation received never exceeds 1200 BTU/hr/sq ft. A typical drum and vent gas collection system is shown in Figure 5.40a.

The knock out drum is usually sized to provide adequate disengaging space while being able to hold the total expected liquid load at maximum relief conditions. As an added safety precaution, the recovered slop oil should be automatically removed from the drum as it accumulates. This can be achieved by using steam-driven pumps which are actuated by liquid level switches on the drum.

The header should be sized to prevent excessive back pressure on the relief valves, yet the vapor velocity should be high enough to avoid excessive piping costs. Usually a design figure of about 200 ft/sec is satisfactory at flowing pressure and temperature.[1] All piping is sloped for drainage to the knock out drum.

Ignition and Pilot Systems

If the exit gas rate from a flare is gradually increased, a velocity will be reached where the flame physically separates from the tip. This is known as the "blow-off velocity." As the flowrate is further increased, the flame becomes increasingly unstable until it is eventually extinguished by atmospheric dilution and cooling. The "blow-out velocity" is dependent upon the heating value of the gas being burned and on atmospheric conditions. The optimum velocity for design purposes has been a matter of considerable debate. In many cases, blow-off occurs at about 200 ft/sec. However, the addition of pilot igniters and burners has allowed satisfactory operation at

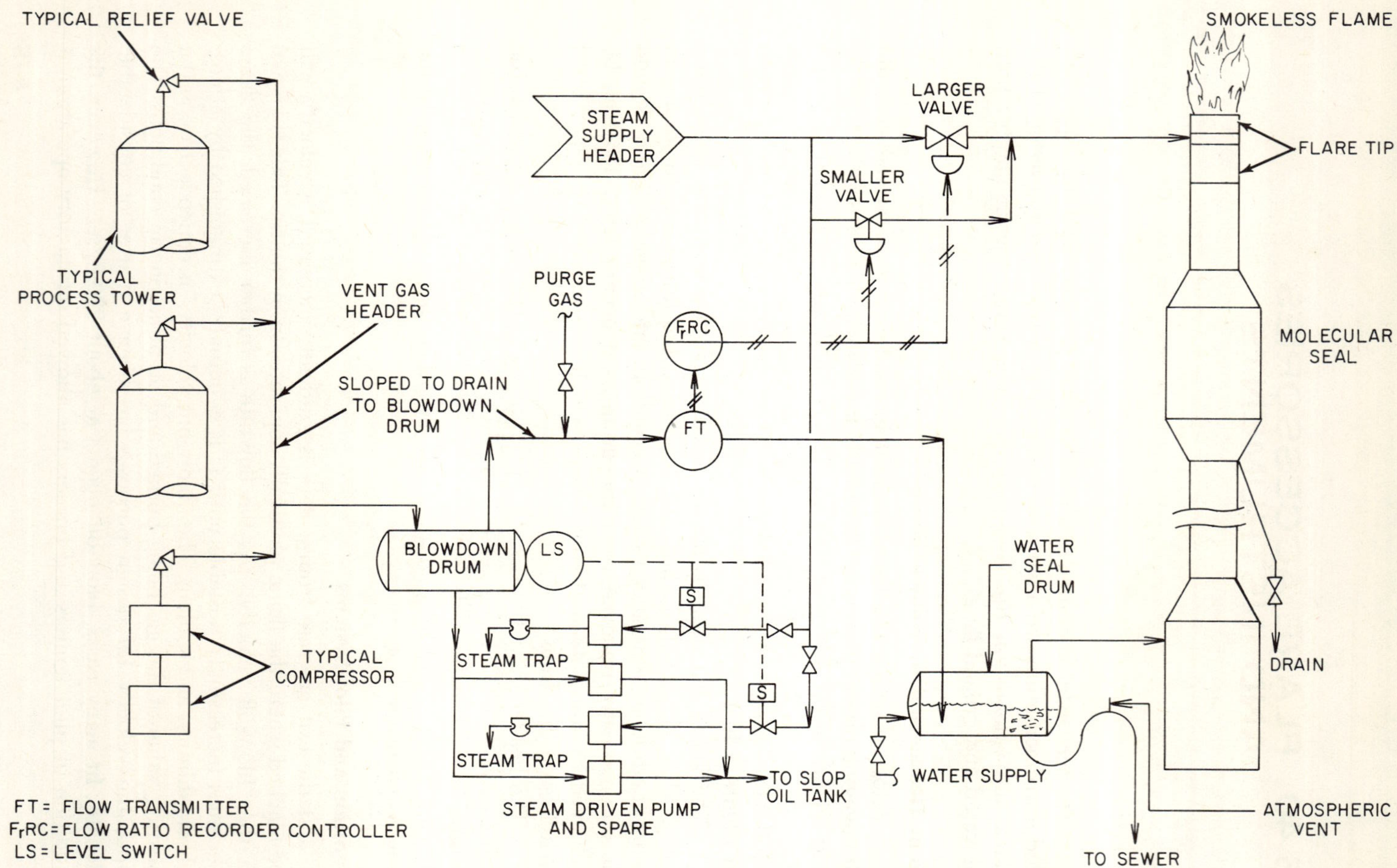

Figure 5.40a Typical vent gas collection and steam control systems.

over 400 ft/sec. The optimum velocity in most cases is probably about 300 ft/sec.

Usually more than one pilot burner is mounted on a flare tip. Each operates with a steady flame produced by the uniform combustion of fuel gas and is relatively unaffected by the variable operating conditions in the main burner. Pilot burners are so positioned that the flare tip remains lighted as long as waste gas is being vented.

A remotely located pilot igniter is used, which is capable of creating a flame and propagating it up to the pilot burners. This is accomplished by mixing air with fuel gas and passing it through a spark generator as shown in Figure 5.40b.

Methods of Preventing Flashback

Low-vent gas rates cause even more concern than high rates, when the material being handled is lighter than air. As zero flow is approached, the static pressure in the base of the flare drops below atmospheric pressure. This causes air to be drawn into the system and the flame flashes back into the stack and vent header, causing violent explosions. At all except very low vent gas rates, sufficient back pressure is developed in the base of the stack to prevent air from entering the system.

An obvious solution to the problem is to intentionally vent sufficient noncondensable inert gas through the system to prevent flashback. However, a velocity of at least 3 ft/sec[1] is usually required; and even if fuel gas is utilized as the purging medium, the operating cost becomes very high for large flares. Consequently, purge gas alone is not an acceptable answer in many cases.

Purge gas requirements can be reduced to only 10% of normal, if a molecular seal is installed just below the flare tip as shown in Figure 5.40a. The seal utilizes the buoyancy of the light purge gas to create a zone of greater than atmospheric pressure which prevents the entrance of air.* Unlike conventional labyrinthine flame arrestors which are subject to plugging, molecular seals contain no obstructions to flow of the vented gases.

Another method commonly utilized is to install a water seal in the vent header between the knock out drum and flare tip. This creates a small back pressure in the header and will work satisfactorily if the vented gases are forced to pass through the water blanket as *discontinuous bubbles*. Care must be taken in design of the seal to prevent surging or pulsations in the gas flow which would affect flame stability.

Automatic Control of the Steam-to-Vent Gas Ratio

Since steam is expensive and the vent gas rate is highly variable, it is desirable in many cases to automatically control the steam addition. This

*This trap prevents the lighter-than-air gas from bleeding out of the system and being displaced by air.

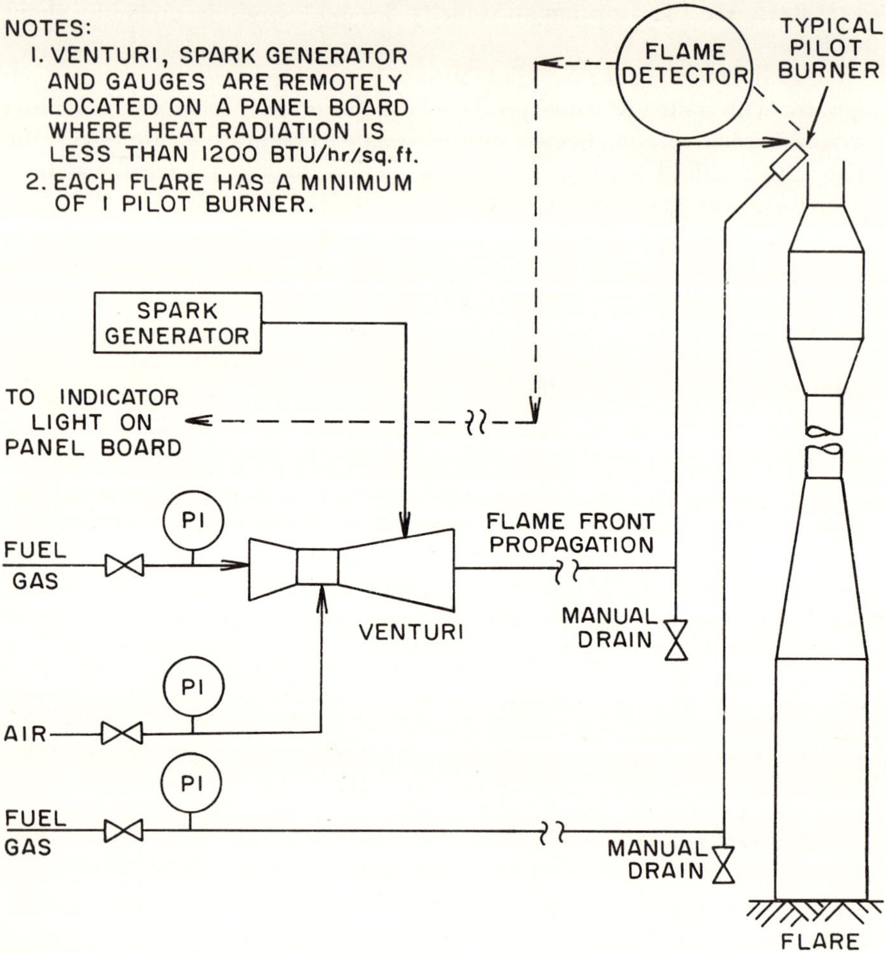

Figure 5.40b Typical pilot igniter and burner.

minimizes steam wastage and achieves smokeless operation at all times (Fig. 5.40a).

Automatic control would be simple if accurate measurement of the vent gas flowrate could be made over the complete flow range. Thermal flow sensing devices[3] recently developed have helped to improve this situation. One of the more accurate conventional measuring devices is an inverted weir located in the vent line inside the water seal drum.[2] Pitot tubes[3] have also been utilized for flare gas flow measurement, but have generally been unsatisfactory at low flowrates.

Flare accessories and instrumentation are required to ensure safe, economical, smokeless operation over the entire vent gas flow range. They add appreciably to the capital investment but significantly reduce operating costs.

REFERENCES

1. Reed, Robert D.: "Design and Operation of Flare Systems," *Chemical Engineering Progress,* vol. 64, no. 6, p. 54, 1968.
2. "Manual on Disposal of Refinery Wastes," Vol. II, "Waste Gases and Particulate Matter," American Petroleum Institute, p. 49, 1957.
3. Lipták, B. G. (ed.): "Instrument Engineers' Handbook," Vol. I, Chap. V, Chilton, Philadelphia, 1969.

5.41 FLARE SELECTION AND APPLICATION

An orientation to assist in the selection of flare systems is provided in Table 5.41a. Most plants have high-vent gas rates during emergency conditions. This usually means that smokeless water spray ground flares are too small. Ground flares with air inspirating venturi burners create high back pressures in the vent header and on the relief valves in the process units. Burning pits are not smokeless and can be used only in the event of a major emergency condition or disaster. Mechanical air blowing usually allows smokeless combustion only in the lower portion of the operating range. Consequently, steam injection elevated flares are the best choice in most cases. The single steam nozzle type is the simplest in design and maintenance; however, the multiple jet and shroud types give more efficient steam utilization.

Ground flares are particularly suited for burning small to intermediate-sized streams of saturated hydrocarbons when steam is unavailable. If the plant is near a residential area or highway, ground flares may also be desirable from a public relations viewpoint, since they emit less light and noise. However, water injection ground flares are prone to smoke unless good atomization can be achieved.

It is virtually impossible to provide meaningful, generalized cost data on flares. Many times, the accessories required for smokeless combustion are more expensive than the flare itself. Ground flares have been reported to be much more expensive than elevated flares; however, the addition of steam lines and elaborate control equipment has changed this situation considerably. The only reliable method of obtaining cost information for a specific case is through actual vendor contacts.

One of the major problems associated with flares today is the noise emitted during emergency venting. However, developments have been reported recently on steam nozzle[1] improvements and noise shields. Additional work will undoubtedly be completed in the near future as attempts are made to solve this problem.

REFERENCE

1. Hersh, Alan S. and James G. Seebold: "Control Flare Steam Noise," *Hydrocarbon Processing*, p. 140, February 1971.

Table 5.41a
ORIENTATION TABLE FOR FLARES
Operating and Design Characteristics

Design Types			Typical Stack Height, ft	Typical Stack Diameter, in	Radiant Heat Effects Per Unit of Vent Gas Burned	Space Requirements	Luminosity	Noise At Ground Level During High Flaring Rates	Smokeless Capability When Burning: Saturated Hydrocarbons	Smokeless Capability When Burning: Unsaturated Hydrocarbons	Automatic Controls For Smokeless Combustion	Typical Vent Gas Capacity, lbm/hr	Maintenance Requirements	Dispersion of Combustion Gases
Elevated Combustion Flares	Air Inspiration With Steam	Multiple Jets[a]	100–300	12–60	M–I	M–I	M–H	I–H	G–E	F–E	A	Less than 1,200,000	I–H	G–E
		Single Nozzle	100–300	12–60	M–I	M–I	M–H	I–H	F–E	P–E	A	Less than 1,200,000	M–I	G–E
		Shroud[a]	100–300	12–60	M–I	M–I	M–H	I–H	G–E	F–E	A	Less than 1,200,000	I–H	G–E
	Mechanical Air Blowing[b]		100–300	12–48	M–I	M–I	I–H	I	F–E[b]	F–G[b]	A[b]	Less than 1,000,000	I–H	F–E
Ground Combustion Flares[c]	Water Injection[d]		15–60	<20 ft[f]	I–H	I–H	M–I	M–I	F–E[g]	F–G[g]	N[g]	Less than	I–H	P–F
	Air Inspirating Burners[e]		15–60 4–6	x[f]	I–H	I–H	M–H	M–H	F–E	F–G	A	100,000[h]	M–I	P–F
Burning Pits[i]			Below grade	Large as required	H	H	H	M–I	P	P	N	Large as required	M	P

E = Excellent; G = Good; F = Fair; P = Poor; N = None; M = Minimal; I = Intermediate; H = High; A = Available; X = Information Not Available

[a] Steam utilization efficiency higher than single nozzle type.
[b] Smokeless combustion generally possible only in the lower portion of the operating range.
[c] Usually attracts less public attention than other type flares since the flame is hidden.
[d] Some manufacturers recommend steam instead of water sprays due to lower maintenance and improved smokeless burning.
[e] Creates up to 7 PSIG back pressure in the flare header since 1 to 4 PSIG is required at the burners.
[f] Outer stack.
[g] Automatic control of the water sprays in proportion to the vent gas rate is very difficult; manual control is used.
[h] Information is not available on the maximum rate; however, it is considerably greater than 100,000 lbm/hr.
[i] Used only in major emergency or disaster.

5.42 STACK TYPES, FEATURES AND OPERATION

Stack Types:	(a) Small; (b) Tall. NOTE: The letters "a" and "b" in the feature summary below refer to the above two categories of stacks.
Stack Diameters:	1 to 3 ft (a); up to 36 ft (b).
Stack Heights:	Under 150 ft (a); over 200 ft (b).
Construction Features:	Refractory lined metal (a); constructed with separate liner (b).
Installed Cost, \$/ft of Height:	\$30 (a), 12-inch ID; \$35 (a), 18-inch ID; \$100 (a), 36-inch ID; See Figure 5.42a (b).
Partial List of Suppliers:	Air Pollution Industries, Inc. (a); BSP Division, Envirotech Corp. (a,b); Buell Engineering Co. (a); The Ceilcote Co. (a); Combustion Equipment Associates, Inc. (a); Joseph Goder Incineraters (a); Monsanto Enviro-Chem Systems, Inc. (b); Peabody Engineering Co. (a); Research-Cottrell, Inc. (a,b); Western Precipitation Division, Joy Manufacturing Company (a); Zurn Industries, Inc. (a,b).

Even with the utilization of the most sophisticated pollution control equipment, small quantities of pollutants will remain in the exhaust air. This cleaned exhaust air is dispersed into the atmosphere through a stack. It discharges the gases at a point sufficiently high that the pollutants are dispersed over a wide area. By this technique, the maximum concentration of odors and toxic components at the ground level is kept within acceptable limits. Stacks can also disperse particulate matter but they must have a considerably greater height than what is required for diluting gases.

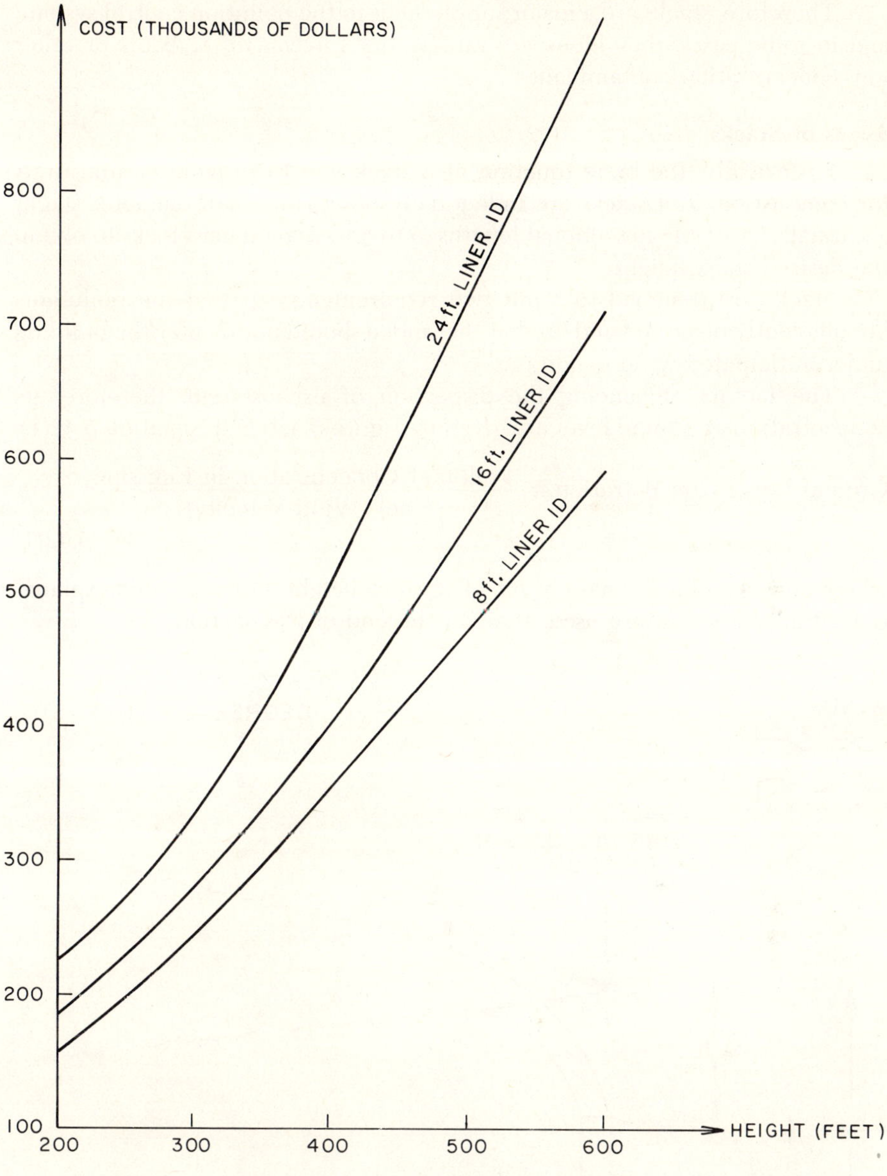

Figure 5.42a Installation costs for tall stacks.

Therefore, stacks are a major supplement to the pollution control system; and in some cases, they alone are satisfactory for control of odors or other low concentration contaminants.

Types of Stacks

Historically, the basic function of a stack was to provide natural draft for combustion. Tall stacks are individually designed. Small diameter stacks are usually available in sectional lengths (5 to 8 ft) as modular blocks to obtain the desired stack height.

Stacks are designed to fulfill two requirements: (1) that the pollutants be sufficiently dispersed and (2) that the smoke should not reenter the building under anticipated wind conditions.

The factors influencing the dispersion of exhaust and, therefore, its concentration at ground level are given in Figure 5.42b and Equation 5.42(1).

$$\text{Ground Level Concentration} \cong \frac{\text{Pollutant Concentration in Emission}}{(\text{he})^2(\text{Wind Velocity})} \qquad 5.42(1)$$

where he is the plume height or effective stack height. For a detailed explanation of all nomenclature used, refer to the end of this section.

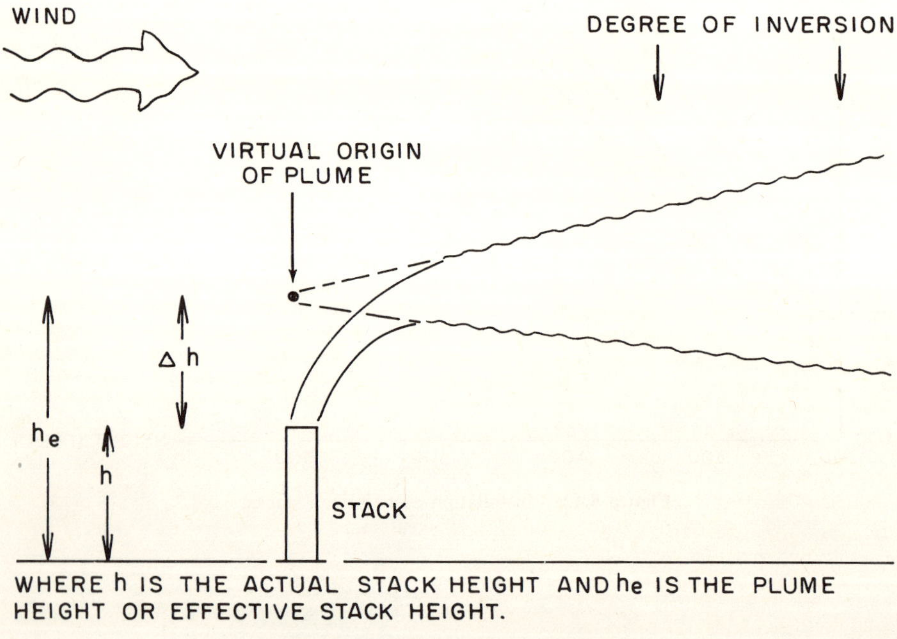

Figure 5.42b Dispersion of exhausts by stacks.

Effective Stack Heights

For stable atmospheres, a semiempirical correlation (developed by Bonasquet, et al.)[3] can be used to predict plume path and to determine effective stack height.

In this method, momentum rise and buoyancy rise (as functions of downwind distance) are calculated separately and added to obtain a total rise.

Momentum rise is given by

$$\Delta h_v = \Delta h_{v,max}\left[1 - 0.8\left(\frac{\Delta h_{v,max}}{x}\right)\right] \quad 5.42(2)$$

where $x > 2\Delta h_{v,max}$

$$\Delta h_{v,max} = \frac{4.77}{(1 + 0.43U/V_s)}\frac{\sqrt{Q_{v1}V_s}}{U} \quad 5.42(3)$$

Buoyancy rise is given by

$$\Delta h_b = \frac{6.37gQ_{v1}\Delta T_1}{U^3T_1}\left(\ln J^2 + \frac{2}{J} - 2\right) \quad 5.42(4)$$

where

$$J = \frac{U^2}{\sqrt{Q_{v1}V_s}}\left[0.43\sqrt{\frac{T_1}{g\left(\frac{d\theta}{dz}\right)}} - 0.28\frac{V_s}{g}\frac{T_1}{\Delta T_1}\right] + 1 \quad 5.42(5)$$

Effective stack height h_e is given by

$$h_e = h + \Delta h_v + \Delta h_b \quad 5.42(6)$$

Equation 5.42(6) is valid for plume rise in a neutral atmosphere and it can be applied for short downwind distance calculations.

Equation 5.42(7), developed by Holland,[4] for plume rise under neutral stability is valid for stack exhausts above 125°F:

$$\Delta h = \frac{V_s d}{U}\left[1.5 + 2.68\,(10^{-3})p\,\frac{T_s - T_a}{T_s}\,d\right] \quad 5.42(7)$$

This relation frequently underestimates effective stack height and, hence, provides a safety factor. Therefore, the result of Equation 5.42(7) can be reduced by 10 to 20% for stable conditions, while for unstable conditions this result should be increased by 10 to 20%.

According to the Briggs correlation,[5] the buoyant plumes follow the "⅔ law" for transitional rise downwind, when there is a wind (regardless of stratification). Plume rise is given by

$$\Delta h = 1.6F^{1/3}U^{-1}X^{2/3} \quad 5.42(8)$$

where

$$F = 0.25\ gV_sd^2\left(\frac{T_s - T_a}{T_a}\right) \qquad 5.42(9)$$

The plume rise ceases when

$$x = 3x' \qquad 5.42(10)$$

where

$$x' = 0.5F^{2/5}\,h^{3/5} \qquad 5.42(11)$$

Refer to Section 5.45 for further discussion of this topic, together with a treatment of diffusion calculations.

Diffusion of Pollutants

After determining the effective stack height, the concentration of pollutants at ground level can be estimated using Pasquill's relations as modified by Gifford.

The concentration C of gas or aerosols (particulates $<20\mu$) at a location defined by the x, y and z coordinates relative to a continuous source is given by

$$C(x, y, z; he) = \frac{Q_h}{2\pi\sigma_y\sigma_zU}\exp\left[-\tfrac{1}{2}(y/\sigma_y)^2\right]\left\{\exp\left(-\tfrac{1}{2}\frac{z - he}{\sigma_z}\right)^2 + \exp\left(-\tfrac{1}{2}\frac{z + he}{\sigma_z}\right)^2\right\} \qquad 5.42(12)$$

For concentrations calculated at ground levels (i.e., $z = 0$) and along the centerline of the plume (i.e., $y = 0$), Equation 5.42(12) simplifies to:

$$c\,(x, o, o; he) = \frac{Q_h}{\pi\sigma_y\sigma_zU}\exp\left[-\tfrac{1}{2}\left(\frac{he}{\sigma_z}\right)^2\right] \qquad 5.42(13)$$

Values of σ_y and σ_z vary with atmospheric conditions. The stability categories assigned to the atmosphere are given in Table 5.42c. Class A is the most unstable and class F is the most stable. For evaluations, the neutral class D is most often used. Having assigned a stability class to the atmosphere, σ_y and σ_z are evaluated using Figures 5.42d and e. These are valid for a sample time of about 10 minutes.

Utilizing these figures, a graphical profile of ground level concentrations (Fig. 5.42f) can be developed. The maximum ground level concentration (C_{max}) should be substantially below the requirements of the ambient air quality standards and the stack height should be increased if this is not so.[1]

The concentration levels of pollutants within the plant must also be within safe limits. It should be assumed that the plant personnel are exposed to the

Table 5.42c
STABILITY CHARACTERISTICS OF THE ATMOSPHERE

Surface Wind Speed, m per sec	*Day*			*Night*	
	Incoming Solar Radiation			*Thinly Overcast*	
	Strong	*Moderate*	*Slight*	*4/8 Cloud*	*3/8 Cloud*
2	A	A–B	B		
2–3	A–B	B	C	E	F
3–5	B	B–C	C	D	E
5–6	C	C–D	D	D	D
6	C	D	D	D	D

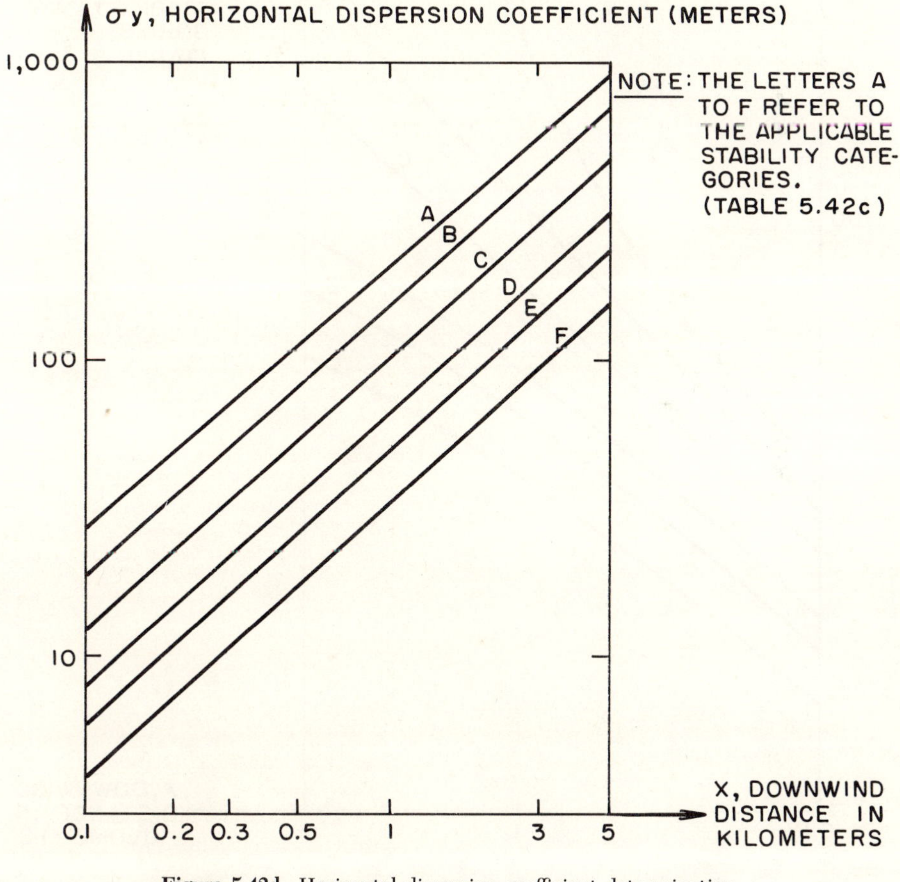

Figure 5.42d Horizontal dispersion coefficient determination.

concentrations of pollutants corresponding to their work stations for 40 hours per week. Occupational Health Standards[2] specify the permissible levels of various pollutants.

For other methods of estimating ground level pollution concentrations, refer to Section 5.45.

Legal Standards

For an overall treatment of air pollution laws and regulations, refer to Chapter 1. Stack height is usually part of the emission standards of the various

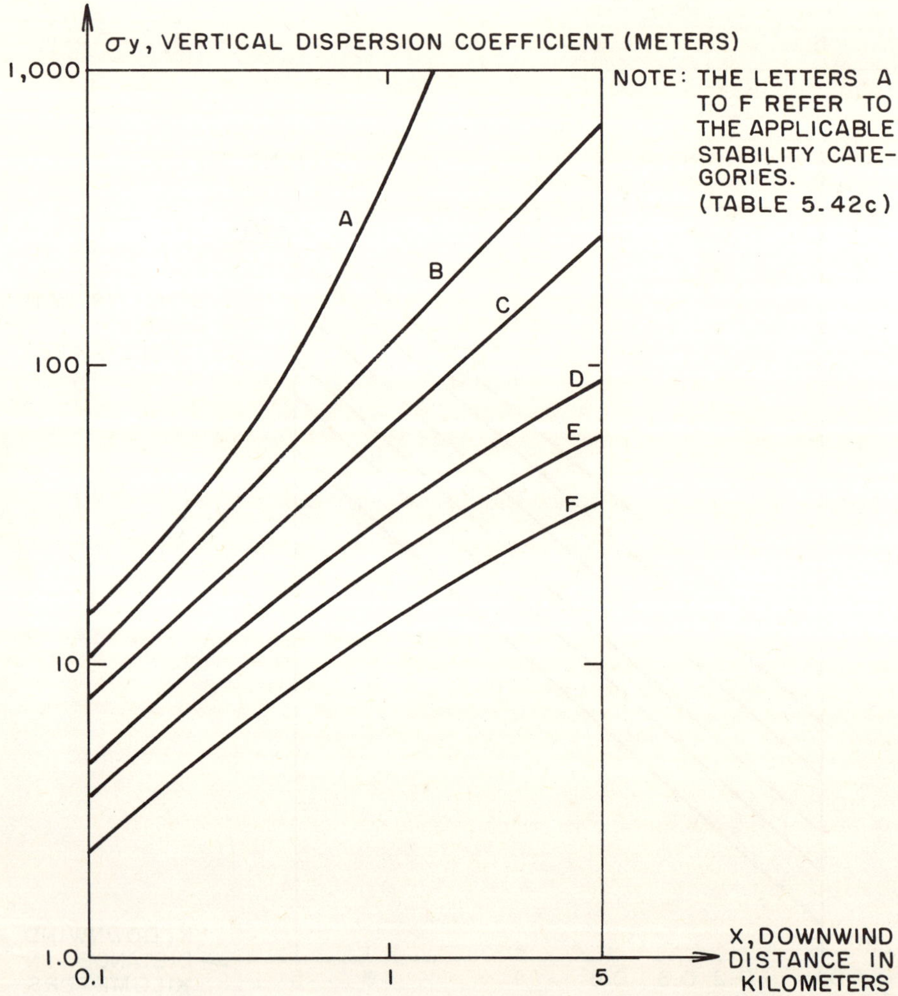

Figure 5.42e Vertical dispersion coefficient determination.

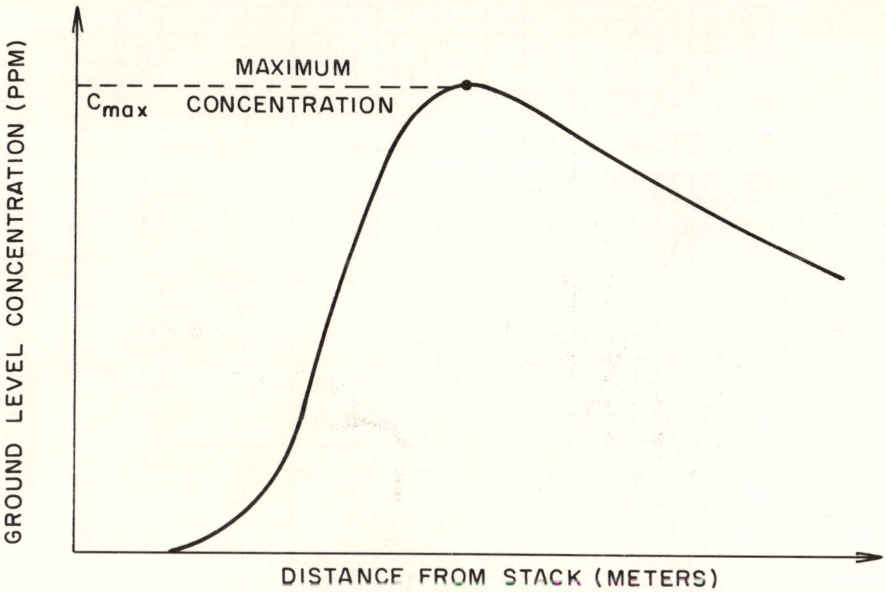

Figure 5.42f Distance-concentration profile.

regulatory agencies. Figure 5.42g and Tables 5.42h and 5.42i are illustrations of such regulations. These are shown only as examples and the applicable local codes should be consulted for design purposes.

Reentry of Stack Exhausts into Buildings

Generally, a stack that meets the requirements of ground level ambient air pollution standards is unlikely to cause reentry of stack exhausts into the buildings. On the other hand, in some cases due to building shape, topography and adjacent structures, extensive analysis by computations, field tests and wind tunnel tests may be required.

Exhaust discharges and equipment should not be enclosed in architectural fences for aesthetic reasons because it tends to trap contaminants and increase chances of reentry.

Because stacks should be designed and located for safe operation at all times, the prevailing winds cannot be relied on to provide safe dispersion.

A knowledge of airflow behavior around buildings is required for proper stack design.

As illustrated in Figure 5.42j, airflow over a building creates a positive pressure zone on the upstream side and a negative pressure zone (cavity) on the roof and downstream side.

The height of the roof cavity is maximum when the wind is perpendicular to the upstream side. This is the design basis condition. Wind tunnel tests

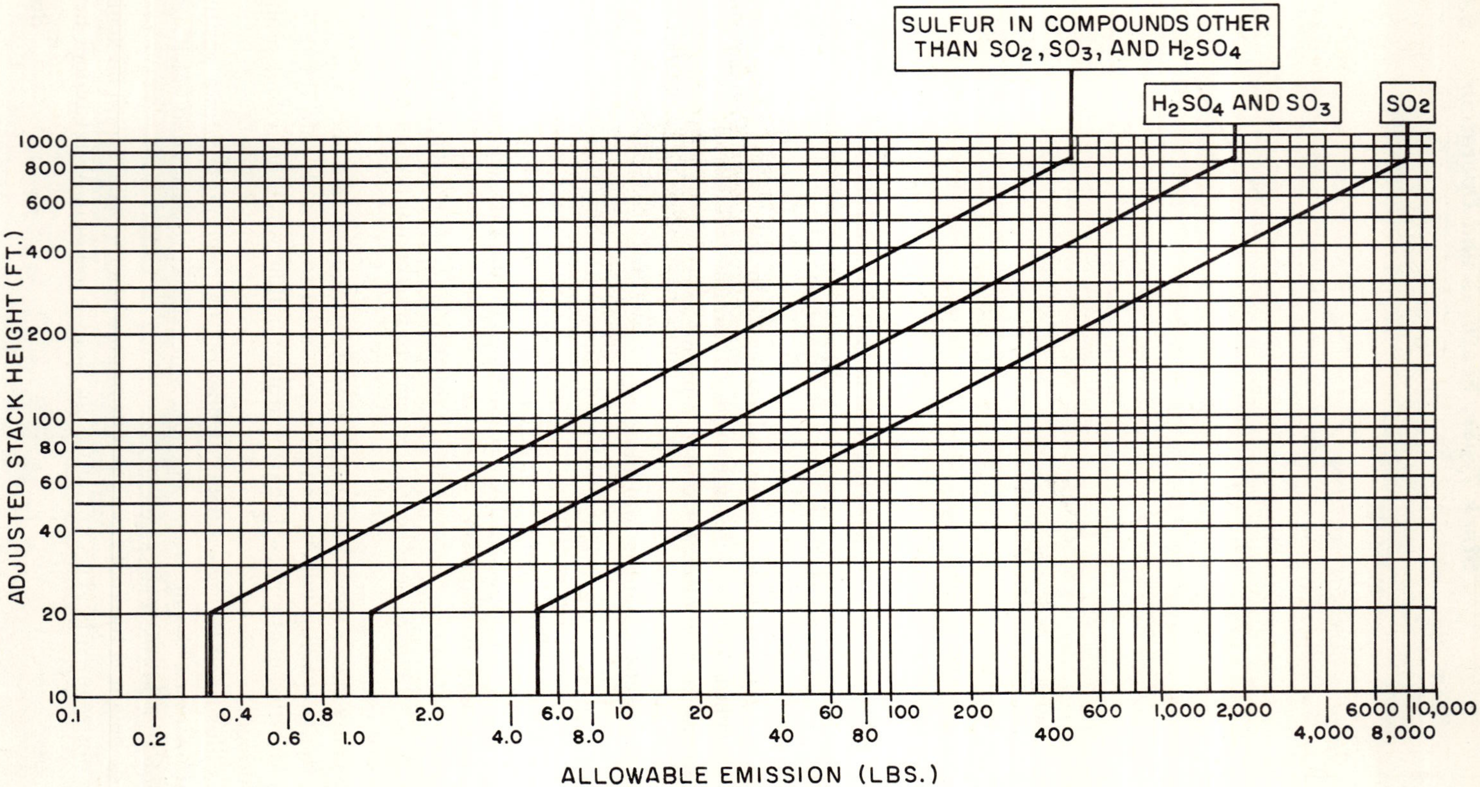

Figure 5.42g Allowable sulfur emissions as a function of stack height.

Table 5.42h
STACK HEIGHTS FOR NITRIC ACID PLANTS

Gas Volume at STP, ft³/min	*Effective Height, ft*	*Plume Rise, ft*	*Basic Chimney Height, ft*
14,000	205	27	180
28,000	287	39	250
42,000	353	47	300
56,000	412	55	350
84,000	468	68	400

Table 5.42i
STACK HEIGHTS FOR CEMENT PLANTS

Clinker Throughput, tons/hr	*Chimney Height, ft*		
	Wet Process	*Semidry Process*	*Dry Process*
30 and less	200	200	200
60	280	260	240
90	340	310	280
120	390	350	310
240	500	460	415
360	550	500	450

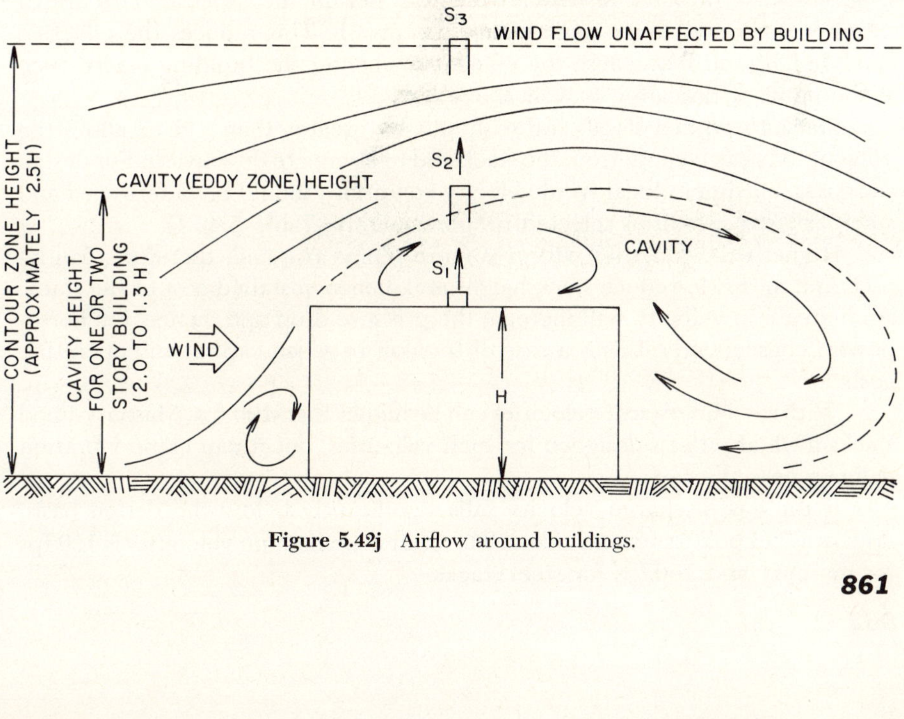

Figure 5.42j Airflow around buildings.

have shown that for perpendicular flow, the height of a roof cavity for a cubical structure will be $1\frac{1}{2}$ times the height of the structure (1.5H). A higher structure of the same width will not increase the cavity height over the roof, but a wider building of the same height (because of greater interference with wind flow) will have a higher cavity dimension.

The contour zone height (2.5H) is that height above a building where there is no further displacement of the airflow. Discharge above the contour zone (S_3 in Fig. 5.42j) is vital for high emission rates, for highly toxic effluents and for nuisance effluents such as mercaptans, which have extremely low odor perception concentrations.

The size and shape of the building airflow contour zone and cavities do not change appreciably with changes in wind velocity, but the pressures do change within these zones.

Topography and nearby structures may disturb the airflow contours around a building and create unfavorable discharge conditions. Contaminants released at roof level (S_1 on Fig. 5.42j) spread over the whole building and reenter the ground on the lee side. Discharge at the cavity height elevation (S_2) still usually results in poor dispersion. Dispersion equations do not apply to S_1 or S_2. Rounding a building outline will decrease the heights of the contour and cavity zones over the roof. Thus, the heights of these zones over the Houston Astrodome, for example, would be relatively low.

Stack Exit Velocities

Wind flow over stacks creates negative pressure zones and eddies (tip and trailing vortices) in the same manner as does airflow around buildings (Fig. 5.42k). Low stack discharge velocities permit the exhaust to be drawn into the stack eddy zone and cause a downwash. This reduces the effective stack height and may cause the effluent to reenter the building cavity even if the point of discharge is well above it.

Stack to wind velocity ratio should be greater than 1.5 to allow the effluent to break cleanly from the stack and to eliminate downwash. For design purposes, maximum local wind velocity conditions should be determined and stack exit gas velocities selected proportionately (Table 5.42 1).

Higher exit velocities will give more plume rise and, therefore, can be used as a means to reduce stack height and, hence, installed cost of the stack. An increase in velocity will increase the pressure drop and, hence, fan horsepower; consequently, this is a case of tradeoff between capital and operating costs.

With steel liners, exit velocities can be higher than 100 fps. Masonry-lined stacks have also been designed for such velocities, but it can cause vibration damage near the top.

If the gas discharge velocity falls significantly below the design basis, downwash of pollutants may occur. The limit on maximum velocities is 120 fps for masonry and 150 fps for steel stacks.

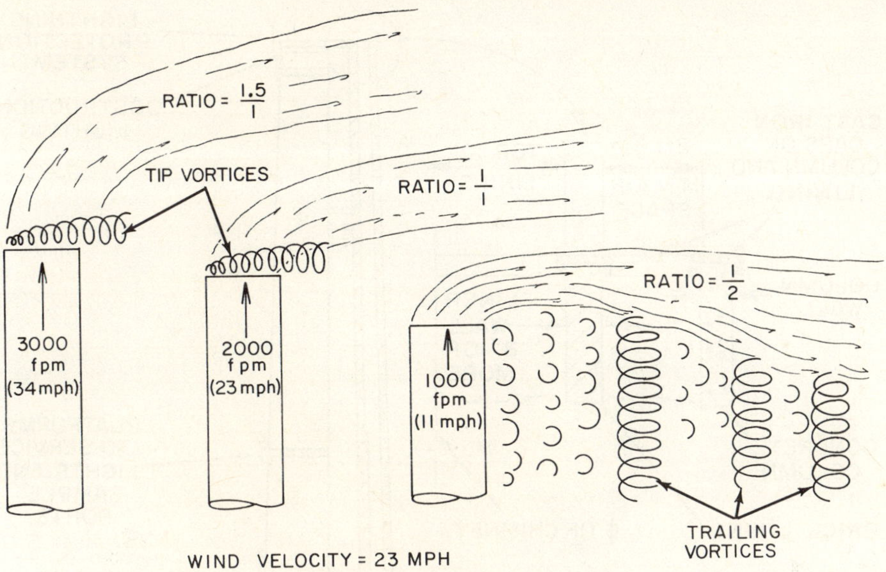

Figure 5.42k Airflow over stacks.

When one stack handles the exhaust from more than one group of equipment, discharge volume variations due to shutdowns may be substantial. In such cases, partitioned liners with separate flues for each source (group of equipment) have been used. If variations in flow are due to process changes, compensating air bleeds can be employed. If the stack is designed for some future capacity, a nozzle at the top of the stack can be used to increase the exit velocity.

Design Considerations

Small stacks are fabricated of steel or refractory-lined steel. These do not involve complicated construction because guy-wires are used to ensure mechanical stability.

Table 5.42 l
RECOMMENDED STACK EXIT VELOCITIES

Wind Velocity, MPH	*Minimum Stack Exit Velocity, ft/min*
10	1,300
15	2,000
20	2,600
25	3,300
30	4,000

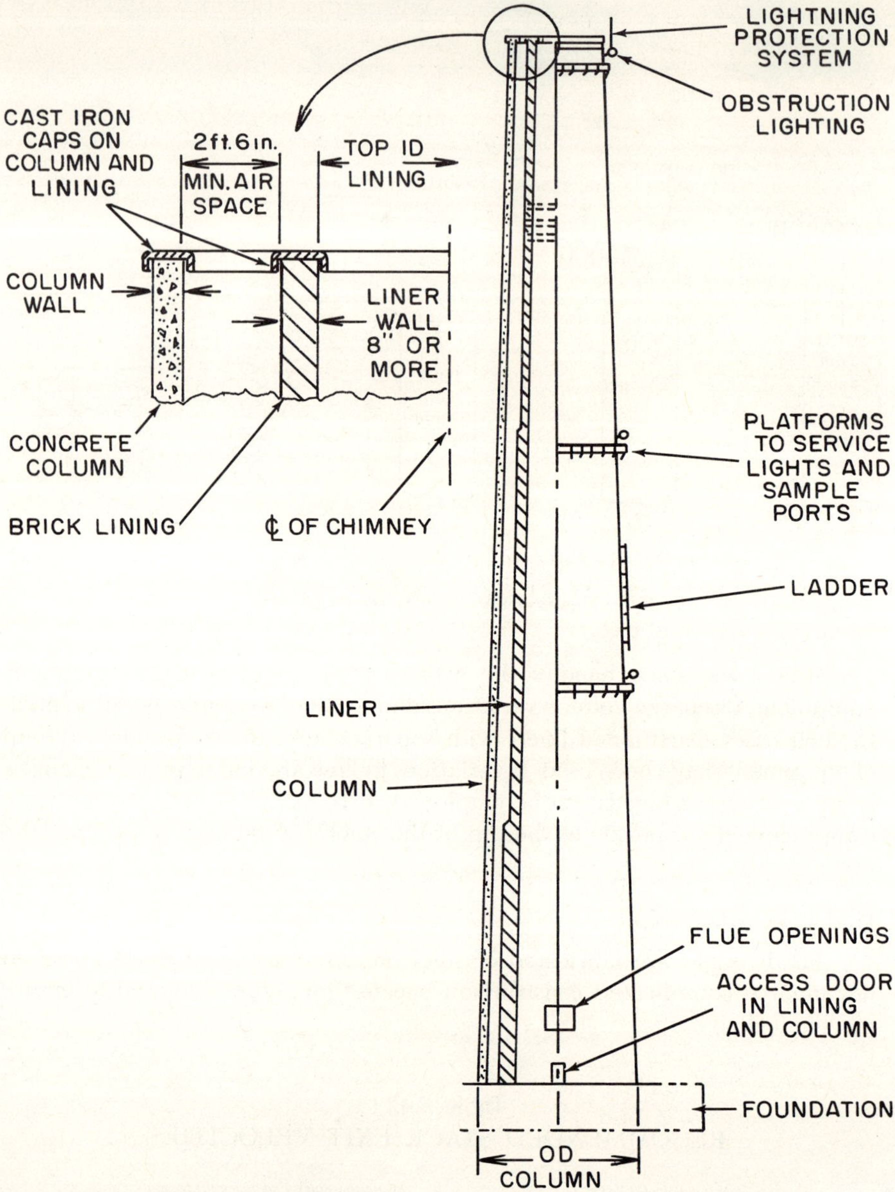

Figure 5.42m Tall stack construction.

In tall stacks (Fig. 5.42m), the liner is designed to withstand the temperature and corrosion of the exhaust gases, while the outside column protects and sometimes supports the liner. Free standing liners are increasingly used where the column serves only to shield the liner from elements and to absorb the wind load.

Liners

Liners may be made of steel or masonry. A steel liner is impervious provided it does not corrode and is more resistant to tremors and vibrations than is masonry. On the other hand, it usually is more expensive except in very tall stacks and requires the temperature of the exit gases to be above the dew point of the corrosive constituents in the gases. When steel liners are used, the top 20 to 50 ft of the liner is made of stainless steel, while the rest is made of a copper-bearing carbon steel.

Masonry liners are more commonly used in the chemical industry. Acid proof brick can withstand chemical attack and the gas temperature does not need to be maintained above the dew point of the corrosive gases. Properties of typical chimney mortars are given in Table 5.42n.

Liners can be constructed for temperatures up to 2000°F. A minimum of 150°F is needed to obtain natural draft and corrosion protection may call for a temperature above the dew point of the corrosive constituents in the exit gas.

In special applications, carbon bricks are employed. Plastic liners are also being developed and will probably be available soon for low-temperature applications. Concrete stacks have also been employed, but corrosion is a problem unless suitable coatings are applied.

Sectional liners are laid up in mortar and supported by the column. In such cases, lining rests on projections from the column, while the space between column and liner is filled with an insulator or is left empty. Sectional liners are not as reliable as free standing liners.

When more than one flue enters a single stack, the angle between them should not be more than 120° and the width of the flue opening should be less than one-half of the inside diameter of the liner.

Table 5.42n
STACK MORTAR PROPERTIES

Type	*pH Resistance Range*	*Maximum Temperature, °F*	*Cost Basis, Compared to Potassium Silicate Cost*
Furfural alcohol	4–12	350	2.0
Potassium silicate	0–5.5	2000	1.0
Colloidal silica	0–7	2000	1.2

Shapes

The best stack shape is a straight cylinder. Converging cones may be used at the top to increase the exit velocity.

The stack should not be enclosed with architectural fences because it has the same effect as if the stack tip was at elevation S_1 in Figure 5.42j.

Stacks should be located on the highest roof of the building and should be tall enough to clear the wake of all nearby structures.

Stack caps should not deflect the exhaust downward or otherwise interfere with the high-velocity vertical discharge.

With the trend toward increasingly stringent air pollution regulations, the use of dispersion as a means of getting rid of toxic gases or pollutants is likely to be more and more restricted in the future. Therefore, stacks will be increasingly supplemented and, in some cases, supplanted by pollution abatement systems which remove the contaminants instead of dispersing them.

Nomenclature

C	pollutant concentration in the plume at a given location (m/l^3)
d	inside diameter of stack (l)
g	acceleration due to gravity (l/t^2)
h	actual height of the stack (l)
h_e	effective stack height (l)
Δh	total plume rise (l)
Δh_v	plume rise due to momentum (l)
$\Delta h_{v,max}$	maximum plume rise due to momentum (l)
Δh_b	plume rise due to buoyancy (l)
p	atmospheric pressure
Q_{vl}	emission rate of stack gases at temperature T_1 (m/t)
Q_h	emission rate of pollutants (m/t)
T_l	absolute temperature at which density of stack gases equals that of the ambient atmosphere (τ)
T_a	absolute temperature of ambient atmosphere (τ)
T_s	absolute temperature of stack gases (τ)
U	wind speed at location (l/t)
v_s	stack gas ejection speed (l/t)
x	downwind distance from stack (l)
y	lateral distance from vertical plane through stack (l)
z	height above ground level (l)
$d\theta/dz$	potential temperature gradient of ambient air (τ/l)
σ_y	horizontal dispersion coefficient (l)
σ_z	vertical dispersion coefficient (l)
Δ	potential temperature gradient of ambient atmosphere (equals T_a in a neutral atmosphere) (τ)

Dimensions of each term given in brackets: 1 = length, t = time, τ = temperature, m = mass. Any consistent set of units can be employed in the equations presented.

All equations in this section are based on metric units, except Equations 5.42(8), (9) and (10) in which English units are employed.

REFERENCES

1. Turner, D. Bruce: "Workbook of Atmospheric Dispersion Estimates," National Air Pollution Control Administration, Cincinnati, Ohio, 1970.
2. Occupational Safety and Health Standards, *Federal Register*, pp. 10466–10714, May 29, 1971.
3. Bosanquet, C. H., W. F. Carey and E. M. Halton: Proc. Inst. Mech. Engrs. (London), vol. 162, p. 355, 1950.
4. Holland, J. Z.: "Meteorology Survey of Oak Ridge Area," U.S. Weather Bureau, Report ORO-99, U.S. Atomic Energy Commission, Oak Ridge, Tennessee, 1953.
5. Briggs, G. A.: "Plume Rise," U.S. Atomic Energy Commission, Division of Technical Information, Oak Ridge, Tennessee, 1969.

BIBLIOGRAPHY

Brink, J. A.: *J. Air Pollution Control Association*, vol. 14, p. 449, November 1964.

Clarke, J. H.: *Heating, Piping and Air Conditioning*, vol. 40, p. 125, March 1968.

Somers, E. V.: Dispersion of Pollutants Emitted Into the Atmosphere, in W. Strauss (ed.), "Air Pollution Control—Part I," Wiley Interscience, New York, 1971.

Stern, A. C.: Air Pollution Standards, Vol. III, Chap. 51, in A. C. Stern (ed.), "Air Pollution," Academic Press, New York, 1968.

Strom, G. H.: Atmospheric Dispersion of Stack Effluents, Vol. I, Chap. 7, in A. C. Stern (ed.), "Air Pollution," Academic Press, New York, 1968.

5.43 STACK ACCESSORIES AND INSTRUMENTS

Types of Stack Accessories: Blowers and instruments. Other accessories such as doors, linings, platforms, lightning arrestors or lighting fixtures are furnished with the stack.

Costs: For blowers with 12″ H_2O ΔP, \$8,000 for 30,000 ACFM, \$10,000 for 65,000 ACFM and \$13,000 for 80,000 ACFM. For monitoring instrument costs, refer to Chapter 4.

Partial List of Suppliers: *Blower Suppliers:* Buffalo Forge Co.; New York Blower Co.; Northern Blower Div., Buell Engineering Co. *Instrument Suppliers* are listed in Chapter 4.

Smoke Detectors: Atlantic Research Corp., Bailey Meter Co., Beckman Instruments, Inc., Climet Instruments, Inc., Combustion Equipment Assoc. Inc., Edison Industries, Thomas A., Environmental Research Corp., Fluid Data, Inc., General Electric Co., Jacoby-Tarbox Corp., Kidde & Co., Inc., Leeds & Northrup Co., Lock & Co., Ltd., Oldham, Lancashire, England. Mine Safety Appliances Co., Monsanto Enviro-Chem Systems Inc., New York Testing Laboratories, Inc., Panametrics Inc., Particle Technology, Inc., Phoenix Precision Instr. Co., Photomation, Inc., Research Appliance Co., Royco Instruments, Inc., Testing Machines Inc., Unico Environmental Instruments, Inc.

Blowers

In the push-through type blowers used in stack exhaust applications, the volumetric capacity varies directly with fan speed while the discharge pressure and power consumption vary with the square and cube, respectively, of the fan speed. The axial propeller type is for very low pressure services only. The axial vane design can develop higher pressures with low power consumption but is limited to clean gas services. The centrifugal types are used in large stacks to develop higher discharge pressures and are described in Section 5.6.

Direct driven fans are used on constant load applications because they are more compact and eliminate belt slippage. Belt-driven units provide more flexibility and are more commonly used in ventilation and exhaust applications. Noise (Chapter III.4) can be an important factor in fan selection.

Foundations

Octagonal foundations are quite common. For heights of up to 300 feet and for normal soil loading of 4,000 lb/ft^2, the maximum octagonal foundation diameter will be one-tenth of the height plus the shaft inside diameter at the top. If resistance to earth tremors is required, this may be increased by up to 15%.

Drainage

Use of cone type stack caps to prevent the rain from entering the stack is undesirable. If a stack operates continuously at an exit velocity of 2,000 fpm or above, no rain will enter it. If the system is operated intermittently, rain protection is required (Fig. 5.43a).

Lightning Arrestors

Stacks being the tallest structures, lightning is prone to strike them in preference to other nearby structures. Steel stacks need no lightning protection

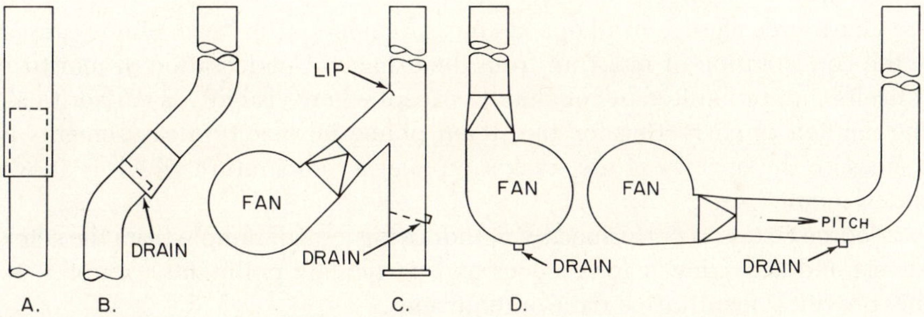

Figure 5.43a Drain stacks and fan drains. *A*, Vertical; *B*, Offset; *C*, Powerhouse; *D*, Stack fan drains. NOTE: The verical type is not fully dry but offers better protection than does a cone cap.

as the frame itself provides an adequate path for discharging the lightning to earth.

Masonry stacks do require lightning arrestors consisting of metal conductors to provide low resistance path to earth. These are usually pointed rods mounted on the top of the stack which are interwired for better conduction.

Lighting

Stacks are lighted at the top in accordance with local regulations. Particularly if the stack is located near an air strip or if it is over 200 ft in height, the lighting standards are to be studied in detail.

Lighting for the rest of the stack including ladders and access platforms follows standard industrial practice.

Instrumentation

Continuous monitoring devices (see Chapter 4) are not yet commonly used.

Most frequently, the observer only looks at the smoke plume and at a Ringlemann chart (Section 1.5) and assigns a smoke number to the plume based on its resemblance to a particular panel.

This procedure is time-consuming and there is no direct correlation between the opacity of the plume and the quantity of pollutants released to the atmosphere. On the other hand, this method can be used to detect a change in the appearance of the plume and, thus, signal a trend which can be analyzed by more precise methods.

The Smoke Inspection Guide of the U.S. Public Health Service suggests that the observer sights the plume and compares it with film strips of varying shades of gray. A further improvement of this technique is the sighting and comparing of the smoke with a reference film which appears in the same field of view. Such methods greatly increase the reproducibility of readings. The automatic, continuous smoke density and color monitors are described in Section 4.15.

One source of error in all opacity measurements is the steam plume caused by the condensation of moisture from the exhaust. Condensation of moisture by itself is not prohibited by regulations except where visibility is very critical. One method of correcting for the steam plume in opacity measurements is to measure the opacity of the stack sample at a temperature which is above its dew point.

Source testing is performed by withdrawing a small sample from the stack exhaust and analyzing it for gaseous and particulate pollutants (Chapter 4). This provides quantitative data on emissions.

5.44 STACK SELECTION AND APPLICATION

To allow one to properly size and design a stack, the following design basis information must be provided:

1. Exhaust gas quantity and possible fluctuations in flowrate.
2. Exhaust gas temperature and pressure range.
3. Draft required ($-'' H_2O$) at the flue opening to the stack. (If this has to be calculated, the number and location of flues, height of openings and elevation of site over sea level need to be known.)
4. Height of nearest buildings and topographical map of the area.
5. Maximum wind velocity, force and prevailing wind direction.
6. Gas composition in detail.

Once the design data information is furnished, the stack diameter can be determined based on the exhaust gas flowrate and on the exit velocity required. The height of the stack is so selected as to guarantee that the ground level concentration of the emitted pollutants will be below the applicable standards.

The next step is to calculate the overall pressure drop in the system; and if a fan is used, determine the required horsepower for its motor. Materials of construction for all components are selected on the basis of exhaust gas composition, including the choice of brick and mortar in tall stacks. The selection of monitoring instrumentation and the approval of the overall design are usually made after consultation with the regulatory agencies.

Some of the stack design rules can be summarized as follows:

1. Ground level concentration varies inversely with the square of the stack height, it is maximum at a distance of 5 to 10 stack heights from the stack and its level is usually 0.1 to 1% of the stack exhaust concentration.
2. Stack heights should be at least 2.5 times the height of the tallest structure in the area and their exit velocity should exceed 60 fps. Emissions from small diameter stacks (under 5 feet) or short stacks (under 200 feet in height) will occasionally hit the ground and increase the ground level concentration of pollutants to undesirable levels.

5.45 METEOROLOGICAL DIFFUSION CALCULATIONS

Pollutant dispersion calculations provide the basis for determining the height of stacks (Section 5.42) and for assessing the effects of pollutant discharges (such as emission rate) and meteorological factors (such as wind speed) on dispersion of pollutants from stacks.

Fundamental Equations

For an explanation of nomenclature used, refer to the end of this section. Dispersion of final exhaust from stacks is described by the mathematics of diffusion. The fundamental equation is based on Fick's second law:

$$\frac{\delta C}{\delta t} = DV^2C \qquad 5.45(1)$$

where D is the diffusion coefficient and C is concentration.

Assuming the wind velocity u to be constant and independent of elevation and neglecting $\delta^2C/\delta x^2$, the diffusion along the wind direction (Fig. 5.45a), we obtain from Equation 5.45(1) at steady-state:

$$u\frac{\delta C}{\delta x} = D_y\frac{\delta^2 C}{\delta y^2} + D_z\frac{\delta^2 C}{\delta z^2} \qquad 5.45(2)$$

Equation 5.45(2) describes diffusion of gaseous or particular dispersoid under the influence of random, homogeneous turbulence.

Practical Dispersion Relations

The two dispersion equations used most frequently are those of Sutton and of Bosanquet-Pearson. They are partially based on Equation 5.45(2) with empirical corrections on diffusion coefficients to account for atmospheric variables, such as atmospheric stability and turbulence. These two equations predict the average ground-level concentration of gaseous pollutant (C_0 in ppm) from a continuous source at downward distance (x in Fig. 5.45a) as follows

Bosanquet-Pearson's Equation:

$$C_0 = \frac{Q\,10^6}{\sqrt{2\pi}\,pqux^2}\exp\left[-\left(\frac{H}{px} + \frac{y^2}{2q^2x^2}\right)\right] \qquad 5.45(3)$$

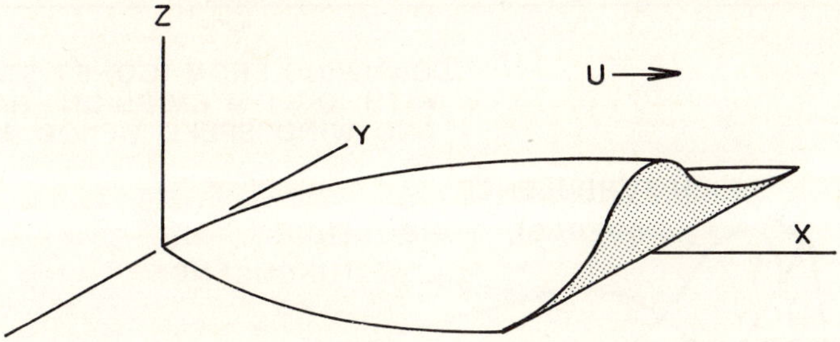

Figure 5.45a Coordinate system for a continuous point source.

Sutton's Equation:

$$C_0 = \frac{2Q\,10^6}{\pi C_y C_z u x^{(2-n)}} \exp\left[-\frac{1}{x^{(2-n)}}\left(\frac{y^2}{C_y{}^2} + \frac{H^2}{C_z{}^2}\right)\right] \qquad 5.45(4)$$

Both of these equations predict a maximum ground-level concentration at some downwind distance. Bosanquet-Pearson's equation yields:

$$C_{max} = \frac{4Q\,10^6}{\sqrt{2\pi}\,e^2uH^2}\frac{p}{q} = 2.15\frac{Q\,10^5}{uH^2}\frac{p}{q} \qquad 5.45(5)$$

at a distance

$$x_{max} = \frac{H}{2p} \qquad 5.45(6)$$

Sutton's equation yields:

$$C_{max} = \frac{2Q\,10^6}{e\pi uH^2}\frac{C_z}{C_y} = 2.35\frac{Q\,10^5}{uH^2}\frac{C_z}{C_y} \qquad 5.45(7)$$

at a distance

$$x_{max} = \left(\frac{H^2}{C_z{}^2}\right)^{\frac{1}{(2-n)}} \approx \frac{H}{C_z} \qquad 5.45(8)$$

Figure 5.45b shows the effect of atmospheric stability on the shape of concentration-distance curves. As can be seen, the average concentration is low close to the stack, increases to a maximum and then tapers off.

Values of diffusion coefficients for the two dispersion equations are available in references 1 and 2.

Reliability of Working Formulas

Montgomery and Corn[1] have evaluated the appropriateness of the Sutton, the Bosanquet-Pearson and the four USPHS-TVA equations for predicting ground-level concentrations of SO_2 from emission and meteorological data.

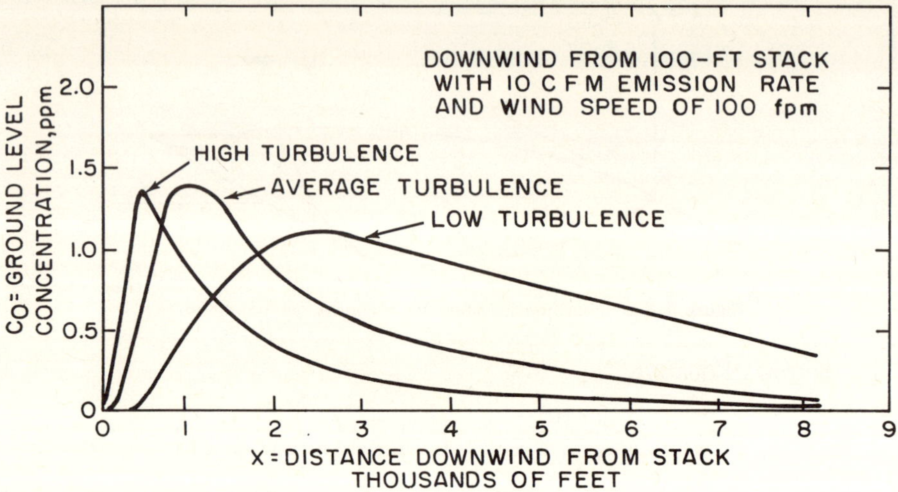

Figure 5.45b Effect of atmospheric stability on diffusion from a stack. Low turbulence is associated with a stable atmosphere in which air temperature increases with height; moderate turbulence is associated with an unstable atmosphere in which the air temperature decreases relatively rapidly with height.

They used the air monitoring data based on half-hourly average SO_2 concentrations for a calendar year at a TVA power plant. Results from their study showed that the Bosanquet-Pearson's equation was reliable 90% of the time. Unsuccessful predictions were primary overestimates for ground-level concentrations closest to the plant. The reliability of the Sutton's model varied from 66 to 93% of the time depending on the diffusion coefficients used.

Gosline[2] has studied the time variations of downwind ground-level concentration. He stated that, close to the stack, 1 to 3-min average maximum concentrations were 6 to 13 times the 30-min or longer average concentrations; 1 to 10 sec maximums were about 50 times the 30-min or longer average. In air pollution control, one should take into account the time-average concentration, the average maximum concentration and the maximum concentration at a point. These ground-level concentrations should be evaluated for ½ to 1-hour, or more; 1 to 5-min; and for 1 to 10-sec periods. The selection of a particular ground-level concentration as a design criterion depends on the type of problem envisioned and on the nature of the pollutant. A sample calculation illustrating this is given in reference 2.

The assumption of constant wind velocity and constant diffusivities with elevation may be an oversimplification in some cases. Exact solutions incorporating the variation of these two with elevation are available in reference 3.

Studies on the reliability of the two dispersion equations were performed on a specific terrain. To solve practical problems, one might substitute one's own test data into the dispersion equation and calculate the diffusion parameters for studies concerning the area of interest. The two dispersion equations are based on smooth, level terrain and they should not be used in calculations for plants in valleys. When a pollution problem is expected, consultation with a meteorologist can be very useful.

Deposition of Particulates

Simultaneous dispersion and deposition are formulated by adding a vertical advective term, $v_f(\delta c/\delta z)$, to Equation 5.45(2):

$$u\frac{\delta C}{\delta x} = D_y\frac{\delta^2 C}{\delta y^2} + D_z\frac{\delta^2 C}{\delta z^2} + v_f\frac{\delta C}{\delta z} \qquad 5.45(9)$$

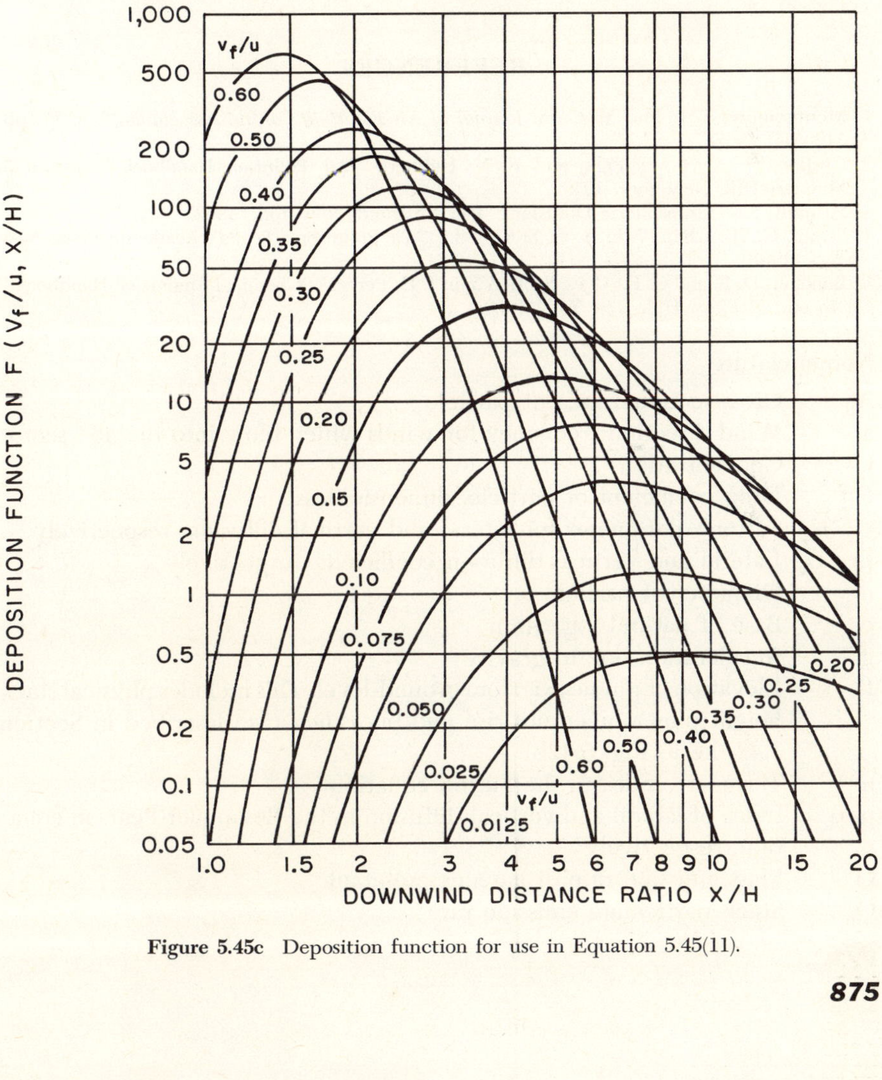

Figure 5.45c Deposition function for use in Equation 5.45(11).

The free settling velocities v_f may be calculated using Equation 5.45(10) and curves relating C_D with Reynolds number given in reference 5.

$$v_f = \frac{2V_p g(\rho_p - \rho)d_p}{\mu A_p R_e C_D} \qquad 5.45(10)$$

The following equation is used with the graph given in Figure 5.45c to calculate the average deposition under average meteorological conditions.[4]

$$\begin{array}{c}\text{Average rate of deposition}\\ (\text{mg/cm}^2\text{/month})\end{array} = \frac{0.31aQ_p}{H^2} F\left(\frac{v_f}{u}, \frac{x}{H}\right) \qquad 5.45(11)$$

Bosanquet, et al. estimated that Equation 5.45(11) gave results accurate within a factor of two when compared with field data and considered the equation suitable for use in the stack design.

REFERENCES

1. Montgomery, T. L. and M. Corn: *Journal of Air Pollution Control Association,* vol. 17, pp. 512–517, 1967.
2. Gosline, C. A., L. L. Falk, and E. N. Helmers: "Air Pollution Handbook," Section 5, McGraw-Hill, New York, 1956.
3. Pasquill, F.: "Atmospheric Diffusion," Van Nostrand, New York, 1962.
4. Strom, G. H.: Chap. 7, in A. C. Stern (ed.), "Air Pollution," 2d ed., Academic Press, New York, 1968.
5. Boucher, D. F. and G. E. Alves: Section 5, in J. H. Perry's, "Chemical Engineers' Handbook," 4th ed., McGraw-Hill, New York, 1963.

Nomenclature

A_p	Cross-sectional area of particle
a	Wind direction frequency for winds which blow into the 45° sector
C	Concentration
C_D	Drag coefficient of particle, dimensionless
C_y, C_z	Sutton's parameter for lateral and vertical diffusion, respectively
D_y, D_z	Lateral and vertical diffusion coefficient, respectively
d_p	Diameter of particle
e	Base of natural logarithm
g	Acceleration due to gravity
H	Elevation of plume axis from ground-level. This includes physical stack height plus momentum rise and buoyancy rise described in Section 5.42 (he in Fig. 5.42b)
n	Turbulent index in the Sutton equation
p, q	Index of lateral and vertical diffusion in the Bosanquet-Pearson equation, respectively
Q	Mass emission rate of gaseous pollutant
Q_p	Stack particulate emission rate

Re	Reynolds number of particle, dimensionless
t	Time
u	Wind velocity
V_p	Volume of particle
v_f	Free settling velocity of particle
x	Distance downwind of stack or source
y	Lateral distance from the vertical plane through the stack or source and parallel to mean wind direction
z	Height above ground
μ	Dynamic viscosity of air
ρ, ρ_p	Mass density of ambient atmosphere and particle, respectively

Mobile Sources of Air Pollution

(Sections 5.46 to 5.51 are devoted to this discussion.)

5.46 THE CASE FOR SMALLER CARS AND FOR IMPROVED MASS TRANSPORTATION

Most people like to use an individual car because it is private and it takes them exactly where they want to go. Man, like many of nature's creatures, has a sense of territorial right; and a large, private car provides security against the violation of this right.

A big car can also be a status symbol and can symbolize a person's individuality, his or her defiance of convention; or it may be adorned in such a manner as to be identified with a particular group of owners.

Since ancient Rome and the chariot races with four-horse teams, circa 550 B.C.,[1] individuals have known the spirit of road competition. Although the machines and technology have changed a great deal, particularly in the last 50 years, the nature of the people has not.

The owner of today's average car is commanding a 4,000 pound machine of 300 potential horsepower with an innate feeling of indestructability. This car is expensive in terms of depreciation, maintenance, insurance and fuel. On the other hand, another individual may drive an auto of half the weight, one-fourth the power, one-fourth the maintenance cost and one-half the insurance cost.

In 1927, Charles Lindbergh flew an aircraft weighing 5,900 pounds and carrying 425 gallons of gasoline a distance of 3,300 miles at an average speed of 108 miles per hour. The wing of this monoplane was 46 feet in span. The engine, designed by Charles L. Lawrence,[2] could muster a maximum of 220 horsepower, but was generally throttled to 150 horsepower to conserve fuel.

A modern private auto weighs half as much as this 1927 aircraft, its average speed is half as great, it is not flying nor is it dragging a 46-foot wing to make it fly; yet, its engine is larger.

Today, the number of people needing transportation, private or public, is increasing, as is the environmental cost of energy for the propulsion of vehicles.

Man is the only one of nature's creatures who utilizes nonmuscular energy other than his own for transporting himself and his cargo.

Animals, in general, do not move very fast because the energy requirement for speed on land, in water and in the air at sea level is an exponential function. To go twice as fast requires eight times the energy.[3]

Mass Transportation of the Past

The chief advantage of railway transport is that when using a steel wheel on a steel rail, very large loads can be moved with comparatively little effort. As early as 1904, electric locomotives pulling trains of passenger cars (each train weighing up to 500 tons) were capable of 94 miles per hour on level track.[4]

In 1933, the Boeing model 247 airplane with two engines of 550 horsepower each transported ten passengers at a cruising speed of 155 miles per hour.[5]

Steam and electric buses and public conveyances[6] as well as those driven by compressed air[7] were in use before 1910. The pneumatic tube rail system had been tried in the 1840s, holding some promise for subterranian high-efficiency rail systems.[8]

Mass Transportation Today

At present, United States electric railways do not excel as a high-speed, convenient method of smog-free transportation. The Japanese National Railways system, on the other hand, utilizes computer controlled electric trains which carry thousands of passengers at speeds up to 150 miles per hour.[9]

Today's passenger aircraft, although still very smoky and noisy, is superior to earlier designs in terms of fuel consumed and smog produced per passenger mile. The Boeing 747 consumes 60 gallons of fuel and 1,000 pounds of oxygen every minute. This, in terms of passenger miles per gallon of fuel consumed, represents a two-fold improvement over the model 247 of 1933. In terms of passenger miles (useful work done) per gallon of fuel (smog produced), the Boeing 747 is less efficient than a small 90-horsepower car carrying four people; but it is four times more efficient than a medium or large American car carrying one person.

Mass Transportation of the Future

One of the possible new methods of railway propulsion is the linear induction motor. In this design, the motor can be within the locomotive, enveloped by a stationary reactor keel alongside of, or between, the rails. But because the power must be brought to the locomotive via trolley, third rail or other means, a conventional traction motor or motors may as well be used, since their electromechanical efficiency far exceeds that of the linear induction motor with its inherently large air gap.

Another possibility is the hovercraft which is a very low altitude aircraft that must operate within a few feet of the surface, be it land or water. They are of modest efficiency and speed and need a relatively flat surface. Since they do not have an immediate grip on this surface, as does a conventional boat, train or auto, their maneuverability, acceleration and stopping power are low.

In the aerotrain, propulsion is via a shrouded air propeller. The effectiveness of this method is exceptionally poor at low speeds and during acceleration, owing to propeller slip. It improves at high speeds, and its potential top speed is above that of the conventional electric train which does not work well above 150 miles per hour due to wheel slip.

Other new propulsion methods are limited only by the imagination of man, but only those designs which require substantially less energy per passenger mile, both public and private, should be considered in the future. For a discussion of the electric car and of other alternatives to the internal combustion engine, refer to Sections 5.47 and 5.48.

REFERENCES

1. Classical Games *Encyclopaedia Britannica,* p. 10, vol. X, 1952.
2. Roseberry, C. R.: "The Challenging Skies," p. 137 Doubleday, New York, 1966.
3. Ferguson, Eugene: Measurement of the 'Man-Day', *Scientific American,* p. 98, vol. 225, no. 4, October 1971.
4. "Transactions of the International Electrical Congress," J. B. Lyon Co., St. Louis, p. 199, vol. III, 1904.
5. Johnson, Robert: The Lessons of 50,000,000 Miles of Flying, *Scientific American,* vol. 149, no. 3, p. 114, September 1933.
6. *Cyclopedia of Automobile Engineering,* American Technical Society, Chicago, vol. I, 1912.
7. Hiscox, Gardner: "Compressed Air and Its Applications," 4th ed., Munn & Co., New York, pp. 573–592, 1903.
8. Edwards, L. K.: High-Speed Tube Transportation, *Scientific American,* p. 35, vol. 213, no. 2, August 1965.
9. Sperry Rand advertisement for Univac, *Scientific American,* p. 83, vol. 220, no. 2, February 1969.

BIBLIOGRAPHY

Allen, Richard: "Revolution in the Sky," Stephen Greene Press, Brattleboro, Vermont, 1964.

Alpers, Antony: "Dolphins, The myth and the mammal," Riverside Press, Cambridge, Massachusetts, 1961.

Hedgpeth, Joel: "Seashore Life," University of California Press, Berkeley, 1962.

Jones, Bradley: "Elements of Practical Aerodynamics," 3d ed., John Wiley & Sons, New York, 1945.

Lusk, Hilton F.: "General Aeronautics," rev. ed., Ronald Press, New York, 1947.

Mansfield, Harold: "Vision," Duell, Sloan and Pierce, New York, 1956.

Norman, J. R.: "A History of Fishes," 2d ed., Hill and Wang, New York, 1963.

Norris, Kenneth: "Whales, Dolphins and Porpoises," University of California Press, Berkeley, 1966.

Page, Victor: "Modern Aircraft," Norman W. Henley Pub. Co., New York, 1929.

Villard, Henry: "Contact," McGraw-Hill, New York, 1968.

5.47 THE ELECTRIC CAR

By the year 1888, three makers of electric automobiles had their wares operating on public roads.[1] Long before records were being set by steam and gasoline driven vehicles, the lowly lead-acid storage battery, invented by R. L. G. Planté in 1859, improved by Fauré, Volkmar (1881), Swan and Correns (1888),[2] provided the electric power for such early designs.

The first official automobile speed record was set in a Jeantaud in France, (38 mph, 1898). Later in 1899, a new record went to another French electric, "Le Jamais Contente," the first car in the world to exceed 60 mph. Walter Baker, an American, completed the "Baker Electric Torpedo" in 1901 (Fig. 5.47a). This fully streamlined wooden-framed car did better than 78 mph in 1902 on Staten Island. Later, the car topped 80 mph but lost control and ran through a crowd of bystanders, killing two of them. This brought infamy to the electric car and was an obvious setback to Baker and his team. After a time, they were back in the racing business and succeeded in reaching 120 mph with their car.[3]

The most notable of the steam-driven cars were the machines of Leon Serpollet of France (1900). The world's speed record following Baker's "Torpedo" was set by the Stanley brothers' streamlined steamer driven by Fred Meriott at Ormond Beach, Florida. In 1906, it reached 127.66 miles per hour.[4,5] Present electric car records exceed 150 mph.[6]

Speed and Range

If the electric cars go fast, they do not go far; and if they go far, they do not go fast. This is due to the small energy content of the batteries.

Battery-driven car designs and research[7] prototypes use the lead-acid battery or some minor modification of it since its cost is far below that of any other battery. The performance that can be expected includes a top speed of 45 mph on level roads and a range per charge of 50 miles if driven conservatively.[8]

Their docile nature is characterized by strong acceleration and hill-climbing ability, combined with regenerative braking in which the energy of the moving vehicle, when slowing or descending hills, recharges the batteries rather than being uselessly dissipated in the brakes as in ordinary cars.

Figure 5.47a Walter Baker's "Electric Torpedo" which exceeded 120 miles per hour in 1905. *(From Floyd Clymer's book "Historical Motor Scrapbook," Clymer Motors, New York, 1944)*

The electric car is inherently reliable and long-lived. The total operating cost is about the same as for a small conventional car. The cost is largely due to the deterioration of the batteries, because the expense of recharging is only about $3 a month for general town use.

Some electric cars will go farther and faster than those mentioned, but their battery weight is so great that their maneuverability is reduced.

Recharging the batteries is done by plugging a cord on the car into a convenient wall socket and leaving it there for 1 to 6 hours, depending on how low the batteries were when the charging began (Fig. 5.47b).

The controls of the car are very simple. There is an accelerator pedal and a brake pedal. In most prototypes, there is no clutch or gearshift. The accelerator controls speed and the degree of regenerative braking. A switch is provided for reverse driving. The brakes are independent of all other systems.

Improvements that might come about through future technology would include a battery of greater energy content per unit weight and a control system which combines smooth control from zero to maximum, low cost and high efficiency. Present controls lack one or another of these features.

The Future of Electric Cars

A common myth is blazing acceleration and power to be obtained from super batteries in the future. This is not likely to occur.[9,10]

Figure 5.47b Modern electric car and motorcycle.

Compared to the lead-acid cell, the nickel-cadmium-alkaline battery has slightly more energy, five times the life and twelve times the cost. The silver-zinc-alkaline system has six times the energy per pound; but the cost per car is $20 thousand and it is worn out in 8 months of service. The zinc-air battery has high-energy content but is easily poisoned by impurities and requires elaborate pumping and filtering systems.

Many improvements have recently been made in flashlight, transistor radio, recorder and other small batteries; but their cost is high because 2,400 size-D flashlight cells would be required to power an electric car for 60 miles.

There are several "hot" batteries under development which operate at high temperatures (500°C), such as the sodium-sulfur cell or others that contain lithium, chlorine and other highly reactive elements. It is probable that a battery may eventually be developed that has twenty times the capacity of an equal weight lead-acid battery. The design problems are several including (1) the containers must be elaborate and expensive to ensure safety, (2) thermal insulation must be effective to keep the battery hot while not in use since it cannot be recharged unless it is hot.

Nonbattery Type Electric Cars

In fuel cells, electric current is directly generated from the oxidation of a fuel such as alcohol or hydrogen. Their momentary current rate is low, rendering them sluggish, so that a vehicle propelled by them would be slow.

Solar cells are not strong enough; if the entire surface of the car was utilized as the cell surface, it would provide less than ½ horsepower.

Pollution and the Electric Car

A popular notion is that the electric car will cause as much pollution as the gasoline car because its batteries must be recharged by a power station which burns fossil fuel and is itself a polluter. This is not correct.

The steam turbo-electric generating plant is many times better than the gasoline engine in terms of energy made available versus smog produced.[11]

Electric autos, at their present performance levels, can be powered by existing electric energy sources without additional research and development work.

If batteries were substantially improved, the increased demand on electric power facilities for charging electric cars would present a new threat to the environment but this threat would probably be less serious than that of today's gasoline-driven automobiles.

SUPPLIERS

Corbin Centry Inc. of Sommersville, Conn. offers an electric bicycle for $1395. Electric Fuel Propulsion of Ferndale, Mich. and Sebring-Vanguard of Sebrin, Florida offer electric cars from $2,068. Battronic Truck Corp. of Boyertown, Pa. and Electrobus of Studio City, California offer electric buses. Operating costs are generally under $.02 per mile and recharging periods vary from three to twelve hours.

The Electric Vehicle Council of 90 Park Avenue, New York City, and Electric Vehicle News (P.O. Box 533, Westport, Conn. 06880) can furnish further information on new products.

REFERENCES

1. Williams, Alice: The Golden Age of the Electrics, *Product Engineering*, p. 130, vol. 41, no. 20, September 14, 1970.
2. Accumulator (Storage Cell), *Encyclopaedia Britannica*, vol. 1, pp. 106–113, 1952.
3. Information from Baker Division Otis Elevator Co., 8000 Baker Avenue, Cleveland, Ohio, 44102.

4. Derr, Thomas S.: "The Modern Steam Car," Floyd Clymer Publications, Los Angeles, p. 57, 1932.
5. The Development of Serpollet Engine, *Light Steam Power*, pp. 127–138, vol. XX, no. 3, May-June 1971.
6. Information from Eagle-Picher Battery Co., American Building, Cincinnati, Ohio, 45202.
7. The Electric Car, a Design Challenge, *Electro-Technology*, p. 57–72, vol. 81, no. 5, May 1968.
8. Consortium's Voltswagon Charges into Production, *Industrial Research*, p. 24, vol. 13, no. 5, May 1971.
9. Howard, John: The Sleepy Battery Business Wakes Up, *New Scientist and Science Journal*, p. 530, vol. 50, no. 753, May 27, 1971.
10. Lindgren, Nilo: Electric Cars—Hope Springs Eternal, *IEEE Spectrum*, pp. 48–60, vol. 4, no. 4, April 1967.
11. Starr, Chauncey: Energy and Power, *Scientific American*, p. 40, vol. 224, no. 3, September 1971.

BIBLIOGRAPHY

Barnard, William: "Heat-power engineering," 3d ed., Part I, Wiley, New York, 1926.
Christie, Clarence: "Electrical Engineering," 2d ed., McGraw-Hill, New York, 1917.
Gray, Alexander: "Electrical Machine Design," McGraw-Hill, New York, 1913.
Kent, R. T.: "Mechanical Engineers Handbook, Power," 11th ed., Wiley, New York, 1936.
Still, Alfred: "Elements of Electrical Design," McGraw-Hill, New York, 1924.
Vinal, George: "Storage Batteries," 3d ed., Wiley, New York, 1940.

5.48 OTHER ALTERNATIVES TO THE INTERNAL COMBUSTION ENGINE

The present automobile engine is the result of roughly 75 years of evolution. It is inexpensive to produce, relatively reliable and easy to manage by inexperienced operators. It is easy to start and requires little warmup; it is highly energetic and docile; and the tooling for its production has been largely paid for.

A modern internal combustion engine at high-output attains combustion temperatures of 5000°F and combustion pressures of 1,000 PSIG. Under these conditions, nitric oxide (NO) is formed; and, due to the extremely rapid cooling rate during the expansion process of this engine (1,000,000°F per second), the nitric oxide molecule does not have time to dissociate and recombine in less noxious forms, is largely fixed as NO (stable at atmospheric conditions) and is rejected from the exhaust pipe as such.

Other chemical compounds rejected in some quantity are carbon monoxide (CO), sulfur dioxide (SO_2), unburned hydrocarbons (due mostly to the "quenching" process), particulate carbon (soot) and compounds of lead.

Smog chemistry is extremely complex so that the finer unpleasantries of engine exhaust will not be dealt with in this article; and indeed, much is yet to be learned of their effect on man and animal.

Due to photocatalytic reactions (sunlight), NO reforms into several other oxides of nitrogen, some of which combine with moisture in the air and in human tissue to form nitric acid (HNO_3). Carbon monoxide is toxic. Sulfur dioxide (SO_2) combines to some extent with moisture in the air and in human tissue to form sulfuric acid (H_2SO_4) and sulfurous acid (H_2SO_3). Large volumes of carbon dioxide (CO_2) and water vapor (H_2O) are also exhausted, some of which form carbonic acid (H_2CO_3).[1]

The internal combustion gasoline-burning engine is the principal source of smog throughout the world and particularly in the United States (Section 2.10). By adding elaborate and expensive afterburners (Sections 5.25 and 5.33), air pumps and other abatement devices, the smog output of the gasoline engine may be reduced but not eliminated.

The Steam Car

The smog characteristics of the steam car are 10 to 100 times better than those of the internal combustion engine. Steam cars have external combustion engines because combustion takes place in a very hot chamber (eliminating quench, the periodic extinguishing of combustion), at very low pressures (eliminating the need for tetraethyl lead or other antiknock compounds) and at low rates of temperature drop (nearly eliminating nitric oxide formation). The steam car (although capable of great bursts of speed) has a modest maximum sustained speed of about 80 mph. It has a complex mechanism and requires more maintenance and understanding than does the equivalent internal combustion car. Its fuel economy at highway speeds is comparable to the conventional car, but its economy in heavy traffic may far exceed that of the ordinary car.

The Electric Car

With present battery designs, this car is limited to in-town use (Section 5.47). As such, it is economical and smog free. It has good, low-speed acceleration, it is nimble and easy to park. It has not been produced in great numbers by any manufacturer.

The Thermoelectric Car

In this car, electric current is caused to flow by the application of heat to thermoelectric junctions of two dissimilar metals (such as iron and constantan). It is smog free, being external combustion in its operation, but has not, thus far, attained competitive performance.

The Compressed Air Car

In this car, an engine (similar to a steam engine) runs on compressed air. The best of these[2] carried 20 passengers for a distance of 17 miles at an average speed of 12 mph (circa 1880).

The Cryogenic Gas Car

In this car, an engine (similar to a steam engine) is powered by the boiling and expansion of liquified gases, such as liquid air and liquid nitrogen. They receive the heat needed to vaporize the cryogenic fuel from the atmosphere. Their performance in the past has been similar to that of the compressed air car.

The Stirling Engine

This engine, invented in 1827, utilizes external combustion as a heat source to drive a working piston and a displacer piston in conjunction with a heat exchanger. These engines are fairly expensive due to design features such as the fact that the best working medium is hydrogen, requiring elaborate seals.[3]

All external combustion engines have the same low smog potential as that of the steam engine.

The Wankel Engine

This is an internal combustion engine in which a triangular rotor moves in planetary motion about an output shaft (Section 5.50). The combustion flame moves around in the chamber always being swept into a relatively cool confinement which results in worse quenching than in conventional engines.[4]

The Wankel can operate at high speeds and is inherently well-balanced and compact. Its fuel consumption is poor (the best achieved economy being 0.55 pound of fuel per brake horsepower hour as compared with 0.40 for conventional internal combustion engines). Being a dirty but small engine, it could be compatible with catalytic pollution abatement devices which require a fair amount of space under the hood and a "rich" effluent gas for best performance.

The Still Engine

This engine employs conventional internal combustion processes, but the cylinders are jacketed by steam which absorbs the heat otherwise wasted in conventional engines. It uses this steam to do work on the underside of the double-acting pistons. This engine is somewhat complex but has proven to yield substantially higher fuel economy than ordinary internal combustion engines. It has not, thus far, been used in automobiles.[5]

The Diesel Engine

This machine, invented and improved by Rudolph Diesel in 1897, burns low-cost fuel. It is inherently reliable and long-lived. Somewhat heavier and more expensive than the gasoline engine, it exhibits better fuel economy at partial loads. It produces less smog but discharges more oil vapors than the gasoline engine.[6]

The Hybrid Car

The engine-electric drive is very efficient and has been used extensively in steamships and locomotives. It is inherently expensive but enables a high-speed prime mover to transmit low-speed, high-torque power to a vehicle. If used with batteries to drive a car, the best features of each system are eliminated by the compromise.

The Momentum (Flywheel) Car

In this system, a motor-generator is coupled to a flywheel and accelerates this wheel, while the vehicle is stationary and power is being supplied to the motor. When the flywheel has reached high speed, it drives the motor as a generator which supplies electric power to a motor geared to the wheels of the vehicle.

This system was ingeniously exploited by the Oerlikon Works in Switzerland some 25 years ago. Their busses carried passengers at speeds up to 30 miles per hour for distances of up to 1.5 miles between reenergizing.

Other Stored Energy Systems

The steel spring (clock spring) and other elastic deformation devices cannot store enough energy to be useful in driving even a small town vehicle.

REFERENCES

1. Obert, E. F.: "Internal Combustion Engines," 3d ed., International Textbook Co., Scranton, Pennsylvania, pp. 363–379, 1968.
2. Hiscox, Gardner: "Compressed Air and Its Applications," 4th ed., Munn & Co., New York, pp. 573–592, 1903.
3. Barnard, William: "Heat-power Engineering," 3d ed., Wiley, New York, pp. 166–168, vol. I, 1926.
4. Obert, E. F.: "Internal Combustion Engines," pp. 511–516, Intext, Scranton, 1968.
5. Marks, Lionel S.: "Mechanical Engineers Handbook," 6th ed., McGraw-Hill, New York, pp. 9–107, 1958.
6. Bryant, Lynwood: Rudolph Diesel and His Rational Engine, *Scientific American*, vol. 221, no. 2, pp. 108–117.

BIBLIOGRAPHY

Anderson, Edwin: "Marine Engineers Hand Book," Theo Audel & Co., New York, 1943.

Barnard, William: "Heat-power Engineering," Part I, 3d ed., Wiley, New York, 1926.

Cummins, Clessie: *"My Days with the Diesel,"* Chilton, Philadelphia, 1967.

Cyclopedia of Automobile Engineering, vol. I, American Technical Society, Chicago, 1912.

Dyke's Automobile Encyclopedia, 17th ed., Goodheart-Willcox, St. Louis, 1935.

Hiscox, Gardner: "Compressed Air and Its Applications, Munn, New York, 1903.

Judge, Arthur W.: "Automobile Engines," 4th ed., Chapman & Hall Ltd., London, 1942.

Kent, R. T.: "Mechanical Engineers Handbook," 11th ed., Wiley, New York, 1936.

Lichty, Lester C.: "Internal Combustion Engines," 5th ed., McGraw-Hill, New York, 1939.

Marks, Lionel S.: "Mechanical Engineers Handbook," 6th ed., McGraw-Hill, New York, 1958.

Obert, Edward F.: "Internal Combustion Engines," 3d ed., Intext, Scranton, Pennsylvania, 1968.

Smith, D. H.: "The Modern Diesel," 11th ed., Clymer Publications, Los Angeles, 1952.

Walton, J. N.: "Doble Steam Cars," 2d ed., Kirk Michael, Isle of Man, 1970.

5.49 THE ROLE OF FUEL COMPOSITION

Pollutants emitted by internal combustion engines consist of vapors escaping from the fuel system and combustion products discharging in the exhaust mixture. Such combustion products—hydrocarbons, oxygenates, nitrogen oxides (NO_x), carbon monoxide (CO), smoke and odorous components—contribute to the air pollution problem to a degree, depending on their levels and chemical character as discussed in Sections 5.48 and 5.51. Level and chemical character of emissions, in turn, depend on several factors, including fuel composition. Dependence of emission on fuel composition is of interest because it provides an approach to emission control. That is, fuel composition can be optimized to reduce pollution from emissions. The potential of this fuel optimization approach for emission reduction depends on how strongly fuel composition influences emissions. According to existing information[1,2] (for fuels not containing tetraalkyl lead or similar anti-knock additives), optimization of fuel formula can effect only modest emission reduction. While this suggestion may be largely correct, it is also true that existing information has not revealed the full potential of the fuel composition factor. On the other hand, it is known that light hydrocarbon fuels create less pollution than gasoline. This section discusses the current understanding of the effect that fuel composition has on emissions and includes some speculation regarding the role of the fuel factor in the problem of emission pollution. (For a discussion on the effects and sources of lead in the atmosphere, refer to Section 3.13.)

Fuel Composition, Definition and Range

Traditionally, fuel composition has been expressed in terms of three component classes—paraffins, olefins and aromatics. Such compositions of gasoline fuels in commercial use in the United States range as follows:[3]

Component Class	*Concentration Range, vol %*
Paraffins	50 to 80
Olefins	2 to 25
Aromatics	7 to 40

Detailed analysis[3] of a typical gasoline fuel is given in Table 5.49a.

Table 5.49a
GASOLINE FUEL COMPOSITION

Component	Concentration, mole %	Component	Concentration, mole %
Paraffins		Olefins	
Isobutane	1	2-Methyl-1-butene	1
n-Butane	7	2-Pentene	1
Isopentane	10	2-Methyl-2-butene	2
n-Pentane	4	2-Methyl-2-pentene	1
2,3-Dimethylbutane	2	Other olefins	6
2-Methylpentane	3	Aromatics	
3-Methylpentane	2	Benzene	1
n-Hexane	2	Toluene	6
Methylcyclopentane	1	Ethylbenzene	1
2,4-Dimethylpentane	2	*m* and *p*-Xylene	5
Cyclohexane	1	*o*-Xylene	2
2-Methylhexane	5	n-Propylbenzene	1
2,2,4-Trimethylpentane	6	1-Methyl-4-ethylbenzene	1
n-Heptane	1	1,3,5-Trimethylbenzene	1
Methylcyclohexane	1	1-Methyl-2-ethylbenzene	1
2,4-Dimethylhexane	1	1,2,4-Trimethylbenzene	3
2,3,4-Trimethylpentane	2	1,2,3-Trimethylbenzene	1
2,3,3-Trimethylpentane	1	Other aromatics	4
2-Methyl-3-ethylpentane	1		
3,4-Dimethylhexane	1		
2,2,5-Trimethylhexane	1		
n-Octane	1		
Other saturates	6		

Defining fuel composition in terms of paraffins, olefins and aromatics is inappropriate from a pollution standpoint. Such a classification of the fuel components, probably originating from early methods for fuel analysis, does not reflect properly the differences in pollution potential among the fuel components. Thus, fuels with identical class-to-class compositions may differ in pollution potential because of differences in within-class composition. A more appropriate definition of fuel composition would be based on a classification in which pollution potential would differ from group to group but would be similar for the components within each group. Based on combustion theory and reasonable speculation, such a classification might be as follows: straight chain paraffins, isoparaffins, cycloparaffins, olefins, benzene, nonalkylbenzenes, polyalkylbenzenes. Further classification of isoparaffins and olefins based on molecular structure may also be necessary.

Effect of Fuel Composition on Emissions

Basic studies on hydrocarbon combustion and automotive engine design suggest that contrary to some recent studies,[1] fuel composition could influence strongly both level and composition of emissions. Such an influence would result from both direct and indirect fuel effects.

The effect of fuel composition on level and composition of evaporative emissions is direct and easily understood. Evaporative emissions are, in themselves, a part (the most volatile) of the fuel. In fact, levels and composition of evaporative emissions can be calculated with reasonable accuracy from fuel composition and relative hydrocarbon volatility data.[4]

The effect of fuel composition on exhaust emissions is extremely complex and represents the net result of several direct and indirect influences. Fuel composition effect on the levels of NO_x, CO and organics in the exhaust is indirect and probably is the composite of effects of air-fuel ratio (A/F) and combustion temperature.[1] Gasolines that differ in composition may also differ in viscosity, density and carbon-hydrogen ratio (C/H). However, variation in fuel viscosity and density will also cause variation in fuel flow through the fuel jets[5] and, hence, variation in A/F and combustion temperature. A/F and combustion temperature, in turn, are known to affect emission levels significantly.[6,7] Further, changing the fuel's C/H will also change the stoichiometric oxygen demand of the fuel, resulting in an effectively richer or leaner air-fuel mixture. Therefore, because engine conditions are not optimized for each fuel separately, the variations in levels of exhaust emissions from fuels of different composition can be explained by the A/F effect.

The effect of fuel composition on the character of exhaust organics is of interest because it determines their reactivity or photochemical smog forming potential.[8] (See Chapter 3 for a detailed discussion on this topic.) One such effect is obvious because some exhaust organics are in themselves unburned fuel. The remaining exhaust organics are fuel combustion products, the composition of which depends on the structure of the fuel molecule, although this dependence has not been fully defined. Nevertheless, reported information[3,7,9–11] suggests that patterns do exist in the association of fuel structure with exhaust organics. Thus, relative to aliphatics, *aromatic fuels* are more resistant to combustion, and their combustion products are at a higher level of unsaturation and include fewer organic radicals. Therefore, exhaust from aromatic fuels contains more unburned fuel and acetylenes. Further, since oxygenates are products from reactions of organic radicals with oxygen, aromatic fuel exhaust contains less oxygenated organic material. From the aliphatic fuels, there are indications that relative to olefins, paraffins may yield higher levels of exhaust olefins.[3,11] Information on the more specific effects of fuel hydrocarbon structure on exhaust organics is scarce. Additional studies of the automotive combustion of hydrocarbons of various structures are much needed.

Effect of Fuel Composition on Evaporative Emissions

Level and reactivity of evaporative emissions depend on the amount and composition of the relatively volatile fuel components. This strong effect of front-end and midrange fuel volatility and composition on evaporative emissions also depends on ambient temperature.[12,13] As a result, reduction of fuel volatility or modification of the fuel formula to replace volatile reactive olefins with unreactive paraffins can *drastically* reduce pollution levels from evaporative emissions.[12-14]

While fuel composition effect on CO and NO_x emissions is small, its effect is not small on the emissions of exhaust organics. The design of the studies in this area was deficient in that the independent variable (fuel composition) was defined in terms of three component classes (paraffins, olefins, aromatics), disregarding the variations in within-class composition. Use of test fuels of varying within-class composition might have yielded a stronger fuel composition effect on exhaust organics.

Effects of Fuel Additives on Emissions

Fuel additives[15] are intended primarily to improve the performance of the engine-fuel system, rather than to reduce pollutant emissions. For this reason, information on the effect of additives on emissions is scarce and consists largely of unreported or informally reported data. Fuel additives have two generally defined functions.[16,17] First, they modify the combustion process directly through participation in the combustion-related chemical reaction. Second, they modify the combustion indirectly by influencing deposits in the combustion chamber and in the induction system. With the exception of the *tetraalkyl-lead* antiknock additive (Section 3.13), the participation of the additives in the combustion process does not seem to affect the emissions significantly.[17] One study even suggested that use of the lead additives decreases the levels of exhaust particulates.[18] The demonstrated effects of additives on emissions are related mainly to formation, modification or reduction of deposits. For example, the formation of lead-salt deposits in the combustion chamber causes an increase in exhaust hydrocarbons but has no effect on CO.[19] The effects of deposit-modifying additives are uncertain.[17] Detergent-dispersant additives, intended to maintain induction-system cleanliness, have a generally beneficial effect, the magnitude of which may depend on the history of engine usage and on the quality of gasoline and lubricant used.[17]

Two other aspects of the effect of additives on emissions remain largely unexplored. One aspect relates to the possibility that additive-derived exhaust components pose heretofore unsuspected pollution problems. Thus, excepting the lead antiknock additive, additive-derived products have not received significant attention. The other aspect relates to the fuel reformulation that may be needed as a substitute for an additive. This case is illustrated by the present trend toward reduced levels of lead additive in gasoline with concurrent reformulation of gasoline to provide the octane quality required. Such reformulations may increase the emission of pollutants.[14]

Emissions from Light Hydrocarbon Fuels

Light hydrocarbon mixtures, including natural gas, LP gas (propane) and Synthane® (a coal-derived mixture consisting of methane with about 10% hydrogen), were found to be low-pollution automotive fuels.[20-22] From a pollution standpoint, these fuels were found to have the following distinct advantages over gasoline:

1. Their organic combustion products are considerably less reactive in the atmosphere.
2. They provide uniform air-fuel mixtures regardless of ambient temperature, thus maintaining low-emission performance even during engine cold-start.
3. They can be used by the engine at leaner air-fuel mixtures, thus resulting in lower levels of NO_x, while maintaining low levels of exhaust organics.

Diesel and Aircraft Engines

Of the pollutants emitted by diesel engines, NO_x, smoke and odor are of most concern. Hydrocarbons (including oxygenates) and CO are within the present specifications for maximum levels imposed on gasoline engines.[23] From studies on fuel and engine factors affecting diesel emissions, it appears that engine design and operating conditions, especially F/A, have an overriding influence on emission levels.[23-25] Fuel composition affects emissions only to the extent that it affects the physical characteristics of the fuel, thereby affecting fuel injection and F/A. Such a fuel effect is negligible. Fuel additives are known to reduce the levels of smoke.[23]

Regarding aircraft engines, the most critical emission, presently at least, is particulate material, emitted mostly as smoke.[25] Studies of the aircraft smoke problem established that fuel composition does influence smoke output.[26] Specifically, fuels with higher C/H, for example, aromatic rich, are conducive to smoke formation, which can be reduced by additives. Nevertheless, neither the fuel C/H nor the additives provide a practical method for smoke control.

REFERENCES

1. Dishart, K. T. and W. C. Harris: "The Effect of Gasoline Hydrocarbon Composition on Automotive Exhaust Emissions," pp. 612–642, vol. 48, Proc. American Petroleum Institute, Div. of Refining, 1968.
2. Dishart, K. T.: "Exhaust Hydrocarbon Composition. Its Relation to Gasoline Composition," presented at 35th Midyear Meeting, API Div. of Refining, Houston, Texas, May 14, 1970.
3. Morris, W. E. and K. T. Dishart: The Influence of Vehicle Emission Control Systems on the Relationship Between Gasoline and Vehicle Exhaust Hydrocarbon Composition, "Effect of Automotive Emission Requirements on Gasoline Characteristics," ASTM Special Technical Publication 487, pp. 63–93, July 1971.

4. Koehl, W. J., Jr.: Mathematical Models for Prediction of Fuel Tank and Carburetor Evaporation Losses, SAE Paper 690506, New York, May 1969.
5. Obert, E. F.: "Internal Combustion Engines," International Textbook Company, Scranton, Pennsylvania, 1968.
6. Jackson, M. W.: Effects of Some Engine Variables and Control Systems on Composition and Reactivity of Exhaust Hydrocarbons, *SAE Vehicle Emissions, Part II,* vol. 12 (selected papers 1963–1966), pp. 241–267, New York, 1967.
7. Ninomiya, J. S. and A. Golovoy: Effects of Air-Fuel Ratio on Composition of Hydrocarbon Exhuast from Toluene, Toleune-*n*-Heptane Mixture and Isooctane, *J. of the Air Pollu. Contr. Ass.*, pp. 342–346, vol. 19, no. 5, May 1969.
8. Altshuller, A. P.: An Evaluation of Techniques for the Determination of the Photochemical Reactivity of Organic Emissions, *J. of the Air Pollu. Contr. Ass.*, pp. 257–260, vol. 16, no. 5, May 1966.
9. Ninomiya, J. S. and A. Golovoy: Use of Diisobutylene Fuel in a Single-Cylinder Engine: Effects of Equivalence Ratio on Exhaust Composition, *J. of the Air Pollu. Contr. Ass.*, pp. 879–883, vol. 19, no. 11, November 1969.
10. Ninomiya, J. S. and B. Biggers: Effects of Toluene Content in Fuel on Aromatic Emissions in the Exhaust, *J. of the Air Pollu. Contr. Ass.*, pp. 609–611, vol. 20, no. 9, September 1970.
11. Fleming, R. D.: "Effect of Fuel Composition on Exhaust Emissions from a Spark-Ignition Engine," U.S. Bureau of Mines Rep. of Investigations 7423, September 1970.
12. Eccleston, D. B., B. F. Noble and R. W. Hurn: "Influence of Volatile Fuel Components on Vehicle Emissions," U.S. Bureau of Mines Rep. of Investigations 7291, February 1970.
13. Jackson, M. W. and R. L. Everett: Effect of Fuel Composition on Amount and Reactivity of Evaporative Emissions, SAE Paper 690088, New York, January 1969.
14. Dimitriades, B., B. H. Eccleston and R. W. Hurn: An Evaluation of the Fuel Factor Through Direct Measurement of Photochemical Reactivity of Emissions," *J. of the Air Pollu. Contr. Ass.*, pp. 150–160, vol. 20, no. 3, March 1970.
15. Barusch, M. R. and J. H. MacPherson: Engine Fuel Additives, in J. John McKetta, Jr. (ed.), "Advances in Petroleum Chemistry and Refining," pp. 457–546, Interscience, New York, 1965.
16. Jones, B. A., B. H. Weil and M. H. Graham: Effect of Additives on the Combustion of Petroleum-Derived Fuels, ACS Advances in Chemistry Series No. 20, "Literature of Combustion of Petroleum," pp. 235–242, 1958.
17. Fleming, R. D., W. F. Marshall and J. R. Allsup: "Fuel Additives and Automobile Exhaust Emissions," U.S. Bureau of Mines Technical Progress Rep. 40, August 1971.
18. Pierrard, J. M. and R. A. Crane: "The Effect of Gasoline Compositional Changes on Atmospheric Visibility and Soiling," presented at the Air Pollution Control Association, Annual Meeting, Atlantic City, New Jersey, June 29, 1971.
19. Coordinating Research Council, Inc., "The Effects of Leaded and Unleaded Gasolines on Exhaust Emissions Caused by Combustion Chamber Deposits," CRC Report 441, June 1970.
20. Eccleston, D. B. and R. D. Fleming: "Engine Emissions Using Natural Gas, Hydrogen-Enriched Natural Gas and Gas Manufactured from Coal (Synthane)," U.S. Bureau of Mines Technical Progress Rep. (in press).
21. Fleming, R. D. and J. R. Allsup: "Emission Characteristics of Natural Gas as an Automotive Fuel," SAE Paper 710833, New York, October 1971.
22. Fleming, R. D., J. R. Allsup, T. R. French and D. B. Eccleston: "Propane as an Engine Fuel for Clean Air Requirements," presented at 64th Annual Meeting of Air Pollution Control Association, Atlantic City, New Jersey, APCA Paper 71–173, June 27–July 2, 1971.
23. Marshall, W. F. and R. D. Fleming: "Diesel Emissions Reinventoried," U.S. Bureau of Mines Rep. of Investigations 7530, July 1971.
24. Fleming, R. D., W. F. Marshall and R. W. Hurn: "The Fuel Factor in Diesel Emissions," presented at National Midyear Meeting of Society of Automotive Engineers, Detroit, Michigan, May 18–22, 1970.
25. George, R. E., J. A. Verssen and R. L. Class: "Jet Aircraft: A Growing Pollution Source," *J. of the Air Pollu. Contr. Ass.*, pp. 847–855, vol. 19, no. 11, November 1969.
26. Linden, L. H. and J. B. Heywood: "Smoke Emission from Jet Engines," Publication 70–12, Dept. of Mechanical Engineering, Fluid Mechanics Lab, Massachusetts Institute of Technology, October 1970.

5.50 THE CASE FOR THE WANKEL ENGINE

In 1926, the German engineer, Felix Wankel began to systematically investigate rotary engines. His efforts culminated in a working internal combustion engine in 1956. Shortly thereafter, he sold a major part of his invention to NSU of Germany. Dr. Froede of NSU kinematically inverted the original Wankel concept to a stationary casing which was feasible to produce. NSU introduced the first Wankel powered vehicle, NSU Spyder, in 1964. Toyo Kogyo of Japan and NSU are both currently manufacturing Wankel powered vehicles.

Curtiss-Wright purchased North American rights in 1958. General Motors and Outboard Marine Corporation are additional United States licensees and practically every major engine manufacturer in the world either is a licensee or is negotiating for a license.

Engine Operation

The basic cycle is illustrated in Figure 5.50a.[1-4] Thermodynamically it follows the 4-stroke Otto cycle. The curve defined by the inner surface of the rotor housing is an epitrochoid and contains an enveloped rotor. A fixed gear on the housing and an internal gear on the rotor maintain the proper relative geometry. Torque is delivered through a crank shaft which is in sliding contact with the rotor. The rotor rotates at $\frac{1}{3}$ crank speed and one power stroke is delivered on each crank shaft revolution.

Engine Features

The Wankel engine is approximately one-half the size and weight of a conventional automotive engine of equivalent horsepower. It is simple and potentially of lower cost. There is no valve train and, with no reciprocating unbalance, the engine is almost vibrationless. Because it inducts fresh mixture during approximately 270 degrees of crank shaft rotation versus 180 degrees in a piston engine, it exhibits superior breathing. Torque is developed for about $\frac{2}{3}$ of the cycle rather than $\frac{1}{4}$ the cycle with the conventional 4-stroke engine and results in a single rotor Wankel being approximately equivalent to a 3-cylinder, 4-cycle engine. The fuel octane requirement appears to be low for a given compression ratio and *unleaded fuel causes no difficulty.*

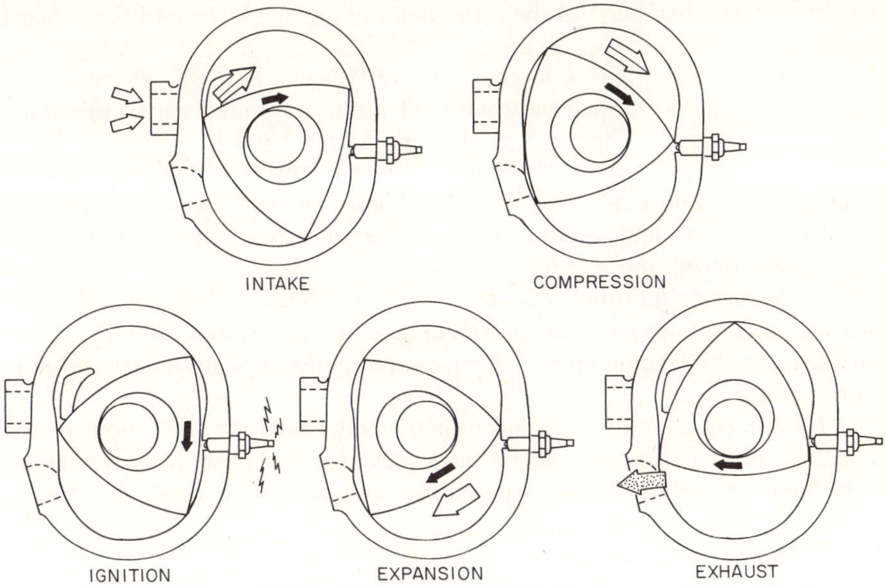

Figure 5.50a Wankel engine.

Deficiencies in early designs caused short apex seal life. Modern Wankel powered vehicles, however, exhibit better than 50,000 mile life. Thermal efficiency was a problem because of leakage, slow burning and the high combustion chamber surface/volume ratio. Again modern designs have brought marked improvement. Oil injection or a fuel-oil mixture is required to lubricate the rotor, rotor housing and seal contacts. Sophisticated seal and rotor housing coating materials may be necessary.

Emission Formation

Unburned Hydrocarbons (HC)

The exhaust HC emissions of the engine are a result of several processes which are also important in the conventional 4-stroke engine.[6]

Wall Quenching. As the propagating flame front approaches the relatively cool chamber wall, it is extinguished or quenched a finite distance from the wall. Raw fuel/air mixture and partially burned and/or cracked fuel exist in high concentrations in this "quench zone." During the exhaust process, some of the HC compounds are scavenged from the chamber walls and are released to the atmosphere while some remain to be partially burned on a succeeding cycle.

This HC pollution source is significant in the Wankel because of a high-combustion chamber surface/volume ratio and because the trailing apex seal of a given chamber "scrapes" the housing quench layer into the exhaust port.

Crevice Effect. This is another form of "quench process" where a combustible mixture is trapped between cool surfaces through which the flame will not propagate. The mass contained in this quench region or "dead volume" consists of a high percentage of fuel/air mixture. In current Wankel designs, the volume above the side seal and between the rotor and side housing is relatively large and vented (unburned or only partially burned) to the atmosphere during the exhaust process.

Blow-by of Unburned Charge. Even a minute leakage of unburned mixture to the atmosphere affects HC emissions significantly. Blow-by of raw mixture past the leading apex seal can pass directly through the exhaust port to the atmosphere.

Incomplete Combustion. Incomplete combustion in the reaction zone is a minor additional source of HC emissions. This problem is intensified at light-load conditions where exhaust dilution is more significant.

CO, NO_x and Particulates

The carbon monoxide emissions are principally caused by poor combustion quality associated with rich mixture ratios and the partial combustion in the quench regions.

The oxides of nitrogen (principally NO) are created[7,8] in or near the high-temperature flame front. Their concentration is a function of the maximum cycle temperature and the equilibrium composition of NO and residence time at these high temperatures.

With leaded gasoline, more than half of the particulate content is due to lead salts. The remaining particulates are principally polynuclear aromatics.

Adaptability to Emission Control

Because of higher combustion chamber surface/volume ratio, slower combustion and internal exhaust gas recirculation, cycle temperatures are reduced and, therefore, the NO emissions are low.

Exhaust HC emissions, which are generally higher than with a conventional engine, and CO emissions can be almost completely eliminated with the use of either a thermal and/or catalytic reactor. Reactors are effective because of both high exhaust gas temperature and rapid warm-up. (For discussions of afterburners, refer to Sections 5.25, 26, 27, 32, and 33.)

High exhaust gas temperature (entering the afterburner) insures a rapid oxidation reaction if excess oxygen is present, and the rapid warm-up decreases the lag time from engine start until the exhaust treatment device becomes effective. The Wankel engine exhaust temperatures are high for several reasons, including:

1. Higher frequency of exhaust through a given exhaust port.

2. Minimum of cooling in exhaust passage.
3. Close proximity of exhaust ports permitting the use of a small exhaust collector which reduces heat transfer.

Rapid warm-up is a result of the foregoing factors and the low thermal inertia of the engines.

Operating Variables

The most influential variables are air/fuel ratio and spark advances.[5] Hydrocarbon and CO concentrations decrease with a reduction in fuel/air ratio to the point where misfiring begins to increase the HC emissions and the NO is near a maximum with slightly leaner than a stoichiometric mixture ratio. (For standard engine, see Fig. 5.51a.)

All important pollutant emissions decrease with spark retardation. Other variables have a relatively small effect on emissions and follow the same trends observed with piston engines.

Design Variables

The influence of design variables on emissions has not been investigated thoroughly. The research efforts in this area have been by the various licensees and the results are, therefore, highly proprietary. However, several observations are appropriate based on conventional engine experience.

The chamber shape can be only slightly modified by changing the shape of the rotor cut-out. A small reduction in emissions is possible at best for HC and CO. Reduction of chamber dead volume and improvement of the apex sealing will reduce the HC emissions.

Modification of the intake and exhaust port size and shape can change the effective valve timing and overlap period and can be used to increase the exhaust residual in the fresh charge for NO_x control.

Thermal Reactor

The thermal reactor[5,9] effectively oxidizes the HC and CO if the residence time and mixing of oxygen with the exhaust gases are adequate along with high exhaust temperatures. Current Wankel powered vehicles manufactured for sale in the United States utilize such thermal reactors. A recirculating type of reactor is shown schematically in Figure 5.50b. Hydrocarbon emission data in Figure 5.50c shows both the effect of the thermal reactor on HC emissions and the quick warm-up of the reactor as indicated by the rapid decrease in HC concentration. At higher speeds with optimized air/fuel ratio and/or air-injection, the HC emissions can be reduced to less than 50 ppm. The NO emissions are not measurably influenced by the thermal reactor.

Catalytic Reactor

The oxidizing catalytic reactor (Section 5.33) appears to be at least as effective as a thermal reactor for CO and HC removal, and it has the added advantage of being able to accommodate minor engine maladjustments without serious degradation of performance. Warm-up is very rapid but over-temperature problems are more likely than with conventional engines. A reducing catalyst would be highly desirable for reduction of NO. Unfortunately, careful control of air/fuel mixture ratio is generally required and a noxious by-product, NH_3, is often produced.

Exhaust Gas Recirculation (EGR)

Exhaust gas recirculation which reduces cycle temperature is currently the only feasible method for retarding NO formation unless it becomes possible to use extremely lean mixtures (Section 5.51). Unfortunately, performance and fuel economy are significantly reduced. Some EGR is available internally during the port overlap period and may be sufficient to achieve the desired control.

A lesser performance penalty is paid by the Wankel to achieve a given level of control because the initial NO emission concentration is less.

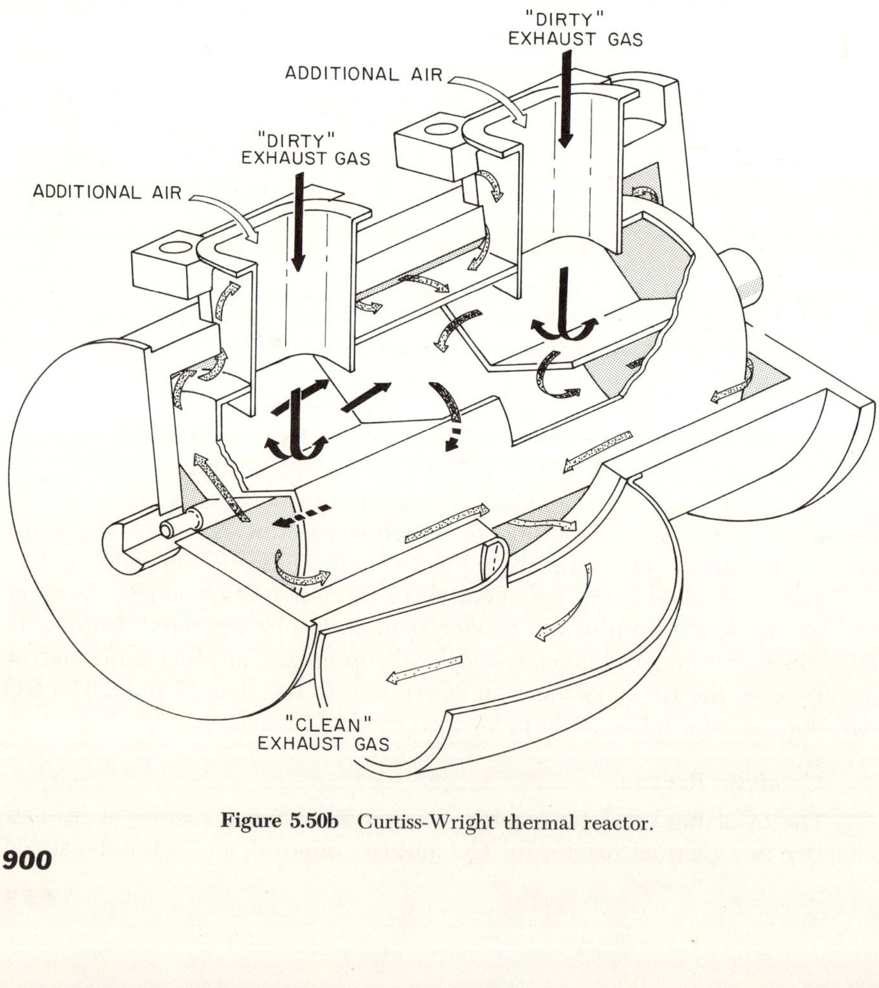

Figure 5.50b Curtiss-Wright thermal reactor.

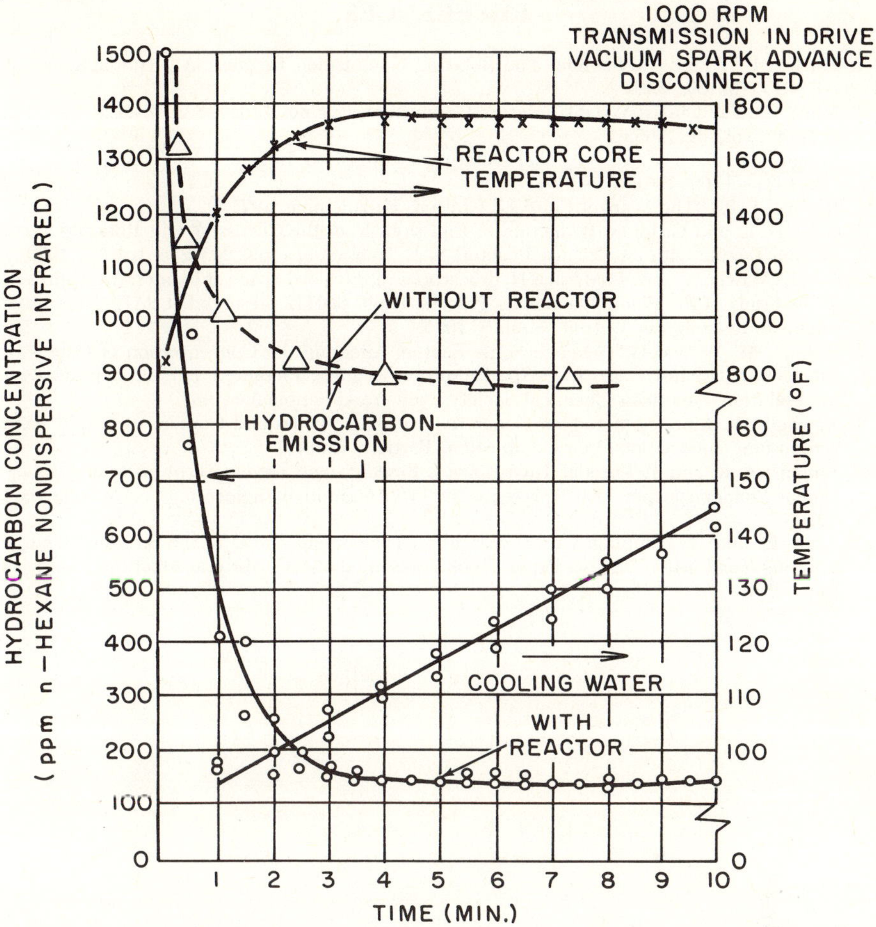

Figure 5.50c Cold-start HC emission performance of the thermal reactor on the Wankel engine.

Stratified Charge

Curtiss-Wright has operated the Wankel as a fuel-injected stratified charge (nonhomogeneous mixture) engine[10] and observed a significant reduction in emissions with catalytic treatment of the exhaust while maintaining acceptable performance and economy at part load. However, full throttle performance was reduced by 20% compared to the carbureted engine.

In summary, it seems that because the Wankel engine requires only half the space for the same horsepower as does a conventional engine and because its exhaust is both rich and hot, it is compatible with the available emission control devices and, in combination with them, can represent an intermediate solution to the present automobile emission problem.

REFERENCES

1. Hurley, R. T. and M. Bentele: The Rotating Combustion Engine, SAE Transactions, pp. 641–649, vol. 69, 1961.
2. Jones, C.: The Curtiss-Wright Rotating Combustion Engines Today, SAE Transactions, pp. 127–147, vol. 73, 1965.
3. Froede, W. G.: The NSU-Wankel Rotating Combustion Engine, SAE Transactions, pp. 179–193, vol. 69, 1961.
4. Norbye, J. P.: "The Wankel Engine," Chilton, Philadelphia, 1971.
5. Cole, D. E. and C. Jones: Reduction of Emissions from the Curtiss-Wright Rotating Combustion Engine with an Exhaust Reactor, SAE Transactions, pp. 205–215, vol. 79, 1970.
6. Gad El-Mawla, A. and W. Mirsky: Hydrocarbons in the Dead Space of Flames: An Approach to the Study of the Flame Quenching Process, Paper 660112 presented at SAE Automotive Engineering Congress, Detroit, January 1966.
7. Jackson, M. W. and G. J. Nebel: Some Factors Affecting the Concentration of Oxides of Nitrogen in Exhaust Gases from Spark Ignition Engines, Presented at the Symposium on Air Pollution, American Chemical Society, New York, September 1957.
8. Newhall, H.: Kinetics of Engine Generated Nitrogen Oxides and Carbon Monoxide, 12th Symposium (International) on Combustion, Portiers, France.
9. Yamamoto, K. and T. Kuroda: Toyo Kogyo's Research and Development on Major Rotary Engine Problems, paper 700079 presented at SAE Automotive Engineering Congress, January 1970.
10. Jones, C. and H. Lamping: Curtiss-Wright's Development Status of the Stratified Charge Rotating Combustion Engine, Paper 710582 presented at SAE Midyear Meeting, Montreal, June 1971.

5.51 THE LEAN REACTOR SYSTEM

As the air-fuel mixture ratio in a gasoline engine is enriched from the stoichiometric (less oxygen present than what is needed for complete combustion), the concentrations of carbon monoxide and unburned hydrocarbons (HC) emitted in the exhaust gas become progressively greater, but the concentration of NO is reduced (Fig. 5.51a). As the mixture is made leaner than stoichiometric, CO is minimized, unburned hydrocarbons pass through a minimum concentration and the concentration of NO increases to a peak but then diminishes rapidly. Brake specific fuel consumption passes through a minimum point on the lean side of stoichiometric point.

Conventional commercial carburetors atomize and vaporize the fuel imperfectly, and often wide variations exist in the air/fuel ratios of the mixture entering the various cylinders of a multicylinder engine. This causes the concentration of unburned hydrocarbons and the brake specific fuel consumption to rise, at overall air/fuel ratios below that of stoichiometric.

The lean reactor system for emission control involves the development of carburetion that produces greatly improved homogeneity of the fuel/air mixture and permits the advantages of lean operation to be realized. When operated so lean that the exhaust contains 2 to 4% of excess oxygen, the engine will exhaust very low concentrations of CO, unburned hydrocarbons and NO. Good fuel economy is an additional worthwhile benefit. The exhaust contains sufficient oxygen (without an added air pump) to oxidize the small amounts of CO and hydrocarbons remaining if time and temperature permit. To enhance oxidation, all possible sensible heat is conserved in the gases leaving each cylinder; and an enlarged, insulated exhaust reactor is used to provide residence time. Hence, this approach is known as the lean reactor system. Such a system can be operated on high octane gasolines containing lead antiknocks, to achieve efficient high-compression ratios.

Experimental cars modified to incorporate the lean reactor system have shown hydrocarbon emissions reduced by 96%, carbon monoxide by 95% and nitrogen oxides by 88%, as compared to the automobiles of the early 1960s without emission controls. Further improvements appear possible.

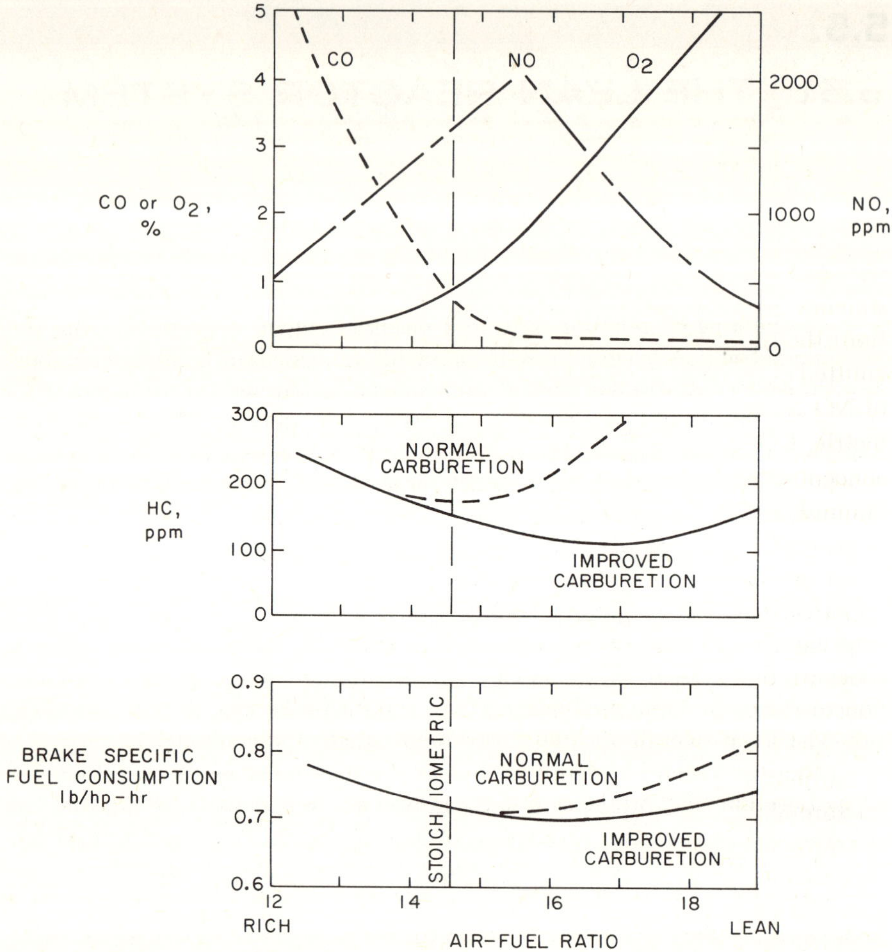

Figure 5.51a Effect of air-fuel ratio and mixture preparation on part-throttle emissions and fuel consumption.

Carburetor

An experimental high-velocity carburetor is used to provide the mixing that permits unusually lean operation. It combines one small-diameter, plain-tube primary carburetor for light-load conditions with two large, variable-venturi secondary throats for medium and high-load conditions. Thus, high air velocities prevail under all operating conditions without restriction of air-breating capacity. The high air velocities in the small primary venturi permit fuel to be supplied under idle and light-load conditions through the main nozzle where it is surrounded by the airstream and is well mixed, rather than through the small wall opening used in conventional carburetors. This improvement eliminates the separate idle system (which is a major source of calibration and maintenance difficulties in conventional carburetors) and also provides greatly improved mixing.

Exhaust Reactors

An important part of the reactor system is the use of liners in cylinder head exhaust ports. These are tubes of stainless steel that prevent cooling of the exhaust gases by the water-cooled cast iron of the exhaust ports. This promotes oxidation at the point of highest temperature.

Gases then discharge into reactor manifolds which consist of an open-tube liner of stainless steel. The liner is surrounded by high-temperature insulation and an outer shell of lower quality sheet steel. Design of the liner permits thermal expansion relative to the cylinder head. Since very little combustible material remains in the exhaust gas from the lean engine, reactor temperatures are not appreciably increased by combustion in the reactor, and temperatures (1400 to 1600°F) do not exceed the capabilities of high-quality stainless steels.

Other Elements of the System

A modulating system of exhaust gas recycle is used to permit different amounts of exhaust gas to be recirculated under different driving conditions for reduction in NO emissions. The recirculated exhaust gas is cooled because cooler gas is more effective than hot in lowering NO; but in order to prevent condensation and operational problems, the gas is maintained above the dew point (about 180°F). Metered gas is well mixed with the fuel/air charge between the primary carburetor and the inlet manifold. The valve and mixing areas are coated with Teflon to prevent deposits.

Ignition system advance characteristics are tailored to give the best compromise among driveability, performance, fuel economy and low emissions.

Comparisons with Other Approaches

A high-compression ratio engine will burn a leaner air/fuel mixture than a low-compression engine. Thus, the lean reactor system benefits from the use of high-compression ratios, which also contribute to high-efficiency and good operating economies. Costs of the engine modifications required for the lean reactor system are believed to be no more, and probably less, than those of alternate systems of comparable effectiveness.

Systems that oxidize emissions downstream from the engine often are susceptible to destructively high temperatures caused by temporary engine malfunctions, such as a misfiring spark plug. In the lean reactor system, so little combustion heat is released that deliberately causing a spark plug to misfire in a car at 70 miles per hour has been found to result in a reduction of the reactor temperature and no harm to the system.

The lean reactor system avoids problems that may be associated with catalysts, such as sensitivity to gasoline additives, the cost and availability of precious metals, deterioration of catalyst activity, catalyst attrition and over-temperature protection. The ability of the lean reactor system to use conventional low-cost gasolines containing lead antiknocks contributes to low operating costs and also prevents problems of exhaust valve seat wear, which otherwise can cause high-maintenance costs and, in some cases, valve leakage with greatly increased emissions. Furthermore, such fuels avoid any necessity for increased use of aromatic materials in gasoline to substitute for lead antiknocks in providing octane quality. This is of value since aromatics have been shown to increase the reactivity and smog-forming potential of exhaust gases and to increase the emission of polynuclear aromatics.

Chapter 6

REMOVING SPECIFIC CONTAMINANTS FROM THE AIR

S. B. Alpert L. J. Bollyky
N. P. Chopey W. J. Hatcher, Jr.
B. I. MacDonald M. J. Matteson P. G. Maurin
I. S. Shah D. J. Sibbett A. Turk
W. C. Zegel

CONTENTS OF CHAPTER 6*

*For a glossary of terms, refer to the front portion of this volume and for a list of abbreviations used in this chapter refer to the listing given after these contents, preceding Section 6.1.

Abbreviations

Atm	Atmosphere
B	Barrel
BTU	British Thermal Unit
cc	Cubic Centimeter
cfm	Cubic Foot Per Minute
cm	Centimeter
D	Day
fps	Foot Per Second
ft	Foot
g, gm	Gram
HP	Horsepower
hr	Hour
ID	Inside Diameter
KN	Kilonewton
KV	Kilovolt (10^3 Volts)
KW	Kilowatt (10^3 Watts)
KWH	Kilowatt Hour
lb	Pound
LEL	Lower Explosive Limit
LFL	Lower Flammability Limit
lpm	Liter Per Minute
M	Meter
MB/D	Million Barrels Per Day
mill	0.1 cent
mm	Millimeter
Mw	Molecular Weight
MW	Megawatt (10^6 Watt)
NAPCA	National Air Pollution Control Administration
Nm^3	Normal Cubic Meter
NPC	National Petroleum Council
ppm	Parts Per Million
PSIA	Pound per Square Inch Absolute
PSIG	Pound per Square Inch Gauge
SCF	Standard Cubic Foot
SCFM	Standard Cubic Feet per Minute
Sec	Second
SG	Specific Gravity
STP	Standard Temperature and Pressure
TVA	Tennessee Valley Authority
W	Watt
Wt	Weight
μ, μm	Micron (10^{-6} Meter)

6.1 CATALYTIC HYDRODESULFURIZATION OF RESIDUAL OIL

Residual Oil Needs in the United States, million barrels/day:	2.2 in 1970, 4.3 in 1985.
Residual Oil Usage Growth Rate:	5% per year or more.
Residual Oil Yield:	7% to 20% of total crude oil; presently closer to 7%.
Fuel Oil Cost Increase in the Last 20 Years:	50% to 200%
Cost of Heavy (Residual) Fuel Oil, ¢/million BTU's:	Low sulfur (0.3%): $1.50; high sulfur: 90¢.
Fuel Oil Cost as Function of Sulfur Content, $/barrel:	3%, $7.00; 2%, $8.50; 0.3%, $10.00
Sulfur Content of Fuels, %:	Distillate oil (0.2-1.0), residual oil (0.7-3.5), coal (2.5-3.5).
Sulfur Standards for Fuels, %:	Distillate oil (0.1), residual oil (0.3), coal (0.7).

Until recently, the United States refiner looked upon fuel oil that is used for power generation and industrial fuel as an unattractive product, since its price was lower than domestic crude oil. Factors which resulted in higher prices for fuel oil and increased demand for this product include:

1. Environmental concerns and setting of standards that preclude the use of high sulfur coal in populated regions of the country.
2. Disappointments in the pace of installing nuclear plants to generate power.
3. Shortages of supply and increasing prices of competing fuels, i.e., natural gas and low sulfur coal.
4. Inability of the domestic United States petroleum industry to site refining plants on the East or West Coast because of public concern about environmental impact.

5. Pressure by the oil producing and exporting countries for higher prices for exported crude.
6. The growing worldwide demand for residual fuel oil.
7. Disappointments in the availability of processing technology to reduce contaminant levels in fuel oil to the desired levels at low cost.

Fuel Oil Costs

Table 6.1a compares price indexes of residual fuel oil in United States markets. During the 1950s and most of the 1960s, the residual fuel price was relatively constant and priced below crude oil. There was an incentive for the United States refiner to convert low valued heavy fuel to gasoline and light distillates. In order to satisfy the growing demand for gasoline, the typical United States refinery installed coking operations. During the last two decades, the yield of heavy fuel oil decreased from about 20% of crude to 7% of the crude oil.

The Eastern United States market for residual fuel became largely dependent on low-cost foreign imports that are brought in without import restrictions. About 68% of the total residual oil used in the United States is derived from foreign sources, primarily from Venezuela. Some 86% of these imports are derived from Caribbean sources. In areas near the United States, i.e., the Caribbean and Canada, foreign crude is processed in relatively simple refineries. The heavy residual fraction representing one-half of the crude is imported to the United States, and low boiling fractions are exported to Europe or used to manufacture petrochemicals in adjacent facilities.

There is concern about loss of investment from the United States to the adjacent foreign regions. Various proposals are under consideration to entice and encourage investment in new plants to be made in the United States because of considerations such as the balance of payments.[1]

Historically, heavy fuel oil commanded low prices, equivalent to about

Table 6.1a
COMPARISON OF FUEL PRICE INDEXES, base: 1950 = 100

Year	*All Commodities*	*Industrial Commodities*	*No. 6 Fuel Oil*	*No. 2 Fuel Oil*	*Coal*
1950	100	100	100	100	100
1955	107	111	117	118	104
1960	115	116	126	114	98
1965	118	118	121	114	94
1969	130	130	98	128	104
1970	135	122	149	137	(150)

Table 6.1b
U.S. DEMAND FOR RESIDUAL FUEL OIL

Year	Millions of bbl daily
1970	2.2
1975	3.1
1980	3.8
1985	4.3

$.30 per million BTUs. Recent pressures have escalated prices sharply for this commodity. A price structure has emerged for low and high sulfur residual fuel. The price for 0.3% sulfur fuel oil is about $10 per barrel, corresponding to about $1.50/million BTU, whereas high sulfur fuel is priced at about $.90/million BTUs.

The tight situation with respect to fuel oil supply is likely to worsen. Table 6.1b shows estimates prepared by the National Petroleum Council.[2] They anticipate that residual fuel demand will increase from 2.2 million barrels in 1970 to 4.3 million barrels by 1985. The historic growth rate for residual oil has been at less than 2% per year, but the NPC forecasts an average growth rate over the next 15 years of about 5% per year. As usual, forecasts by economic prognosticators tend to underestimate what actually occurs. On the other hand, it is probable that the consequences of the 1974 energy crisis—more emphasis on mass transit and on smaller cars—will reduce this growth rate.

Sulfur Standards

Table 6.1c shows suggested standards for sulfur content in fuel oil proposed by the Environmental Protection Agency. In certain populated regions such as New Jersey, New York, Boston and California, such standards have already been promulgated by authorities. These air pollution authorities are forcing industry to comply with the legislated standards (Section 3.20)—albeit with some dislocations and with pleas for time extensions to line up new sources of supply of low sulfur, high-cost fuels.

Table 6.1c
SUGGESTED SULFUR STANDARDS FOR FUELS

Fuel	Typical Sulfur Content of Fuel, %	Suggested "Standard," %
Distillate fuel oil	0.2–1.0	0.1
Residual fuel oil	0.7–3.5	0.3
Bituminous coal	2.5–3.5	0.7

Table 6.1d
PRICES OF FUEL OIL

Sulfur Content, wt%	$/Barrel*
2.9	3.35
2.2	3.42
2.0	3.90
1.7	3.95
0.4	4.90
0.3	5.25

* Approximately doubled in 1974.

Table 6.1d shows the posted price of fuel oil as a function of sulfur content. Over the indicated range from 2.9 to 0.3 wt%, the incremental value of lower sulfur content fuel oil corresponds to an incremental increase of $1.90 per barrel.

Technology for Desulfurization of Residual Fuels

The petroleum refining industry has developed technology for accomplishing desulfurization of the entire residual fuel. Table 6.1e lists nine de-

Table 6.1e
PROCESSES FOR DIRECTLY DESULFURIZING HEAVY FUEL OIL

	Large Size Pilot Plant	Number of Commercial Plants		
		Operating	Shutdown	Under Construction
Hydrocarbon Research Inc., Cities Service	Yes	2	1	1
UOP[a]	Yes	2	2	2
American Oil Co.	Yes			
ER and E[b]	Yes			
Gulf Oil Co.	Yes	2		1
IFP[c]	Yes			
Shell Oil Co.	Under Construction			
British Petroleum, Ltd.	Yes?			
Chevron Research	Yes	1?		

[a]UOP—Universal Oil Products
[b]ER and E—Esso Research and Engineering
[c]IFP—Institut Francais du Petrole

sulfurization processes for residual feeds offered for license or being developed by engineering licensing organizations and by major oil companies that own high sulfur crude.

In these processes, topped crude is processed in the presence of hydrogen, using catalysts that are suitable for the particular feedstock.

If the entire topped material is processed directly without further separation, the process is referred to as a *direct* residual processing step. On the other hand, if further separation is practiced, such as vacuum distillation, or if the feed is subjected to solvent decarbonization to exclude asphaltene high sulfur and high metals containing constituents from the feed, then the processing step is referred to as an *indirect* technique.

Figures 6.1f, g and h show mild and severe indirect and direct processing of Kuwait crude oil.* The mild indirect technique consists of distilling the vacuum gas oil fraction and desulfurizing this fraction in the presence of hydrogen using catalysts. The desulfurized low sulfur gas oil product is then blended back with the virgin vacuum residue.

This processing route accomplishes about 40% lower sulfur content in the overall refinery fuel oil pool. The direct processing of topped crude (shown in Figure 6.1g) can achieve about 75% sulfur removal of the refinery fuel oil pool, reducing the pool to a 1% sulfur level.

*Domestic supply of crude oil probably will not expand and crude oil from the Middle East represents the source of future supply for the United States for the next decade.

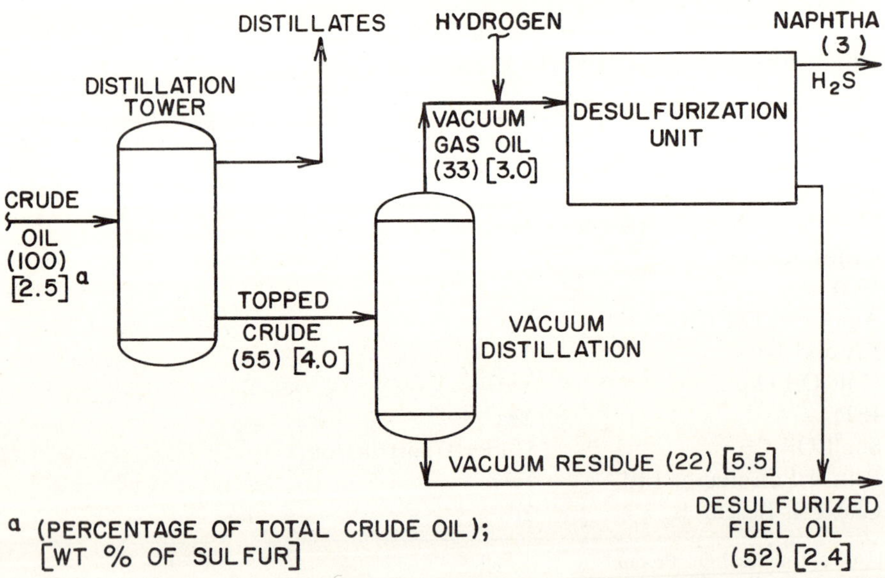

Figure 6.1f Indirect mild desulfurization of Kuwait crude oil.

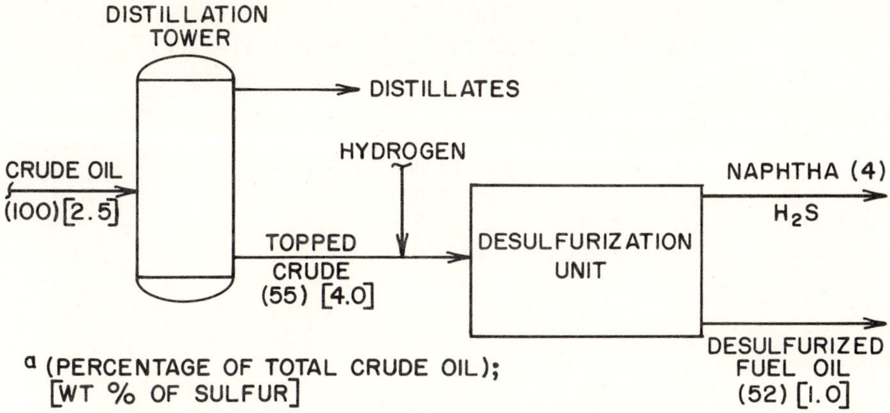

Figure 6.1g Direct desulfurization of Kuwait crude oil.

The indirect route to desulfurization is the current practice in Japan and Venezuela. Applying the indirect technique to 2 to 3% sulfur fuel oil from Venezuela can result in heavy fuel oil with a sulfur content of 1 to 1.5%. The indirect technique is not capable of reducing sulfur levels much further.

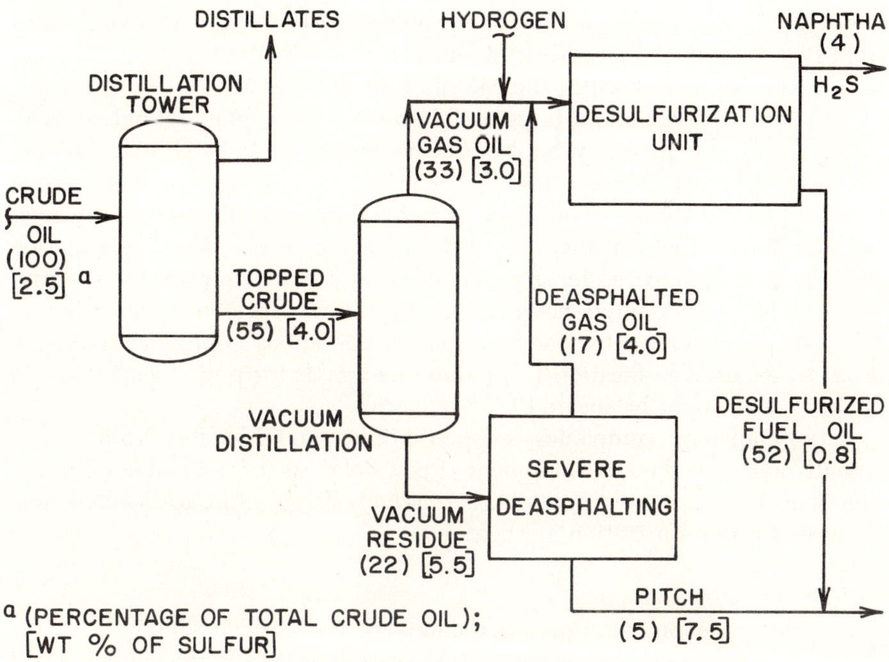

Figure 6.1h Indirect severe desulfurization of Kuwait crude oil.

Table 6.1i
SUMMARY OF DESULFURIZATION CAPACITY

Region	Direct		Indirect	
	No. of Plants	Capacity, MB/D[a]	No. of Plants	Capacity MB/D
Japan	5	193	15	343
Middle East	2	63	0	0
Caribbean	0	0	7	465
Mexico	1	18	0	0
U.S.	3	45	1	20[a]

[a] Excludes indirect units for cat cracking feedstocks.

The direct technique is practiced in the Middle East, Japan and at one United States location (two other operations have been abandoned in the United States at Signal Oil's Texas plant and Humble's Bayway refinery). The severe indirect technique is not yet practiced (Figure 6.1h), but a number of facilities are being studied for applying this indirect technique that uses solvent decarbonization.

Table 6.1i summarizes the regional capacity of all plants in place and under construction that use direct and indirect processes for removing sulfur from heavy residual fuels. Most of the plants are indirect plants, and most of the capacity is located in the Middle East and Japan.

In time, the 0.3% sulfur target will become the acceptable standard. Such a target will also probably be required in Japan. Since catalyst techniques are unavailable to reduce sulfur in the entire fuel oil pool to this level, it is likely that the number 6 oil grade of residual fuel will disappear. However, some high sulfur fuel will always continue to be used in low density population areas in the United States, for ship bunkering, and in areas outside the country.

The high sulfur pitch, however, is surplus and represents a feedstock of relatively low value (1) for partial oxidation operations to produce hydrogen or (2) as feed for gasification to produce fuel gases for power generation in facilities that might be located near refineries.[3]

Two Japanese companies, Nippon Mining and Idemitsu Kosan, have recently announced the development of new catalysts for residual desulfurization. The Japanese have mounted a relatively large effort to develop new catalysts for desulfurization applications.

Reactor Designs

In the development of process technology, two basically different reactor designs have been applied. Figure 6.1j shows schematically the ebullated bed concept featured by the H-Oil process that is licensed by Cities Service Oil

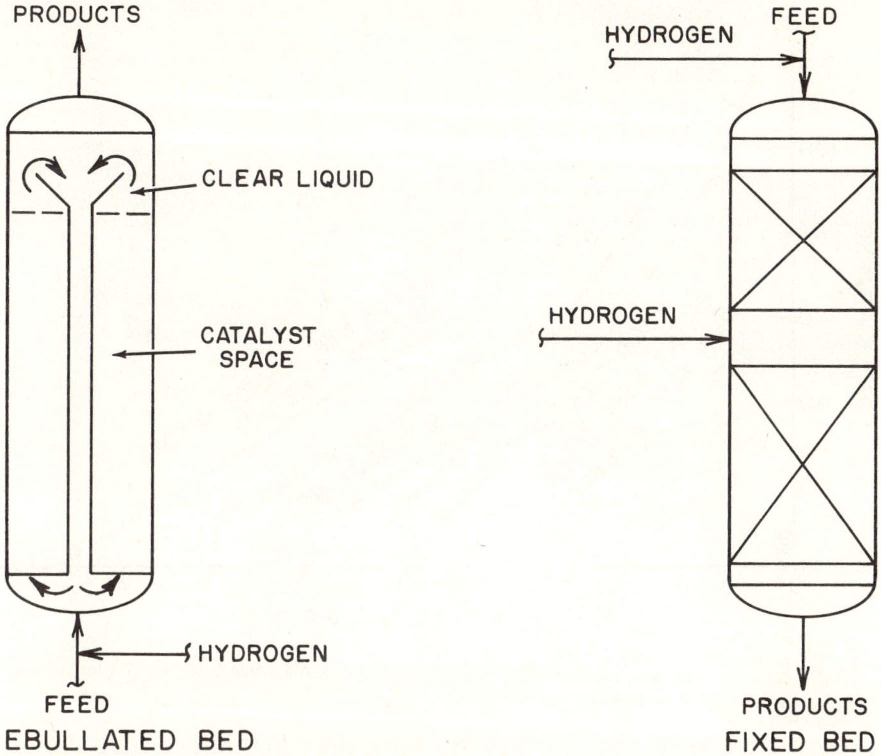

Figure 6.1j Sketch of reactor systems.

Co. and Hydrocarbon Research, Inc. This figure also shows the fixed-bed reactor concept that is used by all other licensors of desulfurization technology. Two operating plants located in the United States and in Kuwait use the ebullating bed technique and all other plants use fixed-bed reactor systems. The features of various processes and reviews of the desulfurization technology are available.[4]

Desulfurization Problems

In applying the desulfurization technology to remove sulfur from topped crude, it would be desirable to directly desulfurize the high sulfur oils to meet projected standards. However, the following limitations exist for direct processing of feeds.

1. Plants need to operate at high pressures, and operating such units is considered hazardous by a number of refiners.
2. Except for relatively easy feeds such as Kuwait, Iranian light

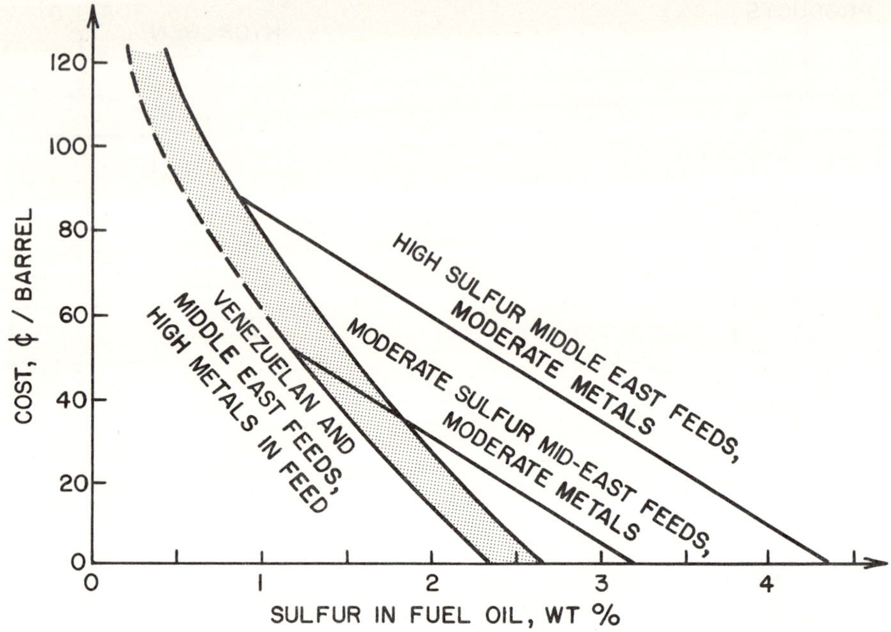

Figure 6.1k Direct and indirect fuel oil desulfurization costs (50,000 bbl daily plant size).

and West Texas crude fractions, the sulfur in residual feedstocks is relatively difficult to remove. Actually there is little evidence that direct processing can achieve sulfur levels of 0.5% in the product oil, except on selected feeds.

3. Some of the more difficult charge stocks contain chelated metallic constituents that poison the catalyst and deleteriously effect performance. The metals are deposited in the pores of the catalyst and preclude regeneration of the spent catalyst. Examples of such feeds include Gach Saran and Cyrus from the Middle East, Bachequero and Tia Juana from Venezuela and the crude oils from the Wilmington Basin in California. Catalyst cost for such feeds is high.
4. Refiners in the United States have been reluctant to commit large investments for desulfurization units because of other needs (e.g., lead removal from gasoline) for capital. They much prefer to locate plants outside the United States and invest capital in stages to meet the required sulfur levels. However, political uncertainties in foreign countries, as well as the lack of firm environmental schedules for low sulfur fuel thus far, have discouraged large capital outlays for new facilities. Attempts to site new facilities

in the United States have also been resisted by environmental groups.

Costs

Figure 6.1k shows in a general way the cost of desulfurizing residual oils by direct and indirect processing. These costs assume utility type economics (20 year plant lives, interest rates of 8 ¾%, 65% debt financing and 13% return on investment). If the more usual capital charges of the petroleum industry were used as a basis, the costs would increase at the 1% sulfur in product level, by about $.30 per barrel. At concentrations below 0.5% sulfur, the cost of desulfurization increases rapidly.

Most of the processes developed cannot handle Venezuelan stocks. Rather

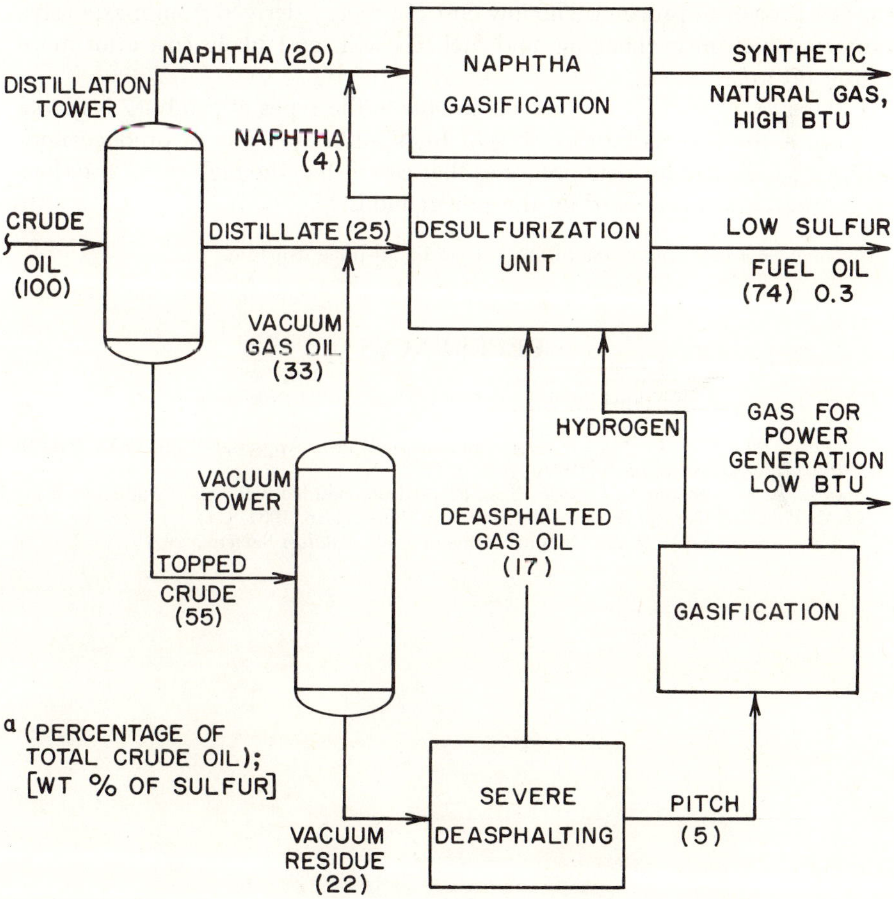

Figure 6.1 l Conceptual clean fuel refinery of Kuwait crude oil.

the developers of these processes look upon the technology as an aid to selling their Middle East crude oil which can be handled by the available technology.

Combination Processing for Low Sulfur Oil and Gas

New concepts of what type of refineries may be built during the next decade are just beginning to emerge. One example of a possible scheme is shown in Figure 6.1 l. This plant would produce (1) 0.3% sulfur fuel oil, (2) synthetic natural gas and (3) low-heating-value gas for use in power generation. It can be shown that with low sulfur fuel valued at a price of $4.50 to $5 per barrel, the cost of gas from such plants can be competitive or lower in cost than gas from other processes.* Although few details are available, Crown Central Petroleum has proposed to build a refinery of this type to process imported oil. A change in import regulations would be required for this project to proceed. The low cost for energy derived from inexpensive imported fuel oil is changing, and fuel oils will most likely cost a lot more in the future.

There is a need for new developments in the types of catalysts available to permit the low cost processing of high sulfur fuel oils to produce nonpolluting fuels and for new processes that can utilize the high sulfur fractions to generate gas to be used by the power industry.

*This might not hold true at the $10 to $12 per barrel price structure.

REFERENCES

1. Cross, J. S.: More Resid Fuel Problems to Come, *Hydrocarbon Processing*, p. 127, September 1971.
2. McLean, John G. (ed.): "U.S. Energy Outlook: An Initial Appraisal 1971–1985," Vol. II, National Petroleum Council, Washington, D.C.
3. Squires, A. M.: "System to Provide Clean Power from Residual Oil," 62nd Annual Meeting of Air Pollution Control Association, New York, June 22–26, 1964.
4. Schuman, S. C. and H. Shalit: Desulfurization of Fuels, *Catalyst Reviews*, vol. IV, no. 2, 1970.

6.2 COAL GASIFICATION

Fuel Gas Heating Values:	Pipeline quality is 900 to 1000 BTU/ft^3 and low-heating-value, utility gas is 150 to 500 BTU/ft^3.
Typical Coal Composition:	75% carbon; 5% hydrogen; oxygen, sulfur and ash.
Status of Plants:	Low-heating-value gas production is in operation.* Pipeline quality gas production is in pilot plant state with commercial operation projected for the late 1970s.**

Coal gasification, broadly defined, is the manufacture of gaseous fuel from coal. The product fuel can be of various heating values and can be used for chemical syntheses as well as for combustion. A typical gasification operation includes coal-handling and/or pretreatment steps, a step for actual conversion of coal to gaseous compounds and subsequent steps to upgrade this mixed gas stream.

Environmental Implications

Coal gasification is drawing much attention in the early 1970s as a fuel-supply method which is likely to become widely used in the United States. This strong attention is due to two national problems: shortage of fuels and power sources other than coal, and concern about the environment.

The shortage problem has various causes, such as dwindling reserves of domestic natural gas, uncertainties in supply of crude petroleum and delays in expansion of nuclear-power capacity. The coal industry itself is not immune from problems such as higher mining costs; but the difficulties associated with

*The first generation Lurgi process can produce power plant fuel gas at about 60¢ per 10^6 BTU in the gas. The gas energy content is 77% of that of the feed coal. The total annual capital plus operating cost of the gas plant (coal cost and sulfur credit included) is about 0.5¢ per KWH generated. The flue gases contain about 550 ppm of SO_2 if the power plant uses this fuel gas.
**As a function of inflation and other factors the capital cost of a 0.25 billion ft^3/day commercial plant is $150 to 300 million and the cost of gas produced is 60¢ to 120¢ per 10^6 BTU.

natural gas, petroleum and nuclear energy have tended to enhance United States coal reserves, which are abundant.

However, coal's sulfur content causes air pollution problems when coal is burned in its natural state. Gasification forestalls this drawback because the gasification operation also removes the sulfur. Thus, it yields fuel that not only capitalizes on coal reserves but is also clean-burning.

The attractiveness of coal gasification will show up most fully in gasification plants which are producing a high-heating-value (e.g., 900 to 1,000 BTU/ft^3), pipeline-quality fuel (hereinafter called pipeline gas) that is fully interchangeable with natural gas since the United States is strongly geared to natural-gas consumption. This pipeline-gas approach is drawing much attention and government support today. Meanwhile, however, utility companies and others are also looking into gasification operations that produce a sulfur-free fuel of lower heating value (e.g., 150 to 500 BTU/ft^3) for burning at utility plants themselves.

History and Current Status

Coal gasification is not new. Plants to make low-heating-value gas from coal have long been used in Europe, where natural gas has been scarce until quite recently. Such gas was also commonly made in the United States until the 1950s when new pipeline networks ushered in the era of natural gas. Coal has also been widely gasified to produce low-heating-value gas for chemical synthesis. However, no commercial operations have produced pipeline gas from coal.

Accordingly, the present status of coal gasification is as follows: (1) commercial manufacture of low-heating-value gas already exists and improved processes for making such gas are being developed; (2) much effort is under way in the United States to develop and commercialize the manufacture of pipeline gas and several projects are expected to be commercially operating by the end of this decade.

Chemistry of Coal Gasification

The chemistry of coal gasification can be put into perspective by considering the difference between coal and methane, because a methane-rich stream is the "target product" of projects aimed at making pipeline gas.

Coal consists mainly of complex organic molecules. Though its carbon content varies, an average figure of 75% is often cited. Similarly, hydrogen content can be taken as around 5%. Among the other significant constituents of the organic molecules in coal are oxygen and sulfur. Pure methane, on the other hand, contains 75% (by weight) of carbon and 25% hydrogen. Thus, the basic chemistry of making pipeline gas calls for breaking down the coal molecules, linking the thus-freed carbon atoms with hydrogen (including hydrogen furnished externally, in elemental form and/or in a compound such as water) to form methane and rejecting the released sulfur and oxygen.

The chemistry behind making low-heating-value gas cannot be set out so simply, because such gases are mixtures of numerous compounds. Even so, the same principles apply: coal molecules are broken down, the carbon is rebuilt into simpler gas molecules and sulfur (though not necessarily oxygen) is rejected.

In either case, however, this simplified chemistry is not only idealistic but also the net result of numerous actual reactions. Five such reactions, for instance, are explicitly gasification reactions:

$$\text{Volatile matter in coal} = CH_4 + C\ (\text{char}) \qquad 6.2(1)$$

$$C + H_2O = CO + H_2 \qquad 6.2(2)$$

$$C + 2\,H_2 = CH_4 \qquad 6.2(3)$$

$$C + O_2 = CO_2 \qquad 6.2(4)$$

$$C + CO_2 = 2\,CO \qquad 6.2(5)$$

Some or all of these reactions, plus others, may take place in a given gasification operation. A reaction also likely to occur during the actual gasification step (even though all reactants and products are gaseous), and which can also be employed downstream from it, is the so-called water-gas shift reaction:

$$CO + H_2O = H_2 + CO_2 \qquad 6.2(6)$$

The downstream use of the reaction in Equation 6.2(6) adjusts the in-process gas stream's H_2/CO ratio, as a prelude to the following methane-producing (methanation) reaction:

$$3\,H_2 + CO = CH_4 + H_2O \qquad 6.2(7)$$

Equation 6.2(7) is not used when making low-heating-value gas. On the other hand, it will be employed in most projects that have thus far been announced for making pipeline gas.

The coal's sulfur content becomes converted to gaseous sulfur compounds during the gasification step. Downstream auxiliary reactions remove these compounds, before the gas stream reaches the methanation step. Similarly, other reactions remove the carbon dioxide produced during gasification and during the water-gas shift reaction.

Chemical Engineering of Coal Gasification

It is no simple matter to physically handle the coal input into the coal gasification plant, especially those varieties that tend to cake. Also, the individual reactions discussed earlier proceed most favorably at temperatures that are quite varied. Reaction 6.2(1) takes place best at around 1100 to 1500°F. Reactions 6.2(2) and 6.2(3) at 1700°F or over, and Reaction 6.2(7) at about 700°F. Reaction 6.2(2) requires a large input of heat, whereas in 6.2(1), 6.2(3) and 6.2(7) heat is released. Indeed the heat release of Reaction 6.2(7) is so vigorous that process design engineers will be hard pressed to forestall a reaction-temperature rise that could harm the reaction catalyst. (Because this methanation reaction is associated with making pipeline gas rather than with the production of low-heating-value gas, there has been no commercial experience with it yet in gasification plants.)

Another design factor is the desirability of "economizing" on carbon: carbon rejected from the overall process as carbon dioxide or as (ungasified) char is wasted, unless the char can be marketed or burned onsite as fuel. Another problem is how to minimize the net input heat required and how best to supply this heat.

In light of such factors and problems, it is not surprising that diverse process designs for coal gasification plants have been and are being developed.

In one new plant designed to make low-heating-value gas in Germany, lump coal reacts with a steam-air mixture in gasifier vessels that operate at 1800 to 2200°F and are equipped with rotary grates for ash removal and with overhead locks for coal feed. Gas made in the vessels goes to a scrubber-cooler system that removes coal dust, tar, alkali salts and other impurities. (This plant needs no sulfur-removal step, because it feeds on low sulfur coal.) Thus cleaned, the gas is ready for onsite use as turbine fuel.

The basic technology that was selected for this German plant is already quite established. Meanwhile, newer approaches are being studied for making low-heating-value gas from coal. Several of these, for instance, employ fluidized-bed gasifiers, wherein particles of an inert material are supported in a fluidlike state by upflowing air. Such beds promise numerous advantages over fixed-bed gasifiers, such as intimate contact between the reactants, and ability to cope with fine particles of coal. Researchers are also studying the moving or entrained-bed gasifiers.

Pipeline Gas Production

Turning now to pipeline gas, one straightforward way of making it is to use established technology to produce a gas stream with low heating value, then use the water-gas shift reaction and the methanation reaction [Reactions 6.2(6) and 6.2(7)] to boost the heating value further. Several pipeline-gas projects that are intended to start commercial operation in the late 1970s will use this approach. Meanwhile, however, there is much effort to develop more advanced overall systems geared explicitly or mainly for pipeline gas. (The upstream portions of some of these, of course, could also be used for making low-heating-value gas.)

In one such approach, now at the pilot-plant stage (Fig. 6.2a), ground-up coal is pretreated with a hot airstream to prevent agglomeration. The coal is then slurried in oil, and the slurry is fed to the top of a three-stage gasifier, in which it flows downward. Meanwhile, a mixture of hydrogen and steam enters the bottom of the gasifier and flows upward, reacting with the coal and picking up the resulting gases. After leaving the vessel, the gas stream

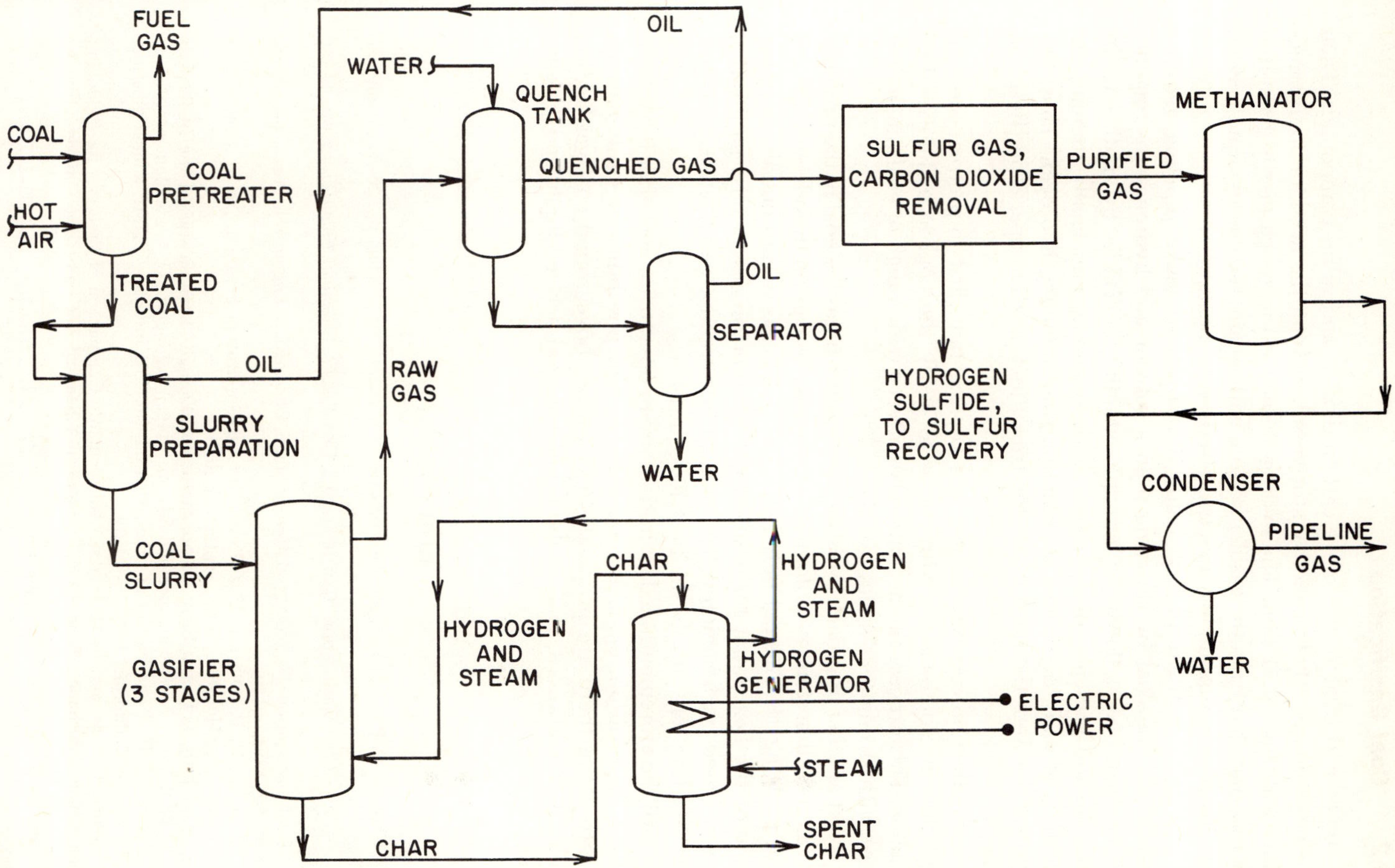

Figure 6.2a One process for making pipeline gas from coal.

is cooled and treated to remove carbon dioxide and sulfur compounds. Then the gases go directly to a methanator vessel, without the need for an intermediate water-gas-shift step. The hydrogen needed for the gasifier is generated by reacting char from the gasifier bottom with steam; the energy input needed for this step is supplied by electrical resistance heating.

A more unusual process being developed to make pipeline gas first "dissolves" the coal by slurrying it in a solvent and hydrogen-treating the slurry, then breaks down the carbon-containing molecules by another treatment in the presence of hydrogen (hydrocracking) and finally submits the resulting stream to yet another reaction step involving hydrogen. Thus, this process sidesteps the need to use the tricky methanation operation [Reaction 6.2(7)].

Another process uses oxygen as a key reactant: first to pretreat the coal and then in combination with steam to gasify the coal in a fluidized-bed reactor. This process can accommodate a very wide range of coals, including ones with strong tendencies to cake.

Yet another approach is distinctive because it gasifies coal in the presence of calcined limestone or dolomite, which absorbs carbon dioxide and releases heat in doing so. This not only helps purify the gas stream but also supplies some of the heat input for gasification. Another process, whose gasification step is based on Reaction 6.2(2), uses molten sodium carbonate to catalyze that reaction and supply the heat it requires.

All the aforementioned operations are to operate on coal after it has been mined. In addition, engineers are studying the possibility of gasifying coal in situ. Also, there are proposals to combine aboveground coal gasification with other operations, such as also producing coal-derived liquid and/or solid fuels.

BIBLIOGRAPHY

Chopey, N. P.: Coal Fuel for Gas Turbines?, *Chemical Engineering*, pp. 36–38, August 24, 1970.

———: Coal Gasification: Can It Stage a Comeback?, *Chemical Engineering*, pp. 44–46, April 3, 1972.

Coal Gasification Technology at Center Stage, *Chemical & Engineering News*, pp. 36–38, January 10, 1972.

Cochran, N. P.: Office of Coal Research Coal Gasification Program, *AIChE* South Texas Section meeting, Houston, Texas, September 1970.

Farnsworth, J. F. and R. A. Glenn: Status and Design Characteristics of the BCR/OCR Bi-Gas Pilot Plant, American Chemical Soc. meeting, Washington, D. C., September 1971.

Garvey, J. R.: "The Impact of Conversion on the Coal Industry," AIME annual meeting, San Francisco, California, February, 1972.

Henry, J. P., Jr. and B. M. Louks: "An Economic Comparison of Processes for Producing Pipeline Gas (Methane) From Coal," American Chemical Soc. meeting, Chicago, Illinois, September 1970.

Hill, G. R.: Some Aspects of Coal Research, *Chemical Technology*, pp. 292–297, May 1972.

Klingman, G. E. and R. P. Schaaf: Make SNG From Coal?, *Hydrocarbon Processing*, pp. 97–101, April 1972.

Office of Coal Research, 1974 Annual Report, United States Department of Interior.

Proceedings of Fifth Synthetic Pipeline Gas Symposium, American Gas Association, Chicago, October, 1974.

Sheldrick, M. G.: Coal Gasification Warms Up, *Chemical Engineering*, pp. 59–61, July 12, 1971.

Shreve, R. N.: "Chemical Process Industries," McGraw-Hill Book Company, New York, 1967.

Squires, A. M.: Keeping Sulfur Out of the Stack, *Chemical Engineering*, pp. 181–189, April 27, 1970.

Symposium on Coal Combustion in Present and Future Power Cycles, American Chemical Soc. meeting, Toronto, Ontario, May 1970.

6.3 PHYSICAL REMOVAL OF PYRITES FROM COAL

Types of Cleaning Operations:	(a) *Wet cleaning,* (a1) Jig washing, (a2) Dense media separation, (a3) Hydrocyclones, (a4) Heavy medium cyclone, (a5) Wet or concentrating tables, (a6) Froth flotation; (b) *Dry cleaning;* (c) *Advanced techniques,* (c1) Magnetic separation, (c2) Electrostatic separation, (c3) Electrophoresis, (c4) Chemical methods, (c5) Biological beneficiation, (c6) Two stage air classification.
Sulfur Concentration of Bituminous Coals:	1 to 7%.
Low Sulfur Coal Availability:	For general use, only 9% of the total 1970 production was available with 1% or less sulfur.
Sulfur Forms in Coal:	Organic (20 to 60%), pyritic (40 to 80%) and sulfate (under 0.05%).
Cost of Coal Washing:	\$.25 to \$.75 per ton of cleaned product.
Effectiveness of Physical Coal Cleaning:	Removes about 50% of pyritic sulfur.
Specific Gravity of Coal Constituents:	Coal: 1.3, Shale: 2.5, Pyrites: 4.9.

The total amount of sulfur present in bituminous coals in the United States varies from less than 1 to 7%. One-third of the coal reserves are east of the Mississippi and two-thirds west. Large deposits of lower-rank subbituminous coal and lignite, which constitute 83% of the relatively low sulfur (~1% by weight sulfur) reserves, lie in the West and are remote from the large urban markets. The low sulfur coal produced in the East amounts to only about 36% of the coal mined. Of this, 26% is exported and 49% sold for metallurgical use, leaving only 25% available for general use. This represents only 9% of the present (1970) total coal production. Because of the air pollution potential of the higher sulfur coals, many urban centers have set limits on sulfur content permitted in coal used for power generation. But the supply of less than 1.0% sulfur coal is limited; therefore, several studies have been made to find methods of more effectively removing the sulfur from coal.

Coal Cleaning

Sulfur occurs in coal in three principal forms: organic, pyritic and sulfate. The amount of sulfate is small (less than 0.05%) and is considered of little significance. The organic sulfur fraction (20 to 60% of the total sulfur) is distributed quite uniformly throughout the coal in a molecular combination with the coal substance and withstands separation unless the coal itself is chemically altered. The pyritic sulfur, however, exists as discrete particles in the coal; the size of these particles varying from seam to seam and from mine to mine. This fraction constitutes from 40 to 80% of the total sulfur in coal. The pyritic sulfur which is finely disseminated as small veins and crystals is the most difficult to remove by coal cleaning techniques; while the pyritic sulfur that is segregated as nodules on cleat faces, as lenses and bands, or as balls or joints, can be removed quite effectively.

Sulfur can be partially removed from coal by means of coal preparation methods now available. NAPCA has estimated[1] that about 25% of the steam-coal production could be cleaned to produce coal with a maximum of 1% sulfur; refinements and broader application of coal cleaning technology might result in an average reduction in sulfur content of the remaining 75% of steam-coal by as much as 40%. Coal washing costs may range from \$.25 to \$.75 per ton of cleaned product.

Separation Methods

Coal preparation or cleaning involves the mechanical removal of waste products such as shale, pyrite and roof-slate by exploiting the differences in physical properties—usually specific gravity—of the coal and the foreign substance. The difference in specific gravity between coal and impurities forms the basis for 99% of all conventional mechanical cleaning processes.[2] The specific gravity of coal is about 1.3, of shale about 2.5 and of pyrites over 4.9.

Mechanical coal cleaning equipment is divided into two general categories: wet and pneumatic. Sometimes elements of these two types of equipment are used in combination. Figure 6.3a illustrates a hypothetical flow scheme for a coal cleaning process. Run-of-mine coal is fed into primary and secondary crushers and is screened until a size range of 6 mesh through 3 inches is attained. A typical wash cycle may consist initially of a jig-washer. In jig-washing, the raw-coal feed is mechanically agitated in a fluid, forming

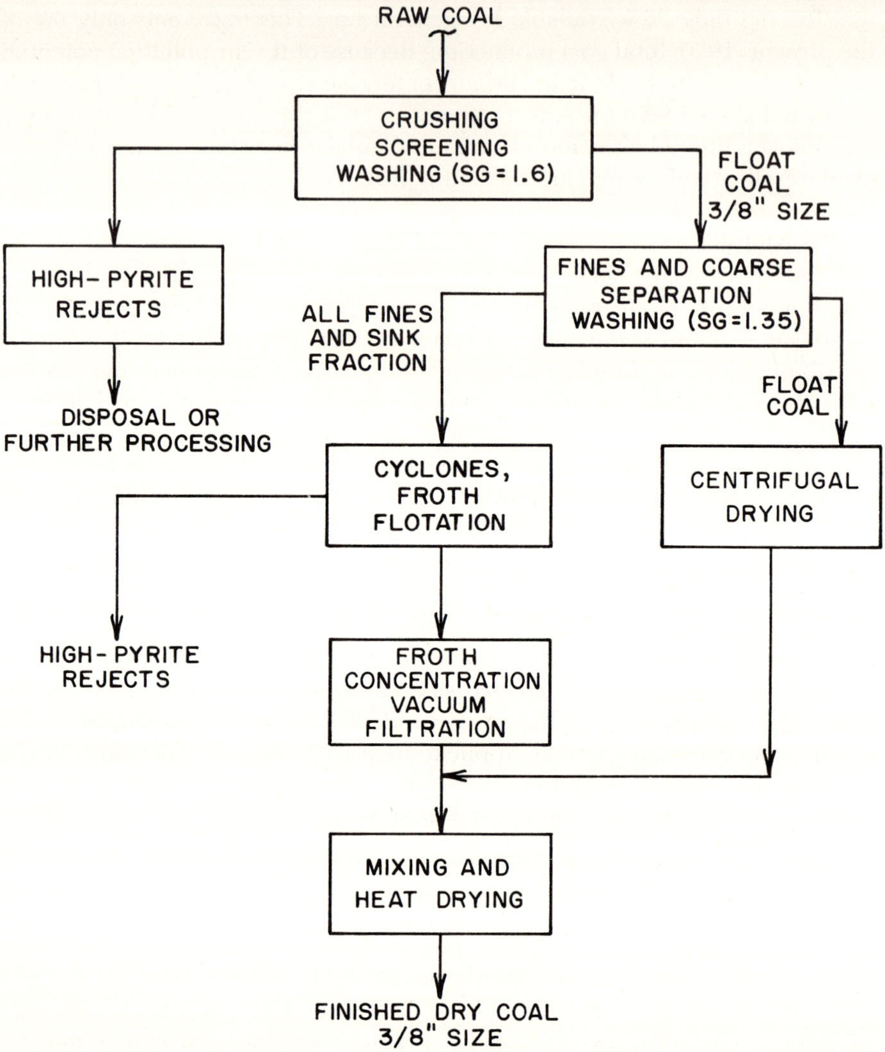

Figure 6.3a Proposed coal cleaning scheme.

a stratification of particles with the lighter specific gravity material (coal) on top and the denser minerals below. These two layers are then sliced into float and sink fractions.

Wet Cleaning

A second type of wet cleaning operation is the dense-medium technique where a size range of 6 mesh through 8 inches pieces of coal are immersed in a high specific gravity liquid. This is usually a bath of finely divided solids in an aqueous solution of electrolytes to simulate a liquid with a specific gravity greater than coal but less than shale or pyrite. About half of the sulfur present as pyrite may be removed by crushing coal to $\frac{3}{8}$ inch and floating it in a liquid of 1.6 specific gravity. The floating portion from this (SG-1.6) bath may vary from a low of 0.88% to a high of 2.67% sulfur; the organic sulfur may vary from a low of 0.718% to a high of 1.58%.[3] Float coal of $\frac{3}{8}$ inch size from the 1.6 SG washing may again be washed at 1.35 SG, from which the floats in turn might be sent to a centrifugal dryer. These dryers might be any of the following types: (1) vertical units with transporting facilities or relying on water or vertical vibration to move the coal, (2) horizontal units with vibration or (3) horizontal solid bowl type (Section I.5.21).

The high-pyrite reject fraction from the 1.35 SG wash is sent to further processing in dense-medium cyclones. The dense-medium cyclones have the advantage of high-cleaning efficiencies at the lower specific gravities. Also the plant construction costs are lower because of less space requirements. Cyclones are able to maintain high-cleaning efficiencies with wide variations in volume and quality of the feed. Hydrocyclones, using water only, have been recently introduced in the United States as wet coal cleaning devices. Though the hydrocyclone is not as effective as the heavy-medium cyclone in cleaning coal in the $\frac{3}{4}$-inch to $\frac{1}{2}$-mm size range, it compares favorably with wet tables for the under $\frac{3}{4}$-inch size range and, with flotation cells for the under $\frac{1}{2}$-mm size range. The hydrocyclone has the advantage over tables and flotation cells in that it has no moving parts, it is compact and requires less structural support.

Wet or concentrating tables separate the coal by use of a differential motion and a riffled deck with water flowing across it. The differential motion provides a sideways conveying action along the tilted deck and the wash water imparts a downward motion on the sloped surface. In many conventional coal washing plants in the United States, the coal is washed by a dense-medium process and the fines and sink fraction are treated on wet tables.

A final step in a wash process could be froth flotation which is based on a difference in surface properties of the particles for pyrite separation. The surface property differences are usually enhanced by the use of chemical additives. Flotation cells may be used to treat coal particles up to 30 mesh in size. The froth is sent to concentrating units and to vacuum filtration, while the sink fraction, high-pyrite reject is sent to further processing. After filtration, the coal is fed to a dryer where it is heated and prepared for shipment.

Dry Cleaning

Pneumatic processes may be used instead of wet cleaning devices; the main advantage is the elimination of the need for drying and water clarification equipment. Air cleaners are used only for coal which is finer than about $\frac{3}{4}$ inch in size, but the cleaning efficiency is low, affected mainly by variations in surface moisture content of the raw coal. This often necessitates the thermal drying of coal prior to dry cleaning. Dust release to the atmosphere also becomes a problem in pneumatic processes.

Advanced Techniques

Tests have indicated that some coals cannot be cleaned to less than 2% sulfur by the conventional means. Advanced methods, still in the experimental stages, may offer solutions to the beneficiation of these coals.

Electrical methods generally employ magnetic or electrostatic separators to dry-clean the coal. A majority of the investigations into the feasibility of magnetic separation of pyrite have involved a type of pretreatment. Thermal steam-air treatment of raw coal imparts magnetic properties of a different nature to pure coal and to pyrite. The formation of magnetic Fe_3O_4 and of hematite Fe_2O_4 on the surface of the pyrite grains supplies the desired magnetic properties, allowing separation of the pyrite from coal. However, extensive work on four West Virginia coals[4] demonstrated that sulfur could be reduced to a greater extent by utilizing a high-intensity superconducting magnetic separation of the raw, untreated coal without the thermal steam-air pretreatment. Magnetic treatment is, however, often unfavorably affected by difficult-to-control variables such as particle size and shape, surface moisture and type of coal. Research at the United States Bureau of Mines involves enhancing the weak magnetism of pyrite by means of microwave radiation.

Electrostatic cleaning employs an electric field acting upon mineral grains flowing through the field. Because of variations in the electrical resistance, a difference in velocity of the grains results. Application of this method to a stream in which the pyrites have already been concentrated may be economically attractive.

A laboratory electrophoresis column has been designed[5] to separate pyrite from coal by taking advantage of an observed difference in the electrophoretic mobilities* of the two minerals in water. The column successfully fractionated a high sulfur raw coal sample, removing both the pyritic and siliceous refuse as well.

Chemical methods using aqueous solutions of nitric acid, chlorine, hydrofluoric acid and caustic for leaching mineral matter from coal have been reported as successful.[4] Molten caustic (sodium/potassium hydroxide ratio 1:1) was used to dissolve pyrite out of coal at various temperatures. Biological beneficiation has involved immersing coal in an acidic (pH 3.5) solution of ferrobacillus-ferrooxidans bacteria at room temperature for four days. Approximately 50% of the pyrite was removed.

The most advanced dry process is the two-stage air classification technique

*Electrophoretic mobility is the velocity of particles in liquid suspension when placed in an electric field of 1.0 volt per cm.

developed by Bituminous Coal Research, Inc. Air classification is a gravity separation method requiring the fine crushing of coal to about 200 mesh. In two-stage cleaning, half of the coal, which is completely pulverized in one pass through the mill, is partially cleaned by air classification. The coarser material rejected by the classifier is more thoroughly cleaned and recycled.

Summary

At present, about 60% of the bituminous coal used in the United States is mechanically cleaned, primarily to remove extraneous impurities and to lower the ash content. These methods are generally effective in the removal of about 50% of pyritic sulfur. Because of the chemical and physical forms in which sulfur occurs in coal, the remaining half of the total sulfur is extremely difficult to liberate. Extremely fine crushing is necessary to break the coal particles apart to permit separation. Fine crushing is prohibitive at the mine because conventional coal transportation methods are not adequate for the fine coal powder. Several advanced experimental approaches to pyrite removal, involving chemical and biological treatment, have shown some promise; however, with existing technology, these would be quite expensive. Magnetic or electrostatic cleaning or air classification seems to hold the greatest possibilities for future treatment.

REFERENCES

1. "Abatement of Sulfur Oxide Emissions from Stationary Combustion Sources," Committee on Air Quality Management, Division of Engineering, National Research Council, Washington, D. C., 1970.
2. Anderson, R. L.: "World-Wide Coal Preparation and Coal Consumption Trends," presented at the 5th International Coal Preparation Congress, Pittsburgh, Pennsylvania, October 3-7, 1966.
3. Fraser, T. and W. L. Crentz: "Washing Characteristics of the Pittsburgh Coal in a High-Sulfur Area in Greene County, Pa.," U.S. Bureau of Mines, Technical Paper 689, Washington, D.C., 1946.
4. Leonard, J. W. and C. F. Cockrell: *Mining Congr. J.*, vol. 56, p. 65, 1970.
5. Miller, K. J. and A. F. Baker: "Electro-Static—Specific Gravity Separation of Pyrite from Coal," U.S. Bureau of Mines, No. 7440 (PB-196-594), 1970.

6.4 HYDRODESULFURIZATION OF COAL

Methods of Hydrodesulfurization:	(a) Solvent refining; (b) H-coal process.
Coal Property Changes Due to Solvent Refining:	Ash content reduced from 7 to 0.14%, sulfur content reduced from 3.3 to 0.95% and heat content (BTU/lb) increased from 14,000 to 16,000.
Status:	Processes discussed are in the pilot stage.

Efforts to convert coal into liquids and gases have had as their primary goal the beneficiation of coal to more valuable products. Sulfur does appear in some of these products. Coal liquefaction has been technically feasible for many years, however, the economics of this process to yield a low sulfur fuel have, until recently, been unfavorable.

Solvent Refining

In the solvent refining process (Section 6.1), finely divided coal particles and anthracene oil are mixed; hydrogen is added to avoid repolymerization; and the slurry is heated to 840°F (Fig. 6.4a). After the coal is dissolved, the ash residue consisting of pyritic sulfur and other minerals is removed by filtration. The light fraction is then removed from the coal solution filtrate by flash evaporation, enabling the solvent to be recovered in a distillation step. The evaporator residue is cooled to a solid, low sulfur, brittle fuel which can be easily pulverized. The product liquefies at 430°F so that it can be burned either in solid or liquid forms. Table 6.4b lists some properties of feed and solvent refined coal.[1]

A recent invention provides for two-stage solvent extraction (Fig. 6.4c).[2] Coal is crushed to an average size of 14 Tyler® screen* at 600 to 900°F and 70 PSIG in the presence of an equal amount of solvent. The coal particles are first contacted with an aromatic hydrocarbon solvent yielding hydrogen-rich and hydrogen-lean liquid coal extract components, and this mixture is filtered to remove the unwanted ash and minerals. In a second extraction step,

*A Tyler mesh 14 has an opening 0.046 inch (1.168 mm) and a wire diameter 0.025 inch.

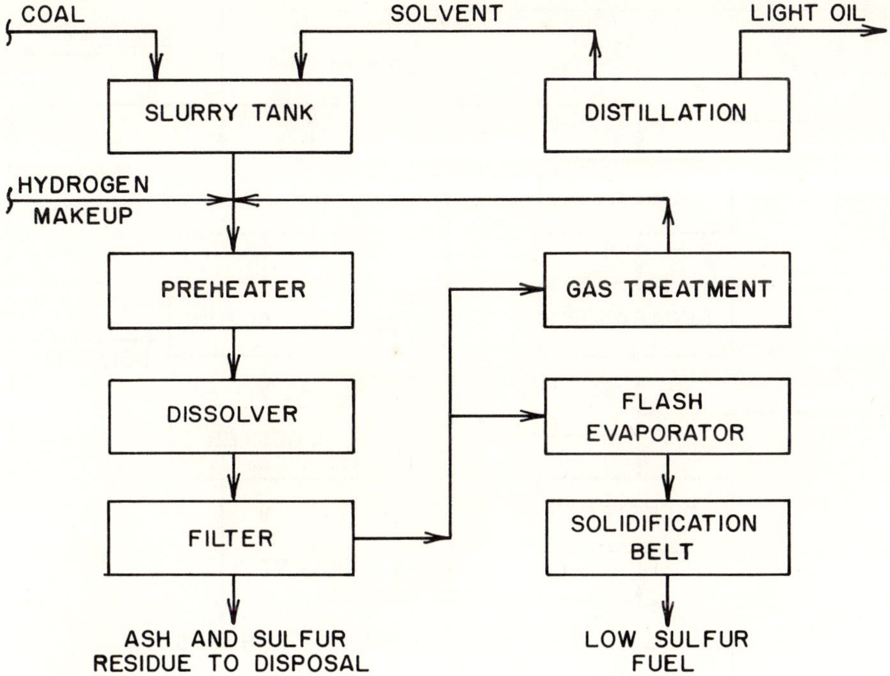

Figure 6.4a Solvent refining.

the liquid coal in the presence of hydrogen is contacted with a ketone selective for the hydrogen-rich fractions.* The preferred operating conditions for this step are 212 to 570°F and 350 to 700 PSIG. The solvent-to-feed ratio is between 0.5 and 5.0 weight basis, with a residence time of between 0.5 and 5.0 LHSV.† The hydrogen-rich components are thereby obtained as a separate product from the hydrogen-lean components contained in the basic liquid coal extract.

* A ketone selective for the hydrogen rich fractions would be one which extracted only the hydrogen-rich portion of the mixture and not the hydrogen-lean portion.
† LHSV is liquid hourly space velocity.

Table 6.4b
PROPERTIES OF SOLVENT REFINED COAL

	Kentucky No. 11 Raw Coal	*Refined Product*
Ash, wt %	6.91	0.14
Sulfur, wt %	3.27	0.95
Heat Content, Btu/lb	13,978	15,956

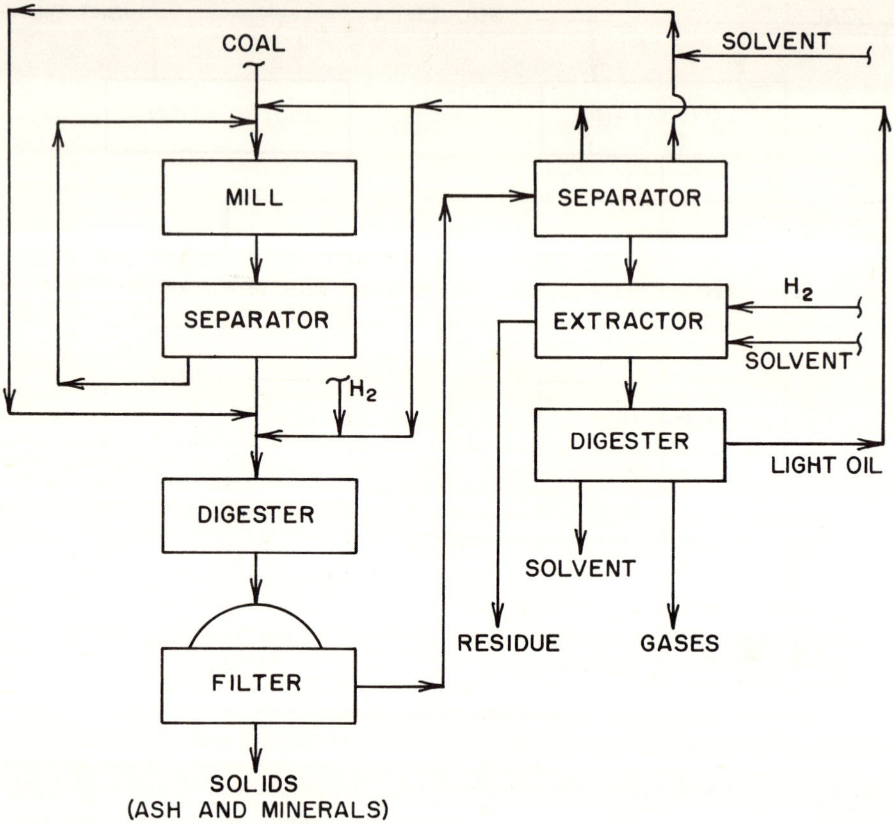

Figure 6.4c Sunagel method for liquefying coal.

H-Coal Process

The H-Coal process recovers via hydrogenation a light crude oil suitable for refining to gasoline. The coal is crushed and mixed with an oily coal derivative. Hydrogenation takes place in catalytic fluidized-bed reactors. The coal is hydrogenated and converted to refinery gases, naphtha, middle distillate and heavy gas oil. The heavy gas oil undergoes catalytic hydrocracking to yield middle distillate, naptha and refinery gas. Naphtha is processed and reformed to gasoline.

The H-Coal process, is shown in Figure 6.4d. Aqueous ammonium sulfides recovered from the coal and from heavy gas-oil hydrogenation steps are converted into H_2S and ammonia

$$(NH_4)_2S \rightleftharpoons H_2S + 2\,NH_3 \qquad 6.4(1)$$

$$NH_4HS \rightleftharpoons H_2S + NH_3 \qquad 6.4(2)$$

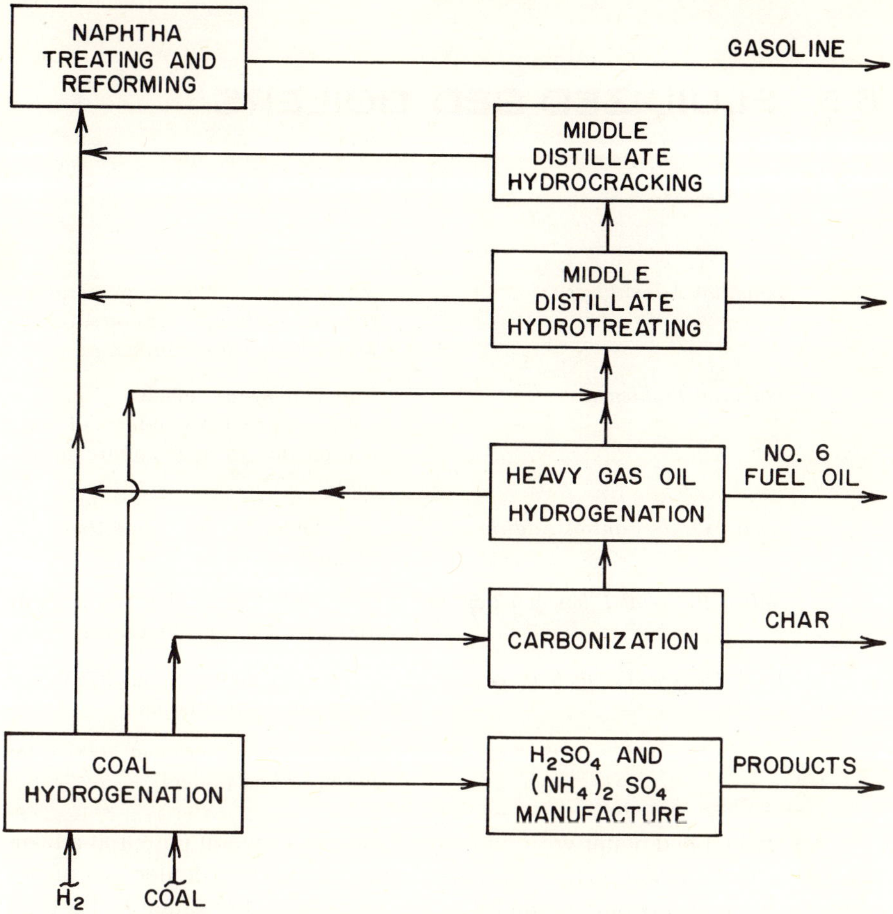

Figure 6.4d H-coal process. *(Developed by Hydrocarbon Research, Inc.)*

and the H_2S is further oxidized to sulfuric acid. The ammonia is reacted with the acid to yield ammonium sulfate.

$$H_2S + 2\,O_2 \rightleftharpoons H_2SO_4 \qquad 6.4(3)$$

$$H_2SO_4 + 2\,NH_3 \rightleftharpoons (NH_4)_2SO_4 \qquad 6.4(4)$$

REFERENCES

1. Jimeson, R. M.: *Chem. Eng. Progr.*, vol. 62, p. 53, 1966.
2. Sunagel, G. R.: U.S. Patent 3,598,717, August 10, 1971.

6.5 FLUIDIZED-BED BOILERS

Potential Advantages:	Smaller size, lower temperature, reduced fouling, corrosion and pollutant (SO_x, NO_x) discharge.
Potential Problems:	Low combustion efficiency, gas distribution problems and loss of fluidization due to agglomerate formation.
Sulfur Oxide Control Additives:	(a) Dolomite; (b) Limestone; (c) Magnesite.
Sulfur Removal Efficiency Reports:	90% with 0.7mm dolomite, nearly 100% with 0.1mm limestone.
Heat Transfer Coefficients in Boilers, $W/m^2 - °C$:	27 to 81 for conventional, 272 to 495 for fluidized-bed boilers.
Gas Temperatures in Boilers, °C:	1500°C for conventional, 1000°C for fluidized bed boilers.

A fluidized-bed boiler with submerged tubes has several potential advantages over the conventional fossil-fuel burning radiant heat boiler as a means of producing high-pressure steam for power generation. For example, the high heat transfer rates achieved within fluidized beds should reduce the surface area required for heat exchange. It also should be possible to inject additives into the bed to reduce or eliminate the emissions of sulfur oxides into the atmosphere. Lower combustion temperatures in the fluidized bed should lower the amount of nitrogen oxides formed by the nitrogen fixation reaction, as well as reduce the fouling and corrosion of the boiler tubes.

Because of the potential of fluid-bed boilers, they are under active investigation in several parts of the world.[1-3] Research in this area has been active for several years;[4,5] however, there have been no large scale applications to date.

Fluidized-Bed Operations

Fluidization, one of the newer processing techniques, is now employed in more than 40 industrial processes where gases and particulate solids take part in chemical reactions or where close temperature control, rapid heat exchange and movement of solids are important factors. Fluidization is achieved when gases passing upward through a bed of particulate solids experience a loss in static pressure equal to the weight of the bed. The fluidized particles are highly mobile in the bed and are circulated through the bed by the rising gas. The bed itself looks very much like a boiling liquid and has properties similar in many respects to liquids. For example, objects of lesser density will float on the surface of the bed, and the surface of the bed will seek a horizontal level. In addition, the fluidized bed has liquid-like flow properties.

The onset of fluidization occurs when the drag force by upward moving gas equals the weight of the solids bed:

$$(\Delta P)_f A = (AL)(1 - \varepsilon)(\rho_s - \rho)g/gc \tag{6.5(1)}$$

where $(\Delta P)_f$ is the frictional pressure drop, A is the bed cross-sectional area, L is the bed height, ε is the porosity of the bed, ρ_s is the solids density, ρ is the gas density, g is the local acceleration due to gravity and gc the mass-force conversion constant. Using Equation 6.5(1) and the Ergun equation[6] for flow through packed beds, the following expression can be derived for the minimum fluidization velocity:

$$v_{mf} = \frac{g(\rho_s - \rho)\, Dp^2}{1{,}650\, \mu} \tag{6.5(2)}$$

where v_{mf} is the minimum fluidization velocity, Dp is the solids particle size and μ is the absolute viscosity of the gas.

The gas flowrate through a fluidized bed then is limited on one hand by v_{mf} and on the other by entrainment of the gas by the solids. This upper limit to the gas flowrate can be estimated by the terminal or free-fall velocity of the particles. Since most solids of interest would have very small particle sizes, most systems will be in the laminar flow regime and Stokes' law applies:

$$v_t = \frac{g(\rho_s - \rho)Dp^2}{18\, \mu} \tag{6.5(3)}$$

The range of velocities that will produce a fluidized bed is illustrated in Figure 6.5a.

Other important features in the design of a fluidized bed include the inlet gas distributor design, the amount of empty vessel or free board that should be included above the fluidized bed in order to minimize entrainment, separation of dust from effluent gas, and solids feeder and discharge system.

Sulfur Oxides Control

Several processing schemes using the fluidized bed for sulfur oxide control have been under study. In these processes, solid particles of dolomite

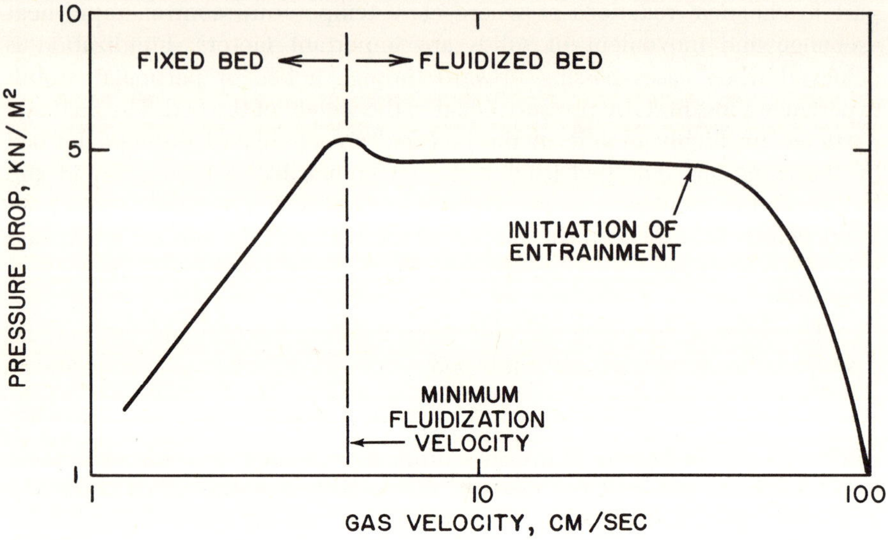

Figure 6.5a Effect of gas velocity on pressure drop through a bed of solid particles.

($CaCO_3$–$MgCO_3$ mixture), limestone ($CaCO_3$) or magnesite ($MgCO_3$) are fed to a high-temperature fluidized bed where the sulfur oxides produced during combustion are fixed as $MgSO_4$ or $CaSO_4$ by chemical reaction. The fluidized solids are a mixture of coal particles and oxides, carbonates and sulfates of calcium or magnesium. Inert sand also has been added to the solids. Air is injected into the bottom of the bed to provide the fluidizing medium. As the air rises through the bed, oxygen reacts with the coal particles to form combustion products and to liberate the heat of combustion.

The reactions of the solids include the calcination reaction that occurs very quickly by:

$$CaCO_3 \longrightarrow CaO + CO_2 \tag{6.5(4)}$$

$$MgCO_3 \longrightarrow MgO + CO_2 \tag{6.5(5)}$$

The oxides then react with sulfur dioxide in accordance with Reactions 6.5(6) and (7).

$$CaO + SO_2 + \tfrac{1}{2}O_2 \longrightarrow CaSO_4 \tag{6.5(6)}$$

$$MgO + SO_2 + \tfrac{1}{2}O_2 \longrightarrow MgSO_4 \tag{6.5(7)}$$

The sulfation reaction occurs on the solid surface; therefore, the amount of surface available for reaction influences the rate of reaction. For example, large-dense oxide particles will have their exterior surface sulfated rapidly, but this sulfate layer on the particle exterior will form a barrier to further reaction.

One solids reactivity was reported to be such that over 90% sulfur removal was achieved at 900°C with a Ca-to-S mole ratio of 2 and with dolomite particle sizes averaging 0.7 mm in diameter.[2] Other studies indicated that 0.1 mm limestone particles gave close to 100% sulfur removal.[7] The difference in solids reactivity is due, at least in part, to the porosity of the particles. Other variables that affect the sulfur removal are the temperature, the gas-solid contact time, the concentration of sulfur oxides in the gas and the amount of calcium in the bed.

The sulfated solids can be withdrawn from the bed continuously and discarded, or they can be regenerated and recycled. Studies of the regeneration reaction have been made at 1050°C using carbon monoxide as the regenerating agent. The reaction occurs in accordance with Equation 6.5(8):

$$CaSO_4 + CO \longrightarrow CO_2 + SO_2 + CaO \qquad 6.5(8)$$

It is visualized that in a large scale operation, regeneration could be carried out by burning coal with a deficiency of air as illustrated by Reaction 6.5(9):

$$CaSO_4 + C + \tfrac{1}{2}O_2 \longrightarrow CaO + SO_2 + CO_2 \qquad 6.5(9)$$

Regeneration permits the recovery of elemental sulfur by processing the SO_2-rich regenerator off-gas. A schematic of a system with a regenerator is shown in Figure 6.5b.

Heat Transfer Considerations

A number of experimental studies have dealt with heat transfer from fluidized beds to immersed vertical tubes.[8] The effects of tube diameter, particle size, shape, density and gas velocity on heat transfer coefficients in horizontal immersed tubes, including data on large-scale beds have also been investigated.[9] Measurements in small-scale fluidized-bed boilers operating in the range of 600 to 1,000°C have also been made. Overall heat transfer coefficients on the order of 272 to 495 W/M^2–°C (50–88 Btu/hr-ft^2-°F) were recorded in these experiments which compare favorably with those predicted by the theoretical correlations.[8,9] Overall heat transfer coefficients in conventional fuel-fired boilers are in the range of 27 to 81 W/M^2–°C. In addition to the high heat-transfer capability, the fluidized-bed system can operate at much lower temperatures, about 1000°C, than the conventional furnace which has gas temperatures of about 1500°C. At the lower operating temperatures, it should be possible to solve the fouling and clinkering problems associated with the mineral matter in coal.

Operational Problems

One of the potential operational problems with the fluidized-bed boiler concerns the combustion efficiency. With fluidized beds of coal and with very low excess air, the combustion efficiency can also be low. In some studies,

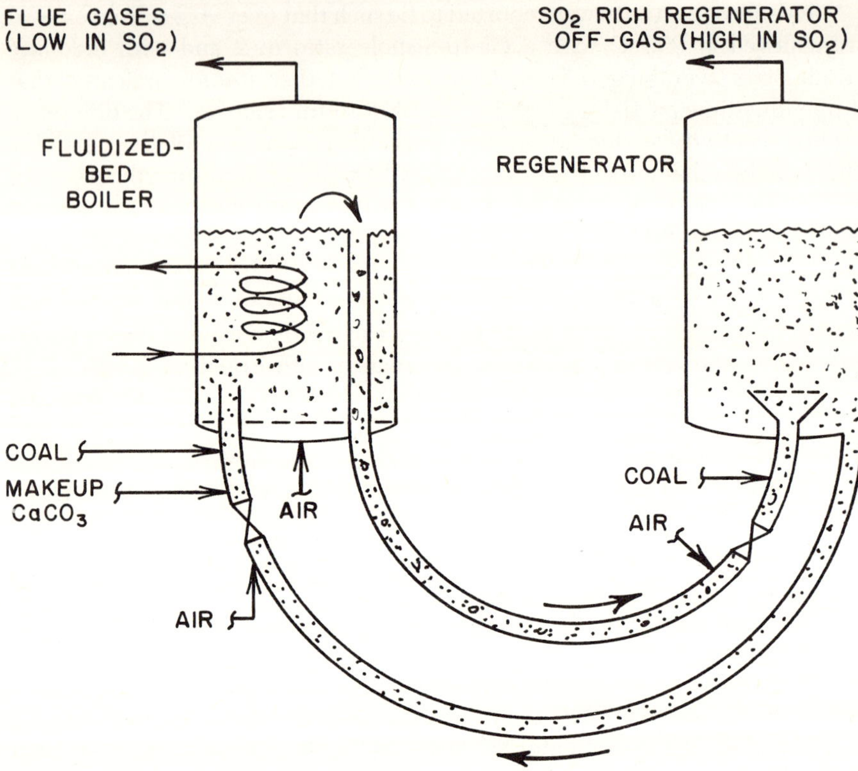

Figure 6.5b Fluidized-bed boiler with regenerator.

inert sand has been added to the bed to improve contacting between the coal and gas.[1] In these studies by the addition of inert solids, almost complete combustion was obtained even at a small percentage of excess air.

Another potential problem is gas distribution. The bed height of a commercial scale operation should be limited to about 1 meter to minimize the air blower power required to overcome the back-pressure caused by the bed. The low height-to-diameter ratio of the bed increases the possibility of maldistribution of the fluidizing gas. A proper grid design can minimize this problem.

Some coals with relatively low ash-fusion points have shown a tendency for the ash to form a soft bond between particles. The agglomerates thus formed are too large to fluidize. This loss of fluidization can be prevented to some extent by higher fluidization velocities which tend to break up the soft agglomerates or by selecting operating temperatures away from the range where agglomerates form.

REFERENCES

1. Bowling, K. McG. and P. L. Waters: *British Chem. Eng.*, vol. 13, no. 8, p. 1127, 1968.
2. Zielke, C. W. et al.: *J. Air Pollu. Contr. Ass.*, vol. 20, no. 3, p. 164, 1970.
3. Godel, A. A.: *Rev. Gen. de Termique* (France), vol. 5, no. 52, p. 349, 1966.
4. Blaskowski, H. J.: Steam Teneration, U.S. Patent 2,842,102, 1958.
5. Johannsen, A. and W. Danz: Combustion of Carbon-Containing Materials and the Recovery of Heat, German Patents 973, 248 (1959) and 973, 344 (1960).
6. Ergun, S.: *Chem. Eng. Progr.*, vol. 48, p. 89, 1952.
7. Williams, D. F.: "Proceedings of the First International Conference on Fluidized Combustion," Oxford, Ohio, November 18–22, 1968.
8. Wender, L. and G. T. Cooper: *AIChE Journal*, vol. 4, p. 15, 1958.
9. Vreendenberg, H. A.: *J. Appl. Chem.*, Supplementary Issue No. 1 S26-S33, 1952.

6.6 SULFUR OXIDE CONTROL— DRY ADDITIVE INJECTION, LIMESTONE

Additives:	Limestone or dolomite.
System Components:	Additive grinders, injection system, dust collection, storage or disposal facilities.
Temperature Ranges:	Boiler temperature from bottom to top: 3000 to 1800°F, temperature required for CaO to react with SO_2: 1200 to 2250°F. At over 2100°F, sintering occurs and best performance is at around 1800°F. The upper limit for MgO is 1550°F.
Velocities:	Flue gas velocity: 60 fps, additive injection velocity 33 to 67 fps.
By-product of the Process:	Sulfur is not recovered in salable elemental form but as gypsum.
Quantity of Additive Required:	For dolomite hydrate see Figure 6.6c.
Sulfate Content of Reacted Solid Additive:	Up to 50% with dolomite hydrate.
Residence Time Requirements:	About 1 to 5 seconds.
Advantages:	Noninterference with boiler operation, low equipment cost, no corrosion of metal surfaces and no loss in plume buoyancy.
Problems:	Control of particle size, dispersion, temperature and residence time.
Disadvantages:	Increased load on precipitator and problems of gypsum disposal.
Costs:	Limestone cost is about $2.05 per ton. In a 200 MW power plant burning coal with 3.5% sulfur, the total cost is about $1.05 per ton of coal burned.

The dry limestone injection process grew out of the need for eliminating sulfur dioxide from flue gases in a simple and direct fashion with as little interference to the operation of the boiler as possible. In principle, the process involves feeding pulverized raw stone to a fuel-fired furnace at a location where maximum distribution of the particles with the hot furnace gases can take place. The limestone or dolomite is calcined to CaO and MgO by the heat of combustion and these react with oxides of sulfur in the flue gas to form sulfites and sulfates:

$$CaCO_3 \underset{\text{heat}}{\rightleftharpoons} CaO + CO_2 \qquad 6.6(1)$$

$$CaO + SO_2 + \tfrac{1}{2}O_2 \rightleftharpoons CaSO_4 \qquad 6.6(2)$$

The solid reaction products are carried by the combustion gases, along with the flyash, to the electrical precipitators, bag filters or other ash collection equipment. The major facilities required are equipment for receiving, handling and grinding limestone or dolomite, an adequate injection system to provide an even distribution of the particles in the combustion gas at the high-temperature zone of the furnace, sufficient collection capacity to handle the increased dust loading and increased capacity for dust disposal or storage.

The dry limestone injection process is relatively easy to operate and is expected to require a low capital investment, especially with existing power stations. The major disadvantage is that the limestone does not completely react with SO_2 to produce a stoichiometric yield of $CaSO_4$. The success of the conversion depends on carrying out the heating and calcining of the particles and the reaction with SO_2 in the few seconds while the particles remain in the flue gas at the correct temperature. From thermochemical considerations, CaO is capable of reacting with SO_2 at temperatures between 1200 and 2300°F if sufficient time is available. In the open, radiant furnace of a typical boiler, the temperature near the bottom is about 3000°F or higher, while at the top it ranges around 1800°F. Since the limestone must be added near the top to satisfy the reaction temperature requirements, and since the residence time of the gas in this temperature zone is very short, the limestone must be well dispersed in the combustion gas early enough so that there is maximum gas-sorbent contact.

A typical gas velocity in this zone is 60 ft/sec or greater, therefore the dynamics of the powder dispersion process become critical. Another complication is that the temperature must be high enough to calcine the limestone without sintering. Furthermore, slagging reactions between the limestone and flyash may remove part of the added lime which might otherwise react with the SO_2. Another variable, particle size, influences the location of injection since the rate and degree of calcination is a function of the size distribution.

Another disadvantage is that this is a "throw-away" process. The sulfur is not recovered in salable form and, therefore, an air pollution problem is converted to a solid waste problem. Attempts are being made, however, to develop successful processes to extract elemental sulfur from gypsum. On the other hand, the abundance of relatively inexpensive limestone and dolomite make the dry additive injection process attractive as a stop-gap measure until a more economical recovery type process can be developed.

Process Chemistry

It is possible to estimate the effectiveness of limestone and dolomite in removing SO_2 from flue gases by calculating the thermochemical equilibrium constants required to support a given mole fraction of SO_2 in the flue gas atmosphere. The CO_2 present as a product of combustion suppresses the release of CO_2 from heated limestone.[1] The calcination temperature must increase in order to make CaO available in equilibrium with higher concentrations of CO_2 (Fig. 6.6a). Also, any CaO formed at a higher temperature will revert back to $CaCO_3$ if the flue gas temperature drops.

Figure 6.6b demonstrates the temperature range where CaO and MgO are effective as SO_2 scavengers. Almost no SO_2 removal is possible with CaO above 2250°F whereas the upper limit for MgO is about 1550°F. Practically all SO_2 is theoretically removed at 1800°F by CaO; however, the equilibrium

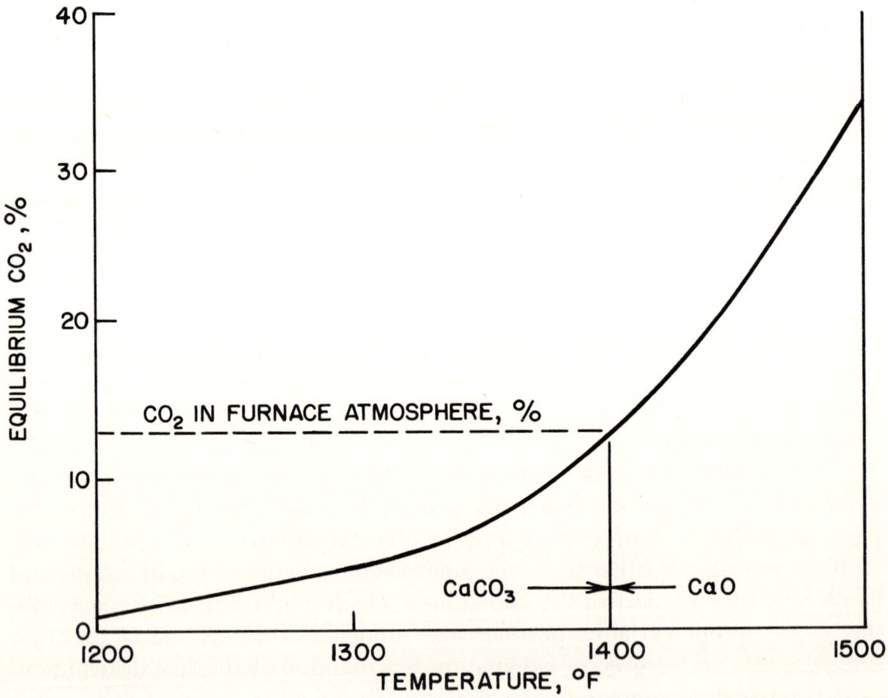

Figure 6.6a Influence of furnace atmosphere on dissociation of $CaCO_3$.

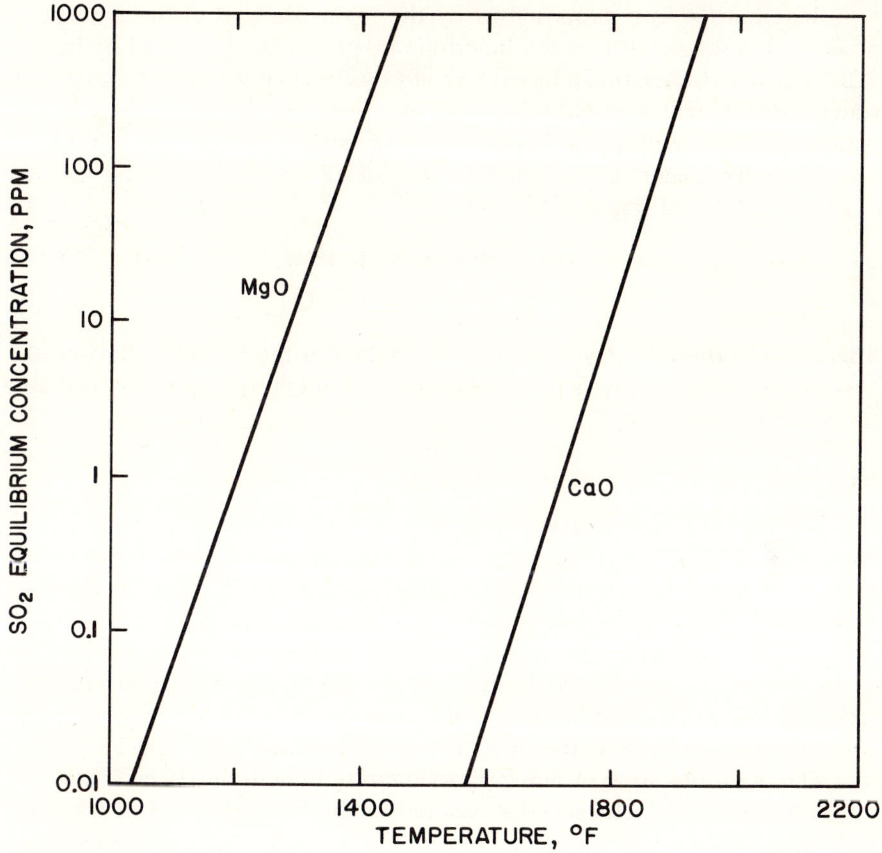

Figure 6.6b Sulfur dioxide concentrations in equilibrium with CaO and MgO at furnace temperatures.

temperature for complete removal (~1 ppm) of SO_2 in the presence of MgO is 1200°F.

Dry solid CaO is a poor absorber of SO_2 and only when water is available at the surface will sufficient absorption and reaction take place.[2] This is illustrated by the following sequence of reactions:

$$CaCO_3 \rightleftharpoons CaO + CO_2 \tag{6.6(3)}$$

$$CaO + H_2O \rightleftharpoons Ca(OH)_2 \tag{6.6(4)}$$

$$Ca(OH)_2 + SO_2 \rightleftharpoons CaSO_3 + H_2O \tag{6.6(5)}$$

$$CaO + SO_2 \rightleftharpoons CaSO_3 \tag{6.6(6)}$$

$$CaSO_3 + \tfrac{1}{2}O_2 \rightleftharpoons CaSO_4 \tag{6.6(7)}$$

$$CaSO_3 + CO_2 \rightleftharpoons CaCO_3 + SO_2 \tag{6.6(8)}$$

Reaction 6.6(5) is proposed to be the main reaction in the absorption process. It is suggested that the limestone be precalcined.[2] Partial hydrolysis at the particle dissociation temperature of $Ca(OH)_2$ in flue gas containing 7.1% water is 682°F, which would rule out its use; however, the rate of dehydration is slow enough to allow the hydrate to exist during the reaction period.

Lime and magnesia are quite active fluxing agents in coal ash and act to lower the fusion point of the ash:

$$2\,CaO + SiO_2 \rightleftharpoons Ca_2SiO_4 \qquad 6.6(9)$$

$$CaO + SiO_2 \rightleftharpoons CaSiO_3 \qquad 6.6(10)$$

This leads to possible slagging and tube fouling in the furnace. Silicates also have the effect to remove some of the lime that might otherwise be available for reaction with SO_2.

Pilot Tests

The idea of injecting an alkaline reagent, derived from dolomite or limestone, directly into a furnace was pioneered by Wickert at Wolfsburg, West Germany.[3] He reported 70 to 90% removal of SO_2 by injecting an approximately stoichiometric amount of slaked calcined dolomite. Later, it was found[4] that the temperature at the point of injection was critical. If the temperature was above 2100°F, the calcium and magnesium oxides sintered, reducing the surface area to less than 0.5 m^2/g. Best results were obtained when the temperature at the point of injection was around 1800°F.

The desulfurization of flue gas by dolomite hydrate has been studied on a bench scale unit.[5] The powder was fed in the bottom of a 4-in ID, 23-ft high vertical reactor. The mixture flowed upward and samples of gas and solid were withdrawn at ports along the axis of flow. The results are shown in Figures 6.6c and 6.6d. Approximately four times as much dolomite as SO_2 is required to remove 100% SO_2 when dispersed as powder smaller than 60 μm. If the size distribution is larger than 60 μm, only about 50% of the SO_2 is removed at the 4:1 feed ratio. An upper limit of 50% of the solid reacts to yield the sulfate; this suggests that, after a surface layer of $CaSO_4$ is formed, there is a rapid decrease in reaction rate.

The application of dry additives to smaller than utility-size industrial installations was also explored.[6] The effects of variations in feed rate, particle size, temperature and residence time on calcination, calcium oxide utilization and additive efficiency were measured.

In a 450 lb/hr coal-firing capacity unit, dry additives were injected at two points: one over the fire bed, just below the reactor exhaust; and the other a few feet toward the front of the furnace and directly over the fire bed. The additives were injected with a modified sandblasting nozzle, using preheated compressed air. A screw type feeder conveyed the powder to the injector

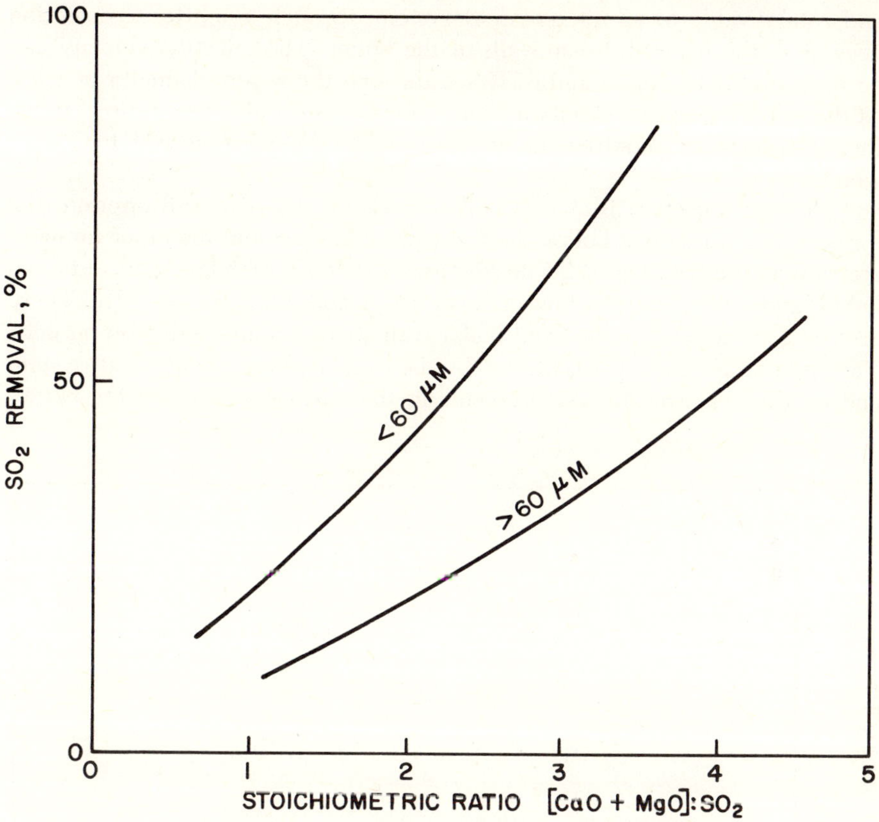

Figure 6.6c Desulfurization with dolomite hydrate.

from the pulverizer and screener. Maximum SO_2 removal was 65% at additive feed rates of 1.5 to 2.0 times the stoichiometric (300 to 400 lb 96% $CaCO_3$ per ton of coal). Tests indicated that very little additional desulfurization occurred after 1-sec residence time at 1800°F and after 0.6 sec at 2400°F.

Sulfur dioxide removal efficiency increased three-fold in going from 50 mesh to 200 mesh at a feed rate of 200 lb per ton of coal burned. Injection air velocity demonstrates a 12% increase in SO_2 removal in going from 150 to 300 ft/sec. Calcination is evidently not a limiting factor for SO_2 sorption.

Plant scale tests at TVA were based on a 20 tons/hr limestone feed (Fig. 6.6e).[7] Crushed limestone in the size range of 0 to $\frac{1}{2}$ inch is dried and pulverized in a ball mill so that 70% is less than 200 mesh in size. The powder is transported pneumatically to a cyclone separator and an air separator; the air separator acts as a classifier to discharge the pulverized material of the appropriate size into a storage silo. The powder, when ready for use, is transported pneumatically to a feed tank, then it is carried by a screw conveyer and rotary valve to an air transport system, through a splitter to injection nozzles in the front and rear walls of the boiler. The injection velocity can be regulated between 33 and 67 ft/sec through the $2\frac{1}{2}$-in. diameter nozzles. At the upper injection velocity and at a limestone-to-sulfur feed ratio of twice the stoichiometric requirement, the ratio of air-to-limestone is 1 to 3 on a weight basis.

Primary injection is at two boiler levels on the rear wall opposite the burners: one a few feet below the top row of burners and the other immediately under the boiler nose. Six nozzles are installed at each level; the distances below and above the top burner level are 4 and 12 ft, respectively. There are also four nozzles on the front boiler wall about 6 ft above the top burners. The amount of carrier gas is limited by the capacity of the boiler draft system and by the heat transfer requirements in the various sections of the boiler.

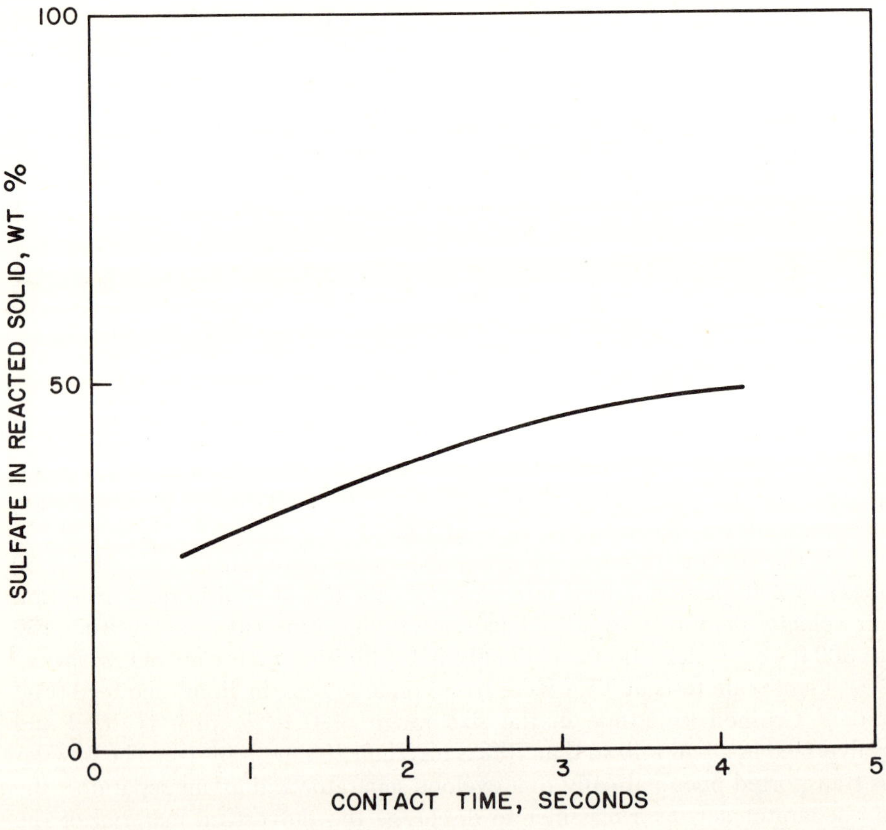

Figure 6.6d Residence times required for conversion of SO_2 by dolomite hydrate.

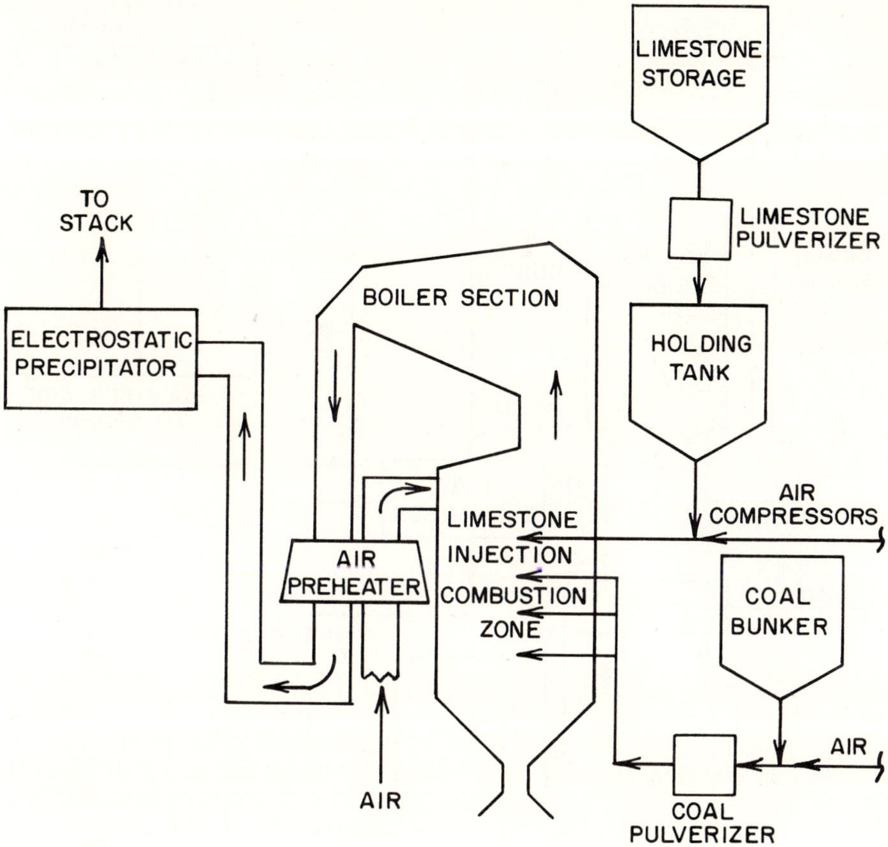

Figure 6.6e Dry limestone injection.

Collection and Disposal of Products

The volume of material to be collected by electrical precipitators (Sections 5.22 to 5.24) in the dry limestone injection process is several times what it would be with only flyash present. In eliminating SO_3 from the flue gas, the collection efficiency of the precipitator is also lowered. A small percentage of SO_3 was formerly an advantage since it would be absorbed by the flyash, thereby reducing its electrical resistivity. It appears, therefore, that existing precipitators will have to be increased in capacity before adopting the limestone injection process. A major drawback to the process lies in the need for the disposal of large amounts of gypsum. This has prompted research into processes for converting the gypsum into a salable product.

Results of an investigation of the thermodynamics and decomposition

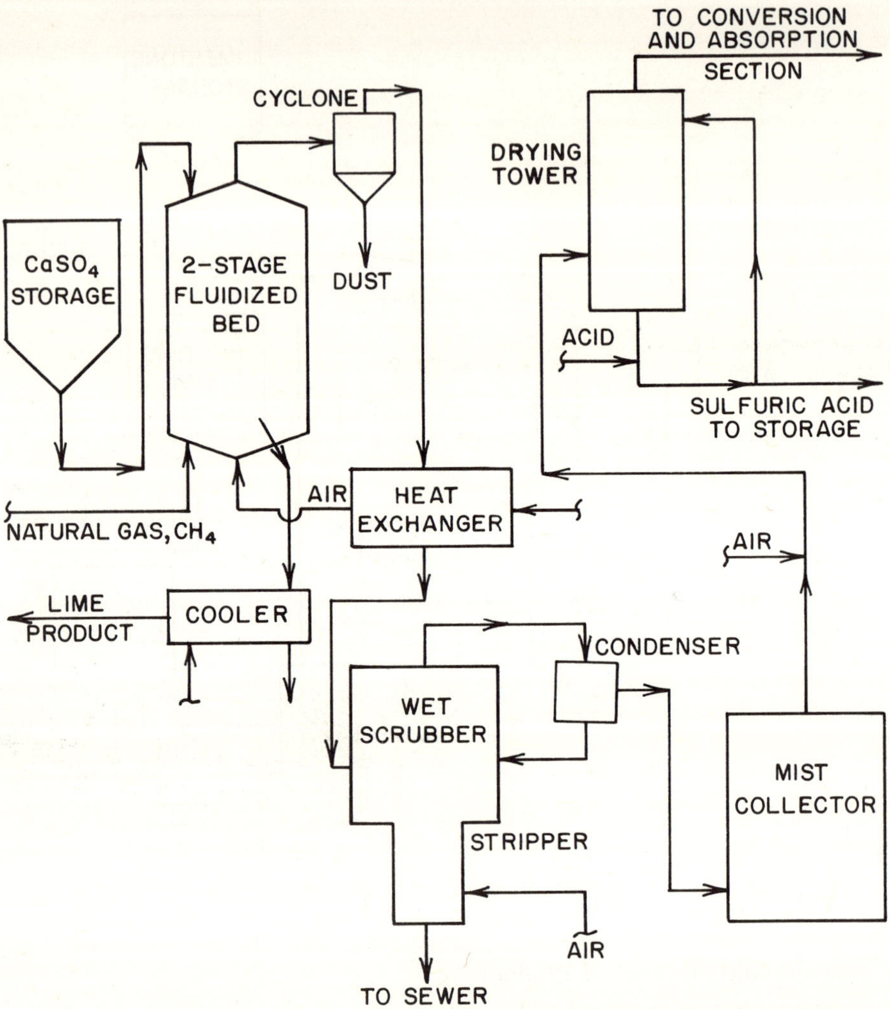

Figure 6.6f Calcium sulfate decomposition and gas purification.

kinetics of calcium sulfate have been reported.[8] Based on these results, a method has been proposed for the large-scale decomposition of $CaSO_4$, involving the heating of solid particles of gypsum or anhydrite in the presence of a reducing gas mixture consisting of unburned hydrocarbons in air (Fig. 6.6f). A feed of 450 tons $CaSO_4$ per day is conveyed to a fluidized-bed reactor and exposed to 1600 cfm of partially oxidized methane. The bed is in two stages. The first stage is used for the preheating of the anhydrite at about 1565°F. The feed then descends to the second stage where the high temperature (2200°F) allows the gypsum to react:

$$4\,CaSO_4 + CH_4 \longrightarrow 4\,CaO + 4\,SO_2 + CO_2 + 2\,H_2O \qquad 6.6(11)$$

Partial combustion of the gas within the fluidized bed supplies the reducing agent and the heat necessary for the reaction. The lime produced is then conveyed to storage for reuse in the flue gas, while the concentrated SO_2 is converted catalytically to sulfuric acid.

Summary

The prospect of a sulfur dioxide removal process based on dry additive injection is attractive from the standpoint of noninterference with boiler operation. The investment in extra equipment is low since the boiler itself is used as the reactor. The flue gas temperature and velocity are not significantly reduced, allowing sufficient plume buoyancy from the stack. Because the process is dry, there is little corrosion in superheaters, air heaters and other metal surfaces in contact with the flue gas. Technical problems which are yet to be solved include achieving maximum dispersion of particles of appropriate size in the correct temperature zone for a residence time sufficient to provide complete reaction of the limestone with the SO_2.

Early estimates of the cost for a 200 MW pulverized-coal-fired unit burning 3.5% sulfur coal, based on a stone additive cost of $2.05 per ton delivered, were $1.05 per ton of coal burned. This is likely to be higher if limestone delivery and crushing costs increase, and expanded dust collection facilities are required.

The limestone process may be applicable to many existing power plants. Since there is no salable product, it would appeal to utilities that do not wish to get into the chemical by-product business.

REFERENCES

1. Reid, W. T.: *Trans. ASME, Series A,* vol. 92, p. 11 1970.
2. Meyer, B. and C. Carlson: "Mechanism on SO_2 Absorption on Metal Oxides, Carbonates and Hydroxides," presented to Second International Clean Air Congress, Washington, D.C., December 6–11, 1970.
3. Wickert, K.: *Mitteilungen der VGB,* vol. 83, p. 74, 1963.
4. Wahnschaffe, E.: *V.J.K.-Berichte,* vol. 155, p. 20, 1964.
5. Kruel, M. and H. Juntgen: *Chemie Ingenieur Techik,* vol. 39, p. 607, 1967.
6. Whitten, C. M. and R. G. Hagstrom: *Trans. ASME, Series A,* vol. 92, p. 1, 1970.
7. Slack, A. V. and H. L. Falkenberry: *Trans. ASME, Series A,* vol. 92, p. 5, 1970.
8. Wheelock, T. D. and D. R. Boylan: "Sulfuric Acid from Calcium Sulfate," *AIChE Technical Manual,* T-37, p. 41, 1971.

6.7 SULFUR OXIDE CONTROL—DRY ABSORPTION PROCESSES

Types of Processes:	(a) Alkalized alumina process (Bureau of Mines); (b) Activated manganese oxide (Mitsubishi DAP-Mm process); (c) Still process; (d) Sodium bicarbonate absorption (Precipitator Corp. and Dow Chemical Co.); (e) Uranyl oxide scrubbing. NOTE: The letters (a) to (e) in the summary below refer to the above listed processes.
SO_2 Absorption Temperature Ranges, °F:	265 to 625 (a); 200 to 900 (c); 32 to 360 (d); over 550 (e).
SO_2 Removal Efficiency:	95% (c); 90% (b); 65% (d).
Advantages:	No loss of plume buoyancy, little corrosion of metal surfaces and, in most cases, no pressure loss in the flue system.
Disadvantages:	Absorbent regeneration is difficult and they lose their activity after some regeneration cycles. The load on the precipitator is increased.
Process Status:	(a) Research discontinued due to high cost of excessive attrition; (b) Plant scale tests with 90% removal efficiency were successfully conducted; (c) Temporarily discontinued; (d) Testing not fully completed; (e) Laboratory studies performed.

A number of dry absorption processes have been proposed wherein the spent absorbent is treated to recover sulfur or sulfur dioxide and the regenerated sorbent is returned to the feed.

The U.S. Bureau of Mines investigated[1] 26 different metal oxides to determine their capacity to absorb SO_2. Gas flow, grain size and gas composition were kept uniform in all tests (Table 6.7a).

Table 6.7a

SULFUR DIOXIDE ABSORPTION TESTS AT 130°C (266°F)[a,b] (90% REMOVAL; HOURLY SPACE VELOCITY OF GAS, 1,050 HR^{-1}; MESH SIZE OF ABSORBENT, 8–24)

Absorbent	*Crystalline Phase, X-ray analysis*	*Purity (wt %)*	*Bulk Density, g/cc*	*SO_2 Absorbed (g/100 g absorbent)*	*Preparation*
Manganese oxide	$MnO_{1.88}$	90	0.14	33	$MnSO_4 \xrightarrow[NH_3]{(NH_4)_2S_2O_8}$ Precipitate washed and dried at 130°C
Cobalt oxide	Co_3O_4	97	0.46	25	$CoSO_4 \xrightarrow{Na_2CO_3}$ Precipitate washed, dried at 130°C and heated in vacua at 300°–340°C for 20 hr
Aluminum-sodium oxide	Al_2O_3 Na_2O	64 31	0.73	24	$Al_2(SO_4)_3 \xrightarrow{Na_2CO_3}$ Precipitate washed, dried at 130°C and heated with H_2 at 600°–640°C for 10 hr
Manganese oxide	$MnO_{1.88}$	94	0.50	23	$MnSO_4 \xrightarrow{Electrolysis}$ Precipitate washed, dried at 130°C and heated in vacua at 300°–340°C for 20 hr
Manganese oxide	γ–Mn_2O_3	96	0.67	19	$MnSO_4 \xrightarrow{Na_2CO_3}$ Precipitate washed, dried at 130°C and heated in vacua at 300°–340°C for 20 hr
Hopcalite	CuO MnO_2	11 79	0.93	13	Dried at 130°C
Aluminum potassium oxide	Al_2O_3 K_2O	61 35	0 63	12	$Al_2(SO_4)_3 \xrightarrow{K_2CO_3}$ Precipitate washed, dried at 130°C and heated with H_2 at 600°–640°C for 10 hr

Table 6.7a (Continued)

Absorbent	Crystalline Phase, X-ray analysis	Purity (wt %)	Bulk Density, g/cc	SO_2 Absorbed (g/100 g absorbent)	Preparation
Cobalt oxide	Co_3O_4	100	0.66	12	$CoSO_4 \xrightarrow[NaOH]{NaOCl}$ Precipitate washed, dried at 130°C and heated in vacua at 300°–340°C for 20 hr
Chromium-sodium oxide	Cr_2O_3 Na_2O	70 26	0.91	12	$Cr_2(SO_4)_3 \xrightarrow{Na_2CO_3}$ Precipitate washed, dried at 130°C and heated with H_2 at 600°–640°C for 20 hr
Nickel oxide	NiO	91	0.74	9	$NiSO_4 \xrightarrow{Na_2CO_3}$ Precipitate washed, dried at 130°C and heated in vacua at 300°–340°C for 20 hr
Nickel oxide	NiO	90	1.49	6	$NiSO_4 \xrightarrow[NaOH]{NaOCl}$ Precipitate washed, dried at 130°C and heated in vacua at 300°–340°C for 20 hr
Sodium carbonate	Na_2CO_3	99	0.98	5	Solution of sodium carbonate dried at 130°C and heated in vacua at 300°–340°C for 20 hr
Sodium stannate	Na_2SnO_3	95	0.91	4	Sodium stannate dried at 130°C and heated in vacua at 300°–340°C for 20 hr
Aluminum-lithium oxide	Al_2O_3 Li_2O	87 10	0.73	4	$Al_2(SO_4)_3 \xrightarrow{(NH_4)_2CO_3}$ Li_2SO_4 Precipitate washed, dried at 130°C and heated with H_2 at 600°–640°C for 10 hr
Iron oxide	α–Fe_2O_3	93	0.98	3	$Fe(NO_3)_3 \xrightarrow{Na_2CO_3}$ Precipitate washed, dried at 130°C and heated at 300°–340°C in a stream of nitrogen for 3 hr

Sodium aluminate	$NaAlO_2$	96	0.90	3	Solution of sodium aluminate dried at 130°C and heated in vacua at 300°–340°C for 20 hr
Cadmium oxide	CdO	97	1.13	1	$CdSO_4 \xrightarrow{Na_2CO_3}$ Precipitate washed, dried at 130°C and heated in vacua at 370°–400°C for 20 hr
Copper oxide	CuO	99	0.39	1	$CuSO_4 \xrightarrow{Na_2CO_3}$ Precipitate washed, compressed at 4000 psi, dried at 130°C and heated in vacua at 300°–340°C for 20 hr
Potassium carbonate		98	0.39	1	Solution of potassium carbonate dried at 130°C and heated in vacua at 300°–340°C for 20 hr
Manganese oxide	$MnO_{1.88}$	94	0.13	61	$MnSO_4 \xrightarrow{(NH_4)_2S_2O_8}$ Precipitate washed, dried at 130°C and heated in vacua at 300°–340°C for 20 hr
Manganese oxide	γ–Mn_2O_3	96	0.57	58	$MnSO_4 \xrightarrow{Na_2CO_3}$ Precipitate washed, dried at 130°C and heated in vacua at 300°–340°C for 20 hr
Hopcalite	MnO_2 CuO	79 11	0.92	57	Dried in vacuo at 300°–340°C
Copper oxide	CuO	99	0.39	56	$CuSO_4 \xrightarrow{Na_2CO_3}$ Precipitate washed, compressed at 4000 psi, dried at 130°C and heated in vacua at 300°–340°C for 20 hr
Manganese oxide	$MnO_{1.88}$	94	0.50	53	$MnSO_4 \xrightarrow{Electrolysis}$ Precipitate washed, dried at 130°C and heated in vacua at 300°–340°C for 20 hr
Cobalt oxide	Co_3O_4	97	0.46	47	$CoSO_4 \xrightarrow{Na_2CO_3}$ Precipitate washed, dried at 130°C and heated in vacua at 300°–340°C for 20 hr

Table 6.7a (Continued)

Absorbent	*Crystalline Phase, X-ray analysis*	*Purity (wt %)*	*Bulk Density, g/cc*	*SO_2 Absorbed (g/100 g absorbent)*	*Preparation*
Cobalt oxide	Co_3O_4	100	0.66	44	$CoSO_4 \xrightarrow[NaOH]{NaOCl}$ Precipitate washed, dried at 130°C and heated in vacua at 300°–340°C for 20 hr
Aluminum-sodium oxide	Al_2O_3 Na_2O	64 31	0.73	19	$Al_2(SO_4)_3 \xrightarrow{Na_2CO_3}$ Precipitate washed, dried at 130°C and heated with H_2 at 600°–640°C for 20 hr
Lead oxide	PbO	99	1.23	18	$Pb(NO_3)_2 \xrightarrow{Na_2CO_3}$ Precipitate washed, dried at 130°C and heated in vacua at 300°–340°C for 20 hr
Chromium-sodium oxide	Cr_2O_3 Na_2O	70 26	0.91	16	$Cr_2(SO_4)_3 \xrightarrow{Na_2CO_3}$ Precipitate washed, dried at 130°C and heated with H_2 at 600°–640°C for 20 hr
Aluminum-potassium oxide	Al_2O_3 K_2O	61 35	0.63	15	$Al_2(SO_4)_3 \xrightarrow{K_2CO_3}$ Precipitate washed, dried at 130°C and heated with H_2 at 600°–640°C for 10 hr
Sodium aluminate	$NaAlO_2$	96	0.90	10	Solution of sodium aluminate dried at 130°C and heated in vacua at 300°–340°C for 20 hr
Nickel oxide	NiO	91	0.74	9	$NiSO_4 \xrightarrow{Na_2CO_3}$ Precipitate washed, dried at 130°C and heated in vacua at 300°–340°C for 20 hr
Nickel oxide	NiO	90	1.49	7	$NiSO_4 \xrightarrow[NaOH]{NaOCl}$

					Precipitate washed, dried at 130°C and heated in vacua at 300°–340°C for 20 hr
Cadmium oxide	CdO	97	1.13	5	$CdSO_4 \xrightarrow{Na_2CO_3}$ Precipitate washed, dried at 130°C and heated in vacua at 370°–400°C for 20 hr
Sodium stannate	Na_2SnO_3	95	0.91	5	Sodium stannate dried at 130°C and heated in vacua at 300°–340°C for 20 hr
Sodium carbonate	Na_2CO_3	99	0.98	4	Solution of sodium carbonate dried at 130°C and heated in vacua at 300°–340°C for 20 hr
Aluminum-lithium oxide	Al_2O_3, Li_2O	87 10	0.73	4	$Al_2(SO_4)_3 \xrightarrow{(NH_4)_2CO_3}$ Li_2SO_4 Precipitate washed, dried at 130°C and heated with H_2 at 600°–640°C for 10 hr
Iron oxide	α–Fe_2O_3	93	0.98	3	$Fe(NO_3)_3 \xrightarrow{Na_2CO_3}$ Precipitate washed, dried at 130°C and heated at 300°–340°C in a stream of nitrogen for 3 hr
Calcium hydroxide	$Ca(OH)_2$, $Ca(OH)_2$ CaO	81 19	0.36	2	$Ca(NO_3)_2 \xrightarrow{NaOH}$ Precipitate washed, dried at 130°C and heated in vacua at 300°–340°C for 20 hr

[a]Bismuth oxide (α–Bi_2O_3), molybdenum oxide (MoO_3), lead oxide (PbO), zinc oxide (ZnO) and calcium hydroxide ($Ca[OH]_2$) absorbed less than 1 g sulfur dioxide for 100 g charge.
[b]Aluminum oxide (γ–Al_2O_3), bismuth oxide (α–Bi_2O_3), calcium oxide (CaO), magnesium oxide (MgO), molybdenum oxide (MoO_3), zinc oxide (ZnO) and potassium carbonate (K_2CO_3) absorbed less than 1 g sulfur dioxide for 100 g charge.

6.7 Dry Absorption Processes

Dry processes, in general, avoid the problem of loss in plume buoyancy, incur little corrosion on the metal surfaces and produce almost no pressure loss in the flue system.

The disadvantages involved in dry absorption include difficulties in regenerating the absorbent, increased load on the precipitator and loss of absorbent activity after repeated cycles of adsorption and regeneration.

The Alkalized Alumina Process

The U.S. Bureau of Mines conducted research and development on dry processes operating at elevated temperatures for removing sulfur dioxide from waste flue gases. Investigations at the Pittsburgh Coal Research Center revealed that a coprecipitating of sodium and aluminum oxides exhibited favorable absorbent characteristics for use in SO_2 removal. These characteristics are:

1. The absorbent performed well at temperatures in the 265 to 625°F range.
2. It could absorb SO_2 in relatively high amounts: 15 to 20 g SO_2/100 g absorbent.
3. The absorbent can be regenerated repeatedly in cyclic operation.
4. Regeneration produces H_2S which is convertible to elemental sulfur in a conventional Claus process.

The process is shown schematically in Figure 6.7b. Flue gases from the boiler are treated in an electrical precipitator or baghouse to remove the flyash and dust. The dust-free gases are fed to a reactor wherein the alkalized alumina absorbs SO_2 at 600°F. The purified gas is preheated before discharge from the stack. The absorbent regeneration is performed by contacting the spent sorbent to about 1250°F with a gas mixture of H_2, CO and CO_2. The sodium aluminate pellets are recycled and the hydrogen sulfide desorbed from the pellets is converted to elemental sulfur in a Claus unit.

During the absorption step, the SO_2 and O_2 in the flue gas react with the absorbent:

$$SO_2 + \tfrac{1}{2} O_2 + Na_2O \rightleftharpoons Na_2SO_4 \qquad 6.7(1)$$

The extent of conversion of SO_2 to sulfate is, of course, limited by the amount of Na_2O in the absorbent. During regeneration, the sulfate is reduced:

$$Na_2SO_4 + 4\,H_2 \rightleftharpoons 3\,H_2O + H_2S + Na_2O \qquad 6.7(2)$$

$$Na_2SO_4 + 4\,CO + H_2O \rightleftharpoons 4\,CO_2 + H_2S + Na_2O \qquad 6.7(3)$$

Alkalized Alumina Preparation

In preparing alkalized alumina absorbent, a slurry is formed by mixing a 17% solution of Na_2CO_3 with a 16% solution of $Al_2(SO_4)_3$ at 190°F. The

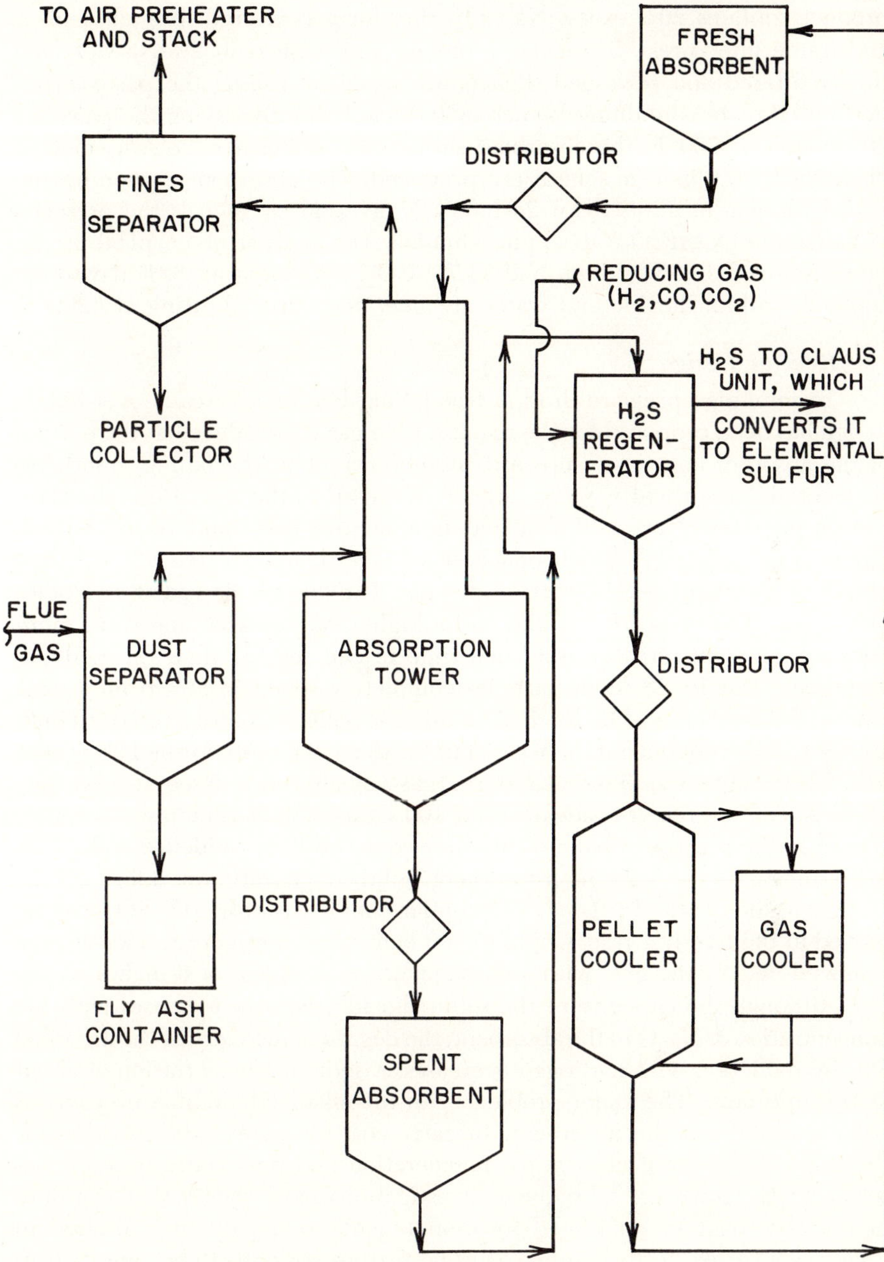

Figure 6.7b Alkalized alumina process.

product contains 20% excess Na_2CO_3; the slurry is filtered through a plate and frame filter press. The cake is broken and washed in 2% Na_2CO_3 and then refiltered and rewashed. Pellets are made by rolling the paste into a perforated plate containing $\frac{1}{8}$-inch cylindrical holes. By drying the paste for 20 minutes at 250°F, the absorbent shrinks and uniform pellets $\frac{1}{16}$ inch in diameter by $\frac{1}{16}$ inch in length are produced. The absorbent is then heated with hydrogen at 1200°F for 24 hours. X-ray analysis reveals the presence of dawsonite [$NaAlCO_3(OH)_2$] plus alumina. The analysis of a typical preparation showed 13.3% sodium and 24.7% CO_2, indicating an 82% dawsonite content. Carbon dioxide and water are liberated during heating at 1200°F.

Reactor Design

To minimize pressure drop, a free-falling bed or a cascade-over-baffles design has been proposed for the reactor. Flue gas is introduced at the bottom of the absorber and passes upward countercurrent to the falling absorbent. Regenerated absorbent is pneumatically returned to the top of the absorber.

In pilot tests,[2] crushed alkalized alumina of 8 to 24 and 16 to 28 mesh size as well as $\frac{1}{16} \times \frac{1}{16}$-inch pellets were investigated in free fall at a rate of 3 lb of absorbent per 1,000 ft^3 of flue gas. The flue gas flowed at a velocity of 8 ft/sec. The crushed particles had a higher surface area and were more buoyant, allowing greater residence time in the reactor than the pelleted absorbent. The 16–28 mesh particles completely removed SO_2 from a 22 ft length of the reactor while the 8–24 mesh and pelletized forms removed only 10% SO_2. The residence time in free fall for the 8–24 mesh crushed absorbent in a 22-ft length was 5 sec at 8 ft/sec gas velocity and 15 sec at 15 ft/sec.

To achieve greater residence time, 100 stainless steel 4-mesh screens were placed in the reactor as baffles; this allowed a 1-minute residence and a 64% SO_2 removal for the 8–24 mesh absorbent. Solids-to-gas ratio was 3 lb/1,000 ft^3 and gas velocity was 12 ft/sec. For similar conditions only 31% SO_2 was removed in the free-fall case. Going to 300 baffles and 2-minute residence time removed 98% of the SO_2 with a consequent pressure loss of 6 inches water.

Although the capacity of the sulfur dioxide removal increases with the concentration of Na_2O in the absorbent, the degree of recovery of the absorbed SO_2 by reduction with hydrogen decreases. A sodium concentration of about 20% is optimum. The major problem with the alkalized alumina process has been in sustaining the absorbent. In early cost estimates,[2] the attrition for the overall process of absorption, regeneration and pneumatic transfer was set at 0.1%. However, subsequent investigations have shown that exposure of the absorbent to continued sorption–regeneration recycling resulted in excessive attrition of the alumina, causing the process costs to become exorbitant. As a consequence, all further research by the Bureau of Mines on this process has been stopped.

Activated Manganese Oxide Process

A dry process called the DAP-Mn process has been developed for SO_2 removal in Japan by Mitsubishi Heavy Industries, Ltd. Plant scale tests sponsored by the Japanese government have been conducted at the Yokkaichi power station, Chubu Electric Power Co., Ltd., treating 5.3×10^6 SCF/hr flue gas. The process was able to remove more than 90% of the SO_2 from flue gas containing 1,100 ppm SO_2. The absorbent is activated manganese oxide which reacts to produce manganese sulfate:

$$MnO_x \cdot i\,H_2O + SO_2 + \tfrac{1}{2}(2 - X)O_2 \rightleftharpoons MnSO_4 + i\,H_2O \qquad 6.7(4)$$

$$MnO_x \cdot i\,H_2O + SO_3 \rightleftharpoons MnSO_4 + \tfrac{1}{2}(X - 1)O_2 + i\,H_2O \qquad 6.7(5)$$

where i and n are about 0.5 to 0.8 and 0.1 to 1.0, respectively. In the regeneration step, ammonium sulfate is recovered as a by-product:

$$MnSO_4 + 2\,NH_4OH \rightleftharpoons Mn(OH)_2 + (NH_4)_2\,SO_4 \qquad 6.7(6)$$

$$Mn(OH)_2 + \tfrac{1}{2}(X - 1)O_2 + (i - 1)H_2O \rightleftharpoons MnO_x \cdot i\,H_2O \qquad 6.7(7)$$

The process is shown in Figure 6.7c. The powdered absorbent is fed to the absorber where it is dispersed in the flue gas and reacts with the oxides of sulfur. Manganese sulfate and excess absorbent are collected in a cyclone and in an electrostatic precipitator in series. A recycle-to-fresh absorbent feed ratio of 6:1 is used. Regeneration of spent absorbent is performed by blowing compressed air at 5 atm into the manganese sulfate solution containing excess ammonium hydroxide at ambient temperature. Fresh $MnO_x \cdot H_2O$ is filtered from the $(NH_4)_2SO_4$ solution and the filtrate is fed to a crystallizer.

The absorber is actually a moving bed or a dilute fluidized bed allowing for a low pressure drop operation. A mean particle size of 40 μm at a feed rate of 4.3 to 7.1 g/SCF is used. The amount of sulfur dioxide recovered increases with the amount of absorbent feed; however, 100% removal is reached at stoichiometric ratios of between 10 and 15 moles of Mn per mole SO_2. The particle size and specific gravity (3.3) of the absorbent facilitates recovery by cyclone. Also the electrical resistivity (3×10^7 to 8×10^7 ohm-cm at 203 to 257°F) permits collection of the finer particle fraction by electrostatic precipitation with little difficulty.

The Still Process

The Still process is a dry additive process based on a CaO containing lignite filter ash as a desulfurizing agent. Hydrated lime, $Ca(OH)_2$, reacts with SO_2 to produce calcium sulfite ($CaSO_3$) exclusively. Absorption takes place below 900°F down to 200°F, allowing sulfur dioxide removal equipment to

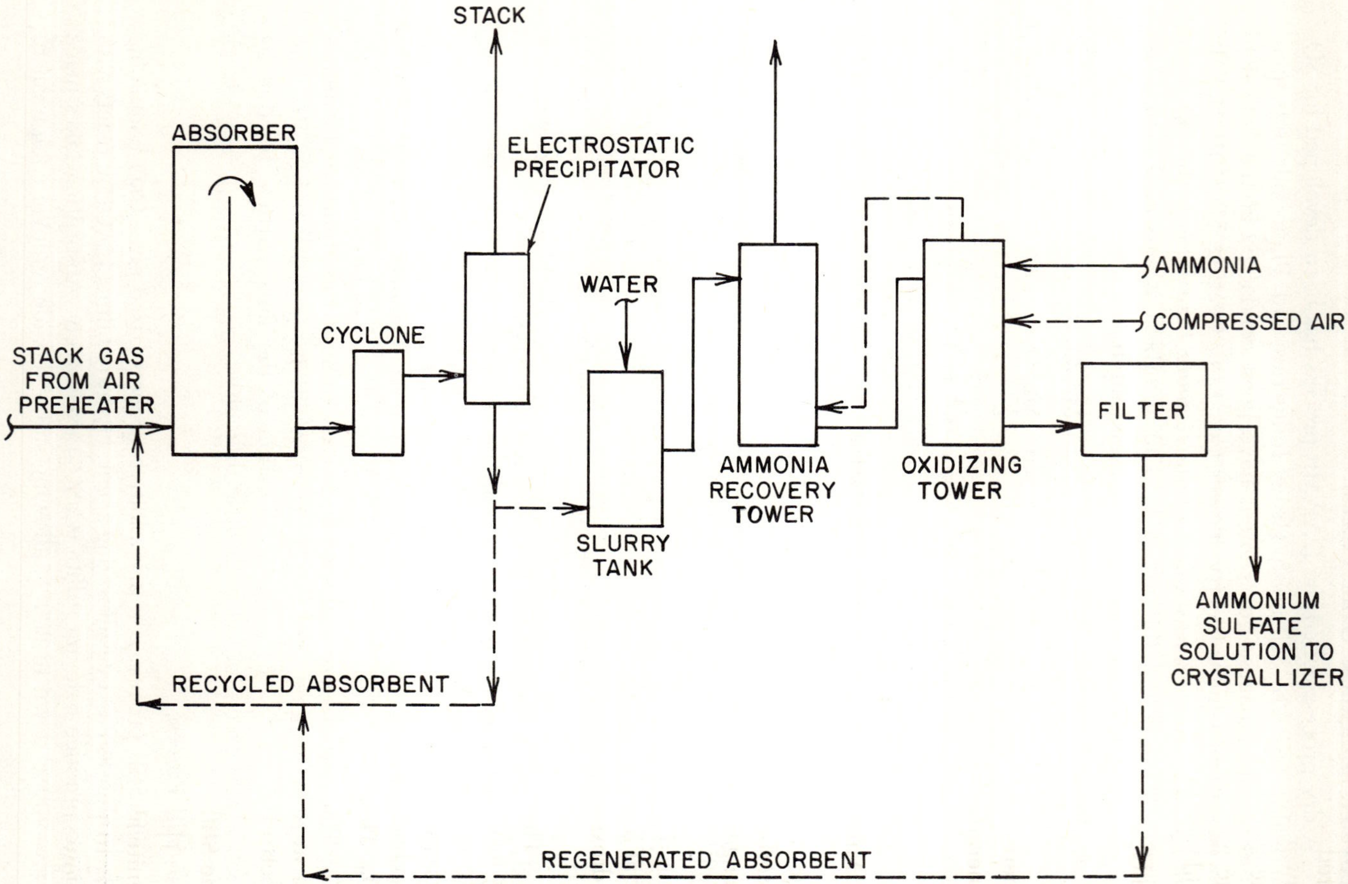

Figure 6.7c Mitsubishi's DAP-Mn process.

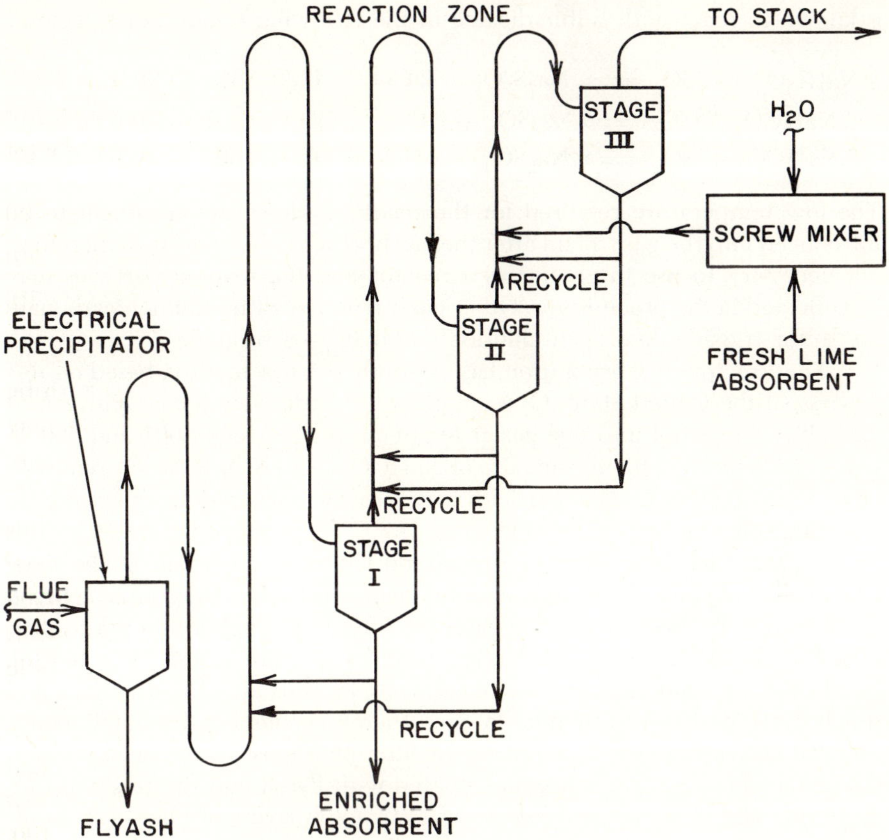

Figure 6.7d Carl Still process.

be placed at the boiler exit (Fig. 6.7d). After removal of the flyash by a cyclone, the flue gas is conducted through three reaction stages. These stages are in the form of vertical, inverted U tubes with a cyclone located at the base of each. Absorbent is introduced at the inlet of each stage and separated at the outlet. Fresh lime is hydrated in a screw mixer with an appropriate water feed rate to produce a dry absorbent for injection at stage III. The Still process* is designed for a flue gas capacity of 700,000 to 1,000,000 SCF/hr and a desulfurization of 95%.

Sodium Bicarbonate Absorption

Sodium bicarbonate or nahcolite is a naturally occurring mineral found in deposits of dawsonite and oil shale in the Piceance Basin in Colorado. It

* Work has been temporarily discontinued on the development of this process.

is known to react with sulfur dioxide in the following temperature regions:

$$2\ NaHCO_3 + 2\ SO_2 \rightleftharpoons Na_2S_2O_5 + 2\ CO_2 + H_2O \quad 32° \text{ to } 90°F \qquad 6.7(8)$$

$$2\ NaHCO_3 + SO_2 \rightleftharpoons Na_2SO_3 + 2\ CO_2 + H_2O \quad 90° \text{ to } 160°F \qquad 6.7(9)$$

$$2\ NaHCO_3 + SO_2 \rightleftharpoons Na_2SO_4 + 2\ CO_2 + H_2 \quad 160° \text{ to } 260°F \qquad 6.7(10)$$

The low temperature required for the reaction allows the absorbent to be introduced into the gas stream after the last heat recovery unit. It is, therefore, not necessary to use the boiler as a reactor and the reacted particles may be collected in the precipitator. Other gases reactive with sodium bicarbonate are sulfur trioxide, hydrogen chloride and hydrogen bromide.

The Precipitair Corporation has established a test facility, based on this process, at the United States Gypsum plant at Clark, New Jersey. Pulverized nacholite is injected into flue gas at temperatures between 300° and 360°F. Using twice the stoichiometric ratio of $NaHCO_3/SO_2$, over 65% SO_2 removals have been achieved. The particles were easily collected by electrostatic precipitation.

A patent has been filed by Frevel and Kressley[3] on behalf of the Dow Chemical Company, which introduces a novel approach to SO_2 removal based on $NaHCO_3$. In this process, flue gases are filtered through a thin porous bed of sodium bicarbonate crystals having a mesh size from about 40 up to 100. In addition to removing SO_2, the bicarbonate crystals are able to attract and retain flyash particles in submicron sizes, thereby achieving the simultaneous removal of flyash. During operation, the filter beds must be replaced because the surface of the crystals becomes coated with flyash and the reaction rate drops off. The filter support consists of compacted layers of fiber glass, compressed to approximately $^3/_8$ inch between wire screens. In a series of pilot tests, a filter medium of 24 mm depth was exposed to between 50 and 60 lpm flue gas for 8 hours. Sulfur dioxide was completely removed and the pressure drop encountered was 7.5 cm H_2O increasing to 15.8 cm at the end of the test. The flyash was substantially contained within the first 1 to 2 mm of the bed. Removal of this layer restored the original pressure drop. Sulfur dioxide was effectively trapped in the first 10 mm of the bed.

In full-scale operation, the number and area of beds would depend on the volume of gas to be filtered. For a large power plant burning 1,500 tons of coal per day, about 10 fiber beds with areas up to 1,000 square meters would be required. A cyclical recharging schedule would be necessary for the continuous operation of the beds.

Uranyl Oxide Scrubbing

Uranyl oxide will remove sulfur dioxide by absorption and reaction at temperatures above 550°F. In the regeneration step, sulfur will reduce

UO_2SO_4 to UO_2 at about 900°F. Laboratory studies[4] demonstrated that between 0.125 and 0.199 gm of SO_2 were absorbed per gm of UO_2.

Summary

The major drawback in recycled dry absorption processes lies with the regeneration of the absorbent. This step is critical since most effective absorbents of SO_2 are costly and, therefore, recovery of this material can make or break the economics of a process.

Further work on the alkalized alumina process has been abandoned because of excessive attrition of the sorbent. The major obstacles in the Carl Still process appear to be that an adequate lignite is not widely available and formation of the calcium sulfate would interfere with the activity of the basic ash in the recycle.

The Mitsubishi process based on an activated manganese dioxide absorbent demonstrates several advantages including a long absorbent life. Sodium bicarbonate has not been fully tested; however, its regeneration is a simple operation and this process may prove competitive in the near future.

REFERENCES

1. Bienstock, D., J. H. Field and J. G. Myers: "Process Development in Removing Sulfur Dioxide from Hot Flue Gases," Part 1, U.S. Bureau of Mines, RI 5735, 1961.
2. Ibid., Part 3, Bureau of Mines, RI 7021, 1967.
3. Frevel, L. K. and L. J. Kressley: U.S. Patent 3,589,863, June 29, 1971.
4. Winsche, W. E., E. Wirsing, Jr. and R. H. Wiswall, Jr.: *J. Air Pollu. Contr. Ass.*, vol. 21, p. 20, 1971.

6.8 SULFUR OXIDE CONTROL—WET ABSORPTION PROCESSES

Types of Processes: (a) *Limestone wet scrubbing,* (a1) Battersea, (a2) Meramec; (b) *Venturi wet scrubbing* (*Chemico-Basic*); (c) *Metal sulfite scubbing* (*Wellman-Lord*); (d) *Grillo desulfurization process;* (e) *Ammonia scrubbing,* (e1) Fulham-Simon Carves, (e2) COMIN-CO, (e3) Mitsubishi, (e4) Two-step treatment; (f) *Caustic soda treatment.* (g) Liquid phase Claus*

Advantages: Not restricted by temperatures or residence times within the furnace. They can be added to existing units without boiler modification.

Limitations: Heat loss due to scrubbing, which reduces plume buoyancy, adds steam to it and necessitates the use of a reheater. Regeneration systems tend to be complex. The acidity of some waste products can be a problem.

Process Status: (a) Battersea installation was expensive, reduced plume buoyancy and required high maintenance. Combustion Engineering improved the process and made two full-scale installations (Section 6.12); (b) In operation; (c) In operation; (d) In operation; (e) In operation, some versions abandoned; (f) In operation.

Cost: See Table 6.12c.

Limestone Wet Scrubbing

Wet scrubbing with limestone was originally studied during the 1930s at Battersea in London, England, and at the Tir John Power Plant in Swansea,

*Citrate by Bu Mines and Powerclaus by Stauffer.

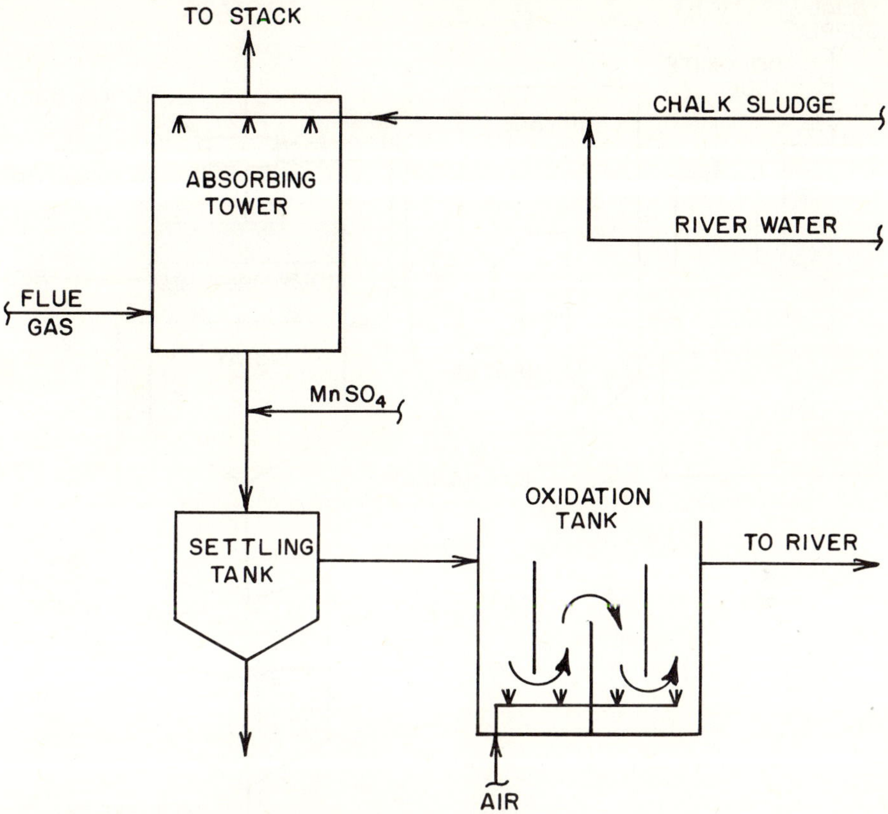

Figure 6.8a Battersea process.

Wales. The experimental plant at Battersea required an investment of over \$5 million to scrub effluent gases from the 120-Mw plant. The installation (Fig. 6.8a) was quite large and was capable of removing 95% of the sulfur dioxide. Thames river water with an added chalk slurry was used as a scrubbing liquor. The absorption tower consisted of timber-lined cast iron with grids of red cedar and teak. The waste water containing gypsum and sulfite was oxidized with manganese sulfate in aerating tanks and the unpurified sludge and cooling water were released into the Thames.

In addition to high initial investment, these plants incurred problems associated with high-maintenance and operating costs, corrosion, abrasion, scale deposit, solid waste disposal and loss of plume buoyancy.

Current studies in the United States are attempting to overcome these problems by improved process and equipment design, closer control of operating parameters, improved materials, systems optimization and reliability.

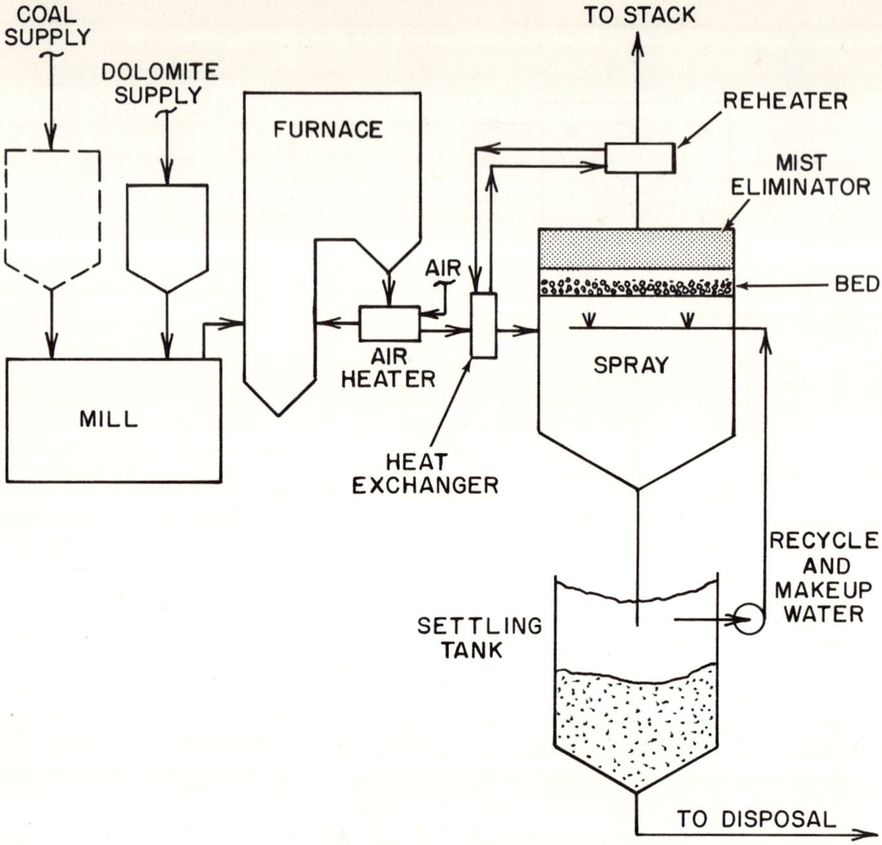

Figure 6.8b Combustion Engineering process.

The Combustion Engineering Company has demonstrated successful sulfur dioxide removal in two full-scale installations at the Meramec Plant of the Union Electric Company in St. Louis and at the Lawrence Plant of the Kansas Power and Light Company. Each unit is designed to scrub all the flue gas from a 125-Mw boiler (Table 6.12a). The Meramec system (Fig. 6.8b) has been in operation since February 1970.* Sulfur dioxide removal efficiency has been about 60%. The system combines dry additive limestone injection with wet scrubbing. Raw limestone rock is pulverized and injected into the boiler through nozzles located above the coal burners. As in the dry limestone process, the rock is calcined at 1500°F to produce the calcium and magnesium oxides and carbon dioxide:

$$CaCO_3 + MgCO_3 + \text{heat} \rightleftharpoons CaO + MgO + 2\,CO_2 \qquad 6.8(1)$$

*Presently it is down for repairs.

The sulfur trioxide in the flue gas is captured:

$$CaO + SO_3 \longrightarrow CaSO_4 \quad 6.8(2)$$

$$MgO + SO_3 \longrightarrow MgSO_4 \quad 6.8(3)$$

and 20 to 30% of the sulfur dioxide is removed:

$$CaO + SO_2 + O_2 \rightleftharpoons CaSO_4 \quad 6.8(4)$$

$$MgO + SO_2 + O_2 \rightleftharpoons MgSO_4 \quad 6.8(5)$$

The reactions are completed in packed marble-bed scrubbers located between the air heaters and the stack.

$$CaO + H_2O \rightleftharpoons Ca(OH)_2 \quad 6.8(6)$$

$$MgO + H_2O \rightleftharpoons Mg(OH)_2 \quad 6.8(7)$$

$$2\,Ca(OH)_2 + 2\,SO_2 + \tfrac{1}{2}\,O_2 \rightleftharpoons CaSO_3 + CaSO_4 + 2\,H_2O \quad 6.8(8)$$

$$2\,Mg(OH)_2 + SO_2 + \tfrac{1}{2}\,O_2 \rightleftharpoons MgSO_3 + MgSO_4 + 2\,H_2O \quad 6.8(9)$$

The wet scrubbers perform the additional duty of removing over 99% of the flyash from the gases.

The calcined dolomite minimizes high-temperature corrosion in the boiler by preventing formation of complex alkali-iron-trisulfates. The calcium and magnesium sulfates formed in reactions with SO_3 combine with alkali sulfates in the ash to form double salts, e.g., $(K_2SO_4)(2\,CaSO_4)$. Also, since SO_3 is removed from the flue gas, sulfuric acid formation is minimized, thereby reducing corrosion of low-temperature heat transfer surfaces, ductwork and stacks. An objectionable feature of this process is the loss of heat by the flue gas in the wet scrubber. This adds to the steam in the plume and reduces the plume buoyancy. To minimize this, a stack gas reheater is located above the demister. This unit is operated in a closed-cycle with a heat exchanger at the furnace exit.

Venturi Wet Scrubbing

An alternative approach in wet scrubbing (Sections 5.12 to 5.21) utilizes a high-velocity venturi rather than the packed-bed type of scrubber. Two types of units, the wet-approach venturi and the Pease-Anthony venturi scrubber have been developed by the Chemical Construction Company for use in sulfur dioxide scrubbing. The wet-approach scrubber features a tangential liquid feed in the convergent section of the venturi. The swirling liquid is brought toward the center of the throat and is impacted by the high-velocity gas stream; the atomized liquid presents a high liquid-gas contact area for mass transfer (Fig. 5.17i). The Pease-Anthony type scrubber employs a radial liquid feed at the throat which creates a curtain of scrubbing liquor for impaction by the gas (Fig. 5.17b). The aerosol, in both systems after leaving the divergent section

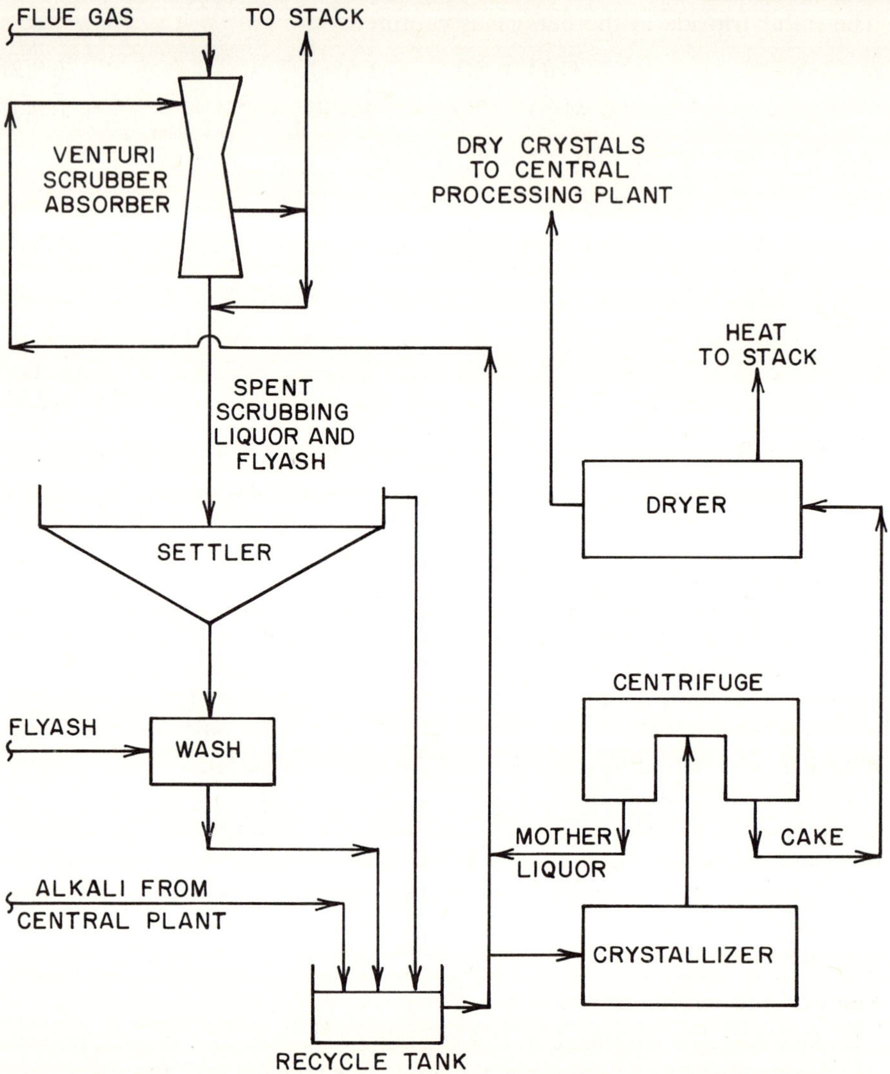

Figure 6.8c Chemico-Basic recovery system.

of the venturi, enters a cyclonic separator (Section 5.15) where the liquor is removed from the gas by both centrifugal and gravitational forces.

The Chemico-Basic recovery process (Fig. 6.8c) uses an absorbing liquor of magnesium sulfite, magnesium sulfate and magnesium oxide in a venturi which absorbs SO_2 according to the following reactions:

$$MgO + SO_2 + 6\,H_2O \rightleftharpoons MgSO_3 \cdot 6\,H_2O \qquad 6.8(10)$$

$$MgSO_3 + SO_2 + H_2O \rightleftharpoons Mg(HSO_3)_2 \quad 6.8(11)$$

$$Mg(HSO_3)_2 + MgO \rightleftharpoons 2\,MgSO_3 + H_2O \quad 6.8(12)$$

$$MgO + SO_3 + 7\,H_2O \rightleftharpoons MgSO_4 \cdot 7\,H_2O \quad 6.8(13)$$

$$MgSO_3 + \tfrac{1}{2}\,O_2 + 7\,H_2O \rightleftharpoons MgSO_4 \cdot 7\,H_2O \quad 6.8(14)$$

The bleed from the absorber slurry passes through a crystallizer* into a centrifuge where $MgSO_3$ and $MgSO_4$ crystals, along with unreacted MgO, are separated from the liquid phase. The mother liquor reenters the absorber feed, and the centrifuged wet cake is dried in a direct fixed dryer to remove surface and bound moisture. Hot gas from the dryer provides the superheat for the flue gas leaving the absorber. The dry crystals are further heated in a direct-fired calciner:

$$MgSO_3 \xrightarrow[\Delta]{\text{heat}} MgO + SO_2 \quad 6.8(15)$$

$$MgSO_4 + \tfrac{1}{2}\,C \xrightarrow[\Delta]{\text{heat}} MgO + SO_2 + \tfrac{1}{2}\,CO_2 \quad 6.8(16)$$

to produce a 15 to 16% SO_2 flue gas for further processing. Make-up water and MgO are added to the recycle tank.

Metal Sulfite Scrubbing

Aqueous solutions of metal sulfites are known to act as strong absorbents for SO_2. The absorbed sulfur dioxide forms the metal bisulfite, which when cooled is precipitated out as the pyrosulfite, $S_2O_5^{2-}$. The pyrosulfite is then heated to release sulfur dioxide and produce the sulfite for recycle.

$$SO_2 + SO_3^{2-} + H_2O \rightleftharpoons 2\,HSO_3^- \quad 6.8(17)$$

$$2\,HSO_3^- \longrightarrow S_2O_5^{2-}\downarrow + H_2O \quad 6.8(18)$$

When SO_3 is present, the sulfate is formed:

$$2\,SO_3 + SO_3^{2-} \rightleftharpoons SO_4 + S_2O_5^{2-} \quad 6.8(19)$$

and, in the presence of oxygen,

$$SO_3^{2-} + \tfrac{1}{2}\,O_2 \rightleftharpoons SO_4^{2-} \quad 6.8(20)$$

The Wellman-Lord sulfur dioxide recovery process (Fig. 6.8d) was originally based on potassium sulfite; however, early indications of high-energy requirements prompted a shift to sodium sulfite. This was found to reduce utility requirements significantly and offer lower capital costs. It is a wet regenerative system wherein the flue gas is contacted countercurrently in an absorber with a sodium sulfite–sodium bisulfite solution. The flue gas is satu-

*Obtaining good crystal growth can be difficult.

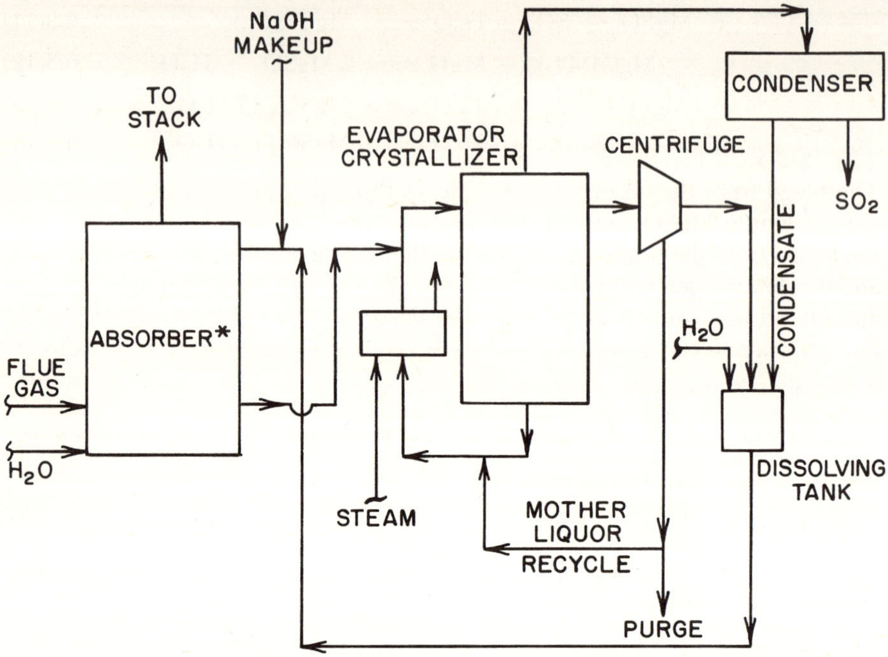

* THE ABSORBER RECYCLE IS INTERNAL TO THE DISC-DONUT TYPE ABSORBER. THIS DISC-DONUT ARRANGEMENT, HOWEVER, HAS BEEN REPLACED BY THREE BEDS OF INTALOX SADDLES, IN THE UPPER SECTIONS OF THE TOP, MIDDLE AND BOTTOM STAGES, WHICH IMPROVES LIQUID-GAS CONTACT AND BRINGS FINAL SO_2 LEVELS TO 200 PPM.

Figure 6.8d Wellman-Lord sulfur dioxide recovery system.

rated in the bottom section of the absorber which consists of a 1-foot packed bed of 3-inch porcelain Intalox® saddles (Fig. 5.14b). Water is sprayed above and below the bed. The upper section consists of three disc and donut sections, each stage with two discs, one donut ring and a collector tray. Lean absorbent solution from the dissolving tank is mixed with absorber recycle and fed to the topmost disc. The solution flows out of the top collector tray and onto the upper disc of the next stage.

The spent absorbent is then sent to an evaporator-crystallizer, where it is heated with low pressure steam (35 PSIG) to about 230°F at 15.7 PSIA, using a shell and tube calandria. The bisulfite is thereby decomposed:

$$2\ NaHSO_3 \longrightarrow Na_2SO_3\downarrow + SO_2\uparrow + H_2O\uparrow \qquad 6.8(21)$$

The Na_2SO_3 slurry is centrifuged and the solids discharged into a dissolving tank for reuse in the absorber. The wet SO_2 gas from the evaporator goes to a condenser and the condensate is sent to the dissolving tank for dissolution of the Na_2SO_3 crystals. The SO_2 rich gas is processed for recovery as sulfuric acid.

Sodium sulfate formed from sulfur trioxide or by oxidation of the sulfite cannot be regenerated and, therefore, is purged for control. This also serves to control small quantities of complex thionates produced in the desorber section. An equivalent amount of caustic soda is added to maintain sodium balance in the system.

The Wellman-Lord SO_2 recovery process has been commercially applied at Olin Corporation's sulfuric acid regeneration plant in Paulsboro, New Jersey. The plant has been designed to handle a 45,000 SCFM gas flow containing 6,000 ppm SO_2 and to discharge less than 500 ppm to the atmosphere. The steam consumption amounts to 30,000 lb/hr, and the caustic make-up is 2.5 STPD* of 100% NaOH.

The Grillo Desulfurization Process

The Grillo process is presently in operation and utilizes a complex mixture of magnesium and manganese oxides in an aqueous suspension as the absorbing-oxidizing agent for sulfur dioxide removal. The oxides may be represented as $(MnO_2)(X\ MgO)$, where X is usually between 3 and 6. The absorption step is performed by the magnesium component, while the manganese donates the oxygen for the chemical reaction. In aqueous solution, the complex is easily hydrated and reacts quite readily with sulfur dioxide, and trioxide. The pH of the suspension before reaction is between 8 and 8.5 and drops to about 4 when spent. Absorption proceeds unabated above pH 7 but drops to between 90 and 95% at pH 7. When spent, a typical absorbent mass is between 17 and 21% sulfur, 50% of which is in the sulfite-bisulfite form, the rest mainly as sulfate in the combined state.

In the Grillo SO_2 removal process (Fig. 6.8e), fresh absorbent is continuously fed to a suspension storage unit where the hydration takes place. From this unit, the 30% suspension is pumped into the absorber-reactor where it contacts the flue gas in a parallel flow arrangement. The flue gas first flows through a spray dryer where it is used to dry the spent slurry. The gas then issues through two absorber-reactors, the first at a temperature between 250 and 300°F and the second ranging from 100 to 175°F, and then into a cyclone where the solids are separated, and the gas exits through the stack.

The regeneration step is accomplished by roasting the spent absorbent in the presence of coal dust or fuel oil between 1500 and 1600°F. The roasted material is hydrated and classified, removing the heavier flyash particles. The hydrate is filtered and returned to the spray dryer. The roasting gas, rich in SO_2, is sent to a sulfur or sulfuric acid recovery system.

*STPD is standard tons per day (2,000 lb/day).

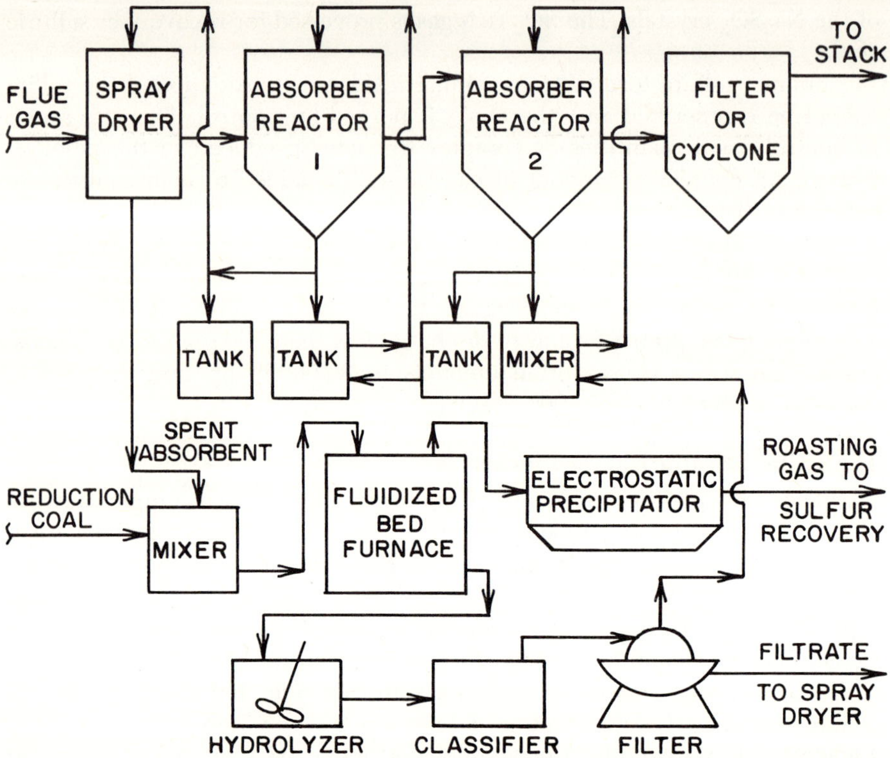

Figure 6.8e Grillo desulfurization process.

Ammonia Scrubbing

An early attempt at sulfur dioxide removal using weak ammonia liquor was made with the Fulham-Simon Carves process (Fig. 6.8f) at the North Wilford power station near Nottingham, England. This plant was commissioned in 1957 to treat 56,000 CFM of combustion waste gases from the burning of 120 tons of coal containing 3% sulfur. The product was 11 tons of ammonium sulfate as well as 2,000 lb of elemental sulfur per day. The basic reaction in the scrubbing tower is the following:

$$NH_3 + H_2O \rightleftharpoons NH_4OH \qquad 6.8(22)$$

$$NH_4OH + NH_3 + 2\,SO_2 \rightleftharpoons (NH_4)_2S_2O_3 \qquad 6.8(23)$$

$$NH_4OH + SO_2 \rightleftharpoons NH_4HSO_3 \qquad 6.8(24)$$

The spent sulfite from the scrubber is autoclaved at 200 psi and 350°F for 3 hours to produce ammonium sulfate and elemental sulfur:

$$2\,NH_4HSO_3 + (NH_4)_2S_2O_3 \longrightarrow 2\,(NH_4)_2SO_4 + 2\,S + H_2O \qquad 6.8(25)$$

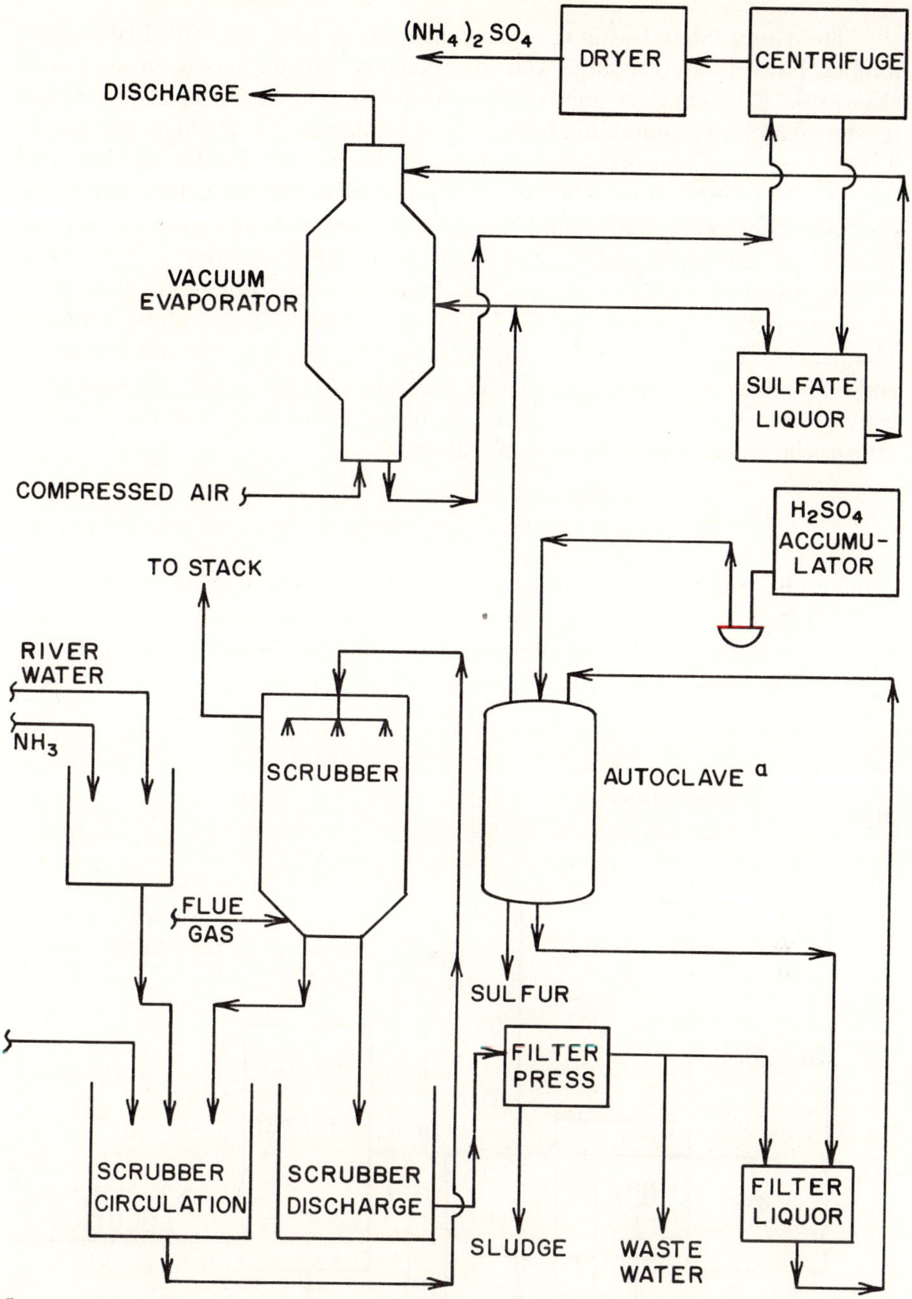

[a] AN AUTOCLAVE IS A PRESSURIZED VESSEL WHICH SUBJECTS MATERIAL TO SUPERHEATED STEAM, IN THIS PARTICULAR CASE 170 °C AT 200 PSI.

Figure 6.8f Fulham-Simon Carves ammonia scrubber.

6.8 Wet Absorption Processes

The Consolidated Mining and Smelting Co., Ltd., at Trail, British Columbia, has operated a sulfur dioxide recovery process based on ammonia absorption (COMINCO) since 1936. The process has been applied to zinc roaster plant gases containing 5.5% SO_2. A reduction to 0.2% SO_2 is claimed. A 30% solution of aqueous ammonia contacts the flue gas producing ammonia bisulfite, NH_4HSO_3 (Fig. 6.8g) In a stripper, 93% sulfuric acid is added to produce ammonium sulfate and free SO_2. A removal efficiency for SO_2 of 97% on an acid plant and 85% on a zinc roaster is claimed.

An ammoniacal liquor process developed by Mitsubishi Heavy Industries of Japan produces, as a by-product, high-purity ammonium sulfate crystals. The process (Fig. 6.8h) involves the reaction of ammonia with sulfur dioxide and water in a cross-flow scrubber to yield ammonium sulfite, subsequently converted to the bisulfite. The spent absorbent is fed to an oxidation unit to reform the sulfite and oxidize it to sulfate:

$$NH_4\,HSO_3 + NH_3 \rightleftharpoons (NH_4)_2SO_3 \qquad 6.8(26)$$

$$(NH_4)_2SO_3 + \tfrac{1}{2}\,O_2 \rightleftharpoons (NH_4)_2SO_4 \qquad 6.8(27)$$

Ammonium sulfate solution can be supplied to a crystallizer in the form of a 45 Wt % solution by using waste heat derived from initial cooling of the flue gas. A variation of this same process, called the Showa Denka process, has recently been abandoned in spite of 90% and higher SO_2 removal rates.

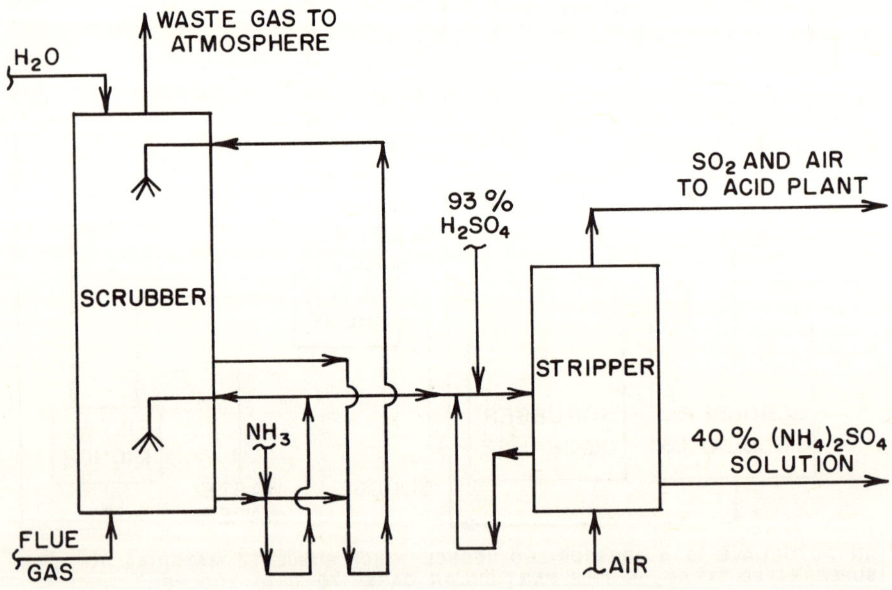

Figure 6.8g Cominco process.

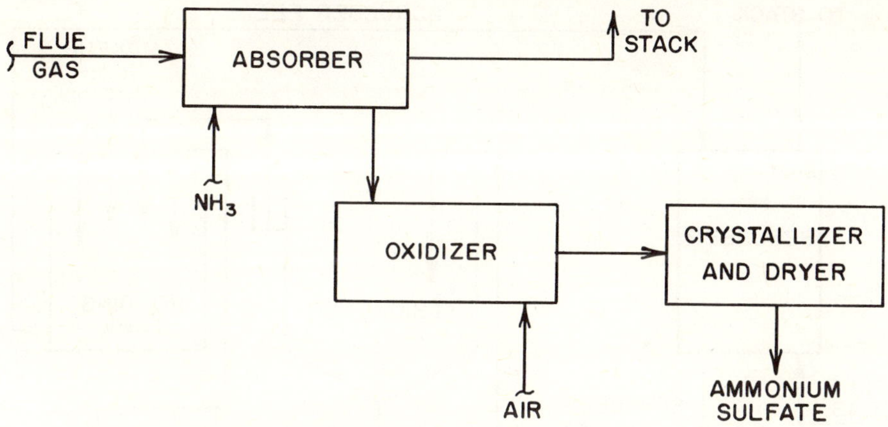

Figure 6.8h Mitsubishi ammonia process.

A two-step treatment for sulfur dioxide removal in flue gases has also been proposed.[1] Ammonia is injected into the gas stream in the presence of water vapor to form the bisulfite-sulfite mixture. The mixture is then brought into contact with either zinc oxide or with an aqueous alkaline earth metal oxide, which liberates NH_3 and forms an insoluble metal sulfite:

$$ZnO + (NH_4)_2SO_3 \longrightarrow ZnSO_3 \cdot H_2O\downarrow + 2\,NH_3 \qquad 6.8(28)$$

The ammonia gas is recycled to the scrubber, while the metal sulfite is dried, dehydrated and thermally decomposed to give off a gas which contains approximately 40% water and 60% sulfur dioxide, by weight.

$$ZnSO_3 \cdot H_2O \xrightarrow[\Delta]{\text{heat}} ZnO + SO_2\uparrow + H_2O\uparrow \qquad 6.8(29)$$

The regenerated metal oxide is then ready for reuse.

The ammonium sulfite is held as a suspended solid in the contacting vessel.

Caustic Soda Treatment

Scrubbing with sodium hydroxide solution with aeration and disposal of the spent absorbent offers a simple, inexpensive treatment in the case of small boilers. Over 90% SO_2 removal has been reported by the process (Fig. 6.8i).[2]

$$2\,NaOH + SO_2 \rightleftharpoons Na_2SO_3 + H_2O \qquad 6.8(30)$$

$$Na_2SO_3 + H_2O + SO_2 \rightleftharpoons 2\,NaHSO_3 \qquad 6.8(31)$$

If only the first step is carried out, aeration of the liquid effluent produces the sulfate which is diluted with other innocuous liquid wastes, yielding a neutral pH and no chemical oxidation demand. Handling the bisulfite is a bit more difficult because of the higher acidity of the waste product.

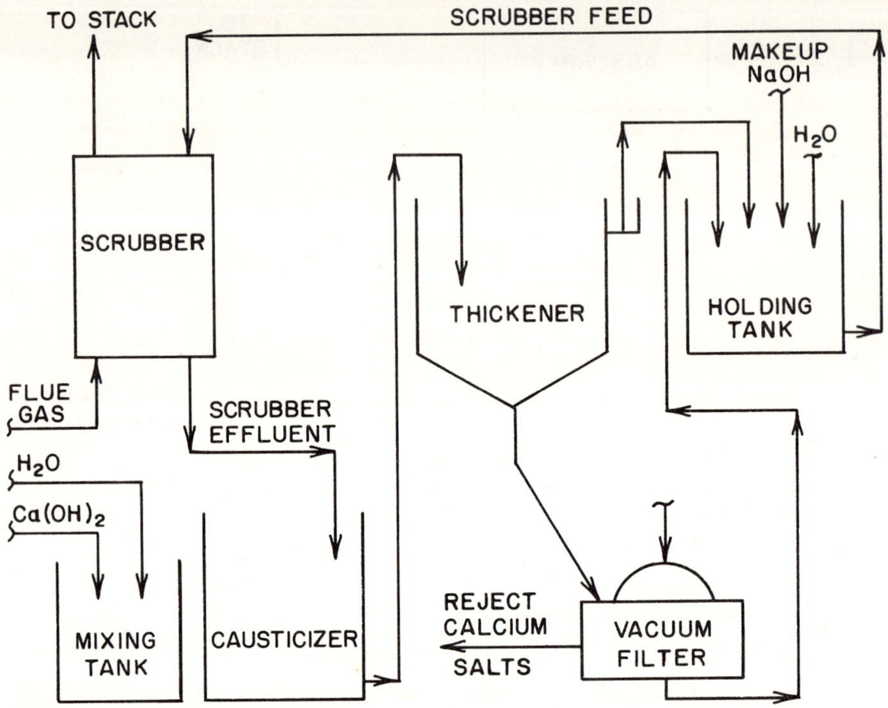

Figure 6.8i Sodium hydroxide scrubbing with lime regeneration.

The sulfite solution may be recovered by reaction with lime to precipitate sulfite and regenerate the hydroxide:

$$Ca(OH)_2 + Na_2SO_3 \longrightarrow 2\,NaOH + CaSO_3\downarrow \qquad 6.8(32)$$

This offers an alternate caustic system in the event disposal of the sulfite is not permissible.

Summary

Wet scrubbing techniques offer a wide possibility of approaches from the standpoint of the type of scrubbing solution used. Absorbents have a capacity for heavy loading but require complex regeneration unless a throw-away system is acceptable. The necessity of reheating the effluent gas stream to maintain plume buoyancy will add to the overall cost. Most systems also must incorporate mist eliminators to avoid excess plume opacity. The wet scrubbing systems are usually operated from the boiler and, therefore, are not restricted by the temperature and by the residence times imposed inside the furnace. Probably the greatest single advantage to wet scrubbing is that these systems may be added on to existing utilities with a minimum amount of boiler modification.*

*In 1974 two liquid phase Claus processes were successfully piloted under the names of Citrate and Powerclaus. They generate elemental sulfur and practically no sludge.

REFERENCES

1. Cann, E. D.: U.S. Patent No. 3,579,296, May 18, 1971.
2. LaMantia, C. R., E. R. Banel, R. Phillips and T. J. Lamb: "Emission Control for Small Scale Facilities," *AIChE Technical Manual,* T-37, p. 142, 1971.

6.9 SULFUR OXIDE CONTROL—ADSORPTION BY CHARCOAL

Types of Processes:	(a) *Dry adsorption (Reinluft) process;* (b) *Wet adsorption processes,* (b1) Lurgi sulfacid process, (b2) Hitachi process; (c) *Elemental sulfur recovery (Westvaco process).*
Types of Charcoals Used:	Activated carbon from black coal, peat coke and oxycoke. Oxycoke is the most abrasion resistant.
Adsorption Temperature Range:	100 to 300°F.
Charcoal Regeneration Methods:	Thermal, scrubbing with sulfate solution or reduction by H_2S.
Residence Times:	4 to 13 seconds.
Charcoal Loss in Dry Adsorption:	10% of adsorbed SO_2 weight plus 0.5% of total coke throughput.
Advantages:	Small absorber-desorber units required due to the short retention periods.
Limitations:	Charcoal abrasion, problems with regeneration including loss of carbon due to CO and CO_2 formation if thermal regeneration is used. Wet processes require added equipment and corrosion resistant construction.
Process Status:	Wet adsorption processes are in operation, dry adsorption has not been particularly successful and is in the pilot stage together with elemental sulfur recovery.
Cost:	See Table 6.12c.

Sulfur dioxide may be removed from flue gases by adsorption techniques if it is brought into contact with activated charcoal at temperatures between

Table 6.9a
SO_2 REMOVAL FROM SIMULATED COMBUSTION GASES BY DIFFERENT CARBONS AFTER PASSAGE OF 12,700 FT^3, STP

Carbon	*Temperature, °C*	*Residence Time, sec*	*% SO_2 Removed*
Activated carbon from black coal with 50% oxidation	80	13.0	60
	120	13.0	45
	160	13.0	35
	80	5.6	30
	120	5.6	20
	160	5.6	15
	80	3.8	
	120	3.8	10–5
	160	3.8	
Peat coke after 3 regenerations	80	13.0	70
	120	13.0	65
	160	13.0	57
Peat coke after 2 regenerations	80	13.0	100
	120	13.0	100
	160	13.0	100
Oxycoke	80	5.6	97
	120	5.6	94
	160	5.6	75
	80	3.8	90
	120	3.8	73
	160	3.8	65

100 and 300°F. The SO_2 is first adsorbed on the surface of the porous structure of the coke and then it is oxidized to sulfuric acid in the presence of water vapor and oxygen in the flue gas. The charcoal is regenerated thermally or by scrubbing with sulfate solution or by reduction with H_2S. In order to meet flue gas exit temperature requirements, it is desirable to use charcoal that will adsorb at a temperature above 200°F, which necessitates the use of a charcoal with a low tendency to spontaneous ignition. Adsorption cokes produced from hard coal meet these requirements particularly well.[1]

The continuous movement of the charcoal material in an adsorber-regenerator system demands a coke of high-abrasion resistance. Investigations have shown[2] that pelletized charcoal is harder than similar sizes obtained by crushing and screening. Charcoal derived from black coal with 50% oxidation has been investigated[3] together with oxycokes obtained from coking air-oxidized hard coals and peat coke after several regenerations, for physical and SO_2 adsorptive properties. The data are shown in Table 6.9a. The oxy-carbons, although absorbing somewhat less than the peat, proved to be much more abrasive resistant and require less resident time.

A favorable aspect of the charcoal adsorption processes, in general, is the possibility of using small adsorber-desorber units. This is due to the brief retention periods necessary to attain the rapid SO_2 adsorption and high SO_2 equilibrium loading of charcoal.

The regeneration step is a key factor in controlling the economics of this class of sulfur dioxide control processes. If the adsorbed material is treated in an inert atmosphere at 600°F, substantial amounts of carbon may be lost by reaction with the bonded sulfuric acid:

$$H_2SO_4 + C \xrightleftharpoons{\text{heat}} CO + H_2O + SO_2 \qquad 6.9(1)$$

$$2\,H_2SO_4 + C \xrightleftharpoons{\text{heat}} CO_2 + 2\,H_2O + 2\,SO_2 \qquad 6.9(2)$$

This also leads to a lowering of the abrasion resistance of the coke. One remedy is to control regeneration in such a way that only SO_3 is produced.

$$H_2SO_4 \xrightleftharpoons{\text{heat}} SO_3 + H_2O \qquad 6.9(3)$$

Regeneration by scrubbing with aqueous ammonia solution results in a dilute product which must be concentrated. Also a buildup of surface oxides reduces the adsorptive activity of the coke.

Dry Adsorption

In the Reinluft process (Fig. 6.9b), the adsorbing charcoal is channeled downward into the reactor through a set of guide vanes. The flue gas flows counter-currently to the coke through the baffles up to the stack. The most favorable temperature range for the adsorption reaction process is between 200 and 250°F.

The flue gas enters the lower portion of the adsorber at about 300 to 400°F; and after partial adsorption of the SO_2, it moves through a heat exchanger where it is cooled to 220°F and returned to the upper section of the adsorber. The exit gas temperature is about 215°F.

The spent charcoal falls by gravity to a desorber where the SO_2 is continuously driven off by thermal regeneration with a scavenging gas at between 650 and 850°F. The charcoal at this temperature reacts with the sulfuric acid producing CO, CO_2, SO_2 and water vapor. This acts to renew the surface of the coke, thereby reactivating it. The regenerated coke is recycled to the reactor after screening and dust separation.

Operating costs are determined to a large extent by the consumption of the adsorption material. A loss of about 0.1 lb per lb of adsorbed SO_2 and

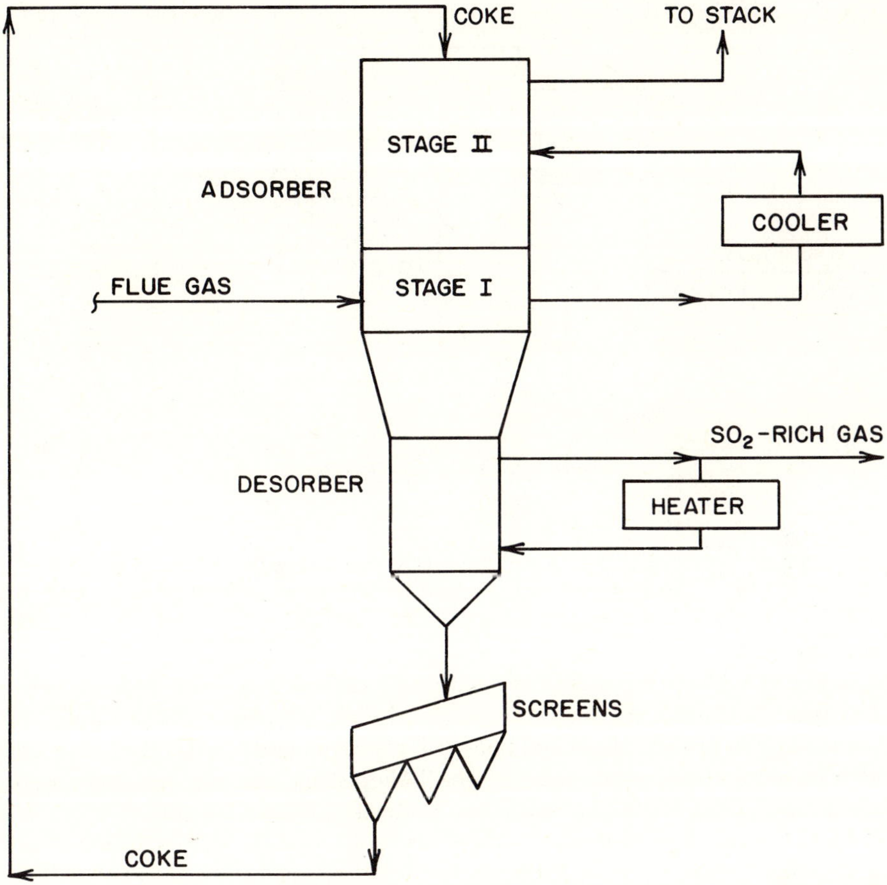

Figure 6.9b The Reinluft process.

a loss, by abrasion, of 0.5% of the coke throughout has been reported.[4] These results are based on a cheap semi-coke of peat carbonized under vacuum at 1100°F in a pilot operation at the Volkswagenwerk at Wolfsburg, Germany. The degree of desulfurization was estimated at 83%.

Wet Adsorption Processes

The Lurgi Sulfacid Process (Fig. 6.9c) is a wet adsorption process involving continuous water spraying of the charcoal. Flue gas is first cooled to 100 to 120°F by contacting the 10 to 20% acid solution elutriated from the charcoal adsorber. A beneficial side-effect is the enrichment of H_2SO_4 in the scrubbing solution by absorption of SO_2 from the gas phase. With flue gas temperatures ranging between 300 and 400°F, acid concentrations of 60

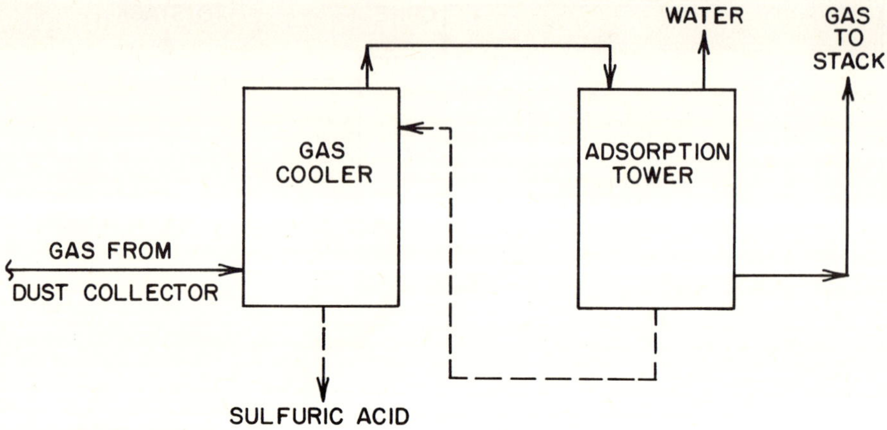

Figure 6.9c Lurgi sulfacid process.

to 75% have been achieved. Further processing of the acid yields a final concentration of 98%. The total pressure drop of the system varies between 6 and 8″ H_2O with a water consumption of 0.5 to 0.75 gallon per 1,000 ft^3 STP flue gas treated.

The Hitachi process also uses activated carbon in a series of fixed beds. The process consists of two basic stages: drying/adsorbing and washing/regenerating. In one set of packed beds, the wet regenerated activated charcoal is dried with the sensible heat of the boiler stack gas and simultaneously adsorbs SO_2 from the boiler stack gas. Water is flushed through the other set of fixed beds and extracts the sulfuric acid adsorbed by the carbon. A final acid concentration of 20 to 25% is achieved after several washings. The flow of flue gas can be controlled in a manifold arrangement by changing dampers to the various beds, permitting continuous operation. An operation cycle for a 5-bed unit would consist of a 60-hour period, 12 hours of which is used for washing each one of the beds. By staggering the washing periods, it is possible for four of the beds to be adsorbing at all times.

The use of water for acid desorption avoids chemical consumption of the carbon and also removes the dust deposited on the carbon. By avoiding direct contact with the washing stream, the flue gas is cooled only 20 to 40°F in the drying portion of the cycle. This low temperature drop is also partially due to the heats of adsorption and reaction on the carbon surface. The Hitachi process uses a carbon that is dry-distilled and is activated with high-temperature steam, yielding an ignition temperature of over 750°F and, thereby, avoiding combustion problems.

The dilute acid can be concentrated to recover sulfuric acid or treated with ammonia or limestone to recover ammonium sulfate or gypsum.

Elemental Sulfur Recovery

A novel charcoal regeneration process which simultaneously recovers the adsorbed sulfur oxide products as elemental sulfur has been developed by Westvaco (Fig. 6.9d). Flue Gas is contacted with activated carbon in the adsorber unit. The acid-laden spent carbon is then fed to a recovery unit (sulfur generator) where it is contacted with H_2S:

$$H_2SO_4 + 3\,H_2S \longrightarrow 4\,S + 4\,H_2O \qquad 6.9(4)$$

As can be determined by the above reaction, one-fourth of the final elemental sulfur came from the acid and is, therefore, the desired by-product. The sulfur is recovered by vaporization in a fluidized sulfur stripper. The fraction of by-product sulfur is stripped from the carbon by heating to 1000°F with inert gases. The sulfur is then condensed and sent to storage. The remaining three-

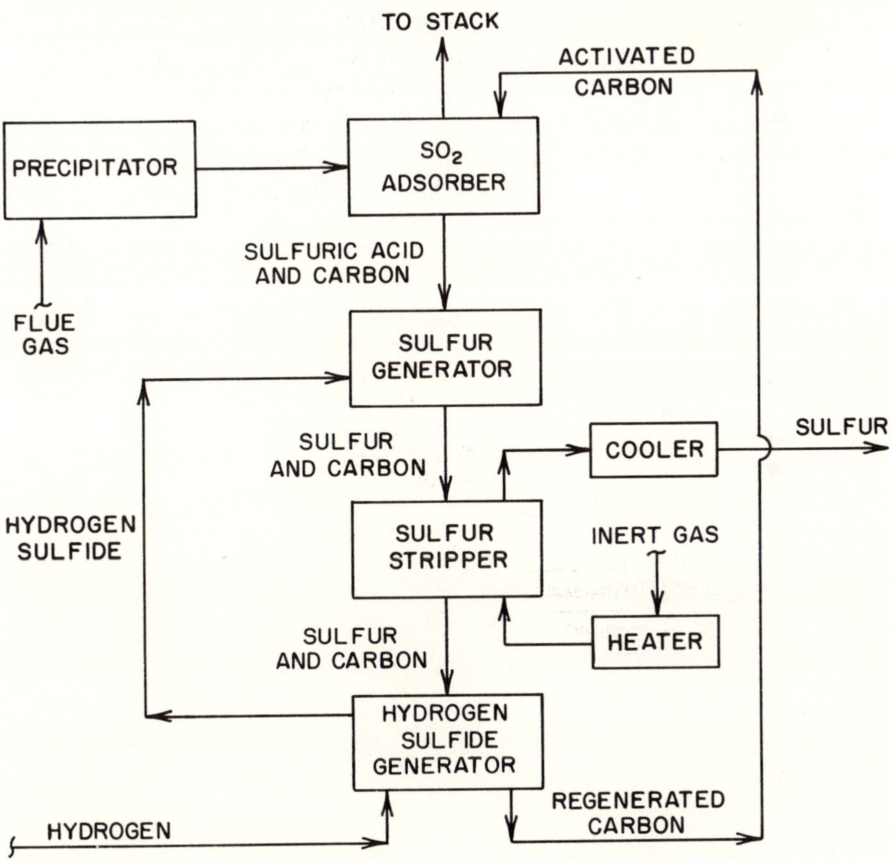

Figure 6.9d Westvaco activated carbon process.

fourths of the adsorbed sulfur reacts with hydrogen, from a coal gas producer or natural gas reformer, to regenerate the hydrogen sulfide:

$$H_2 + S \longrightarrow H_2S \qquad 6.9(5)$$

Finally, the carbon is cooled and returned to the adsorber.

Summary

Activated carbon with an adsorption area in the range 3.5×10^6 to 5.0×10^6 ft^2 per lb, in the presence of oxygen and water vapor, will chemisorb SO_2 and catalyze the conversion to sulfuric acid. The sulfuric acid can be removed thermally, reactivating the carbon in the process, as in the Reinluft process. The Reinluft process depends on large supplies of peat coke, which are not available in central Europe, and tests are being conducted to find a superior coke with high abrasion resistance, ignition temperature and adsorptive capacity.

Wet processes involve much added equipment in the form of spray systems, collection and concentration units. Cooling of the treated flue gas and the need for acid-resistant materials must be taken into account. The Hitachi process avoids gas flow during washing, offering lower temperature drops, but requires higher capital costs for its multibed arrangement. If adequate supplies of H_2S acid producer gas are available, the Westvaco process offers an effective system of elemental sulfur recovery.

REFERENCES

1. Jüntgen, H., and W. Peters: *Staub-Reinhalt. Luft,* vol. 28, p. 1, 1968.
2. Jüntgen, H.: *Habilitationsschrift,* Heidelberg, 1966.
3. Dratwa, H. and H. Jüntgen: *Staub-Reinhalt. Luft,* vol. 27, p. 301, 1967.
4. Zentgraf, K. M.: *Staub-Reinhalt. Luft,* vol. 28, p. 6, 1968.

6.10 SULFUR OXIDE CONTROL—CATALYTIC CONVERSION OF SO_2

Types of Processes:	(a) *Catalytic oxidation,* (a1) Pennsylvania Electric Co., (a2) Air Preheater Co., (a3) Monsanto Co., (a4) Research-Cottrell, Inc.; (b) *Ammonia injection* (*Kiyoura-TIT*); (c) *Catalytic reduction* (*Princeton Chemical Research*).
Catalytic Oxidation Temperatures:	700 to 900°F. For existing power generating stations, this requires the use of auxiliary fuel in a flue gas reheater.
Cat-Ox Catalyst Losses During Regeneration:	2.5% per cleaning.
Catalytic Reduction Temperatures:	250 to 300°F.
Advantages of Catalytic Oxidation:	It is a simple process with no catalyst recycling required. It involves no heat loss and maintains stack gas buoyancy.
Disadvantages of Catalytic Oxidation:	It requires corrosion resistant materials of construction and its economy is affected by the uncertain sulfuric acid market. Ammonia injection reduces the corrosion problem and produces a potential fertilizer component.
Cost:	See Table 6.12c.
Process Status:	(a) In operation; (b) and (c) In pilot stage.

Table 6.10a
EQUILIBRIUM CONSTANTS FOR THE REACTION
$SO_2 + \frac{1}{2} O_2 \rightleftharpoons SO_3$

Temperature, °F	K_p
620	4780.000
980	53.000
1340	3.670
1700	0.630
2240	0.114

Catalytic oxidation of gaseous sulfur dioxide to the trioxide in the presence of a vanadium catalyst occurs according to Equation 6.10(1):

$$SO_2 + \frac{1}{2} O_2 \rightleftharpoons SO_3 \qquad 6.10(1)$$

where the equilibrium constant is

$$K_p = \frac{P_{SO_3}}{P_{SO_2}(P_{O_2})^{1/2}} \qquad 6.10(2)$$

where P_{SO_3} = partial pressure SO_3
P_{SO_2} = partial pressure SO_2
P_{O_2} = partial pressure O_2

Table 6.10a shows that for temperatures below 1340°F the equilibrium favors the formation of SO_3; however, at temperatures above 1340°F, SO_2 is favored.[1]

Several commercial catalysts have been tested in the oxidation of 0.35% simulated flue gas.[2] It was found that vanadia on silica, K_2O promoted Harshaw catalyst VO204 converted up to 85% of the SO_2 at 716°F. An alkalized vanadia catalyst consisting of 39.6% S_iO_2, 16.5% K_2O, 27.3% SO_3, 7.1% V_2O and 9.5% H_2O was dried at 806°F, and prepared in granular form, oxidized 97% of the SO_2 at 689°F. A process of this type was developed by the Pennsylvania Electric Company, the Air Preheater Company, the Monsanto Company and Research-Cottrell Incorporated.

The Monsanto Cat-Ox Process

The Cat-Ox Processes of Monsanto Company are a direct translation of the contact sulfuric acid process (Section 7.16). The integrated Cat-Ox system is intended for new power generating stations since the converter uses gas at temperatures of 700°F to 900°F. For a new power plant, equipment that would normally be included in the plant design is adapted to meet the requirements of the SO_2 control systems. Economizers and air heaters must be built to Cat-Ox specifications and designed into the Cat-Ox system. The reheat Cat-Ox system for existing power generating stations performs the same air pollution control service as the integrated system. However, the flue gas must be reheated with an auxiliary fuel to approximately 850°F for conversion. In most cases, dust loading is too high for optimum Cat-Ox performance and additional dust removal equipment is needed. When in operation, the reheat Cat-Ox system functions independently of the steam generation equipment and does not interfere with the power plant heat cycle.

The Integrated Cat-Ox System

The integrated Cat-Ox system involves removing all flyash (>99.6%) in a combination mechanical separator and electrostatic precipitator and catalytically oxidizing SO_2 to SO_3. The SO_3 reacts with water vapor in the flue gas and is condensed as sulfuric acid at an average concentration of 78% (Fig. 6.10b).

Flue gas is admitted to the dust collector at 950°F—the required initiation temperature for the catalyst. The hot flyash flows freely from the collecting hoppers under the dust collector and is carried away from the unit hydraulically in a united conveyer system.

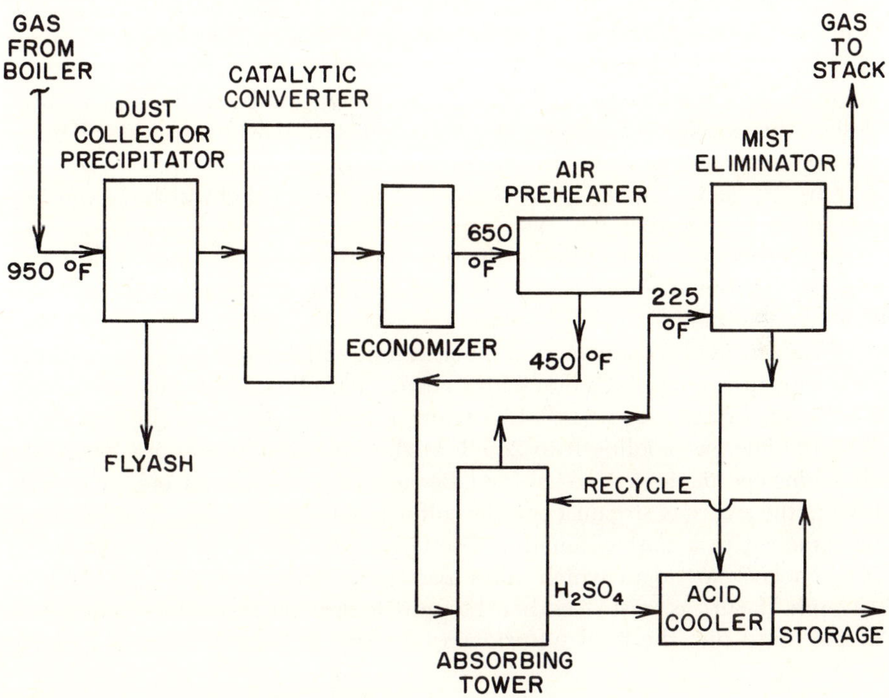

Figure 6.10b Catalytic oxidation process of Monsanto.

Table 6.10c
PRESSURE LOSSES IN CAT-OX SYSTEM EQUIPMENT

Unit	Pressure Drop (in H_2O)	
	Clean	Dirty
Precipitator	3.5	3.5
Converter	0.5	3.0
High-level economizer	4.0	4.0
Air preheater	5.0	5.0
Cooling tower	5.0	5.0
Mist eliminator	13.0	18.0
Ducts	3.0	3.0
Control	7.5	0.0
Total	41.5	41.5

The converter system provides oxidation of SO_2 to SO_3 in excess of 90% so long as the required temperature levels are maintained.[3] The catalyst is a proprietary material, Cat-Ox A, which may be cleaned while the converter is in service. Handling and screening results in about a 2.5% loss of mass per cleaning. A critical feature of SO_2 removal systems is their flexibility in adjusting to variations in boiler loads. In cutting the boiler load to 50% by design, the exit flue gas temperature may drop 175°F, which would still allow for 80% conversion of SO_2 to SO_3 in the Cat-Ox system.

In the heat exchanger, the flue gas is cooled to slightly above the SO_3 dew point in two steps. The first step cools the gas to 650°F in a finned-tube exchanger used to preheat boiler feed water; the second cooling is to 450°F in a Ljungstrom air preheater,* operated such that all surfaces are above the flue gas dew point.

Sulfuric acid is condensed in the next unit which consists of a packed-bed absorbing tower in conjunction with an external shell-and-tube heat exchanger. The tower brings a cool (110°F) stream of sulfuric acid into direct contact with the flue gas, cooling it to 225°F at the exit. Design pressure loss in the absorbing cooling tower is 5″ of H_2O. Because of entrainment from the cooling tower, the exit gas stream contains sulfuric acid mist particles which must be removed in a mist eliminator (Section 5.19). This is an arrangement of fiber-packed cartridges which allow passage of only 1.0 mg H_2SO_4/SCF or 10 ppm H_2SO_4 as mist. Another 11 ppm is derived from the vapor phase at 225°F, so that the total sulfuric acid loading in the exit stack is approxi-

*A Ljungstrom preheater is a gas-gas heat exchanger where hot air from the converter is cooled by incoming, countercurrent flue gases.

mately 20 ppm. The mist eliminator accounts for the major fraction of the system pressure loss as shown in Table 6.10c.

Temperature controllers are required at the air preheater and the cooling tower. These ensure that condensation is avoided and that the flue gas temperature leaving the preheater is controlled. In the cooling tower, it is necessary to ensure that the acid is adequately cooled and that the flowrate of acid is satisfactory. An infrared analyzer monitors the quantity of SO_2 in the stack gases.

The Reheat Cat-Ox System

This system contains the same elements as the integrated system with the addition of an oil or gas-fired furnace between the precipitator and the converter. The flue gas leaving the electrostatic precipitator in most existing units is relatively cool (250 to 350°F) and must be reheated to approximately 850°F before entering the converter. The gas is brought through a heat exchanger and up to the desired temperature by direct addition of hot gas from the reheat furnace, designed to maintain proper conversion temperature regardless of boiler load.

Ammonia Injection

The Kiyoura-T.I.T. process[4] employs a completely dry method to recover flue gas sulfur as ammonium sulfate or phosphate. The sulfur dioxide in the flue gas is oxidized to SO_3 in the presence of vanadium oxide in the temperature range 700 to 800°F (Fig. 6.10d). Ammonia is then injected in the gas in a venturi mixer between two stages of an air preheater. The preheater is necessary to hold the temperature above the sulfuric acid dew point. By controlling the temperature at 400 to 500°F, the formation of ammonium

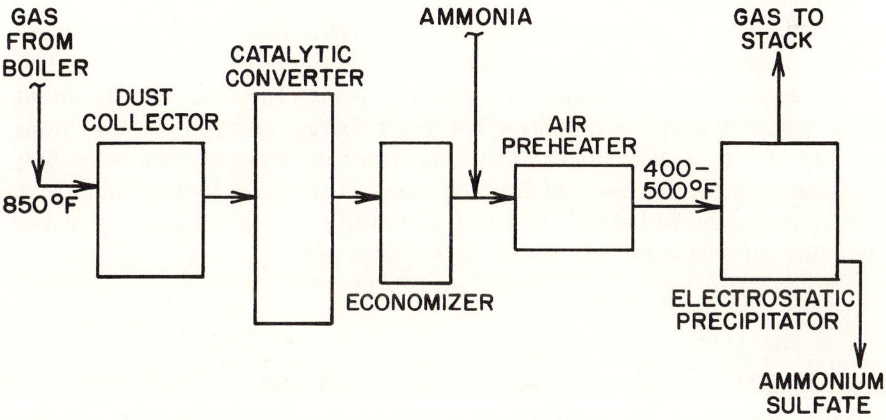

Figure 6.10d Kiyoura-T.I.T. version of catalytic oxidation.

bisulfate, amides, imides and corrosion problems in general is avoided. This temperature also allows the formation of ammonium sulfate aggregate particles of sizes suitable for removal by an electrostatic precipitator. The crystal size of ammonium sulfate is on the order of 1 to 3 μ and the aggregates reach 100 μ.

In a 300 cfm pilot plant operating on fuel oil exhaust with 3.5% sulfur, conversions of 91 to 93% were achieved. The gas was passed through a fixed bed of vanadium pentoxide catalyst placed in three layers at temperatures of 790 to 845°F and with a space velocity of 50 ft^3/ft^2-min. The ammonium sulfate was precipitated electrostatically at 59 to 63 kV at an efficiency of 97.5%.

Catalytic Reduction

In a typical Claus process, hydrogen sulfide is extracted from sour gas by absorbing it in ethanolamine and then by heating. The H_2S then may be converted to elemental sulfur by contacting one part SO_2 with two parts H_2S over a ferric oxide catalyst.

$$SO_2 + 2\,H_2S \underset{Fe_2O_3}{\rightleftharpoons} 2\,H_2O + 2\,S \qquad 6.10(3)$$

Princeton Chemical Research, Inc., utilizes the latter step to react the SO_2 in stack gases at 250 to 300°F. Some of the sulfur is recovered for catalytic reaction with methane at high temperatures to produce hydrogen sulfide for recycling and for further reaction with stack gases.

$$4\,S + CH_4 + 2\,H_2O \underset{cat}{\rightleftharpoons} 4\,H_2S + CO_2 \qquad 6.10(4)$$

Metal Salt Catalysis

Certain metal salts and oxides act as catalytic agents for the oxidation of SO_2. It was found that manganese sulfate is the most efficient and does not produce dithionic acid in a side reaction as the others do.[5] The resulting formation of acid, however, slows the reaction rate and the reaction stops at a pH of 2 to 3. The kinetics of the heterogeneous reaction between SO_2 in air and aqueous aerosols of $MnSO_4$ was also studied.[6] It was found that oxidation was controlled by a four-step chemical reaction in the liquid phase, involving the formation of intermediate complexes:

$$SO_2 + Mn^{2+} \rightleftharpoons Mn \cdot SO_2^{2+} \qquad 6.10(5)$$

$$2\,Mn \cdot SO_2^{2+} + O_2 \rightleftharpoons [(Mn \cdot SO_2^{2+})_2 \cdot O_2] \rightleftharpoons 2\,Mn \cdot SO_3^{2+} \qquad 6.10(6)$$

$$Mn \cdot SO_2^{2+} + H_2O \rightleftharpoons Mn^{2+} + HSO_4^- + H^+ \qquad 6.10(7)$$

$$HSO_4^- + H^+ \rightleftharpoons H_2SO_4 \qquad 6.10(8)$$

This reaction is inhibited easily by the presence of phenols which apparently interfere with the complex formation. When SO_2 was oxidized in solutions of $MnCl_2$, $CaCl_2$, $FeCl_2$ and $CoCl_2$, it was found that the potency of the catalysts decreased in the order given.[7] The general kinetic behavior of SO_2 oxidation with all these salts suggests a mechanism similar to that proposed for manganese sulfate. Also the number of sites available for attachment of the SO_2 molecule to the metal ion to form the complex $(M \cdot SO_2)^{2+}$ decreases in approximately the same order as does the catalyst potency.

Summary

The catalytic oxidation method of SO_2 removal is simple and requires no recycling of the vanadium base catalyst. Since the reaction is exothermic and no heat is lost to the scrubbing media, effluent stack gas buoyancy is maintained. The disadvantages of the process are the need for expensive corrosion resistant materials and the uncertain market for sulfuric acid. The Kiyoura process partially avoids the corrosion problems to produce a possible fertilizer component, ammonium sulfate. Metal salt catalysts have not been applied directly to stack gas SO_2 reclamation because of difficulties in maintaining a high pH in the scrubbing liquid.*

* Sodium phosphate and citric acid are used as buffers in newer processes.

REFERENCES

1. Juntgen, H.: *Erdoel Kohle,* vol. 16, p. 119, 1963.
2. Bienstock, D., J. H. Field and J. G. Meyers: "Process Development in Removing Sulfur Dioxide from Hot Flue Gases," Part 1, U.S. Bureau of Mines RI 5735, 1961.
3. Stites, J. G., W. R. Horlacher, Jr., J. L. Bachofer, Jr. and J. S. Bartman: *Chem. Eng. Prog.* vol. 65, p. 63, 1969.
4. Kiyoura, R., M. Munidasa, H. Kuronuma and G. Uwanishi: *Bull. Tokyo Inst. Tech.,* vol. 81, p. 1, 1967.
5. Bassett, H. and W. G. Parker: *J. Chem. Soc.,* vol. 1951, p. 1540, 1951.
6. Matteson, M. J., W. Stokes and H. Luther: *Ind. Eng. Chem. Fund.,* vol. 8, p. 677, 1969.
7. Junge, C. E. and T. G. Ryan: *Quart. J. Roy. Meteor. Soc.,* vol. 84, p. 46, 1958.

6.11 OTHER SO_2 REMOVAL PROCESSES

Types of Processes: (a) Caustic scrubbing with electrolytic regeneration (Stone and Webster/Ionics); (b) Submerged combustion (Black, Sivalls & Bryson); (c) Molten carbonate process (Atomics International); (d) Chamber process adaptation (TYCO® Process); (e) Wade process (Gilbert station); (f) New concepts, (f1) Gas diffusion, (f2) Centrifugation, (f3) Mass spectrometry, (f4) Condensation, (f5) Corona discharge catalysis.

Process Status: All processes are in the pilot stage.

Caustic Scrubbing with Electrolytic Regeneration

Advanced technology has been applied in a number of approaches to eliminate sulfur dioxide from combustion gas products. The Stone and Webster/Ionics process was originally designed to remove CO_2 from submarine and spacecraft atmospheres. It is a wet caustic scrubbing system with a closed-circuit electrolytic regeneration of the absorbing liquor. The recovered sulfur dioxide is fed to a sulfuric acid processing plant (Fig. 6.11a).

Hot flue gases are passed through a precipitator and are introduced to the bottom of the absorber tower where they are quenched to between 130 and 150°F. Caustic solution enters the upper portion of the absorber. The sodium sulfite product moves to the lower stage where a large fraction is converted to the bisulfite using hydrogen generated in the electrolytic cell:

$$Na_2SO_3 + SO_2 + H_2O \underset{H_2}{\rightleftharpoons} 2\ NaHSO_3 \qquad 6.11(1)$$

The resulting solution is reacted with dilute H_2SO_4 to yield sodium sulfate and concentrated SO_2 gas.

$$Na_2SO_3 + 2\ NaHSO_3 + 2\ H_2SO_4 \rightleftharpoons 3\ SO_2\uparrow + 2\ Na_2SO_4 + 3\ H_2O \qquad 6.11(2)$$

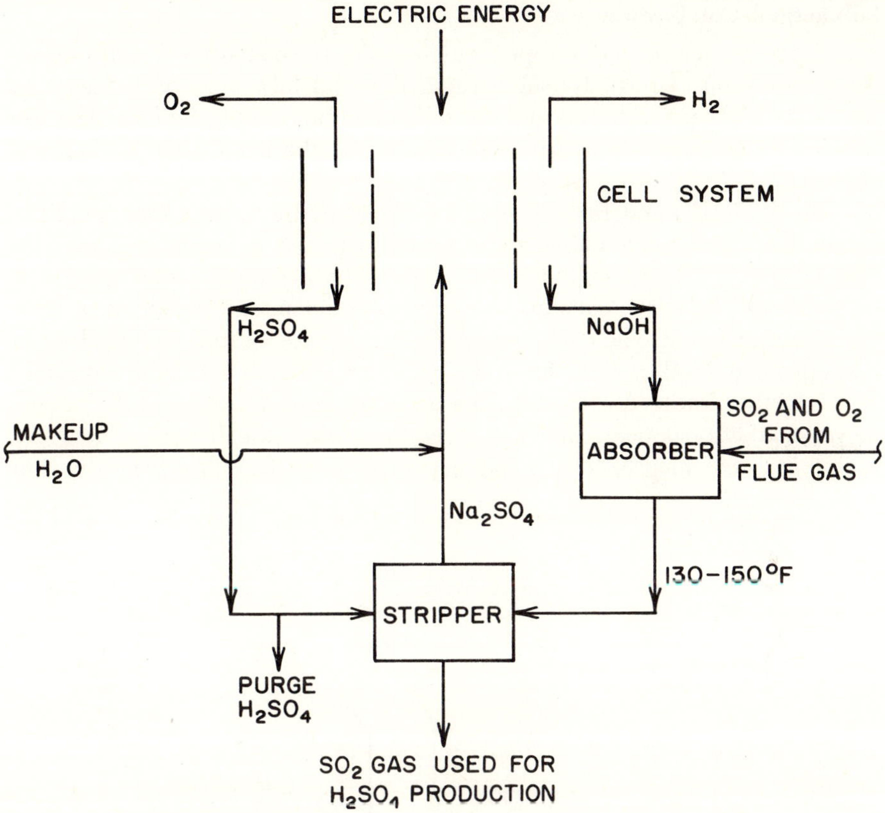

Figure 6.11a Stone and Webster/Ionics SO_2 process chemistry.

The SO_2 is stripped from the mixture, dried and converted to concentrated H_2SO_4.

A key factor in the process is the electrolytic regeneration unit. Sodium sulfate from the stripper is pumped into the electrolytic cell between a membrane and a diaphragm. The electrode reactions at the cathode are:

$$2\,Na^+ + 2\,H_2O + 2\,e^- \rightleftharpoons 2\,NaOH + H_2\uparrow \qquad 6.11(3)$$

and at the anode,

$$2\,SO_4^{2-} + 2\,H_2O \rightleftharpoons 2\,H_2SO_4 + O_2\uparrow + 4\,e^- \qquad 6.11(4)$$

The cell thereby regenerates caustic for the absorber and sulfuric acid for use in releasing SO_2 from the sulfite/bisulfite solution.

6.11 Other SO_2 Removal Processes

Submerged Combustion

A process has been developed which avoids the oxidation of sulfur during the combustion of coal. By submerging the coal in a mass of molten iron, the coal dissolves and burns, and the sulfur combines with the iron. The iron sulfide then migrates to a slag layer floating on the iron and is periodically removed.

Black, Sivalls and Bryson, Inc., have designed a furnace that would be compatible with conventional boilers used in power generating plants. The furnace (Fig. 6.11b) is a cylindrical-shaped, refractory-lined, metal vessel with parts for slag and iron exits and air lances for pneumatic injection of limestone/slag and coal. The coal is injected into the molten iron and combustion takes place at 2500 to 2700°F, producing an off gas rich in carbon monoxide. This is further oxidized with secondary air in the conical section of the furnace to yield additional heat. The hot gases then enter the boiler section. The limestone is calcined to form a slag which removes the sulfur from the iron:

$$CaO + FeS \rightleftharpoons CaS + FeO \qquad 6.11(5)$$

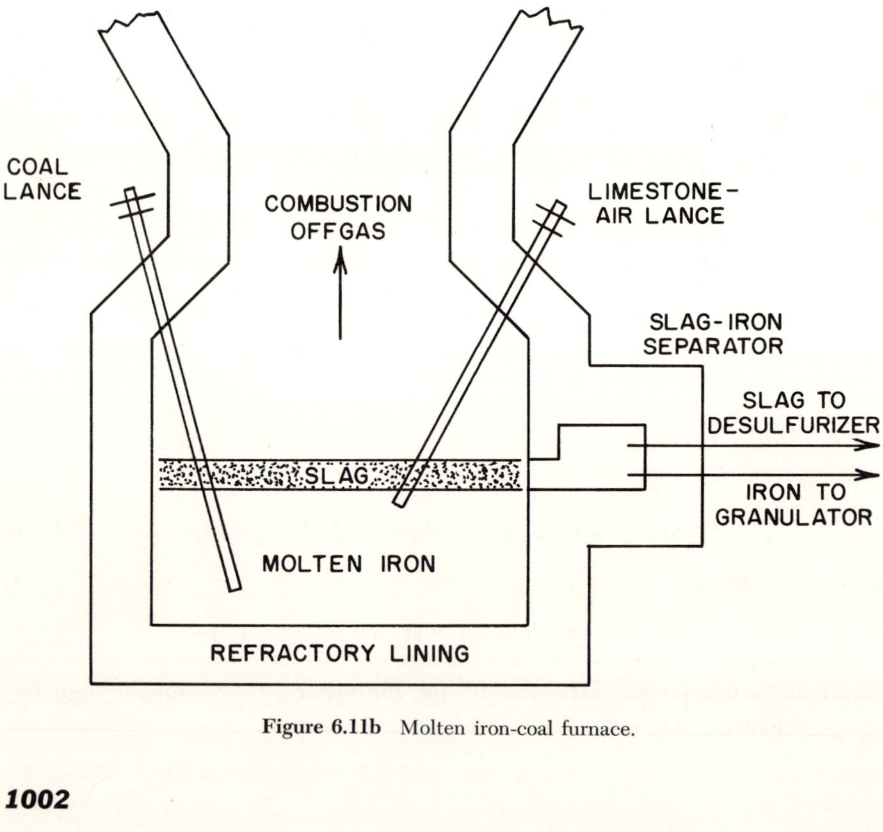

Figure 6.11b Molten iron-coal furnace.

The slag is periodically withdrawn from the furnace and contacted with steam and air, producing a sulfur rich gas:

$$2\,CaS + O_2 \xrightarrow[\text{heat}]{} 2\,CaO + 2\,S \qquad 6.11(6)$$

The sulfur is condensed from the gas and deposited in pits, while the desulfurized slag is recycled. The iron is removed in a slag/iron separator, quenched and granulated.

Molten Carbonate Process

In the molten carbonate process, a molten eutectic* mixture of 33 wt% sodium, 35 wt% potassium and 32 wt% lithium carbonate is used to remove sulfur dioxide from powerplant waste gas streams. The SO_2 reacts with the carbonates to yield sulfites and sulfates that remain in the excess unreacted carbonate mixture. This mixture is regenerated in two steps. First the sulfite and sulfate are reduced thermally to sulfide at 1500°F with excess petroleum coke; then the carbonate-sulfite melt is contacted with a mixture of carbon dioxide and steam which yields the carbonate plus gaseous H_2S (Fig. 6.11c).

Flue gas is removed from the boiler at 850°F and flyash is extracted in an electrical precipitator. The gas then enters the scrubber contacting a spray of the molten carbonate. The scrubbed gases are returned to the boiler for heat recovery and exhausted out the stack. The spent absorbent melt is filtered to remove ash, heated and fed into the reducer. Here the sulfite and sulfate are completely reduced to the sulfide in about 20 minutes. The reduced melt is quenched to 950°F by mixing with melt at 850°F; and the effluent is filtered, cooled further to 850°F and sent to the regenerator.

In the regenerator, the reduced melt reacts with CO_2 and steam in a multistage, countercurrent sieve-tray column. The resulting hydrogen sulfide is passed to a Claus plant, where it is converted to elemental sulfur. The regenerated melt is recycled to the scrubber.

The process has been under development at Atomics International since 1966, during which time bench-scale engineering tests have been performed on the scrubbing and filtration tests. A pilot plant to process a sidestream of stack gases from a 172 Mw oil-fired boiler is proposed.

Chamber Process Adaptation

Some flue gas cleaning processes have been proposed wherein both the oxides of sulfur and nitrogen are scrubbed. In the traditional chamber process, oxides of nitrogen serve to catalyze the oxidation of sulfur dioxide by oxygen to form nitrososulfuric acid:

$$2\,SO_2 + NO + NO_2 + O_2 + H_2O \longrightarrow 2\,HSO_4NO \qquad 6.11(7)$$

*Having the lowest possible melting point.

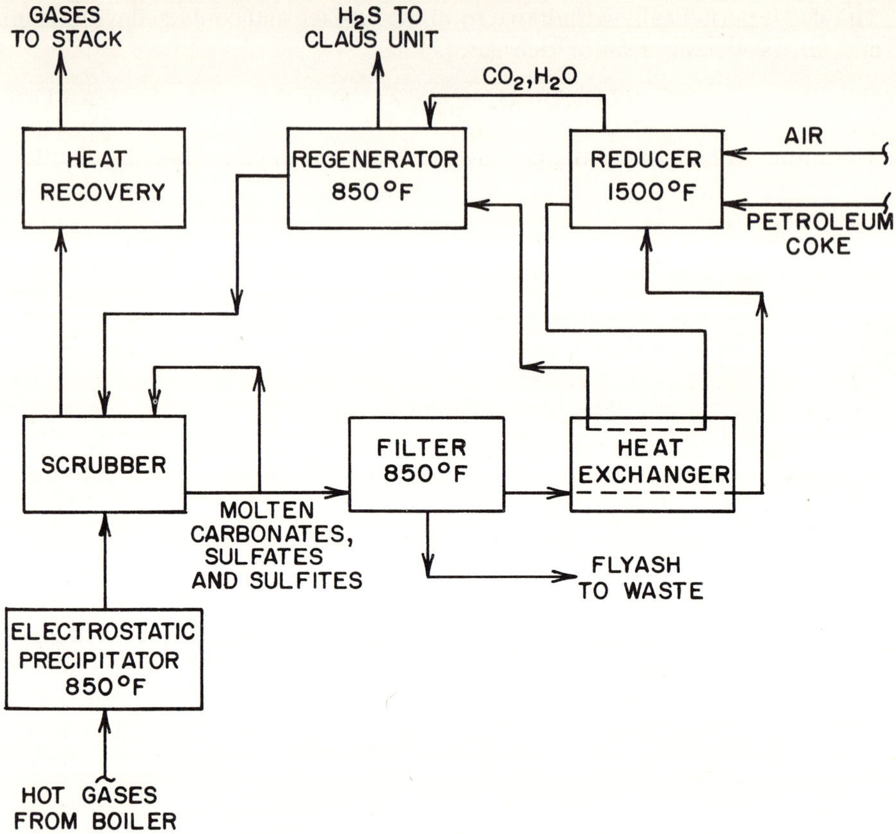

Figure 6.11c Molten carbonate SO_2 removal.

Steam is injected and droplets of sulfuric acid mist are formed, while the oxides of nitrogen are released:

$$2\,HSO_4NO + H_2O \longrightarrow 2\,H_2SO_4 + NO\uparrow + NO_2\uparrow \qquad 6.11(8)$$

The TYCO Process is an adaptation of the chamber process wherein flyash is first removed from a flue gas in a cyclone; and then the gas is mixed with recirculated oxides of nitrogen in a reactor in quantities sufficient to give a $NO_x:SO_2$ mole ratio of 2:1. A retention time of two seconds is adequate to completely oxidize the sulfur dioxide (Fig. 6.11d). After the reactor, the gases go to an absorber where they are washed with a recycled stream of 80% sulfuric acid. The absorber is a packed column fitted at the top with a mist collector. The droplet-free gas stream is then sent to the stack. The absorbent solution strips both the oxides of nitrogen and the sulfuric acid from the gas. This solution is then sprayed countercurrent to a hot sidestream of flue gases.

The oxides of nitrogen in the flue gas stream are enriched by those from the spent scrubber solution and thereby a gradual accumulation of nitrogen oxides occurs, resulting in the production of nitric acid.

The gas stream from the stripper is converted to nitric acid by aqueous scrubbing while a portion of the gases are recycled to the main reactor. The sulfuric acid leaving the stripper is concentrated to 80%, cooled, filtered and partially recycled before storage.

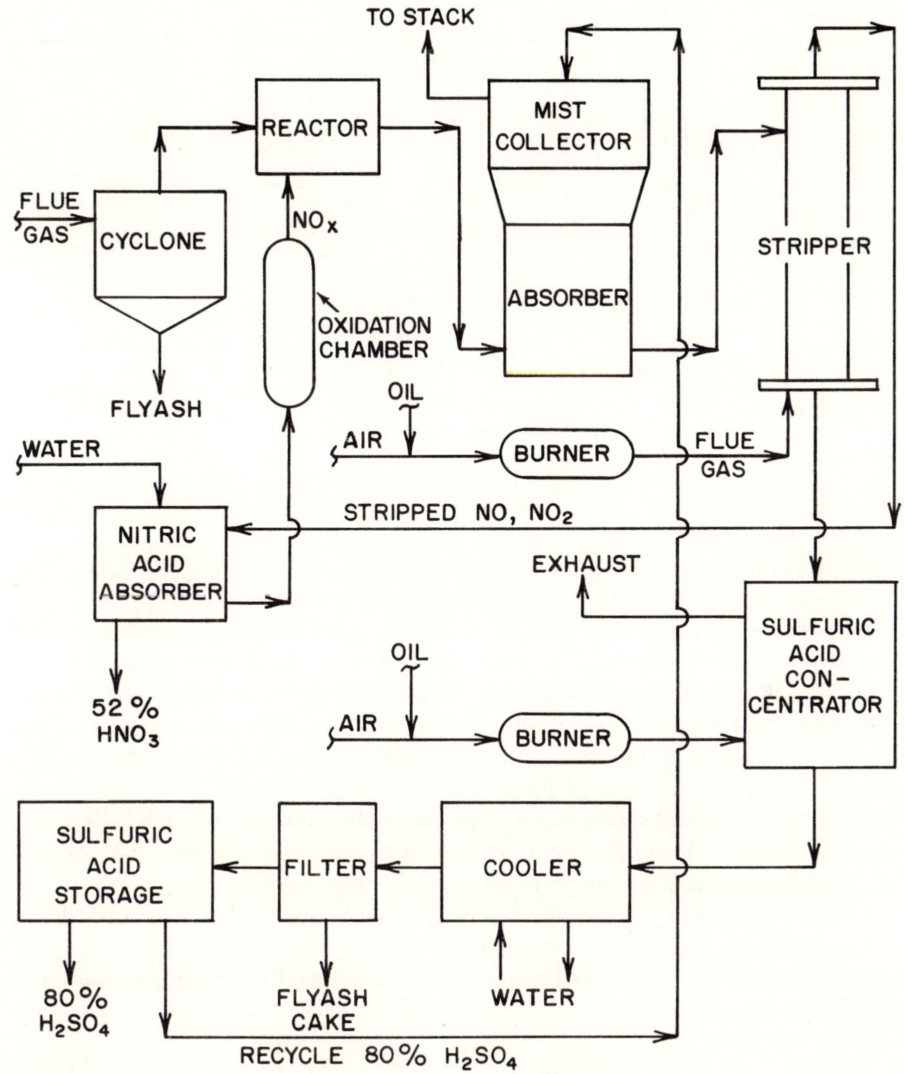

Figure 6.11d TYCO process.

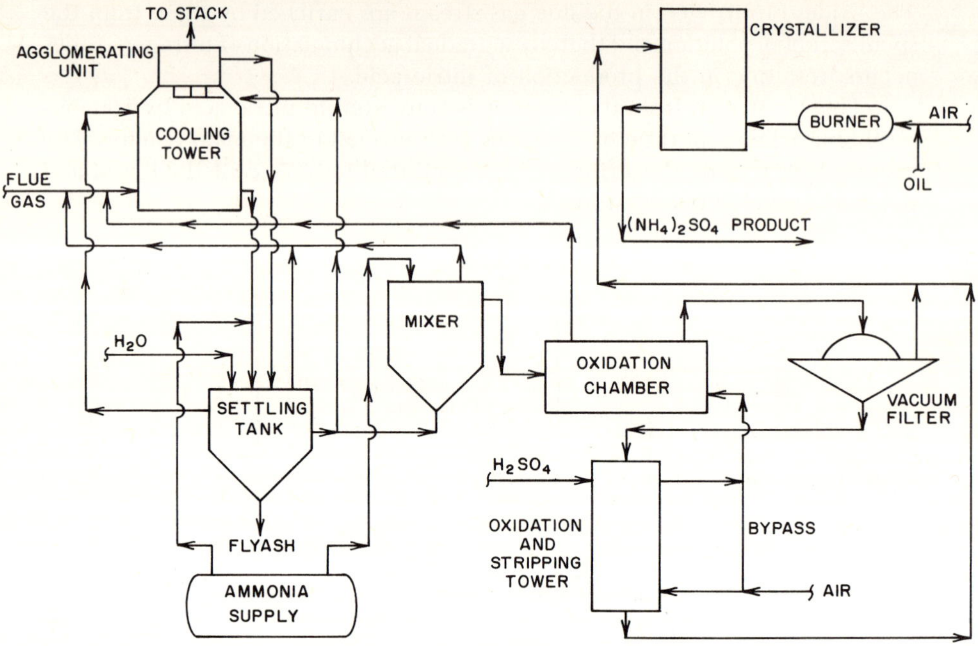

Figure 6.11e Wade process.

The Wade Process

A method for simultaneous removal of SO_2 and flyash is based on rapid acceleration and deceleration of the flue gas in a stack of polypropylene discs. The process, developed by Wade Company, is shown in Figure 6.11e. The hot flue gases are cooled in a tower packed with Raschig rings (Fig. 5.14b) and rise to the agglomerating unit. This unit consists of elements of rings chamfered on all sides and held apart by spacers. The gas flows through a venturi-like opening, being accelerated radially outward and then decelerated to produce a coagulation of flyash particles. The particles and gas are contacted with a second scrubbing stream containing ammonia which falls to the cooling section. The pH of this stream is maintained between 5.5 and 6.5 by adding ammonia as the solution becomes more acidic.

Concentrations of ammonium sulfite, ammonium bisulfite and flyash particles build up in the settling tank, which is tapped periodically. The suspension may be transferred to a mixing tank, where the bisulfite is converted to sulfite:

$$NH_4HSO_3 + NH_3 \rightleftharpoons (NH_4)_2SO_3 \qquad 6.11(9)$$

Then the sulfite is oxidized to the sulfate, using air:

$$(NH_4)_2SO_3 + \tfrac{1}{2}\,O_2 \rightleftharpoons (NH_4)_2SO_4 \qquad 6.11(10)$$

This reaction is catalyzed by trace metal oxides in the flyash.

The flyash is removed by vacuum filtration and the solution is further treated with sulfuric acid to ensure complete oxidation.

$$(NH_4)_2SO_3 + H_2SO_4 \rightleftharpoons (NH_4)_2SO_4 + SO_2\uparrow + H_2O \qquad 6.11(11)$$

Sulfur dioxide liberated in this stage is recycled to the main scrubber. A final crystallization unit consists of hot air flowing through a bed of $(NH_4)_2SO_4$ crystals.

The final product, ammonium sulfate, has a market as a nitrogenous fertilizer. Wade Co. has demonstrated the scrubbing unit with 3,000 cfm from boiler flue gases at the Gilbert generating station of the New Jersey Power and Light Co. The effluent contained about 1,000 ppm SO_2 and 5.7 grams per cubic meter flyash and removal of 98% SO_2 and flyash was obtained.

New Concepts in Sulfur Dioxide Control

The feasibility of using a number of physical methods for separating SO_2 from flue gas was investigated by the Westinghouse Electric Corporation Research Laboratories.[1] Gaseous diffusion, centrifugation, mass spectrometry and condensation techniques were considered for directly separating SO_2 from the flue gas. Electrical, electromagnetic and nuclear radiation excitation were studied as means of converting SO_2 to the more easily removable SO_3.

Gaseous diffusion has been applied to the separation of isotopes of uranium and depends on the molecular effusion through a number of porous barriers or stages. The gas mixture is forced through a barrier with pore sizes of the order of the mean free path of molecules. The average velocity of the lighter molecules is higher, resulting in a preferential flow of these molecules through the barrier. The enriched fraction is fed to the next stage while the lean fraction is returned to the previous barrier. Even with barriers of high permeabilities, the number of stages required for SO_2 removal would be between 50 and 60, based on 10^6 SCFM corresponding to a 500-Mw powerplant. This flowrate would require compressor power costs in excess of \$500 million per year.

Gas separation by centrifugation was also found to be economically prohibitive at costs of \$100 per KW. Mass spectrometers are unattractive for large-scale separation of mixtures of gases because they operate on a batch basis, involving start-up, production and cool down. Also, the operating pressure is in the order of 2×10^{-5} mm mercury. Separation by condensation requires temperatures lower than $-190°F$, the saturation temperature of SO_2 in the flue gas. Since CO_2 solidifies at $-147°F$, the need for removing the CO_2 from the condenser makes this method even more difficult.

Recently, there have been reports on the corona discharge catalysis of sulfur dioxide. This treatment involves passing a flue gas through an electrical field where the SO_2 gas mixture is exposed to the electrical discharge from a point or wire maintained at high voltages. The SO_2 is oxidized to SO_3 and, in the presence of water vapor, sulfuric acid mist is generated. The force of the electrical field drives the acid mist to the grounded electrode where it is collected by condensation. It has been shown[2] that a damped or pulsed, high-frequency current is more effective than direct current for SO_2 conversion. In the tests, the reactor was placed in an oven at 266°F; and a conversion efficiency of 96% was obtained with a power expenditure of 9 watts over a period of 30 minutes. The reaction kinetics of SO_2 oxidation was studied using a direct current corona discharge.[3] It was concluded that the rate controlling step was the generation of atomic oxygen by the corona.

REFERENCES

1. Hamm, J. R.: "Study of the Applicability of Physical Methods of Separation to the Development of New Processes for the Control of Sulfur Dioxide Pollution," U.S. Dept. HEW, Nat. Ctr. for Air Poll. Contr., Contract No. PH-22-68-34, January 31, 1969.
2. Palumbo, F. J. and F. Fraas: *J. Air Pollu. Contr. Ass.*, vol. 21, p. 143, 1971.
3. Matteson, M. J., H. L. Stringer and W. L. Busbee: "The Corona Discharge Catalysis of Sulfur Dioxide," presented to AIChE 72nd National Meeting, St. Louis, Missouri, May 21–24, 1972.

6.12 RELATIVE MERITS OF SO_2 REMOVAL PROCESSES

Operating SO_2 Removal Installations:	See Tables 6.12a and 6.12b.
*SO_2 Removal Costs**:	See Table 6.12c.
Cost of Sulfur:	$8 to $15 per ton.
List of System Suppliers:	See Table 6.12c.

Sections 6.6 through 6.11 were devoted to the description of processes for the removal of sulfur oxides—and, coincidentally in some cases, particulates—from stack gases. This section summarizes the current status of these systems. A listing of representative commercial installations is given in Tables 6.12a (electric utilities) and 6.12b (industrial applications).

When evaluating alternative processes for the removal of sulfur oxides from a particular waste gas stream, the following factors must be considered:

* According to the National Environmental Research Center report titled: "A Critical Evaluation of Processes for the Removal of SO_2" presented at the June 1973 meeting of APCA:

1. The lime scrubbing and the Wellman-Lord processes are the best proven.
2. Regenerable processes are less polluting while the throwaway processes are less expensive if waste disposal is cheap.
3. Annualized costs of reasonably waste free processes are or will soon be in the range of 15–30¢/10^6 BTU. Based on the use of coal with 3.5% sulfur content in 500 MW plants and assuming $3/ton waste disposal cost and $15/ton sulfur credit, the cost estimates are as follows:

	Capital Cost (S/KW)	*Annualized Cost (mills/KWH)*
Throwaway Processes	24–30	2.0–2.25
Regenerable Processes	30–41	1.7–2.0

Throwaway processes are the Double Alkali, Lime Scrubbing and Limestone Scrubbing processes, while the Citrate, Wellman-Lord, MgO Scrubbing and the Cat-Ox processes are considered to be regenerable.

Operating pilot plants with minimum sludge generation and with elemental sulfur production include the Wellman-Lord unit at Northern Indiana Public Service, the Citrate installation at Pfizer's Terre Haute, Indiana pilot plant and the Stauffer Powerclaus installation (a liquid phase Claus process with sodium phosphate buffer solution) at Norwalk Harbor, Conn. Total costs are reported to be under 10% of delivered 1974 electrical costs.

Table 6.12a

SO_2 REMOVAL SYSTEM INSTALLATIONS[a] IN DOMESTIC[m] AND FOREIGN ELECTRIC UTILITIES

Domestic Utility	*Predicted Removal Efficiency, % (Sulfur Oxides/ Particulate Matter)*	*Plant Size, mw*	*Fuel*	*Process*	*Designer or Manufacturer*	*Initial Operation*
Union Electric Company Meramec #2 Meramec, Missouri	70/99[b]	125	Coal	Furnace injection of limestone followed by wet scrubbing[c]	Combustion Engineering, Inc.	Sept. '68[d]
Kansas Power & Light Co. Lawrence #4 Lawrence, Kansas	70/99[b]	125	Coal/Natural Gas	Furnace injection of limestone followed by wet scrubbing[c]	Combustion Engineering, Inc.	Nov. '68
Kansas Power & Light Co. Lawrence #5 Lawrence, Kansas	80/99	420	Coal/Natural Gas	Furnace injection of limestone followed by wet scrubbing[c]	Combustion Engineering, Inc.	Nov. '71
Commonwealth Edison Co. Will County #1 Will County, Illinois	76/98	200	Coal	Limestone scrubbing[c]	Universal Oil Products/Babcock & Wilcox	Early '72
Boston Edison Company Mystic #6 Everett, Massachusetts	90/—	150	Oil	MgO scrubbing[c]	Chemical Construction Corporation	Early '72
City of Key West Stock Island #1 Key West, Florida	85/98	37	Oil	Limestone scrubbing[c]	Zurn Industries	Early '72
Louisville Gas & Electric Co. Paddy's Run #6 Louisville, Kentucky	80/99	70	Coal	$Ca(OH)_2$ scrubbing[c]	Combustion Engineering, Inc.	Mid '72

Detroit Edison Company River Rouge #1 Detroit, Michigan	[e]	260	Coal	Limestone scrubbing[ch]	Peabody-Lurgi	Late '72
Kansas City Power & Light Co. Hawthorne #3 & #4 Kansas City, Missouri	70/99	200 (total)	Coal	Furnace injection of limestone followed by wet scrubbing[c]	Combustion Engineering, Inc.	Late '72
Arizona Public Service Cholla Station Joseph City, Arizona	[f]	115	Coal	Limestone scrubbing[ci]	Research-Cottrell	Late '72
Kansas City Power & Light Co. LaCygne #1 LaCygne, Kansas	80/99	800	Coal	Limestone scrubbing[c]	Universal Oil Products/Babcock & Wilcox	Late '72
Duquesne Light Company Phillips #1–#6 Pittsburgh, Pennsylvania	80/99.3	387[g] (total)	Coal	CaO scrubbing[c]	Chemical Construction Company	Early '73
Nevada Power Company Reid Gardner Station Moapa, Nevada	—	250	Coal	Na(OH) scrubbing[c]	Combustion Equipment Associates	Mid '73
Illinois Power Company Wood River #4 Wood River, Illinois	85/—	100	Coal	Catalytic[i]	Monsanto Co.	Mid '73
Potomac Electric Company Dickerson #3 Dickerson, Maryland	90/99.5	195[j]	Coal	MgO scrubbing[c]	Chemical Construction Corporation	Late '73
Northern States Power Sherburne County #1 & #2 Sherburne County, Minnesota	50/99[k]	1,400	Coal	Limestone scrubbing[c]	Combustion Engineering, Inc.	Early '75

Table 6.12a (Continued)

Foreign Utility	Predicted Sulfur Oxides Removal Efficiency, %	Plant Size, mw	Fuel	Process	Designer or Manufacturer	Initial Operation
Japan Synthetic Rubber Chiba Factory Power Plant	90	75	Oil	Na_2SO_3 scrubbing[c]	Wellman Power Gas[m]/ Mitsubishi Heavy Industries	July '71
Showa Denko Kawasaki Mill Power Plant	98	15	Oil	NaOH scrubbing[c]	Showa Denko	Aug. '71
Kansai Power Company Amahigashi #2	—	20	Oil	$Ca(OH)_2$ scrubbing[c]	MHI/Kansai Power	March '72
Kansai Power Company Sakai #8	80	35	Oil	Adsorption on activated charcoal[h]	Sumitomo Shipbuilding & Machinery/Kansai Power Company	Early '72
Chubu Electric Power Company Yokkaichi Station	—	110	Oil	DAP-Mn[l]	Mitsubishi Heavy Industries/Chubu Electric Power Company	Mid '72
Tokyo Electric Power Company Kashima #3	—	80	Oil	Adsorption on activated charcoal[h]	Tokyo Electric Power Company/Hitachi Ltd.	Late '72
Steag	80	45	Coal	Adsorption on activated carbon[h]	Bergbau-Forschung	1973

[a] All coal-fired applications also have particulate matter removal capability.
[b] Actual operating results.
[c] Detailed description of noted process given in Section 6.6 or 6.8.
[d] In 1972, it was announced that this installation had been shut down. The possibility of conversion to limestone scrubbing is being considered.
[e] Expected emission of 300 ppm SO_2 and 0.03 lb particulate matter per 1,000 lb gas.
[f] Burning 0.5% sulfur coal.
[g] SO_2 removal system will treat only 25% of resulting flue gas.
[h] Detailed description of noted process given in Section 6.9.
[i] Detailed description of noted process given in Section 6.10.
[j] SO_2 removal system will treat only 50% of resulting flue gas.
[k] Burning 0.8% sulfur coal; provisions for additional scrubbing stage for future increase in SO_2 removal.
[l] Detailed description of noted process given in Section 6.7.
[m] Wellman Power Gas (name recently changed to Davy Power Gas) will start up its 115 mw coal fueled installation at Northern Indiana Public Service Co. in Gary, Indiana, in 1975.

Table 6.12b

SO_2 REMOVAL SYSTEMS IN DOMESTIC AND FOREIGN INDUSTRIAL APPLICATIONS

Installations	Unit Description	SO_2 Removal Process[a]	Designer or Manufacturer	Initial Operation
Domestic				
Olin-Mathieson Corp.[b] Paulsboro, N.J.	700 ton/day (equivalent to 65,000 kw) Sulfuric Acid Plant	Na_2SO_3-$NaHSO_3$ scrubbing	Wellman Power Gas[b]	June '70
Standard Oil of California El Segundo Refinery El Segundo, California	Claus Tail Gases	Na_2SO_3-$NaHSO_3$ scrubbing	Wellman Power Gas	Late '71
ASARCO El Paso, Texas	Copper Smelter	Na_2SO_3-$NaHSO_3$ scrubbing	Wellman Power Gas	Late '71
General Motors Corp.	25,000 kw Coal-fired boiler	NaOH Scrubbing; Removal of sulfur by external treatment with lime.	—	Mid '72
Foreign				
Showa-Hokkaichi Oil Co. Yokkaichi Refinery	70,000 cfm	Shell Flue Gas Desulfurization Process	Japan Shell Technical Co., Ltd.	
Toa Fuel Oil Company Kawasaki Refinery	42,000 cfm Claus Tail Gas	Na_2SO_3 Scrubbing	W-L/Sumitomo	
Sodersjukhuset Hospital Stockholm, Sweden	50,000 cfm (total) from 3 boilers	$Ca(OH)_2$ Scrubbing	A. B. Bacho, Sweden	
Yahagi Iron Works	50,000 cfm from 1 boiler	$Ca(OH)_2$ Scrubbing	A. B. Bacho, Sweden	
Japanese Industrial Sources	350,000 cfm from 9 boilers	NaOH Scrubbing	A. B. Bacho, Sweden	

[a] Detailed description of noted process given in Section 6.8.
[b] Names changed to Olin Corp. and Davy Power Gas respectively.

1. The extent to which the recovery process has been demonstrated to be effective under industrial operating conditions.
2. The ability of the recovery process to respond to load changes and other anticipated operating upsets and still meet the requirements of state air quality regulations.
3. The avoidance of creating water pollution or solid waste disposal problems.
4. Capital cost.
5. Operating cost, with particular attention to possible revenues from sale of by-products, or costs to cover the safe disposal of by-products that cannot be sold.

Capital and Operating Costs

Some estimated capital and operating cost information has been published, although much of it is based on extrapolation of laboratory or pilot plant data. Representative data are given in Table 6.12c.

A great many process concepts for the recovery of SO_2 from waste gas streams have been developed and a good number of these have been tested in the laboratory. Some processes have also been studied on the pilot plant scale and relatively few have been installed on full-scale commercial plants.

Those which have been tested in pilot plant scale or larger equipment belong to the following categories:

1. Dry additive injection, limestone
2. Dry absorption
3. Wet absorption, scrubbing
4. Adsorption by charcoal
5. Catalytic conversion of SO_2

Lime and Limestone Processes

Much recent work on sulfur oxide removal from stack gases has centered on limestone processes in which the sulfur is captured as calcium sulfite or calcium sulfate and discarded. Limestone is abundantly available and cheap; however, limestone which is low in its magnesium content is preferred since magnesium salts are relatively more water soluble than are calcium salts.

Addition of dry limestone to existing fossil-fuel-fired utility boilers followed by dry collection has the advantage of relatively low capital and operating cost; but sulfur removal efficiencies vary widely with the type of limestone used and generally are disappointingly low (Section 6.6).

When adding dry limestone during coal combustion, the sulfur recovery efficiencies can be improved by the use of a fluidized-bed boiler (Section 6.5); but this technology has not been developed past the pilot plant stage, and even when it has, it will not benefit existing power plants.

Table 6.12c
ESTIMATED COST OF SO_2 REMOVAL PROCESSES[f]

SO_2 Removal Process Description	*Capital Cost, \$/kw*	*Operating Cost, mills/kw-hr*	*Manufacturer*
Caustic soda treatment[a]	15–20	1.0	A. B. Bacho, Enköping, Sweden
Limestone wet scrubbing[a]	15–25	0.2–0.5	Combustion Engineering, Inc.
Limestone wet scrubbing[a]	20–25	0.7	Zurn Industries
Caustic soda treatment[a]	25	Not available	Showa Denko
Metal sulfite scrubbing (Na salt reagent)[a]	30	Not available	Wellman Power Gas/Mitsubishi
Venturi wet scrubbing (MgO reagent)[a]	35–40	1.2[d]	Chemical Construction Co.
Limestone wet scrubbing[a]	35–40	Not available	Research Cottrell
Adsorption by charcoal[b]	40	Not available	Sumitomo/Kansai
Limestone wet scrubbing[a]	40–45	2.0	Peabody Engineering
Limestone wet scrubbing[a]	40–60	1.5–25	Universal Oil Products/ Babcock & Wilcox
Hydrated lime ($Ca[OH]_2$) scrubbing[a]	45	2.0	Mitsubishi/Kansai
Catalytic oxidation[c]	60–70	[e]	Monsanto Co.

[a]Detailed process description in Section 6.8.
[b]Detailed process description in Section 6.9.
[c]Detailed process description in Section 6.10.
[d]Less income from sales of sulfur.
[e]Depends on by-product acid sales.
[f]EPA lists all limestone scrubbing processes at about the same cost.

Wet scrubbing of waste gas streams with water slurry of lime or limestone today is regarded as the basic process for control of SO_2 emissions from powerplants, even though major uncertainties exist. A number of commercial installations are now planned or are in construction as shown in Table 6.12a. In addition, dry limestone injection followed by wet scrubbing (Fig. 6.8b) has been studied on a large scale for several years, although Union Electric Company announced that their Meramec #2 unit has been shut down due to plugging both in the boiler and in the scrubber system.

The "double alkali" concept has been proposed as a method for avoiding scrubber plugging difficulties (Figs. 6.8h and 6.8i). In this concept, the stack gas is scrubbed with ammonia or caustic solution and the sulfur removed by precipitation with lime in a separate circuit. Pilot plant work has been undertaken by several companies, most notably by General Motors and by the (Copper) Smelter Control Research Association. The concept has not yet been developed to the point of demonstrated commercial feasibility.

Dry Absorption Processes

High losses of absorbent (considerably in excess of the 0.1%) override the advantage of sulfur recovery, particularly with current depressed sulfur prices in the alkalized alumina process (Fig. 6.7b). The system has never advanced beyond the size of a 1,000-cfm pilot unit on a coal-fired boiler.

The DAP-Mn process (Fig. 6.7c), developed by Mitsubishi Heavy Industries, recovers ammonium sulfate (following the reaction of manganese sulfate with ammonia, water and air) and manganese dioxide for recycle. The ammonium sulfate can be sold as a by-product or can be treated with lime to recover the ammonia for recycle and producing gypsum. However, preliminary dust removal is necessary to maintain absorbent integrity. Emission of manganese compounds could present a health hazard. Japanese health authorities are said to be investigating the problem.

Copper oxide on an activated alumina carrier absorbs sulfur dioxide to form copper sulfate. Reduction of the copper sulfate yields CuO for recycle and recovers SO_2 for conversion to elemental sulfur or sulfuric acid.

Shell Oil Company and the Bureau of Mines, working independently, are the principal developers. Although the processes are similar, Shell uses a static-packed bed (flue gas passes alongside absorbent mass) for absorption while the Bureau of Mines utilizes a fluidized-bed. A static bed requires less pressure drop than a fluidized bed but may require a larger size to obtain optimum contact time.

Large-scale installations of the copper oxide process are planned; one using the Shell process will be in a Japanese oil refinery.

Table 6.12d
REAGENTS USED TO CONCENTRATE THE SO_2 STREAM IN SCRUBBING PROCESSES

Reagent	*Developer*
Magnesium Oxide	Chemico-Basic Corporation
Sodium Salts	Wellman-Power Gas
Sodium Citrate	U.S. Bureau of Mines
Potassium Formate	Consolidation Coal Company
Sodium Hydroxide	S&W—Ionics
DMA	ASARCO
Sodium Phosphate	Stauffer

Table 6.12e
BY-PRODUCTS OF SO_2
REMOVAL PROCESSES

Reagent	*By-product*	*Developer*
Sodium Hydroxide	Na_2SO_4	Various
Ammonia	$(NH_4)_2$ SO_4 and S	Cominco (and others)

Wet Absorption (Scrubbing) Processes

Many wet scrubbing processes have been proposed (Section 6.8). Some of these act as SO_2 "enrichers" by removing SO_2 from a dilute waste gas stream and giving up a gas stream relatively rich in SO_2 when the sorbent is regenerated. Other processes produce a by-product such as ammonium sulfate or sodium sulfate.

Examples of processes that "enrich" an effluent SO_2 stream are listed in Table 6.12d.

Examples of processes that produce a sulfur-containing by-product are summarized in Table 6.12e.

Demonstration plants and commercial installations utilizing some of the foregoing processes are listed in Tables 6.12a and 6.12b.

Adsorption by Charcoal

Several processes are under investigation in which SO_2 is adsorbed on carbon and is oxidized to SO_3. Subsequently, the SO_3 can be removed by washing, yielding dilute sulfuric acid, or by high temperature regeneration in which SO_3 is reduced back to SO_2 by reaction with the carbon itself (Section 6.9).

Principal development work has been done in Europe and Japan, including, for example, Lurgi (Sulfacid Process) (Fig. 6.9c), Bergbau-Forschung and Hitachi. An elemental sulfur recovery process is being developed in the United States by Westvaco (Fig. 6.9d).

In the past, autoignition of the carbon has been a problem, particularly with the Reinluft process (Fig. 6.9a). Loss of sorbent by attrition and decreasing reactivity with time is also a problem. However, the process concept is sufficiently attractive to warrant construction of demonstration plants as shown in Table 6.12a.

Catalytic Oxidation

Oxidation of SO_2 to SO_3 followed by absorption in water to produce sulfuric acid is an obvious route to consider (Section 6.10). Contact process sulfuric acid technology is well established, but conversion of this technology to recovery of SO_2 from stack gases has proven difficult, primarily because of the need for high-temperature particulate removal ahead of the catalyst bed which is difficult. Monsanto has tested their Cat-Ox process (Fig. 6.10b) extensively, and a 100 megawatt test installation is now being built in Illinois. The economics of the process is adversely affected by the fact that the by-product acid contains more water and dissolved impurities than does commercial grade sulfuric acid.

Work on catalytic oxidation has been done by others such as Kiyoura in Japan (Fig. 6.10d) and Tyco Laboratories (Fig. 6.11d).

Markets for Recovered By-products

Sulfur demand on the North American continent by 1975 is estimated to be approximately 15 million long tons per year. Total production capability in 1975 in Canada, Mexico and the United States combined will be between 20 and 25 million long tons per year. Thus, sulfur probably will be in a surplus supply for a number of years ahead and, consequently, sulfur and sulfur products will continue to experience depressed prices.*

The most important use for sulfur in the United States is in the manufacture of sulfuric acid. A major use for sulfuric acid is in the production of phosphate fertilizer.

Frasch† sulfur is produced near the Gulf of Mexico in both the United States and Mexico. Lesser quantities are produced in West Texas. Total Frasch sulfur production in 1972 was approximately 7½ million long tons. Frasch deposits are not unlimited but are expected to be adequate for many years to come.

Sulfur is also recovered from "sour" natural gas. Production in Western Canada might be as much as 7½ million long tons per year by 1975.

Sulfur is produced in Eastern Canada from various by-product sources, including smelter gases. Production in 1972 was approximately 2 million long tons, although potential production is even greater.

Other sources of sulfur in North America are from desulfurization of sour crudes and from other by-product recovery operations.

Sulfur dioxide can be recovered from SO_2-containing stack gases in several forms, depending on the stack-gas cleaning process selected.

Principal forms in which sulfur may be recovered are:

1. Sulfuric acid
2. Sulfur dioxide (Gas or Liquid)
3. Elemental sulfur
4. Sulfur containing salts

* The trend as of 1974 seems to be in conflict with this prediction (editor).

† "Frasch" sulfur is a term used in the trade to identify sulfur produced from underground deposits using the Frasch process. The implication is that this is "virgin" sulfur as opposed to sulfur produced as a by-product.

Sulfuric acid is a widely used commodity for which many markets exist. However, freight cost is of major concern, since sulfuric acid is only about one-third the sulfur content of elemental sulfur. Also, in view of the relatively low selling price of sulfuric acid, investment in storage facilities is not always justified.

The market for *sulfur dioxide* is relatively small. While sulfur dioxide contains a higher sulfur content than does sulfuric acid, it rarely is economic to ship it, except possibly by pipeline when very large volumes are involved.

Elemental sulfur is easily handled and stored and is more cheaply transported than are the sulfur-containing compounds.

Special markets including export markets do exist in certain locations for *sulfur-containing salts* such as ammonium sulfate, sodium sulfate and potassium sulfate. Careful market research is suggested, however, because these materials usually are readily available from other sources.

BIBLIOGRAPHY

Environmental Briefs, *Power*, September 1970–August 1971.

Maurin, P. G. and J. Jonakin: Removing Sulfur Oxides from Stacks, *Chemical Engineering*, April 27, 1970.

Nippen Kogyo and *Yomiuri Newspapers*, February to July 1971.

Oil Paint and Drug Reporter, August–November 1971.

Proceedings, American Institute of Chemical Engineers, 68th National Meeting, Houston, Texas, February 28–March 4, 1971; 70th National Meeting, Atlantic City, New Jersey, August 29–September 1, 1971.

Proceedings, Second International Clean Air Congress of the International Union of Air Pollution Prevention Associations, Washington, D.C., December 6–11, 1970.

Proceedings, Second International Lime/Limestone Wet Scrubbing Symposium, Sponsored by the Environmental Protection Agency, New Orleans, Louisiana, November 8–12, 1971.

Slack, A. V.: Sulfur Dioxide Removal from Waste Gases, Noyes Data Corporation, Park Ridge, New Jersey, 1971.

Sulfur from Gas is Glutting the Market, *Business Week*, July 24, 1971.

6.13 DIRECT FLAME INCINERATION

Capacity:	250 to 56,000 SCFM.
Fan Motors:	2 to 150 HP.
Materials of Construction:	Refractory-lined steel.
Operating Temperature:	1000 to 2200°F.
Residence Time:	0.3 to 1.2 seconds.
Partial List of Manufacturers:	Bigelow-Liptak Corp.; The Air Preheater Co., A Division of Combustion Engineering; Combustion Equipment Associates, Inc.; Coppers Engineering Corp.; Gas Processors, Inc.; Granco Equipment, Inc.; Hirt Combustion Engineers; Mahon Technology Group, Inc.; Maxon Corp.; North American Manufacturing Co.; Ross Engineering Machinery Division, A Division of Midland-Ross Corporation; Thermal Research & Engineering Corp.

Direct-flame incineration is a control technique for combustible organic air pollutants. The purpose of the process is to oxidize the organic gases and vapors completely into carbon dioxide and water. This is accomplished in the presence of a flame and sufficient oxygen by raising the temperature of the gases above their ignition temperature and then maintaining this temperature until the oxidation reactions are complete.

Direct-flame incineration can be used both for organic-rich fumes and for air-rich fumes. The first is used where the organic concentration is at or above the limits of flammability; the second, where the organic concentrations are below the limits of flammability. The equipment used for these two types of streams is similar in fundamental operation but different in design.

Incineration of Organic-Rich Fumes

Organic rich fumes, such as those generated by petrochemical plants, are destroyed in a flare system (Fig. 6.13a). The exact design of a flare system depends upon gas composition, the type of unit generating the waste gas and environmental factors of the location. In general, the flare is designed to thoroughly mix the waste gas with air, ensure its ignition at various waste gas flowrates and meteorological conditions, safely reject the heat generated to the atmosphere and operate without generating smoke. These require a turbulent flame, a reliable pilot light and placing the flare away from, or at least shielded from, other units.

The design of a flare system is a fairly complex subject (see Sections 5.39 to 5.41) particularly when the flare is used as a safety valve in the event that an emergency requires the immediate venting of a large quantity of organic gases.

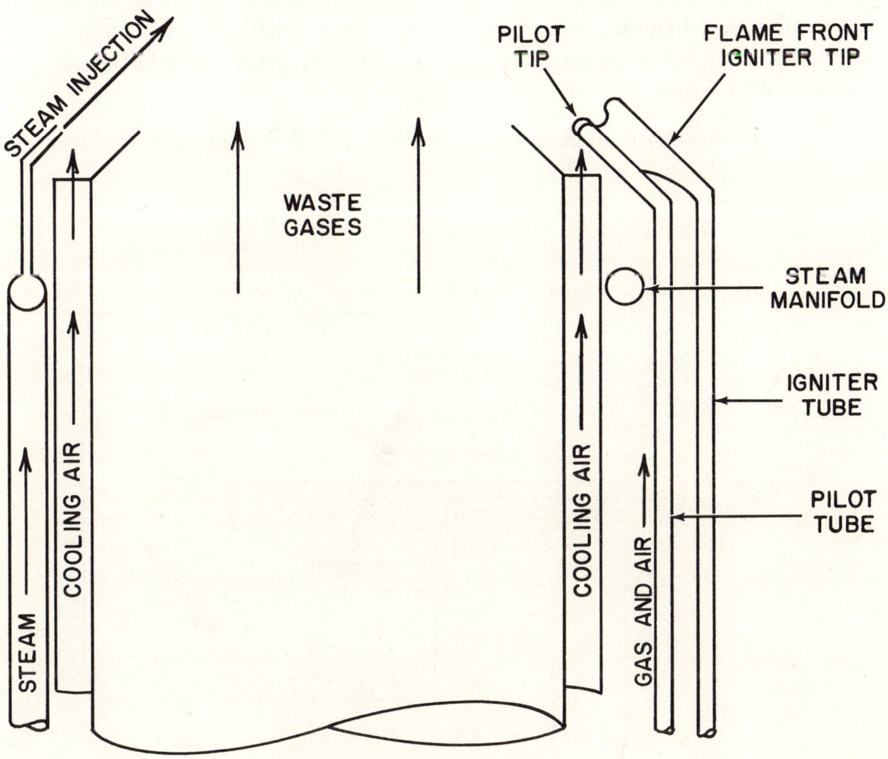

Figure 6.13a Flare tip components.

6.13 Direct Flame Incineration

Incineration of Air-Rich Fumes

In most air pollution situations, the concentrations of the organic materials in the gas stream are below the lower limits of flammability. Processes in which hydrocarbons or organic solvents are released, usually add large quantities of dilution air to maintain the organic concentrations at about one quarter of the lower limits of flammability. This is done as a safety measure, to eliminate the possibility of an explosion within the equipment. The dilution of the organics with air complicates the problem of incineration (Sections 5.32, 35, 36, 37 and 38). The heat available from the combustion of the organics is insufficient to maintain the combustion temperature. In order to provide the additional energy needed, an auxiliary fuel is mixed with the oxygen-containing fumes before they are burned (Fig. 6.13b).

For proper oxidation of the organics, there must be contact with the flame, sufficient average temperature and time at that temperature to complete combustion. Thus, an intimate mixing of flame and effluent gases will aid the oxidation of the organics by rapidly bringing all the fumes to a high temperature, giving a higher average temperature over the residence time of the unit and bringing all the fumes in contact with the flame. A well-designed direct flame incinerator will have three features in addition to the presence of a flame:

1. Sufficient average temperature for the oxidation reaction to proceed rapidly.

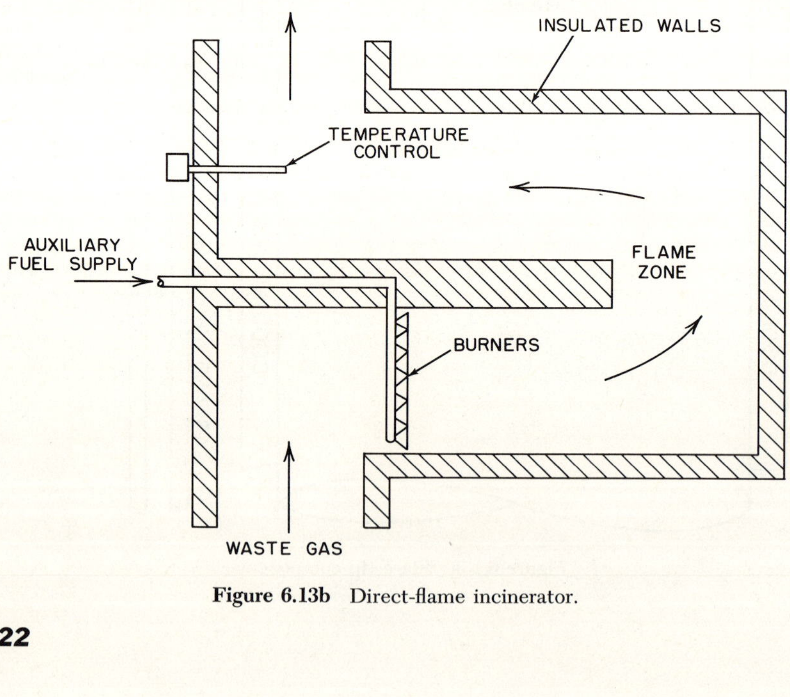

Figure 6.13b Direct-flame incinerator.

2. Sufficient time at that temperature for the oxidation reaction to reach completion.
3. Sufficient turbulence to promote good heat transfer between the flame and the gases and to minimize temperature gradients that may quench the oxidation reactions.

Increasing any of these allows for a reduction of the others, within some limits.

The temperature at which the unit operates can be increased by increasing the quantity of fuel being burned in the unit or by reducing the quantity of gases passing through the unit. The dwell time of the gases within the unit can also be increased by reducing the flowrate of the gases passing through the unit, as well as by increasing the volume of the unit. Turbulence is influenced by the design of the burner and by the design of the inside of the incinerator. A tightly profiled multiple burner that directs the fumes over, around and through a number of small gas flames coming from a line burner efficiently promotes contact between the flames and the fumes. Because of the high temperatures to which the baffles in such a burner are exposed, this design requires more maintenance than do large single burners. Baffles installed downstream of the burner also help to promote mixing and to increase the level of turbulence.

Results of Improper Design or Operation

The optimization of an incinerator design requires a detailed knowledge of the composition and flow of the stream to be controlled and the auxiliary fuels available, as well as a thorough knowledge of combustion principles and an understanding of the incineration process. In many cases, a direct-flame incinerator is poorly designed for a particular application as a result of incorrect information on the stream to be controlled or due to inexperience on the part of the design engineers. In other cases, a unit is properly designed; however, after installation, the quantity of material to be treated increases beyond the design capacity of the unit because of process changes.

As a result of forcing more gases through a unit than its design allows, the oxidation reaction may be arrested by insufficient temperature and/or insufficient dwell time. This can result in the emission of large amounts of toxic carbon monoxide and highly odorous aldehydes with other products of incomplete combustion. Thus, the result of overloading a direct-flame incinerator may well be an intensification of the problem rather than a solution of it.

Heat Recovery

Because direct flame incineration involves the heating of exhaust gases to high temperatures, heat-recovery equipment is usually justified in larger installations to reduce fuel costs. The level of heat recovery selected must be weighed against the installation and operating costs (Section I.8.9) that are acceptable to the individual company.

There are many different ways in which heat recovery equipment can be applied to direct-flame incineration systems. Figure 6.13c illustrates some of them. The most direct heat-recovery scheme is to simply mix a portion of the incinerated gases with cool ambient air and recirculate them back into a process where hot air is required, such as in a baking oven. In this manner, the fuel required to heat the ambient air normally used in the process will be reduced. Another direct approach is to recirculate some of the gases to the inlet of the incinerator. This acts to preheat the gases going to the incinerator and to reduce its fuel requirements. This also acts to increase the quantity of gases the incinerator must handle and, hence, its size and capital costs. It is also subject to an oxygen limitation, because between 16 to 18% oxygen is needed in the inlet gases of the incinerator for its proper operation. Heat can be transferred to the incinerator gases from the outlet gases by use of gas-gas heat exchangers to recover heat without reducing the oxygen content, but this will substantially increase the overall unit cost.

An intermediate fluid such as a heating oil can also be used to transfer heat from the outlet gases to other points in the plant where heat is required.

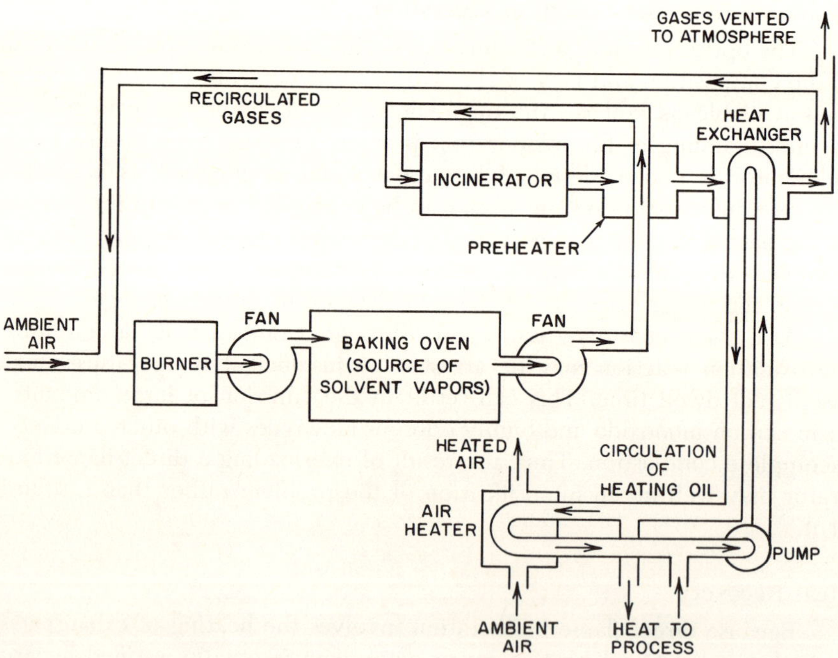

Figure 6.13c Techniques for recovery of heat from a fume incinerator.

The incinerator can also be used as a source of low-pressure steam for use within the plant (Section I.8.9). All of these features will increase equipment costs but reduce the fuel requirements and, hence, the operating costs. The overall system must be optimized to give the best air pollution control operation for the least expense, within the constraints provided by a particular installation.

BIBLIOGRAPHY

Danielson, J. A. (ed.): "Air Pollution Engineering Manual," U.S. DHEW, PHS, National Center for Air Pollution Control, Cincinnati, Ohio, PHS Publication No. 999-AP-40, 1967.

Rubin, F. L. et al.: Heat Transfer Equipment, in J. H. Perry (ed.), "Chemical Engineer's Handbook," 4th ed., McGraw-Hill Book Co., New York, 1963.

U.S. DHEW, PHS, "Control Techniques for Hydrocarbon and Organic Solvent Emissions from Stationary Sources," National Air Pollution Control Administration Publication No. AP-68, 1970.

6.14 HYDROCARBON CONTROL—CATALYTIC INCINERATON

Capacity:	1,000 to 12,000 SCFM.
Fan Motor:	2 to 50 HP.
Materials of Construction:	Refractory-lined steel.
Operating Temperature:	600 to 900°F.
Residence Time:	0.01 to 0.3 seconds.
Frequency of Catalyst Replacement:	3 to 5 years or less.
Partial List of Manufacturers:	Engelhard Industries; Maithey Bishop, Inc.; Oxy-Catalyst, Inc.

Catalytic incineration (Sections 5.25 and 5.33) is a control technique for the removal of dilute concentrations of combustible organic air pollutants. The purpose of the process is to oxidize the organic gases and vapors to carbon dioxide and water completely. This is accomplished by exposing the mixture of dilute organic gases and oxygen to a catalytic surface at a temperature which is high enough for the oxidation to occur, for a length of time sufficient for the oxidation to go to completion.

Catalysts

Catalysts are usually solids that are neither reactants nor products of a reaction, yet alter the rate of chemical reactions. The catalyst reduces the potential energy barrier over which the reactants must pass to form products. Reducing this activation energy for the reaction increases the rate of the reaction, sometimes dramatically. To be effective, a catalyst need not be present in large amounts, but it must be accessible to the reactants; thus, it is generally distributed over a large surface area. Because it is generally a solid and distributed over a large surface area, usually as a porous material, there are several steps necessary for the completion of a reaction (Fig. 6.14a).

1. Reactants must diffuse from the main body of the gases to the exterior surface of the porous material containing the catalytic sites.

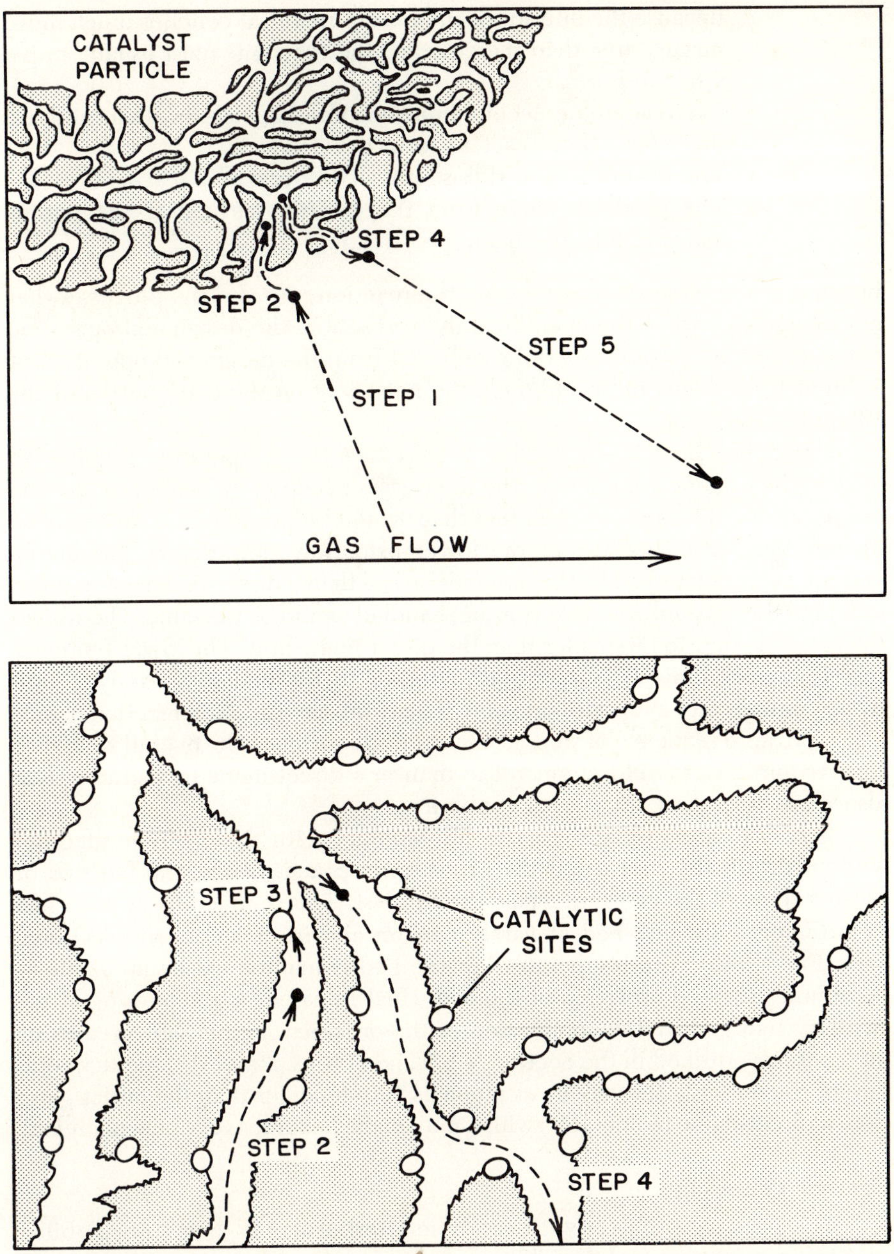

Figure 6.14a Five steps the reactants and products must undergo in a catalytic reaction. Step 1 is diffusion from the bulk flow to the catalyst. Step 2 is diffusion into the catalyst pores. Step 3 is association with a catalytic site and reaction. Step 4 is diffusion out of the catalyst pores. Step 5 is diffusion to the bulk flow.

2. Because the interior of the porous material contains much more surface area than the exterior, the reactants must usually move into the pores.
3. The reactant molecules must become associated with a catalytic site where they react to form the products.
4. The products then diffuse out of the pores.
5. The products move from the exterior surface of the porous material into the main gas stream.

Because of the effect of a catalyst on the reaction rate and the processes that must occur as part of the overall catalytic reaction, the design and operation of a catalytic incinerator are quite different from the design and operation of a direct-flame incinerator, although the net effect on the composition of the effluent is the same.

Since the effect of the catalyst is to reduce the temperature required to oxidize the organic compounds, the inlet gases need not be heated to ignition temperatures. This requires less fuel than would be needed for a direct-flame incineration. Also, because of the lower oxidation temperature, the actual volume of gases handled by the unit will be less than a direct-flame incinerator, although the mass (or standard volume) handled would be the same. This allows the catalytic unit to be smaller than the direct-flame unit. The lower temperature may also allow less-expensive materials to be used in a catalytic unit when compared to a direct-flame unit. The catalytic oxidation reaction occurs as a very rapid sequence of steps at the catalytic surface. The overall residence time required is, therefore, much less than in a direct-flame incinerator. This also tends to reduce the size of the unit.

It is desired to force the gases into contact with the catalytic material without drastically increasing the pressure drop and, hence, the fan size or causing some of the gases to bypass the catalyst. This is generally approached by utilizing a shallow bed of catalyst material with a large cross-sectional area (Fig. 6.14b). The temperature of the bed is maintained by an auxiliary fuel burner located upstream of the bed. Heat released by the burning of the auxiliary fuel and by the combustion of the organics can be recovered using the methods outlined in the Sections I.8.9 and II.6.13. However, the temperature of the outlet gases may be as much as 700°F cooler than the outlet gases from a direct-flame incinerator, which may make heat recovery uneconomical.

Catalyst Poisoning

The primary problem with catalytic afterburners is their susceptability to catalyst poisoning and the additional maintenance this requires as cleaning or replacement of the catalyst is needed to maintain activity. Poisoning may be caused by several different mechanisms:

1. Blinding of the pores with dust, dirt or process particulates.
2. Chemical poisoning of the catalytic sites.
3. Erosion of the catalytic surfaces by physical or chemical degradation.
4. Changing the crystal or physical structure of the catalyst by heat.

The blinding of the pores by particulate matter renders the largest portion of the catalytic surface unavailable to the waste gases. This can be caused by ordinary atmospheric dirt or by particulate matter originating from the process. For example, in some paint-baking ovens, evaporation of the paint occurs so rapidly that pigments, binders and resin particles are entrained in

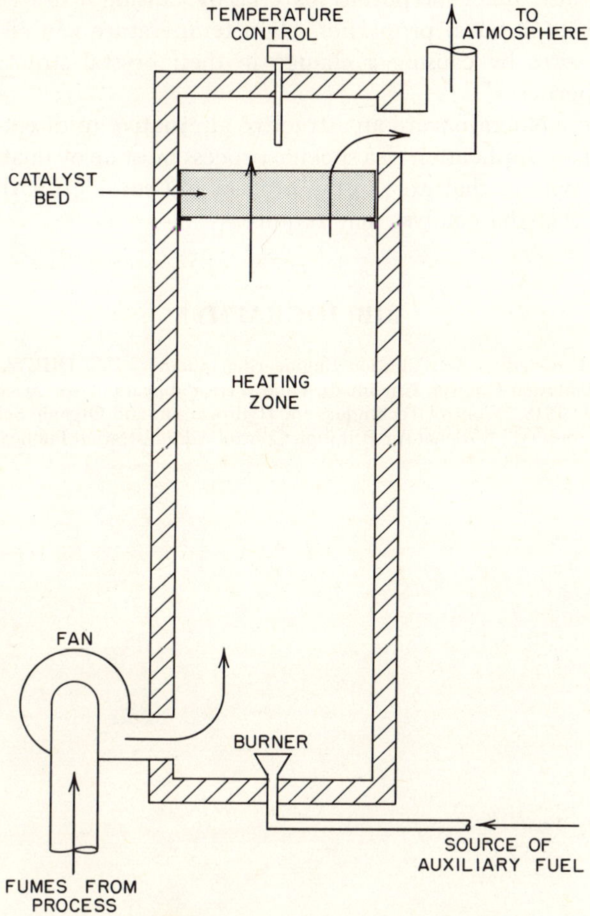

Figure 6.14b Catalytic incinerator.

the waste gases and are deposited on the catalyst, as well as on other areas of the outlet system.

Chemical poisoning occurs when the catalytic sites are deactivated by a chemical reaction. Materials such as halogens, halogen compounds, heavy metals, sulfur, silicon and phosphorus have poisoned various catalyst systems. A specific catalyst system may resist poisoning by some of these compounds but may be poisoned by other compounds.

Erosion of the catalytic surfaces may occur as a result of the movement of catalyst within the catalyst bed. It may also occur as a result of a chemical degradation of the porous material used to distribute the catalyst over a large surface area. Excessive temperature caused by a slug of high-concentration organics may also affect the porous material by causing it to lose its porosity or to change its physical properties. High temperature can also deactivate the catalytic sites by causing a change in their crystal structure or other physical properties.

Catalytic incinerators are an attractive alternative to direct-flame incineration, but their application to a specific process must be evaluated regarding the potential poisons that exist in the process and considering the additional maintenance that the catalyst may require.

BIBLIOGRAPHY

Danielson, J. A. (ed.): "Air Pollution Engineering Manual," U.S. DHEW, PHS, National Center for Air Pollution Control, Cincinnati, Ohio, PHS Publication No. 999-AP-40, 1967.

U.S. DHEW, PHS, "Control Techniques for Hydrocarbon and Organic Solvent Emissions from Stationary Sources," National Air Pollution Control Administration Publication No. AP-68, 1970.

6.15 HYDROCARBON CONTROL—ADSORPTION

Capacity:	500 to 450,000 SCFM.
Fan Motor:	2 to 250 HP.
Materials of Construction:	Carbon steel (plastic or glass lining provided where necessary).
Partial List of Manufacturers:	American Norit Co.; Barneby-Cheney Co.; Connor Engineering Corp., Div. Jenn-Air Corp.; Courtaulds Engineering, Ltd.; Pittsburgh Activated Carbon; D. T. Melsheimer Co., Inc.; Vic Manufacturing Company; Vulcan-Cincinnati, Inc.

Adsorption is a process in which a special solid surface collects selected gases or vapors. In adsorption, the molecules of the gas or vapor adsorbed contact and adhere to the surface of the solid in an extremely thin layer. Depending upon the nature of the gas or vapor and of the solid, this layer will range in thickness from one to many molecules.

The amount of adsorption that occurs on most solids is rather limited because of the small surface area available for adsorption. Certain materials (e.g., silica gel, activated alumina, molecular sieves and activated carbon) have a porous, sponge-like structure that vastly increases their surface area and enables them to adsorb substantial quantities of gases and vapors. Adsorption of a gas or vapor is similar to condensation in that the molecules of the adsorbed components are brought into close proximity and remain loosely bound to each other under the influence of the solid. For this reason, the adsorption is more efficient near the dew point of the adsorbed gases and vapors. Therefore, an adsorption system should operate at the highest pressure and lowest temperature within the limitations of the process.

In any adsorption process, three steps are necessary:

1. Intimate contact between the gases and vapors and intimate contact between the gases and the solid adsorbent. During this

step certain compounds in the gas phase are adsorbed. The adsorbed components are termed adsorbates.

2. Separation of the unadsorbed gases and vapors from the adsorbent and adsorbate.
3. Regeneration of the adsorbent by removing the adsorbate or discarding the used adsorbent and replacing it with fresh material.

The first two steps are carried out in the adsorption equipment. In air pollution control applications, this may take the form of a stationary bed of adsorbent or a continuous flow of the solid phase through a series of fluidized beds. The third step may be performed in a number of ways, depending upon the constraints of the overall system. (See Section I.6.14.)

Regeneration of the Adsorbent

In an adsorption process, the quantity of gases or vapors that can be adsorbed is finite. After some period of use, the adsorbent becomes saturated with the adsorbate and will no longer adsorb. When this occurs, it must be either regenerated or replaced. In most installations, regeneration is favored. Regeneration may be accomplished in a number of ways:

1. The temperature of the adsorbent may be increased until the vapor pressure of the adsorbate exceeds atmospheric pressure.
2. The pressure surrounding the adsorbent may be decreased until the vapor pressure of the adsorbate exceeds the ambient pressure.
3. An inert fluid containing none of the components of the absorbate can be passed through the bed of adsorbent, stripping the adsorbate from the adsorbent.
4. In the case of a hydrocarbon adsorbate, the bed of adsorbent can be heated and a stream of air passed through it. The hot air will then oxidize the hydrocarbons and, in that way, remove them from the adsorbent.

In most applications in which recovery of the adsorbate is desired, a stream of an easily condensed fluid, such as steam, is passed through the bed, combining the effects of methods (1) and (3). The stripped adsorbate can then be recovered by condensing and separating the mixture. After regeneration, the adsorbent is restored to its activity and the adsorbate is removed for disposal, reuse or sale.

Adsorption Equipment

The primary function of the adsorption unit is to bring the gases and solid adsorbent into direct contact. In order to accomplish this, a bed of adsorbent with a large cross-sectional area and relatively small depth is

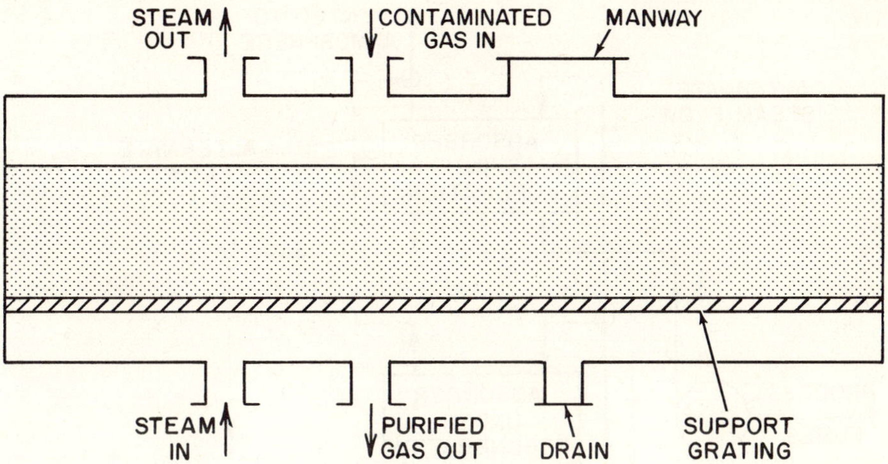

Figure 6.15a Fixed-bed adsorber.

generally used as illustrated in Figure 6.15a. To achieve continuous operation with such fixed-bed equipment, two or more adsorbers must be used, with one or more being in operation while the other is being regenerated as shown in Figure 6.15b.

Moving-bed equipment is also available, particularly for large-scale, continuous operation. In such a unit (Fig. 6.15c), solvent-laden gases are drawn from the process by a fan. They pass through a heat exchanger, to control their temperature, and are fed to the base of the adsorber vessel. The adsorber vessel contains a number of shallow fluidized beds of adsorbent. The gases flow upward through these beds where the solvent is progressively adsorbed onto the adsorbent. Solvent-free gases are finally discharged to the atmosphere through special dust collectors at the top of the adsorber vessel. Fresh adsorbent is fed onto the top tray and is fluidized. It flows across the tray, over a weir and down a downcomer to the tray beneath, progressively becoming loaded with solvent. The adsorbent is discharged at the bottom of the adsorber where it is regenerated and transported back to the top of the adsorber.

Adsorbents

The most important characteristics of an adsorbent are its large surface-to-volume ratio and its preferential affinity for specific substances. Almost all adsorbents are capable of adsorbing organic impurities and water from gas streams, but each has a particular affinity for either polar or nonpolar compounds. Water vapor is a polar compound while organic vapors are nonpolar. Molecular sieves, silica gel and aluminum oxide adsorbents have an affinity for polar compounds. Activated carbon is the most commonly used nonpolar adsorbent.

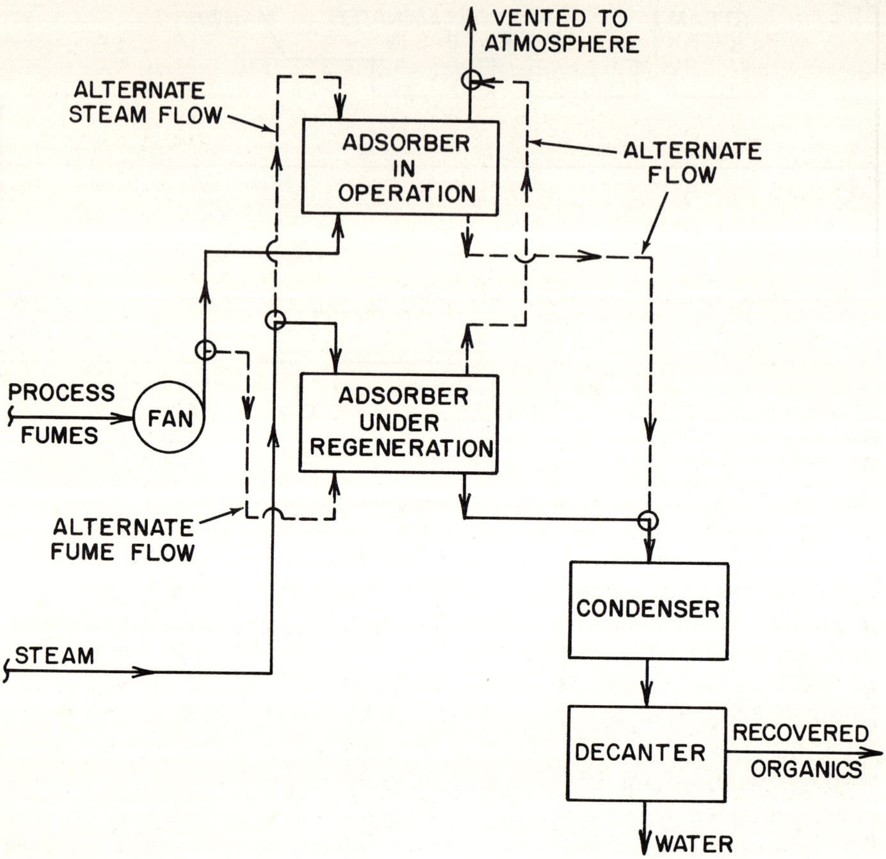

Figure 6.15b Two-bed adsorption-regeneration system showing one bed in operation and the other under regeneration.

Activated carbons adsorb organic vapors from gases selectively, even in the presence of water vapor. They are generally used for the control of organic emissions because of this property and the fact that they can recover organic materials from waste streams in which the organics are present at low concentrations. Their use is limited by the fact that they are inefficient in treating high-temperature streams and they are also susceptible to fouling by particulate matter in the stream to be treated. The particulate matter coats the carbon particles, closing its pores and significantly reducing the available surface for adsorption.

Thus, the adsorption of organic material on activated carbon is an attractive control technique for low-temperature, low-concentration streams (Sections I.6.15 and II.6.9). The problems of regeneration or disposal of the spent

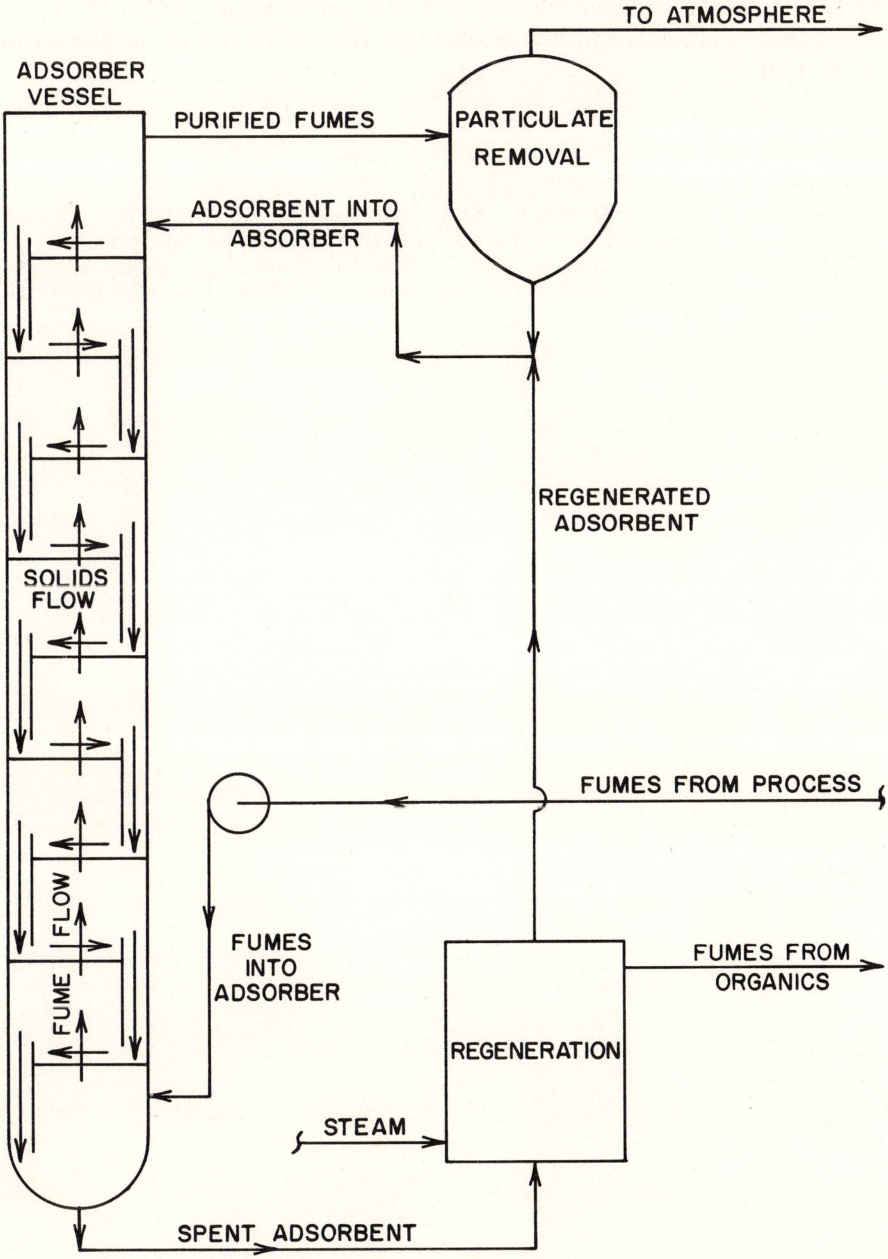

Figure 6.15c Continuous fluidized-bed adsorber.

carbon should be considered in any potential application (Section I.6.16), as should the requirement of low levels of particulate matter in the stream to be treated.

BIBLIOGRAPHY

Danielson, J. A. (ed.): "Air Pollution Engineering Manual," U.S. DHEW, PHS, National Center for Air Pollution Control, Cincinnati, Ohio, PHS Publication No. 999-Ap-40, 1967.

U.S. DHEW, PHS, "Control Techniques for Hydrocarbon and Organic Solvent Emissions from Stationary Sources," National Air Pollution Control Administration Publication No. AP-68, 1970.

6.16 HYDROCARBON CONTROL—CONDENSATION AND ABSORPTION

Capacity:	300 to 70,000 SCFM.
Materials of Construction:	Carbon steel, coated steel, lined steel, stainless steel, depending on application.
Partial List of Manufacturers:	ARCO Industries Corp., Process Equipment Division; Ceilcote Co.; Combustion Equipment Associates, Inc.; Heil Process Equipment Corporation; Joy Manufacturing Company, Western Precipitation Division; Koch Engineering Co., Inc.; Schutte & Koerting Co.; Slick Corporation, Pulverizing Machinery Division; UOP Air Correction Division; John Zink Company.

Condensation

The condensation of higher boiling compounds from a waste stream by cooling is an important but limited method for the control of organic emissions (Section 5.28). Condensation has two important applications:

1. Reducing the volume of an emission by removing high concentrations of condensable materials.
2. Controlling the emissions of high-molecular weight organics such as fats and oils.

Both of these have limited application and depend upon special conditions for their effectiveness.

When a waste stream contains a large amount of condensable material, such as steam, along with various organic compounds, condensation will remove some of the organics with the steam; and it will significantly reduce the volume of the emissions. This reduction in volume will allow a smaller control unit to handle the emission more effectively and at a lower cost. An example of this type of application is the control of the emissions from the cooking operation in a rendering plant. The emissions of the operation contain large amounts of steam and organic malodors. Condensation of the steam removes much of the odorous components with the steam. The remaining small volume of uncondensed gases can be effectively incinerated in a small afterburner or boiler firebox. The effluent wastewater must be properly disposed of to prevent a water pollution problem.

Operations which emit high-molecular weight organics, such as paint resin manufacture or deep fat frying of food or very high concentrations of condensable materials such as gasoline in petroleum refining or styrene in petrochemical manufacturing, can utilize condensation to simultaneously retain these materials in the process and to reduce the emission from the process. Since the condensation process is limited by the equilibrium partial pressure of the component condensed, complete removal is not possible. In the case of a high molecular weight fat or oil, the removal is almost complete. In the case of a relatively volatile material, such as gasoline or many solvents, the levels may be only reduced to 1,000 to 10,000 ppm. Thus, in these cases, condensers are usually followed by a secondary air pollution control unit such as an adsorber or an incinerator.

Condensation Equipment

Condensers may be classified into two types: surface condensers and contact condensers or spray chambers. In a surface condenser (Fig. 6.16a), the cooling medium and the vapors are separated by a wall. In a contact condenser (Fig. 6.16b), the cooling medium, usually water, is sprayed into

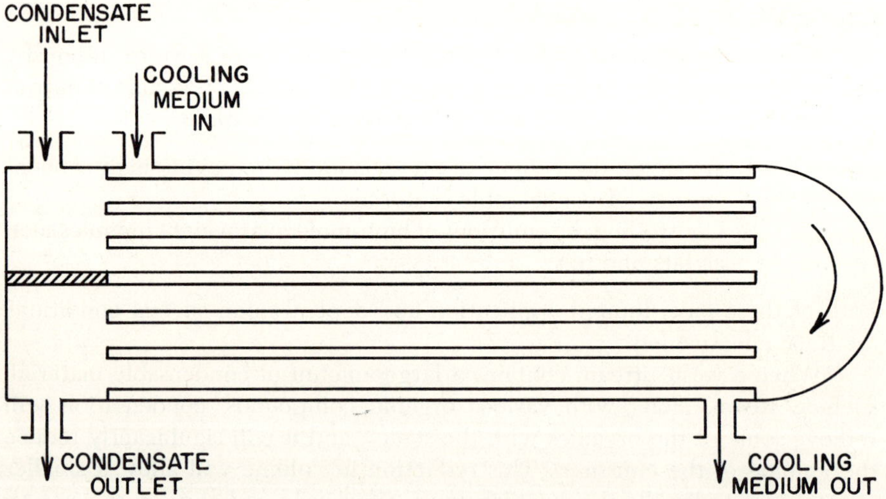

Figure 6.16a Surface condenser (shell and tube type).

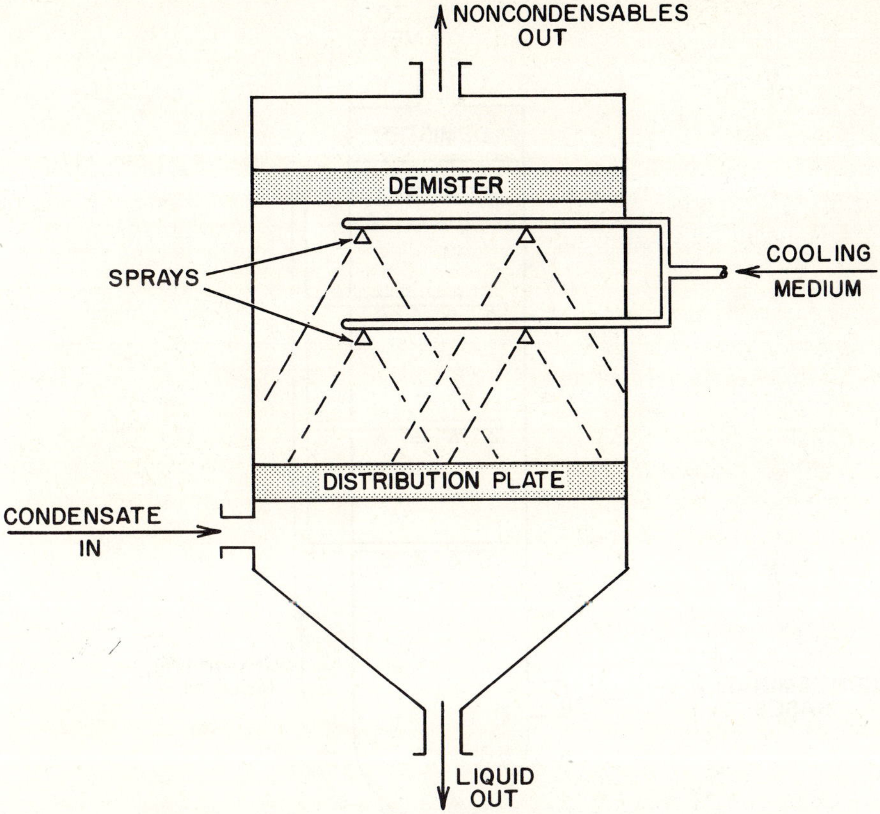

Figure 6.16b Contact condenser (spray type).

the contaminated gases and vapors and may perform scrubbing as well as condensation of the vapors. In general, contact condensers are less expensive, more flexible and more efficient in removing organic compounds than surface condensers. However, they require large amounts of water to be effective and can produce considerable quantities of contaminated water that must be disposed of. Surface condensers, on the other hand, allow reuse of the cooling medium and produce no such waste disposal problem.

The design basis for a condenser is the quantity of heat to be removed from the condensing vapors. This includes the latent heat of the condensable vapors, the additional cooling of the condensed liquid and the cooling of the noncondensing gases. In a contact condenser, this heat must be absorbed by the water or other cooling medium which is sprayed into the unit. The quantity of water needed is equal to the heat load divided by the allowable difference between outlet and inlet water temperatures. In a surface condenser, the

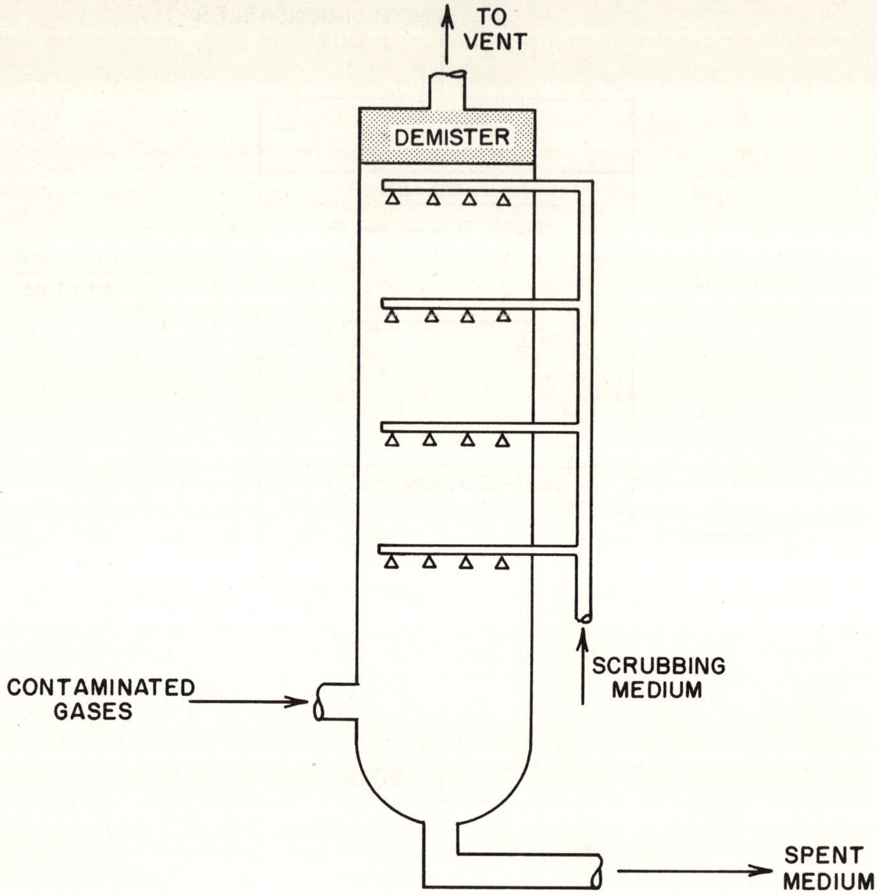

Figure 6.16c Spray tower.

existence of a wall between the condensing vapors and the cooling medium limits the transfer of heat between the two fluids. Thus, the area of the wall and the overall heat transfer coefficient that can be attained are the critical factors in the design of a surface condenser.

Condensers have found many applications in controlling organic emissions. In considering their use for any specific situation, allowance must be made for the type and quantity of coolant available, for the quantity of material to be condensed and for the possible liquid disposal problems which are created. Condensers are best used as primary devices for the control of organic emissions, to be followed by more efficient devices such as adsorbers or afterburners.

Absorption or Scrubbing

Absorption is a process in which a soluble component of a gas stream is transferred into a liquid. The component may become dissolved in the liquid or it may chemically react with the liquid or a compound dissolved in the liquid. The process depends upon an exchange of a component between a gas phase and a liquid phase. This exchange is limited by the area of the interface between the phases and by the solubility of the component in the liquid phase. The area of the interface can be increased by:

1. Dispersing the liquid phase within the gas phase.
2. Dispersing the gas phase within the liquid phase.
3. Spreading the liquid phase over a large surface in contact with the gas phase.

The first can be accomplished by spraying the liquid into the gas, either under pressure or by use of a mechanical device, or by moving the gas phase past the liquid phase rapidly enough to entrain the liquid. Examples of this type of equipment are spray towers (Fig. 6.16c) and venturi scrubbers (Fig. 6.16d), or flooded disc scrubbers and cyclone spray chambers (see Sections 5.12 to 5.21).

The dispersion of the gas phase within the liquid phase can be accomplished by breaking the gas phase into bubbles within the liquid phase either

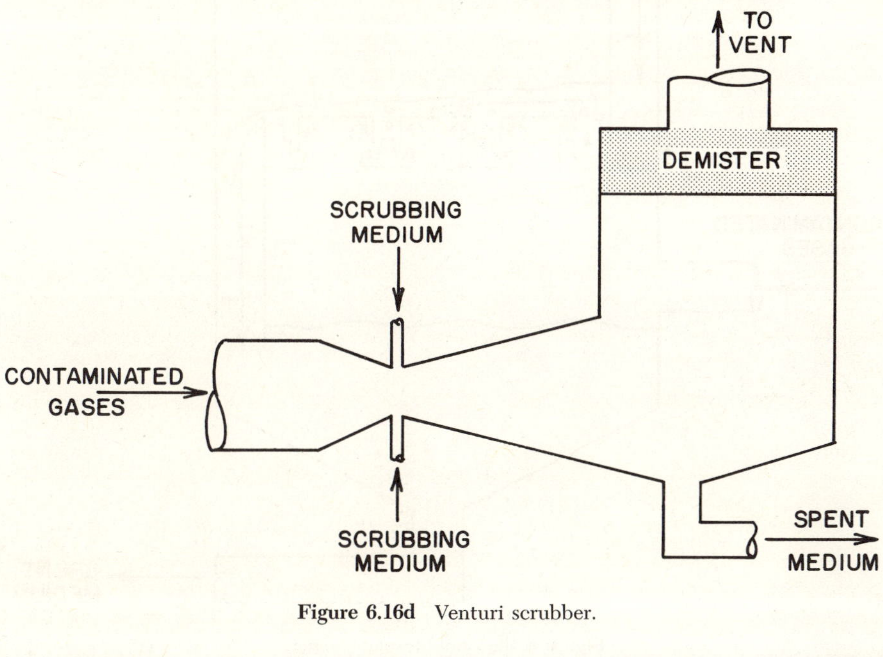

Figure 6.16d Venturi scrubber.

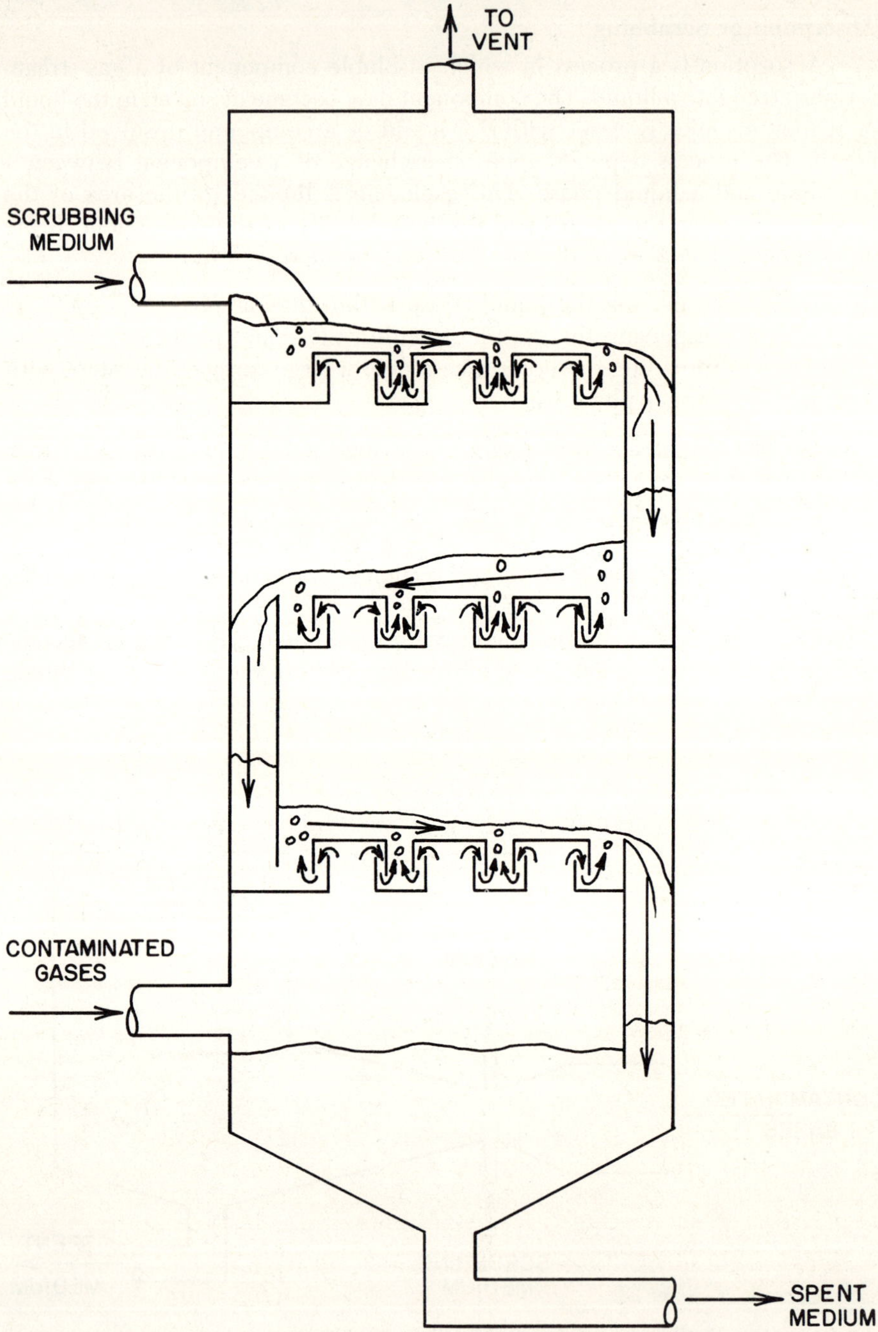

Figure 6.16e Bubble plate tower.

by supplying the gas under pressure to spargers or bubble plates or by using some form of mechanical agitation. Examples would include bubble plate towers (Fig. 6.16e), sparging tanks and various agitated vessels.

The last approach to increasing the area of the interface is probably the most commonly used in air pollution control because the units are simple and have relatively low pressure losses. The latter is important where large volumes of gases with low concentrations of contaminants are being treated. Examples

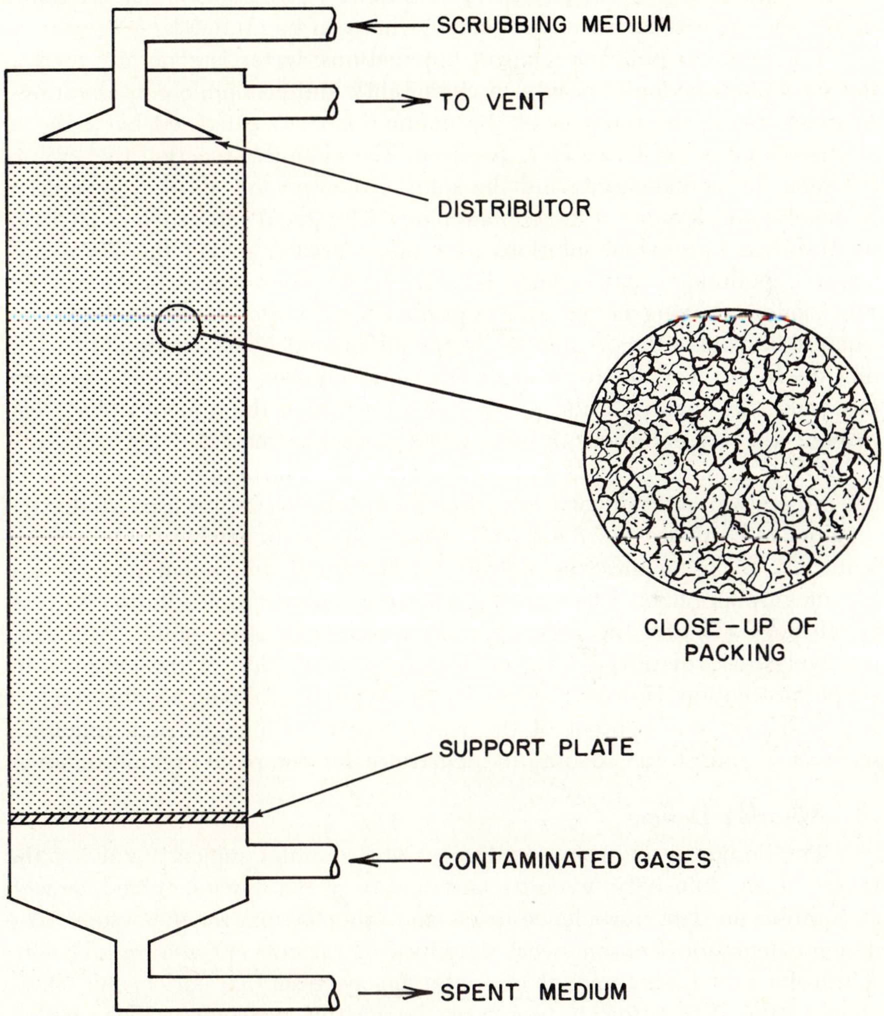

Figure 6.16f Packed-tower scrubber (with close-up of random packing).

of these types of units are wetted wall towers and the popular packed towers (Fig. 6.16f).

Scrubbing Media

Absorption of a gaseous contaminant by a liquid occurs because the liquid is not saturated with the contaminant at the conditions existing in the absorber. The difference between the concentration of the contaminant in the liquid if the liquid were saturated and the actual concentration provides the driving force for the absorption. Thus, the more soluble a contaminant is in the liquid phase, the greater the overall efficiency that can be attained.

For most air pollution control applications, water is generally used as the scrubbing medium because of availability and economic considerations. In many cases, the removal of the contaminants is enhanced by using an aqueous solution as a scrubbing medium. The chemical reaction that occurs between the contaminants and the solution component allows the solution to dissolve much more of the contaminants. Thus, acid gases such as HCl can be absorbed in alkaline solutions to a much greater extent than in water. Organic pollutants, particularly odorous compounds, can be removed by solutions of oxidizing agents, such as permanganate salts, chlorine and oxygen. These will oxidize the organics to water soluble aldehydes, acids and carbon dioxide. In this way, the organics can be removed from the gaseous emissions. However, the incorporation of the organics within the aqueous phase can present a water pollution problem if not properly treated.

Other scrubbing media that are used for the control of organic emissions are mineral oil and nonvolatile hydrocarbon oils. These are used where the organics to be removed from a waste stream are particularly soluble in these materials. By using these media, the recovery and subsequent reuse of the organics are simplified. The absorption mechanism is a simple solution process, which does not alter the organic as an aqueous oxidizing solution will. The relatively low volatility of the media allows separation of the organics by simple distillation. However, generally this type of scrubbing is only effective where the organic content of the waste stream is high. It may require a secondary control unit such as an incinerator for complete organic removal.

Absorber Design

The design of absorbing equipment is a complex subject involving the temperature, diffusivity, viscosity and density of the fluids involved, as well as equilibrium data, turbulence levels and other factors for the system. The design calculations are somewhat simplified by the concept of a transfer unit. A transfer unit is the amount of contacting necessary to enrich one phase, equal to the average driving force producing this enrichment. The number of transfer units needed to achieve a certain amount of scrubbing is then the change in concentration desired for one phase, divided by the transfer unit for that phase. The height of a transfer unit is the length of scrubbing tower needed to effect one transfer unit. Multiplication of the number of transfer units required and the height of a transfer unit will yield an estimate of the height of tower needed to accomplish the desired amount of scrubbing. There are established procedures for carrying out these calculations.[1]

In the control of organic emissions, absorption has limited applicability because of the low concentrations and high gas volumes which are normally encountered. It is generally best used in conjunction with other control techniques to achieve the required degree of organics removal.

REFERENCE

1. Danielson, J. A. (ed.): "Air Pollution Engineering Manual," U.S. DHEW, PHS, National Center for Air Pollution Control, Cincinnati, Ohio, PHS Publication No. 999-AP-40, 1967.

BIBLIOGRAPHY

Foust, A. S. et al.: "Principles of Unit Operations," John Wiley, New York, 1960.

Rubin, F. L. et al.: Heat Transfer Equipment, in J. H. Perry (ed.), "Chemical Engineer's Handbook," 4th ed., McGraw-Hill Book Co., New York, 1963.

U.S. DHEW, PHS, "Control Techniques for Hydrocarbon and Organic Solvent Emissions from Stationary Sources," National Air Pollution Control Administration Publication No. AP-68, 1970.

6.17 HYDROCARBON CONTROL—SALABLE BY-PRODUCTS

Cost Aspects: See Figure 6.17a

Air pollution control does not generally have an immediate monetary return on investment associated with it. It is something that is done to maintain good relations with neighbors or is done in response to local, state or federal laws. In some cases, the economics of a control system can be significantly improved by considering the salable by-products of the process. These may vary from the recovery of a valuable material for sale to the recovery of heat for use in some other area of the plant.

Incineration By-products

In both direct-flame and catalytic incineration, the result of the process is the oxidation of the hydrocarbons to carbon dioxide and water with the release of heat. This heat results from the oxidation of the hydrocarbons and from the burning of the auxiliary fuel used in the unit. Because the hydrocarbons are destroyed and the carbon dioxide and water produced are not particularly valuable materials, the heat contained in the outlet gases represents the only salable by-product of the control process (Section I.8.9). Recovery of this energy for use either by the control unit to preheat the inlet gases or by some other process in the plant can result in significant fuel savings. The resulting decrease in operating expenses must be weighed against the increased cost of equipment needed for the recovery of the heat. The more heat recovered the lower the overall operating expenses; but, at the same time, the recovery equipment becomes more expensive (Fig. 6.17a). The overall control system must be optimized for each individual situation, since factors such as uses for the recovered heat, amortization periods and fuel costs will vary from location to location.

Adsorption

By its nature, an adsorption system retains the hydrocarbons by adsorbing them onto the activated carbon matrix. If the retained hydrocarbons have sufficient value, they can be recovered from the activated carbon by various processes for reuse or sale (Section I.6.15). The cost of recovery depends upon

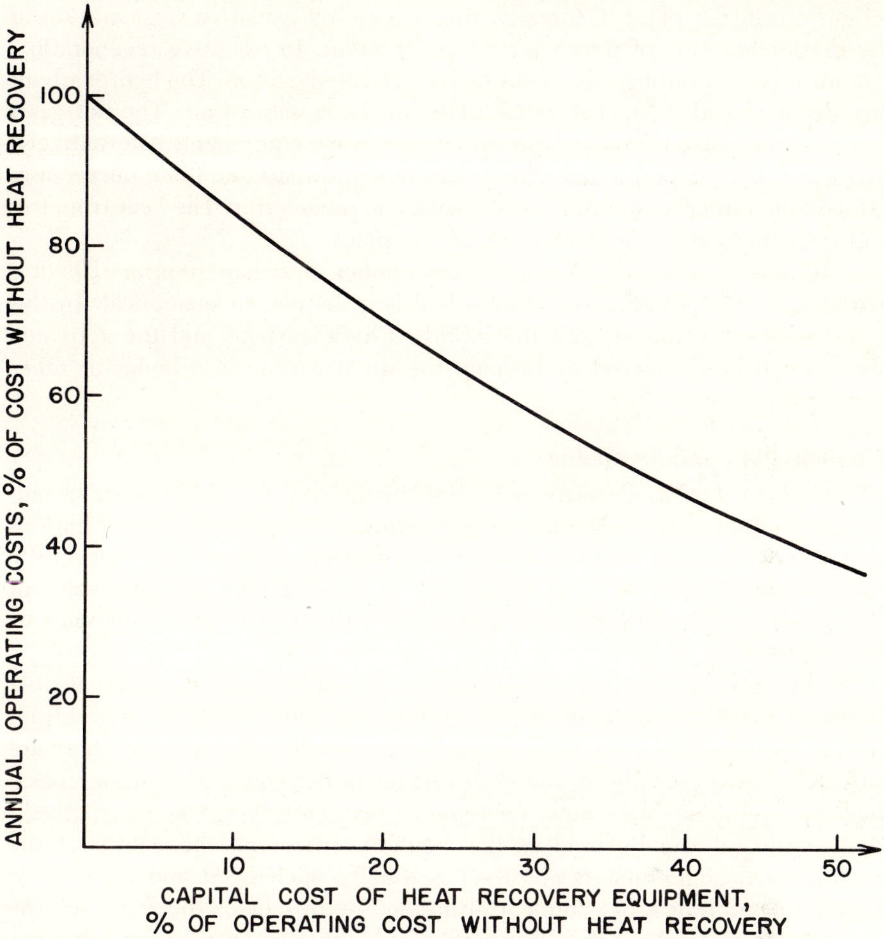

Figure 6.17a Relationship between annual operating costs and capital costs of heat recovery equipment.

the concentration of the hydrocarbons retained on the activated carbon, which, in turn, depends upon the concentration of the hydrocarbons in the waste stream. Generally, if the waste stream concentration is greater than about 1,000 ppm, recovery of the hydrocarbons is economically feasible. Recovery is generally effected by purging the bed with steam, condensing the steam and hydrocarbon vapors and separating and distilling the hydrocarbons to purify them for reuse or sale.

If the concentration of the hydrocarbons in the waste stream is below 1,000 ppm but above about 10 ppm, recovery of the hydrocarbons may not be advantageous, but their heat value may be recovered to reduce fuel costs elsewhere in the plant. Generally, this is done by oxidative regeneration or by the incineration of purge gases from the unit. In oxidative regeneration, hot air is passed through the spent bed of activated carbon. The hydrocarbons are oxidized and their heat value added to the stream of air. The hot gases can then be passed through appropriate recovery equipment. Alternatively, the spent bed can be purged with a stream of warm air and these purge gases oxidized in either a direct-flame or a catalytic incinerator. The heat from this unit can then be recovered for use in the plant.

In situations where the waste gases contain less than 10 ppm of hydrocarbons, regeneration of the carbon bed is generally uneconomical. In this case, the heat value of both the adsorbed hydrocarbons and the activated carbon can be recovered by burning the spent carbon in a boiler or other similar equipment.

Condensation and Scrubbing

Condensation as a control method is effective only when the waste stream contains either high molecular weight hydrocarbons or high concentrations of hydrocarbons. In either case, the hydrocarbons collected are generally rather pure and can be recycled directly or collected for sale. The value of these collected hydrocarbons is extremely important to the economics of many processes.

The spent liquid from absorption or scrubbing may contain valuable hydrocarbons that can be recovered for reuse or sale. Usually the liquid from an aqueous scrubber contains only small amounts of hydrocarbons. These are usually at least partially oxidized to remain in the aqueous solution; consequently, they have little value for reuse or sale. However, when a relatively nonvolatile oil can be used as the scrubbing medium, the nature of the hydrocarbon contaminants absorbed is usually unchanged and recovery is easily accomplished by simple distillation. In this case, the value of the recovered hydrocarbons should be considered in the choice of a control system.

Summary

In any situation requiring control of hydrocarbon emissions, the evaluation of control alternatives should include the following factors:

1. Technical feasibility of the control process
2. Physical constraints of the existing plant
3. Capital costs of the control process
4. Operating costs of the control process
5. Value of by-products from the control process
6. Cost of recovering these by-products

In many cases, the value of the by-products will determine which overall control strategy is optimum for a particular situation.

BIBLIOGRAPHY

Chilton, C. H.: "Cost Engineering in the Process Industries," McGraw-Hill Book Co., New York, 1960.

Danielson, J. A. (ed.): "Air Pollution Engineering Manual," U.S. DHEW, PHS, National Center for Air Pollution Control, Cincinnati, Ohio, PHS Publication No. 999-AP-40, 1967.

U.S. DHEW, PHS, "Control Techniques for Hydrocarbon and Organic Solvent Emissions from Stationary Sources," National Air Pollution Control Administration Publication No. AP-68, 1970.

6.18 HYDROCARBON CONTROL—RELATIVE MERITS OF THE VARIOUS PROCESSES

The various hydrocarbon control processes presented in the preceding sections have advantages and disadvantages which depend upon the properties of the emission to be controlled and on the reasons for controlling the emission (Table 6.18a).

Direct Flame Incineration

Direct-flame incineration has the outstanding virtue of being a simple process that is easy to operate. Once it is installed, it needs little maintenance. It is applicable to the control of streams containing almost any concentration of hydrocarbons, although the fuel required to incinerate low concentrations may make it uneconomical. In operation, the concentration of the hydrocarbons may vary significantly, within limits, without altering its control capabilities, thereby allowing it to be used to control variable emission sources. A direct-flame incinerator is also capable of handling almost any type of hydrocarbon and will, in addition, eliminate some carbonaceous particulates.

The primary disadvantage of a direct-flame incinerator is the high temperature that must be maintained in the unit. Generally, auxiliary fuel is needed to maintain this temperature and special materials are needed to handle the hot gases. The cost and availability of the auxiliary fuel generally limit the use of direct incineration, thereby restricting it to emissions containing a relatively high concentration of hydrocarbons. This relationship is affected by the factors such as the temperature of the emissions and the amount of heat that can be recovered from the outlet gases. The higher the temperature of the emissions, the less auxiliary fuel is needed and the lower is the concentration of hydrocarbons that can be handled economically. The cost of the auxiliary fuel can be reduced by recovering some of the heat in the outlet gases of the incinerator and using it in place of burning more fuel. In this way, the overall fuel consumed by a plant may increase only slightly with the addition of an incinerator.

A disadvantage of a direct-flame incinerator which is a result of the combustion occurring within the unit is that it will emit air pollutants. At the flame temperature within the unit, the nitrogen and oxygen in the air

Table 6.18a
ADVANTAGES AND DISADVANTAGES OF MAJOR HYDROCARBON CONTROL PROCESSES

Process	*Advantages*	*Disadvantages*
Adsorption	Can recover hydrocarbon; will remove low concentrations of hydrocarbons at low temperatures; does not require heat	Limited to low temperatures; limited to low concentrations; must be regenerated with steam or disposed of; must remove particulates
Direct flame incineration	Simple to operate; destroys organics; little maintenance	Emits nitrogen oxides; may require expensive heat recovery equipment; may emit carbon monoxide; generally requires auxiliary fuel
Catalytic incinerators	Smaller than a flame unit; uses less fuel than a direct-flame unit; less nitrogen oxide emissions than a direct-flame unit	Efficiency may decline with time; susceptible to poisons including particulates; requires regular maintenance
Condensation and scrubbing	Can recover hydrocarbons; will also remove particulate matter; conditions stream for other control units	Limited to high-concentration of hydrocarbons; potential disposal problem; requires second control unit

will combine to form nitrogen oxides. The unit may also emit carbon monoxide, as a function of the combustion conditions and depending upon the type of hydrocarbon being burned.

A direct-flame incinerator is best for a stream containing relatively high levels of hydrocarbons. Streams with high temperatures and streams from processes which could utilize the heat generated by an incinerator also favor their use. Direct-flame incinerators generally require an auxiliary fuel supply and produce nitrogen oxides and sometimes carbon monoxide as a result of their operation.

Catalytic Incineration

Catalytic incinerators have several advantages over direct-flame incinerators. Since they are operated at lower temperatures than direct-flame incinerators, they can be smaller, lighter, use less fuel and emit less nitrogen oxides than a direct-flame incinerator. Because of the nature of catalytic oxidation, they are also more effective when the hydrocarbon concentrations are low.

The problems of catalyst poisoning and the somewhat higher maintenance are the biggest disadvantages when compared to a direct-flame incinerator. The bed must be replaced at least every 3 to 5 years, and more frequently if there is a poisoning problem. The stream to be controlled by a catalytic incinerator must not only be free of chemical poisons, but also of particulate matter which can blind the catalyst pores.

A catalytic incinerator requires more maintenance than a direct-flame incinerator; however, in many cases, it is less expensive to operate.

Adsorption

The adsorption process, unlike incineration, does not destroy the hydrocarbons. Hence, they may be recovered and reused or sold if desired. Activated carbon is particularly effective in adsorbing hydrocarbons from low concentration and low temperature sources. For these reasons, it is able to control sources of hydrocarbon emissions that other methods cannot control economically.

The major disadvantage of adsorption is that it is inherently a batch process and requires either regeneration or disposal of the spent adsorbent. In either case, additional equipment and facilities are required. Adsorption, like catalytic incineration, also requires low levels of particulate matter in the stream to be controlled to prevent blinding of the pores.

Thus, activated carbon adsorption is best applied to the control of cool, clean sources of low-concentration hydrocarbons. Provisions must be made for regeneration or disposal of the spent carbon.

Condensation and Scrubbing

Both condensation and scrubbing equipment are useful as preliminary units for the control of high-concentration hydrocarbon emission sources. The hydrocarbons removed from the stream may be recovered for reuse or sale. To meet most control requirements, this equipment must be followed by a more efficient unit such as an incinerator or an adsorber.

In the special cases of an emission containing a high boiling point hydrocarbon or a hydrocarbon that is particularly soluble in a scrubbing medium, condensation or scrubbing, respectively, can be used as the only control process. These situations are relatively rare, however.

Both scrubbing and condensation are generally used as a primary process for the control of high-concentration sources of hydrocarbons. Since most hydrocarbon emissions have a low concentration and high volume, this limits their use. In the case of condensation, the cooling requirement may also be a limiting factor. A disadvantage of the scrubbing process is that disposal or treatment of the spent scrubbing medium may be difficult and expensive.

Other Hydrocarbon Control Methods

Significant quantities of hydrocarbons are released to the atmosphere from storage tanks and through the transfer of hydrocarbons from one tank to another. These releases represent substantial losses of valuable hydrocarbons. Their recovery not only eliminates one major source of hydrocarbon pollution, but also represents a saving of a valuable product.

The emissions of storage tanks are due to mechanisms which occur simultaneously as a tank becomes warmer (Fig. 6.18b). The vapor within the tank expands and is released to the atmosphere, carrying hydrocarbons with it. The higher temperature also raises the equilibrium partial pressure of the

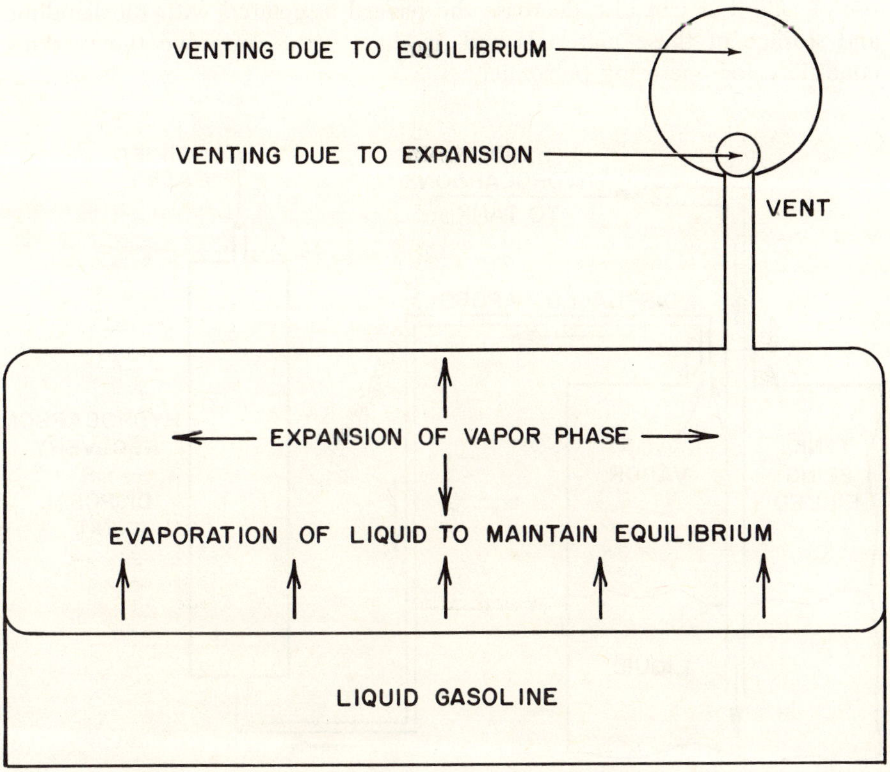

Figure 6.18b Fuel tank evaporation processes.

hydrocarbons. In an effort to maintain equilibrium, more hydrocarbons are evaporated from the liquid phase. These evaporated hydrocarbons displace some of the vapor phase, causing further venting.

These types of emissions can be controlled by:

1. Eliminating the vent and building a tank which is strong enough to withstand the expected pressure.
2. Installation of a floating roof, thereby minimizing the vapor phase and allowing for changes in the volume of the stored hydrocarbons with temperature.
3. Passing the vented hydrocarbons through a control unit such as an adsorber.

Transfer losses are primarily due to the displacement of hydrocarbon vapors from a storage tank as the tank is filled. These emissions are best controlled by first sealing the tank with a vapor-tight closure (Fig. 6.18c) and then directing the vented vapors into a vapor recovery or disposal system.

These control methods can not only recover valuable hydrocarbons for use or sale, but can also decrease the hazard associated with the handling and storage of these materials and, in many cases, improve the working conditions for operating personnel.

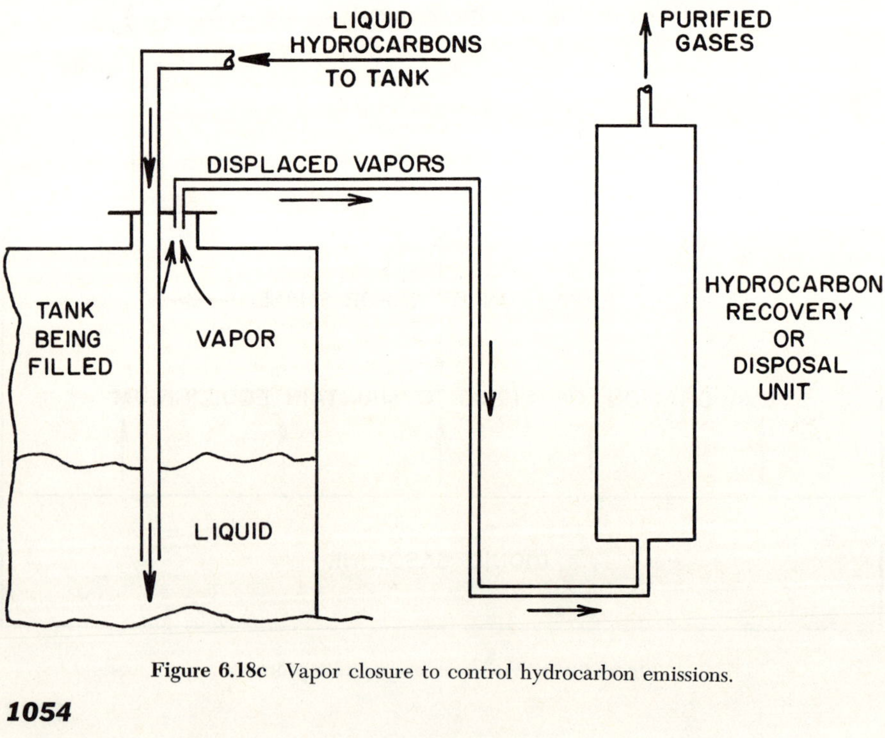

Figure 6.18c Vapor closure to control hydrocarbon emissions.

BIBLIOGRAPHY

American Industrial Hygiene Assn.: "Air Pollution Manual, Part II," Detroit, Michigan, 1968.

Danielson, J. A. (ed.): "Air Pollution Engineering Manual," U.S. DHEW, PHS, National Center for Air Pollution Control, Cincinnati, Ohio, PHS Publication No. 999-AP-40, 1967.

U.S. DHEW, PHS: "Control Techniques for Hydrocarbon and Organic Solvent Emissions from Stationary Sources," National Air Pollution Control Administration Publication No. AP-68, 1970.

6.19 ODOR CONTROL— ODOR, ITS SOURCES AND CHARACTERISTICS

Odor is the sensation of smell perceived as a result of olfactory stimulation by an odorous material. Odorous materials can range from inorganic gases and vapors, such as ammonia, hydrogen sulfide, and hydrogen chloride, across the full spectrum of organic compounds, such as mercaptans, amines, acids, aldehydes and ketones. No instrument has yet been developed with the capability and the sensitivity of the human olfactory system for detecting the presence of this wide range of materials (Section 4.13). The human nose can respond to thousands of different materials, simultaneously detecting both high and low concentrations.

In the life of man, odor plays an important role which varies from person to person, depending upon his background and environment. The odor of fresh bread may remind someone of home and/or start his digestive juices flowing. A nausating fecal odor may force him to leave the area or may make him physically sick. Offensive odors are capable of producing nausea, vomiting and headache, as well as disturbing sleep, imparing nutrition and generally interfering with the enjoyment of life and property.

Odor Characteristics

Odor has two basic characteristics which determine its effect on humans: intensity and quality. The intensity of an odor is a quantization, either through words or numbers, of the strength of an odor. The sensation of intensity of an odor varies logarithmically with the concentration of the odorous compound, according to the Weber-Fechner psychophysical law:[1]

$$I = k \ln C \qquad 6.19(1)$$

where I = the intensity of the odor

k = a constant related to the pervasiveness of the odor

C = the concentration of the odorous compound

The threshold concentration of an odorous compound gives an odor intensity of zero. Since all people do not have the same sensitivity for the detection of odors, the threshold concentration will vary from person to person. For precise work, the threshold concentration is generally taken as the concentration at which 50% of a trained odor-panel will be able to detect the odor (Section 4.13). Because of problems with the purity of the odorous compounds, the presentation system or the panel, threshold values in the literature show a great variation and care should be taken in their use.

The intensity of a mixture of compounds may be additive, counteractive, synergistic or independent, depending upon the compounds and their concentration. This is determined by a comparison of the values for the actual component concentrations and the effective concentration of the mixture in the Weber-Fechner law. For an additive mixture, the effective concentration is equal to the sum of the component concentrations; whereas in an independent mixture, it is equal to the dominant component concentration. In a counteractive mixture, it is less than either component concentration; and in a synergistic mixture, it is more than the sum of the component concentrations.

The quality of an odor is a description of the sensation experienced by an observer in terms of the type of odor rather than its intensity. Generally, this is done by associating an unknown odor with familiar odors. Several such classification schemes have been proposed such as the seven basic odors:[2]

1. Camphoraceous, such as moth repellent
2. Pungent, such as vinegar
3. Etheral, such as dry-cleaning fluid
4. Floral, such as roses
5. Pepperminty, such as mint candy
6. Musky, such as angelica root oil
7. Putrid, such as a bad egg

By use of this basic vocabulary and reference, an observer can describe the quality of any unknown odor. For example, the odor of acetylene tetrabromide can be described as smelling both camphoraceous and etheral.

The quality of an odor may change with dilution for reasons that are not fully understood. In mixtures of odorous compounds, this may be due to differences in the pervasiveness of the compounds. The pervasiveness refers to the ability of an odor to maintain a detectable intensity in spite of dilution with large volumes of air. As previously mentioned, the pervasiveness is indicated by the slope of the Weber-Fechner equation. The pervasiveness increases as the value of this slope [k in Equation 6.19(1)] decreases.

Odor Sources

Almost any human activity produces some odorous compounds. Not all are unpleasant odors; however, the unpleasant ones tend to be noticed and result in complaints. The most frequently reported resources of obnoxious odors in the ambient air are listed in Table 6.19(a).[3] The sources of odor indicated can be either natural or man-made.

Natural odors generally result from the bacterial decomposition of animal

TABLE 6.19a
MOST FREQUENTLY REPORTED ODOR SOURCES

Source of Odor	*No. Reported*
Animals	
Meat packing and rendering plants	12
Fish oil odors from manufacturing plants	5
Poultry ranches and processing	4
Combustion processes	
Gasoline and diesel engine exhaust	10
Coke-oven and coal-gas odors (steel mills)	8
Poorly adjusted heating systems	3
Food processing	
Coffee roasting plants	8
Restaurants	4
Bakeries	3
Paint and related industries	
Manufacture of paint, lacquer and varnish	8
Paint spraying	4
Commercial solvents	3
General chemicals	
Hydrogen sulfide	7
Sulfur dioxide	4
Ammonia	3
General industries	
Burning rubber from smelting and debonding	5
Odors from dry-cleaning shops	5
Fertilizer plants	4
Asphalt odors (roofing and street paving)	4
Asphalt odors (manufacturing)	3
Plastic manufacturing	3
Foundries	
Core-oven odors	4
Heat treating, oil quenching, and pickling	3
Open-dump fires	2
Combustion of waste	
Home incinerators and backyard trash fires	4
City incinerators burning garbage	3
Open-dump fires	2
Refineries	
Mercaptans	3
Crude oil and gasoline	3
Sulfur	1

TABLE 6.19a (Continued)

Source of Odor	No. Reported
Decomposition of waste	
Putrefaction and oxidation (organic acids[a])	3
Organic nitrogen compounds (decomposition of protein[a])	2
Decomposition of lignite (plant cells)	1
Sewage	
City sewers carrying industrial waste	3
Sewage treatment plants	2

[a]Probably related to meat processing plants.

and vegetable matter. Such odors develop primarily in stagnant water, such as in swamps, or as a result of algae blooms in lakes, bays or the ocean. An extreme example of natural odor pollution occurred in the Los Angeles area in 1964. Ocean water temperatures remained over 70°F for several days. The extended high ocean temperature stimulated the growth of the plankton responsible for the "red tide." Consequently, millions of fish were killed by the toxins released by these plankton, which were washed ashore. The fish decomposing on the beaches produced a considerable stench, resulting in the closing of the beaches for several weeks.

Other natural sources of unpleasant odors include volcanoes, sulfur springs and other sources of gases from the earth's interior. Fires such as brush and forest fires also release a significant amount of odorous compounds. More intimately, the human body also releases highly odorous materials that can be a problem in a confined space such as a spacecraft, submarine or odor measurement room.

Industry produces many odorous compounds, generally as an unwanted by-product of their operations. Some of the more significant industrial sources include food processing, the various aspects of the meat industry, coke ovens, petroleum production and refining, petrochemical production, various chemical plants, pulp and paper mills, sewage plants and portions of the iron and steel industry, as well as many mobile and stationary combustion sources.[4]

The most unpleasant odors caused by industry and, consequently, those causing the most complaints result from the release of sulfur and nitrogen-containing compounds and from oxygenated organic materials. For example, one of the major sources of strong odors is the meat packing and rendering plants. In many cases, this is due in part to the presence of a significant amount of decaying flesh. This releases various amines, such as the very strong odorants putricine and cadaverine, ammonia and hydrogen sulfide. Other strong odorants released by industry include mercaptans, organic sulfides, phenolics, aldehydes, napthenic acids, alcohols and several forms of halogens. Any process dealing with such materials, or with easily decayed material such as flesh, has the potential for being a strong odor source. It is only through careful design of storage and materials-handling facilities, as well as controlling any release of material from the process, that such odor sources can be controlled.

REFERENCES

1. Greyton, Arthur C.: "Textbook of Medical Physiology."
2. Amoore, J. E., J. W., and M. Rubin: *Scientific American*, vol. 210, no. 2, p. 42, 1964.
3. Kerka, W. F. and R. R. Kaiser: *J. Air Pollu. Contr. Ass.*, vol. 7, p. 4, February 1958.
4. Sullivan, R. J.: "Preliminary Air Pollution Survey of Odorous Compounds," A literature review prepared under contract No. PH22-68-25, National Air Pollution Control Administration No. APTD 69–42, 1969.

6.20 ODOR CONTROL— AIR OXIDATION

Related Sections:	On afterburners: 5.25, 26 and 27; on direct-flame incineration: 5.32 and 6.13; on catalytic incineration: 5.33 and 6.14; and on odor control: I.6.33 and I.7.45.
Oxidation Method:	(a) Flame; (b) Thermal; (c) Catalytic. NOTE: The letters "a" to "c" in the feature summary below refer to the above listed oxidation methods.
Operating Temperatures, °F:	600 to 900 (c); 1,000 to 1,500 (b); from 1,500 to over 2,500 (a).
Equipment Cost, \$/SCFM of Capacity:	1.75 to 5 (c); 1.75 to 10 (b); 5 to 50 (a).
Annual Fuel Cost, \$/1,000 SCFM Treated:	0 to 4.5 (c); 0 to 7.5 (b); 0 to 20 (a).
Suppliers:	See Sections 6.13 and 14.

Most odorants are organic compounds and their complete oxidation to carbon dioxide, water and nitrogen is one way to eliminate odor. Complete oxidation is essential because partially oxidized organics can be worse offenders than their parent compounds. The few inorganic odorants, such as hydrogen sulfide, can also be effectively treated by oxidation because their oxidation products usually have a comparatively higher odor threshhold. The air oxidation can be carried out by direct-flame incineration in a combustion chamber or by catalytic incineration over precious metal such as a platinum catalyst surface. Variations of these methods include the direct combustion in the fire boxes of boilers or the indirect-fired processes such as kilns and thermal incineration on hot ceramic or metallic surfaces.

Direct-Flame Incineration

This method operates at the relatively high temperature of at least 1100 to 1500°F. The temperatures required for the destruction of some odorous fatty acids released by sewage sludge are compiled in Table 6.20a.

Table 6.20a
COMBUSTION TEMPERATURES OF ODOROUS FATTY ACIDS

Fatty Acid	Temperature °F
Acetic acid	1370
Propionic acid	1370
Butyric acid	1425
Caproic acid	1425

The operation is more economical when the combustible odorant concentration is relatively high, requiring less fuel to maintain the operating temperature. The effect of combustible vapor concentration on the temperature rise is shown in Table 6.20b.[1]

Besides the sufficiently high temperature, the incinerator must provide adequate oxygen, turbulence for good mixing and sufficient residence time for complete combustion. A residence time of 0.3 to 0.5 second is satisfactory for most organic odorants except benzene and aliphatic acids which need 1 second.

The hot exhaust gases can be used to preheat the entering gases through a single or double pass heat exchanger (Fig. 6.13c). A substantial portion of

Table 6.20b
THE EFFECT OF COMBUSTIBLE VAPOR CONCENTRATION ON TEMPERATURE RISE

Vapor	Molecular Weight	Heating Value, Btu/lb	Vapor Concentration by Volume, ppm	Vapor Concentration by Weight, $lb/10^6 ft^3$	Heat Value of Vapor-air Mixture, BTU/ft^3	Theoretical Temperature Rise of, from Initial 500°F
Gasoline	114	20,750	1	0.3	0.0062	0.62
			10	3.0	0.0620	6.20
			100	30.0	0.6200	62.00
			1,000	300.0	6.2000	620.00
Ethyl Alcohol	46	11,620	1	0.12	0.0014	0.14
			10	1.2	0.0140	1.40
			100	12.0	0.1400	14.00
			1,000	120.0	1.4000	140.00
Kerosene	170	19,810	1	0.44	0.0087	0.87
			10	4.4	0.0870	8.70
			100	44.0	0.8700	87.00
			1,000	440.0	8.7000	870.00

Table 6.20c
DIRECT-FLAME INCINERATION COSTS WITH VARYING DEGREES OF HEAT RECOVERY

Cost Items	*System Types*		
	Without Heat Recovery $	*With a Single-Pass Primary and a Single-Pass Secondary Heat Exchanger, $*	*With a Double-Pass Primary and a Single-Pass Secondary Heat Exchanger, $*
Initial equipment cost	16,750	42,300	58,800
Approximate erection cost	3,800	4,600	5,400
Hourly operating cost	4.33	2.58	1.84
Annual operating cost = fuel cost minus makeup-air savings	25,900	13,510	9,630
Annual fuel savings for makeup air	None	1,960	1,370
Total system cost	20,550	46,900	64,200
Annual cost, including initial cost plus operating cost, based on 15-year life	27,270	16,630	13,910
Operating cost/hr/1,000 cfm	0.64	0.38	0.27

NOTES: Fuel cost $\$.50/10^6$ BTU;* annual operating hours, 6,000; erection of incinerator only is included.

the fuel cost can be saved by this mode of operation. However, the capital cost becomes much higher. Comparative cost data is given in Table 6.20c.[2]

Catalytic Incineration

The catalytic combustion occurs faster and at a lower temperature (500 to 800°F). The reaction is exothermic, liberating a substantial amount of heat which can be used to preheat the entering gas or can be recovered as waste heat steam (Fig. 6.20d).

The catalyst surface may accumulate dirt and ashes and the like, which usually can be removed by periodic washing. A more serious problem is catalyst poisoning and physical deterioration of the catalyst due to inadvertant over heating (Section 6.14).

Catalytic incineration is usually applied to lean air mixtures. When the air mixture has a heat value of at least 10 BTU per cubic foot, the resulting temperature rise can sustain combustion (Table 6.20b). Insurance regulations require that air mixture not exceed one-fourth lower explosive limit (LEL).

The costs of catalytic and direct-flame incineration (Table 6.20e) indicate that the catalytic incineration is more economical.

* Approximately doubled as of 1974.

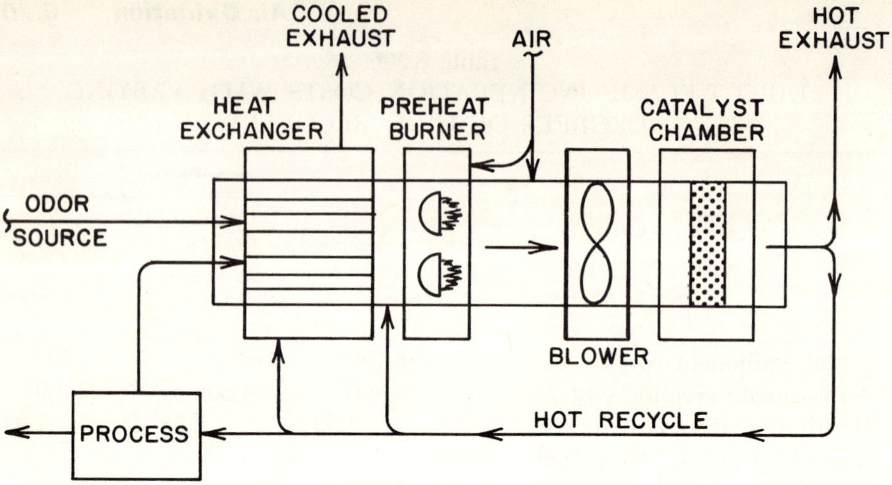

Figure 6.20d Catalytic incineration.

Table 6.20e
COSTS OF CATALYTIC AND DIRECT-FLAME INCINERATION

Operating Condition			*Unit Costs, \$/hr per 1,000 cfm*		
Extent of Waste Heat Recovered	*Calorific Value of Odor Source Gas*	*Cost Item*	*Catalytic Method*	*Direct-Flame Method*	*Cost Ratio Flame/Catalytic*
None	None	Fixed charges	0.15	0.10	
		Burner fuel	0.51	1.06	
		Maintenance	0.09	0.06	
		Labor	0.03	0.03	
		TOTAL	0.78	1.25	1.60
None	¼ LFL[a]	Fixed charges	0.15	0.10	
		Burner fuel	0.03	0.58	
		Maintenance	0.09	0.06	
		Labor	0.03	0.03	
		TOTAL	0.30	0.77	2.45
½ of input	None	Fixed charges	0.19	0.17	
		Burner fuel	0.27	0.58	
		Maintenance	0.11	0.10	
		Labor	0.03	0.03	
		TOTAL	0.60	0.88	1.33
½ of input	¼ LFL[a]	Fixed charges	0.19	0.17	
		Burner fuel	(−0.20)	0.10	
		Maintenance	0.11	0.10	
		Labor	0.03	0.03	
		TOTAL	0.13	0.40	3.10

[a] At concentrations of one-quarter of the LFL, many hydrocarbon gases have a calorific value of about 12 BTU per cu. foot. It is usually sufficient to sustain catalytic combustion for hydrocarbons with catalytic ignition temperatures below 660°F.

REFERENCES

1. Cromwell, W. E.: Ozone in Air Pollution Abatement, *Ind. Eng. Chem.*, vol. 51, p. 83A, June 1959.
2. Yocom, J. E. and R. A. Duffee: Controlling Industrial Odors, *Chemical Engineering*, June 15, 1970.

BIBLIOGRAPHY

Corey, R. C.: "Principles and Practices of Incineration," Wiley, New York, 1969.
Mueller, J. H.: What It Costs to Control Process Odor, *Food Engineering*, April 1971.
Summer, W.: "Odor Pollution of Air, Causes and Control," CRC Press, London, 1971.
Turk, A.: "Industrial Odor Control," *Chem. Eng.*, p. 199, April 27, 1970.

6.21 ODOR CONTROL—OXIDATION BY SUBSTANCES OTHER THAN AIR

Related Sections:	On properties of ozone and other oxidants and use in odor control, Sections I.6.33, I.7.45 and II.3.14.
Ozone Generator Data:	Designs are plate or tube types with water or air cooling. On air as feed gas, the power requirement is 10 to 11 KWH to generate 1 lb of ozone in the outlet gas, concentrated to 1 wt% of O_3. With oxygen enriched feed gas, the output concentration is 2 wt% O_3 and the power required is 3.5 to 4 KWH/lb of O_3.
Odor Control Ozone Dosages:	From 1 to 50 volumetric ppm is required as a function of the application (Table 6.21b).
Residence Time for Ozonation:	5 to 45 seconds are required, depending on application.
Ozonation Equipment Cost:	\$200 to \$1,000 per lb/day ozone generating capacity (Fig. 6.21e).
Partial List of Ozonator Suppliers:	Crane Corp.; Davison Chemical; Ozone Research & Equipment Corp.; Pollution Control Industries, Inc.; The Welsbach Corp.
Deodorizing Cost Using Permanganate:	For scrubber costs, see Fig. 6.23b. The permanganate cost in 100 lb lots is \$.85/lb. Purafil* is \$1.50/lb.
Partial List of Permanganate Suppliers:	Carus Chemical Co.

Oxidizing agents such as ozone and permanganates can oxidize certain odorants in the air at ambient temperatures. However, the oxidation is not

*Permanganate impregnated alumina available under the trade name of Purafil from H. E. Burroughs and Assoc.

always complete and often leads to incomplete oxidation products other than carbon dioxide and water, which are not necessarily odorless nor have a high-odor threshold. The contact between the oxidizing agent and the odorant can be gas-gas, gas-liquid and gas-solid, depending on the character of the oxidizer and on the odor-producing process. The oxidizing power of an oxidizer is proportional to its oxidation-reduction potential (Table 6.21a).

Ozonation

Ozone (O_3) is a triatomic allotrope of oxygen which oxidizes most oxidizable organic and inorganic compounds. The sulfur and nitrogen containing odorants of sewage treatment plants and food and chemical processing plants are converted to sulfoxides, sulfones and amine oxides that have a substantially higher odor threshold.

The ozone dosage required for odor control* can be less than equimolar because the ozone can initiate a free radical reaction and thereby "catalyze" the oxidation of the odorant by the oxygen in the air. When the odorant concentration is unknown, use the ozone dosage shown in Table 6.21b.

Gas-Gas Contact

The odorant-containing air is discharged into a duct or mixing chamber. The ozone is introduced into the intake of the fan which transports the emission to promote thorough mixing.

*In Nagoya, Japan, a 56,000 ACFM sewage digestion gas stream requires 0.37 lb/hr of ozone to reduce ammonia concentration by 60% and eliminate all sewage odors after a 4 second residence time.

Table 6.21a

OXIDATION POTENTIALS OF CHEMICAL DISINFECTANTS

		Oxidation Potentials, E°_{ox} Volts
Ozone	$O_3 + 2H^+ + 2e^- \longrightarrow O_2 + H_2O$	2.07
Permanganate	$MnO_4^- + 4H^+ + 3e^- \longrightarrow MnO_2 + 2H_2O$	1.67
Hypobromous acid	$HOBr + H^+ + e^- \longrightarrow \frac{1}{2}Br_2 + H_2O$	1.59
Chlorine dioxide	$ClO_2 + e^- \longrightarrow ClO_2^-$	1.50
Hypochlorous acid	$HOCl + H^+ + 2e^- \longrightarrow Cl^- + H_2O$	1.49
Hypoiodous acid	$HOI + H^+ + e^- \longrightarrow \frac{1}{2}I_2 + H_2O$	1.45
Chlorine gas	$Cl_2 + 2e^- \longrightarrow 2Cl^-$	1.36
Oxygen	$O_2 + 4H^+ + 4e^- \longrightarrow 2H_2O$	1.23
Bromine	$Br_2 + 2e^- \longrightarrow 2Br^-$	1.09
Hypochlorite	$ClO^- + H_2O + 2e^- \longrightarrow Cl^- + 2OH^-$	0.94
Chlorite	$ClO_2^- + 2H_2O + 4e^- \longrightarrow Cl^- + 4OH^-$	0.76
Iodine	$I_2 + 2e^- \longrightarrow 2I^-$	0.54

Table 6.21b
OZONE DOSAGES REQUIRED FOR ODOR CONTROL

Odor Control Application	*Ozone Dosage, ppm by Volume*
Restaurants	1
Sewage plants, general	1
Morgues	3
Phenol plants	3–10
Rubber plants	3–10
Fish processing plants	10
Sludge storage and vacuum filter units	10
Rendering plants	10
Fermentation processes (manufacture of antibiotics in pharmaceutical)	0–50

The contact chamber should be of sufficient size to allow for 5 to 45 seconds residence time, depending on the oxidizability of the odorant. The ozonated air exhaust should be discharged through a tall stack. The ozone dosage should be adjusted so that no large excess of unreacted ozone remains in the treated air. An excess ozone concentration of 0.1 ppm is considered acceptable at the stack discharge (Fig. 6.21c).

Overozonation, Deozonation

This technique is similar to the gas-gas contact method. A two-three fold excess of ozone dosage is used to accelerate the oxidation and shorten residence time. After contacting, the ozonated air is passed through a catalyst layer (about 1-inch thick) to decompose the excess ozone to oxygen. The catalyst selected can be manganese dioxide or potassium permanganate (Purafil) on activated alumina support, iron oxide or another ozone decomposer catalyst.

The role of the catalyst is often more than just an ozone decomposer. It can also accelerate the oxidation process by adsorbing on its surface both the unreacted odorant and the excess ozone. Manganese dioxide is an excellent catalyst for this purpose. A typical overozonation-deozonation system is shown in Figure 6.21d. Ozone has also been used to control the growth of odor-producing microorganisms in cellars and cold rooms. Care should be taken that the ozone concentration does not exceed the 0.1 ppm limit when workers are present (Fig. 6.21c). The cost of ozone-generating equipment is shown in Figure 6.21e.

Permanganate

Potassium permanganate can oxidize odorants containing sulfur into their less odorous sulfone and sulfoxide forms or can oxidize offensive nitrogen-

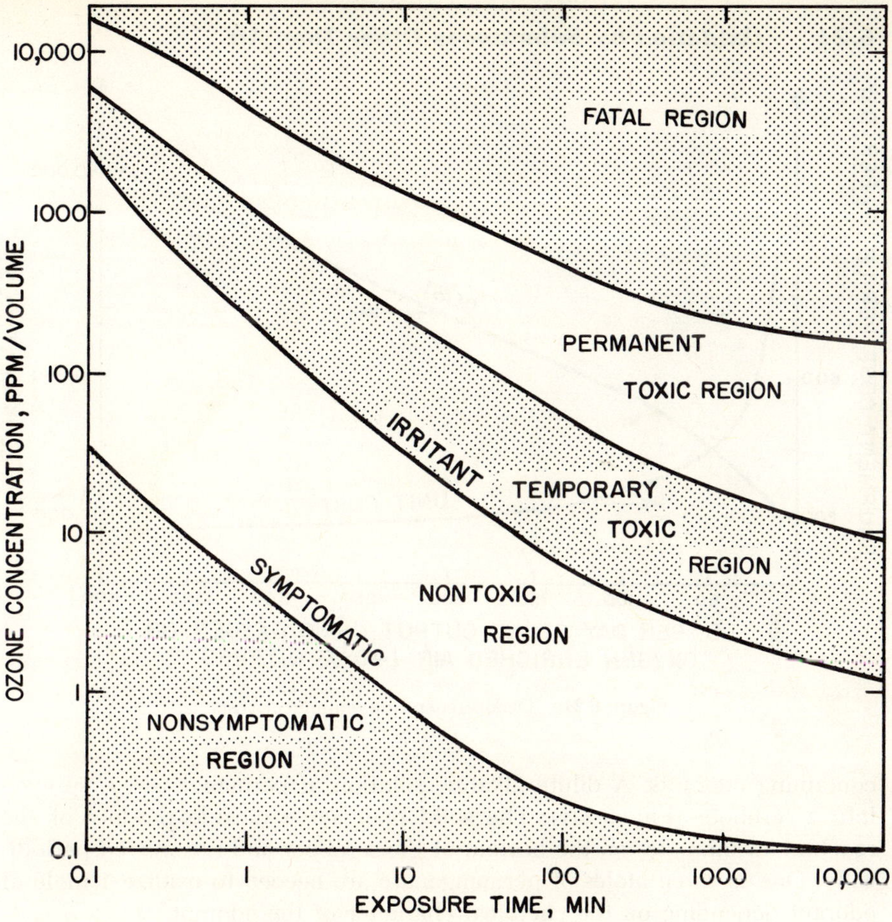

Figure 6.21c Human tolerance for ozone.

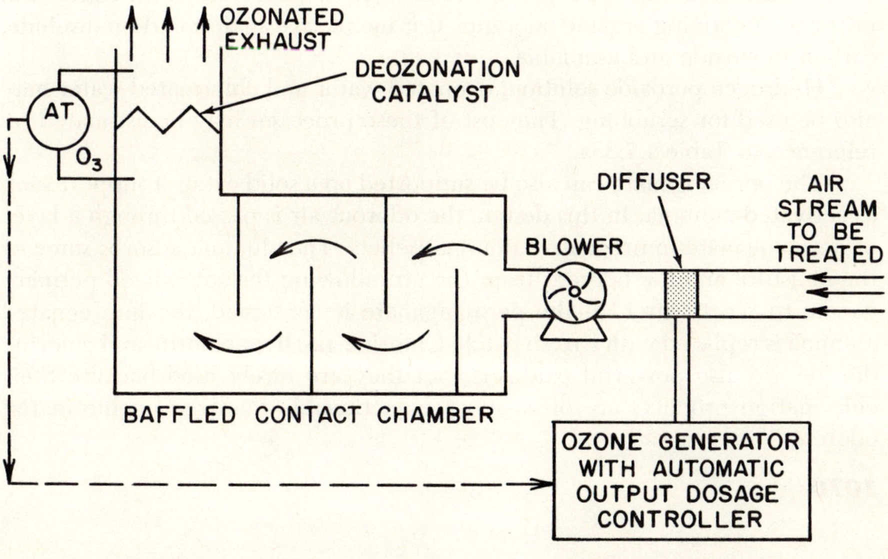

Figure 6.21d Odor control by overozonation followed by deozonation.

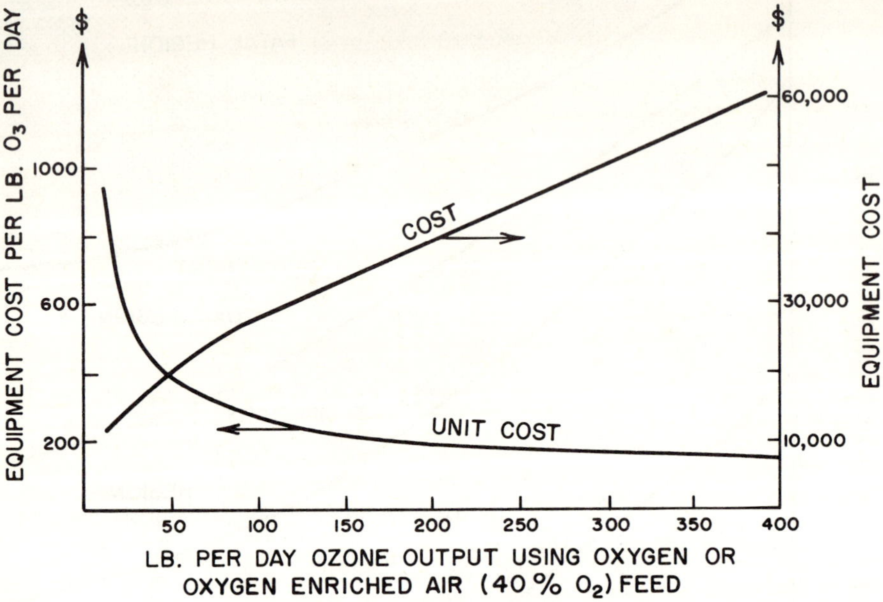

Figure 6.21e Ozone-generating equipment costs.

containing odorants. A dilute alkaline solution of permanganate is sprayed into a scrubber (Fig. 6.21f). The solid manganese dioxide product of the reaction accumulates on the bottom of the scrubber and is removed periodically. One to three moles of permanganate are needed to oxidize 1 mole of odorant depending on the chemical character of the odorant.

Permanganate scrubbing[1] is usually carried out with 1 to 2% $KMnO_4$ solution at pH 8 to 9.5. The pH is maintained by the addition of sodium bicarbonate or borax. This process is used to control odor from sulfur and nitrogen-containing organic odorants. It is ineffective against carbon disulfide, carbon monoxide and ammonia.

Hydrogen peroxide solution, ozonated water and chlorinated water may also be used for scrubbing. The cost of these processes may be estimated by reference to Table I.7.33a.

The permanganate can also be supported on a solid catalyst support such as activated alumina. In this design, the odorous air is passed through a layer of permanganate-impregnated alumina pellets. The alumina adsorbs some of the moisture and the odorant from the air, allowing the solubilized permanganate to react. After all the permanganate is consumed, the impregnated alumina is replaced with a fresh batch. Chlorine gas, hypochlorite and chlorine dioxide are also powerful oxidizers, but they are rarely used because their chlorination products are often more objectionable than the unchlorinated odorant.

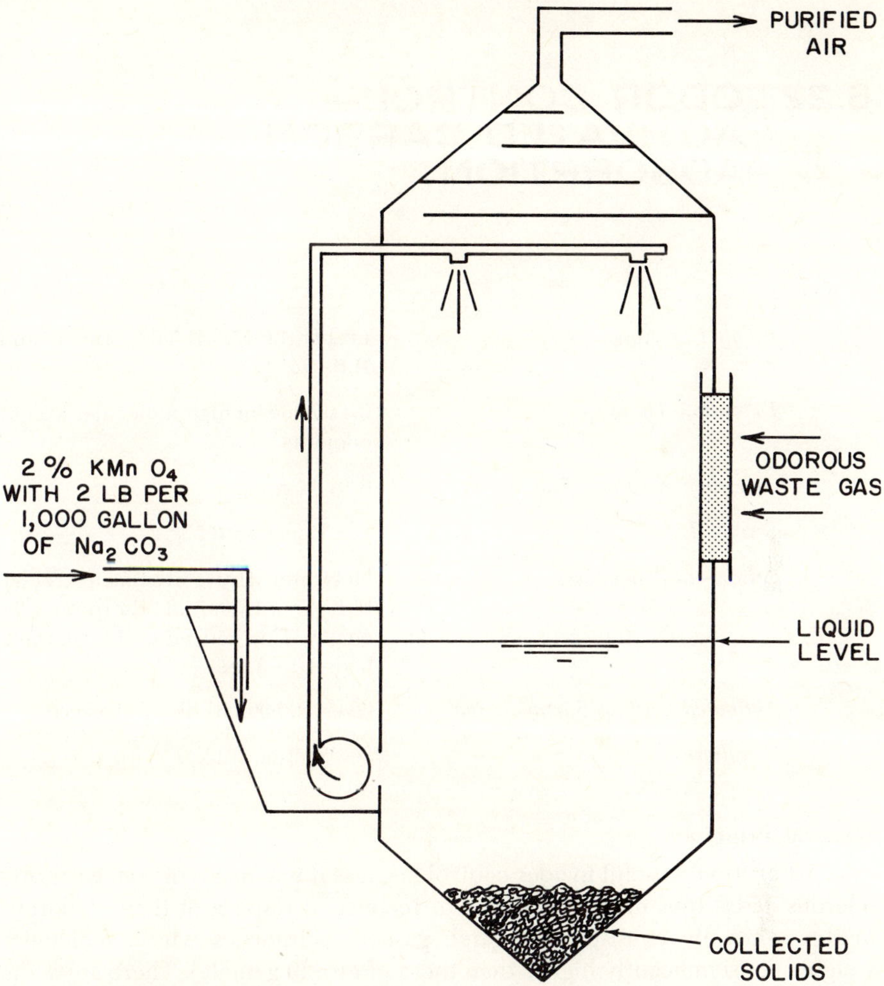

Figure 6.21f Odor control by use of a permanganate scrubber.

REFERENCE

1. Posselt, H. S. and A. H. Reidies: "Odor Abatement with Potassium Permanganate Solution," I & E. C. Product Research and Development, vol. 4, p. 48–50, March 1965.

BIBLIOGRAPHY

Summer, W.: "Odor Pollution of Air, Causes and Control," CRC Press, London, 1971.

Turk, A.: Odor Control, "Kirk Othmer Encyclopedia of Chemical Technology," 2d ed., vol. 14, p. 170, Interscience, New York, 1967.

Yocom, J. E. and R. A. Duffee: Controlling Industrial Odors, *Chem. Eng.*, vol. 160, June 15, 1970.

6.22 ODOR CONTROL—ACTIVATED CARBON ADSORPTION

Related Sections:	I.6.14, I.6.15, II.5.29, II.6.9 and II.6.15.
Detention Time:	0.1 second for high-molecular weight odorants.
Adsorbent Granule Size Range:	6 to 14 mesh.
Service Life:	See Equation 6.22(1).
Adsorbent Thickness:	Determined by total capacity (retentivity) not by initial adsorption efficiency. Thin bed: under 1 inch; thick bed: 1 to 3 feet.
Activated Carbon Surface Area:	Over 100,000 ft^2/lb of charcoal.
Suppliers:	See Section 6.15.

General Principles

Adsorption is useful in odor control because it is a means of concentrating odorous gases, thus making it easier to recover or dispose of them. Odorous matter generally consists of organic gaseous substances whose molecular weights are significantly higher than those of air (29 g/mole). There are some exceptions, however, notably ammonia (molecular weight 17), formaldehyde (30), hydrogen sulfide (34) and methylamine (31). The high molecular weight odorants can generally be concentrated by physical adsorption. The lighter odorants, on the other hand, require some form of chemical conversion on the adsorbent surface if they are to be effectively separated from air.

The capacity of a given adsorbent bed to deodorize a flowing gas stream depends on the total quantity of adsorbable matter to which it is exposed, not on the total quantity of odorants alone. Thus, a stream containing a small concentration of a highly odorous sulfide and a large concentration of almost odorless paraffinic hydrocarbons will saturate an adsorbent bed at a rate that is determined by its hydrocarbon content, not by its odor. The odor control performance, however, is judged solely on the odorous properties of the effluent stream after it emerges from the adsorbent. So long as the effluent is odorless, the odor removal is complete.

When the mass transfer zone of odorous matter approaches the downstream end of the adsorbent bed, effluent odor will begin to be detectable ("odor breakthrough"). When the odorous matter consists of various components, some separation usually takes place on the adsorbent, and the first downstream odors will therefore be different from the upstream odor. In addition, some chemical conversions, typically oxidation, may takes place on the adsorbent bed, possibly generating odors of a different quality when breakthrough occurs.

Given a fresh adsorbent surface and an odorous gas of relatively high molecular weight whose molecules are readily adsorbed (for example, butyl mercaptan, 90, or methyl methacrylate, 100), the proportion of molecules reaching the surface that are retained on it approaches 100%. Therefore, the rate of adsorption depends on those factors that determine the rate at which odorant molecules are brought to the adsorbent surface. Using a well-packed bed of adsorbent granules in the 6 to 14 mesh size range, adsorption efficiencies up to 97% can be attained within a detention time of less than 0.1 second. Such performance is equivalent to 30 dilutions of an odorant within 0.1 second of detention time, or about 300 dilutions per second, and implies that the reduction of odorant concentrations to values below the detection thresholds can be achieved with relatively thin adsorbent beds operating at low to modest pressure drops. Indeed, the designed thickness of adsorbent beds for odor control is almost always determined by the requirements for total capacity rather than for initial efficiency.

Adsorbed odorants must be disposed of. The choice of procedure will depend on the relative values of recovered adsorbate and adsorbent, and on the relative costs of recovery and disposal. When both adsorbate and adsorbent are recovered, ("solvent recovery") they are recycled into the process and do not constitute wastes. When only the adsorbent is to be recovered ("reactivation"), the adsorbate is removed, usually by steam desorption, and is either disposed of as a water-borne waste or is burned. When neither adsorbate nor absorbent is to be recovered ("throw-away" adsorbers), the saturated adsorbent constitutes a solid waste.

Properties of Activated Carbon

Activated carbon (also called activated charcoal or active carbon) consists of particles of carbon of varying degrees of purity which have a large surface area per unit weight, usually in the order of several hundred thousand square feet per pound. Raw material sources include coal, wood, petroleum residues, nut shells and fruit pits. Activation is effected by driving out the volatiles by heating (charring or coking), followed by a high-temperature reaction with steam which creates an extensive network of micropores that provides the desired large surface area. The pores, whose radii may approach molecular dimensions, can serve as sites for capillary condensation of adsorbed matter, and thus enhance the overall capacity of the adsorbent.

Activated carbon is the adsorbent of choice because it is nonpolar and, therefore, does not selectively adsorb polar substances such as water. Instead, it tends to retain the larger, less polar organic molecules which typically comprise the odorous components of a gas stream. Therefore, the deodorizing capacity of activated carbon is not vitiated by the presence of moisture. By contrast, polar adsorbents such as silica gel, activated alumina and synthetic zeolites selectively adsorb and retain water and, therefore, lose their deodorizing function in even moderately moist atmospheres.

The total adsorptive capacity of a sample of activated carbon may be measured by its activity or retentivity for a standard vapor. The activity is the maximum amount of a vapor which can be adsorbed by a given weight of carbon under specified conditions of temperature, concentration of the vapor and concentration of other vapors (usually water). The retentivity is the maximum amount of adsorbed vapor which can be retained by the carbon after the vapor concentration in the ambient air or gas stream has been reduced to zero. Because an adsorbent may be required to retain its adsorbate even in pure air, the retentivity represents the practical capacity of the carbon in service.

Typical specifications for activated carbon to be used for air purification are given in Table 6.22a.

The service life of an activated carbon system is given by the general formula:

$$t = \frac{kWS}{ECQ} \tag{6.22(1)}$$

where t = service life, time in hours

k = a constant that depends on the choice of units and on the average molecular weight of the gas = $6.43 \times 10^6/M$ where M is the average gram molecular weight

W = mass of carbon, pounds

S = proportionate maximum saturation of carbon, fractional; typical values are 0.10 to 0.25

E = adsorption efficiency, fractional

C = entering concentration of adsorbable vapors, ppm by volume

Q = volumetric rate of airflow, ft^3/min

Activated carbon can be impregnated with various reagents that will react with specific odorants to convert them to inodorous products or to render them more easily retained by the carbon. Such measures are usually reserved for low molecular weight odorants for which unimpregnated carbon has a poor retentivity. An example is the use of lead acetate as an impregnant to

Table 6.22a
TYPICAL SPECIFICATIONS FOR ACTIVATED CARBON USED FOR AIR PURIFICATION

Property	*Specification*
Activity for CCl_4[a]	At least 50%
Retentivity for CCl_4[b]	At least 30%
Apparent density	At least 0.4 g/ml
Hardness (ball abrasion)[c]	At least 80%
Mesh distribution	6–14 range (Tyler Sieve Series)

[a] Maximum saturation of carbon at 20 °C and 760 mm Hg in an airstream equilibrated with CCl_4 at 0 °C.
[b] Maximum weight of adsorbed CCl_4 retained by carbon on exposure to pure air at 20 °C and 760 mm Hg.
[c] Percent of 6 to 8 mesh carbon which remains on a 14-mesh screen after shaking with 30 steel balls of 0.25 to 0.37-inch diameter per 50 g carbon for 30 minutes in a vibrating or tapping machine.

help retain hydrogen sulfide. Activated carbon can also be impregnated with catalysts that will promote the oxidation of adsorbed matter. Transition metals or metal oxides are typically used.

Equipment and Systems

Activated carbon equipment must be so designed that it displays a uniform exposure of the adsorbent bed to the contaminant gases and provides a satisfactory efficiency and capacity for deodorization without imposing a prohibitive resistance to airflow. When low resistance is critical, thin-bed adsorbers are used. Typical units, with bed thicknesses of about 2 cm, are illustrated in Figures 6.22b and 6.22c.

Thick-bed adsorbers (Fig. 6.22d) are used when a high-adsorbing capacity is needed. Typical bed depths are 1 to 3 feet.

Activated carbon adsorption may be applied to deodorize either (1) an exhaust gas before it is released to the atmosphere, (2) an odorous outside atmosphere before it is used to supply inside air for a building or (3) an indoor atmosphere that is to be recirculated inside an enclosure such as a building or a room.

In the first instance (exhaust air purification), odorant concentrations are typically high and thick-bed adsorbers are used. If the adsorbate is valuable, it may be recovered by a regenerative ("solvent recovery") system such as is shown in Figure 6.22d. If it is not valuable, it may be desorbed and incinerated on a catalytic grid or by other means. Odorant removal need not necessarily be completely effective, because some additional dilution will occur as the gases disperse into the outside atmosphere.

In case of air intake purification, odorant concentrations are typically low and a long service life (months) may be realized even with relatively thin-bed adsorbers. High efficiency is needed because the intake air constitutes the total supply for human occupancy.

Recirculation involves a gradual reduction of odor level by infusion of deodorized air into an odorous space. If the initial concentration of odorant is C_O, then the concentration C after any time t, during which the air is recirculated through a carbon adsorber, is:

$$C = C_O\, e^{-EQt/V} \quad 6.22(2)$$

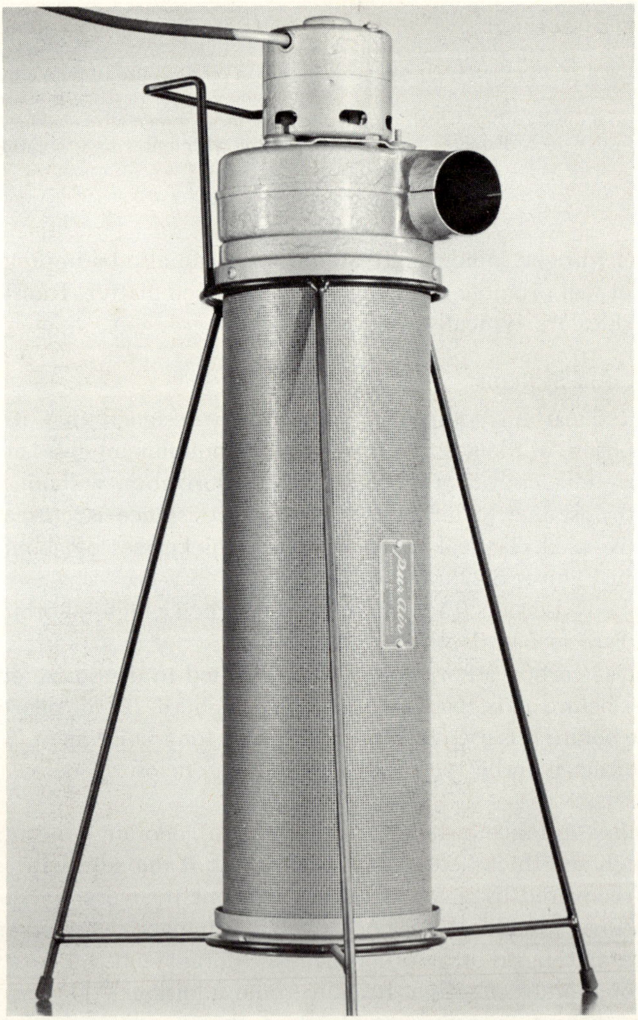

Figure 6.22b Cylindrical thin-bed canister adsorber. *(Courtesy of Barnebey-Cheney Company)*

Figure 6.22c Pleated thin-bed adsorber. The small test element on the face of the adsorber is removed when needed for analysis to establish the degree of saturation of the entire cell. (*Courtesy of Connor Engineering Division, Jenn-Air Company*)

where E is the adsorption efficiency, Q is the rate of airflow through the adsorber and V is the volume of the space. (The term Qt/V is the number of air changes.) This relationship means that the adsorption efficiency does not independently control the deodorizing performance, which cannot be instantaneously effective in any case. Thus, a 100 cfm unit at 50% efficiency is equivalent to a 50 cfm unit at 100% efficiency. Thin-bed adsorbers are usual for recirculation applications, because low pressure drops are advantageous (especially where fan noise might otherwise be a problem) and high efficiency is not critical. The sources of odorous adsorbable vapors associated with human occupancy (bodies, cosmetics, food and tobacco) usually contribute so little mass that a typical thin-bed adsorber of the type used with air conditioning systems may last for a half-year or more before reactivation is needed. However, an extraneous massive source of vapor, such as a solvent spillage, may rapidly saturate such a system.

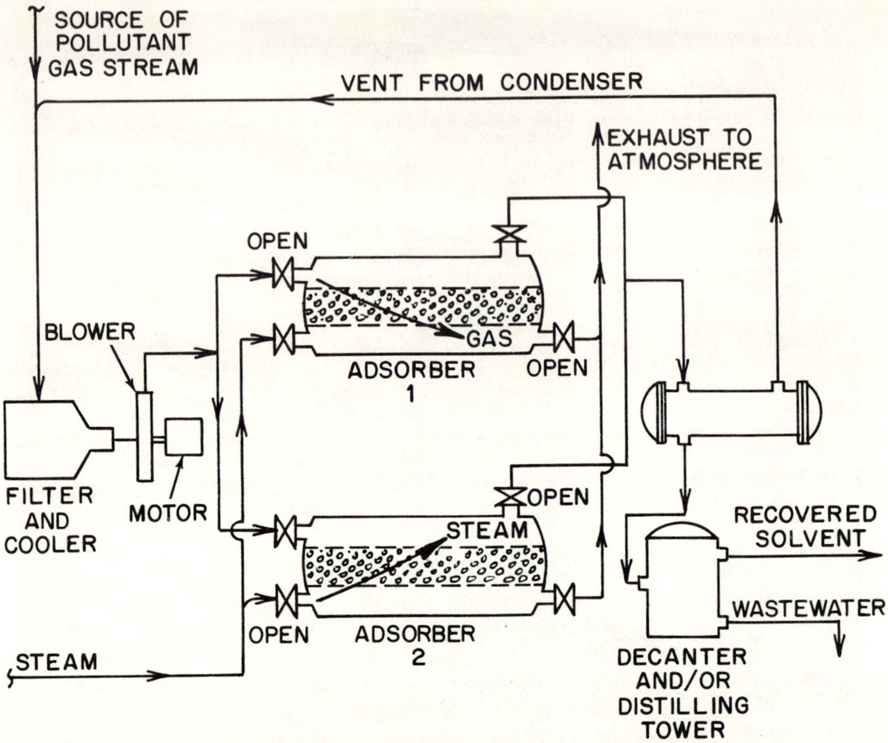

Figure 6.22d Two-stage regenerative system. Adsorber 1 is adsorbing; adsorber 2 is steaming. After No. 1 is saturated and No. 2 is clean, their functions are reversed.

The capacity of activated carbon decreases with increase of temperature. Temperatures above about 40°C should be avoided. When necessary, cooling can be utilized by dilution with ambient air.

Saturated thin-bed carbon adsorbers are generally removed from the system and replaced with fresh units. The spent carbon is removed from its canister or cell and is reactivated as part of a bulk batch with saturated carbon from other units. Thick-bed adsorbers, on the other hand, are usually reactivated in place by superheated steam or through air oxidation.

Typical Applications

Exhaust air deodorization with activated carbon is best applied to highly odorous materials of relatively high molecular weight which appear in moderately low concentrations in the effluent to be treated. For example, the air discharge from a scrubber used to treat animal waste effluents will generally retain the intense odors of the heavier, less soluble but fairly dilute nitrogenous vapors, which will be very effectively deodorized by activated carbon. For effluents with relatively high concentrations of adsorbable vapors, activated carbon is applicable when it can be used to recover valuable solvents, such as methylethyl ketone or cumene. Activated carbon adsorbers are also used as standby protection against the accidental release of small quantities of highly odorous matter, such as mercaptans or sulfides used for gas odorization.

Air intake purification is used whenever the outside air quality is unsatisfactory. In some cases, the odor source may be quite close to the intake; in fact, it may even be an odorous exhaust from the same building. In such cases, it is much cheaper to purify the intake rather than the exhaust air, because only a small fraction of the discharged odorants need then to be adsorbed.

When the outside air quality is satisfactory, deodorization of recirculated air may be viewed as a conservation of the energy expended in heating or cooling the air for human occupancy. Such treatment is therefore most appropriate, and used most often, in air conditioned buildings where energy savings can be effected during both cooling and heating seasons.

BIBLIOGRAPHY

Barnebey, H. L.: *J. Air Pollu. Contr. Ass.*, vol. 15, p. 422, 1965.

Hassler, J. W.: "Activated Carbon," Chem. Publ. Co., New York, 1963.

Sleik, H. and A. Turk: "Air Conservation Engineering," 2d ed., Connor Engr. Corp., Danbury, Connecticut, 1953.

Turk, A.: Source Control by Gas-Solid Adsorption and Related Processes, Vol. III, Chap. 47, in A. Stern, "Air Pollution," pp. 497–519, Academic Press, New York, 1968.

6.23 ODOR CONTROL— SCRUBBING AND OTHER PROCESSES

Related Sections:	On scrubbing: II.5.11 to 5.21, 5.31, 6.8 and 6.16; on stacks and dispersion: II.2.6, 5.42 to 5.45.
Odor Control Methods:	(a) Scrubbing; (b) Dispersing; (c) Neutralization; (d) Masking.
Scrubber Equipment Cost:	\$.10 to \$.40 per SCFM of airflow capacity.
Cost of Stacks:	See Figure 5.42a.
Cost of Odor Neutralization:	An 800 ft long "vapor fence" has a chemical consumption cost of \$.40/hr.

Scrubbing

Scrubbers used for odor-control are usually the spray tower, packed tower or wetted porous media type. They operate by dissolving, condensing or chemically reacting the odorant with the scrubbing liquid.

The odorant in the air exhaust can be present as vapor or as fine droplets or can be absorbed on the surface of particulate contaminants. It can usually be removed by liquid washing of the air. The same equipment can be used to control the temperature and the humidity of the air or to recover valuable vapors (e.g., solvents) from the air at the same time.

The odorant vapor may condense from the air when it contacts the cold scrubbing liquid. The scrubbers utilizing this principle are also called contact condensers. This equipment can be effectively used for the removal of odorous contaminants from air exhausts which are rich in water vapors such as are the rendering plant cooker exhausts. Contact condensers are also commonly used for dehumidifying the air.

The odorant vapor may have sufficient solubility in the scrubbing liquid to be removed without chemical action. Inorganic gases such as HCl and NH_3 are usually removed by water scrubbing. The choice of scrubbing liquid is dependent on the odorant. Glycol solutions remove a large variety of odorants, but they are too expensive to throw away and have to be regenerated usually by distillation.

The addition of chemicals to the scrubbing solution is often effective in odor removal. Aqueous acid solutions effectively remove basic odorants such as amines or pyridine and aqueous base solutions remove acidic odorants such as hydrogen sulfide. The addition of oxidizing agents such as potassium permanganate can destroy numerous organic odorants (Fig. 6.21f). The scrubbing liquid must be maintained in relatively odor-free condition to avoid recontamination of the air. Therefore, either fresh water should be added continuously or the organic solvent should be regenerated continuously to assure this requirement. The discarded water is likely to be sufficiently contaminated to require treatment prior to being discharged.

A spray tower and a packed tower type scrubber are shown in Figure 6.23a. Cost data is provided in Figure 6.23b.

Dispersion

This method accomplishes odor control by dispersing the odorant and diluting it to a concentration which is below the odor threshold level. The odorous gas is usually dispersed from a stack. (Sections 5.42 to 5.45). In general, the higher the stack, the better the dispersion.

Ground level odorant concentration can be determined by Equations

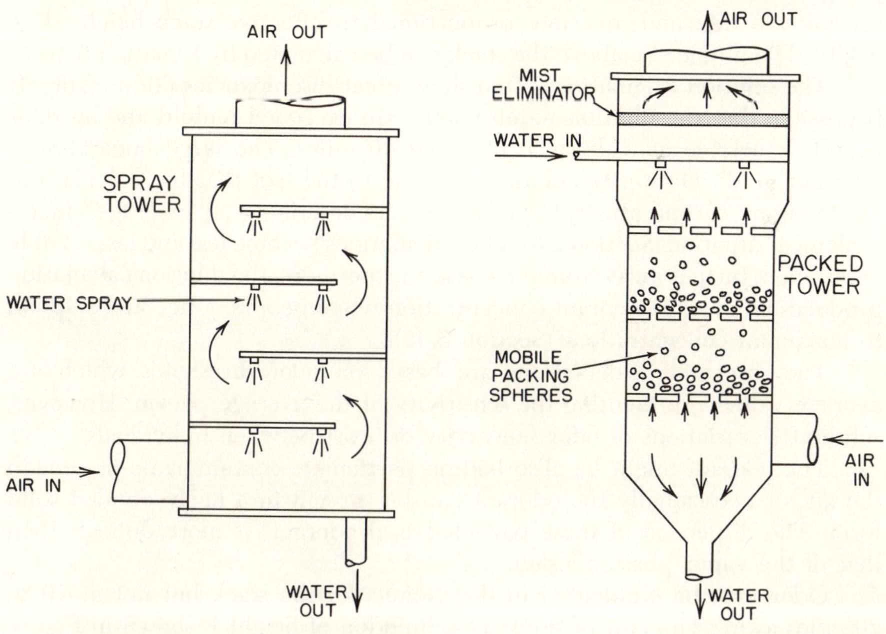

Figure 6.23a Scrubber designs.

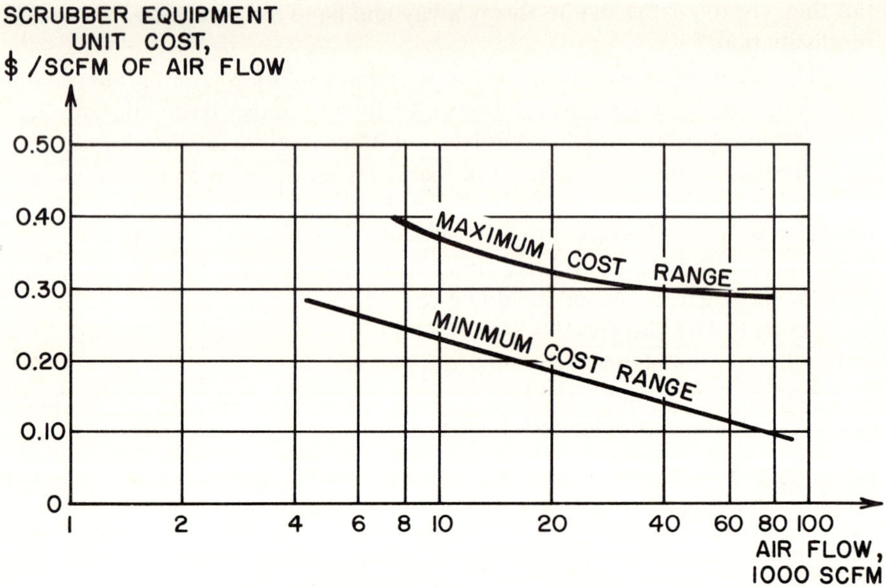

Figure 6.23b Scrubber costs.

5.45(3) and (4). The ground level odorant concentration is directly proportional to emission rate and inversely proportional to effective stack height (Fig. 5.42b). The plume rise above the stack can be calculated by Equation 5.42(7).

The dilution calculations often show great discrepancies (Table 2.6f). It is possible that the dilution requirements are exceeded tenfold and an odor is still detectable at a distance of 10 to 40 miles. The large uncertainties inherent in all dilution calculations are due to the fact that the calculations are based on certain atmospheric parameters describing an "average" meteorological situation (Section 2.6). The calculations become less and less reliable at distances further away from the stack. Furthermore, the dilution calculation produces an average odorant concentration while people notice and respond to maximum concentrations (Section 5.45).

The dispersion calculations are based on odor thresholds which are average values representing the sensitivity of the average person. However, substantial variations of odor sensitivity do exist between individuals.

The odorant might be absorbed on particulate contaminants present in the air, or occasionally the odorant can be present in a finely divided solid form. The dispersion of these particle-based odorants is more difficult than that of the vapor phase odorants.

Odor may be eliminated in the vicinity of the stack but not at 10 to 20 miles away. The cost of stacks as a function of height is shown in Figure 5.42a.

Neutralization

This method consists of interposing an odor-neutralizing vapor wall between the odor source and the surrounding area. The neutralizing agent is selected on the basis of odor counteraction.[1] A combination of two odors can increase or decrease the sensation of odor (Section 6.19). The odorant that decreases the odor of another one is called a counteractant. There are known odor pairs that serve as counteractants for each other. These odor pairs include rubber-cedarwood, musk–bitter almond, skatole-coumarin and butyric acid–oil of juniper.

Malodors are usually mixtures of various odors and require different counteractants or counteractant mixtures. The counteractants and their concentrations needed for neutralization are selected empirically. The success of neutralization is largely dependent on the talent and experience of the formulator.

The neutralization of odor at its source is costly and not always possible. Neutralization can be achieved much more readily at some distance from the source, usually at the property line where the odorant is diluted substantially and is distributed more evenly in the air.

The counteractant can be applied through a compressed air or steam-driven spray nozzle or through a vaporizer-blower that assures complete evaporation. The output of the vaporizer-blower can be funneled into a ductwork which is running along the property line. The counteractant is released through small orifices to form a "vapor fence."

The equipment can be automatically operated, taking wind direction, wind velocity and the time of the day into account. An 800-foot long unit consumes chemicals at the rate of about $.40/hr.

This method is commonly used for odor control of lagoons, ditches and other large, open odor sources to which alternate techniques cannot be readily applied.

Masking

With this method, a second odorant is injected into the air to obscure the odor of the first one already present. The mode of application is very similar to that of neutralization. This method is also used for odor control of lagoons and ditches.

The concentration of the masking agent is usually in the range of 1 to 10 ppm. The processes which commonly use the masking technique include the scrap metal, bone and waste sludge processing.

REFERENCE

1. Zwaademaker, H.: *Die Physiologie des Geruchs*, Leipzig, Germany, 1895.

BIBLIOGRAPHY

Air Pollution Control, Basic Technology, *Chem Eng. Deskbook*, April 27, 1970.

Leonardos, G., D. Kendall and N. Barnard: Odor Threshold Determinations of 53 Odorant Chemicals, *Journal Air Pollution Control Assoc.*, vol. 19, pp. 91–95, 1969.

Summer, W.: "Odor Pollution of Air, Causes and Control," CRC Press, London, 1971.

Turk, A.: Industrial Odor Control, *Chemical Eng. Deskbook*, April 27, 1970.

Yocom, J. E. and R. E. Duffee: Controlling Industrial Odors, *Chemical Eng.*, June 15, 1970.

6.24 ODOR CONTROL— RELATIVE MERITS OF PROCESSES

The Degree of Odor Control

Adverse effects of odors include annoyance, interference with rest or relaxation, loss of sleep or appetite, interference with the use of property, social detriments such as reluctance to entertain one's friends in an odorous environment and the depression of property values. It is impractical to monitor all of these effects in order to judge the performance of an odor control system. It is usually assumed that the reduction of odorant concentration to below the threshold level eliminates the problem and, therefore, represents complete control. According to this concept, the degree of control can be measured by the additional dilution required to reach the threshold concentration. (See Section 4.13.)

Alternatively, odor control effectiveness may be monitored by some direct psychophysical scaling of odor intensity, by an evaluation of odor character, by some estimate of its acceptability or by an appropriate chemical analysis.

When odor masking or counteraction is used (Section 6.23), the sensory evaluation of intensity or of threshold levels, or the chemical analysis of any specific components, is inappropriate. Here, the control objective is to make the odors less objectionable by the simple admixture of the malodor with the control agents, and no chemical conversions or physical separations are involved. Therefore, the criterion of effectiveness must be some measured decrease of objectionability, such as the reduction of complaints.

Some odor control ordinances specify a maximum permissible dilution ratio, such as, for example, "not to exceed the threshold odor level after dilution of 1 volume of the ambient air with 3 volumes of odorous air at any point outside the plant property." Under such conditions, a threshold dilution value is the mandatory criterion of odor control.

Effect of Degree of Confinement

Adsorption, scrubbing and incineration all involve the expenditure of energy in moving an airstream against a resistance. Incineration also demands thermal energy. All of these odor control techniques, therefore, require the confinement of the gas into as small a volumetric discharge rate as possible.

In practice, these techniques are, therefore, applied to odorant streams that are discharged to the atmosphere through ducts, vents or stacks.

Unconfined sources, such as settling ponds, open storage areas, drainage ditches and the like, are amenable to treatment by masking or counteraction methods.

Nature and Concentration of Pollutants

Of the various methods of controlling odors by concentration or conversion, the least selective with regard to chemical composition of the odorant are incineration and activated carbon adsorption. These methods are, therefore, favored for the control of odorants that consist of many kinds of chemical components. Control by activated carbon is most attractive when both concentrations and temperatures are low, as in air intake purification (see Section 6.22). Incineration is favored for emissions with high temperatures and concentrations, as in solvent exhausts from baking or curing ovens. Intermediate ranges of temperature and concentration may make either method very costly. Catalytic combustion is designed to save energy costs by reducing the required oxidation temperature by about 200 to 300°C. However, when catalyst activity gradually deteriorates, oxidation becomes less complete and some intermediate odorous oxidation products may be released.

The various ambient-temperature chemical conversions of odorants are much more specific in their actions than are adsorption or incineration. Most of these methods use oxidants, such as potassium permanganate, ozone, chlorine and hydrogen peroxide. These agents are relatively slow in their reactions, requiring detention times of seconds or even minutes rather than fractions of a second. Furthermore, they do not generally provide complete oxidation to carbon dioxide and water. Permanganates, for example, typically produce glycols, aldehydes and organic acids; ozone also produces aldehydes and acids; chlorine yields chlorinated products. These various intermediates may be almost odorless (glycols) or may have pleasant odors (aromatic aldehydes) or highly unpleasant ones (chlorinated phenols). Therefore, any such specific chemical conversions must be rigorously evaluated by sensory as well as chemical methods before they are used in a full-scale control system.

Scrubbing without chemical reaction is most advantageous when the odorant stream contains relatively high concentrations of soluble matter, such as ammonia or low molecular weight amines, which are both soluble in water. Caution must be exercised, however, lest a small insoluble residue of highly odorous matter carries most of the odor past the scrubber into the atmosphere.

Multistage Control Systems

Odor control objectives may best be realized by abatement systems that comprise two or more stages.

A cycling adsorption-oxidation system comprises a bed of activated

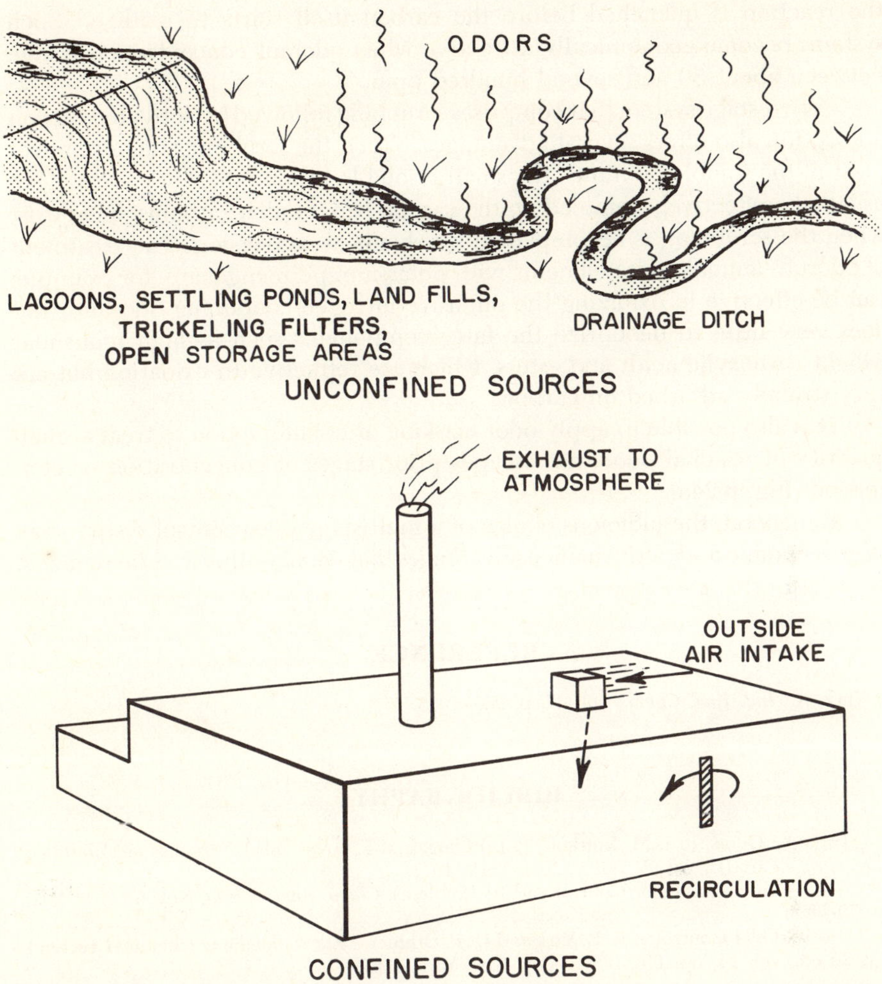

Figure 6.24a Choice and application of odor control methods. Methods for controlling unconfined sources include masking or counteraction by aerosol sprays or evaporative systems. Methods for controlling confined sources include (1) for exhausts: incineration; catalytic combustion; thick-bed activated carbon, with "solvent recovery"; scrubbers—water, hydrocarbon oil, oxidants such as $KMnO_4$, H_2O_2 and Cl_2 or acid-base neutralization, as with caustic soda; ozone injected into the stack; masking or counteraction and (2) for intake and recirculation: thin-bed activated carbon. Methods for controlling unconfined sources include masking and counteraction using aerosol sprays or evaporative systems.

carbon which functions at ambient temperature until it is saturated. The odorous gas stream is then diverted to a fresh bed while the contaminants on the saturated bed are desorbed onto a catalytic grid where they are oxidized. Alternatively, the adsorbate may be oxidized directly on the carbon bed if the reaction is quenched before the carbon itself starts to oxidize.[1] Such systems become economically attractive when odorant concentrations range between about 50 and several hundred ppm.

A two-stage system that comprises scrubbing followed by activated carbon adsorption also offers several advantages. When the scrubbing stage removes most of the soluble odorants, the small insoluble residue that may escape can usually be effectively retained on the carbon bed. Similar circumstances exist when the scrubbing is accompanied by chemical conversion. The treatment of odorous animal waste effluents with potassium permanganate, for example, can be effective in oxidizing the offensive nitrogen-containing odorants, but does very little to deodorize the fatty components such as high molecular weight carboxylic acids and esters, which are refractive to oxidation but are very strongly adsorbed on carbon.

It is also possible to apply odor masking or counteraction to treat a small quantity of residual odor that escapes prior stages of concentration or conversion (Fig. 6.24a).

In general, the judicious choice of a multistage odor control system may offer economic and performance advantages that cannot otherwise be realized.

REFERENCE

1. Turk, A.: *Ind. Eng. Chem.*, vol. 47, p. 966, 1955.

BIBLIOGRAPHY

Turk, A.: Odors, in I. M. Kolthoff, P. J. Elving, and F. Stress (eds.), "*Treatise on Analytical Chemistry,*" Part III, vol. 2, Wiley, New York, 1971.

———: Industrial Odor Control and Its Problems, *Chem. Engineering,* November 3, 1969, pp. 70–78.

———: Odor Control in R. E. Kirk and D. F. Othmer, "Encyclopedia of Chemical Technology, 2d ed., vol. 14, pp. 170–178, Wiley, New York, 1967.

6.25 FURNACE MODIFICATIONS

Furnace Modification Methods:	(*a*) *Operating condition modifications,* (a1) Low excess air combustion, (a2) Two-stage combustion, (a3) Flue gas recirculation, (a4) Reduced air preheat, (a5) Steam or water injection; (*b*) *Design feature modifications,* (b1) Burner configuration, location and spacing, (b2) Tangential firing, (b3) Steam temperature control, (b4) Air and fuel flow pattern, (b5) Fluid bed combustion.
Typical NO_x Concentrations:	100 to 700 ppm by volume in boiler emissions, depending on operating conditions.
NO_x Control Effectiveness:	% reduction in NO_x concentration in emissions, utilizing the different methods: 20 to 30% (a1); 40 to 50% (a2); 90% (a1 plus a2); 30 to 35% (a3); 40 to 80% (a1 plus a3); 20% (a4); 10% (a5); 50 to 60% (b2).

The various sources of emissions of nitrogen oxides and their effects are discussed in Section 3.12. The instrumentation for the detection of nitrogen oxides is presented in Section 4.12. In Sections 6.25 and 6.26, the various methods to control nitrogen oxide emissions are described. Section 6.25 deals with furnace modification methods which prevent the formation of nitrogen oxides, and Section 6.26 deals with combustion flue gas treatment methods which attempt to reduce the quantity of nitrogen oxides once they are formed.

Combustion modification techniques consist of modifying operating and/or design features to obtain optimum control of nitrogen oxides emission.

Operating Condition Modifications

Low Excess Air Combustion

Nitrogen oxides concentrations can be lowered 20 to 30% by reducing excess air rates. The effectiveness of low excess air combustion is proven for gas and oil combustion, whereas no test data are yet available for coal-fired equipment (Fig. 6.25a).

Reduction in combustion air limits the availability of oxygen throughout the combustion equipment and increases the burner air-fuel mixture ratio to some extent. Very low NO emissions with oil fuel have been attained with excess air values as low as 2% (equivalent to 0.5% O_2) in the flue gas. The minimum possible oxygen content is about 0.2% equivalent to approximately 1% excess air. Nevertheless, simple lowering of excess air does not appear as effective nor as damage proof as two-stage combustion. In coal firing, serious imbalances in fuel/air distribution may result from low excess air, and problems of unburned fuel or carbon monoxide emissions may limit the utility of this approach. By this method, a reduction in emissions of one pollutant may result in an increase of another.

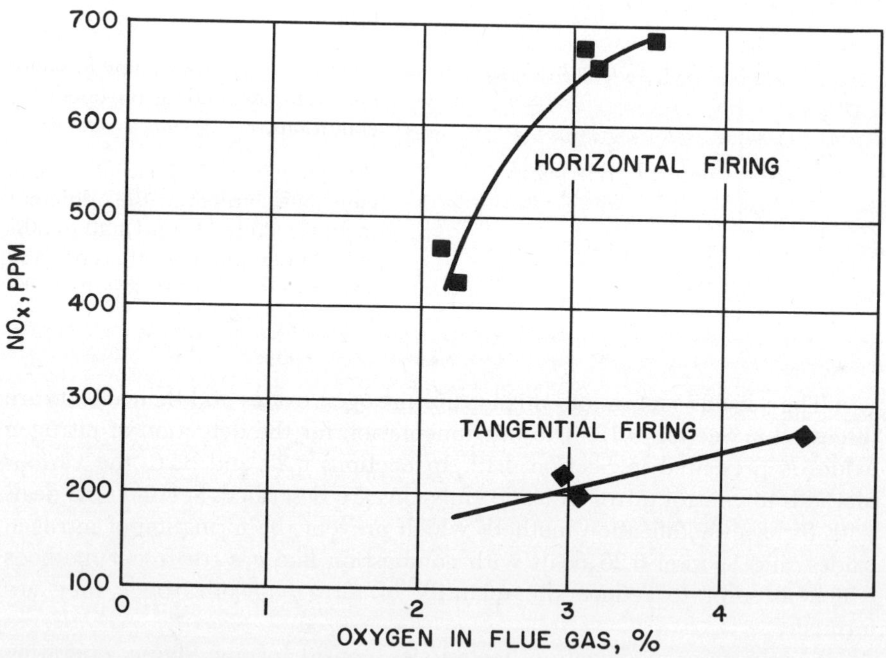

Figure 6.25a. Effect of excess air on NO_x emissions from oil-fired boilers.

Two-stage Combustion

This is one of the most effective control methods for nitrogen oxides. In this method of combustion, only 90 to 100% of the theoretical combustion air requirement is injected into the burner, whereas the remaining air is introduced at a few feet downstream of the burner to complete combustion over a longer zone. With this arrangement, the total excess air is held to the same value as what is used during normal firing. By supplying substoichiometric quantities of primary air to the burners in oil or gas-fired combustion, substantial reductions of 40 to 50% in nitrogen oxides emissions may be achieved. Large steam generators have been observed to operate on two-stage combustion without any measurable loss in combustion efficiency with either gas or oil firing. Complete burnout of the fuel is accomplished by injecting secondary air at lower temperatures, where nitrogen oxide formation is limited by kinetics. This technique is best used in conjunction with low excess air firing and reductions as high as 90% have been achieved in NO_x concentration in large gas-fired power plants, when both techniques were applied. Figure 6.25b shows a typical front-fired boiler using the two-stage combustion technique. The data in Table 6.25c show the effect of two-stage combustion when natural gas and residual fuel oil are fired. On a percentage basis, the reduction attained is higher in case of oil firing than with gas firing, and this is due to the differential in flame temperature between the two fuels. No full-scale performance data are available for coal-fired generators.

Flue Gas Recirculation

The recirculation of cool flue gas and its injection through the burner is found to be quite effective in reducing nitrogen oxides formation. This technique lowers the peak flame temperatures by diluting the primary flame zone with recirculated combustion flue gases. The injection of flue gas into the base of the boiler is found to be ineffective. A direct reduction of maximum combustion temperature is realized with flue gas recirculation in addition to a reduction of oxygen concentration. This technique of nitrogen oxides emission control has been (almost entirely) used in tangentially fired units and is definitely applicable to wall-fired designs.

Figure 6.25d shows typical reductions in nitrogen oxides for increasing percentages of flue gas recirculation at full load with natural gas. Emission concentration levels below 100 ppm are possible with over 25% flue gas recirculation. Similar results have been achieved with fuel oil combustion in the same units. Most existing units are presently limited to approximately 20% recirculation, due to superheat and reheat steam temperature elevation, and would have to be modified to attain a higher percentage of gas recirculation.

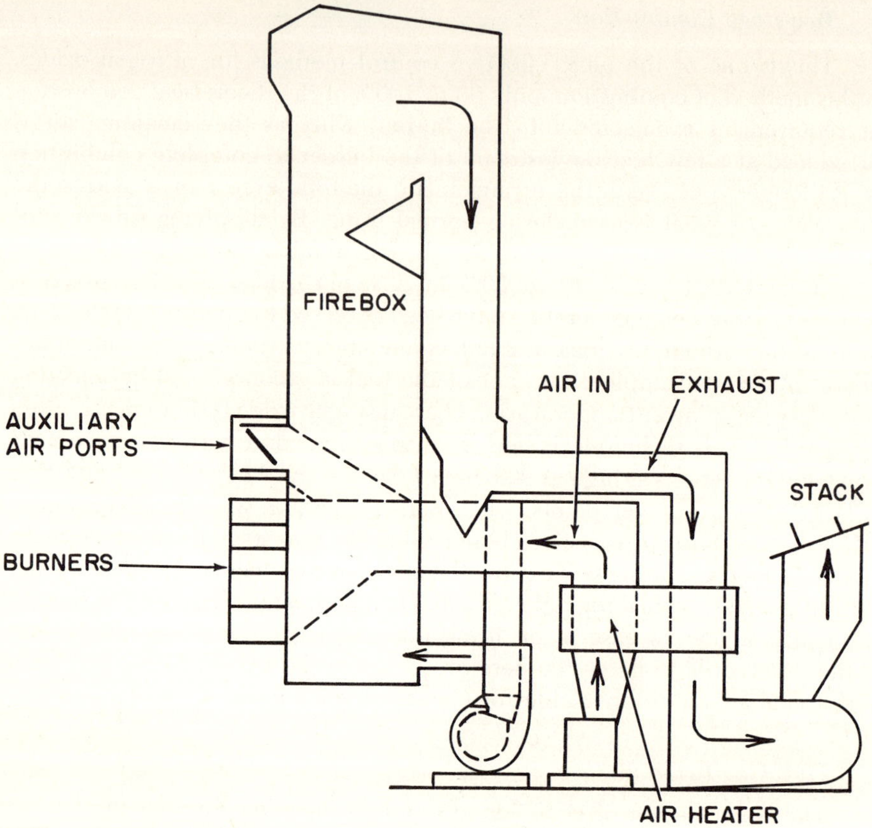

Figure 6.25b Typical front-fired boiler with two-stage combustion.

Table 6.25c
EFFECT OF TWO-STAGE COMBUSTION ON EMISSION OF NITROGEN OXIDES FROM A LARGE STEAM GENERATOR AT FULL LOAD

Fuel	*Nitrogen Oxides Concentration When All Air Enters Through Burners, ppm by Volume*	*Two-stage Combustion*			
		Air Through Burners, % Theoretical	*Total Air, % Excess*	*Nitrogen Oxides, ppm by Volume*	*% Reduction in NO_x Concentration*
Oil and gas combined	525	95	7–10	385	27
Oil only	580	90	7–10	305	47

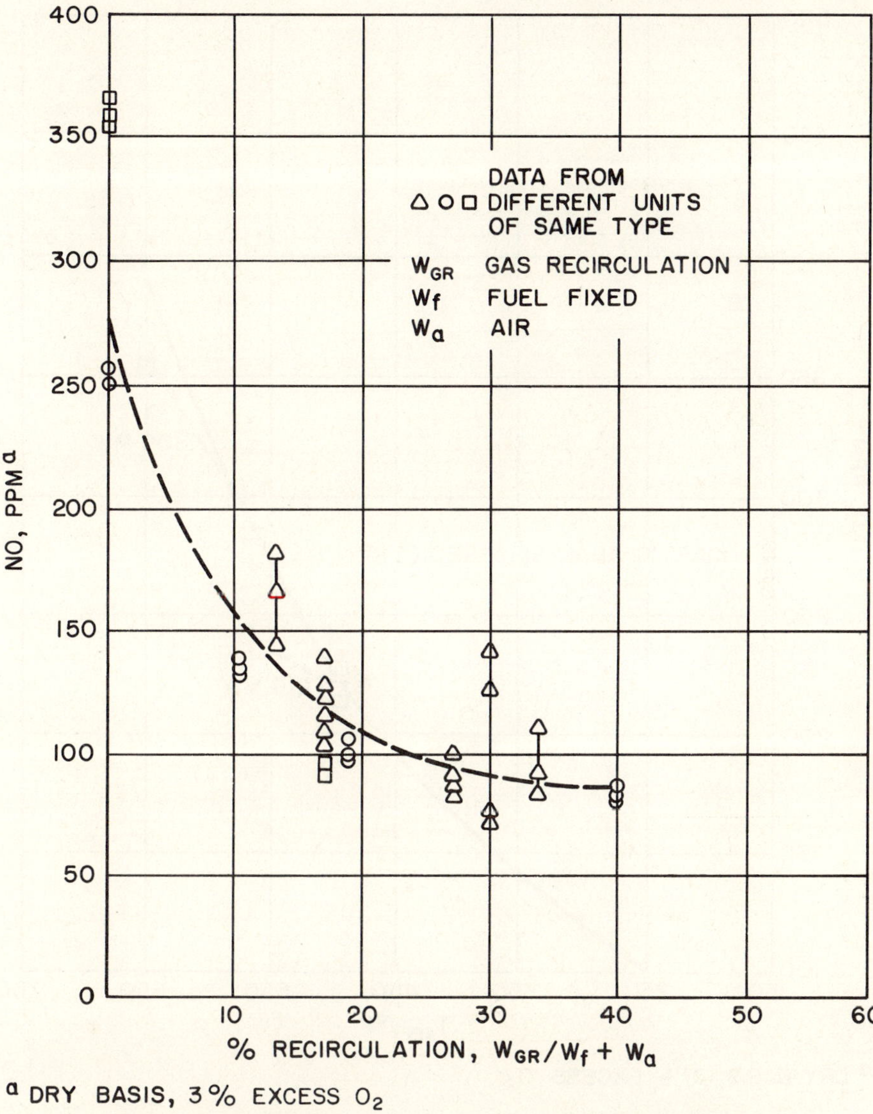

Figure 6.25d Gas recirculation with natural gas firing: 320MW corner-fired unit.

Reduced Air Preheat

Reducing the air preheat is a direct means of lowering primary combustion zone peak temperatures and nitrogen oxides formation. Limited data for natural gas tangential firing indicates a drop in nitrogen oxides concentration in the order of 20% with a 100% reduction in air preheat (Fig. 6.25e).

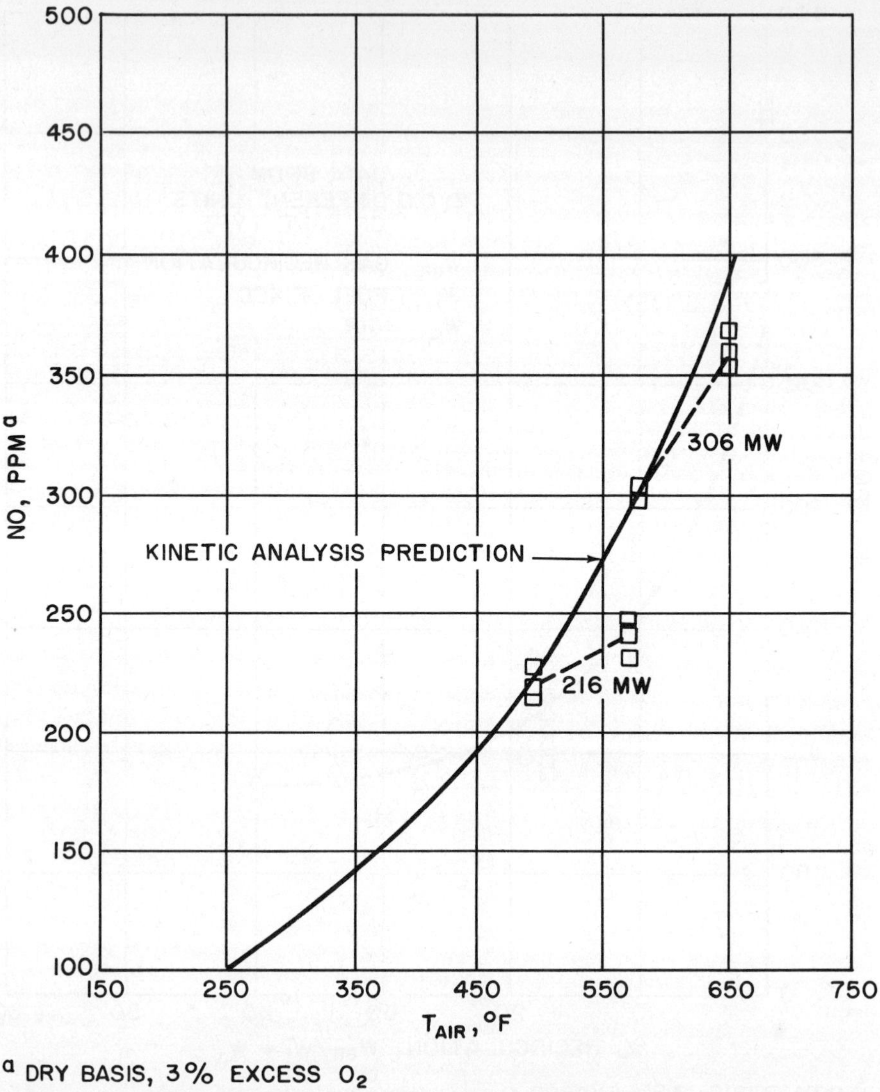

Figure 6.25e Reduced air preheat with natural gas firing: 320 MW corner-fired unit.

Operational limitations with this technique have not been established, but there is a basic disadvantage in that the thermal efficiency of the unit is reduced. This is undesirable not only from an operational viewpoint, but also because more fuel is consumed to generate the same electrical power.

Steam or Water Injection

Injection of low temperature water or steam is similar in concept to flue gas recirculation; thermal dilution of the flame can be achieved by this method, resulting in reduced emissions. This technique appears to have limited utility for furnace control because of operating debits. However, it has been shown in laboratory-scale studies of internal combustion engines that this technique is very effective for control of nitrogen oxide emissions, apparently without a loss of efficiency.

Combination of Techniques

The techniques of off-stoichiometric combustion, flue gas recirculation and reduced air preheat can be effectively utilized in combination. Off-stoichiometric combustion techniques, namely low excess air and two-stage combustion, reduce the quantities of combustion gases reacting at maximum temperatures while flue gas recirculation and reduced air preheat directly influence the maximum level of combustion temperatures.

Limited tests combining off-stoichiometric combustion with gas recirculation on a 320 MW corner-fired unit have shown that nitrogen oxide emissions are reduced from 330 to 110 ppm, when using only 75% of the theoretical combustion air requirement; and when combined with 30 to 35% flue gas recirculation, the oxides emissions were furthur reduced by 50%. It has also been found that the combined conditions necessary to achieve such low levels of oxides are not compatible with operational procedures. The reductions obtained by using individual techniques cannot be added when utilizing combined techniques. Table 6.25f summarizes estimated nitrogen oxides reductions achieved by combustion modification techniques.

Design Feature Modifications

The effectiveness of the different techniques in reducing emissions of nitrogen oxides, and their ease of application varies from boiler to boiler, as the specific boiler designs and operating characteristics do play an important role in determining the final emissions of nitrogen oxides.

Burner Configuration, Location and Spacing

The concentrations of nitrogen oxides have been found to vary with different burner types, spacings and their locations. The cyclone burners are known for their highly turbulent operation, and this results in high-level emissions of nitrogen oxides in coal-fired generating plants.

The amount of heat release in the burner zone, calculated by considering the total energy input into the furnace and the amount of energy lost to the surface area of the water walls near the burners, seems to have a direct effect on oxides concentration (Fig. 6.25g). The reduction techniques of off-stoich-

Table 6.25f
ESTIMATED PERCENT REDUCTIONS OF NITROGEN OXIDES BY COMBUSTION MODIFICATIONS
NO_x Control Method

Boiler Size and Fuel	*Low Excess Air*	*Two-Stage Combustion*	*Low Excess Air and Two-stage Combustion*	*Flue Gas Recirculation*	*Low Excess Air and Flue Gas Recirculation*	*Steam or Water Injection*
1,000 MW						
Gas	33	50	90	33	80	10
Oil	33	40	73	33	70	10
Coal	25	35	60	33	55	10
750 MW						
Gas	33	50	85	33	80	10
Oil	33	40	73	33	70	10
Coal	25	35	60	33	55	10
500 MW						
Gas	33	50	70	33	70	10
Oil	33	40	65	33	65	10
Coal	25	35	55	33	55	10
250 MW						
Gas	33	45	60	30	60	10
Oil	30	35	55	30	55	10
Coal	25	30	50	30	50	10
100 MW						
Gas	33	40	53	30	53	10
Oil	30	30	45	30	45	10
Coal	25	25	40	30	40	10

iometric combustion have been found to be most effective when applied to larger generating units with larger burners. This is true as essentially all the nitrogen oxides are formed in the primary combination zone; and as the burner size increases, the primary zone tends to become less efficient.

The difference in burner spacing has essentially no effect on nitrogen oxides concentration in the boiler emissions, when the boilers are at full load operation. However, at reduced loads, a close burner spacing would result in higher nitrogen oxide release as the close burner spacing inhibits effective bulk gas recirculation into the primary combustion zone.

The distribution of airflow, through the primary air ducts located imme-

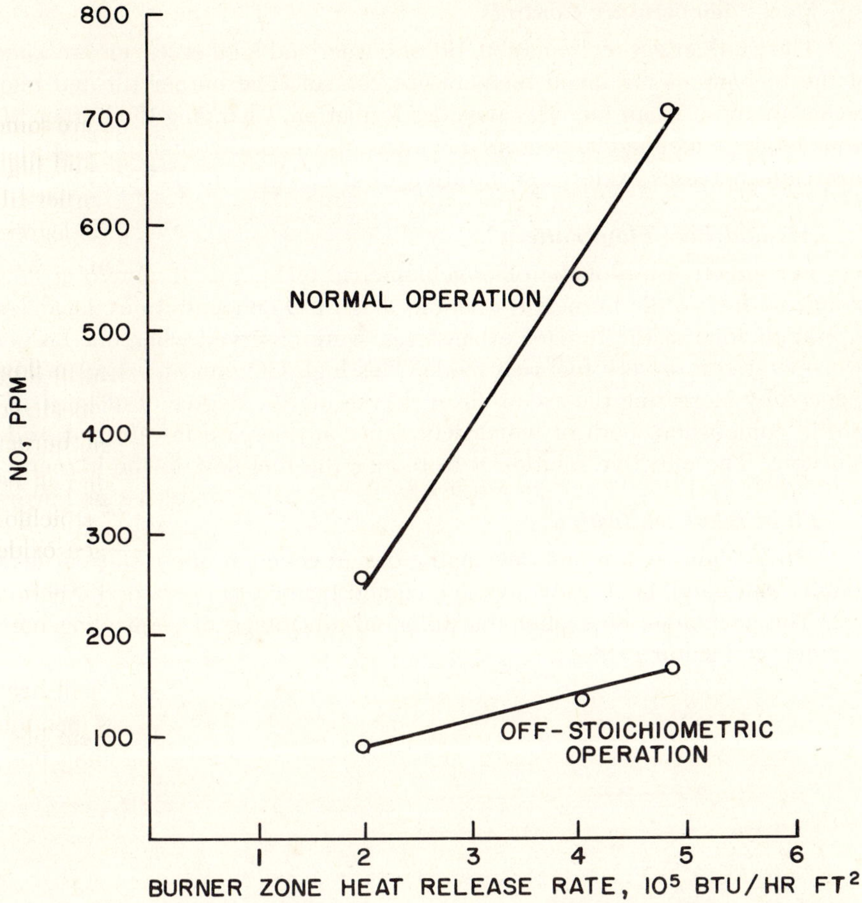

Figure 6.25g Nitric oxide increases with increasing burner zone heat release rate: 480 MW unit, natural gas fuel.

diately above and below the fuel nozzles and through secondary air ducts located immediately above and below the primary ducts, can be used as an effective means to reduce nitrogen oxides emissions.

Tangential Firing

In tangential firing, the furnace itself is used as the burner, thus resulting in lower peak flame temperatures and in simultaneous reductions (as much as 50 to 60% compared with conventional methods of firing) in nitrogen oxides emissions.

Steam Temperature Control

The product gas recirculation, tilt of burner and high excess air are some of the techniques for steam temperature control. The burner tilt and high excess air result in an increase in oxides formation. Changing the burner tilt from an angle inclined 30 degrees downstream from horizontal to 10 degrees upstream increases oxides concentration from 225 ppm to 335 ppm.

Air and Fuel Flow Pattern

For effective use of the off-stoichiometric technique, the uniform flow of air and fuel to the burners is essential. High CO concentrations, localized in a small zone of the furnace exhaust gas, were observed when one burner was operated in a very fuel-rich mode. This high CO concentration can be reduced by increasing the excess air or decreasing the degree of off-stoichiometric combustion, both of which will cause an increase in nitrogen oxide emissions. The effective solution is to reduce the fuel flow to the burner.

Fluid Bed Combustion

This technique, a useful developing design concept, offers efficient heat transfer rates and, hence, low average combustor bed temperatures (Section 6.5). This technique does offer the potential advantage of controlling both nitrogen and sulfur oxides.

6.26 NITROGEN OXIDE CONTROL—COMBUSTION FLUE GAS TREATMENT

Methods of Treatment: (a) *Catalytic decomposition;* (b) *Catalytic reduction;* (c) *Physical separation;* (d) *Adsorption by solids;* (e) *Absorption by liquids,* e1) Sulfuric acid scrubbing, e2) Lime scrubbing, e3) Magnesium hydroxide scrubbing, e4) Modified lime scrubbing, e5) Urea scrubbing.

The furnace modification (combustion modification) techniques prevent the formation of nitrogen oxides in the first place; whereas the combustion flue gas treatment techniques provide potential alternate methods to control or reduce the quantity of nitrogen oxides, once they are formed.

The concentrations of nitrogen oxides in the flue gases are low (200 to 500 ppm), but the large volumes of flue gases emitted by generating stations make it necessary to further reduce the concentration of these contaminants before their release into the atmosphere. The processing of such large volumes of gas with low nitrogen oxide concentrations presents formidable engineering problems in efficient gas contacting, equipment size and operation. The presence of moisture, carbon dioxide, sulfur oxides and oxygen do interfere with the removal processes.

There are no removal processes or systems which are commercially available to reduce the concentration of oxides of nitrogen in emissions from power plants. The known adsorbents and absorbents have low capacity, and the known catalytic decomposition and reduction methods are inadequate to cope with dilute stack gases. For a chemical reaction method to be efficient, the less reactive NO has to be converted into NO_2 and/or N_2O_3 forms.

The flue gas treatment processes can be categorized as follows:

1. Catalytic decomposition
2. Catalytic reduction
3. Adsorption by solids
4. Absorption by liquids
5. Physical separation

Catalytic Decomposition

Several attempts have been made to decompose NO and NO_2 to the elements of nitrogen and oxygen. The decomposition is extremely slow in the absence of a catalyst. While the thermodynamically favored decomposition would provide a comparatively simple method of controlling emissions, no commercially available catalysts have been found which would be sufficiently active at reasonably low temperatures. The indications are that when and if a catalyst is found, it may have to be used at temperatures above 1000°F, which necessitates the installation ahead of the air heater.

Catalytic Reduction

For the control of nitric acid tail gas emissions, nonselective reduction under net reducing conditions has been used with some success. The requirement for net reducing conditions means that either the combustion be carried out under substoichiometric air conditions or reducing agents such as CO, H_2 or CH_4 be added to the flue gas upstream of the catalytic reactor. Both approaches would result in emissions of these gases and/or in overall reduction of thermal efficiency.

Ammonia is known to be capable of reducing NO selectively in an oxygen-containing atmosphere. This technique, if found effective, has the potential benefit that it might also simultaneously control sulfur oxide emissions. This reaction requires temperature control and a catalyst which is still not available.

Physical Separation

The removal of nitrogen oxides from flue gases based on differences in such physical properties as molecular size, condensation temperature and magnetic susceptibility has been investigated; but, this technique is ineffective as the physical properties of NO are not sufficiently different from the properties of the other components of the flue gas.

Adsorption by Solids

Some investigations on laboratory scale have shown that both activated carbon and silica gel adsorb measurable amounts of NO_2 and that activated carbon offers more promise in the control of emissions from combustion sources. The other solid sorbents which show potential are certain metal oxides, particularly manganese and alkalized ferric oxides. However, severe attrition of the sorbent under flue gas treatment conditions is a major technological problem yet to be solved.

The common adsorbents such as silica gel, alumina, molecular sieves, char and ion exchange resins do show some capacity for oxidizing NO to NO_2 and then for adsorbing the NO_2. These capacities are quite low at the NO concentrations of the typical flue gas.

Absorption by Liquids

The absorption systems using alkaline solutions seem to have the most potential of all the control techniques. For this method to be effective, equimolar concentrations of NO and NO_2 (N_2O_3) are essential in the flue gas. The most feasible and promising possibility is the recycle of NO_2 formed by oxidizing the concentrated NO produced by thermal, catalytic or chemical regeneration of the spent absorbent. None of the potential absorption systems are proven, technically or economically, nor are they commercially available. The potential absorption systems are described in Figures 6.26a through 6.26e.

Sulfuric Acid Scrubbing Process

TYCO's process[3] is a modified lead chamber sulfuric acid process in which NO_2 is recycled to oxidize SO_2 to sulfuric acid mist and to form the N_2O_3

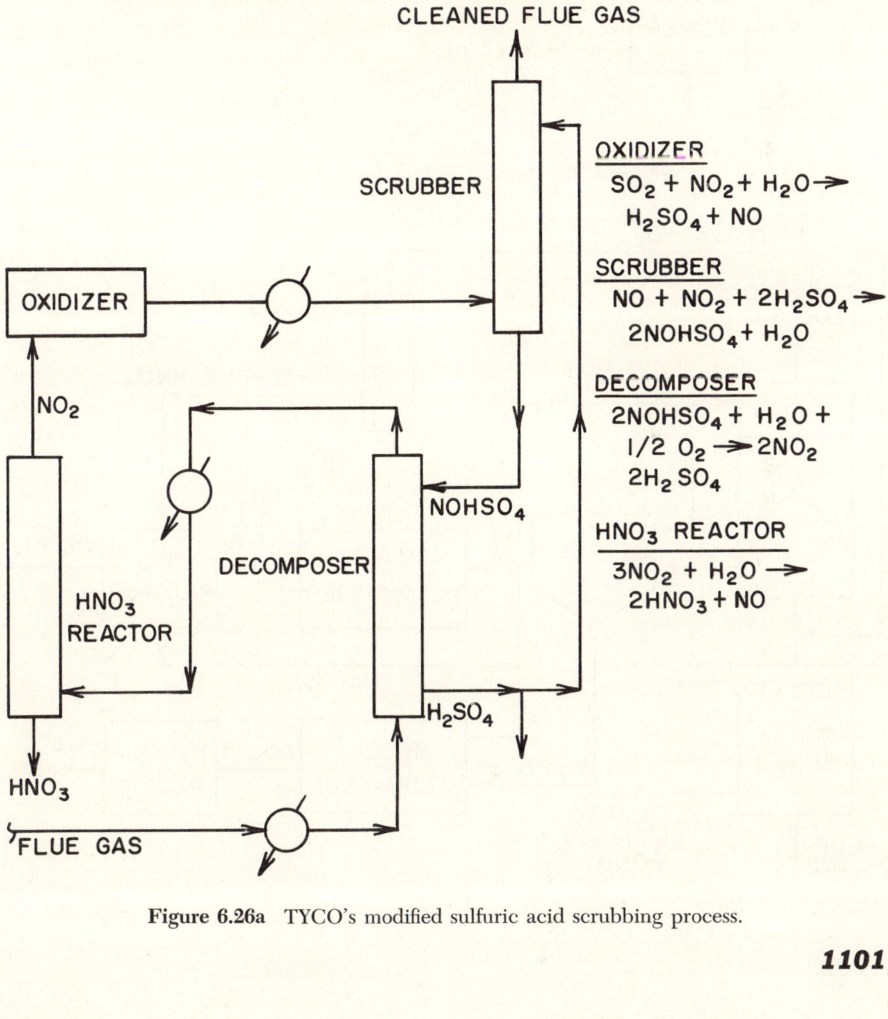

Figure 6.26a TYCO's modified sulfuric acid scrubbing process.

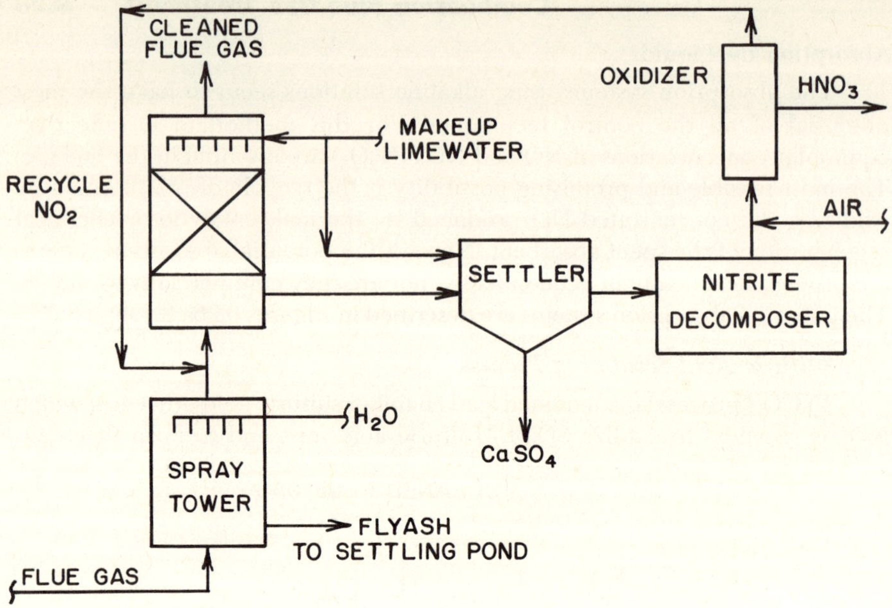

Figure 6.26b Lime scrubbing for NO_x removal.

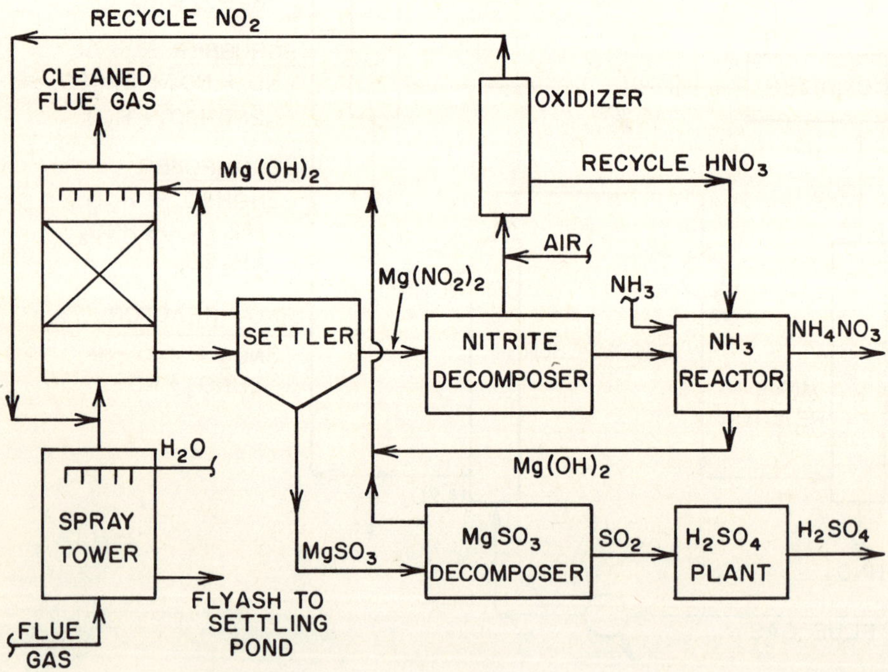

Figure 6.26c Magnesium hydroxide scrubbing of NO_x.

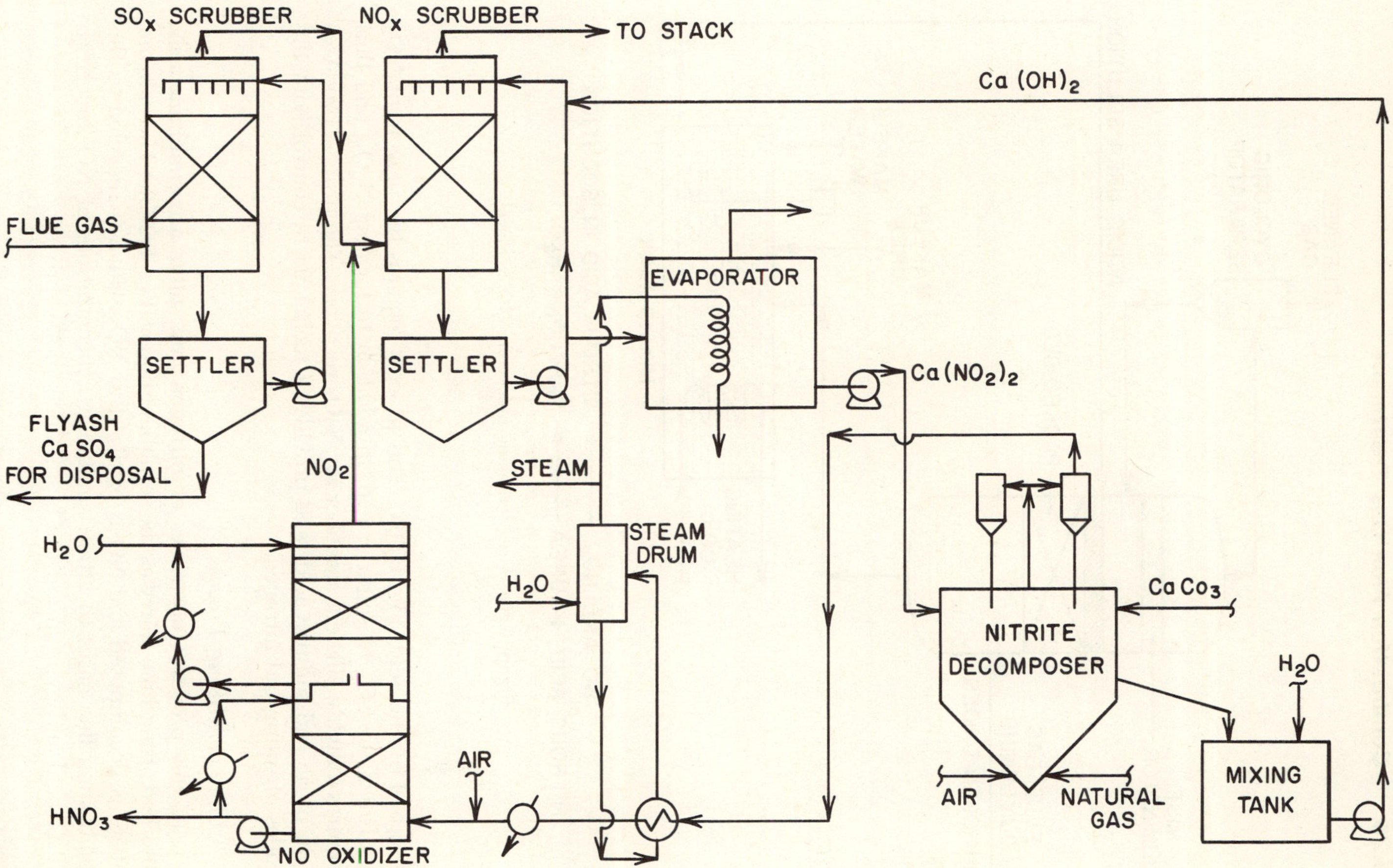

Figure 6.26d Modified lime scrubbing process for NO_x control.

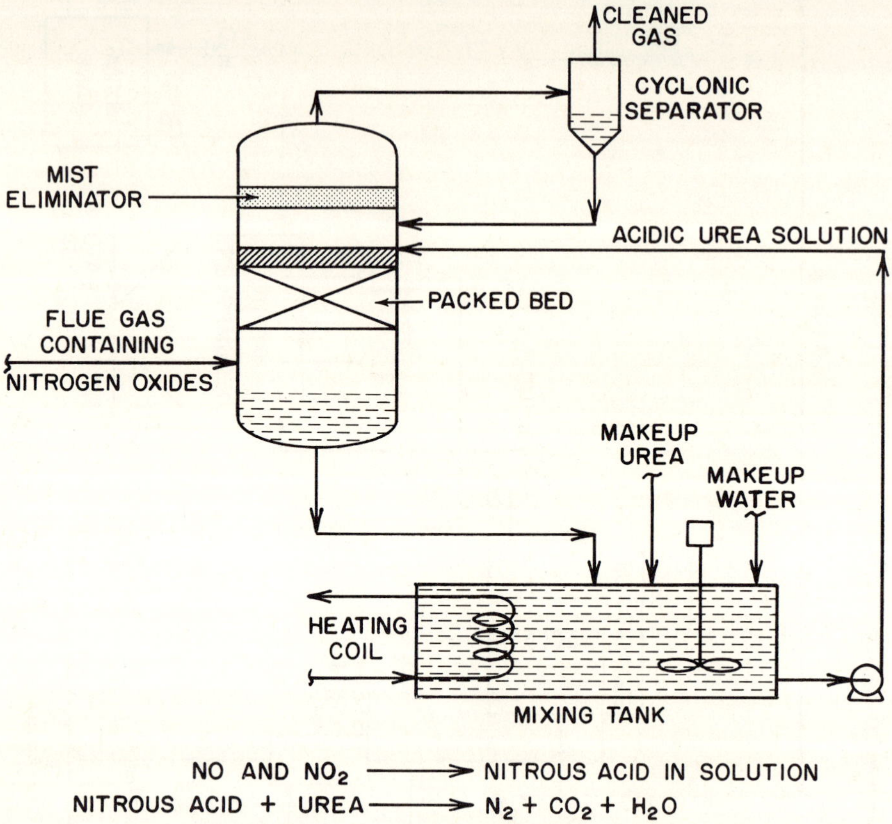

Figure 6.26e Urea scrubbing process for NO_x removal.

absorbed by the acid (Fig. 6.26a). This process requires the recycle of NO_2 to oxidize the SO_2 to sulfuric acid, in addition to forming N_2O_3, and thus requires a highly efficient scrubber for N_2O_3 recovery. The process produces both nitric and sulfuric acids and provides a technique to control both the oxides of sulfur and nitrogen.

Lime Scrubbing Process

In this process, the oxides of nitrogen and sulfur are simultaneously absorbed, using lime slurry as the scrubbing liquor (Fig. 6.26b).[3] The calcium nitrite is decomposed and NO is oxidized to NO_2 which is recycled to the combustion flue gases to attain equimolar concentrations of NO and NO_2. Nitric acid and gypsum are the by-products of this process.

Magnesium Hydroxide Scrubbing Process

In this process,[3] both the oxides of nitrogen and sulfur are simultaneously absorbed, using magnesium hydroxide as the scrubbing liquor (Fig. 6.26c). A settling tank is used to separate the insoluble magnesium sulfite. The overflow containing soluble magnesium nitrite is then evaporated and decomposed. The NO so formed is oxidized to NO_2 which is recycled into the combustion flue gases to attain equimolar concentrations of NO and NO_2. The magnesium sulfite is decomposed to form MgO and SO_2. This process results in production of sulfuric acid and ammonium nitrate.

Modified Lime Scrubbing Process

This process[4] is similar to the lime scrubbing process described in Figure 6.26b except that the sulfur oxides and nitrogen oxides are absorbed separately (Fig. 6.26d).

Urea Scrubbing Process

In this process, the flue gas containing nitrogen oxides is contacted in a countercurrent contactor with acidic urea solution (Fig. 6.26e).[1,2] The nitrogen oxides dissolve to form nitrous acid in solution. The resulting nitrous acid reacts with dissolved urea to form the products of nitrogen, carbon dioxide and water. After efficient separation of any entrained liquor, the cleaned flue gas is discharged to the atmosphere. The elevated temperature of the scrubbing liquor increases the reaction rates between nitrogen oxides and the urea solution. The resulting liquor from the tower is sent to the mixing tank where makeup urea and water are added and then recycled to the contacting section. This technique has been proven only on the pilot scale. The cost of urea is the major operating expense.

REFERENCES

1. Bagwell, F. A., et al.: *Journal of APCA*, vol. 21, no. 11, pp. 702–708, November 1971.
2. "Air Pollution Engineering Manual," 999-AP-40, pp. 554–557.
3. "Systems Study of Nitrogen Oxide Control Methods for Stationary Sources," Final Report, vol. I, Contract No. PH-22-68-55.
4. Hall, J. H., and William Bartok: *Environmental Science and Technology*, vol. 5, no. 4, pp. 320–326, April 1971.

6.27 MERCURY CONTROL

Related Sections:	On the properties of mercury: III.5.19; on the sources and uses of mercury: III.5.20; on the detection of mercury: I.4.12 and II.4.11; on mercury as an air pollutant: II.3.11; and on mercury removal from water: I.7.28.
Removal Processes:	(a) Cordero Mining absorption process, (b) Pittsburgh Activated Carbon Company absorbents, (c) Kokkola process of Hg recovery from zinc smelter, (d) Mercury removal from hydrogen, (e) Allied Chemical Corporation process, (f) Monsanto Enviro-Chem demisters.
Mercury Absorbent Performance:	1 ounce is required for each 1,000 ft^3 of gas saturated with Hg. Absorbents pick up Hg in the ratio of 3% by weight.

The technology available for the removal of mercury vapor from exhaust gases is not extensive. Considerable development work will be required if incinerators, smelters and other ore processing plants and fossil fuel combustion systems are to be equipped with efficient and economic mercury removal devices.

Cordero Mining Absorption Process

The Cordero Mining Company has patented a procedure[1] for removing mercury vapor from airstreams. The process utilizes a chamber containing a porous absorbent coated with a mercury-wettable metal. The absorbent metal is preferably gold, silver or alloys of these elements. The support may be glass wool, nickel wool, granular alumina or porous ceramic pellets. Gold, silver or combinations of these metals are reduced onto the support surfaces to form a very thin film with relatively high-surface areas.

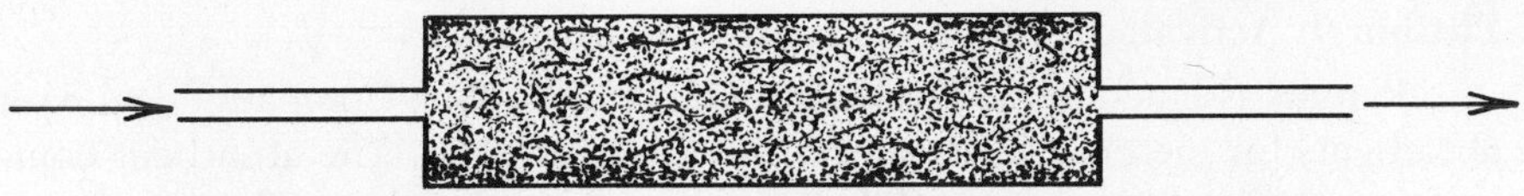

Figure 6.27a Device for absorbing mercury vapor from a gas stream.

Devices which utilize these absorbents are shown in Figures 6.27a and b. The chamber illustrated in Figure 6.27a is filled with the supported absorbent which may be regenerated by heating to drive off and recover the mercury. For economic reasons in relatively large columns, silver is preferred as the absorbent, particularly where H_2S and Cl_2 are absent in the gas stream. Figure 6.27b shows a typical cartridge installation which may be used for the removal of mercury vapor from large airflows. For use in purifying the air, the cartridge of absorbent may be replaced or a regenerable system can be devised.

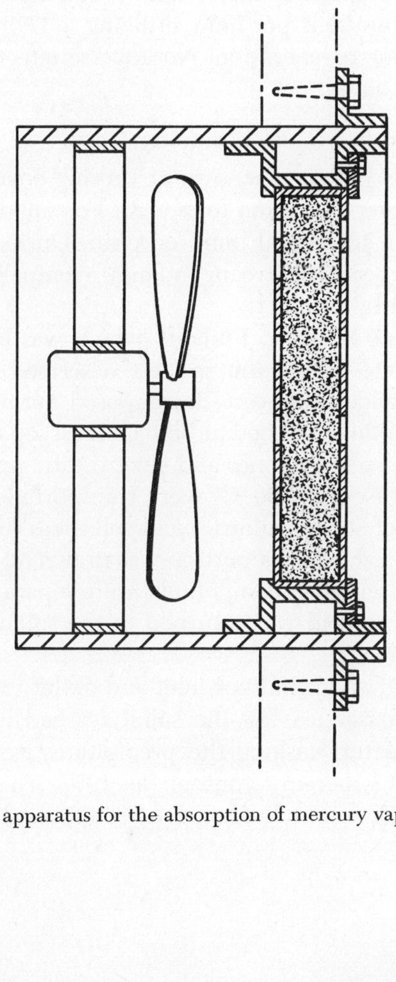

Figure 6.27b Assembled apparatus for the absorption of mercury vapor from air or other gases.

Pittsburgh Activated Carbon Company Absorbents

Patents assigned to the Pittsburgh Activated Carbon Company[2] deal with absorbents for mercury in streams of a variety of gases. The absorbent compositions utilized are specifically silver on activated carbon. The absorbents are completely regenerable at 300 to 350°C when the mercury is recovered. The absorbent contains between 5 and 50% silver on activated carbon sized between 4 and 60 mesh. These absorbents pick up in excess of 3% by weight of mercury and can be utilized in rates of 1 ounce of absorbent to 1,000 cubic feet of gas saturated with mercury vapor.

To be most reactive in mercury absorption, the silver-charcoal preparation requires that the silver crystallite sizes be not greater than 250 Å. This is obtained by reduction of a silver thiosulfate complex following impregnation on the carbon support. Reduction to the metal is obtained in hydrogen at approximately 400°C. The resultant absorbent will remove mercury from hydrogen, carbon dioxide, air, oxygen and nitrogen. It is also regenerable with absorbent life being in excess of one year without failure. This absorbent has been tested with hydrogen systems at space velocities to 11,000 volumes of gas per volume of absorbent per hour utilizing a 1.5 hour absorption cycle followed by a 0.5 hour regeneration. No adverse effects caused by water in the gas stream were noted.

Mercury Recovery from SO_2-Rich Smelter Gases

Most sulfidic ores contain mercury at varying concentrations above the environmental average ($<$0.1 ppm for rocks). For sulfidic zinc ores, this may vary between 10 ppm to several hundred ppm. During ore processing, this mercury enters the process gas stream and may eventually be partially carried into the sulfuric acid by-product.

In a zinc plant at Kokkola, Finland (which was brought on stream by Outokumpu Oy in late 1969), the process described in Figure 6.27c was utilized. Mercuric sulfide ores were decomposed during the roasting of the zinc concentrates in a fluidized bed at 950°C. Dust separation from the gases was obtained by the use of cyclones and electrostatic precipitators. From the dust separators, the gases (at 350°C) were transported to a sulfatizer where countercurrent flow of strong sulfuric acid sulfatizes the mercury. The bulk of this acid was recycled, while a portion was drawn off to a thickener where mercuric sulfate (and selenium compounds) were separated from the solution. The thickener overflow acid was returned to the sulfatizer storage tank for reuse.

Using 90% H_2SO_4 and sulfatizer inlet and outlet temperatures of 40 and 180°C respectively, gases leaving the sulfatizer had less than 0.2 mg/Nm^3 elemental mercury. After washing the precipitates from the thickener, the residue had about 50% mercury content (Se 10%). This filtered residue was mixed with lime and charged to a furnace at 650°C, and the metallic mercury

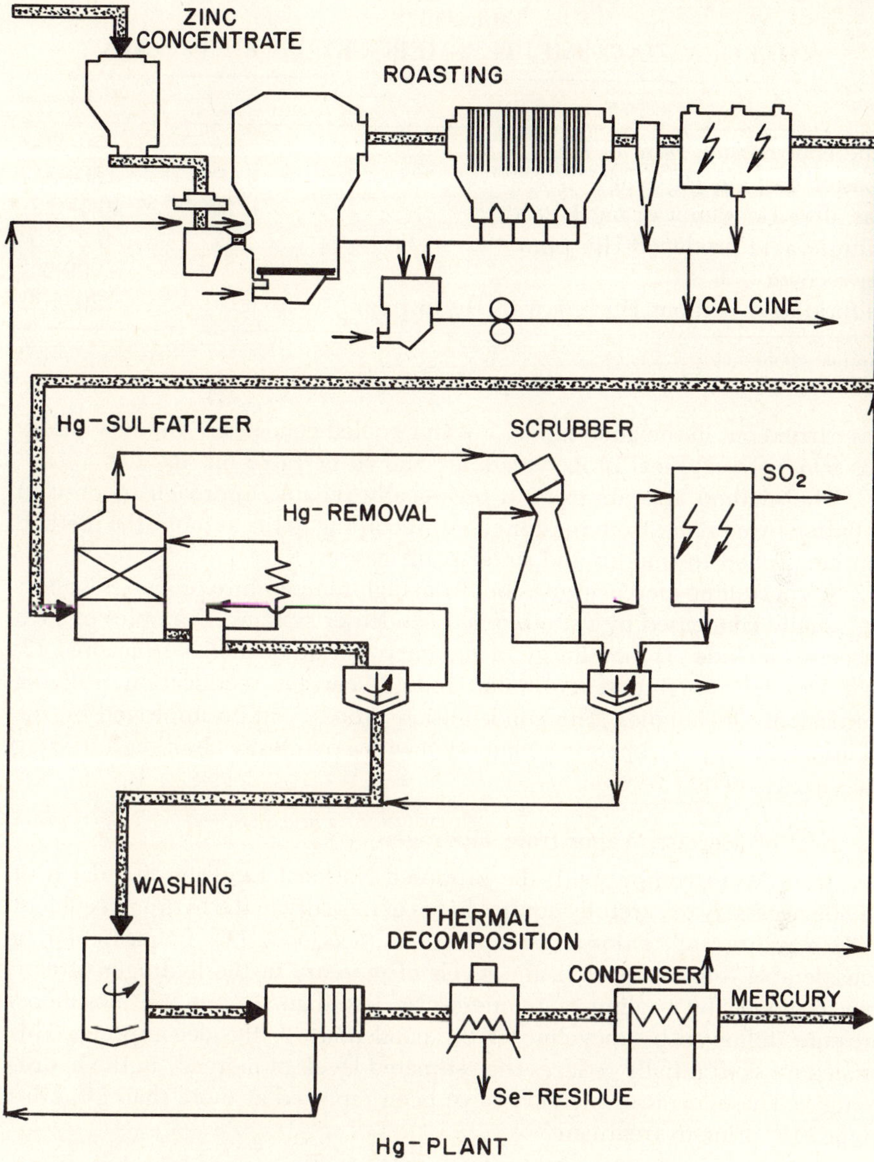

Figure 6.27c Mercury removal process, zinc smelter in Kokkola, Finland.

Table 6.27d
KOKKOLA ZINC SMELTER MERCURY PROCESS DATA

Stream	Concentration
Zinc concentrate, ppm of Hg	50–200
Roaster gas, mg Hg/Nm^3	20–80
Gas after Hg removal, mg Hg/Nm^{3a}	0.1–0.2
Sulfuric acid produced Hg, ppm	0.2–0.5
Hg recovery, %	99.5
Sulfuric acid without Hg removal, Hg in ppm	20–100

[a]Normal cubic meter.

was carried out in the airstream to a water cooled condenser. Mercury purity was 99.99+%. Typical process data are shown in Table 6.27d.

The method appears to be a technically reliable approach to prevent pollution by mercury from roaster gases. According to the authors, the method can be applied to smelter and sinter-plant gases.[3]

In controlling mercury emissions from high-temperature processes, vapors are usually condensed by water cooled condenser systems. Examples of such processes include (1) metallurgy of mercury and other metals from ores, (2) heat transfer using mercury boiler and (3) current rectification utilizing mercury arc techniques. The condensation process can be improved by the addition of water scrubbers in conjunction with a pyrolusite absorber following the final condenser section.

Removal of Mercury Vapor from Hydrogen

In order to comply with the proposed National Emission Standards, it will be necessary to carefully control losses of mercury in the hydrogen effluent from mercury cell, chlor-alkali operations (Section 3.11). In particular, a considerable improvement in the levels of mercury in the hydrogen stream may be made by cooling to temperatures less than 0°C at 2 atmospheres pressure, followed by recycling of the condensate to the decomposer. This procedure substantially reduces the estimated levels of mercury in the hydrogen gas. These levels of mercury have been reported at more than 20 milligrams/M^3 prior to treatment.

Allied Chemical Corporation Process

The Allied Chemical Corporation has developed a patented process whereby mercury vapor levels in hydrogen streams are reduced by passage through beds of absorbent consisting of silver on fused alumina supports.[4]

The process has been used on hydrogen effluents flowing at an average rate of 303 SCFM and 80 PSIA. The gas contained about 20 milligrams/M^3

of mercury before processing. Initially, the stream was cooled from 25°C to 5°C and the condensate was removed by a cyclone. The cooled hydrogen gas was then passed to an absorption tower packed with two absorbents arranged in succession. The initial absorbent section consists of fused alumina chips (4 to 6 mesh) arranged in a bed measuring 18 inches in diameter and 12 inches deep. The mercury absorbent section consists of a silver on alumina bed measuring 18 by 36 inches. The latter absorbent was comprised of 5/16-inch diameter spheres, on which 5 to 12% by weight of metallic silver had been deposited by impregnation followed by reduction to the metallic state. This absorption process reduced the mercury content of the effluent hydrogen gas to $<10\ \mu g/M^3$.

In practice, two parallel absorbent towers were utilized, each operating separately. After approximately 30 days operation, the tower in use was shut down and the hydrogen stream diverted to the second unit. The first tower was regenerated by passing dry nitrogen at 240°C countercurrent to the hydrogen flow for 1.5 hours. The mercury in the nitrogen stream was recovered by cooling, recovering the condensate and passing it through a similar absorbent column.

It was reported that the process is relatively insensitive to variations in space velocity through the beds; ratios from 0.0016 to 16 volumes of hydrogen per volume of bed per second were utilized. Temperatures of operation should be maintained in the range from 0 to 50°C. Davison Catalyst SMR-55 may be employed as the absorbent.

Monsanto Enviro-Chem Demisters

The mercury content of hydrogen off-gases from chlor-alkali operations may be considerably decreased by a demisting process marketed by Monsanto-Enviro-Chem Systems. In this process, the hot gases (at 80 to 120°C) are cooled to 5°C and passed through fiber beds packed between parallel flat screens or concentric cylindrical screens. This procedure removes 90 to 95% of the particulate mercury leaving the vapor concentration of mercury at approximately $3.7\ \mu g/M^3$.

REFERENCES

1. Williston, S. H. and M. H. Morris: U.S. Patent No. 3,232,033, February 1, 1966.
2. Manes, M. and R. J. Grant: U.S. Patent No. 3,193,987, July 13, 1965; and M. Manes, U.S. Patent No. 3,374,608, March 26, 1968.
3. Rastas, J., E. Myholm and J. Kangas: *Engineering/Mining Journal,* p. 123, April 1971, Finnish Patent Application No. 1901/70.
4. Park, J. C. and L. O. Winstrom: U.S. Patent No. 3,257,776, 1966.

6.28 FLUORIDE CONTROL

Limits on Fluoride Emissions:	In phosphate fertilizers: 0.4 lb per ton of production rate.
Fluoride Emissions:	From uncontrolled processes: 75 to 85 lb per ton of aluminum produced, 50 to 90 lb of fluorine per ton of phosphoric acid that is concentrated.
Fluoride Removal Efficiency:	Up to 95% in the steel industry, 80 to 99% in the aluminum industry.

Emphasis on recovery of fluorides from plant effluents has been intensified in recent years, mainly because of the destructive effects of these emissions, even in very low concentrations, on crops and livestock. Thus, governments have imposed codes restricting the quantity of emissions from a single source as well as limiting the fluorine concentrations in foliage (Section 3.8). The recent escalation of restrictions from a maximum of 0.6 lb of fluoride (as F) per ton of P_2O_5 production rate to a new maximum of 0.4 lb for new plants has been accompanied by significant improvements in design and by innovations in fluoride recovery systems.

Sources of Fluoride Emissions

The fluoride emissions have been reported in the following industries: brick, tile and ceramic;[1,2] steel;[3] aluminum;[4-6] and phosphate fertilizer.[7-11]

In the *brick, tile and ceramic industry,* the source of fluorides is the loam or clay from which the fluorides are emitted during the firing process. The fluorides are emitted from the tunnel furnace to a greater extent than they are from the ring oven. In the operation of a frit furnace, fluorides of sodium and calcium are released both as fume and dust.

Emissions from the glass-melting operation consist primarily of particulates and fluorides, if fluoride-containing fluxes are used in the process. Fluoride emissions can be calculated on the basis of 20% of the input fluoride being emitted.[12]

In the *steel industry,* fluorspar, used as a flux for shaping slag during the final stage of the heat cycle, and the iron ore itself are the main sources of fluoride emissions. In sintering plants, the fluoride evolution is reduced to nearly half by the addition of ground limestone to the sintering mix. The ⅛-inch size limestone is added to ore piles as they are stacked and it is then blended in the ore bedding area. Fluorides in the waste flue gases are in both the gaseous and particulate state. By addition of ground limestone to the gas stream, the gaseous fluorides are converted to the particulate state. Hydrated lime is added to the flue gas in case of open hearth process, and ground limestone is added in case of the sintering process.

In the *aluminum industry,* fluorides are emitted as particulates and as gaseous HF from the electrolytic cell where the electrolytic reduction of alumina takes place in the presence of molten cryolite and fluorspar. In Soderberg® pots, the fluoride particulates range in size from approximately 0.05 to 0.75 micron, with the majority being under 0.25 micron. In the particulates, the water soluble portion is mainly hydrated aluminum fluoride; the insoluble portion contains cryolite, calcium fluoride and substantial quantities of alumina.

About 50% of the total fluorides are in gaseous and 50% in particulate form. No actual data on fluoride emissions are available; but from the consumption data on cryolite and other fluoride-containing ingredients, an estimated 75 to 85 pounds of fluoride, as fluorine, is emitted per ton of aluminum produced from an uncontrolled process.

The gaseous pollutants emitted from *phosphate fertilizer operations* consist primarily of hydrogen fluoride and silicon tetrafluoride. Small quantities of ammonium silicofluoride and ammonium bifluoride are emitted in diammonium phosphate production. Solid wastes containing fluorides are also emitted to the atmosphere as a suspension of particulates. The various process sources, the types of contaminants and their concentrations are summarized in Table 6.28a.

Silicon fluoride (SiF_4) evolution occurs in various operations in fertilizer manufacturing. The recorded values of SiF_4 emitted in various process operations are summarized in Table 6.28b.

Control Methods and Equipment

Most of the fluorides both in the gaseous and particulate form are water soluble and easy to collect. Table 6.28c gives a summary of types of equipment used to combat air pollution resulting from emissions of fluorides. Dry collectors, electrostatic precipitators and wet scrubbers are used for the collection of particulate matter, whereas different types of wet scrubbers, namely, spray and packed towers, cyclonic and venturi scrubbers (Chapter 5), are used for gaseous fluorides.

In the ceramic, brick and tile industry, essentially no information is available on the recovery and control of the emitted fluorides. In the steel

Table 6.28a
POLLUTANTS FROM PHOSPHATE FERTILIZER OPERATIONS

Process Source	Contaminant	Concentrations
Drying and grinding of phosphate rock	Solid fluoride only	
Calcining of phosphate rock	Gaseous SiF_4 and HF	66% of total fluoride in rock; 500–2,000 mg/ft^3
	Solid fluoride	100 mg/ft^3
Normal superphosphate manufacture	Gaseous SiF_4	11–42% of total fluorine in the rock fed; 2,000–5,000 mg/ft^3
Phosphoric acid manufacture	Gaseous SiF_4 and HF	2–12% of total F in rock; 2–8 lb of F per ton of rock fed when 26–30% P_2O_5 is produced; for air cooled reactor 40–80 mg/ft^3 of F; for vacuum flash cooling, 200–500 mg/ft^3 of F
Phosphoric acid concentration from 26–30% to 53–58% P_2O_5	Gaseous SiF_4	25–35% of F present in the rock; 1st stage, 0.83–2.58%; 2nd stage, 2.96–4.80%
Phosphoric acid neutralization		
diammonium phosphate production	Gaseous HF	20–60 mg/ft^3
triple superphosphate production	Gaseous SiF_4	From reactor-dryer 25 lb of F/ton of P_2O_5, 60–100 mg/ft^3
	Dust	15 g/ft^3
	Gaseous SiF_4	From cooler 15–30 mg/ft^3
	Dust	15 g/ft^3
Storage	Gaseous SiF_4	From storage piles 0.2–0.6 lb of F per day per ton of P_2O_5 3–15 mg/ft^3

industry, a combination method of injecting pulverized limestone and utilizing the final dust collection equipment, including cyclones and an electrostatic precipitator, resulted in 96% reduction of fluorides.

In the *aluminum industry,* particulates are removed by dry collectors or electrostatic precipitators. The exhaust from this primary control equipment

Table 6.28b
FLUORIDE EMISSIONS IN VARIOUS PHOSPHATE FERTILIZER PROCESSES

Process	Concentration, mg/SCF as F
Rock-acidulation (H_2SO_4)	
batch	300–600 maximum
continuous	60–150
Triple superphosphate (H_3PO_4)	
granulation	50–100
dryer	50–100
cooler	10
Phosphoric acid	
digestion	10–40
Superphosphoric acid	200–500
Diammonium phosphate (NH_3 recovery with H_3PO_4)	6–9 lb F per ton produced

Table 6.28c
FLUORIDE REMOVAL EQUIPMENT FOR VARIOUS INDUSTRIES

Brick tile and ceramic industry	Venturi scrubber, spray tower
Steel industry	Cyclone, precipitator, venturi and cyclonic scrubbers
Phosphoric acid production	Venturi scrubber, cyclonic scrubber, spray and packed towers and combinations thereof
Phosphoric acid concentration	Surface condenser, venturi, cyclonic and packed scrubbers
Drying and calcination of rocks	Dry cyclones, spray tower and packed tower
Diammonium phosphate production	Venturi and cyclonic scrubbers
Triple superphosphate	Venturi and cyclonic scrubbers, turbulent contact absorbers and packed scrubbers
Aluminum industry	Dry cyclone, precipitator, spray towers, tray and sieve towers, cyclonic and venturi scrubbers and turbulent contact absorber

is discharged into wet scrubbers of tray, spray and sieve tower, venturi and cyclonic designs. In some cases, filament mats or screens wetted by sprays are used to attain 80 to 85% removal efficiency of fluorides. The main objection to this system is that it requires continuous repair and maintenance of the sprays and manual washing of the screens to remove accumulated sludge. A sieve-plate gas absorber column is reported[13] to obtain 99% absorption of the fluorides. A floating bed type of scrubber has been reported[6] to overcome the tar fouling problem and to have an efficiency of 95% removal of fluorides and up to 90% removal of solids. In one recent installation, cyclonic wet scrubbers were capable of reducing HF concentration to approximately 0.03 mg/SCF, using a weak caustic solution for scrubbing.

In the *phosphate fertilizer industry,* the main air pollutants are HF and SiF_4. Both are relatively soluble gases; but due to some unique difficulties, effective recovery has only been achieved recently.

Some of these unique difficulties are (1) low pressure drop requirement (as 1-inch pressure drop will cost $3,000/year at $0.01/KWH, for a 500 tons/day plant producing triple superphosphate); (2) need to neutralize or reuse scrubbing liquid because the presence of pollutant is undesirable in the discharged water; (3) high unit costs for materials of construction due to high corrosiveness of fluorine-containing gas or solution. (Suitable materials are rubber-lined steel, fiberglass-reinforced polyesters or epoxies, the Hastelloys,® some monels and, in some cases, 316 ELC* stainless steel; (4) gelatinous form of precipitated silica, formed as a result of absorption.

Recovery of fluorine compounds from effluent gases is normally achieved by scrubbing with water. Scrubbing systems used in phosphoric acid plants include venturis, wet cyclones, sprays, packed sections and combinations of two or more of these. Packed sections are not used for scrubbing digester gases because of the requirement for high solids-handling capacity. Recently a combination system, consisting of spray, cyclonic or venturi scrubber for initial contact with the digester gases to remove solids and followed by a packed section, is being used.

During the concentration of phosphoric acid, from 50 to 90 lb of fluorine per ton of P_2O_5 is released to atmosphere. A surface condenser is used effectively for the recovery of high-concentration fluosilicic acid in the vacuum evaporation process, whereas venturi, cyclonic or packed scrubbers are used for submerged combustion systems. An alternative approach is to add sodium carbonate to precipitate sodium fluosilicate and provide 80% recovery of the fluorine in the acid prior to concentration.

The flue gas resulting from thermal defluorination of phosphate rock contains 2 to 4% HF by volume and 3 g/ft^3 of dust. A scrubbing efficiency of 99.5%+ is required to meet present emission code requirements. Use of

* Extra-low-carbon.

dry cyclones and combination of spray and packed sections can provide this high removal efficiency.

In diammonium phosphate production, the primary pollutant is fluorine in the dust. However, stripping of fluorine during recovery of ammonia by scrubbing with phosphoric acid results in a secondary pollution problem. Venturi and cyclonic scrubbers are used for dust collection with 25 to 32% P_2O_5 acid as the scrubbing liquor. For efficient recovery of stripped fluorine, a packed scrubber or a spray tower is used.

The major gaseous fluorine effluents from granular triple superphosphate production result from the drying and cooling steps. The fluorine content approximates 0.5 lb/ton of P_2O_5, and the scrubbing systems used are combinations of venturi and cyclonic scrubbers with packed towers used as tail gas scrubber. Outlet loadings in the range of 0.04 to 0.15 lb of fluorine per ton of P_2O_5 have been achieved with this combined system.

REFERENCES

1. Bohne, H.: *Staub,* vol. 24, pp. 261–265, July 1964.
2. Wentzel, K. F.: *Staub,* vol. 25, pp. 121–125, March 1965.
3. Purvance, W. T.: *Chem. Eng. Progr.,* vol. 55, pp. 49–53, July 1959.
4. Ott, R. R. and R. E. Hatchard: *J. Air Pollu. Contr. Ass.,* vol. 13, pp. 141–148, March 1966.
5. "Extractive Metallurgy of Aluminum," vol. 2, Interscience, New York, 1963.
6. Kielback, A. W.: *Chem. Eng. Progr. Symposium Ser.,* vol. 57, no. 35, pp. 51–54, 1961.
7. "Phosphoric Acid," vol. I, part II, Marcel Dekker, New York, 1968.
8. National Air Pollution Control Administration Publication No. AP-48.
9. Specht, R. C.: *J. Water Pollu. Contr. Fed.,* vol. 32, no. 9, pp. 964–974, September 1960.
10. Huffstutler, K. K.: Paper presented at 59th AIChE National Meeting, Columbus, Ohio, May 15–18, 1966.
11. Huffstutler, K. K. and W. C. Starnes: Paper presented at meeting of APCA, San Francisco, June 20–24, 1966.
12. Semrou, K. T.: *J. of Air Pollu. Contr. Ass.,* vol. 7, pp. 92–108, August 1957.
13. French Patent no. 1, 244, 078, assigned to Elektrokemisk, September 12, 1960.

Chapter 7

CONTROL OF SPECIFIC AIR POLLUTION SOURCES

R. L. Bennett E. C. Bingham L. W. Canter
J. W. T. Ferretti W. C. Gardiner
J. L. Gilliland B. I. MacDonald J. D. Rushton
I. S. Shah F. G. Smith, Jr. G. P. Sutton

In Chapter 5, the general air pollution abatement techniques for the removal of air contaminants were discussed. Chapter 6 covers the removal of specific air pollutants from odor to SO_2 and from hydrocarbons to NO_x. The topic of this chapter is the control of specific industrial and municipal sources of air pollution. The authors of this last chapter eliminate double coverage by giving references only to the previous chapters where the pieces of equipment or processing steps that are applicable to the control of the specific pollution source under discussion are described.

CONTENTS OF CHAPTER 7*

*For a glossary of terms used, refer to the back portion of the volume and for a list of abbreviations used in this chapter, refer to the listing given after these contents, preceding the first section.

ABBREVIATIONS

BOF	Basic oxygen furnace
BTU	British thermal unit
CFM, cfm	Cubic feet per minute
ID	Induced draft
MBTU	Million British thermal units
mg	Milligram
NO_x	Various oxides of nitrogen (NO and NO_2)
NPK	Nitrogen, phosphorous, potassium product
ppb	Parts per billion
ppm	Parts per million
PSIG, psig	Pounds per square inch gauge
SCF, scf	Standard cubic feet
SCFM, scfm	Standard cubic feet per minute
SO_x	Various oxides of sulfur (SO_2 and SO_3)
stp	Standard temperature and pressure

7.1 ASPHALT PLANTS

Hot-mix asphalt is formed from uniformly mixed aggregates and asphalt cement coating. As shown in Figure 7.1a, the basic components of a batch plant include a rotary dryer, a screening system, a mixer and conveying equipment consisting of bucket and belt elevators.[1] Aggregate is conveyed from storage piles to the rotary dryer by means of conveyors. The aggregates, ranging in size up to 2½ inches, are heated to temperatures of 250 to 350°F in the dryer. Mineral filler, which is commonly crushed rock, limestone, hydrated lime or Portland Cement, is used in some mixes. At least 65% of this filler must pass a 200-mesh (74 μ) sieve. From the dryer, the aggregate is conveyed by a bucket elevator to screens where it is classified by size into elevated storage bins. Specified amounts of the sized aggregates are then dropped to the weigh hopper and, thence, into a mixer along with 3 to 12% by weight of hot asphalt cement. Batch plants may vary considerably in size. Many plants generate a 2 to 3 ton batch and have production rates of 100 to 150 tons per hour.

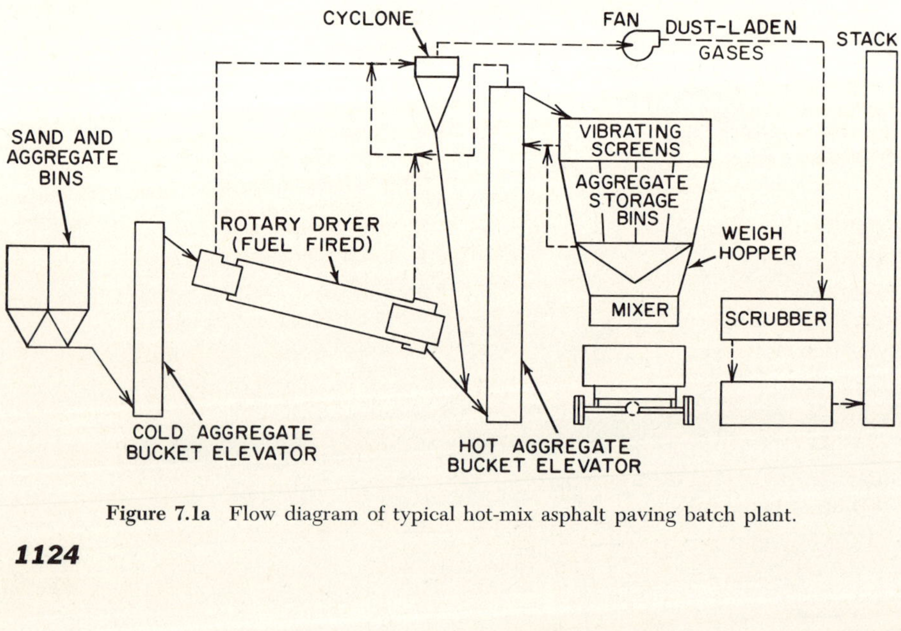

Figure 7.1a Flow diagram of typical hot-mix asphalt paving batch plant.

Sources of Air Pollutants

Gaseous and particulate air pollutants are generated as a result of heating of component materials, grading, mixing and transporting operations.[1] The gaseous emissions originate from combustion of fuel in the dryer, from asphaltic cement in the mixer and from trucks hauling the mix from the plant. Gaseous air pollutants include carbon monoxide, oxides of nitrogen and oxides of sulfur. Carbon monoxide and oxides of nitrogen emissions depend on the completeness of fuel combustion. Sulfur oxide emissions are a function of the sulfur content of the fuel. Gaseous emissions can be minimized by the selection of low-sulfur-bearing fuels, by the use of natural gas and by proper combustion.

Particulate emissions are composed of stone dust, flyash, soot, unburned droplets of fuel oil and carryover of mineral fillers from the rotary dryer. Stone dust refers to dust released from aggregate surfaces during heating as well as from aggregate fracturing during transport. Flyash originates from impurities in fuel oil which create solid rather than gaseous combustion products. Soot refers to unburned carbon resulting from less than optimum combustion. Mineral filler carryover may be as high as 55% of the minus 200 mesh fraction.

Particulates are the major air pollutants released, and stone dust and mineral filler carryover are the main components. Particle sizes range from 1 to 300 μ.[2] The major particulate source is the rotary dryer, with 80 to 90% of the total emissions originating from this unit. The total particulate emissions from a plant may range from 1 to 3% of the process weight rate. "Fugitive" dust losses of 10 to 20% of the total may occur from the cold aggregate conveyance system, hot aggregate elevator, screens, storage bins, weigh hopper and mixer. Also, airborne dusts can be generated from aggregate stockpiles, plant grounds, roadways and mix discharges.

Control of Particulate Air Pollutants

Dust pickup is necessary at all sources of dust discharge; therefore, the dryer and the other, secondary dust sources should be enclosed. The total ventilation requirement is a function of plant size. A 3 ton per batch plant usually requires about 22,000 SCFM total airflow, 80 to 85% of which is used in transporting dryer emissions.

At the present time, centrifugal type dry collectors (cyclones), wet washers (scrubbers) and baghouses (filters) are the collectors used in asphalt plant particulate control. (See Chapter 5 for hardware description.) One combination consists of a "dry-wet" system composed of a cyclone and scrubber. A "dry-dry-wet" system includes a cyclone, multiple-tube cyclone and scrubber. Another possibility is for a "dry" system composed of a cyclone followed by filters.[3]

Most batch plants have cyclones, but a few plants still operate without them. A conventional cyclone will remove 98% of the 30-plus μ particles and 70% or more of the particles down to 10 μ.

Two of the principal types of wet washers (scrubbers) which are used are the multiple centrifugal type spray chamber and the baffled type spray chamber. Of the two, the multiple centrifugal type spray chamber has proven to be the most efficient, yielding efficiencies of greater than 90% for particles of 5 μ and greater. A water use rate of at least 6 to 10 gal/1,000 SCF of airflow is required for optimum particulate collection.[1] The water used in the wet collectors may be fresh or recirculated. Settling pits or concrete tanks of sufficient capacity to allow the collected dust to settle are required for a recirculating water system.

Filter baghouses are increasingly used in asphalt plants due to the necessity for meeting the more stringent particulate emission regulations. Due to the temperature of the emissions (usually around 250°F), Nomex® bags are utilized. Actually, the temperature must be maintained near 250°F in order to prevent mudding, and auxiliary heaters may be required for this purpose. Reverse flow is the technique utilized for cleaning the bags. The collections from the baghouse can be used as mineral fillers and, therefore, there is no problem with dust disposal.

REFERENCES

1. Danielson, J. P.: "Air Pollution Engineering Manual," Public Health Service Publication No. 999-AP-40, pp. 325–334, 1973.
2. "Environmental Pollution Control at Hot-Mix Asphalt Plants," Information Series 27, The National Asphalt Pavement Association, Riverdale, Maryland, 1970.
3. "Guide for Air Pollution Control of Hot Mix Asphalt Plants," Information Series 17, The National Asphalt Pavement Association, Riverdale, Maryland, 1970.

7.2 BOILERS AND HEATERS

Boilers, heaters, steam generators and similar combustion equipment fired with fossil fuels are used in commerce and industry for the transfer of heat. In 1968, an estimated 8,900,000 tons of particulate, 24,000,000 tons of sulfur oxides and 10,000,000 tons of nitrogen oxides (NO_x expressed as NO_2) were emitted from stationary fossil fuel combustion sources in the United States. The majority of the particulate emissions result from the combustion of coal and oil in the form of ash. Sulfur oxide emissions are produced from the oxidation of sulfur in the combustion of fossil fuels. The bulk of the nitrogen oxide emissions are generated by high-temperature fixation of combustion air and fuel constituent nitrogen. (For the Federal emission standards applicable to steam generators, refer to Table 5 of the Preface to this Volume.)

A boiler or heater consists of a burner, firebox, heat exchanger and a means of creating and directing a flow of gases through the unit. Larger combustion equipment also includes flame safety devices, soot blowers, air preheaters and fuel heaters.

Types of Boilers and Heaters

The vast majority of combustion equipment is used to heat or vaporize water. Industrial boilers, water heaters, oil heaters and boilers, refinery heaters and power plant steam generators represent some of the main types.

Industrial boilers and water heaters are of the fire tube, water tube or sectional designs. Fire tube boilers constitute the largest share of small and medium-size industrial units. In fire tube boilers, the products of combustion pass through the heat exchanger tubes, while water, steam or other fluids are outside the tubes. The water tube design is used in both small and large industrial units, and all large boilers are of this type. In water tube boilers, the water, steam, or heat transfer medium is circulated through the tubes, while hot products of combustion pass outside the tubes. Sectional boilers use irregularly shaped heat exchangers. Hot combustion gases are directed through some of these passages, transferring heat through metal walls to water or steam in the other passages.

In hot oil heaters and boilers, a stable heat transfer oil is heated or vaporized. Some of these units are simple water tube boilers in which the heat transfer oil merely replaces water. These boilers, most often found in the chemical process industries, are used to transfer heat to other fluids in heat exchangers.

Refinery heaters include large combustion units and are fired with a wide variety of refinery by-product fuels, notably heavy residual oils and high sulfur-bearing gas streams. In most refinery heaters, an oil or other petroleum product flows inside the heat exchange tubes.

The largest boilers are located at steam power plants where high-pressure, superheated steam is used to drive turboelectric generators.

Air Pollution Sources

The combustion equipment emits both visible and nonvisible air contaminants. The steam generation facilities burning fossil fuels produce 29% of particulate matter, 60% of sulfur oxides and 29% of nitrogen oxides emitted by all sources of these pollutants. Steam generators also emit small quantities of carbon monoxide, hydrocarbons, sulfur trioxide and trace metallic compounds.

Particulate Emissions

The ash content of the fuel is emitted as particulate matter in the flue gas. Typical flue gas particulate concentrations and estimates of annual emissions from steam generators are given in Table 7.2a for the year 1968. Both the uncontrolled and actual emissions are tabulated for coal, oil and natural gas fuels. The particulate emissions from oil and natural gas steam generation units are relatively small when compared with those from coal-fired units. The use of coal-fired boilers in the power industry has doubled during the last 10 years, in terms of capacity.

The maximum flyash particle size is about 20 μ. Measureable quantities of particulates in exhaust flue gas cover a size range of from 1 to 300 μ. The peak of the size distribution is about 10 μ when using a mechanical collector and much lower when passed through an electrostatic precipitator.

Table 7.2a
ESTIMATED PARTICULATE EMISSIONS FROM STEAM GENERATORS FOR THE YEAR 1968

Type of Fuel	*Emissions, lb/MBTU*	*Dust Loading in Exhaust Flue Gas, grains/SCF*	*Dust Loading in Flue Gas at Collector Inlet, grains/SCF*	*Annual Controlled Emissions, 10^6 tons/year*	*Annual Uncontrolled Emissions, 10^6 tons/year*
Coal	0.6–2.0	0.36–1.2	3.6–12.0	8.1	60.0
Fuel Oil	0.05	0.03	0.03	0.04	0.04
Natural Gas	0.015	0.008	0.008	0.03	0.03

Table 7.2b
ESTIMATED SULFUR DIOXIDE EMISSIONS
FROM STEAM GENERATORS FOR THE YEAR 1968

Type of Fuel	*Emissions, lb/MBTU*	*Sulfur Dioxide Content in Exhaust Flue Gas, ppm*	*Annual Uncontrolled Emissions, 10^6 tons/year*
Coal	5.0	2600	17.9
Fuel Oil	3.0	1500	1.8
Natural Gas	Nil	Nil	Nil

Sulfur Oxides

The sulfur content of the fuel is converted to sulfur oxides, mainly sulfur dioxide, in combustion equipment. Sulfur trioxide comprises only 1 to 5% of the total sulfur oxides emission, with the balance being sulfur dioxide. The average sulfur dioxide concentrations and total emission estimates for the year 1968 are given in Table 7.2b. The emissions are based on 3% sulfur content in coal and 2.5% sulfur content in oil.

Nitrogen Oxides

The nitrogen oxides result from the fixation of combustion air nitrogen and oxygen at combustion temperatures. Nitric oxide (NO) and nitrogen dioxide (NO_2) are present in the exhaust flue gases. The term NO_x is generally used to represent NO and NO_2, both of which play important roles in smog formation. Table 7.2c is a summary of estimated nitrogen oxide emissions during the year 1968 in the United States. Nitrogen oxides produced by coal combustion are higher than those from fuel oil or natural gas.

Other Emissions

Carbon monoxide, hydrocarbons and trace metallic compounds are also emitted by steam generation units. The present designs result in complete

Table 7.2c
ESTIMATED NITROGEN OXIDES EMISSIONS
FROM STEAM GENERATORS FOR THE YEAR 1968

Type of Fuel	*Emissions, lb/MBTU*	*NO_x Content in Exhaust Flue Gas, ppm*	*Annual Uncontrolled Emissions, 10^6 tons/year*
Coal	1.4	1000	4.4
Fuel oil	0.7	530	0.7
Natural Gas	0.4	330	0.8

combustion, which usually maintains carbon monoxide emissions below 50 ppm. Hydrocarbon and trace metallic compound emissions are negligible.

Air Pollution Controls

Particulate Emissions

Generally, mechanical collectors and electrostatic precipitators are used for control of particulate matter from exhaust flue gases leaving coal and oil-fired generating units. The most widely used are electrostatic precipitators, wet scrubbers, fabric filters or bag houses and mechanical collectors. (See Chapter 5.)

Electrostatic precipitators are normally used in combination with mechanical collectors because mechanical collectors alone are not efficient enough to satisfy existing and proposed particulate regulations. Fabric filters can achieve the desired collection efficiencies but they have not been used on large fuel-fired equipment because of cost. When emission regulations become sufficiently restrictive to make fabric filters competitive with electrostatic precipitators, fabric filters will be widely used.

The wet scrubbers, which have been extensively used for dust collection in other industries (i.e., steel, pulp and paper), were never used for control of particulates from steam-generating units. Considerable research and development, directed toward utilizing wet scrubbers, has proven the feasibility of their application.

Burning of a low ash fuel and use of a higher BTU fuel can also help in reducing particulate emissions.

Sulfur Dioxide

The two basic approaches to the reduction of SO_2 emissions are the use of low sulfur fuels and the desulfurization of the flue gases. Low sulfur fuel supplies are relatively limited and isolated and are unlikely to be an adequate or universal solution to sulfur dioxide emission problems. Substantial progress has been made during the last 5 years toward developing the technology for the removal of sulfur oxides from the combustion product gases. Progress has also been made in fuel desulfurization prior to combustion. With the use of the various processes, now under demonstration and development, SO_2 concentrations can be reduced by at least 90%. These various processes are described in detail in the starting sections of Chapter 6. The particular process utilized depends upon plant location, proximity to alkali supplies, by-product marketability and transportation, reliability considerations and cost factors.

Nitrogen Oxides

Although a great deal of laboratory development work is in progress, there are no adequately demonstrated nitrogen oxide absorption processes in existence. The combustion modification methods known to reduce NO_x emissions include:

1. Low stoichiometric combustion—selective burner service and two stage combustion.
2. Flue gas recirculation.
3. Reduced combustion air preheat temperature.
4. Low excess air.
5. Reduced load operation.

The methods for removal of NO_x are also described in Chapter 6.

BIBLIOGRAPHY

"Air Pollution Engineering Manual," PHS Publication No. 999-AP-40.
"Combustion Engineering," rev. ed., Combustion Engineering Inc.
"Steam: Its Generation and Use," 37th ed., Babcock and Wilcox Co.
"Control Techniques for Particulate Air Pollutants," NAPCA Publication No. AP-51.
"Emissions from Coal-Fired Power Plants," PHS Publication No. 999-AP-35.
"Atmospheric Emissions from Coal Combustion," NAPCA Publication No. AP-24.
"Emissions from Fuel Oil Combustion," PHS Publication No. 999-AP-2.

7.3 CEMENT MANUFACTURING (Portland)

In the United States, portland cement is manufactured by 50 companies, operating 181 plants in 41 states. Effective January 1, 1972, the United States industry converted the unit of production from the barrel (376 pounds) to the short ton. For 1969, total production of finished portland (including masonry) cement was 77 million tons, utilizing about 80% of the estimated available production capacity. The individual plants vary widely in size from less than 200,000 to over 3,000,000 tons per year.

Portland cement consists primarily of calcium silicates, aluminates and alumino ferrites. The raw materials include limestone, marl, cement rock, shell, chalk, blast furnace slag, sand, iron ore, shale and clay.

The Process

The four main steps in the manufacture of portland cement are shown in Figure 7.3a. Step 1 illustrates the quarrying, rock crushing and raw material storage operations. Step 2 presents the alternative dry and wet processes used for the preparation of the kiln feed. In the dry process, the materials are predried (usually in rotary dryers) or are dried in the grinding circuit. In the wet process, the raw materials are ground with approximately 30 to 40% water. Except for the method of feeding the kiln, the burning process, illustrated in step 3, is similar for both dry and wet process plants. Some kilns include special heat recovery equipment, such as chains, crosses or preheaters. The kilns are direct-fired with coal, oil or gas. In the burning zone, temperatures of 2700 to 2800°F are reached. Gas discharge temperatures range up to 1400°F for dry process kilns and up to 700°F for wet process kilns. The clinker discharged from the kiln falls into an air cooler, which serves to preheat the secondary combustion air, as well as to cool the clinker. In step 4, the clinker and added gypsum are ground to the desired product fineness. Cement is shipped in bulk or in 94-pound sacks.

Principal Atmospheric Emissions

Particulates from the kilns, clinker coolers, dry grinding circuits, material dryers and the various material-handling operations are the principal emissions from cement manufacturing. Health hazards have not been associated with these emissions.

Water vapor, carbon dioxide and nitrogen are the principal gaseous emissions from the kilns. Sulfur oxides may be generated if sulfur is present in the fuel and raw materials. Nitrogen oxides may be generated in the burning zone of the kiln. Significant quantities of both sulfur oxides and nitrogen oxides combine with the alkaline dusts, which are caught in the particulate collector, so that the theoretical quantities are not emitted in the kiln gases.

Applications of Control Equipment

Cyclones. Older installations have used cyclones for particulate controls on kilns and clinker coolers. Cyclones still find some use as precleaners for bag collectors or electrostatic precipitators. (See Section 5.9.)

Scrubbers. Very few cement plants have utilized scrubbers for the control of kilns, due to the hydraulic nature of the kiln dust. (See Section 5.12.)

Bag Collectors. The application of bag collectors in cement plants is very general, with the choice of fabric dependent upon the operating temperature. High-temperature synthetic fabrics are used for dryers and clinker coolers. Glass fabrics are used for clinker coolers and kilns. Bag collectors have had limited application on wet process kilns because of problems with high pressure drop due to fabric blinding. (See Section 5.5.)

Electrostatic Precipitators. While the electrostatic precipitators are principally applied in controlling kilns, they are also used for clinker coolers, raw material dryers and dry grinding circuits. Gases from dry process kilns are usually moisture-conditioned before they enter the precipitator to decrease their electrical resistivity. (See Section 5.22.)

Emission Standards

Particulate emission standards vary among the state and local control agencies. In some cases, exit gas dust loading limits apply, although the process weight type of regulation is more common. It should be noted that 1 ton of dry raw material is required to produce approximately 1,300 pounds of clinker. For wet process plants, the slurry feed rate for 1 ton of dry raw material is approximately 3,300 pounds. These factors must be taken into account when calculating allowable emissions based on process weight. The federal "new source standards" for cement plant particulate emissions refer to the allowable emissions per ton of dry raw feed, for both the kiln and the clinker cooler. Opacity at the kiln stack, the clinker cooler stack and at all other sources is also limited. No standards have yet been set for gaseous emissions from cement plants.*

Table 7.3b provides an orientation to the applications of air pollution control hardware in cement plants.

*See Table 5 in the Preface for the new source performance standard of EPA.

STEPS IN THE MANUFACTURE

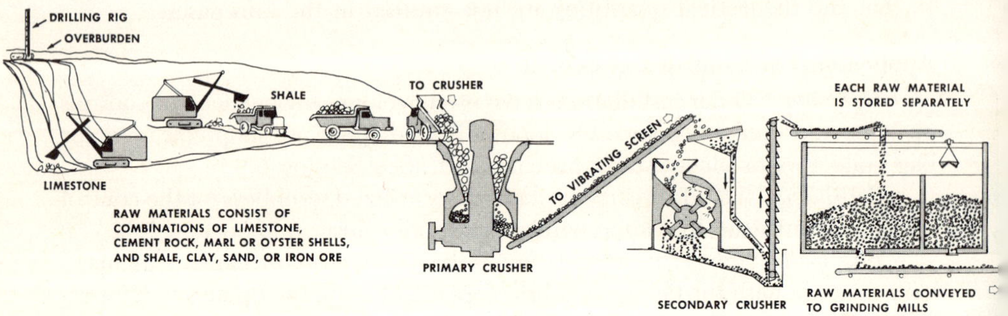

1 Stone is first reduced to 5-in. size, then ¾ in., and stored

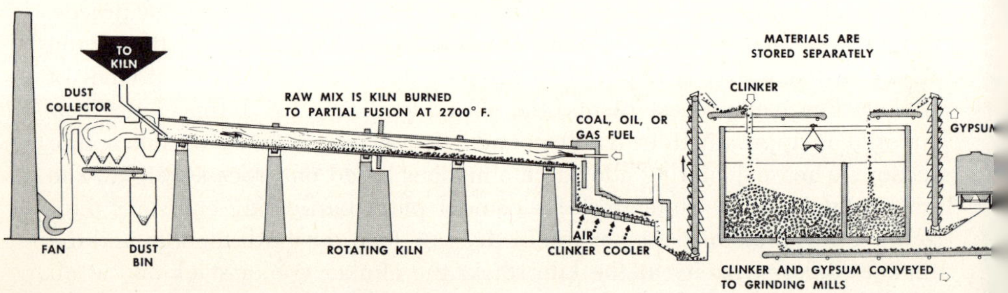

3 Burning changes raw mix chemically into cement clinker

Figure 7.3a The manufacture of portland cement. (*Courtesy of Portland Cement Association*)

OF PORTLAND CEMENT

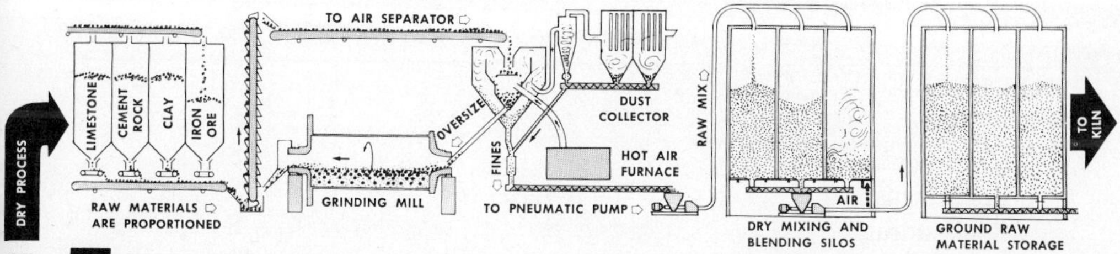

OR

2 Raw materials are ground to powder and blended, or

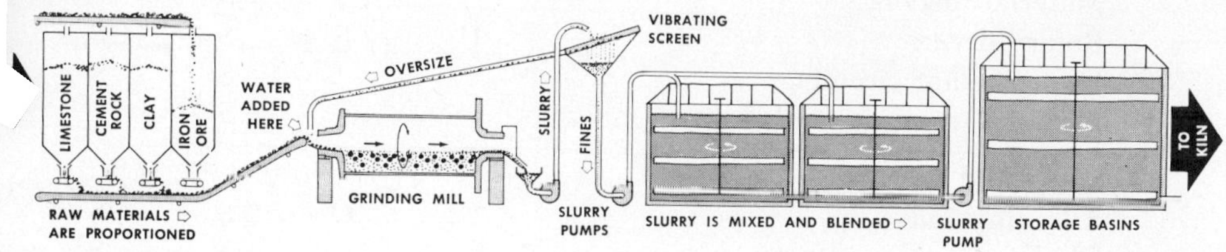

2 Raw materials are ground, mixed with water to form slurry, and blended

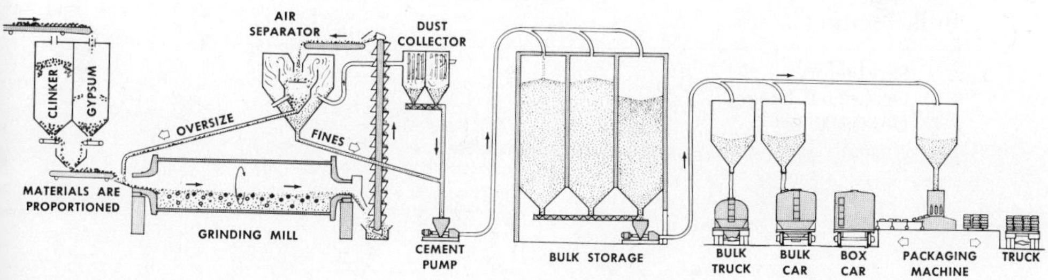

4 Clinker with gypsum added is ground into portland cement and shipped

Table 7.3b
APPLICATIONS OF CONTROL EQUIPMENT

Processing Operation	*Available Control Equipment*				
	Water Sprays	*Cyclones*	*Scrubbers*	*Fabric Filters*	*Precipitators*
Quarry					
Drilling	S			S	
Crushing	S		S	S	
Rock transport	S			S	
Rock storage	S			S	
Raw Feed Preparation					
Material feeders				S	
Rotary dryers		U	S	S	S
Dry grinding circuits				S	S
Material transport and storage				S	
Burning and Cooling					
Kiln emissions		U	O	S	S
Clinker cooler emissions		S[a]	U	S	S
Clinker transport and storage			U	S	
Finish Grinding					
Material feeders			U	S	
Grinding circuit			U	S	S
Material transport and storage			U		
Shipping					
Sacking			U	S	
Bulk loading			U	S	

KEY: S Satisfactory
O Occasional Usage
U Unsatisfactory

[a] Opacity usually satisfactory, except during kiln upset.

7.4 CHLORINE PLANTS

Control methods are discussed for chlorine, hydrogen chloride and hydrochloric acid, salt dust, lead, chlorinated hydrocarbon, asbestos and mercury in the diaphragm cell and mercury cell processes for chlorine.

The Process

Solutions of NaCl or KCl can be electrolyzed by one of two processes to produce chlorine and caustic soda or caustic potash. Only the electrolysis of NaCl need be considered because the methods are the same for both NaCl and KCl. The sodium, magnesium and lithium industries also produce chlorine as a by-product, but they have no additional air contaminants and therefore will not be considered here.

In the United States, 70% of the chlorine is made by the diaphragm cell process (Fig. 7.4a) and 28% by the mercury cell process (Fig. 7.4b).

In the diaphragm cell process, an asbestos diaphragm separates the anode and the cathode. Brine flows slowly through the diaphragm to the cathode where H_2 and NaOH are formed. Cell liquor leaving the cell contains approximately 14% NaCl and 15% NaOH which is evaporated to 50% NaOH and the NaCl is crystallized and recycled.

In the mercury cell process, brine is electrolyzed between the anode and a flowing mercury cathode forming chlorine and dilute sodium amalgam. The sodium amalgam reacts with pure water in a decomposer forming H_2 and 50% NaOH. The denuded mercury is recycled.

Air contamination from the diaphragm cell process can occur due to salt dust, chlorine, hydrochloric acid, lead, chlorinated hydrocarbons and asbestos.

Air contamination from the mercury cell process can occur due to salt dust, chlorine, hydrochloric acid and mercury.

Salt Dust

Salt dust in low concentrations is not unhealthy but it corrodes metal structures and kills plants in the surrounding areas. It can be controlled by protecting operations from wind action. Unloading areas and conveyors can be enclosed and kept under a slight vacuum. Steam heat can also be used to reduce corrosion in enclosed areas. Storage piles can be enclosed in plastic

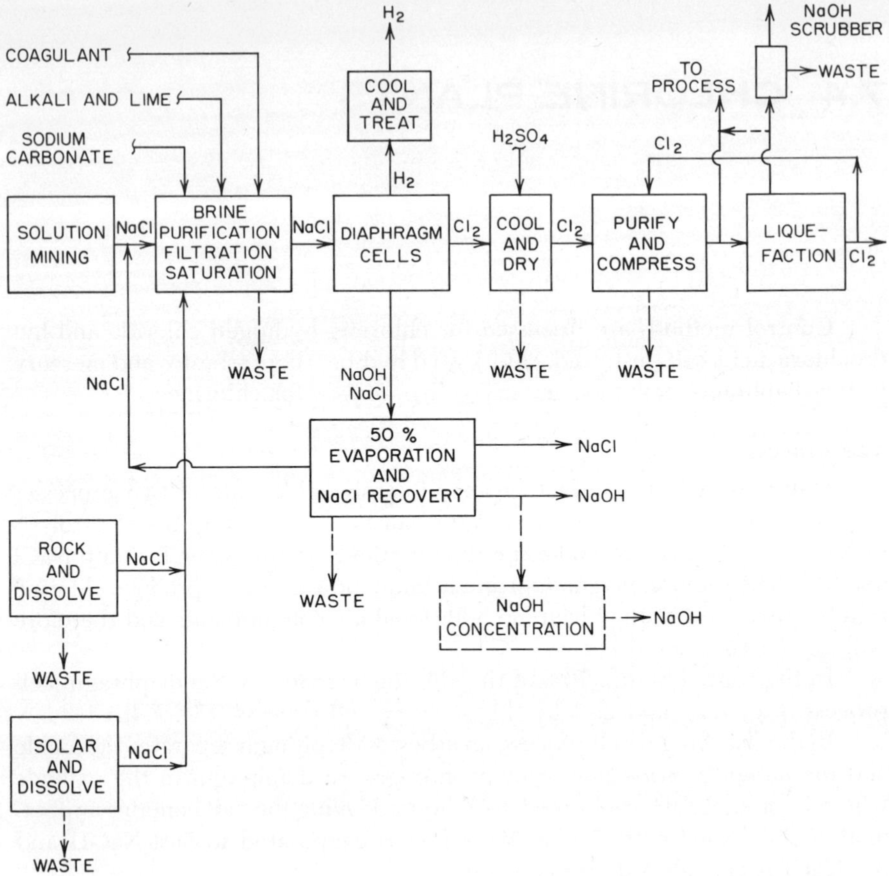

Figure 7.4a Chlorine-caustic soda flow diagram—diaphragm cell type process.

tents or in permanent buildings. Vent systems can be exhausted to a water scrubber.

Chlorine

Chlorine is collected from the cells under a slight vacuum, cooled to condense the water, dried with sulfuric acid, compressed, usually liquefied by refrigeration and shipped in cylinders and tanks. Accidental leaks may occur through the piping, packing and seals due to back-pressure on the cells, equipment or instrument failures and explosions.

A well-designed plant includes a vent system which connects all sources of dilute chlorine and exhausts into an alkali scrubber system. The resulting hypochlorite may be used as such, or decomposed by adding a trace of cobalt, nickel or copper salt and heating. Automatic suction controls should be provided at the compressors, together with back-pressure relief to the scrubbing system. Power failure to the compressors should also shut down the electricity to the cells.

Respirators should be provided for escape and self-contained breathing equipment for men who take remedial action.

Hydrogen chloride is frequently made by burning hydrogen and chlorine together to make hydrochloric acid at chlorine plants. It should be vented to the alkali scrubber system together with the chlorine.

Chlorinated hydrocarbons from chlorine purification are generally collected and incinerated with precautions to avoid emissions of HCl or are buried in drums in a hazardous materials site.

Lead

Lead is used to seal the graphite anodes into the copper contact of many diaphragm cell designs. At the end of the life of the anodes, the lead is melted out. Emissions from the lead melting furnace must be controlled and hoods must be provided at the furnace openings. Pouring temperature should be as low as practical. Preheating the set-up table before pouring the lead permits a lower pouring temperature. Area ventilation of five air changes per hour, or more, should be provided. Lead content of the working area should be monitored and so should the blood and urine of operators.

Asbestos

Asbestos may be applied to the cathode in the form of sheet or cloth but is usually applied as a suspension of fibers in cell liquor. A hazard exists in handling the dry material which can be controlled by providing a hood with adequate suction. It should exhaust through a filter so as not to contaminate the community.

Mercury

Mercury is emitted to the atmosphere from mercury cells in three ways:

1. With the hydrogen (H_2) stream
2. Through the ventilation of end-boxes
3. With the cell room ventilation air

H_2 leaves the decomposer (Fig. 7.4b) at temperatures up to 120°C. It can be treated by one of the following four methods, in order to protect against the escape of mercury. The resulting concentration of Hg in the H_2 vented is noted for each method:

1. By indirect cooling to 30 to 40°C with mist elimination [29 mg/m^3 (stp)].
2. By indirect cooling to 4°C with mist elimination [3 mg/m^3 (stp)].

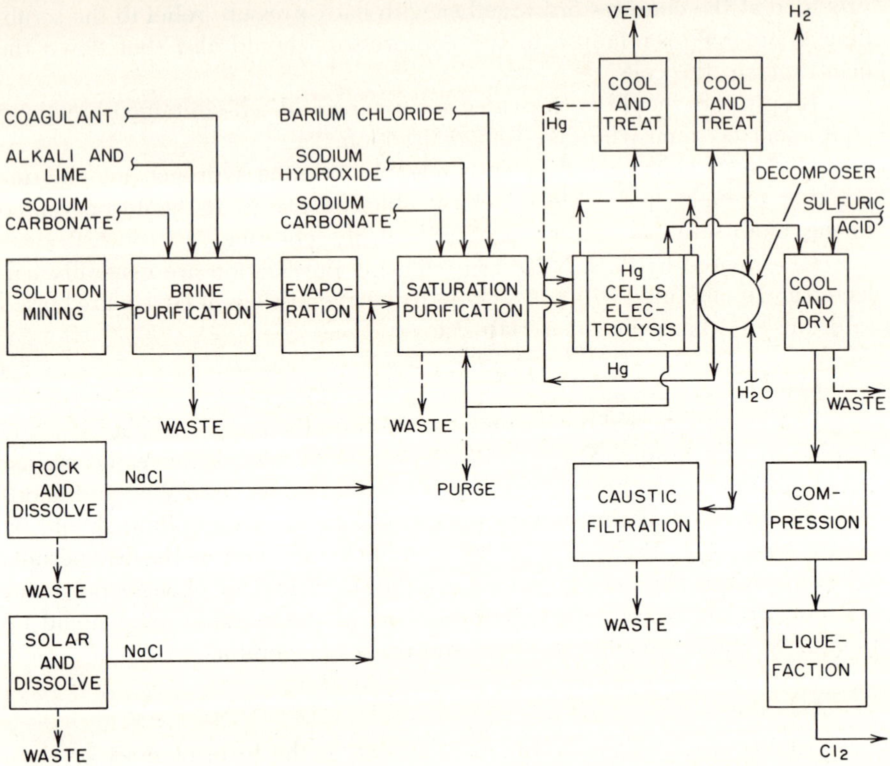

Figure 7.4b Chlorine-caustic soda flow diagram—mercury cell type process.

3. By scrubbing with chlorinated brine or hypochlorite and then with NaOH [1 mg/m^3 (stp)].
4. By contacting impregnated activated carbon or chelating resin [0.01 mg/m^3 (stp)].

Ventilation gas from the end-boxes can be treated the same way as hydrogen. Ventilation air requirements amount to 100,000 to 1,000,000 cfm per 100 daily tons of chlorine capacity. There is no known way to treat this quantity of air, so mercury emissions must presently be controlled at their source by improved equipment design and operating procedures.

BIBLIOGRAPHY

"Control Techniques for Mercury Emissions," Environmental Protection Agency, Bureau of Stationary Source Pollution Control, Division of Compliance, Durham, North Carolina.

"Industrial Waste Study of Inorganic Chemicals, Alkalies and Chlorine," Final Report to Environmental Protection Agency, General Technologies Corp., Reston, Virginia, May 1971.

Sconce, J. S.: "Chlorine—Its Manufacture and Uses," Reinhold Publishing Co., New York, 1962.

7.5 COMMERCIAL FACILITIES

Commercial facilities can be loosely defined as those business establishments where the public is admitted and/or present. This would include, but not be limited to:

Office buildings	Theaters
Multiple residences	Stores
Restaurants	Hospitals

All of the above have various types of heating plants which are discussed elsewhere. In all commercial facilities, there is the subject of human consumption of air which would be a matter for air purification rather than air pollution control. In this section, however, commercial facilities as air polluters are discussed.

Hospitals

Sources of possible air pollution include the pathological and other incinerators (Sections 7.2 and 7.10).

In hospitals with asthmatic and other respiratory patients, a specific area is set aside to provide the extreme cleansing of air required. A complete system designed for such use includes a consumer-size electrostatic precipitator, a wet scrubber and an activated carbon absorption unit, all integrally supported by temperature and humidity control systems.

Apartment and Office Buildings

Air pollution caused by heating and incinerator units has been discussed in Sections 7.2 and 7.10.

The quality of building air supplies is another area that requires attention. It used to be axiomatic that when designing air conditioning or ventilation systems, the quality of the surrounding air can be assumed to be good for human use. This concept has been altered in many densely populated areas of the United States where SO_2, NO and particulate concentrations have increased appreciably in the last decade or two.

The quality of outside air also contributes to material deterioration. The life expectancy, for example, of commercial air conditioning cooling towers has dropped from about 10 years to as low as 3 years, due to the corrosion caused by the abundance of SO_2 and SO_3 in the ambient air of metropolitan areas.

Restaurants

The stoves and fryers of restaurants are a source of fumes, odors and other cooking emissions.

The use of a duct type scrubber installed immediately adjacent to the fume hood for fire protection purposes with an activated carbon bed (self-regenerating) followed by an exhaust fan can be a solution for a rural restaurant.

Urban building codes require exhaust emissions of this nature to be vented a specified height above surrounding structures.

Metropolitan laws usually require the use of fume hoods with continuously welded exhaust stacks. These are vented higher than the height of structures within a 50 to 250-foot radius. It is also required that the stack be scraped clean (by contract) every 90 to 180 days.

These requirements have a considerable economic impact upon the restaurants. The hazard of fire is always present (even if alleviated by CO_2 systems inserted in the duct work) and the maintenance is costly enough for many owners to attempt to circumvent the regulations.

Duct work 7 to 9 floors high is more costly to install and maintain than pollution control systems that can also cool, clean and greaseproof the gases before discharge. If such systems are installed, the restaurant can seek a variance from the code requirement to allow these cleansed and cooled gases to be vented at the street level, eliminating the need for ducts to the roof.

7.6 FERTILIZER MANUFACTURING

The United States fertilizer manufacturing industry is composed of about 100 basic producers of nitrogen, phosphate and potash materials used within the industry and more than 10,000 mixers, granulators and blenders that formulate the basic products into final fertilizer products.

The Process

Basic phosphate producers, principally in Florida and North Carolina, mine and process phosphate rock into wet process phosphoric acid and/or single and triple superphosphate and/or mono and diammonium phosphate. Producers of superphosphate frequently manufacture the sulfuric acid needed. Likewise, producers of ammonium phosphate frequently manufacture the ammonia required.

Basic potash producers merely mine and process the mineral "potash" which is usually in the form of muriate (KCl).

Basic nitrogen producers fix atmospheric nitrogen into ammonia. Some of the ammonia is converted into nitric acid which is combined with other ammonia to form ammonium nitrate. Some of the ammonium nitrate is combined with ammonia to produce an ammoniating solution for mix plants. Certain producers make urea from ammonia and from by-product carbon dioxide generated by the ammonia plant. Urea is processed into solid prills and/or added in solution form to ammonium nitrate to produce nitrogen fertilizer solutions. Some ammonium nitrate is usually processed into solid form.

In granulator or mixed fertilizer plants, the three basic plant nutrients are combined to form an N-P-K (Nitrogen, Phosphorous, Potassium) product, usually in granulated form. Some mix plants manufacture their triple superphosphate by reacting phosphate rock with sulfuric acid which usually is a reprocessed spent acid from another industry. Some plants use phosphoric acid directly in place of a superphosphate. Mix plants ammoniate their product either by use of ammoniating solutions or by the use of ammonia itself. In any event, the three plant food elements are mixed together in such plants, usually in a granulating drum. Heat from the acid-ammonia reaction causes the moistened bed of material to form granules in the rotating drum. The resultant granulated material is dried and screened to size.

In liquid mix plants, which are located in farming areas, liquids such as phosphoric acid and ammonia are combined or high analysis dry fertilizers are dissolved to make a base solution. Other ingredients, such as nitrogen solutions, potash and micronutrients are added to produce the desired formulation.

Using dry nitrogen, phosphate and potash fertilizers, together with micronutrients, the blending plants mix the dry granular fertilizers together in various proportions to make custom formulations.

Pollutants and Pollutant Sources

Air pollutants and their sources in fertilizer manufacturing are listed in Table 7.6a.

Sulfur Dioxide

Some phosphate fertilizer complexes include sulfuric acid manufacturing plants. Most of these plants are of the single absorption type which emit a concentration of 1,500 to 2,500 ppm sulfur dioxide (SO_2). Many plants have installed electrostatic precipitators and mist eliminators and/or water scrubbers to reduce the visible emissions. The more modern plants are also using the double conversion process which reduces the SO_2 concentration in the emissions to 500 ppm.

Nitrogen Oxides

Essentially all basic nitrogen fertilizer producers possess captive nitric acid plants. Most such plants are relatively new and include catalytic type

Table 7.6a
POLLUTANTS AND POLLUTANT SOURCES
IN FERTILIZER MANUFACTURING

Type of Pollutant	*Source*
Sulfur Dioxide	Sulfuric acid plants
Nitrogen Oxides	Nitric acid plants
Ammonia	Ammonia, urea, mono and diammonium phosphate, granulation and ammoniation plants
Fluorine compounds	Wet process phosphoric acid, mono and diammonium phosphate, single and triple superphosphate and granulation plants
Particulates	Phosphate rock grinding, wet process phosphoric acid, ammoniation and all dry fertilizer processing plants, and ammonium nitrate plants

tail gas combustors (afterburners), which reduce the orange-colored nitrogen dioxide NO_2 in the stack to colorless nitric oxide NO. Newer units now being installed are equipped with improved combustors capable of eliminating virtually all the oxides of nitrogen (NO_x). New emission standards promulgated by the Environmental Protection Agency restrict any plants now being put into service from emitting more than 3 pounds of NO_x (expressed as NO_2) for each ton of 100% nitric acid produced. Standards have not yet been established for the existing plants. Nitrogen oxides from all industrial processes, including those of the fertilizer industry, contribute less than 1% of the total nation wide NO_x emissions.

Ammonia

Ammonia is a gas with a very strong odor and is readily captured by acidic scrubbing. Newer ammonia plants record losses of about 0.5 pound of nitrogen (N) per ton of ammonia produced, which is about 5% of the losses experienced in plants built 25 years ago. Ammonia loss can be kept to a minimum in most urea plants. Once-through urea plant off-gas is frequently channeled through an ammonium nitrate neutralizer for recovery of ammonia. Excess ammonia in the neutralizer vent gas can be effectively scrubbed with a weak acidic ammonium nitrate solution that is recycled into the process. Ammonia losses from mono and diammonium phosphate, granulation and ammoniation plants can be kept to very low levels by wet scrubbing.

Fluorine Compounds

Emissions of fluorine compounds as gases and in particulates are, by far, the most difficult and expensive to control of all air pollutants arising from the manufacture of fertilizer materials. Fluorine emissions from wet process phosphoric acid plants, from mono and diammonium phosphate and from granulation plants can be kept to a low level by wet scrubbing of the exit gases. This is a procedure that is generally followed throughout the domestic industry. The fluorine compounds produced from mixing single and triple superphosphate are collected in most instances so that emissions to the atmosphere from these operations are low. However, large amounts of fluorine compounds are discharged from the curing sheds ("dens") because few sheds have adequate fluorine controls. To adequately control the fluorine emissions from curing superphosphates, a large ventilation system is required, with wet scrubbing of all exhausted air and proper treatment and disposal of the resultant scrubber liquor effluent.

Particulates

Major potential sources of particulate emissions in the fertilizer industry are the phosphate rock drying and grinding operations. Electrostatic precipitators and bag collectors currently in use have been *unable* to meet existing removal standards. Extensive development is required to further reduce such emissions.

Particulate emissions from wet process phosphoric acid and ammoniation plants equipped with scrubbers are at an acceptably low level. Dry fertilizer processing and sizing plants generate significant particulate emissions from their stacks and from each piece of materials-handling equipment. Many of these plants are equipped with dust collection systems to control particulate emissions, but poor operation and breakdowns are common and emissions during such periods can be significant.

In the ammonium nitrate manufacturing segment of the fertilizer industry, the recent switch to prilling a high-density prill has created severe particulate emissions, the control of which is yet unsolved. To produce a high-density prill, the ammonium nitrate solution is concentrated to 99.8%. Such concentration requires extensive evaporation of the 83% feedstock. The most efficient evaporators for this service are the air-swept falling film type. Use of such evaporators results in significant emissions. Likewise, significant emissions arise due to the spraying of the concentrated solution in the prill tower for the formation of the prills. Many of the emissions are ordinary dust size particles while others are submicron in size and behave like smoke. Extensive research is being conducted within the industry to effect positive control of this serious particulate emission problem.

BIBLIOGRAPHY

"Atmospheric Emissions from Nitric Acid Manufacturing Processes," Public Health Service Publication No. 999-AP-27, U.S. Department of Health, Education and Welfare, Cincinnati, Ohio, 1966.

"Atmospheric Emissions from Sulfuric Acid Manufacturing Processes," Public Health Service Publication No. 999-AP-13, U.S. Department of Health, Education and Welfare, Cincinnati, Ohio, 1965.

"Atmospheric Emissions from Wet-Process Phosphoric Acid Manufacture," AP-57, National Air Pollution Control Administration, U.S. Department of Health, Education and Welfare, April, 1970.

"Cleaning Our Environment, the Chemical Basis for Action," A Report by the Subcommittee on Environmental Improvement, Committee on Chemistry and Public Affairs, American Chemical Society, Washington, D.C., 1969.

"Fertilizers and Agricultural Chemicals," Subcouncil Report, National Industrial Pollution Control Council, Washington, D.C., October 1971.

7.7 IRON FOUNDRIES

Control of iron foundry emissions has been hampered by their complex nature, by a general lack of experience in controlling them and by the severe mechanical and thermal environments to which control devices are often subjected.

Cupola Furnaces

The cupola furnace is the basic melting unit of most iron foundries and it is the most important single source of emissions. This furnace, which produces gray cast iron from pig iron, iron and steel scrap, coke and limestone, emits mixed gases (typically, 10 to 15% carbon monoxide, 10 to 15% carbon dioxide, small amounts of sulfur dioxide with the balance being nitrogen and water vapor) at temperatures of 1000°F and up. This gas stream also carries volatilized oil and grease and a wide range of particulates, including iron and other metal oxides, limestone dust, coke breeze and flyash.

In normal operation, the flow of cupola gases up the stack draws atmospheric air in through charging openings. This air mixes with the carbon monoxide in the cupola gases causing them to spontaneously ignite and burn. The resulting fire also consumes some of the volatilized hydrocarbons. To assure continuity of this fire and to aid in the mixing of air with the cupola gases, afterburners are placed on the cupola stack which fire into the gas stream.

Particulate control is more difficult, due to the high temperature of the carrier gases and to the wide range of particle sizes emitted (typically, 0.01 to over 40 μ). Modern, successful cupola emission control systems usually employ baghouse filters, high-energy scrubbers or, less frequently, electrostatic precipitators. Well-engineered installations employing these units have reduced cupola emissions to extremely low levels, so selection of one unit over another is usually a matter of economics, operator preference or factors not directly related to emission control. Figures 7.7a and 7.7b show two typical control systems.

Most baghouse, scrubber or precipitator-based installations retain the afterburners, either on the cupola stack or on a separate chamber, to incinerate carbon monoxide and thus reduce the hazard of explosions in control equip-

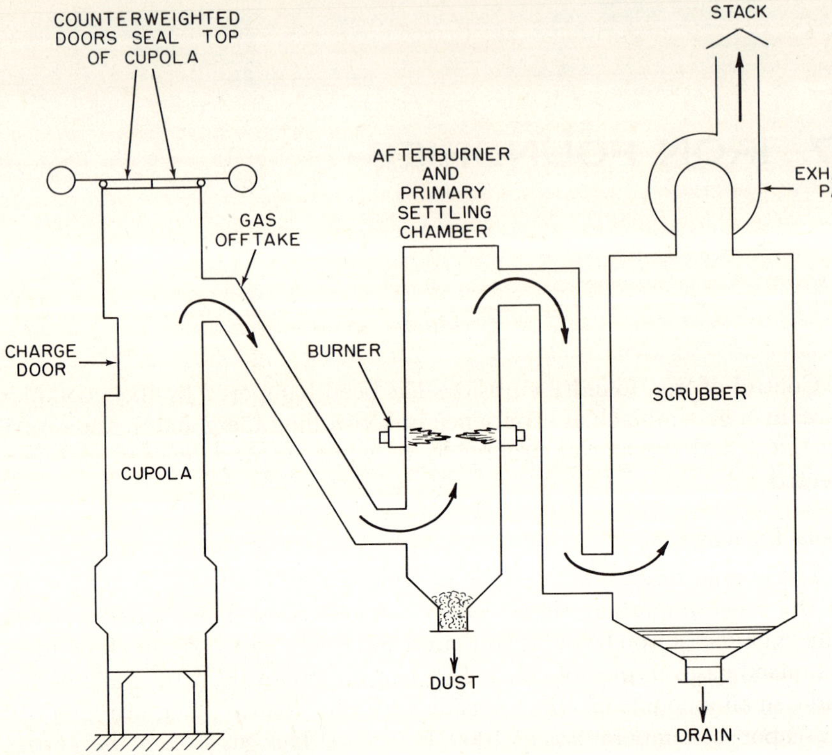

Figure 7.7a Cupola control system with remote afterburners and wet scrubber.

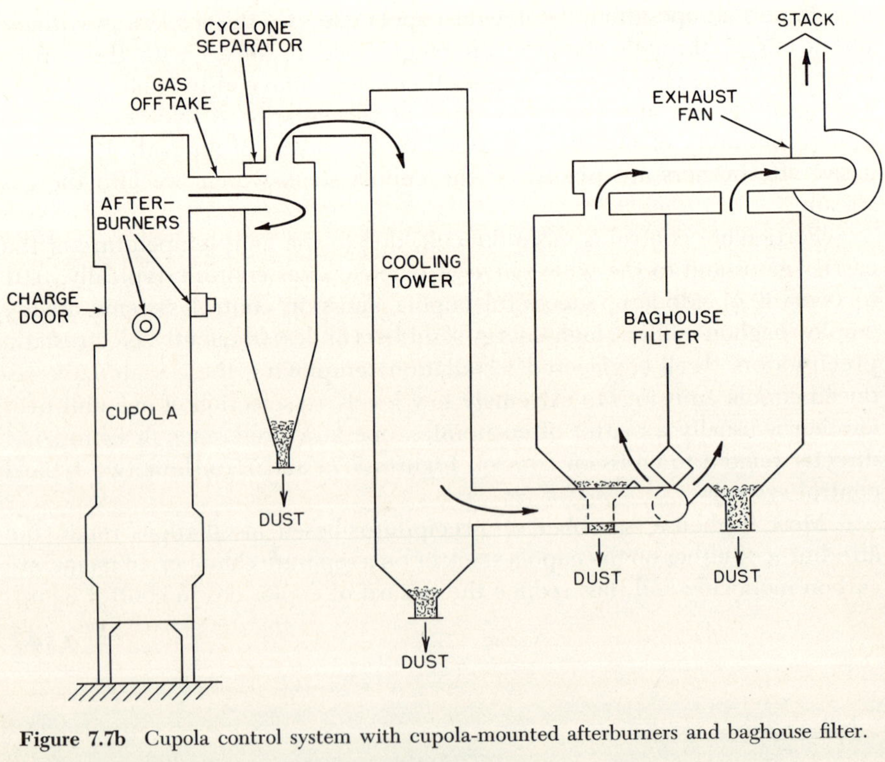

Figure 7.7b Cupola control system with cupola-mounted afterburners and baghouse filter.

ment. Afterburners also dispose of oil and grease vapors that could foul or plug the control unit.

Temperatures of exhaust gases leaving the cupola are too high to permit direct venting to control equipment, so cooling is necessary. Dilution with cold ambient air has been used, but it adds to the volume of gases to be treated by the final control unit.

Preheating of the incoming blast air to the cupola with the exhaust gases has economic advantages, because the use of hot blast air permits reduction of the amount of coke (fuel) charged into the cupola. High dust loads in the cupola gases, however, cause fouling of heat exchanger surfaces and require frequent cleaning and maintenance. Most current techniques make no effort to recover the heat, but settle instead for cooling the gases in water spray quenchers. This also removes some of the coarser particulates from the gas stream, reducing their burden on the final control unit. Cyclones or other low efficiency mechanical collectors are also occasionally used for this purpose.

Reverberatory and Air Furnaces

Reverberatory or air furnaces are used to a limited extent by foundries producing malleable iron or mill rolls. These furnaces are charged with pig iron, scrap and alloying elements, which are then melted by burning powdered coal or other fuels. Emissions include flyash, carbon, metal oxides and sulfur dioxide. Substituting fuel oil for coal and fitting the furnace with an afterburner will eliminate virtually all flyash and carbon. This approach has some metallurgical drawbacks, however.

Sulfur dioxide levels are proportional to the sulfur content of the fuel and can be reduced by burning low sulfur oil. Metallic oxides make up about 10% of the particulates emitted by a coal-fired furnace. Substituting oil for coal will not appreciably change their level. Baghouses and electrostatic precipitators have been used to a limited degree, but the advanced age of most air furnaces makes their economic justification difficult. Many foundries have chosen instead to install inherently cleaner induction melting furnaces.

Induction Furnaces (Electric)

Induction furnaces employ electricity to induce secondary electrical currents in the metal charge. The metal's resistance to these currents causes it to heat and melt. Some induction furnaces are charged with molten metal and are used as holding and refining units.

Induction furnace operation is inherently clean. In furnaces melting scrap metal, volatilized oil and grease on the charge is responsible for the bulk of emissions. These can be combatted either by precleaning the charge or by installing a fume collection system over the furnace and venting the collected vapors to an afterburner. Small amounts of fine metal oxides are also emitted. Generally, their levels are too low to require control measures. In extreme cases, however, venting the furnace to a baghouse collector or scrubber should achieve the desired degree of control.

Core Ovens

Ovens for baking sand cores bonded with oil emit volatilized hydrocarbons as smoke. Afterburners will control these emissions.

Sources of Dust

The average foundry has many potential sources of dust emissions, including:

1. Furnace charge preparation (coke fines and limestone dust)
2. Foundry sand preparation and reclamation (silica dust, cereal and wood flour dust and fine clay particles)
3. Cleaning and finishing operations (sand, metal powder and abrasive dust)

Effective control of these emissions requires enclosure of storage bins and conveying equipment; removal (by screening or other suitable methods) of objectionable fines; and collection, through hoods and exhausters, of airborne particles. Exhausted particles are frequently gathered in some type of wet or dry collector.

When molten metal is poured into sand molds, organic additives in the sand volatilize and decompose to form smoke and odors. Many of these materials can be incinerated in afterburners while others must be collected by scrubbers. Control efforts have been hampered by the difficulty of collecting the emissions as they escape from the mold into the foundry building. Close-fitting fume hoods, where feasible, have the best chance of success.

7.8 NONFERROUS FOUNDRIES

Emissions common to all types of nonferrous foundries include the dust raised in the handling and preparation of sand; bentonite (clay), wood and cereal flours; and other powdered additives used in the preparation of molding sand. The most effective control measure is to enclose the storage bins and transporting equipment. Enclosing the mixing equipment is difficult and a more practical approach is the use of collection hoods vented to exhausters, which in turn discharge to filters or water baths. This approach has merit as long as duct inlet velocities are low enough to avoid entrainment of particles that would otherwise remain undisturbed.

In foundries where sand cores are used, the chief source of emissions is the oven used for the baking of cores which were bonded with oil. Volatilized hydrocarbon oils make up the bulk of emissions, which usually appear as bluish smoke. This condition can be corrected by equipping the oven with an afterburner or by changing the method of bonding cores to one that does not require baking of oil-base binders. On occasion, core ovens emit the particularly acrid, offensive smell of "burnt oil." This odor is caused by decomposition of the core oil at excessive temperatures. Closer control of oven operation, particularly temperatures, should cure this problem.

Emissions from the foundry floor include smoke and the characteristic odor of "burnt sand." Both result from the hot metal decomposing the various binders and additives in the sand mold and from the dust liberated when castings are shaken out of their molds.

Complete entrapment of these emissions within the large open expanse of most foundry buildings is virtually impossible. The fume hood-exhauster-collector system described for sand preparation facilities may serve to control the extremely heavy concentrations in the immediate vicinity of the smoking molds and shakeout area.

Cleaning and finishing operations generate relatively coarse metal and abrasive dusts. If source equipment is vented to filters or to water baths, this usually achieves the desired degree of dust abatement.

Heat treating operations are inherently clean, unless oil or grease has been left on castings charged into the heat treating furnace, in which case volatilized hydrocarbons will result in the furnace exhaust. Precleaning the castings is the most practical cure.

Melting room emissions vary to a great degree with the alloy being melted. In aluminum foundries, typical emissions include chlorine gas, which is used for purifying the metal, oxides of zinc and magnesium from alloys containing those metals and metal chlorides and fluorides from chemical reactions between the alloy and flux compounds. Chlorine, choride and fluoride emissions can be eliminated by substituting alternative fluxing compounds (e.g., nitrogen gas) or by passing the metal through an inert filtering media. If such process changes are not feasible, scrubbers or filters can be used for at least partial control. Metal oxides are usually at such low levels that control measures are not necessary.

Major emissions from brass and bronze melting operations are metallic oxides, chiefly those of zinc and lead. Control techniques include exhausting furnace emissions to filters or placing cover fluxes on the metal to retard the formation and loss of oxides.

Zinc melting operations generate zinc oxide fumes. Exhausting them to filters is usually sufficient to control them.

If the furnace is an indirect-fired or induction unit (where metal is melted in a pot or crucible), fumes are usually collected by a hood over the furnace. On so-called reverberatory furnaces, burners fire directly into a refractory-lined enclosure containing the metal. Here fumes mingle with spent burner gases and leave the furnace through the stack, necessitating the treatment of the entire flow of stack gases.

Nonferrous melting furnaces handling scrap and machine chips often generate smoke and vapors from volatilized oil and grease on the charge. The most practical method of control is the predrying of the metal in an oven or kiln equipped with an afterburner.

7.9 PHOSPHORIC ACID PLANTS

Phosphoric acid is produced by one of two processes: (1) thermal process and (2) wet process. The thermal process accounts for approximately one-third of the annual production of phosphoric acid. Acid mist, in the form of orthophosphoric acid (H_3PO_4) is the principal atmospheric emission from this process. Loadings as high as 6,000 mg/SCF of dry stack gas have been reported. Hydrogen sulfide is emitted in the preparation of food-grade phosphoric acid. In the wet process, fluorides, namely hydrogen fluoride and silicon tetrafluoride, are the principal atmospheric emissions. A small quantity of phosphate rock dust may also become airborne during the handling and feeding of the digester. Packed towers, venturi scrubbers, electrostatic precipitators and mist eliminators (Chapter 5) are successfully used to reduce the atmospheric emissions.

The Processes

Thermal Process

The thermal process[1] for phosphoric acid manufacture (Fig. 7.9a) consists of three steps:

1. Burning the liquid phosphorus in a combustion chamber to produce phosphorus pentoxide.
2. Hydrating the phosphorus pentoxide with dilute acid or water to produce phosphoric acid liquid and mist.
3. Removal of the phosphoric acid mist from the gas stream.

In the preparation of food-grade phosphoric acid, the raw acid is treated with a sulfide (hydrogen sulfide, sodium hydrosulfide or sodium sulfide) to precipitate arsenic and other heavy metal ions. Hydrogen sulfide is highly toxic and flammable and its recommended maximum allowable concentration is 20 ppm for an 8-hour period.

Figure 7.9a shows the emission sources and the type and quantity of pollutants. The phosphoric acid mist and hydrogen sulfide are the pollutants emitted to the atmosphere.

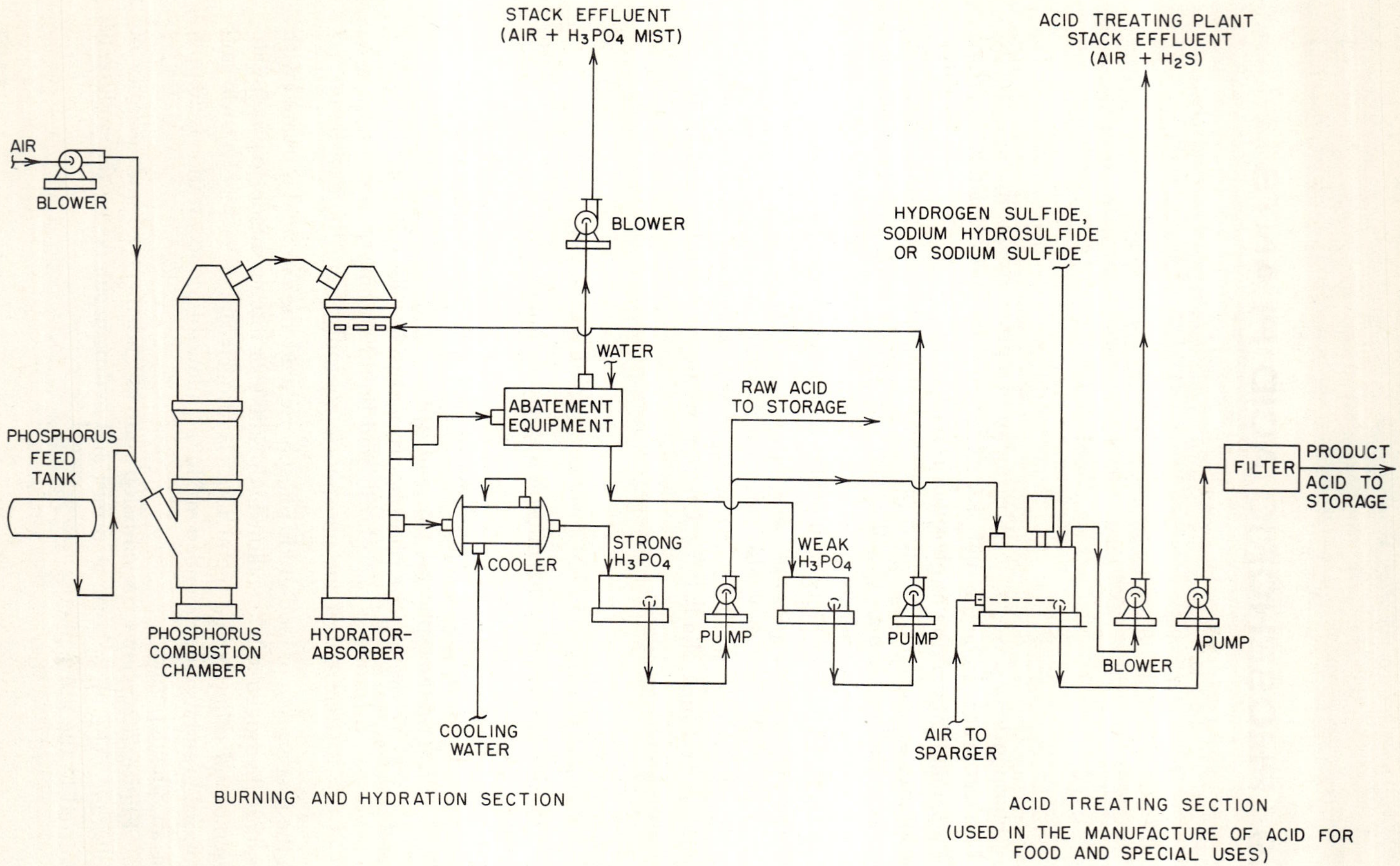

Figure 7.9a Typical thermal-process phosphoric acid plant.

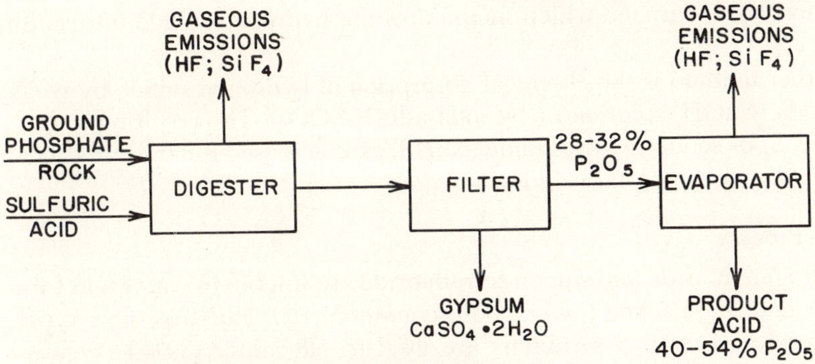

DIGESTER REACTIONS

7.9 (1) $3\,Ca_{10}(PO_4)_6F_2 + 30\,H_2SO_4 + 60\,H_2O \rightarrow 30\,CaSO_4 \cdot 2H_2O + 18H_3PO_4 + 6HF\,(aq.)$

7.9 (2) $6\,HF_{(aq.)} + SiO_{2\,(c)} \rightarrow H_2SiF_6 \cdot 2H_2O$

7.9 (3) $H_2SiF_{6\,(aq.)} \longrightarrow 2HF_{(g)}\uparrow + SiF_{4\,(g)}\uparrow$

Figure 7.9b Wet process for phosphoric acid manufacture.

Wet Process

Figure 7.9b is a simplified flow diagram for the wet process, showing the emission sources and the type and quantity of pollutants. The ground phosphate rock is digested with strong sulfuric acid in a digester to produce phosphoric acid and gypsum slurry. About 60% of the fluorine in the rock is converted to hydrofluoric acid which in turn reacts with the silica in the rock to form fluosilicic acid. The fluosilicic acid readily vaporizes from the hot solution and decomposes to gaseous hydrogen fluoride and silicon tetrafluoride. These are the only gaseous fluorides emitted from rock processing.

The main evolution of fluorides during the processing takes place in the evaporator where phosphoric acid strength is increased from 30% P_2O_5 to 40 to 54% P_2O_5. Approximately 13 to 40% of the fluorine fed with the rock is released.

Control of Emissions

Thermal Process

The acid mist[1] is the major air pollutant emitted to atmosphere from the thermal process. Packed and spray towers, venturi scrubbers, cyclonic separators with wire-mesh mist eliminators, fiber mist eliminators, high-energy wire-mesh contactors and electrostatic precipitators (Chapter 5) have been successfully used to collect the acid mist. Collection efficiency in the range of 95 to 99.9% can be easily attained.[1]

In many plants, no attempts are made to remove hydrogen sulfide[1,3] from the acid-treating tank vent gas. In some cases, the effluent gases are vented to the phosphorus furnace which incinerates the hydrogen sulfide into sulfur dioxide.

Another method is the chemical absorption of hydrogen sulfide by weak caustic soda (NaOH) solution or by soda ash (Na_2CO_3). The resulting sodium sulfide (Na_2S) or sodium hydrosulfide (NaHS) can be reused for acid treatment to precipitate heavy metal sulfides.

Wet Process

Hydrogen fluoride and silicon tetrafluoride are soluble in water and form fluosilicic acid.[2,4,5] Packed towers, both countercurrent and cross-flow types, spray towers and venturi scrubbers are used to efficiently (99%+) remove these pollutants. Water or phosphoric acid are generally used as scrubbing liquors. Fluorine is also recovered as a marketable 15 to 25% solution of fluosilicic acid, which is an agent for fluoridation of water or a raw material for synthetic cryolite used in the manufacture of aluminum.

A small quantity of phosphate rock dust may become airborne during handling and feeding the digester. Whenever the physical arrangement of the rock grinding and acid plant permits, the conveying and weighing equipment is exhausted into the mill vent air system. Otherwise, a dust collector, cyclone or bag filter is installed on the rock bin to collect this dust before discharge.

REFERENCES

1. "Atmospheric Emissions from Thermal Process Phosphoric Acid Manufacture," NAPCA Publication No. 48.
2. "Phosphoric Acid," vol. I., part II., chap. 9, Marcel Dekker, New York, 1968.
3. "Air Pollution Engineering Manual," Public Health Service Publication No. 999-AP-40.
4. "Control Techniques for Particulate Air Pollutants," NAPCA Publication No. AP-51.
5. Craig, John M.: "Fluoride Removal from Wet Process Phosphoric Acid Reactor Gases," Ph.D. Dissertation, University of Florida, 1970.

7.10 POWER PLANTS, INCINERATORS, HEATERS

Boilers, steam generators and liquid heaters are basically furnaces in which fuel is burned to heat or vaporize a liquid heat transfer medium. The heated liquid or vapor is piped to the point of use, where the heat is extracted. The cooled liquid or condensate is then either discarded or recycled to the heating unit. Water is the most common heat transfer medium, although organic fluids are also used on many installations. Organic fluids are always recycled, therefore the only emissions associated with handling of the heat transfer medium are steam and water vapor.

All objectionable emissions associated with incinerators, steam generators and other fired heaters are the result of the combustion process. They can be traced back to the composition of the fuel, deviations from optimum firing practice, inability of combustion equipment to properly burn the fuel or to chemical reactions in the combustion process that produce undesirable by-products.

From the standpoint of combustion, the ideal situation is the complete oxidation of a pure carbon or hydrocarbon fuel in pure oxygen. Combustion products in this case are solely water and carbon dioxide. In practice, this is not attainable because:

1. Available fuels are contaminated with impurities, usually sulfur, sulfur compounds and ash. In case of incinerators, the fuel composition is unpredictable.
2. The use of pure oxygen in large-scale combustion applications is not yet commercially feasible. Air, containing about 23% oxygen and 76% nitrogen by weight, is the most economical oxygen source.
3. Even the best available combustion equipment is not able to achieve 100% complete reaction between fuel and oxygen.

Flyash

Flyash is airborne particulate matter, chiefly silica and other metal oxides. Noncombustible minerals, present in all coal and, to a lesser extent, in most fuel oils, are liberated as fine particles during the combustion process and are carried out of the furnace by the flow of hot, buoyant combustion gases. Other factors being equal, flyash emissions from boilers and steam generators increase with higher ash content of the fuel, finer ash particle sizes and increasing stack gas velocity (a function of boiler firing rate and stack draft).

Many grades of coal contain 10 to 15% ash. No commercially feasible process can yet remove this ash,* so control efforts have been directed instead at removing it from the stack gases.

Boiler firing methods have a great deal of influence on flyash emissions. They tend to be low on hand and stoker-fired boilers, where relatively large ash particles are trapped in the fuel bed and can easily resist entrainment by slow-moving flue gases. High-emission levels are characteristic of pulverized coal-fired boilers (Fig. 7.10a), where finely ground coal is burned in suspension in a turbulent, high-velocity airstream. In short, coal-burning systems designed for rapid combustion and high-fuel efficiency tend to be the worst emitters of flyash.

Early efforts at flyash control concentrated on manipulating combustion chamber designs, gas flow patterns in the boiler and methods of feeding coal into the boiler. Many of these were moderately successful but still could not achieve sufficiently low stack emission levels without compromising the boiler's primary function.

The use of tall stacks for wide dispersion of particulates is a time-honored practice, but it fails to reduce the total weight of flyash released to the atmosphere.

Mechanical collectors have been used, with some success, to trap flyash; but their collection efficiencies are generally too low to satisfy present air quality codes, especially on large steam generators. Consequently, modern electrostatic precipitators (Chapter 5), with collection efficiencies of 98% and

*The Solvent Refined Coal (SRC) process for the removal of sulfur and ash from coal has been successfully piloted by the Interior Department's Office of Coal Research.

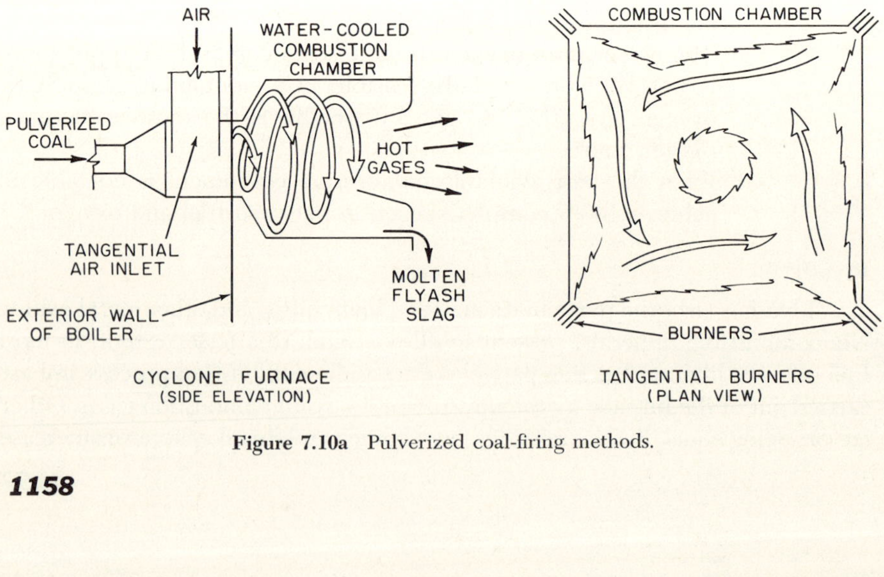

Figure 7.10a Pulverized coal-firing methods.

up, are almost universally used by powerplants and other large installations. The high initial and operating costs of electrostatic precipitators have discouraged their use on smaller installations because operators usually find conversion to oil or gaseous fuels more practical.

Sulfur Oxides

Whenever coal or oil containing sulfur or its compounds is burned, sulfur dioxide (SO_2) and sulfur trioxide (SO_3) are formed. Sulfur dioxide forms the bulk of these emissions, although it subsequently oxidizes to SO_3 which is the more offensive of the two pollutants. Since the level of sulfur oxide emissions varies with the sulfur content of the fuel, sulfur-free and low sulfur grades of oil and coal are in great demand. Naturally occurring low sulfur grades are scarce, for which reason many coal plants and oil refineries desulfurize their products.

Desulfurization of coal is, at best, only partially successful. Much of the sulfur borne in coal cannot be removed by any known practical method, and complete desulfurization will probably become practical only when a process is developed for the converting of coal to synthetic liquid or gaseous fuels. See Chapter 6.

The technology for desulfurizing oil is well-developed (Chapter 6). The existing refinery capacity is insufficient to meet the demand for desulfurized oil, but additional facilities are being put on stream.

Virtually all natural gas entering the domestic market is either naturally sulfur-free or is completely desulfurized. Its use causes no sulfur oxide emissions.

For the short term, at least, supplies of low sulfur and sulfur-free fuels are likely to remain inadequate. The most promising solutions to the sulfur oxide problem revolve around the use of fuel additives and flue gas treatment. Numerous methods have entered pilot plant or limited operational stages. Several of the more important ones are described in the early sections of Chapter 6.

Nitrogen Oxides

Nitric oxide (NO) and nitrogen dioxide (NO_2) form at combustion temperatures from nitrogen and oxygen in the combustion air supply. Chemical reactions resulting in the formation of these oxides are largely independent of the type of fuel burned but are accelerated by combustion temperatures over 2000°F and by increasing amounts of excess air. (In combustion reaction, excess air is that air which is supplied beyond the amount stoichiometrically required for complete oxidation of the fuel.) Since exclusion of nitrogen from the combustion reaction is a practical impossibility, control of the combustion chamber reactions and of flue gas treatment is receiving thorough investigation (Chapter 6).

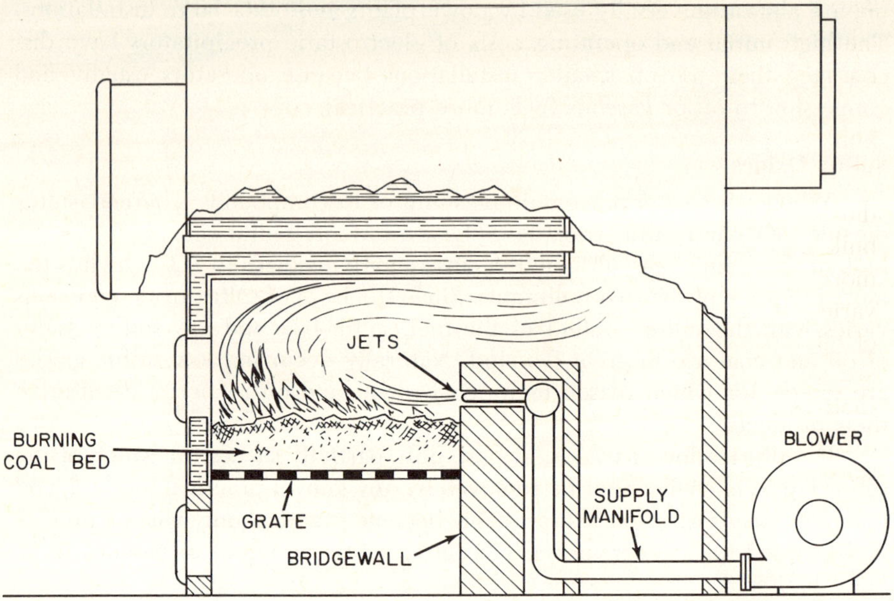

Figure 7.10b Application of overfire air jets to a coal-fired boiler.

Products of Incomplete Combustion

Furnishing a carbonaceous or hydrocarbon fuel with insufficient combustion air or failing to mix the fuel intimately with air results in the formation of incomplete combustion products. Incomplete burning of coal and oil produces a variety of complex hydrocarbons and carbon monoxide. In aggravated cases, some of the fuel may escape the stack completely unchanged. Smoke, ranging in color from bluish white to black, usually results. Incomplete combustion of natural gas usually yields carbon monoxide and, in extremely poor situations, free hydrogen.

Complete combustion, desirable for optimum fuel economy as well as for clean air, can be accomplished by the use of modern, efficient combustion equipment and control accessories and by keeping them well maintained and adjusted.

Mixing of fuel and air is influenced by many factors, but the most important one is the size of the fuel particles. The ease of mixing the fuel and air increases with increasing surface area-to-volume ratio of the fuel. From this, it follows that gaseous fuels, which can be intimately mixed with air on a molecular scale, are easiest to burn quickly. Liquid fuels, which are generally atomized into fine droplets to prepare them for combustion, are somewhat more difficult to burn. The same holds true for finely pulverized coal.

The most difficult fuel to burn completely is lump coal. With some grades, the problem is aggravated by a high level of volatile constituents. The heat of combustion drives these volatiles out of the coal before the carbonaceous material has burned completely. Unless sufficient air and an ignition source are present to consume them, they escape unburned. The use of coals with low volatile content will alleviate this problem, as will the use of overfire air jets (Fig. 7.10b). These nozzles assist combustion by injecting secondary air at high velocity into the furnace space above the burning fuel bed.

Oil atomization is generally accomplished with liquid pressure atomizers, in which a high pressure (about 76 psig) stream of oil is forced through a small orifice, or with viscous friction atomizers in which a stream of pressurized air or steam is used to impinge on the oil, tearing it into fine droplets. Modern atomizers, which are integral parts of the oil burners they serve, are capable of dividing oil into extremely fine droplets, provided that:

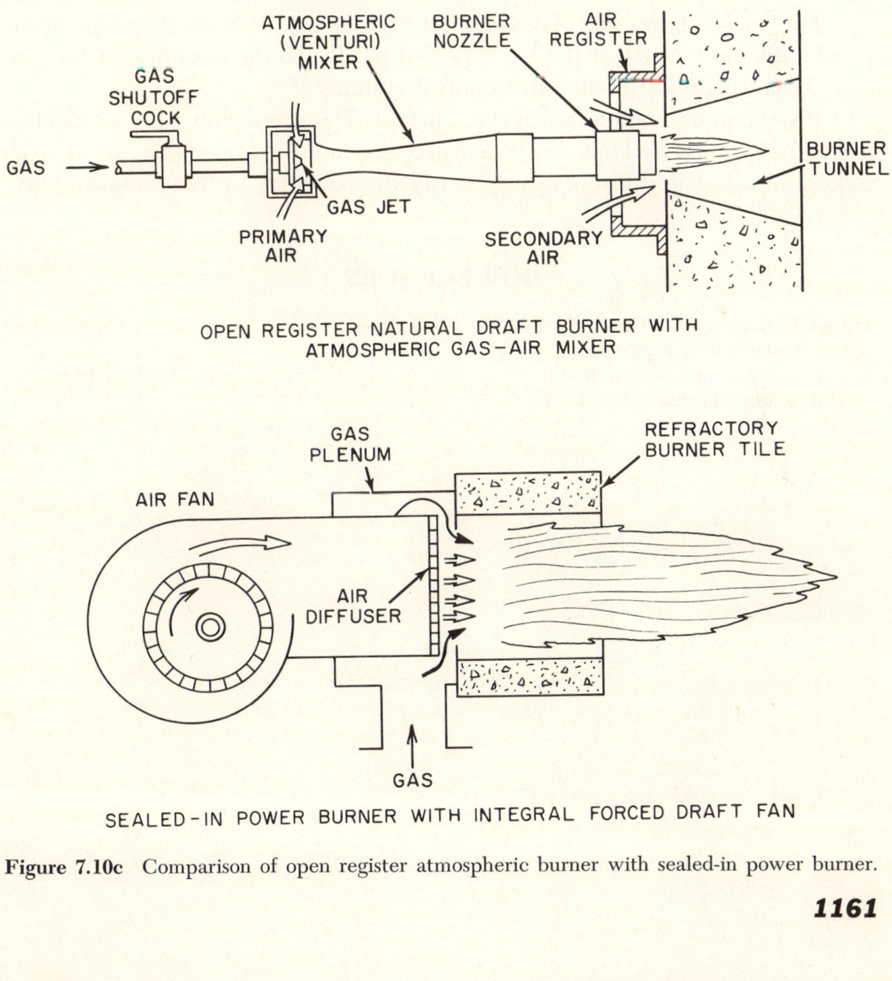

Figure 7.10c Comparison of open register atmospheric burner with sealed-in power burner.

1. The oil pressure exceeds the required minimum.
2. The atomizing medium pressure (for viscous friction atomizers) exceeds required minimum.
3. The oil viscosity is sufficiently low (this requires preheating of heavier grades).
4. The atomizer passages and orifices are clean.

Gas and oil burners must be capable of injecting combustion air and fuel into the boiler in a highly turbulent fashion. This ensures fast mixing and completion of combustion reactions before the hot gases are quenched on cold heat transfer surfaces. Two areas for improvement in this field are the replacement of atmospheric (blowerless) burners with those having a forced air supply and the replacement of open-register burners with sealed-in types. (With sealed-in burners, all combustion air passes through the burner orifice and can therefore be metered.) Figure 7.10c illustrates two extremes of firing practices.

Fuel and airflow control devices must be capable of maintaining accurate, predictable flow ratios at the burner. This has led to the adoption of increasingly sophisticated fuel-air ratio control systems.[1]

Proper maintenance is very important. Dampers, flow control devices and burners must be kept clean and properly adjusted, and flue gas analysis should be monitored frequently to verify the accuracy of these adjustments.[2]

REFERENCES

1. Lipták, B. G. (ed.): "Instrument Engineers' Handbook," Vol. II, Section 10.6, Chilton Book Co., Philadelphia, Pennsylvania, 1969.
2. ———: "Instrumentation in the Processing Industries," Chapter IV, Chilton Book Co., Philadelphia, Pennsylvania, 1973.

7.11 PROTECTIVE COATINGS MANUFACTURE

Protective coatings include paints, lacquers, varnishes and synthetic resins which are used for decorative and protective purposes. Coatings are scientifically formulated and engineered to meet specific end-use requirements in both the pigmented and nonpigmented form.[1]

Raw materials include pigments and extenders which are finely sized solids and are usually dusty to handle. The other materials, including binders (varnishes and resins), plasticizers, solvents and thinners, are organic compounds having a degree of volatility. The nature of these raw materials makes the coating manufacturer a potential source of atmospheric pollution for particulate matter, hydrocarbons and odor. The quantity of pollution will depend upon the materials used and the operational conditions.[2]

The best method of emission control is based on economics, i.e., the reduction or elimination of costly material losses by changes in handling methods and processes. (See Section 5.4.)

Another approach to reduce emissions is to substitute some of the process materials: use of less volatile and less photochemically reactive solvents; reformulation to aqueous base systems; conversion to high solids coatings, powdered coatings and the like.

The last approach involves the installation of control equipment. The control of dust emissions can be accomplished with centrifugal collectors (Section 5.9), bag filters (Section 5.7) and scrubbers (Section 5.12). Hydrocarbon emissions can be controlled with condensers, afterburners, scrubbers, absorbers and adsorbers or combinations thereof (Chapter 6).

The type and extent of control needed can be determined by a thorough system evaluation of governing regulations and industry responsibilities to the community.

With respect to hydrocarbons and organic solvents, state regulations may set quantitative limits on emissions. However, in most localities, it will be necessary to control the emissions because of the odor nuisance and not because of the volume of hydrocarbons. The emissions losses resulting from the manufacture of coatings are much lower than the volume emitted during their application.[3]

Raw Material Handling

The handling of organic solvents in stationary bulk systems is a potential source of air pollution involving hydrocarbons and odors (Chapter 6). Emissions can be minimized by the use of submerged fill pipes, conservation vents[4] and closed circuit vapor return lines. Vapor emissions during filling and tank breathing require a collection system leading to a control device (Section 5.39).

Chemicals and pigments used in the powdered or flaked form and in large volume may be handled in stationary bulk systems for both economic reasons and to minimize air pollution control problems. The movement of these materials can cause a dusty situation emitting particulate matter and heavy metals into the atmosphere unless properly controlled.

Paint Manufacturing

Paint represents the pigmented form of protective coatings.[1] It is usually manufactured in a batch process because of the great number of custom formulated products. The manufacturing process includes mixing, grinding, thinning and filling (Fig. 7.11a).

The pigment is mixed in the high-speed dispenser with resin and/or solvent to make a paste of proper grinding consistency. The loading operation emits a small amount of pigment fines which can be controlled by a collection system exhausting to a bag filter (Section 5.7). Hydrocarbon emissions also occur because of displacement and evaporation. Mixing is often done in open tanks where the collection of fumes is inefficient resulting in large volumes of air with low concentrations of hydrocarbons.

"Grinding" the paste to break down the pigment aggregates can be accomplished with the sand grinder, which represents a continuous step in an otherwise batch process. Both open and closed units are available. The elevated operating temperature increases the hydrocarbon evaporation. For this reason, closed units offer an advantage in reducing solvent emissions. Emissions from sand grinders are easily collected and can be exhausted to a pollution control device.

Ball or pebble mills combine the mixing and grinding operations. Once loaded, the mill is closed and there are no emissions.

Two, three and five-roll mills, also used for grinding, present a large open surface for evaporation of solvents and thus require a refined vapor pick-up system for handling the hydrocarbon emissions.

The ground paste is reduced with the remaining resins and solvents to yield the finished product which is tinted to the required color. This operation is logically done in enclosed tanks to prevent the loss of solvent. When smaller batches are thinned and tinted in open tanks, large evaporation losses can result and control of emissions becomes necessary.

After meeting quality control tests, the batches are strained and filled into containers. Straining can be done in an enclosed, pressure type unit to

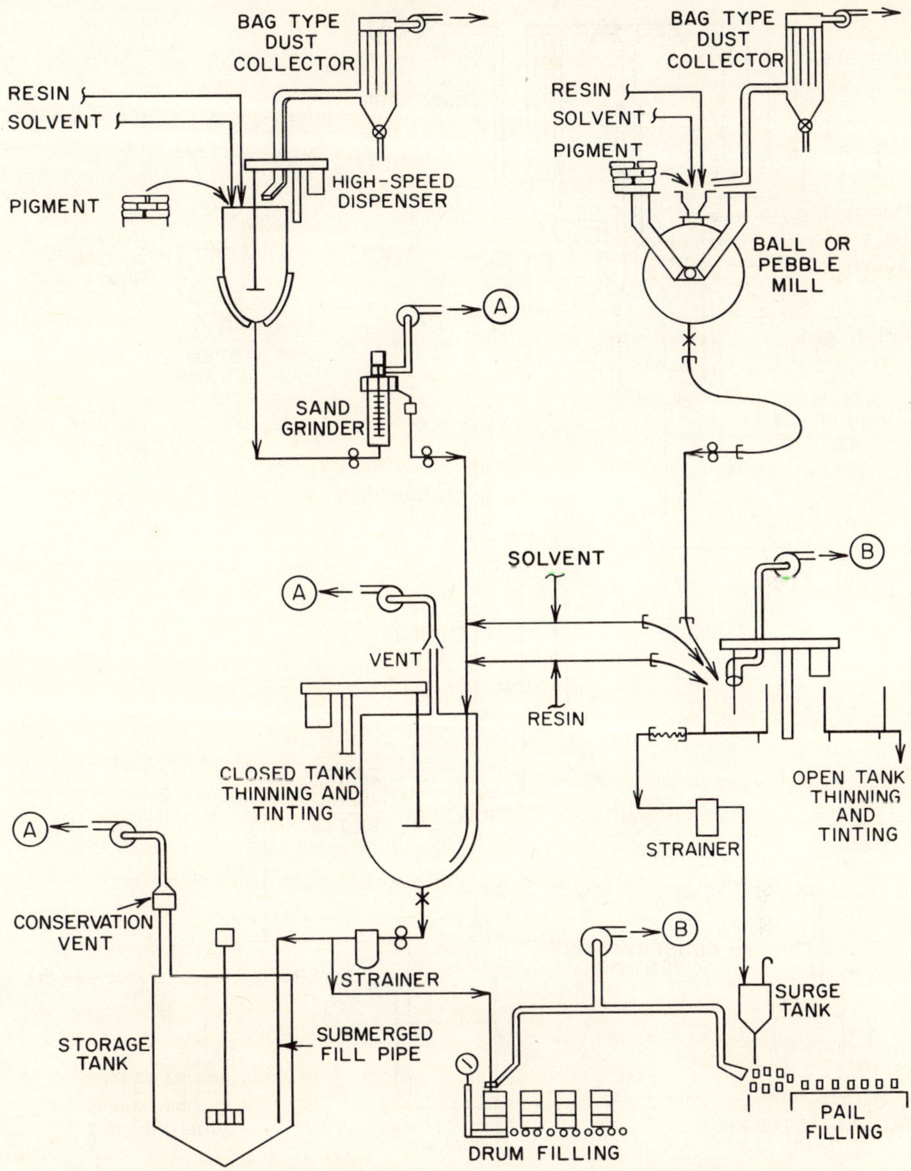

Ⓐ A POLLUTION CONTROL DEVICE SUCH AS AN INCINERATOR, SCRUBBER, OXIDIZER OR ADSORBER.

Ⓑ A POLLUTION CONTROL DEVICE SUCH AS A SCRUBBER, OXIDIZER OR ADSORBER.

Figure 7.11a Paint manufacturing processes.

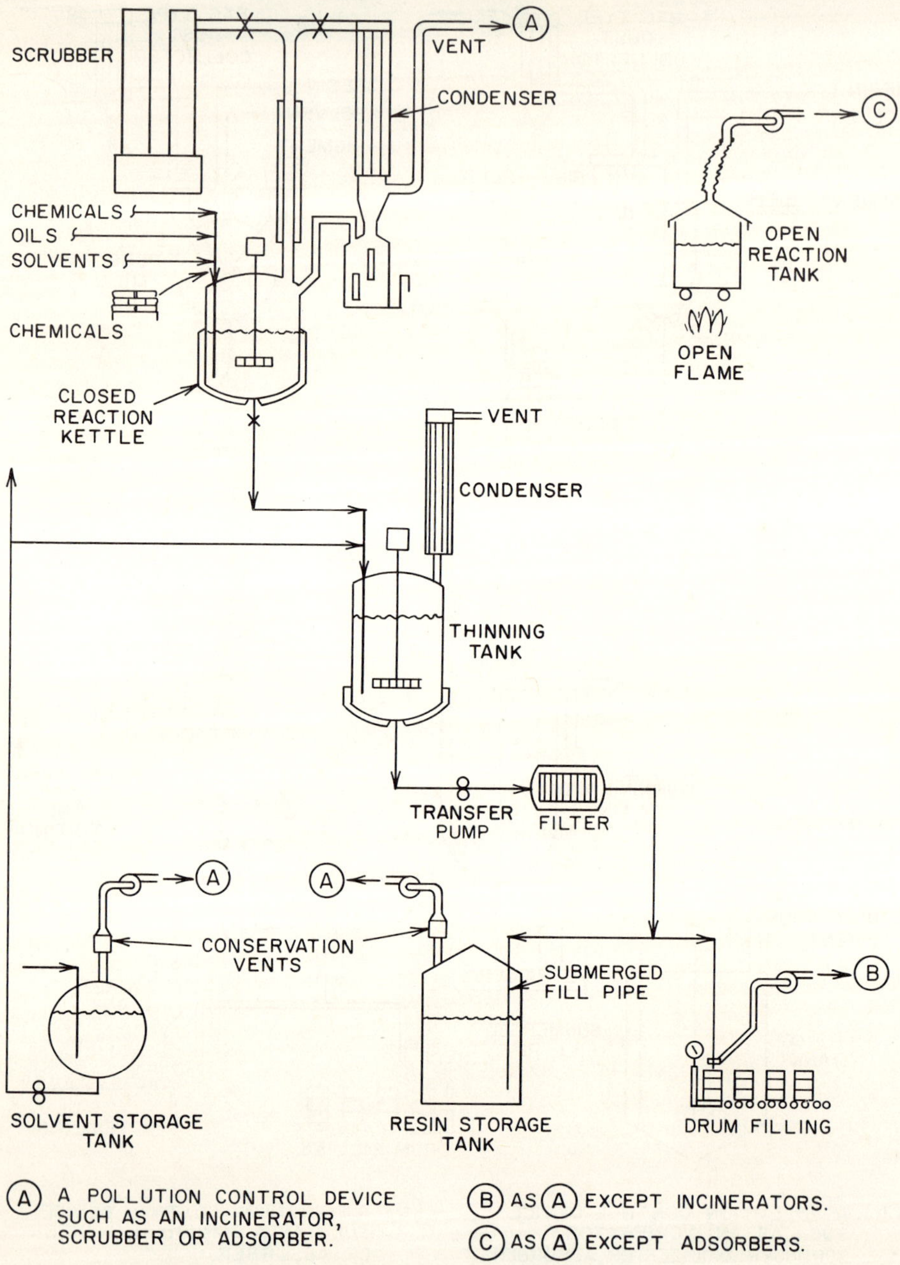

Figure 7.11b Varnish and synthetic resin manufacturing steps.

eliminate hydrocarbon losses. Filling operations make use of integrated vapor collection systems to handle the displaced vapors.

Varnish and Synthetic Resin Manufacture

Varnish and synthetic resins represent the nonpigmented form of protective coatings. Their manufacture (Fig. 7.11b) includes the operations of chemical reaction, thinning, mixing, filtering and storing.[5]

The reaction may be done in open or totally enclosed kettles. The high-reaction temperature (200 to 600°F) causes evaporation of the various constituents and by-products, with many of the latter having disagreeable and irritating odors.[3]

Losses to the atmosphere are substantial from open kettles if vapor collection systems, with control devices, are not used.

Enclosed kettles minimize the material losses. Control of released vapors is accomplished by condensers in the azeotropic distillation process and by scrubbers in the blow-hard process. Secondary control devices may also be required, especially to control the undesirable odorous materials.

Solvent is added to the resin to provide a viscosity suitable for efficient handling. Thinning is preferably done in enclosed vessels, either by adding the solvent to the hot mass of resin or by adding the hot resin to the colder solvent. In either case, solvent loss is controlled with condensers.

After thinning and mixing, the resin batches are filtered to remove foreign matter and gel particles. Filtration is best carried out in an enclosed unit to minimize solvent loss and emissions to the atmosphere. Cleaning the filter requires disassembly of the unit, at which time residual solvent on the exposed surfaces can evaporate.

After filtration, the resin is pumped to storage tanks or to shipping containers. The emission of hydrocarbons during these operations occurs as vapors are displaced.

REFERENCES

1. Martens, C. R. (ed.): "Technology of Paints, Varnishes and Lacquers," Reinhold Publishing Corp., New York, 1968.
2. Stenburg, R. L.: Atmospheric Emissions from Paint and Varnish Operations, Part 1, *Paint Varn. Prod.*, vol. 49, no. 10, pp. 61–65, September 1959.
3. Control Techniques for Hydrocarbon and Organic Solvent Emissions from Stationary Sources," U.S. Dept. of HEW, AP-68.
4. Lipták, B. G. (ed.): "Instrument Engineers' Handbook," Vol. I, Section 10.4, Chilton Book Co., Philadelphia, Pennsylvania, 1969.
5. Mattiello, J. J. (ed.): "Protective and Decorative Coatings," Vol. III, John Wiley, London, 1943.

7.12 PULP AND PAPER MILLS

The pulp and paper industry is a principal polluter of the atmosphere. Although there is still a long way to go before pulp and paper mills will go unnoticed by their neighbors, significant efforts have been made in the pollution abatement area.

The major polluters (*kraft* or *sulfite* mills), can be generally classified according to the active chemical used in the digestion of the wood to produce pulp.[1] Since the number of kraft mills far exceeds the number of sulfite mills, this discussion will concentrate on the kraft mills. This discussion will also be confined to those systems which are the most significant sources of air pollution.

Kraft Mill Digester Systems

Kraft mills, and all other chemical process pulp mills, use either a continuous or a batch digestion system to convert wood into a pulp suitable for making paper. In a continuous digester, wood, chemicals and steam are continuously introduced; and pulp, spent chemicals and air pollutants are continuously discharged. The batch digester is loaded with wood and chemicals; and after being sealed, it is *cooked* with live steam for a specified period of time. During the batch cycle, some air pollutants are relieved as excess gases, but the majority of the pollutants are released at the end of the cycle when the digester is "blown" to discharge the cooked pulp.

The major air pollutants from both continuous and batch digesters include hydrogen sulfide, methyl mercaptan, dimethyl sulfide and dimethyl disulfide.[2] These compounds constitute what is generally referred to as *kraft odor*. While they are usually not present in sufficient quantities in the atmosphere to be toxic and normally are not associated with smog formations, the compounds have an extremely low odor threshold (0.4 to 5 ppb for hydrogen sulfide) and are easily detected by our sense of smell.

Although methods of treatment are available to reduce the emissions, their odor threshold is so low that pulp mills are not likely to be completely *odor free*. At the present time, the gaseous pollutants are usually treated by chlorination or by incineration. However, due to the air pollution laws now being enforced, particularly in the Northwest region of the United States, incineration in an existing lime kiln, recovery boiler or a specially designed incinerator is the most preferred method. The major problem with chlorination or incineration is the collection of the gases since even a small leak can be detected at relatively great distances (i.e., outside the mill's property line).

In a mill using only a continuous digester, the problem of collection is greatly simplified because the gases come off the digester at a controlled rate and usually can be treated directly. When batch digesters are used, the bulk of the gases is exhausted at the end of a batch when the digester is blown. To take care of these surges, a large holding tank with a diaphragm or a floating head must be included in the system. The gases can then be metered out of the holding tank at a controlled rate for treatment. In mills using a softwood which contains turpentine, the gases first go to a turpentine system to collect the by-product for sale and also to remove the highly explosive turpentine from the gases before incineration.

In addition to the batch and continuous digesters, odorous gases must also be collected from vents off black liquor evaporator hot wells and from black liquor flash tank vents (if any). Figure 7.12a is a schematic flow diagram of a gas collection and incineration system for a mill having both continuous and batch digesters in addition to an evaporator.[3]

Kraft Mill Recovery Boiler

Weak black liquor separated from the pulp by washers is concentrated by black liquor evaporators and is sent to a recovery boiler. Here it is burned to recover its chemical content and also to dispose of the wood organics dissolved in the liquor which are burned to carbon dioxide and water vapor. The chemicals are recovered in the form of a smelt containing sodium sulfide and sodium carbonate with a small amount of sodium hydroxide. Makeup chemical in the form of sodium sulfate is also added to the recovery boiler. This is reduced to sodium sulfide which is also included in the smelt (Fig. 7.12b).

The recovery boiler, because of the volume of flue gas it emits, is the largest single source of air pollution in a pulp and paper mill. The flue gas does not present an insurmountable problem since it theoretically contains only combustion products of carbon dioxide and water vapor. (Depending on furnace operation, some sulfur dioxide and nitrogen oxides might also be present.) However, until recently, most mills used a direct contact evaporator, where the hot flue gases were contacted with the incoming black liquor feed to concentrate the liquor to 60 to 65% solids before burning. The carbon dioxide and sulfur dioxide in the flue gases reacted with the sodium sulfide in the black liquor to produce hydrogen sulfide in a sufficient quantity to cause a serious odor problem and, sometimes, a health hazard.

The most widely practiced method of preventing the emission of hydrogen sulfide from recovery boiler stacks includes the oxidization of the strong

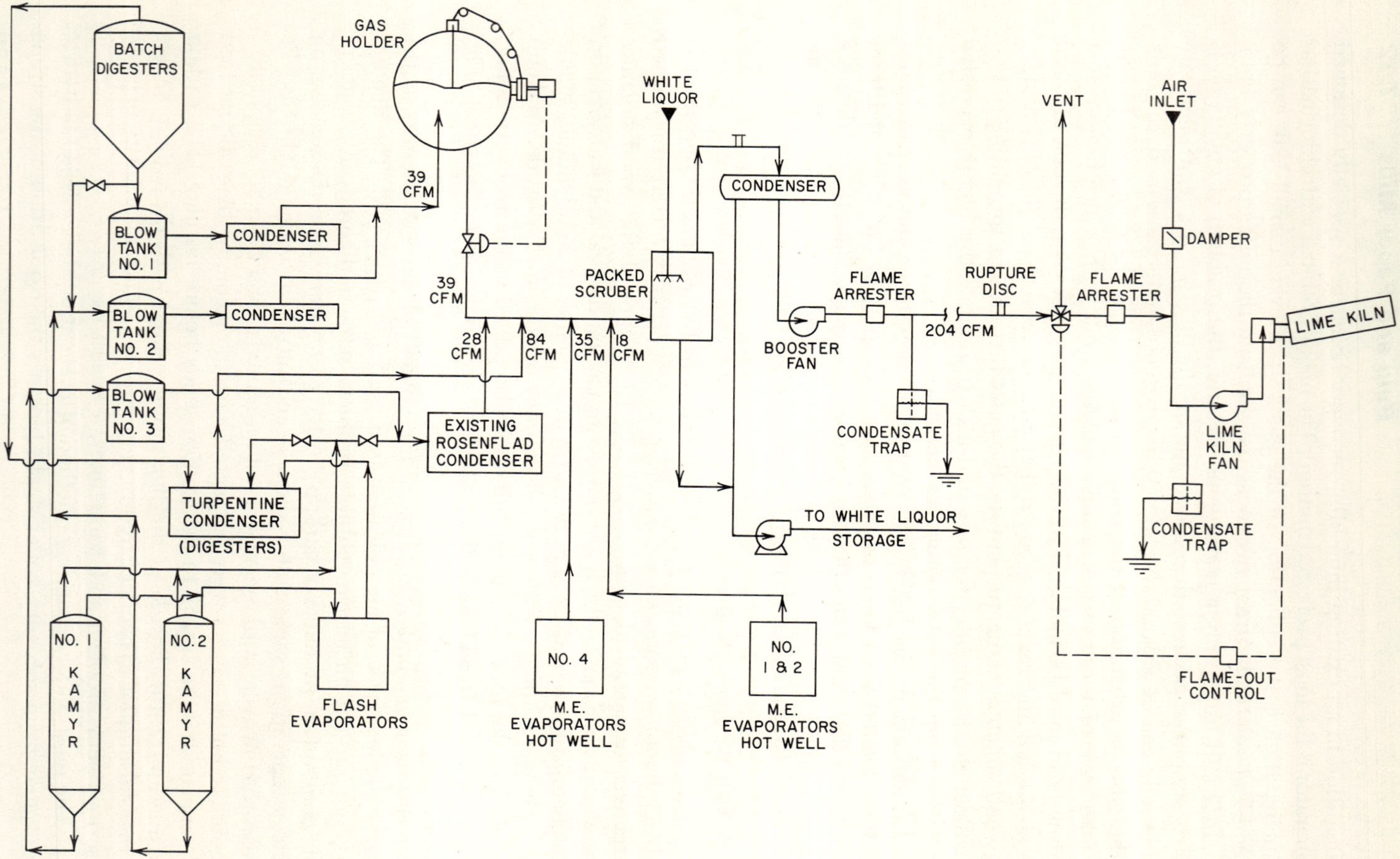

Figure 7.12a System for incineration of odorous gases from a kraft mill.

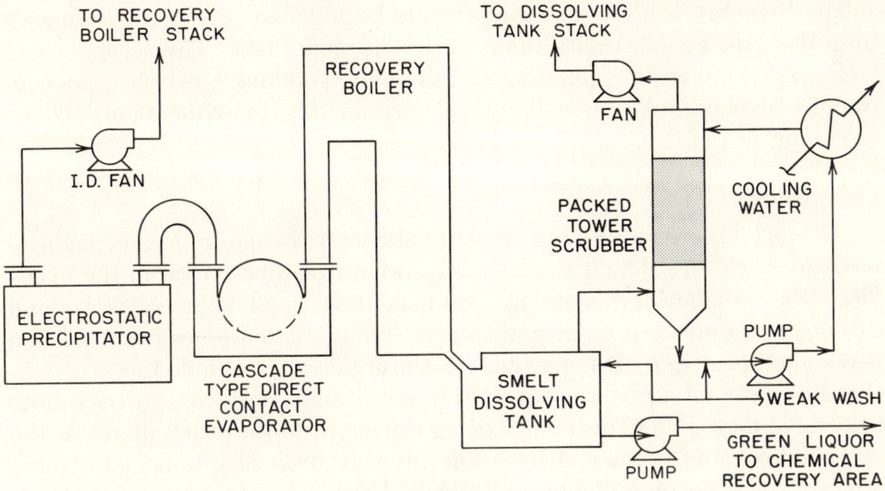

Figure 7.12b Conventional recovery boiler system.

or weak black liquor, by converting the sodium sulfide to sodium thiosulfate or to sodium sulfate before it reaches the boiler. Black liquor oxidation has been only partly successful, and scrubbers had to be installed to reduce hydrogen sulfide emissions. Scrubbers present a problem in the selection, cost and disposal of the scrubbing solutions; and this further complicates the emission problem. Perhaps the most significant new development is the advent of *odor free* recovery boilers. These boilers eliminate the need for contacting the flue gases and black liquor by one of the following methods:

1. The use of multiple-effect evaporators upstream of the boiler to concentrate the feed liquor to 60 to 65%, thus eliminating the direct contact evaporator.
2. The use of air heaters to heat the combustion air for the boiler with the hot flue gases. The heated combustion air is then used in the direct contact evaporator, replacing the flue gas.

The flue gases also contain particulates and, regardless of the prior treatment, they must be passed through an electrostatic precipitator to remove the particulates.

Another air pollution problem can be attributed to the chemical smelt which is discharged from the recovery boiler and is dissolved in water before being sent to the chemical recovery area. The hot smelt reacts violently with the water. In addition, a considerable volume of pressurized steam is also used to break up the smelt stream. This combination produces a large exhaust of steam from the dissolving tank which also carries some quantity of chemicals and particulates. Therefore a scrubber must be added to remove the chemicals from this exhaust and return them to the dissolving tank. This scrubber can be a packed tower designed to use wash liquor, containing very low concentrations of chemicals (same liquor as used in the dissolving tank), as the scrubbing media.

Lime Kilns

Lime kilns are required to convert calcium carbonate from the chemical recovery cycle to calcium oxide (lime) so that it can be reused in the cycle. Since the calcium carbonate may also contain small amounts of sodium carbonate, sodium hydroxide and sodium sulfide, the off gases from the kiln may contain calcium oxide particulate, sulfur dioxide and soda fumes. Traditionally, these off gases are passed through a high-energy venturi scrubber to remove these pollutants before being exhausted to the atmosphere. As the air pollution laws become stricter, the pressure drop requirement for these scrubbers will become higher and higher. Even now, the pressure drop requirements are high enough that alternates such as electrostatic precipitators are being studied, although no electrostatic precipitators have been installed

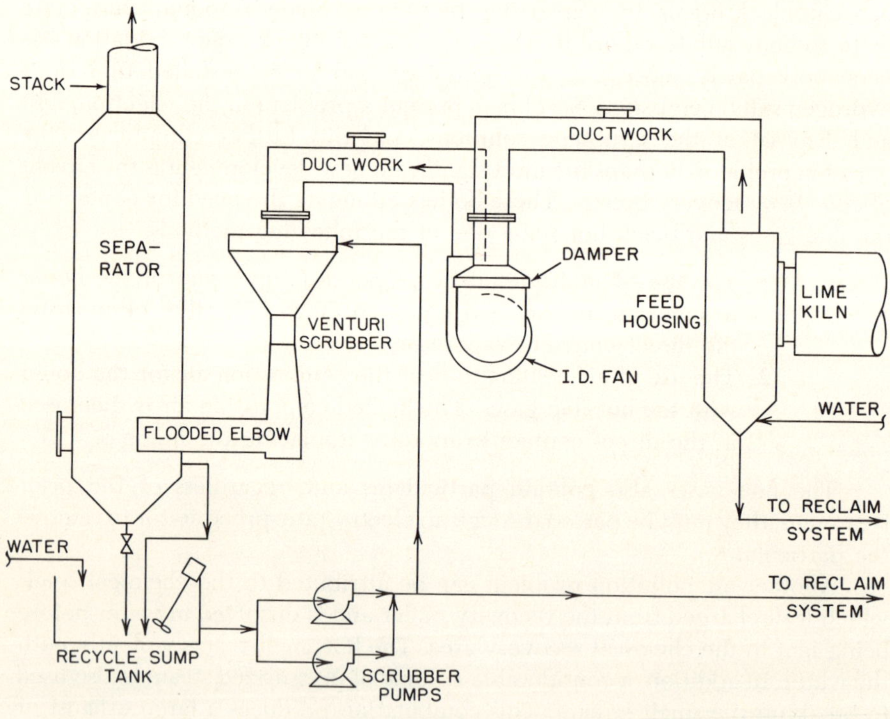

Figure 7.12c Lime kiln installation equipped with high-energy venturi scrubber.

in this service as of this writing. Figure 7.12c illustrates a lime kiln installation equipped with a high-energy venturi scrubber.

REFERENCES

1. Stephenson, J. Newell (ed.): "Pulp and Paper Manufacture, Vol. I, McGraw-Hill, New York, 1950.
2. Sarkanen, Hrutfiord, Johanson and Gardner: Kraft Odor, *TAPPI*, vol. 53, no. 5, pp. 766–782, May 1970.
3. Lipták, B. G. (ed.): "Instrumentation in the Processing Industries," Chapter IX, Chilton Book Co., Philadelphia, Pennsylvania, 1973.

BIBLIOGRAPHY

Hough, Gerald W. and Lyle J. Gross: "Air Emission Control in a Modern Pulp and Paper Mill," American Paper Industry, pp. 36–44, February 1969.

Personal Communication between S. L. McCleskey, Jr., and the author.

Lewis, C. J. and B. B. Crocker: The Lime Industry's Problem of Airborne Dust, *J. of the Air Pollu. Contr. Ass.*, vol. 19, pp. 31–39, January 1969.

Libby, C. Earl: Pulp and Paper Science and Technology, Vol. I, "Pulp," McGraw-Hill, New York, 1962.

National Council for Stream Improvement: "Current Practices in Thermal Oxidation of Noncondensable Gases in the Kraft Industry," Atmospheric Pollution Technical Bulletin No. 34, NCSI, New York, 1967.

Sarkanen, Kyosti V., Bjorn F. Hrutfiord, Lennart N. Johanson and Howard S. Gardner: Kraft Odor, *Tappi*, vol. 53, pp. 766–783, May 1970.

Stephenson, J. Newell: "Pulp and Paper Manufacture," Vol. I, McGraw-Hill, New York, 1950.

Thoen, G. N., G. G. Dehaas, R. G. Tallent and A. S. Davis: Effect of Combustion Variables On The Release of Odorous Compounds from a Kraft Recovery Furnace, *Tappi*, vol. 51, pp. 329–333, August 1968.

"Control of Atmospheric Emissions in the Wood Pulping Industry," Vol. I, II and III, U.S. Department of Health, Education and Welfare, Consumer Protection and Environmental Health Service, National Air Pollution Control Administration, J. E. Sirrine Company, Greenville, March 15, 1970.

Wenz, F. J. Hermann: *Sulphite Pulping Technology*. Lockwood Trade Journal Company, New York, 1965.

7.13 SEWAGE TREATMENT PLANTS

Odor control in municipal sewage treatment is the subject of increasing public concern, tightening local regulations, tougher legislation and more frequent use of the legal recourse by offended parties. Careful monitoring, identification, treatment and control of odors is therefore required.

The odorous gases emanating from wastewaters can be irritating, but fortunately they are mostly nontoxic. The toxic gases, such as ammonia or hydrogen sulfide, are usually in such dilute concentrations as to be well below the danger threshold.

It has been known for over 100 years (since John Snow traced a cholera epidemic to a sewage contaminated well in London) that pathogenic organisms (brucellosis, encephalitis, hepatitis, cholera, poliomyelitis or tuberculosis) are associated with sewage. It is possible that biological material may enter the air in some of the processes used in the treatment of waste waters and may be transferred by the wind to adjacent locations.

No measurements are as yet available to detect the air transport of virus from sewage plants, but virus is much smaller than bacteria and more capable of withstanding environmental changes. They too are likely to be readily air transported. The spreading of disease by aerosol transport is difficult to prove and the chances for infection are probably small, but potentially serious.

Sanitary facilities usually have an unfavorable image, causing many local complaints including that they are a bad influence on adjacent real estate values and, in general, a difficult public relations problem.

Identification of Odors

Many of the odor producing chemicals have not yet been accurately identified or measured (See Chapter 6). Hydrogen sulfide is suspected to be one of the main contributors of sewage plant odor. There are other inorganic odors (ammonia, sulfur dioxide or carbon disulfide). There are mercaptans, proteins degraded by bacteria, often resulting in various amines. The odor threshold of concentrations of mercaptans, certain amines or hydrogen sulfide is about 10 times lower than that of sulfur dioxide and it in turn is 10 times lower than the threshold for ammonia.

When several odor-producing chemicals are emitted simultaneously, there are synergistic effects and accurate determination is difficult.

Table 7.13a

SOURCES OF ODORS IN MUNICIPAL WASTEWATER TREATMENT PLANTS

Code	Location	Odor Source	Comment
1A	Sewage entering treatment plant	Gas produced by anaerobic reaction in sewer line	Certain industrial wastes can intensify the odor (food processing)
1B	Weirs	Volatile gases escape	Minimize weir drops elevation
1C	Screening, grit removal	Evaporation from grit, screenings or wash water	
2	Aeration tanks or basins	Evaporation of small droplets or the collapse of bubbles causes formation of aerosols	
3	Heat processing of sludge	Volatile products from digestors or pressure cooking off-gases	Proper design can eliminate these odors
		Supernatant hot liquors and warm sludges	Cooling reduces odor
4	Ponds or holding tanks	Usually caused by anaerobic reaction	Aeration will prevent odor
5	Sludge handling and solid/liquid separation	Conveyors, filters, trucks, landfills and drying beds	Proper sludge treatment can eliminate odor
6	Sludge incineration	Incomplete combustion	Proper design and operation will eliminate these odors

7.13 Sewage Treatment Plants

Sources of Odor

There are many places or unit processes in the treatment of sewage where annoying odors can be generated. Table 7.13a gives some typical or potential sources. The specific source locations are identified on the flowsheets for two typical sewage treatment plants (Figs. 7.13b and c). Figure 7.13b shows a relatively conventional biological treatment process with trickling filters to remove some of the suspended solids and organics and with sludge digestion equipment. Figure 7.13c illustrates a physical chemical treatment process, where phosphorus and dissolved organics are also largely removed and where activated carbon is recycled and reused. These diagrams are not explained here because a detailed description of sewage treatment processes is given at the end of Chapters I.5 and I.6. The numbers in these figures refer to locations for potential odor sources corresponding to the numbered items in Table 7.13a.

Odor Prevention or Removal

The technology for the prevention and/or removal of odors from sewage treatment processes is an empirical art (Chapter 6). There are three ways in which odors can be controlled: (1) changes to the treatment process to prevent the generation of odors; (2) removing the odorous chemicals and destroying or changing the nature of the odor-producing gases or vapors; and (3) emitting a counteracting odor to mask or ameliorate the action of the irritating odor.

Shown in Table 7.13d are the techniques for the control of odors in

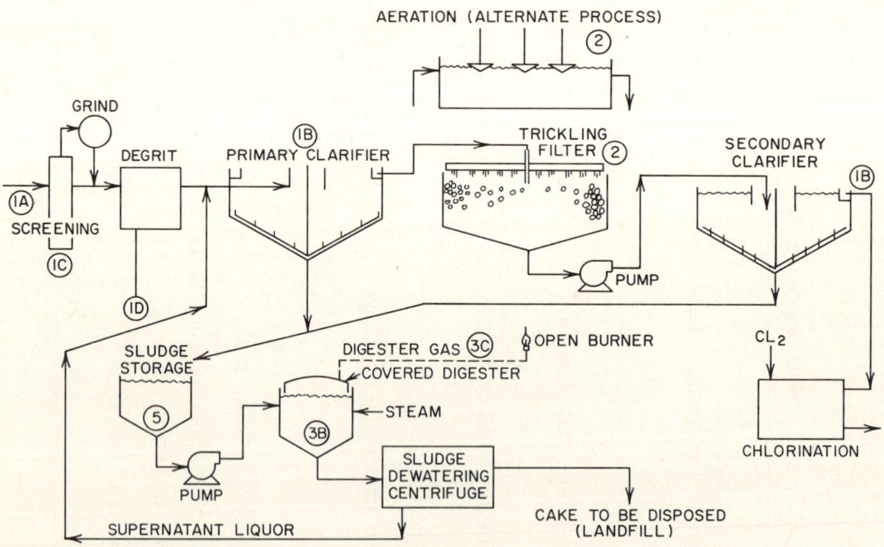

Figure 7.13b Primary and secondary treatment with sludge digestion.

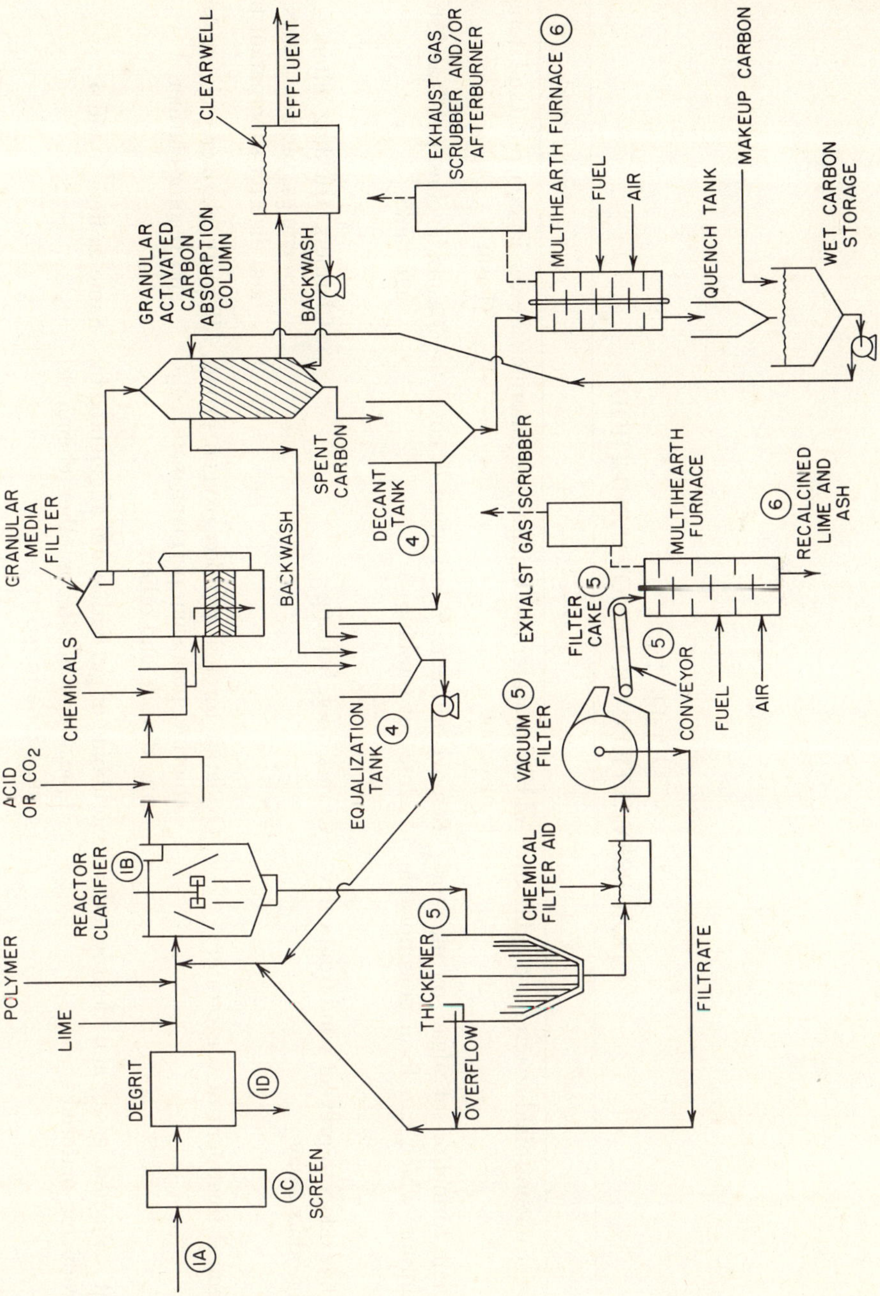

Figure 7.13c Physicochemical treatment of wastewater.

Table 7.13d
ODOR PREVENTION OR REMOVAL METHODS

Code	Method	Suggested Action	Typical Location
1	Prevent uncontrolled anaerobic conditions	Add air to water and agitate; add chlorine or ozone	Sewer, lagoon, pond and storage basin
2	Prevent formation of algae or microorganisms, which can form objectionable odors	Remove nutrients; chlorination, or cover to close out sunshine	Basins, lagoons, ponds and storage tanks
3	Confine odors to prevent wind from carrying odors away	Covers, roof, housing	Clarifiers, trickling filters, screenings, vacuum filters and digester supernatant tanks
4	Chemical treatment of wastewater	Oxidation through aeration, oxygen, ozone, chlorine, potassium permanganate, chlorine dioxide and others	Primary clarifier, raw sewage and reactor-clarifier
5	Treat collected gases and destroy odor-producing material	Scrubbing of gas with water augmented by lime, hypochlorite or other chemicals; add ozone to gas	Same as item 3 and 4

6	Radiation treatment	Ultraviolet or gamma radiation—usually to augment chemical treatment	
7	Adsorb on activated carbon	Use granular carbon column or powdered carbon	Add to treatment flowsheet
8	Biological treatment of odorous gas (proven in a few experimental installations)	Pass gas through packed column with living biological culture	Need separate unit process for gas treatment
9	Combustion	Add wet scrubber	Incinerator exhaust gas
10	Electrolytic treatment	Partial oxidation and transformation of odorous chemicals	Digester supernatant
11	Masking or counteracting	Spraying of special chemicals to emit strong, but nonirritating odor to cover or mask the objectionable odor	Downwind of plant
12	Improve plant operation	Minimize spills, upset biota, improper sampling procedures, bypassing of key process steps, water standing in storage tanks, poor housekeeping; train operating personnel	

municipal wastewater treatment. The selection of the best technique for a given odor problem will depend on many factors: the type, intensity, frequency, duration of the odor, the specific chemicals (if known) that cause the odors, their quantity, the local weather conditions, the plant process design, the skill and experience of the operators, the prior experience with the specific odor control process, the capital and operating costs and the background and experience of the people doing the selecting.

Conclusions

The exact mechanisms for generation of specific odorous chemicals and the variation of the odor intensities with treatment process parameters or with raw waste characteristics are not really understood. Once the techniques for identifying, characterizing and measuring specific odors become available, it will be possible to readily select the most effective methods of odor prevention and abatement for each particular application.

There is scientific evidence that disease can be spread by carrying aerosolized bacteria and virus from wastewater treatment plants. There are no hard data indicating the extent of the danger of the spreading of diseases in this manner. By proper plant design and good operating practices, it is believed that this danger can be essentially eliminated.

Many municipal wastewater treatment plants have had a bad public image, mostly because of their odors. As plant designs and operations improve, sewage plants will operate without irritating odors.

Many of the irritating odors occur only during relatively short periods of time. These are believed to be caused by occasional overloading, low flow, warm summer flow and/or by the temporary occurrence of toxicants or special industrial wastes which are mixed into the sewage. The optimum remedy, therefore, is one that can be applied on a demand basis when an odor problem develops. As of today, good, reliable, reproducible odor-sensing instruments are not available.

In this section (see Table 7.13d), twelve methods of preventing, abating or controlling malodors from municipal wastewater treatment plants were listed. Generally speaking, one or more of these techniques are usually effective for treating any one or a combination of several odors.

BIBLIOGRAPHY

Adams, A. P. and J. C. Spendlove: "Coliform Aerosols Emitted by Sewage Treatment Plants," *Science*, vol. 169, no. 18, pp. 1218–1220, September 1970.

Harris, S. M.: "Incineration—Multiple Hearth Furnaces," *Water and Sewage Works*, August 1967.

Hicks, Ronald: "Odor Control in Wastewater Treatment," paper presented at Second International Clean Air Congress, Washington, D.C., December 1970.

Mackie, W. C.: "Odor Control in Sewage Treatment Plants," *Water and Sewage Works*, p. 488, September 1966.

"Research on Tastes and Odor," Report of the American Water Works Association Committee on Tastes and Odors, *Journal of AWWA*, vol. 62, no. 1, January 1970.

Sutton, G. P.: "Odors and Air Pollution from the Treatment of Municipal Wastewater," paper presented at Air Pollution Control Association Meeting, Atlantic City, July 1971.

7.14 SMELTING OF ORES OF NONFERROUS METALS

Air pollutants are evolved in the high-temperature processes employed to win nonferrous metals from their ores. These high-temperature processes are correctly identified as *pyrometallurgy;* the term *smelting,* while less correct, is more widely employed and will be used here.

Pollutants arise not only from the generation of volatile compounds during smelting, but also by the entrainment of solid particles in gas streams employed in or arising from the process. Pollutants commonly found in nonferrous smelting include:

Particulates:	Metal containing dusts; ash from fuels used to furnish process heat; mists, such as hydrated sulfur trioxide.
Gases:	Sulfur dioxide; fluorides; volatile metal oxides or other metallic compounds; carbon monoxide.

Chemical Nature of Nonferrous Ores

Nonferrous metals of commercial importance most often occur in the natural state as sulfides or oxides. Common ores of a number of metals are listed in Table 7.14a.

Oxide Ores Processes

Metals are recovered from oxide ores by reducing the oxide, usually using carbon, although other reductants such as hydrogen are sometimes employed. Energy is supplied by carrying out the process either in an electric furnace or in a blast furnace. Other more complex processes are also employed for winning certain specialty metals such as beryllium.[1,2]

Control of Air Pollutants

Blast furnace control techniques are discussed in Section 7.15. In electric furnaces, when used for metals other than aluminum, gases are generated by reaction of the oxygen in the ore with carbon. Sulfur, which usually is present in small amounts in the electrodes or in the reductant, will also react, forming volatile sulfur compounds. Gases leaving the furnace entrain particulates, principally from fresh feed material. These gases are collected by a hood where

Table 7.14a
PRINCIPAL ORE TYPES
OF NONFERROUS METALS

Oxide Ores	*Sulfide Ores*
Aluminum	Cadmium[a]
Beryllium	Cobalt[b]
Chromium	Copper
Manganese	Lead
Tin	Mercury
Titanium	Molybdenum
Tungsten	Nickel
Zirconium	Selenium
	Zinc

[a] Usually associated with lead and zinc and recovered as a by-product.
[b] Almost always associated with iron, copper, nickel or silver.

they are mixed with air drawn in from the room. The configuration of equipment involved is shown in Figure 7.14b.

Solids Recovery. This is accomplished either by dry collection or wet collection. A description of various types of equipment available for this function is given in Sections 5.7, 5.11, 5.21 and 5.24. The recovered solids usually are recycled to the process.

Wet collection systems have the advantage of accomplishing virtually complete solids removal, which can be important where the gas is to be burned as a fuel.

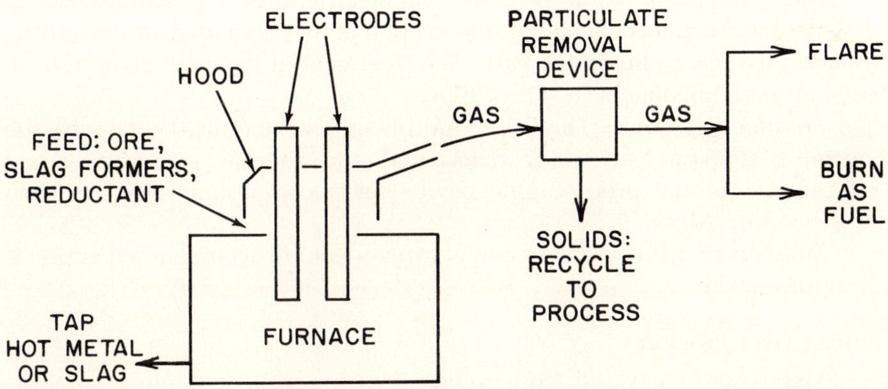

Figure 7.14b Pollution control techniques. Electric furnace reduction of oxide ores other than aluminum.

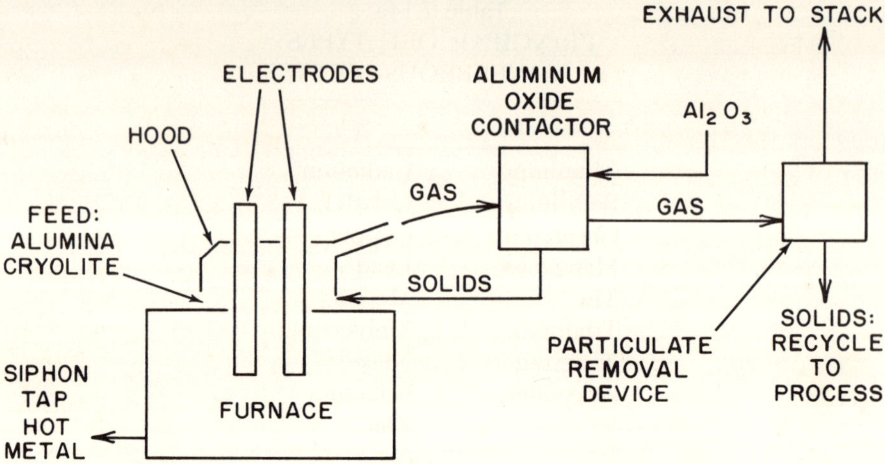

Figure 7.14c Pollution control techniques applied to the electric furnace reduction of aluminum oxide.

Gas Treatment. This usually follows solids recovery and involves combustion in those instances where carbon monoxide is present in appreciable amounts. The gas may be burned under controlled conditions and the heat recovered or it may be sent to an afterburner or flared. A discussion of afterburners is given in Section 5.27. Treatment to remove sulfur dioxide may also be required, depending on its concentration and on applicable air quality regulations. If so, refer to the discussion of sulfur oxides control methods later in this section. (See also Section 6.12.)

Aluminum Furnace

Gases are generated in the aluminum electric furnace process exactly as described in the preceding paragraph except that they contain fluorides, arising from the use of cryolite in the bath. The treatment of the gases evolved from an aluminum pot line is shown in Figure 7.14c.

Fluoride Recovery. This is accomplished by contacting the gas with the incoming alumina feed which reacts with the fluorine, capturing it and returning it to the process. One device for accomplishing this has been developed by Alcoa.[3]

Subsequent solids recovery and gas treatment are accomplished as previously described.

Sulfide Ores Processes

Metals are recovered from sulfide ores by removing sulfur as sulfur dioxide, utilizing oxygen in the form of air. Metal is recovered either as the metal itself (as, for example, with copper) or as an oxide, in which case a subsequent reduction step is employed to win the metal (as, for example, with molybdenum).

While the sulfur oxidation reaction is exothermic, high temperatures are required to initiate the reaction and, also, to permit liquid phase separation between metal and slag. Consequently, the overall smelting process consists of various reaction vessels in which chemical reactions or physical separations take place, all at elevated temperatures.

A certain amount of the sulfur dioxide generated in the smelting process becomes further oxidized to sulfur trioxide. This complicates the air pollution control problem because as the sulfur-containing gases are cooled, the sulfur trioxide absorbs moisture and becomes a mist. This mist is extremely finely dispersed and is difficult to remove; it also imparts visibility to the stack exhaust.

A number of smelter processes are listed in Table 7.14d. Those most commonly applied industrially are illustrated in Figures 7.14e, f and g.[4,5]

Control of Particulates

Particulates arising from smelter processes can consist of dust particles from a variety of sources such as the ores themselves, slag-forming additives and ash from fuels burned to generate heat for the process. The composition of the particulates can influence the selection of the control process, since in some instances hazardous materials may be present, such as compounds of beryllium, cadmium or selenium.

Electrostatic precipitators are preferred for particulate control in many instances. Since they can be operated at reasonably high temperatures, serious corrosion problems from "dewing out" sulfuric acid mist can be minimized. Also, the reclaimed solids contain recoverable values and recycling is facilitated if it is not required to separate the solids from water as one would have to do with a wet scrubbing system. However, the electrostatic precipitator will not always achieve the degree of control required. (See Section 5.24.)

Other devices which should be considered are bag filters, centrifugal collectors and wet collectors. See Sections 5.7, 5.11 and 5.21.

It may be necessary to consider treatment of the air exhausted from an entire smelter building. Such treatment can be extremely costly since the quantities of air involved may be very large. Therefore, the alternative of reducing particulate loading by the use of hoods and localized dust collection systems should be considered.

Control of Sulfur Oxides

Selection of the preferred process to control sulfur oxides emitted from an existing smelting process is complicated by the following important considerations.

Differing Gas Concentrations. Some smelting process steps emit sulfur

Table 7.14d

SMELTING PROCESS METHODS USED IN THE RECOVERY OF METALS FROM SULFIDE ORES

Process Method	*Purpose*	*Continuous (C) or Batch (B)*	*Relative SO_2 Content of Gas Stream*	*Used in the Recovery of (Metal)*	*Notes*	*Figure*
Multiple-hearth roaster	Roasting	C	Rich	Cu, Mo, Ni and Zn		7.14e
Fluidized-bed roaster	Roasting	C	Rich	Cu, Mo, Ni and Zn		7.14e
Flash smelter	Roasting	C	Rich	Cu	Relatively new technology	7.14e
Reverberatory furnace	Roasting; slag-matte separation	C	Weak	Cu and Ni		7.14f
Converter	Conversion of matte to blister	B	Rich	Cu and Ni		7.14g
Sintering machine (downdraft)	Roasting	C	Weak	Pb and Zn		7.14f
Sintering machine (updraft)	Roasting	C	Rich	Pb and Zn		7.14e
Blast furnace	Reduction	C	Extremely weak	Pb		
Retort	Reduction	C	Extremely weak	Zn	Control of zinc oxide fume may be a problem	

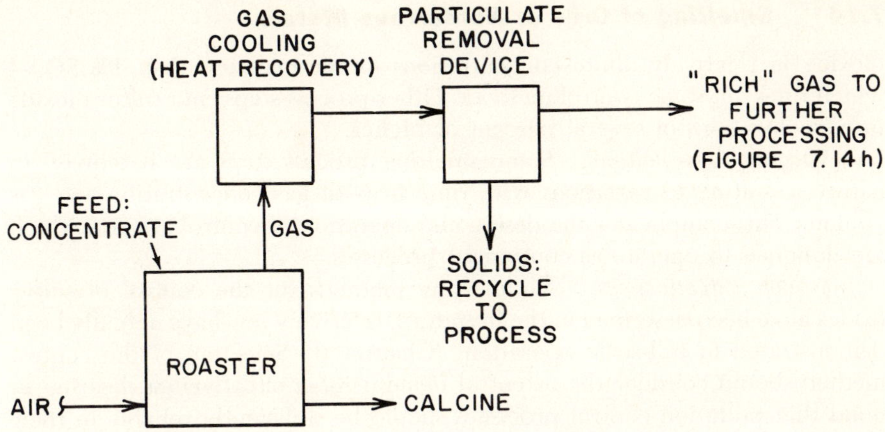

Figure 7.14e Pollution control techniques applicable to nonferrous smelting roasters (also applicable to flash smelters and updraft sintering machines).

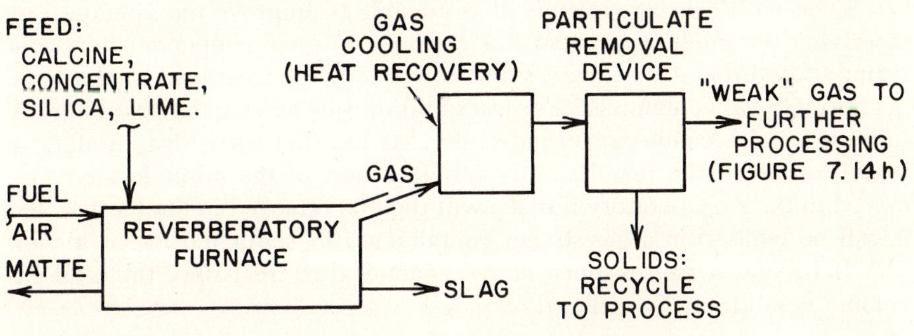

Figure 7.14f Pollution control techniques applicable to nonferrous smelting reverberatory furnaces (also applicable to downdraft sintering machines).

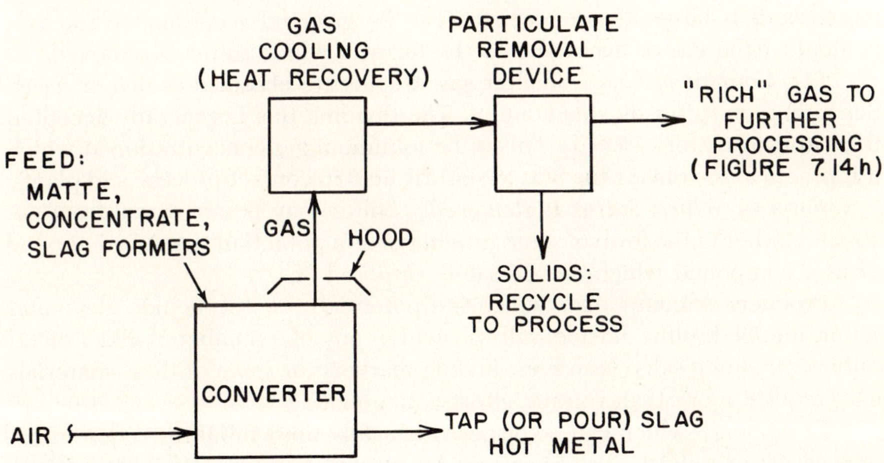

Figure 7.14g Pollution control techniques applicable to nonferrous smelting converters.

dioxide in relatively dilute concentrations—sometimes less than 1% SO_2—which rules out some control methods. Other process steps emit sulfur dioxide in concentrations of several percent or higher.

Varying Gas Volume. Some smelting process steps are batchwise in nature, resulting in variations with time in both gas concentration and gas volume. This complicates the design and operation of control systems which are designed to operate as continuous processes.

Proven Effectiveness. While many methods for the control of sulfur oxides have been described in the literature, relatively few have actually been demonstrated in full-scale operations (Chapter 6). Selection of the control method should consider the extent of *demonstrated* effectiveness, bearing in mind that pollution control processes should be sufficiently reliable in their operation so that they do not interfere with smelter operations.

Smelting Process Modifications

Even with existing smelters, it is possible to improve the operation by modifying the smelting process. For example, in some copper smelters consisting of reverberatory furnaces and converters, the treatment of the large quantities of gas containing 2% or less sulfur dioxide leaving the reverberatory furnace can be cumbersome and costly. As an alternative, by installing a roaster ahead of the reverberatory furnace, most of the sulfur formerly removed in the reverberatory furnace will then be removed in the roaster; and it will be emitted in a gas stream containing 10% sulfur dioxide or higher. This richer gas stream is much more economical to treat since the total gas volume is substantially reduced.

Removal from Gas Streams

The number of alternative processes for recovering sulfur dioxide from gas streams is large. These processes can be grouped according to the SO_2 content of the gas or according to the form in which sulfur is removed.

SO_2 Content of Gas. Smelter gas streams are classified as *rich* or *weak*, depending on sulfur dioxide content. The dividing line is generally accepted to be approximately 4% SO_2. This is the minimum gas concentration at which it is practical to convert the SO_2 to sulfuric acid in contact process acid plants.

Form in Which Sulfur Is Removed. Sulfur may be recovered from gas streams either in the form of some product or as a potentially useful compound or as a compound which can be safely disposed of.

Products or compounds which are potentially useful include elemental sulfur, liquified sulfur dioxide, sulfuric acid or any of a number of alkali metal sulfur-containing salts. However, finding markets for some of these materials may require more than routine effort (Chapter 6).

Compounds which can be disposed of safely must fulfill the requirement that no water pollution problem will be created by them. While gypsum is

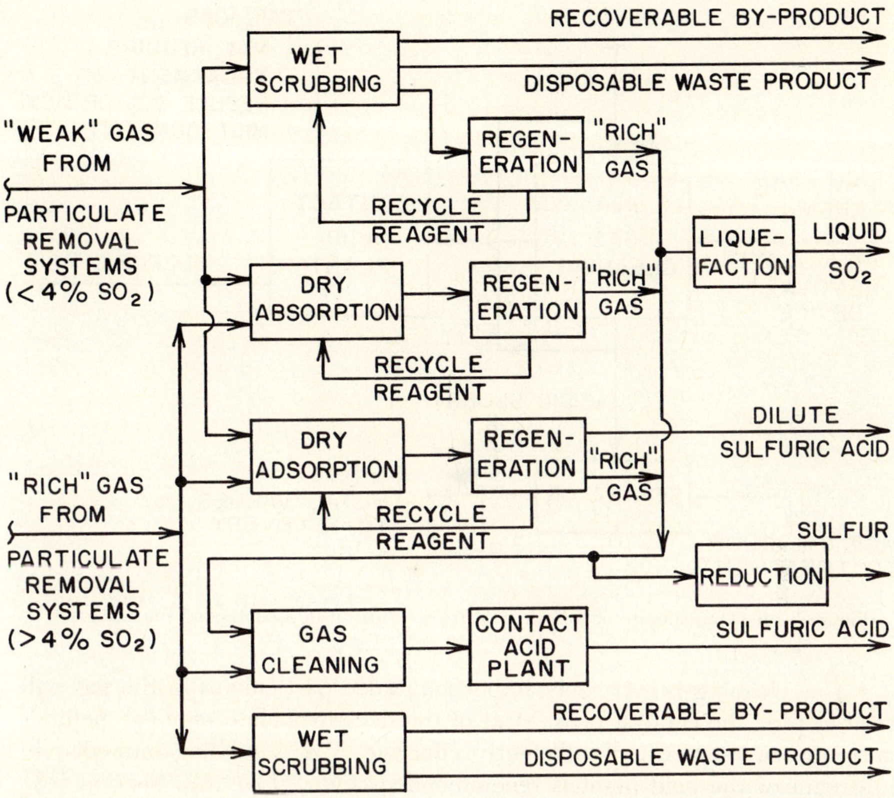

Figure 7.14h Alternative methods for treating sulfur dioxide–containing gases from nonferrous smelters.

generally regarded as meeting this requirement, care must be taken to see that only limestone containing very low quantities of magnesium is used for neutralization, since magnesium sulfate is appreciably more soluble in water than calcium sulfate.

SO_2 control technique alternatives, grouped by the type of process, are shown in Figure 7.14h, which also indicates their interrelationships where one process can or must precede another. In Figure 7.14h and in the paragraphs that follow, it has been assumed that the gas has been subjected to particulate removal prior to sulfur oxide recovery.

Sulfuric Acid Manufacture

Manufacture of sulfuric acid from sulfur dioxide–rich gas streams by the contact process has been practiced commercially for years and may be considered as proven technology.[7] When considering this control technique, the following should be considered.

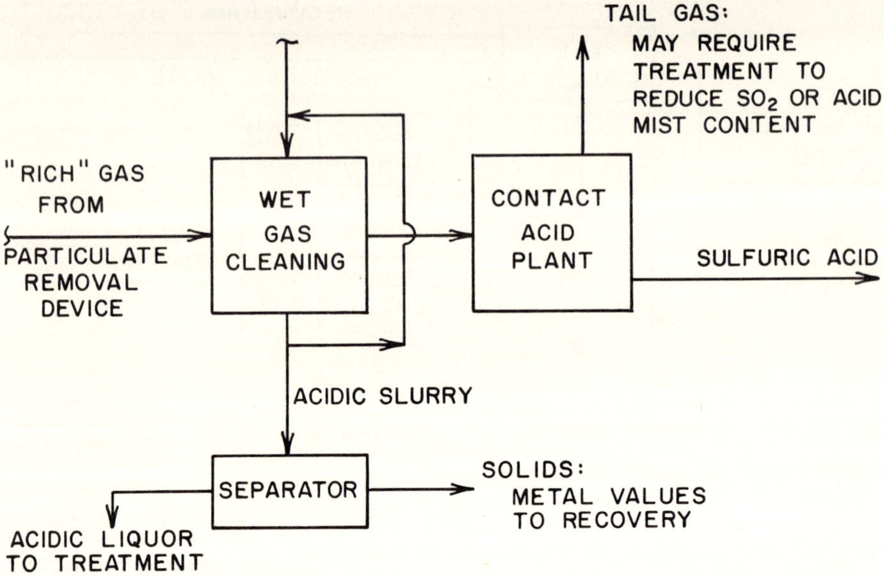

Figure 7.14i Manufacture of sulfuric acid from sulfur dioxide–containing smelter gases.

Gas cleaning is extremely important, since particulates in the gas will tend to plug the catalyst beds. Most of the dry particulate removal methods currently in use are not sufficiently effective. Wet scrubbing immediately upstream of the acid plant is recommended (Fig. 7.14i). (See Section 5.21 for a discussion of wet scrubbing equipment.)

Adjustment of the SO_2 content of the gas may be necessary to ensure that sufficient oxygen is present to oxidize the SO_2 to SO_3. Treatment of the "tail gas" leaving the acid plant may be required to meet applicable air quality standards. Sulfuric acid plant emission control techniques are discussed in Section 7.16. It may also be necessary to treat the tail gas to control the emission of sulfuric acid mist. See Section 5.19 for a discussion of mist eliminators.

Other techniques have been developed to convert SO_2 in stack gases to sulfuric acid, although none of these have been practiced commercially. Catalytic oxidation methods (the use of activated carbon and other catalytic processes) are covered in Chapter 6. These processes usually produce relatively dilute acid. The handling and disposition of this acid must be considered carefully.

Wet Scrubbing

Wet scrubbing of gas streams has been widely practiced for many years, although it has been used to only a limited extent in connection with nonferrous smelting. A discussion of the relative merits of various types of wet

Table 7.14j
IMPORTANT FACTORS TO BE CONSIDERED IN EVALUATING WET SCRUBBING OF SO_2 CONTAINING GASES

Factor	*Special Considerations*
Will air quality requirements be met?	Scrubbing effectiveness given expected range of variability in gas flowrate and SO_2 content
In what form will sulfur be removed from the process?	Marketing problems; capital and operating costs; avoidance of water pollution problems
Will it be necessary to reheat the effluent gas before sending it to the stack?	Plume buoyancy (see Section 5.45)
Is the process sufficiently reliable for plant operation?	Demonstrated commercial performance; possible plugging of equipment; anticipated losses of expensive reagents

scrubbers is given in Section 5.21. A number of factors which should be considered in evaluating wet scrubbing control methods are summarized in Tables 7.14j and k.

Dry Absorption

Dry absorption processes, in which the SO_2 is chemically absorbed on a dry solid, have been studied extensively in the laboratory but have not been used extensively on a commercial scale. Their advantage, potentially, is that the sulfur oxides can be absorbed at elevated temperatures, making subsequent reheating of the tail gas unnecessary. Also, SO_2 is regenerated as a relatively *rich* gas suitable for further processing.

In the dry absorption process, the solids are allowed to contact the stack gas stream; and after reaction with sulfur oxides, the solids are transported to a separate vessel where SO_2 is removed, usually by heating. One of the major problems to be solved in the continuing development work is reduction of loss of expensive solid reagent by attrition during transport. A number of dry absorption processes are discussed in Chapter 6.

Dry Adsorption

Dry adsorption processes, in which the SO_2 is adsorbed on the surface of dry solids such as activated carbon, have been studied in the laboratory and in pilot plants but have not been used extensively on a commercial scale.

Table 7.14k
WET SCRUBBING PROCESSES GROUPED ACCORDING TO FORM IN WHICH SULFUR IS REMOVED

Process Type	*Example*[a]
Sulfur converted to marketable by-product; scrubbing reagent consumed	Cominco process; (by-product $[NH_4]_2SO_4$)
Sulfur regenerated as "rich" SO_2 for further processing[b]; scrubbing reagent recycled	Magnesium oxide scrubbing
Sulfur converted to "throwaway" product; scrubbing reagent consumed	Lime scrubbing

[a] See Chapter 6.
[b] Further processing may include manufacture of sulfuric acid, reduction to elemental sulfur or liquefaction for transport.

In this process, adsorbed SO_2 is oxidized to SO_3 in situ using oxygen in the flue gas, and the SO_2 can be removed from the carbon by washing with water. Unfortunately, this usually produces sulfuric acid which is diluted below commercial strength.

Alternatively, the SO_3 can be removed from the carbon by heating. During this step, the carbon reacts, reducing the SO_3 to SO_2, which is swept out of the carbon bed by a flow of inert gas. Development of this process concept has been hampered by the fact that activated carbon has an auto ignition temperature which can be reached within the normal operating temperature ranges of smelter flue gases.

Tall Stacks

National ambient air quality standards have established maximum permissible ground level concentrations for various pollutants. (See Chapter 1.) In many cases, ground level or ambient pollutant concentrations can be maintained at or below the required levels by controlling a number of physical variables related to the stack exit conditions. The interrelationship among these physical variables and ambient pollutant concentrations can be expressed in various diffusion equations which are discussed in Chapter 2.

Where state regulations permit, ambient air quality requirements for single source emitters such as nonferrous smelters sometimes can be met by a combination of (1) increased stack height, (2) increased stack gas exit velocity and (3) reduced pollutant concentration in the exit gas.[6] To be effective, this control method requires that ground level air quality monitoring stations be operated. In addition, it may be necessary to curtail smelter operations on certain days when meteorologic conditions are particularly unfavorable. Meteorological considerations are discussed in Section 5.45.

REFERENCES

1. Hayward, Carle Reed: "An Outline of Metallurgical Practice," 3d ed., New York, 1952.
2. Kirk-Othmer: "Encyclopedia of Chemical Technology," 2d ed., New York, 1968.
3. Cook, C. C., G. R. Swany and J. W. Colpitts: Operating Experience with the Alcoa 398 Process of Fluoride Recovery, *J. of the Air Pollu. Contr. Ass.*, vol. 21, no. 8, pp. 479–483, August 1971.
4. Staff, U.S. Bureau of Mines: "Control of Sulfur Oxide Emissions in Copper, Lead and Zinc Smelting," Circular No. 8527, U.S. Department of the Interior, Washington, D.C., 1971.
5. Bolt, Joseph R., Jr. and Paul Queneau: "The Winning of Nickel," Canada, 1967.
6. Leavitt, J. M., S. B. Carpenter, J. P. Blackwell and T. L. Montgomery: Meteorological Program for Limiting Power Plant Stack Emissions, *J. of the Air Pollu. Contr. Ass.*, vol. 21, no. 7, pp. 400–405, July 1971.
7. Duecker, Werner W. and James R. West: "The Manufacture of Sulfuric Acid," American Chemical Society Monograph No. 144, New York, 1959.

7.15 STEEL MILLS

Ore Preparation—Agglomeration Processes

Sintering

Sintering machines are continuous, traveling grate furnaces which fuse ore fines or iron-bearing blast furnace flue dust, or both, into more easily handled lumps called *sinter.* The sinter charge contains coal or coke as a fuel. The mixture is ignited and burns as fans draw air downward through it.

Sinter plant emissions include iron oxide dust and sulfur oxides (from the coal or coke) drawn from the burning bed and exhausted by the air fans. Oxide dust is generally controlled by electrostatic precipitators, less often by baghouse filters. High-energy wet scrubbers are also suitable, especially if capture of sulfur oxides is desired. (See Chapter 5 for hardware descriptions.)

Pelletizing

Another agglomeration method is pelletizing, in which finely ground iron ore and binders are rolled into balls and fired to hardness in furnaces or kilns. The process emits ore and rock dust in the material preparation stages and emits nitrogen oxides from the furnaces. Control efforts are hampered by the sheer size of the equipment and the severe environment in which it operates.

Coke Manufacture

Iron is produced by the reduction of oxide ores with coke. Coke, a carbonaceous product of the destructive distillation of coal, is produced in batteries of narrow, rectangular retorts called by-product coke ovens. Coke oven emissions have been difficult to suppress and, in some cases, continue to defy control.

Coke ovens are charged by gravity-feeding the coal from a charging car into circular ports in the roof. The large mass of coal, dropping suddenly into the hot oven, begins to burn, adding smoke and volatilized hydrocarbons to oven gases it has displaced. These gases, carrying fine coal dust, shoot out the charging ports. Inserting retractable charging sleeves into the ports, placing the oven under negative pressure during charging and exhausting the ports to wet scrubbers have reduced, but not eliminated, the emissions. Modified charging systems, including one using afterburners for incinerating combustible gases, are being tested.

When coking of the coal is completed, the oven is emptied by opening doors at both ends, inserting a ram into one end and pushing the incandescent coke out the other onto a rail car. The hot coke at this point begins to burn, generating smoke until it is quenched by water sprays. Gaseous hydrocarbons are also emitted if the coke is pushed before all volatile constituents are driven off. Improved oven control has reduced the frequency of premature pushes, but normal smoke emissions have evaded control. Movable hoods, enclosing the coke as it emerges from the oven, have been tested. Smoke evolved under the hood in that case is exhausted to a wet collection system. Initial results using this technique were disappointing, but tests continue.

The ultimate solution to coke oven emissions may come as a result of an entirely different method of producing coke. Development of coke pellet-making processes is under way, and this or some as-yet unknown method may eventually replace by-product ovens.

Ironmaking—The Blast Furnace

Pig iron, a basic ingredient of cast iron and steel, is produced by the reduction of iron oxides (ores, pellets and sinter) by coke in a blast furnace. These raw materials, plus limestone or dolomite, are charged into the top of the furnace and slide slowly downward as molten iron and slag are tapped from the bottom. Heated air, blown into the furnace, sustains combustion reactions.

The production of 1 ton of pig iron generates about 3 tons of blast furnace gas and up to 100 pounds of flue dust. The gas, about 55% N_2, 20% CO, 20% CO_2, 4% H_2 and a trace of methane, leaves the furnace at 300 to 700°F, carrying the flue dust. Flue dust contains up to 30% iron oxides. Both emissions have economic value, the gas as a fuel, the iron-bearing dust as blast furnace charge material; so collecting them is an established practice. A typical collection system is shown in Figure 7.15a. Some installations omit the cooling tower. After leaving the precipitator, blast furnace gas is clean enough to be used as a fuel. The collected dust is agglomerated in a sinter plant.

This highly effective control system is sometimes, of necessity, bypassed. Ideally, charge materials descend slowly and uniformly through the furnace. Occasionally, however, they drop suddenly, displacing large volumes of gas. To prevent the resulting pressure buildup from damaging the furnace, relief valves are used to discharge the gas and any entrained dust to the atmosphere. These occurrences have become less frequent with the advent of more uniformly sized charge materials (pellets and sinter) and due to closer operating control of furnaces.

Steelmaking Furnaces

Open Hearth Furnace

The open hearth furnace is a shallow, rectangular, refractory chamber in which gas, oil or tar is burned with heated air to melt a charge of pig

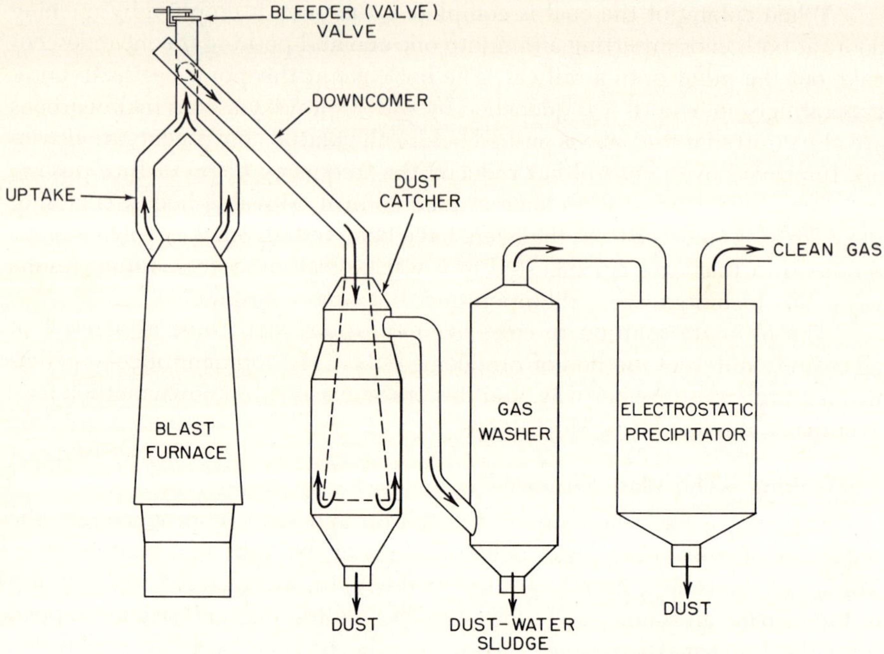

Figure 7.15a Blast furnace emission control system.

iron and steel scrap and refine it into steel. Oxygen is sometimes injected into the furnace to speed the process. Major emissions include iron oxide dust and smaller amounts of other metal oxides. Passage of flue gases through regenerative brick checkers, and occasionally through waste heat boilers (to recover heat), allows much of the coarse dust to settle out. Further heat extraction may be required before the gases are sent to a baghouse, electrostatic precipitator or high-energy wet scrubber for final treatment.

Electric Arc Furnace

Electric arc furnaces are bowl-shaped, refractory-lined vessels in which steel scrap is melted and refined. Heat is generated by electric arcs between carbon or graphite electrodes and the metal. The electrodes are suspended through openings in a removable furnace roof. Action of the intense, high-temperature electric arcs on the metal forms large quantities of submicron oxides, mostly of iron. These escape the furnace through the electrode openings and access doors in the sides of the furnace.

Emission control requires gathering the furnace gases carrying the particles and exhausting them to a collector. Baghouse filters are most widely used. Electrostatic precipitators have also been employed on occasion, but the high resistivity of electric furnace fumes has discouraged wide use. High-energy wet scrubbers are theoretically suitable but have been applied only occasionally.

The most difficult problem associated with arc furnace control is the collection of fumes. Gathering hoods and exhaust ducts must not interfere with removal and replacement of the furnace roof, with the addition of sections to electrodes as they are consumed and with the tilting of the furnace as it is tapped. Exhausting the entire building area around the furnace permits placing the fume hood out of the way but requires collection of air volumes which are several times those associated with close-fitting hoods. Consequently, the size of the fume handling system must be increased accordingly. For this reason, close-fitting hoods are preferred. One highly successful approach uses several small hoods adjacent to electrode openings and access doors. These hoods are manifolded into a common duct fixed to the structural work supporting the furnace roof. A slip joint between this duct and the one feeding the dust collector permits the tilting of the furnace. The most recent approach involves using the furnace roof as a collection hood (Fig. 7.15b). An exhaust fan maintains a slightly negative pressure in the furnace, drawing all emissions through a duct in the furnace roof. Gas volumes handled by this system are at a minimum, because little ambient air is entrained; but the high temperatures force the use of heat resistant materials and extensive cooling equipment in the exhaust system.

Basic Oxygen Furnace

The basic oxygen furnace (BOF) is a pear-shaped, refractory-lined vessel in which scrap and molten iron are melted and refined by injected oxygen. Vigorous oxidation reactions produce a heavy, reddish brown plume of fine iron oxides, other metal oxides and carbon monoxide. A typical BOF control system is shown in Figure 7.15c. Secondary air is admitted to the water-cooled collection hood to burn the carbon monoxide before the dust-laden gases are passed to the final control unit, usually an electrostatic precipitator.

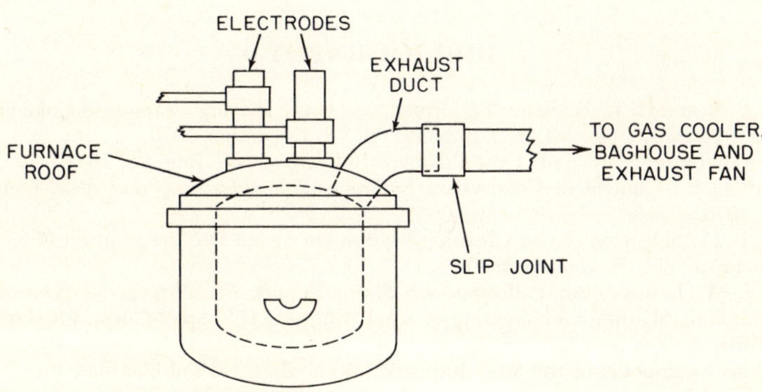

Figure 7.15b Direct evacuation of arc furnace emissions.

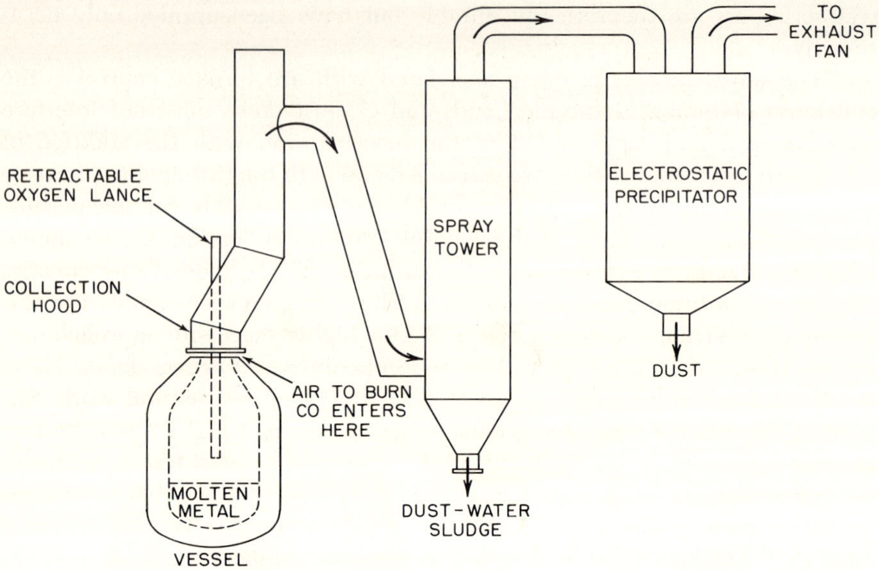

Figure 7.15c Basic oxygen furnace with emission control system.

Heating Furnaces and Boilers

A typical, large steel mill has numerous fuel-fired furnaces for heat treating, annealing and reheating steel, as well as a power-generating station. Emissions from these sources are rare; they are usually the result of improper operation of fuel-burning equipment or (in the case of boilers) the use of coal. Sections 7.2 and 7.10 describe these emissions and their control. In general, operating recommendations pertaining to gas and oil-fired boilers can be extended to steel-heating furnaces.

BIBLIOGRAPHY

Balla, P. A. and G. E. Wieland: Performance of Gas Cleaning Systems on Coke-Oven Larry Car at Burns Harbor, *Iron and Steel Engineer,* p. 78, December 1970.

Coke Pelletizing Plant Sans Dust Planned, *Industry Week,* June 28, 1971.

Dancy, T. E.: Control of Coke Oven Emissions, *Blast Furnace and Steel Plant,* p. 811, November 1970.

Egley, B. D.: Selection of Gas Cleaning Equipment for an Ore Preparation Plant, *Iron and Steel Engineer,* p. 111, November 1970.

Kane, J. M.: Foundry Air Pollution. . .A Status Report, *Foundry,* p. 50, November 1968.

"The Making, Shaping and Treating of Steel," 9th ed., U.S. Steel Corp., Pittsburgh, Pennsylvania, 1970.

"Watkins Cyclopedia of the Steel Industry," 13th ed., Steel Publications, Inc., Pittsburgh, Pennsylvania, 1971.

7.16 SULFURIC ACID PLANTS

The two main processes for the manufacture of sulfuric acid* are the chamber process and the contact process. The essential steps in the production of sulfuric acid are (1) generation of sulfur dioxide (SO_2); (2) its oxidation to sulfur trioxide (SO_3); and (3) the hydration of the SO_3 to form sulfuric acid. In the chamber process, the oxidizing mechanism to convert the SO_2 to SO_3 is the reduction of nitrogen oxide to nitric oxide whereas, in the contact process, a catalyst is used to oxidize the SO_2 to SO_3.

The main air pollutants in the exit gases from chamber process plants are sulfur dioxide, nitrogen oxides and acid mist. The sulfur dioxide emissions vary from 1,000 to 2,000 ppm in the case of sulfur burning plants, whereas the emissions may be as high as 2,000 to 4,000 ppm when other raw materials are used. The nitrogen oxides concentrations may vary from 1,000 to 2,000 ppm; and the acid mist, composed of sulfuric acid and dissolved nitrogen oxides, may be as high as 30 to 40 mg/scf of stack gas.

Since the individual plants are relatively small and the total acid produced by the chamber process is also small, a very low air pollution potential exists due to emissions from chamber plants.

The waste gas from the absorber exit stack of contact process plants contains unreacted sulfur dioxide, unabsorbed sulfur trioxide and sulfuric acid mist as atmospheric pollutants. Sulfur dioxide concentrations as high as 5,400 ppm and as low as 1,300 ppm have been observed. The acid mist content in the case of a contact plant may vary from 1.0 to 50 mg/scf of stack gas, whereas the acid mist concentrations range from 50 to 200 mg/scf of gas, leaving the drum type acid concentrator. The contact process plants represent a very high air pollution potential.

Emission Sources

Chamber Process

Less than 10% of sulfuric acid in the United States is produced by the chamber process.[1-3] The acid produced being more dilute (60°Be′, 77.7%) than from the contact process; and therefore, the transportation costs per unit of H_2SO_4 being higher, the chamber plants are captive and of low capacity. A simplified flow diagram for the chamber process is given in Figure 7.16a, which

*For some additional discussion of sulfuric acid plant emissions refer to Sections 7.6 and 7.14.

also shows the type of air pollutants and their concentrations emitted from the chamber process. The main air pollutants in the exit gases are sulfur dioxide, nitrogen oxides and acid mist.

Contact Process

The contact process,[3-5] using a catalyst to oxidize the SO_2 to SO_3, is the more modern and the more commonly encountered method. Figure 7.16b is a typical flow diagram of a sulfur-burning contact plant, while Figure 7.16c is a typical flow diagram of a metallurgical type contact plant. Sulfur dioxide (1,300 to 5,400 ppm), sulfur trioxide and acid mist (1.0 to 50 mg/scf) are the main pollutants in the waste gas from the absorber exit stack.

Acid Concentrators

Two types of concentrators,[6,7] vacuum and drum types, are used for the concentration of dilute acid, for reuse purposes or to reduce shipping costs. The concentration step is accompanied by the evolution of acid mist and sulfur dioxide. The acid mist emissions are minor in the vacuum process; whereas in the drum type concentrator (Fig. 7.16d), flashing occurs due to direct contact with hot combustion gases and results in mist formation. The acid mist concentrations range from 50 to 200 mg/scf of gas; and approximately 80 to 85% of this mist is 3 μ and smaller in size.

Control of Emissions

Chamber Process

The emissions from chamber plants[8,9] represent a low air pollution potential, as less than 10% of the sulfuric acid is produced this way and the plants are relatively small. Recovery equipment is rarely employed for pollution abatement in the exit gases from the final Gay-Lussac tower. Concentrations of sulfur dioxide in the exit gases are lower in the average chamber plant than in the average contact plant. The future air pollution potential of the chamber process is even lower because the use of this process is diminishing.

Contact Process

Sulfur dioxide can be absorbed by using alkaline materials. Various processes, some in commercial operations and some under construction or development (see Chapter 6), are available for control of sulfur dioxide from stack flue gases.

The two processes which have been adequately demonstrated* and which increase acid production include the (1) dual absorption process and (2) the Wellman–Power Gas process.

*See page 1009 for EPA evaluation.

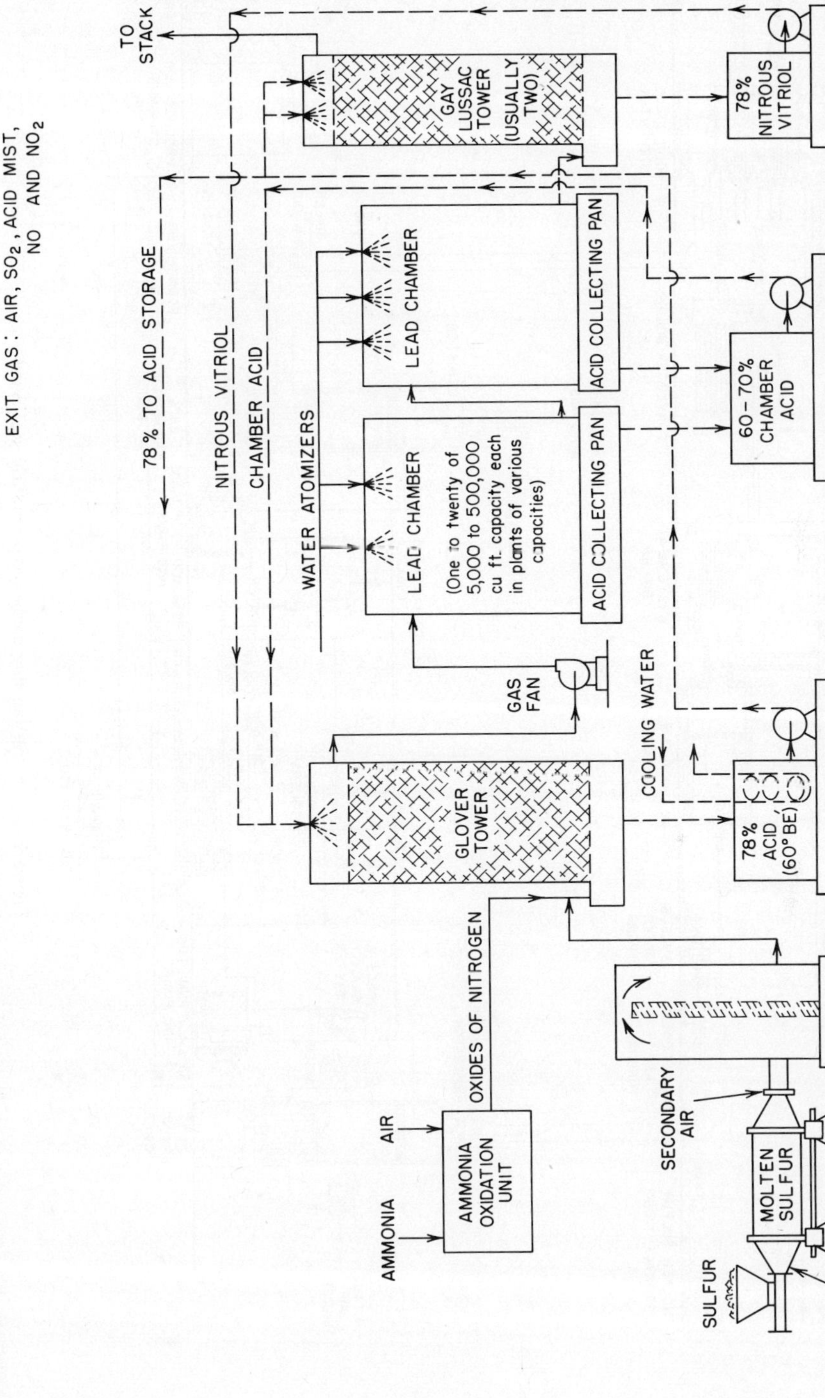

Figure 7.16a Lead chamber process for sulfuric acid manufacture (based on use of elemental sulfur as the raw material).

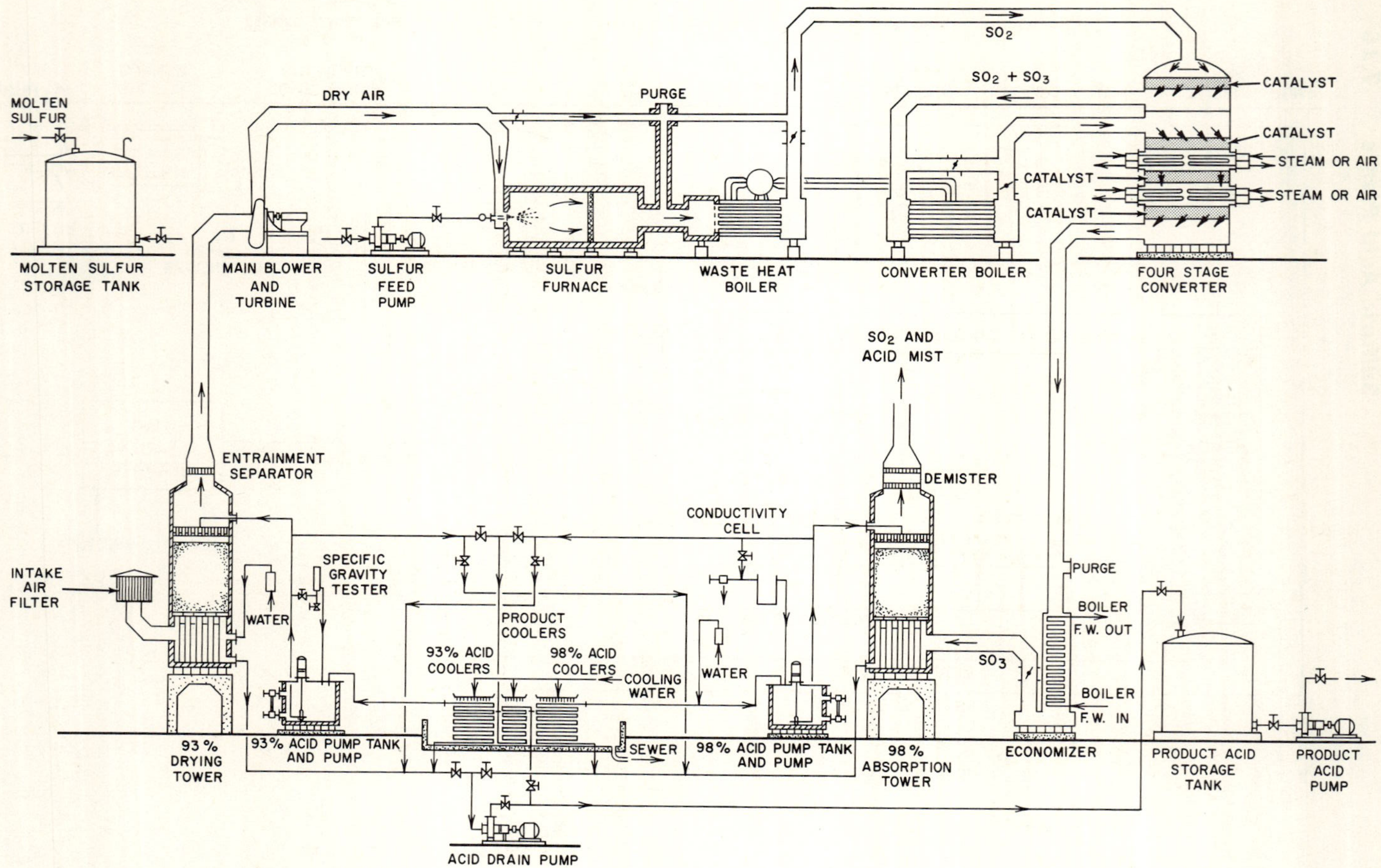

Figure 7.16b Sulfur-burning contact type sulfuric acid plant.

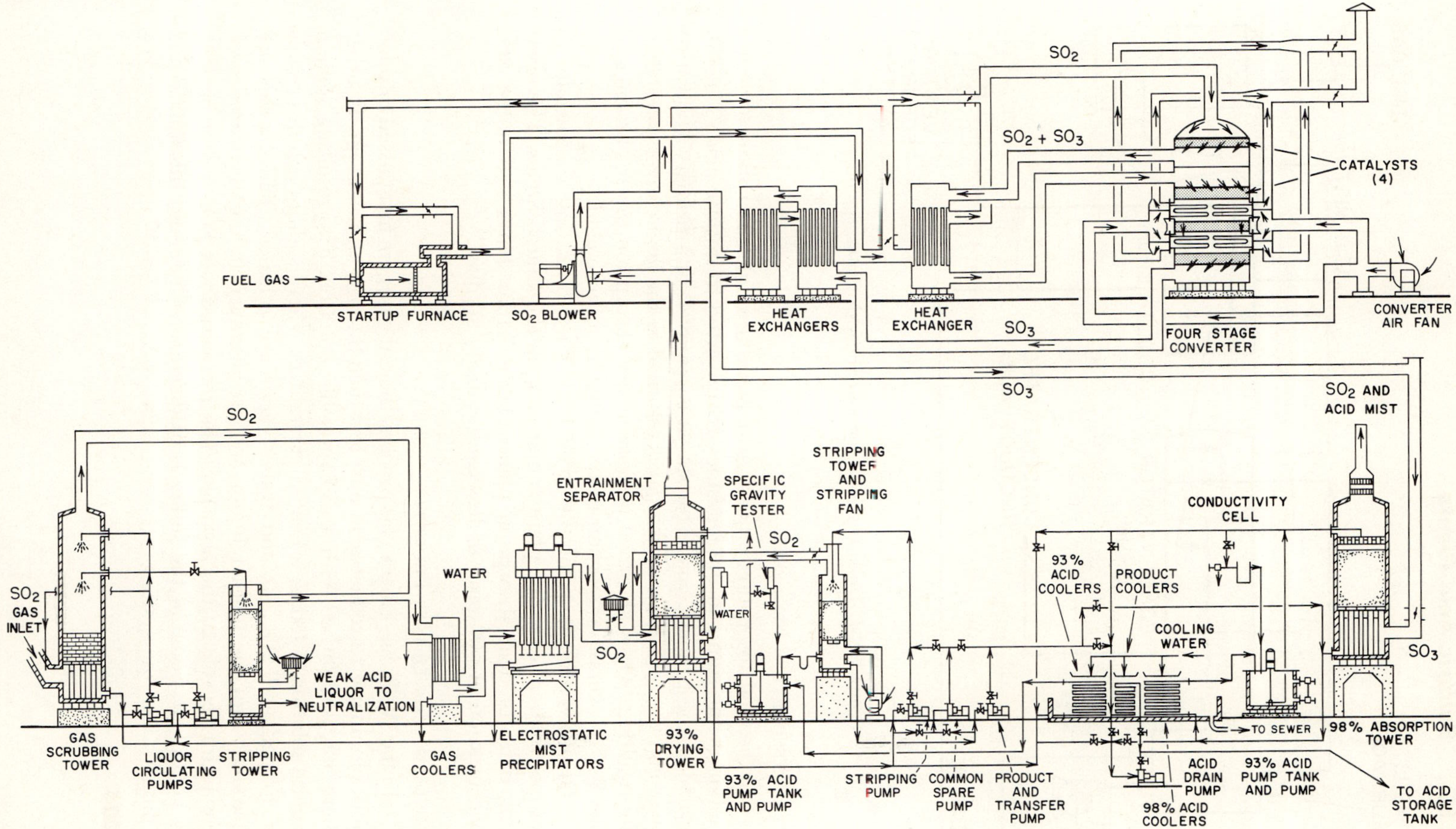

Figure 7.16c Metallurgical contact type sulfuric acid plant.

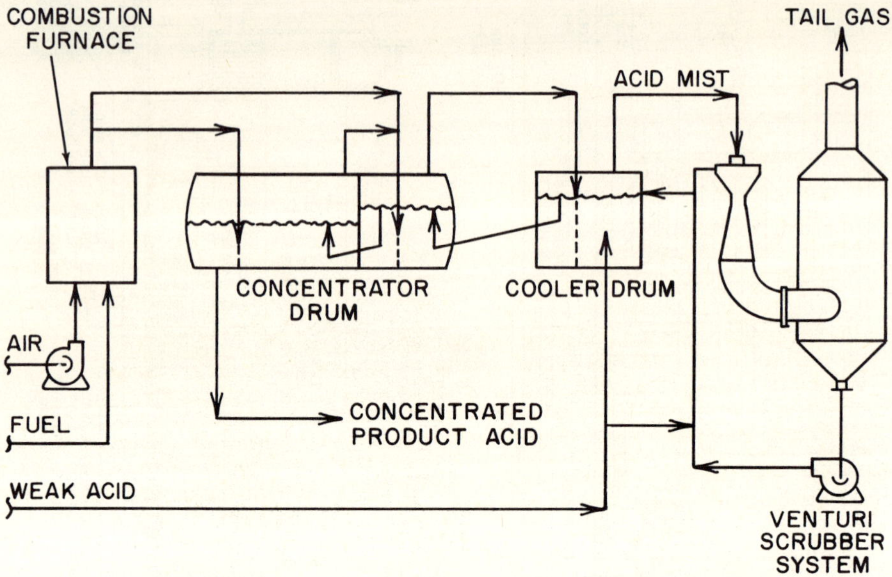

Figure 7.16d Drum type acid concentrator with venturi scrubber.

Dual Absorption Process

In the dual absorption process, as opposed to single absorption, a greater fraction of the sulfur in the feedstock is converted to sulfuric acid. The SO_3 formed in the primary converter stages is removed in a primary absorption tower and the remainder of the gas is returned to the final conversion stages. The resulting SO_3 is absorbed in a secondary absorption tower obtaining at least 99.7% overall conversion of the sulfur to sulfuric acid.

Over twenty plants of this type have been operating successfully in Europe, whereas only one plant is operating in the United States. The operating data for dual absorption plants (Table 7.16e) shows that outlet SO_2 concentration of 500 ppm or less can be easily attained.

Wellman–Power Gas Process

In the Wellman–Power Gas process, the tail gases are first passed through a mist eliminator to reduce acid mist and then through a three-stage absorber where SO_2 is absorbed with a sodium sulfite solution. The resulting sodium bisulfite solution is fed to a heated crystallizer where sodium sulfite crystals are formed and SO_2 gas and water vapor are released. The crystals are separated from the mother liquor and are dissolved in the recovered condensate for recycle to the absorber. The recovered wet SO_2 is sent back to the acid plant. Some of the sulfite is oxidized to sulfate and thiosulfate and is purged from the system.

Table 7.16e
OPERATING DATA FOR EUROPEAN DUAL ABSORPTION PLANTS

Plant Location	*SO_2 Source*	*Capacity, tons per day H_2SO_4*	*Gas Flow, SCFM*	*SO_2 Inlet Concentration Range, %*	*Conversion Efficiency Range, %*	*Outlet SO_2 Concentration, ppm*	*Range, lb. per ton*
Farbenfabriken Bayer AG Leverkusen Werke, Germany							
Plant 1	Roasting sulfide ore	550	28,700	8.2–9.3	99.71–99.83	139–260	2.0–3.1
Plant 2	Elemental sulfur	550	27,700	8.5–9.2	99.78–99.83	148–202	2.0–2.5
Farbenfabriken Bayer Dormagen Werke, Germany	Roasting of sulfide ores	550	—	9.0–9.7 (avg. 9.4)	99.27–99.5	510–790	6.5–9.2
Berzelius Metallhutten AG, Homburg Zinkhutte Duisberg, Germany	Roasting of zinc ore	220	15,600	5.9–6.4	99.60–99.73	190–300	3.7–5.5
Sud-Chemie AG Kehleim Werke, Munich, Germany	Elemental sulfur	110	6,900	9.09–9.23	99.90–99.92	74–91	1.0–1.2
Saurefabrik Schweizerhalle, Basel, Switzerland	Elemental sulfur	220	—	9.9–10.5	99.48–99.81	220–600	2.4–7.0

The first commercial plant handles 35,000 to 36,000 SCFM gas flow from an acid plant producing 730 tons/day of sulfuric acid. The outlet SO_2 concentration was reduced to 227 to 250 ppm which is equivalent to 2.59 to 2.85 lb SO_2/ton of acid produced.

Acid Mist Control

Mist eliminators, electrostatic precipitators and venturi scrubbers are all used in the removal of acid mist.

The *wire-mesh mist eliminator* effectively removes particles larger than about 3 μ in diameter. Particle size and corrosion from concentrated acid mist must be considered when selecting a wire-mesh eliminator. The eliminator is commonly constructed with two beds in series and operates with pressure drops of 1 to 3″ H_2O. Efficiency in the range of 92 to 96% can be attained.

The high-efficiency *glass-fiber mist eliminator* is capable of operating with acid-mist collection efficiencies of over 99%. The pressure drop is usually between 5 and 10″ H_2O, and this mist eliminator is capable of maintaining a high-collection efficiency at varying gas flowrates.

Electrostatic precipitators are highly efficient (99%+) regardless of size of the acid mist particles. Precipitators operate with pressure drops less than 1″ H_2O and may be either of the wet or dry type. Power requirements range from 25 to 90 KVA for 50 to 1,500 tons per day plants. Space requirements are large, ranging from 5 × 11 × 25 ft high for a 150 tons/day plant to 30 × 30 × 40 ft high for a 1,500 tons/day plant.

The *venturi scrubbers* are capable of operating with high acid-mist collection efficiency, 99%+, but at the expense of high pressure drops. The scrubbers also remove unabsorbed sulfur trioxide. Exit acid-mist loadings ranging from 0.5 to 5 mg/scf have been reported.

EPA Emission Standards

Standards of performance have been established only for sulfur dioxide and acid mist. The standards are:

Sulfur Dioxide — No more than 4.0 lb of SO_2 per ton of acid (100% H_2SO_4) produced or 2.0 kilograms SO_2 per metric ton.

Acid Mist
1. No more than 0.15 lb of acid mist (measured as H_2SO_4) per ton of acid (100% H_2SO_4) produced or 0.075 kilograms acid mist per metric ton.
2. No visible air pollutants shall be released to the atmosphere.

The standard allows about 0.3 to 1% of the feedstock sulfur to be released to the atmosphere as SO_2. The range 180 to 350 ppm is applicable to most contact acid plants. The standard for acid mist emission is equivalent to a concentration of 0.8 mg/scf. The standard of performance applies only to contact acid plants burning elemental sulfur, alkylation acid, hydrogen sulfide, organic sulfides and mercaptans, or acid sludge. It does not apply to metallurgical plants using pyrites or smelter gas, to chamber process plants or to acid concentrators. These performance standards apply to any facility, the construction or modification of which is commenced after August 17, 1971.

REFERENCES

1. Jones, E. M.: Chamber Process Manufacture of Sulfuric Acid, *Eng. Chem.*, vol. 42, pp. 2208–2210, November 1950.
2. Fairlee, Andrew: "Sulfuric Acid Manufacture," A.C.S. Mono. 69, Reinhold Pub. Co., New York, 1936.
3. Duecker, W. W. and J. J. West: "Manufacture of Sulfuric Acid," A.C.S. Mono. 144, Reinhold Pub. Co., New York, 1959.
4. Monsanto Designed Sulfuric Acid Plants, *Tech. Bull. Monsanto Co.*, St. Louis, Missouri, 1961.
5. "Modern Contact Sulfuric Acid Processes," Bulletin No. 116, Chemical Const. Corp. New York.
6. Berger, J. H. and A. J. Gloster: *Chem. Eng. Prog.*, vol. 43, pp. 225–236, May 1947.
7. Shah, I. S.: *Chem. Eng. Prog.*, vol. 67, pp. 51–56, May 1971.
8. "Atmospheric Emissions from Sulfuric Acid Manufacturing Processes," PHS publication No. 999-AP-13, 1965.
9. Chemical Construction Corporation, "Engineering Analysis of Emission Control Technology for Sulfuric Acid Manufacturing Processes," vols. I and II, NAPCA Publication No. PB-190-393, 1970.

7.17 TEXTILE MANUFACTURING

The pollution problem originates in the fiber processing steps illustrated by Figure 7.17a.

The light hydrocarbon solvents and plasticizers used in the manufacture of fibers include DOP, MEK and THF.*

In the weaving or knitting of synthetics, oil, that later becomes one of the pollutants, is used for lubrication purposes.

In the dyeing process, scouring is usually the first step. This is repeated

*DOP, Di-ochtyl phythalates; MEK, Methyl-ethyl ketones; THF, Tetra hydro furan.

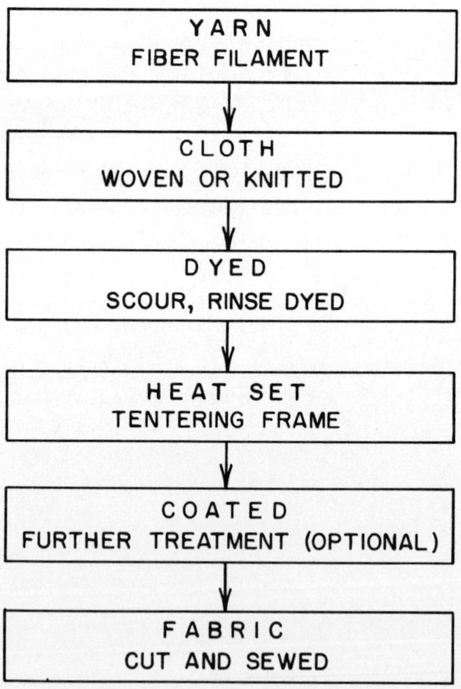

Figure 7.17a Textile processing steps.

as many times as is necessary to remove most of the lubricating oils. Fixatives and dye-setting agents are also applied and can become sources of pollution.

Fibers are heat-set which means that the cloth is brought up to its temperature of plasticity and is kept there for a specified time (0.1 to 6.0 seconds) until the desired heat-setting of the cloth is accomplished.

During the heat-setting step, a liberation of the solvents, plasticizers and all other substances, such as lubrication oil residuals, fixatives and dye carriers, takes place. This emission is a light blue or white smoke that hangs in the atmosphere and does not easily dissipate.

The air pollutants released by the textile industry are mostly high BTU content hydrocarbons. These tend to leave a residue in the ducts which represents a potential fire hazard. The pollution abatement techniques used are those discussed in Chapter 6 under "Hydrocarbon Control." The most successful installations are those which use a combination of oxidation, adsorption and scrubbing techniques. In small installations where heat recovery is not an important consideration, afterburners have been successfully used.

7.18 SECONDARY METAL PROCESSING

The reclamation of copper from scrap electrical wire and steel from junked automobiles involves burning, either to destroy nonmetallic components or to decompose them so they can be more easily separated from the metal. To prevent excessive oxidation of the recoverable metal, burning must be carried out at temperatures too low for complete combustion of the nonmetallics. Heavy smoke, chiefly partially burned materials and some ash, results. The problem is especially acute with open pit burning, since no effort is made to control either combustion or smoke evolution. More stringent air quality codes are quickly eradicating these sources of pollution.

Minimum control of burning operations requires enclosing the burning scrap in a combustion chamber and venting all smoke to an afterburner or to a secondary combustion chamber. This is usually satisfactory for wire reclamation or batch burning of a few automobiles at a time; residual levels of noncombustible particulates are tolerably low. Large, continuous auto-burning furnaces emit significant amounts of particulates. Therefore, scrubbers, baghouses or electrostatic precipitators are required for final control (Chapter 5). The three methods are competitive; selection usually depends on cost and on relative compatability of the control unit to the furnace.

The melting of aluminum alloys results in the generation of dross, a mixture of metal oxides and other impurities. When this dross is removed from a furnace, some molten aluminum is invariably trapped and carried along. Recovery of aluminum from the dross is often economically feasible. Processes vary but most involve controlled melting of the aluminum-dross mixture and treatment with chlorine gas or chloride salts to facilitate separation. Emissions include fine oxide dust, metallic chloride fumes and, sometimes, chlorine gas. Baghouse collectors will control dust and fumes; if elimination of chlorine is also desired, scrubbers are preferred.

Fuel-fired or induction furnaces are used to remelt and refine aluminum, copper, zinc and lead scrap. Typical emissions include volatilized hydrocarbons from grease or oil on the scrap, metal oxide dust, volatilized metal (particularly zinc) and flux residues (chiefly chlorine and chlorides from the refining of aluminum). Induction furnaces emit little more than hydrocarbon fumes and can usually be controlled by venting the fumes to an afterburner. Fuel-fired furnaces generate larger amounts of oxides and metal vapors; baghouses or scrubbers are usually selected for their control.

GLOSSARY

AA Spectroscopy An analytical method in which an element is atomized in order to permit the observation and the measurement of its atomic absorption spectrum.

Abatement The effort or measures taken to reduce the degree or intensity of pollution.

Absorber Apparatus used in the process in which one material is employed to retain another. Absorbers are used to remove a gaseous or liquid material selectively from another gas or liquid.

Absorption A process in which a soluble component of a gas stream is transferred into a liquid. The component may become dissolved in the liquid, or it may chemically react with the liquid or a compound dissolved in the liquid. The process depends upon an exchange of a component between a gas phase and a liquid phase. This exchange is limited by the area of the interface between the phases and by the solubility of the component in the liquid phase.

Acclimation Ability of organism to adapt to change in environment.

Accretion A process that increases the size of particles by external additions; a form of agglomeration.

Acetaldehyde See Aldehyde.

Acetylene Gaseous hydrocarbon made by action of water on calcium carbide. ($HC\equiv CH$)

Acid A substance that in aqueous solution turns blue litmus red, furnishes hydronium ions and contains hydrogen, which can be replaced by a metal.

Acid, Organic See Organic Acid.

Activated Carbon Any form of carbon characterized by high adsorptive capacity for gases, vapors and colloidal solids. Carbon must usually be activated to develop adsorptive power; achieved by heating to 800 or 900°C with steam or CO_2 which produces a porous particle structure.

Acute Toxicity Short term toxicity.

Additives Agents used to improve and control air pollutant emissions from various types of fuels such as coal, gasoline, petroleum. Additives such as finely powdered solids ($MgCO_3$, Zn metal, ZnO), sprayed liquids (Na_2CO_3) and NH_3 gas, which reacts with SO_3 in flue gases to form solid sulfates, have been used to control acid smut formation.

Adiabatic Lapse Rate A theoretical thermodynamic property describing a fluid's temperature loss (due to expansion in a contained volume), without a corresponding loss of heat. Rising air will cool due to adiabatic expansion, while subsiding air warms in response to adiabatic compression. Generally, mechanically produced turbulence will approximate the adiabatic lapse rate, which is 5.5°F/1,000 feet. When the real atmospheric lapse rate exceeds this theoretical magnitude, the lapse rate is superadiabatic; while during the evening hours, downward convection of the cooling earth's surface causes a subadiabatic, isothermal or even inverted lapse rate. The lapse rate is g/Cpd, where g is the acceleration of gravity and Cpd is the specific heat of dry air at constant pressure, equal to 9.8°C/Km or about 5.5°F/1,000 ft.

Adiabatic Process A process during which no heat is transferred between initial and final states.

Adsorbate Molecules of gas, dissolved substances or liquids which adhere in an extremely thin layer to the surface of a solid body.

Adsorbent The solid material which collects the adsorbate.

Adsorber Apparatus used for a process in which the molecules of either a gas or liquid are captured by solid material. Usually the capturing solid is activated carbon, which has a large surface area of capillary form.

Adsorption A process in which a special solid surface is able to collect selected gases or vapors. In adsorption, the molecules of the gas or vapor adsorbed contact and adhere to the surface of the solid in an extremely thin layer. Depending upon the nature of the gas or vapor and of the solid (silica gel, activated alumina, molecular sieves, activated carbon), this layer will range from one to many molecules in thickness.

Advection The transportation of an atmospheric property solely by the mass motion of the atmosphere; the term is more particularly applied to the transfer of heat by horizontal motion of air. The transfer of heat from high to low altitudes is the most obvious example of advection. Fog drifts from one place to another by advection. Cold air moves from polar regions southward.

Aeration Bringing about intimate contact between air and a liquid by such methods as spraying the liquid in the air, bubbling air through the liquid or agitating the liquid to promote surface absorption of air. In general, any process whereby a substance becomes permeated with air or other gas.

Aerometer Air or gas density detector.

Aerometric Measurements The detection of atmospheric temperatures and currents.

Aerosol A cloud of solid particles and/or liquid droplets smaller than 100 μ in diameter, suspended in a gas. (Smoke, fog, mist.)

Afterburner A burner located so that the combustion gases are made to pass through its flame in order to remove smoke and odors; it may be attached to or be separated from an incinerator proper.

Agglomeration The process by which precipitation particles grow larger by collision or contact with cloud particles or other precipitation particles.

Agglutination The process of uniting solid particles coated with a thin layer of adhesive material or of arresting solid particles by impact on a surface coated with an adhesive.

Air Ions Small particles in the atmosphere (molecules or microscopic dust particles) which have an induced electrical charge acquired through loss or gain of an electron, or through the adsorption of a molecule which has lost or gained an electron.

Air Mass A widespread body of air established while that air was either situated over a particular region of the earth's surface or undergoing specific modification while in transit away from the source region.

Air Monitoring Station System to monitor continuously the concentrations of gaseous pollutants. In conjunction with the Continuous Air Monitoring Program (CAMP), these have been located in or near the principal business district in each city where the program established such a station. Data collected by these stations provide continuous, comparable, concurrent information on significant gaseous air pollutants (SO_2, NO, NO_2, CO, hydrocarbons, O_3 and oxidants) from a number of urban areas.

Air Pollutant Substances added in sufficient concentration to produce a measurable effect on man or other animals, vegetation or material. Pollutants may thus include almost any natural or artificial composition of matter capable of being airborne. They may occur as solid particles, liquid droplets or gases or in various admixtures of these forms. Exclusive of pollen, fog and dust, which are of natural origin, about 100 contaminants have been identified and fall into the following categories: solids, sulfur compounds, organic compounds, nitrogen compounds, oxygen compounds, halogen compounds, radioactive compounds, odors. Solids include carbon, flyash, ZnO, PbCl; gaseous and other airborne compounds include SO_2, SO_3, H_2S, mercaptans, aldehydes, hydrocarbons, tars, NO, NO_2, NH_3, O_3, CO, CO_2, HF, HCl; radioactive compounds include radioactive gases and aerosols.

Air Pollution Control District District created and authorized by state legislation to adopt comprehensive control programs and establish the limits of air pollution contaminants which may be emitted from sources within their jurisdiction.

Air Pollution Index An arbitrarily derived, mathematical combination of air pollutants which give a single number attempting to describe the ambient air quality. The formula $API = 20 [SO_2 + 1 (CO) + 2 (\text{smoke shade})]$ has no scientifically derived basis. Experience indicates that the average value of this index is 12.0. An index reading of 50.0 is considered adverse and a cause for alarm.

Air Pollution Source Types

Continuous Point Source—Strong powerplant.

Continuous Line Source—Main urban traffic artery.

Continuous Area Source—An entire urban complex.

Continuous Volume Source—An entire urban complex.

Virtual Point Source—A series of powerplants whose emissions are entrained within the plant volume and can be extrapolated behind the emission complex to a virtual point.

Instantaneous Point Source—Explosion, volcanic eruption.

Instantaneous Line Source—A plane emitting pesticides.

Air Pollution Survey A critical examination of a given geographical area for the purpose of determining the nature, sources, extent and effects of air pollution.

Albedo Reflective power indicating the fraction of incident light or electromagnetic radiation that is reflected by a surface. Black or totally absorbent bodies (such as the oceans) have low values, while white or totally reflective bodies (such as freshly fallen snow) have high values. Expressed as percentage.

Alcohol Hydroxyl derivatives of hydrocarbons, which are analogous to ethyl alcohol: C_2H_5OH.

Aldehyde Highly reactive family of compounds typified by acetaldehyde and characterized by the group CHO.

Algorism (Algorithm) In modern mathematics, any method or procedure for computation.

Aliphatic Organic compounds having an open-chain structure, consisting of the paraffin, olefin and acetylene hydrocarbons and their derivatives.

Alkali A water-soluble metallic hydroxide that ionizes strongly.

Alkoxyl Any univalent substituent group of a compound of the general composition, $—O—C_nH_{2n} + 1$ or $—O—C_6H_4C_nH_{2n+1}$ or similar molecular structures.

Alkyl Any univalent aliphatic substituent group of a general composition, C_nH_{2n+1}.

Alpha Radiation Atomic particles arising from the nucleus of radioactive material, consisting of two protons and two neutrons.

Alumina Aluminum oxide (Al_2O_3) occurring native as corundum and in hydrated forms as bauxite.

Alveolar Air Air which has mixed and equilibrated with pulmonary capillary blood while dead space air is air within conducting tubes which has not equilibrated with blood and is unchanged from ambient air. During exhalation, dead space air is exhaled first and may be discarded. The terminal portions of the exhalation may be collected and considered to represent alveolar air.

Alveolus A small, sac-like dilation at the innermost end of the airway in the lungs through whose walls gaseous exchange takes place.

Amalgam An alloy of mercury with another metal.

Ambient Air Surrounding atmospheric air.

Amine Ammonia derivate resulting from the replacement of hydrogen by one or more univalent hydrocarbon radicals. They are generated by protein decomposition during putrefaction.

Ammonia A colorless gaseous alkaline compound which is water soluble and can be easily condensed into a liquid. (NH_3)

Anemia A condition in which the number of red blood cells per cubic centimeter of blood is below normal.

Anode The positive electrode at which oxidation occurs.

Anoxia A deficiency of the amount of oxygen reaching the body tissues.

Anthropogenic Involving man's impact on nature.

Antiparticles These have the same mass and spin as particles but opposite electromagnetic properties such as charge and magnetic moment.

Arithmetic Mean The sum of numerical values divided by the number of the numerical values, such as $\frac{1 + 2 + 3}{3}$.

Aromatic Cyclic hydrocarbons and their derivatives characterized by the presence of at least one benzene or other ring structure.

Aryl Any univalent aromatic substituent group of compound having the general composition C_nH_{2n+1}—C_6H_4—.

Asbestos Although defined by Webster as "a grayish mineral, a silicate of calcium and magnesium," in reality the term asbestos is not limited to any one substance and, instead, refers to a class of minerals having certain chemical and physical features in common. They are all mineral silicates and are fibrous and noncombustible carcinogens.

Astronomical Unit 1.5×10^{13} centimeter distance.

Atmosphere The whole mass of air surrounding the earth composed largely of oxygen and nitrogen.

Atmospheric Assimilation Photochemical reactions whereby chemical substances in the atmosphere are mixed, diffused and chemically reacted, thereby maintaining concentrations reasonably constant within defined atmospheric regions.

Atomization The dividing of a liquid into extremely minute particles, either by impact with a jet of steam or compressed air or by passage through some mechanical device.

Attenuation Any process in which the flux density (or power, amplitude, intensity, illuminance) of a "parallel beam" of energy decreases with increasing distance from the energy source. A more general usage of this word is any reduction in strength, density, effect, or amplitude.

Autotroph An organism capable of manufacturing its own food and releasing oxygen in the process.

AZO Compounds Chemicals containing the group —N=N— united at both ends to carbon.

Background Concentration The concentration of constituents in normal ambient air. It varies by geographic area, by season and by altitude. It is considered to be the average composition of air.

Baffle A plate, grating or refractory wall used especially to block, hinder or divert a flow or to hinder the passage of a substance.

Baffle Chamber In incinerator design, a chamber designed to promote the settling of flyash and coarse particulate matter by changing the direction and/or reducing the velocity of the gases produced by combustion.

Bag Filter Fabric filter used in industrial operations to recover valuable matter as well as to control atmospheric pollution at the source. Such a filter is commonly of a tubular form or similar to an envelope and can be slipped over a wire frame.

Barometer A detector for sensing atmospheric pressure.

Base A substance that in aqueous solution turns red litmus blue, furnishes hydroxyl ions and reacts with an acid to form a salt and water only.

Beta Radiation Electrons emitted due to disintegration of the atomic nucleus in a radioactive substance.

Binary Arithmetic A system of calculation in which the only numerals used are 0 and 1, and all real numbers are represented as powers of 2.

Biosphere Living beings together with their environment.

Bit Condensation of binary digit. In computer and information theory, a single binary digit, either 0 or 1.

Bivane A sensitive wind vane used in turbulence studies to obtain a record of the horizontal and vertical components of wind.

Black Liquor Spent cooking liquor from a digester. Contains dissolved organic wood materials and residual active alkali compounds left over from the cook.

Boolean Algebra An algebra obeying special laws that make it suitable for such uses as the study of logic and planning switching circuits.

Breeching A passage or conduit to conduct products of combustion to the stack or chimney. A high-temperature duct lined with refractory material that conveys combustion gas from the furnace to the chimney. In large incinerators, the breechings are refractory-lined masonry construction. In smaller furnaces, the breeching may be refractory-lined steel duct.

Breeze Very fine particles of coke.

Bronchoconstriction A diminution in the size of the lumen of a bronchus.

Bronchodilatation An increase in the size of the lumen of a bronchus.

Bronchus One of the larger air passages in the lung.

Bubbler Absorption apparatus used in gas sampling. Usually U-tube absorbers filled with a specific amount of reagent and fitted with a glass partition, so that the air or gas led into them passes through the reagent solution in the form of very fine bubbles.

Buffer A solution containing either a weak acid and its salt or a weak base and its salt, thereby resisting changes in acidity or basicity.

Capacity, Forced Vital The largest amount of gas which can be forcibly expired from the lungs following a maximal inspiration.

Capacity, Vital The maximum volume of gas which can be expired from the lungs following a maximum inspiration.

Carbohydrate A compound of carbon, hydrogen and oxygen, usually having hydrogen and oxygen in the proportion of 2 to 1.

Carbonaceous Rich in carbon.

Carbon Black Finely divided forms of carbon made by the incomplete combustion or thermal decomposition of natural gas or liquid hydrocarbons. Principal types, according to the method of production, are channel black, furnace black and thermal black.

Carbon Dioxide Colorless, odorless gas or heavy, volatile, colorless liquid or white, snowlike solid.

Carbon Monoxide Colorless gas, exceedingly faint metallic color and taste; highly poisonous, inducing asphyxiation; 0.2% in air is poisonous and 0.43% will induce asphyxiation. It is found in trace quantities in the natural atmosphere but also is produced by the incomplete combustion of carbonaceous fuels.

Carcinogenesis The production of cancer.

Carina A ridgelike structure; in anatomy of the respiratory tract, it is a prominence on the lowest tracheal cartilage and is situated between the orifices of the two main-stem bronchi.

Cascade Impactor Sampling device in which air is drawn through a series of jets against a series of slides. Air particles then adhere to the microscope slides, which are coated with an adsorbing medium. Jet openings are sized to obtain a size distribution of particles.

Catalyst A substance capable of increasing the velocity of a reaction without chemically or physically changing itself.

Cathode The electrode at which reduction occurs: negative electrode.

Caustic Solid caustic is NaO. Caustic soda is its hydrated form, also called sodium hydroxide (NaOH). Soda ash is sodium carbonate (Na_2CO_3).

Ceilometer An automatic, recording, cloud-height indicator.

Ceramic Filter Component of a stack sampling system. These are suitable for high-temperature (1000°F) use.

Charcoal A product of the destructive distillation of wood used for absorbent, filtering and decolorizing media. It can absorb various gases and vapors, showing preferential absorption for vapors of liquids that are volatile at ordinary temperatures. One volume of carbon can absorb about 16 volumes of O_2 or N_2.

Checkwork A pattern of multiple openings in a refractory wall to promote turbulent mixing of products of combustion.

Chemisorption Chemical adsorption which depends on chemical bond formation between the adsorbent and adsorbate, but which is distinct from a chemical reaction in that it takes place only in a monolayer on the surface of the adsorbent.

Chimney Effect A phenomenon consisting of a vertical movement of a localized mass of air or other gases due to temperature differences.

Chromatographic Analysis A method of separation based on selective adsorption. A solution of the substance or substances desired is allowed to flow slowly through a column of adsorbent. Different substances will pass with different speeds down the column and will eventually be separated into zones. The column core can then be pushed out and the zones of material cut apart and analyzed.

Chromatography/Gas An analytical technique for separating mixtures of volatile substances. The procedure consists of introducing the sample mixture into a separating column and washing it down with an inert gas. The column is packed with absorbent materials which selectively retard the components of the sample. The surface is usually coated with a relatively nonvolatile liquid designated the stationary phase. This gives rise to the term gas-liquid chromatography. If the liquid is not present, the process is gas-solid chromatography.

Chronic Toxicity Long-term toxicity.

Climate See Weather.

Coal Desulfurization Process of removing sulfur from coal. Iron pyrites, which account for about one-half of the total sulfur content of coal, can be removed mechanically.

Coal Gasification The manufacture of gaseous fuel from coal. The product fuel can be of various heating-values and can be used for chemical syntheses as well as for combustion. A typical gasification operation includes coal-handling and/or pretreatment steps, a step for actual conversion of coal to gaseous compounds and subsequent steps to upgrade this mixed-gas stream.

Coefficient of Haze Units are based on the transmission of light through a soiled filter. Coefficient of haze units are defined as 100 times the optical density, determined by transmittance.

Coke Plant Chemical plant for the purification of gases evolved in the carbonization of coal and where commercial products such as ammonia, tar, benzene and naphtha are recovered.

Collector A device for removing and retaining contaminants from air or other gases. Usually this term is applied to cleaning devices in exhaust systems.

Combustion[1] Any chemical change that produces heat and, usually, light. Most commonly, the combination of the oxygen of the air with a substance: burning.

Compaction Compressing metallic or nonmetallic powders, as a briquette in a die.

Concentration The total mass of the suspended or dissolved particles contained in a unit volume at a given temperature and pressure.

Condensation The physical process of converting a material from a gaseous phase to a liquid or solid state by decreasing temperature, increasing pressure or both. Usually in air sampling, only cooling is used.

Condensation Sampling Gas sampling technique in which the gas is drawn through trapping tubes immersed in refrigerants.

Coning On cloudy days or nights with strong winds, mechanical turbulence prevails and "normal" dispersion occurs, forming a *cone* to the eye of an observer looking upwind into the plume.

Conjunctivitis An inflammation of the delicate membrane lining the eyelids and covering the exposed surface of the eyeball.

Consolidation The process by which a diseased lung passes from an aerated collapsible state to one of an airless solid consistency because of accumulation of exudate.

Contacting Power Rule A relationship which states that the efficiency of a scrubber on a given aerosol is essentially dependent only upon the power per unit of volumetric gas flow rate that is dissipated in the gas-liquid contacting process, and not upon the geometry or size of the unit or the manner in which the power is applied to the contacting process.

Continuous Air Monitoring Program Program undertaken by the U.S. Public Health Service to increase knowledge about the nature of air pollution by providing information in the form of comparable, continuous, concurrent data on significant gaseous pollutants from a number of urban areas. Monitoring stations have been established in or near the principal business district of such cities as Chicago, Cincinnati, Philadelphia, San Francisco, St. Louis and Washington to monitor gaseous pollutants, SO_2, NO, NO_2, CO, hydrocarbons, O_3 and oxidants. Measurements comparable with the CAMP operations are also made in cities at stations operated by local groups (such as Detroit, New York, Los Angeles).

Control Agencies The term "air pollution control agency" means any of the following: (1) a single state agency designated by the governor of that state as the official state air pollution control agency for the purposes of the Clean Air Act; (2) an agency established by two or more states and having substantial powers or duties pertaining to the prevention and control of air pollution; (3) a city, county or other local government health authority where an agency other than the health authority is charged with responsibility for enforcing ordinances or laws relating to the prevention and control of air pollution; or (4) an agency of two or more municipalities located in the same state or different states and having substantial powers or duties pertaining to the prevention and control of air pollution. The term "interstate air pollution control agency" means (1) an air pollution control agency established by two or more states, or (2) an air pollution control agency of two or more municipalities located in different states.

Controlled Atmosphere Atmospheres which may be used as primary standards for the calibration of analytical techniques or as simulated environments for the study of biological responses or of the resistance of materials.

Cooling Rate Rate of cooling of a body, substance or gas environment obtained by plotting time against temperature under constant conditions.

Cosmic Rays Primary cosmic rays (galactic, solar and Van Allen belt) are those rays which reach the earth's surface without interaction with atomic nuclei present in the earth's atmosphere. Secondary cosmic rays are primary rays which have interacted with atomic nuclei, mostly nitrogen and oxygen, in the earth's atmosphere. The major cosmic ray particles in the earth's atmosphere are nucleons, mesons, leptons and photons. The dose at the earth's surface from the ionizing component of cosmic rays is highly latitude and altitude dependent and varies with the solar cycle. The dose rate at sea level at mid-latitude, which does not underestimate the dose to the gonads, is 34 mrad/yr.

Counter Instrument for measuring the size and number of dust particles in the atmosphere.

Counting To determine the number of dust particles per unit area of a standard cell in the case of optical microscope or per representative portion of the filter in the case of a membrane filter. The total concentration of dust can then be calculated and expressed in millions of particles per cubic foot of air sampled.

Count Median Size The median measurement of particle diameters of particulate matter of a gas sample.

Cracking A method of separating the constituents of petroleum, using pressure, heat and catalysts, in which the hydrocarbons of high molecular weight are broken down into those of lower molecular weight.

Criteria, Air Quality A compilation of the scientific knowledge of the relationship between various concentrations of pollutants in the air and their adverse effects.

Cumulative Sample A sample obtained over a period of time with the collected atmosphere being retained in a single vessel or a separated component accumulating into a single whole.

Curie See Roentgen.

Curtain Wall A partition wall between chambers in an incinerator under which combustion gases pass.

Cut-size In a cyclone is the particle size which can be collected at 50% efficiency.

Cyclone A structure without moving parts in which the velocity of an inlet gas stream is transformed into a confined vortex, from which centrifugal forces tend to drive the suspended particles to the wall of the cyclone body.

Cyclone Separator A separator which utilizes centrifugal force to arrest all or part of the particles suspended in a current of gas.

Cyclonic Spray Tower Liquid scrubbing apparatus where sprays are introduced countercurrent to gases for removal of contaminants.

Decay Rate Disintegrations A unit quantity of radioactive material per unit time.

Dedusting Operation or process to arrest all or part of the dust present in gases.

Defoliation Removal of leaves from growing plants due to the effect of liquid or solid chemical agents (herbicides as well as some pesticides).

Degree Day (Often referred to as heating degree day) is the departure of one degree of the daily mean temperature from a standard. Here 18°C (65°F) is meant, which is the threshold for space heating requirements. Usually the daily departures below this value are cumulated by months to judge fuel needs.

Deliquescent Substance A substance that absorbs moisture from the air and dissolves in the water so absorbed.

Demister Apparatus made of wire mesh or glass fiber and used to eliminate the acid mist as in the manufacture of H_2SO_4.

Deodorizer Equipment for the removal of noxious gases and odors. It may consist of a combustion, an absorption or an adsorption unit.

Deposit Gauge General name for air pollution instruments used to measure the amount of material deposited on a given area during a given time.

Desulfurization In coal processing, desulfurization involves elutriation, froth flotation, laundering, magnetic separation; in petroleum refining, removing sulfur or sulfur compounds from a charge stock (oil that is to be treated in a particular unit). Removing sulfur from iron, metal or an ore.

Detectors Device which indicates the presence of air pollutants.

Deterioration Degradation of quality. For example, degradation of materials due to the chemical activity of air pollutants, especially evident in cases of corrosion and oxidation.

Dew Point Temperature of saturation for air at which level it deposits excess moisture.

Diameter, Count Median The geometric median size of a distribution of particles, based on a numerical count.

Diameter, Mass Median The geometric median size of a distribution of particles, based upon a weight.

Diffraction Analysis Application of diffraction technique (of x-rays, electrons, neutrons) to study the structure of matter, gases and solids. This may involve diffraction of electromagnetic radiation or particle beams.

Diffuser A porous plate or tube through which air is forced and divided into minute bubbles for diffusion in liquids. Commonly made of carborundum, alundum or silica sand.

Diffusion The spreading or scattering of a gaseous or liquid material. (1) Eddy diffusion: the most important mixing process in the atmosphere. It causes a parcel of polluted air to occupy larger and larger volume. (2) Molecular diffusion: a process of spontaneous intermixing of different substances, attributable to molecular motion and tending to produce uniformity of concentration.

Discomfort Threshold A value at which a stimulus just produces a discomfort sensation or comes just within the limits of perception.

Disinfection The killing of the larger portion (but not necessarily all) of the harmful and objectionable microorganisms in, or on, a medium by means of chemicals, heat, ultraviolet light.

Dispersion The dilution of a pollutant by diffusion or turbulent action or the like. Technically, a two-phase system of two substances, one of which (the dispersed phase) is uniformly distributed in a finely divided state through the second substance (the dispersion medium). The dispersion medium may be a gas, liquid or solid; and the dispersed phase may be any one of these.

Dispersion Forecasts Prediction of gas plume dispersions.

Distillation Vaporization of a liquid, followed by condensation of the vapor.

Diurnal Recurring daily.

Doppler Effect Apparent change in frequency of sound waves or light waves caused by relative motion of the source toward or away from an observer.

Downwash Occurs when chimney gases are drawn downward by a system of vortices or eddies in the lee of a chimney when wind is blowing.

Draft A gas flow resulting from the pressure difference between the incinerator, or any component part, and the atmosphere, which moves the products of combustion from the incinerator to the atmosphere.

Drag Coefficient A dimensionless ratio of the component of force parallel to the direction of flow exerted on a body by a fluid to the kinetic energy of the fluid multiplied by a characteristic surface area on the body.

Droplet A small spherical liquid particle that may fall free under still conditions but which may remain suspended in the atmosphere under turbulent conditions.

Dry Impingement The process that pushes particulate matter carried by a gas stream against a retaining surface. The retaining or collecting surface may be coated with an adhesive film.

Dust Solid particles larger than colloidal size and capable of suspension in air and in other gases. Also see Particulate Matter.

Dust Collector Device for monitoring dust emissions.

Dust Counter An instrument for counting the dust particles in a known volume of air. In Aitken's dust counter, condensation is made to occur on the nuclei present by adiabatic expansion of air, and the number of drops is counted. In Owen's dust counter, a jet of damp air is forced through a narrow slit in front of a microscope coverglass. The fall of pressure due to expansion of the air after passing the slit causes the formation of a film of moisture on the glass, to which dust adheres, forming a record which can be studied under a microscope.

Dust Flow Rate at which dust is carried in a gaseous medium.

Dust Horizon The top of a dust layer which is confined by a low-level temperature inversion in such a way as to give the appearance of the horizon when viewed from above against the sky.

Dyspnea Difficult or labored breathing.

Eductors Device for mixing air with water. A liquid pump operating under a jet principle, using liquid under pressure as the operating medium.

Electron Capture Detector (ECD) The technique is based on the principle of the electron affinity (energy liberated in the capture of a free electron by a neutral species). ECD measures the loss of signal due to recombination phenomena. The carrier gas flowing through the chromatograph is ionized by a source of radiation (such as radium or tritium) to form slow electrons which migrate to the anode under an applied voltage. These electrons are captured by the compounds injected into the chromatograph. The result is a decrease of incurrent flow. The ECD is very sensitive to some molecules such as organic and inorganic halogen-containing compounds and metallorganic compounds.

Electron Charge The smallest possible negative electric charge (4.80298×10^{-10} electrostatic units).

Electrostatic Filter Filter where electrostatic charge is applied to the filter element.

Electrostatic Precipitators They operate on the general principle of passing air between charged surfaces, imparting a charge to particles in the air and collecting the particles on an oppositely charged surface or plate.

Elutriation Separating the lighter particles of a powder from the heavier by means of an upward stream of fluid. It is especially useful for very fine particles below the usual screen sizes and is used for pigments, clay dressing and ore flotation.

Emission The total amount of a solid, liquid or gaseous pollutant emitted into the atmosphere from a given source in a given time and indicated in grams per cubic meter of gas or by a relative measure (e.g., smoke density) upon discharge from the source. Primary emissions may be characterized as follows: fine solids (less than 100 μ in diameter); coarse particles (greater than 100 μ in diameter): sulfur compounds, organic compounds, nitrogen compounds, oxygen compounds, halogen compounds, radioactive compounds. Secondary emissions may be characterized as the products of polluted air reactants, such as those which occur in atmospheric photochemical reactions. Secondary pollutants include 0_3, formaldehyde, organic hydroperoxides, free radicals, NO and O. Emissions are commonly reported in terms of weight of pollutant per unit of time.

Emission Inventories Systematic collections and collations of detailed information concerning the air pollution emissions in a given area.

Emittor/Emitter Any technical equipment which discharges matter foreign to air into the open air.

Energy Absorption Process whereby some or all the kinetic energy of electromagnetic radiation is lost to another substance.

Enthalpy A thermodynamic property defined as the sum of the internal energy and the product of the pressure and specific volume (BTU/1bm or Cal/Kg units).

Entrainment The mist or fog droplets of liquid carried off by the vapor of a boiling liquid or more frequently from a liquid through which bubbles of gas or vapor are passing rapidly.

Eutrophic Water which is rich in dissolved nutrients which support excessive algal growth. Due to the oxygen demand of the decaying algal remains, such waters are deficient in oxygen on a seasonal basis. This reduces their animal life supporting capacity.

Excess Air In combustion reactions, excess air is that air supplied beyond the amount stoichiometrically required for complete oxidation of the fuel.

Exposure Subjecting a person, animal, plant or material to an environment containing a harmful concentration of an air pollutant or air pollutants.

Exposure Chamber A chamber for exposing human, plant, animal or material to a particular environment. Various parameters, such as pollutants, humidity, temperature, pressure, noise, movement, radiation and fluid contents, may be controlled.

Eye Irritation Caused by air pollutants, such as gases, contained in smog and automotive exhausts (SO_2, SO_3, NO_2, O_3, olefins, formic acid, oil, gasoline); secondary pollutants caused by the photochemical reactions of NO or NO_2 and hydrocarbons such as PAN (peroxyacetyl nitrate), formaldehyde or acrolein, alone, or in combination with aerosols (such as NaCl and silica dust).

Fabric Filter A collector designed to remove particles from a carrier gas by filtration of the gas through a porous medium.

Fanning As radiational cooling progresses, the inversion reaches up to the plume height. Since most of the vertical dispersion is suppressed by the extremely stable conditions, the plume *fans out* in the horizontal. Strong concentrations at plume height are exhibited downwind of the stack.

Fanning (Mechanical) In blast furnace operation, it is the idling period between the blowing periods when the blast pressure is reduced to a minimum. During this period, no charging or tapping of slag or metal is carried out.

Fiber Glass Generic name for a manufactured nonflammable fiber fabricated from glass. A continuous filament or staple fiber having unusual resistance to heat and chemicals. The strongest fiber known and perfectly elastic up to its ultimate strength. It is attacked by hydrofluoric acid and alkalies but resistant to most other chemicals and solvents.

Filter A porous material used for the removal of solid particles from liquid or gas.

Filtration The process of separating suspended solids from a liquid or gas, usually by forcing a carrier gas or liquid through a porous medium.

Fixation of Nitrogen The combination of atmospheric nitrogen with other elements to form chemical compounds.

Flame Photometry Emission spectroscopy in the ultraviolet and visible regions which makes use of flame sources. Some seventy-odd elements can be qualitatively or quantitatively determined based on the emission spectra of these elements when they are excited in an arc or high-voltage spark.

Flash Point The lowest temperature at which a substance, such as fuel oil, will give off a vapor that will flash or burn momentarily when ignited.

Float The floating layer from a wet gravity separation operation such as coal cleaning.

Flue A passage for conducting combustion gases in an incinerator installation. Also, used synonymously with chimney (buildings).

Flue Gas Waste gas from combustion processes.

Fluidization Achieved when gases passing upward through a bed of particulate solids experience a loss in static pressure equal to the weight of the bed. The fluidized particles are highly mobile in the bed and are circulated through the bed by the rising gas. The bed itself looks very much like a boiling liquid and has properties similar in many respects to liquids. For example, objects of lesser density will float on the surface of the bed, and the surface of the bed will seek a horizontal level. In addition, the fluidized bed has liquid-like flow properties.

Fluorescent Particle Technique Use of fluorescent particles as visual traces to study diffusion in the atmosphere.

Fluorine A nonmetallic univalent halogen element which normally is in the gas form and is both irritating and toxic. It forms fluorides with other elements or radicals.

Fluorometer An instrument to measure the intensity and color of fluorescent radiations.

Flyash Finely divided particles of ash entrained in flue gases resulting from the combustion of fuel. The particles of ash may contain incompletely burned fuel and minerals.

Fog Visible aerosols in which the dispersed phase is liquid. In meteorology, fog is a dispersion of water or ice.

Fog Chamber A confined space in which air or other gas is supersaturated by reduction of pressure or cooling; used to study the movement and interaction of electrified particles by the condensation (fog tracks) they produce.

Fog Horizon The top of a fog layer which is confined by a low-level temperature inversion in such a way as to give the appearance of the horizon when viewed from above against the sky.

Fourdrinier The name given to paper machines using conventional moving wire screens for wet-end paper formation.

Fractionation Separation in successive stages, each removing from a mixture some proportion of one of the substances. The operation may be precipitation, crystallization or distillation.

Free Radical See Radical.

Freezing Out A procedure used to separate flue gases through progressively lowered temperatures for the purpose of analysis.

Frequency Distribution An arrangement of statistical data which exhibits the frequency of the occurrence of the values of a variable.

Fume Fine solid particle predominantly less than 1 μ in diameter, resulting from condensation, sublimation or chemical reaction, e.g., condensed metal oxides.

Fumigation Sunrise initiates solar heating; and the gradual warming of atmosphere from the ground upwards "burns off" the inversion. As the strong convective mixing reaches the fanned plume, it immediately mixes the large pollutant concentrations toward ground level, *fumigating* that area.

Gas Black Finely divided carbon produced by incomplete combustion or thermal decomposition of natural gas. Other names used are carbon black, channel black, furnace black.

Gas Plume Stack effluent composed of gases alone or gases and particulates. The geometrical form and concentration distribution of gas plumes are dependent on turbulence. Because of distinctly different atmospheric conditions, gas plumes may be classified into five types: looping, coning, fanning, fumigating and lofting plumes.

Gaseous Diffusion Separator Instrumentation to separate a gas mixture into its components, as when a gas is an isotopic mixture and is allowed to diffuse through another (preferably heavier) gas. Separation of the various isotopes is based on the principle that the lighter isotopic molecules diffuse through the heavier gas more readily than the heavier isotopic molecules.

Geometric Mean The product of numerical values divided by the number of the numerical values used, such as $\frac{1 \times 2 \times 3}{3}$. It can also be defined as the antilog of the mean of the logarithms of the individual readings.

Geometric Standard Deviation The standard deviation of the logarithms of the individual readings.

Gob Piles Fires which result when air can flow readily through the interstices of large piles of low-combustible refuse from coal mine preparation plants. Slow oxidation occurs, heat is liberated and higher temperatures develop until ignition occurs.

Gonadal Dose Absorbed dose by the gonads due to external plus internal radiation from nuclides within the gonads or within radiation range of the gonads. It has an important effect on the genetic fertility of children.

Gradient Either the rate of change of a quantity (such as temperature, pressure) or a diagram or curve representing this rate.

Grain 15.4 grains is a gram and 437.5 grains is an ounce. One grain is 65 milligrams.

Green Liquor The name given to the mixture of recovery smelt and weakwash formed in the dissolving tank.

Grit Solid particles of natural or man-made origin, retained on a 200 mesh British Standard sieve, i.e., 76 μ or greater in diameter.

Ground Level Concentration The amount of solid, liquid or gaseous material per unit volume of air, from 0 to 2 meters above the ground.

Gypsum Hydrous calcium sulfate mineral used as a soil conditioner and in making plaster. It is also a by-product of some of the limestone additive processes for SO_2 removal from boiler exhausts ($CaSO_4 \cdot 2H_2O$).

Halide Any binary compound of a halogen.

Halogen Any one of the chemically related elements fluorine, chlorine, bromine, iodine and astatine.

Haze A state of atmospheric obscurity due to the presence of fine dust solid particles in suspension; visibility exceeds 1 but is less than 2 km. The particles are so small that they cannot be felt or individually seen with the naked eye.

Haze Horizon The top of a haze layer which is confined by a low-level temperature inversion in such a way as to give the appearance of the horizon when viewed from above against the sky.

Hemoglobin An iron-containing protein respiratory pigment found in red blood cells, responsible for oxygen transport to all parts of the body.

Hydrocarbon A compound containing only hydrogen and carbon. This group is subdivided into alicyclic, aliphatic and aromatic hydrocarbons, according to the arrangement of the atoms and the chemical properties of the compounds.

Hydrocarbon Cracking Decomposition by heat, with or without catalysis, of petroleum or heavy petroleum fractions, with production of lower-boiling materials useful as motor fuels, domestic fuel oil or other needed products.

Hydrogen Sulfide A colorless gas that has a characteristic, disagreeable odor described as that of rotten eggs. It can be detected by the sense of smell at concentrations as low as 0.1 ppm. Exposures for short times can result in a fatigue of the sense of smell because it can become densensitized after 2 to 15 minutes of exposure to 100 ppm or more. Hydrogen sulfide can also be an irritant to the eyes and respiratory system of humans at concentrations considerably above those that cause detectable odors. Unlike what is known of photochemical eye irritants, hydrogen sulfide causes inflammation and edema of the conjunctiva. At still higher concentrations, it may produce systemic effects and can even result in death.

Hydrosphere Aqueous envelope of the earth.

Hygrometer An instrument which measures the water vapor content of the atmosphere.

Hygroscopic Readily absorbing and retaining moisture.

Hν Symbol for one quantum of radiant energy (Planck's constant multiplied by the vibration frequency of the radiant energy). One quantum is the amount of energy needed to convert one molecule photolytically to a higher or more energetic state.

Immission The occurrence of solid, liquid and gaseous substances polluting the air close to the ground.

Impaction A forceable contact of particles of matter; synonymous with impingement.

Impactor, Cascade An instrument which employs several impactions in series to collect successively smaller sizes of particles.

Incidence The rate at which a certain event or disease occurs.

Incinerator Equipment in which solid, semisolid, liquid or gaseous combustible wastes are ignited and burned, leaving solid residues containing little or no combustible material.

Indicator Plants Vegetation used to detect air pollution.

Individual Perception Threshold The lowest concentration of a particular odor at which a subject indicates both an initial positive response and a repeated response.

Inertial Separator Includes all dry type collectors utilizing the relatively greater inertia of the particle to effect particulate gas separation.

Infrared Gas Analyzer Infrared spectrometer with two equivalent calcium fluoride infrared window sources, one for the comparison cell and the other for the sample cell. It functions on the principle that the sample air will absorb infrared radiation at a different rate than the gas in the comparison cell. Thus, with proper instrumentation, the CO concentration in the sample gas may be measured.

Inversion Hot gases generated by burning, being warmer than surrounding air, rise into the upper atmosphere. However, since the pressure at higher levels is lower, these gases expand and cool themselves. Under normal circumstances, the upper layers of the atmosphere are colder than those nearer the surface and the gases continue to rise. Occasionally, however, a warm air mass moves in over a cool one. It is higher than the cooler air; so this condition is quite stable and may remain for a considerable time. Under these circumstances, hot gases rising through the cold air become heavier than the warm upper air and cannot penetrate the warmer layer but form a layer between the two air masses. This condition is called an *inversion.* It can be subsidence or radiative inversion, which are also defined in this glossary.

Inversion Height A term commonly used to denote the height of a temperature inversion which is a layer of atmosphere in which the temperature increases with height.

Ion An atom or a group of atoms with an electric charge.

Ionization The loss of one or more outer electrons of an atom, causing the atom to become electrically charged positively, or the gain of one or more extra electrons, causing the atom to become electrically charged negatively.

Ionization Chamber An apparatus used to study the production of small ions in the atmosphere by cosmic ray and radioactive bombardment of air molecules. The chamber is an airtight container usually cylindrical in shape and 25 to 50 liters in volume. An insulated electrode is centrally located in the chamber. In operation, a potential is applied between the electrode and the chamber wall. The ions produced in the chamber are collected by the electrode system and measured by an electrometer.

Isoconic Line A line joining points of equal dust precipitation per unit area and time.

Isokinetic A line in a given surface connecting points with equal wind speed. (British term for isotach) (also called isovel).

Isokinetic Sampling Defined as removal of the sample, usually through a probe or sampling system, by utilizing the same linear velocity of particulates (and gases) in sampler as is found in the stream or exhaust at the sampling point. Particulate matter of diameter greater than 3 μ tends to be selectively removed if the linear velocity is changed during the sampling procedure.

Isopleth A line on a map connecting points at which a given variable has a specified constant value.

Konimeter An instrument for determining the dust content of a sample of air.

Kraft The term used to describe the sulfate or alkaline pulping process.

Kytoon A captive balloon used to maintain meteorological equipment aloft at approximately a constant height.

Labile Unstable.

Lacrimation Tear formation, especially in excess.

Langley A unit of radiation intensity. One langley equals 1.0 gcal/cm^2/min = 221 BTU/min/ft^2 = 69.7 milliwatts/cm^2.

Lapse Rate The rate at which temperature decreases or lapses with altitude.

Leaching The term applied to the process of extracting a water soluble material from a solid by filtering the water through the solid. An excellent example of the process is a coffee pot. In a coffee pot, the soluble coffee is dissolved from the insoluble grounds by the water, and the resultant solution is separated from the grounds. In a similar manner, soluble organic matter is extracted from landfills by the percolation of rainwater through the soil. This organic contaminated water is a particularly potent source of water pollution when it finds its way into streams and wells around the landfill.

Lidar Laser-radar.

Light Attenuation Reduction in the intensity of light due to the action of the transmitting medium (absorption and scattering).

Light Scattering Reflection or scattering of light in all directions when it enters a body of matter, however transparent.

Lime *Limestone* is an accumulation of organic remains consisting mostly of calcium carbonate ($CaCO_3$). When burned, it yields *lime* which is a solid (CaO). The hydrated form of chemical lime is calcium hydroxide [$Ca(OH)_2$].

Line Formula A way of presenting a chemical formula in one line rather than as a two dimensional structure: R—COOH (line),

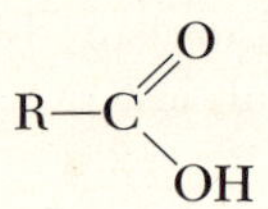 (2-dimensional)

Lithosphere Solid part of the earth.

Lofting Over rural areas, just after sunset, a superadiabatic layer exists over the newly forming ground-based inversion. The elevated plume cannot disperse downward due to the inversion, but strongly disperses above the inversion. Urban areas do not have inversion conditions at this type due to heat release from buildings (the "heat island" effect).

Longitudinal Studies Mean studies where test groups are followed over a period of time.

Looping In the afternoon, strong solar insolation, producing an excessive amount of convective turbulence, causes the plume to *loop* in the vertical in response to the strong eddy motion.

Lower Explosive Limit The minimum concentration of a material which can be ignited to cause an explosion.

Luminescence Any emission of light below temperatures required for incandescence. Frequently, a body becomes luminescent because of previous exposure to some form of radiation. If the body is luminous during the time it is being irradiated, the phenomenon is called fluorescence. In phosphorescence, the emission of light continues after irradiation.

Macrometeorology Study of the largest-scale aspects of the atmosphere, such as the general circulation, weather types and the Grosswetterlagen. There is a wide gap between this and the relatively small scale of mesometeorology. The gap is bridged by those atmospheric characteristics referred to as cyclonic scale. (*Grosswetterlag* is defined as the mean pressure distribution for a time interval during which the positions of the stationary cyclones and anticyclones remain unchanged.)

Manometer An instrument for measuring pressure differences usually consisting of a U-shaped tube containing a liquid.

Mass Concentration Concentration expressed in terms of mass of substance per unit volume of gas or liquid.

Mass Median Size The median measurement of particle size for samples of particulate matter related to diameter size and thus mass.

Maximum Allowable Concentration The concentration of a pollutant considered harmless to healthy adults during their working hours, assuming that they breathe uncontaminated air for the remainder of the time.

Maximum Emission Concentration Standards for maximum concentration of air pollutant emission from stationary or moving sources. These include opacity standards, such as the Ringelmann chart gradations often applied with time specification; gravimetric emission standards (expressed as weight of emitted pollutant per volume or unit weight of carrier gas); volumetric emission standards (percent of gaseous pollutant by volume in the emitted gas or as parts per volume in specific parts of effluent gas).

Membrane Filters Combine extremely high collection efficiencies with relatively low resistance to flow. Such filters are not made up of a fibrous mat but instead are usually composed of gels of cellulose esters or other polymeric substances in such fashion that a smooth surface of predictable characteristics is formed. Such filters contain many small holes, or pores, and may be made to exacting specifications in this regard so that their performance characteristics may be predicted. In addition, the filters are usually of high chemical purity and are well suited to trace metal analyses. Some of the membrane filters can also be rendered transparent and thus permit direct observation of collected particles with a microscope. Alternatively, the filters can be dissolved in an organic solvent, and the particles may be isolated and studied. Most membrane filters are not affected greatly by relative humidity changes and can be weighed before and after use.

Mercaptans Aliphatic compounds that have structures similar to alcohols except that the oxygen is replaced by sulfur; hence, they are called thioalcohols.

Mesh Numbers U.S. sieve sizes are applied to screens and they designate the number of holes per inch.

Mesh Number	Corresponding Opening, Inches
8	0.093
12	0.066
14	0.055
20	0.0328
30	0.023
40	0.0165
100	0.0058
200	0.0029
325	0.0017

Methane Hydrocarbon gas (CH_4) produced by the decomposition of organic matter in marshes, mines and elsewhere.

Micrometeorology The portion of the science of meteorology which deals with the observation and explanation of the smallest scale physical and dynamic occurrences within the atmosphere.

Mist Low-concentration dispersion of relatively small liquid droplets. In meteorology, the term mist applies to a light dispersion of water droplets of sufficient size to fall from the air.

Mixing Depth The atmospheric layer between the surface and the base of the inversion layer.

Mixing Height Height at which stack effluent begins mixing with the atmosphere as it leaves the stack.

Molecular Weight Relative weight of a molecule compared to the weight of an atom of carbon taken as exactly 12.00; the sum of the atomic weights of the atoms in a molecule.

Monitoring Instrumentation and procedures for the continuous measurement of air pollutants and the application of regulatory or control measures when an established standard has been exceeded.

Monochromatic Light A beam of light having some desired, narrow range of wavelengths.

Mortality The ratio of the total number of deaths to the total population, or the ratio of the number of deaths from a given disease to the total number of people having that disease.

Nasopharynx The part of the pharynx (throat) lying above the level of the soft palate.

National Air Sampling Network Network of air sampling stations established in the United States and urban and nonurban communities. The objectives of this network include determination of the extent and nature of air pollution as well as the study of trends in the levels of atmospheric contamination; and investigation of relationships between air pollution and socioeconomic, geographic, topographic, meteorological and other factors.

Nebulize To reduce to a fine spray.

Nocturnal Occurring at night.

Nonvolatile Hydrocarbons Hydrocarbons without sufficient vapor pressure (volatility) to be found ordinarily in the atmosphere in the gas phase, but which may possibly be found in solid airborne particles.

Noxious Gases Gases which have been proved to cause ill effects in human beings.

Nucleus A particle in the size range from 0.1 to 1 μ which serves as a nidus on which water or other vapors in the air can condense to form liquid droplets.

Nuisance Threshold Standard for the concentration of an air pollutant that is considered objectionable. In the case of an odor, that concentration which can be detected by a human being.

Odor The sensation of smell perceived as a result of olfactory stimulation by an odorous material. Odorous materials can range from inorganic gases and vapors, such as ammonia, hydrogen sulfide and hydrogen chloride, across the full spectrum of organic compounds, such as mercaptans, amines, acids, aldehydes and ketones. No instrument has yet been developed with the capability and the sensitivity of the human olfactory system for detecting the presence of this wide range of materials. The human nose can respond to thousands of different materials, simultaneously detecting both high and low concentrations.

Odor Threshold Concentration of an odor in air that can be detected by a human being.

Odor Unit The quantity of any single odorous substance or a combination of substances which, when completely dispersed in 1 ft^3 of odor-free air, is detectable by a median number of observers in a panel of eight or more persons. The odor concentration is determined by a modification of the ASTM D1391-57 observer panel technique which, in its present form, allows optional approaches to sampling and dilution.

Olefin An unstable open-chain hydrocarbon containing at least one double bond.

Oleum An oily, corrosive solution of sulfur trioxide in anhydrous sulfuric acid.

Olfactory Pertaining to the sense of smell.

Opacity Rating A measurement of the opacity of emissions, defined as the apparent obscuration of an observer's vision to a degree equal to the apparent obscuration of smoke of a given rating on the Ringelmann chart.

Open Hearth Furnace Reverberatory furnace, containing a basin-shaped hearth, for melting and refining suitable types of pig iron, iron ore and scrap for steel production. In all open hearth furnaces, a large amount of dust from ore and other materials and splashings from slag are carried away by the waste gases. This is especially serious in the basic process.

To prevent choking of the regenerators and the fluxing and glazing of the checker-bricks, a supplementary chamber is used for collecting slag and dust.

Organic Acids Carbon compounds that contain a carboxyl group as the functional group (formula R—COOH, where R = group containing C attached to carboxyl group).

Organoleptic Employing one or more of the organs of special sense.

Overfire Air Jet Streams of high-velocity air issuing from nozzles in a furnace enclosure to provide turbulence and oxygen to aid combustion or to provide cooling air.

Oxidant Any substance with an oxidation potential greater than oxygen.

Oxidation A process in which an atom or group of atoms loses electrons; the combustion of a substance with oxygen.

Oxidation Potential A single electrode potential related to the total efm (electromotive force) of a cell, as shown by the following equation:

$$E_{cell} = E_{ox} + E_{red}$$

The standard oxidation potential E°ox refers to the oxidation potential when each substance involved in the electrode reaction is at a unit "activity."

Oxides of Nitrogen These include nitric anhydride or nitrogen pentoxide (N_2O_5), nitrogen tetroxide (N_2O_4), nitrous anhydride or nitrogen sesquioxide (N_2O_3), nitrogen trioxide (NO_3), nitrogen dioxide (NO_2), nitrous oxide (NO).

Oxides of Sulfur These include sulfur dioxide (SO_2), sulfur trioxide (SO_3).

Ozonosphere General stratum of the upper atmosphere in which there is an appreciable ozone concentration and in which ozone plays an important part in the radiative balance of the atmosphere. The region lies between 10 and 50 km, with maximum ozone concentration at about 20 to 25 km (also called ozone layer).

Packed Bed Consists of mesh solid support used in gas chromatography. Components of a mixture are separated by volatizing a sample into a carrier gas stream which is passing through and over the bed of packing.

Packed Column A vertical column used for distillation, absorption and extraction.

PAN Peroxyacetyl nitrate. The class of compounds of which PAN is one is referred to as peroxyacyl nitrates, which are a noxious and unstable product of photochemical smog. Its acetyl derivative has the following form:

$$\underset{\displaystyle \text{O}}{\overset{}{\text{CH}_3\underset{\|}{\text{C}}\text{OONO}_2}}$$

PAN has been used to stand for both peroxyacetyl nitrate and peroxyacyl nitrates. Peroxyacetyl nitrate is a member of a family of peroxyacyl nitrates.

Particle A small discrete mass of solid or liquid matter such as aerosols, dusts, fumes, mists, smoke and sprays.

Particle Concentration Concentration expressed in terms of number of particles per unit volume of air or other gas.

Particle Fall A measurement of air contamination consisting of the mass rate at which solid particles deposit from the atmosphere. A term used in the same sense as the older terms "dust fall" and "soot fall" but without any implication as to nature and source of the particles.

Particle Size Distribution The relative percentage by weight or number of each of the different size fractions of particulate matter.

Particulate Matter Any dispersed matter, solid or liquid, in which the individual aggregates are larger than single small molecules (about 0.0002 μ in diameter) but smaller than about 500 μ. Particles in this size range have a lifetime in the suspended state in air ranging from a few seconds to several months. Additional terms used to further describe particulate matter include dust, flyash, smoke, soot, aerosols, droplets, mist, fog and fumes.

Pathogenic Disease causing.

Peak Concentrations During a period of monitoring an air pollutant or pollutants, the highest concentration of the pollutant or pollutants during that period.

Periodic Variations Hourly, daily, seasonal or annual variations in the concentrations of air pollutants due to meteorological, technological and/or sociological factors.

Permeability The ability to permit penetration or passage.

Pharynx The upper expanded portion of the alimentary canal lying between the mouth, the nasal cavities, the beginning of the esophagus and the throat.

Phenols Hydroxybenzenes, C_6H_5OH, where one or more ring hydrogens may be substituted by another group.

Photochemical Reaction A chemical reaction which involves either the absorption or emission of light radiation.

Photochemical Smog Results from the action of sunlight with oxides of nitrogen and vaporized organics to produce ozone and a variety of noxious products. While the whole system of reactions is complex and not yet completely understood, its major causes and effects are well known.

Photoelectric Effect Emission of electrons by certain substances when struck by light or other radiation.

Photoelectric Particle Counter Instrumentation which detects ionizing particles or radiation by means of the light flash emitted when the radiation is absorbed in the phosphors.

Photolysis Chemical decomposition due to the action of radiant energy.

Photon Elementary particle which is a quantum of radiation.

Photosynthesis The formation of carbohydrate from carbon dioxide and water in the presence of chlorophyll and light, in plant tissues.

Phytoplankton Plant life of the plankton.

Phytotoxic Poisonous to plant life.

Plankton Floating or weakly swimming plant or animal life forms in water.

Plant Damage Damage to living plants caused by air pollutants such as SO_2, fluorides, oxidants such as O_3 and peroxacyl nitrates, nitrogen oxides, ethylene, chlorine and hydrogen sulfide. Visible symptoms include leaf tissue collapse, chlorosis or other color changes and growth alterations.

Plume The path taken by the continuous discharges of products from a chimney. The shape of the path and the concentration distribution of gas plumes is dependent on turbulence. Looping, coning, fanning, fumigating and lofting are among the designations of various shaped plumes.

Pneumonitis A general term for inflammation of the lung.

Pneumotachograph An instrument used to determine the force and velocity of respired air.

Polynuclear Hydrocarbons Hydrocarbons with two or more usually hexagonal carbon rings in their structure.

Potash Potassium compounds used in agriculture and industry. Potassium carbonate (K_2CO_3) can be obtained from wood ashes. The mineral potash is usually a muriate (KCl). Caustic potash is its hydrated form (KOH).

Potentiation Synergism, as between two agents which together have a greater effect than the sum of their effects when acting separately.

Precipitation The formation of solid particles in a solution. Also, the settling out of small particles in either a liquid or gaseous medium. Precipitation may be due to such forces as chemical, electric, electrostatic, thermal and ultrasonic.

Precombustion Preliminary stage of combustion.

Precursor Chemical compounds which are gaseous air pollutants and which react with other substances in the atmosphere to produce different pollutants, such as the case where photochemical reactions producing ozone involve preliminary reactions of NO and NO_2 with the oxygen of the air.

Prilling To convert a solid into spherical pellets.

Primary Air Air admitted to a furnace during the first part of the firing cycle, i.e., for firing fresh fuel.

Primary Emission Pollutants emitted directly into the air from identifiable sources.

Primary Standards Ambient air quality levels, based on the criteria and allowing an adequate margin of safety, which, in the judgment of the Administrator of EPA, are necessary to protect the public health.

Probe A tube used for sampling or for measuring pressures at a distance from the actual collection or measuring apparatus. It is commonly used for reaching inside stacks and ducts.

Psychrometric Chart A plot of the properties of air–water vapor mixtures in a graphical form.

Pure Air Since the composition of the atmosphere is slightly variable with respect to certain components, the term "pure air" has no precise meaning; but it is commonly air free of nongaseous suspensoids (dust, hydrometeors) and free of such gaseous contaminants as industrial effluents. By far the most important gas in air, from the meteorological standpoint, is water vapor. "Dry air" denotes air from which all water vapor has been removed. Pure dry air has a density of 1.2923 g/cu cm at a pressure of 1,013.25 mb and a temperature of 0°C; a specific heat at constant volume of 0.1707 cal/(g)(°C); a specific heat at constant pressure of 0.2396 cal/(g)(°C); and its gas constant (per gram) is 2,870,400 ergs/(g)(°C). The percent by volume of those gases found in relatively constant amount in dry air is very nearly as follows: N_2, 78.084; O_2, 20.946; A, 0.934; CO_2, 0.033; Ne, 0.0018; He, 0.000524; CH_4, 0.0002; Kr, 0.000114; H_2, 0.00005; N_2O, 0.00005; Xe, 0.0000087.

Pyrite Iron disulfite mineral (FeS_2). Sulfuric acid is produced by burning it. It is the main impurity of coal.

Pyrolysis Chemical change brought about by the action of heat.

Quadrupole Electric Field The field is produced by four rectangular hyperbola electrodes, the opposite pairs of electrodes are connected together electrically. Although the electrodes usually have a uniform hyperbolic cross section, cylindrical rods can be used if they are properly spaced.

Quantum See Hν.

Quantum Yield The number of molecules (or the fraction of a molecule) formed in a photochemical reaction per quantum of light absorbed by the primary reactant.

Radar Electronic device used to detect distant objects by means of transmitted high-frequency radio waves and their reflection from the objects.

Radiation Units See Roentgen.

Radiative Inversion Radiative cooling of the ground cools the air in contact with the surface which makes the upper warmer air to act as an "inversion lid," preventing the polluted surface air from rising.

Radical Atom or group of atoms replaceable by a single atom which is capable of remaining unchanged during a series of reactions. It is a portion of an organic molecule which in many respects behaves as a unit. Such a radical can be described as a "free" radical when split off from its parent compound. Such free radicals are chemically extremely reactive.

Rale An abnormal respiratory sound heard in auscultation of the chest.

Rate Constant Second order rate constants are constants which, when multiplied by the concentrations of the two reactants, yield the rate of reaction.

Reinjection Method of disposing of flyash where it is reinjected in a furnace for further burning.

Reinluft Process A dry scrubbing (German) process for the adsorptive separation and recovery of gaseous oxides of sulfur, nitrogen and phosphorus from industrial gases.

REM See Roentgen.

Retentivity The capacity of an adsorber to hold some amount of adsorbent usually expressed as, for example, weight of adsorbate per weight of dry carbon.

Residual Fuel(s) Liquid or semiliquid products obtained as residues from the distillation of petroleum and used as fuel.

Residual Oil When crude oil is extracted from the ground, it is a combination of many organic molecules. In the refining process, the smaller and lighter molecules are evaporated and sold as petroleum gas. The slightly heavier molecules become gasoline and diesel oil. Intermediate molecules are "cracked" under high temperature and pressure to yield additional gasoline. Still heavier oils become lubricants. Paraffin and asphalt are also extracted from the crude oil. The remaining residual oil, which is not economically usable for other industrial purposes, is sold for burning.

Respirable Fit for breathing.

Rhonchus A dry, coarse sound usually originating from partial obstruction in a bronchial tube.

Ringelmann Cards A series of five cards that present graduated shades of gray from white to black for the evaluation of smoke density or opacity. It utilizes a rectangular grill of black lines of definite width and spacing on a white background, which can be accurately reproduced. The rule by which the chart may be reproduced is as follows:

Card 0 All white.
Card 1 Black lines 1 mm thick, 10 mm apart, leaving white spaces 9 mm square.
Card 2 Lines 2.3 mm thick, 10 mm apart, leaving white spaces 7.7 mm square.
Card 3 Lines 3.7 mm thick, 10 mm apart, leaving white spaces 6.3 mm square.
Card 4 Lines 5.5 mm thick, 10 mm apart, leaving white spaces 4.5 mm square.
Card 5 All black.

Ring Oven The operating principle of the ring oven is based on the addition of solutions and solvents to the center of a filter paper. The paper is positioned on a ring oven which consists of a small annular block of aluminum, which is heated to about 110°C. When the solutions reach the heated oven, the solvents evaporate and a ring is left on the paper. Soluble substances in this ring can be identified by spot tests. By comparing the test ring with standard rings prepared in the same way, the quantitative determination of traces can be made. The method will detect metals in nanogram to microgram quantities with an accuracy about 95%.

Roentgen and Other Units of Radiation The measure of radioactive dose, exposure and, in short, the intensity of radiation in a certain area, is the Roentgen. To give an indication of how much radiation intensity one roentgen represents, it can be said that a person being 3 feet from 1 gram of radium for an hour will receive a radioactive dose of 1 roentgen. Another way to describe this unit would be to say that a person should not receive a dose of higher than 250 roentgen over an entire lifetime. In industrial measurements, the radiation field intensity is normally expressed in milliroentgens per hour units. Exposure to external radiation is referred to in units of "rem" (roentgen + equivalent + man). A person receives the dose of 1 rem when exposed to 1 roentgen of radiation in any time period.[2] A person should not receive more than 250 rems over his entire life. The rate at which this exposure is accumulated is also important. It is desirable to keep the yearly dose below 5 rems and it should definitely not exceed 12 rems per year or 3 rems per quarter. The normal industrial practice is to limit the weekly exposure to between 100 and 200 millirems per week.

The unit rate by which a radioactive material generates radiation is a "curie." The radioactivity of 1 gram of radium is a curie. One curie of radioactivity is generated by any material which is disintegrating at the rate of 37 billion atoms per second. In order to compare some of the commonly used radioactive materials, their masses which will generate 1 curie of radioactivity are listed below:

Radium:	1,000 milligrams
Cobalt 60:	0.88 milligrams
Cesium 137:	0.115 milligrams

The same amount of radioactivity is generated by substantially different masses of materials. The curie is not a measure of radiation intensity, only an indication of the number of disintegrations per second. The more frequently used unit of radioactivity is the millicurie, representing one thousandth of a curie or the radioactivity produced by one milligram of radium.

The highest incidence of lung cancer (which has a latent period of 10 to 20 years) is occurring in the group of miners who have worked in mines in which the average concentration of radon daughters was usually higher than 10 working levels (any combination of radon daughters in 1 liter of air that will result in the emission of 1.3×10^5 MEV of potential alpha energy), whose cumulative exposure is greater than 1,000 WLM (working level months) and who are moderate to heavy smokers.

Ruds Reflectance units of dirt shade, units based on the reflection of light. Zero ruds are defined as absolutely black. The reflectance unit of dirt shade is defined as 100 times the optical density determined by reflectance. Ruds are expressed either in terms of 10,000 lineal feet of air (American Society for Testing and Materials unit) or 1,000 lineal feet of air.

Sample Blanks These are the tests applied to the solutions or the filter medium in the absence of samples. The tests employ the same chemical or analytical procedures as for the sample determination. The results obtained in a blank test are used to correct for the normal background contaminants and to correct for and to discover possible operator error.

Saturated Hydrocarbon A compound having the formula $Cn\ H_{2n+2}$. If two of the adjacent hydrogen atoms of a saturated hydrocarbon are removed, allowing the carbon electrons which were formerly shared with the hydrogen to be shared internally, the molecule becomes unsaturated. One or more double or triple bonds between carbon atoms can be present in linear, branched or cyclic unsaturated hydrocarbons.

Screen A plate, sheet or a woven cloth, or other device, with regularly spaced apertures of uniform size, mounted in a suitable frame or holder for use in separating material according to size.

Scrubber Apparatus used in sampling and in gas cleaning in which the gas is passed through a space containing wetted "packing" or spray. Devices which utilize a liquid to achieve or assist in removal of solid or liquid particles from a carrier gas stream.

Secondary Emission Substances that are secondary products of pollutants such as O^3, formaldehyde, organic hydroperoxides, O and NO.

Secondary Standards The levels required to protect the public welfare from any known or anticipated adverse effects associated with the presence of the pollutants in the ambient air. Public welfare is defined as ". . . includes, but is not limited to, effects on soils, water, crops, vegetation, man-made materials, animals, wildlife, weather, visibility and climate, damage to and deterioration of property, and hazards to transportation, as well as effects on economic values and on personal comfort and well-being."

Septic The anaerobic condition of water having no dissolved oxygen, where putrefactive conditions prevail.

Settling Chamber Chamber designed to reduce the velocity of gases in order to permit the settling out of flyash.

Sintering To cause to become a coherent, nonporous mass by heating but not melting. Flyash accumulated in the smokestack is a salable item for use in the manufacture of cinder blocks and other ceramic products. However, it must first be sintered, or baked, at a very high temperature.

Sizing Classifing dust particles based on the measurements of dimensions (sizes). For the case of spherical particles, the size is completely defined by the particle diameter. For the general case of nonspherical particles, the size may be defined in terms of equivalent diameter, depending on various geometrical or physical properties of the particles.

Smelt The molten ash formed in the bed of the recovery furnace as a result of burning concentrated black liquor. Consists mainly of $NaCO_3$ and Na_2S which is mixed with weakwash in the dissolving tank to form green liquor.

Smelter The plant where smelting, melting and refining of metals, is accomplished.

Smog A term to describe a combination of smoke and fog, but widely used in the United States to describe objectionable air pollution. Also see Photochemical Smog. Two basic types of smog occur. "Reducing" smog results from the combination of sulfur-containing fuels which lead to high levels of sulfur oxides and particulate matter in air. The second "oxidizing" type smog requires sunlight along with oxides of nitrogen and vapor phase hydrocarbons or other organics to be evident. Its severity is commonly estimated by its capacity to oxidize iodide.

Smog Alert Conditions indicated by meterological forecasts and air sampling where authorities responsible for air pollution control in a given area recommend or impose reduction of pollutant output by industry and other sources which contribute to smog development.

Smoke Small gas borne particles resulting from incomplete combustion. Such particles consist predominantly of carbon and other combustible material and are present in sufficient quantity to be observable independently of other solids.

Smoke Candle Apparatus designed for the collection of acid mist.

Smoke Density The amount of solid matter contained in smoke; often, measured systems that relate the grayness of the smoke to an established standard.

Smoke Horizon The top of a smoke layer which is confined by a low-level temperature inversion in such a way as to give the appearance of the horizon when viewed from above against the sky.

Smoke Plume Stack effluent, the geometrical form and concentration of which depend on meteorological factors, such as atmospheric turbulence and wind. Five types due to distinctly different atmospheric conditions are looping, coning, fanning, fumigating and lofting.

Smoke Shade A method for measuring the particulate matter in the air. A clean, white piece of filter paper is placed in an apparatus which forces a known quantity of air through it. The particulates which adhere to the paper darken it. A "smoke shade" reading may be obtained by comparing the discolored paper with known standard shades.

Social Classes Social Class I are professionals; social class II, persons in managerial positions; social class III, white collar workers and skilled blue collar workers; social class IV, semiskilled blue collar workers; and social class V, unskilled and unemployed persons.

Soiling Index Results are expressed in coefficients of haze per 1,000 lineal feet (COHs/1,000 LF) for transmittance measurements and in reflectance units of dirt shade (Ruds) for reflectance measurements.

Solar Energy Application of sunlight as a special purpose energy source, as in the case of a solar furnace.

Solar Radiation The total electromagnetic radiation emitted by the sun.

Solute Adsorbate.

Soot An agglomeration of carbon particles impregnated with "tar," formed by the incomplete combustion of carbonaceous material. These particles are larger than those present in smoke.

Soxhlet Extraction An analytical operation where a given amount of solvent is continuously distilled and condensed over the sample so that clean solvent is always in contact with the sample and the volume of the extract is kept constant.

Specific Gravity The ratio between the density of a substance at a given temperature and the density of some substance assumed as standard.

Spectroscope Optical instrument for the study of spectra.

Spectrum Band of electromagnetic radiation in which the constituent parts are arranged according to their wave-lengths or their frequencies.

Stack Effluent Gaseous and particulate waste products discharged to the atmosphere through stacks of some form.

Stack Height Selection Various limitations prevent the construction of stacks as high as might be desired from the pollution viewpoint. Among these are cost, air traffic considerations and hazards, aesthetic considerations, aerodynamic factors, vibration, materials. Some stacks are now reaching the 700-ft level, and the 1,000-ft level is anticipated.

Stagnant Inversion During cloudless evenings and nights, particularly in fall and winter, the earth's surface loses heat by radiation and in consequence the air adjacent to the ground is cooled and reduced to some temperature below that of the air at some heights above. Thus, air pollutants are trapped in this layer of air which, being colder than the layer above it, will not rise.

Staining The visible deleterious change that arises from a chemical reaction between the affected object and the pollutant.

Standard Condition 0°C and 1 atmosphere (760 mm or 29.92 in of mercury) pressure.

Stochastic Sampling Random sampling.

Stoker A machine for feeding coal into a furnace and supporting it there during the period of combustion.

Stratosphere The upper layer of the atmosphere, 6 to 7 miles above sea level. Here heating is mostly by the absorption of solar radiation by ozone.

Subsidence Inversion A large-scale phenomena where the adiabatic compression of an air layer at about 2,000 feet causes it to warm up and, thereby, prevents the heavily polluted lower and cooler air layer from rising.

Sulfate The final decomposition product of the organic sulfur compounds ($SO_4^=$).

Sulfate Pulping Describes the alkaline pulping process using sodium sulfate (Na_2SO_4) or "saltcake." Also see "Kraft."

Sulfhydryl The chemical structure corresponding to the substituent group, $-SH$.

Sulfide A salt or ester of H_2S in high pH waters or a H_2S gas in acidic waters.

Sulfite Salt or ester of sulfurous acid.

Sulfite Pulping The name given to the acid or neutral base pulping process.

Sulfur Content Both oil and coal are the result of decomposition of vegetation which grew on the earth millions of years ago. These products consist of many complex organic structures, the characteristics of which vary, depending upon the vegetation from which they were formed. If the vegetation contained sulfur, then the resultant oil or coal will also contain sulfur. Such sulfur may be separated from the fuel by either chemical or physical means.

Sulfur Dioxide The result of combustion of sulfur and oxygen. It may also be generated by the combustion of some compound which contains sulfur.

Temperature Gradient Rate of change of temperature or a diagram or curve representing this.

Temperature Inversion A layer in which temperature increases with altitude. The principal characteristic of an inversion layer is its marked static stability, so that very little turbulent exchange can occur within it.

Terpene Cyclic, nonaromatic hydrocarbon found in essential oils, pine tar and other wood sources. $[(C_5H_8)n]$

Thermal Inversion See Inversion.

Thermal Precipitation Whenever a strong temperature gradient exists between two adjacent surfaces, there will be a tendency for particles to be deposited on the colder of these surfaces. Collection of aerosols by this means is termed "thermal precipitation"; and, in practice, several commercial devices, which collect by means of thermal precipitation, are available.

Thermosphere Atmospheric shell extending from the top of the mesosphere to outer space. It is a region of more or less steadily increasing temperature with height, starting at 70 or 80 km. It, therefore, includes the exosphere and most or all of the ionosphere.

Thioalcohols See Mercaptans.

Threshold Limit Values Refers to airborne concentrations of substances and represents conditions under which it is believed that nearly all workers may be repeatedly exposed, day after day, without adverse effect. Because of wide variation in individual susceptibility, exposure of an occasional individual at or even below the threshold limit may not prevent discomfort, aggravation of a preexisting condition or occupational illness.

Topped Crude The material remaining after light petroleum fractions are removed by distillation at atmospheric pressure. It roughly represents that portion of crude oil boiling higher than about 600°F.

Toxicity The quality of being poisonous.

Tracer A foreign substance mixed with or attached to a given substance to enable the distribution or location of the latter to be determined subsequently.

Transfer Unit The amount of contacting necessary to enrich one phase equal to the average driving force producing this enrichment. The number of transfer units needed to achieve a certain amount of scrubbing is then the change in concentration desired for one phase, divided by the transfer unit for that phase. The height of a transfer unit is the length of scrubbing tower needed to effect one transfer unit. Multiplication of the number of transfer units required and the height of a transfer unit will yield an estimate of the height of tower needed to accomplish the desired amount of scrubbing.

Transmission The process by which radiant flux is propagated through a medium or body.

Transmissometer An instrument for measuring the extinction coefficient of the atmosphere for the determination of visual range. Also called telephotometer, transmittance meter, hazemeter.

Trapping Usually occurs during the daylight hours. Under subsidence and warm frontal inversion conditions, pollutants are trapped below the inversion level.

Troposphere From sea level to 12 kilometers, at which point the temperature has dropped to about −60°C. It contains 90% of the mass of the atmospheric air and it is heated mostly from the ground, not from the sun.

Turbulence Atmospheric eddies classified as mechanical or thermal as they are produced primarily by shearing stress or by convection. Turbulence or turbulent energy promotes atmospheric dilution of pollutants.

Ultrasonic Having a frequency above that audible to human beings; about 20,000 cycles per second.

Ultraviolet Electromagnetic radiation with wavelengths shorter than visible violet but longer than x-rays.

Underfeed A stoker consisting of a trough or pot-shaped retort into which coal is forced by an endless screw or ram. Coal is fed to the fire zone by being pushed up from underneath.

Unsaturated Hydrocarbon See Saturated Hydrocarbon.

Valence A number that represents the combining power of an atom or radical referred to hydrogen as a standard.

Vapor Recovery System System used in petroleum refining for separating a mixed charge of miscellaneous gases and gasolines into desired intermediates for further processing.

Visual Acuity Acuteness or clearness of vision.

Visual Threshold Value of a visual stimulus that just produces a sensation, is just appreciable or comes just within the limits of perception.

Volatile Vaporizable at low temperatures. Combustible at or below 600°C.

Weather Designates the instantaneous state of the atmosphere over a place or area. In contrast, climate encompasses the statistical ensemble of all weather conditions over a long interval of time (generally comprising at least several decades), over a place, area or the whole globe. Weather and climate are usually characterized by measurable elements, such as atmospheric pressure, temperature, wind, precipitation and events such as fog, thunderstorms, freezes. However, climate can also be qualitatively established by its effects on fauna and flora, glaciers and icesheets and geologic formations.

White Liquor The main cooking liquor used in a digester. Consists of active alkalis: NaOH and Na_2S.

Wind Rose Any one of a class of diagrams designed to show the distribution of wind direction experienced at a given location over a considerable period; it thus shows the prevailing wind direction. The most common form consists of a circle from which 8 or 16 lines emanate, one for each compass point. The length of each line is proportional to the frequency of wind from that direction; the frequency of calm conditions is entered in the center.

Zooplankton Animal life of the plankton.

[1] A liquid is called "combustible" if its flash point is above 100°F (37.8°C) and is called "flammable" if its flash point is below 100°F.

[2] One "rad" is the dose corresponding to the absorption of 100 ergs per gram of exposed body tissue. The dose of one rad due to X, gamma, or beta radiation or 0.1 rad due to neutrons and protons is equivalent to the dose of one "rem."

[illegible]

APPENDIX

CONTENTS OF APPENDIX

Table A.1

PERIODIC TABLE

This is a common long form of the table. Roman numerals and letters heading the vertical columns indicate the groups (there are differences of opinion regarding the letter designations, those given here being probably the most generally used). The horizontal rows represent the periods, with two series removed from the two very long periods and represented below the main table. Atomic numbers are given above the symbols for the elements and atomic weights when known are given below.

IA	IIA	IIIB	IVB	VB	VIB	VIIB	VIII			IB	IIB	IIIA	IVA	VA	VIA	VIIA	ZERO
1 H 1.008																1 H 1.008	2 He 4.003
3 Li 6.940	4 Be 9.013											5 B 10.82	6 C 12.011	7 N 14.008	8 O 16.000	9 F 19.00	10 Ne 20.183
11 Na 22.991	12 Mg 24.32											13 Al 26.98	14 Si 28.09	15 P 30.975	16 S 32.066	17 Cl 35.457	18 Ar 39.944
19 K 39.100	20 Ca 40.08	21 Sc 44.96	22 Ti 47.90	23 V 50.95	24 Cr 52.01	25 Mn 54.94	26 Fe 55.85	27 Co 58.94	28 Ni 58.71	29 Cu 63.54	30 Zn 65.38	31 Ga 69.72	32 Ge 72.60	33 As 74.91	34 Se 78.96	35 Br 79.916	36 Kr 83.80
37 Rb 85.48	38 Sr 87.63	39 Y 88.92	40 Zr 91.22	41 Nb 92.91	42 Mo 95.95	43 Tc	44 Ru 101.1	45 Rh 102.91	46 Pd 106.4	47 Ag 107.88	48 Cd 112.41	49 In 114.82	50 Sn 118.70	51 Sb 121.76	52 Te 127.61	53 I 126.91	54 Xe 131.30
55 Cs 132.91	56 Ba 137.36	57 *La 138.92	72 Hf 178.50	73 Ta 180.95	74 W 183.86	75 Re 186.22	76 Os 190.2	77 Ir 192.2	78 Pt 195.09	79 Au 197.0	80 Hg 200.61	81 Tl 204.39	82 Pb 207.21	83 Bi 209.00	84 Po	85 At	86 Rn
87 Fr	88 Ra	89 ‡Ac															

* LANTHANIDE SERIES	58 Ce 140.13	59 Pr 140.92	60 Nd 144.27	61 Pm	62 Sm 150.35	63 Eu 152.0	64 Gd 157.26	65 Tb 158.93	66 Dy 162.51	67 Ho 164.94	68 Er 167.27	69 Tm 168.94	70 Yb 173.04	71 Lu 174.99
‡ ACTINIDE SERIES	90 Th 232.05	91 Pa	92 U 238.07	93 Np	94 Pu	95 Am	96 Cm	97 Bk	98 Cf	99 Es	100 Fm	101 Md	102 No	103 Lw

Table A.2
CHEMICAL ELEMENTS
with international atomic weights

Element & Symbol	*Atomic Number*	*Atomic Weight* (O = 16)	(C = 12)
actinium (Ac)	89		
aluminum (Al)	13	26.98	26.9815
americium (Am)	95		
antimony (Sb)	51	121.76	121.75
argon (Ar)	18	39.944	39.948
arsenic (As)	33	74.91	74.9216
astatine (At)	85		
barium (Ba)	56	137.36	137.34
berkelium (Bk)	97		
beryllium (Be)	4	9.013	9.0122
bismuth (Bi)	83	209.00	208.980
boron (B)	5	10.82	10.811
bromine (Br)	35	79.916	79.909
cadmium (Cd)	48	112.41	112.40
calcium (Ca)	20	40.08	40.08
californium (Cf)	98		
carbon (C)	6	12.011	12.01115
cerium (Cc)	58	140.13	140.12
cesium (Cs)	55	132.91	132.905
chlorine (Cl)	17	35.457	35.453
chromium (Cr)	24	52.01	51.996
cobalt (Co)	27	58.94	58.9332
columbium (Cb)	(see niobium)		
copper (Cu)	29	63.54	63.54
curium (Cm)	96		
dysprosium (Dy)	66	162.51	162.50
einsteinium (Es)	99		
erbium (Er)	68	167.27	167.26
europium (Eu)	63	152.0	151.96
fermium (Fm)	100		
fluorine (F)	9	19.00	18.9984
francium (Fr)	87		
gadolinium (Gd)	64	157.26	157.25
gallium (Ga)	31	69.72	69.72
germanium (Ge)	32	72.60	72.59
gold (Au)	79	197.0	196.967
hafnium (Hf)	72	178.50	178.49
helium (He)	2	4.003	4.0026
holmium (Ho)	67	164.94	164.930

Table A.2
CHEMICAL ELEMENTS (Continued)

Element & Symbol	Atomic Number	Atomic Weight (O = 16)	Atomic Weight (C = 12)
hydrogen (H)	1	1.008	1.00797
indium (In)	49	114.82	114.82
iodine (I)	53	126.91	126.9044
iridium (Ir)	77	192.2	192.2
iron (Fe)	26	55.85	55.847
krypton (Kr)	36	83.80	83.80
lanthanum (La)	57	138.92	138.91
lawrencium (Lw)	103		
lead (Pb)	82	207.21	207.19
lithium (Li)	3	6.940	6.939
lutetium (Lu)	71	174.99	174.97
magnesium (Mg)	12	24.32	24.312
manganese (Mn)	25	54.94	54.9380
mendelevium (Md)	101		
mercury (Hg)	80	200.61	200.59
molybdenum (Mo)	42	95.95	95.94
neodymium (Nd)	60	144.27	144.24
neon (Ne)	10	20.183	20.183
neptunium (Np)	93		
nickel (Ni)	28	58.71	58.71
niobium (Nb)	41	92.91	92.906
nitrogen (N)	7	14.008	14.0067
nobelium (No)	102		
osmium (Os)	76	190.2	190.2
oxygen (O)	8	16.000	15.9994
palladium (Pd)	46	106.4	106.4
phosphorus (P)	15	30.975	30.9738
platinum (Pt)	78	195.09	195.09
plutonium (Pu)	94		
polonium (Po)	84		
potassium (K)	19	39.100	39.102
praseodymium (Pr)	59	140.92	140.907
promethium (Pm)	61		
protactinium (Pa)	91		
radium (Ra)	88		
radon (Rn)	86		
rhenium (Re)	75	186.22	186.2
rhodium (Rh)	45	102.91	102.905
rubidium (Rb)	37	85.48	85.47

Table A.2
CHEMICAL ELEMENTS (Continued)

Element & Symbol	*Atomic Number*	*Atomic Weight* (O = 16)	(C = 12)
ruthenium (Ru)	44	101.1	101.07
samarium (Sm)	62	150.35	150.35
scandium (Sc)	21	44.96	44.956
selenium (Se)	34	78.96	78.96
silicon (Si)	14	28.09	28.086
silver (Ag)	47	107.880	107.870
sodium (Na)	11	22.991	22.9898
strontium (Sr)	38	87.63	87.62
sulfur (S)	16	32.066	32.064
tantalum (Ta)	73	180.95	180.948
technetium (Tc)	43		
tellurium (Te)	52	127.61	127.60
terbium (Tb)	65	158.93	158.924
thallium (Tl)	81	204.39	204.37
thorium (Th)	90	232.05	232.038
thulium (Tm)	69	168.94	168.934
tin (Sn)	50	118.70	118.69
titanium (Ti)	22	47.90	47.90
tungsten (W)	74	183.86	183.85
uranium (U)	92	238.07	238.03
vanadium (V)	23	50.95	50.942
wolfram (W)	(see tungsten)		
xenon (Xe)	54	131.30	131.30
ytterbium (Yb)	70	173.04	173.04
yttrium (Y)	39	88.92	88.905
zinc (Zn)	30	65.38	65.37
zirconium (Zr)	40	91.22	91.22

Table A.3
TECHNICAL SOCIETIES

Aerospace Industries Association (AIA)
Air Conditioning and Refrigeration Institute (ARI)
American Gas Association, Inc. (AGA)
American Leather Chemists Association (ALCA)
American National Standards Institute (ANSI)
American Petroleum Institute (API)
American Public Health Association (APHA)
American Society for Testing and Materials (ASTM)
American Society of Agricultural Engineers (ASAE)
American Society of Heating, Refrigerating, and Air Conditioning Engineers, Inc. (ASHRAE)
American Society of Mechanical Engineers, The (ASME)
American Vacuum Society (AVS)
American Water Works Association (AWWA)
Association of American Railroads (AAR)
Chlorine Institute, The (CI)
Compressed Gas Association (CGA)
Contractor's Pump Bureau (CPB)
Cooling Tower Institute (CTI)
Edison Electric Institute (EEI)
Electronic Industries Association (EIA)
Factory Insurance Association (FIA)
Factory Mutual System (FM)
Fluid Controls Institute, Inc. (FCI)
Illuminating Engineering Society (IES)
Industrial Gas Cleaning Institute, Inc. (IGCI)
Institute of Electrical and Electronics Engineers (IEEE)
Instrument Society of America (ISA)
International Association of Plumbing and Mechanical Officials (IAPMO)
International Conference of Building Officials (ICBO)
Manufacturers Standardization Society of the Valve and Fittings Industry (MSS)
Metal Powder Industries Federation (MPIF)
National Association of Relay Manufacturers (NARM)
National Bureau of Standards (NBS)
National Electrical Manufacturers Association (NEMA)
National Fire Protection Association (NFPA)
National Fluid Power Association (NFPA)
National Printing Ink Research Institute (NPIRI)
National Standards Association (NSA)
Natural Gas Processors Association (NGPA)
Pipe Fabrication Institute (PFI)
Radio Technical Commission for Aeronautics (RTCA)
Range Commanders Council (RCC)
Scientific Apparatus Makers Association (SAMA)
Sheet Metal and Air Conditioning Contractors National Association, Inc. (SMACC)
Society of Automotive Engineers, Inc. (SAE)
Society of Motion Picture and Television Engineers (SMPTE)
Society of Naval Architects and Marine Engineers (SNAME)
Society of the Plastic Industry, Inc. (SPI)
Technical Association of the Pulp and Paper Industry (TAPPI)
Ultrasonic Manufacturers Association, Inc. (UMA)
Underwriters' Laboratories, Inc. (UL)
Variable Resistive Components Institute (VRCI)
Yacht Safety Bureau, Inc. (YSB)
International Standardization
- International Organization for Standardization (ISO)
- International Electrotechnical Commission (IEC)
- Pan American Standards Commission (COPANT)

Table A.4***

CHEMICAL RESISTANCE OF MATERIALS

MATERIALS
× — Very Good Service
+ — Moderate Service
− − — Limited or Variable Service
○ — Unsatisfactory
Blank — No Information

CHEMICALS
Solids Assumed in Sol'n.
Room Temperatures Assumed
Unless Otherwise Stated

	METALS																												CARBONS & CERAMICS								
CHEMICALS	*Carbon Steel; Fe*	*Cast Iron & Ductile Iron; Fe*	*304 Stainless Steel; Fe, 18Cr, 8Ni*	*316 Stainless Steel; Fe, 16Cr, 10Ni, 2Mo*	*347 Stainless Steel; Fe, 17Cr, 9Ni; (Cx10)Cb*	*Ni-Resist Iron; Fe, 14Ni, 2Cr, 2Si*	*Durimet 20; Carpenter 20; Fe; 4Cu, 20Cr, 29Ni, 2Mo, 1Si*	*Worthite; 3Mo, 2Cu, Fe, 20Cr, 24Ni, 3Si*	*Duriron; Fe, 14Si; Durichlor, Fe, 14Si, 3Mo**	*Copper; Brass; Bronzes; Everdur*	*Aluminum; Al (and Alloys)*	*Lead; Pb*	*Monel; 67Ni, 30Cu, 1.4 Fe*	*Nickel; Ni*	*Inconel; 76Ni, 15Cr, 8Fe*	*Hastelloy B; Ni, 26Mo, 4Fe*	*Hastelloy C; Ni, 16Mo, 4Fe, 14Cr, 4W*	*Hastelloy D; Ni, 8Si, 3Cu*	*Chlorimet 3; 3Fe, 1Si, 60Ni, 18Mo, 18Cr*	*Chlorimet 2; 63Ni, 32Mo, 3Fe, 1Si*	*Stellite; Co, 28Cr, 4W*	*Zirconium; Zr*	*Tantalum; Ta*	*Silver; Ag*	*Platinum; Pt*	*Dowmetal; (Mg alloys)*	*Titanium; Ti*	*Molybdenum; Mo*	*Carbon & Graphite*	*Glass, "Pyrex" brand*	*Silicaware*	*Silicate Cements*	*Chemical Stoneware*	*Transite (asbestos & cement)*	*Chemical Porcelain*	*Concrete-Unbonded*	*Concrete-Mortar Bonded*
Acetic Acid, 100%, CH_3COOH	○	○	+	×	−	○	×	×	×	○	×	○	−	+	−	×	×	×	×	×	×	×	×	×	×		×		×	×	×	×	×		×		
Acetic Acid, Dilute	○	○	×	×	−	○	×	×	×	×	−	−	−	○	+	×	×	+	×	×	×	×	×	×	×		×	×	×	×	×		×		×		
Acetic Anhydride, $(CH_3\ CO)_2O$	−	+	×	×	×	−	×	×	×	−	−	○	−	−	+	+	×	+	×	×	×		×	×	×		×		×	×	×	×	×		×		
Acetone, $CH_3\ COCH_3$	×		×	×	×	×	×	×	×	×	×	×	×	×	×	×	×	×	×	×		×		×			×		×	×	×		×	×	×		
Acetyl Chloride, $CH_3\ COCl$	○		−	×			×		×	○			+						×	×										×					×		
Aluminum Chloride, $AlCl_3$	○	○	○	○	○	−	×	○	×	−	○	○	○	−	−	×	+	×	×	×	−	×	×	×	×		×		×	×	×	×	×		×		
Aluminum Hydroxide, $Al\ (OH)_3$			×	×			×	×	+	×	×		+	+					×	×			×							×					×		
Aluminum Sulfate, $Al_2\ (SO_4)_3$	○		−	−	×	−	×	×	×	−	×	×	−	−	−	×	×	×	×	×	×	×	×	×	×		×		×	×	×		×	×	×		
Alums, Conc., $Al_2(SO_4)_3 \cdot K_2\ SO_4$, etc.	○		○	−	−		+		×	−	−	×							×	×			×		×				×	×					×		
Alums, Dilute	○	○	×	−	−	−	×	×	×	−	×	+	+	+		+	+		×	×		×	×	×	×				×	×	×	×	×	×	×		
Amines, various	×	×	×	×	×		×	×	×	−	−					×	×	×	×	×	×		−	×	×				×	×			×		×		
Ammonia (Gas), Moist, NH_3	×		×	×	×	×	×		×	○	○	−	−	−	−	×	×	×	×	×	×		−	×	×	×	−		×	×	×	○	×	×	×		
Ammonium Carbonate, $(NH_4)_2\ CO_3$	+	+	+	+	×		×		×	−	−		+	×	+	+	+	+	×	×			×							×		×			×		
Ammonium Chloride, $NH_4\ Cl$	○	○	○	−	−	×	×	×	×	○	○	−	+	+	−	+	+	×	×	×	−	×	×	×	×		×	+	×	×	×	×	×	×	×	×	
Ammonium Hydroxide, $NH_4\ OH$	×	×	×	×	×	×	×		+	○	−	+	○	○	×	+	+	+	×	×	×	×	−	×	×	×	×	−	×	−			×		×		
Ammonium Nitrate, $NH_4\ NO_3$	−		×	×	×	−	×	×	×	○	○		×		−	○	×	○	×	×			×						×	×	×		×		×	×	
Ammonium Persulfate, $(NH_4)_2\ S_2\ O_8$	○		×	×	×		×	×	×	−	−		○	○	−	○	×	○	×	×			×							×							
Ammonium Phosphate, $(NH_4)_1H_2\ PO_4$	○		×	×		−	×		×	○									×	×										×					×		
Ammonium Phosphate, $(NH_4)_2H\ PO_4$	−		×	×		×	×		×	−	−								×	×										×					×		
Ammonium Phosphate, $(NH_4)_3PO_4$	×	−	×	×	×	×	×	×	×	○	−	×	×	×	×	×	×	×	×	×	×		×						×	×			×		×		
Ammonium Sulfate, $(NH_4)_2SO_4$	×		−	−	+	×	×	×	×	−	−	−	+	+	×	×	×	×	×	×	×	×	×	×			×		×	×	×		×		×		
Amyl Acetate, $C_5H_{11}COOCH_3$	−		×	×	×	×	×	×	×	−	×		+	+	+	×	×	×	×	×			×						×	×	×		×		×	×	
Amyl Alcohol, $C_5H_{11}OH$	×		×	×			×		×	×	×								×	×									×	×	×		×	×	×		

* NOTE: Duriron is as shown. Durichlor is also satisfactory on chlorides and HCl.

** Durcon 5 would be the preferred formula.

*** Reproduced by permission of Chem Flow Corporation.

Table A.4
CHEMICAL RESISTANCE OF MATERIALS (Continued)

RUBBERS					THERMOPLASTICS															THERMOSETTING PLASTICS																			WOODS					
Hard Rubber (Natural)	*Soft Rubber (Natural)*	*Neoprene*	*Butadiene Derivatives*	*Nitrile Rubber (Chemigum)*	*Viton*	*Asphaltic, Bitumastic*	*Cellulose Acetate*	*Cellulose Acetatebutyrate*	*Ethyl Cellulose (Ethocel)*	*Cellulose Nitrate*	*Acrylic (Lucite, Plexiglas)*	*Coumarone Resins*	*Polyethylene*	*Polyvinyl Chloride, Rigid or Unplasticized*	*Tygon (P. V. C. & Copolymers)*	*Saran (Vinyl chloride, vinyldene chloride)*	*Kel-F (Polytrifluorochloroethylene)*	*Teflon (Polytetrafluoroethylene)*	*Uscolite CP (Styrene-acrylonitrile-butadiene)*	*Penton (Chlorinated Polyether)*	*Shellac Compounds*	*Organic Polysulfides*	*Polystyrene (Styron)*	*Vinylidene Chlorides*	*Vinyl Chloride Acetates*	*Cast Phenol Formaldehyde*	*Haveg 41 (Phenolic w. asbestos)*	*Heresite (Phenol formaldehyde)*	*Molded Phenolformald, (Durez)*	*Phenol Furfural Plastics*	*Urea Formaldehyde*	*Casein Plastics*	*Epoxy Resins*	*Furane Resins (Haveg 61, Duralon)*	*Silicone Resins*	*Permanite (Furan, Glass Fiber)*	*Nylon (Adipic Acid—Hexameth, Diamine)*	*Durcon 6 (Modified Epoxy)*	*Cypress*	*Fir*	*Maple*	*Oak*	*Pine*	*Redwood*
×	—	○	○	○	○	○	○		×	○	○		○	○	○	—	×	×	○	+		×	○	○	○	—	—	×	×	—			—	×	○	—	—	×	—	—	×	×	×	—
×	—	○	×	○		×	×	×	×	×	×	×	×	—	—	×	×	×	—	×	×	×	○	—	—	×	×	×	×	×		×	×	×	×	×	○	×	×	×	×	×	×	×
×		○	—	○	○	○	○		○	○	○			○	○	—		×		+			○			—	○	×	×	×				○		—		×						
○	○	○	×	—	○	○	○	○	○	○	○	○	○	○		○		×	○	+	○	×	—	○	○		○		—	—	×	×	×			×	×							
																	×	×																				×						
×	×	×	×	×	×	×				×			×	×	×	×	×	×	×	×	×		×	×	×	×	×	×	×	×			×	×		×	×	×						
					×								×	×			×	×		×																	×	×						
×	×	×	×	—	×	×				×	×	×	×	×	×	×	×	×	×	×	×	×	×	×	×	—	×	×	×			×		×		×	×	×	×	×	×	—	×	×
		×											×	×	×	×	×	×	×	×							×	×					×					×						
×	×	×	×	×		×				×	×	×	×	×	×	×	×	×	×	×	×	×	×	×	×	—	×	×	×			×	×	×				×	×	×	×	—	×	×
—					○									○	—		×	×									×	×									×	×						
—	—	×	×		○		—	×	×		×		×	—		○	×	×	—	×	×		○	×	×	×			×	—			×	×	—			×						
×				○		×							×	×		×	×	×	×	×			×						×	×			×	×			×	×						
×	×	×	×	—	×	×	×	×		×	×	×	×	×	×	×	×	×	×	×	×		×	×	×	×	×	×	×	×		×	×	×		×	○	+°°						
—	×	—		×	+	—					×		×	×	—	○	×	×	×	×			×	○			×	×	○	○			×	×	—	×	×	×						
—	—	—	×		+	×	×			×	×		×	×		×	×	×	×	×	—		×	×	×	×			×	×		×	×				×	×						
						×								×	×		×	×		×													×	×				×						
														×			×	×	×	×																		×						
														×			×	×	×	×																		×						
×													×	×	×		×	×	×	×							×	×										×						
×	×	×	×			×				×	×	×		×	×	×	×	×	×	×	×	×		×	×	×	×	×	×	×		×	×			×		○						
○	○	○	○		○	○	○	○	○	○	○	○	—	○		○	×	×		×	○		○	○	×	—			—	—		×	×			×	×	×						
—	—	×	×		×	○	×	○	○	○	×	—	—	×		×	×	×	×	×	○		×	—	—	×			×	×		×	×				×	×						

Table A.4
CHEMICAL RESISTANCE OF MATERIALS (Continued)

MATERIALS
× — Very Good Service
+ — Moderate Service
− — Limited or Variable Service
○ — Unsatisfactory
Blank — No Information

CHEMICALS
Solids Assumed in Sol'n.
Room Temperatures Assumed
Unless Otherwise Stated

CHEMICALS	METALS																												CARBONS & CERAMICS								
	Carbon Steel; Fe	*Cast Iron & Ductile Iron; Fe*	*304 Stainless Steel; Fe, 18Cr, 8Ni*	*316 Stainless Steel; Fe, 16Cr, 10Ni, 2Mo*	*347 Stainless Steel; Fe, 17Cr, 9Ni; (Cx10)Cb*	*Ni-Resist Iron; Fe, 14Ni, 2Cr, 2Si*	*Durimet 20; Carpenter 20; Fe; 4Cu, 20Cr, 29Ni, 2Mo, 1Si*	*Worthite; 3Mo, 2Cu, Fe, 20Cr, 24Ni, 3Si*	*Duriron; Fe, 14Si; Durichlor, Fe, 14Si, 3Mo°*	*Copper; Brass; Bronzes; Everdur*	*Aluminum; Al (and Alloys)*	*Lead; Pb*	*Monel; 67Ni, 30Cu, 1.4 Fe*	*Nickel; Ni*	*Inconel; 76Ni, 15Cr, 8Fe*	*Hastelloy B; Ni, 26Mo, 4Fe*	*Hastelloy C; Ni, 16Mo, 4Fe, 14Cr, 4W*	*Hastelloy D; Ni, 8Si, 3Cu*	*Chlorimet 3; 3Fe, 1Si, 60Ni, 18Mo, 18Cr*	*Chlorimet 2; 63Ni, 32Mo, 3Fe, 1Si*	*Stellite; Co, 28Cr, 4W*	*Zirconium; Zr*	*Tantalum; Ta*	*Silver; Ag*	*Platinum; Pt*	*Dowmetal; (Mg alloys)*	*Titanium; Ti*	*Molybdenum; Mo*	*Carbon & Graphite*	*Glass, "Pyrex" brand*	*Silicaware*	*Silicate Cements*	*Chemical Stoneware*	*Transite (asbestos & cement)*	*Chemical Porcelain*	*Concrete-Unbonded*	*Concrete-Mortar Bonded*
Amyl Chloride, $C_5H_{11}Cl$	×		−	−	−		×	−	×	×	○		−	+	−	×	×	×	×	×			×						×	×	×		×		×		
Antimony Trichloride, $SbCl_3$	○		○	○	○				×	−	○	×	×	×	×	×	×		×	×			×							×	×		×		×		
Arsenic Acid, $HAsO_8$	○		−	−		×	×	×	×	−		−			−				×	×			×	×					×	×					×		
Barium Carbonate, $BaCO_3$			×	×	×		×	×	×	×	○		+	×	×	×	×	×	×	×		×	×				×			×					×		
Barium Hydroxide, $Ba(OH)_2$	×			×		×	×	×	+	○	○		×	×		+	+	+	×	×		×	×				+			×					×		
Barium Sulfide, BaS	○		×	×		×	×	×	×	○	○		−						×	×				○						−					×		
Benzaldehyde, C_6H_5 CHO	×		×	×			×		×		−	○							×	×									×	×	×		×	×	×	×	
Benzene, C_6 H_6	+	+	×	×	×	×	×	×	×	×	×	+	+	+	×	+	+	+	×	×	×		×	×	×	×			×	×		×	×		×		
Benzoic Acid, C_6 H_5 COOH	○		×	×	+		×	×	×	+	×	○	+	+	+	×	×	×	×	×			×	×						×					×		
Borax, $Na_2B_4O_7$	×	−	×	×	×	×	×		×	−	−	×	×	×					×	×				○					×	×	×		×	×	×	×	
Boric Acid, H_3BO_3	○	○	+	+	+	−	×	×	×	+	+	+	+	+	×	×	×	×	×	×	×		×	×	×		×		×	×		×	×		×		
Bromine, Wet, Br_2	○		○	○	○	○	○	○	○	○	○	−	○	○	○	○	×	○	×	○	−		×	○	−		○		○	×	×		×		×	×	
Butanol, C_4H_9OH	×		×	×			×		×	×	×		×	×	×	×	×	×	×	×										×					×		
Butyl Acetate, $C_4H_9COOCH_3$	−			×			×	×	×	×	×	×	+			×	×	×	×	×			×			−			×	×	×		×	×	×	×	×
Butyric Acid, C_3 H_7 COOH	○	○	○	×	○		×	×	×	−	−	○	+	+	+	×	×	×	×	×			×				×			×					×		
Calcium Bisulfate, $CaHSO_4$	○		○	×		○	×			○	○		○	○					×	×										×					×		
Calcium Bisulfite, Ca HSO_3	×		○	×		○	×	×	○	−	−	×	○						×	×							×			×	×		×		×		
Calcium Carbonate, $CaCO_3$	×		×	×		×	×	×	×	×	○	×	×	×	×	×	×	×	×	×			×				×			×					×		
Calcium Chlorate, $CaClO_3$			−	×	○		+	−	+	−			+	+	+		+	+	+	+			×						×	×					×		
Calcium Chloride, $CaCl_2$	+	+	−	−	−	−	+	×	×	+	−	○	+	+	+	+	×	+	×	×	−	×	×				×		×	×	×	×	×	×	×	×	
Calcium Hydroxide, Ca $(OH)_2$	×		×	×	×	×	×	×	+	+	−		×	×	×	×	×	×	×	×			×				×			×					×		
Calcium Hypochlorite, Ca $(OCl)_2$	−	○	○	−	−	−	−	−	×	○	○	○	○	○	○	○	−	○	−	○	○	×	×		×		×		×	×	×	○	×		×	×	
Calcium Sulfate, Ca SO_4	−		×	×			×	×	×	×	−		+			+	+	+	×	×			×							×					×		
Carbon Dioxide (Dry), CO_2	×		×	×	×	×	×		×	×	×	×	×	×	×	×	×		×	×				×				×		×					×		
Carbon Dioxide (Wet or H_2 CO_3)	−	−	×	×	×	−	×	×	×	−	−	−	−	×	×	×	×	×	×	×	×		×	×					×	×	×		×	×	×	×	×
Carbon Disulfide, CS_2	×	+	+	+	×	×	×	×	×	−	−	+	+	+	+				×	×			×	−					×	×	×	×	×		×		

° NOTE: Duriron is as shown. Durichlor is also satisfactory on chlorides and HCl.
°° Durcon 5 would be the preferred formula.

Table A.4
CHEMICAL RESISTANCE OF MATERIALS (Continued)

RUBBERS						THERMOPLASTICS															THERMOSETTING PLASTICS																		WOODS					
Hard Rubber (Natural)	*Soft Rubber (Natural)*	*Neoprene*	*Butadiene Derivatives*	*Nitrile Rubber (Chemigum)*	*Viton*	*Asphaltic, Bitumastic*	*Cellulose Acetate*	*Cellulose Acetatebutyrate*	*Ethyl Cellulose (Ethocel)*	*Cellulose Nitrate*	*Acrylic (Lucite, Plexiglas)*	*Coumarone Resins*	*Polyethylene*	*Polyvinyl Chloride, Rigid or Unplasticized*	*Tygon (P. V. C. & Copolymers)*	*Saran (Vinyl chloride, vinyldene chloride)*	*Kel-F (Polytrifluorochloroethylene)*	*Teflon (Polytetrafluoroethylene)*	*Uscolite CP (Styrene-acrylonitrile-butadiene)*	*Penton (Chlorinated Polyether)*	*Shellac Compounds*	*Organic Polysulfides*	*Polystyrene (Styron)*	*Vinylidene Chlorides*	*Vinyl Chloride Acetates*	*Cast Phenol Formaldehyde*	*Haveg 41 (Phenolic w. asbestos)*	*Heresite (Phenol formaldehyde)*	*Molded Phenolformald, (Durez)*	*Phenol Furfural Plastics*	*Urea Formaldehyde*	*Casein Plastics*	*Epoxy Resins*	*Furane Resins (Haveg 61, Duralon)*	*Silicone Resins*	*Permanite (Furan, Glass Fiber)*	*Nylon (Adipic Acid—Hexameth, Diamine)*	*Durcon 6 (Modified Epoxy)*	*Cypress*	*Fir*	*Maple*	*Oak*	*Pine*	*Redwood*
–	○	–	×		–	○								○		×	×	×		×				×	×	–			–	–		×	×					×						
			×		×				–					×		×	×	×	×	×													×				○	×						
		×			×								×	–			×	×		×																	×	×						
					×									×			×	×	×	×													×				×	×						
					×	×							×	×			×	×	×	×													×				×	×						
×					×								×	–	×	×	×	×	×	×																	×	×						
○	○	○	×			○							○	○		–	×	×		–	○			×		×			×	×			–					×	×					
○	○	○		○	–	○	–	○	○		○		–	○	○	–	×	×		+			○		○		–	×	×	×			×	×		×	×	×°°						
						×								×	×	×	×	×		×													×				○	+°°						
×	×	×	×		×	×	×	×	×	×	×	×		×			×	×	×	×	×		×	×	×	×			×	×							×	×	×	–	–	–	×	–
×	×	×			×									–	×	×	×	×	×	×							×	×	×	×				×		×	×	×						
–	○		–		–	○							○	○	–	○	×	×		–			–				○	×	–	–	–		○	○		○	○	×°°						
–	–	–									×			×		×	×	×	×	×													–			×		×						
○	○	○	○		○	○	○	○	○	○	○			○		○	×	×		+			○	○	○	–			–	–	○	×	×					×						
×	–	○			○	○								○		–	×	×		×													–				○	×						
																	×	×																				×						
×	×	×	×											×		×	×	×	×	×						×			×	×								×						
×	×	–			×									×		×	×	×		×																	×	×						
						×								×			×	×	×	×													×					×						
×	×	×	×	×	×	×	×	×	×	×	×	×	×	×	×	×	×	×	×	×	×	×	×	×	×	–	×	×	–	–			×	×		×	○	×	×	×	×	×	×	–
×	×	×			×	×								×		×	×	×	×	×													×			×	×	×						
×	×	○	×	–		–							×	×	×	×	×	×	×	×		×	×		×	○	–	–	○	○			×	×			○	×	○	○	○	○	–	○
×	×	×			×	×								×		×	×	×	×	×													×				×	×						
×	×	×				×								×		×		×	×	×													×					○						
×	×	×	×			×	×	×	×	×	×	×	×	×	×	×		×	×	×	×	×	×	×	×	×	×	×	×	×	×	×						×	×	×	×	×	×	×
○	○	○	×		×	○	○	○	○	×	×		○	○		–	×	×		–	–		○	○		×			×	×			○	×		×	×	×						

Table A.4
CHEMICAL RESISTANCE OF MATERIALS (Continued)

MATERIALS
× — Very Good Service
\+ — Moderate Service
– — Limited or Variable Service
○ — Unsatisfactory
Blank — No Information

CHEMICALS
Solids Assumed in Sol'n.
Room Temperatures Assumed
Unless Otherwise Stated

METALS

CHEMICALS	Carbon Steel; Fe	Cast Iron & Ductile Iron; Fe	304 Stainless Steel; Fe, 18Cr, 8Ni	316 Stainless Steel; Fe, 16Cr, 10Ni, 2Mo	347 Stainless Steel; Fe, 17Cr, 9Ni; (Cx10)Cb	Ni-Resist Iron; Fe, 14Ni, 2Cr, 2Si	Durimet 20; Carpenter 20; Fe; 4Cu, 20Cr, 29Ni, 2Mo, 1Si	Worthite; 3Mo, 2Cu, Fe, 20Cr, 24Ni, 3Si	Duriron; Fe, 14Si; Durichlor, Fe, 14Si, 3Mo*	Copper; Brass; Bronzes; Everdur	Aluminum; Al (and Alloys)	Lead; Pb	Monel; 67Ni, 30Cu, 1.4 Fe	Nickel; Ni	Inconel; 76Ni, 15Cr, 8Fe
Carbon Tetrachloride (Moist) CCl_4	–	–	○	–	–	–	×	×	×	–	–	+	+	+	×
Chloracetic acid, $ClCH_2CO_2H$	○	○	○	○	○		–	○	×	○	○	○	–	–	+
Chloric Acid, $HClO_3$			○	○											
Chlorine (Dry), Cl_2	×	×	×	×	×		×	×	×	×	○	×	×	×	×
Chlorine (Wet), Cl_2	○	○	○	○	○	○	○	○	–	○	○	–	○	–	○
Chlorobenzene, C_6H_5Cl	×		×	×	×		×	×	×	+			–	–	–
Chloroform, $CHCl_3$	×		×	×	×		×	×	×	×	–	×	×	×	×
Chromic Acid, CrO_3 solution	–	+	○	–	○	○	+	×	×	○	+	+	○	○	–
Copper Chloride, $CuCl_2$	○		○	○	○		○	×	○	○	○		○	○	○
Copper Cyanide, $Cu(CN)_2$			×	×	×		×	×	×	○	○		○	○	○
Copper Nitrate, $Cu(NO_3)_2$	○		×	×	–		×	×	×	○	○		○	○	○
Copper Sulfate, $CuSO_4$	○	○	×	×	×	–	×	×	×	○	○	–	–	–	–
Cresylic acid	○		×	×	×		×	×	×	○	–		–	×	–
Dichlorethane, $C_2H_4Cl_2$	○		×	×	×		×		×				×	×	×
Diethylamine, $(C_2H_5)_2NH$	×			×			×								
Diphenyl, $C_6H_5C_6H_5$	×		×				×		×					×	
Ethers, Various	×		×	×	×		×	×	×	×	×	×	+	+	+
Ethyl Acetate, $C_2H_5COOCH_3$	×		×	×		×	×	×	×	×	×		×	×	×
Ethyl Alcohol, C_2H_5OH	+	+	×	×	×	×	×	×	×	+	–	+	+	+	+
Ethyl Chloride, C_2H_5Cl	×		×	×	×	○	×	–	×	–	–		+	+	×
Ethylene Chlorohydrin, $Cl(C_2H_4)OH$	×		×				×		×			×			
Ethylene Dichloride, $C_2H_4Cl_2$	×		×	×	×		×		×	–	○	×	+	+	–
Ethylene Glycol, CH_2OHCH_2OH	–	+	×	×	×	×	×	×	×	+	×	+	+	+	+
Ethylene Oxide, CH_2OCH_2	×		×				×		×						
Fatty Acids, Various	○	○	×	×	×	–	×	×	×	–	×	○	+	×	×
Ferric Chloride, $FeCl_3$	○	○	○	○	○	○	○	○	–	○	○	○	○	○	○
Ferric Nitrate, $Fe(NO_3)_3$	○		×	×	×		×	×	×	○	○		○	○	○
Ferric Sulfate, $Fe_2(SO_4)_3$	○		–	×	×	○	×	×	×	○	○	×	○	○	○

METALS (continued)

CHEMICALS	Hastelloy B; Ni, 26Mo, 4Fe	Hastelloy C; Ni, 16Mo, 4Fe, 14Cr, 4W	Hastelloy D; Ni, 8Si, 3Cu	Chlorimet 3; 3Fe, 1Si, 60Ni, 18Mo, 18Cr	Chlorimet 2; 63Ni, 32Mo, 3Fe, 1Si	Stellite; Co, 28Cr, 4W	Zirconium; Zr	Tantalum; Ta	Silver; Ag	Platinum; Pt	Dowmetal; (Mg alloys)	Titanium; Ti	Molybdenum; Mo
Carbon Tetrachloride (Moist) CCl_4	+	×	+	×	×	×	×	×	×			×	
Chloracetic acid, $ClCH_2CO_2H$	×	×	–	×	×	–		×	×	×		×	
Chloric Acid, $HClO_3$									○				
Chlorine (Dry), Cl_2	×	×	×	×	×			×	×	×		○	+
Chlorine (Wet), Cl_2	○	×	○	×	○		○	×	×	×		–	○
Chlorobenzene, C_6H_5Cl				×	×			×					
Chloroform, $CHCl_3$	×	×	×	×	×			×				×	
Chromic Acid, CrO_3 solution	○	+	○	+	○	×	×	×	○	×	–	×	+
Copper Chloride, $CuCl_2$	×	×	×	–	○		–	×	–			×	
Copper Cyanide, $Cu(CN)_2$	×	×	×	×	×			×					
Copper Nitrate, $Cu(NO_3)_2$	○	×	○	×	×			×	○				
Copper Sulfate, $CuSO_4$	×	×	×	×	×	×	×	×	–	×			
Cresylic acid	–	–	–	×	×								
Dichlorethane, $C_2H_4Cl_2$	×	×		×	×								
Diethylamine, $(C_2H_5)_2NH$				×	×								
Diphenyl, $C_6H_5C_6H_5$				×	×								
Ethers, Various	×	×	×	×	×								
Ethyl Acetate, $C_2H_5COOCH_3$	×	×	×	×	×			×					
Ethyl Alcohol, C_2H_5OH	×	×	+	×	×		×	×			×		
Ethyl Chloride, C_2H_5Cl	×	×	×	×	×			×					
Ethylene Chlorohydrin, $Cl(C_2H_4)OH$				×	×								
Ethylene Dichloride, $C_2H_4Cl_2$	×			×	×			×				×	
Ethylene Glycol, CH_2OHCH_2OH				×	×								
Ethylene Oxide, CH_2OCH_2				×	×								
Fatty Acids, Various	×	×	×	×	×	×		×	×	×			
Ferric Chloride, $FeCl_3$	○	+	○	–	○	–	○	×	○			×	○
Ferric Nitrate, $Fe(NO_3)_3$	○	+	○	×				×	○				
Ferric Sulfate, $Fe_2(SO_4)_3$	○	+	○	×		×		×	○			×	

CARBONS & CERAMICS

CHEMICALS	Carbon & Graphite	Glass, "Pyrex" brand	Silicaware	Silicate Cements	Chemical Stoneware	Transite (asbestos & cement)	Chemical Porcelain	Concrete-Unbonded	Concrete-Mortar Bonded
Carbon Tetrachloride (Moist) CCl_4	×	×	×	×	×	×	×	×	
Chloracetic acid, $ClCH_2CO_2H$		×			×		×		
Chloric Acid, $HClO_3$		×							
Chlorine (Dry), Cl_2	×	×	×		×		×		
Chlorine (Wet), Cl_2	–	×	×		×		×		
Chlorobenzene, C_6H_5Cl		×	×		×		×		
Chloroform, $CHCl_3$	×	×	×		×		×		
Chromic Acid, CrO_3 solution	○	×	×	×	×		×		
Copper Chloride, $CuCl_2$		×	×		×		×		
Copper Cyanide, $Cu(CN)_2$		+					×		
Copper Nitrate, $Cu(NO_3)_2$		×					×		
Copper Sulfate, $CuSO_4$	×	×	×	×	×		×		
Cresylic acid		×					×		
Dichlorethane, $C_2H_4Cl_2$		×					×		
Diethylamine, $(C_2H_5)_2NH$	×	×					×		
Diphenyl, $C_6H_5C_6H_5$		×					×		
Ethers, Various	×	×	×		×	×	×	×	×
Ethyl Acetate, $C_2H_5COOCH_3$	×	×	×		×	×	×	×	×
Ethyl Alcohol, C_2H_5OH	×	×	×	×	×	×	×		×
Ethyl Chloride, C_2H_5Cl	×	×					×		
Ethylene Chlorohydrin, $Cl(C_2H_4)OH$	–	×	×		×		×		
Ethylene Dichloride, $C_2H_4Cl_2$	×	×	×		×		×		
Ethylene Glycol, CH_2OHCH_2OH	×	×	×	×	×		×		
Ethylene Oxide, CH_2OCH_2		×					×		
Fatty Acids, Various	×	×	×	×	×	×	×	×	×
Ferric Chloride, $FeCl_3$	×	×	×	×	×		×	×	
Ferric Nitrate, $Fe(NO_3)_3$		×					×		
Ferric Sulfate, $Fe_2(SO_4)_3$	–	×	×		×		×		

* NOTE: Duriron is as shown. Durichlor is also satisfactory on chlorides and HCl.

** Durcon 5 would be the preferred formula.

Table A.4
CHEMICAL RESISTANCE OF MATERIALS (Continued)

RUBBERS					THERMOPLASTICS															THERMOSETTING PLASTICS																			WOODS					
Hard Rubber (Natural)	*Soft Rubber (Natural)*	*Neoprene*	*Butadiene Derivatives*	*Nitrile Rubber (Chemigum)*	*Viton*	*Asphaltic, Bitumastic*	*Cellulose Acetate*	*Cellulose Acetatebutyrate*	*Ethyl Cellulose (Ethocel)*	*Cellulose Nitrate*	*Acrylic (Lucite, Plexiglas)*	*Coumarone Resins*	*Polyethylene*	*Polyvinyl Chloride, Rigid or Unplasticized*	*Tygon (P. V. C. & Copolymers)*	*Saran (Vinyl chloride, vinyldene chloride)*	*Kel-F (Polytrifluorochloroethylene)*	*Teflon (Polytetrafluoroethylene)*	*Uscolite CP (Styrene-acrylonitrile-butadiene)*	*Penton (Chlorinated Polyether)*	*Shellac Compounds*	*Organic Polysulfides*	*Polystyrene (Styron)*	*Vinylidene Chlorides*	*Vinyl Chloride Acetates*	*Cast Phenol Formaldehyde*	*Haveg 41 (Phenolic w. asbestos)*	*Heresite (Phenol formaldehyde)*	*Molded Phenolformald. (Durez)*	*Phenol Furfural Plastics*	*Urea Formaldehyde*	*Casein Plastics*	*Epoxy Resins*	*Furane Resins (Haveg 61, Duralon)*	*Silicone Resins*	*Permanite (Furan, Glass Fiber)*	*Nylon (Adipic Acid—Hexameth. Diamine)*	*Durcon 6 (Modified Epoxy)*	*Cypress*	*Fir*	*Maple*	*Oak*	*Pine*	*Redwood*
○	○	○	×	–	×	○	×	–		×	×		○	–	○	–	×	×		×	○	–	○	×	×	×	×	×	×	×			×	×		×	×	×						
–	–	–	×		×	○			–					×	○	–	×	×		+		×	×				×	×	×	×							○	×						
																	×	×																				○						
×	–	○	–		×	○	–	–					×	–	×	–	×	×	×	+			×	×	×	–	–	–	–	–					–	–	○	○						
×	×	○	×			–	–	–					×	–	×	–	×	×	×	+			×	×	×	–	–		–	–			○		–	–	○	+						
○	○	○			–	○					–		○	○		–	×	×		+		○	○	○			–						–	×		×		–°°						
○	○	○			–	○	○	○	○		○		○	○		–	×	×		+													×				×	×°°						
○	○	○	○	○	–	○	○	○	○		○	○	×	–	×	×	×	×	–	–			○	×		○	○	×	○	○			○	○		○	○	×						
–	×	×	×		×	×							×	×		×	×	×	×	×	–						×						×	×			○	×						
						×								×				×	×	×													×					×						
×		×			×								×	×		×	×	×	×	×																	×	×						
×	×	×	×	○	×	×	×	×		×	○		×	×	×	×	×	×	×	×	×		×			×	×	×	×	×			×	×		×	○	×						
○	○	○			–									–			+	×		×																	○	×						
																	–	×																				×						
		○												○			×	×		×																		×						
																	×	×																				×						
○	○	○	○		○	○	○	○	○	○	○		○	○		–	×	×		×			○	○	○	–	–		–	–		×	–	×		×		×						
○	○	○	○		○	○	○	○	○	○	○		○	○		–	×	×		+			○	○	○	–	○		–	–		×	×	×		×	×	×						
×	×	×	×	×	×	×	×	–	–		–		○	×	–	×	×	×	×	×	○	–	–	×	×		×		×	×	×		×	–			×	×						
○		–			×	○								○		–	×	×		×													×				×	×						
–	–	–	–			○	○	○						○		○	×	×		×				–		–			–	–			×					×	×					×
○	○	○	○		–	○	○				○		–	○		–	×	×		+			○	○	○		×						×			×	×	×						
–	–	–	×	×	×	×	×	×		×	×		○	×	–		×	×	×	×			×	–	–	–	×		–	–			×	×			×	×				×		
					–									○		○		×		×																	×	×						
○	○	○	×		×	○	×	×		×			○	–	×	×	×	×	×	×			○	×	–	–	×	×	×	–	×		×	×	–	×	×	×	×	×	×	×	×	–
×	×	–	×		×	×			–	×			×	×	×	–	×	×	×	×	×		×	×	×	×	×	×	×	×	×		×	×			○	+°°						
○		×			×	×								×		×	×	×	×	×							×						×	×		×		×						
×	×	×	×		×	×			–	×				×	×	×	×	×	×	×	×		×	×	×	×	×	×	×	×	×		×	×		×	×	×						

Table A.4
CHEMICAL RESISTANCE OF MATERIALS (Continued)

MATERIALS
× — Very Good Service
+ — Moderate Service
– — Limited or Variable Service
○ — Unsatisfactory
Blank — No Information

CHEMICALS
Solids Assumed in Sol'n.
Room Temperatures Assumed Unless Otherwise Stated

METALS

CHEMICALS	*Carbon Steel; Fe*	*Cast Iron & Ductile Iron; Fe*	*304 Stainless Steel; Fe, 18Cr, 8Ni*	*316 Stainless Steel; Fe, 16Cr, 10Ni, 2Mo*	*347 Stainless Steel; Fe, 17Cr, 9Ni; (Cx10)Cb*	*Ni-Resist Iron; Fe, 14Ni, 2Cr, 2Si*	*Durimet 20; Carpenter 20; Fe; 4Cu, 20Cr, 29Ni, 2Mo, 1Si*	*Worthite; 3Mo, 2Cu, Fe, 20Cr, 24Ni, 3Si*	*Duriron; Fe, 14Si; Durichlor, Fe, 14Si, 3Mo°*	*Copper; Brass; Bronzes; Everdur*	*Aluminum; Al (and Alloys)*	*Lead; Pb*	*Monel; 67Ni, 30Cu, 1.4 Fe*	*Nickel; Ni*	*Inconel; 76Ni, 15Cr, 8Fe*	*Hastelloy B; Ni, 26Mo, 4Fe*	*Hastelloy C; Ni, 16Mo, 4Fe, 14Cr, 4W*	*Hastelloy D; Ni, 8Si, 3Cu*	*Chlorimet 3; 3Fe, 1Si, 60Ni, 18Mo, 18Cr*	*Chlorimet 2; 63Ni, 32Mo, 3Fe, 1Si*	*Stellite; Co, 28Cr, 4W*	*Zirconium; Zr*	*Tantalum; Ta*	*Silver; Ag*	*Platinum; Pt*	*Dowmetal; (Mg alloys)*	*Titanium; Ti*	*Molybdenum; Mo*
Ferrous Chloride, $Fe\ Cl_2$	○	○	○	○	○	×	–	○	+	○	○	○	○	○	○	–	+	+	+	–			×	○				
Ferrous Sulfate, $FeSO_4$	○	○	+	○	×	–	×	×	×	–	+	–	○	○	○	+	+	+	×	×	–		×	○			×	
Fluorine, F_2			○	○			–		○	○	○	–	○	×	×	+	×		+				○	×			○	○
Formaldehyde, CH_2O	–	–	–	–	–	–	×	×	×	×	–	○	–	×	–	×	×	×	×	×	×		×	×		–	×	
Formic Acid, HCOOH	○	○	+	+	–		×	×	×	○	–	○	+	+	–	–	×	–	×	×	×	×	×				○	+
Fuel Oil	×		–	×	×	×	×		×	×	×		×	×	+	×	×	×	×	×			×					
Gallic Acid, $(OH)_3\ C_6H_2COOH$	○		×	×	×		×		×	×	×		–	+	+	×	×	×	×	×								
Gasoline, Refined	×		×	×	×	×	×		×	×	×		+	+	+	×	×	×	×	×			×					
Glycerol, $CH_2\ OH\ CHOH\ CH_2\ OH$	+		+	+	+	×	×		×	×	×	–	+	×	×	×	×	×	×	×				×			×	
Hydrobromic Acid, HBr	○		○	○	○		○	○	–	–	○	○	○	○	○	×	–	–	–	+			×				×	
Hydrochloric Acid, (Conc.), HCl	○	○	○	○	○	○	○	○	+	○	○	○	○	○	○	+	–	○	+	×	×	×	×	×	×		○	×
Hydrochloric Acid, (Dilute)	○	○	○	○	○	–	○	○	×	○	○	–	–	–	–	+	–	–	×	×	–	×	×	×	×		–	
Hydrochloric Acid (Dry Gas)	×		–	×			×		×		○	–	–	–	–	×	×		×	×		×	×					
Hydrocyanic Acid, (Conc.), HCN	○		×	×	×	×	×	×	×	○	×	–	×	–	–	×	×	×	×	×			×					
Hydrocyanic Acid, (Dilute & Gas)			○	○																			×					
Hydrofluoric Acid, (Conc.), HF	–	○	○	○	○	○	+	–	○	○	○	○	+	–	–	+	+	+	×	×	–		○	×	×	×	○	
Hydrofluoric Acid, (Dilute)	○	○	○	○	○	–	+	○	○	–	○	+	+	+	–	+	+	+	×	×	–		○		×	○	○	
Hydrofluosilicic Acid, H_2SiF_6	○		○	○	○	○	+		○	○	–		–	–	–	–	–	–	×	×			○					
Hydrocarbons (Aliphatic)	×	×	×	×	×	×	×		×	×	×		×	×	×	×	×	×	×	×	×		×	×	×	×		+
Hydrocarbons (Aromatic)	×	×	×	×	×	×	×		×	×	×		×	×	×	×	×	×	×	×	×		×	×	×	×		
Hydrogen Gas, H_2	×						×		×	×	×	×	×	×					×	×				×				×
Hydrogen Peroxide (Conc.), $H_2\ O_2$	○	○	+	○	–	○	+	–	+	○	×	–	–	–		–	×	–	×	×	×		×	○	×		×	
Hydrogen Peroxide (Dilute)	○	○	+	+			×	×	×	○	×	○	+	+	+	×	×	–	×	×	×	×	×	×	×		–	
Hydrogen Sulfide (Dry) $H_2\ S$	×		–	×		–	×		×	○	○	×	–	×	+	–	×	–	×	+			×	○				+
Hydrogen Sulfide (Wet)	–		–	×	–	–	×	×	+	○	○	○	○	×	+	–	+	–	×	+				○				
Iodine, I_4 Wet	○	○	○	○	○		○		○	○	○	○	○	○	–	○	–	○	+	○	–		×	–	×	–		
Iodoform, CHI_3	○		○	×	×		×		×		×								×	×								
Kerosene	×		×	×	×		×		×	×	×		×	×	×	×	×	×	×	×			×					
Ketones, Various	×		×				×		×		–	×							×	×								
Lactic Acid, $CH_3\ CHOHCOOH$	○	○	–	–	×	–	×	×	×	–	–	○	–	–	–	×	×	×	×	×	×	×	×	×	×		×	×
Lead Acetate, $Pb(CH_3\ COO)_2$	○		×	×	+		×	×	×	–	○		×	+		×	×	×	×	×			×				×	

CARBONS & CERAMICS

CHEMICALS	*Carbon & Graphite*	*Glass, "Pyrex" brand*	*Silicaware*	*Silicate Cements*	*Chemical Stoneware*	*Transite (asbestos & cement)*	*Chemical Porcelain*	*Concrete-Unbonded*	*Concrete-Mortar Bonded*
Ferrous Chloride, $Fe\ Cl_2$	×	×	×	×	×		×		
Ferrous Sulfate, $FeSO_4$	×	×	×	×	×		×		
Fluorine, F_2	○	○					○		
Formaldehyde, CH_2O	×	×	×		×		×	×	
Formic Acid, HCOOH	×	×	×	×	×		×		
Fuel Oil		×					×		
Gallic Acid, $(OH)_3\ C_6H_2COOH$		×					×		
Gasoline, Refined	×	×	×		×		×		
Glycerol, $CH_2\ OH\ CHOH\ CH_2\ OH$	×	×	×		×		×	×	
Hydrobromic Acid, HBr	×	×	×		×		×		
Hydrochloric Acid, (Conc.), HCl	×	×	×	×	×		×		
Hydrochloric Acid, (Dilute)	×	×	×		×		×		
Hydrochloric Acid (Dry Gas)	×	×					×		
Hydrocyanic Acid, (Conc.), HCN		×					×		
Hydrocyanic Acid, (Dilute & Gas)		×					×		
Hydrofluoric Acid, (Conc.), HF	○	○			○		○		
Hydrofluoric Acid, (Dilute)	×	○		○	○		○		
Hydrofluosilicic Acid, H_2SiF_6	×	–					○		
Hydrocarbons (Aliphatic)	×	×	×		×		×		
Hydrocarbons (Aromatic)	×	×	×		×		×		
Hydrogen Gas, H_2		×					×		
Hydrogen Peroxide (Conc.), $H_2\ O_2$		×	×		×		×		
Hydrogen Peroxide (Dilute)	×	×	×	×	×		×		
Hydrogen Sulfide (Dry) $H_2\ S$		×	×		×		×	×	
Hydrogen Sulfide (Wet)		×					×		
Iodine, I_4 Wet	○	×	–				×	×	
Iodoform, CHI_3		×					×		
Kerosene	×	×					×		
Ketones, Various	–	×					×		
Lactic Acid, $CH_3\ CHOHCOOH$	×	×	×		×		×		
Lead Acetate, $Pb(CH_3\ COO)_2$		×					×		

° NOTE: Duriron is as shown. Durichlor is also satisfactory on chlorides and HCl.

°° Durcon 5 would be the preferred formula.

Table A.4
CHEMICAL RESISTANCE OF MATERIALS (Continued)

RUBBERS					THERMOPLASTICS																				THERMOSETTING PLASTICS															WOODS					
Hard Rubber (Natural)	*Soft Rubber (Natural)*	*Neoprene*	*Butadiene Derivatives*	*Nitrile Rubber (Chemigum)*	*Viton*	*Asphaltic, Bitumastic*	*Cellulose Acetate*	*Cellulose Acetatebutyrate*	*Ethyl Cellulose (Ethocel)*	*Cellulose Nitrate*	*Acrylic (Lucite, Plexiglas)*	*Coumarone Resins*	*Polyethylene*	*Polyvinyl Chloride, Rigid or Unplasticized*	*Tygon (P. V. C. & Copolymers)*	*Saran (Vinyl chloride, vinyldene chloride)*	*Kel-F (Polytrifluorochloroethylene)*	*Teflon (Polytetrafluoroethylene)*	*Uscolite CP (Styrene-acrylonitrile-butadiene)*	*Penton (Chlorinated Polyether)*	*Shellac Compounds*	*Organic Polysulfides*	*Polystyrene (Styron)*	*Vinylidene Chlorides*	*Vinyl Chloride Acetates*	*Cast Phenol Formaldehyde*	*Haveg 41 (Phenolic w. asbestos)*	*Heresite (Phenol formaldehyde)*	*Molded Phenolformald, (Durez)*	*Phenol Furfural Plastics*	*Urea Formaldehyde*	*Casein Plastics*	*Epoxy Resins*	*Furane Resins (Haveg 61, Duralon)*	*Silicone Resins*	*Permanite (Furan, Glass Fiber)*	*Nylon (Adipic Acid—Hexameth, Diamine)*	*Durcon 6 (Modified Epoxy)*	*Cypress*	*Fir*	*Maple*	*Oak*	*Pine*	*Redwood*
×	×	×	×		×	×				×				×		–	×	×	×	×	×		×	×	×	×	×		×	×	×		×	×			○	×						
×	×	×	×	×	×	×				×			×	×	×	–	×	×	×	×	×		×	×	×	×	×	×	×	×	×		×	×		×	○	+						
						○							–	–				–	○	○							○						○	○			○							
–	–	–	×		○	×	○	×		×	×	×	×	–	–	×	×	×	–	×	×			×	×		×	×			×	×	×				×	×						
–	○		×	×	+	○			–		○		×	–	×	–	×	×	–	×	–		×	×	×	×	×	×	×	×	×	×	×	×			○	×						
								×					○	×			×	×					–	×			×											+°°						
					+									×	×		×	×		×																	×	×						
–	○	–	×		×	○	×	×	–	×	×			–	–	×	×	×	–	×			–	–	–	–	×		–	–		×	×	×		×	×	×						
×	×	–	–	×	×	×	×	×		×	×		×	×	×	×	×	×	×	×			×	–	–	–			–	–			×			×	×	×						
–	×	–	×		+	○	×						×	×	–	×	×	×	–	×	×	×	×	×	×	×	×		×	×			×	–		×	○	×						
×	×	–	×	○	–	×	○	○	○	○	×	×	×	×	×	×	×	×	×	×	×	○	○	○		–	×	×	–	–			×	–	–	×	○	×	○	–	○	○	○	○
×	×	–	×	–		×	×	×	–	×	×	×	×	×	×	×	×	×	×	×	×	×	×	×	×	×	×	×	–	–		×	×		–	×	○	×	×	×	×	×	×	–
						○								×			×	×	×														×					×						
						×								×			×	×	–														×					×						
–		–												×		×		×		–													×					×						
○	○	○		○							×		×	–	○	○	×	×		+			×			–	○	○	–	–		–	×	○	○	–		×						
×	×	–		×	+	○					×		×	–	–	–	×	×	×	×			×				○	–	×	×			×	×	○	×	○	×						
		×													–			×		×																×		×						
○	○	×	×					–	–	×	×				–		×	×			–	–	×	–	–	–	×	×	–	–	×	×						+°°						
○	○		○				×	–	–		–				○		+	×			○	○	○	–	–	–	×	×	–	–		×						×						
					×	×											×	×		×													×				×	×						
–	–	○	×			–	–				×			×	–		×	×	–	+		○	○	○		–	×	×	×	–							○	×						
–	–	○	×	×		–	–				×		×	×	×	×	×	×	×			○	○	–		×	×	×	–	–			○	○			○	×						
×	×		×		×	×								×			×	×	×	×		–		×		–	×		×	–			×	×		×	○	×						
					×									×			×	×		×																		×						
–	–	○	–		×	○	–						–	○		–	×	×									○	×					○	○		○	○	×						
					–													×																			×	×						
○	○	○			×						×			×		×	×	×	×	×																	×	×						
○		○			○	○		○			–		○	○	○	○	+	×	○	+			○	–			×						–			×	×	×						
×	–	×			+		–	×	–				×	×	×	×	×	×	–	×			×				×	×					×	×		×	–	×						×
						×								×			×	×	×	×							×						×	×		×		×						

Table A.4
CHEMICAL RESISTANCE OF MATERIALS (Continued)

MATERIALS
× — Very Good Service
\+ — Moderate Service
– — Limited or Variable Service
○ — Unsatisfactory
Blank — No Information

CHEMICALS
Solids Assumed in Sol'n.
Room Temperatures Assumed
Unless Otherwise Stated

METALS: columns Carbon Steel through Molybdenum; CARBONS & CERAMICS: columns Carbon & Graphite through Concrete-Mortar Bonded

Chemicals	*Carbon Steel; Fe*	*Cast Iron & Ductile Iron; Fe*	*304 Stainless Steel; Fe, 18Cr, 8Ni*	*316 Stainless Steel; Fe, 16Cr, 10Ni, 2Mo*	*347 Stainless Steel; Fe, 17Cr, 9Ni; (Cx10)Cb*	*Ni-Resist Iron; Fe, 14Ni, 2Cr, 2Si*	*Durimet 20; Carpenter 20; Fe; 4Cu, 20Cr, 29Ni, 2Mo, 1Si*	*Worthite; 3Mo, 2Cu, Fe, 20Cr, 24Ni, 3Si*	*Duriron; Fe, 14Si; Durichlor, Fe, 14Si, 3Mo°*	*Copper; Brass; Bronzes; Everdur*	*Aluminum; Al (and Alloys)*	*Lead; Pb*	*Monel; 67Ni, 30Cu, 1.4 Fe*	*Nickel; Ni*	*Inconel; 76Ni, 15Cr, 8Fe*	*Hastelloy B; Ni, 26Mo, 4Fe*	*Hastelloy C; Ni, 16Mo, 4Fe, 14Cr, 4W*	*Hastelloy D; Ni, 8Si, 3Cu*	*Chlorimet 3; 3Fe, 1Si, 60Ni, 18Mo, 18Cr*	*Chlorimet 2; 63Ni, 32Mo, 3Fe, 1Si*	*Stellite; Co, 28Cr, 4W*	*Zirconium; Zr*	*Tantalum; Ta*	*Silver; Ag*	*Platinum; Pt*	*Dowmetal; (Mg alloys)*	*Titanium; Ti*	*Molybdenum; Mo*	*Carbon & Graphite*	*Glass, "Pyrex" brand*	*Silicaware*	*Silicate Cements*	*Chemical Stoneware*	*Transite (asbestos & cement)*	*Chemical Porcelain*	*Concrete-Unbonded*	*Concrete-Mortar Bonded*
Magnesium Chloride, Mg Cl_2	–	–	–	+	–	×	×	×	×	+	–	○	+	+	+	×	×	×	×	×	–		×	×			×			×	×	×	×	×	×	×	×
Magnesium Hydroxide, Mg $(OH)_2$	×		×	×	+		×	×	+	–	○		×	×	×	×	×	×	×	×			×	×			×			×					×		
Magnesium Sulfate, Mg SO_4	+	○	+	+	+	×	×	×	×	+	×	+	+	+	+	×	+	×	×	×		×	×	×			×			×	×	×	×	×	×	×	×
Maleic Acid, CO_2 H C_2 H_2CO_2 H	○		×	×			×	×	×							×			×	×				×						×					×		
Malic Acid, CO_2 H CH_2 CHOH CO_2 H	○		×	×	×		+		×	○	–		+	+	+	×	×	×	×				×							×	×		×		×		
Mercuric Chloride, Hg Cl_2	○		○	○	○		–	–	+	○	○	–	–		+	○	×	○	+	○		×	×	○			×			×					×		
Mercury, Hg	×		×	×	×		×		×	○	○	○	+	+	×	×	×	×	×	×				○			–			×					×		
Methanol, (Conc.), CH_3OH	+	+	+	+	+	×	×	×	×	+	–	+	+	+	+	×	×	+	×	×		×	×	×						×		×			×		
Methanol, (Dilute)	–						×		×	–	–					×	×	–	×	×			×							×					×		
Methyl Chloride, CH_3 Cl	○		–	–	+		×	×	×	×		×	×	×	×				×	×			×							×					×		
Naphtha, Petroleum	×		×	×	×	×	×		×	×	×		+	+	+	×	×	×	×	×									×	×	×		×		×		
Nickel Chloride, Ni Cl_2	○		–	–			+	○	+	–	○		–	×		○	×	○	×	×		×	×				×		×	×					×		
Nickel Sulfate, Ni SO_4	○	○	–	–			×	×	×	+	–	+	+	–	–	○	+	○	×	×		×	×						×	×					×		
Nitrating Acid ($>$15% H_2SO_4)	×	×	–	–	–		×	×	×	○	○		○	○		○	×	○	×	×	×		–	×	×		–		○	×			×		×		
Nitrating Acid ($<$15% H_2 SO_4)	○		–	–	–		×	×	×	○	○		○	○		○	×	○	×	×	×		–	○	×		×		○	×			×		×		
Nitrating Acid ($<$15% HNO_3)	○		–	○	–		×		×	○	○		○	○		○	×	○	×	○	×	○	–	○	×		–		○	×			×		×		
Nitrating Acid ($<$1% Acid)	○		–	×	–		×		×	○	○		○	○		○	×	○	×		×	○	–	○	×				○	×			×		×		
Nitric Acid (Conc.), HNO_3	○	○	+	+	×		×	×	×	○	–	–	○	○	–	○	+	○	×	○	×	×	×	○	×		×	+	○	×	×	×	×		×		
Nitric Acid, Dilute	○	○	×	×	×	○	×	×	×	○	○	○	○	○	○	○	×	○	×	○	×	×	×	○	×		×	○	–	×	×		×		×		
Nitrobenzene, C_6 N_5 NO_2	×		×	×			×	×	×			○							×	×			×							×					×		
Nitrous Acid, HNO_2	○		×	×	–		×	×	×	×			×	○					×				×							×	×		×		×		
Oleic Acid, C_8 H_{17} CH:CH$(CH_2)_7CO_2H$	–	–	–	×	×	–	×	×	×	–	×	○	×	×	×	×	×	×	×	×	×		×	×	×				×	×	×		×	×	×	×	×
Oxalic Acid, CO_2 H CO_2 H	○	○	–	–	–	–	×	×	×	+	–	○	+	–	×	+	+	+	×	×	×	×	×	×	×		–	+	×	×	×	×	×	×	×	×	
Phenol (Conc.), C_6 H_5 OH	+	+	–	×	–	×	×	×	×	○	–	–	+	+	×	+	+	+	×	×	×	×	×	×	×	×	–		×	×	×		×	×	×	×	
Phenol (Dilute)	+	+	×	×	×		×		×	–	×	–	+	+	–	+	+	+	×	×		×								×					×		
Phosphoric Acid (100%), H_3PO_4	○	○	–	–	×		×		×		○	+	–	○	–	+	×	×	×	×	×	×	×	+	×			×	+	–	×		×	×	×	×	×
Phosphoric Acid ($>$45% Hot)	○	○	○	○	×	○	×	×	×	○	○	+	○	○	○	–	–	–	×	×	×	–	×		×				×	–		×			○		

°NOTE: Duriron is as shown. Durichlor is also satisfactory on chlorides and HCl.

°°Durcon 5 would be the preferred formula.

Table A.4
CHEMICAL RESISTANCE OF MATERIALS (Continued)

RUBBERS					THERMOPLASTICS															THERMOSETTING PLASTICS																			WOODS					
Hard Rubber (Natural)	*Soft Rubber (Natural)*	*Neoprene*	*Butadiene Derivatives*	*Nitrile Rubber (Chemigum)*	*Viton*	*Asphaltic, Bitumastic*	*Cellulose Acetate*	*Cellulose Acetatebutyrate*	*Ethyl Cellulose (Ethocel)*	*Cellulose Nitrate*	*Acrylic (Lucite, Plexiglas)*	*Coumarone Resins*	*Polyethylene*	*Polyvinyl Chloride, Rigid or Unplasticized*	*Tygon (P. V. C. & Copolymers)*	*Saran (Vinyl chloride, vinyldene chloride)*	*Kel-F (Polytrifluorochloroethylene)*	*Teflon (Polytetrafluoroethylene)*	*Uscolite CP (Styrene-acrylonitrile-butadiene)*	*Penton (Chlorinated Polyether)*	*Shellac Compounds*	*Organic Polysulfides*	*Polystyrene (Styron)*	*Vinylidene Chlorides*	*Vinyl Chloride Acetates*	*Cast Phenol Formaldehyde*	*Haveg 41 (Phenolic w. asbestos)*	*Heresite (Phenol formaldehyde)*	*Molded Phenolformald, (Durez)*	*Phenol Furfural Plastics*	*Urea Formaldehyde*	*Casein Plastics*	*Epoxy Resins*	*Furane Resins (Haveg 61, Duralon)*	*Silicone Resins*	*Permanite (Furan, Glass Fiber)*	*Nylon (Adipic Acid—Hexameth, Diamine)*	*Durcon 6 (Modified Epoxy)*	*Cypress*	*Fir*	*Maple*	*Oak*	*Pine*	*Redwood*
×	×	×	×		×	×	×	×		×		×	×	×	×	×	×	×	×	×	×	×	×	×	×	×	×	×	×	×	×		×	×			×	×						
×		×			×									×		×	×	×	×	×																	×	×						
×	×	×	×		×	×	×	×		×		×	×	×		×	×	×	×	×	×	×	×	×	×	×			×	×	×		×	×			×	×						
		×			+	○								–		–	×	×	×	×													×				×	+°°						
×	×	×	×		–									×			×	×		×			×														×	×						
×		–			×	×	×							–	×	×	×	×	×	×													×					×						
					×	×								–		×		×		×													×				×	×						
–	×	–		×	–	×					○		–	×		×	×	×	×	×			×				–		×	×			×	×			×	×						
×	×													×			×	×	×	×			×						×	×			×			×		×						
○		○			–	○								○		–	×	×		×													×				×	×						
–	○	×	×		+	○				×	×			×			×	×		×			–	–	–	–			–	–		×	×				×	×						
×		×			×	×							×	×	×	×	×	×	×	×													×	×			×							
×	×	×			×	×							×	×	×	×	×	×	×	×			×						×	×			×	×			×	×						
–	–					○									○		×	×									○	×					○					×°°						
–	–					○									–		×	×									–	×										×						
–	–					○									–		×	×									–	×										×						
–	–					○									–		×	×									–	×										+°°						
○	○	○	○	○	○				○				○	○	○	–	×	×	○	○			○	○	○	○	○	×	○	○			○	○	–	○	○	×						
–	○	○	×	○	–	–			–		×		–	×	–	×	×	×	×	+			×	–	–	–	–	×	–	–			○		–	○		×						
○	○	○			○	○							–	○		–	×	×		+													–					×						
–	○	○	×			×											×	×		×		×																×						
○	○	○	×		+	○	×	×		×			○	×	×	×	×	×	×	×			×	×	×	–	×	×	×	–	×		×				×	×	×	×	×	×		–
×	×	–	×		–	×							×	×	×	×	×	×	×	×			×			×	×	×	×	×			×	×		×	○	×						
–	○	–	×		×	○	○	○	–		×			–	–	–	×	×	○	+			○			–	–	×	–	–				×		×	○	×						
–		–				–								×	–	–	×	×	○				○						×	×			–	×				×						
×	×	×	×			×	×		○				×	×	×		×	×	×					×	×		×	×						–	–	–	–	×						
–	–	–		×	+								–	×	×		×	×	×								×	×	○	○			×	×	○			×						

Table A.4
CHEMICAL RESISTANCE OF MATERIALS (Continued)

MATERIALS
× — Very Good Service
+ — Moderate Service
– — Limited or Variable Service
○ — Unsatisfactory
Blank — No Information

CHEMICALS
Solids Assumed in Sol'n.
Room Temperatures Assumed
Unless Otherwise Stated

	METALS																												CARBONS & CERAMICS								
CHEMICALS	*Carbon Steel; Fe*	*Cast Iron & Ductile Iron; Fe*	*304 Stainless Steel; Fe, 18Cr, 8Ni*	*316 Stainless Steel; Fe, 16Cr, 10Ni, 2Mo*	*347 Stainless Steel; Fe, 17Cr, 9Ni; (Cx10)Cb*	*Ni-Resist Iron; Fe, 14Ni, 2Cr, 2Si*	*Durimet 20; Carpenter 20; Fe; 4Cu, 20Cr, 29Ni, 2Mo, 1Si*	*Worthite; 3Mo, 2Cu, Fe, 20Cr, 24Ni, 3Si*	*Duriron; Fe, 14Si; Durichlor, Fe, 14Si, 3Mo*[*]	*Copper; Brass; Bronzes; Everdur*	*Aluminum; Al (and Alloys)*	*Lead; Pb*	*Monel; 67Ni, 30Cu, 1.4 Fe*	*Nickel; Ni*	*Inconel; 76Ni, 15Cr, 8Fe*	*Hastelloy B; Ni, 26Mo, 4Fe*	*Hastelloy C; Ni, 16Mo, 4Fe, 14Cr, 4W*	*Hastelloy D; Ni, 8Si, 3Cu*	*Chlorimet 3; 3Fe, 1Si, 60Ni, 18Mo, 18Cr*	*Chlorimet 2; 63Ni, 32Mo, 3Fe, 1Si*	*Stellite; Co, 28Cr, 4W*	*Zirconium; Zr*	*Tantalum; Ta*	*Silver; Ag*	*Platinum; Pt*	*Dowmetal; (Mg alloys)*	*Titanium; Ti*	*Molybdenum; Mo*	*Carbon & Graphite*	*Glass, "Pyrex" brand*	*Silicaware*	*Silicate Cements*	*Chemical Stoneware*	*Transite (asbestos & cement)*	*Chemical Porcelain*	*Concrete-Unbonded*	*Concrete-Mortar Bonded*
Phosphoric Acid (>45% Cold)	○	○	–	×	×	–	×	–	×	○	○	+	–	–	–	+	×	+	×	×	×	×	×	×	×		–		×	×			–		–		
Phosphoric Acid (<45% Cold)	○	○	–	×	×	–	×	–	×	○	–	+	–	–	–	+	×	+	×	×	×	×	×	×	×		×		×	×			–		–		
Phosphoric Anhydride, Dry or Moist	○		×	×			×		×										×	×									×	×							
Phosphoric Anhydride Molten, P_2O_5	×		×	×			×		×	○	○		○	○					×	×										○							
Phthalic Anhydride, $C_6H_4(CO)_2O$	×		×				×	×	×							×	×		×	×			×							×					×		
Picric Acid, Sol'n., $HOC_6H_2(NO_3)_3$			×	×	+		×	×	×	○	–	–	○	○	○	×	×	×	×	×										×					×		
Potassium Bromide, KBr	○		–	–			×		×	×	–	×	+	×		×	×	×	×	×			×	○			×			×					×		
Potassium Carbonate, K_2CO_3	–		×	×		–	×	×	×	–	○	×	×	×	+	×	×	×	×	×			–							–					×		
Potassium Chlorate, $KClO_3$	×		×	×			×	–	×	–	+		+	+	+	○	×	○	×	○			×							×					×		
Potassium Chloride, KCl	×		–	–		×	×	×	×	–	–		+	+		×	×	×	×	×			×				×			×					×		
Potassium Cyanide, KCN	×		×	×		×	×		×	○	○		×			×	×	×	×	×			×	○	×					–					×		
Potassium Dichromate, $K_2Cr_2O_7$			×	×	–		×	×	×	×	×		–	–	+	○	×	○	×	×		×	×	×			×			×					×		
Potassium Ferrocyanide, $K_4Fe(CN)_6$	○		×	×			×	×	×	×	–	×	×	×	×	×	×	×	×	×			×	○						×					×		
Potassium Hydroxide, KOH	+	–	+	+	–	–	×	×	+	○	○	○	×	×	×	+	+	+	×	×		×	–	×			–	+		–		○			–		
Potassium Nitrate, KNO_3	×		×	×	×		×	×	×	×	×	–	+	+	+	○	×	○	×	×			×	○						×					×		
Potassium Permanganate, $KMnO_4$	+	+	+	+	+		×	×	×	×	×	○	–	+	+	○	×	○	×	○			×				×			×					×		
Potassium Sulfate, K_2SO_4	×		+	+	+	×	×	×	×	+	×	×	+	+	+	+	+	+	×	×			×				×			×		×			×		
Potassium Sulfide, K_2S	–		×	×		–	×		×										×	×				○						–					×		
Pyrogallol, $C_6H_3(OH)_3$			×	×			×		×	×				+		×	×	×	×	×										×					×		
Silver Nitrate, $AgNO_3$			×	×			×	×	×	○	○	○			+	+	+	+	×	×		×	×							×					×		
Sodium, Molten 210°-400°F.	×	×	×	×	×		×	×	×	×	×		×	×	×	×	×		×	×		×	×			–		×		–					○		
Sodium Acetate, $NaCH_3COO$	○		–	×			×	×	×	×	–	×	+			×	×	×	×	×							×			×					×		
Sodium Bicarbonate, $NaHCO_3$	–		×	×	×	×	×	×	×	–	×	×	×	×	×	×	×	×	×	×		×	×	×					×	×	×		×	×	×	×	×
Sodium Bisulfate, $NaHSO_4$	○	○	–	×	×		×	×	×	–	○	×	+			×	×	×	×	×	×		–		×		×		×	×	×		×		×		
Sodium Bisulfite, $NaHSO_3$	○		×	×			×	×	–	–	×	×							×	×									×	×	×		×		×		

[*] NOTE: Duriron is as shown. Durichlor is also satisfactory on chlorides and HCl.
[**] Durcon 5 would be the preferred formula.

Table A.4
CHEMICAL RESISTANCE OF MATERIALS (Continued)

RUBBERS					THERMOPLASTICS																				THERMOSETTING PLASTICS															WOODS					
Hard Rubber (Natural)	*Soft Rubber (Natural)*	*Neoprene*	*Butadiene Derivatives*	*Nitrile Rubber (Chemigum)*	*Viton*	*Asphaltic, Bitumastic*	*Cellulose Acetate*	*Cellulose Acetatebutyrate*	*Ethyl Cellulose (Ethocel)*	*Cellulose Nitrate*	*Acrylic (Lucite, Plexiglas)*	*Coumarone Resins*	*Polyethylene*	*Polyvinyl Chloride, Rigid or Unplasticized*	*Tygon (P. V. C. & Copolymers)*	*Saran (Vinyl chloride, vinyldene chloride)*	*Kel-F (Polytrifluorochloroethylene)*	*Teflon (Polytetrafluoroethylene)*	*Uscolite CP (Styrene-acrylonitrile-butadiene)*	*Penton (Chlorinated Polyether)*	*Shellac Compounds*	*Organic Polysulfides*	*Polystyrene (Styron)*	*Vinylidene Chlorides*	*Vinyl Chloride Acetates*	*Cast Phenol Formaldehyde*	*Haveg 41 (Phenolic w. asbestos)*	*Heresite (Phenol formaldehyde)*	*Molded Phenolformald, (Durez)*	*Phenol Furfural Plastics*	*Urea Formaldehyde*	*Casein Plastics*	*Epoxy Resins*	*Furane Resins (Haveg 61, Duralon)*	*Silicone Resins*	*Permanite (Furan, Glass Fiber)*	*Nylon (Adipic Acid—Hexameth, Diamine)*	*Durcon 6 (Modified Epoxy)*	*Cypress*	*Fir*	*Maple*	*Oak*	*Pine*	*Redwood*
–	–	–				×							×	×	×	×		×	×	×			×				×	×					×		–			×						
×	×	–											×	×	×	×		×	×	×							×	×							–	×	○	×						
													×					×																				×						
																		×																				×						
×	×					×										×		×															×					×						
		–				○								○	×	○		×		×																		×						
×		×												×		×		×	×	×																		×						
×		×			×	×							×	–		×		×	×	×													×				×	×						
					×									×		×		×	×	×																	○	×						
×		×			×	×							×	×		×		×	×	×							×						×	×		×	×	×						
×		×			×									×	×	×		×	×	×																	×	×						
–	×	×				×							×	×	×	×		×	×	×													×					×°°						
						×								×		×		×		×													×				×	×						
×	×	×		○	×	×							×	×		–		×	×	×			×				○		○	○			×	×		×	×	×						
×		×				×							×	×		×		×	×	×													×					×						
–	○	○			×	○							×	×		–		×	×	×			×						×	×			×	×			○	×						
×		×		×	×	×								×		×		×	×	×			×				×		×	×			×	×		×	×	×						
														×				×	×																		×	×						
											×				×			×																				×°°						
×	×				×						×		×	×		×		×	×	×						–			–				×				×	×						
								○					○	○	○	○		–		○			○	○			○	○										×						
		×			○	×								×		×		×		×													×				–	×						
×	×	×	×			×	×	×		×	×			×		×		×	×	×	×		×	×	×	×	×		×	×			×	×		×	×	×						
×	×	×	×											×	×	×		×		×			×	×	×	–	×	×	×	–		×	×	×		×		×						
×	×	–	×			×							×	×	×	×		×	×	×									×			×	×				○	×°°	×	×	×	×	×	

Table A.4
CHEMICAL RESISTANCE OF MATERIALS (Continued)

METALS: columns Carbon Steel through Molybdenum. CARBONS & CERAMICS: columns Carbon & Graphite through Concrete-Mortar Bonded.

MATERIALS
× — Very Good Service
+ — Moderate Service
– — Limited or Variable Service
○ — Unsatisfactory
Blank — No Information

CHEMICALS
Solids Assumed in Sol'n.
Room Temperatures Assumed
Unless Otherwise Stated

Chemicals	*Carbon Steel; Fe*	*Cast Iron & Ductile Iron; Fe*	*304 Stainless Steel; Fe, 18Cr, 8Ni*	*316 Stainless Steel; Fe, 16Cr, 10Ni, 2Mo*	*347 Stainless Steel; Fe, 17Cr, 9Ni; (Cx10)Cb*	*Ni-Resist Iron; Fe, 14Ni, 2Cr, 2Si*	*Durimet 20; Carpenter 20; Fe; 4Cu, 20Cr, 29Ni, 2Mo, 1Si*	*Worthite; 3Mo, 2Cu, Fe, 20Cr, 24Ni, 3Si*	*Duriron; Fe, 14Si; Durichlor, Fe, 14Si, 3Mo°*	*Copper; Brass; Bronzes; Everdur*	*Aluminum; Al (and Alloys)*	*Lead; Pb*	*Monel; 67Ni, 30Cu, 1.4 Fe*	*Nickel; Ni*	*Inconel; 76Ni, 15Cr, 8Fe*	*Hastelloy B; Ni, 26Mo, 4Fe*	*Hastelloy C; Ni, 16Mo, 4Fe, 14Cr, 4W*	*Hastelloy D; Ni, 8Si, 3Cu*	*Chlorimet 3; 3Fe, 1Si, 60Ni, 18Mo, 18Cr*	*Chlorimet 2; 63Ni, 32Mo, 3Fe, 1Si*	*Stellite; Co, 28Cr, 4W*	*Zirconium; Zr*	*Tantalum; Ta*	*Silver; Ag*	*Platinum; Pt*	*Dowmetal; (Mg alloys)*	*Titanium; Ti*	*Molybdenum; Mo*	*Carbon & Graphite*	*Glass, "Pyrex" brand*	*Silicaware*	*Silicate Cements*	*Chemical Stoneware*	*Transite (asbestos & cement)*	*Chemical Porcelain*	*Concrete-Unbonded*	*Concrete-Mortar Bonded*
Sodium Borate $NaBO_2$			×	×			×		×	×	–		+			×	×	×	×	×			×							×	×			×	×	×	
Sodium Carbonate, Na_2CO_3	+	+	+	+	×	×	×	×	×	+	○	–	+	+	–	+	×	×	×	×	×		–	×	×	×	×		×	–	–	○	×	×	×	×	
Sodium Chlorate, $NaClO_3$	–		×	×			–	–	×	+	+				+	○	×	○	×	○			×				×			×	×		×	×	×	×	×
Sodium Chloride, $NaCl$	+	+	–	–	–	–	×	×	+	+	–	+	+	+	–	+	+	–	×	×	–	×	×	×			×		×	×		×	×		×		
Sodium Cyanide, $NaCN$	×		×	×			×	×	×	○	○	×	×						×	×				○			×			–					×		
Sodium Fluoride, NaF	○		–	–			+	–	○	–	–		+			×	×	×	×	×		×	○				×			–					×		
Sodium Hydroxide, (Conc.), $NaOH$	–	–	–	–	×	×	×	×	–	–	○	○	×	×	×	×	+	×	×	×	×	×	○	×	×		×		×	–	○	○	○		○		
Sodium Hydroxide, (Dilute)	×	+	+	+	×		×	×	+	×	○	○	–	–	–	×	+	×	×	×	×	×	–		×	–	×	+	×	–	○	○	–		–		
Sodium Hydrosulfite							×		×	–	×		○	○	○	–	×	–	×	×										×	×		×		×		
Sodium Hypochlorite, $NaOCl$	○	○	–	–	–	–	–	–	×	○	○	○	○	○	○	○	–	○	+	○	–		–	×	×		×		○	×			×		×		
Sodium Hyposulfate	–		×	×			×		×	○	○		×						×	×				○						×					×		
Sodium Nitrate, $NaNO_3$	×	+	×	×	×	–	×	×	×	+	×	○	+	+	×	○	+	○	×	○	×		–	–	×				×	×	×	×	×		×		
Sodium Peroxide, Na_2O_2	–		×	×	+		×		×	–	–		×	+	+	+	×	+	×	×			○	○						–					×		
Sodium Phosphate, (Tri) Na_3PO_4	×		×	×	+		×		×	–	○		+	+		×	×	×	×	×			×	×			×			–	–		×	×	×		
Sodium Silicate, Na_2SiO_3	×		×	×	+	×	×		×	–	–		+	×	+	×	×	×	×	×			×							–					×	×	
Sodium Sulfate, Na_2SO_4	×	×	–	–	×	×	×	×	×	×	×	–	+	×	×	×	×	×	×	×	–		–	×			×		×	×			×		×		
Sodium Sulfide, Na_2S	+	×	–	×	×	×	×	×	×	○	○	–	+	+	+	×	×	×	×	×	×		–	○			×		×	–	–		×	×	×		
Sodium Sulfite, Na_2SO_3	+	×	–	–	–	×	×	×	○	–	–	×	–	–	–	○	×	○	×	×	×		–				×		×	×	×		×		×		
Stannic Chloride, $SnCl_4$	○		○	○	○		×	–	×	○	○	×	○	○	○	○	×	○	–	○			×				×			×					×		
Stannous Chloride, $SnCl_2$	○		○	–			×	–	×	○	○		○	○	○	×	×	×	×	×		×	×				×			×					×		
Stearic Acid, $CH_3(CH_2)_{16}COOH$	–	–	×	×	×	–	×	×	×	–	–	–	×	×	×	×	×	×	×	×	×		×	×			×		×	×			×		×		
Sulfur, Molten, S	–	×	–	–	–	–	×	×	×	○	×	○	–	–	×	○	○	○	×	–	○		×	○			×	+	–	×				–		–	
Sulfur Chloride, (Wet), S_2Cl_2	–		○	–		×	–	–	×	○	○		–	–					×					○						×					×		
Sulfur Dioxide (Dry), SO_2	×	×	×	×	–	×	×	×	×	×	×	×	○	○	○	○	×	○	×	–	×		×	○		×			×	×	×		×		×		
Sulfur Dioxide (Wet)	○		–	×			×		–	○		×	○	○	○	○	×	○	×				×				+		×	×	×				×		
Sulfur Trioxide, SO_3	×		×	×		×	×	×	–	–	×		○	○	○				×				○	○						×			×		×		
Sulfuric Acid (Fuming to 98%)	–	–	○	○		–	×	×	○	○	–	○	○	○	○	×	○	×	×	×		○	○	○			○		○	×	×	×					
Sulfuric Acid (Hot Conc.) H_2SO_4	○		○	○			+		×	○	○	–	○	○	○	○	○	×	+	+	○	○	–				○			×							

°NOTE: Duriron is as shown. Durichlor is also satisfactory on chlorides and HCl.

°°Durcon 5 would be the preferred formula.

Table A.4
CHEMICAL RESISTANCE OF MATERIALS (Continued)

RUBBERS					THERMOPLASTICS															THERMOSETTING PLASTICS																			WOODS					
Hard Rubber (Natural)	*Soft Rubber (Natural)*	*Neoprene*	*Butadiene Derivatives*	*Nitrile Rubber (Chemigum)*	*Viton*	*Asphaltic, Bitumastic*	*Cellulose Acetate*	*Cellulose Acetatebutyrate*	*Ethyl Cellulose (Ethocel)*	*Cellulose Nitrate*	*Acrylic (Lucite, Plexiglas)*	*Coumarone Resins*	*Polyethylene*	*Polyvinyl Chloride, Rigid or Unplasticized*	*Tygon (P. V. C. & Copolymers)*	*Saran (Vinyl chloride, vinyldene chloride)*	*Kel-F (Polytrifluorochloroethylene)*	*Teflon (Polytetrafluoroethylene)*	*Uscolite CP (Styrene-acrylonitrile-butadiene)*	*Penton (Chlorinated Polyether)*	*Shellac Compounds*	*Organic Polysulfides*	*Polystyrene (Styron)*	*Vinylidene Chlorides*	*Vinyl Chloride Acetates*	*Cast Phenol Formaldehyde*	*Haveg 41 (Phenolic w. asbestos)*	*Heresite (Phenol formaldehyde)*	*Molded Phenolformald, (Durez)*	*Phenol Furfural Plastics*	*Urea Formaldehyde*	*Casein Plastics*	*Epoxy Resins*	*Furane Resins (Haveg 61, Duralon)*	*Silicone Resins*	*Permanite (Furan, Glass Fiber)*	*Nylon (Adipic Acid—Hexameth, Diamine)*	*Durcon 6 (Modified Epoxy)*	*Cypress*	*Fir*	*Maple*	*Oak*	*Pine*	*Redwood*
×	×	×	×		×	×	×	×		×	×			×	×			×	×		×		×	×	×	–			–	–							×	×°°						
×	×	×	×	×	×	×	–	×		–	○		×	×	×	×	×	×	×	×			×			×	×	×	×	×		○	×	×		×	×	×°°	+	+	+	×	+	+
×	×	×	×		×	×	×	×		×	×		×	×		×		×	×	×	×	×	×	×	×	×			×	×	×	×					○	×	×	×	×	×	×	
×	×	×		×	×	×	×	×			×		×	×	×	×	×	×	×	×			×				×	×	×	×			×	×	×	×	×	×	×	×	×	×	+	+
×	×	×			×	×								×	×	–		×	×	×													×				–	×						
×	×				×									×		×		×	×	×																	×	×						
×	×	×	×	×	–	×	○	–	×	○	×		×	×	–	○	×	×	×	×			×			○	○	○	○	○	○		×	×	–	×	×	×	–	–	○	○	○	
×	×	×	×	×		×	–	×	×	–	×		×	×	×	–	×	×	×	×		×	×	×	×	–	○	–	○	○	×		×	×	–		×	×	–	–	○	○	–	–
×	–		×															×								×			×	×								×						
×	–	○	×		–	–	–			×				×	–	×		×	×	×			×	×	×	–	○	×	–	–			○	○		○	○							
																		×																										
×	–	×	×	×	×	×	×	×		×				×	×	×		×	×	×			×			×	×	×	×	×			×	×			○	×						
														×				×	×																			×						
×	×	–	×			×			×		×			×	×			×	×				–			–	–		×	–			×	×		×	×	×						
×	×	×	×		×	×								×				×	×				×			×			×	×							×	×						
×	×	×			×	×							×	×	×	×		×	×	×							×	×					×	×		×	×	×						
×	–	×	×		×	×							×	×	×	×		×	×				×				×	×				–	×				×	+°°	–	–	○	○	–	–
×	×	○	×		×	×								×	×	×		×	×	×						–	×	×	–	–			×	×		×	×	×						
					×	○								×	×		×	×	×	×							×									×	○	×						
×	×	–				×							×	–		×		×	×	×													×					×						
–		○			×	○							×	×	×	×		×	×	×							×	×					×				×	×						
														○	○			×		○							×	○										×						
																		×																				×						
–	–	○	–			×			–					×	–	–		×		×				×	×		×						×				–	×						
–		○			×									–		–		×		×																	○	×						
×	×	○	×		×	○			○					×				×						×	×								×				×	×						
		○		○		×	○	○	○	○	○			○		–		×	○	○			○			○			○	○			○	○	○			×	○	○	○	○	○	○
○	○	○			○													×	○	–							○						○	○				×	○	○	○	○	○	○

Table A.4
CHEMICAL RESISTANCE OF MATERIALS (Continued)

MATERIALS
× — Very Good Service
+ — Moderate Service
− — Limited or Variable Service
○ — Unsatisfactory
Blank — No Information

CHEMICALS
Solids Assumed in Sol'n. Room Temperatures Assumed Unless Otherwise Stated

METALS

CHEMICALS	*Carbon Steel; Fe*	*Cast Iron & Ductile Iron; Fe*	*304 Stainless Steel; Fe, 18Cr, 8Ni*	*316 Stainless Steel; Fe, 16Cr, 10Ni, 2Mo*	*347 Stainless Steel; Fe, 17Cr, 9Ni; (Cx10)Cb*	*Ni-Resist Iron; Fe, 14Ni, 2Cr, 2Si*	*Durimet 20; Carpenter 20; Fe; 4Cu, 20Cr, 29Ni, 2Mo, 1Si*	*Worthite; 3Mo, 2Cu, Fe, 20Cr, 24Ni, 3Si*	*Duriron; Fe, 14Si; Durichlor, Fe, 14Si, 3Mo°*	*Copper; Brass; Bronzes; Everdur*	*Aluminum; Al (and Alloys)*	*Lead; Pb*	*Monel; 67Ni, 30Cu, 1.4 Fe*	*Nickel; Ni*	*Inconel; 76Ni, 15Cr, 8Fe*
Sulfuric Acid (Cold Conc.)	+	+	×	×	−		×	×	×	○	−	×	+		
Sulfuric Acid (75%–95%)	+	−	○	○	−	−	×	×	×	○	○	×	○	○	−
Sulfuric Acid (10%–75%)	○	○	○	○	−	−	×	×	×	○	○	×	−	○	−
Sulfuric Acid (<10%)	○	○	○	−	−	−	−	×	×	○	−	×	−	−	−
Sulfurous Acid, H_2SO_3	○	○	−	×	−	○	×	×	○	○	−	−	○	○	○
Sulfuryl Chloride, SO_2Cl_2							+		×						
Tannic Acid	−		×	×			×	×	×	−	×		+	+	+
Tartaric Acid, $(CHOH\ COOH)_2$	−		×	×			×	×	×	−	○	×	+	+	
Toluene, $CH_3C_6H_5$	×		×				×		×	×	×	×	×		
Trichlorethylene, Dry, Cl_2CCHCl	−	−	−	×	−	−	×	×	×	−	−		×	×	×
Water, Fresh, H_2O	×		×	×			×		×	×	×		×	×	
Water, Distilled Lab.	○		×	×		○	×		×	×	○		○	−	
Zinc Chloride, $ZnCl_2$	○	○	○	−	○	−	×	×	×	○	○	−	+	+	
Zinc Sulfate, $ZnSO_4$	○	○	+	+	+	×	×	×	×	+	−	+	+	+	+

METALS (continued); CARBONS & CERAMICS

CHEMICALS	*Hastelloy B; Ni, 26Mo, 4Fe*	*Hastelloy C; Ni, 16Mo, 4Fe, 14Cr, 4W*	*Hastelloy D; Ni, 8Si, 3Cu*	*Chlorimet 3; 3Fe, 1Si, 60Ni, 18Mo, 18Cr*	*Chlorimet 2; 63Ni, 32Mo, 3Fe, 1Si*	*Stellite; Co, 28Cr, 4W*	*Zirconium; Zr*	*Tantalum; Ta*	*Silver; Ag*	*Platinum; Pt*	*Dowmetal; (Mg alloys)*	*Titanium; Ti*	*Molybdenum; Mo*	*Carbon & Graphite*	*Glass, "Pyrex" brand*	*Silicaware*	*Silicate Cements*	*Chemical Stoneware*	*Transite (asbestos & cement)*	*Chemical Porcelain*	*Concrete-Unbonded*	*Concrete-Mortar Bonded*
Sulfuric Acid (Cold Conc.)	×	×	×	×	×	−	○	−		×		○	−	○	×					×		
Sulfuric Acid (75%–95%)	×	×	−	×	×	−	−	−	○	×		○		×	×			×		×		
Sulfuric Acid (10%–75%)	×	×	×	×	×	−	×	−	×	×		−	×	×	×	×		×		×		
Sulfuric Acid (<10%)	×	×	×	×	×	−	×	×	×	×		×		×	×	×		×		×		
Sulfurous Acid, H_2SO_3	○	+	+	×	×	×	×	×	×			×		×	×	×	×	×		×		
Sulfuryl Chloride, SO_2Cl_2				+	+			×							×	×		×		×		
Tannic Acid	×	×	×	×	×		×	×	×			×		×	×			×		×	×	
Tartaric Acid, $(CHOH\ COOH)_2$	×	×	×	×	×		×	×	×			×	+	×	×	×		×		×		
Toluene, $CH_3C_6H_5$				×	×		×							×	×	×		×		×		
Trichlorethylene, Dry, Cl_2CCHCl	×	×	×	×	×	×		×			×	×		×	×			×		×		
Water, Fresh, H_2O				×	×			×	×				×	×	×	×		×	×	×	×	×
Water, Distilled Lab.				×	×				×						+					×		
Zinc Chloride, $ZnCl_2$	+	+	+	×	×		×	×	×			×		×	×		×			×	×	
Zinc Sulfate, $ZnSO_4$	+	+	+	×	×			×	×			×		×	×					×		

° NOTE: Duriron is as shown. Durichlor is also satisfactory on chlorides and HCl.
°° Durcon 5 would be the preferred formula.

Table A.4
CHEMICAL RESISTANCE OF MATERIALS (Continued)

RUBBERS					THERMOPLASTICS																				THERMOSETTING PLASTICS															WOODS				
Hard Rubber (Natural)	*Soft Rubber (Natural)*	*Neoprene*	*Butadiene Derivatives*	*Nitrile Rubber (Chemigum)*	*Viton*	*Asphaltic, Bitumastic*	*Cellulose Acetate*	*Cellulose Acetatebutyrate*	*Ethyl Cellulose (Ethocel)*	*Cellulose Nitrate*	*Acrylic (Lucite, Plexiglas)*	*Coumarone Resins*	*Polyethylene*	*Polyvinyl Chloride, Rigid or Unplasticized*	*Tygon (P. V. C. & Copolymers)*	*Saran (Vinyl chloride, vinyldene chloride)*	*Kel-F (Polytrifluorochloroethylene)*	*Teflon (Polytetrafluoroethylene)*	*Uscolite CP (Styrene-acrylonitrile-butadiene)*	*Penton (Chlorinated Polyether)*	*Shellac Compounds*	*Organic Polysulfides*	*Polystyrene (Styron)*	*Vinylidene Chlorides*	*Vinyl Chloride Acetates*	*Cast Phenol Formaldehyde*	*Haveg 41 (Phenolic w. asbestos)*	*Heresite (Phenol formaldehyde)*	*Molded Phenolformald. (Durez)*	*Phenol Furfural Plastics*	*Urea Formaldehyde*	*Casein Plastics*	*Epoxy Resins*	*Furane Resins (Haveg 61, Duralon)*	*Silicone Resins*	*Permanite (Furan, Glass Fiber)*	*Nylon (Adipic Acid—Hexameth, Diamine)*	*Durcon 6 (Modified Epoxy)*	*Cypress*	*Fir*	*Maple*	*Oak*	*Pine*	*Redwood*
○	○	○			+						–			×	○	–		×	○	–							–						○			○	○	×	○	○	○	○	○	○
		○		○		×							×	×	○	–		×	○	+			–				–		–	–			–	–	–			×						
×	×	–	×	–	×	×	×	×	○	○	×		×	×	–	×	×	×	–	×	×	×	×	×	×	–	×		–	–		○	×	×	–		–	×	–	–	–	–	–	–
×	×	×	×	×	×	×	×	×	–	–	×		×	×	×	×	–	×	×	×	×	×	×	×	×	–	×		×	×	×			×	–	×	–	×	–	–	×	–	–	–
–	–	×				×							○	×	–	–		×	×	×			×			×	×		×	×		×	×	×	–	×	○	+°°				×	×	
													○					×											×									×						
×	–	×				×			–				×	×		×		×	×	×			×			×	×		×	×	×	×	×	×		×		×				×	×	
×	×	×			×	×	×	×						×		×		×	×	×						–	×		–	–	×		×	×		×	×	×						
○	○	○	○		–		–	–	○		○		–	○	○	–	–	×		+	○	○	○	–	–	–	–		–	–		×		×		×	×	×	–	–	×	×	×	
○	○	○			–	○					○		○	○	○	–	○	×		+							×						×				×	×						
×	×	×	×		×	×	×	×	×	×	×	×		×		×		×	×	×	×	×	×	×	×	×			×	×	×	○						×	×	×	×	×	×	×
×		×									×			×				×	×	×																		×						
×	×	×	×	×	×	×			×				×	×		×		×	×	×			×			–	×		×	–			×	×		×	–	×						
×	×	×			×	×							×	×	×	×		×	×	×			×				×		×	×			×	×		×	×	×						

Table A.5[1]
CONVERSION FACTORS

To Convert	Into	Multiply by
	A	
Abcoulomb	Statcoulombs	2.998×10^{10}
Acre	Sq. chain (Gunters)	10
Acre	Rods	160
Acre	Square links (Gunters)	1×10^{5}
Acre	Hectare or sq. hectometer	.4047
acres	sq feet	43,560.0
acres	sq meters	4,047.
acres	sq miles	1.562×10^{-3}
acres	sq yards	4,840.
acre-feet	cu feet	43,560.0
acre-feet	gallons	3.259×10^{5}
amperes/sq cm	amps/sq in.	6.452
amperes/sq cm	amps/sq meter	10^{4}
amperes/sq in.	amps/sq cm	0.1550
amperes/sq in.	amps/sq meter	1,550.0
amperes/sq meter	amps/sq cm	10^{-4}
amperes/sq meter	amps/sq in.	6.452×10^{-4}
ampere-hours	coulombs	3,600.0
ampere-hours	faradays	0.03731
ampere-turns	gilberts	1.257
ampere-turns/cm	amp-turns/in.	2.540
ampere-turns/cm	amp-turns/meter	100.0
ampere-turns/cm	gilberts/cm	1.257
ampere-turns/in.	amp-turns/cm	0.3937
ampere-turns/in.	amp-turns/meter	39.37
ampere-turns/in.	gilberts/cm	0.4950
ampere-turns/meter	amp/turns/cm	0.01
ampere-turns/meter	amp-turns/in.	0.0254
ampere-turns/meter	gilberts/cm	0.01257
Angstrom unit	Inch	3937×10^{-9}
Angstrom unit	Meter	1×10^{-10}
Angstrom unit	Micron or (Mu)	1×10^{-4}
Are	Acre (US)	.02471
Ares	sq. yards	119.60
ares	acres	0.02471
ares	sq meters	100.0
Astronomical Unit	Kilometers	1.495×10^{8}
Atmospheres	Ton/sq. inch	.007348
atmospheres	cms of mercury	76.0
atmospheres	ft of water (at 4°C)	33.90
atmospheres	in. of mercury (at 0°C)	29.92
atmospheres	kgs/sq cm	1.0333
atmospheres	kgs/sq meter	10,332.
atmospheres	pounds/sq in.	14.70
atmospheres	tons/sq ft	1.058
	B	
Barrels (U.S., dry)	cu. inches	7056.
Barrels (U.S., dry)	quarts (dry)	105.0
Barrels (U.S., liquid)	gallons	31.5
barrels (oil)	gallons (oil)	42.0
bars	atmospheres	0.9869

[1]Reproduced by permission of Jerguson Gauge and Valve Company.

Table A.5
CONVERSION FACTORS (Continued)

To Convert	Into	Multiply by
bars	dynes/sq cm	10^6
bars	kgs/sq meter	1.020×10^4
bars	pounds/sq ft	2,089.
bars	pounds/sq in.	14.50
Baryl	Dyne/sq. cm.	1.000
Bolt (US Cloth)	Meters	36.576
BTU	Liter—Atmosphere	10.409
Btu	ergs	1.0550×10^{10}
Btu	foot-lbs	778.3
Btu	gram-calories	252.0
Btu	horsepower-hrs	3.931×10^{-4}
Btu	joules	1,054.8
Btu	kilogram-calories	0.2520
Btu	kilogram-meters	107.5
Btu	kilowatt-hrs	2.928×10^{-4}
Btu/hr	foot-pounds/sec	0.2162
Btu/hr	gram-cal/sec	0.0700
Btu/hr	horsepower-hrs	3.929×10^{-4}
Btu/hr	watts	0.2931
Btu/min	foot-lbs/sec	12.96
Btu/min	horsepower	0.02356
Btu/min	kilowatts	0.01757
Btu/min	watts	17.57
Btu/sq ft/min	watts/sq in.	0.1221
Bucket (Br. dry)	Cubic Cm.	1.818×10^4
bushels	cu ft	1.2445
bushels	cu in.	2,150.4

To Convert	Into	Multiply by
bushels	cu meters	0.03524
bushels	liters	35.24
bushels	pecks	4.0
bushels	pints (dry)	64.0
bushels	quarts (dry)	32.0
	C	
Calories, gram (mean)	B.T.U. (mean)	3.9685×10^{-3}
Candle/sq. cm	Lamberts	3.142
Candle/sq. inch	Lamberts	.4870
centares (centiares)	sq meters	1.0
Centigrade	Fahrenheit	$(C° \times 9/5) + 32$
centigrams	grams	0.01
Centiliter	Ounce fluid (US)	.3382
Centiliter	Cubic inch	.6103
Centiliter	drams	2.705
centiliters	liters	0.01
centimeters	feet	3.281×10^{-2}
centimeters	inches	0.3937
centimeters	kilometers	10^{-5}
centimeters	meters	0.01
centimeters	miles	6.214×10^{-6}
centimeters	millimeters	10.0
centimeters	mils	393.7
centimeters	yards	1.094×10^{-2}
centimeter-dynes	cm-grams	1.020×10^{-3}
centimeter-dynes	meter-kgs	1.020×10^{-8}

Table A.5
CONVERSION FACTORS (Continued)

To Convert	*Into*	*Multiply by*	*To Convert*	*Into*	*Multiply By*
centimeter-dynes	pound-feet	7.376×10^{-8}	circular mils	sq inches	7.854×10^{-7}
centimeter-grams	cm-dynes	980.7	Cords	cord feet	8
centimeter-grams	meter-kgs	10^{-5}	Cord feet	cu. feet	16
centimeter-grams	pound-feet	7.233×10^{-5}	Coulomb	Statcoulombs	2.998×10^{9}
centimeters of mercury	atmospheres	0.01316	coulombs	faradays	1.036×10^{-5}
centimeters of mercury	feet of water	0.4461	coulombs/sq cm	coulombs/sq in.	64.52
centimeters of mercury	kgs/sq meter	136.0	coulombs/sq cm	coulombs/sq meter	10^{4}
centimeters of mercury	pounds/sq ft	27.85	coulombs/sq in.	coulombs/sq cm	0.1550
centimeters of mercury	pounds/sq in.	0.1934	coulombs/sq in.	coulombs/sq meter	1,550.
centimeters/sec	feet/min	1.1969	coulombs/sq meter	coulombs/sq cm	10^{-4}
centimeters/sec	feet/sec	0.03281	coulombs/sq meter	coulombs/sq in.	6.452×10^{-4}
centimeters/sec	kilometers/hr	0.036	cubic centimeters	cu feet	3.531×10^{-5}
centimeters/sec	knots	0.1943	cubic centimeters	cu inches	0.06102
centimeters/sec	meters/min	0.6	cubic centimeters	cu meters	10^{-6}
centimeters/sec	miles/hr	0.02237	cubic centimeters	cu yards	1.308×10^{-6}
centimeters/sec	miles/min	3.728×10^{-4}	cubic centimeters	gallons (U.S. liq.)	2.642×10^{-4}
centimeters/sec/sec	feet/sec/sec	0.03281	cubic centimeters	liters	0.001
centimeters/sec/sec	kms/hr/sec	0.036	cubic centimeters	pints (U.S. liq.)	2.113×10^{-3}
centimeters/sec/sec	meters/sec/sec	0.01	cubic centimeters	quarts (U.S. liq.)	1.057×10^{-3}
centimeters/sec/sec	miles/hr/sec	0.02237	cubic feet	bushels (dry)	0.8036
Chain	Inches	792.00	cubic feet	cu cms	28,320.0
Chain	meters	20.12	cubic feet	cu inches	1,728.0
Chains (surveyors' or Gunter's)	yards	22.00	cubic feet	cu meters	0.02832
circular mils	sq cms	5.067×10^{-6}	cubic feet	cu yards	0.03704
circular mils	sq mils	0.7854	cubic feet	gallons (U.S. liq.)	7.48052
Circumference	Radians	6.283	cubic feet	liters	28.32
			cubic feet	pints (U.S. liq.)	59.84

Table A.5
CONVERSION FACTORS (Continued)

To Convert	Into	Multiply by
cubic feet	quarts (U.S. liq.)	29.92
cubic feet/min	cu cms/sec	472.0
cubic feet/min	gallons/sec	0.1247
cubic feet/min	liters/sec	0.4720
cubic feet/min	pounds of water/min	62.43
cubic feet/sec	million gals/day	0.646317
cubic feet/sec	gallons/min	448.831
cubic inches	cu cms	16.39
cubic inches	cu feet	5.787×10^{-4}
cubic inches	cu meters	1.639×10^{-5}
cubic inches	cu yards	2.143×10^{-5}
cubic inches	gallons	4.329×10^{-3}
cubic inches	liters	0.01639
cubic inches	mil-feet	1.061×10^{5}
cubic inches	pints (U.S. liq.)	0.03463
cubic inches	quarts (U.S. liq.)	0.01732
cubic meters	bushels (dry)	28.38
cubic meters	cu cms	10^{6}
cubic meters	cu feet	35.31
cubic meters	cu inches	61,023.0
cubic meters	cu yards	1.308
cubic meters	gallons (U.S. liq.)	264.2
cubic meters	liters	1,000.0
cubic meters	pints (U.S. liq.)	2,113.0
cubic meters	quarts (U.S. liq.)	1,057.
cubic yards	cu cms	7.646×10^{5}
cubic yards	cu feet	27.0
cubic yards	cu inches	46,656.0
cubic yards	cu meters	0.7646
cubic yards	gallons (U.S. liq.)	202.0
cubic yards	liters	764.6
cubic yards	pints (U.S. liq.)	1,615.9
cubic yards	quarts (U.S. liq.)	807.9
cubic yards/min	cubic ft/sec	0.45
cubic yards/min	gallons/sec	3.367
cubic yards/min	liters/sec	12.74
	D	
Dalton	Gram	1.650×10^{-24}
days	seconds	86,400.0
decigrams	grams	0.1
deciliters	liters	0.1
decimeters	meters	0.1
degrees (angle)	quadrants	0.01111
degrees (angle)	radians	0.01745
degrees (angle)	seconds	3,600.0
degrees/sec	radians/sec	0.01745
degrees/sec	revolutions/min	0.1667
degrees/sec	revolutions/sec	2.778×10^{-3}
dekagrams	grams	10.0
dekaliters	liters	10.0
dekameters	meters	10.0
Drams (apothecaries' or troy)	ounces (avoidupois)	0.1371429
Drams (apothecaries' or troy)	ounces (troy)	0.125

Table A.5

CONVERSION FACTORS (Continued)

To Convert	Into	Multiply by
Drams (U.S., fluid or apothecaries)	cubic cm.	3.6967
drams	grams	1.7718
drams	grains	27.3437
drams	ounces	0.0625
Dyne/cm	Erg/sq. millimeter	.01
Dyne/sq. cm.	Atmospheres	9.869×10^{-7}
Dyne/sq cm.	Inch of Mercury at 0°C	2.953×10^{-5}
Dyne/sq. cm.	Inch of Water at 4°C	4.015×10^{-4}
dynes	grams	1.020×10^{-3}
dynes	joules/cm	10^{-7}
dynes	joules/meter (newtons)	10^{-5}
dynes	kilograms	1.020×10^{-6}
dynes	poundals	7.233×10^{-5}
dynes	pounds	2.248×10^{-6}
dynes/sq cm	bars	10^{-6}
	E	
Ell	Cm.	114.30
Ell	Inches	45
Em, Pica	Inch	.167
Em, Pica	Cm.	.4233
Erg/sec	Dyne—cm/sec	1.000
ergs	Btu	9.480×10^{-11}
ergs	dyne-centimeters	1.0
ergs	foot-pounds	7.367×10^{-8}
ergs	gram-calories	0.2389×10^{-7}
ergs	gram-cms	1.020×10^{-3}
ergs	horsepower-hrs	3.7250×10^{-14}
ergs	joules	10^{-7}
ergs	kg-calories	2.389×10^{-11}
ergs	kg-meters	1.020×10^{-8}
ergs	kilowatt-hrs	0.2778×10^{-13}
ergs	watt-hours	0.2778×10^{-10}
ergs/sec	Btu/min	$5{,}688 \times 10^{-9}$
ergs/sec	ft-lbs/min	4.427×10^{-6}
ergs/sec	ft-lbs/sec	7.3756×10^{-8}
ergs/sec	horsepower	1.341×10^{-10}
ergs/sec	kg-calories/min	1.433×10^{-9}
ergs/sec	kilowatts	10^{-10}
	F	
farads	microfarads	10^{6}
Faraday/sec	Ampere (absolute)	9.6500×10^{4}
faradays	ampere-hours	26.80
faradays	coulombs	9.649×10^{4}
Fathom	Meter	1.828804
fathoms	feet	6.0
feet	centimeters	30.48
feet	kilometers	3.048×10^{-4}
feet	meters	0.3048
feet	miles (naut.)	1.645×10^{-4}
feet	miles (stat.)	1.894×10^{-4}
feet	millimeters	304.8
feet	mils	1.2×10^{4}

Table A.5
CONVERSION FACTORS (Continued)

To Convert	Into	Multiply by
feet of water	atmospheres	0.02950
feet of water	in. of mercury	0.8826
feet of water	kgs/sq cm	0.03048
feet of water	kgs/sq meter	304.8
feet of water	pounds/sq ft	62.43
feet of water	pounds/sq in.	0.4335
feet/min	cms/sec	0.5080
feet/min	feet/sec	0.01667
feet/min	kms/hr	0.01829
feet/min	meters/min	0.3048
feet/min	miles/hr	0.01136
feet/sec	cms/sec	30.48
feet/sec	kms/hr	1.097
feet/sec	knots	0.5921
feet/sec	meters/min	18.29
feet/sec	miles/hr	0.6818
feet/sec	miles/min	0.01136
feet/sec/sec	cms/sec/sec	30.48
feet/sec/sec	kms/hr/sec	1.097
feet/sec/sec	meters/sec/sec	0.3048
feet/sec/sec	miles/hr/sec	0.6818
feet/100 feet	per cent grade	1.0
Foot—candle	Lumen/sq. meter	10.764
foot-pounds	Btu	1.286×10^{-3}
foot-pounds	ergs	1.356×10^{7}
foot-pounds	gram-calories	0.3238
foot-pounds	hp-hrs	5.050×10^{-7}

To Convert	Into	Multiply by
foot-pounds	joules	1.356
foot-pounds	kg-calories	3.24×10^{-4}
foot-pounds	kg-meters	0.1383
foot-pounds	kilowatt-hrs	3.766×10^{-7}
foot-pounds/min	Btu/min	1.286×10^{-3}
foot-pounds/min	foot-pounds/sec	0.01667
foot-pounds/min	horsepower	3.030×10^{-5}
foot-pounds/min	kg-calories/min	3.24×10^{-4}
foot-pounds/min	kilowatts	2.260×10^{-5}
foot-pounds/sec	Btu/hr	4.6263
foot-pounds/sec	Btu/min	0.07717
foot-pounds/sec	horsepower	1.818×10^{-3}
foot-pounds/sec	kg-calories/min	0.01945
foot-pounds/sec	kilowatts	1.356×10^{-3}
Furlongs	miles (U.S.)	0.125
furlongs	rods	40.0
furlongs	feet	660.0
	G	
gallons	cu cms	3,785.0
gallons	cu feet	0.1337
gallons	cu inches	231.0
gallons	cu meters	3.785×10^{-3}
gallons	cu yards	4.951×10^{-3}
gallons	liters	3.785
gallons (liq. Br. Imp.)	gallons (U.S. liq.)	1.20095
gallons (U.S.)	gallons (Imp.)	0.83267
gallons of water	pounds of water	8.3453

Table A.5
CONVERSION FACTORS (Continued)

To Convert	Into	Multiply by
gallons/min	cu ft/sec	2.228×10^{-3}
gallons/min	liters/sec	0.06308
gallons/min	cu ft/hr	8.0208
gausses	lines/sq in.	6.452
gausses	webers/sq cm	10^{-8}
gausses	webers/sq in.	6.452×10^{-8}
gausses	webers/sq meter	10^{-4}
gilberts	ampere-turns	0.7958
gilberts/cm	amp-turns/cm	0.7958
gilberts/cm	amp-turns/in	2.021
gilberts/cm	amp-turns/meter	79.58
Gills (British)	cubic cm.	142.07
gills	liters	0.1183
gills	pints (liq.)	0.25
Grade	Radian	.01571
Grains	drams (avoirdupois)	0.03657143
grains (troy)	grains (avdp)	1.0
grains (troy)	grams	0.06480
grains (troy)	ounces (avdp)	2.0833×10^{-3}
grains (troy)	pennyweight (troy)	0.04167
grains/U.S. gal	parts/million	17.118
grains/U.S. gal	pounds/million gal	142.86
grains/Imp. gal	parts/million	14.286
grams	dynes	980.7
grams	grains	15.43
grams	joules/cm	9.807×10^{-5}
grams	joules/meter (newtons)	9.807×10^{-3}
grams	kilograms	0.001
grams	milligrams	1,000.
grams	ounces (avdp)	0.03527
grams	ounces (troy)	0.03215
grams	poundals	0.07093
grams	pounds	2.205×10^{-3}
grams/cm	pounds/inch	5.600×10^{-3}
grams/cu cm	pounds/cu ft	62.43
grams/cu cm	pounds/cu in	0.03613
grams/cu cm	pounds/mil-foot	3.405×10^{-7}
grams/liter	grains/gal	58.417
grams/liter	pounds/1,000 gal	8.345
grams/liter	pounds/cu ft	0.062427
grams/liter	parts/million	1,000.0
grams/sq cm	pounds/sq ft	2.0481
gram-calories	Btu	3.9683×10^{-3}
gram-calories	ergs	4.1868×10^{7}
gram-calories	foot-pounds	3.0880
gram-calories	horsepower-hrs	1.5596×10^{-6}
gram-calories	kilowatt-hrs	1.1630×10^{-6}
gram-calories	watt-hrs	1.1630×10^{-3}
gram-calories/sec	Btu/hr	14.286
gram-centimeters	Btu	9.297×10^{-8}
gram-centimeters	ergs	980.7
gram-centimeters	joules	9.807×10^{-5}
gram-centimeters	kg-cal	2.343×10^{-8}
gram-centimeters	kg-meters	10^{-5}

Table A.5
CONVERSION FACTORS (Continued)

To Convert	Into	Multiply by
	H	
Hand	Cm.	10.16
hectares	acres	2.471
hectares	sq feet	1.076×10^{5}
hectograms	grams	100.0
hectoliters	liters	100.0
hectometers	meters	100.0
hectowatts	watts	100.0
henries	millihenries	1,000.0
Hogsheads (British)	cubic ft.	10.114
Hogsheads (U.S.)	cubic ft.	8.42184
Hogsheads (U.S.)	gallons (U.S.)	63
horsepower	Btu/min	42.44
horsepower	foot-lbs/min	33,000.
horsepower	foot-lbs/sec	550.0
horsepower (metric) (542.5 ft lb/sec)	horsepower (550 ft lb/sec)	0.9863
horsepower (550 ft lb/sec)	horsepower (metric) (542.5 ft lb/sec)	1.014
horsepower	kg-calories/min	10.68
horsepower	kilowatts	0.7457
horsepower	watts	745.7
horsepower (boiler)	Btu/hr	33.479
horsepower (boiler)	kilowatts	9.803
horsepower-hrs	Btu	2,547.
horsepower-hrs	ergs	2.6845×10^{13}
horsepower-hrs	foot-lbs	1.98×10^{6}
horsepower-hrs	gram-calories	641,190.
horsepower-hrs	joules	2.684×10^{6}
horsepower-hrs	kg-calories	641.1
horsepower-hrs	kg-meters	2.737×10^{5}
horsepower-hrs	kilowatt-hrs	0.7457
hours	days	4.167×10^{-2}
hours	weeks	5.952×10^{-3}
Hundredweights (long)	pounds	112
Hundredweights (long)	tons (long)	0.05
Hundredweights (short)	ounces (avoirdupois)	1600
Hundredweights (short)	pounds	100
Hundredweights (short)	tons (metric)	0.0453592
Hundredweights (short)	tons (long)	0.0446429
	I	
inches	centimeters	2.540
inches	meters	2.540×10^{-2}
inches	miles	1.578×10^{-5}
inches	millimeters	25.40
inches	mils	1,000.0
inches	yards	2.778×10^{-2}
inches of mercury	atmospheres	0.03342
inches of mercury	feet of water	1.133
inches of mercury	kgs/sq cm	0.03453
inches of mercury	kgs/sq meter	345.3
inches of mercury	pounds/sq ft	70.73
inches of mercury	pounds/sq in.	0.4912
inches of water (at 4°C)	atmospheres	2.458×10^{-3}
inches of water (at 4°C)	inches of mercury	0.07355
inches of water (at 4°C)	kgs/sq cm	2.540×10^{-3}

Table A.5
CONVERSION FACTORS (Continued)

To Convert	Into	Multiply by
inches of water (at 4°C)	ounces/sq in	0.5781
inches of water (at 4°C)	pounds/sq ft	5.204
inches of water (at 4°C)	pounds/sq in.	0.03613
International Ampere	Ampere (absolute)	.9998
International Volt	Volts (absolute)	1.0003
	J	
joules	Btu	9.480×10^{-4}
joules	ergs	10^7
joules	foot-pounds	0.7376
joules	kg-calories	2.389×10^{-4}
joules	kg-meters	0.1020
joules	watt-hrs	2.778×10^{-4}
joules/cm	grams	1.020×10^4
joules/cm	dynes	10^7
joules/cm	joules/meter (newtons)	100.0
joules/cm	poundals	723.3
joules/cm	pounds	22.48
	K	
kilograms	dynes	980,665.
kilograms	grams	1,000.0
kilograms	joules/cm	0.09807
kilograms	joules/meter (newtons)	9.807
kilograms	poundals	70.93
kilograms	pounds	2.205
kilograms	tons (long)	9.842×10^{-4}
kilograms	tons (short)	1.102×10^{-3}
kilograms/cu meter	grams/cu cm	0.001
kilograms/cu meter	pounds/cu ft	0.06243
kilograms/cu meter	pounds/cu in.	3.613×10^{-5}
kilograms/cu meter	pounds/mil-foot	3.405×10^{-10}
kilograms/meter	pounds/ft	0.6720
Kilogram/sq. cm.	Dynes	980,665
kilograms/sq cm	atmospheres	0.9678
kilograms/sq cm	feet of water	32.81
kilograms/sq cm	inches of mercury	28.96
kilograms/sq cm	pounds/sq ft	2,048.
kilograms/sq cm	pounds/sq in.	14.22
kilograms/sq meter	atmospheres	9.678×10^{-5}
kilograms/sq meter	bars	98.07×10^{-6}
kilograms/sq meter	feet of water	3.281×10^{-3}
kilograms/sq meter	inches of mercury	2.896×10^{-3}
kilograms/sq meter	pounds/sq ft	0.2048
kilograms/sq meter	pounds/sq in.	1.422×10^{-3}
kilograms/sq mm	kgs/sq meter	10^6
kilogram-calories	Btu	3.968
kilogram-calories	foot-pounds	3,088.
kilogram-calories	hp-hrs	1.560×10^{-3}
kilogram-calories	joules	4,186.
kilogram-calories	kg-meters	426.9
kilogram-calories	kilojoules	4.186
kilogram-calories	kilowatt-hrs	1.163×10^{-3}

Table A.5
CONVERSION FACTORS (Continued)

To Convert	Into	Multiply by
kilogram meters	Btu	9.294×10^{-3}
kilogram meters	ergs	9.804×10^{7}
kilogram meters	foot-pounds	7.233
kilogram meters	joules	9.804
kilogram meters	kg-calories	2.342×10^{-3}
kilogram meters	kilowatt-hrs	2.723×10^{-6}
kilolines	maxwells	1,000.0
kiloliters	liters	1,000.0
kilometers	centimeters	10^{5}
kilometers	feet	3,281.
kilometers	inches	3.937×10^{4}
kilometers	meters	1,000.0
kilometers	miles	0.6214
kilometers	millimeters	10^{6}
kilometers	yards	1,094.
kilometers/hr	cms/sec	27.78
kilometers/hr	feet/min	54.68
kilometers/hr	feet/sec	0.9113
kilometers/hr	knots	0.5396
kilometers/hr	meters/min	16.67
kilometers/hr	miles/hr	0.6214
kilometers/hr/sec	cms/sec/sec	27.78
kilometers/hr/sec	ft/sec/sec	0.9113
kilometers/hr/sec	meters/sec/sec	0.2778
kilometers/hr/sec	miles/hr/sec	0.6214
kilowatts	Btu/min	56.92
kilowatts	foot-lbs/min	4.426×10^{4}
kilowatts	foot-lbs/sec	737.6
kilowatts	horsepower	1.341
kilowatts	kg-calories/min	14.34
kilowatts	watts	1,000.0
kilowatt-hrs	Btu	3,413.
kilowatt-hrs	ergs	3.600×10^{13}
kilowatt-hrs	foot-lbs	2.655×10^{6}
kilowatt-hrs	gram-calories	859,850.
kilowatt-hrs	horsepower-hrs	1.341
kilowatt-hrs	joules	3.6×10^{6}
kilowatt-hrs	kg-calories	860.5
kilowatt-hrs	kg-meters	3.671×10^{5}
kilowatt-hrs	pounds of water evaporated from and at 212°F.	3.53
kilowatt-hrs	pounds of water raised from 62° to 212°F.	22.75
knots	feet/hr	6,080.
knots	kilometers/hr	1.8532
knots	nautical miles/hr	1.0
knots	statute miles/hr	1.151
knots	yards/hr	2,027.
knots	feet/sec	1.689
	L	
league	miles (approx.)	3.0
Light year	Miles	5.9×10^{12}
Light year	Kilometers	9.46091×10^{12}

Table A.5
CONVERSION FACTORS (Continued)

To Convert	Into	Multiply by
lines/sq cm	gausses	1.0
lines/sq in.	gausses	0.1550
lines/sq in.	webers/sq cm	1.550×10^{-9}
lines/sq in.	webers/sq in.	10^{-8}
lines/sq in.	webers/sq meter	1.550×10^{-5}
links (engineer's)	inches	12.0
links (surveyor's)	inches	7.92
liters	bushels (U.S. dry)	0.02833
liters	cu cm	1,000.0
liters	cu feet	0.03531
liters	cu inches	61.02
liters	cu meters	0.001
liters	cu yards	1.308×10^{-3}
liters	gallons (U.S. liq.)	0.2642
liters	pints (U.S. liq.)	2.113
liters	quarts (U.S. liq.)	1.057
liters/min	cu ft/sec	5.886×10^{-4}
liters/min	gals/sec	4.403×10^{-3}
lumens/sq ft	foot-candles	1.0
Lumen	Spherical candle power	.07958
Lumen	Watt	.001496
Lumen/sq. ft.	Lumen/sq. meter	10.76
lux	foot-candles	0.0929
	M	
maxwells	kilolines	0.001
maxwells	webers	10^{-8}
megalines	maxwells	10^{6}

To Convert	Into	Multiply by
megohms	microhms	10^{12}
megohms	ohms	10^{6}
meters	centimeters	100.0
meters	feet	3.281
meters	inches	39.37
meters	kilometers	0.001
meters	miles (naut.)	5.396×10^{-4}
meters	miles (stat.)	6.214×10^{-4}
meters	millimeters	1,000.0
meters	yards	1.094
meters	varas	1.179
meters/min	cms/sec	1.667
meters/min	feet/min	3.281
meters/min	feet/sec	0.05468
meters/min	kms/hr	0.06
meters/min	knots	0.03238
meters/min	miles/hr	0.03728
meters/sec	feet/min	196.8
meters/sec	feet/sec	3.281
meters/sec	kilometers/hr	3.6
meters/sec	kilometers/min	0.06
meters/sec	miles/hr	2.237
meters/sec	miles/min	0.03728
meters/sec/sec	cms/sec/sec	100.0
meters/sec/sec	ft/sec/sec	3.281
meters/sec/sec	kms/hr/sec	3.6
meters/sec/sec	miles/hr/sec	2.237

Table A.5
CONVERSION FACTORS (Continued)

To Convert	Into	Multiply by
meter-kilograms	cm-dynes	9.807×10^{7}
meter-kilograms	cm-grams	10^{5}
meter-kilograms	pound-feet	7.233
microfarad	farads	10^{-6}
micrograms	grams	10^{-6}
microhms	megohms	10^{-12}
microhms	ohms	10^{-6}
microliters	liters	10^{-6}
Microns	meters	1×10^{-6}
miles (naut.)	miles (statute)	1.1516
miles (naut.)	yards	2,027.
miles (statute)	centimeters	1.609×10^{5}
miles (statute)	feet	5,280.
miles (statute)	inches	6.336×10^{4}
miles (statute)	kilometers	1.609
miles (statute)	meters	1,609.
miles (statute)	miles (naut.)	0.8684
miles (statute)	yards	1,760.
miles/hr	cms/sec	44.70
miles/hr	feet/min	88.
miles/hr	feet/sec	1.467
miles/hr	kms/hr	1.609
miles/hr	kms/min	0.02682
miles/hr	knots	0.8684
miles/hr	meters/min	26.82
miles/hr	miles/min	0.1667
miles/hr/sec	cms/sec/sec	44.70

To Convert	Into	Multiply by
miles/hr/sec	feet/sec/sec	1.467
miles/hr/sec	kms/hr/sec	1.609
miles/hr/sec	meters/sec/sec	0.4470
miles/min	cms/sec	2,682.
miles/min	feet/sec	88.
miles/min	kms/min	1.609
miles/min	knots/min	0.8684
miles/min	miles/hr	60.0
mil-feet	cu inches	9.425×10^{-6}
milliers	kilograms	1,000.
Millimicrons	meters	1×10^{-9}
Milligrams	grains	0.01543236
milligrams	grams	0.001
milligrams/liter	parts/million	1.0
millihenries	henries	0.001
milliliters	liters	0.001
millimeters	centimeters	0.1
millimeters	feet	3.281×10^{-3}
millimeters	inches	0.03937
millimeters	kilometers	10^{-6}
millimeters	meters	0.001
millimeters	miles	6.214×10^{-7}
millimeters	mils	39.37
millimeters	yards	1.094×10^{-3}
million gals/day	cu ft/sec	1.54723
mils	centimeters	2.540×10^{-3}
mils	feet	8.333×10^{-5}

Table A.5
CONVERSION FACTORS (Continued)

To Convert	Into	Multiply by
mils	inches	0.001
mils	kilometers	2.540×10^{-8}
mils	yards	2.778×10^{-5}
miner's inches	cu ft/min	1.5
Minims (British)	cubic cm.	0.059192
Minims (U.S., fluid)	cubic cm.	0.061612
minutes (angles)	degrees	0.01667
miles (naut.)	feet	6,080.27
miles (naut.)	kilometers	1.853
miles (naut.)	meters	1,853.
minutes (angles)	quadrants	1.852×10^{-4}
minutes (angles)	radians	2.909×10^{-4}
minutes (angles)	seconds	60.0
myriagrams	kilograms	10.0
myriameters	kilometers	10.0
myriawatts	kilowatts	10.0
	N	
nepers	decibels	8.686
Newton	Dynes	1×10^{5}
	O	
OHM (International)	OHM (absolute)	1.0005
ohms	megohms	10^{-6}
ohms	microhms	10^{6}
ounces	drams	16.0
ounces	grains	437.5
ounces	grams	28.349527
ounces	pounds	0.0625
ounces	ounces (troy)	0.9115
ounces	tons (long)	2.790×10^{-5}
ounces	tons (metric)	2.835×10^{-5}
ounces (fluid)	cu inches	1.805
ounces (fluid)	liters	0.02957
ounces (troy)	grains	480.0
ounces (troy)	grams	31.103481
ounces (troy)	ounces (avdp.)	1.09714
ounces (troy)	pennyweights (troy)	20.0
ounces (troy)	pounds (troy)	0.08333
Ounce/sq. inch	Dynes/sq. cm.	4309
ounces/sq in.	pounds/sq in.	0.0625
	P	
Parsec	Miles	19×10^{12}
Parsec	Kilometers	3.084×10^{13}
parts/million	grains/U.S. gal	0.0584
parts/million	grains/Imp. gal	0.07016
parts/million	pounds/million gal	8.345
Pecks (British)	cubic inches	554.6
Pecks (British)	liters	9.091901
Pecks (U.S.)	bushels	0.25
Pecks (U.S.)	cubic inches	537.605
Pecks (U.S.)	liters	8.809582
Pecks (U.S.)	quarts (dry)	8
pennyweights (troy)	grains	24.0

Table A.5
CONVERSION FACTORS (Continued)

To Convert	Into	Multiply by
pennyweights (troy)	ounces (troy)	0.05
pennyweights (troy)	grams	1.55517
pennyweights (troy)	pounds (troy)	4.1667×10^{-3}
pints (dry)	cu inches	33.60
pints (liq.)	cu cms.	473.2
pints (liq.)	cu feet	0.01671
pints (liq.)	cu inches	28.87
pints (liq.)	cu meters	4.732×10^{-4}
pints (liq.)	cu yards	6.189×10^{-4}
pints (liq.)	gallons	0.125
pints (liq.)	liters	0.4732
pints (liq.)	quarts (liq.)	0.5
Planck's quantum	Erg—second	6.624×10^{-27}
Poise	Gram/cm. sec.	1.00
Pounds (avoirdupois)	ounces (troy)	14.5833
poundals	dynes	13,826.
poundals	grams	14.10
poundals	joules/cm	1.383×10^{-3}
poundals	joules/meter (newtons)	0.1383
poundals	kilograms	0.01410
poundals	pounds	0.03108
pounds	drams	256.
pounds	dynes	44.4823×10^{4}
pounds	grains	7,000.
pounds	grams	453.5924
pounds	joules/cm	0.04448
pounds	joules/meter (newtons)	4.448
pounds	kilograms	0.4536

To Convert	Into	Multiply by
pounds	ounces	16.0
pounds	ounces (troy)	14.5833
pounds	poundals	32.17
pounds	pounds (troy)	1.21528
pounds	tons (short)	0.0005
pounds (troy)	grains	5,760.
pounds (troy)	grams	373.24177
pounds (troy)	ounces (avdp.)	13.1657
pounds (troy)	ounces (troy)	12.0
pounds (troy)	pennyweights (troy)	240.0
pounds (troy)	pounds (avdp.)	0.822857
pounds (troy)	tons (long)	3.6735×10^{-4}
pounds (troy)	tons (metric)	3.7324×10^{-4}
pounds (troy)	tons (short)	4.1143×10^{-4}
pounds of water	cu feet	0.01602
pounds of water	cu inches	27.68
pounds of water	gallons	0.1198
pounds of water/min	cu ft/sec	2.670×10^{-4}
pound-feet	cm-dynes	1.356×10^{7}
pound-feet	cm-grams	13,825.
pound-feet	meter-kgs	0.1383
pounds/cu ft	grams/cu cm	0.01602
pounds/cu ft	kgs/cu meter	16.02
pounds/cu ft	pounds/cu in.	5.787×10^{-4}
pounds/cu ft	pounds/mil-foot	5.456×10^{-9}
pounds/cu in.	gms/cu cm	27.68
pounds/cu in.	kgs/cu meter	2.768×10^{4}

Table A.5
CONVERSION FACTORS (Continued)

To Convert	Into	Multiply by
pounds/cu in.	pounds/cu ft	1,728.
pounds/cu in.	pounds/mil-foot	9.425×10^{-6}
pounds/ft	kgs/meter	1.488
pounds/in.	gms/cm	178.6
pounds/mil-foot	gms/cu cm	2.306×10^{6}
pounds/sq ft	atmospheres	4.725×10^{-4}
pounds/sq ft	feet of water	0.01602
pounds/sq ft	inches of mercury	0.01414
pounds/sq ft	kgs/sq meter	4.882
pounds/sq ft	pounds/sq in.	6.944×10^{-3}
pounds/sq in.	atmospheres	0.06804
pounds/sq in.	feet of water	2.307
pounds/sq in.	inches of mercury	2.036
pounds/sq in.	kgs/sq meter	703.1
pounds/sq in.	pounds/sq ft	144.0
	Q	
quadrants (angle)	degrees	90.0
quadrants (angle)	minutes	5,400.0
quadrants (angle)	radians	1.571
quadrants (angle)	seconds	3.24×10^{5}
quarts (dry)	cu inches	67.20
quarts (liq.)	cu cms	946.4
quarts (liq.)	cu feet	0.03342
quarts (liq.)	cu inches	57.75
quarts (liq.)	cu meters	9.464×10^{-4}
quarts (liq.)	cu yards	1.238×10^{-3}
quarts (liq.)	gallons	0.25
quarts (liq.)	liters	0.9463

To Convert	Into	Multiply by
	R	
radians	degrees	57.30
radians	minutes	3,438.
radians	quadrants	0.6366
radians	seconds	2.063×10^{5}
radians/sec	degrees/sec	57.30
radians/sec	revolutions/min	9.549
radians/sec	revolutions/sec	0.1592
radians/sec/sec	revs/min/min	573.0
radians/sec/sec	revs/min/sec	9.549
radians/sec/sec	revs/sec/sec	0.1592
revolutions	degrees	360.0
revolutions	quadrants	4.0
revolutions	radians	6.283
revolutions/min	degrees/sec	6.0
revolutions/min	radians/sec	0.1047
revolutions/min	revs/sec	0.01667
revolutions/min/min	radians/sec/sec	1.745×10^{-3}
revolutions/min/min	revs/min/sec	0.01667
revolutions/min/min	revs/sec/sec	2.778×10^{-4}
revolutions/sec	degrees/sec	360.0
revolutions/sec	radians/sec	6.283
revolutions/sec	revs/min	60.0
revolutions/sec/sec	radians/sec/sec	6.283
revolutions/sec/sec	revs/min/min	3,600.0
revolutions/sec/sec	revs/min/sec	60.0
Rod	Chain (Gunters)	.25

Table A.5
CONVERSION FACTORS (Continued)

To Convert	Into	Multiply by
Rod	Meters	5.029
Rods (Surveyors' meas.)	yards	5.5
rods	feet	16.5
	S	
Scruples	grains	20
seconds (angle)	degrees	2.778×10^{-4}
seconds (angle)	minutes	0.01667
seconds (angle)	quadrants	3.087×10^{-6}
seconds (angle)	radians	4.848×10^{-6}
Slug	Kilogram	14.59
Slug	Pounds	32.17
Sphere	Steradians	12.57
square centimeters	circular mils	1.973×10^{5}
square centimeters	sq feet	1.076×10^{-3}
square centimeters	sq inches	0.1550
square centimeters	sq meters	0.0001
square centimeters	sq miles	3.861×10^{-11}
square centimeters	sq millimeters	100.0
square centimeters	sq yards	1.196×10^{-4}
square feet	acres	2.296×10^{-5}
square feet	circular mils	1.833×10^{8}
square feet	sq cms	929.0
square feet	sq inches	144.0
square feet	sq meters	0.09290
square feet	sq miles	3.587×10^{-8}
square feet	sq millimeters	9.290×10^{4}
square feet	sq yards	0.1111
square inches	circular mils	1.273×10^{6}
square inches	sq cms	6.452
square inches	sq feet	6.944×10^{-3}
square inches	sq millimeters	645.2
square inches	sq mils	10^{6}
square inches	sq yards	7.716×10^{-4}
square kilometers	acres	247.1
square kilometers	sq cms	10^{10}
square kilometers	sq ft	10.76×10^{6}
square kilometers	sq inches	1.550×10^{9}
square kilometers	sq meters	10^{6}
square kilometers	sq miles	0.3861
square kilometers	sq yards	1.196×10^{6}
square meters	acres	2.471×10^{-4}
square meters	sq cms	10^{4}
square meters	sq feet	10.76
square meters	sq inches	1,550.
square meters	sq miles	3.861×10^{-7}
square meters	sq millimeters	10^{6}
square meters	sq yards	1.196
square miles	acres	640.0
square miles	sq feet	27.88×10^{6}
square miles	sq kms	2.590
square miles	sq meters	2.590×10^{6}
square miles	sq yards	3.098×10^{6}

Table A.5
CONVERSION FACTORS (Continued)

To Convert	Into	Multiply by
square millimeters	circular mils	1,973.
square millimeters	sq cms	0.01
square millimeters	sq feet	1.076×10^{-5}
square millimeters	sq inches	1.550×10^{-3}
square mils	circular mils	1.273
square mils	sq cms	6.452×10^{-6}
square mils	sq inches	10^{-6}
square yards	acres	2.066×10^{-4}
square yards	sq cms	8,361.
square yards	sq feet	9.0
square yards	sq inches	1,296.
square yards	sq meters	0.8361
square yards	sq miles	3.228×10^{-7}
square yards	sq millimeters	8.361×10^{5}
	T	
temperature (°C) +273	absolute temperature (°K)	1.0
temperature (°C) +17.78	temperature (°F)	1.8
temperature (°F) + 460	absolute temperature (°R)	1.0
temperature (°F) −32	temperature (°C)	5/9
tons (long)	kilograms	1,016.
tons (long)	pounds	2,240.
tons (long)	tons (short)	1.120
tons (metric)	kilograms	1,000.
tons (metric)	pounds	2,205
tons (short)	kilograms	907.1848
tons (short)	ounces	32,000.
tons (short)	ounces (troy)	29,166.66
tons (short)	pounds	2,000.
tons (short)	pounds (troy)	2,430.56
tons (short)	tons (long)	0.89287
tons (short)	tons (metric)	0.9078
tons (short)/sq ft	kgs/sq meter	9,765.
tons (short)/sq ft	pounds/sq in.	2,000.
tons of water/24 hrs	pounds of water/hr	83.333
tons of water/24 hrs	gallons/min	0.16643
tons of water/24 hrs	cu ft/hr	1.3349
	V	
Volt/inch	Volt/cm.	.39370
Volt (absolute)	Statvolts	.003336
	W	
watts	Btu/hr	3.4129
watts	Btu/min	0.05688
watts	ergs/sec	107.
watts	foot-lbs/min	44.27
watts	foot-lbs/sec	0.7378
watts	horsepower	1.341×10^{-3}
watts	horsepower (metric)	1.360×10^{-3}
watts	kg-calories/min	0.01433
watts	kilowatts	0.001
Watts (Abs.)	B.T.U. (mean)/min.	0.056884

Table A.5

CONVERSION FACTORS (Continued)

To Convert	Into	Multiply by	To Convert	Into	Multiply by
Watts (Abs.)	joules/sec.	1	webers/sq in.	webers/sq cm	0.1550
watt-hours	Btu	3.413	webers/sq in.	webers/sq meter	1,550.
watt-hours	ergs	3.60×10^{10}	webers/sq meter	gausses	10^4
watt-hours	foot-pounds	2,656.	webers/sq meter	lines/sq in.	6.452×10^4
watt-hours	gram-calories	859.85	webers/sq meter	webers/sq cm	10^{-4}
watt-hours	horsepower-hrs	1.341×10^{-3}	webers/sq meter	webers/sq in.	6.452×10^{-4}
watt-hours	kilogram-calories	0.8605			
watt-hours	kilogram-meters	367.2		Y	
watt-hours	kilowatt-hrs	0.001	yards	centimeters	91.44
Watt (International)	Watt (absolute)	1.0002	yards	kilometers	9.144×10^{-4}
webers	maxwells	10^8	yards	meters	0.9144
webers	kilolines	10^5	yards	miles (naut.)	4.934×10^{-4}
webers/sq in.	gausses	1.550×10^7	yards	miles (stat.)	5.682×10^{-4}
webers/sq in.	lines/sq in.	10^8	yards	millimeters	914.4

Footnotes to Figure A.6

Headquarter's mailing address:
U.S. Environmental Protection Agency Washington, D.C. 20460

For information on grant applications see publication:
"Catalog of Federal Domestic Assistance," Public Documents Department, Government Printing Office, Washington, D.C. 20402

For information on survey and demonstration grants:
Water 703-557-9338
Air 919-688-8146/345
Solid waste 513-684-4333

For information on research:
703-557-9338

Figure A.6 Organization of the United States Environmental Protective Agency.

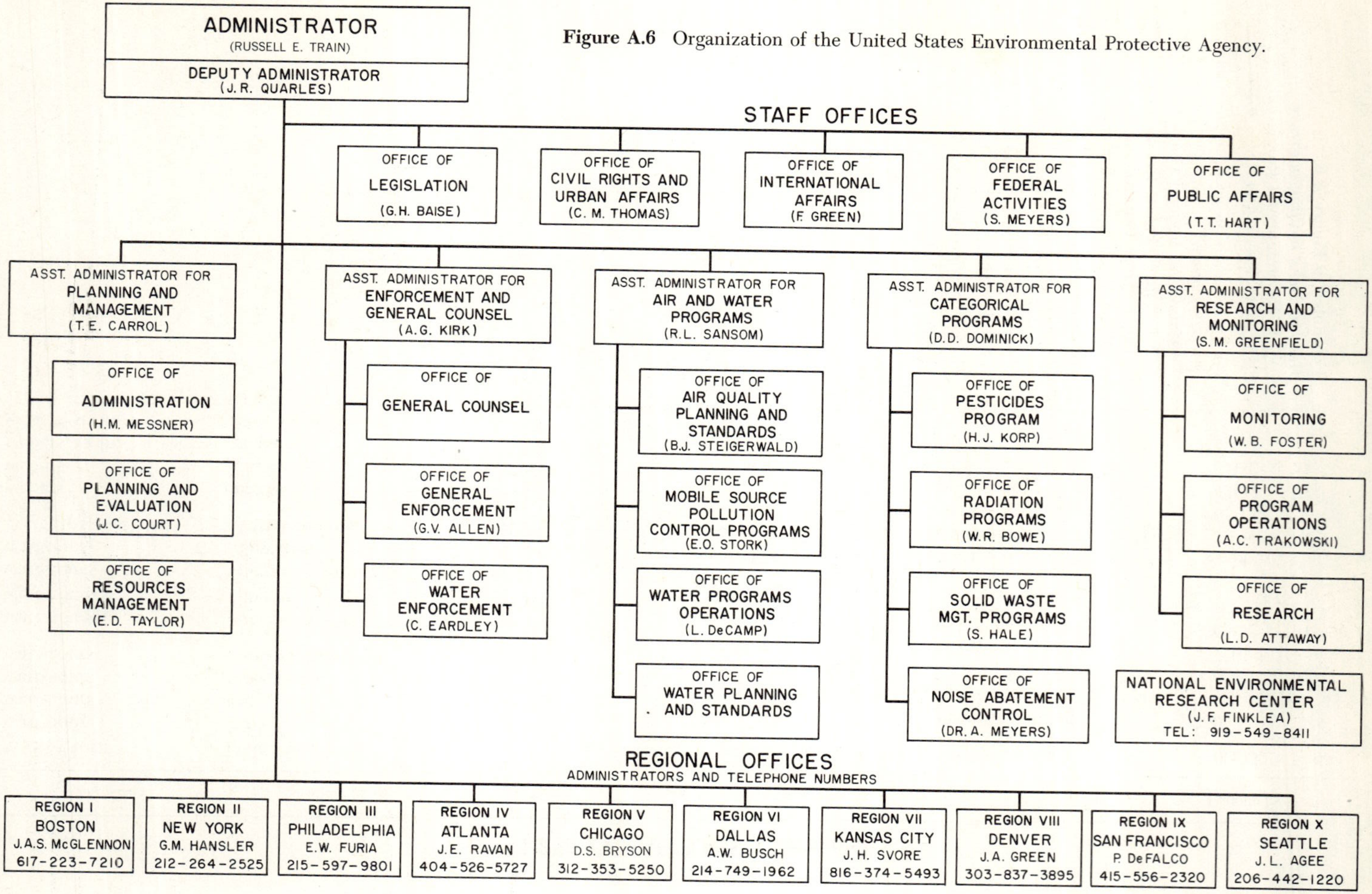

Table A.7
INTERNATIONAL SYSTEM OF UNITS

Quantity	*Unit*	*SI Symbol*	*Formula*
BASE UNITS			
length	metre	m	...
mass	kilogram	kg	...
time	second	s	...
electric current	ampere	A	...
thermodynamic temperature	kelvin	K	...
amount of substance	mole	mol	...
luminous intensity	candela	cd	...
SUPPLEMENTARY UNITS:			
plane angle	radian	rad	...
solid angle	steradian	sr	...
DERIVED UNITS:			
acceleration	metre per second squared	...	m/s^2
activity (of a radioactive source)	disintegration per second	...	(disintegration)/s
angular acceleration	radian per second squared	...	rad/s^2
angular velocity	radian per second	...	rad/s
area	square metre	...	m^2
density	kilogram per cubic metre	...	kg/m^3
electric capacitance	farad	F	$A \cdot s/V$
electrical conductance	siemens	S	A/V
electric field strength	volt per metre	...	V/m
electric inductance	henry	H	$V \cdot s/A$
electric potential difference	volt	V	W/A
electric resistance	ohm	Ω	V/A
electromotive force	volt	V	W/A
energy	joule	J	$N \cdot m$
entropy	joule per kelvin	...	J/K
force	newton	N	$kg \cdot m/s^2$
frequency	hertz	Hz	(cycle)/s
illuminance	lux	lx	lm/m^2
luminance	candela per square metre	...	cd/m^2
luminous flux	lumen	lm	$cd \cdot sr$
magnetic field strength	ampere per metre	...	A/m
magnetic flux	weber	Wb	$V \cdot s$
magnetic flux density	tesla	T	Wb/m^2
magnetomotive force	ampere	A	...

Table A.7
INTERNATIONAL SYSTEM OF UNITS (Continued)

Quantity	*Unit*	*SI Symbol*	*Formula*
Derived Units:			
power	watt	W	J/s
pressure	pascal	Pa	N/m^2
quantity of electricity	coulomb	C	A · s
quantity of heat	joule	J	N · m
radiant intensity	watt per steradian	...	W/sr
specific heat	joule per kilogram-kelvin	...	J/kg · K
stress	pascal	Pa	N/m^2
thermal conductivity	watt per metre-kelvin	...	W/m · K
velocity	metre per second	...	m/s
viscosity, dynamic	pascal-second	...	Pa · s
viscosity, kinematic	square metre per second	...	m^2/s
voltage	volt	V	W/A
volume	cubic metre	...	m^3
wavenumber	reciprocal metre	...	(wave)/m
work	joule	J	N · m

Table A.7
INTERNATIONAL SYSTEM OF UNITS (Continued)

Si Prefixes

Multiplication Factors	*Prefix*	*SI Symbol*
1 000 000 000 000 = 10^{12}	tera	T
1 000 000 000 = 10^{9}	giga	G
1 000 000 = 10^{6}	mega	M
1 000 = 10^{3}	kilo	k
100 = 10^{2}	hecto[a]	h
10 = 10^{1}	deka[a]	da
0.1 = 10^{-1}	deci[a]	d
0.01 = 10^{-2}	centi[a]	c
0.001 = 10^{-3}	milli	m
0.000 001 = 10^{-6}	micro	μ
0.000 000 001 = 10^{-9}	nano	n
0.000 000 000 001 = 10^{-12}	pico	p
0.000 000 000 000 001 = 10^{-15}	femto	f
0.000 000 000 000 000 001 = 10^{-18}	atto	a

[a] To be avoided where possible.

INDEX